Fibers

Dieter Veit

Fibers

History, Production, Properties, Market

Springer

Dieter Veit
Institut für Textiltechnik
RWTH Aachen University
Aachen, Germany

ISBN 978-3-031-15308-2 ISBN 978-3-031-15309-9 (eBook)
https://doi.org/10.1007/978-3-031-15309-9

This Springer imprint is published by the registered company Springer Nature Switzerland AG
The registered company address is: Gewerbestrasse 11, 6330 Cham, Switzerland

Preface

Today, life without fibers is unimaginable. For thousands of years, fibers have determined human existence, be it in the form of clothing, floor coverings, seat covers of all kinds, technical and medical textiles, hygiene articles, and much more. Naturally occurring fibers have been used since prehistoric times, e.g., wool, flax, hemp, and possibly cotton, later silk and asbestos. Today, about 75% of all fibers are produced artificially from renewable raw materials (e.g., cellulose), from petroleum (e.g., polyester) and from inorganic materials (e.g., glass fibers).

This book is an attempt to summarize the current knowledge about all relevant fibers in a compact form. It is based on my lectures over the past 20 years on fiber production as well as on numerous sources mentioned in the respective chapters.

The book is aimed at students of textile technology with and without prior knowledge as well as at professional users in the textile and apparel industry who are looking for information on the production and the properties of fibers. An extensive bibliography at the end of each chapter contains further sources, most of which are available online. Numerous Internet links lead to websites with further information.

The book is divided into 12 parts. In Part I, after an introduction to the history of fiber production and the markets for fiber materials, their internal structure is explained. Then, Part II describes all the important natural fibers, their history, their extraction, their properties and typical applications, and the development of the respective markets. In the case of synthetic and cellulosic man-made fibers in the following parts, the basics of their production, the machinery and equipment, and the main processes are also explained. Part VI covers biopolymers, detailing both early developments and the current status. Part VII covers inorganic fibers, and Part VIII deals with metallic fibers. The most important test methods for fibers including the relevant standards are explained in Part IX. An overview of the properties of the most important fibers and corresponding comparisons are the subject of Part X. An extensive list of selected manufacturers and trade names of fibers can be found in Part XI. In Part XII, an outlook on the future of fiber production is attempted.

All persons who have contributed to the success of this book are acknowledged in the front matter.

If, despite carefully checking all the facts, errors have crept in, I would be grateful for any comments.

Aachen, Germany Dieter Veit

Acknowledgments

This book could not have been written without the support of many kind people who provided me with documents, pictures, fiber samples, and other information or helped me in other ways.

In alphabetical order of the companies, these are the following people, to whom I would like to express my sincere thanks for their support and uncomplicated help:

Angus Ireland, Cian King, Australian Wool Innovation Limited, https://www.wool.com

Curd Schneider, Jörg Alexander, BB Engineering GmbH, https://www.bbeng.de

Michiel Van Hooreweder, Bekaert Fiber Technologies https://www.bekaert.com

Elke Hortmeyer, Bremen Cotton Exchange, https://baumwollboerse.de/

Eckhard Sonntag, Emil BRÖLL GmbH & Co.KG, https://www.broell.com

Hans-Peter Welzel, ChemieFreunde Erkner e.V., http://www.chemieforum-erkner.de

Isabelle Thoele, Cotton made in Africa, https://cottonmadeinafrica.org

Marie Luise Möller, Georgi Gogoladze, Deutsche Basalt Faser GmbH, www.deutsche-basalt-faser.de

Sachin Doshi, Doshi Group, http://www.doshi-group.com

Khosrow Rahimi Harchegani, DWI - Leibniz Institute for Interactive Materials e.V. , https://www.dwi.rwth-aachen.de

Karolína Pittnerová, David Brokl, ELMARCO s.r.o., https://www.elmarco.com

Janusz Berger, Evonik Industries AG, httsp://www.evonik.de

Renate Fourné, FOURNÉ Maschinenbau GmbH, https://fourne-maschinenbau.de

Volker Röhring, Freudenberg Performance Materials, https://www.freudenberg-pm.com

Sabine Reimer, GKD - Gebr. Kufferath AG, Metal Weaving & Wire Weaving, https://www.gkd-group.com

Eva-Maria Hartmann, General Association Linen e.V.

Peggy Thiele, P-D Glasseiden GmbH Oschatz, https://www.oschatz-glas.com

Andrea Kossmann, Oliver Schmidt, Gneuss Kunststofftechnik GmbH, https://www.gneuss.com

Marina Stettler, Heberlein AG, https://www.heberlein.com
David Chiles, Hills, Inc, https://www.hillsinc.net
David Schmelzeisen, Impact Products GmbH, https://upperhand.co
Erika Flores, International Alpaca Association, http://aia.org.pe/
Jessica Schnell, John Deere Walldorf GmbH & Co. KG, https://www.johndeere.com
Stefanie Heigl, Kelheim Fibres GmbH, https://www.kelheim-fibres.com
Stefan Steinbrecht, Kermel, http://www.kermel.com
Carina Schauer, KnitsByCari, https://www.knitsbycari.de
Sarah Koller, Lenzing Aktiengesellschaft, https://www.lenzing.com
Andreas Roye, Lucem GmbH, https://lucem.com
Annika Belloff, Maag Germany GmbH , https://www.maag.com
Michel Taglang, Mahogany, https://www.mahogany-cashmere.com
Ilka Haase, Mahr Metering Systems GmbH, https://www.mahr.de
Dirk Weber, Ing. A. Maurer SA, https://www.maurer-sa.ch
Kazunori Kawahara, Karl Brentrup, Mitsubishi Chemical Europe, https://eu.mitsubishi-chemical.com
Monika Meier, Neptu GmbH, https://neptugmbh.de
Philipp Jungbecker, Susanne Beyer, Ralf Buchholz, Oerlikon Barmag, Branch of Oerlikon Textile GmbH & Co. KG, https://www.oerlikon.com/polymer-processing/de/
Claudia Henkel, Oerlikon Neumag, branch of Oerlikon Textile GmbH & Co KG, https://www.oerlikon.com/polymer-processing/
Olga Leontieva, OJSC Svetlogorsk Khimvolokno, http://www.sohim.by
Susanne Brenk-Umhauer, Pappella, https://www.pappella.de
Margaret Rye, Qiveut Designs, https://qiveut.com
Walter Notter, Qiviuk Europe, https://www.qiviut.ch
Dietmar Hutmacher, Queensland University of Technology, https://www.qut.edu.au/institute-of-health-and-biomedical-innovation
Hans-Georg Geus, Martin Neuenhofer, Reifenhäuser GmbH & Co. KG Maschinenfabrik, https://reifenhauser.com
Michaela Friemer, Saurer Technologies GmbH & Co KG Elastomer Components & Engineered Bearing Solutions, https://www.saurer.com
Georg Tetzlaff, Saurer Technologie GmbH & Co KG Twisting Solutions, https://www.saurer.com
Rebekka Glogger, Steffen Janetzko, SGL TECHNOLOGIES GmbH, http://www.sglcarbon.com
Martina Finken, smartfiber AG, https://www.smartfiber.de
Thomas Reiser, Sossna GmbH, https://sossna.de
Thomas Elsener, Silvan Schäfli, SSM Schärer Schweiter Mettler AG, https://www.ssm.ch
David Amrein, Stamixco AG, https://www.stamixco.com
Sigrid Eder-Ince, Starlinger & Co. Gesellschaft m.b.H., http://www.starlinger.com
Malte Giesenow, Steppenstrolch GmbH, https://www.happy-camel.de
Ulrich Mörschel, Textechno Herbert Stein GmbH & Co. KG, https://www.textechno.com

Wolfram Soell, Uster Technologies, https://www.uster.com
Sally Fox, Vreseis, Ltd, https://www.vreseis.com
Julian Eichhoff, Marc Felske, Fanny Habermeyer, Tobias Löffler, Birgit Reiter, Steffen Reppich, Jan Serode, and Lother Tallner have also provided me with photos from their private collections, for which I would like to thank them all.

In addition, over many years, my student coworkers Julia Webler, Inke Eley, Lena Gerten, Amrei Becker, Kutlu Öztaş, and Henrike Bensiek collected a large amount of data that was used in many chapters. My colleague Roswitha Jacobs not only created some of the more complex illustrations, but also patiently gave me useful tips so that I could do most of the drawings myself. Marita Rohs picked out many SEM images from her large collection and also made some new ones, and Sabine Weber photographed many of our lecture exhibits, which added the finishing touch to the book.

My special thanks go to my former colleague Gabriele Satlow, who tirelessly updated our lectures on all fibrous materials over decades and whose thorough research work on fiber materials laid the foundation for many of the chapters in this book, especially those on natural fibers.

Finally, my biggest thanks go to my wife Sabine and my children Markus, Marion, and Michael, who have patiently put up with me working on the book in the evenings and on weekends for the last 3 years. Without your support, I could not have written this book. And special thanks go to my parents Rolf and Hedwig Veit, without whom I would not have come into existence and, as a consequence, neither the book.

Contents

Part I Fundamentals

1 Definitions . . . 3
1.1 Definition of Fibers and Classification . . . 3
1.2 Systematic Taxonomy . . . 4
1.3 Abbreviation . . . 5
References . . . 8

2 History . . . 9
2.1 Natural Fibers . . . 9
2.2 Cellulosic Man-Made Fibers . . . 12
2.3 Protein-Based Man-Made Fibers . . . 22
2.4 Synthetic Man-Made Fibers . . . 22
2.5 Inorganic Man-Made Fibers . . . 36
2.6 Metal Fibers . . . 38
References . . . 38

3 Market . . . 41
3.1 World Fiber Production . . . 41
3.2 Regions Producing Natural Fibers . . . 45
3.3 Regions and Countries Producing Man-Made Fibers . . . 46
3.4 Prices . . . 61
References . . . 61

4 Internal Structure of Fibers . . . 63
4.1 Chemical Structure of Fibers . . . 63
4.2 Structural Models . . . 63
4.3 Crystallinity . . . 63
4.4 Model Conceptions . . . 69
4.5 Crystal Structures . . . 70
4.6 Length of the Macromolecules . . . 83
4.7 Orientation . . . 85
4.8 Drawing Process . . . 86
4.9 Summary . . . 88
References . . . 89

Part II Natural Fibers

5 Cotton . . . 93
5.1 History . . . 93
5.2 Cultivation, Growth, Varieties . . . 97
5.3 Harvest . . . 106
5.4 Ginning . . . 110
5.5 Composition and Fine Structure . . . 121
5.6 Properties . . . 124
5.7 Classification . . . 132
5.8 Production, Export, Import . . . 133
5.9 Important Cotton Producing and Processing Countries . . . 143
5.10 Processing in Germany . . . 167
5.11 Pests . . . 169
5.12 Genetic Engineering (“Engineered Cotton”) . . . 173
5.13 Organic Cotton . . . 178
5.14 Trade . . . 181
5.15 Areas of Application . . . 185
5.16 Trends . . . 185
References . . . 187

6 Bast Fibers . . . 189
6.1 Flax (Linen) . . . 189
6.2 Hemp . . . 213
6.3 Jute . . . 223
6.4 Ramie . . . 230
6.5 Kenaf . . . 238
6.6 Nettle Fiber . . . 241
6.7 Bamboo . . . 243
6.8 Property Comparison . . . 246
References . . . 247

7 Hard Fibers . . . 249
7.1 Agave Fibers (Sisal, Henequen) . . . 249
7.2 Musa Textilis (Manila Hemp, Abacá) . . . 255
7.3 Pineapple . . . 257
7.4 Lily Fibers . . . 260
7.5 Grass Fibers . . . 260
7.6 Palm Fibers . . . 261
References . . . 262

8 Fruit Fibers . . . 263
8.1 Coir Fibers . . . 263
8.2 Kapok (Wool Tree) . . . 267
8.3 Poplar Fibers . . . 268
8.4 Betel Nut Fibers . . . 269
8.5 Cactus Fibers . . . 270
References . . . 271

9 Wool . . . 273
9.1 Terms . . . 273
9.2 History . . . 274
9.3 Sheep Breeds and Breeding . . . 279
9.4 Wool Extraction . . . 282
9.5 Structure of the Wool Fiber . . . 286
9.6 Features . . . 292
9.7 Classification . . . 296
9.8 Production . . . 301
9.9 Trade . . . 317
9.10 Wool Research . . . 322
9.11 Areas of Application . . . 322
9.12 Wool Seals . . . 325
References . . . 325

10 Fine Animal Hair . . . 327
10.1 Camel . . . 327
10.2 Goat . . . 338
10.3 Angora Rabbit . . . 344
10.4 Yak . . . 344
10.5 Horsehair . . . 345
10.6 Dog Hair . . . 347
10.7 Fur Hair . . . 347
10.8 Comparison of the Most Important Fine Animal Hair . . . 349
References . . . 349

11 Silk . . . 351
11.1 Terms . . . 351
11.2 History . . . 351
11.3 Breeding . . . 354
11.4 Structure . . . 357
11.5 Classification . . . 359
11.6 Reeling . . . 360
11.7 Terms for Silk Products . . . 362
11.8 Spinning Mill . . . 363
11.9 Production . . . 364
11.10 Properties . . . 364
11.11 Finishing . . . 368
11.12 Areas of Application . . . 372
11.13 Tussah Silk (Tussar or Tasar Silk) . . . 373
11.14 Other Silks . . . 374
References . . . 380

12 Asbestos . . . 381
12.1 Terms and Definitions . . . 381
12.2 History . . . 381
12.3 Origin, Geology . . . 382
12.4 Deposits . . . 383
12.5 Generation . . . 384

12.6 Classification . . . 385
12.7 Chemical Composition and Structure . . . 386
12.8 Properties . . . 388
12.9 Production and Consumption . . . 391
12.10 Environmental Behavior . . . 392
12.11 Processing . . . 394
12.12 Areas of Application . . . 395
12.13 Health Hazards and Damage . . . 396
12.14 Refurbishment of Buildings Containing Asbestos . . . 397
12.15 Substitute Materials . . . 399
References . . . 401

13 Other Natural Fibers . . . 403
13.1 Cellulosic Fibers . . . 403
13.2 Animal Fibers . . . 406
References . . . 408

Part III Chemical Fibers: Processes and Machines

14 Man-Made Fibers: Polymer Formation Processes . . . 413
14.1 Extraction of Raw Materials from Oil, Coal and Natural Gas . . . 413
14.2 Raw Material Extraction from Cellulose . . . 413
14.3 Raw Material Extraction from Protein . . . 417
14.4 Polymer Formation . . . 418
14.5 Formation of Polymers from Several Components: Copolymers . . . 427
14.6 Recycling . . . 428
References . . . 429

15 Aggregates for Polymer Formation . . . 431
15.1 Reactor Types . . . 431
15.2 Autoclave . . . 433
15.3 Disk Ring Reactor . . . 434
15.4 VK Tube . . . 435
15.5 Discharge and Booster Pumps . . . 436
15.6 Tube Flows . . . 437
15.7 Heat Exchanger . . . 439
15.8 Static Mixers . . . 443
15.9 Granulator . . . 445
15.10 Dryer . . . 446
References . . . 452

16 Processes for the Production of Man-Made Fibers . . . 453
16.1 Process Steps . . . 453
16.2 Solution Spinning . . . 453
16.3 Melt Spinning . . . 461
16.4 Spinnability . . . 461

16.5 Orientation of Macromolecules and Mechanical Properties . . . 465
16.6 Theory of Drawing . . . 468
16.7 Force-Elongation Curve During Drafting . . . 470
16.8 Spinning Orientation and Take-Up Speed . . . 472
16.9 Spin Dyeing . . . 474
References . . . 475

17 Aggregates for the Production of Synthetic Filaments . . . 477
17.1 Extruder . . . 477
17.2 Filter . . . 478
17.3 Spin Pump . . . 482
17.4 Melt Spinneret . . . 484
17.5 Spinning Packages . . . 489
17.6 Spinning Beam . . . 491
17.7 Quench Duct . . . 492
17.8 Spin Preparation . . . 496
17.9 Drafting . . . 498
17.10 Entangling . . . 500
17.11 Winder . . . 500
References . . . 505

18 Aggregates for the Production of Solution Spun Man-Made Fibers . . . 507
18.1 Reactor . . . 507
18.2 Filter . . . 507
18.3 Spinning Pump . . . 508
18.4 Spinneret . . . 508
18.5 Winder . . . 509
References . . . 511

19 Processes for the Production of Textile Filament Yarns . . . 513
19.1 POY Process . . . 513
19.2 FDY Process . . . 514
19.3 FOY Process . . . 516
19.4 Performance Increase in POY Spinning of Polyester . . . 516
References . . . 519

20 Processes for the Production of Industrial Yarns . . . 521
20.1 Reactor . . . 521
20.2 FDY Process . . . 521
20.3 Draw Twisting . . . 521
20.4 Cabling . . . 523
References . . . 525

21 Process for the Production of Monofilaments . . . 527
21.1 Production . . . 527
21.2 Properties and Typical Applications . . . 527
References . . . 529

22 Process for the Production of Carpet Yarns . . . 531
22.1 Spinneret . . . 531
22.2 Process . . . 531
Reference . . . 535

23 Process for the Production of Film Yarns . . . 537
23.1 Basics . . . 537
23.2 Technical Data . . . 537
23.3 Aggregates . . . 537
23.4 Products . . . 537
Reference . . . 541

24 Processes for the Production of Staple Fibers . . . 543
24.1 Market . . . 543
24.2 Production . . . 543
24.3 Conversion of Fiber Cables . . . 547
24.4 Features . . . 553
References . . . 556

25 Processes for the Production of Nonwovens . . . 557
25.1 Spunbond . . . 557
25.2 Meltblown Process . . . 560
25.3 Products . . . 561
25.4 Market . . . 562
References . . . 567

26 Bicomponent Fibers . . . 569
26.1 Terms . . . 569
26.2 History . . . 570
26.3 Source Materials . . . 571
26.4 Manufacturing Technologies . . . 571
26.5 Fine Fibers . . . 576
26.6 Properties . . . 576
26.7 Areas of Application . . . 579
26.8 Market . . . 582
References . . . 582

27 Fine Fibers . . . 585
27.1 Electrospinning . . . 585
27.2 Melt Electrospinning . . . 599
27.3 Flash Spinning . . . 600
27.4 Centrifugal Spinning . . . 601
27.5 Areas of Application . . . 602
References . . . 603

28 Aerogels . . . 605
28.1 Manufacturing Principle . . . 605
28.2 Aerogel Fibers . . . 605
28.3 Features . . . 607
28.4 Applications . . . 607
References . . . 608

29 Texturing . . . 609
29.1 False Twist Texturing . . . 609
29.2 Air-Jet Texturing . . . 617
29.3 Other Processes . . . 621
References . . . 622

Part IV Synthetic Chemical Fibers: Production and Properties

30 Polyester . . . 625
30.1 Definition . . . 625
30.2 History . . . 625
30.3 Raw Materials . . . 627
30.4 Production of Polymers . . . 628
30.5 Fiber Modification . . . 631
30.6 Chemical Structure . . . 632
30.7 Properties of Polyester Fibers . . . 632
30.8 Market and Trade Names . . . 641
30.9 Processing Properties and Areas of Application . . . 643
References . . . 648

31 Polyamide . . . 649
31.1 Definition . . . 649
31.2 History . . . 649
31.3 Chemical Constitution and Raw Materials . . . 652
31.4 Process for the Preparation of Polyamides . . . 660
31.5 Process for the Production of Polyamide Fibers . . . 663
31.6 Features . . . 669
31.7 Yarn Types . . . 674
31.8 Market and Manufacturers . . . 675
31.9 Processing Properties and Areas of Application . . . 676
References . . . 680

32 Aramid . . . 681
32.1 History . . . 681
32.2 Raw Materials, Formation of the Polymer . . . 682
32.3 Production . . . 683
32.4 Properties . . . 685
32.5 Products and Applications . . . 690
32.6 Market . . . 691
References . . . 691

33 Polyolefins . . . 693
33.1 Polyethylene (PE) . . . 693
33.2 Polypropylene (PP) . . . 696
33.3 Polyvinyl Chloride (PVC) . . . 714
33.4 Polyvinyl Dichloride (PVDC) . . . 716
33.5 Polyvinyl Difluoride (PVDF) . . . 717
33.6 Polytetrafluoroethylene (PTFE) . . . 718
33.7 Polyacrylic (PAN) . . . 718

33.8 Polyvinyl Alcohol (PVA) . . . 718
References . . . 719

34 Polyacrylonitrile . . . 721
34.1 Definition . . . 721
34.2 Inventors, Development . . . 721
34.3 Raw Material, Formation of the Polymer . . . 722
34.4 Production Method . . . 725
34.5 After-treatment . . . 727
34.6 Recovery of the Solvent . . . 730
34.7 Special Fiber Types . . . 730
34.8 Properties . . . 733
34.9 Applications and Fields of Use . . . 734
34.10 Market and Trade Names . . . 735
References . . . 737

35 Polylactic Acid . . . 739
35.1 Definitions . . . 739
35.2 Inventors, History of Development . . . 739
35.3 Raw Materials and Synthesis . . . 740
35.4 Chemical Composition . . . 741
35.5 Production Method . . . 741
35.6 Properties . . . 743
35.7 Areas of Application . . . 746
35.8 Market and Trade Names . . . 747
References . . . 747

36 Elastane . . . 749
36.1 History . . . 749
36.2 Starting Materials, Formation of the Polymer . . . 749
36.3 Production . . . 752
36.4 Features . . . 753
36.5 Products and Applications . . . 756
36.6 Market . . . 756
References . . . 759

37 High-performance Fibers . . . 761
37.1 Definition . . . 761
37.2 Polytetrafluoroethylene Fibers (PTFE) . . . 761
37.3 Polyimide Fibers (PI) . . . 762
37.4 Polyetherketone Fibers (PEK, PEEK) . . . 764
37.5 Polysulfone Fibers (PSU, PES) . . . 767
37.6 Polyphenylene Sulfide Fibers (PPS) . . . 767
37.7 Melamine Resin Fibers (MF) . . . 768
37.8 Novoloid Fibers . . . 769
37.9 Polyazole Fibers (PBO) . . . 770
37.10 Polybenzimidazole Fibers (PBI) . . . 773
37.11 Polyoxadiazole Fibers (POD) . . . 773
37.12 Liquid Crystal Polymers (LCP) . . . 774

37.13 Properties 776
37.14 Manufacturers 776
References 777

38 Carbon Fibers 779
38.1 Development and History 779
38.2 Raw Materials 779
38.3 Fiber and Yarn Production 779
38.4 Inner Structure 790
38.5 Features 791
38.6 Further Processing 792
38.7 Areas of Application 793
38.8 Market and Manufacturers 794
References 798

39 Carbon Nanotubes and Graphene 801
39.1 Definition and Types 801
39.2 History 801
39.3 Structure 802
39.4 Production 804
39.5 Properties 807
39.6 Functionalization 808
39.7 Production of Yarns from Carbon Nanotubes 809
39.8 Production of Composite Materials with Nanotubes 809
39.9 Areas of Application 811
39.10 Health Effects 813
39.11 Market and Manufacturers 813
39.12 Graphene 814
References 816

40 Optical Fibers 819
40.1 History 819
40.2 Principle 819
40.3 Manufacture of Optical Fibers 821
40.4 Types of Polymer Optical Fibers 821
40.5 Production of Polymer Optical Fibers 822
40.6 Properties 824
40.7 Products and Applications 825
40.8 Market 826
References 827

Part V Cellulosic Chemical Fibers: Production and Properties

41 Cellulosic Man-Made Fibers 831
41.1 Basics 831
41.2 Regenerated Fibers 837
41.3 Lyocell 858
41.4 Other Direct Solution Methods 866
41.5 Production 867

41.6 Derivative Fibers 870
41.7 Nanocellulose 877
References 879

Part VI Biopolymers

42 Biopolymers 883
42.1 Basics 883
42.2 Protein Fibers 883
42.3 Polysaccharide Fibers 891
42.4 Other Biopolymers 893
42.5 Interim Summary 896
42.6 BioPET and BioPA Fibers 897
42.7 Degradability 898
42.8 Market and Manufacturers 900
References 900

Part VII Inorganic Fibers

43 Glass Fibers 905
43.1 Development and History 905
43.2 Raw Materials 905
43.3 Chemical Composition and Structure 905
43.4 Glass Production 908
43.5 Fiber and Yarn Production 911
43.6 Fiber Optic Types 914
43.7 Properties 916
43.8 Further Processing 918
43.9 Products 920
43.10 Market and Manufacturers 921
References 922

44 Basalt Fibers 923
44.1 Basalt Occurrence 923
44.2 Chemical Composition and Structure 923
44.3 Fiber and Yarn Production 923
44.4 Fiber Types 923
44.5 Properties 923
44.6 Further Processing 924
44.7 Products 924
44.8 Market and Manufacturers 924
References 926

45 Ceramic Fibers 927
45.1 Development and History 927
45.2 Raw Materials and Structure 927
45.3 Fiber and Yarn Production 931
45.4 Properties 934
45.5 Further Processing 934

45.6 Products 935
45.7 Market and Manufacturers 936
References 937

Part VIII Metal Fibers

46 Metal Fibers 941
46.1 Basics 941
46.2 Development and History 941
46.3 Metal Extraction 942
46.4 Metal Fiber Types 943
46.5 Production 944
46.6 Processing of Metal Fibers 951
46.7 Properties 953
46.8 Applications 954
46.9 Market and Manufacturers 956
References 956

Part IX Test Methods

47 Test Methods 959
47.1 Determination of the Fiber Content 959
47.2 Mechanical Properties 961
47.3 Geometric Properties 968
47.4 Moisture and Water Absorption 971
47.5 Standards 972
References 972

Part X Properties of Natural and Chemical Fibers

48 Properties 977
48.1 Fineness 977
48.2 Strength and Modulus of Elasticity 979
48.3 Degree of Polymerization and Length of the Polymers . . . 982
48.4 Composition 982
48.5 Density 982
48.6 Flammability and Use Temperature 984
48.7 Resistance to Light and Weather 987
48.8 Resistance to Acids and Alkalis 987
48.9 Resistance to Microorganisms 988
48.10 General Overview 988
References 1000

Part XI Producers and Trade Names

49 Producers and Trade Names 1003
49.1 Producers 1003
49.2 Trade names 1003

Part XII Outlook and Trends

50 Outlook and Trends 1023
50.1 Natural Fibers 1023
50.2 Man-Made Fibers 1025
50.3 Inorganic and Carbon Fibers 1025
50.4 The World in 2050 1026
References 1027

Index 1029

Part I

Fundamentals

1 Definitions

The word "fiber" is probably derived from the Proto Indo European word **gwhī-* which means "thread, tendon". In addition to textile fibers, there are also fibers in plants and in animal and human bodies, e.g. muscle fibers. Corresponding expressions, such as "with every fiber of his being" refer to this and also exist in other languages, e.g. in French "avec chaque fiber de son corps" and German "mit jeder Faser seines Körpers".

Textile fibers determine our lives and without fibers our existence would be unthinkable. Over the millennia, a wide variety of fiber materials have been used: vegetable, animal, mineral, and, since around 1890, increasingly also the so-called man-made fibers, initially on a cellulosic basis and, since the 1930s, also from petroleum. Figure 1.1 shows well-known examples.

1.1 Definition of Fibers and Classification

According to the German standard DIN 60000, fibers are "textile fibrous materials [...] which can be processed into textiles. They have the common characteristic of being long in relation to their cross-section and sufficiently strong and flexible" (DIN 60000, 1969). Raw spaghetti therefore do not count as 'fibers' because they are thin but not pliable, nor do they count as 'fibers' when cooked because they are not significantly longer than thick. The length-to-diameter ratio must normally be at least about 2000:1, which corresponds to a diameter of 10 µm for a fiber length of 20 mm, e.g. for short cotton fibers. Only then can fibers be processed into a yarn. However, there are also shorter and coarser fibers from which, for example, nonwovens or bristles can be produced.

Generally, a distinction is made between:

- fibers of finite length: staple fibers from natural resources and man-made staple fibers
- short, coarse fibers: bristles
- very short fibers: flock
- "continuous" fibers: filaments (monofilaments, multifilaments)
- coarse monofilaments: wires
- fiber composites: roving, tape, cable.

The transition between the individual areas is sometimes fluent, e.g. between "monofilaments" and "wires" in the case of metal fibers. Also "tape" and "cable" are not exactly delimited from each other.

The US standard (ASTM D7641-21, 2021) lists fibers used to manufacture textile products, gives their biological and commercial names and defines specific groups.

Textile fibers are divided into natural fibers (vegetable, animal, mineral, Fig. 1.2) and man-made fibers based on natural or synthetic polymers as well as industrially produced inorganic fibers. The two latter groups are often combined as in Fig. 1.3.

D. Veit, *Fibers*, https://doi.org/10.1007/978-3-031-15309-9_1

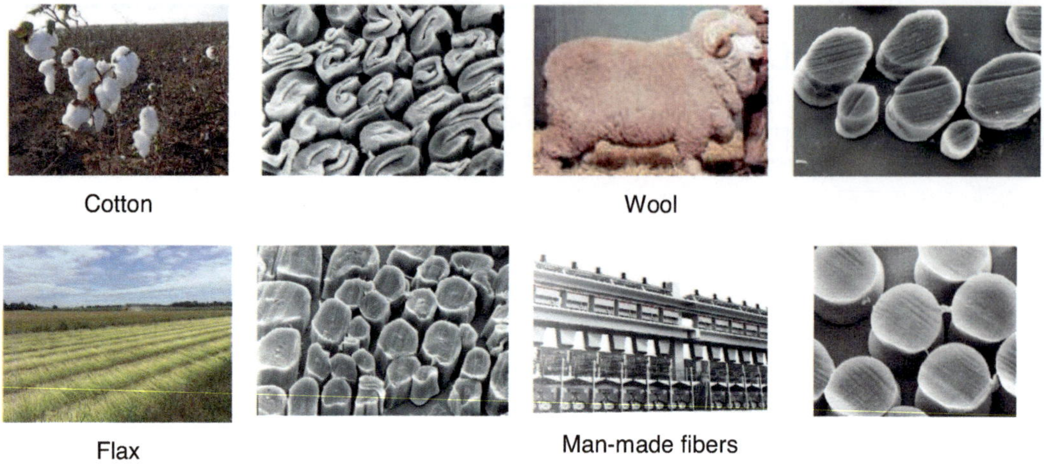

Fig. 1.1 Selected fibers and their producers (cotton: Lothar Tallner; wool; flax: Mark Felske; man-made fibers: Oerlikon Barmag)

The designation "hard fibers" in Fig. 1.2 is mainly used by textile manufacturers, the official designation according to the standard is "leaf fibers" or "fruit fibers". In general throughout this book, the fiber materials are always abbreviated in the overviews if this abbreviation is also usually used colloquially, e.g. "PET" for polyester made from polyethylene terephthalate. In both classification schemes, only typical representatives of the respective fiber group are named.

The generic names of natural fibers are defined in (DIN EN ISO 6938, 2015).

1.2 Systematic Taxonomy

There are various approaches to the classification (systematics) of plants and animals. In the following, the classical approach according to the systematics of angiosperms is used for natural fibers.

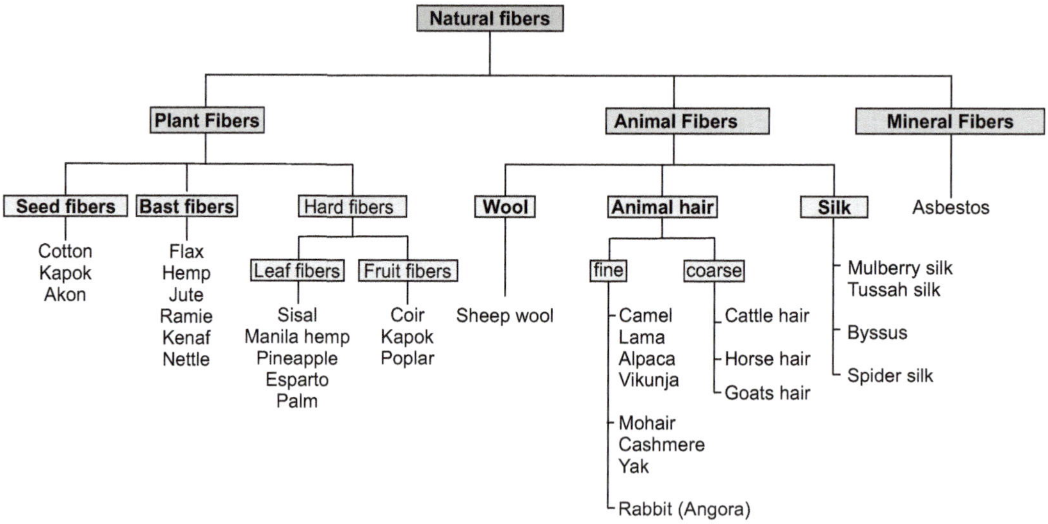

Fig. 1.2 Classification of natural fibers

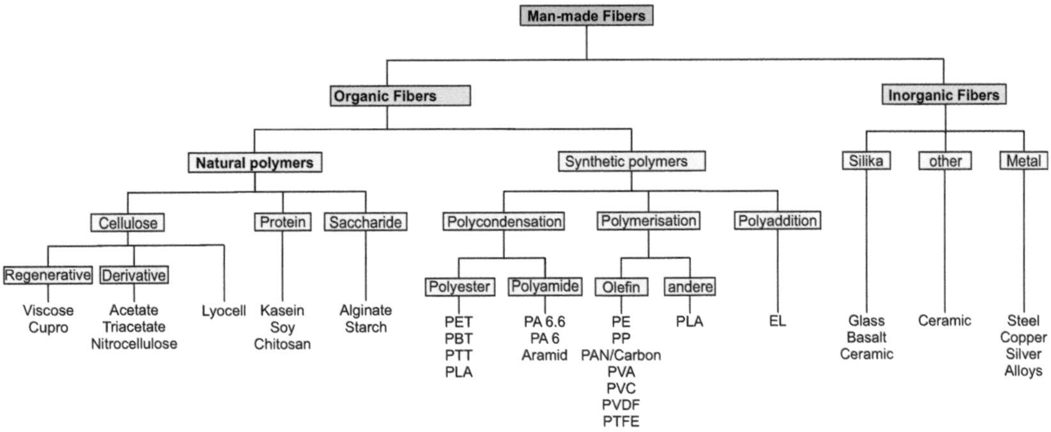

Fig. 1.3 Classification of man-made fibers

1.2.1 Natural Vegetable Fibers

Figure 1.4 shows the systematic classification of selected cellulosic natural fibers.

1.2.2 Animal Natural Fibers

Figure 1.5 shows the classification of important animal fiber producers. With the exception of angora rabbits and silk worms, the animal fiber producers all belong to the order of even-toed ungulates. Sheep can usually be interbred, for example a Corriedale is produced from a combination of Merino and Lincoln. Hybrids can also be produced in goats, for example by crossing cashmere and mohair goats, resulting in cashgora goats, which are used for fiber production.

1.3 Abbreviation

Table 1.1 lists the abbreviations for selected fibers (DIN 60001-3 (1991) et al.).

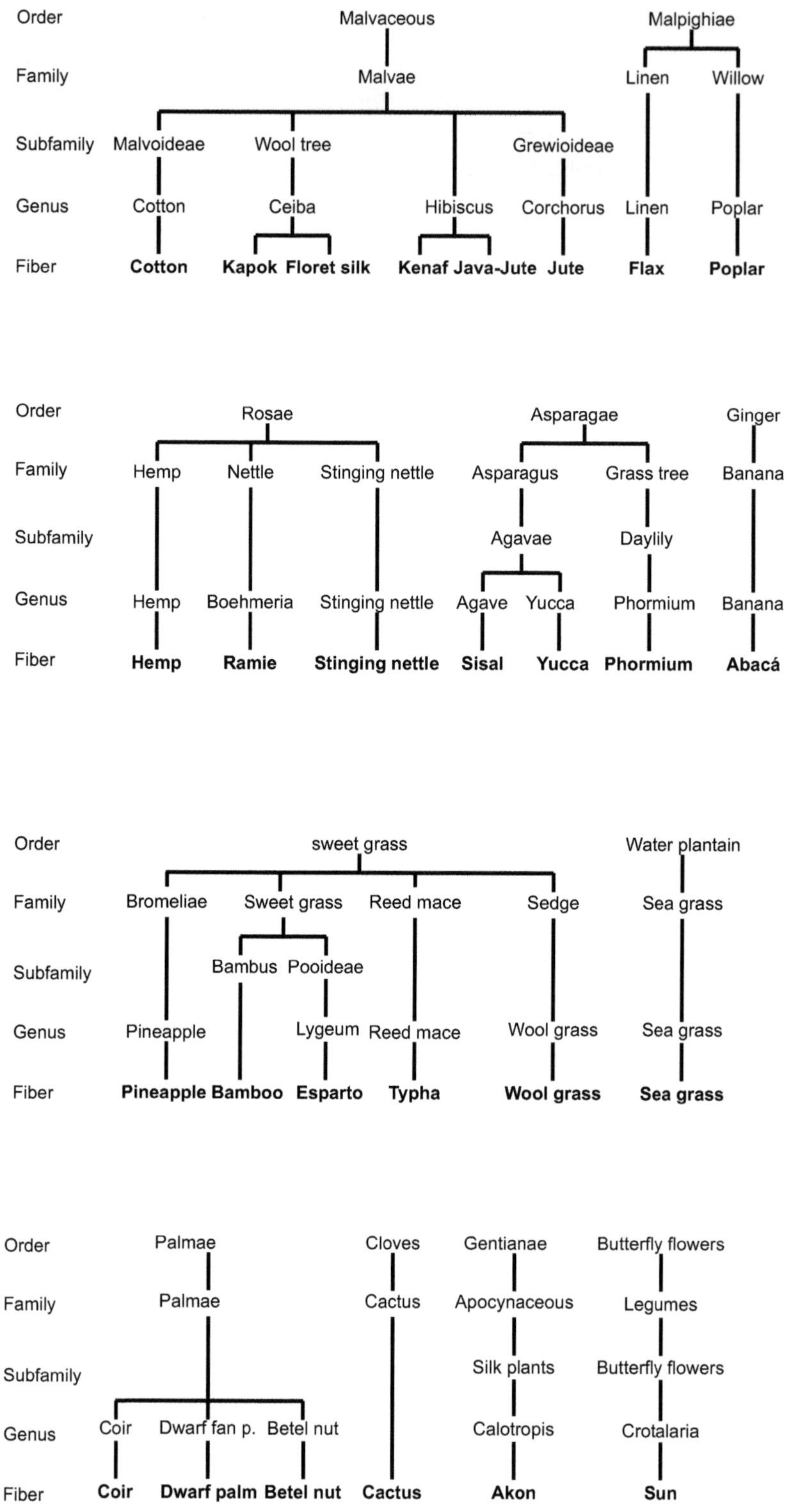

Fig. 1.4 Systematics of selected natural plant fibers

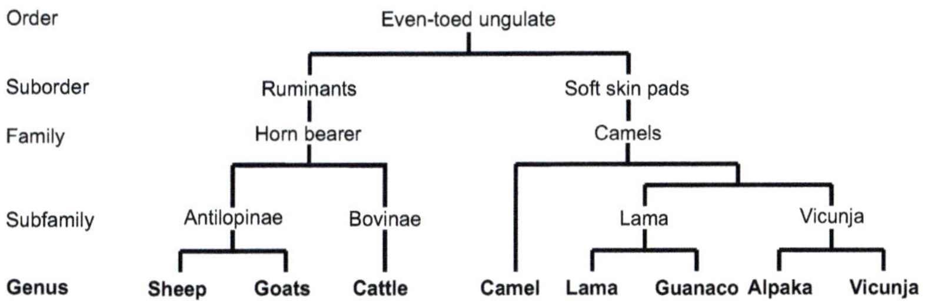

Fig. 1.5 Systematics of selected animal natural fiber producers

Table 1.1 Abbreviations of selected fibers

Seed fibers		**Man-made fibers from natural polymers**	
Cotton	CO	Acetate	CA
Kapok	KP	Alginate	ALG
Bast fibers		Cupro	CUP
Flax, linen	LI	Rubber	LA
Hemp	HE	Modal	CMD
Jute	JU	Triacetate	CTA
Kenaf	KE	Viscose	CV
Ramie	RA	**Man-made fibers from synthetic polymers**	
Rosella	JS	Aramid	AR
Sun	SN	Elastane	EL
Urena	JR	Melanin resin fiber	MF
Hard fibers		Modacrylic	MAC
Abacá (Manila hemp)	AB	Polyacrylic	PAN
Esparto	AL	Polyamide	PAN
Fiqué	FI	Polyazole	PBO
Henequen	HE	Polybenzimidazole	PBI
Coir	CC	Polybutylenterephthalate	PBT
Phormium (New Zealand flax)	NF	Polyester	PES
Sisal	SI	Polyetheretherketone	PEEK
Wool		Polyetherimid	PEI
Wool	WO	Polyetherketone	PEK
Fine animal hair		Polyethylene	PE
Alpaka	WP	Polyethylenterephthalate	PET
Angora	WA	Polyimide	PI
Guanako	WK	Polyoxadiazole	POD
Camel	WN	Polyphenylensulfide	PPS
Cashmere	WS	Polypropylene	PP
Lama	WL	Polysulfon	PSU, PES
Mohair	WM	Polytetrafluorethylene	PTFE
Vicunja	WG	Polytrimethylenterephthalate	PTT
Yak	WY	Polyvinylalcohol	PVA
Coarse animal hair		Polyvinylchloride	PVC
Cattle hair	HR	Polyvinyldichloride	PVDC
Horse hair	HS	Polyvinyldifluoride	PVDF
Goat hair	HZ	**Man-made fibers from inorganic materials**	
Silk		Glass	GF
Mulberry silk	SE	Carbon	CF
Tussah silk	ST	Metal	MTF

References

ASTM D7641-21. (2021). *Standard guide for textile fibers*. ASTM International.

DIN 60000. (1969). *Textilien – Grundbegriffe*. Beuth-Verlag. Retrieved from https://www.beuth.de

DIN 60001-3. (1991). *Textile Faserstoffe – Chemiefasern*. Beuth-Verlag. Retrieved from https://www.beuth.de

DIN EN ISO 6938. (2015). *Textilien – Naturfasern – Gattungsnamen und Definitionen*. Beuth-Verlag. Retrieved from https://www.beuth.de

Oerlikon Barmag. *Remscheid*. Retrieved from https://www.oerlikon.com/polymer-processing/de/

2 History

In this chapter, we look at the most important historical developments in fiber extraction and production. For many fibers, detailed explanations are given in the respective chapters. The history of man-made fiber production is explained in more detail here because many developments build upon each other and took place within a relatively short period of about 70 years. Hence, the description is largely in chronological order.

Figure 2.1 shows the development of the fiber production from 1900 to the present day. Until 1930, almost exclusively natural fibers were produced, the production of cellulosic man-made fibers on a larger industrial scale began around 1950, and from 1960 onwards, synthetic fibers were added. Since 1970, the amount of polyester produced from PET in particular has increased rapidly. In 2000, man-made fiber production overtook the amount of natural fibers produced and today, man-made fibers have a market share of around 70%.

2.1 Natural Fibers

Natural fibers have been used to produce textile structures for thousands of years. These include in particular the bast fibers flax and hemp as well as wool and cotton. Details on the historical significance of the individual natural fibers are covered in the respective chapter (Fig. 2.2).

2.1.1 Flax

Flax was already cultivated more than 6000 years ago in Egypt and in Mesopotamia (today's Iraq) and came to southern Central Europe in the late Stone Age (third millennium BC). Finds of fabrics come from ancient Egyptian graves (e.g. mummy bandages) and from lakeside settlements ("pile dwellings") in Switzerland from the time around 5000–3000 BC. In addition to clothing, belts, cords, strings, lamp wicks and sails were made from flax (Fig. 2.3).

2.1.2 Hemp

Hemp is already known around 1500–1000 BC in China and India. It is first used for the production of intoxicating drinks and medicines from the petals and seeds and only later cultivated as a fiber plant. From India, hemp culture first reached southern Russia via the Scythians (1000–500 B.C.) and from there northwards to the Slavs and Germanic tribes. A southern way led the hemp of the Black Sea-countries over Asia Minor to the Greeks and Romans and is mentioned by (Plinius, 50) as fiber-plant and cure. From there, hemp also reached Gaul and has since been cultivated in northern France and Belgium. In the Middle Ages, clothing made of hemp was often considered a sign of poverty because of its hard textile handle. Hemp's main use was in

D. Veit, *Fibers*, https://doi.org/10.1007/978-3-031-15309-9_2

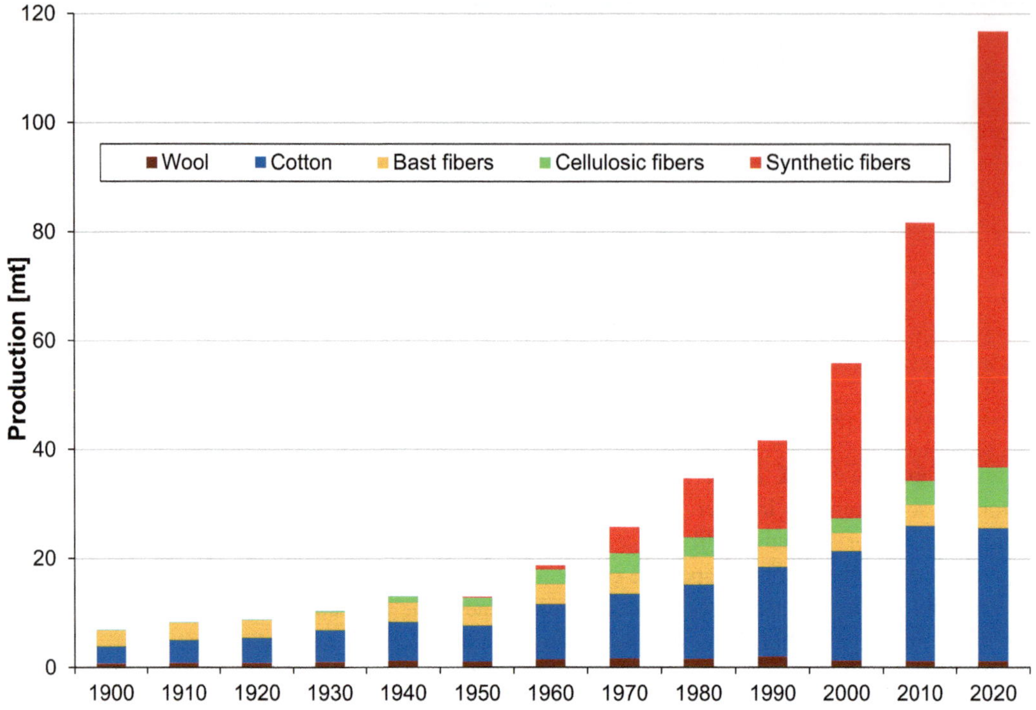

Fig. 2.1 Development of fiber production since 1900 (organic fibers)

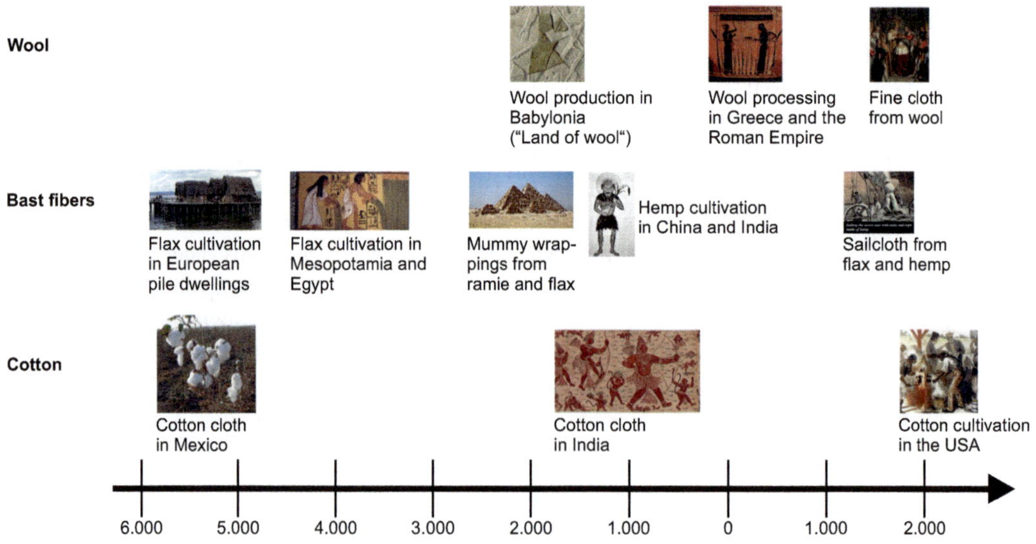

Fig. 2.2 Overview of the history of important natural fibers

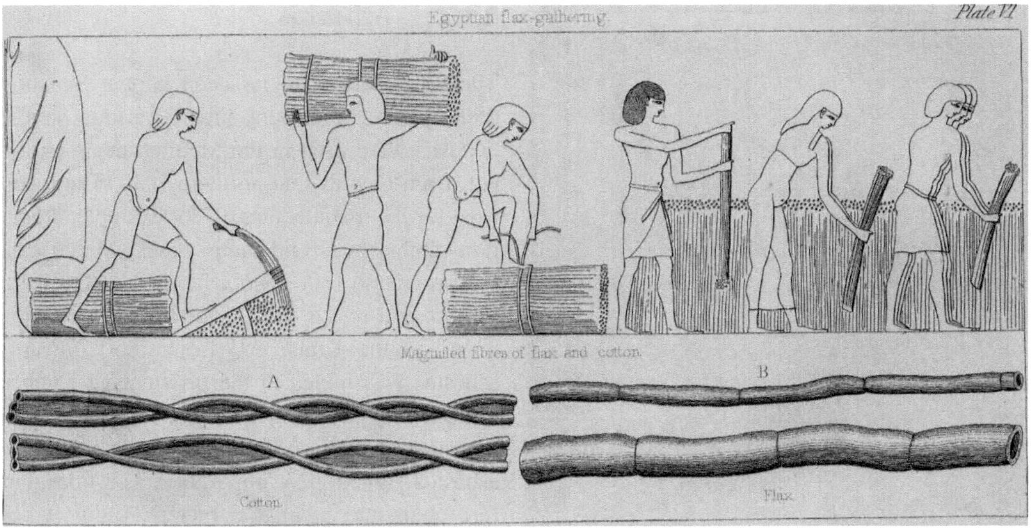

Fig. 2.3 Flax in ancient Egypt (Yates, 1873)

shipbuilding for sails and ropes because of its great wet strength.

2.1.3 Other Bast Fibers

Jute has been grown in India for several thousand years for fibers, as a vegetable crop and for medicinal purposes. Ramie has also been used as a source of fibers for more than 4500 years. The oldest finds come from Egypt (mummy bandages), the first written mention from China (1797 BC, as perennial hemp). At the beginning of the twentieth century, ramie reached Europe from India and was processed into yarns in England.

2.1.4 Cotton

The oldest cotton textiles date back to around 5800 BC and were discovered in Mexico. Textile finds from the third and second millennium BC are also known from Peru. In Pakistan and especially in India, cotton was cultivated on a large scale from 1500 BC onwards. Herodotus mentions cotton fabrics several times in his Histories around 445 BC, including so-called cotton tank tops worn by Indian soldiers. India was the main supplier of cotton fabrics until the sixteenth century. In Europe, cotton was largely unknown before the Middle Ages. In the eighteenth century, the USA were the largest producer of cotton, mainly with the help of slaves from Africa who harvested and ginned the cotton. Since then, cotton has been by far the most important of all natural fibers.

2.1.5 Wool

Sheep's wool is probably the first textile raw material used by humans to make clothing, because the fibers are comparatively easy to obtain. The oldest written reference is on clay tablets from Babylonia ("land of wool") from around 2000 BC. Initially, felts were made from plucked or shorn hair. With the development of spinning and weaving techniques, woven fabrics ("cloth") were also produced. Fine wool was a luxury article in Europe for a long time. White wool in particular was highly prized because it could be dyed well in many shades. Owing to its excellent color absorption, the expression "dyed in the wool" became synonymous with thoroughgoing and uncompromising.

Fig. 2.4 Silk production in Japan in the nineteenth century

2.1.6 Silk

The silk threads of the mulberry silk worm were already obtained in China 5000 years ago. However, the production and wearing of silk fabrics was initially reserved for the emperor and the nobility. Later, silk was exported to Europe via the "Silk Road", first mentioned by Ferdinand v. Richthofen at the end of the nineteenth century. Due to the artificial scarcity of silk on the part of the Chinese and the long distances over which it needed to be transported, it was often weighed in gold. Dyed with purple, silk fabrics were worn by Roman emperors as a sign of their power. Apart from China, silk was mainly produced in Japan and remained a luxury good until modern times (Fig. 2.4).

2.1.7 Asbestos

The use of asbestos has been known for about 2500 years. In ancient times, various smaller deposits were used in the Mediterranean region, e.g. from Cyprus and northern Italy. They were used for the manufacture of, for example, cremation cloths, mats and lamp wicks. Marco Polo reports around 1300 about a fabric from the Tartar Empire that resists fire. Around 1710, extraction began in the Urals, and from 1850 onwards, asbestos was mined in the province of Quebec, Canada. This was also the beginning of industrial asbestos production. In the twentieth century, asbestos gained great importance as a fiber reinforcement and insulation material until it was recognized that asbestos can cause fatal lung diseases. Asbestos has therefore been banned in the EU since 2005, but is still used in other regions of the world.

2.2 Cellulosic Man-Made Fibers

Silk is generally regarded as the model for the manufacture of artificial threads, especially filament yarns. It is believed that as early as 1000 BC, attempts were made in China to artificially produce threads from the secretion of the silkworm by dipping a needle into it.

In 1665, Robert Hooke (1635–1703) was the first historically documented researcher to describe a process for "artificially producing a glue-like mass similar to the substance from which the silk worm spins its balls of thread". He also already formulated the idea of making textiles from threads of an artificially produced liquid, "if only very quick ways of drawing it out into small wires for use could be found" (Hooke, 1665). Due to the insufficient technical possibilities at the time, however, his idea fell into oblivion and was not taken up again until about 200 years later. Therefore, Robert Hooke is usually only known for "Hooke's law", which

Fig. 2.5 Hooke (left) and R.A.F. Réaumur (Pizzetta, 1893)

describes the relationship between stress, strain and Young's modulus.

René-Antoine Ferchau de Réaumur (1683–1757) postulated that silk is a formerly liquid gum, which is formed from it by drying. It should therefore be possible to draw threads from lacquers and process these into fabrics whose sheen and durability resemble silk (De Réaumur, 1734). His idea, too, was not pursued at first, and so he is remembered mostly only for the temperature scale named after him (Fig. 2.5).

In 1839, Anselme Payen (1795–1871) succeeded in isolating cellulose from wood. He gave it its present name. However, the way to the production of artificial cellulose-based fibers was only cleared by the discovery of how cellulose can be dissolved.

2.2.1 Nitrocellulose

In 1846, Christian Friedrich Schönbein (1799–1868) invented a process for producing cellulose nitrate, the starting material for nitrate silk, from nitrated cotton (so-called gun cotton) by dissolving it in alcohol ether. Schönbein's research, however, focused on the use of the solution in bullets. As Paul Schlack remarked at a lecture at the Deutsches Museum in Munich in 1967, Schönbein "was not particularly interested in filaments. To him, banging and shooting were more important" (Klare, 1985).

It was the Swiss chemist George Philippe Audemars who was the first to dissolve cellulose nitrate in alcohol and ether in 1855. He patented his process and proposed a fiber production with a needle and a winding device (Audemars, 1855). However, the process was unsuitable for industrial production. Joseph W. Swan (1828–1914), who received a patent for an incandescent lamp in England in 1878, made lamp filaments from nitrocellulose dissolved in acetic acid in 1883. He called the filaments "artificial silk" (Swan, 1884). His wife made them into home textiles, which he presented to the Society of Chemical Industry in London in 1884 (Fig. 2.6).

From about 1905 onwards, it became apparent that the nitrocellulose process could not compete with the copper and viscose processes, neither economically nor in terms of quality (Becker, 1912).

2.2.2 Process (Copper Process)

Nitrocellulose is explosive and therefore the fiber production was dangerous and the textiles made from it were highly flammable. Hence, an alternative solvent was required. Matthias Eduard Schweizer (1818–1860) succeeded in 1857 in dissolving cellulose obtained from cotton with copper oxide and ammonia (Schweizer's reagent, "Cuoxam"). His wife may also have played a decisive role in this by bringing his cotton laboratory coat accidentally into contact with Cuoxam, thus creating a hole by dissolving the cellulose. In 1881, William Crookes (1832–1919) produced incandescent lamp filaments from

Fig. 2.6 Payen (Wiki, 1870), Schönbein (Hanfstängl, 1857), Swan (Wiki, 1907)

Cuoxam. However, it took until 1890 (Louis Henry Despaissis) and 1892 (Max Fremery and Johann Urban) to develop processes for the production of cellulose threads for textiles. It was possible to work with lower temperatures than with nitrocellulose and the costly denitration step was not necessary because the copper silk produced could be used immediately. The corresponding patent was not registered by Fremery and Urban, but by Hermann Pauly, a chemistry professor in Würzburg. This was presumably to prevent competitors from becoming aware of the new process and to buy time for industrial production. The copper oxide ammonia process therefore later became known as the "Pauly patent". In 1935, cupro fibers accounted for about 9% of the market for cellulosic man-made fibers. IG Farben manufactured them under the brand name Cuprama®. Until the 1990s, such fibers were used for lining materials and for membranes of dialyzers for extracorporeal hemodialysis. Nowadays, cupro has been largely replaced by viscose (for linings) and synthetic fibers.

2.2.3 Cellulose Acetate

Paul Schützenberger (1829–1897) dissolved cotton in acetic anhydride and thus produced cellulose acetate in 1865. Charles F. Cross (1855–1935) and Edward J. Bevan (1856–1921) developed a process for dissolving cellulose acetate in chloroform. This was the starting point for later fiber production. Again, it became apparent that chemists such as Schweizer and Schützenberger were not able or willing to translate their discoveries into industrial fiber production. This was only achieved by engineers after a considerable delay. The need to produce durable and inexpensive filaments for the incandescent lamps developed by Edison in 1880 created a market demand for cellulose filaments that could not be met by conventional natural fibers. In 1904, Artur Eichengrün (1867–1949) succeeded in resaponifying triacetate to secondary acetate, from which filaments could be produced by dry-spinning (Fig. 2.7).

2.2.4 Viscose

In 1892, Charles F. Cross, Edward J. Bevan and Clayton Beadle developed viscose, a spinning solution of cellulose sodium xanthate (Cross et al., 1893). They formed Viscose Syndicate, Ltd., in 1894, which industrialized the patents in the United States, Germany, and France. Courtoulds produced according to these patents in Great Britain. In 1900, yarns and fabrics made from viscose rayon were first shown at the World Exhibition in Paris. About the same time, C. H. Stearn and C. F. Topham in England discovered that viscose requires a "maturing process" before

Fig. 2.7 Schweizer, Schützenberger (Wiki, 1863) and Eichengrün (Wiki, 1900)

it can be spun into usable threads. The degree of polymerization of viscose is then around 200–350. Once the appropriate process conditions had been found (including the addition of weak acids), nothing stood in the way for the industrial production of viscose. As a result, viscose achieved a market share of 88% of all man-made fibers as early as 1935.

Leon Lilienfeld developed a high-strength viscose fiber at the end of the 1920s that had a tenacity of up to 50 cN/tex (compared to "normal" viscose with 11–15 cN/tex). For this purpose, he used highly concentrated sulfuric acid as a precipitation bath, which led to a special internal structure of the viscose fiber, so that a high stretching of up to 100% was possible. However, this process did not catch on due to the dangerous production conditions as well as the very low elongation of the "Lilienfeld silk" and was discontinued after a few years.

2.2.5 Modal

Modal fibers are viscose fibers with a modified internal structure. In their manufacture, the spinning conditions are changed, for example by adding more zinc sulfate to the precipitation bath, in order to increase their strength and reduce elongation by changing the internal structure of the fibers. It reaches less than 15% in the wet state at a tensile load of 22.5 cN/tex. In addition, the fibers have higher abrasion resistance than viscose. Their degree of polymerization is 700–1100.

Polynosic fibers (from French "polymer non-synthétique"), which can be mercerized, belong to the modal fibers. In addition, there are HWM types (high wet modulus), which have only a very low elongation when wet. They are thus similar to cotton fibers. The strength of modal fibers (35–45 cN/tex) is about twice that of viscose fibers. The elongation, on the other hand, is only half at 8–12%. Today, HWM fibers are mainly produced in Europe, polynosic fibers almost exclusively in China.

2.2.6 Industrial Production of Filament Yarns

It was decisive for the industrial production of artificial fibers that it was possible to isolate cellulose not only from cotton but also from lignified fiber cells. Benjamin C. Tilghman (1821–1901) succeeded in doing this on a technical scale in 1866 by digestion with sulfurous acid. From 1875 onwards, cellulose was produced industrially using the sulfite process developed from this by Alexander Mitscherlich (1836–1918).

Count Hilaire de Chardonnet (1839–1924) was granted a patent for the production of artificial silk from nitrocellulose in 1885 (de Chardonnet, 1891). Fabrics made from this

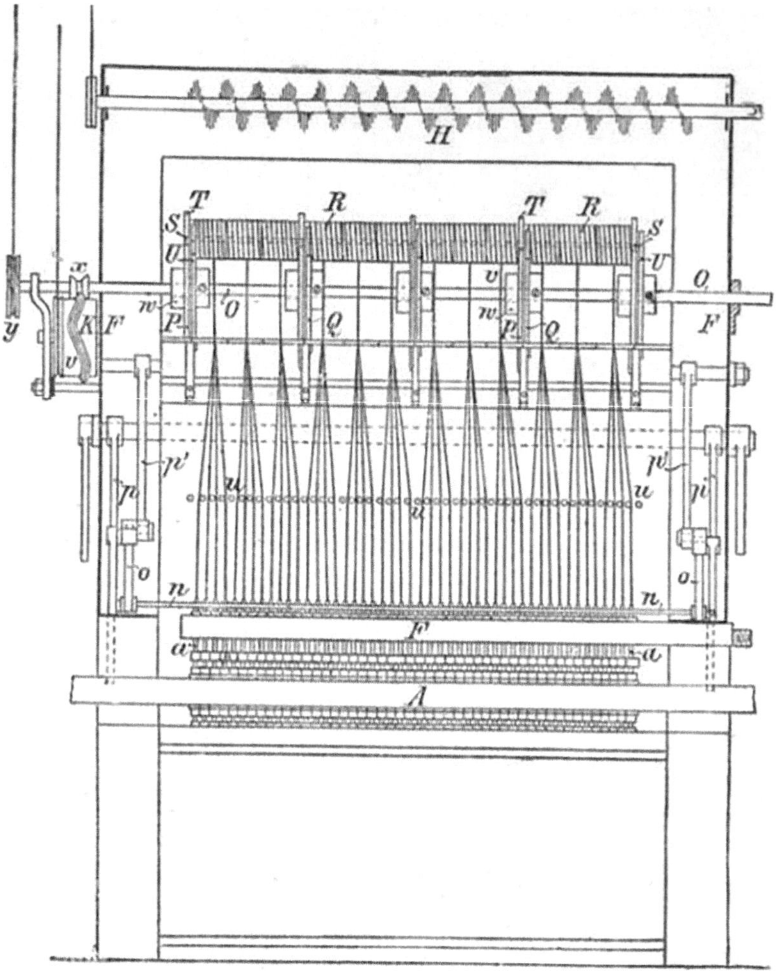

Fig. 2.8 Spinning machine by Count H. de Chardonnet (1890)

and also his artificial silk spinning machine were presented at the World Exhibition in Paris in 1889. The front view of the machine is shown in Fig. 2.8. The spinning direction is from the bottom (A: spinnerets) to the top (T: take-up).

In 1891, he began producing nitrate silk in his first factory in Besançon. He succeeded in reducing the explosive tendency of nitrocellulose by using a new denitrating process. However, as a result, the strength and elongation values also deteriorated. Commercially, his venture was not a success and so Chardonnet died impoverished in 1924, having spent his entire fortune on the industrial production of nitrocellulose. By drawing out the still plastic thread, F. Lehner succeeded in refining it and increasing its strength, and other factories were founded in Europe, including one by Hugo Küttner in Pirna in 1908. But they also were not commercially successful. Küttner switched to the production of viscose as early as 1910 for economic reasons. In 1935, there was no longer any significant production of nitrocellulose in the world (Fig. 2.9).

Table 2.1 shows the development of the world production of nitrocellulose and its market share. It is obvious that although the production volume increased more than 10 times from 1896 to 1928, at the same time its market share fell from 100% to 4.5%. Viscose, on the other hand, quickly became the dominant fiber, with 250,000 t produced as early as 1930.

In 1891 M. Fremery and J. Urban founded the Rheinische Glühlampenfabrik (Rhenish Incandescent Lamp Factory) in Aachen and in 1892

Fig. 2.9 Alexander Mitscherlich (Richter, 2012) and Count Hilaire de Chardonnet (Wiki, 1899)

began manufacturing cellulose filaments for carbon filament incandescent lamps in Oberbruch. These were made from a solution of cellulose in copper oxide ammonia. The company became Vereinigte Glanzstoff-Fabriken AG in 1899 with its headquarters in Aachen, which was moved to Elberfeld, now Wuppertal, in 1901. In 1903 a factory was built in Oberbruch, district of Heinsberg (Fig. 2.10).

This was followed by the founding of Glanzstoff in St. Pölten (1904) and in Liverpool (1908). In 1906 there were already a total of 22 artificial silk factories. These either produced directly using the viscose process or later converted their production to the viscose process (Glanzstoff, 1916). J.P. Bemberg AG in Barmen, founded in 1900, continued to produce artificial copper silk for decades. In 1901, the copper process was significantly improved by the development of draw spinning with the spinning funnel invented by Edmund Thiele (Fig. 2.11). In this process, the thread emerging from the spinneret is drawn by a precipitating bath flowing from top to bottom in the spinning direction. Its application requires a high viscosity of the spinning solution, which is provided by Cuoxam. In the 1920s, "Bemberg stockings" were in high demand because the cupro fibers showed a high tenacity. In the 1940s, cupro fibers were mainly used to make parachutes. After the end of World War II, in which the production facilities were largely destroyed, the production of copper silk was almost completely discontinued.

Asahi Kasai, Japan, has been producing copper silk since 1929, at that time as a cooperation with Vereinigte Glanzstoff-Fabriken AG or Bemberg AG, and to this day under the brand name Bemberg™, making it the only manufacturer in the world with an annual production of 16,000 tonnes. Cotton linters are used as the starting material.

C. F. Topham and the watchmaker Friedrich Eilfeld (1868–1942) invented multi-hole spinnerets made of platinum, which replaced the previous "spinnerets", which consisted of individual glass capillaries and were thus complex and expensive to manufacture. Platinum was required because of its high corrosion resistance. Due to its high material price, alternatives were soon sought and F. Eilfeld then developed

Table 2.1 Production of nitrocellulose (Klare, 1985)

	Nitrocellulose [t]	Market share [%]	Cellulosics total [t]
1896	600	100	600
1909	2400	48	5000, thereof Viskose: 800
1924	4900	7.8	62,790, therof Viskose: 54,000
1928	7040	4.4	158,460

Fig. 2.10 Glanzstoff factory in Oberbruch (1903)

spinnerets made of tantalum, which were successfully used from the 1930s onwards. By the 1940s they already had up to 15,000 capillaries. As early as 1901, C. F. Topham developed a regulating device for the uniform supply of the spinning mass to the nozzle, the forerunner of the spinning pump. Figure 2.12 shows the two processes in use at that time. On the left is the centrifugal spinning process according to Topham, on the right the continuous process according to E. Boos. Here, all process steps from spinning to winding of the washed yarn are combined in a single process. It still forms the basis of viscose spinning today.

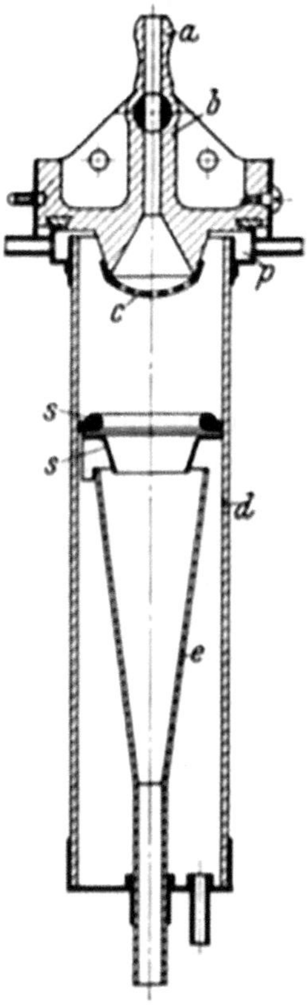

Fig. 2.11 Thiele's spinning funnel

The viscose process was modified several times. As early as 1898, C. H. Stearn obtained a patent for a spinning process whereby viscose was coagulated in solutions of ammonium salts, especially ammonium chloride. Despite all efforts, the mechanical properties of viscose were still insufficient, especially the low wet strength was a major problem. Typical values for copper silk were 13 cN/tex, viscose only reached about 8 cN/tex. Wet strength for both was about 3 cN/tex. In contrast, mulberry silk reached values of 40 cN/tex (dry) and 35 cN/tex (wet). Shrinkage was also still much too high, which made uniform dyeing difficult, among other things.

The patent registered in 1905 by P. Koppe—a spinning bath (Müllerbad I) consisting of sulfuric acid and sodium sulfate and originating from M. Müller—made it possible to produce viscose with improved properties economically. The first viscose factory founded in Germany by Henkel von Donnersmark near Stettin worked according to this process. It was improved in 1912 by V. Hottenroth with the Müllerbad II. Nevertheless, it was not until E. Bronnert succeeded in

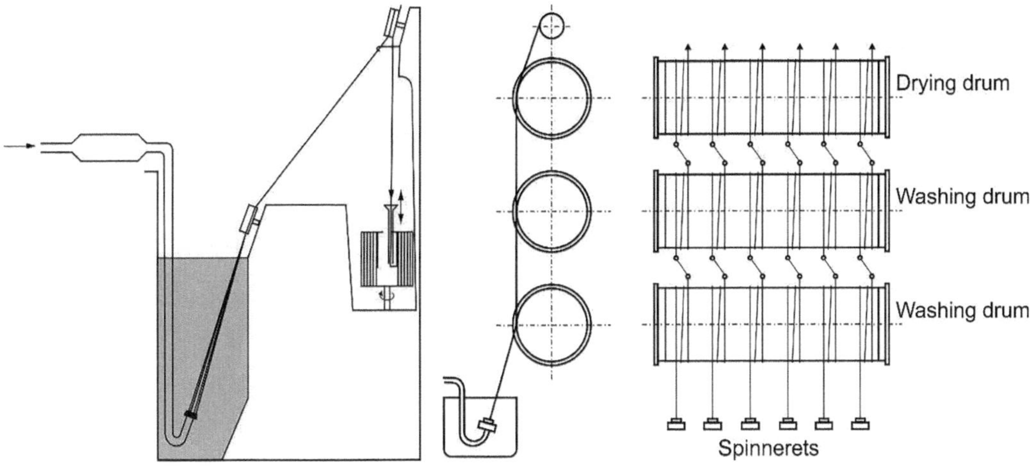

Fig. 2.12 Topham box (centrifugal spinning process) and Boos contouring process

spinning not only coarse fibers (7–12 dtex) but also fine fibers (2–3 dtex) by varying the acid concentration in the spinning bath, which had already been possible for some time with the Cuoxam process using Thiele's spinning funnel.

The production costs of viscose in 1926 were approx. 12 Reichsmark/kg, for nitrocellulose the costs were over 13 Reichsmark/kg. The average hourly wage for factory workers in 1910 was approx. 0.50–0.80 RM. With selling prices of 10–15 RM/kg, the profit margin for viscose was large, but for nitrocellulose it was very small, if it existed at all (Bodenbender, 1929). Consequently, nitrocellulose virtually disappeared from the market in the following period and industrial artificial silk production concentrated on the viscose process. The demand for ladies' stockings had risen sharply as a result of women's knee-length fashions being introduced for the first time in 1925, which led to a rapid increase in production. Thus, the production of artificial silk opened up a broad market, which is reflected, among other things, in the book "Das kunstseidene Mädchen" (1932) by Irmgard Keun. The production of dull artificial silk began on a larger scale in the 1920s, when fashions changed and ultra-bright fibers were no longer in demand. TiO_2 was relatively quickly used as a delustering agent because it was cheap and easy to process. It was now also possible to produce spun-dyed artificial silk.

Despite all the technical successes, the theoretical principles of cellulose spinning were insufficiently known. For reasons of secrecy, little was published in general and there were also only a few patent applications. Thus, most technological advances were based on empiricism. This did not change until the 1920s and 1930s. In 1919, for example, a research institute was founded in Teltow near Berlin that was specifically concerned with the development of artificial silk. In particular, the uneven dyeability was a problem that was only solved after it was recognized that uniform dyeing was crucially dependent on stretching.

In 1937, a special artificial silk (tire cord) was developed for technical applications, e.g. fabric inserts for car tires. From about 1948, the continuous production of viscose fibers also reached operational maturity on a larger scale.

The term artificial silk was followed by the term rayon in the USA in 1924 (rayonne in France) and "rayon" in Germany in 1948. Initially they were used as generic terms for all three types of artificial silk (viscose, copper and acetate rayon), later only for filament yarns produced by the viscose process.

2.2.7 Industrial Production of Staple Fibers

Emil Bronner operated a small experimental plant for the production of staple fibers in 1905. In 1916 a first staple fiber spinning machine went into operation in Oberbruch and in 1917 Vereinigte Bronner-Fabriken AG built a plant in Sydowsaue near Stettin for a production of 10 t/day. These spun fibers were very uneven and stiff because only artificial silk filaments were cut. In Premnitz, Köln-Rottweil AG started production of the viscose staple fiber Vistra® in 1920, which was developed as a staple fiber from the outset. After initial successes, when viscose staple fibers were used as a substitute for cotton, especially during World War I, production declined again because the attainable properties did not meet the customers' expectations. It was only through further research efforts that a competitive viscose staple fiber could be developed. In 1935, the name rayon was introduced for all chemical staple fibers based on cellulose. In 1937, wool-like special types were developed for cloth and clothing fabrics, and in 1938, high-wet-strength types for cotton spinning. This was followed by spun-dyed rayon, special types for upholstery fabrics and carpets as well as coarse, highly crimped core/sheath fibers and sheathless round fibers (Fig. 2.13).

Fig. 2.13 M. Fremery, J. Urban and E. Bronnert (from left)

In 1935, with a total production of 489,000 t, around 87% of viscose was processed into filaments and only 13% into staple fibers. Just 5 years later rayon had overtaken filaments before the ratio was reversed again in 1950 (Fig. 2.14).

Figure 2.15 shows the typical development of man-made fiber production from cellulose, taking Germany as an example, and how cellulosic man-made fibers were increasingly replaced by synthetic ones, especially polyester, from about 1970 onwards. A similar development also took place in the other Western European countries as well as in the USA and Japan.

Today, 6.5 million tons of staple fibers are produced annually compared to only 0.5 million tons of filaments. This is mainly due to the significantly cheaper competition from polyester for filament yarns. Staple fibers, on the other hand, are often processed in blends with cotton and offer dyeing advantages over petroleum-based synthetic fibers due to their similar cellulosic structure.

2.2.8 Lyocell

Lyocell fibers are cellulosic man-made fibers obtained by regenerating the cellulose in fiber form from a solution of cellulose in an organic solvent ("Lyo" derived from Greek: lyein = to dissolve, "Cell" from cellulose). This is a purely physical process in which the cellulose is not chemically altered beforehand. N-methylmorpholine-N-oxide (NMMO) is used as the solvent. The NMMO process is based on the ability of amine oxide to dissolve cellulose under certain conditions. Cellulose can be regenerated from these solutions. A first patent application was filed for it in 1969.

The main problems were initially the high investment costs and the difficulty of recovering the expensive solvent as completely as possible. After this was achieved, the process has been used by Courtaulds plc in Great Britain since 1987 (from 1991 also in the USA) and by

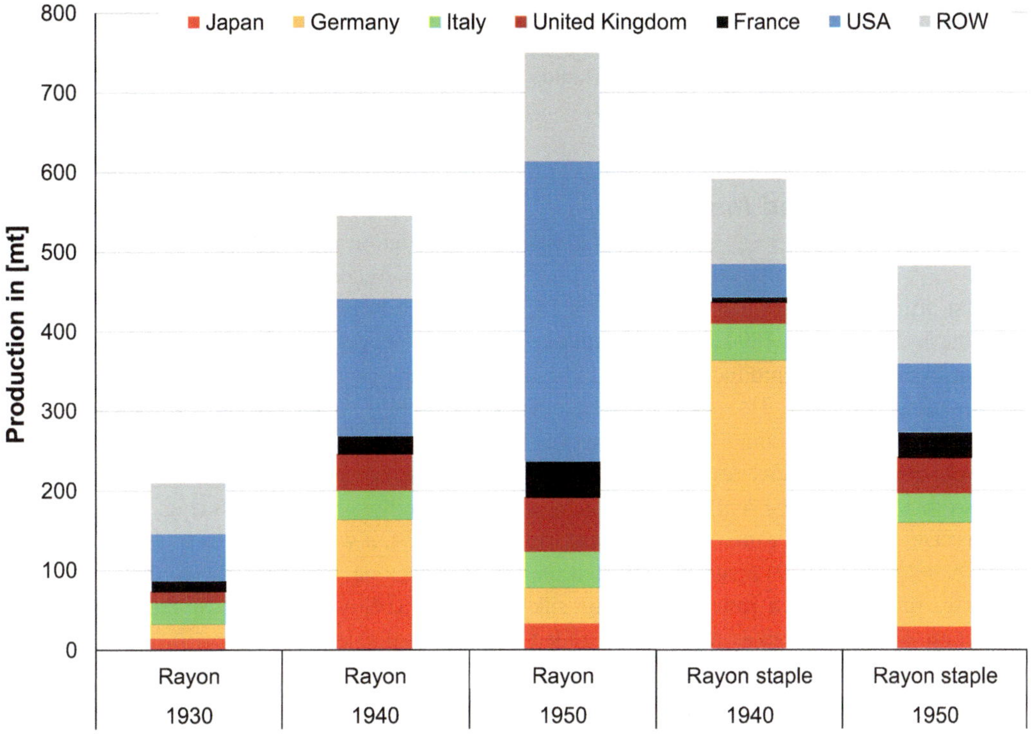

Fig. 2.14 Production of filament yarns (rayon) and staple fibers (rayon staple) from viscose

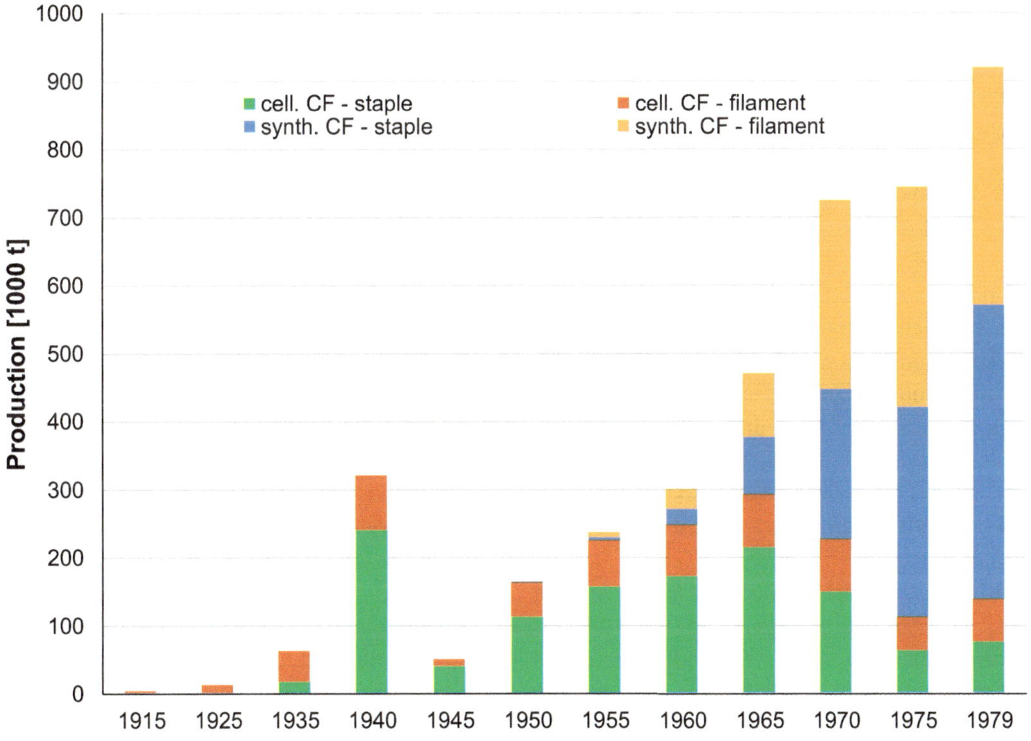

Fig. 2.15 Development of man-made fiber production in Germany acc. to Wagner (1981)

Lenzing AG in Austria since 1997. Lyocell is marketed under the brand name Tencel®, among others, originally the brand name of Courtaulds plc, which was bought by Lenzing in 2004.

2.3 Protein-Based Man-Made Fibers

The first artificial protein fiber was patented by A. Millar in England in 1894. He used gelatine solution from which he produced threads by the dry spinning process. He also carried out experiments with casein and albumin. However, both wet strength and water resistance were very poor, so that the so-called Vandura® silk was a commercial failure. From 1904, Friedrich Todtenhaupt (1873–1919) again investigated the suitability of casein as a raw material for silk production. However, like Millar, he failed because casein does not contain any sulfur, as it is found, for example, in the cystine of wool. The resulting sulfur bridges are apparently crucial for thread formation. Antonio Ferretti's attempts to first extract casein from skimmed milk (about 3 g/l) and produce a fiber, which he marketed under the name Lanital™ (from Lana italiana: Italian wool), also failed. A major problem with his process was the extremely long production time of about 20 h from spinning to the finished dried fiber. R. Wormell continued the experiments and used peanuts, castor and soybean protein, wool keratin and fish protein, among others, as starting materials. He too failed due to the unsuitability of the proteins used for fiber production. The same was experienced by a Polish manufacturer who marketed its fibers under the name Wipolan™. Q-Milk GmbH, Hanover, has been trying again for some years to produce artificial protein fibers.

2.4 Synthetic Man-Made Fibers

In Klare (1985) the history of man-made fiber research from its beginnings to the 1980s is presented in great detail. A good overview of the most important developments in synthetic-based man-made fibers from 1950–2000 is given in Anonymous (2000).

2.4.1 Polyvinyl Chloride

The chemist Fritz Klatte (1880–1934) patented the production of fibers from polyvinyl chloride in 1913. According to many researchers, this marks the beginning of the history of synthetic fibers. However, it was not until 1930 that the first fibers were produced from PVC by Emil Hubert (1887–1945) in the Wolfen film factory using solution spinning. Therefore, the year 1930 could also be called the beginning of the production of fully synthetic fibers (Fig. 2.16).

In 1934, it was possible to spin out PVC fibers wet in acetone. However, due to their low softening point (below 80 °C), the so-called PeCe fibers were unsuitable for clothing. Instead, they were used for fishing nets, acid-proof clothing and filter cloths. They were also used as material for rheuma blankets. They were also processed in the form of dyed bristles, as Fig. 2.17 shows.

2.4.2 Polyamide

The history of polyamide fibers begins in the early twentieth century. In 1899, 1902 and 1904, S. Gabriel, Th. M. Maass, A. Manasse and J. von Braun worked on the ring formation of ω-aminocaproic acid, 20–30% of which could be converted into caprolactam, the basic material for polyamide 6. However, the importance for industrial use was not recognised at that time.

In 1929, Wallace Hume Carothers began working at the chemical company DuPont de Nemours & Co., USA, on polycondensation and ring formation of bifunctional compounds. He was familiar with Hermann Staudinger's work on the theory of macromolecules from the second half of the 1920s and intended to prove his findings by synthesizing very large molecules. Carothers found that polycondensation of a variety of bifunctional compounds produced meltable, filament-forming polymers that had been

Fig. 2.16 F. Klatte, E. Hubert (Wiki, 1944) and advertising brochure for PeCe fibers from VEB Filmfabrik AGFA in Wolfen (PeCe, 1934)

unknown until then. With the publication of "An Introduction of the General Theory of Condensation Polymers" by Carothers in August 1929, a new chapter in man-made fiber research began (Carothers, 1929).

In 1932, Carothers and J.W. Hill found that molecular distillation of molten polyesters produces so-called superpolyesters from which threads can be drawn that solidify in air. Thus the principle for the production of synthetic fibrous materials was found, even if the first superpolyesters were unsuitable for textile purposes due to their low melting point of less than 100 °C.

2.4.2.1 Polyamide 66 and Polyamide 6

As early as 1930, Carothers succeeded in producing a polyamide from ε-aminocaproic acid. However, he apparently did not attempt to produce filaments from it. On February 28, 1935, he succeeded in synthesizing polyamide from hexamethylene diamine and adipic acid. When Carothers succeeded in significantly increasing the strength of the polyamide by cold drawing, this was the breakthrough for the production of synthetic fibers. Due to the practically unlimited raw material base (benzene and phenol), this polyamide was selected for further development on an industrial scale and a patent application was

Fig. 2.17 Sample box with bristles made of PeCe fibers from Filmfabrik Wolfen

filed on April 9, 1937. It became known as PA 6.6 or PA 66. DuPont first published the results of their research in 1938, and small-scale production began that same year. Carothers did not live to see this. After the further development of PA 66 had been transferred to another research group at DuPont, he turned his attention to other polyamides. It turned out that PA 66, which had originally been chosen rather at random, was by far the most suitable of all the polyamides. In 1937, Carothers committed suicide aged only 41.

Without knowing anything about Carothers' work, Paul Schlack had been working on the production of a synthetic fiber, including the polymerization of caprolactam, at IG Farben AG in Berlin since 1928. Previously, Schlack had tried to procure the basic materials for the production of PA 66. However, he was unable to do so because of the poor raw material situation in Germany. In his work published in 1930, Carothers had ruled out the polymerization of ε-caprolactam with or without a catalyst. His experiments had only resulted in a polyamide with a low degree of polymerization, from which no filaments could be spun. Schlack, however, continued to work in this field. On 29 January 1938, he succeeded in splitting the lactam ring using aminocaproic acid hydrochloride as a catalyst and converting it into a linear polymer. This was the birthday of polyamide 6 (Fig. 2.18).

A patent application for the caprolactam process was filed on June 11, 1938 (Schlack, 1938). The first stocking made of PA 6 was produced in the same year, initially under the name "Perluran" (Fig. 2.19).

In July 1938, a delegation from IG Farben set out for the United States to demonstrate the new fiber to Dupont and to negotiate licenses. Already under way on the Atlantic, the delegation learned that a DuPont delegation was on its way to Germany to visit IG Farben and negotiate licenses for the production of a new synthetic fiber. To their dismay, they found out that IG Farben was not interested in the new synthetic fiber polyamide 66, but had its own polyamide 6, which was based on a different monomer and had not been patented by DuPont.

Fig. 2.18 Polyamide spinning machine by Paul Schlack (1938)

An amicable agreement was then reached, the respective polyamide patents were exchanged and the worldwide sales markets were defined. Eastern Europe and Asia went to IG Farben, while Western Europe and America went to DuPont. This division of the world market can still be seen today: In Eastern Europe and Asia, mainly PA 6 and only very little PA 66 is produced, while in the USA and South America, mainly PA 66 is manufactured.

The fact that both companies knew nothing about the competitor's development was due to the fact that the results of their research departments were practically not published. Patents were only filed for the essential process steps, but the other company apparently had no knowledge of this.

In 1939, DuPont began the industrial production of Nylon™, as the fiber was now called. The first nylon stockings came onto the market in the

Fig. 2.19 W.H. Carothers (left), P. Schlack (center) and the first PA 6 fibers produced by Paul Schlack (right)

USA in 1940. However, production soon focused exclusively on military applications, such as parachute fabrics and high-strength threads as reinforcement for tires (tire cord). It was not until after 1945 that it was expanded for civilian and industrial applications. By the end of the war, the production of Nylon™ already amounted to 15,000 t/year.

The coveted nylon stockings quickly became a means of payment for American soldiers in Germany and soon spread to Europe as well (Fig. 2.20).

IG Farben began producing PA 6 in 1939, and toward the end of the year the first monofilaments came on the market under the name Perluran, later designated Perlon L. Large-scale production of perlon filament yarn began in 1943 at the Landsberg a. d. Warthe plant. The plant under construction in Premnitz (Westhavelland) was unable to start its planned large-scale production of perlon filament yarn because of the war. In 1946, the production of PA 6 filament yarn was resumed at the VEB Thüringisches Kunstfaserwerk "Wilhelm Pieck" plant in Schwarza near Rudolstadt, after smaller quantities had already been produced there in 1945. In 1949, production of perlon filament yarns began at the Bobingen artificial silk factory near Augsburg, followed in 1950 by the Oberbruch Vereinigte Glanzstoff plant.

By 1953, the West German man-made fiber industry had the trade name Perlon® protected, so the East German polyamide manufacturers were forced to create a new name: Dederon® (after "DDR", the German abbreviation of the "German Democratic Republic"). This product, based on PA 6, was successful in eastern Germany until German reunification in 1990.

From the 1950s onwards, outerwear was also made from polyamide instead of cotton. The so-called "Nyltest" shirt became particularly famous. It was made of 100% PA 66 and was the first easy-care outer shirt. The water absorption of PA is similar to that of cotton, but polyamide fibers store the absorbed moisture. This leads to the wearer of such a shirt starting to sweat very quickly and also very strongly and then expiring bad smells after some time. Therefore, this article was not very successful in the market did in fact define the bad image of synthetic textiles for a long time (Fig. 2.21).

2.4.2.2 Discovery of Further Polyamide Types

Another method of obtaining a polyamide was found by W. Zeltner and M. Genas. This polyamide 11 is prepared on the basis of castor oil, from which the 11-aminoundecanoic acid is synthesized. The patent for this was filed on April 21, 1944. Initial trials were carried out in

Fig. 2.20 Queue in front of a nylon stockings store in the USA (left) and happy consumer (right)

France in the mid-1940s, and large-scale production began in 1955. This fiber was marketed under the brand name Rilsan™ (Genas, 1962).

The production of polyamide 7 as an alternative to PA 6 is carried out in different ways. In the USA, 7-aminoheptanoic acid was obtained from caprolactone with subsequent conversion to methyl 7-aminoheptanoate. In the former USSR, in 1951–1955, A. A. Strepicheev and R. CH. Freidlina used a different method. It was based on telomerization, similar to polyamide 11. The starting material is ω-aminoenanthic acid. However, the process was discontinued, despite better properties of PA 7 compared to PA 6, due to inefficiency and lack of raw materials.

Polyamide 12, which is already mentioned in the basic patents of Carothers and Schlack, consists of 12-aminododecanoic acid or λ-laurinlactam. The process for polymerizing laurinlactam was developed by W. Griehl and S. Schaaf and patented in Switzerland in 1966 and 1967, respectively. Polyamide 12 was produced as a monofilament under the brand name Swiss Polyamide Grilon-Monofil D, DN, DS. The monomer is also used for adhesive fibers in copolyamides (Swiss Polyamide Grilon fibers K 115 and K 140).

Fig. 2.21 Stockings production on cotton knitting machines (left) and advertisement for Nyltest shirt (right; translation: His shirt: unsurpassed, easy-care)

Despite partly more favorable fiber properties than other polyamides, attempts to produce fibers from polyamides 3, 4, 8 and 9 did not bring the desired industrial success. The production of PA 3 from β-lactam by wet spinning and of PA 4 fibers from polymerized γ-butyrolactam by the dry spinning process were also unsuccessful. In the former USSR, a fiber called Perlagon, a PA 9 type, was temporarily on the market. However, production was later discontinued (Zaremba et al., 1997).

2.4.2.3 Aramid

The first aramid fibers were developed in 1955–1965 (DuPont, Monsanto). The name is the abbreviation for aromatic amides. H. Mark found in 1964 that the strength and modulus of elasticity of aliphatic polyamides can be specifically increased if aromatic components are inserted into the chain. Starting in 1967, DuPont marketed its m-aramid fiber, first produced as early as 1962, under the trade name Nomex™. Monsanto and companies in the USSR were now also producing such fibers. After Stephanie Kwolek had succeeded in synthesizing p-aramid for the first time in 1966, DuPont began producing p-aramid on an industrial scale under the name Kevlar™ in 1971 (Fig. 2.22).

In Japan, Teijin began producing Conex™, an m-aramid, in 1972, followed by Enka Glanzstoff BV in Arnhem in 1975 with the production of Twaron™, a p-aramid. The first important applications for m-aramids were, in particular, high-temperature resistant textiles, while p-aramids were mainly used to reinforce tire cord and other rubber products.

Fig. 2.22 Stephanie Kwolek (1923–2014) (Wiki, 1986)

2.4.3 Polyester

The preparation of linear, filament-forming polyesters by reaction of aliphatic dicarboxylic acids and glycols had already been studied by W. H. Carothers as part of his work on polymers (and on polycondensation reactions in particular) which he begun in 1928. The first synthetic man-made fiber shown by Carothers and Hill in 1931 was a polyester thread made from sebacic acid and ethylene glycol. Because of the low melting point of this polyester and its solubility in many organic solvents, which he considered a disadvantage, he turned to polyamides, which seemed more promising for commercial use.

Building on Carothers' research, Schlack and Ludewig realized in 1939 that terephthalic acid increases the melting point of the polymer as well as the stiffness and elasticity of the yarn. However, they initially used it for polyamide production. In 1942, Schlack condensed terephthalic acid dihalides and terephthalic acid esters with butylene glycol to form a primary polyester and further condensed this with diisocyanates to form nitrogen-containing polyesters. These have high melting points and can be hot-drawn to three to four times their initial length under orientation.

2.4.3.1 Polyester Terephthalate (PET)

Independently of this, at almost the same time in 1939, the English researcher John R. Whinfield (1902–1966), together with his collaborator James T. Dickson, discovered that when the aliphatic dicarboxylic acid, which had been regarded as unsuitable by Carothers, was replaced by an aromatic dicarboxylic acid, the solubility of the polyester in organic solvents was considerably reduced and its melting point substantially increased (Whinfield & Dickson, 1941). The polyester they discovered from terephthalic acid and ethylene glycol was given the name "terylene". From this polyester, in 1941, Whinfield and Dickson produced usable polyester filaments at the research laboratory of the Calico Printers Association, Ltd. (C.P.A.), in Accrington, England, for the first time. In 1944, laboratory-scale production of polyester filaments was carried out by Imperial Chemical Industries, Ltd. (ICI), produced terylene yarns. In 1947, C.P.A. and ICI reached an agreement on patent rights. ICI acquired the patent rights for terylene outside the United States. Trial production of terylene yarn was initiated in 1947, and large-scale production began in 1953.

The patent rights for the United States were acquired from the C.P.A. in 1946 by E. I. DuPont de Nemours & Co. of Wilmington, USA. C.P.A. had already conducted work with polyterephthalate independently of ICI in 1944. Beginning in 1950, DuPont de Nemours produced small quantities of filament yarns and staple fibers under the name "Fiber V." In 1953, it began production under the trade name "Dacron". In Germany, polyester production began in the mid-1950s under the trade names "Trevira" (Hoechst) and "Diolen" (Akzo).

ICI, for its part, had granted production licenses for polyester fibers using its proprietary process to fiber manufacturers in other countries. From 1954 onwards, these began to produce under various brand names. The basic patents for the production of polyester fibers expired at the end of 1966. Shortly afterwards, large-scale production of polyester began worldwide (Fig. 2.23).

PET has also been used as a base material for beverage bottles since the early 1970s. This has now become a rapidly growing market, with PET replacing glass more and more.

The production volume of PET fibers exceeded that of cotton for the first time in 1992 and PET is now by far the dominant fiber on the world market. Recently, the production of PET microfibers has also gained importance.

Fig. 2.23 John R. Whinfield and James T. Dickson

2.4.3.2 Polytrimethylene Terephthalate (PTT) and Polybutylene Terephthalate (PBT)

Since the 1990s, polyesters made from terephthalic acid and diols other than ethylene glycol have been investigated for their suitability for fiber production. Their production costs are about 20% higher than those of PET due to the more expensive raw materials, but they are only about 90% of the production costs of polyamide 6 and polyamide 66. PTT in particular competes with these fibers due to its high elasticity, its good dyeability already below 100 °C and its high recovery capacity, which is of great importance especially for carpet yarns. PBT made from terephthalic acid and butylene glycol has so far remained a niche product.

2.4.3.3 Other Polyester

In addition to the polyester made from terephthalic acid and ethylene glycol, a polyester has also been developed whose polycondensation uses an aromatic diol instead of ethylene glycol (1,4-dimethylol-cyclohexane). The fibers produced from this ('Kodel' or 'Vestan') have somewhat different physical characteristics.

In order to improve certain properties of polyester fibers—in particular their only moderately good dyeability and the susceptibility of textiles made from staple fiber yarns to pilling—chemical and physical modifications were made which significantly altered the properties of these 'second and third generation' polyester fibers compared with those of the standard types. In the case of chemically modified polyester fibers, in addition to the terephthalic acid, a small proportion of a third or fourth component, e.g. isophthalic acid or p-hydroxybenzoic acid, is incorporated into the polyethylene glycol terephthalate chain. Physical modifications are carried out today by cross-sectional profiling or spinning of bicomponent fibers. Other new developments led to the polyester ether fiber.

Fig. 2.24 Herbert Rein (1899–1955) (Erkner, 2021)

2.4.4 Polyacrylonitrile

In 1913, the German chemist Fritz Klatte proposed the use of synthetic raw materials for the production of fibers. His work on the polymerization of vinyl compounds resulted in a patent specification in which he mentions, among other things, the processability of the resulting "horn-like mass" into threads (Klatte, 1913).

For a long time, acrylonitrile was considered technically insignificant because it decomposed before reaching melting temperature and was insoluble in common solvents. In 1929, H. Fikentscher suggested using acrylonitrile for fiber production (Fikentscher and Heuck, 1937). The first patents for the production of polyacrylonitrile date from 1930. In 1939, Otto Bayer and Peter Kurtz developed the direct synthesis of acrylonitrile from hydrocyanic acid and acetylene, which made industrial production possible, but was still uneconomical at first because of the expensive solvents.

In 1939, IG Farben made the first attempts to produce filaments from polyacrylonitrile under the brand name "PAN-Fasern". In 1941, H. Rein and R.C. Houtz almost simultaneously discovered dimethylformamide (DMF), the most economically suitable solvent. In the USA, DuPont produced a dry-spun polyacrylonitrile fiber under the name Orlon™ in 1950. In Germany, the first industrially produced polyacrylonitrile fibers came on the market in 1952 under the brand names "PAN" and "Redon". In Japan, PAN fibers were also produced by a different process and marketed as Sinsen™ (Fig. 2.24).

In the 1950s and 1960s, the production of polyacrylonitrile fibers boomed. A first fiber crisis in the mid-1970s, caused by the oil shortage, and a second one at the beginning of the 1980s due to the change in consumer buying behavior, which turned to natural fibers and away from man-made fibers, had a serious impact on the polyacrylonitrile market. Polyacrylonitrile fiber production in Europe and North America declined, while production in Asia increased. Nowadays, PAN is almost exclusively used to produce staple fibers, 10% of which are still manufactured in Germany.

2.4.5 Polypropylene

Ethylene, as a representative of the polyolefins, was long considered difficult and only polymerizable under high pressure. In 1953, Karl Ziegler (1898–1973) developed a method for polymerizing ethylene at low temperatures and under normal pressure using transition metal catalysts. The resulting polyethylene was highly crystalline, similar to polyamide, in contrast to the radically polymerized, highly branched high-pressure polyethylene (HD-PE). This discovery laid the foundation for the industrial production of polypropylene fibers. Giulio Natta (1903–1979) succeeded in 1954 in producing the first isotactic polypropylene in laboratory scale. The monofilaments spun by extrusion and subsequent drawing had a very high tenacity of 75 cN/tex. Natta and Ziegler were awarded the Nobel Prize for their work in 1963 (Fig. 2.25).

Industrial production of polypropylene began in Italy as early as 1957. Shortly after its market launch, this new fiber established itself alongside the other man-made fibers produced industrially at the time (polyester, polyamide and polyacrylonitrile). An even faster upswing of the new fiber material was prevented on the one hand by the poor dyeing properties and on the other hand by the patent situation, because the Italian Montecatini group owned almost all basic patents in Europe for melt spinning of polypropylene. Until the early 1970s, there was hardly any possibility for other companies to produce polypropylene fibers on an industrial scale. It was only after the first patents had expired that other manufacturers started production. Due to the low production costs, the market for these fibers expanded very quickly and PP is now the third most important synthetic man-made fiber in terms of volume.

2.4.6 Polyethylene

Polyethylene (PE) was discovered in 1898 by the chemist Hans von Pechmann and first produced industrially in 1933 in the ICI laboratories in England under a pressure of approx. 1400 bar and at a temperature of 170 °C. The first polyethylene was produced in the USA. It was not until 1940 that an economical manufacturing process was developed. With the aid of Ziegler-Natta and metallocene catalysts, PE can also be produced at normal pressure.

One type that is particularly important for technical textiles is UHMW-PE. (UHMW: ultra high molecular weight), which is marketed under the trade names Dyneema™ by DSM (Netherlands) and Spectra™ by Honeywell (USA). It was first synthesized by Albert Pennings in 1968 and patented in 1979. Its production in industrial scale began in 1990.

Fig. 2.25 Karl Ziegler (left) and Giulio Natta (right)

2.4.7 Polylactide

Polylactide (PLA) is produced from lactic acid. Dilactide was produced as early as 1833 (Gay-Lussac) and 1893 (Bischoff and Walden) before Carothers succeeded in producing a polymer from lactic acid for the first time in 1932. Since 1954 DuPont used PLA fibers as suture material and for implants. In Japan, Shimadzu Corp. and others produced small quantities of consumer goods from PLA. Because PLA can be hydrolytically decomposed, it has been used as a binder for cellulosic nonwovens since 1992. Today, NatureWorks LLC in the USA and Toray Ind., Inc., (Japan) are the largest producers of PLA.

2.4.8 Short Time Travel

The following figures show the above in compact form. The industrial production of man-made fibers on a cellulosic basis began in the 1940s, mainly in Europe. In the USA, there were proprietary developments and subsidiaries of European companies, in Japan a separate path was taken with proprietary technologies (Fig. 2.26).

Due to the destruction in Japan during World War II, production there shrank considerably. The same applies to some countries in Europe, where production declined in Germany but increased in France and England (Fig. 2.27).

In the 1960s, the production of synthetic fibers began in the USA, Europe and Japan (Fig. 2.28). At the same time, the volume of cellulosic man-made fibers increased, with Europe taking a leading role. Viscose and related fibers were also produced in large quantities in Eastern Europe. There are gaps in the data for this region, which is reflected in some figures by their exclusion.

Until the 1970s, the production capacity of synthetic fibers was massively expanded and the USA, Europe and Japan shared the world market. At the same time, the amount of man-made cellulosic fibers produced continued to increase (Fig. 2.29).

Until 1980, the production of synthetic man-made fibers continued to grow strongly, especially in the USA and also in Europe. Due to lower demand, the amount of cellulose fibers produced decreased in many countries (Fig. 2.30).

This development continued until the beginning of the 1990s, when China started producing mainly synthetic man-made fibers (Fig. 2.31). In India, large-scale production of cellulosic man-made fibers is started.

In the 1990s and especially the 2000s, production capacities are massively expanded, especially in Asia. China is the largest producer, but large

Fig. 2.26 Main producing countries of man-made fibers 1940 (data in [1000 t])

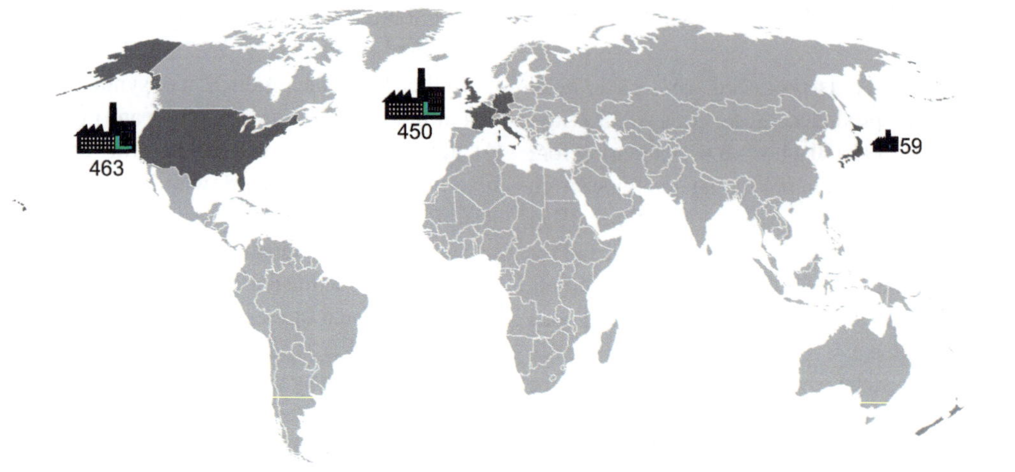

Fig. 2.27 Main producing countries of man-made fibers 1950

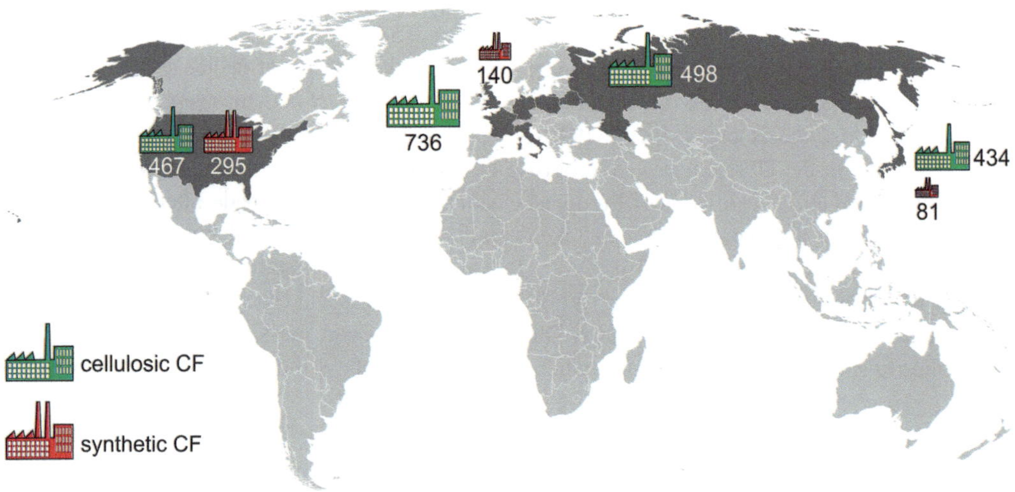

Fig. 2.28 Main producing countries of man-made fibers 1960

production capacities are also being built up in South Korea and Taiwan. In contrast, the quantity produced in Japan, a high-wage country, is declining (Fig. 2.32).

As of 2010, China dominates the world market for synthetic fibers and achieves a market share of 60% (Fig. 2.33). India is also massively expanding its capacities and becomes No. 2 worldwide. The demand for cellulosic fibers increases strongly again and capacities are expanded accordingly, especially in China, Europe and the USA. In Japan, the production of synthetic fibers is being massively reduced.

From 2010 onwards, production facilities are greatly expanded, primarily in China, and to a lesser extent in India (Fig. 2.34). Indonesia rises to become the fourth largest man-made fiber producer in the world. Today, 70% of synthetic fibers and 60% of cellulosic man-made fibers are produced in China. The USA, Europe and South

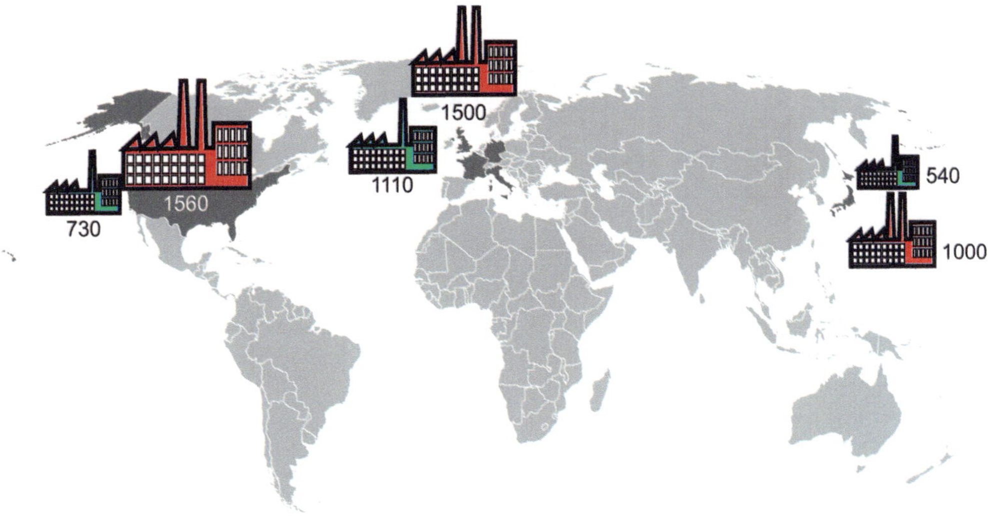

Fig. 2.29 Main producing countries of man-made fibers 1970

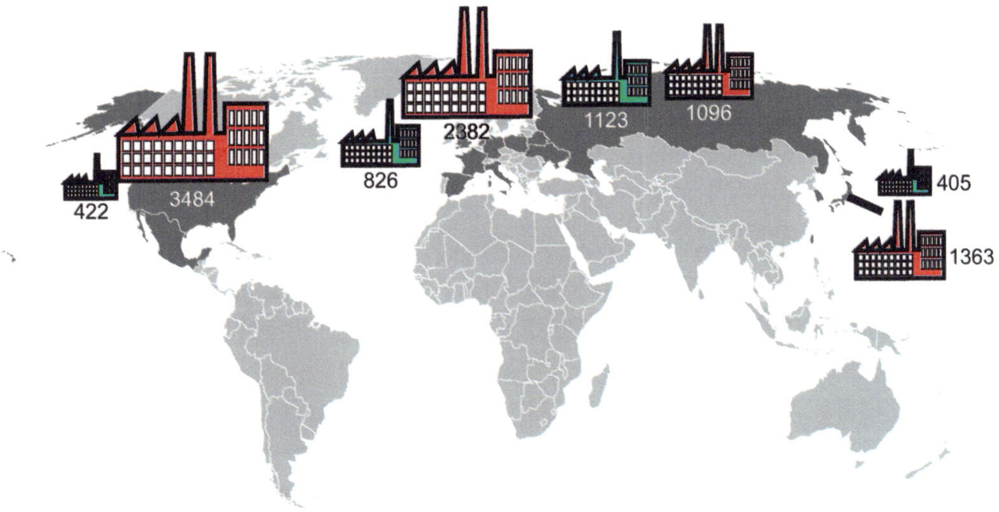

Fig. 2.30 Main producing countries of man-made fibers 1980

Korea as well as Japan continue to be important man-made fiber producing countries, especially for technical fibers.

2.4.9 Carbon Fibers

The history of carbon fiber begins in 1860 when J. W. Swan, the inventor of "artificial silk" develops carbon fibers for incandescent lamps. In 1879, Thomas A. Edison filed a patent for the production of carbon filaments for electric light bulbs. Both used cellulose as the starting material. These incandescent lamp wires were soon replaced by tungsten and so carbon fibers were largely forgotten for the next 50 years.

In 1958, William T. Soltes received a patent for the production of carbon fibers from cotton (Soltes, 1961), William F. Abott also filed a patent for carbon fibers from viscose a year later (Abott, 1962). In 1964, the American company Union Carbide launched cellulose-based carbon

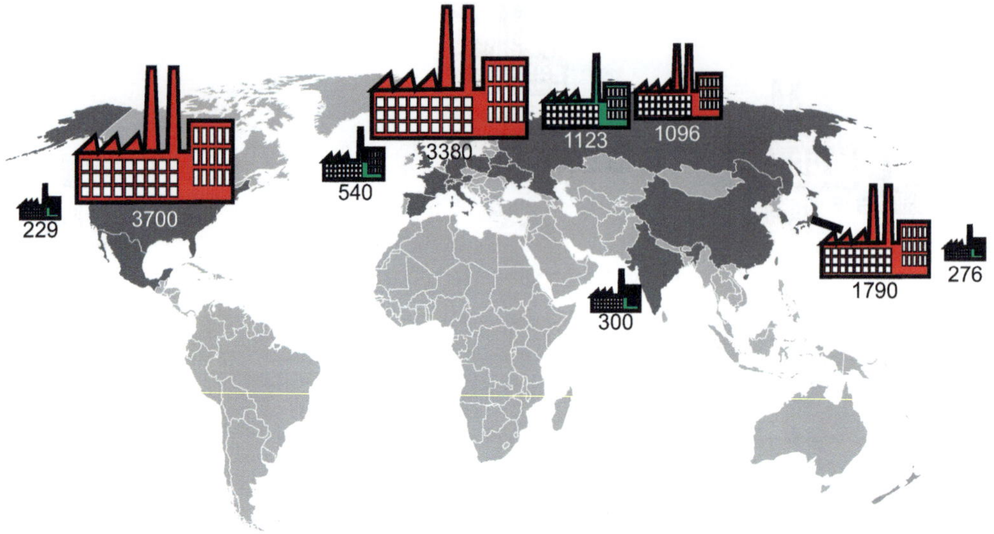

Fig. 2.31 Main producing countries of man-made fibers 1990

fibers on the market under the trade name Thornel™. However, carbon fibers made of cellulose were not able to establish themselves on the market.

In 1965, William Watt and others at Royal Aircraft Establishment (RAE) in Farnborough, UK, succeed in producing carbon fibers from polyacrylonitrile (Watt et al., 1971). In 1966, the process patented by RAE is licensed to a consortium of three UK companies to produce carbon fibers on an industrial scale. In 1967, 1 t of carbon fibers is thus produced. The fibers are given to various companies free of charge on the condition that they identify applications for them. In the USA, the production of carbon fibers according to this patent begins in 1971.

S. Otani began researching pitch as a starting material for carbon fibers in 1965, and in 1970 the first carbon fiber based on polyvinyl chloride pitch (PVC) is launched by Kureha Corporation,

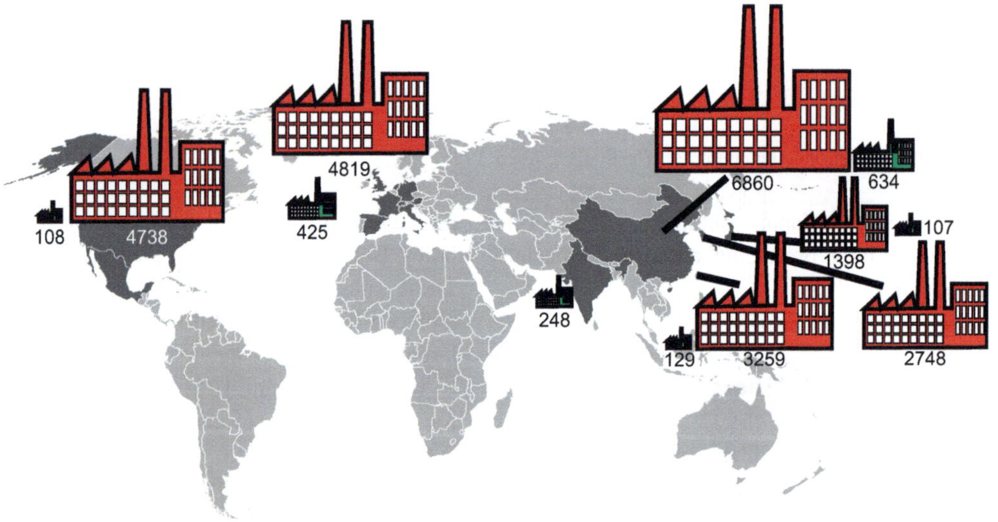

Fig. 2.32 Main producing countries of man-made fibers 2000

Fig. 2.33 Main producing countries of man-made fibers 2010

Japan, under the trade name Kreca™. Subsequently, carbon fibers are used for the first time in sports equipment (e.g. golf clubs, fishing rods). Components for aircraft are now also coming onto the market.

Since the 1980s, more and more new applications for carbon fibers have been developed, especially for the aerospace industry. With the expansion of the wind power industry, carbon fibers are also increasingly being used there to reinforce rotor blades. High-speed machine parts are also increasingly being made from carbon fiber-reinforced plastics for weight reasons, e.g. guide bars in the tufting process and in warp knitting.

Figure 2.35 shows an overview of important milestones in the development of cellulosic and synthetic man-made fibers.

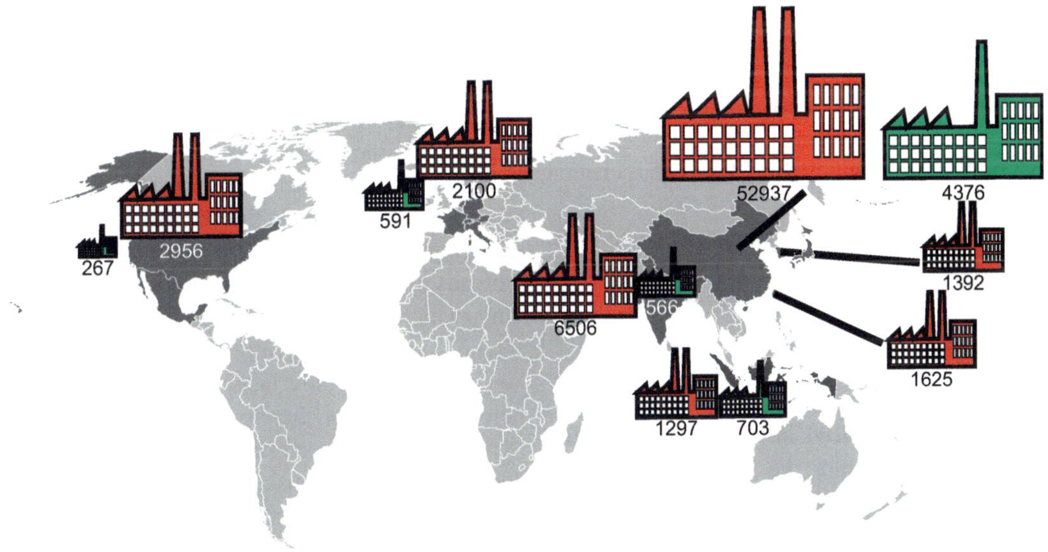

Fig. 2.34 Main producing countries of man-made fibers 2020

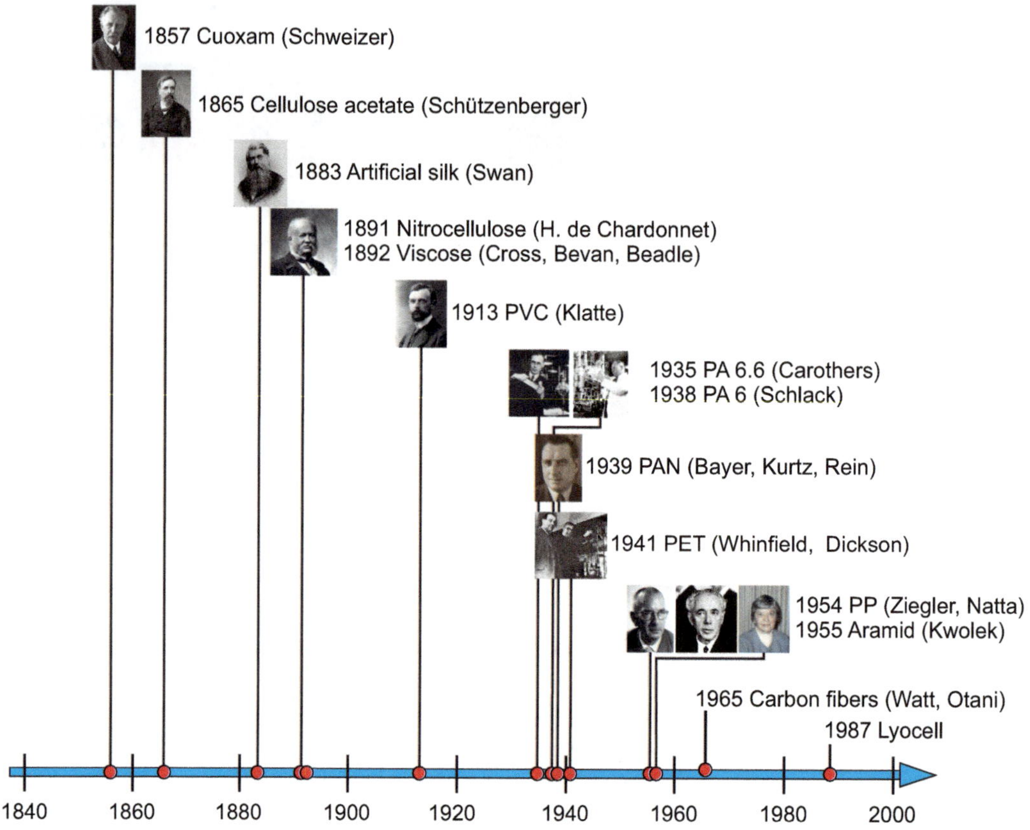

Fig. 2.35 Important milestones in the development of man-made fibers

2.5 Inorganic Man-Made Fibers

2.5.1 Glass Fibers

As early as 1600 BC, glass threads were used in Egypt to decorate vessels. The glass was pulled out of the liquid melt by using tongs, thus creating a thread (Fig. 2.36).

In 1713, R.-A. F. de Réaumur put some thinking into the use of glass fibers for textiles. He postulated that it would soon be possible to spin glass fibers so finely that they could be used to make fabrics. However, he himself did not succeed. From the sixteenth century onwards, glass threads were produced in Venice as decorative elements, and from the eighteenth century onwards, the so-called angel's hair (German: "Engelshaar") was also produced in the Thuringian Forest using the rod drawing process. In 1822, A. and D. Gordon patented lamp wicks made of glass fibers in England, in 1830, a glass fabric was used to cover the coffin of Napoleon I and in 1842 the first woven textiles made of glass fibers were produced on a larger scale (Wulfhorst et al., 1993).

In 1842, the silk manufacturer Louis Schwabe (1798–1845) exhibited glass threads and fabrics made from them at the British Association meeting in Manchester. He had obtained these threads by drawing them out through a thin opening, which thus represents the first documented use of a type of spinneret (Glocker, 1992). However, these glass silk fabrics were unsuitable as clothing. Schwabe therefore called for the development of another substance from which threads could be drawn to make textiles. However, neither he nor others succeeded in this in the next few years and he committed suicide in 1845.

Fig. 2.36 Vessels decorated with glass threads in Egypt, ca. 1600 BC

The first factory for glass fiber production was founded in Vienna in 1866. It produced glass threads with diameters of 6–12 μm for wigs, caps and bridal veils. In 1893, Edward D. Libbey in the USA used glass weft yarns for a festive dress worn by actress Georgia Cayven. When the dress was presented to the public, however, there was great disappointment because the fabric was—of course—not transparent. Presumably, it was also rather uncomfortable to wear. One of three copies of this dress is exhibited in the Deutsches Museum in Munich, Germany (Fig. 2.37).

In 1908, coarse glass filaments were produced by W. v. Paczinsky. Industrial production of glass fibers began in the USA in 1930 and in Germany in 1939. As early as 1932, the Owens-Illinois Glass Company in Newark developed the jet-blowing process purely by accident, when an employee (G. Slayter) directed a jet of compressed air at a molten glass mass and thus produced glass fibers. This process was patented in 1933 and is still used today. From this time onwards, fiber composites were also reinforced with glass fibers, e.g. for tanks in aircraft. In 1938, W. Schuller in Coburg developed the rod-drum drawing process from the rod-drawing process, which was used to manufacture glass staple fibers. Since 1950, the drum drawing process for the production of staple fibers and the jet drawing process for the spinning of filament yarns have become established technologies.

2.5.2 Basalt Fibers

The production of basalt fibers was first patented in the USA in 1921 by Paul Dhé (1921). After World War II, the development of basalt fibers was advanced in the USA, Europe and especially in the Soviet Union. Since the 2000s, basalt fibers have been increasingly used for fiber composites.

2.5.3 Ceramic Fibers

Ceramic fibers consist of inorganic, non-metallic material. A distinction is made between oxidic and non-oxidic fibers. The Welshman E. Parry produced mineral wool from ceramic fibers for the first time in 1840. In 1870 J. Player patented a manufacturing process for ceramic fibers in the USA and in 1871 the first industrial production of ceramic fibers began in Osnabrück, Germany. This involved blowing a rapid stream of air through a curtain of liquid slag. C. C. Hall developed a process to produce fibers from a molten stone in 1897, thus establishing the stone wool industry in the USA. In 1942, ceramic stone wool was first used as an insulation material for high-

Fig. 2.37 Georgie Cayven in her dress made of glass fibers (Wiki, 1893)

temperature applications. However, industrial production did not begin until 1953. Since the 1970s, the production of ceramic stone wool has been steadily increasing. In the meantime, ceramic fibers are mostly produced by pyrolysis of a precursor, which leads to better fiber properties compared to earlier processes.

2.6 Metal Fibers

The first metal fibers were used as early as around 3000 BC by the Egyptians and in the Levant in the form of threads wrapped with thin strips of gold, e.g. from flax and byssus. The wire threads were forged by hammering or cut from thin metal sheets. In the fifth century BC, the drawing iron was developed. In this process, a forged, coarse wire is pulled through a tapering opening in the drawing iron. This makes the wire longer and thinner. The Romans used wound dressings made of silver threads because of their aseptic effect. In the Middle Ages, metal threads from Egypt and China, the so-called "Cyprian gold", were used in ceremonial robes. From the twelfth century onwards, gold and silver threads were produced in Europe (especially in France, Spain and Italy). In this process, a drawn gold wire is rolled flat and wound around a silk thread. For price reasons, pure gold was replaced by gilded silver and later by silver-plated and gilded copper. In Spain, mainly "faux" (leonic) metal threads are produced. From the fifteenth century onwards, pure gold and silver threads were woven into heavy brocade fabrics and trimmings. After the establishment of the first workshop for the production of "leonic threads" in Nuremberg in 1569 by A. Fournier, the production of metal threads began in Germany. In the nineteenth century, the first rolling mills were opened and wire drawing developed from a craft to an industrial production and diamond drawing dies were introduced. Beginning in the 1960s, tire cord made of steel fibers became established in Europe and in 1962 the "shape memory effect" in nickel-titanium alloys (Nitinol) was discovered.

References

Abott, W. F. (1962). *Method for carbonizing fibers*. US Patent Nr. 3,053,775.

Anonymous. (2000). *Chemical Fibers 1950–2000, Chemical Fibers International 50* (pp. 104–105). Deutscher Fachverlag.

Audemars, G. P. (1855). *GB PS-283*.

Becker, F. (1912). *Die Kunstseide (Nachdruck)* (p. 343ff). Salzwasser Verlag.

Bodenbender, H. G. (1929). *Selbstkostenberechnung in der Kunstseidenindustrie* (pp. 494–496). Melliand Textilberichte 10.

Carothers, W. H. (1929). Studies on polymerization and ring formation. I. An introduction to the general theory of condensation polymers. *Journal of the American Chemical Society, 51*, 2548–2559.

Cross, F. C., Bevan, E. J., & Beadle, C. (1893). *Herstellung eines in Wasser löslichen Derivats der Zellulose, genannt "Viskoid". Deutsches Reichspatent 70999*. Reichspatentamt.

de Chardonnet, H. (1890). *Deutsches Reichspatent Nr. 56331, Maschine zur Herstellung künstlicher Seide*. Reichspatentamt.

de Chardonnet, H. (1891). *FR-PS 165349*.

De Réaumur, F. (1734). *Memoires pour servir a l'historie des insectes* (Vol. I). L'imprimerie royale.

Erkner. (2021). *ChemieFreunde Erkner e.V.* Retrieved from http://www.chemieforum-erkner.de/chemie-geschichte/personen/rein_h.htm

Fikentscher, H., & Heuck, C. (1937). *Verfahren zur Herstellung von Polymerisationprodukten, Patent DE654989*. Deutsches Reichspatentamt.

Genas, M. (1962). Rilsan (Polyamid 11), Synthese und Eigenschaften. *Angewandte Chemie, 74*, 535–540.

Glocker, W. (1992). *Glastechnik – Technikgeschichte im Deutschen Museum*. Beck'sche Verlagsbuchhandlung.

Hanfstängl. (1857). Retrieved from https://commons.wikimedia.org/wiki/File:Christian_Friedrich_Sch%C3%B6nbein.jpg

Hooke, R. (1665). *Micrographia*. Royal Society.

Klare, H. (1985). *Geschichte der Chemiefaserforschung*. Akademie-Verlag.

Klatte, F. (1913). *Verfahren zur Herstellung einer auf Hornersatz, Filme, Kunstfäden, Lacke und dergleichen verarbeitbaren Masse, DRP 281877*. Deutsches Reichspatentamt.

Paul, D. (1921). *Filament composed of basalt*. US Patent Nr. 1,438,428.

PeCe. (1934). Retrieved from https://de.wikipedia.org/wiki/Datei:W77_PeCe-Faser_Chemiepark.jpg, CC BY 3.0 DE.

Pizzetta. (1893). Retrieved from https://commons.wikimedia.org/w/index.php?curid=100577

Plinius, D. Ä. *(50), Naturalis Historia 19 and 20. Somewhere in Germania*.

Richter. (2012). Retrieved from https://commons.wikimedia.org/w/index.php?curid=18582552, CC BY-SA 3.0.

Schlack, P. (1938). *Preparation of polyamides*. US Patent Nr. 2,241,321.

Soltes, W. T. (1961). *Electrically conducting fibrous carbon*. US Patent Nr. 3,011,981

Swan, J. W. (1884). *GB-PS 5975*.

Wagner, E. (1981). *Die textilen Rohstoffe*. Deutscher Fachverlag.

Watt, W., Johnson, W., & Moreson, R. (1971). *Production of carbon fibers*. US Patent 3,532,466A.

Whinfield, J. R., & Dickson, J. T. (1941). *Improvements relating to the manufacture of highly polymeric substances*. US Patent 578,079.

Wiki. (1863). Retrieved from https://commons.wikimedia.org/w/index.php?curid=35305608

Wiki. (1870). Retrieved from https://commons.wikimedia.org/w/index.php?curid=2573381

Wiki. (1893). Retrieved from https://commons.wikimedia.org/wiki/File:Georgia_Cayvan_in_her_glass_dress_1893.jpg

Wiki. (1899). Retrieved from https://commons.wikimedia.org/wiki/File:Hilaire_de_Chardonnet.jpg

Wiki. (1900). Retrieved from https://commons.wikimedia.org/w/index.php?curid=29704748

Wiki. (1907). Retrieved from https://commons.wikimedia.org/w/index.php?curid=19124987

Wiki. (1944). Retrieved from https://commons.wikimedia.org/w/index.php?curid=58187936, CC BY 3.0 DE.

Wiki. (1986). Retrieved from https://commons.wikimedia.org/wiki/File:Stephanie_Kwolek_1986.TIF,C CC BY-SA 3.0

Wulfhorst, B., Kaldenhoff, R., & Hörsting, K.-H. (1993). *Glass fibers. Fiber table according to P.-A. Koch of the Institut für Textiltechnik of RWTH*. Aachen University, Deutscher Fachverlag.

Yates, J. (1873). *The textile manufactures of the ancients*. Reprint by Hansebooks GmbH.

Zaremba, S., Steffens, M., Wulfhorst, B., Hirt, P., & Riggert, K. H. (1997). *Polyamide. Fiber table according to P.-A. Koch*. Institut für Textiltechnik of RWTH Aachen University.

3 Market

The world fiber production figures in statistics are somewhat problematic. For some fibers, relatively accurate statistics exist when they are traded across national borders and their quantities are thus recorded. These include cotton and wool. For other fibers, which include most man-made fibers, there is only information on production capacity, but the values of the utilization of the corresponding plants are usually unknown. Therefore, only estimates of the actual quantity of fibers produced are possible. The quantities of fine animal hair are often unreliably recorded and the production of metallic fibers is completely neglected in most fiber statistics because they are usually not counted as fibers. Polypropylene is also often not listed because many fibers are cut from tapes and thus considered films. Glass and carbon fibers, although recorded separately, are nevertheless usually not included in man-made fiber statistics, partly because the industries record their data separately. Therefore, the figures given in this chapter should all be seen with the caveat of a certain degree of imprecision. The world's best and most comprehensive compilation of fiber production statistics is published annually by The Fiber Year GmbH (The Fiber Year, 2021).

3.1 World Fiber Production

At present, around 113 million tons of natural and man-made fibers are produced each year. Added to this are glass and metal fibers, so that the total market amounts to over 130 m tons. In 2020, approximately 32 m tons of natural fibers were produced, of which cotton is by far the most important. This contrasts with 81 m tons of man-made fibers on a synthetic and cellulosic basis. Polyester (PET) accounts by far for the largest share, mainly as filament. A total of around 59 m tons of staple fibers and 54 m tons of filament yarns are currently produced (Fig. 3.1).

Figure 3.2 shows the quantities of important fibers produced since 1900. The production of man-made fibers on a larger scale did not begin until around 1900. Until 1960, man-made cellulosic fibers dominated, followed by polyamides from 1950 onwards, before the production of polyester, especially PET, increased massively since 1970. In 1992, PET overtook cotton as the most important fiber in terms of volume, and in 1997 the volume of man-made fibers produced exceeded the production of natural fibers (Fig. 3.3). After many years of stagnation, since 2010 the production of cellulosic-based fibers, mainly viscose and lyocell, has been rising steeply, by 64% from 2010 to 2020. The relative increase in the production of synthetic fibers during this period is lower, at 57%. It is expected that the importance of cellulosic fibers will continue to rise because they are obtained from renewable raw materials and are therefore considered more sustainable than petroleum-based synthetic fibers. However, this does not mean that polyester will

D. Veit, *Fibers*, https://doi.org/10.1007/978-3-031-15309-9_3

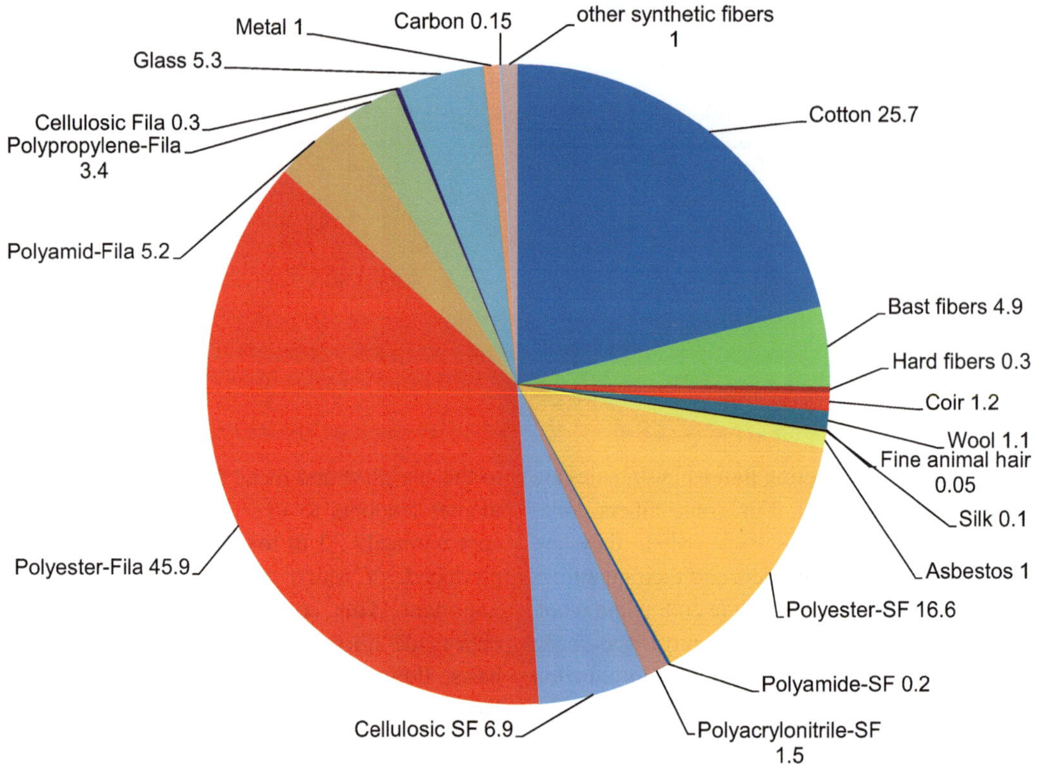

Fig. 3.1 Shares of important fibers in the global market in [m tons] (data: The Fiber Year, 2021; faostat, 2021)

be replaced by viscose or lyocell in the mean run. Apart from the higher price of cellulosic man-made fibers compared to synthetics, the properties of polyester are as yet unrivalled and PET is therefore not easy to replace. It is also conceivable, and there are corresponding efforts, to produce polyester and polyamide on the basis of renewable raw materials (cf. Chap. 42 on biopolymers).

The impact of the Covid-19 pandemic on global fiber consumption is shown in Fig. 3.3. Never in the history of man-made fiber production has the decline in consumption been so large (−14%) in a single year. Although natural fiber production is virtually unchanged, natural fiber production cannot be reduced as easily as man-made fiber production. It is expected that the resulting oversupply of natural fibers will continue to influence the market in the coming years.

Figure 3.4 shows the annual change in fiber production compared to the previous year. The production of natural fibers occasionally decreases due to climatic influences. Nevertheless, there is a slight annual growth on average until the 2010s. Due to the limited agricultural land that can be used for fiber production, the amount of natural fiber production seems to be approaching a limit. This is all the more so as the cultivation of fiber crops is in direct competition with food production. Man-made fibers do not have these limiting, external factors, so the amount of man-made fibers produced grows by about 5% each year. This is expected to continue for the time being, even though the effects of the Covid-19 pandemic are clearly visible.

Against the background of a steadily growing world population and an increase in prosperity, per capita fiber consumption is also rising, as Fig. 3.5 shows. The decline caused by the

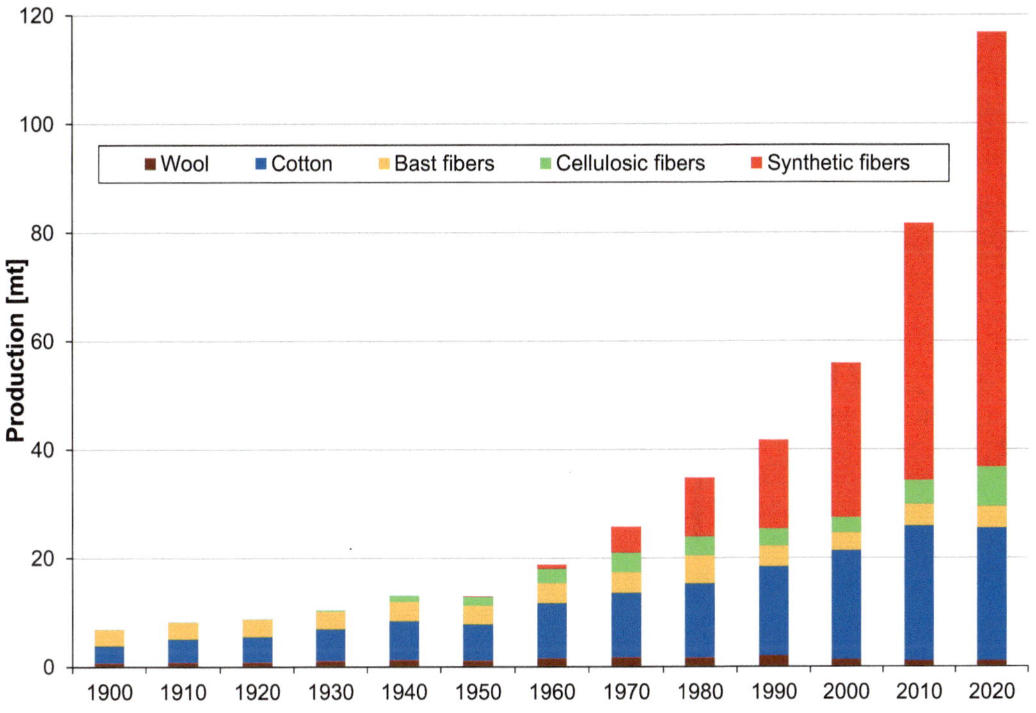

Fig. 3.2 Shares of important fibers in world production

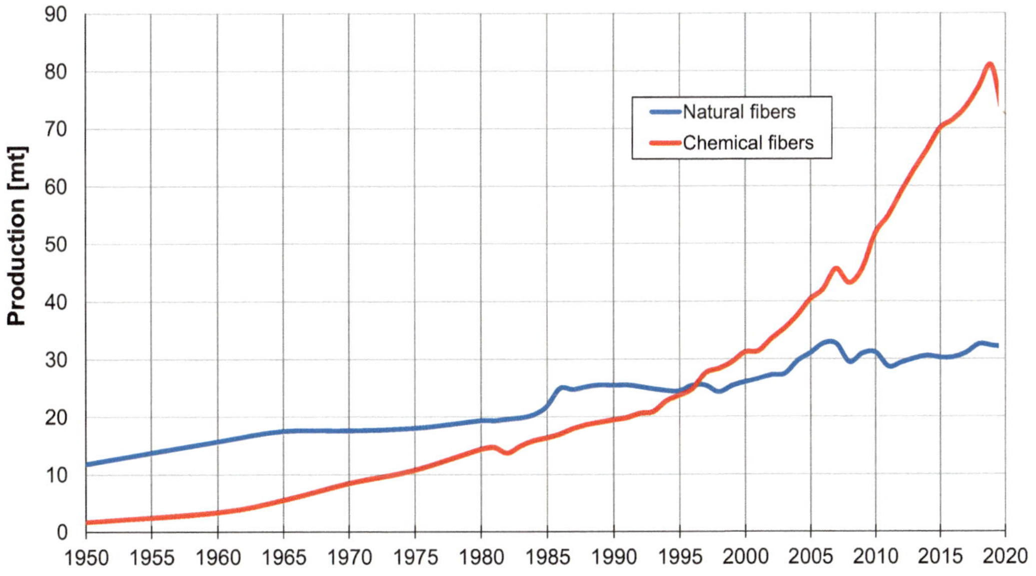

Fig. 3.3 Evolution of consumption of natural and man-made fibers (data: The Fiber Year, 2021; CIRFS, 2021)

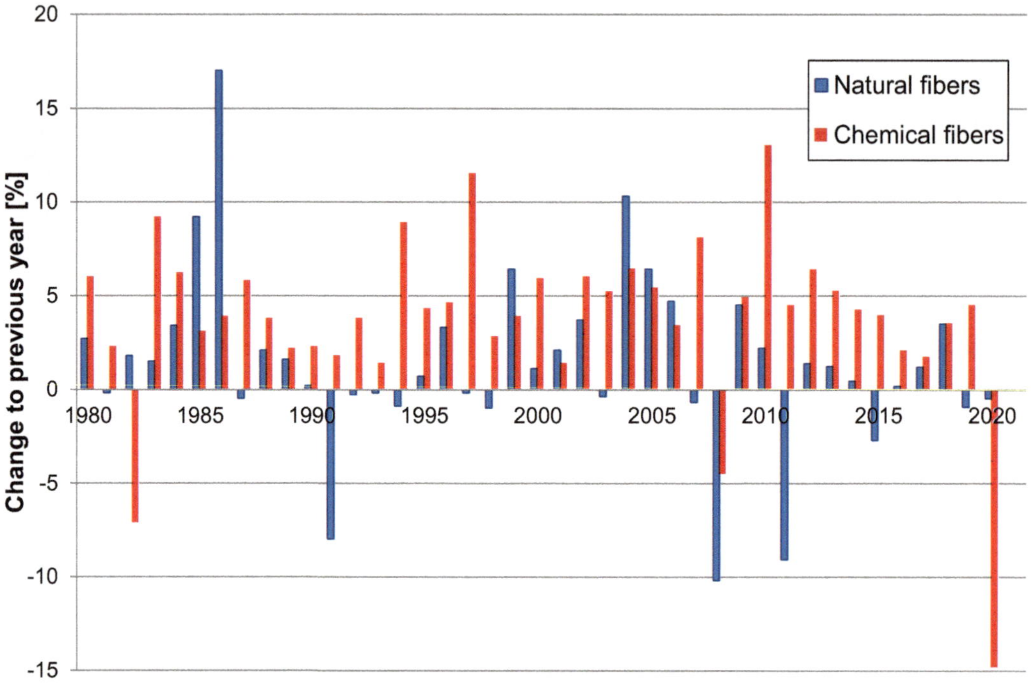

Fig. 3.4 Change in fiber consumption compared to the previous year (data: The Fiber Year, 2021)

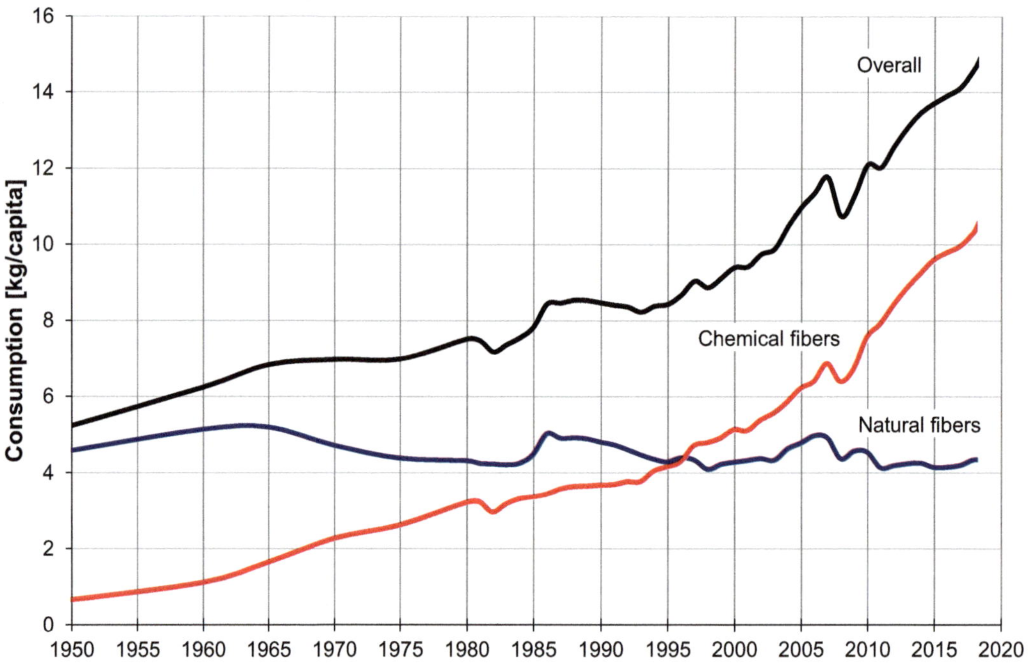

Fig. 3.5 Fiber consumption per capita (data: The Fiber Year, 2021)

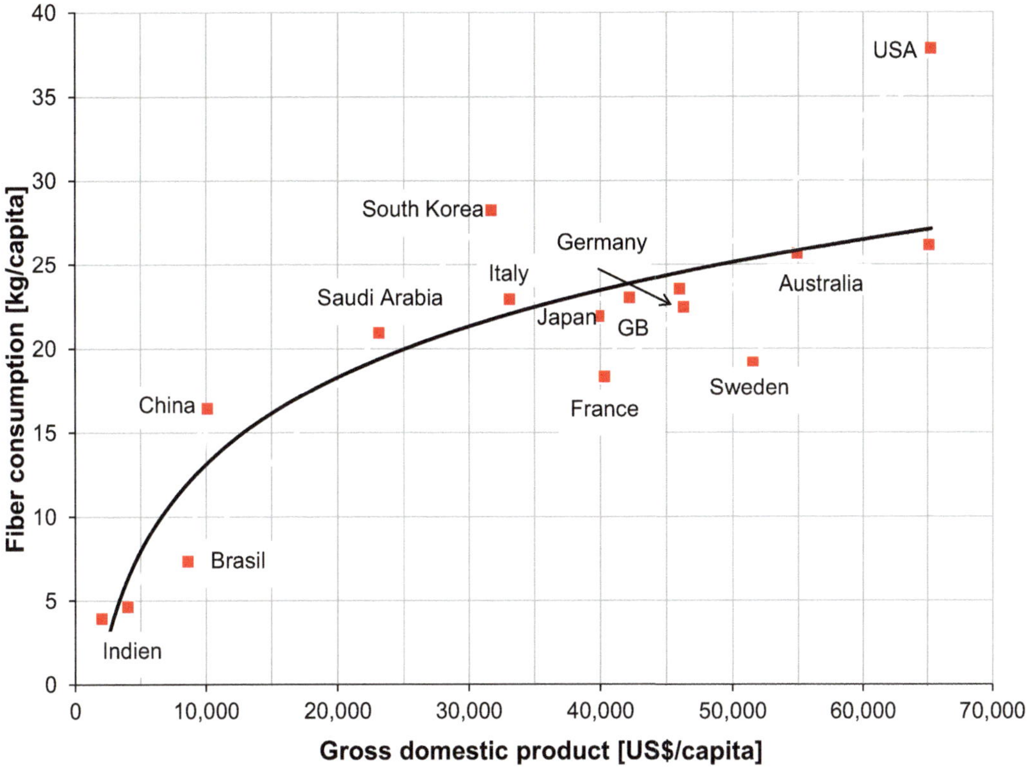

Fig. 3.6 Fiber consumption as a function of gross national product

pandemic is expected to be made up for within 2–3 years, after which further steady growth in fiber consumption can be expected.

There are large regional differences, depending on the wealth of the population (Fig. 3.6). The corresponding numerical values for the fiber consumption of individual countries were determined indirectly because exact figures are not recorded. Therefore, they are only approximate values showing the trend.

3.2 Regions Producing Natural Fibers

Figure 3.7 shows the world regions relevant for the production of the most important natural fibers. Apart from the polar regions, natural fibers are produced in significant quantities on all continents. It is estimated that about 40 m households worldwide produce natural fibers, of which 29 m are involved in cotton production, 6 m in jute and 5 m in wool production. Around 200 m people work in natural fiber production (DNFI, 2021).

Table 3.1 shows the most important producers of selected natural fibers. India dominates cotton and jute production, China is the second largest producer of cotton and wool and the main producer of cashmere. The only European countries of importance for natural fiber production are Russia (asbestos) and France and Belgium for flax. Most wool comes from Australia.

The following figures show the most important producing regions of natural fibers of the respective continents.

In Africa, cotton dominates. Egyptian Maco cotton is considered the finest in the world. However, large quantities of sisal are also grown and sheep are bred, especially in the south (Fig. 3.8).

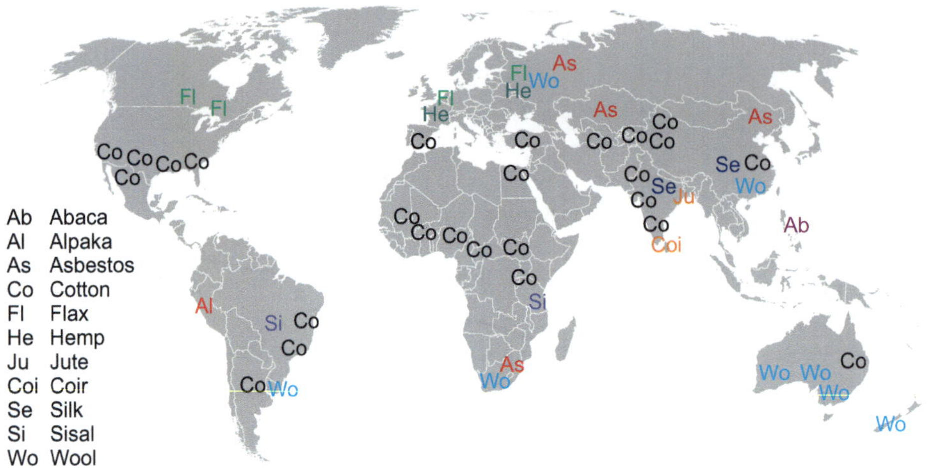

Fig. 3.7 World regions with significant natural fiber production

Table 3.1 The main producers of selected natural fibers

Cotton		**Jute**		**Flax**	
India	6200 t	India	2.0 mt	France	580,000 t
China	5800 t	Bangladesch	1.5 mt	Belgium	94,000 t
USA	4336 t				
Wool		**Cashmere**		**Asbestos**	
Australia	270,000 t	China	15,000 t	Russia	700,000 t
China	179,000 t	Mongolia	3000 t	Kasachstan	200,000 t
New Zealand	105,000 t				

In addition to cotton, numerous other natural fibers are also produced in Asia. These include jute, silk, coir and wool (Fig. 3.9).

In Australia and New Zealand, mainly sheep are bred. In Australia, there are regularly good cotton harvests despite the difficult climatic conditions (Fig. 3.10).

In Europe, in addition to small quantities of cotton, flax and hemp are harvested in the north and coarse wool is produced in southern Mediterranean countries (Fig. 3.11).

In Central and North America, cotton cultivation dominates. In addition, there is flax and sisal production in small quantities (Fig. 3.12).

In South America, sheep are bred as well as large quantities of cotton are grown. Fine animal hair from alpaca and llama is also of great local economic importance. In addition, the production of sisal is worth mentioning (Fig. 3.13).

As Fig. 3.14 shows, it has been possible to significantly increase yields in the cultivation of natural fiber plants over the last 60 years. This is partly due to new breeds and partly to the use of genetic engineering. The amount of wool obtained per sheep also increased during this period.

3.3 Regions and Countries Producing Man-Made Fibers

The market shares of the main producing countries of natural fibers have remained largely unchanged for many years. Natural fibers require certain environmental conditions and therefore production can hardly be relocated to other countries. Moreover, the production of agricultural goods is organized differently from

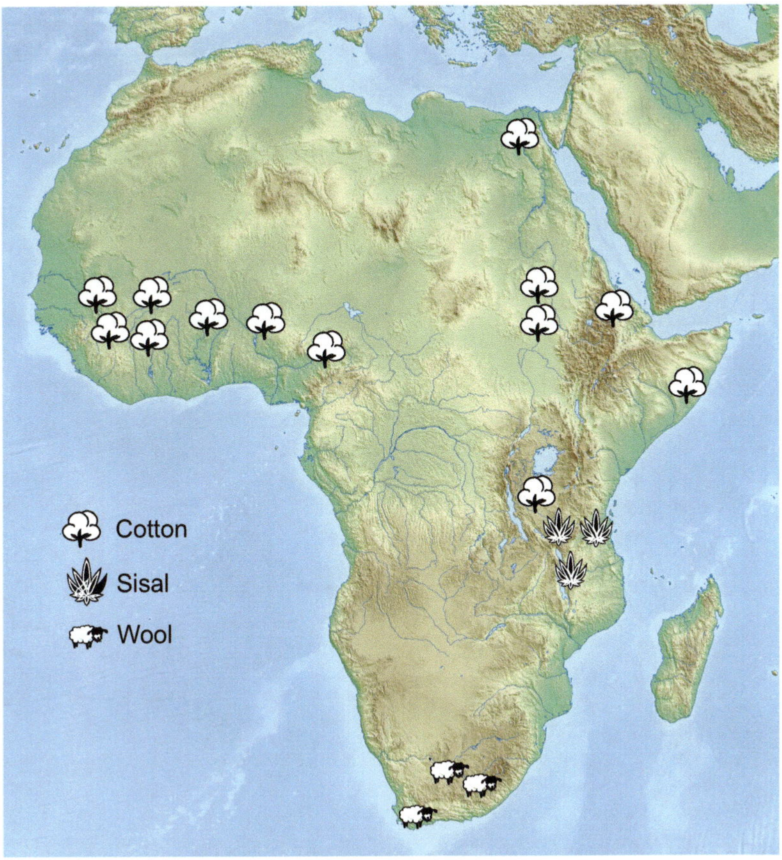

Fig. 3.8 Regions with natural fiber production in Africa (map: Mapswire)

industrial production. Therefore, the situation for producers of man-made fibers is completely different.

3.3.1 World

The growth driver in the production of man-made fibers is polyester (PET). This applies in particular to PET filament yarns, but PET staple fibers also dominate the synthetic staple fiber market. Polyamide clearly focuses on filament production, PA staple fibers are hardly produced. Polypropylene has a stable market with slight growth, while PAN production is declining. Cellulosic staple fibers are experiencing a renaissance, with the highest growth after PET (Fig. 3.15 and Table 3.2). Cellulosic filament yarns do not play a role in the world market, mainly for cost reasons, compared to PET and PA.

Table 3.2 shows the production figures for the most important man-made fibers in terms of volume from 2000 to the present day.

Figure 3.16 shows the development of the production volumes of filaments and staple fibers. In the case of synthetic fibers, filaments dominate with a current growth rate of 6–7%. Staple fibers also show a steady growth of 2–5%. Cellulose fibers are used to 95% as staple fibers; filaments are in competition with the significantly cheaper PET and therefore hardly play a role.

In Fig. 3.17 it can be seen that in 1970 filament yarns made of PA dominated over viscose. Polyester (PET) has gained considerably in importance since 1990 and today more than 80% of all filament yarns are made from it.

Table 3.3 shows the main PET and PA filament producing countries for 2019. The current figures for 2020 and 2021 are distorted for

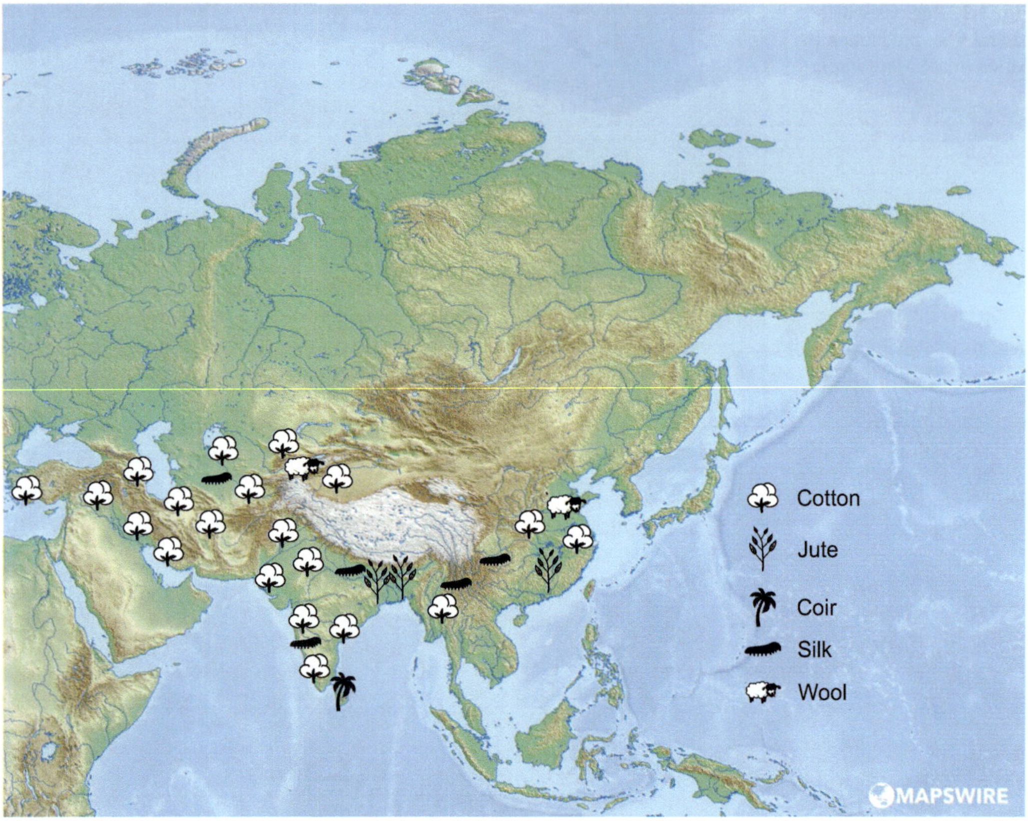

Fig. 3.9 Regions with natural fiber production in Asia (map: Mapswire)

individual countries due to the pandemic and are therefore not included.

Cotton is the most important staple fiber in terms of volume, although PET staple fibers now have a market share of around 25%. The production of cellulosic staple fibers (viscose, lyocell) has also increased considerably in recent years (Fig. 3.18).

Figure 3.19 shows the development of the fiber production in selected countries from 1960 to 2020. While Western Europe and the USA dominated until the 1990s, around 80% of synthetic fibers are now manufactured in Asia, primarily in China. There are two main reasons for this: The costs of producing textiles and clothing are lower in Asia than in Europe and the USA (labor costs, energy costs, environmental regulations). Therefore, textiles and clothing are mainly produced in Asia and therefore, in addition to the lower costs, there are also great logistical advantages to having the fiber production also locally. While Japan has largely discontinued its production of standard polymers and is concentrating on special fibers (e.g. aramid), polyester and polyamide for the mass market are increasingly being produced in India.

Figure 3.20 shows the rise of China in particular, but also of India, as chemical fiber producing countries. Taiwan, Japan and South Korea have significantly reduced their capacities in recent years, while Indonesia and Turkey have seen a slight increase. In the USA, new capacities have been built up on a small scale, mainly for technical fibers.

Figure 3.21 shows this development for selected countries. While China had a market share of only 13% in 1995, this figure has risen

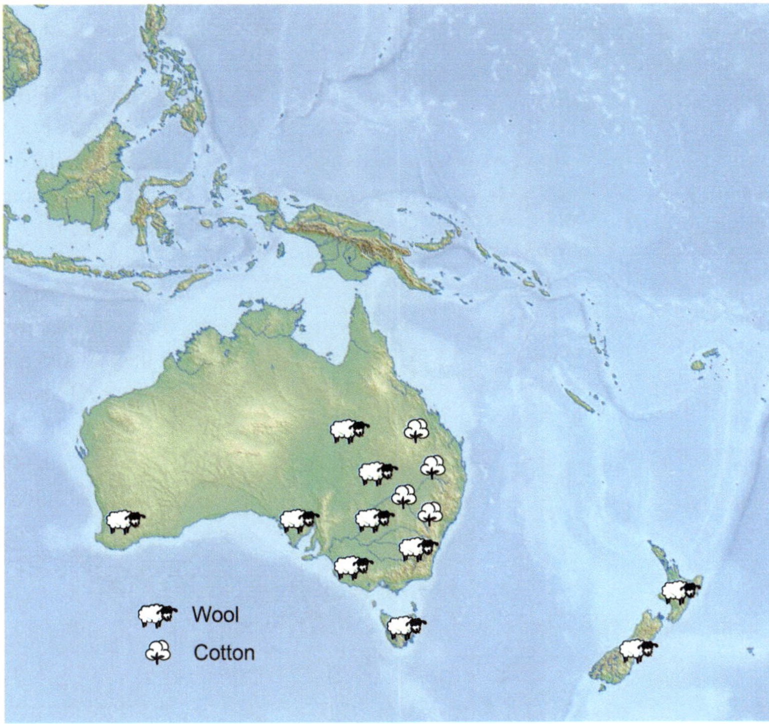

Fig. 3.10 Regions with natural fiber production in Oceania (map: Mapswire)

Fig. 3.11 Regions with natural fiber production in Europe (map: Mapswire)

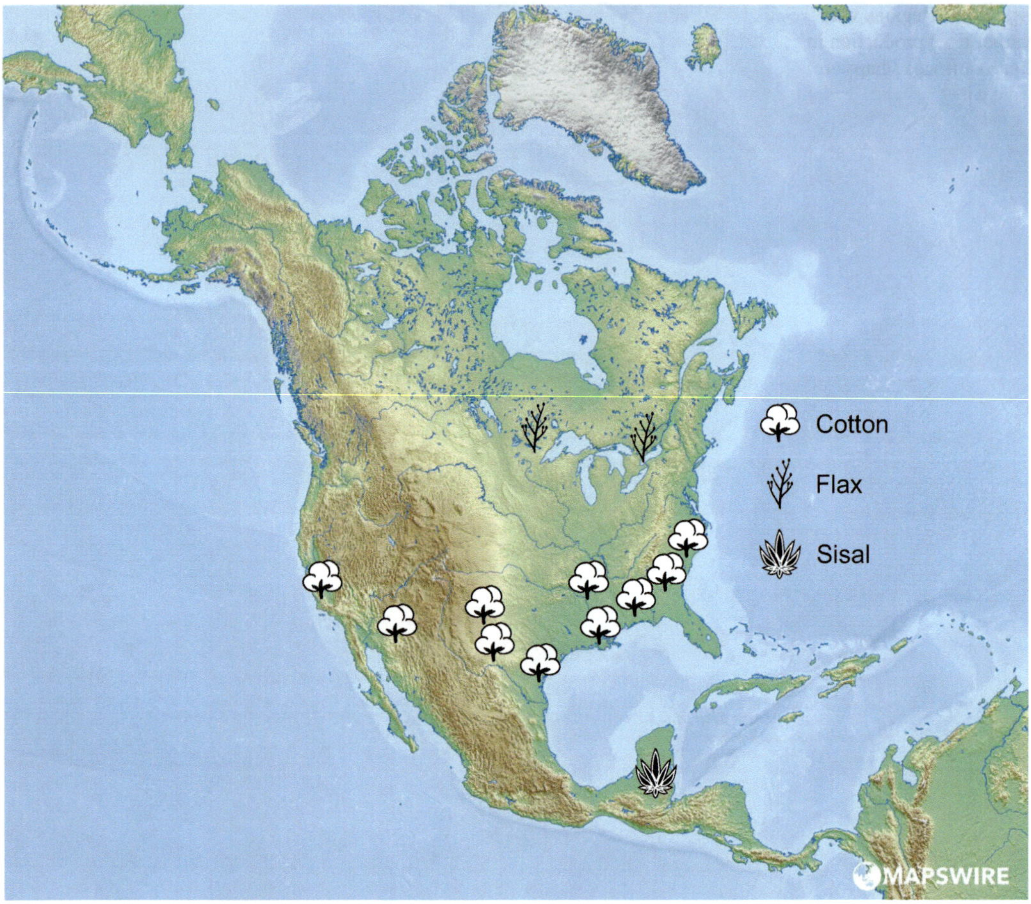

Fig. 3.12 Regions with natural fiber production in Central and North America (map: Mapswire)

to 70% by 2020. India produces as many fibers as China in 2000 and thus has a market share of 8% today.

The most important producing countries for staple fibers made from PET, cellulose, PAN and PA are shown in Table 3.4. While China clearly dominates the production of fibers from PET and cellulose, the production of PAN fibers in Europe is similar to that in China. PA staple fibers are mainly produced in China, the USA and Mexico as well as in Germany.

3.3.2 Germany

Figure 3.22 shows the development of the man-made fiber production in Germany. While the amount of polyamide produced has been steadily decreasing since the mid-1990s, the production of PAN has remained almost constant at 150,000–200,000 tons per year for 40 years. This corresponds to 10% of world production. The production of polyester has also stabilized at around 180,000 t in the last 10 years.

In contrast to many other former chemical fiber producing countries in Europe, the German industry has been able to hold its own in a difficult market. In addition to the use of modern spinning technologies, this is mainly due to a sharp increase in productivity. Whereas in 1970, 53,800 employees produced around 700,000 tons of man-made fibers, in 2020 only 7600 employees produced around 540,000 tons of fibers. This corresponds to a fivefold increase in production per employee (Fig. 3.23).

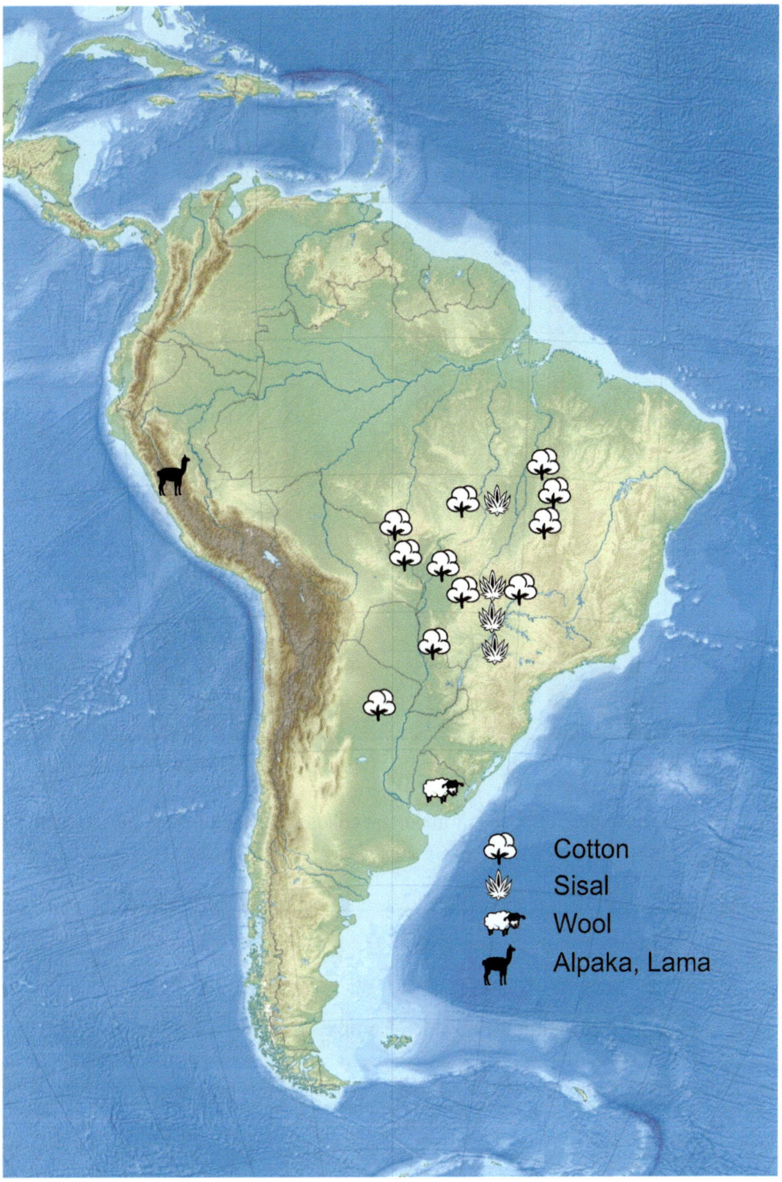

Fig. 3.13 Regions with natural fiber production in South America (map: Mapswire)

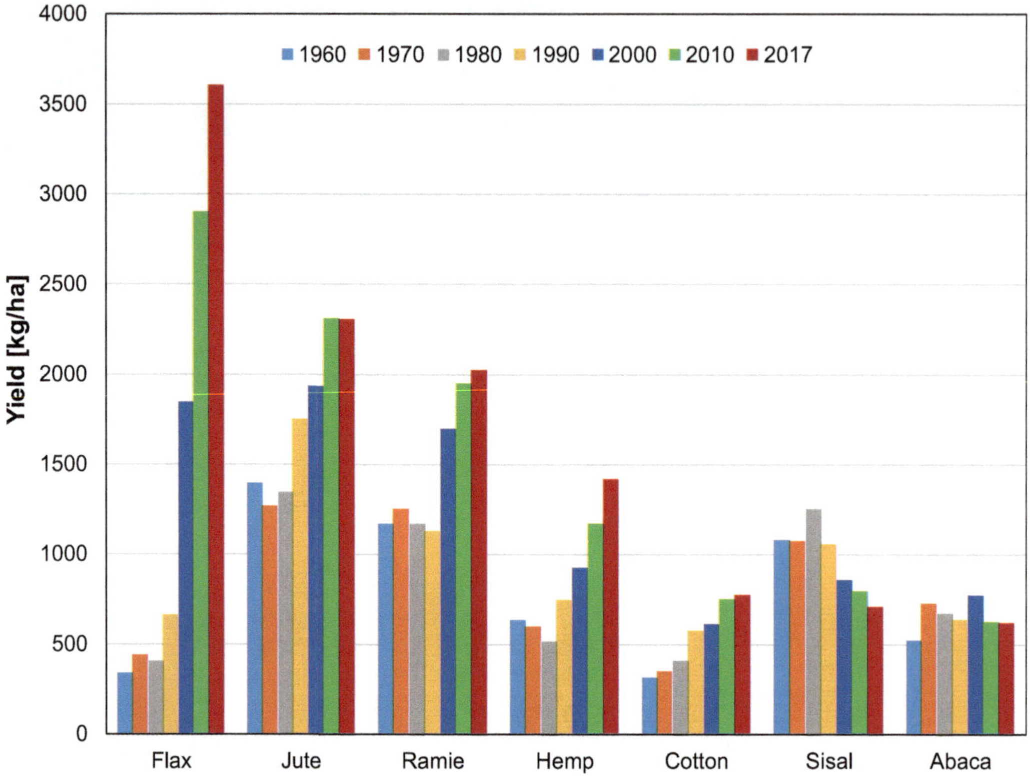

Fig. 3.14 Development of yields of important cellulosic natural fibers (data: faostat, 2021)

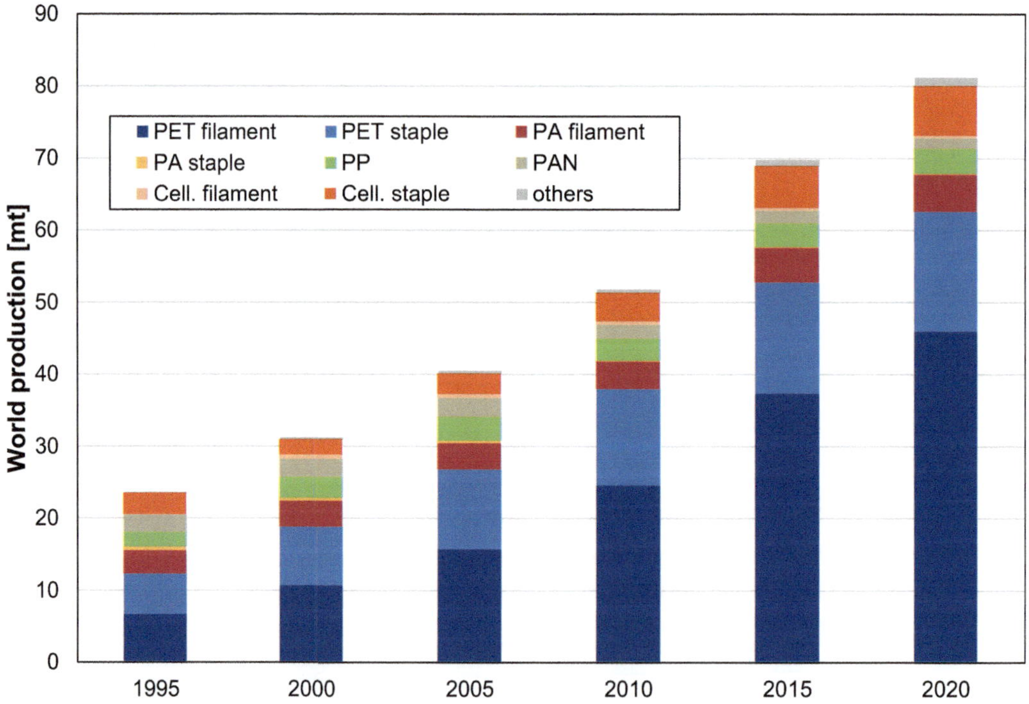

Fig. 3.15 Development of man-made fiber production (data: The Fiber Year, 2021)

Table 3.2 Production volumes of important man-made fibers in [mt] (data: The Fiber Year, 2021)

	2000	2010	2020
PET Fila	10.7	24.5	45.9
PET SF	8.1	13.4	17.0
PA Fila	3.6	3.8	5.1
PA SF	0.5	0.2	0.3
PP	2.8	3.0	3.2
PAN	2.6	2.0	1.3
Cell. SF	2.2	4.0	6.5
Cell. Fila	0.5	0.4	0.3
Total	31.0	51.3	79.6

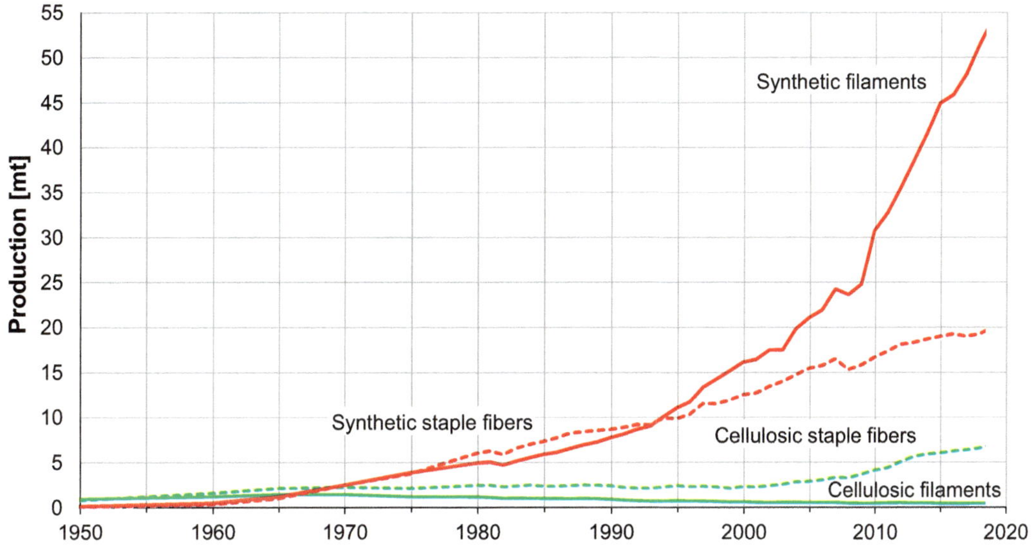

Fig. 3.16 Development of filament and staple fiber production (data: The Fiber Year, 2021)

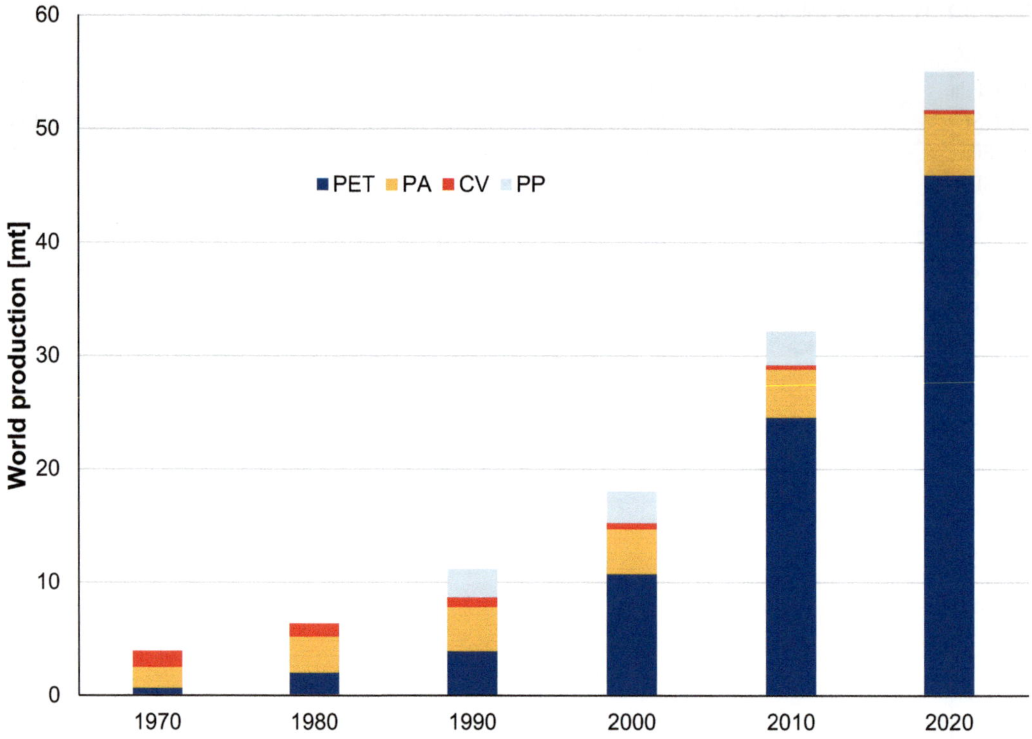

Fig. 3.17 Development of filament yarn production (data: The Fiber Year, 2021)

Table 3.3 Main producing countries of PET and PA filament yarns in [mt] (data: The Fiber Year, 2021)

PET	Textile yarns	Technical yarns	Total
China	35,530	1780	37,310
India	3964	28	3992
Taiwan	702	67	769
Indonesia	575	44	619
South Korea	378	165	543
Europe	312	148	460
USMCA	200	25	225
South America	104	10	114
Japan	72	43	115
ROW	1012	158	1170
Total	42,849	2468	45,317

PA	Textile yarns	Technical yarns	BCF yarns	Total
China	2513	725	75	3313
USMCA	27	107	444	578
Europe	122	161	129	412
Taiwan	206	53	0	259
India	100	70	0	170
South Korea	35	40	4	79
Japan	23	49	4	76
South America	29	33	0	62
Indonesia	13	35	0	48
ROW	132	60	8	200
Total	3200	1333	664	5197

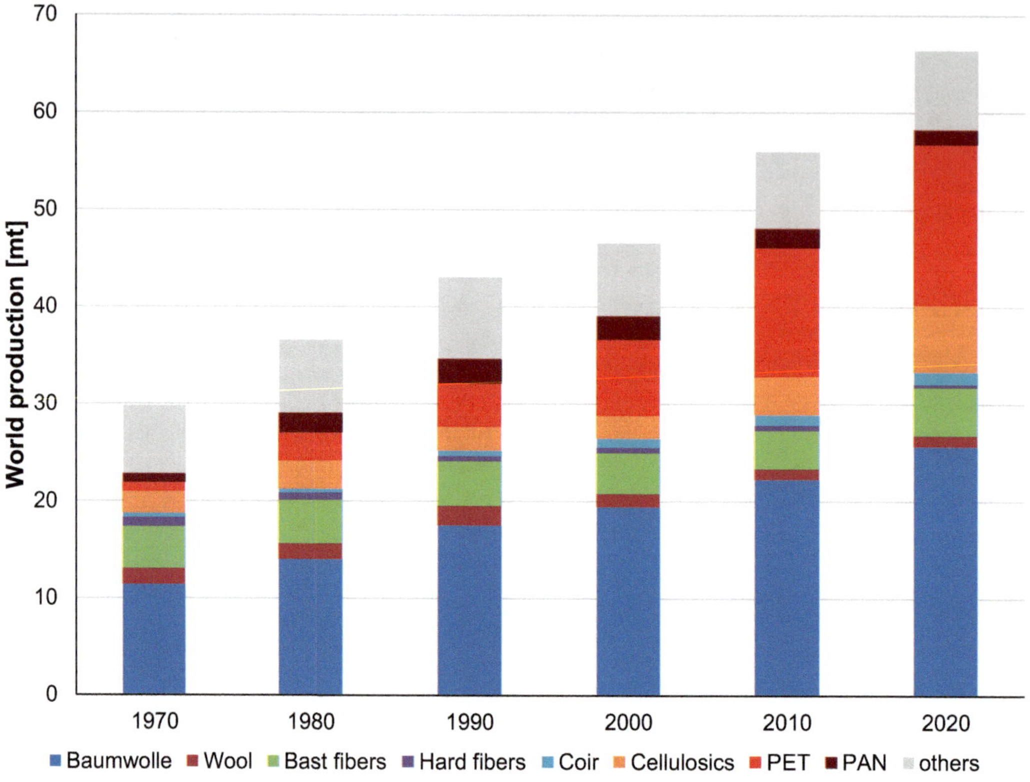

Fig. 3.18 Development of staple fiber production (data: The Fiber Year, 2021; faostat, 2021)

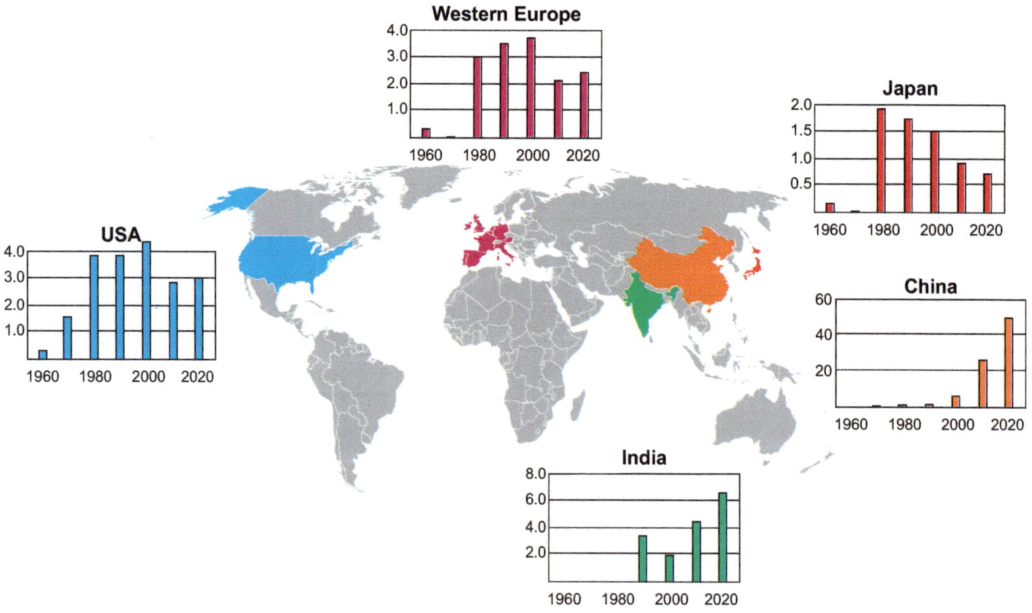

Fig. 3.19 Major man-made fiber producing countries by (data: The Fiber Year, 2021, et al. sources), all figures in [mt]

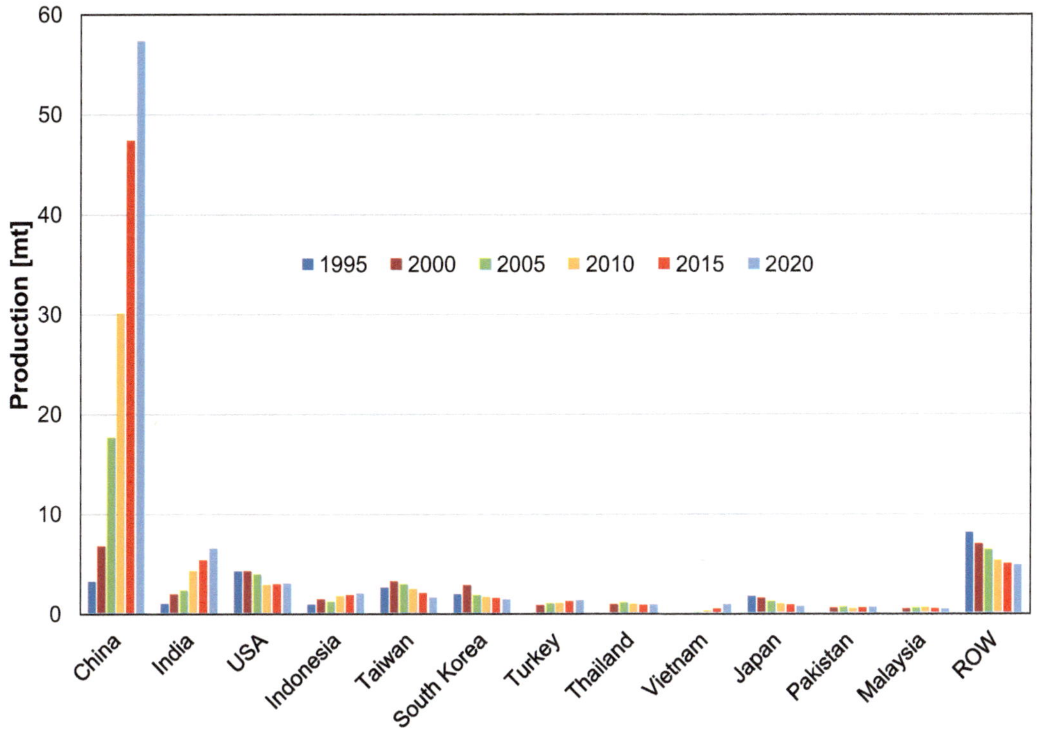

Fig. 3.20 The development of the production of selected producing countries of synthetic man-made fibers according to (data: The Fiber Year, 2021, et al. sources)

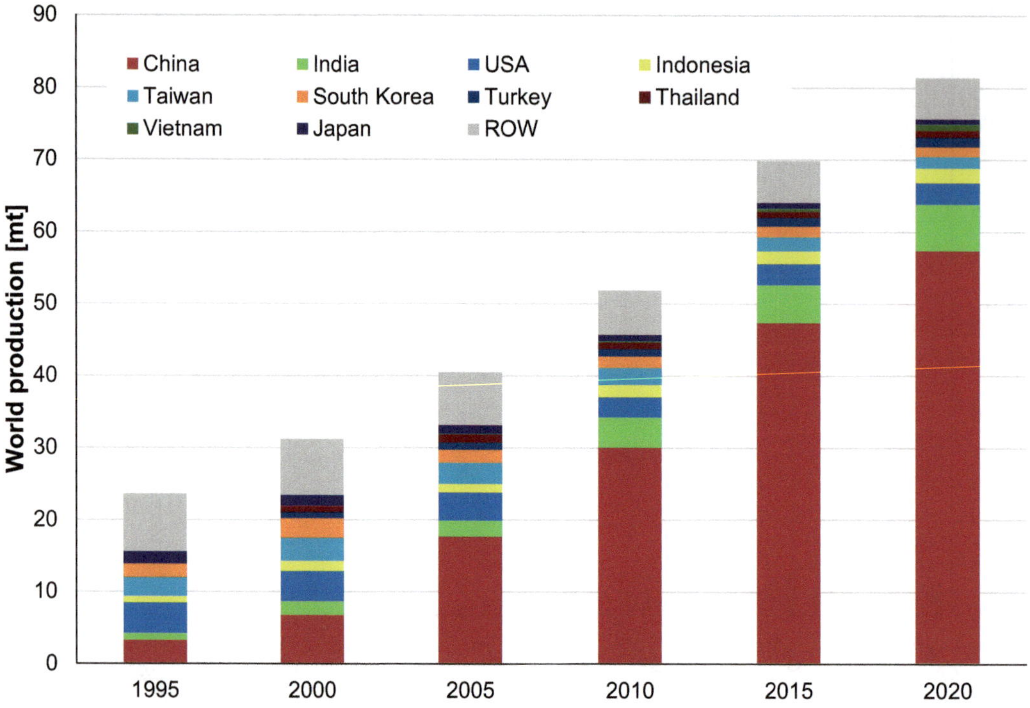

Fig. 3.21 The development of the production of selected producing countries of synthetic man-made fibers according to (data: The Fiber Year, 2021, et al. sources)

Table 3.4 Main producing countries of staple fibers from PET, cellulose, PAN and PA in [million t] (data: The Fiber Year, 2021)

	PET	Cellulosics	PAN	PA	Total
China	10,200	4376	580	187	15,343
India	1590	566	104	0	2260
Europe	828	610	464	28	1930
Indonesia	608	703	0	0	1311
USMCA	751	267	41	33	1092
South Korea	658	26	39	0	723
Taiwan	449	64	0	0	513
Japan	83	142	115	0	340
South America	121	10	24	0	155
ROW	1356	131	110	0	1597
Total	16,644	6895	1477	248	25,264

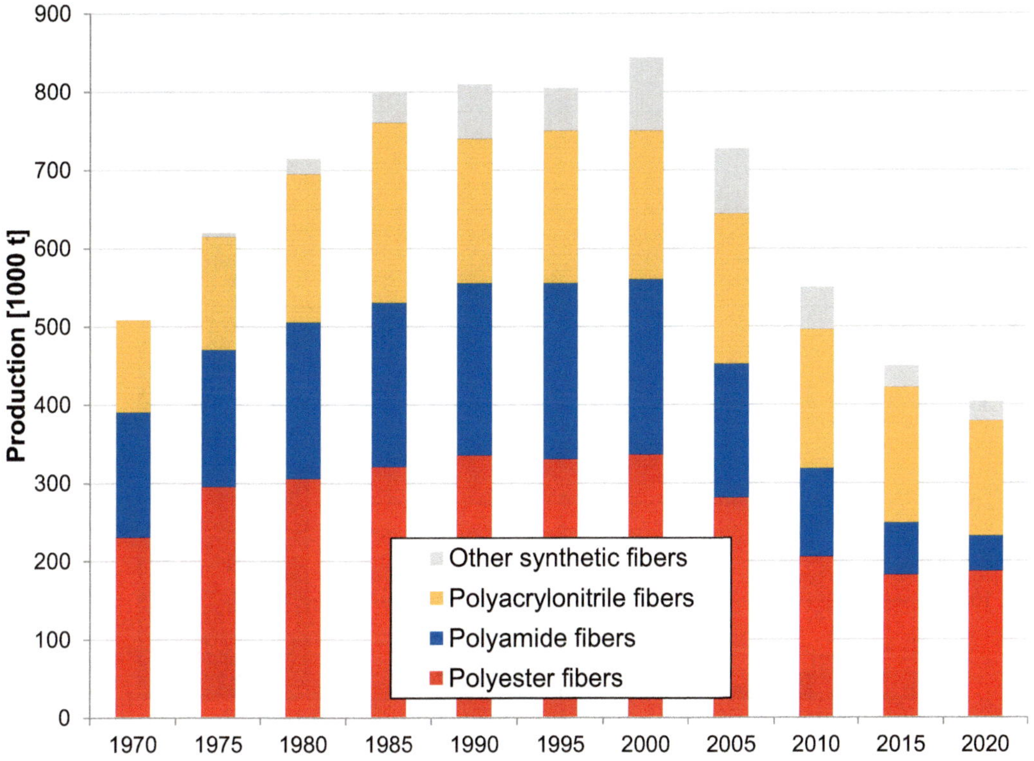

Fig. 3.22 Development of man-made fiber production in Germany (data: statista, 2021)

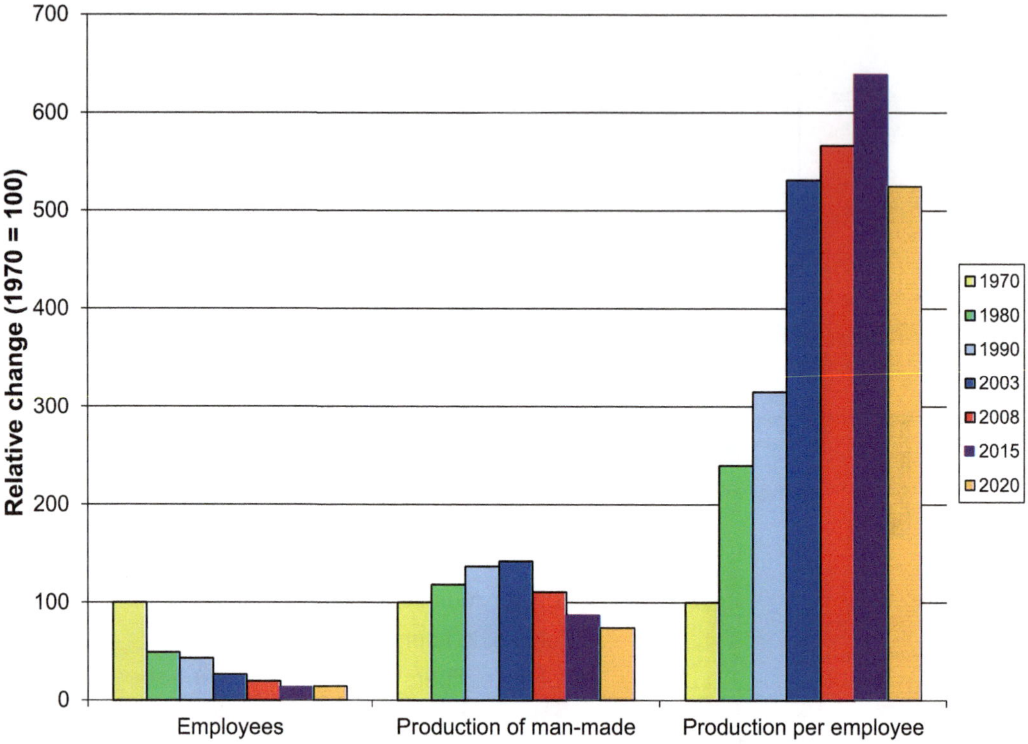

Fig. 3.23 The development of productivity in the German man-made fiber industry (data: statista, 2021; IVC, 2021)

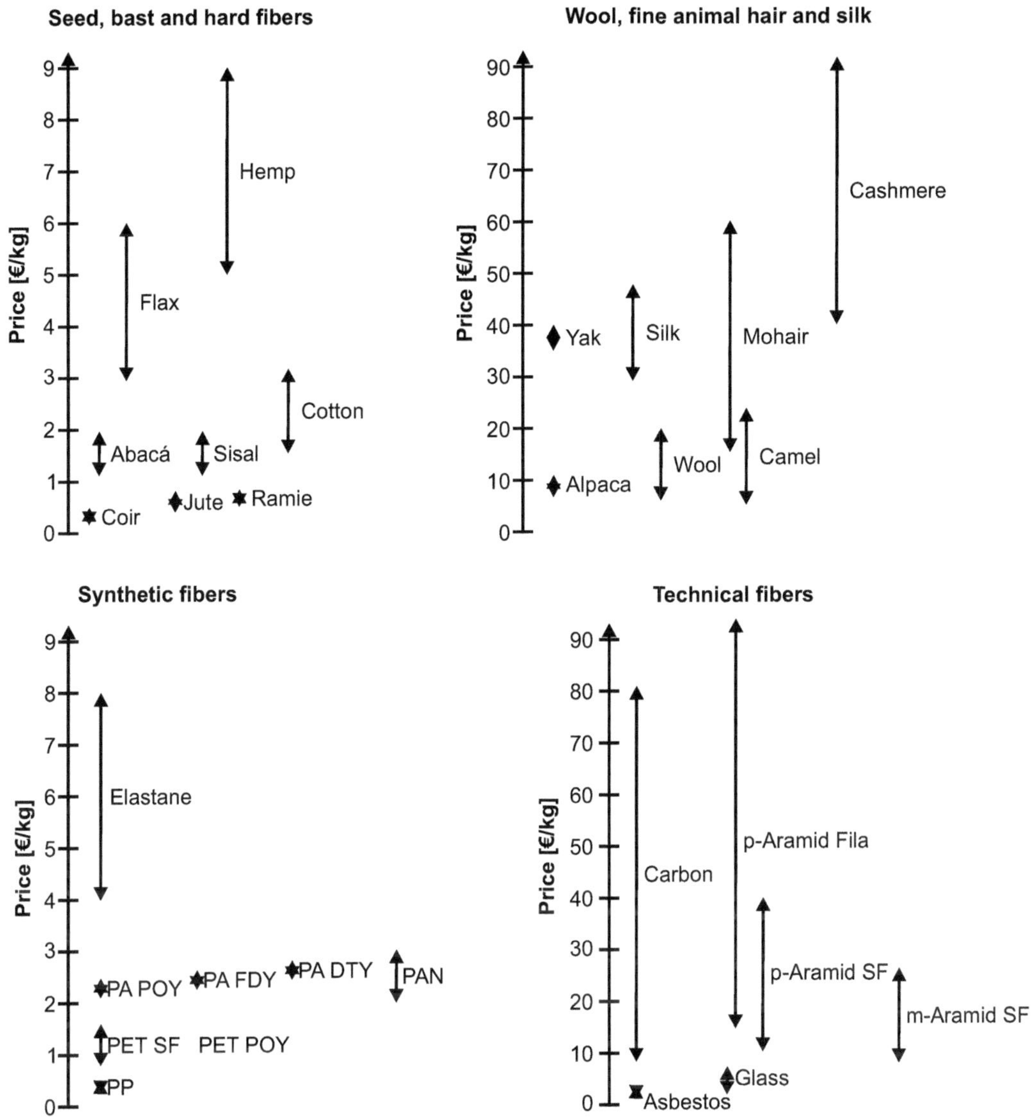

Fig. 3.24 Prices of selected fibers (SF: staple fibers, Fila: filament)

Further information and data on the quantities produced of the various fibers are given in the relevant chapters.

3.4 Prices

In Fig. 3.24, the usual market prices for selected natural and man-made fibers are compiled. The prices for vicuña and other fine animal hair as well as ceramic fibers are not given, as they would exceed the scale at 500–4000 €/kg. Further information on some fibers, in particular on price trends and influencing factors, is given in the chapters on cotton, wool and fine animal hair, for example.

References

CIRFS. (2021). Retrieved from http://www.cirfs.org
DNFI. (2021). Retrieved from https://dnfi.org/abaca/40-million-households-produce-natural-fibers_22590/

faostat. (2021). Retrieved from http://faostat.fao.org

IVC. (2021). *Die Chemiefaserindustrie in der Bundesrepublik Deutschland*. Retrieved from http://www.ivc-ev.de

Mapswire. Retrieved from https://mapswire.com

statista. (2021). Retrieved from http://www.statista.d

The Fiber Year. (2021). *The Fiber Year GmbH, Roggwil*. Retrieved from https://www.thefiberyear.com

4 Internal Structure of Fibers

This chapter deals with the internal structure of synthetic and cellulosic fibers. It differs fundamentally from the structure of inorganic fibers (e.g. glass) and metal fibers. Their structures are considered separately in the corresponding chapters.

4.1 Chemical Structure of Fibers

Synthetic and cellulosic fibers are composed of macromolecules. These usually have long (aliphatic or aromatic) chains of carbon atoms and often polar, functional groups (Fig. 4.1). Polyolefins (e.g. polypropylene) obtain their mechanical properties from long segments of carbon atoms, which can be arranged parallel to each other. The resulting weak bonds, the so-called van der Waals forces, lead to good cohesion of the macromolecules. If polar groups are present, electrostatic forces form between them, resulting in more or less strong bonds. As a result, the macromolecules attract each other, which can lead to crystal formation and then increase the modulus of elasticity, for example. Polyamides and polyesters belong to this group.

4.2 Structural Models

Macromolecules are present in the melt or solution in the form of irregularly tangled chains. On cooling, this amorphous structure is retained for many polymers (e.g. atactic polypropylene). If a polymer is already present in the melt in a partially crystalline form, these structures are retained after cooling and crystallites in the order of 15–100 nm are formed. When ordered superstructures form from these, they are called crystals (Fig. 4.2).

4.3 Crystallinity

4.3.1 Kinetics of Crystal Formation

Unbranched molecular chains with few, ideally regularly arranged side groups crystallize best. These include, for example, polyethylene and isotactic polypropylene. In these structures, there are many positions for two macromolecules where they can be closely aligned.

Polyesters and polyamides also crystallize according to this mechanism, although they are not ideally elongated but zigzag-shaped and have polar side groups. Due to the looping of the polymer chains, polymers generally crystallize much more slowly than substances with low molecular weight.

Crystallites can form randomly when macromolecules are present in parallel as a whole or in sections. If a critical minimum size is exceeded, the crystallites continue to grow, otherwise they disintegrate again.

Mostly, the formation of crystals starts from crystallization nuclei which "contaminate" the

D. Veit, *Fibers*, https://doi.org/10.1007/978-3-031-15309-9_4

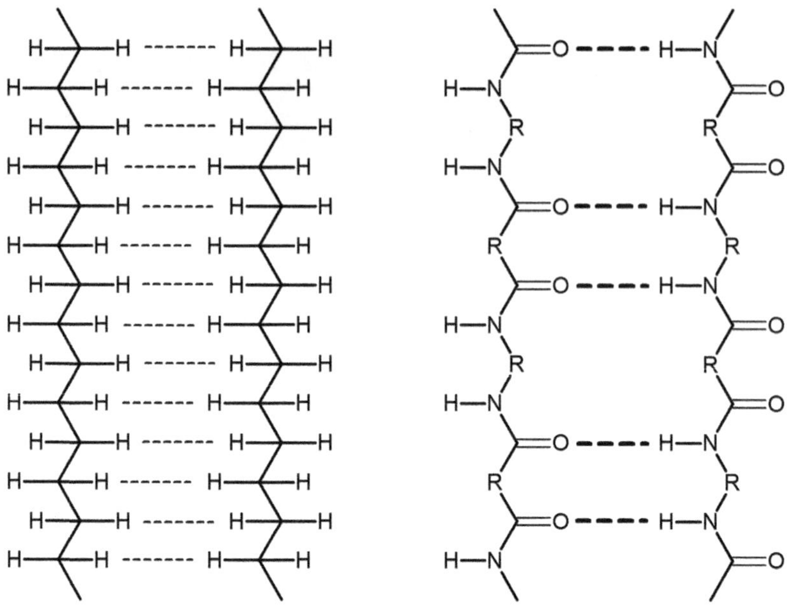

Fig. 4.1 Internal structure of fibers (left: aliphatic chains; right: polar groups)

polymer. Particles with a size of approx. 3 µm are used for this purpose, e.g. metal salts and metal oxides in a concentration of 0.1–0.5%. These materials are polymer-specific because they do not work equally well for all polymers.

The crystallization rate is strongly dependent on the polymer temperature, as Fig. 4.3 shows. Crystal structures can form in a range of about 30 °C above the glass transition temperature and at least 10 °C below the melting temperature. The rate of formation reaches a maximum at a certain temperature and decreases towards lower and towards higher temperature values. This is due to the fact that with increasing temperature above T_{glass} the mobility of the macromolecules increases, thus facilitating crystal formation up to the maximum crystallization rate at T_{max}. Thereafter, this rate decreases again because the macromolecules are too mobile to be incorporated into crystal structures. The crystallization rate continues to increase sharply from the start of crystallization until T_{max} is reached, because the heat released in the process further increases the temperature of the polymer (snowball effect).

The crystallization rate also depends on the cooling rate. If it is high, then many crystallization nuclei are formed and thus many small

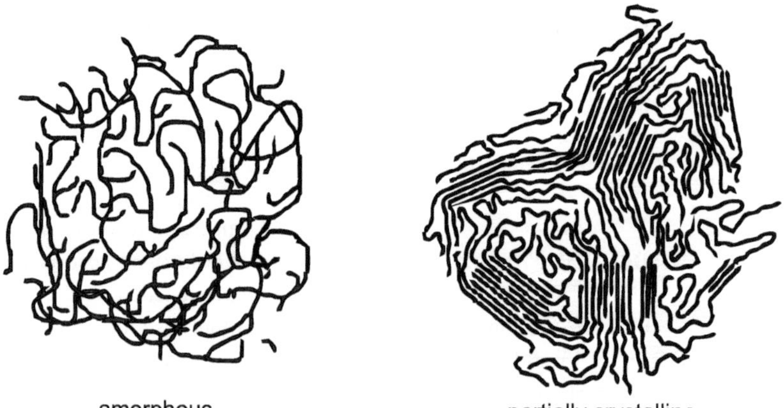

Fig. 4.2 Structure of macromolecules with amorphous and partially crystalline regions

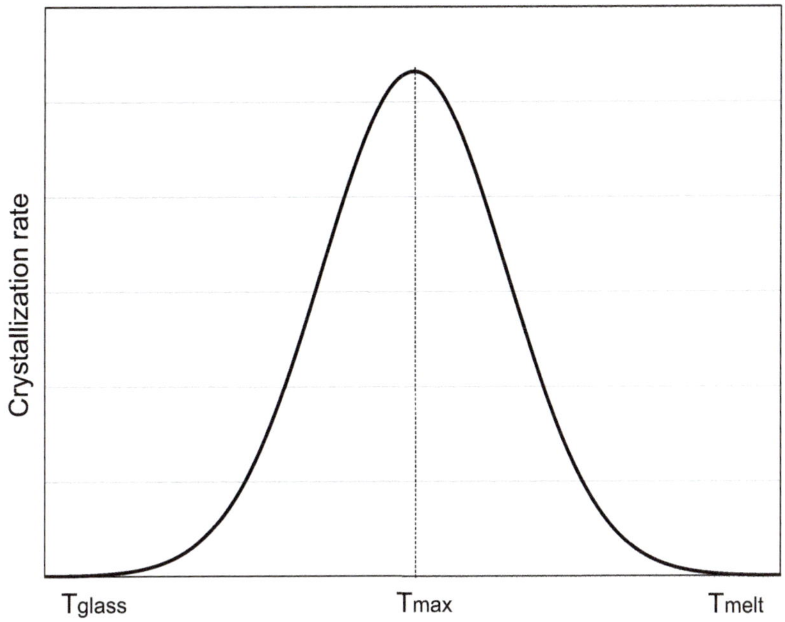

Fig. 4.3 Dependence of the crystallization rate on the polymer temperature acc. to Ziabicki (1974)

crystals. These have a larger relative surface area compared to large crystals and thus a higher interfacial energy. Therefore, they melt earlier than large crystals. Hence, very rapid cooling leads to a decrease in the average melting temperature. The width of the melting temperature range depends on the homogeneity of the crystal size. The more homogeneous the crystal size, the smaller its width.

Table 4.1 shows the characteristic temperatures for some important polymers. The temperature T_{max} is not exactly halfway between T_{glass} and T_{melt}. For maximum crystallization rate, the temperature of the polymer apparently needs to be well above the glass transition temperature. For this reason, additional heating is often used during the spinning process immediately below the spinneret in order to promote crystal formation ("annealing"). Alternatively, this effect can also be achieved during subsequent drawing by heated godets.

The crystallization rate of polymers can be calculated with the Avrami equation. It describes the weight fraction of the non-crystallized melt at time t in relation to the mass of the melt at time t_0:

$$\frac{w_{m,t=t}}{w_{m,t=0}} = e^{-k \cdot t^n}$$

w_m is the function of the weight of the non-crystalline melt as a function of time t, k is a rate constant and n is the Avrami exponent.

For slowly crystallizing fibers (near the melting temperature), the Avrami exponent is $n = 2$, and a normal distribution is obtained according to:

$$\frac{w_{m,t=t}}{w_{m,t=0}} = e^{-k \cdot t^2}$$

Table 4.1 Characteristic temperatures of important polymers acc. to Ziabicki (1976) in (°C)

Polymer	T_{glass}	T_{max}	T_{melt}
PP (isotactic)	−20	65	180
PET	67	190	267
PA 6	45	146	228
PA 66	45	150	264

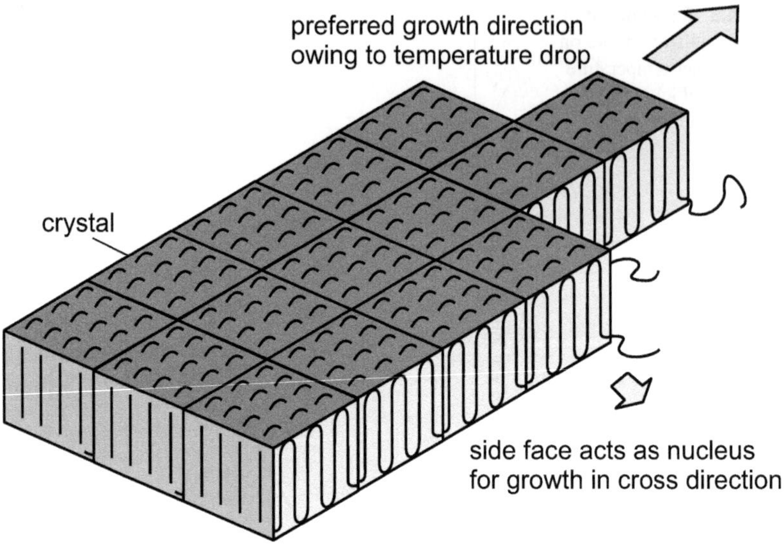

Fig. 4.4 Formation of lamellae from crystals acc. to Menges et al. (2002)

In the case of rapid crystallization (temperature significantly below the melting temperature), n = 1, and thus:

$$\frac{w_{m,t=t}}{w_{m,t=0}} = e^{-k \cdot t}$$

A detailed description of the theoretical relationships can be found in Avrami (1939), and Ziabicki (1976).

4.3.2 Formation of Lamellae and Crystal Growth

During crystallization, the macromolecules are arranged parallel to each other in sections. In the energetically ideal case, the macromolecules would be arranged parallel over their entire length. In reality, this is normally not possible. Macromolecules can be part of several crystallites. Between the crystallites, they exist in amorphous form and therefore one speaks of partially crystalline structures. The molecules are therefore often folded together and the so-called lamellae are formed.

The crystals grow by the addition of further folded molecular chains in the direction of the largest temperature gradient. On the one hand, the temperature must be sufficiently far below the melting temperature so that the attached chains do not detach again through thermally induced movement. On the other hand, the temperature must be sufficiently far above the glass transition temperature so that the mobility of the macromolecules is sufficiently large to be able to attach (Fig. 4.4).

The cohesion of macromolecules is determined by their structure. If polar groups are present (e.g. polyester, polyamide), the intermolecular forces are correspondingly greater due to the formation of hydrogen bonds than in the case of macromolecules without such groups (e.g. polyolefins), in which the cohesion is essentially determined by van der Waals forces.

At the top and bottom of the crystals are the amorphous areas, so that no crystal growth can take place in these directions. The directed growth results in so-called lamellae, i.e. thick bands of interconnected crystals.

4.3.3 Formation of Spherolites

Under isotropic, quasi-stationary conditions, further superstructures form from the lamellae. If these are radially symmetrical, they are called spherolites. Polyamide 6, polypropylene, polyethylene and polybutylene terephthalate form such structures. Spherolites can only form as long as they are surrounded by amorphous

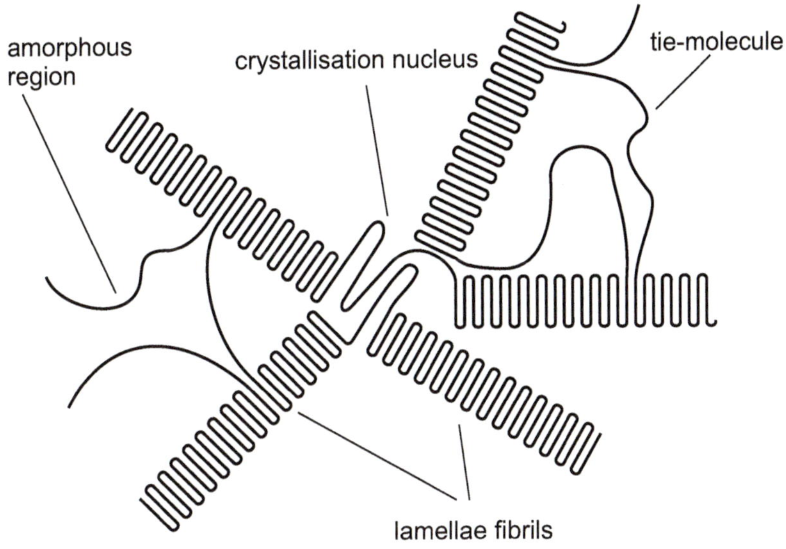

Fig. 4.5 Formation of spherolites

areas. If spherolites meet, flat surfaces form between them (Fig. 4.5).

If a strong temperature gradient is present, dendritic superstructures are formed from lamellae arranged in parallel.

When the polymer flows slowly during cooling, dumbbell-shaped structures are formed, which are also called shish-kebab structure. Their inner part consists of chains arranged in parallel, while on the outside there are folded lamellae. A typical example of this is polyethylene (Fig. 4.6).

The internal energy of large crystals is higher than that of small crystals. If the temperature in the polymer below the spinneret is increased by external heat supply (annealing), the crystal size increases.

4.3.4 Crystallization Due to Stretching

Crystal formation depends largely on the rate of cooling during spinning and on the crystallizers used. If cooling is very fast, this can largely prevent crystallization and an amorphous material is obtained. The arrangement of the molecules or

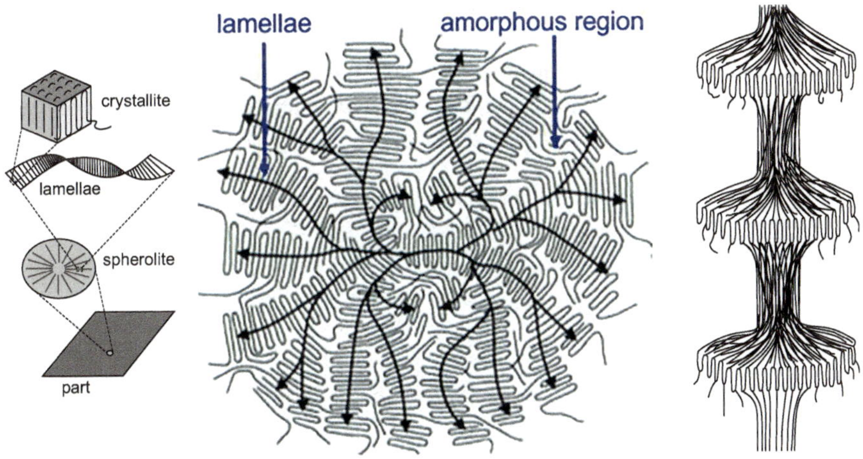

Fig. 4.6 Nomenclature of superstructure and spherolite acc. to Roland (2015), Shish kebab structure

crystals can still be changed by subsequent drawing. In this case, thread tension and heat supply can lead to a transformation of the crystal structure, e.g. in the case of PA 6.

Crystallization occurs much faster when the macromolecules are oriented. This is the case at high take-up speeds in the spinning process. For example, the crystallization of PET at 5000 m/min takes place within 1 ms, whereas it takes approx. 10 s in the polymer at rest at 160 °C (Kikutani et al., 1989).

In principle, it is difficult to significantly change the overall degree of crystallinity after crystal formation. This is due to the fact that a change in the order in one area, due to the low mobility of the macromolecules, inevitably disturbs the order of neighboring areas. Thus, although local changes in the crystal structure can still be achieved, as a result, the overall degree of crystallinity changes only marginally.

During drawing, the macromolecules are oriented in the direction of the fibers. This increases the strength because covalent bonds are very difficult to separate.

4.3.5 Determination of the Degree of Crystallinity

The crystallinity of a polymer can be determined using various methods. These include density measurement with Digital Scanning Calorimetry (DSC), X-ray diffraction, IR spectroscopy and Nuclear Magnetic Resonance (NMR).

From such investigations, the dimensions of the crystalline unit cell, the size of the crystalline areas and the degree of their orientation to the fiber axis, but also typical values for the crystalline portion in the polymer are obtained. Since the specific volumes of the non-crystalline polymer and the crystalline regions differ, the degree of crystallinity can also be estimated from density measurements. However, this should be done with caution, as it is assumed that purely amorphous and crystalline areas occur in the semi-crystalline polymer in a distinguishable and separate manner, which is usually not the case.

4.3.6 Influence on Fiber Properties

In Table 4.2, typical degrees of crystallinity of important polymers are listed. Most values are in the range of 10–30%, in exceptional cases, e.g. after heat treatment, also 30–40%.

Kong and Hay (2003) investigated the dependence of the degree of crystallinity for PET on thermal treatment (temperature, time). They found that higher temperatures and longer exposure time can significantly increase the degree of crystallinity, because the addition of heat allows the macromolecules to move better and thus adhere to each other. This can be achieved, for example, by using hot godets with fibers wrapped around several times before winding.

Table 4.3 shows typical values for selected polymers for tenacity and Young's modulus. Even highly crystalline fiber materials such as PE and aramid achieve only 11% of the theoretically possible strength values, but still 49–65% of the maximum possible modulus of elasticity.

Table 4.2 Typical degrees of crystallization of important polymers acc. to Ehrenstein (1999)

	Crystallinity [%]
Polyamide (PA 6, PA 66)	15–45
Polyester (PET)	10–40
Polyester (PBT)	40–50
Polypropylene (isotactic)	70–80
Polypropylene (syndiotactic)	30–40
Polypropylene (atactic)	0
Polyethylene (HD-PE)	70–85

Table 4.3 Theoretical and actual values for strength and modulus of elasticity acc. to Hearle (2001)

	Tenacity [cN/tex]			E-Modulus [cN/tex]			
	Theoretical	Real	%	Theoretical	Real	%	Crystallinity [%]
PE	3300	370	11	24,700	12,000	49	95
Aramid	2100	230	11	12,700	8300	65	50–75
PA 6	2300	80	3	12,500	500	4	15–45
PET	2000	80	4	9000	1000	11	10–40
PP	2000	60	3	3800	600	16	30–60

4.4 Model Conceptions

Comparison of the chain lengths (for PA 6.6: approx. 200 nm) with the dimensions of the crystalline regions (approx. 5–10 nm) revealed early on that each polymer chain passes through several crystallites, interrupted by amorphous sections. Together with other investigations, this led to a series of increasingly refined model conceptions of the superstructure of partially crystallized polymers, which moved further and further away from the original two-phase model of amorphous and crystalline regions.

One of the earliest models was developed by W. Seifriz (1930) and K.H. Meyer. They assumed that the macromolecules form brick-like structures, the so-called micelles. However, the question of how the micelles hold together remained unsolved. W. Abitz, O. Gerngroß and K. Herrmann (1930) assumed that the macromolecules extend beyond the length of a micelle and form a fringe-like area ("fringe micelle"). Under certain circumstances, the macromolecules can also be incorporated into an adjacent micelle, thus connecting micelles. This model has become accepted for the internal structure of cellulose, but it is not applicable to synthetic polymers. A special feature is that refolding of the macromolecules is only present in the amorphous regions, but not within crystallites (Fig. 4.7).

Electron microscopic studies have shown that synthetic polymers are likely to have crystalline layers consisting of perpendicularly folded polymer chains. Crystalline layers can also be observed in oriented polymers, i.e. in drawn fibers. From X-ray small angle diagrams, their average thickness and spacing can be estimated, and also their average angle to the direction of orientation. Nowadays, the model shown in Fig. 4.8 prevails. Alternating amorphous and crystalline regions are formed, which are connected to each other by macromolecules.

In addition to the models described above, there are a number of other theories published by various researchers (Hearle, 1958; Kargin, 1958; Hindeleh & Hosemann, 1988).

With the constantly evolving model concepts, many properties of the fibers can be explained. They also provided ideas for new processes to achieve special fiber properties. Some of these relationships are briefly explained below by way of typical examples.

The crystallization of a polymer only starts noticeably at a certain distance above the glass

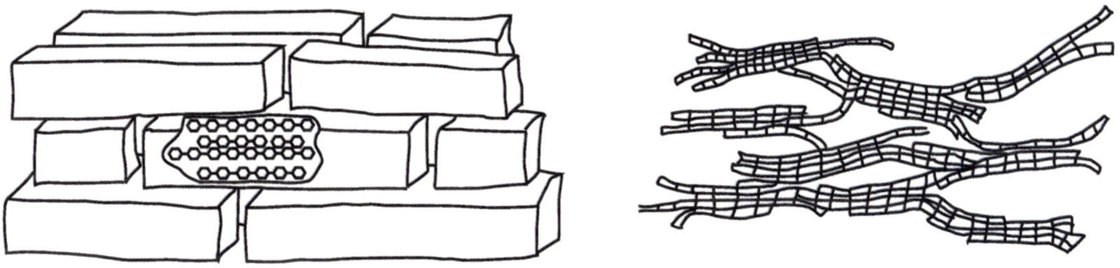

Fig. 4.7 Micellar models acc. to Seifriz (1930) and Abitz et al. (1930)

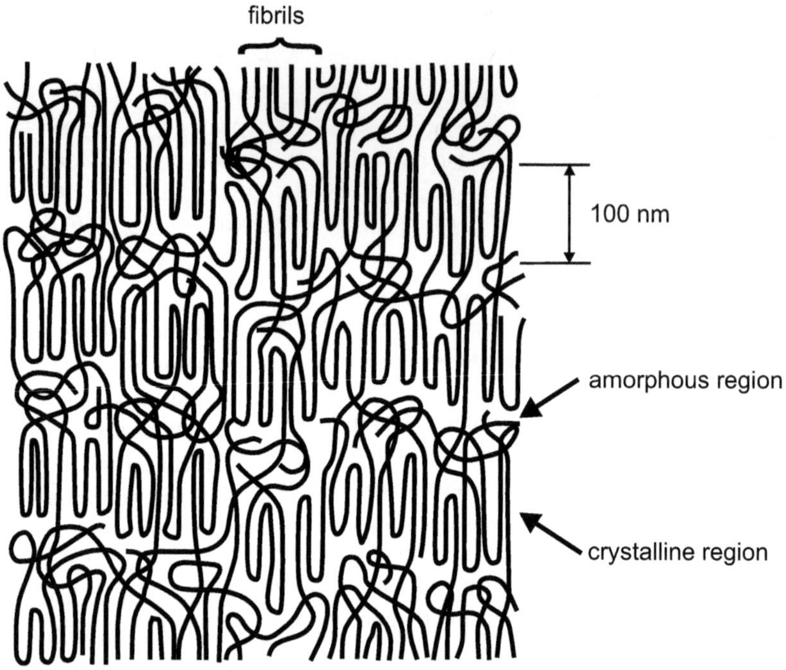

Fig. 4.8 Crystalline and amorphous regions in a polymer acc. to Hearle (1958)

transition point T_G and there is a temperature of maximum crystallization speed. Just below the melting range of the polymer, if present, a crystallite melting point can often be observed. The rate of crystallization and the degree of crystallization reached depend, in addition to temperature, on the chemical nature of the polymer and the spatial structure of the chain molecules. Moreover, they rise significantly with increasing orientation. The position of the crystallization region is essentially determined by the glass transition temperature and thus also depends on the above-mentioned influencing variables which determine this temperature. There are polymers which are already highly crystalline in the almost unoriented spinning filament and others which initially exhibit only a low level of crystallinity, which increases considerably during orientational drawing. And there are polymers in which the filaments are generally amorphous directly after spinning and in which crystallization only begins during drawing, e.g. polyester.

The crystallinity of the fiber end products can be increased by heat treatment in the textile and the inner structure thus formed can be fixed. For example, fibers or yarns in the highly twisted state of a twisted yarn or in the corrugated state in a woven fabric can be "frozen" so that they retain their shape permanently.

The non-crystalline parts provide the drawability of the finished structures, which is often necessary for textiles. Also, only amorphous areas allow diffusion processes. The more amorphous areas a fiber has, the more dye it can absorb.

There are also fiber-forming polymers that do not crystallize or crystallize only very weakly and still yield very useful fibers, e.g. polyacrylonitrile.

Salem (2000) gives a good overview of the currently used models for PET, PA, PE and other homopolymers as well as copolymers.

4.5 Crystal Structures

Morton and Hearle (2008) proposed a selection of parameters that describe the internal structure of different polymers in a sufficiently distinguishable manner. These include

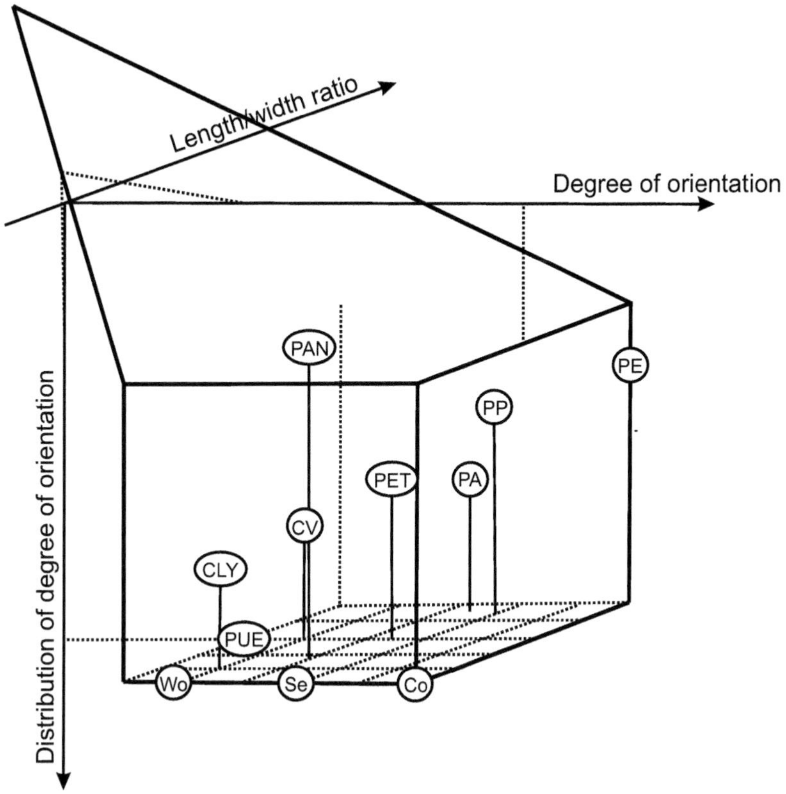

Fig. 4.9 Assignment of important polymers to crystal structure types acc. to Morton and Hearle (2008)

1. degree of order,
2. uniformity index of the internal structure,
3. length/width ratio of crystalline regions.

According to J. Hearle, the degree of order DO can be calculated from the average polymer density and the density of the amorphous or crystalline regions of the polymer:

$$DO = \frac{\rho - \rho_{amorph}}{\rho_{kristallin} - \rho_{amorph}}$$

Accordingly, values between zero (completely amorphous) and one (fully crystalline) result. The respective numerical values can be determined, for example, via X-ray diffraction, IR absorption or nuclear magnetic resonance. The second parameter describes the distribution of the degree of orientation between defined areas within the polymer. A narrow distribution of the degree of orientation (similar to a normal distribution) results in a small value (differences in the OG very large within the polymer). A uniform distribution of the OG results in a medium value and if there are only areas of very low (amorphous) and very high order (crystalline) would result in a high uniformity index. The third parameter as length/width ratio is a measure of the shape of the crystalline regions. For long fibrils, a very high value is obtained, for micelles a value of one and for flat layers values close to zero are obtained. In Fig. 4.9, these 3 parameters form the axes of a cube.

Natural fibers typically have a high internal uniformity because they grow slowly. This also explains the relatively large crystal size of natural fibers, whose macromolecules, however, are usually also significantly longer than those of synthetic materials. Solidification processes during the production of synthetic fibers always lead to a significantly lower internal uniformity. Man-made cellulose-based fibers have amorphous and few crystalline regions, while polyester (PET) and polyamide 6 and 66 are partially crystalline, respectively. HD polyethylene fibers (PE) are extremely oriented (90–95%).

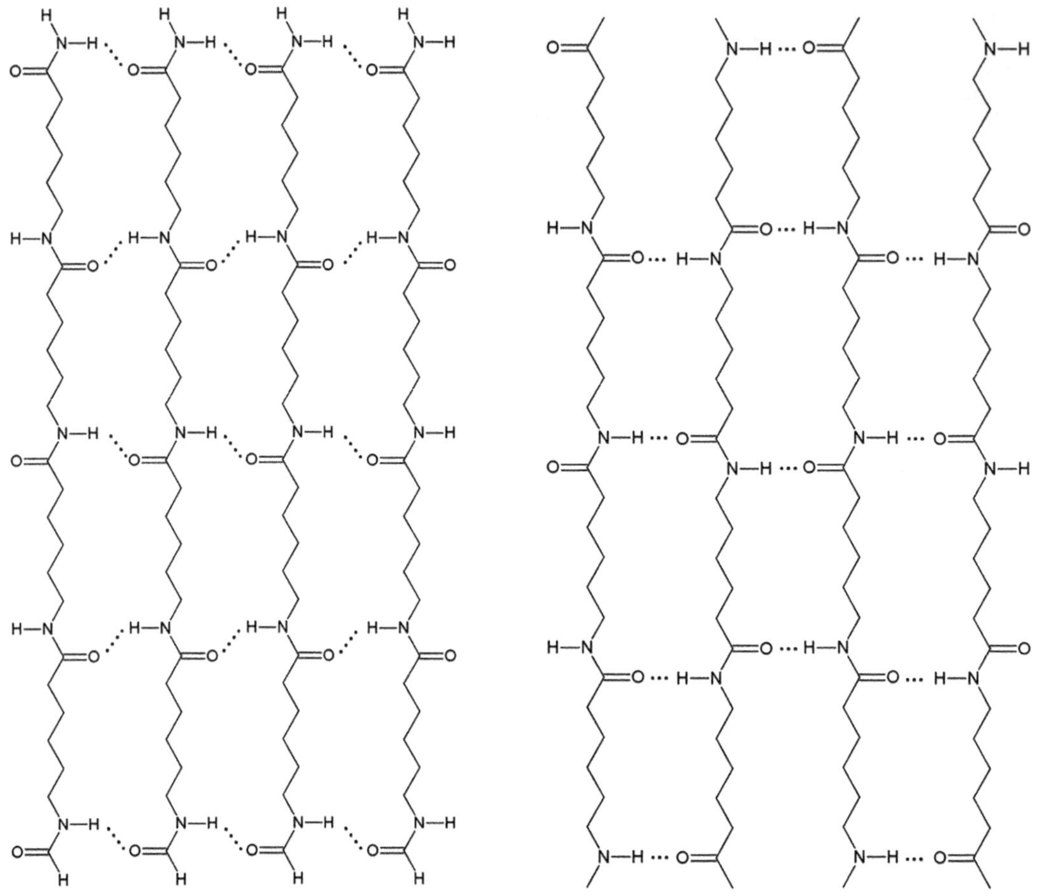

Fig. 4.10 2D crystal structure of PA 6 (left: parallel; right: anti-parallel)

Synthetic fibers are mostly composed of regular macromolecules with only small side groups. This allows a high packing density of the chains. At the same time, most macromolecules have strong dipoles, resulting in large attractive forces between the chains. Furthermore, most macromolecules are relatively flexible, which makes orientation by rearrangement of the chains possible in the first place. This is particularly true for polyester and polyamide, which is why these polymers in particular are used as fiber materials.

4.5.1 Polyamide

The two most important polyamides for fiber materials, polyamide 6 and polyamide 66, basically have a very similar structure. The main differences are the 50% shorter length of the repeating unit in PA 6 compared to PA 66. An error in the arrangement of the PA 66 molecules has a correspondingly stronger effect on the crystal structure than in PA 6 because of the greater length of the repeating unit. Furthermore, PA 6 molecules have two different orientations (depending on the sequence of the amide group in the chain: CO-NH or NH-CO). The effect of this on the crystal structure is shown in Figs. 4.10 and 4.11. Parallel orientation of the chains (same sequence of CO-NH groups) results in higher internal stresses than anti-parallel arrangement (CO-NH different for adjacent chains). In reality, mixed forms exist. It is not possible in the process of fiber formation to reorient PA 6 chains with a

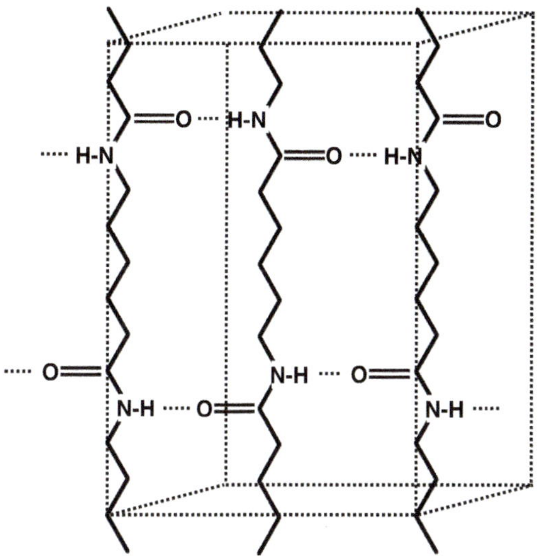

Fig. 4.11 3D crystal structure of anti-parallel PA 6

"wrong" sense of direction. Corresponding defects in the crystal structure are therefore irreparable.

The melting temperature of PA 6 depends linearly on the crystallite length in the chain direction, as Fig. 4.12 illustrates for PA 6, crystallized in the α-structure.

Figure 4.13 (left and center) shows the α-folded sheet structure of PA 66. It has already been demonstrated experimentally in this way by Hearle and Greer (1970). Some macromolecules connect the mycelia via the so-called tie molecules, while others "fray" and turn into amorphous regions. Radiographic studies by Murthy et al. (1990) showed that this ideal structure does indeed exist in reality, albeit with somewhat more variation with respect to the arrangement of the mycelia. Figure 4.13 (right) shows their model for PA 6. The length and width of the mycelia is about 15 nm.

Macromolecules will not normally orient completely, so the degree of orientation of polyamide is about 50%. Furthermore, there are 2 different stable crystal configurations for polyamide,

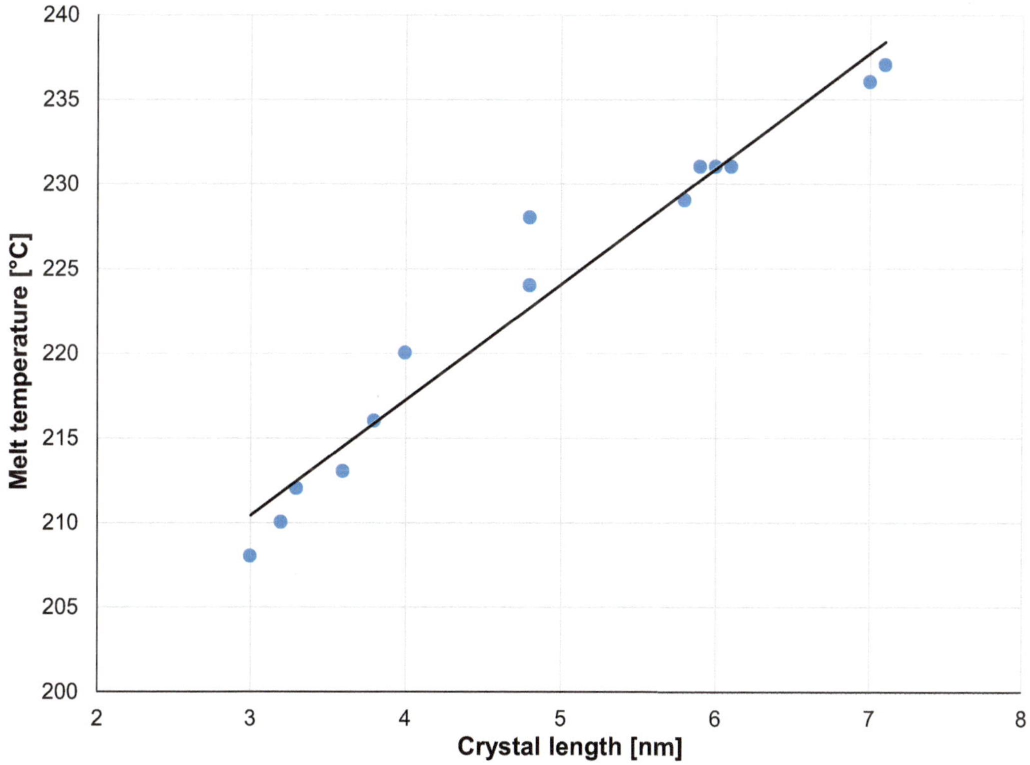

Fig. 4.12 Influence of crystallite length in PA 6 on the melting temperature acc. to Illers and Haberkorn (1971)

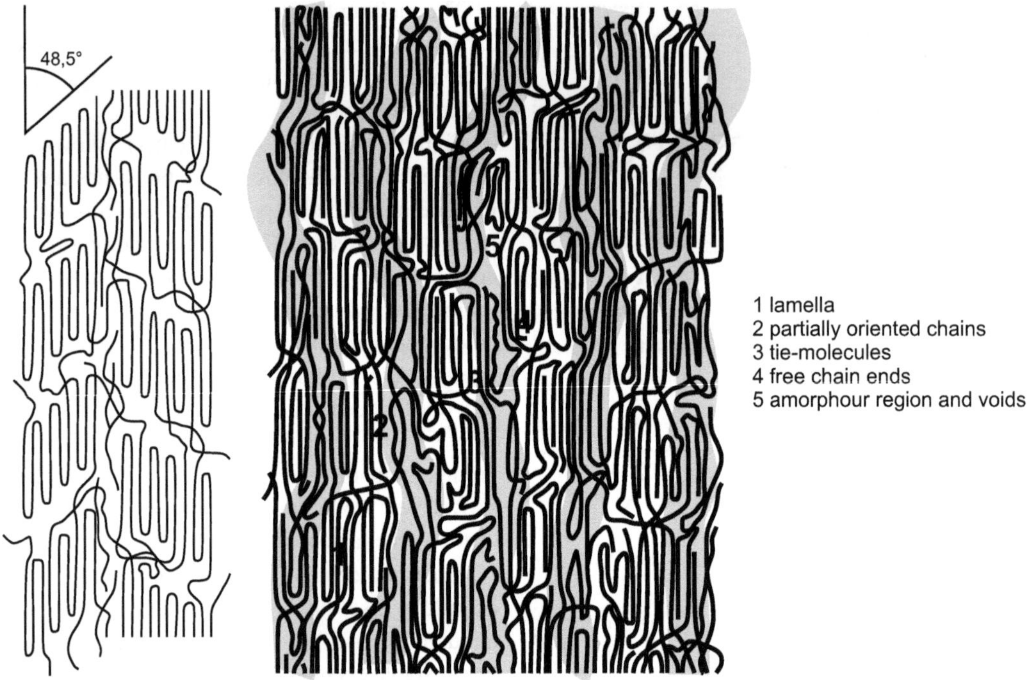

Fig. 4.13 Crystal structure of polyamide 66 (left) acc. to Hearle and Greer (1970) and of polyamide 6 (right) acc. to Murthy et al. (1990)

the α-structure and the γ-structure, and a transitional form, the β-crystals. They are shown in simplified form in Fig. 4.14.

Details of the triclinic crystal structure of PA 66 are depicted in Fig. 4.15. Due to the attracting forces between the polar NH- and CO-groups, respectively, the macromolecules are slightly offset from each other. This results in a slightly shifted structure in the z-axis. In contrast to PA 6, macromolecules with a "wrong sense of direction" cannot be present. A non-ideal crystal structure can therefore be corrected, e.g. by thermal

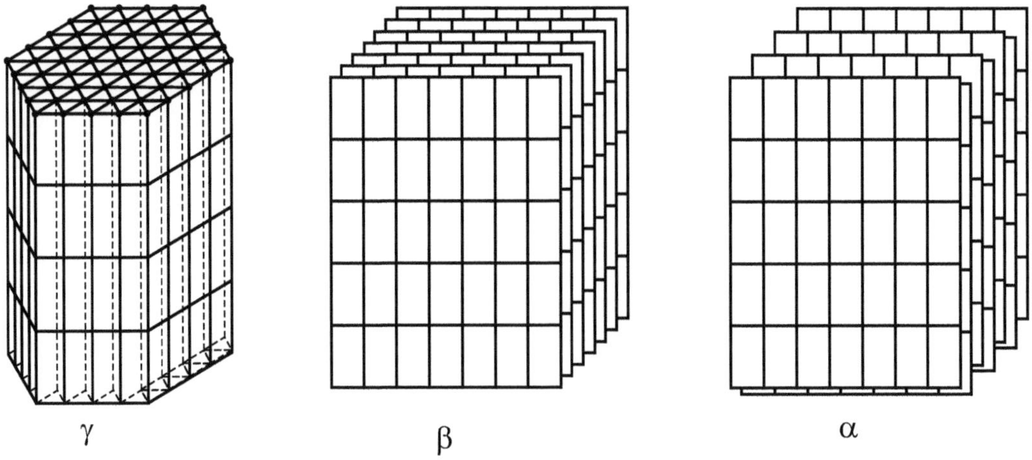

Fig. 4.14 Typical crystal structures of polyamide acc. to Bunn and Garner (1947), Arimoto (1964, 1965)

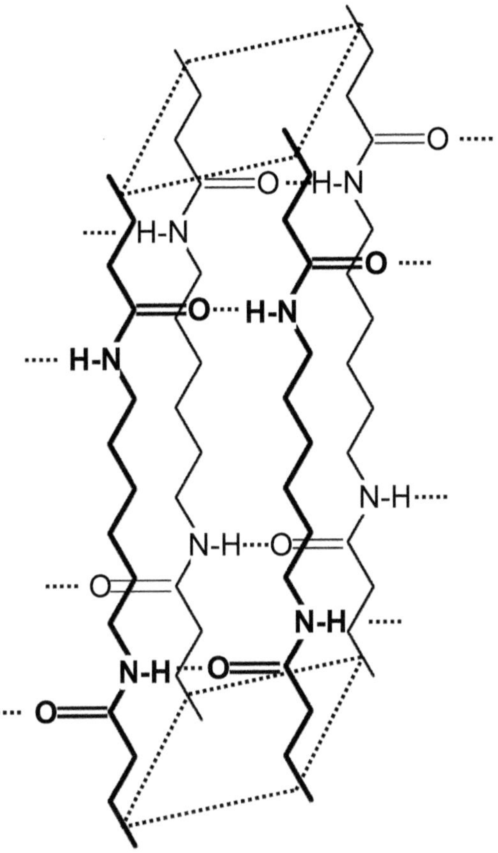

Fig. 4.15 Crystal structure of α-polyamide 66 acc. to Bunn and Garner (1947)

treatment, whereby the macromolecules can be shifted within the crystals, resulting in an energetically more favorable structure.

For PA 6, Heuvel and Huisman (1982) found that relatively large γ-crystals already form directly below the spinneret and their degree of orientation is about 60%. The number of γ-crystals rises with increasing take-off speed (= internal tension) and their thickness initially decreases, reaches a minimum at a speed of approx. 2700 m/min, and then increases again (Fig. 4.16). Furthermore, the closer the crystals are to the fiber axis, the greater the crystal thickness. On the other hand, more α-crystals are formed with increasing winding speed, and their thickness is only slightly dependent on their position relative to the fiber axis.

The approximately hexagonal-tightly packed γ-crystals can be transformed into slate-like α-crystals under the influence of stress and temperature, which considerably facilitates their drawability. At higher temperatures, the crystals become thicker in the process. As Fig. 4.17 shows, the transformation of γ- to α-crystals does not occur abruptly, but in a gradual process. It requires less energy to convert already oriented structures. Therefore, the already oriented γ-crystals are more easily transformed into α-crystals, which reduces the orientation of the

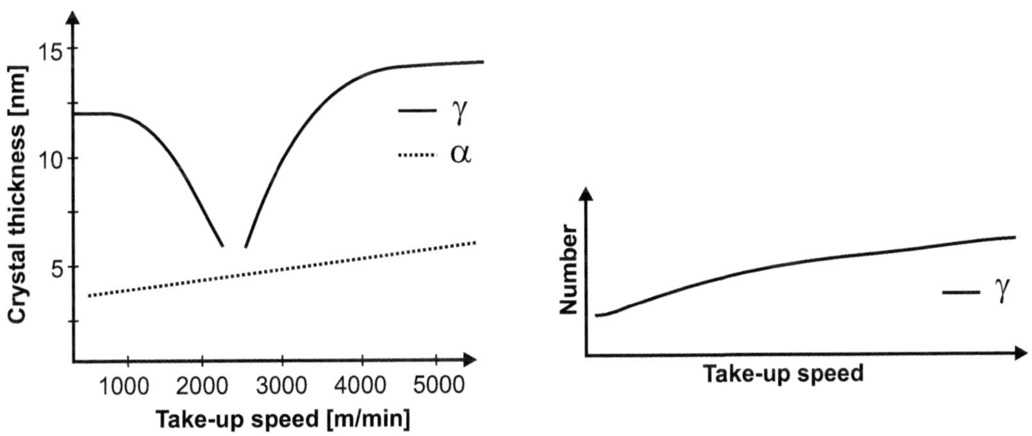

Fig. 4.16 Dependence of the crystal structure in PA on the take-up speed acc. to Heuvel and Huisman (1982)

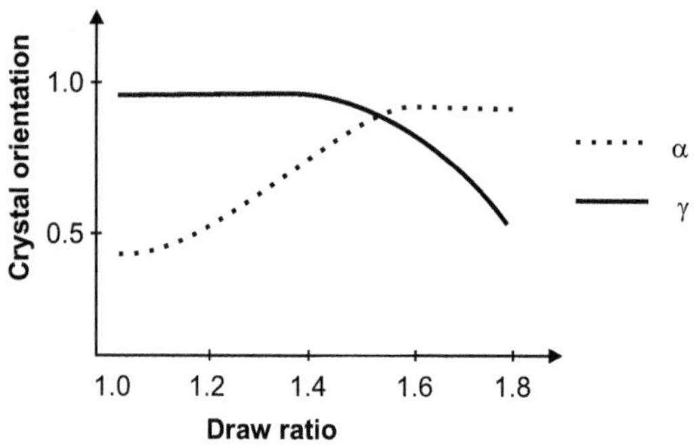

Fig. 4.17 Influence of the drawing ratio on the orientation of the crystals according to Heuvel and Huisman (1982)

remaining γ-crystals accordingly and increases that of the α-crystals overall at the same time.

4.5.2 Polyester

Aliphatic polyesters are not thread-forming, as Carothers already discovered. Only when the chains are sufficiently torsion-resistant through the incorporation of phenyl rings are stability and strength sufficient to produce fibers. By far the most commonly used polyester is polyester terephthalate (PET). It has phenyl rings as a central component and forces develop between the phenyl rings similar to those between the C-chains in polyamide. Only when the phenyl rings are incorporated in a para arrangement (CH_2 groups dock at opposite points in the ring) can crystalline regions form. In contrast to polyamide, polyester is almost completely amorphous at the beginning of fiber formation and crystallizes only slowly. The degree of orientation as well as the crystal thickness increase approximately linearly with rising take-off speed. Up to a take-up speed of approx. 3000 m/min, the relaxation forces of the polymer counteract the stretching forces that orient the macromolecules. From approx. 3000 m/min upwards, the chain molecules are sufficiently oriented in the direction of the fibers and, with simultaneous cooling, crystallization sets in. PET is then in a partially crystalline (pre-oriented) form and is therefore referred to as POY (partially or pre-oriented yarn).

For fully oriented PET yarns, the degree of orientation is about 40%. The aspect ratio is about 2, which is twice that of PA 6 and PA 66, respectively. The triclinic crystal structure of polyester (PET) is depicted in Fig. 4.18.

Longer aliphatic portions in the polyester, such as in polytrimethylene terephthalate (PTT), increase the mobility of the chains and therefore lead to elongation values similar to those of polyamides. Their crystal structure then also corresponds to that of polyamides (cf. chapter on polyester for details).

4.5.3 Aramid

Aramids consist of rigid macromolecules in which amide groups alternate with phenyl groups. This increases the attracting forces between the chains, both through the amide groups and through the mutual attraction of the phenyl rings. The latter forces are of the order of the van der Waals-forces in aliphatic polymers. Basically, two types of aramid are distinguished, meta- and para-aramid. Although they differ only in the position of the functional groups relative to each other, they lead to very different crystal structures.

p-aramid fibers are manufactured using the air-gap spinning process. Already in the spinning solution, small crystalline areas, so-called liquid crystals, are formed due to the strong mutual attraction of the macromolecules (Fig. 4.19). During coagulation of the fibers in the air gap and in the subsequent precipitation bath, these crystals are oriented and large crystalline zones with a high packing density are formed. In between, there are amorphous areas which can also be oriented under tensile stress and heat.

p-aramid displays the superstructure shown in Fig. 4.20. Here, the crystallites form radial, leaflet-like structures. This results not only in high strength in the direction of the fiber axis, but also transversely, which can only be observed to a limited extent in PA 6, PA 66 and PET.

In m-aramid, the NH and CO functional groups point alternately in different directions (Fig. 4.21). This massively hinders crystal

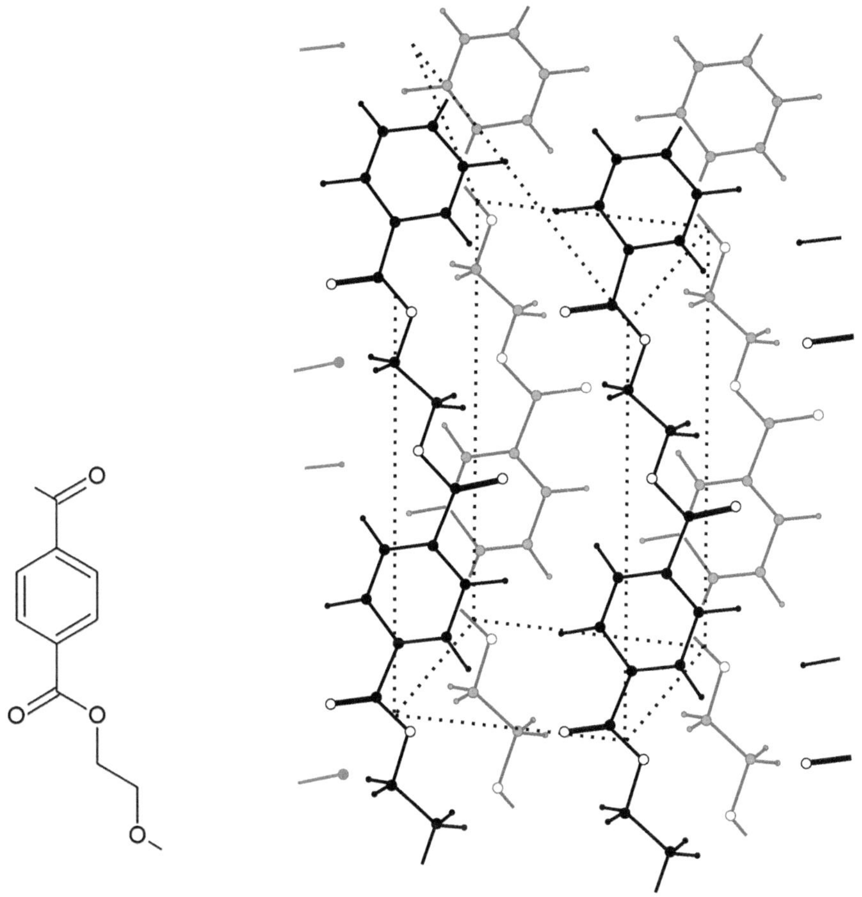

Fig. 4.18 Repeat unit (left) and triclinic crystal structure of PET (right)

formation in solution and there are no liquid crystals as in p-aramid. Therefore, after spinning both, partially crystalline and amorphous areas, are formed, similar to polyamide and polyester. Accordingly, the mechanical properties, in particular the modulus of elasticity, are significantly lower than those of p-aramid.

4.5.4 Polyolefins

Polyethylene is a very simple polymer without side groups. Therefore, it is easy to crystallize and degrees of orientation of 80% and more are achieved.

In polypropylene, there is a methyl group in each monomer unit (-CH_3). If all methyl groups are on one side (isotactic PP), either a monoclinic or an α-helix structure results. This is not possible if the side groups are arranged randomly, as is the case with atactic PP. This polymer cannot be crystallized and is therefore unsuitable as a fiber material. However, the tendency to crystallize and the degree of order can be influenced by deliberate incorporation of atactic segments into isotactic PP. Syndiotactic polyethylene is partially crystalline (Fig. 4.22).

4.5.5 Polyacrylonitrile

The high polarity of the $C \equiv N$-group in the macromolecules causes the polyacrylonitrile chains to form a helix. Several helices accumulate next to each other in parallel to form the crystal structure shown in Fig. 4.23. The electrostatic

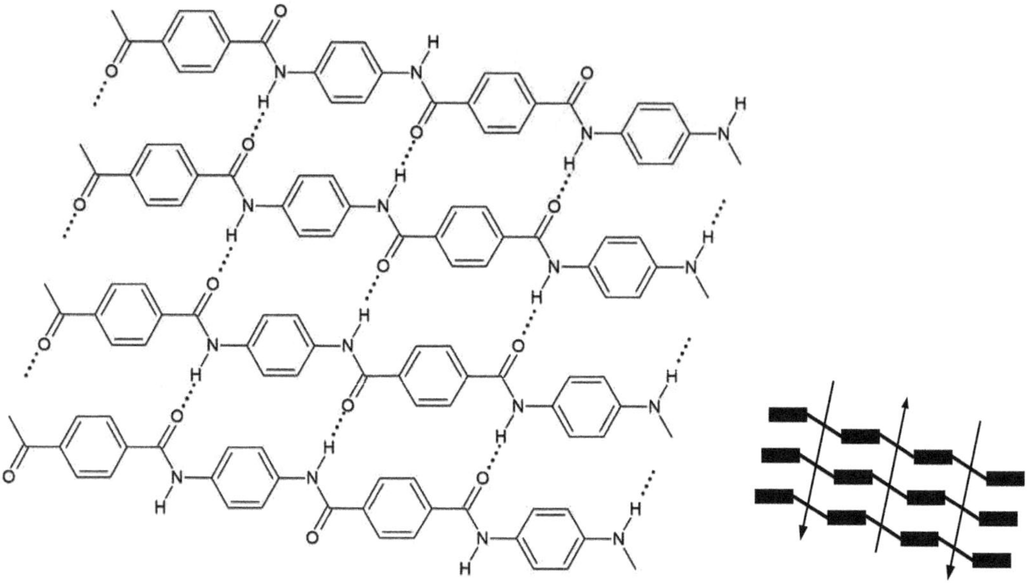

Fig. 4.19 Crystal structure of para-aramid (p-aramid)

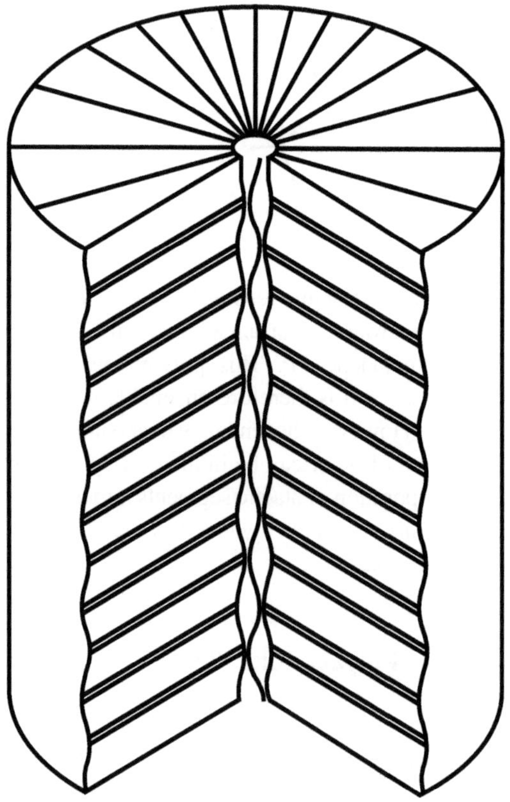

Fig. 4.20 Radial folded sheet structure of para-aramid acc. to Morton and Hearle (2008)

forces between the chain molecules have no preferred orientation. Therefore, one speaks of a pseudo-crystal. These areas make up about 50% within the structure, the rest is amorphous.

PAN fibers tend to fibrillate. This is due to the solution spinning process, where precipitation is not perfect and uniform. Thus, solvent bubbles form in the polymer which are precipitated later. The resulting cavities collapse and, when stretched, and a fibrillar network is formed from this sponge-like structure (Fig. 4.24).

4.5.6 Cellulosic Fibers

The β-D-glucose units of cellulose macromolecules possess numerous polar OH-groups. The OH-group at C atom no. 5 can form a hydrogen bond with the OH-groups of the glucose unit of a neighboring cellulose molecule. This results in very regular and stable crystalline structures (Fig. 4.25).

A detailed model of the cellulose structure was proposed by K. H. Meyer and L. Misch in 1937. It is also known as "Cellulose I" (Fig. 4.26).

Mercerization reduces the lattice height by approx. 10% and the drawn angle is reduced to

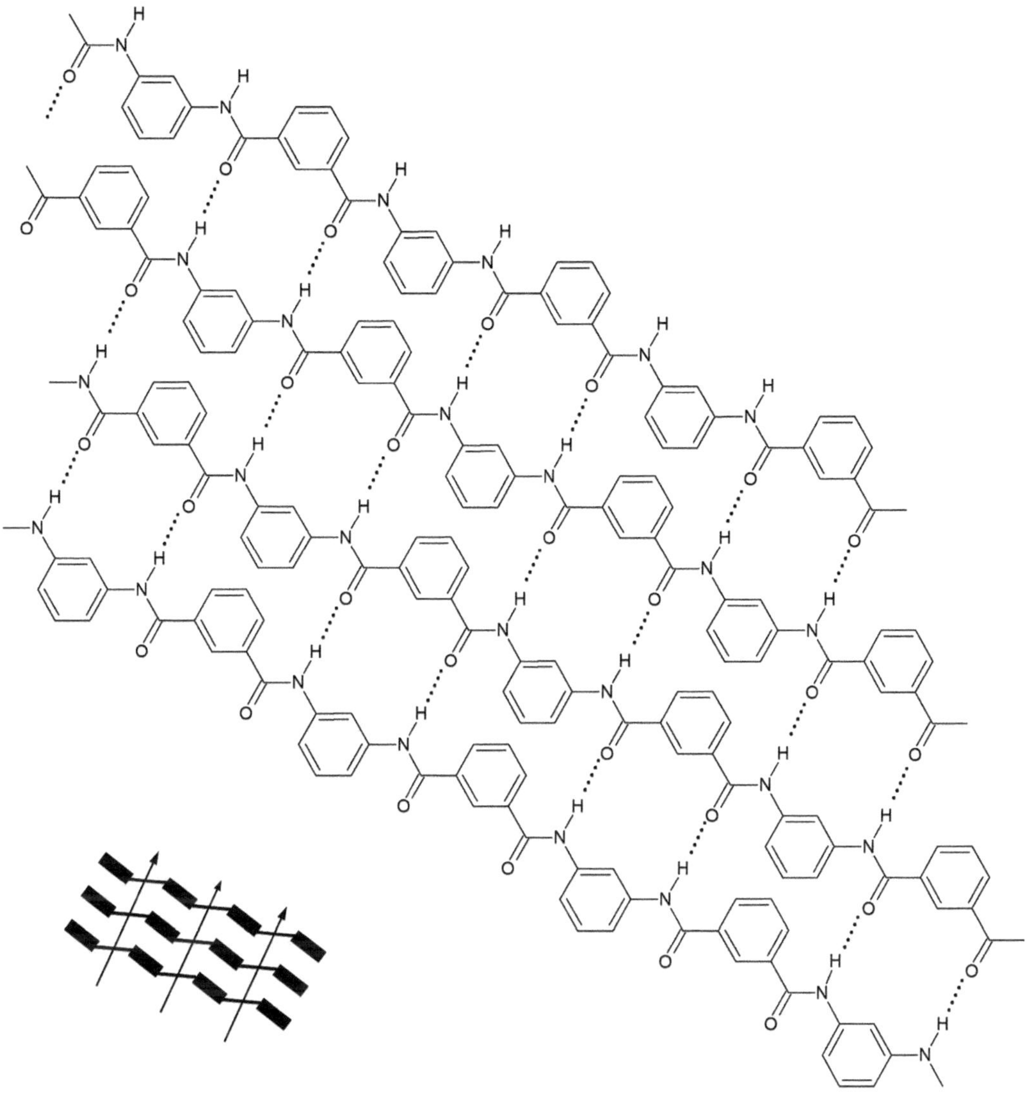

Fig. 4.21 Crystal structure of meta-aramid

62°, resulting in the so-called "cellulose II". It is also present in regenerated cellulose fibers.

Cellulose is almost 100% crystalline in fibers. The measurably detectable "disorder" is caused by small crystallites that are not perfectly arranged to each other. Amorphous areas, however, hardly exist or not at all.

Cellulose chains grow by enzymatic processes. As a result, the cellulose macromolecules are mostly aligned and parallel to each other. Crystallization occurs in such a way that the free energy in the polymer is minimized. The greatest attractive forces between the macromolecules exist along the sides of the chain as shown in Fig. 4.27. This results in block-like or slate-like superstructures known as microfibrils. These in turn can form (concentric) lamellae, which can also be seen in microscopic images.

The microfibrils are helically deposited in layers on the inside of cells in natural cellulosic fibers. This automatically creates tension between the fibrils, which leads to separation. Thus, a single, large cellulose crystal does never occur,

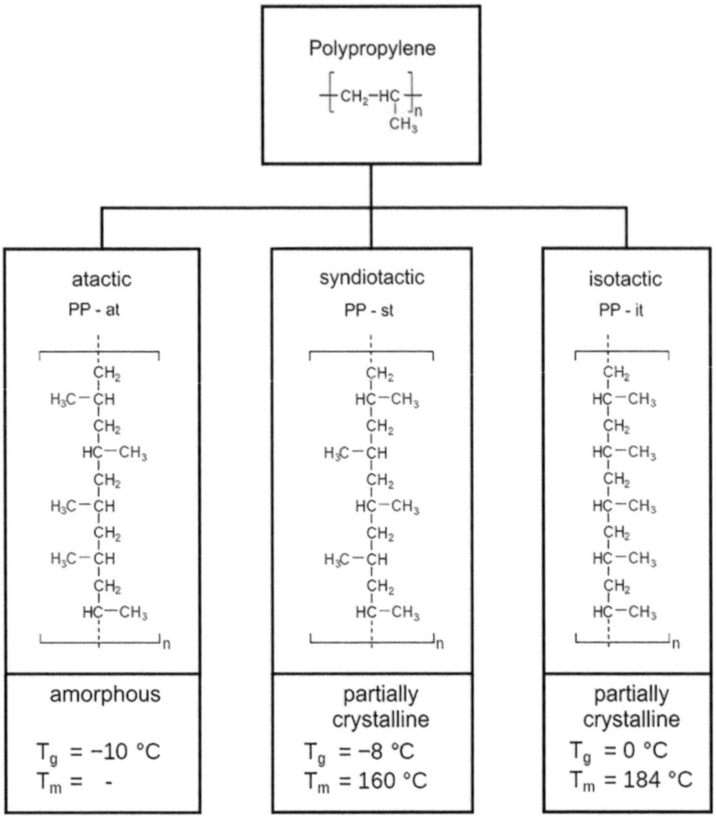

Fig. 4.22 Different structures of polypropylene and their influence on the properties (Minihaa, 2016)

but rather numerous small crystals are formed and thus always a certain "disorder" of the structure.

In crease-resistant finishing, a resin is added to the fibers (usually cotton), which polymerizes in the fiber. It connects the OH-groups of the cellulose chains with each other via covalent bonds. This largely prevents shrinkage and thus considerably reduces the formation of wrinkles.

During the regeneration of cellulose in the viscose process, a core-shell structure is formed. It is assumed that the Zn^{2+} ions form a chemical bond with the cellulose xanthate in the precipitation bath. Polar forces generated by the Zn^{2+} ions quickly lead to the formation of a small-scale network before the zinc cellulose xanthate is regenerated into cellulose. Normally, the Zn^{2+} ions cannot penetrate to the core regions because of their low mobility, so a different crystal structure with larger units is formed there. During the subsequent stretching, the mantle areas are therefore better oriented and thus have a higher strength than the core.

In acetate fibers, the OH-groups of the glucose rings are replaced by acetate groups. In the case of diacetate, there are on average 2.5 out of 3 substituted, and in the case of triacetate all 3. Because the acetate groups are not very polar and do not form hydrogen bonds, the attractive forces between neighboring macromolecules are lower than in normal cellulose. In addition, the acetate groups are relatively large, so the distance between the chains is always greater than in cellulose, and the overall structure is much more irregular than cellulose, making it difficult for crystalline regions to form. Therefore, diacetate fibers display a lower tenacity than viscose, higher elongation, lower density, and absorb less water. Tricacetate fibers have a higher internal order because their structure is more regular. However, they absorb even less water because there are no free OH-groups.

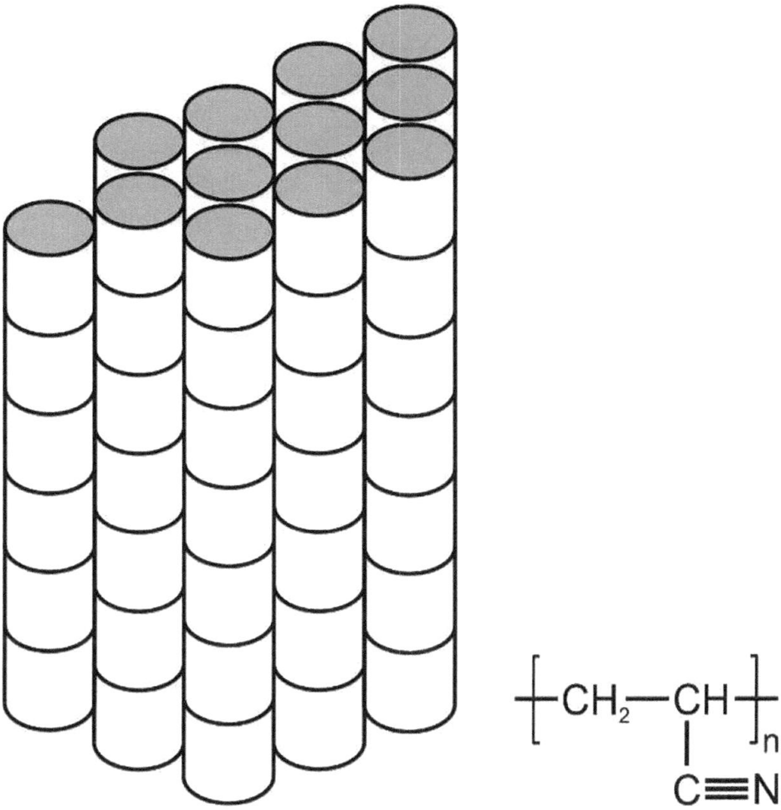

Fig. 4.23 Pseudo-crystal structure of polyacrylonitrile acc. to Morton and Hearle (2008) and repeating unit of PAN

4.5.7 Protein Fibers

Protein-based fibers are made up of amino acid building blocks. This includes all silks, wools and fine animal hair.

Mulberry silk consists of two fibroin filaments surrounded by a sheath of sericin. This is removed during processing (degumming). Four amino side groups are primarily present in fibroin: –H (glycine), $-CH_3$ (alanine), $-CH_2$ OH (serine), and $-CH_2-C_6-H_4-OH$ (tyrosine). Sulfur-containing side groups are hardly present. Therefore, the intermolecular forces are mainly based on attracting forces of the main chains and, to a lesser extent, on hydrogen bonds of the side chains. The crystalline regions form a folded sheet structure (Fig. 4.28).

As X-ray investigations have shown, the macromolecules are mostly fully stretched and highly crystalline. There is no refolding as in

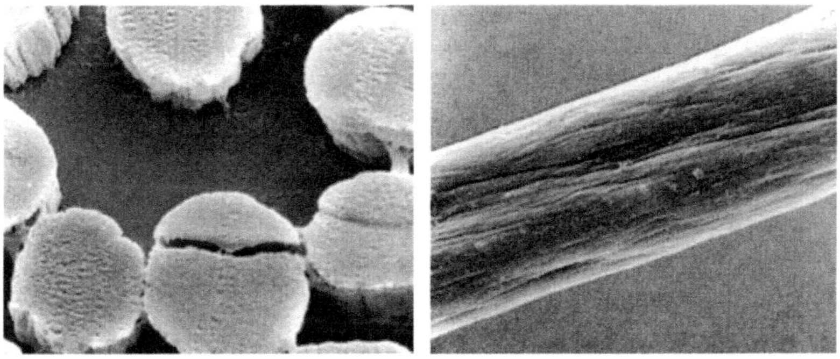

Fig. 4.24 PAN fibers

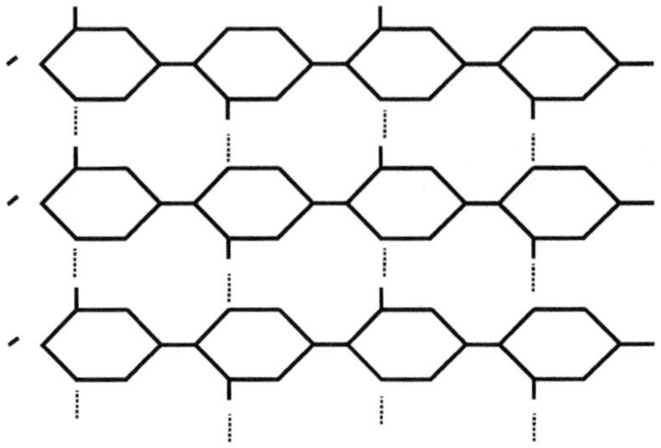

Fig. 4.25 Structure of cellulose (top) and model of cellulose crystal (bottom) acc. to Morton and Hearle (2008)

cellulose. Coleman and Howitt (1947) found that fibroin macromolecules are composed of two segments. A short segment (17 of 140 nm) comprises the amino acids with large side groups (e.g. tyrosine), while the long segment (123 of 140 nm) is composed of glycine, alanine and serine. This forms crystalline regions, while the short segment is largely non-crystalline. Thus, fibroin is a co-polymer.

The peptide chain of wool fibers contains 17 different amino acid side groups, with most of them occurring in similar concentrations. Similar to silk, according to (Pauling and Corey, 1951) the wool proteins consist of crystalline and non-crystalline domains. The former contain the short side chains without sulfur bridges and thus can easily attach parallel to each other. The latter have sulfur bridges and, in particular, glycine and tyrosine side chains. These largely

Fig. 4.26 Model of cellulose I acc. to Meyer and Misch after Morton and Hearle (2008)

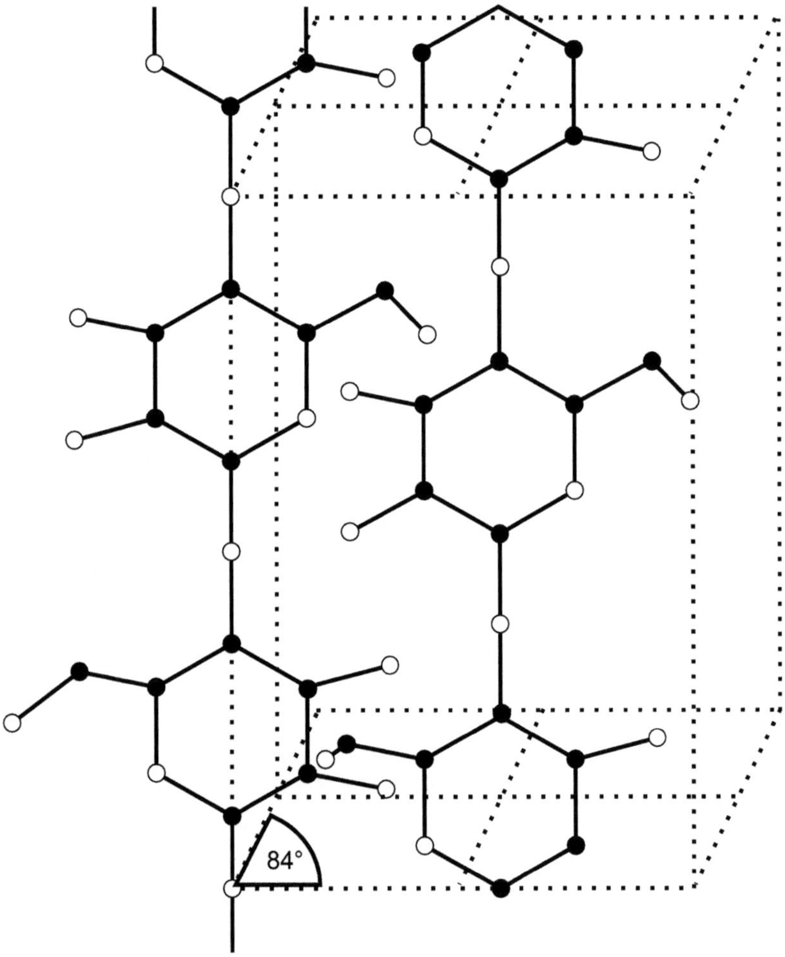

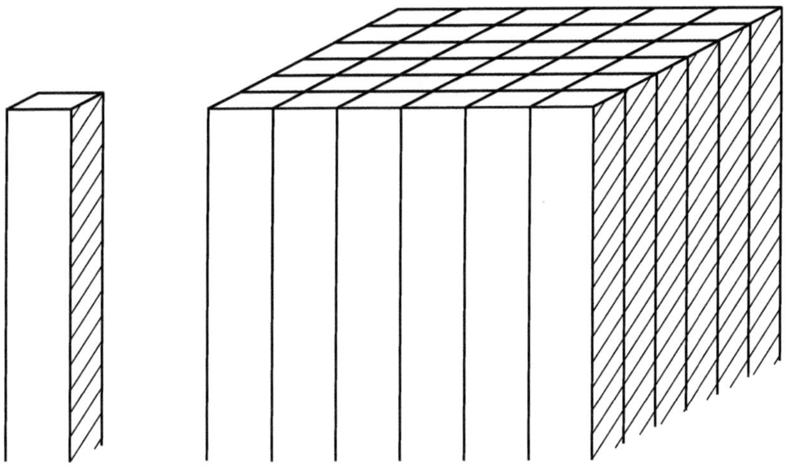

Fig. 4.27 Formation of microfibrils from single cellulose molecules acc. to Morton and Hearle (2008)

amorphous regions are closely linked via numerous sulfur bridges (cystine).

In raw wool, the peptide chains are present in the form of what is known from (Pauling et al., 1951) as the α-helix (α-keratin). If the fiber is stretched, this structure is transformed into β-keratin, the structure of which is similar to that of fibroin. Several protein chains then lie parallel to each other (Fig. 4.29).

4.6 Length of the Macromolecules

Figure 4.30 gives an overview of the chain length of important synthetic polymers. It is evident that most macromolecules have lengths between 100 and 350 nm, while typical crystal lengths are 5 to 10 nm. Thus, each macromolecule is either part of several crystalline domains and in

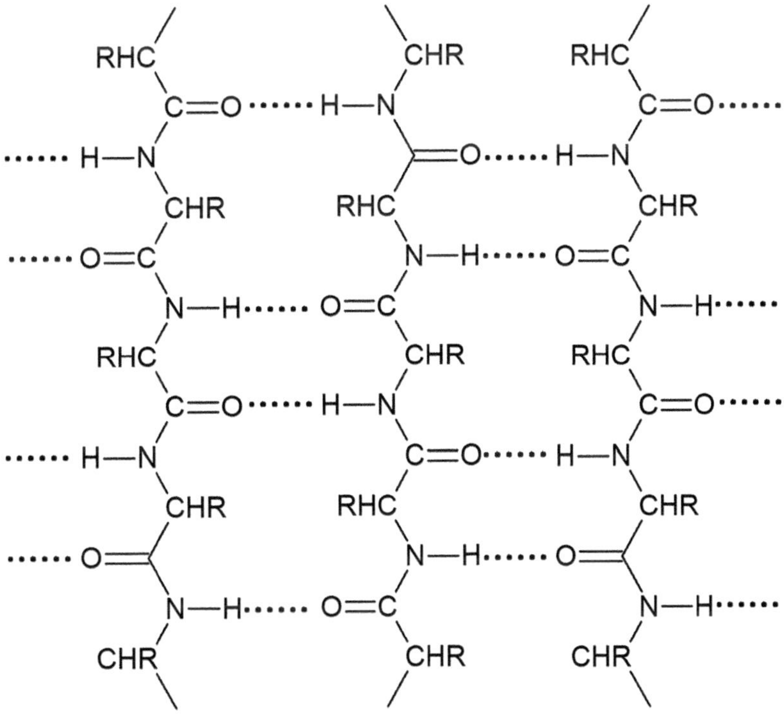

Fig. 4.28 Folded sheet structure of β-keratin (fibroin)

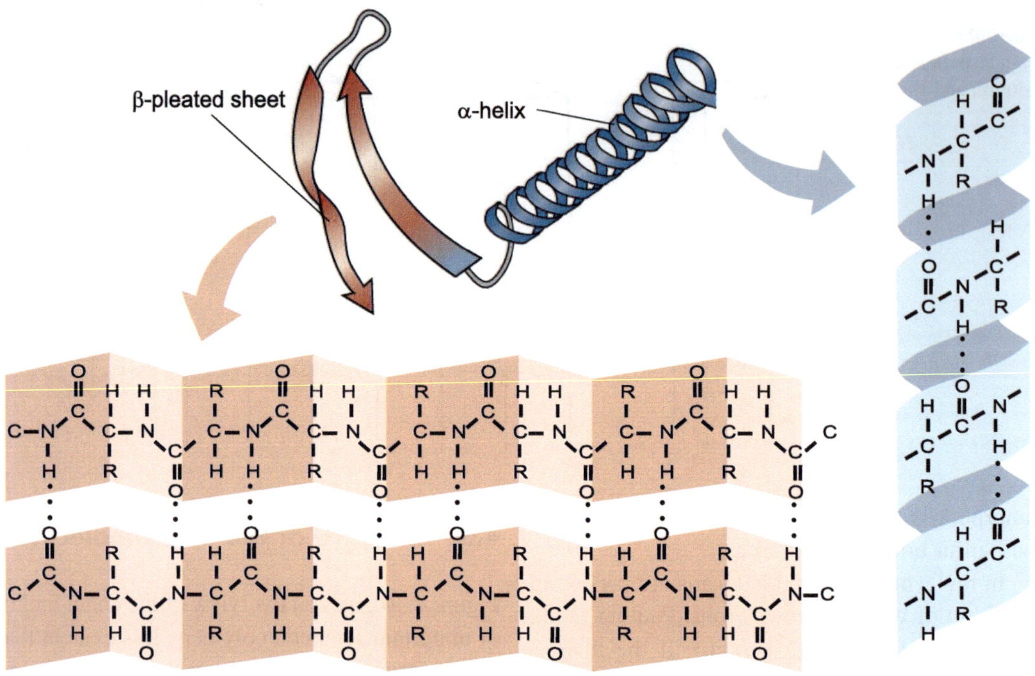

Fig. 4.29 Structural transformation of β-keratin to α-keratin (wool)

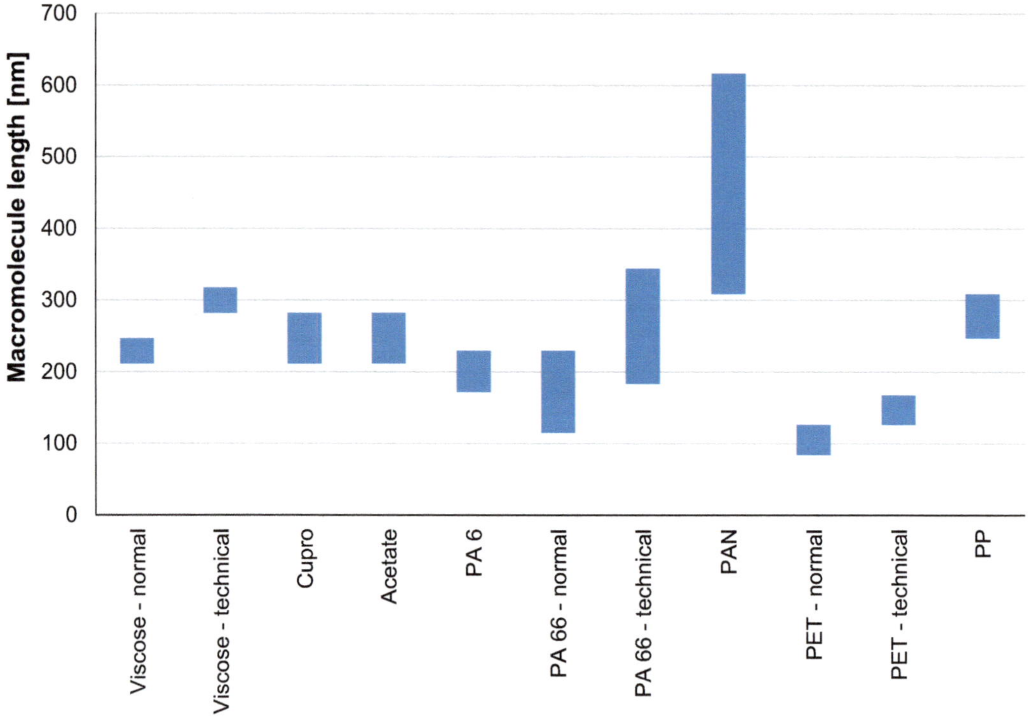

Fig. 4.30 Chain length of important polymers

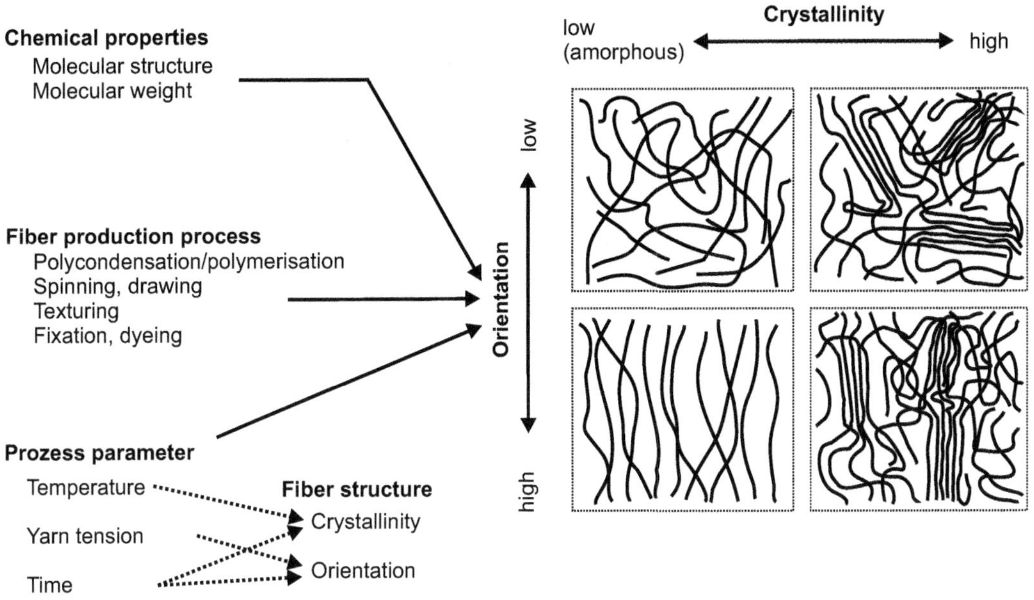

Fig. 4.31 Crystallinity and orientation of polymers acc. to Fourné (1995)

between in an amorphous state or passes through a crystal several times as it folds back.

The chain length of natural cellulosic fibers is significantly higher, with values between 2000 and 3000 nm (cotton) and 3500 and 5600 nm (flax). This is due to the fact that these macromolecules form very slowly and thus reach greater lengths than, for example, viscose, where the viscosity of the polymer at such chain lengths would be much too great for the spinning process. Therefore, the original chain length of the cellulose is reduced to 200 to 300 nm before spinning.

4.7 Orientation

Stiff chains quickly form stable crystal structures, as shown above. This leads to a high Young's modulus. Structures with long aliphatic chain segments tend to tangle. However, they can be easily drawn, which is often facilitated by applying heat, and thus oriented. This leads to fibers with high tenacity.

In the melt or solution of a linear polymer, the chain molecules are generally in a disordered random state. In the solidified state—preferably above the glass transition temperature—they can be oriented by external deformation of the polymer structure. In this process, the molecules or at least parts of them—depending on the degree of orientation—adopt more and more the direction of deformation. Both, the chain parts of the non-crystalline portion and the crystalline regions of ordered chain parts, are oriented. Figure 4.31 depicts the four different states schematically. Normally, there is always a mixture of these states present in a polymer.

Natural fibers of plant origin already grow in an oriented state. In the case of man-made fibers, this must first be brought about by changing the macromolecules orientation under tension, partly during spinning, but for the most part by a separate process step, drawing. The increasing parallelism of chain parts causes a denser packing or allows them to come together to form ordered crystalline areas. Their character usually differs considerably from that of the crystalline structure, which may already have existed in the unoriented state. For these structural changes during orientation, numerous model conceptions have been worked out, which are continuously developed further.

The close arrangement created during stretching prevents the stretched yarn from contracting significantly again when the applied tension is released and it is still above the glass transformation point. This inhibition can be further enhanced by heat treatment.

The birefringence n is characteristic for the orientation. It represents the difference of the refractive indices and thus of the dielectric polarizabilities in the direction of orientation and perpendicular to it. In addition to birefringence and the evaluation of wide-angle X-ray diagrams, there are other methods of measuring orientation, some of which even distinguish between amorphous and crystalline regions. An actual measurement of how far the filaments can be drawn additionally can already give an indication of the internal macromolecular orientation. This is easy to see in a uniaxial tensile test.

As the orientation of a fiber increases, its tenacity and modulus of elasticity generally increase and elongation at break decreases. It is therefore possible to adapt the fiber material to the respective application. In practice, however, it is often not easy to separate the influences of the orientation from those of the crystallinity, which usually increases with it. This also applies to the dyeing behavior. The maximum dye uptake decreases with increasing crystallinity. The diffusion rate and thus the speed of the dyeing process is additionally influenced by the packing density of the amorphous regions. It usually goes up with increasing orientation and slows down the dyeing process accordingly.

4.8 Drawing Process

The mechanical properties of the fibers immediately after spinning are rarely sufficient: Tenacity, modulus of elasticity and thermal stability are usually too low, and plastic ductility is often too high for textile processing.

The main reason is the insufficient orientation of the chain molecules, both in the amorphous and the crystalline areas. By applying an external tension (e.g. by drawing the filaments below the spinneret), the macromolecules are very quickly oriented in the direction of the fiber axis (Fig. 4.32). At the same time, crystalline areas often form from polymer chains arranged in parallel.

Depending on the polymer, this process goes differently: In PAN, the internal structure remains amorphous, but more cross-links are formed between the molecular chains. In PET, a partial transformation from amorphous to crystalline regions takes place. In the case of PA and PP, which are already partially crystalline shortly after leaving the spinneret, the crystallinity increases further. The same applies to fast-spun PET.

The stretching process is illustrated in Fig. 4.33 for a semi-crystalline polymer. Initially, the deformation at small strains is associated with a linear increase in force (1). It is often reversible. In the plateau phase, the amorphous regions are aligned in the fiber axial direction (2). This is associated with a large increase in length and a relatively constant force up to (3). The draft at this

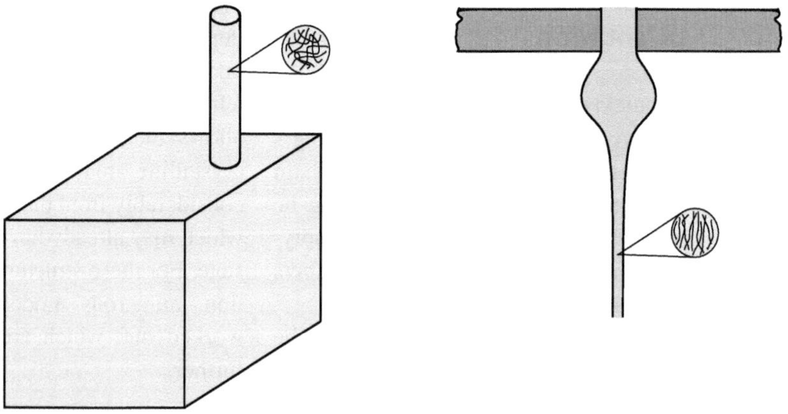

Fig. 4.32 Comparison of the orientation of the macromolecules in undrawn (solid) and drawn polymer acc. to Gries

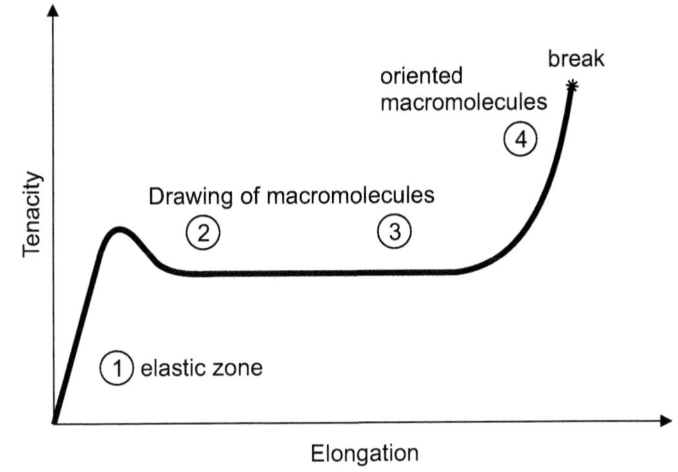

Fig. 4.33 Force-elongation curves during stretching of polymers

point is referred to as the "natural draw ratio". Thereafter, in phase (4), all macromolecules are aligned in the fiber axial direction and as the load increases, the yarn eventually breaks. If phase (4) takes place above the glass transition temperature, the resulting strain is reversible up to a certain point. The forces required are correspondingly lower due to the macromolecules moving more easily as a result of the higher temperature.

In the case of solution-spun yarns, residues of the solvent in the spinning mass can act as a kind of lubricant and thus considerably reduce the drawing forces while simultaneously achieving a high elongation.

The shape of the tenacity-elongation curves depend in particular on the temperature of the polymer, as Fig. 4.34 shows. The temperature increases from curve 1 to 9. At temperatures below T_{glass}, fracture occurs in a brittle manner (curves 1–3). Above this, plastic deformation occurs with a pronounced maximum (4–6), after which the curves increasingly flatten. The maximum fracture work (toughness) results for curves 4–6, because the strain is large and also the force is relatively high. After that, the strain increases further (7–9), but the forces become relatively low, which significantly reduces the area under the curve and thus the fracture work.

Various ideas have been developed for the structural transitions, but these are as certain or uncertain as the model ideas of the structure of semi-crystalline oriented and non-oriented polymers. In 1963, K. Kobayashi assumed that crystals originally oriented transverse to the fiber axis dissolve completely and form new crystal structures in the fiber axial direction (Fig. 4.35, left). Peterlin developed this model further in 1967 on the basis of his own structural investigations (Fig. 4.35, right). He assumed that the originally formed crystals rearrange themselves, twisting and partially unfolding in the process, before forming new structures, oriented both in the axial direction and transversely.

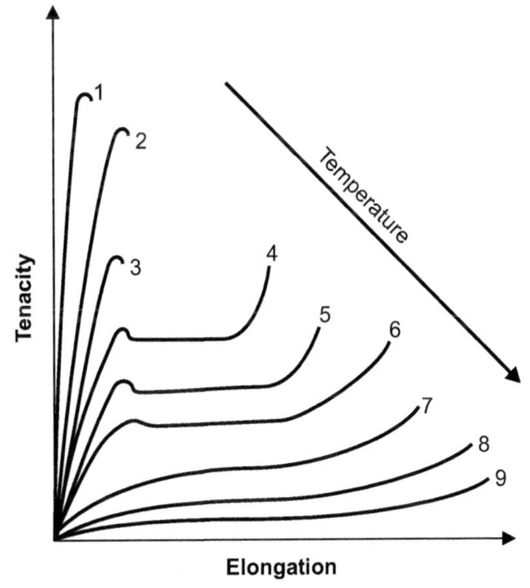

Fig. 4.34 Dependence of the KD curve on the temperature of a polymer acc. to Ziabicki (1976)

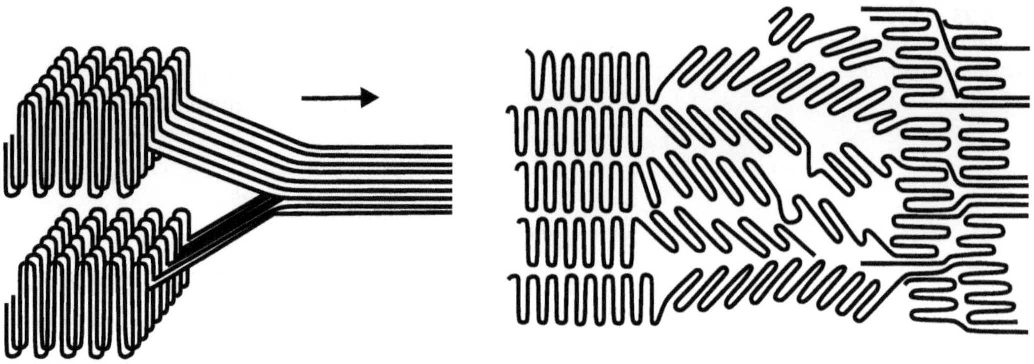

Fig. 4.35 Model representation of plastic deformation acc. to Kobayashi (left) and Peterlin (1965, right)

If one observes the stretching process macroscopically, a so-called stretching shoulder can usually be detected. Elias (2001) formulated from this the structural model shown in Fig. 4.36, which builds on Peterlin's model. It is considered by many to be the model that best explains the phenomena observed in practice. The length of the crystalline regions depends on the polymer and is usually in the range of 100–600 nm.

Figure 4.37 shows the so-called residual stretching for PET, PA 66 and PA 6 as a function of the winding speed. This is the remaining plastic stretching in the spun yarn. A value of e.g. 180% means that the yarn can still be stretched to 1.8 times its length. Due to the higher internal tension, the residual stretching decreases with increasing speed. However, this is not linear, because the rearrangement of the molecular chains not only depends on the winding speed, but is also influenced by the rearrangement processes described above. In the case of PA 6, this becomes clear in comparison with PA 66. The repeating unit is much smaller in PA 6 than in PA 66, therefore the chain molecules are more mobile and there is less residual stretching for the same induced stress. The value of the residual elongation is independent of the yarn count.

4.9 Summary

The internal structure of man-made fibers always depends decisively on the conditions during spinning and during after-treatment. Basically, drawing tension (spinning and winding speed), heat treatment and residence time can significantly change the crystal structure and orientation.

Amorphous regions are oriented by applying a tensile stress, e.g. by a higher spinning speed. By inducing heat, the number of crystallization nuclei can be increased and thus the size of the crystals formed can be reduced. Through storage,

Fig. 4.36 Structural model acc. to Elias (2001)

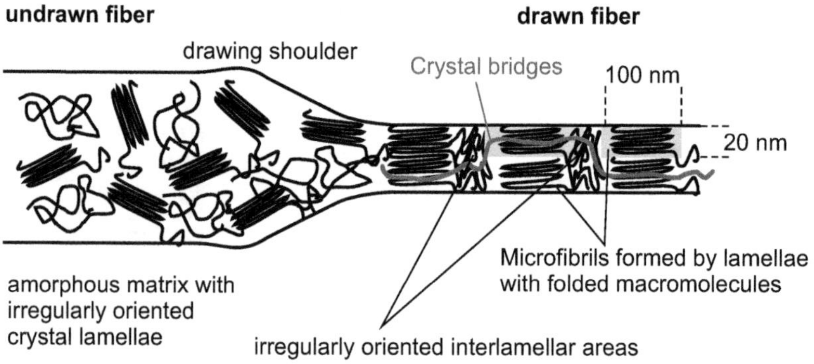

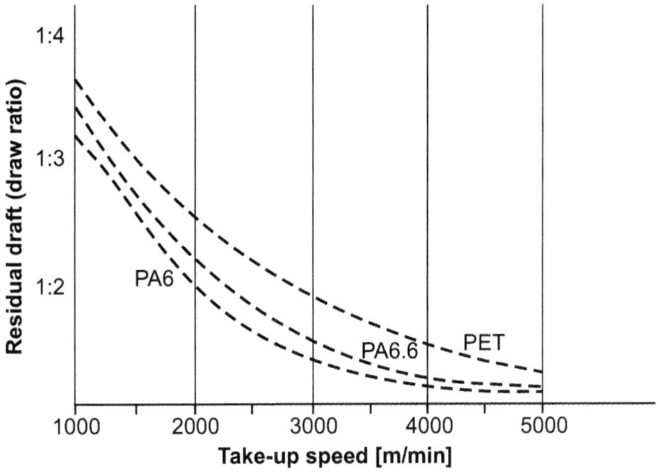

Fig. 4.37 Residual stretching as a function of the winding speed

an internal thermodynamic equilibrium can occur in the fiber material, whereby a new, energetically more stable state is established, which can be accompanied by a rearrangement of the crystals.

References

Abitz, W., Gerngroß, O., & Herrmann, K. (1930). Zur röntgenographischen Strukturerfoschung des Gelatinemicells. *Naturwissenschaften, 18*, 754–755.

Arimoto, H. (1964). Alpha-gamma transition of Nylon 6. *Journal of Polymer Science Part A, 2*, 2283f.

Arimoto, H. (1965). Crystal structure of gamma-form of Nylon 6. *Journal of Polymer Science Part A, 3*, 317f.

Avrami, M. (1939). *Kinetics of phase change I – General theory*. American Institute of Physics.

Bunn, C. W., & Garner, E. V. (1947). The crystal structures oft two polyamides ('nylons'). *Proceedings of the Royal Society A, 189*, 47f.

Coleman, D., & Howitt, F. O. (1947). The properties and constitution of fibroin. The conversion of fibroin into a water-soluble form and its bearing on the phenomenon of denaturation. *Proceedings of the Royal Society A, 190*.

Ehrenstein, G. W. (1999). *Polymer-Werkstoffe*. Hanser.

Elias, H. G. (2001). *Makromoleküle*. Wiley-VCH.

Fourné, F. (1995). *Synthetic fibers*. Hanser.

Hearle, J. W. S. (1958). A fringed fibril theory of structure in crystalline polymers. *Journal of Polymer Science, 28*.

Hearle, J. (2001). *High-performance fibres*. Woodhead Publishing.

Hearle, J. W. S., & Greer, R. (1970). On the form of lamellar crystals in nylon fibres. *Journal of the Textile Institute, 61*, 240f.

Heuvel, H. M., & Huisman, R. (1982). Effects of winding speed, drawing and heating on the crystalline structure of Nylon 6 Yarns. *Journal of Applied Polymer Science, 26*, 713–732.

Hindeleh, A. M., & Hosemann, R. (1988). *Paracrystals representing the physical state of matter*. IOP Publishing.

Illers, K. H., & Haberkorn, H. (1971). Melting behavior, structure and crystallinity of 6-polymide. *Makromolekulare Chemie, 142*.

Kargin, V. A. (1958). Structure and phase state of polymers. *Journal of Polymer Science, 39*.

Kikutani, T., et al. (1989). Measurement of filament temperature and analysis of orientation-induced crystallization behavior in high-speed melt spinning process. *Seikei-Kakou, 1*(3).

Kong, Y., & Hay, J. N. (2003). The enthalpy of fusion and degree of crystallinity of polymers as measured by DSC. *European Polymer Journal, 39*.

Menges, G., Haberstroh, E., Michaeli, W., & Schmachtenberg, E. (2002). *Werkstoffkunde Kunststoffe*. Hanser Verlag.

Minihaa. (2016). Retrieved from https://commons.wikimedia.org/wiki/File:Polypropylene_tacticity_de.svg, CC0 1.0 Universal (CC0 1.0)

Morton, W. E., & Hearle, J. W. S. (2008). *Physical properties of textile fibres*. Woodhead Publishing.

Murthy, N. S., Reimschuessel, A. C., & Kramer, V. (1990). Changes in void content and free-volume in fibers during heat setting and their influence on dye diffusion and mechanical properties. *Journal of Applied Polymer Science, 40*, 249f.

Pauling, L., & Corey, R. B. (1951). The pleated sheet, a new layer configuration of polypeptide chains. *Proceedings of the National Academy of Sciences, 37*.

Pauling, L., Corey, R. B., & Branson, H. R. (1951). The structure of proteins: Two hydrogen-bonded helical configuarions of the polypeptide chain. *Proceedings of the National Academy of Sciences, 37*.

Peterlin, A. (1965). Crystalline character in polymers. *Journal of Polymer Science: Part C, 3*, 61–69.

Roland. (2015). Retrieved from https://commons.wikimedia.org/wiki/File:Spherulite2de.svg, CC BY-SA 3.0.

Salem, D. R. (Ed.). (2000). *Structure formation in polymeric fibers*. Hanser Verlag.

Seifriz, W. (1930). The alveolar structure of protoplasm. *Protoplasma, 9*, 177–280.

Ziabicki, A. (1974). Theoretical analysis of oriented and non isothermal crystallization. *Colloid and Polymer Science, 252*, 3.

Ziabicki, A. (1976). *Fundamentals of fibre formation*. Wiley.

Part II

Natural Fibers

5 Cotton

According to ASTM D7641-21 and the German standard DIN 60001 (part 1), cotton belongs to the group of plant or seed fibers. It is cultivated in monocultures on mostly very large fields on all continents except Antarctica (Fig. 5.1).

The fibers are unicellular and belong to various Gossypium species. After harvesting, seed cotton is obtained with the fibers still attached to the seed. After ginning (separating the fibers from the seed), raw cotton (lint cotton) and linters are derived. Linters are short, non-spinnable fibers with a length of 1 to 6 mm which are used for paper production (e.g. banknotes). Edible oil may be gathered from the seed by pressing. The fibers are slightly twisted and have a typically kidney-shaped cross-section (Fig. 5.2).

5.1 History

The cotton plant (latin: Gossypium) belongs to the Malvaceae (mallow) family. It has been domesticated at least four times: Twice in America (Gossypium hirsutum and Gossypium barbadense), once in Asia (Gossypium arboreum) and once in Africa (Gossypium herbaceum). The oldest finds of cotton textiles date from around 5800 BC from a cave near Tehuacán (Mexico). Remains of cotton bolls and textiles were found there. The oldest finds from South America come from Peru from the third and second millennium BC. From the lower reaches of the Indus in what is now Pakistan come fragments of fabric dating to 3000 BC. In India, cotton cultivation and fiber processing were practiced on a larger scale from 1500 BC. Herodotus mentions protective armor made of cotton fabrics worn by soldiers in India around 445 BC (Herodotus, 3, 106). In the second century BC, cotton fabrics were known as luxury articles in Greece, and later also in the Roman Empire. However, people had a somewhat blurred idea of the origin of cotton, as Fig. 5.3 shows. It was assumed by many until the Middle Ages that cotton fibers were produced by sheep-like animals that grew on trees and died when the grass under the tree was eaten away.

Until the sixteenth century, India was the main supplier of cotton fabrics. It was brought to Spain by the Crusaders and the Arabs around 1000 AD via Sicily and North Africa, where it was also cultivated. Cotton was introduced on a larger scale to Egypt in 1821 by the Khedive Mohamed Ali. The name for cotton used in most European languages, "katun", "coton" and also "cotton", is derived from the Arabic "al qatun".

Systematic production in the USA did not begin until the beginning of the seventeenth century (1621 Virginia, 1733 Carolina, 1734 Georgia). In the southern states of the USA, East Indian cotton varieties were cultivated at the beginning of the eighteenth century (Fig. 5.4). Until the second half of the nineteenth century, the American farmers tried to solve the problem of the enormous manpower required for the manual extraction of cotton fibers (harvesting,

D. Veit, *Fibers*, https://doi.org/10.1007/978-3-031-15309-9_5

Fig. 5.1 Cotton field shortly before harvest (Vardeman, 2009)

ginning) under difficult climatic conditions by importing slaves from Africa.

Cotton obtained through slave labor was exported to Europe for further processing. Some of the cotton was exported back to Africa to pay slave traders for new slaves, who were in turn trafficked to the Americas. This commodity cycle is called the triangular trade (Fig. 5.5).

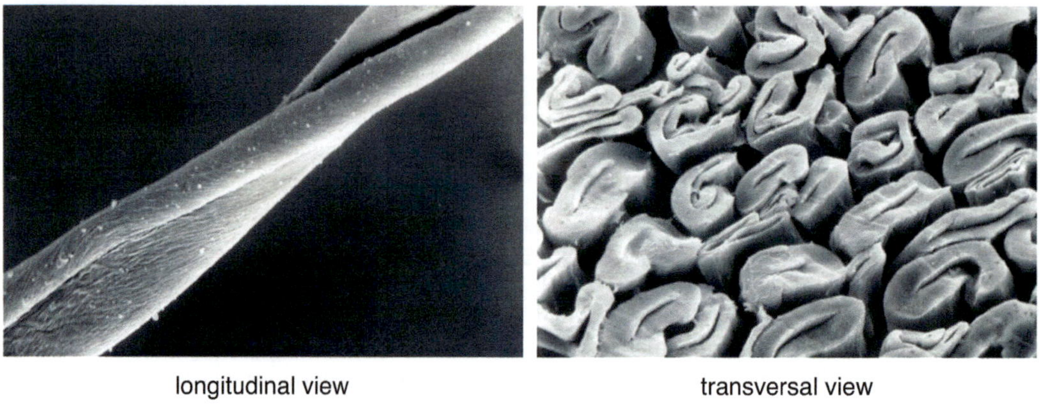

Fig. 5.2 SEM images of cotton fibers

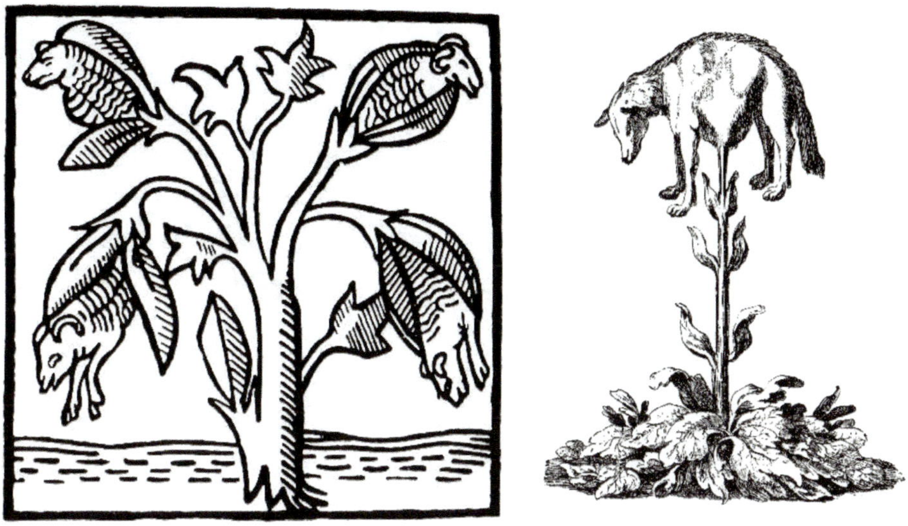

Fig. 5.3 Cotton tree (Mandeville, fourteenth century; Lee, 1887)

That the obverse of many 1861 Confederate States of America banknotes depicts slaves picking, processing, and loading cotton shows how important cotton farming was to the Southern states (Fig. 5.6).

However, partly because slavery contradicted the declared American rights of freedom, the North went to war against the South. With the victory of the North, slavery ended and the textile industry of the Southern states initially fell into a severe crisis.

The processing of cotton in Europe only really took off after the spinning machine (1764: Spinning Jenny by J. Hargreaves) and the

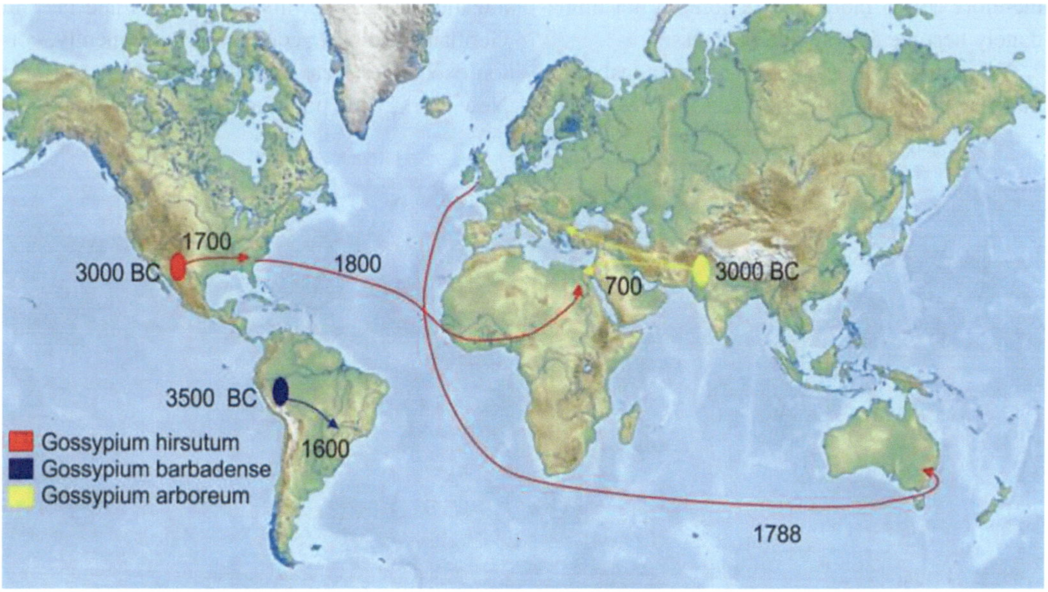

Fig. 5.4 Spread of cotton (map: Mapswire)

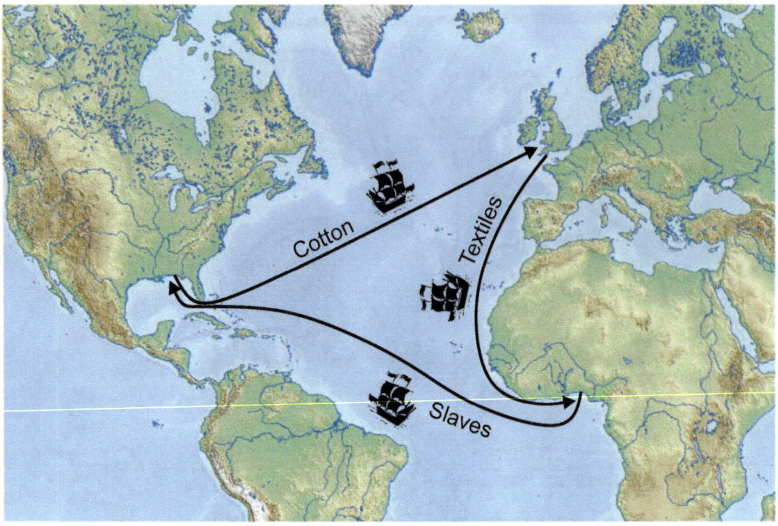

Fig. 5.5 Principle of triangular trade

mechanical loom (1785: E. Cartwright) were invented in England in the second half of the eighteenth century. This enabled cotton from their East Indian colonies to be processed in England into katun—a cotton fabric—which was exported all over the world. Compared to the wool and flax fibers processed until then, however, cotton is much finer and shorter. Therefore, it could not be processed with the usual machines in sufficient quantity and quality and therefore cotton fabric production did not immediately become a profitable business.

When the American Eli Whitney constructed the gin to separate the fibers from the seed in 1793, which was 1000 times faster than the previously applied manual methods (500 g of fiber per day per worker), the triumph of cotton was unstoppable (Fig. 5.7).

In 1806, Napoleon imposed an economic blockade on England, the so-called Continental Blockade, and thus encouraged the development of cotton spinning mills and mechanical weaving mills on the European mainland. The processing of cotton then marked the beginning of industrialization in Europe, first in England and later in Germany and other countries. Subsequently, cotton exchanges were established in New Orleans, New York, Liverpool, and Bremen beginning in

Fig. 5.6 Confederate $100 bill with slaves loading bales of cotton (Godot13, 2015)

Fig. 5.7 Cotton production in the USA in the nineteenth century (www01) and Eli Whitney (Gutenberg, 1919)

1870. By 1900, cotton accounted for 80% of the world fiber market, with an annual production of 4 m tons.

With the development of man-made fibers since the 1930s, cotton's market share has been steadily declining and in 1994, for the first time, the amount of man-made fibers produced annually worldwide was greater than the amount of cotton produced. This trend since continuous.

5.2 Cultivation, Growth, Varieties

According to their height, there are

- annual shrubby (1 to 2 m high) and
- perennial arboreal (up to 6 m high) plants.

5.2.1 Cultivation

The cotton plant thrives in tropical and subtropical climates ("cotton belt") around the equator, between 38° and 43° north and south latitude (Fig. 5.8). It requires fine sandy, loose and nutrient-rich soil and needs moisture during sowing as well as only small fluctuations between day and night temperatures. During maturity (formation of seeds and seed hairs) the plant needs a uniform high air and soil temperature.

Flat fields are best suited for growing cotton. A change in the crop rotation increases the yield, but is rarely applied. The one-sided nutrient consumption by the cotton plant is compensated by fertilizing with nitrogen, phosphate and potash. Sowing is done after harrowing the soil in rows about 1 m apart, about 5 cm deep. The night temperature should then not drop below 10 °C. For one hectare, 17–28 kg of seeds are required, which corresponds to approx. 150,000–250,000 pieces.

It takes between 25 and 32 weeks from sowing to harvesting, depending on the variety. Under ideal conditions, the sprout breaks through the soil after 5–10 days. During this time, the plant has only 2 leaves. After 2 weeks, more leaves develop and after 5 weeks the first flowers appear, which remain on the cotton plant for several weeks. Within 3 days, self-fertilization of the flower occurs and its color changes from white or yellowish to reddish before withering and falling off. A green, spherical fruit then develops, containing up to 32 seeds, which later develop the cotton fibers. This fruit (boll) grows to the size of a fig while the cotton fibers ripen inside (Figs. 5.9 and 5.10). An average boll contains about 500,000 fibers, with up to 100 bolls attached to a single shrub.

Approx. 100 days after planting (5–10 weeks after formation of the boll), the boll gradually begins to break open and dry out. The dried, strip-shaped capsule remnants are known as

Fig. 5.8 Cotton growing areas (map: Mapswire)

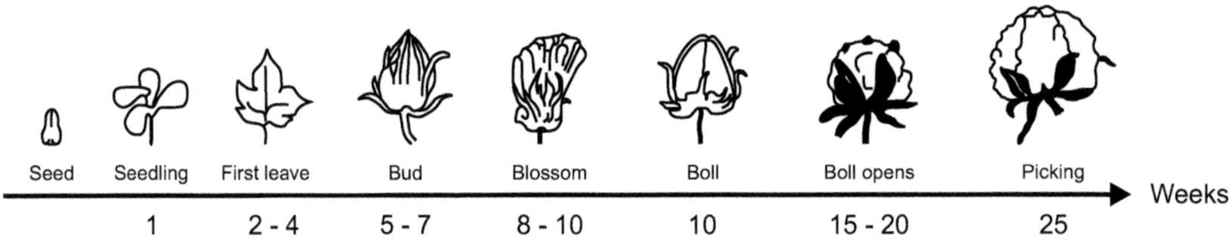

Fig. 5.9 Growth of the cotton plant

Fig. 5.10 Fields of Upland cotton (at flowering: Nichols, 2016; before harvest: Sandlapper, 2015)

"bur" and hold the cotton fibers together until harvesting.

The yield and fiber quality of cotton are decisively influenced by the climatic conditions in which it grows. The best qualities are obtained when the soil is kept moist by rain or artificial irrigation during the growing phase and dry weather prevails during the harvesting period. Cotton is therefore grown in both, dry and semi-humid climates.

Although cotton is in fact a perennial plant, it is usually destroyed after harvesting and the ground prepared for replanting to keep the yield constant.

5.2.2 Irrigation

Cotton plants require different amounts of water during their growth, as Fig. 5.11 shows as an example.

Only about a quarter of the world's cotton is still produced in dryland farming, i.e. without artificial irrigation (rain-grown cotton). This requires an annual rainfall of 500–700 mm. From 800 to 1000 mm per year, high-quality upland varieties can be grown. Cotton is cultivated in this way in Central and South America, in Africa and in parts of Southeast Asia.

If this amount of water is not available at the right time through natural rainfall, artificial irrigation must be used. This affects most cotton growing regions, as Fig. 5.12 shows.

More than half of the world's cultivated area requires therefore artificial irrigation. Around 75% of cotton is grown as irrigated cotton. Irrigation systems consume between 7 m^3 (Israel) and 29 m^3 (Kazakhstan) of water per kilogram of raw cotton, depending on the irrigation technology used (Fig. 5.13). Cotton is grown in this way in Spain, Greece, Egypt, Uzbekistan, Pakistan, the northern regions of India, large parts of China, the USA and Mexico.

In Uzbekistan and Kazakhstan, the extensive use of water from the Aral Sea for cotton cultivation over several decades has led to its almost complete drying up. In India and Pakistan, spring melt water from the mountains is stored in reservoirs, and only a small portion of the water for irrigation comes from springs or groundwater reservoirs. In Egypt, the water of the Nile is dammed (Aswan Dam), so that sufficient water is available all year round. The amount of water used in selected countries is shown in Fig. 5.14.

Compared with other fibers, considerably more water is needed to grow cotton (Fig. 5.15).

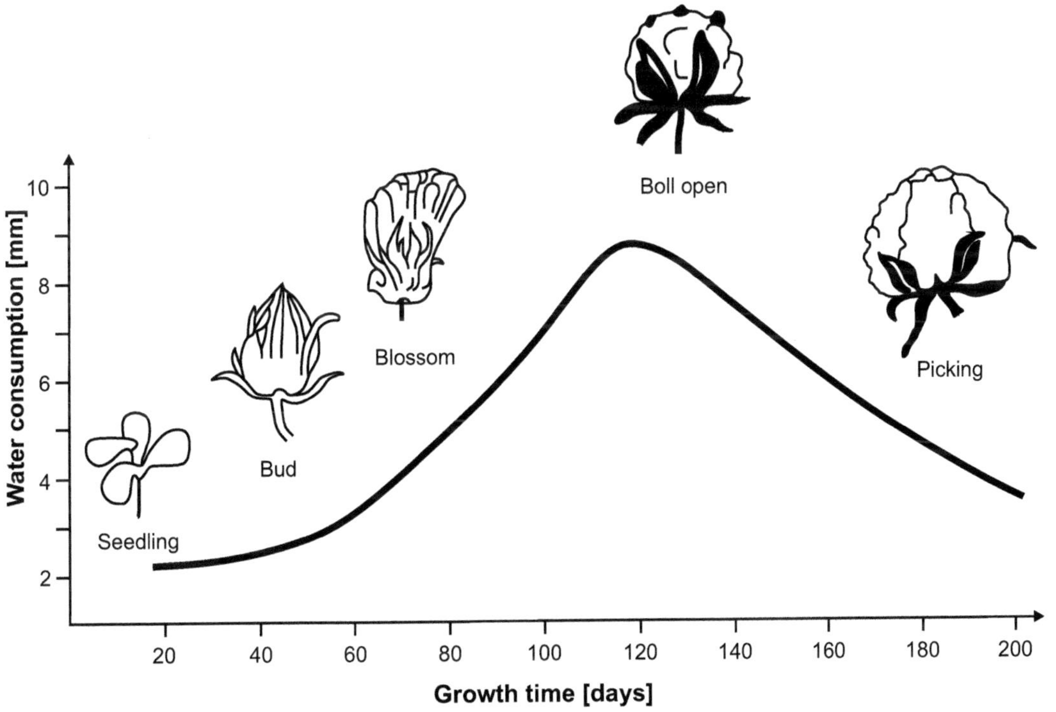

Fig. 5.11 Typical water consumption of Upland cotton

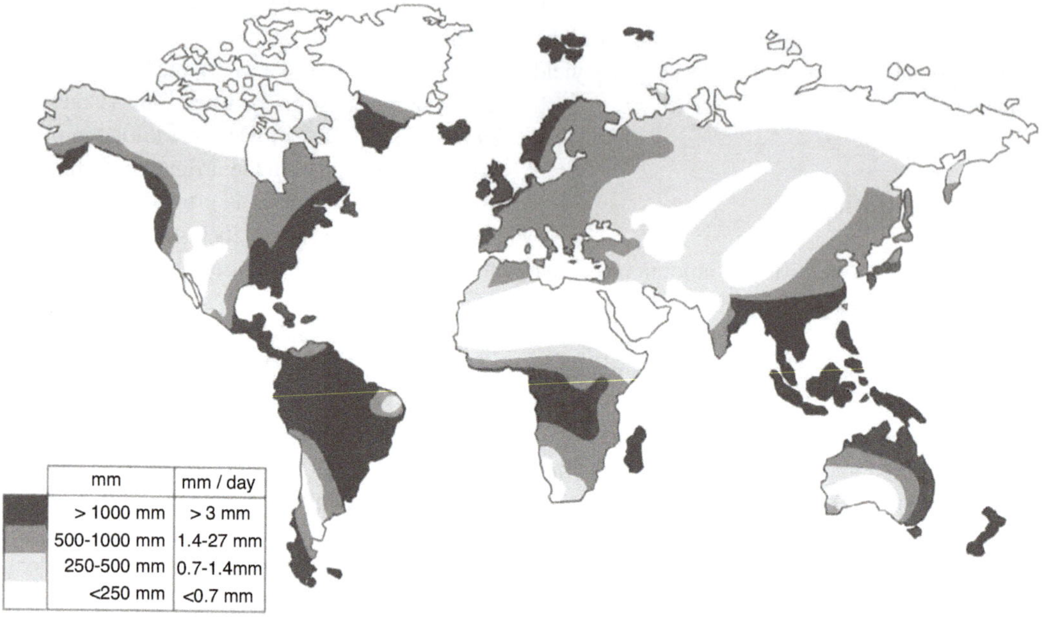

Fig. 5.12 Rainfall worldwide (FAO, 2021)

Worldwide, around 3% of the water consumed in agriculture is used for cotton cultivation (Fig. 5.16).

Figure 5.17 shows the water consumption in the production of selected products. The production of chocolate, beef and biodiesel requires even more water than the cultivation of cotton. However, the share of artificial irrigation is significantly lower in each case.

The use of fertilizers in cotton cultivation and the large differences between individual countries are shown in Fig. 5.18.

Fig. 5.13 Artificial irrigation in the USA (Hullwarren, 2012)

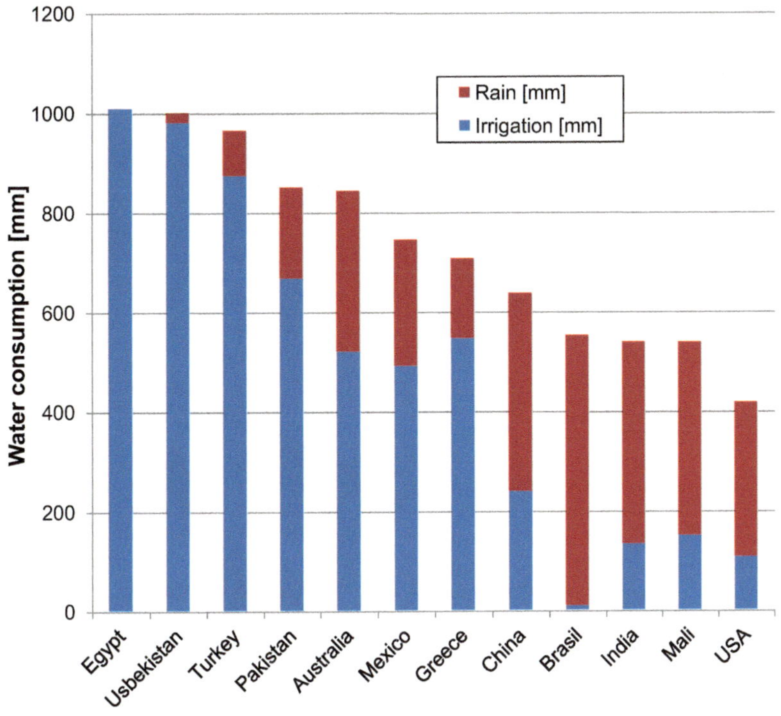

Fig. 5.14 Irrigation data of major cotton growing countries (data: Chapagain et al., 2006)

5.2.3 Breeding

Genetically, the species can be divided according to the number of chromosomes into those of the New World (13 haploid chromosomes) and those of the Old World (26 haploid chromosomes).

Old World species include Gossypium arboreum, G. herbaceum and G. nangking. Gossypium herbaceum and G. arboreum are very closely related and are both descended from the African native G. herbaceum var. africanum. It probably originated in West Africa

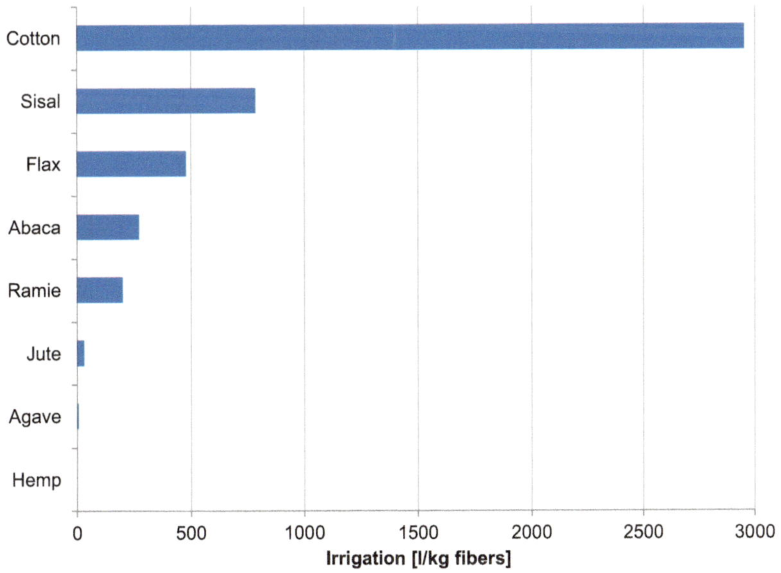

Fig. 5.15 Water consumption of selected fiber crops under artificial irrigation (data: Mekonnen & Hoekstra, 2011)

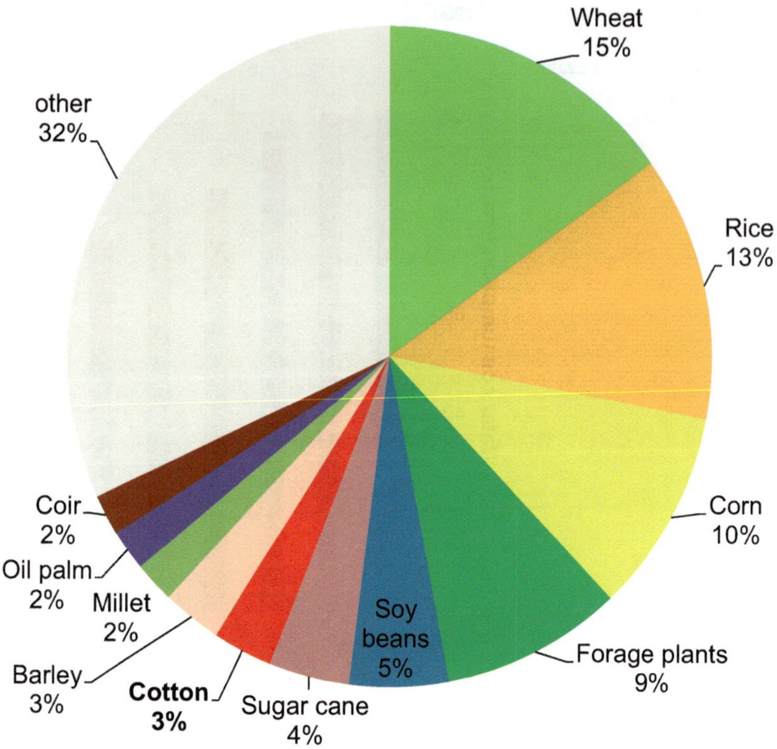

Fig. 5.16 Share of world water consumption of selected crops (data: Mekonnen & Hoekstra, 2011)

and the Arabian Peninsula. As early as 5000 years ago, G. arboreum was cultivated in present-day Gujarat, and it is still cultivated in southern India. Forms of G. herbaceum were developed in the Persian region and spread through Pakistan, India and China. As perennial shrubs, these types of cotton still grow wild in Africa and Asia. In cultivation, however, they are usually planted as annuals. Today, these short-haired varieties (fiber length often $<1''$) are still cultivated in India and

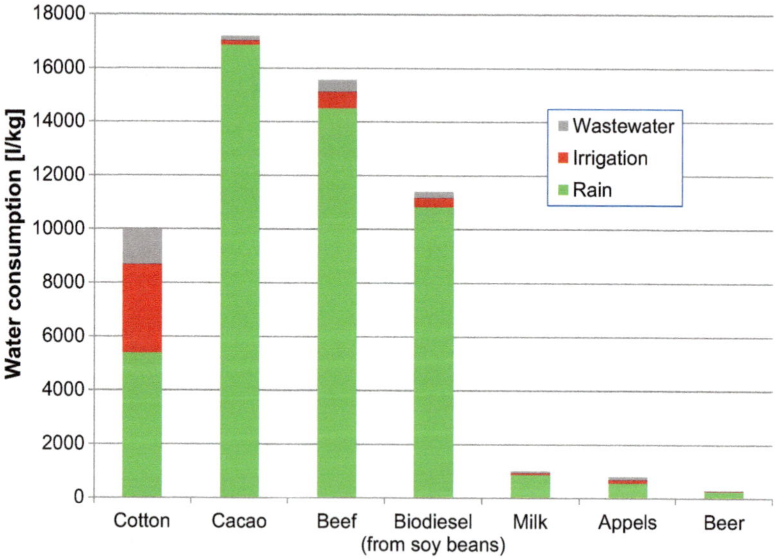

Fig. 5.17 Water consumption for selected products (data: Mekonnen & Hoekstra, 2011)

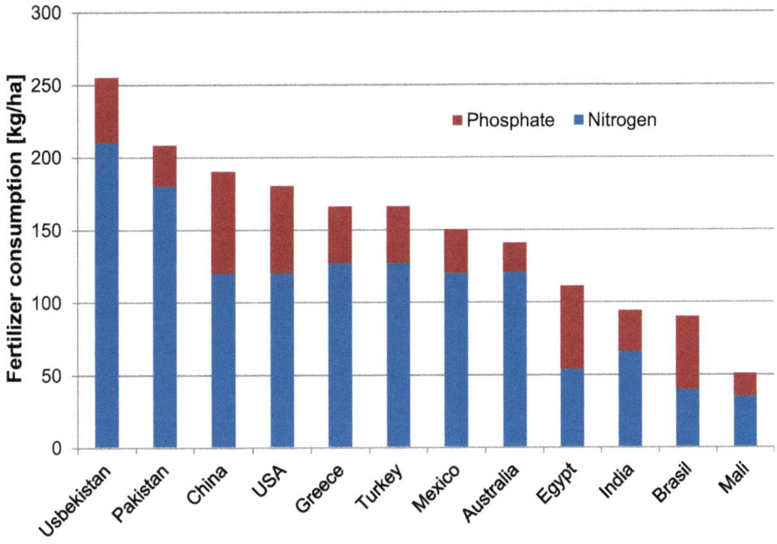

Fig. 5.18 Use of fertilizers of important cotton growing countries (data: Mekonnen & Hoekstra, 2011)

Pakistan, but their production volume is steadily decreasing despite the fibers' high degree of maturity. They can only compete with modern cotton where cultivation can take place without irrigation.

New World species include Gossypium hirsutum, G. barbadense, G. brasiliense, G. purpurascens and G. peruvianum. It is uncertain how these originally African species reached the Americas but it is believed that G. hirsutum and G. barbadense are hybrids of G. herbaceum which somehow managed to cross the Atlantic and cross-bred with local varieties. Gossypium barbadense was domesticated in Peru more than 4000 years ago, Gossypium hirsutum already 1000 years before in Mexico. Because of their good fiber qualities (long staple length, good strength, high fineness) and their high yields, varieties and crossbreds of these two species account for the largest proportion of global cotton cultivation.

Cotton is the product of at least 7000 years of breeding by man. Initially, plants were selected naturally according to their adaptability to the most diverse growing areas with the most diverse soil conditions and climates, and later a crossbreeding of the most diverse varieties took place. Breeding successes in the nineteenth century were mainly in the USA and Egypt. M. Jumel in particular had great success in Egypt (1818–1822) by crossing long-staple cotton (G. barbadense) from South America (Peru: Pima), Central America and the USA (sea island cotton) with native, old Egyptian and Ethiopian cotton varieties. Egyptian cottons thus began their triumphant advance as the best cottons in the world (Mako cotton). Similar successes were achieved in the southern states of the USA, where English settlers first cultivated and further bred Indian cotton varieties. Especially in the Mississippi region south of Memphis, cotton breeding stations were established, the most famous of which is in Stoneville/Mississippi. The new American upland varieties were introduced and further bred in many countries because of their high yield, e.g. in West Africa, Central America, Paraguay and Argentina (chaco), Spain, Greece and in the last thirty years particularly successfully in Australia. Even the former Soviet cotton varieties of Central Asia are partly descended from American upland varieties introduced in Lenin's time. Also worth mentioning are the American Pima cotton varieties of California (San Joaquin Valley, SJV) and cross-breeds between Upland and Pima, such as the Acala cotton varieties.

Cotton with a shortened ripening period and special weather stability was bred in the High Plains of Texas (Lubbock). Such commercially available varieties are used in many cotton-growing countries. In addition, cultivars of special indigenous varieties are often grown.

The development of the scientific basis of heredity by G. Mendel from 1865 onwards brought about decisive progress in systematic cotton breeding. In the USA in particular, in addition to state-supported breeding stations, many private cotton breeding companies were established, to which substantial progress in cotton cultivation can be attributed. This resulted in a doubling of the cotton yields within a century. Comparable increases in production were only achieved by the introduction of artificial fertilization based on the findings of J. Liebig.

Similar progress is expected today from the application of genetic engineering in cotton breeding. The breeding of so-called engineered cotton has gained worldwide acceptance, starting in the USA. Here, effective protection against pests is achieved by inserting a bacillus gene (Bt = Bacillus Thuringiensis) into the genes of the cotton. At present, around 90% of the world's cotton areas are planted with such varieties.

Breeding attempts in the USA with blue cotton (jeans) were not successful. In the USSR, there were attempts to breed red-brown colored cotton fibers, which also was no economic success. On an experimental scale, low-growing plants succeeded in shortening the growing season to 135 days. Other experiments were aimed at varieties in which all the fibers ripened at the same time. So far, none of these cotton varieties has been economically successful.

The current goals of cotton breeding are to increase yields, disease and pest resistance, varietal stability, early maturity and the improvement of those fiber properties for which the cotton farmer obtains a higher price when selling. These include, for example, greater fiber staple length, higher strength and better micronaire value. Seed-fiber bonding varies and depends on the variety. Attempts have therefore been made to breed cottons with lower seed-fiber adhesion (slickseed types) in order to achieve gentler ginning and thus a lower short-fiber content.

The traditional breeding of a new cotton variety often takes four to ten years. The yield of the plant, its soil and climate tolerance, its disease and insect resistance, and its fiber properties are decisive for the success of a breeding. However, these properties can only be assessed on the real result of the plant. Therefore, the cotton plant must be able to grow over several generations. Once a particular variety has been selected, it takes another four to six years to produce enough seed for commercial agricultural cultivation through multiplication. At the same time, field trials must be carried out to test whether the variety is resistant. Thus, the introduction of a new cotton variety takes 7–15 years (Fig. 5.19).

A cotton variety introduced into agriculture degenerates within ten to fifteen years due to inbreeding through self-pollination caused by monoculture. In the process, fiber yield and fiber properties both deteriorate significantly. Therefore, cotton breeding must be seen on a very long-term basis and must be systematically planned and controlled. If long-term cotton breeding for the succession of existing varieties is interrupted by warlike or political disasters, it takes a very long time before new and internationally competitive cotton varieties are again available for cultivation. Examples of this are formerly excellent cotton provenances in Iran, Uganda and Afghanistan, which have now largely disappeared.

5.2.4 Varieties

Worldwide, cotton of the species G. hirsutum is the main crop (87%), followed by G. barbadense (8%), G. arboreum (3%) and G. herbaceum (2%).

The characteristics of the main varieties are:

- **G. hirsutum (87%):** annual shrub form (1–2 m high), in areas with a rainfall of 60 mm ("rain-grown cotton"), most important variety, good yield, medium fineness and staple length, yields American "Upland" and "Acala" cotton.
- **G. barbadense (8%):** annual shrub form, in areas with artificial irrigation ('irrigated cotton'), particularly long staple (up to 50 mm), fine fibers, 'Sea Island' and 'Pima'.
- **G. arboreum (3%):** perennial, tree-like plant (2–3 m high), low yield.

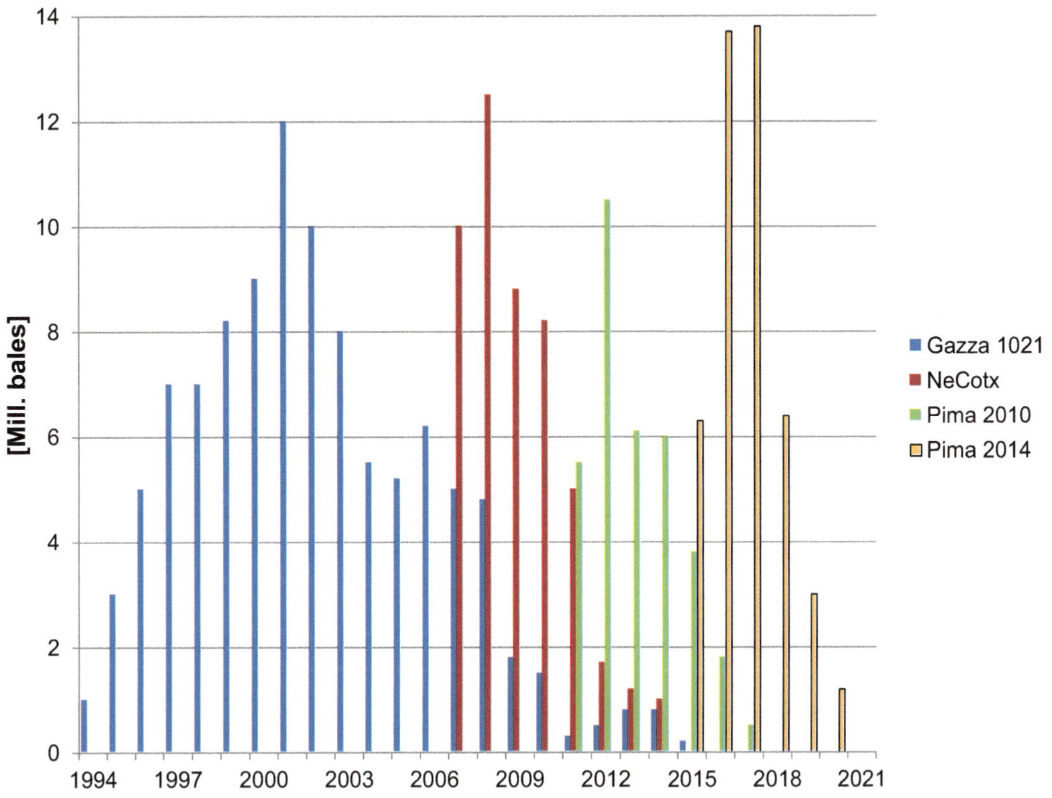

Fig. 5.19 Typical breeding cycles for cotton varieties

- **G. herbaceum (2%):** annual herbaceous plant (0.5–1 m high), medium to low yield, coarse, short fibers.

In most growing areas, only annual, shrubby plants are grown because they are severely damaged during harvesting (defoliation, mechanical harvesting, etc.).

Depending on the country and climatic conditions, different varieties dominate in certain cotton-growing countries. In the USA, for example, these are Upland cotton (market share approx. 97%), followed by Pima cotton (3%, almost exclusively in California). In China, various varieties are grown, mainly Bt cotton, in Uzbekistan 4 varieties dominate the market, in Egypt the particularly long Giza cotton. In some African countries (e.g. Togo and Uganda), only one variety is grown (Table 5.1).

In Fig. 5.20, the fineness (micronaire value) and staple length of important cotton varieties are shown.

This diversity on the international cotton market offers the cotton farmer a wide range of options for selecting the optimum variety for his needs. In addition to yields, other characteristics also play a role, such as

Table 5.1 Typical properties of US cotton

		G. hirsutum (Upland)	G. barbadense (Pima)	G. arboreum
Staple length	**[1/32"]**	35–40	36–48	20–23
	[mm]	27.8–31.8	28.6–38.1	16–18
Micronaire		3.6–4.6	4.1–4.8	3.3
Tenacity	**[cN/tex]**	29.2–34.6	31.9–46.5	36.8

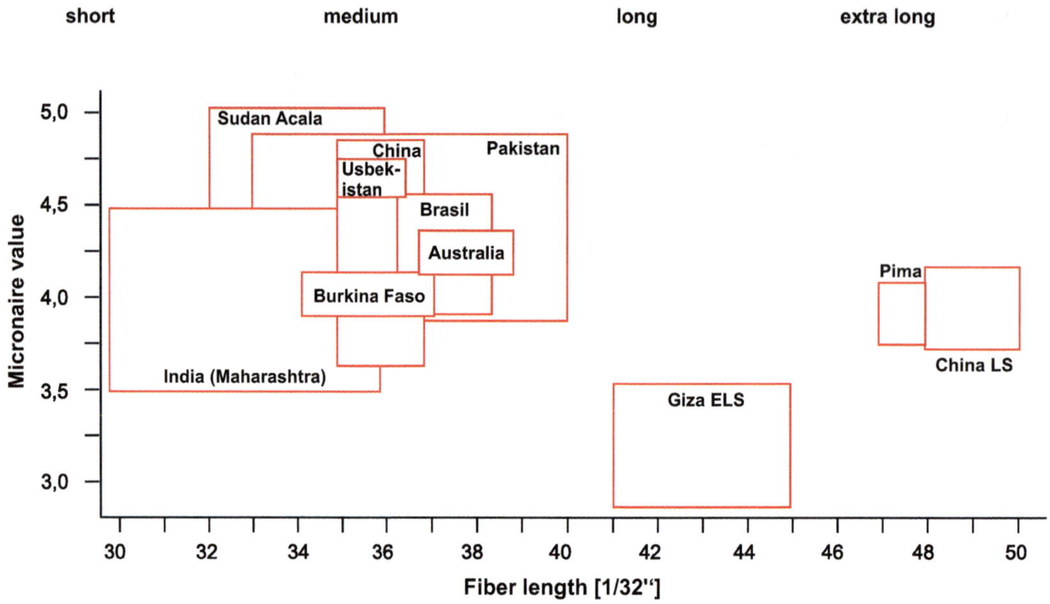

Fig. 5.20 Fineness and fiber length of important cotton varieties

- the response of cotton to soil quality and climate,
- resistance to diseases and pests,
- the response to irrigation and fertilization,
- the reaction to plant protection products,
- the suitability of the plants for mechanical harvesting.

Ultimately, many different factors determine the price of the cotton produced. They can also change over time. For example, if new cleaning processes are better and gentler, the trash content of the fiber material is no longer of decisive importance and other factors come to the fore. The development of new spinning processes can also change the requirements for cotton (e.g. greater staple length required for air-jet spinning).

It is also important whether the cotton is to be grown in a country with traditional agriculture or on a large farm (plant protection, harvesting method).

Figure 5.21 shows the most economically important cotton varieties today.

5.3 Harvest

There are two types of harvesting:

- Hand picking, and
- machine picking.

5.3.1 Hand picking

Because the fields of a cotton plantation are not normally cultivated at the same time, the bolls do not all mature at the same time. With hand harvesting, it is possible to harvest only the fully developed cotton fibers. The harvesting process must then be repeated several times. Therefore, the fields are usually harvested 3–4 times. For small fields, e.g. in China (average 0.1 ha), this method is more economical than mechanical harvesting.

The proportion of trash in the raw cotton (from seed coat fragments, stick and leaf fragments, etc.) is very low. This is a great advantage for further processing.

Fig. 5.21 The most important types of cotton

The disadvantage is the required large work force combined with a low picking yield, which is around 80 to 120 kg of seed cotton per person and day, from which 30–40 kg of cotton fibers can be obtained.

There are two types of hand picking, hand peeling and snapping. Hand peeling is the careful harvesting by hand. Snapping involves less careful harvesting, in which the picker strips the capsules from the plant with a glove-protected hand (Fig. 5.22).

5.3.2 Mechanical harvesting

In most cotton growing countries, cotton is harvested by machine. Depending on the type of machine and yield, 700–1500 kg of seed cotton can be obtained in 1 h. However, not only mature but also immature and dead cotton fibers are harvested. In addition, the proportion of trash particles is often very high. As a rule, the cotton is defoliated before harvesting, which prevents too many sticks and leaves from getting into the cotton.

Fig. 5.22 Manual harvesting in India and Côte d'Ivoire (Renault, 2005; Kidoudeni, 2017)

Two types of harvesters are used:

- Spindle picker (spindle picker) and
- stripper.

5.3.2.1 Spindle Picker

Harvesting with the spindle picker is the dominant method today (Figs. 5.23 and 5.24). It has vertically arranged rotating spindles that twist the fibers off the plants together with adhering seeds, leaves and dirt. The harvested material passes from the spindles into a large transport container where it is collected. This machine works selectively, i.e. it only extracts the fibers from the already opened bolls, so that the immature bolls remain on the bush.

The spindle picker harvests 4–8 rows of cotton simultaneously. Its speed is approx. 4 to 14 km/h.

Generally, the spindle picker passes over the field only once, when all bolls are open. However, harvesting twice is also possible, with 80–85% of the fibers being harvested in the first pass and the remaining 15–20% in the second.

Worldwide, and especially in the USA, spindle pickers are increasingly being used to keep the amount of trash particles as low as possible and thus achieve a higher price for the cotton.

5.3.2.2 Stripper

This machine uses rotating, rubberized beaters and brushes to collect both the mature and unopened bolls from previously defoliated plants. A pickup takes over the harvested cotton bolls and via an air flow, they are sucked off to the rear. This method thus allows a field to be harvested in a single pass. The stripper is therefore particularly suitable for short, early-maturing cotton varieties. Its travel speed is 6–9 km/h. Its production rate is higher than that of the spindle picker, but considerably more trash particles get into the cotton

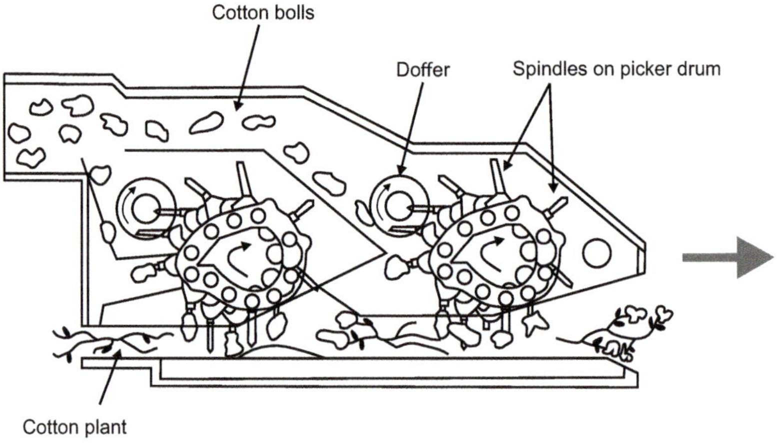

Fig. 5.23 Principle of the spindle picker according to ICAC (2001)

Fig. 5.24 Spindle picker (John Deere, 2021)

because the entire field is harvested at once. Usually the field is defoliated beforehand to reduce the amount of vegetable impurities (Fig. 5.25).

The number of brushes and wipers ("bat") per unit is usually 2 × 6 (Fig. 5.26), although the ratio of brushes to rubberized wipers varies (e.g. 3:3, 5: 1).

5.3.2.3 Comparison of Harvesting Methods

Machine-picked cotton contains more trash particles and more moisture than hand-picked cotton. Therefore, subsequent fiber drying and cleaning must be coordinated with the harvesting method.

Because only one harvesting operation is required when harvesting with the stripper and this machine is simpler in design than the spindle picker, the operating costs are lower for stripper cotton than for spindle picker cotton. This advantage of the stripper is offset by the poorer fiber and yarn characteristics as a disadvantage. However, these differences are only slight in the case of very strong cotton fibers.

Fig. 5.25 Structure of the harvesting unit and stripper (John Deere, 2021)

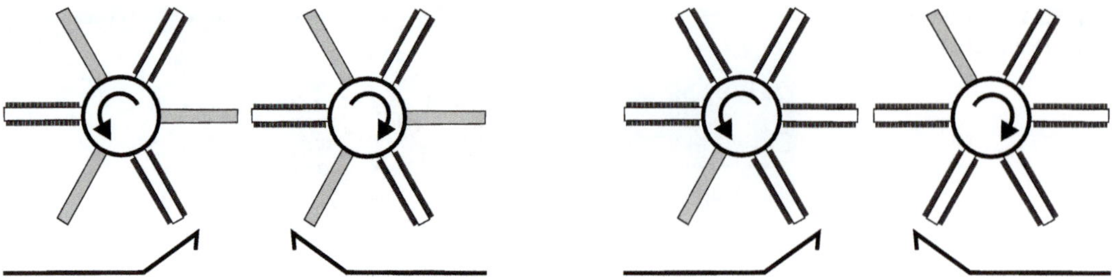

Fig. 5.26 Possible arrangement of brushes and scrapers (3:3, 5:1)

5.3.3 Storage and Transport

After harvesting, the seed cotton is directly packed on modern harvesting machines and deposited in the field as a so-called module (Fig. 5.27). The usual module weights are 2–2.5 t (round) and 3.5 t (rectangular). They remain there until they are transported to the ginning plant for ginning. To prevent damage to the cotton, the moisture content within the bale should not exceed 12%.

In countries with high labor costs, mechanical harvesting dominates, regardless of the quantity produced. In India and China, as well as in almost all countries in Africa, harvesting is almost exclusively manual (Fig. 5.28).

Harvesting depends on the maturity of the cotton, which, depending on the growing region, occurs at different times, as Fig. 5.29 shows.

5.4 Ginning

Ginning is a multi-stage process that consists of

- the drying, cleaning and dusting of the seed cotton,
- the actual ginning process and
- the cleaning of the lint cotton.

5.4.1 History

The ginning of cotton, the separation of the fibers from the seed, was initially carried out by hand. A worker needed about 2 years to gin a bale of 250 kg.

In India, the churka was used early on as an aid to ginning, from which the roller gin was later

Fig. 5.27 Formation of cotton modules in the field after harvesting (John Deere, 2021)

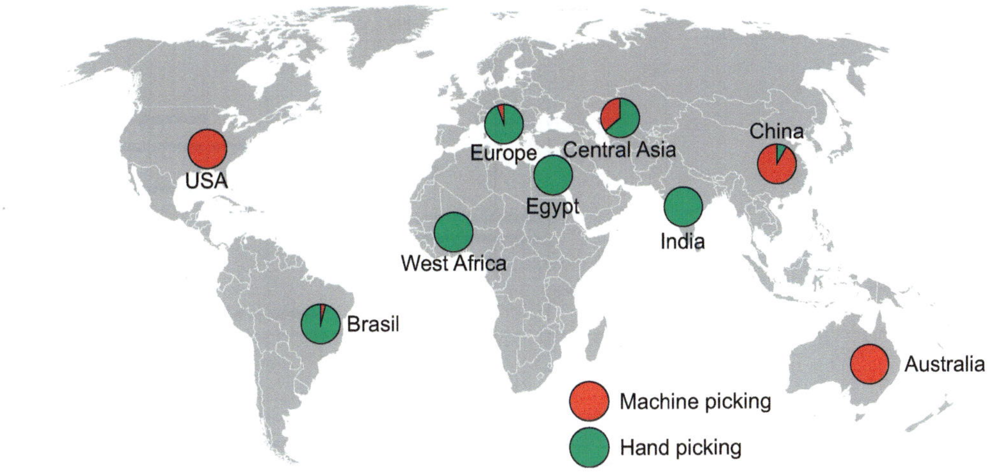

Fig. 5.28 Proportion of manual and mechanized harvesting for selected countries (data: ICAC, 2021)

developed. In 1794, the American Eli Whitney patented the pin roller gin (Fig. 5.30). With the aid of a roller fitted with pins, the cotton is pulled through a narrow slit. This separates the up to 20,000 fibers from the—thick—seed. A rotating brush then strips the fibers from the roller. The machine replaced the labor of 100 workers. After "engine", Whitney named his invention "gin".

The business model of Eli Whitney and his partner, Phineas Miller, was as simple as it was unsuccessful: They first produced a large number of gins and installed them in numerous locations, mainly in Georgia. Then they offered cotton farmers to gin their cotton in exchange for 40% of the profit from the sale of the ginned cotton. Understandably, few farmers took advantage of this offer. Many copied the gin machine instead, which led to considerable patent disputes that were not settled until many years later. Eli Whitney did not become rich with his invention. He only achieved business success through the mass production of firearms, which eventually made him a rich man.

Henry Ogden Holmes received a patent for a saw-tooth roller gin in 1796. This made ginning a continuous process and further increased productivity. Both principles are still in use today (roller gin, saw gin).

The invention of a gin machine made it possible to process cotton on a large scale. It was therefore a decisive step towards the industrialization of textile production.

5.4.2 Gin

Today, almost all cotton worldwide is ginned by machine where the cotton is separated from the seed, and at the same time the fibers are cleaned and dried (Figs. 5.31 and 5.32).

The stored seed cotton is first pneumatically conveyed out of the storage, then cleaned of coarse trash particles and then transported in an air stream to the first dryer.

5.4.3 Tower Dryer

This dryer usually works with an air temperature of approx. 175 °C (350 °F). The cotton is transported downwards by the hot air stream over 16–24 levels in 10–15 s and dried in the process. The air temperature decreases by approx. 60 °C. With decreasing humidity, the strength of the cotton fibers also decreases, but trash particles can then be removed more easily. In most cases, a moisture content of 6–7.5% is set for the lint cotton in the spinning mill.

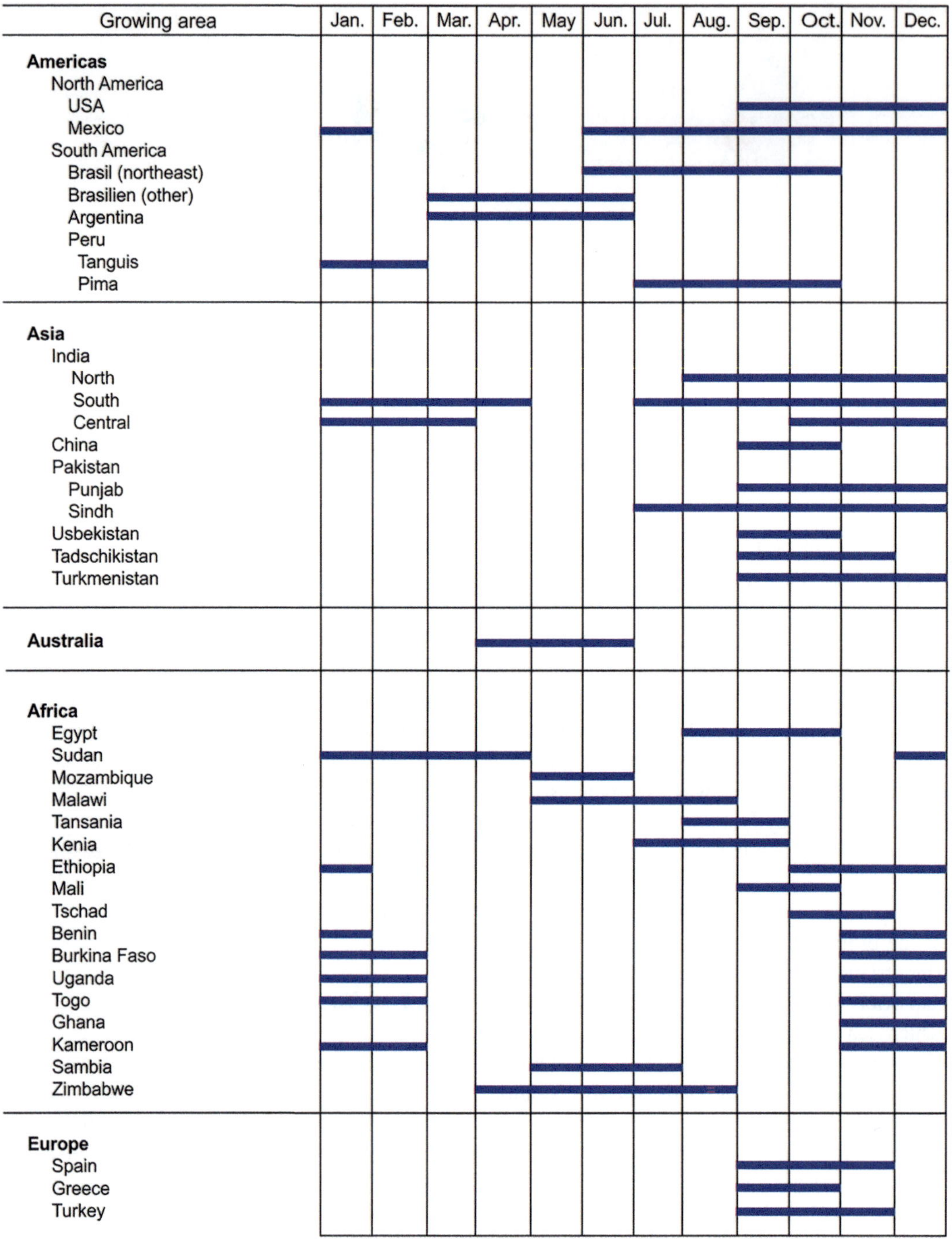

Fig. 5.29 Harvest times in the main growing areas (data: ICA, 2019)

Fig. 5.30 Ginning with churka (Sheppard, 1869) and gin (Betten & Kaufmann, 1919)

5.4.4 Cylinder Cleaner

The drying process is followed by a first cleaner. This is similar to the step cleaners used in the blowroom. It usually consists of 6 pin rollers which remove coarse trash from the cotton flocks at approx. 400 rpm. Coarse and fine trash particles are removed via separating grates. In this way, 6 bales of 250 kg cotton per metre of dryer length can be cleaned per hour.

5.4.5 Stick Machine

The second cleaning process is a saw-tooth cleaner ("stick machine"). This also removes coarse trash particles. The fibers are caught by saw-tooth rollers rotating at 300–400 rpm, and plant and trash particles are separated out by centrifugal force. Per metre of cleaner length, 4–6 bales of 250 kg each can be processed per hour (Fig. 5.33).

In most ginning plants, this process step is followed by a dryer and another saw-tooth cleaner. After the last dryer, the moisture content of the cotton should be 6–7.5%. If the moisture is below 5%, fiber shortening will occur during ginning. If the moisture is above 7.5%, not enough dirt may have been removed and there is a risk that the gin will clog.

Fig. 5.31 Seeds with fibers (left) and gin with operator (right)

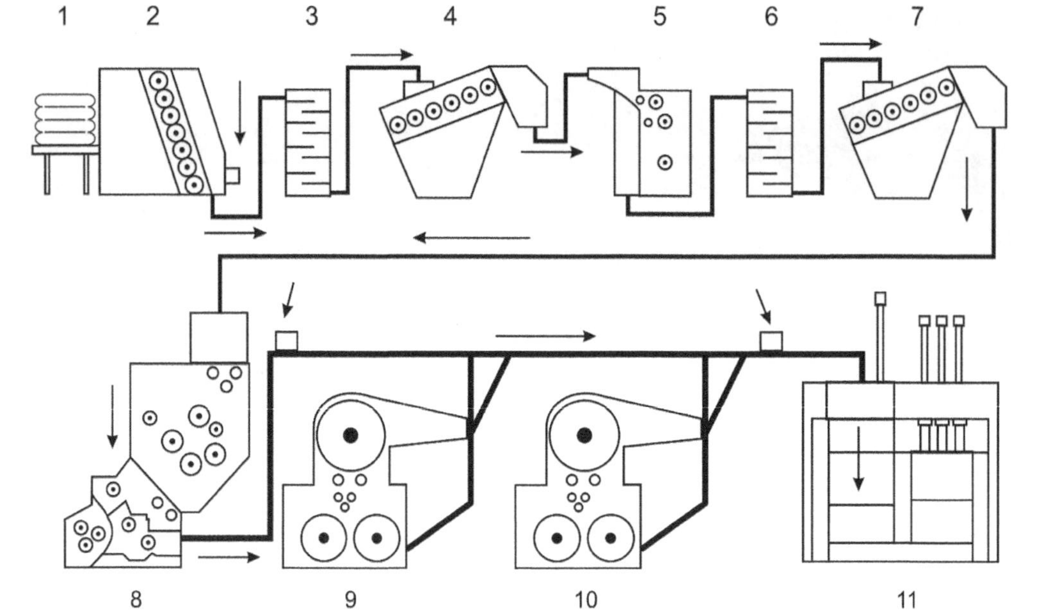

Fig. 5.32 Material flow diagram of a saw gin plant

5.4.6 Gin Machine (Gin Stand)

In ginneries, several gin machines are usually operated simultaneously. A distinction is made between roller gins and saw gins. The latter are more aggressive, but also more productive and cheaper. Roller gins are usually only used for long-staple cotton varieties (e.g. Pima).

5.4.6.1 Roller Gin

The roller gin consists of a rubberized roller and fixed as well as rotating knives. The cotton flocks are first fed into the gusset between the roller and the knives. Ginning then occurs due to the frictional force exerted on the seed cotton between the roller, which moves, and fixed knives. This presses the seed cotton against the knives, which separate the seed from the lint cotton. The fibers stick to the roller and are removed by a rotating brush. A rotating knife cleans the fixed knives from seed parts and adhering cotton fibers. In the roller gin, not only are the seeds separated from the cotton fibers, but also about 50% of the trash particles in the cotton are removed. The most expensive component of a roller gin is the rubber roller, which is reinforced with up to 13 layers of fabric (Fig. 5.34).

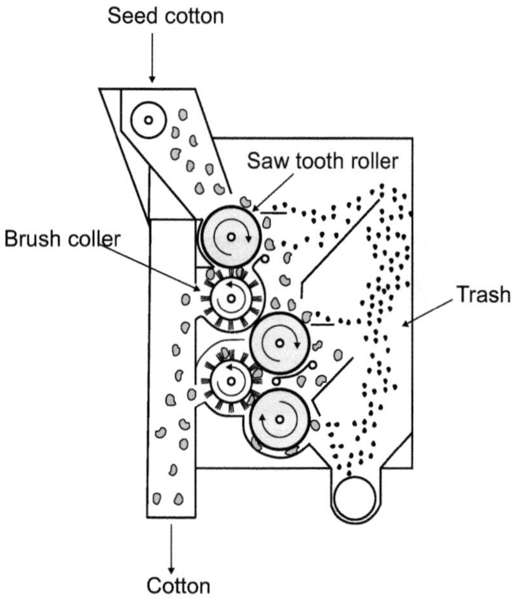

Fig. 5.33 Stick machine

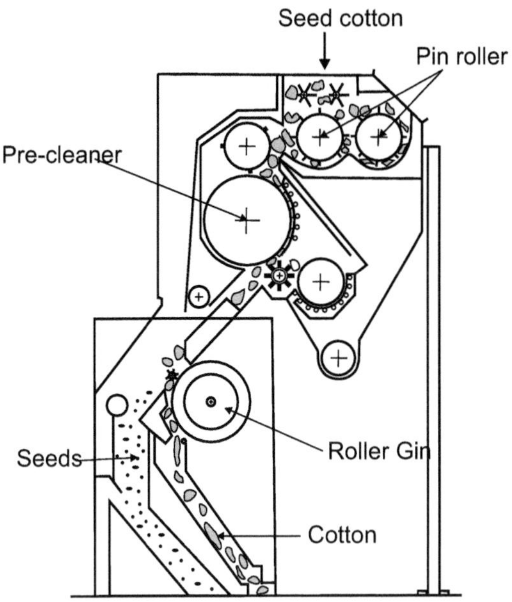

Fig. 5.34 Roller gin

The production of this simply constructed machine with 150–350 kg/h lint cotton is lower than that of the saw gin and requires a high manual labor input. However, the roller gin is very gentle on the fibers. It is therefore mainly used for ginning long-staple varieties, e.g. in Egypt.

5.4.6.2 Saw Gin

In a saw gin, there are up to 234 saw teeth arranged on a roller up to 120 inches wide and rotating at about 700 rpm (sample data from Lummus®). The rotating saw teeth separate the seed from the lint cotton. The saw blades are steel discs with a width of about 1 mm, a diameter of 12 inches and about 200 teeth. The production of these machines reaches up to 3500 kg/h of lint cotton. The ginned cotton is removed from the saw blades with brushes or in an air stream. When using an air stream, the settings of air quantity and air pressure are decisive for fiber shortening and nep formation. Fiber damage and the content of neps in the lint cotton can be influenced by the machine setting. The saw gin is relatively aggressive to the fibers and is mainly used for Upland type cotton with a staple length up to 30 mm (Figs. 5.35 and 5.36).

5.4.7 Lint Cleaner

In the lint cleaner, further trash particles are removed from the cotton. Roller gin cotton is treated differently than saw gin cotton.

Saw gin cotton is first compressed by a condenser and formed into a sliver. This is presented to a saw-tooth roller and thus cleaned out. The dirt particles are clamped between the saw teeth and the grid plates and discharged via the latter. The distance between the grid plates and the saw teeth is therefore decisive for the cleaning effect and the possible damage to the fibers. The cleaned fibers are removed from the saw-tooth roller by a rotating brush.

Fig. 5.35 Saw gin

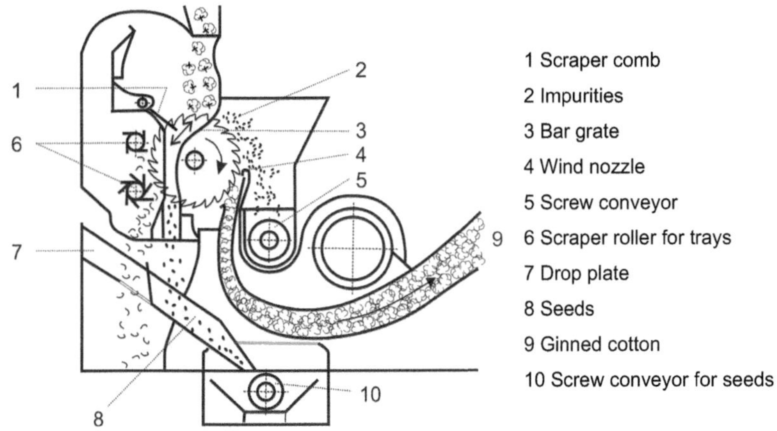

Fig. 5.36 Saw Gin

Roller gin cotton is usually cleaned from trash particles by rotating rollers with the aid of an air stream and by utilizing centrifugal force.

A lint cleaner can be used to improve the quality of a cotton (trash content, color) by cleaning out trash particles and mixing the fiber material. If the cotton is too dry, however, undesirable fiber shortening may occur.

5.4.8 Influence of the Ginning on the Fiber Properties

In modern ginning plants, due to the high production and intensive cleaning, leaves, stems, boll and seed parts are crushed to very small particles, which can hardly be removed in the blowroom. Furthermore, intensive cleaning produces fiber neps. Cotton which has been ginned with moderate cleaning intensity shows better fiber properties and produces fewer neps during carding than cotton which has been treated more aggressively. The values for yarn breakage frequency and yarn evenness are also better.

In general, the average fiber length decreases significantly in some cases during ginning, as the example in Fig. 5.37 shows.

Overall, the roller gin works much more gently than the saw gin. Roller gin cotton contains fewer neps and has a longer staple length. Roller gin cotton also produces a higher fiber yield than saw gin cotton. Roller ginning breaks fewer long and fine fibers than saw tooth ginning. Therefore, with saw tooth ginning, there is a severe decrease in both the 50% and 2.5% span length values. In contrast, saw gin cotton contains less clearly visible impurities, resulting in a higher classification value and thus a higher commercial value. However, the very small trash particles in Saw Gin cotton are difficult to remove in the blowroom, but this is not taken into account in the usual classification.

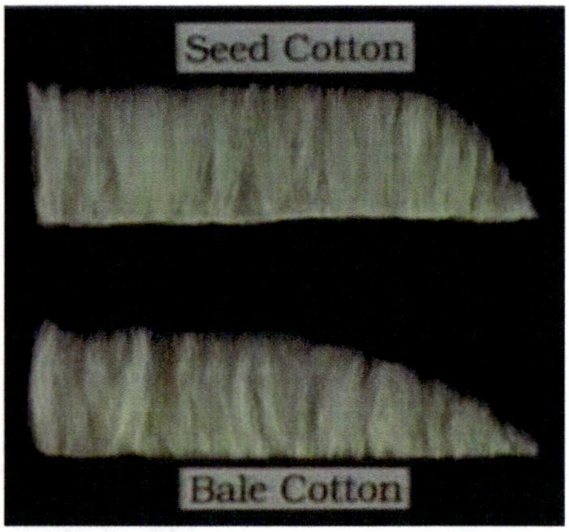

Fig. 5.37 Fiber length distribution before (top) and after (bottom) ginning

In the blowroom, roller gin cotton results in a higher waste percentage, i.e. it has a higher trash particle content. However, it also has a better cleanability than saw gin cotton. The carding influence on the length parameters is higher in roller gin cotton than in saw gin cotton, owing to the higher percentage of long and fine fibers in roller gin cotton, which are more damaged during carding than short fibers. During carding, the 50% span length is more affected than the 2.5% span length. After carding, the quality difference between differently ginned cottons is therefore smaller than before.

The uniformity and purity of the draw frame slivers is better with roller gin cotton than with saw gin cotton. For coarse and medium coarse yarns, there is practically no difference in yarn breakage frequency, but for fine yarns, roller gin cotton is better in this respect. The ginning process has almost no influence on yarn strength and elongation.

With regard to yarn uniformity and appearance (thick, thin places, neps, foreign matter, etc.), roller ginned cottons achieve the better results.

The standard process is the saw gin due to its very high throughput compared to the roller gin, which is mainly used for long staple cotton. In Table 5.2, the most important differences are summarized.

5.4.9 Bale Press

After ginning, the lint cotton is pneumatically conveyed to a bale press (Fig. 5.38). In the process, the moisture content of the cotton is adjusted to approx. 8%. It is important that no moist nests are formed. This can lead to damaged areas after prolonged storage, especially in the case of highly compacted bales.

In modern gins, up to 60 bales per hour and press are processed today. This corresponds to approx. 15 t/h. After the bale has been pressed, packed and tied, samples are taken from each bale and the bale is stored in the warehouse.

For the ginning plants, the costs increase with increasing bale density. For storage, on the other hand, a high-density bale with high stability and stacking safety is ideal. Similarly, high bale density and small dimensions are also sought for long-distance and overseas shipping in containers, because they allow high container filling. In spinning mills, bales are preferred which always have the same density and dimensions after removal of the wrapping layer and straps. A high density of the bales makes the cleaning of the cotton more difficult. When removing the cotton with the bale opener, the bale structure consisting of several individual layers, which has been common up to now, is disadvantageous, since in some cases lumps and

Table 5.2 Comparison of roller gin and saw gin cotton properties

	Roller Gin	Saw Gin
Production	150–350 kg/h	1200 kg/h
Manual labor	High	Low
Fiber types	long staple	normal staple
Fiber yield	High	Low
Number of neps	Low	High
Fiber shortage	Low	High
Span length	Better	Worse
Visible impurities	High	Low
Cleanability	Better	Worse
Yarns down for fine yarns	Lower	High
Yarn tenacity	No influence	No influence
Twist	No influence	No influence
Yarn evenness	Better	Worse
Thick and thin places	Less	More

Fig. 5.38 Bale press

remnants of the individual layers are carried along, which requires more effort when opening the fiber flocks. For these reasons, various bale formats have been standardized.

5.4.10 Bale Size

New transport systems and the increasing automation of spinning mills, combined with higher demands for consistent quality, led to the requirement for uniform bale density and dimensions. However, views on the ideal bale differ in the various industrial sectors.

Bale weight since about 1920 has usually been about 500 American pounds. If packed in jute, deducting about 12 pounds for the emballage and 9 pounds for the steel straps, gives a net cotton weight of 480 pounds (218 kg). If wrapped in PP film, 3 pounds must be deducted for steel straps and only 1 pound for the film, resulting in a net bale weight of 496 pounds (225 kg). Bale densities range from 250 to 650 kg/m^3, with the current standard density being approximately 450 kg/m^3.

In ISO 8115, an attempt was made in 1986 to standardize bale sizes. Two bale sizes were defined for this purpose (Fig. 5.39):

$$L = 1060\ \text{mm},\ W = 530\ \text{mm},\ H = (780\text{–}950)\ \text{mm}$$
$$L = 1400\ \text{mm},\ W = 530\ \text{mm},\ H = (700\text{–}900)\ \text{mm}$$

Because ginners, sliver manufacturers, warehousemen, hauler and spinners have a different understanding of bale height, it has not been possible to specify a uniform height (tie, bulge). Nowadays, typical formats (H/W/L) are in the range of (50–70/50–70/140) cm^3.

Bale sizes are standardized, but because there are different standards, the size and therefore the bale weight are still not uniform. It varies between 100 kg and 330 kg (Egypt), although the international standard bale with a weight of 480 pounds (218 kg) is becoming increasingly popular.

5.4.11 The Wrapping (Emballage)

The cotton bales are packed after pressing and before setting. The wrapping consists of films or

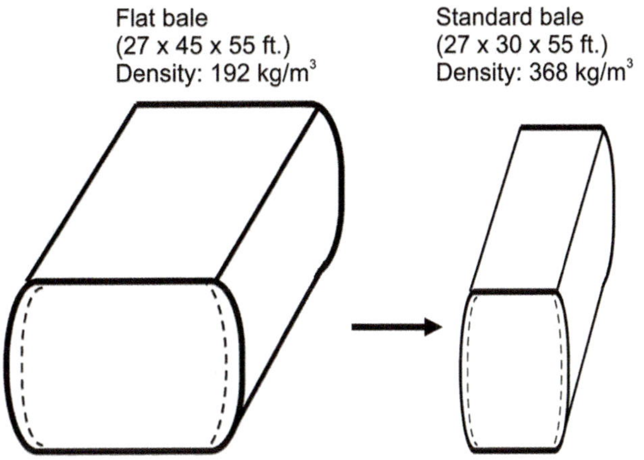

Fig. 5.39 Bale dimensions in accordance with ISO 8115 (left: approx. 1950, right: 2021) and bales

fabrics and is intended to protect the bales from soiling and fiber loss during transport and storage.

Today, polypropylene (PP), polyester (PET) and jute are used as packaging materials, with jute becoming less and less important for cost and weight reasons (Fig. 5.40).

As fiber can get from the wrapping layer into the cotton easily, e.g. during sampling and removal of the wrapping material in the spinning mill, the wrapping material is of great importance. Both jute and polypropylene fibers lead to yarn breaks in the spinning mill. Jute fibers are no longer recognizable after bleaching in cotton and show the same dyeing behavior as cotton fibers. Polypropylene, on the other hand, cannot be dyed and therefore remains visible. Complex foreign

Fig. 5.40 Bales with PP wrapping (Lenzing)

fiber detection systems are therefore required to remove such fibers from the yarn later on.

5.4.12 Yield

From 1000 kg of raw cotton can typically be obtained:

- 420 kg of raw fibers suitable for the production of textile fibers, including 340 kg of high-quality raw fibers (lints), 80 kg of short fibers (linters) used, e.g., for the production of man-made cellulosic fibers such as viscose and paper,
- 540 kg of other natural products (seed cake 200 kg, oil 110 kg, husks 210 kg, retained seed 20 kg),
- 40 kg of foreign matter generated as waste.

This results in a fiber yield for the production of yarns of approx. 34%. These values vary depending on the cotton variety and growing region.

Cotton linters are used, among other things, for the production of banknotes. The oil from the cotton seeds is used to make mayonnaise, margarine, frying oils, explosives, cosmetics, gum, soap, insecticides and other products. Cottonseed oil is cholesterol-free and contains high levels of monosaturated fatty acids. Therefore, it is considered to provide health benefits. About 0.2 kg of oil can be produced from 1 kg of cottonseed. Oil from cottonseed occupies seventh place in the list of the most important vegetable oils, with a world market share of around 2% (equivalent to approx. 4.5 million tons, Fig. 5.41).

With a price of approx. US$ 0.30/kg (2020), the seeds (Fig. 5.42) make a significant contribution to the income of cotton farmers. A few years ago, it was still US$ 0.15/kg, so the price is highly volatile.

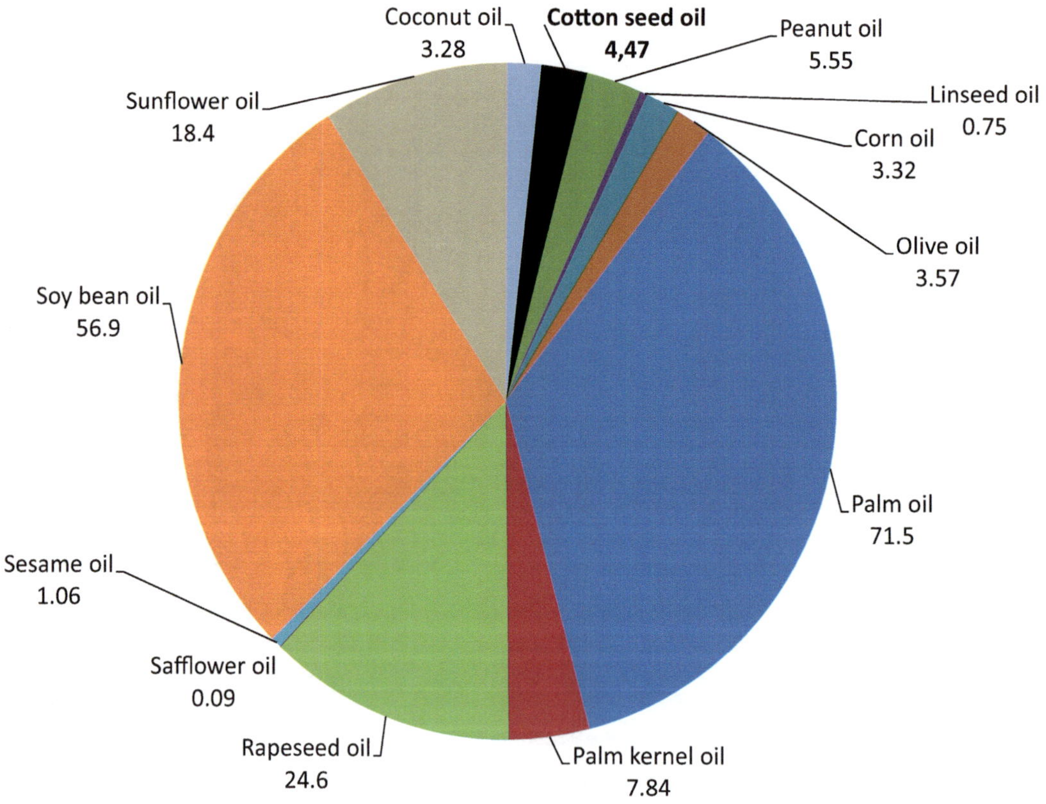

Fig. 5.41 Production volume of the main vegetable oils worldwide [mt] (data: FAO, 2021)

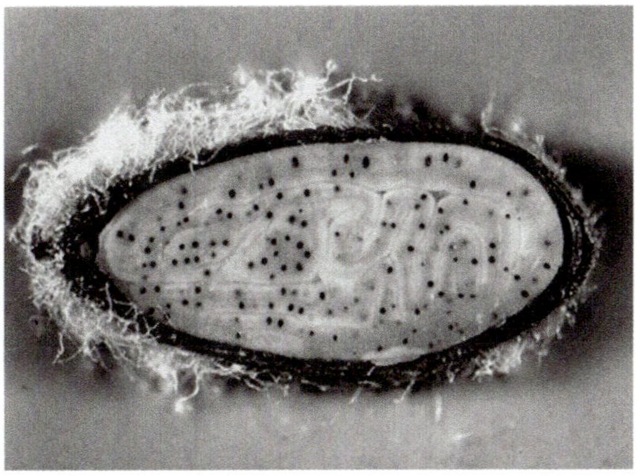

Fig. 5.42 Cotton seed in cross section with linters

5.5 Composition and Fine Structure

Cotton fibers grow in the closed boll first in length and then in thickness. They either develop into long fibers or soon cease to grow and cover the seed as short-fibered linters.

5.5.1 Chemical Structure

Cotton fibers belong to the natural cellulose fibers, the macromolecular cellulose serving as the framework substance. In addition to cellulose, cotton contains other natural accompanying substances. These occur in particular in the primary wall.

Among plant fibers, cotton has one of the highest cellulose contents and is free of wood components. In terms of anhydrous fiber, it consists of around 95% cellulose (Table 5.3).

The fully extended cellulose molecule (Fig. 5.43) has the shape of a flat ribbon with laterally protruding hydroxyl groups that form intra- and intermolecular hydrogen bonds. The surface of the ribbon consists mainly of hydrogen atoms directly linked to carbon atoms and is therefore hydrophobic.

These two characteristics of the cellulose molecule are responsible for the structure of the crystallites. This in turn determines many chemical and physical properties of the cotton fiber. In Fig. 5.44, the crystalline structure is shown. If the cellulose chains are largely stretched (left), the result is a high modulus of elasticity. If the chains form more of a zigzag shape, a higher elongation is obtained (right).

Table 5.3 Main components of cotton fibers

	Weight percent
Cellulose	82–89
Sorbed water	7–10
Pectins	0.6–1.1
Proteins	1.0–1.8
Organic acids	0.5–0.9
Wax	0.4–0.9
Other residue	1
Ash	0.6–1.5

The proportion of hemicellulose and pectin is comparatively low. The average degree of polymerization is 5000–10,000 (Hessler et al., 1948).

5.5.2 Fine Structure

During length growth, the fiber is a thin-walled tube filled with protoplasm and enveloped by a membrane (primary wall). During subsequent thickness growth, the secondary wall is then formed. The formation takes place layer by layer from the primary wall into the fiber interior. Growth layers are formed by the deposition of cellulose in the day-night cycle. Their number therefore corresponds to the number of days from the beginning of thickness growth to maturity (Fig. 5.45, right). The fiber lives until the capsule opens.

The internal structure of cotton fibers was described in detail by Kling and Mahl (1952). According to them, the primary wall consists of only 10% cellulose. It has a smooth surface formed by waxes and pectins (Greek: pektos = coagulated). Underneath, there are two superimposed fibril nets. The outer fibril net is more longitudinally oriented, the inner fibril net has a transverse orientation. The secondary wall is the actual cell wall. It consists mainly of cellulose (about 94%). The arrangement of the microfibrils in relation to each other has a decisive influence on the properties of the cotton fiber.

Fig. 5.43 Structure of the cellulose repeating unit

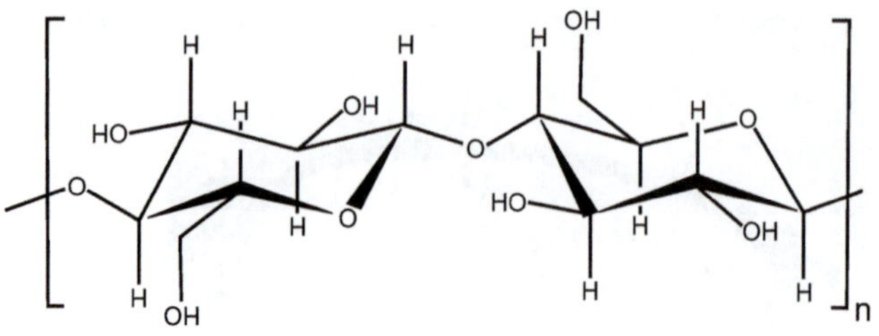

Fig. 5.44 Crystal structure of cellulose (left: high modulus of elasticity; right: high elongation)

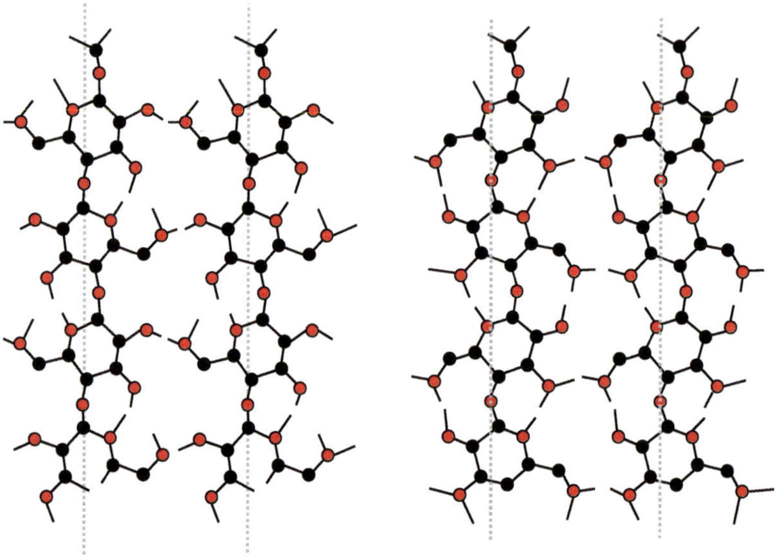

The tertiary wall is the inner layer of the fiber wall. Like the primary wall, it consists of little cellulose and is heavily interspersed with accompanying substances (pectin, fats, waxes). In addition, it forms the closure of the secondary wall against the lumen (lat.: lumen = cavity). The lumen is filled with plasma during growth. After cotton harvesting, the plasma gradually dries out, resulting in a cavity (Fig. 5.46).

Between the elementary fibrils and the microfibrils there are cavities into which chemical compounds consisting of small molecules can

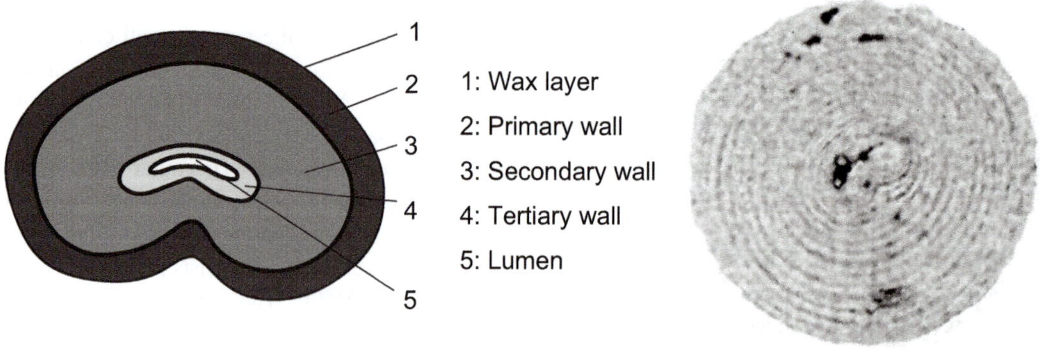

Fig. 5.45 Structure of a cotton fiber (right: after J. Hearle)

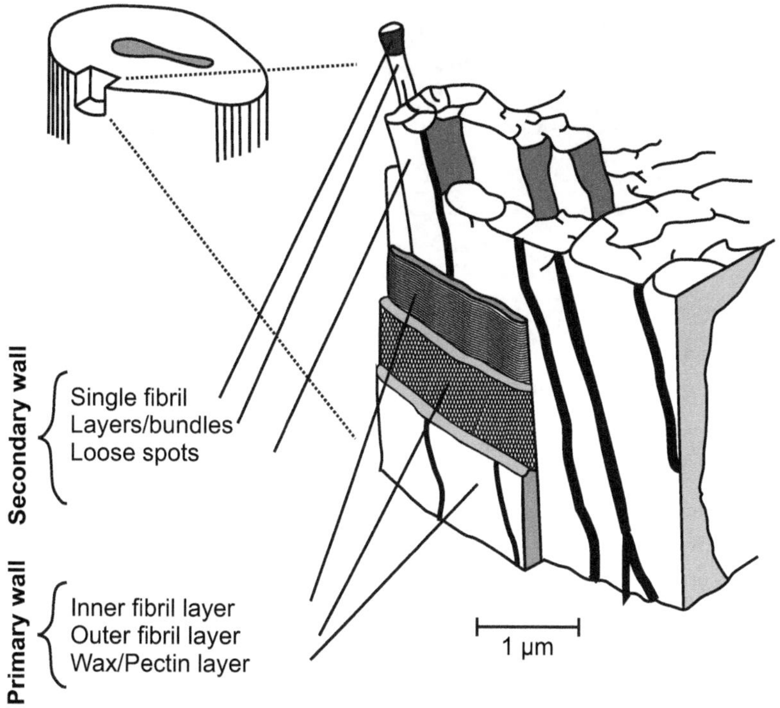

Fig. 5.46 Fine structure of the cotton fiber acc. to Kling and Mahl (1952)

penetrate. These interstices are also referred to as the capillary cavity system (Latin: capillus = hair).

The cotton fibers first grow in length (approx. 20–25 days), after which the secondary wall is formed and the fibers acquire their final structure and fineness (Fig. 5.47).

As a result of the desiccation that follows the opening of the boll, the hitherto cylindrical fiber collapses and becomes a multiply twisted fiber due to the different stress states in the layers of the secondary wall. These twists frequently change their sense of direction (Fig. 5.48).

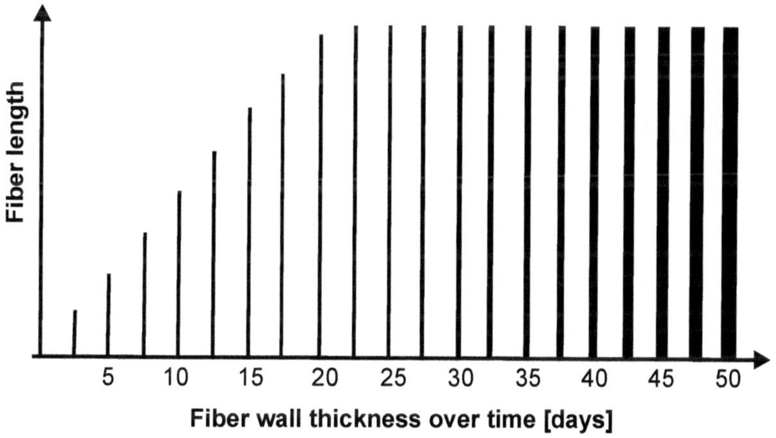

Fig. 5.47 Time course of the growth of the cotton fiber

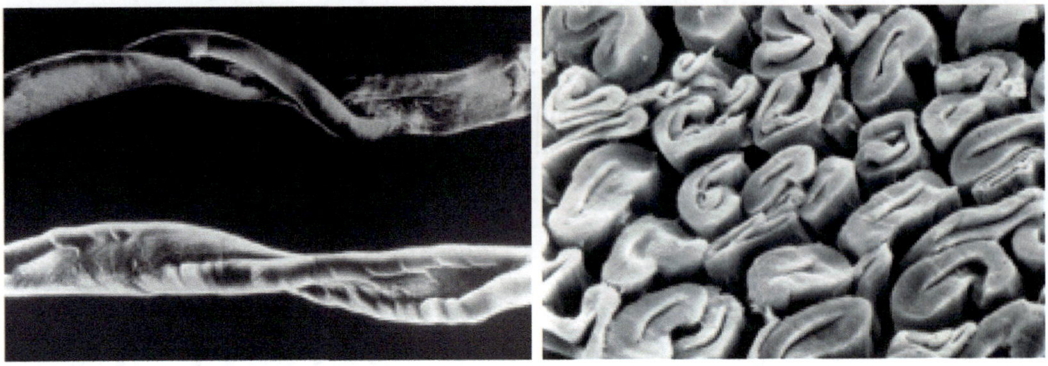

Fig. 5.48 Longitudinal and cross-sectional view of cotton fibers

In one boll there are up to 32 seeds in 3–5 compartments and from each seed 2000 to 7000 fibers grow.

5.6 Properties

5.6.1 Fiber Length

The fiber length of cotton is one of the most important properties, because it essentially determines the minimum yarn count. It lies between 10 and 60 mm, values of 25–30 mm are usually achieved.

The staple length, also known as the classifying staple, is the length of a typical part of the fibers. It denotes the average fiber length without taking into account the short fibers and is required for setting the drafting system in the spinning mill.

In many countries, the following length classes are distinguished (Table 5.4):

A cotton of staple length 32/32 inches is usually referred to as "1" and a cotton of staple length 35/32 inches is referred to as "1 3/32" (Fig. 5.49).

In addition to the mean length, the fiber length distribution and the proportion of short fibers are also important. Figure 5.50 shows typical stacks of different cotton provenances.

Length Uniformity

This value is the ratio of the average fiber length of all fibers and the "upper half mean length" and is given as a percentage. The greater the value, the less the lengths of the individual fibers of the sample differ. If all fibers had the same length, the value would be 100. Typical values today are in the range of 80.5–81.5 for American upland cotton.

Figure 5.51 shows the typical distribution of fiber length for American Upland cotton. Above 35/32″ fiber length, one speaks of "premium quality". Fibers longer than 40 mm are referred to as Extra Long Staple (ELS).

Staple lengths have increased steadily in recent years through selective breeding and are now

Table 5.4 Staple lengths acc. to US standard

Length class	(mm)	(inch)
Short staple S	<20.6	<13/16
Medium staple M	20.6–25.4	13/16–1
Medium long staple ML	26.2–27.8	1 1/32–1 3/32
Long staple L	28.6–33.3	1 1/8–1 5/16
Extra long staple XL	>34.9	>1 3/8

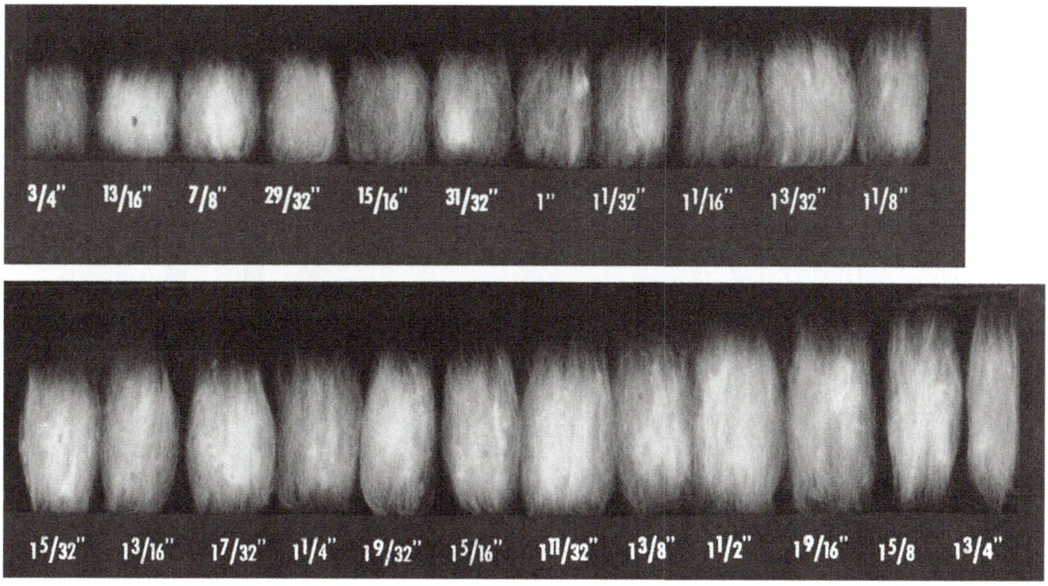

Fig. 5.49 Length classes for cotton (in each case mean fiber length in inches)

around 36/32″ (approx. 29 mm) for US Upland cotton (Fig. 5.52).

5.6.2 Fiber Fineness

The fineness of cotton fibers is usually determined with the so-called "Micronaire device", which was developed as early as 1940. A defined air flow passes through a fiber sample and the pressure drop is measured. This is a direct indicator of the average fineness and degree of maturity of the fibers (Fig. 5.53). The corresponding characteristic value is dimensionless and is referred to simply as "micronaire".

Typical micronaire values:

<3	Very fine
3 to 3.9	Fine
4 to 4.9	Average
5 to 5.9	Slightly coarse
>6	Coarse

Micronaire values can be converted to dtex values according to:

$$[dtex] = \frac{\textit{Micronaire}\ \text{Value}}{2.54}$$

The fineness of cotton fibers in the tex system is between 1 and 4 dtex, values of 1.5–2 dtex are common.

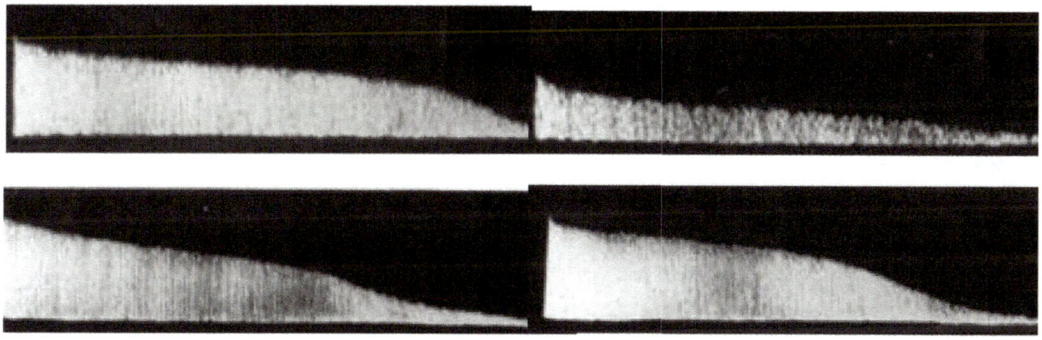

Fig. 5.50 Staple length distribution of different cotton provenances

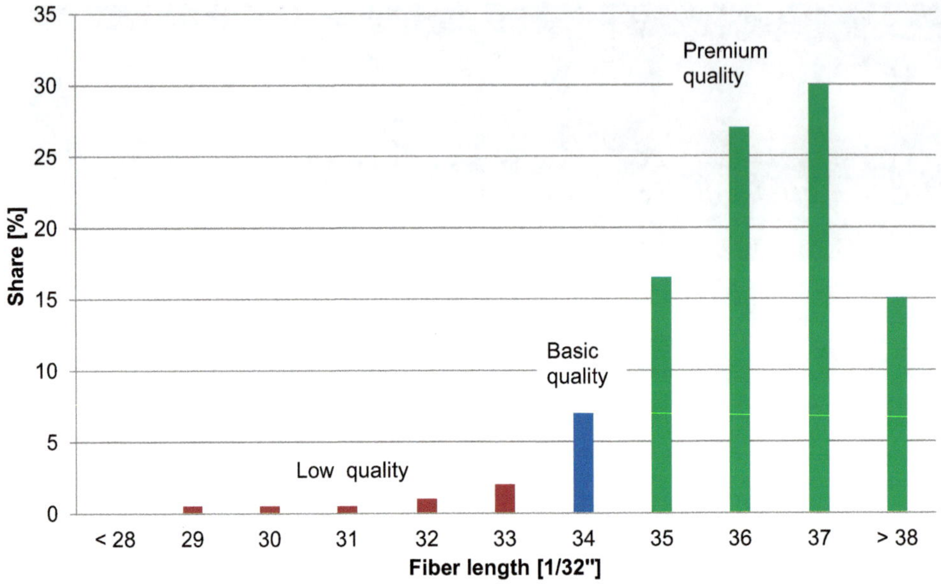

Fig. 5.51 Upland cotton fiber length distribution (data: Cotton Inc., 2021)

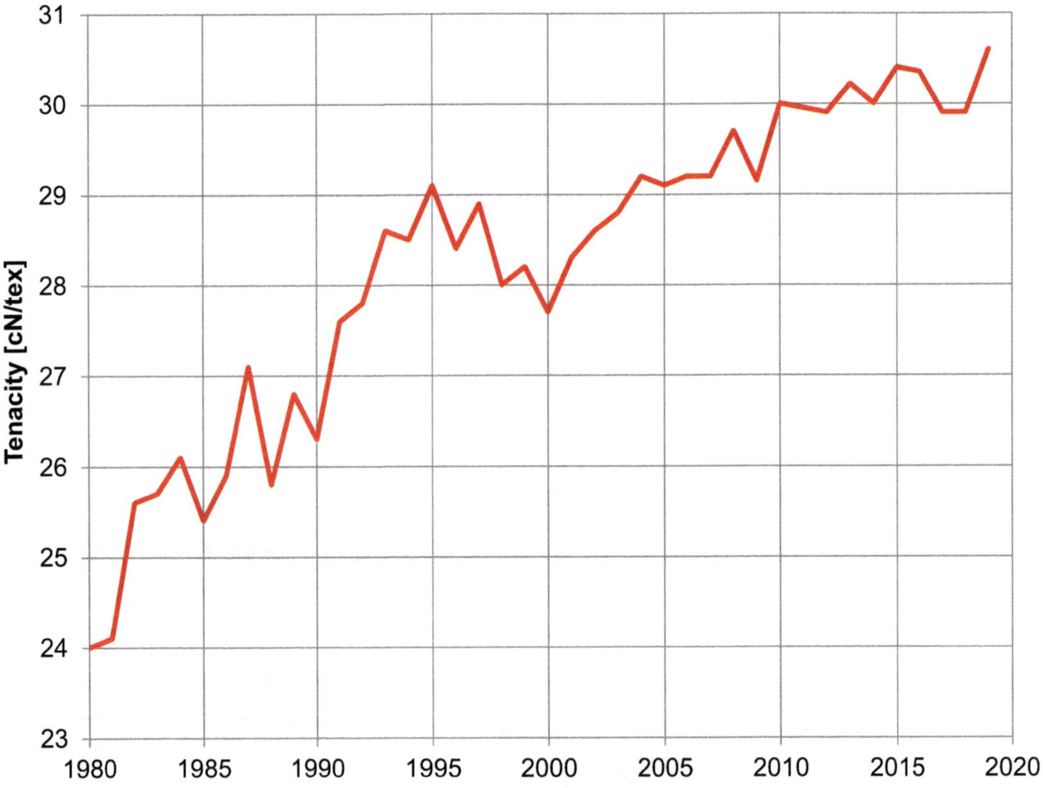

Fig. 5.52 Development of the staple length of US Upland cotton (data: Cotton Inc., 2021)

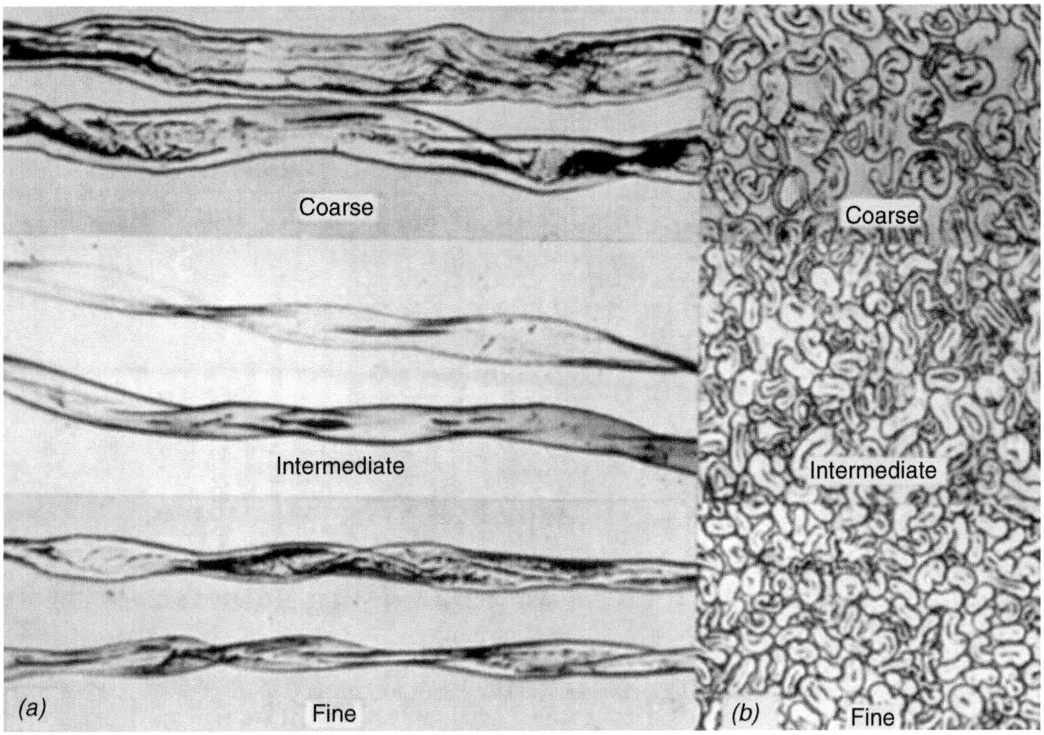

Fig. 5.53 Coarse, medium and fine cotton in longitudinal and cross-sectional view

As Fig. 5.54 shows, the micronaire values for upland cotton wools are within a wide range. Premium grades are 3.7–4.2 micronaire (1.5–1.7 dtex).

As Fig. 5.55 shows, the mean micronaire values have been increasing in recent years, so the fibers are becoming coarser. Typical values for upland cotton are 4.2–4.4.

5.6.3 Strength

This value is usually given in g/tex or cN/tex and rounded to 1/10. It is determined on a fiber bundle and is therefore an averaged value.

The distribution of fiber strengths shows Fig. 5.56. From a value of 30 cN/tex, the base price of the cotton increases.

Typical values for fiber strength according to this method today are often up to 30 cN/tex for American upland cotton wools. The trend is towards greater strengths, because these lead to fewer yarn breaks in downstream processing and thus increase machine efficiency (Fig. 5.57).

5.6.4 Trash and Foreign Matter

An important quality characteristic of cotton is its degree of soiling or purity, i.e. its content of dust and leaves and other plant fragments and of foreign matter (Fig. 5.58). The size of the fragments plays a particularly important role. Therefore, automatic determination of trash particle area, as is done in some test methods, may be misleading because, although many small particles have a smaller total area than a few large ones (making the cotton appear "cleaner"), these particles are much more difficult to remove from the fiber material.

The trash content is measured with a scanner that scans the surface of a fiber sample and calculates the trash content from the ratio of the dark trash area to the total area of the sample.

Foreign materials in particular (remnants of packaging, feathers, plastic, etc.) can lead to serious problems in downstream processing. In the worst case, the contamination is only detected in the yarn, where it then has to be removed at great expense.

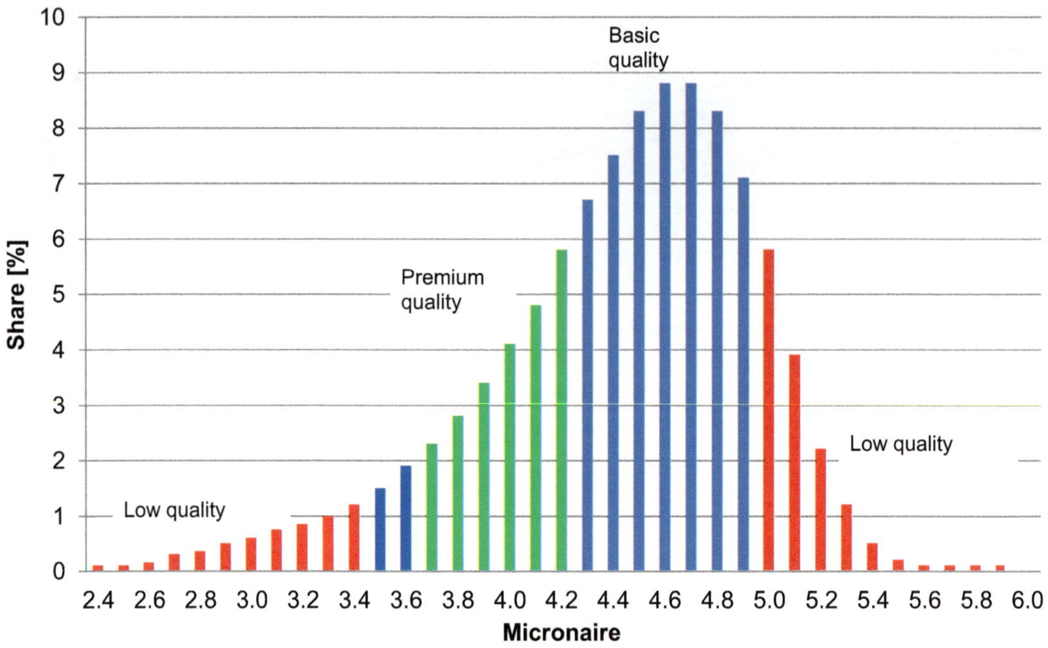

Fig. 5.54 Distribution of fiber fineness for Upland cotton (data: Cotton Inc., 2021)

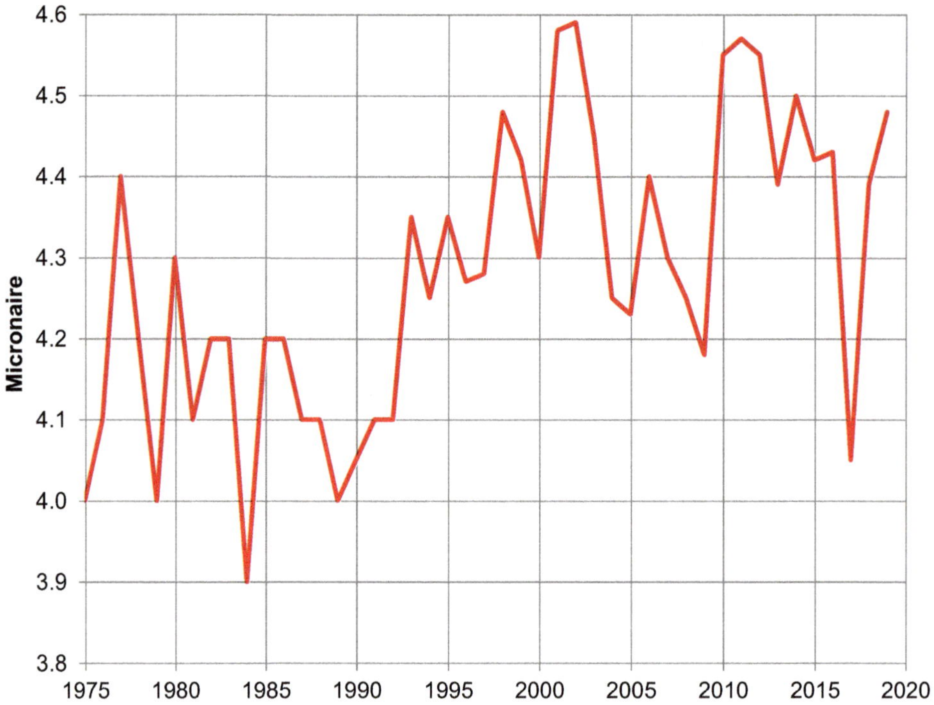

Fig. 5.55 Development of micronaire values for upland cotton (data: Cotton Inc., 2021)

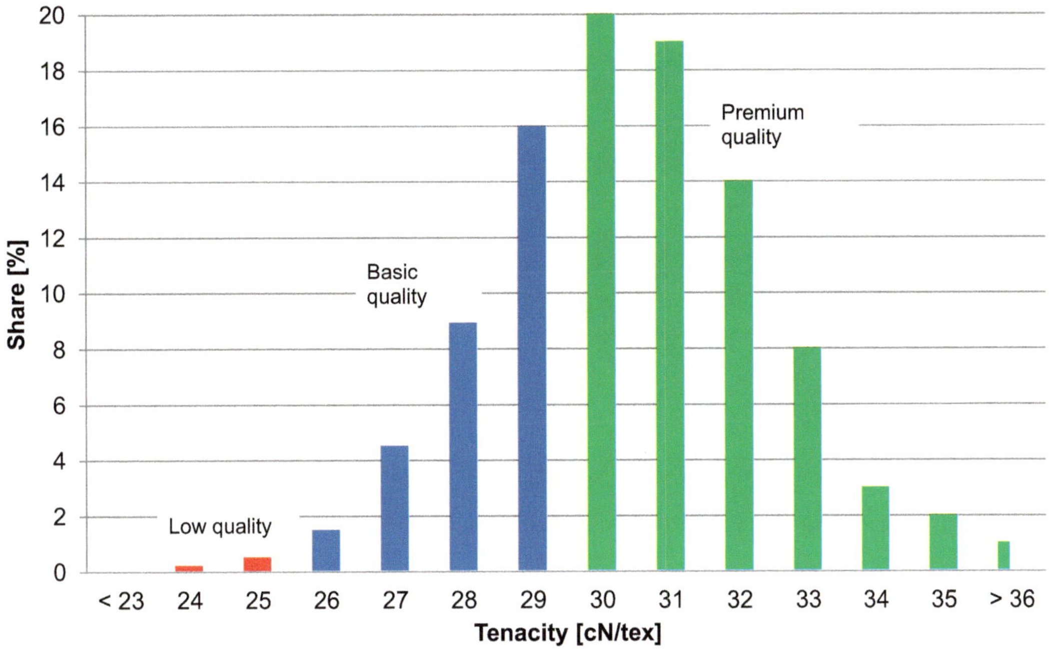

Fig. 5.56 Upland cotton fiber strength distribution (data: Cotton Inc., 2021)

5.6.5 Color

Natural color is an important criterion for classification, because it gives a direct indication of disease attacks and weathering (rain influences). Sick cotton plants produce yellowish fibers, too much rain may result in gray fibers. The color can range from whitish to yellowish/greyish to reddish/brownish. Color variations are standardized and are usually described in words, such as middling spotted or low middling tinged. The best

Fig. 5.57 Development of Upland Cotton Strength (data: Cotton Inc., 2021)

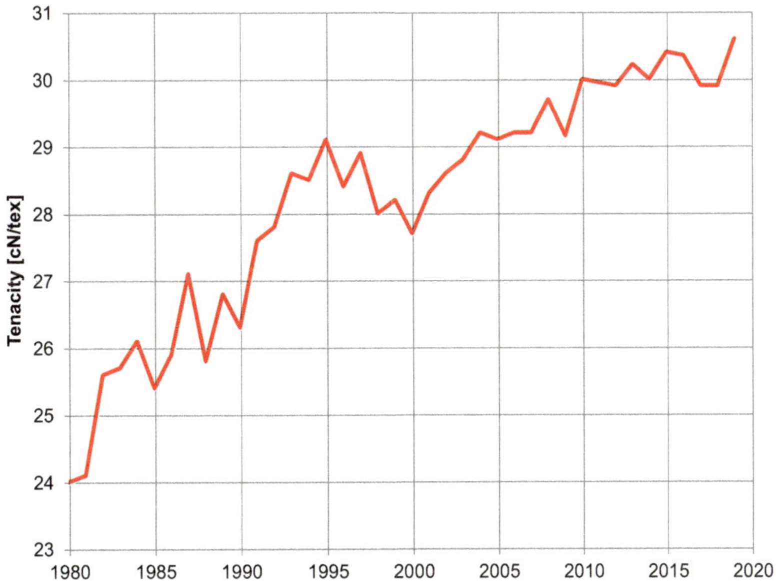

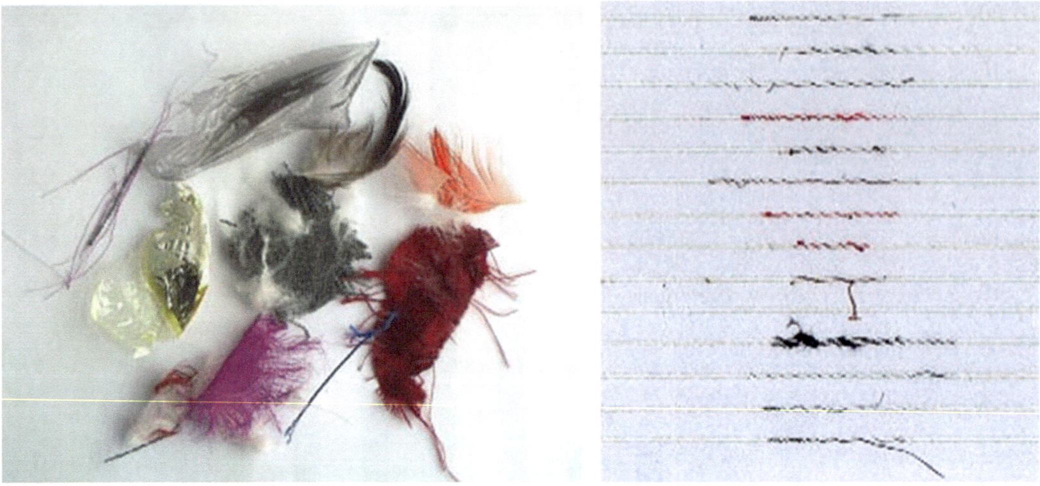

Fig. 5.58 Foreign materials in raw cotton and foreign fibers in yarns

color grade is white. The standard color grades with corresponding price reductions are:

- Light spotted (slightly brownish),
- light grey,
- spotted (brownish spotted),
- grey,
- tinged (strong yellowish to brownish),
- yellow stained (interspersed with strong brown spots).

Normally mature fibers generally have a dull luster, long-staple varieties often have a slightly silky luster.

The color of the cotton is of great importance for further processing (dyeing). Therefore, 25 different colors are distinguished.

Color Code

The color code is determined from the degree of reflectance (Rd) and the yellowness (+b). The Rd. value is a measure of how shiny or dull a sample is, the yellowness depends on the color pigments in the fiber. The color code is given as a combined value of these two characteristics. For example, an Rd. value of 72 combined with a yellowness of 9.0 gives a color code of 41–3 (Fig. 5.59).

5.6.6 Maturity

Depending on the state of development of the cotton (wall thickness of the cell wall), a distinction is made according to DIN 53943, T 1:

Mature fibers:

Fully formed fibers with a strongly thickened cell wall and twists.

Immature fibers:

Fibers harvested before maturity, with a weakly thickened cell wall.

Dead fibers:

Are dead before the wall thickness growth, consist only of the primary wall.

The degree of maturity can be determined by various methods and is usually expressed as a percentage (proportion of mature fibers). It depends on the provenance of the cotton and the year of harvest. Common classifications (%) are:

>82: mature
76–81: medium maturity
70–75: immature
<70: very immature

Often the values are in the range of 75–85% (Fig. 5.60).

Immature and dead fibers are difficult to dye. Table 5.5 summarizes the important physical and chemical properties of cotton fibers.

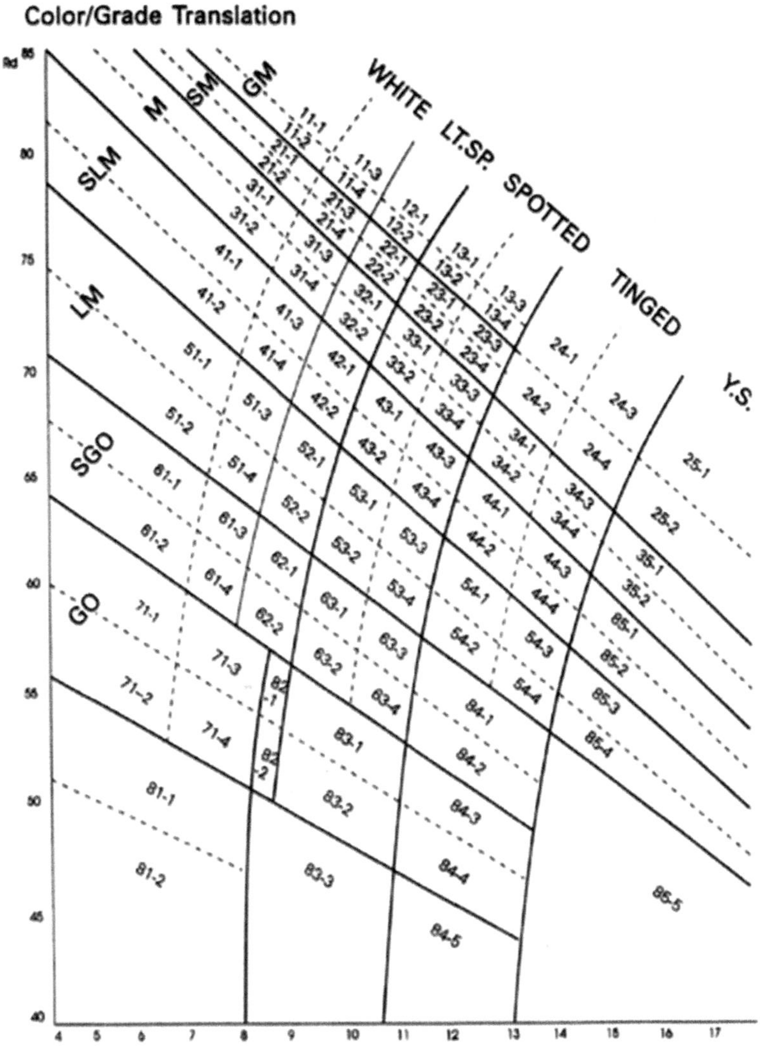

Fig. 5.59 Color code diagram

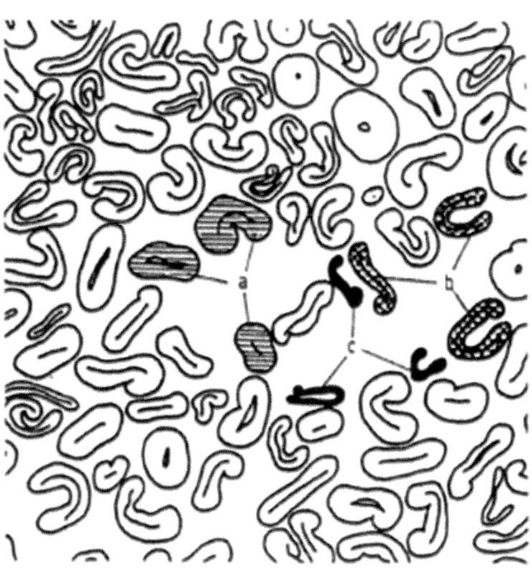

Fig. 5.60 Mature (**a**), immature (**b**) and dead (**c**) cotton fibers in cross-section

5.6.7 Other Properties

Table 5.5 Physical and chemical properties of cotton fibers

Physical properties	
Strength	
Tenacity (cN/tex)	20–50
Wet strength ratio (%)	100–120
Maximum tensile force-elongation (%)	6–10
Wet maximum tensile strength/elongation ratio (%)	100–110
E-modulus (cN/tex)	300–600
Tenacity (cN/tex) at = 2 %	6–8
Tenacity (cN/tex) at = 5 %	10–20
Mercerized cotton may have a single fiber tenacity of more than 70 cN/tex.	
Moisture behavior	
Moisture absorption (%) at 65 % rel. h.	7–9.5
Moisture absorption (%) at 95 % rel. h.	14–18
Water retention capacity (%)	42–53
Other	
Density (g/cm^3)	1.5
Electrostatic charge:	Low
Chemical properties	
Resistance to strong acids	Moderate
Resistance to ammonia and soda solution	Good
Resistance to caustic soda	Poor
Thermal properties	
Temperatures (°C):	
Glass temperature	−5
Staining temperature	120–150
Ironing temperature	180–220
Decomposition temperature	>180
Auto-ignition temperature	400
Charring temperature	430
Usage behavior	
Creasing, dry and boil wash	Large, to be removed by ironing
Crease recovery	Good due to high refinement
Fibrillation and pilling tendency	Low
Boiling water resistance	Very good (hygiene products)
Skin-friendliness, absorbency, air permeability	Very good

5.7 Classification

Classification, i.e. the qualitative grading of cotton, is carried out in most cotton producing countries according to standardized testing procedures.

5.7.1 Manual Classification

In this method, individual characteristic values of the cotton are determined manually, visually or with the aid of simple measuring instruments (Fig. 5.61).

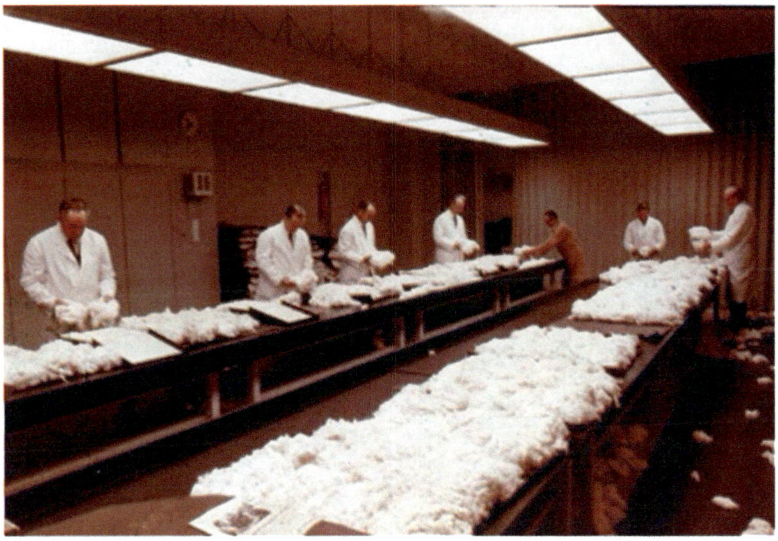

Fig. 5.61 Manual classification

In the USA and many other countries, the "US Universal Cotton Standards", published by the United States Department of Agriculture (USDA), are used for this purpose.

US Universal Cotton Standards
These standards consist of sample boxes containing 6 samples of 300 g fibers for each class (Fig. 5.62). The standards are reviewed in the USA at 3-year intervals at a conference in Memphis, Tennessee, and revised as necessary. Classification by manual-visual comparison with the standards is done by experienced classifiers. This visual classification has all but disappeared and been replaced by automated systems.

5.7.2 High Volume Instruments (HVI) and Fibre Classifying System (FCS)

Alternatively, an objective classification can also be carried out by an automated determination of properties, so-called instrument testing. Here, several instruments are applied to determine fineness (micronaire), strength, color grade, trash content etc. The first such testing system was developed by USTER Technologies AG, Switzerland, and called HVI (High Volume Instruments). An alternative system is the Fibre Classifying System (FCS) by Textechno, Germany (Fig. 5.63). These systems have already largely replaced manual-visual quality testing in the USA, Uzbekistan, China and many other countries. However, different grading standards are used due to the differences between Upland and Pima cotton. These differences are genetic and are also caused by the different ginning processes used for Upland and for Pima cotton. Pima is ginned via roller gins and therefore appears more uniform and softer than Upland varieties. In addition, the color of Pima cotton is creamier.

Calibration cotton is required for instrumental classification in the HVI process. For this purpose, corresponding bales are selected from 7 testing laboratories worldwide and sent in the form of small samples for calibration of the measuring instruments. The Fiber Classifying System (FCS) from Textechno does not require calibration cotton and provides the same parameters as the HVI fiber testing line. However, the values are not directly comparable.

5.8 Production, Export, Import

Cotton is one of the world's most economically important agricultural products, especially for developing countries. There, the cultivation of cotton secures the jobs of an estimated 125 m

Fig. 5.62 US Universal Cotton Standards sample cases

Fig. 5.63 HVI fiber test line from USTER (left) and FCS devices from Textechno (right)

people. The value of the cotton produced there is around US$ 40bn annually. For some African countries in particular, cotton is one of the most important trade products and thus of crucial economic importance.

5.8.1 Producer

Cotton grows in both dry and semi-humid climates. The most important growing areas are China, India, the USA, Pakistan, Uzbekistan, Turkey, Australia, Brazil and Egypt. In Europe, Greece and Spain are note-worthy. In total, cotton is currently grown on around 32 m hectares of arable land in more than 40 countries, which corresponds to 0.7% of the world's arable land (Fig. 5.64).

Figure 5.65 shows the development of cotton production for all continents. Around 1940, America had a world market share of about 60%, Asia 30% and Africa 10%. Around 1980, Asia and the Americas were tied at about 45% each; since then, production has more than tripled in Asia, roughly doubled in the Americas, and today, market shares are 60% for Asia and 30% for the Americas. Africa has been able to maintain its market share of 8–10% to date. Australia has a highly fluctuating share of 2–4%, because record harvests are often followed by droughts. Europe does not play a role with about 2% market share (Fig. 5.65).

Within the last 50 years, improved fertilization, weed removal and pest control have more than doubled yields on almost the same area under cultivation (Fig. 5.66). However, there are major regional differences.

A very high yield is over 2000 kg of fiber per ha (e.g. Australia), a low yield is around 400 kg/ha (e.g. India). The world average is around 760 kg/ha (2020). Yields were increased in particular through the targeted use of modern irrigation methods together with increased use of pesticides.

The yields of the growing countries are very different as Fig. 5.67 shows. Australia, China, Brazil and Turkey each achieve very high yields, while India and the countries of Africa rank at the bottom.

World cotton production has tripled in the last 50 years, but the number of unsold bales (ending stocks) in storage has increased even more, to

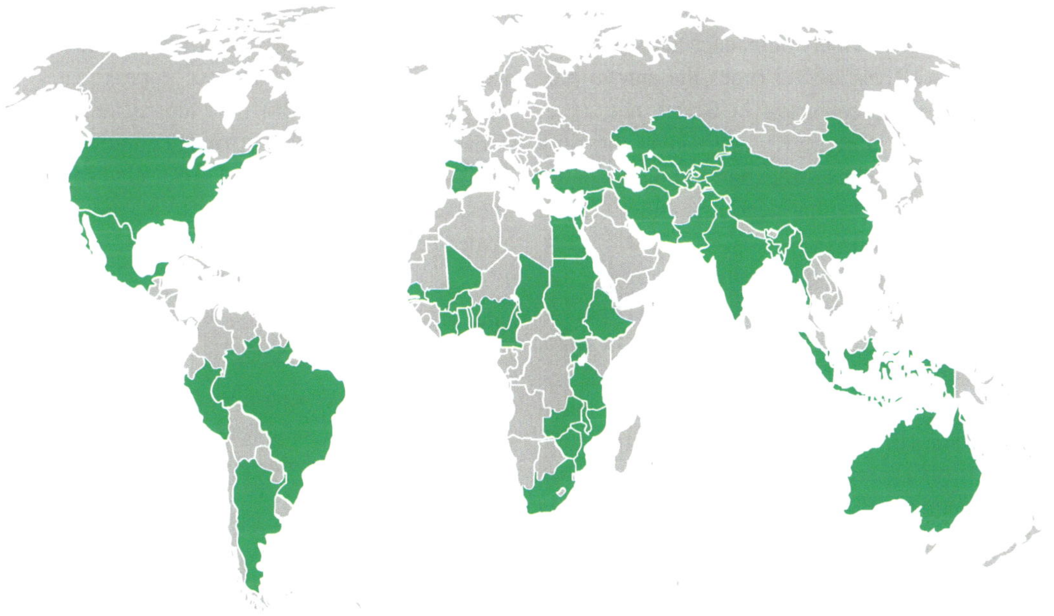

Fig. 5.64 Cotton growing countries

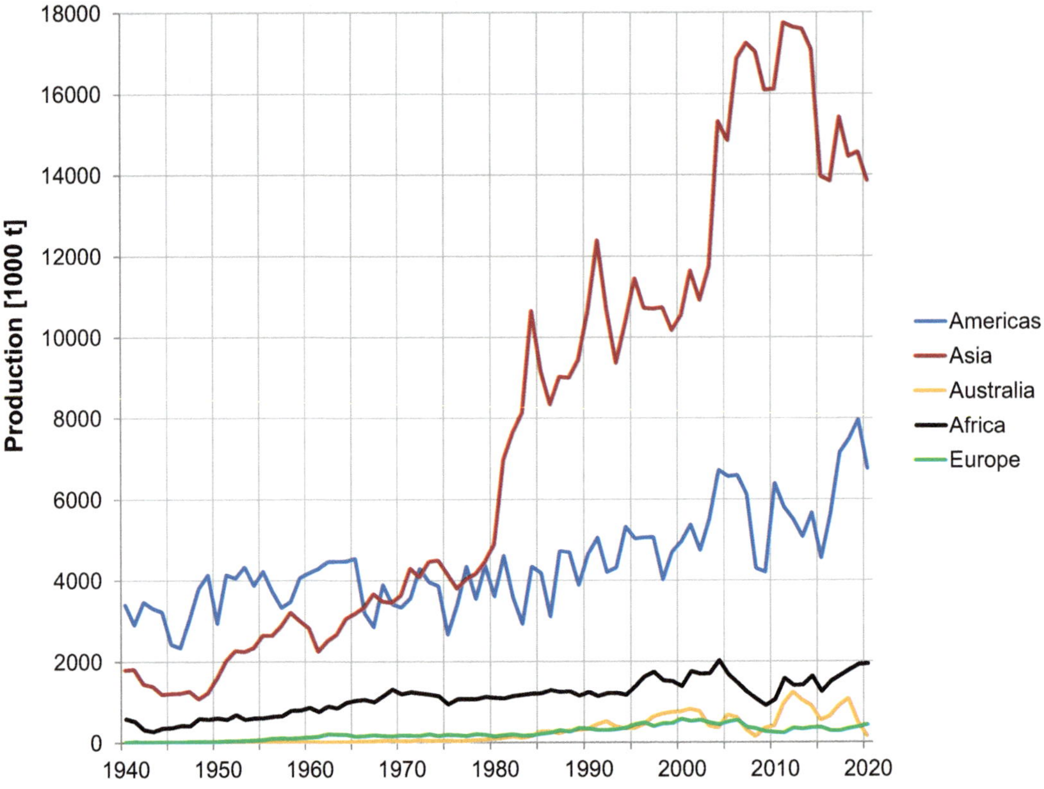

Fig. 5.65 Development of cotton production since 1970 for all continents (data: ICAC, 2021)

almost a year's crop at present (Fig. 5.68). There are large country-specific differences.

Stocks are extremely high, especially in China and Turkey. They are 1.5 times the annual production of these countries. In Brazil, almost an entire crop is in storage. Australia is the only major cotton producer to export almost its entire cotton crop, but also has nearly an entire crop in storage (Fig. 5.69).

The main producing, importing, exporting and consuming countries are shown in Table 5.6. Australia had an extremely poor harvest in 2020, but is well ahead of Mexico in normal years.

The four largest growing countries, India, China, the USA and Brazil, account for more than 70% of world production (Fig. 5.70). While, until the 1980s, the USA and subsequently China were the main producers of cotton, in 2015, for the first time since the nineteenth century, India took first place.

5.8.2 Exporters

Worldwide, around 8.5 m tons of cotton are exported with a total value of approximately € 12bn. The USA and Brazil dominate, far ahead of India, Greece, the West African countries Benin, Mali and Burkina Faso and Australia. Together, these 8 countries export 80% of the world's traded cotton (Fig. 5.71).

5.8.3 Importers

Imports are also dominated by only a few countries. Only 7 countries import 84% of the cotton traded worldwide (Fig. 5.72).

5.8.4 Consumer

Likewise, a strong concentration on only a few countries can be seen in processing cotton. China

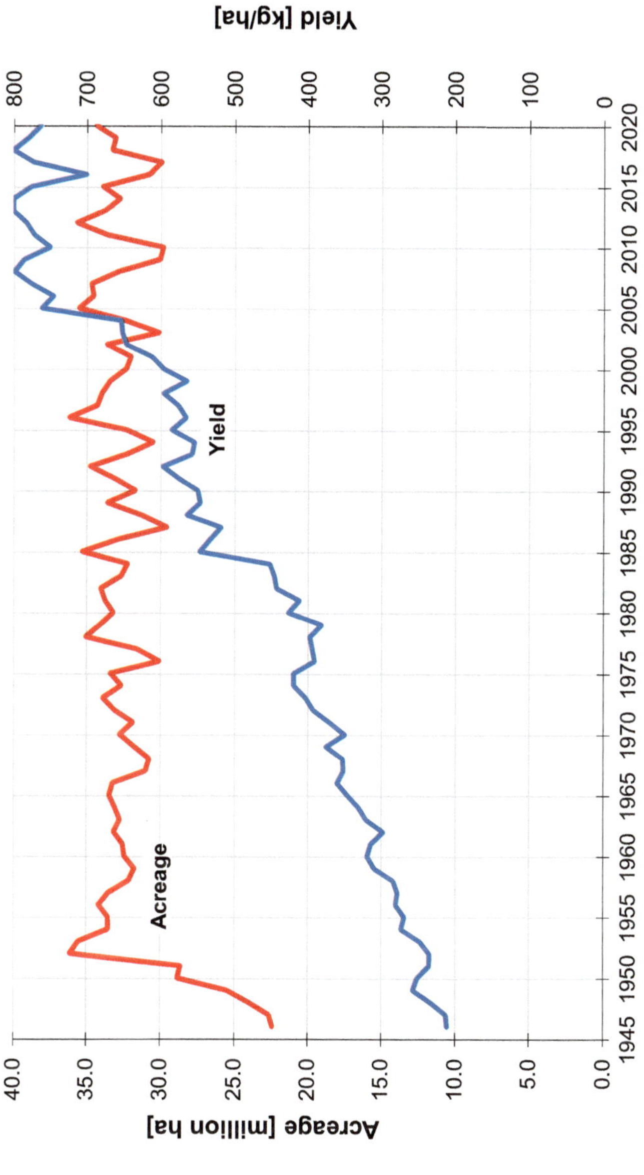

Fig. 5.66 World cultivated areas and average yields (data: ICAC, 2021)

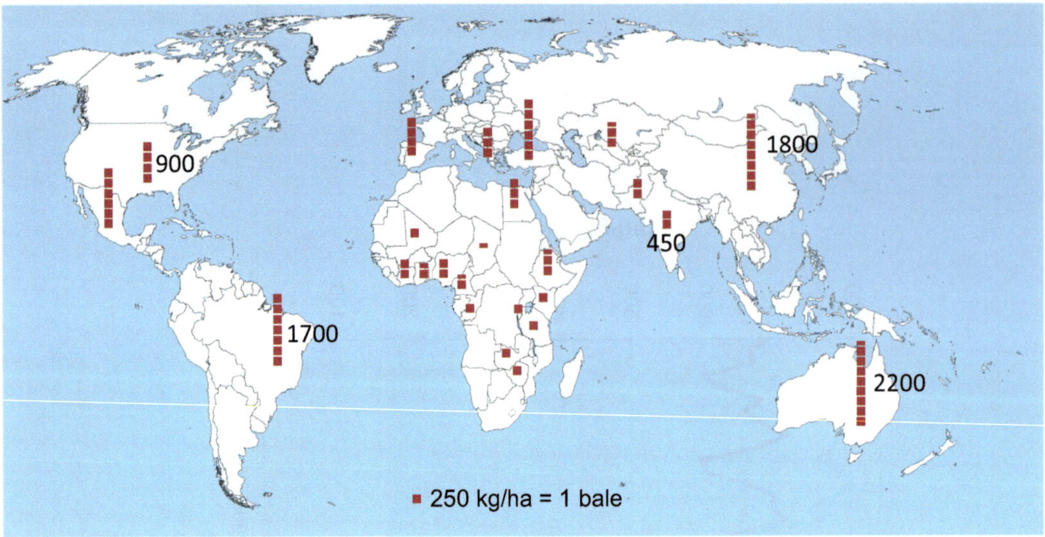

Fig. 5.67 Yields of selected countries in [kg/ha] (data: ICAC, 2021)

and India account for more than half of the world's production of cotton products (Fig. 5.73).

The development of the last decades is depicted in Fig. 5.74. While the USA processed more than 10% of the world's raw cotton until 2000, they now play only a minor role (2%). Cotton processing is declining rapidly in China, while it has been stable at high levels in India, Pakistan, Turkey and Bangladesh since 2010. The strongest growth is in Vietnam (+300% since 2010).

The EU currently processes 150,000 tons (660,000 bales) of cotton per year. Figure 5.75 shows the most important consumers in the EU (2020). In addition, there is Turkey with around 1.5 m tons. It thus ranks fourth in the world.

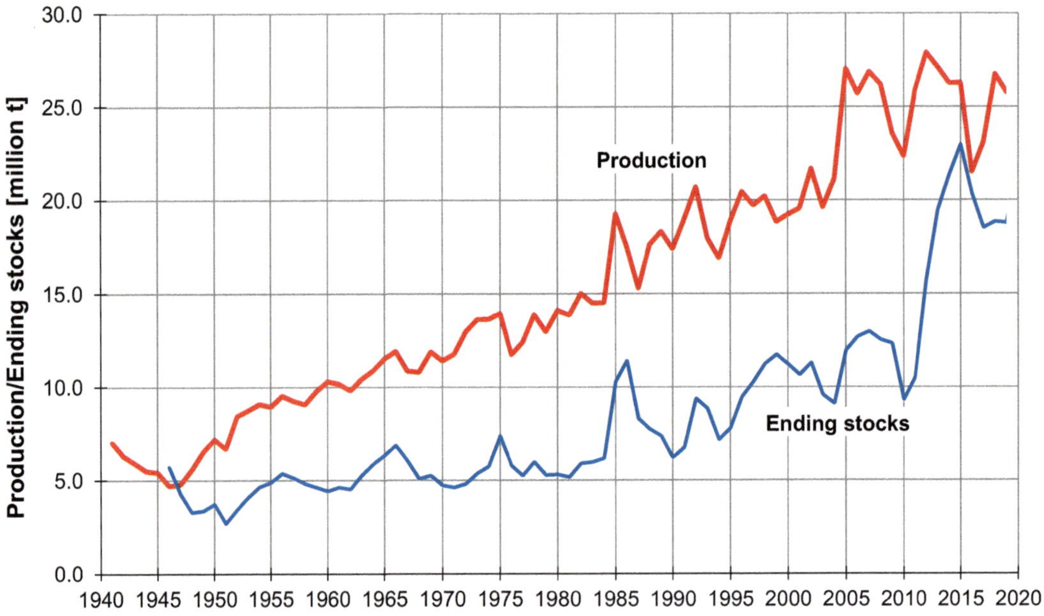

Fig. 5.68 Production and Ending Stocks worldwide (data: ICAC, 2021)

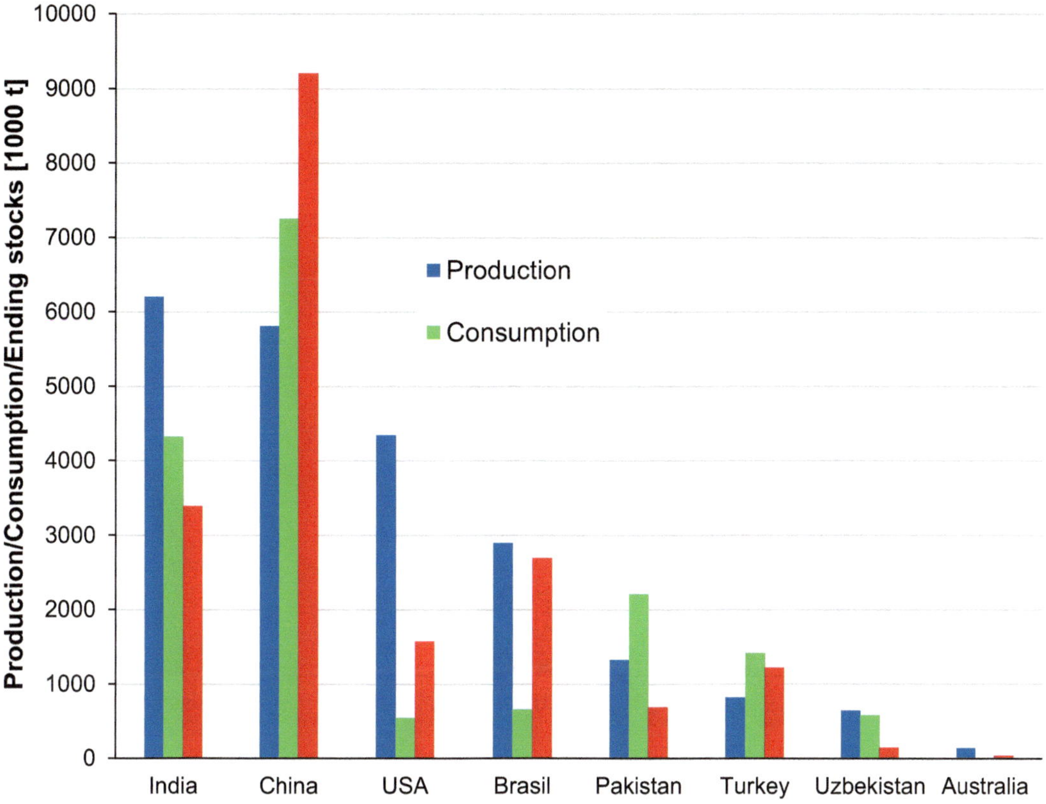

Fig. 5.69 Production, consumption and ending stocks of the TOP 8 cotton producers for 2020 (data: ICAC, 2021)

5.8.5 Extra Long Staple Cotton (ELS)

The picture is different from that for standard cotton for the extra-long staple cotton varieties with a fiber length of more than 35 mm (44/32″) as shown in Fig. 5.76. Until 2011, Egypt ("mako cotton") and the USA were dominant, leading China and India. The total volume was around 360,000 t in 2009, rose to a record 570,000 t by 2011, and has been around 350,000–450,000 t since 2012. Today, the US is the largest producer of long staple cotton (120,000 t) ahead of India (85,000 t) and Egypt (58,000 t). China cut subsidies for long staple cotton in 2017,

Table 5.6 Top 10 producers, importers, exporters, consumers (1000 t, data: ICAC, 2021)

Producer		Importer		Exporter		Consumer	
India	6200	China	1815	USA	3309	China	7250
China	5800	Vietnam	1401	Brasil	1901	India	4320
USA	4336	Bangladesh	1216	India	528	Pakistan	2204
Brasil	2891	Turkey	900	Greece	320	Turkey	1415
Pakistan	1320	Pakistan	823	Benin	263	Vietnam	1401
Turkey	815	Indonesia	630	Mali	262	Bangladesh	1216
Uzbekistan	641	Thailand	211	Australia	235	Brasil	650
Mexico	369	Malaysia	162	Burkina Faso	174	Indonesia	630
Turkmenistan	300	South Korea	136	Turkmenistan	149	Uzbekistan	578
Australia	135	India	40	Ivory Coast	145	USA	537

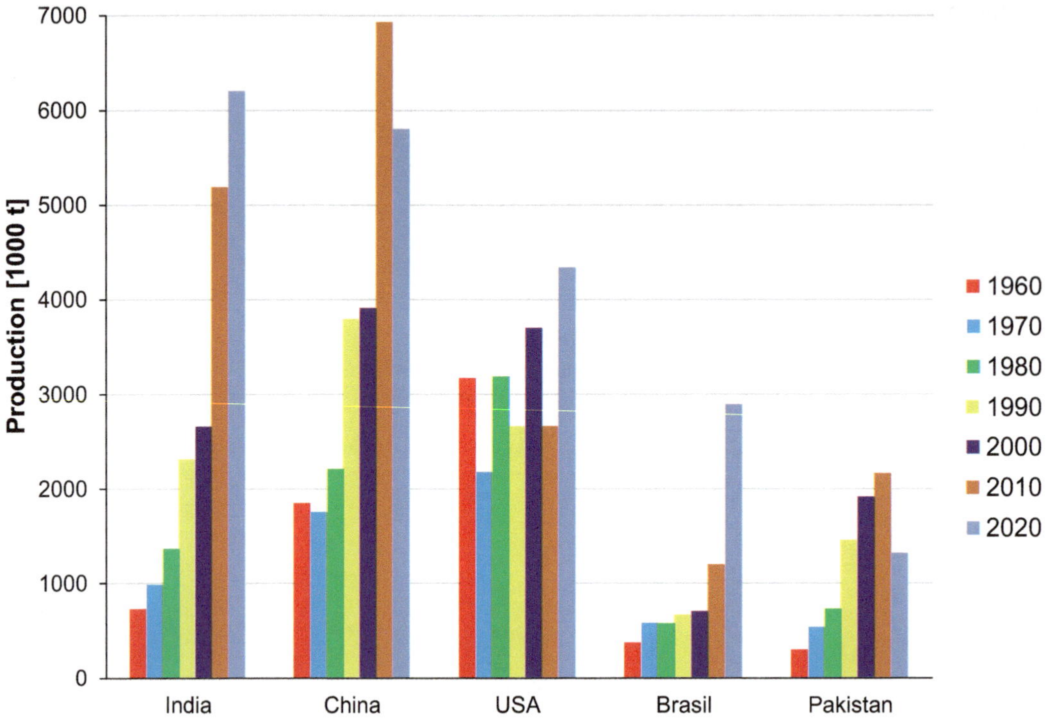

Fig. 5.70 Development of the production of the TOP 5 cotton producers (data: ICAC, 2021)

Fig. 5.71 Main cotton exporting countries (data: ICAC, 2021)

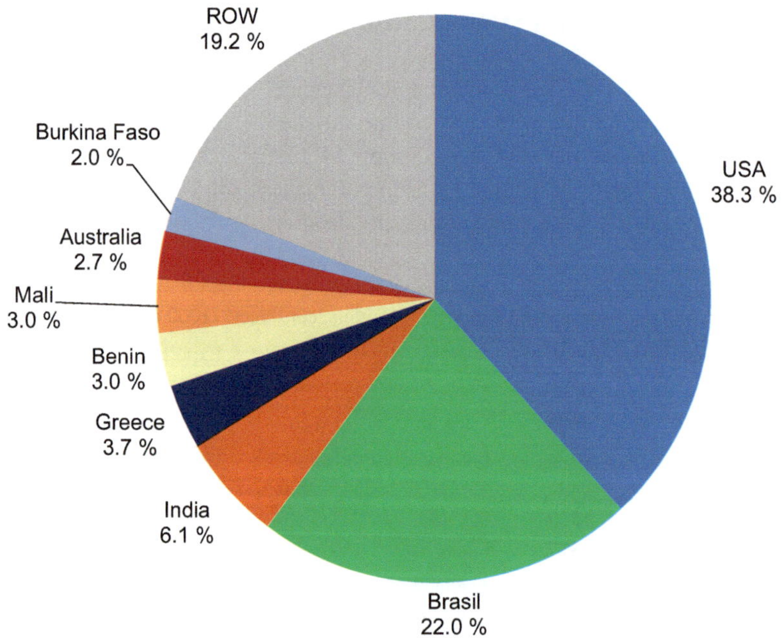

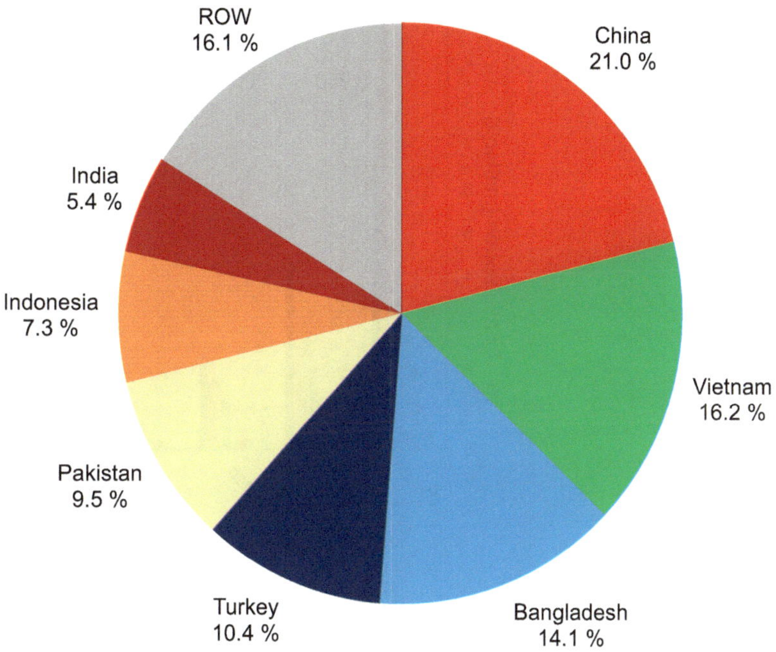

Fig. 5.72 Main cotton importing countries (data: ICAC, 2021)

whereupon production declined drastically (Fig. 5.77). The market price is subject to large fluctuations and is higher than the price for standard qualities (upland) by a factor of 1.5–2.0. Unlike upland varieties, which grow within about 5 months, ELS cotton takes 8–10 months to mature, which increases the cost of fertilizer, pest control and irrigation. For example, on average, insecticides for upland cotton worldwide cost about US$150/ha, compared to about US$570 for

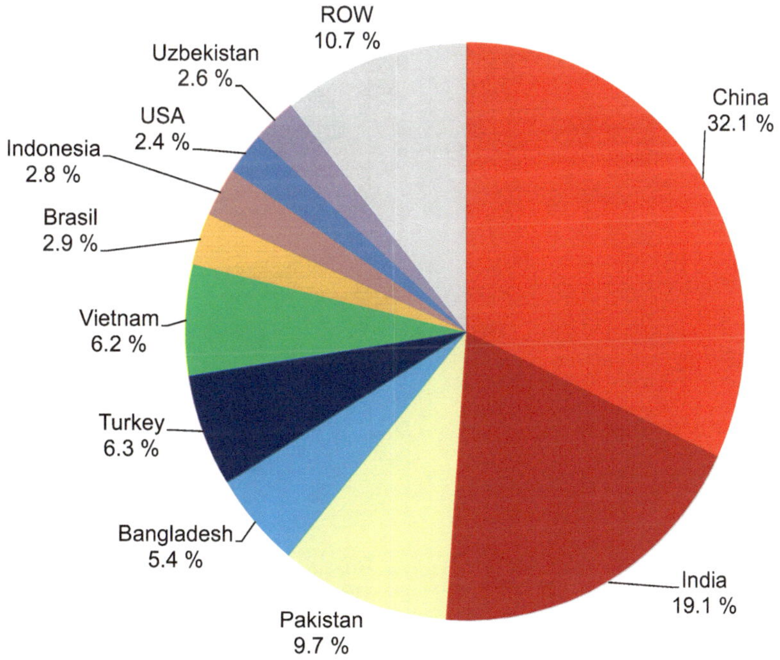

Fig. 5.73 Main consumers of cotton (data: ICAC, 2021)

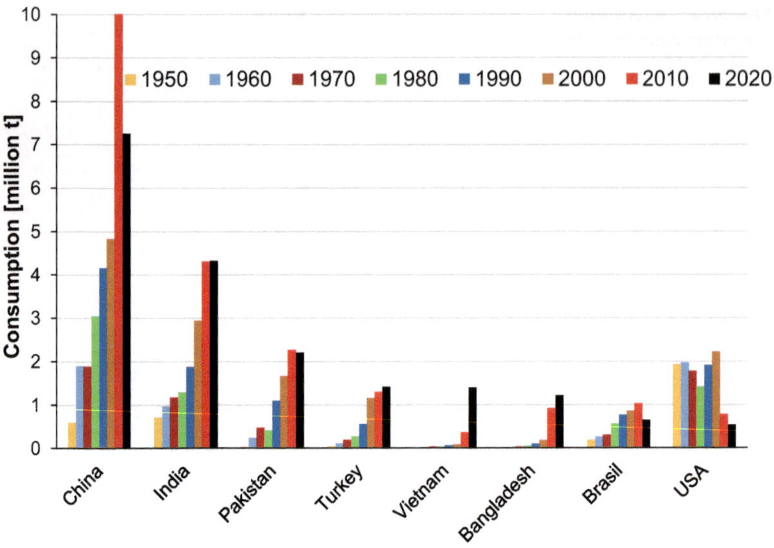

Fig. 5.74 Development of the processing of cotton of important countries (data: ICAC, 2021)

ELS cotton. As yields have also fallen in recent years, ELS production is under considerable cost pressure.

Figure 5.78 shows the typical length distribution for extra long staple (ELS) cotton compared to upland cotton.

ELS cotton has considerably greater strength than upland cotton (Fig. 5.79).

Egypt processes only 15% of the long staple cotton produced in the country itself; in the USA the export share is over 97%. The largest

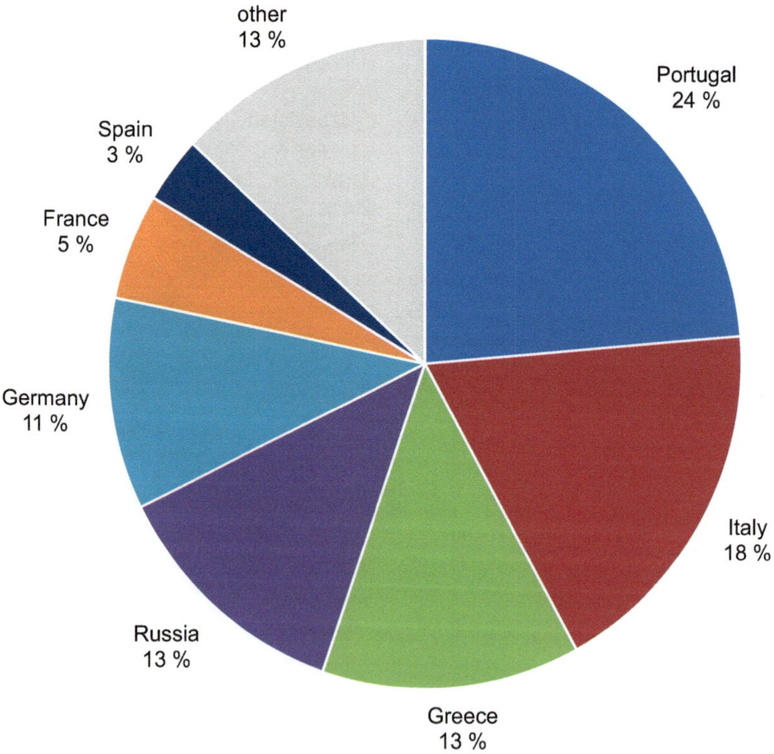

Fig. 5.75 Main cotton consumers in the EU (data: USDA, 2020a)

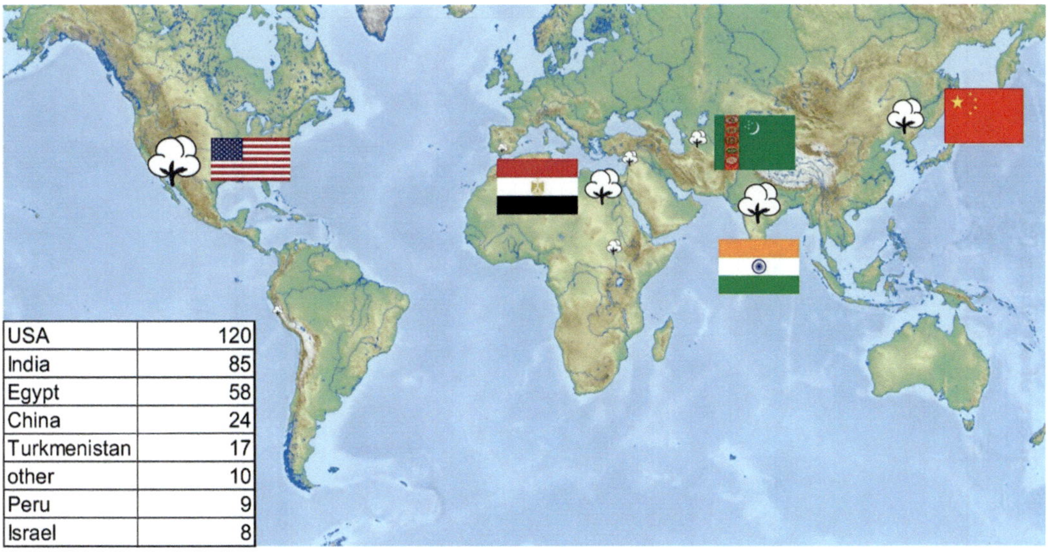

USA	120
India	85
Egypt	58
China	24
Turkmenistan	17
other	10
Peru	9
Israel	8

Fig. 5.76 Major producers of ELS cotton (data: Bremer Cotton Report, 2021)

consumer overall is China (144,000 t), ahead of India (106,000 t) and Peru (13,000 t).

5.9 Important Cotton Producing and Processing Countries

In the following, we look at typical characteristics of selected cotton producing and processing countries and regions.

5.9.1 USA

Cotton production in the USA is concentrated in about a dozen states in the South, the so-called cotton belt (Fig. 5.80).

Cotton production is an important economic factor for the USA, providing work for more than 190,000 people in around 21,000 companies with a turnover of around US$ 8.9bn (2020). Cotton

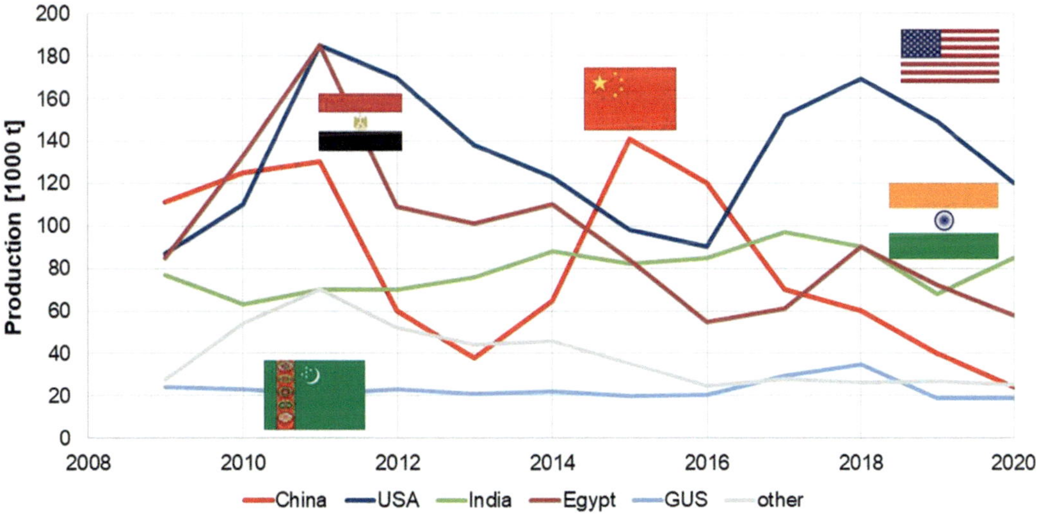

Fig. 5.77 Development of LS and ELS cotton production (data: Bremer Cotton Report, 2021)

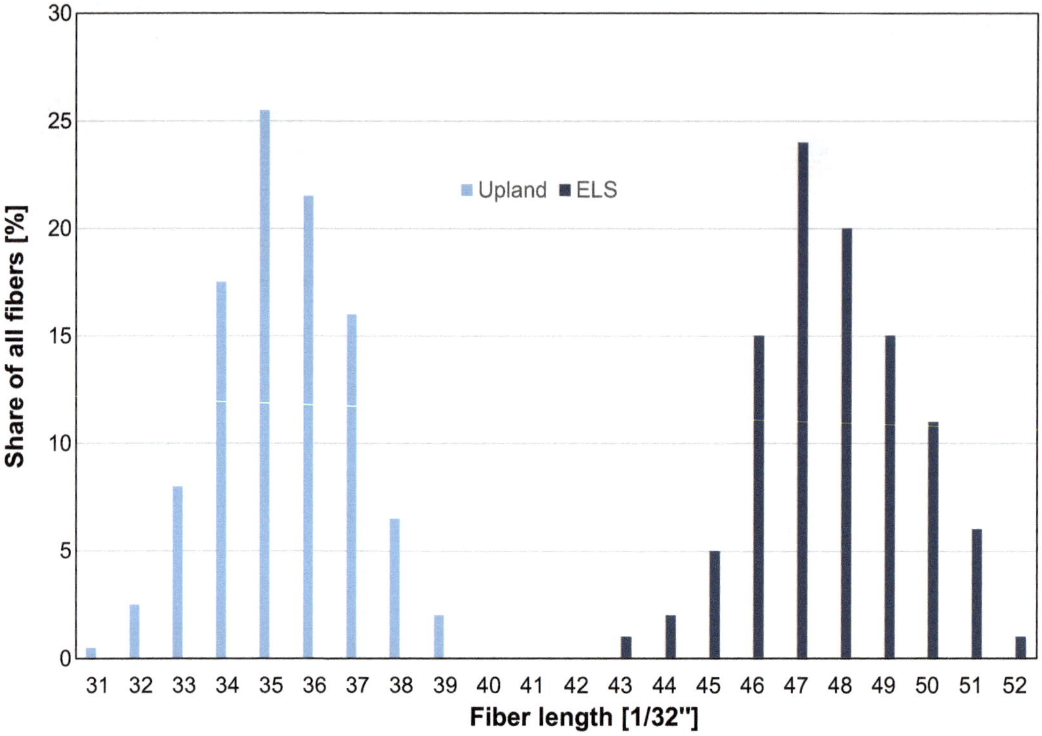

Fig. 5.78 Comparison of fiber length of ELS cotton with Upland varieties (data: intracen, 2020)

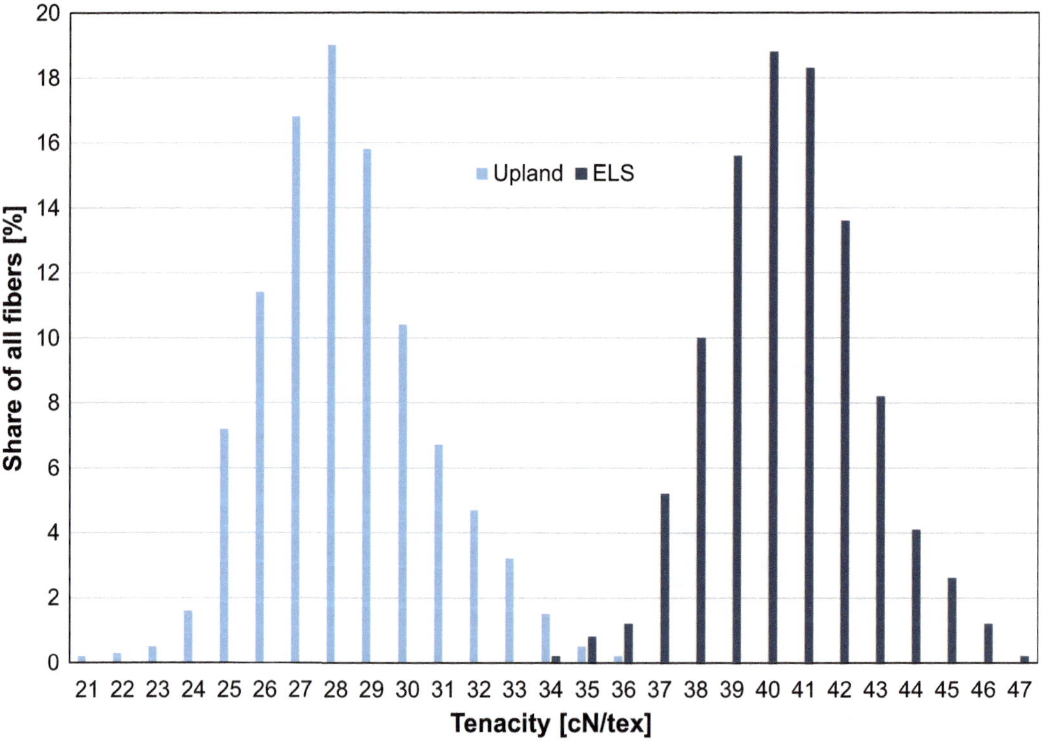

Fig. 5.79 Comparison of the strength of ELS cotton with Upland varieties (data: intracen, 2020)

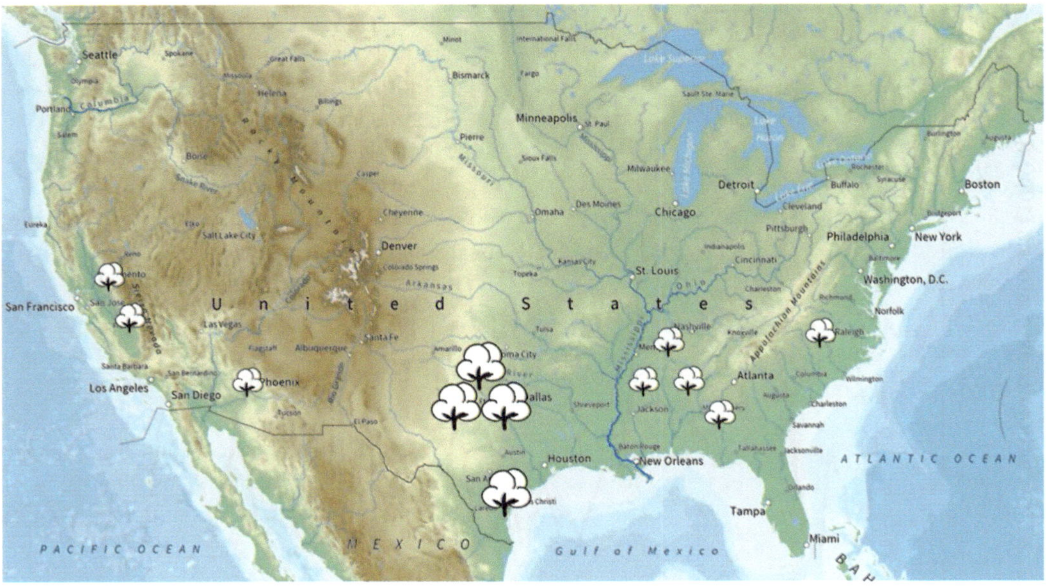

Fig. 5.80 Most important cotton growing areas in the USA (map: Mapswire)

fibers account for 80% of this turnover and cotton seeds for 20%.

The development of the cultivated area as well as the yield since 1990 is shown in Fig. 5.81. The dramatic decline in acreage in the 1950s and 1960s and again from 2006 to 2009 as a result of the global economic crisis can clearly be seen. In 2011, the production level of 2008 was reached

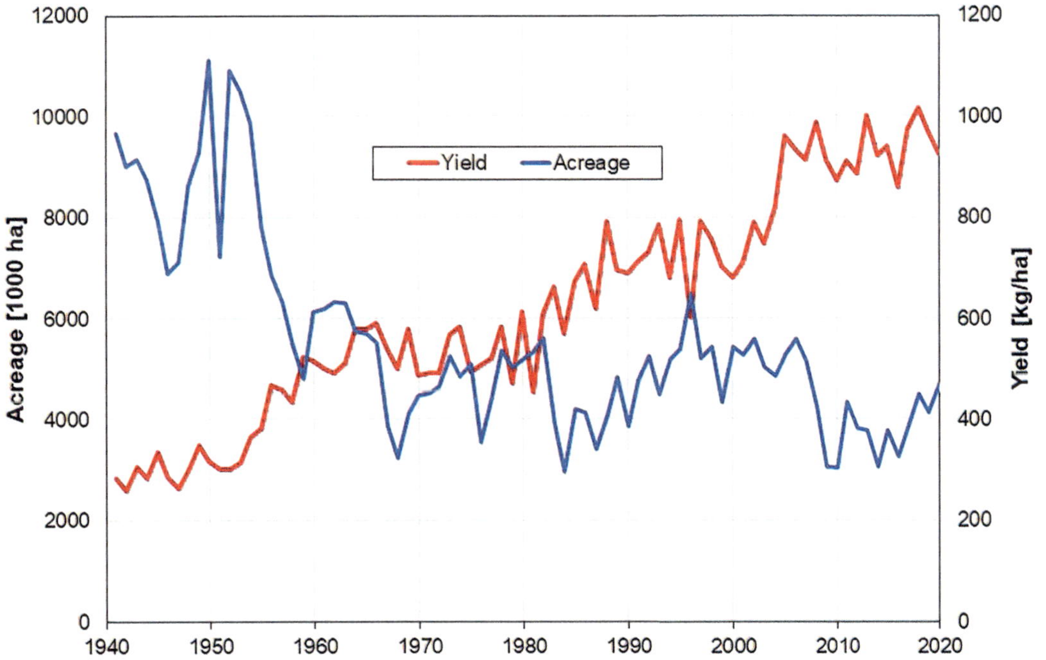

Fig. 5.81 Area under cultivation and yield in the USA (data: ICAC, 2021)

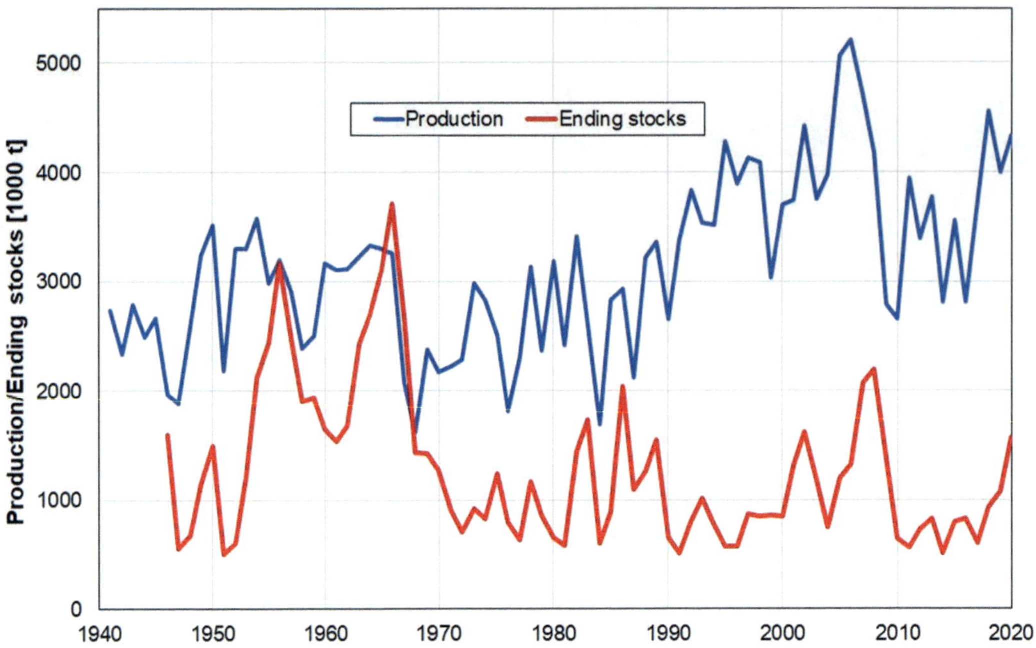

Fig. 5.82 US cotton production (data: ICAC, 2021)

again and since then the area under cultivation has fluctuated within the usual range. The use of genetic engineering has succeeded in increasing yields since 2000 (680 kg/ha) to over 900 kg/ha (2020).

Cotton cultivation competes with the production of biofuels and the reduction of direct payments to cotton farmers in favor of crop insurance has recently made their economic situation more difficult.

This development is also directly reflected in total production, as Fig. 5.82 illustrates. Upland cotton currently accounts for around 96% and Pima cotton for 4% of the production of 4.3 m tons. The quantity of cotton in storage is 1.5 m tons, which corresponds to about 40% of a year's harvest.

In 1950, the world market share of US cotton was still 50%, today it is less than 20%. In the same period, China's share grew from 3% to 25%.

Most cotton is grown in Texas (29%), followed by California (11%) and the eastern states of Georgia (11%), Mississippi, Tennessee and South Carolina (9% each), Louisiana and Missouri account for 5% each. In addition, Arkansas, Arizona and Alabama are of lesser importance.

85% of the cotton produced in the USA is exported, mainly to Mexico and China

5.9.2 China

Until the twelfth century, China imported cotton from India, after which cultivation began under the first Mongol rulers. A cotton field yielded about 10 times more fiber than a comparable field of ramie or hemp. As a result, cotton, also encouraged by government support, gradually replaced these two traditional fiber raw materials over the next centuries.

Today, cotton cultivation in China is mainly concentrated in the province of Xinjiang in northern China, the northernmost cotton growing region in the world. Here, cotton grows only with the help of artificial irrigation. In the traditional cotton growing areas in the south along the Yangtze and Huang Ho (Yellow River), on the other hand, production is gradually being reduced due to the increasingly serious resistance of the bollworm to the usual pesticides. Currently, about 90% of cotton plants are genetically modified to solve this problem. All areas are characterized by moderate winters and prolonged periods of heat, both of which are prerequisites for good harvests (Fig. 5.83).

In 2020, 3.3 m ha of land was planted with cotton. The reduction in cultivated area in recent years, which competes with food production, has been partly offset by a sharp increase in yields per hectare. Production is around 5.8 m tons, equivalent to 25.5 m bales. Around 9 m households in China make their living from cotton cultivation (Fig. 5.84).

A cause for concern is the large amount of cotton currently in storage ("ending stocks"). It has been drastically reduced in recent years by leaving land fallow. But, it still exceeds the amount of cotton produced annually by more than 50% (Fig. 5.85).

5.9.3 India

Around 3000 BC, a civilization had developed along the Indus, in the cities of Mohenjo Daro and Harappa, whose inhabitants, according to archaeological finds, were already familiar with the cotton plant and its use. Cotton gradually spread throughout the subcontinent and its fabrics were exported as an important commodity. Greeks, Romans and Egyptians became acquainted with the soft material. The Greek philosopher Herodotus wrote: "The wild trees there also bear wool in fruit, which in beauty and goodness deserves to be preferred to sheep's wool. From these trees, the Indians make themselves garments." (Herodotus, 3, 106).

Later, the transport of cotton was taken over by European nations, especially England. In 1600, England's Queen Elizabeth I granted the later East India Company a monopoly on trade with India, in which cotton played a prominent role.

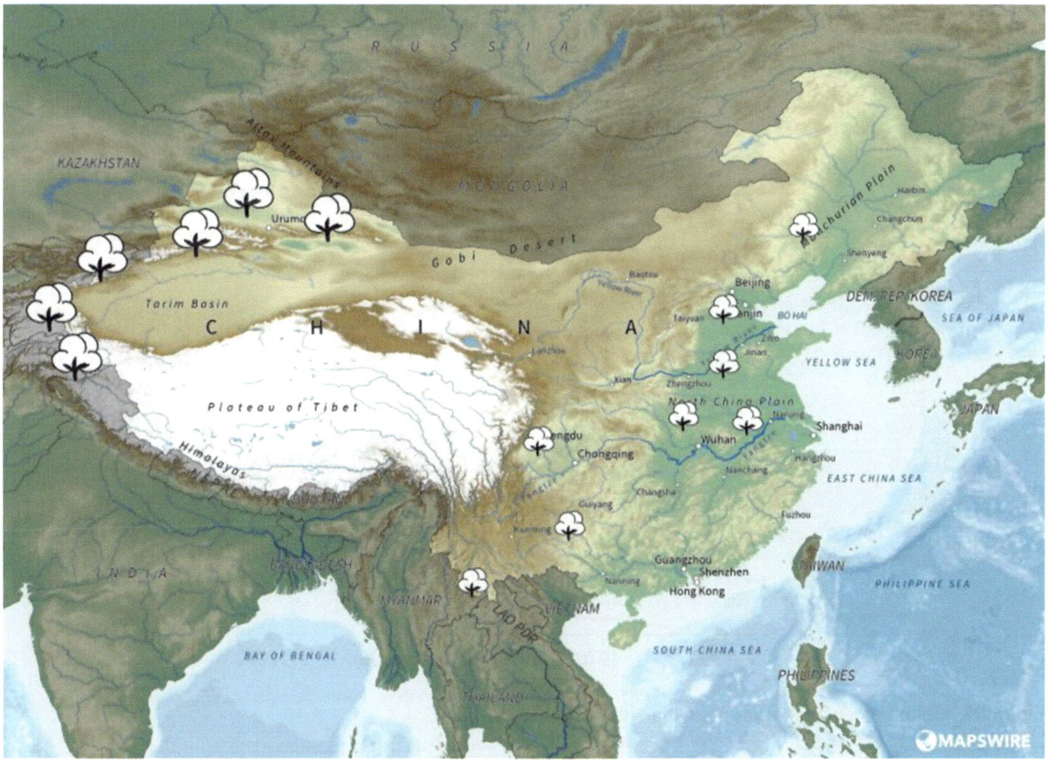

Fig. 5.83 Cotton growing regions of China (map: Mapswire)

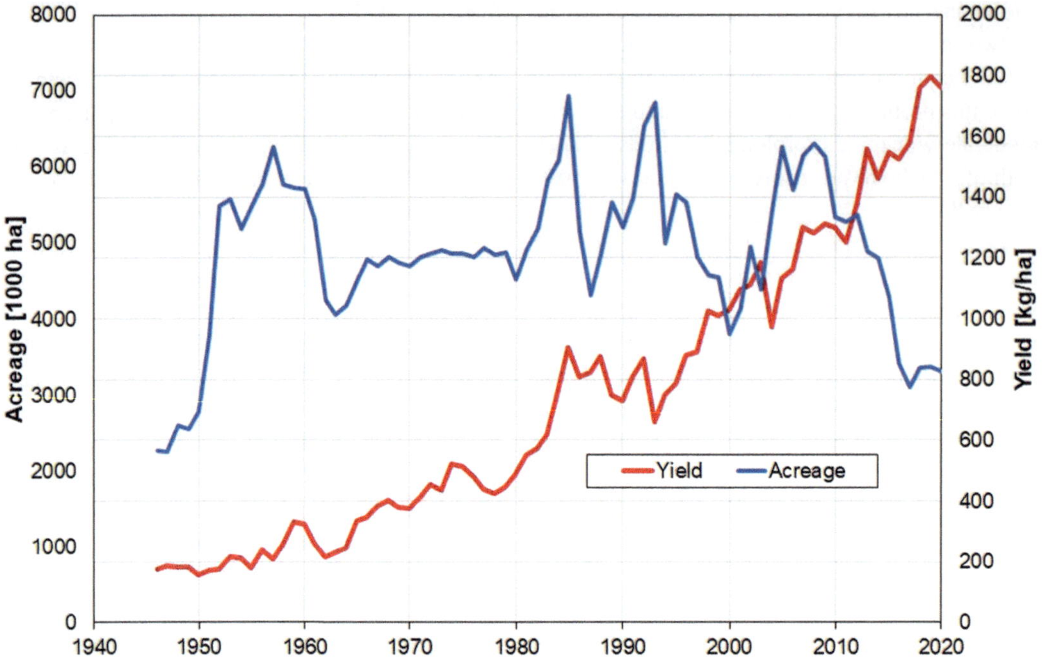

Fig. 5.84 Area under cultivation and yield in the PRC (data: ICAC, 2021)

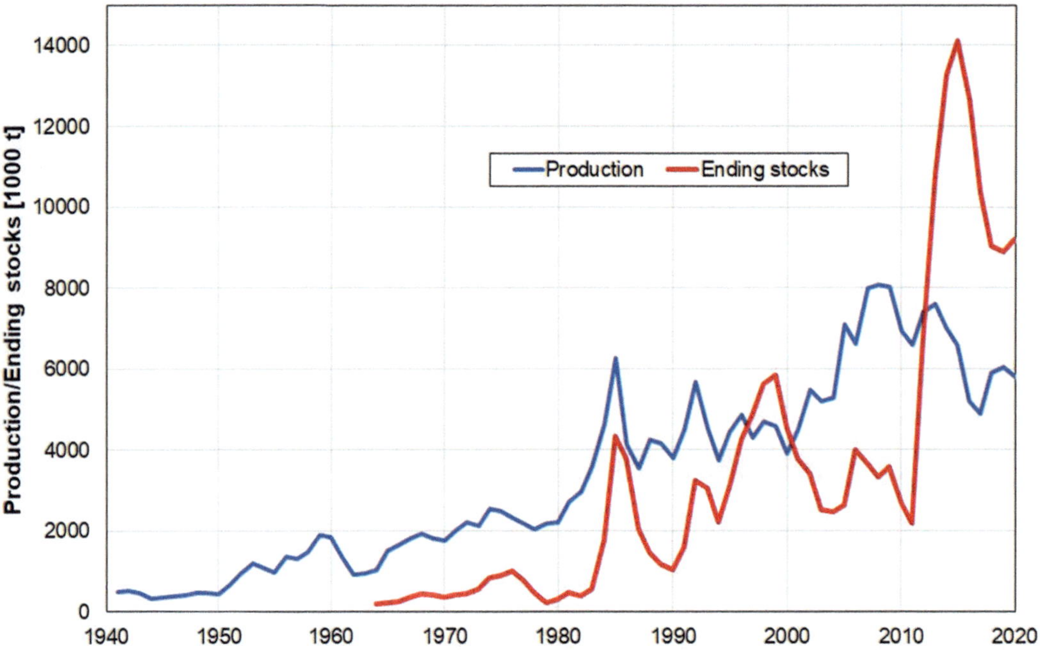

Fig. 5.85 Cotton production of the PR China (data: ICAC, 2021)

India remained the world's most important supplier of cotton and cotton fabrics until the end of the eighteenth century.

Relations between Great Britain and India deteriorated when machines for spinning and weaving were developed in England as part of the industrialization process. As a result, the demand for raw cotton rose sharply and many Indian small businesses suffered from the price pressure of the cheap large competitors. In 1818, the British succeeded in subduing India. The country became a British colony and the new masters had sovereignty over the "white gold". India was converted from a producer of precious cotton fabrics to a mere supplier of raw cotton for the European industry.

This period saw the awakening of resistance against colonial exploitation, later led by Mahatma Gandhi. Among other things, the charismatic pacifist called for a boycott of cheap imported goods. He appealed to the people of India to return to their own identity and wear homespun cotton (Fig. 5.86). In the end, Gandhi overcame colonial rule with passive resistance and in 1947, the last British troops left the subcontinent.

Today, India is not only the largest producer of cotton, but also one of the most important users. About 10 m households live directly from cotton cultivation.

Cotton is grown in India mainly in the southwest and central parts of the country on more than 12 m ha (Fig. 5.87). The production volume in 2020 was around 6.2 m tons (28 m bales). About 18 m bales of cotton, more than half of the annual production, are in stock. The Indian textile industry, which mainly processes cotton, contributes about 8% to the GDP with a turnover of US$ 200 m, about 14% of industrial production is textiles, accounting for 11% of export earnings. After agriculture, the textile industry is the second largest employer for about 51 m people.

The area under cultivation has fluctuated in recent years, but yields have increased considerably since 2000, mainly due to the use of genetically modified varieties (90% market share). This initially resulted in fewer crop failures because the new cotton varieties were less sensitive to pests.

In India, two-thirds of the cotton grown is a hybrid, which is a cross between G. hirsutum (good yields) and G. barbadense (high staple length). In genetic modification, the transgenic traits are transferred only to the upland genome. This results in half of the hybrid genome being non-genetically modified, thus reducing the pest

Fig. 5.86 Mahatma Gandhi at the spinning wheel (www02) and cotton warehouse (Bremen Cotton Exchange, 2021)

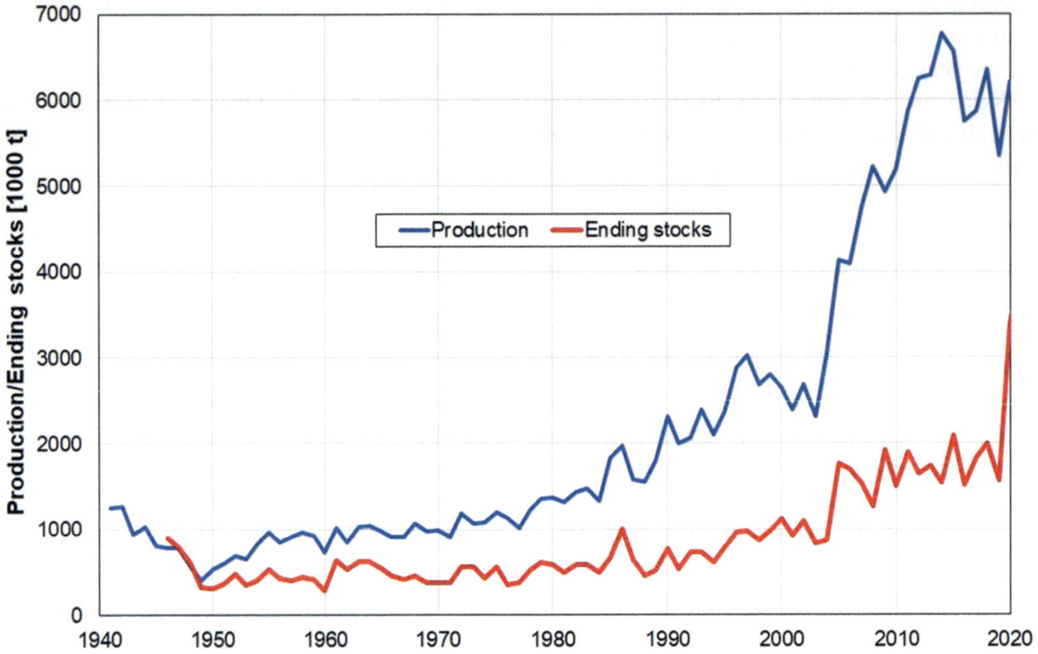

Fig. 5.87 India's cotton production (data: ICAC, 2021)

control effect by 50%. This allowed resistance to develop in the insects, which further reduced the effect of Bt cotton.

Due to the introduction and gradual increase of the minimum support price for cotton ("floor price"), the area under cotton cultivation has risen sharply in recent years. However, yields per hectare of 400–500 kg are still well below those of China and the USA (Fig. 5.88).

India is the only country where all 4 major varieties of cotton are planted. In the north (Punjab, Rajasthan, Haryana) G. hirsutum and G. arboreum are grown. In central India (Gujarat, Madhya, Pradesh and Maharashtra), G. hirsutum, G. arboreum and G. herbaceum are planted. In the south (Andra Pradesh, Karnatska and Tamil Nadu) all varieties are cultivated (Fig. 5.89).

5.9.4 Pakistan

In Pakistan, cultivation is concentrated in the area along the Indus River. It is almost totally irrigated artificially as it only rains from July to September and therefore a continuous supply of water to the plants is not possible. As a result, production increased tenfold from 1950 to 2015 to 2.3 m t. In 2020, after a number of crop failures, annual production was only 1.2 m t, with ending stocks of 0.7 m t (Fig. 5.90). Cotton cultivation on an area of 2.2 m ha gives income to about 2 m households, and harvesting is only done manually. The market share of genetically modified cotton is around 80%.

In contrast to India, both the area under cultivation and the yield, which is 500 to 600 kg/ha, are stagnating in Pakistan. This is due to political reasons, among others, as the government has been mostly unstable in recent years, making economic conditions difficult. Because cotton is produced by about 2 m farmers, each producing only about 1 t of cotton per year, the introduction of new technologies and modern cultivation and harvesting methods is difficult. The collapse of the system of state-controlled variety approval means that poor quality seed is often used. The falling water table and the increasing salinization of the soil make cultivation even more difficult. It

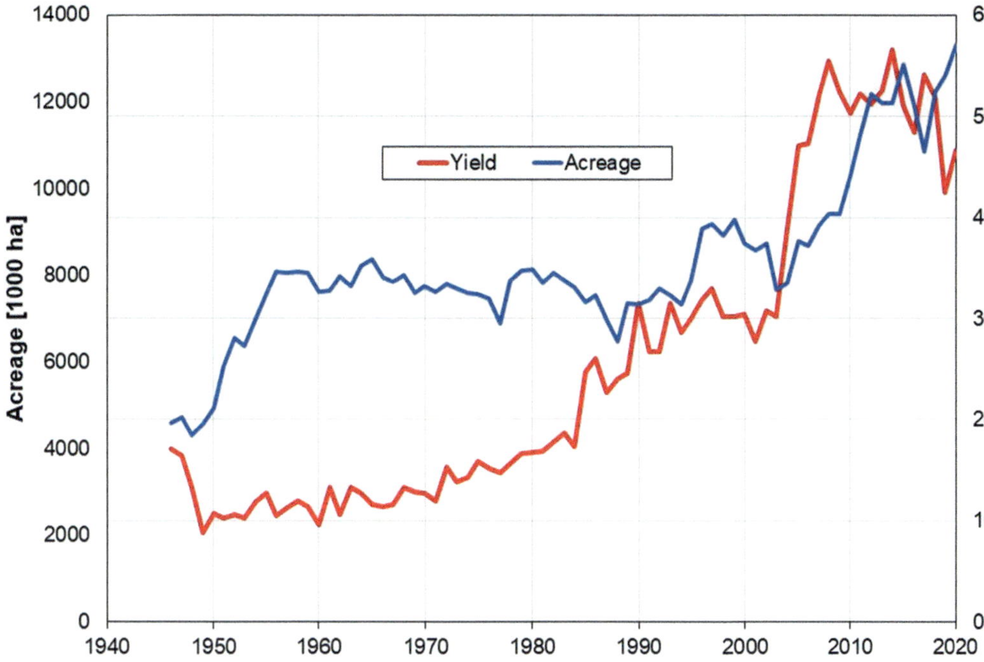

Fig. 5.88 Cultivated area and yield in India (data: ICAC, 2021)

has also not yet been possible to develop a cotton variety that is resistant to the widespread cotton leaf curl virus.

5.9.5 Central Asia

In Central Asia, cotton cultivation is now concentrated mainly in Uzbekistan (especially in the south and east and south of the Aral Sea). While in 1987 the area under cultivation was 2.1 m ha, today it is only 0.9 m ha. Until 2010, about 70–90% of the cotton was exported, meanwhile 90% is processed in the country in about 7000 enterprises. The great importance of cotton for Uzbekistan's economy can be seen in the national coat of arms, which is mainly formed by cotton and wheat ears (Fig. 5.91).

Production is around 640,000 tons (2.8 m bales), with ending stocks of around 140,000 tons (Fig. 5.92).

All cotton production is artificially irrigated, which has led to considerable environmental damage in some cases (extensive drying up of Aral Sea).

Due to major environmental problems, both the total area under cultivation and the yield have declined in recent years. At present, however, 3 m households still live from cotton cultivation (Fig. 5.93).

Other producers of cotton in Central Asia are Turkmenistan (300,000 tons), Tajikistan (100,000 tons), Kazakhstan (80,000 tons) and Kyrgyzstan (12,000 tons) as shown in Fig. 5.94.

5.9.6 Australia

Cotton seeds reached Australia as early as 1788 with James Cook and were successfully cultivated there. In 1810, the first three bales of cotton were exported to England. The cotton industry grew slowly but steadily and by 1934, 17,000 bales were exported. Due to devastating droughts and the collapse of the world market, by 1954 Australia's cotton industry was almost non-existent. The construction of several dams led to a renaissance of cotton cultivation at the end of the twentieth century. Since 2005,

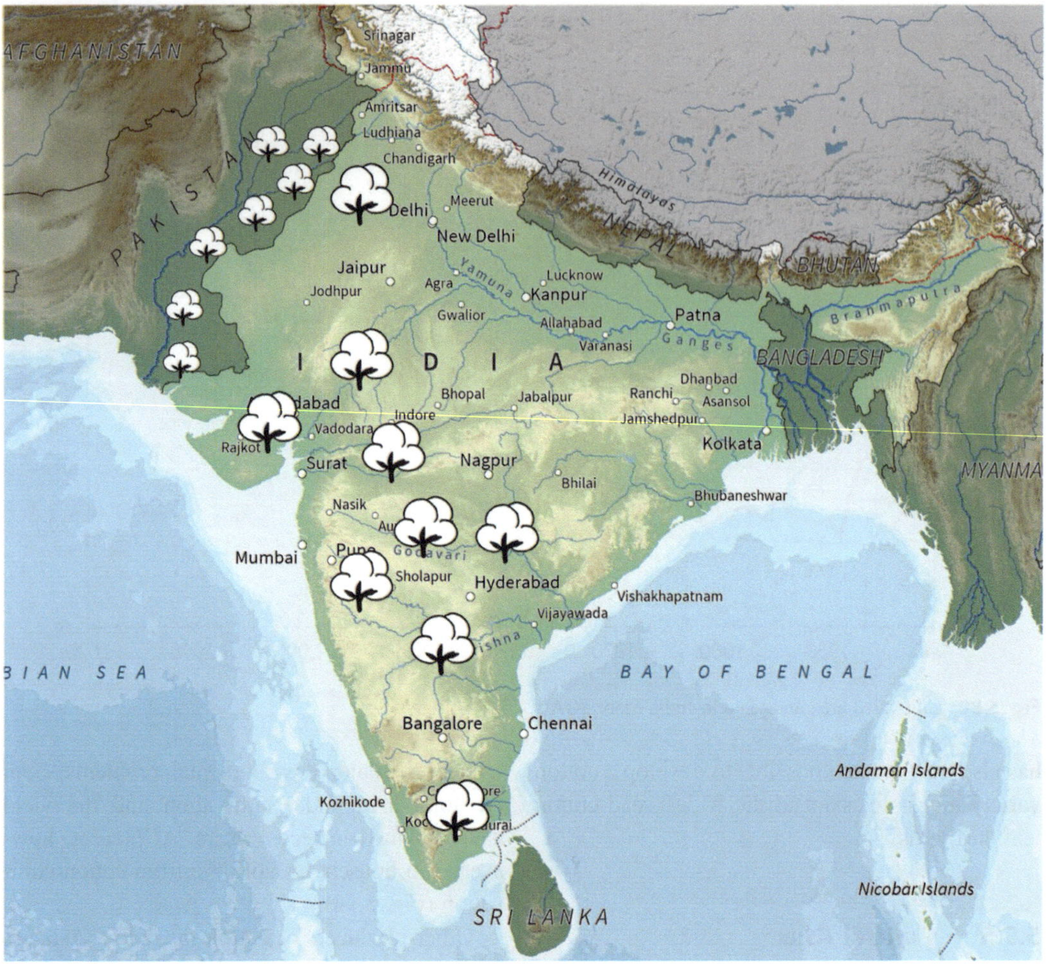

Fig. 5.89 Cotton growing areas in India and Pakistan (Map: Mapswire)

Australia has regularly been among the top 10 producing countries.

In Australia, cotton production is concentrated in the Darling River estuary area in New South Wales (70% of the crop) and some regions of Queensland (Fig. 5.95). All cotton fields are irrigated artificially. This led to major problems in recent years because the natural water resources, which were limited anyway, had almost dried up due to drought.

In 2020, there were approximately 1500 cotton farms, 90% of which were family-owned and planted an average of 300 ha of land with cotton (Cotton Australia, 2021). It is unusual in comparison to other countries that all cotton farmers also grow fruit or keep livestock (e.g. sheep and cattle). In no country in the world does cotton production fluctuate as much as in Australia. Catastrophic droughts in 2006 and 2007 halved cotton production and many farmers gave up growing cotton. Then, from 2009 to 2010, due to favorable climatic conditions, the harvest doubled and almost reached the all-time high of the late 1990s. In the meantime, production fell again significantly, mainly due to a prolonged drought, and after a recovery is now only at 135,000 t (0.5 m bales) in 2020. Ending stocks also fluctuate strongly accordingly, but in recent years have been below the world average at 20–40% of a harvest (Figs. 5.96 and 5.97).

Owing to Australia's extreme climate fluctuations, predicting future development is

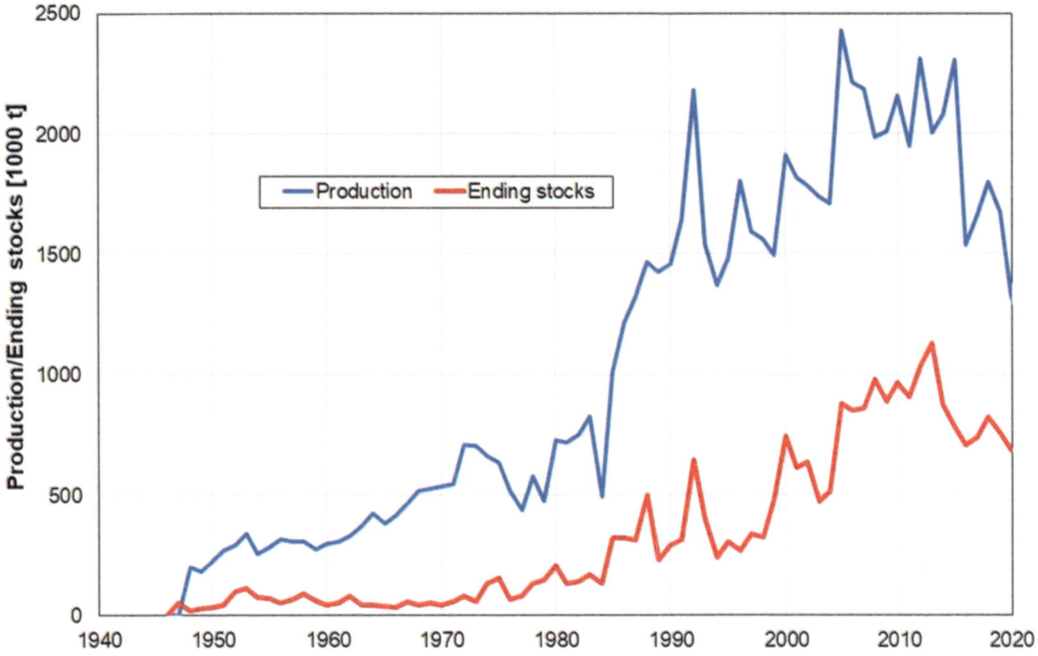

Fig. 5.90 Pakistan's cotton production (data: ICAC, 2021)

extremely difficult and cotton farmers are constantly struggling to survive. Because they receive no subsidies from the government, their economic situation is often difficult. However, they also produce by far the highest yields of any cotton producer in the world, at over 2000 kg/ha. This is achieved mainly through sophisticated water management, a very scarce

Fig. 5.91 National coat of arms of Uzbekistan (www03) and drying up of the Aral Sea (left: 1989; right: 2008; www04)

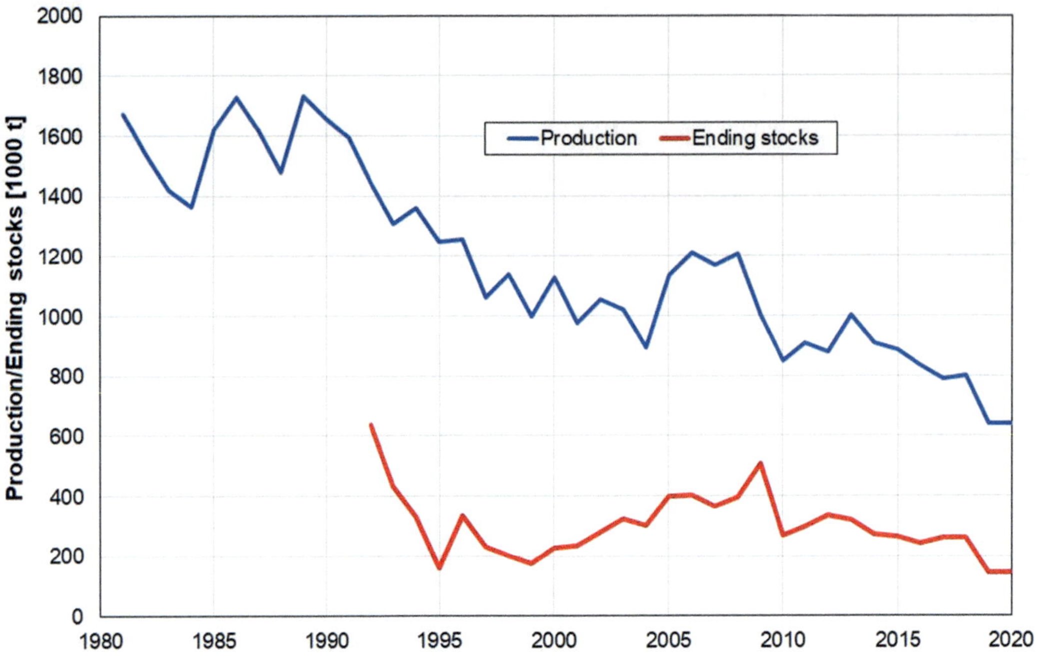

Fig. 5.92 Uzbekistan's cotton production (data: ICAC, 2021)

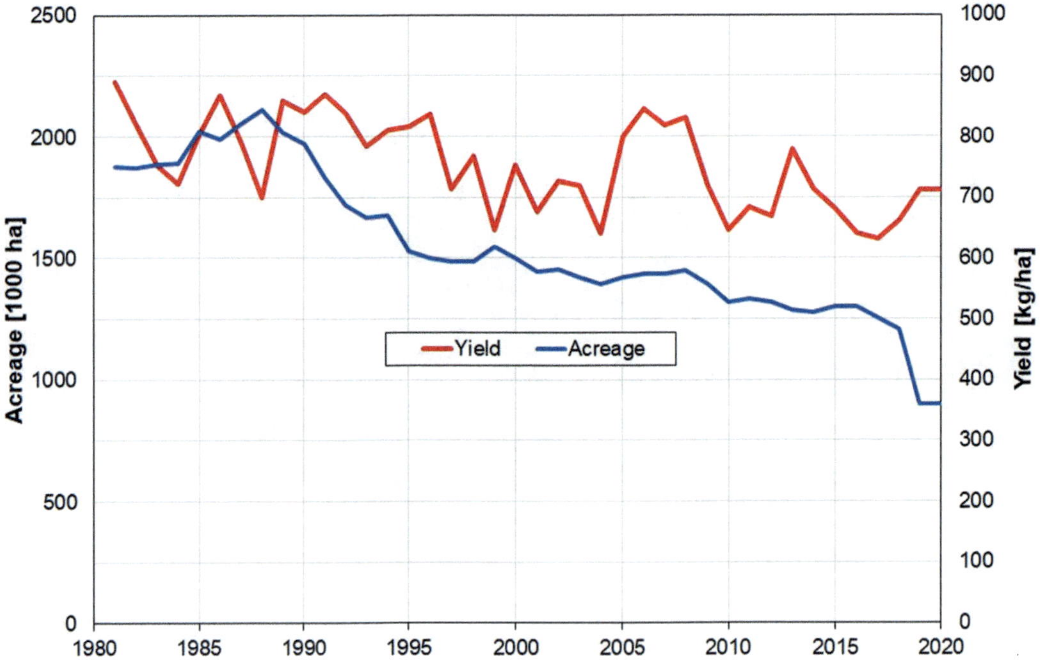

Fig. 5.93 Cultivated area and yield in Uzbekistan (data: ICAC, 2021)

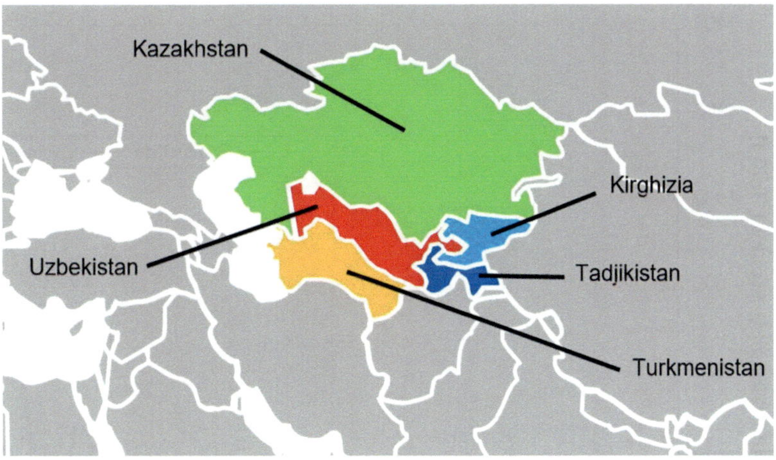

Fig. 5.94 Cotton producing countries in Central Asia

Fig. 5.95 Cotton growing regions in Australia

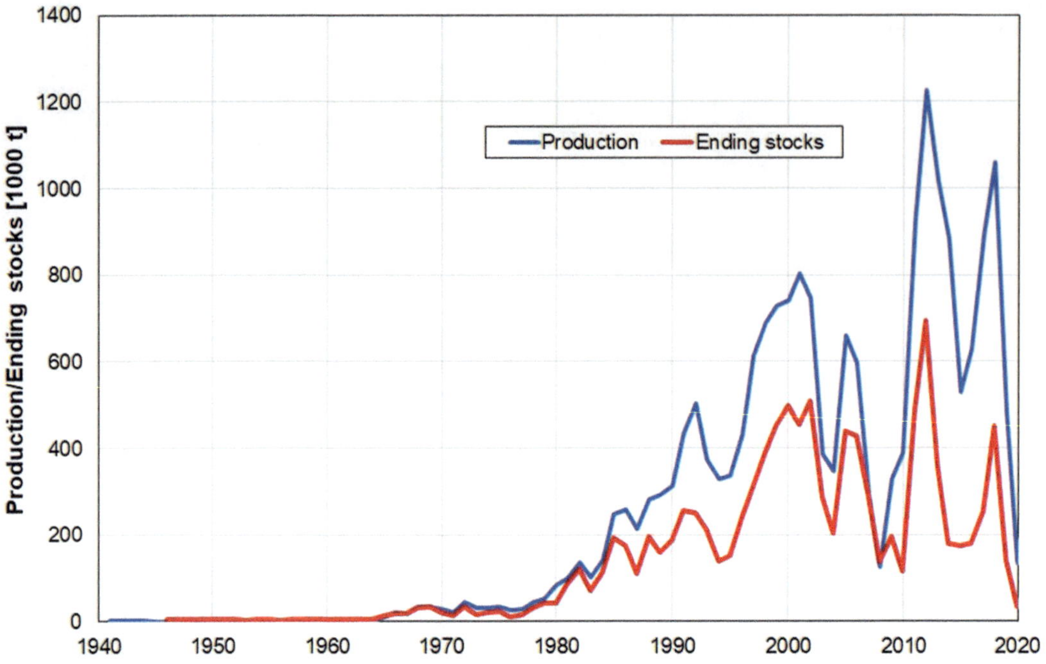

Fig. 5.96 Australian cotton production (data: ICAC, 2021)

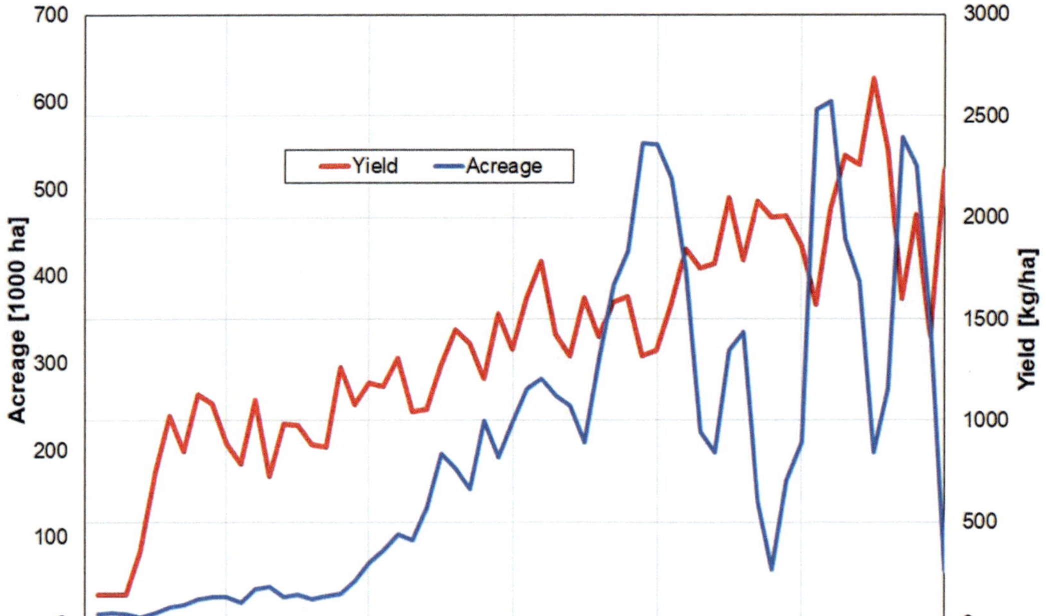

Fig. 5.97 Cultivated area and yield in Australia (data: ICAC, 2021)

Fig. 5.98 Cotton growing regions in Brazil

resource in Australia. As a result, only 3000 litres of water are needed per kg of cotton (world average: 8000 litres).

The cotton produced in Australia is almost exclusively exported, only about 1% is still processed locally.

5.9.7 Brazil

Brazil is the country of origin of the G. barbadense variety. Other varieties are also grown, such as G. hirsutum and G. mustelinum. This genetic diversity is not only used in Brazil, but also in other countries of the world for breeding new, improved varieties.

Brazil is one of the largest cotton producers in the world. The main growing areas are in the northeast, the midwest and the southeast. The northeast is semi-arid and therefore produces only 10% of the country's total yield. The states of Paraná (57%) and Sao Paulo (21%) are the most important growing areas (Figs. 5.98 and 5.99).

The production volume has roughly doubled since the turn of the millennium (Fig. 5.100). However, a complete year's harvest has also been in stock for many years, which keeps the price low. Considering the production volume, the number of Brazilian farmers is very low at 1200. This means that around 2500 tons of cotton are produced per farm, which makes the use of capital-intensive technologies economical and also quickly catches on. However, Brazil also uses a large amount of artificial fertilizers and insecticides, herbicides and fungicides.

Although there has been considerable investment in the expansion of the textile industry in Brazil in recent years, around 2/3 of the harvest is exported.

Fig. 5.99 Cotton field in Brazil (Bremen Cotton Exchange, 2021)

The area under cultivation has decreased considerably in recent years, but the cultivation of new varieties and the use of genetic engineering (market share 90%) have led to a considerable increase in yield. It reaches the highest values worldwide for rain-grown cotton with more than 1700 kg/ha (Fig. 5.101).

5.9.8 Egypt

The cultivation of cotton began under the reign of Pasha Mohamed Ali (1769–1849) by chance: a Frenchman living in Egypt discovered a cotton plant growing wild there and then grew several plants, with which he produced three bales of cotton. He sold these at a high price in France because the staple length of the fibers was exceptionally high. This in turn prompted Mohamed Ali to make the cultivation of cotton a state responsibility. In the early years, therefore, to protect the emerging "cotton industry", the state held a monopoly on cultivation and marketing. After a few years, this was lifted and since then cotton cultivation is mostly in private hands.

In Egypt today, cotton is mainly grown west of the Nile and in the Nile delta (Fig. 5.102). The climate becomes harsher as the distance from the Mediterranean increases. The extra-long staple varieties Giza 45 and Giza 70 are grown in the lower delta (Damiette), the better qualities of long staple varieties in the upper delta and in central Egypt and the most heat-tolerant varieties (poorer qualities) mainly in Dendera and in Upper Egypt.

While Egypt still had a world market share of 7% in 1970, this has since fallen to just 0.2%. However, this includes the highest quality mako cotton (extra long staple, ELS), which is why cultivation is still economical. Cotton is Egypt's most important agricultural export. It is currently grown on 100,000 ha of land, with a yield of about 660 kg/ha (Fig. 5.103).

In 2020, 67,000 t of cotton was produced (295,000 bales) and ending stocks were 1.5 times a year's crop. Egypt imports about 80,000 tons of cotton, processes about 100,000 tons and exports about 50,000 tons.

Due to Egypt's rapidly growing population, many cotton fields have been abandoned in favor of growing food crops that yield a higher

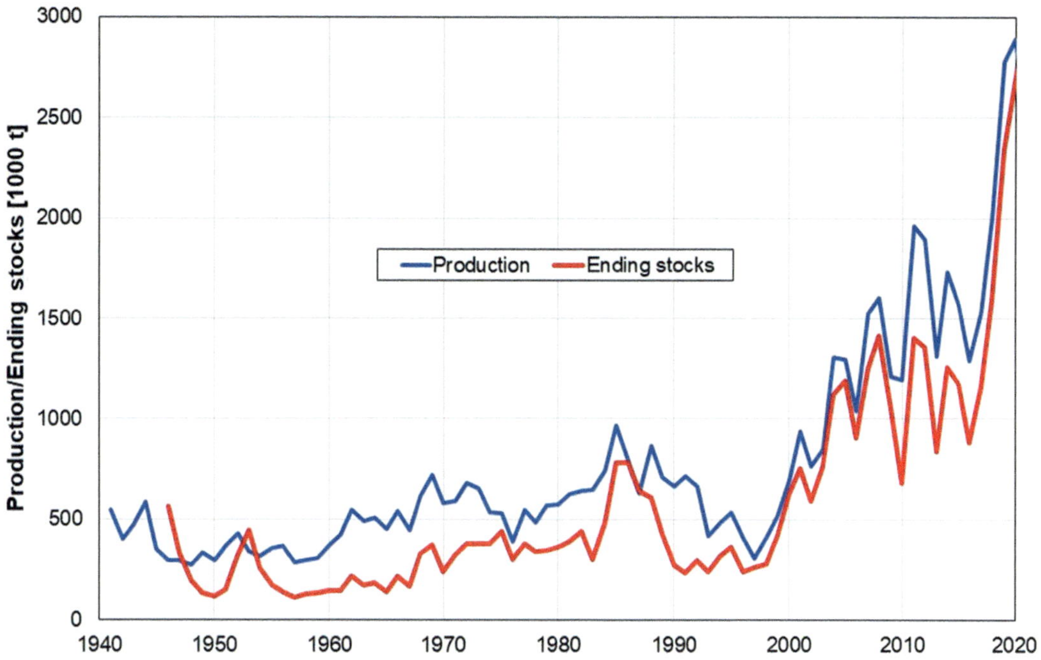

Fig. 5.100 Brazil's cotton production (data: ICAC, 2021)

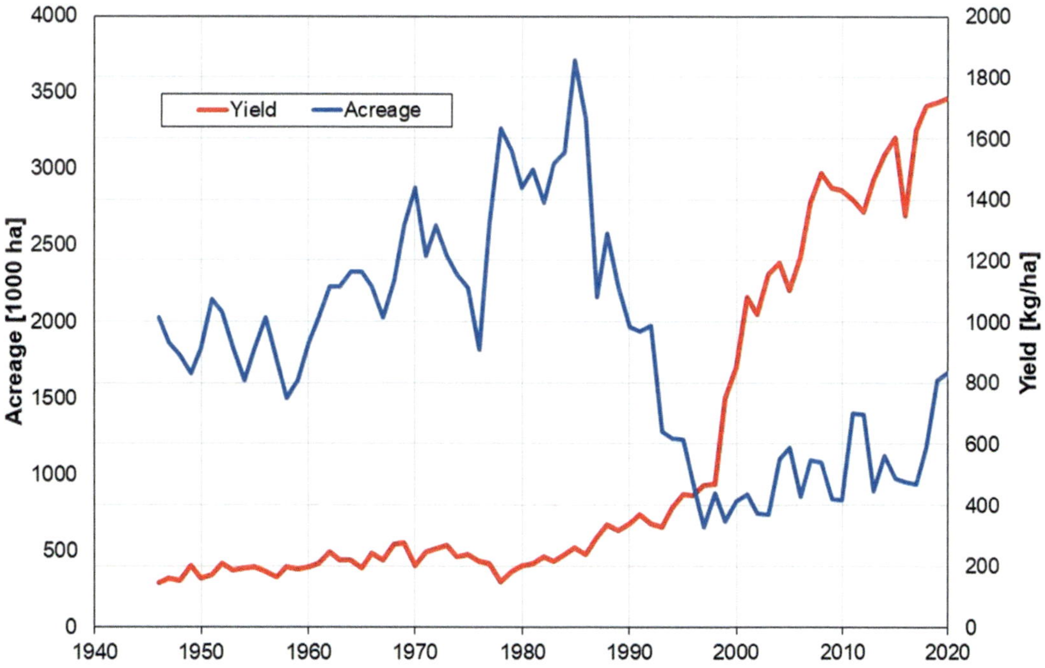

Fig. 5.101 Cultivated area and yield in Brazil (data: ICAC, 2021)

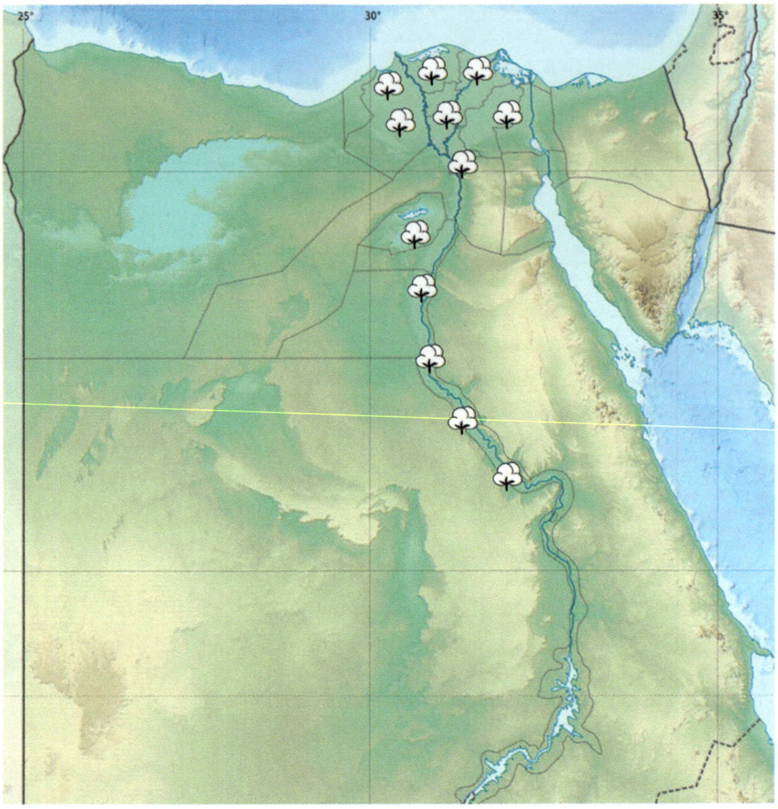

Fig. 5.102 Cotton growing areas in Egypt (map: Mapswire)

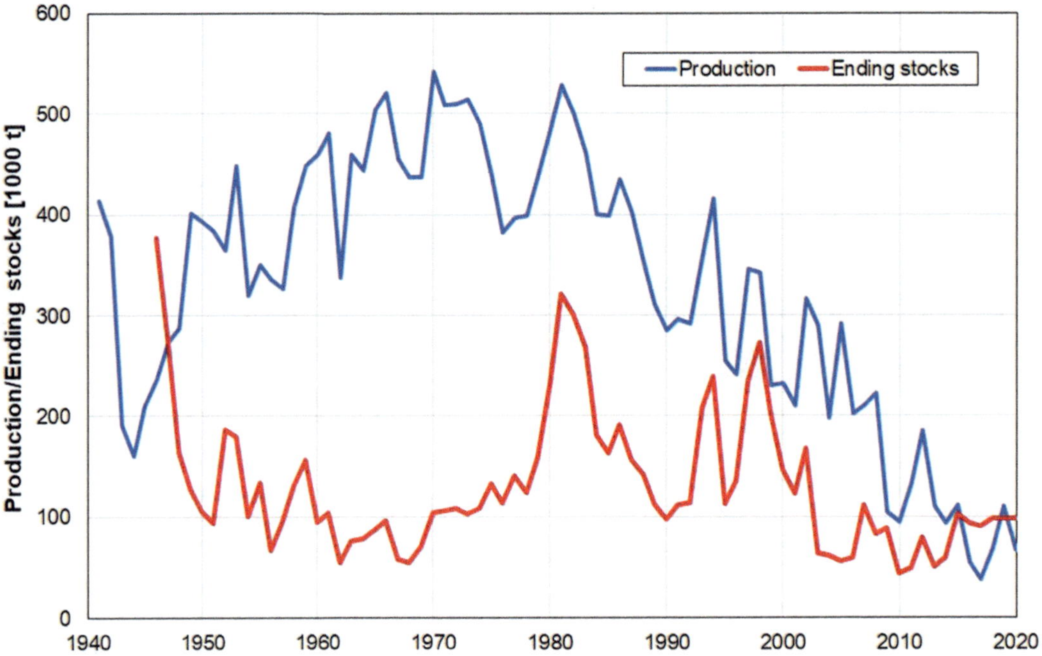

Fig. 5.103 Egypt's cotton production in (1000 t) (data: ICAC, 2021)

return. Therefore, the production volume is not expected to increase in the near future.

5.9.9 Africa

Currently, about 2 m tons (8.8 m bales; 8% of the world's cotton) are produced on an area of 3.3 m ha in Africa, mainly in sub-Saharan and East Africa (Fig. 5.104). More than 1.5 m households earn their living from growing cotton. The countries participating in the Cotton Made in Africa (CmiA) program are marked in dark.

With average yields of 400–600 kg/ha, African cotton farmers are below the world average of around 800 kg/ha. For this reason and also because of the mostly small field sizes, African cotton farmers are not competitive on the world market with countries such as India and China. Added to this is the poorly developed infrastructure, which makes it difficult to transport the cotton to the buyers, and the often uncertain political situation in the growing countries. Figure 5.105 shows the development of the amount of cotton produced in Africa. The interim slump in the mid-2000s is mainly due to reduced production in West Africa, especially in Burkina Faso, Mali and Côte d'Ivoire. Nevertheless, the overall trend is positive, despite the low absolute volumes compared to other world regions.

One possible way out of this difficult situation is the "Cotton made in Africa" initiative of the Aid by Trade Foundation, which was founded by German entrepreneur Michael Otto in 2005. More than 1 m smallholders with over 6 m family members in 10 countries produce around 630,000 tons of sustainable cotton on 1668 m ha under this program (2021).

The cotton is mostly produced rain-grown with no artificial irrigation, with correspondingly low yields per hectare (Fig. 5.106). At the same time, the use of pesticides and fertilizers is kept to a minimum and the fibers are harvested entirely by hand. Furthermore, no genetically modified cotton may be planted. Buyers are primarily consumers in Europe who demand products

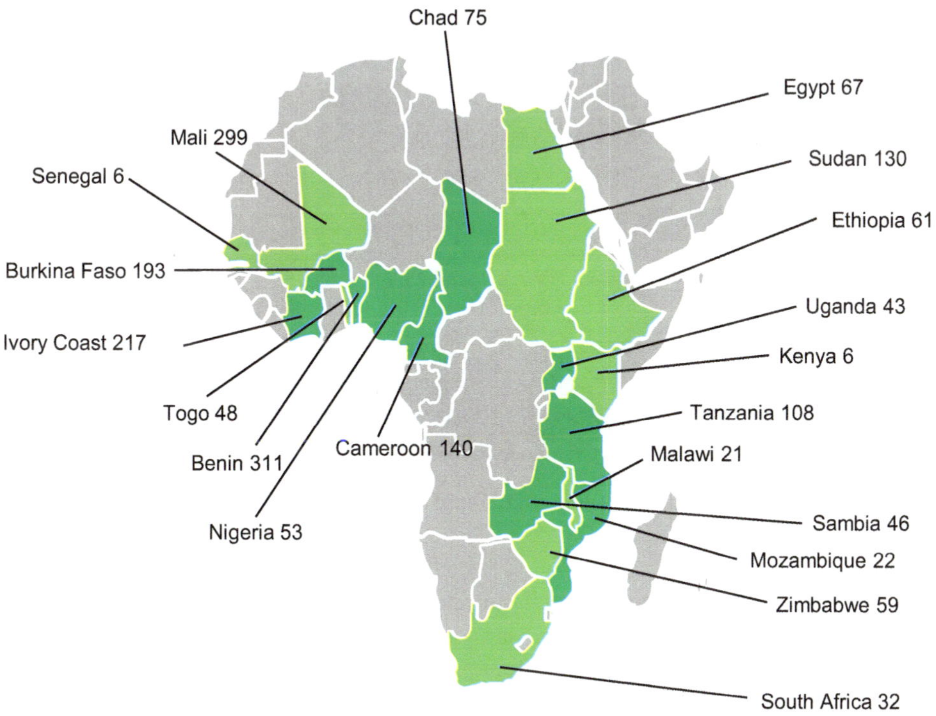

Fig. 5.104 Cotton growing countries in Africa and production (1000 t) (Data: ICAC, 2021)

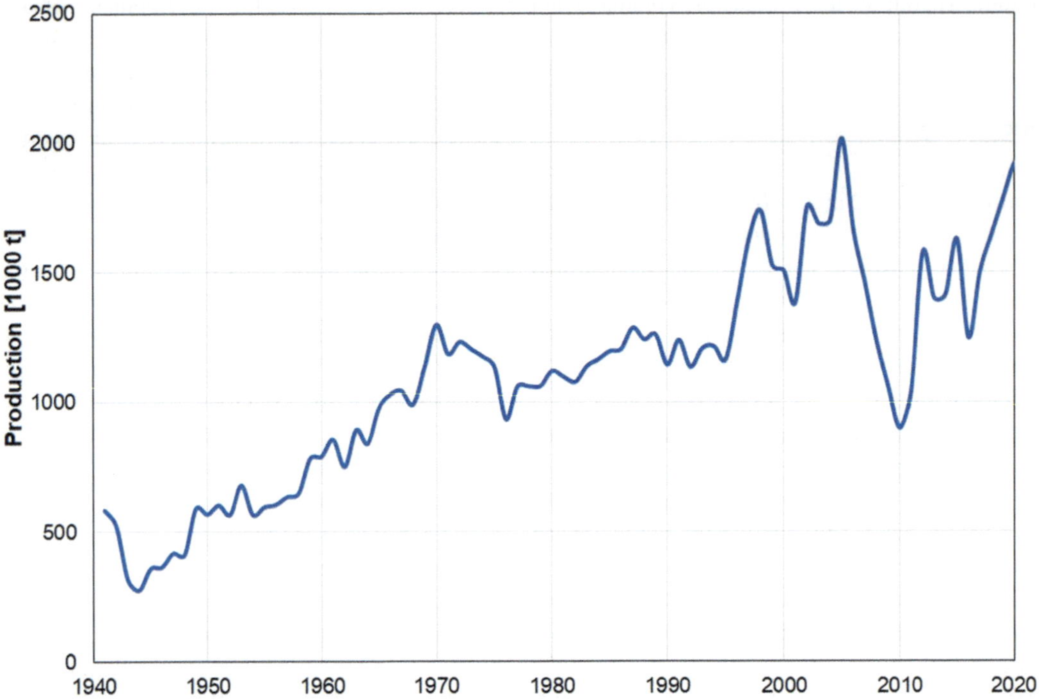

Fig. 5.105 African cotton production (data: ICAC, 2021)

Fig. 5.106 Cotton harvesting in Africa (Kielmann, 2021) and logo of the "Cotton made in Africa" initiative (CmiA, 2021)

made from "organically produced cotton". In 2020, approximately 276 m CmiA textiles were thus brought to market, an increase of 120% from 2019. CmiA also includes 217 spinning and textile mills in 20 countries (8 of which are in Africa), three times as many as in 2016.

It has been shown in recent years that cotton production in those countries that are part of the CmiA initiative has continued to increase, while it has stagnated in countries that do not participate. Moreover, CmiA cotton fetches higher prices on the world market.

Figure 5.107 shows trends in cotton production for West, North, and Southern Africa. It is evident that cotton production in West Africa has increased significantly in the last 10 years, while it has stagnated or completely collapsed in the other regions. In North Africa, Egypt is no longer a major producer, while Sudan grows the most cotton in the region.

In West Africa in particular, cotton production is an important economic factor, as the following figures illustrate. Many, but not all, of these countries are members of the CmiA.

The largest producer of cotton in Africa is Benin (311,000 t), followed by Mali (299,000 t) and Côte d'Ivoire (217,000 t) and Burkina Faso (193,000 t). Of lesser importance are Cameroon, Sudan and Tanzania and a few other countries. In Rwanda and Burundi, formerly important cotton producers, cotton is no longer grown today due to the civil war 20 years ago. In all African countries, most cotton is ginned locally. Table 5.7 shows the share of cotton in all exports for selected countries.

The cultivation of organic cotton has increased sharply in recent years, especially in Tanzania (2% world market share), Uganda, Benin and Burkina Faso, where around 24,000 farmers produce cotton in this way (Fig. 5.108).

Figure 5.109 shows typical impressions from the cotton growing.

5.9.10 Turkey

Cotton is grown in various parts of southern Turkey (Fig. 5.110). The area under cotton

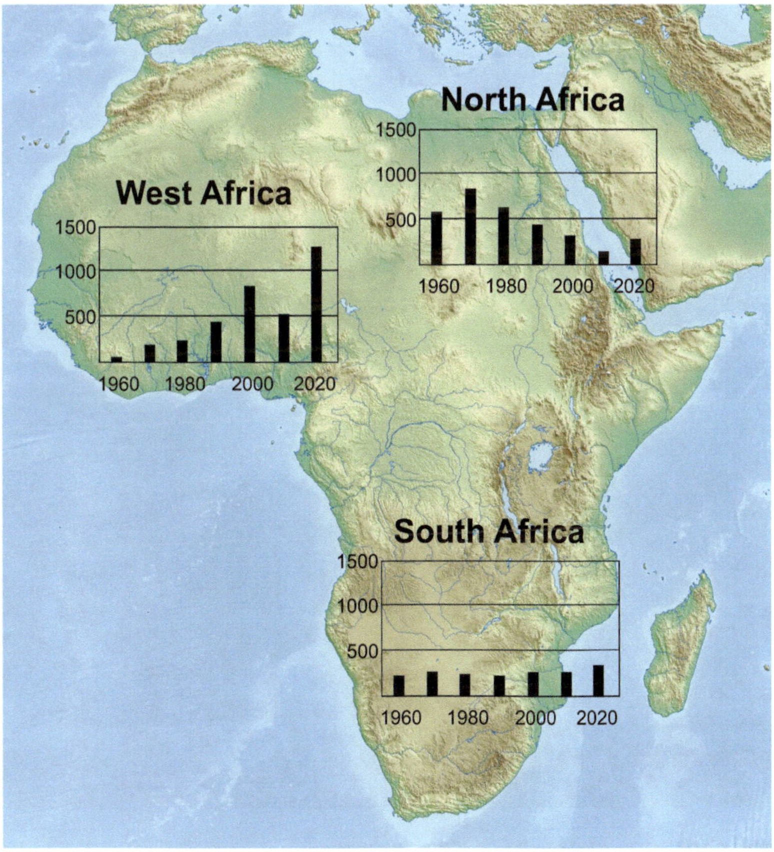

Fig. 5.107 Development of cotton production in important regions in Africa (Data: ICAC, 2021; Map: Mapswire)

Table 5.7 Share of cotton exports in all exports (Trading Economics)

	%
Benin	55
Togo	14
Burkina Faso	10
Mali	7
Cameroon	5
Sudan	4
Ivory Coast	3
Uganda	1.6
Egypt	1.5

cultivation was around 700,000 ha until the year 2000 and has since been reduced to 480,000 ha.

At the same time, however, yields have increased enormously to the current level of 1700 kg/ha, so that cotton production has now returned to the level of the year 2000 with approx. 800,000 t (3.5 m bales). However, the "ending stocks" have also increased accordingly and currently amount to more than a whole year's harvest (Figs. 5.111 and 5.112).

Cotton farmers in Turkey largely shun from using genetically modified cotton. Therefore, pest control is still traditionally done by pesticides and insecticides. This leads to costs of about 550 US $/ha for pesticides, which is 3 times as much as the world average . After China, Turkey imports the most cotton (1/3 of it from the USA) and is one of the largest users. Almost the entire crop is processed in the country. Turkey is one of the largest producers of organic cotton with a market share of 10%, which corresponds to approximately 110,000 tons.

The textile industry, which mainly processes cotton, currently accounts for around 18% of all Turkey's exports. Thus, US$17bn worth of apparel and US$16bn worth of textiles were exported in 2019. With approximately 53,000

Fig. 5.108 Cotton field in Benin (Schmidt, 2004)

Fig. 5.109 Cotton farmers and cotton warehouses in Africa (Kielmann, 2021)

companies employing more than 400,000 people, the textile industry is of enormous importance to Turkey.

Therefore, Turkish cotton farmers are massively supported by the government. Direct subsidies of about 25 ct/kg are paid. At the same time, the government supports dam projects and road construction in the cotton growing areas. Thus, the cultivation area, which is completely irrigated artificially, is to be further expanded. However, these areas are largely located in the south near the border with Syria. The unstable political situation there jeopardizes this goal.

5.9.11 Europe

99% of cotton production within the EU (world market share approx. 1.5%) is concentrated in Greece and Spain (Fig. 5.113). Until 2005, the production of cotton was subsidized with around US$ 700 m per year, which corresponded to approx. 150% of the world market price. Due to the reduction of subsidies and the simultaneous increase in prices for other agricultural products, cotton production in Europe has been halved in recent years. The total area under cultivation is

Fig. 5.110 Main cotton growing regions in Turkey (map: Mapswire)

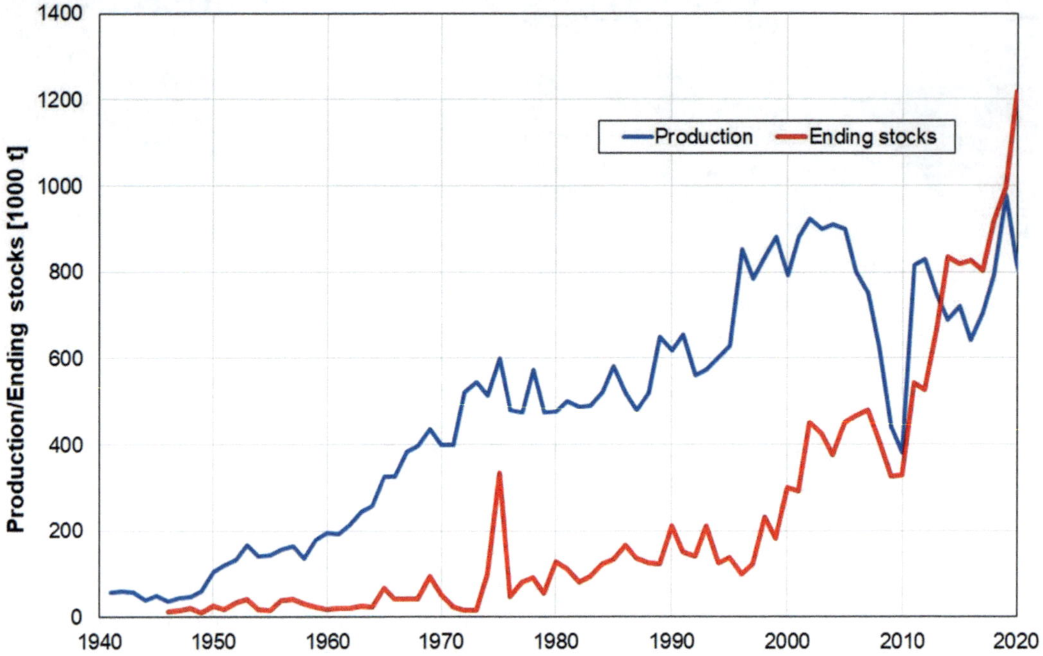

Fig. 5.111 Production and Ending Stocks in Turkey (data: ICAC, 2021)

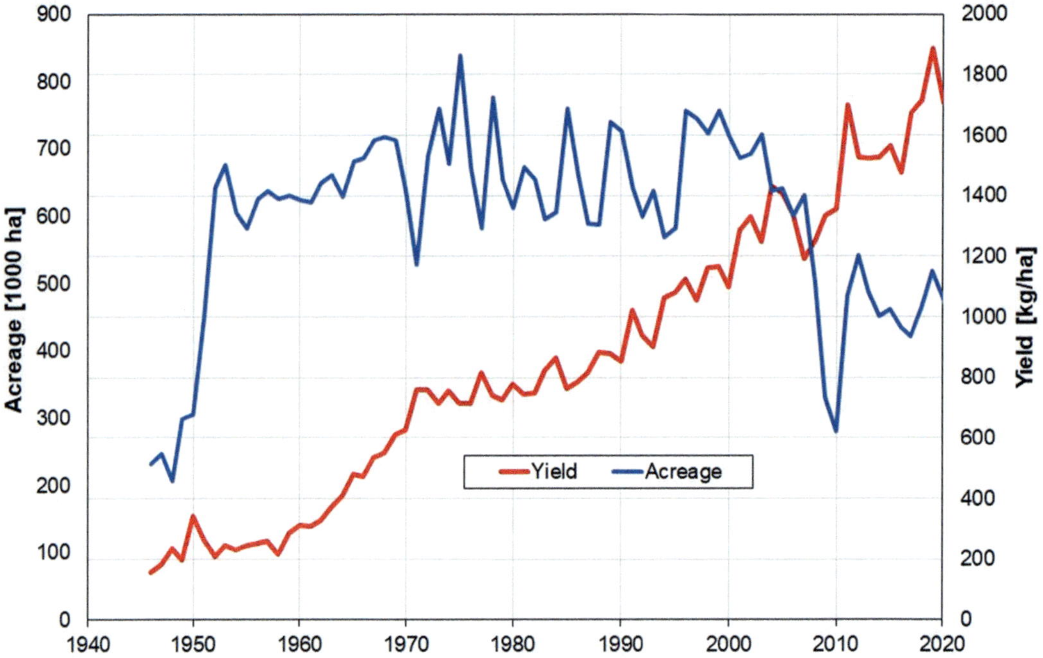

Fig. 5.112 Cultivated area and yield in Turkey (data: ICAC, 2021)

Fig. 5.113 Main cotton growing areas in Greece and Spain (map: Mapswire)

around 340,000 ha (2020) with a slight downward trend, production is around 400,000 t.

5.9.11.1 Greece

The total cultivated area is 277,000 ha with a yield of 1200 kg/ha (2020). Cotton is mainly grown in Thessaly, Central Greece and Thrace.

In 2020, the total production volume was around 325,000 tons (1.4 million bales), while the ending stocks were around 43,000 tons, which is an extremely low proportion compared to other countries (Fig. 5.114).

Whereas at the turn of the millennium, around 30% of domestic cotton was still processed in Greece, today almost the entire harvest is exported. This is essentially due to strong competition from China.

5.9.11.2 Spain

In Spain, cotton production is concentrated in the region around Seville, where 13% of the EU harvest is produced. This area along the Guadalquivir river stretches from Cordoba to the Mediterranean. Cotton production here competes with other irrigation-intensive crops, such as sunflowers, maize, wheat and citrus fruits. A total of 60,000 ha of farmland is planted with cotton with a yield of 1000 kg/ha, which has almost tripled since 2010.

As a result of the financial crisis in 2007, cotton production in Spain fell drastically. Since then, it has recovered somewhat and in 2020 will stand at 70,000 tons or 300,000 bales. Ending stocks amount to around 65% of a harvest (Fig. 5.115).

By 2000, more than half of the harvest was processed in the country. With the opening of the EU internal market and the fall of the Multifiber Arrangement, textile production in Spain quickly became uneconomical and migrated to the countries of Eastern Europe and Asia. Today, almost 100% of the produced cotton is exported.

5.10 Processing in Germany

Cotton processing no longer plays such a major role in Germany as it did until the 1980s, although

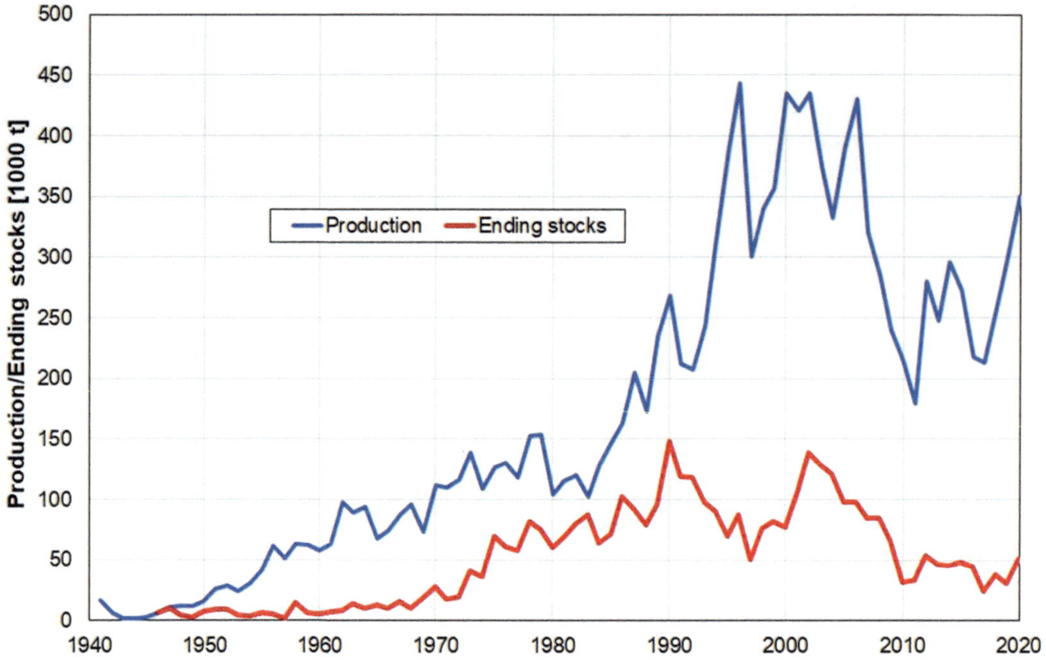

Fig. 5.114 Cotton production in Greece (data: ICAC, 2021)

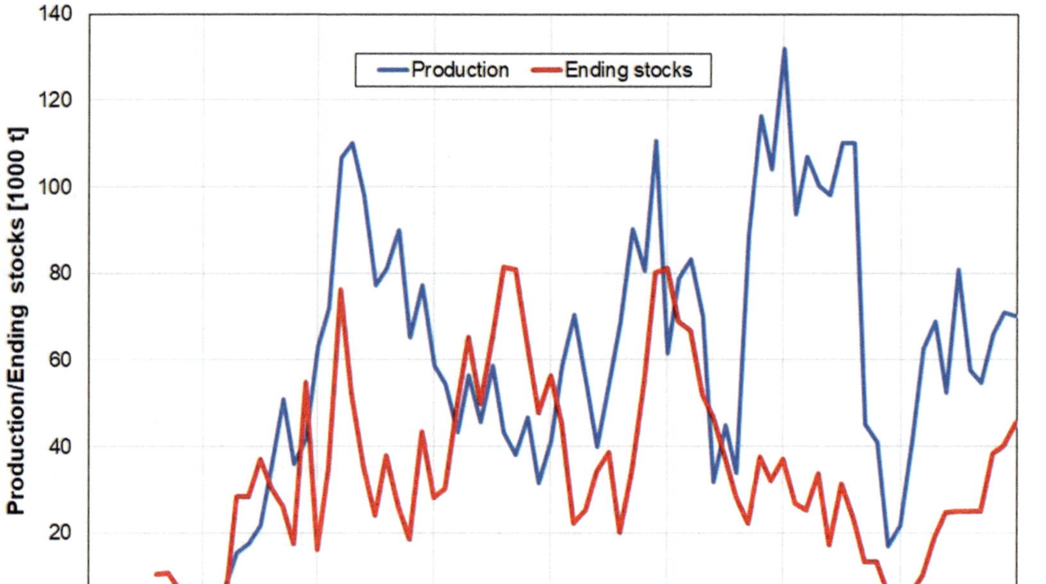

Fig. 5.115 Cotton production in Spain (data: ICAC, 2021)

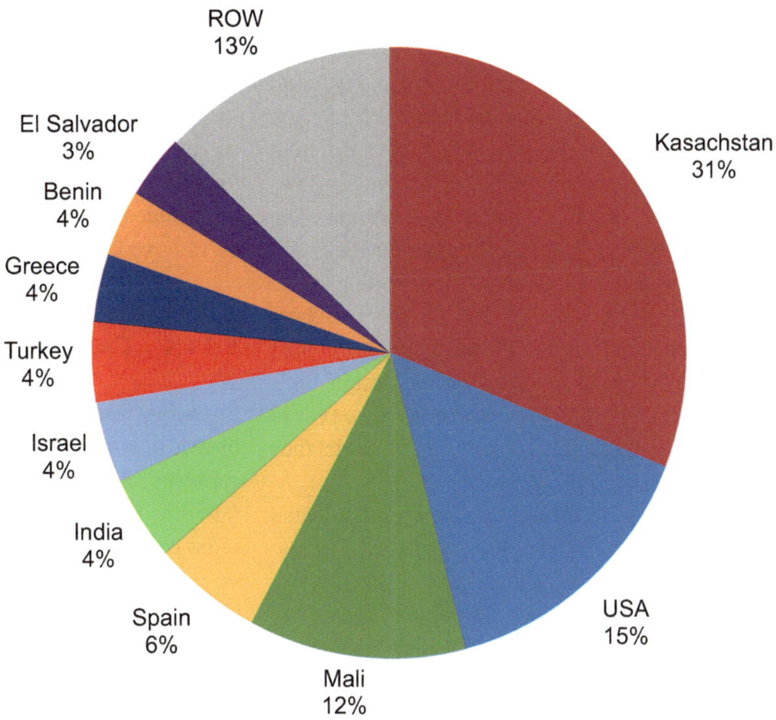

Fig. 5.116 Cotton imports to Germany (data: Bremer Cotton Report, 2021)

around 16,000 tons are still imported annually. Figure 5.116 provides an overview of the most important countries exporting cotton to Germany.

5.11 Pests

Cotton is mostly grown in monocultures and is therefore particularly sensitive to pests. Cotton is attacked by several hundred different types of insects worldwide. The boll offers the pests not only food but also protection against weather and natural enemies as well as against control by humans.

5.11.1 Boll Weevil

It is the most common cotton pest and arrived in the USA from Mexico around 1890 (Fig. 5.117). After mating in spring, the female begins laying eggs in the young flower buds. Three days after egg laying, the larvae hatch. After another 6–12 days, they pupate and 3–5 days later the beetles hatch, which start laying eggs again after about 5 days. A complete development cycle thus takes only 3 weeks.

The damage occurs due to the premature opening and yellowing of the plant's leaves. In addition, the buds fall off and the fiber quality of the bolls still hanging on the bush also deteriorates.

Fig. 5.117 Cotton boll weevil (USDA, 2020b)

5.11.2 Boll Worm

This pest is also called red bollworm and is found in almost all cotton growing regions. There are different species of this pest, which have different importance depending on the region. In the USA, the pink boll worm is the most common. The female moth lays 100 to 500 eggs on the buds, capsules and leaves. After 4–5 days the larvae hatch, which are mostly white at first and later pink. Their length is 12–15 mm. During the next 10–14 days, the larvae feed on the cotton bolls before they pupate and become butterflies (wing-span 15–20 mm). The development cycle thus lasts 4–6 weeks (Fig. 5.118).

The damage is mainly caused by fibers eaten and soiled by the larvae, destroyed seeds, immature and non-opening as well as hollowed bolls. In case of early infestation, the upper parts of the plant may die. Furthermore, pests and bare leaves are widespread problems.

5.11.3 Cotton Louse

This louse is also found in all cotton growing areas. It causes direct damage to the plant by sucking out the plant juices. This causes the leaves to curl up, especially on young shoots (Fig. 5.119). As a result, the plant is stunted and retards its development. Indirectly, considerable damage is also caused by the excretions of the louse ("honeydew"). When these sticky secretions enter the capsules, the fibers stick together, which leads to very serious problems in further processing because honeydew cannot be removed.

5.11.4 Cotton Bug (Stink Bug)

Stink bugs are about 12 mm long and green to brown in color (Fig. 5.120). Stink bugs bore into capsules, causing young capsules in particular to fall off. However, the main damage occurs to older capsules shortly before ripening. The fibers discolor and turn yellowish, and the seeds may also shrivel. Damaged capsules often fail to open and therefore cannot be harvested. Stink bugs may also spread bacteria or fungi, which in turn attack the cotton.

5.11.5 White Fly

This fly also produces honeydew (Fig. 5.121). This can lead to fungal infections that affect the plants. Whiteflies also transmit other plant

Fig. 5.118 Boll worm at work (Matthews, 1962) and infested cotton plants

Fig. 5.119 Cotton aphid on a cotton plant (Smith, 2009)

Fig. 5.120 Stink Bug (Ottens) and damaged cotton boll

diseases. A total of around 1550 different species are known, which damage numerous other crops besides cotton.

5.11.6 Other Diseases

Other common diseases of cotton plants include bacteriosis, fusarium and verticillium wilt, and phymatotrichum root rot.

5.11.7 Pest Control

By regularly controlling pests from the ground or from the air with pesticides (Fig. 5.122), the negative effects can be reduced. However, this requires an estimated 150,000–250,000 tons of pesticides worldwide. In the USA, cotton pesticides account for 11% of all pesticides, and in Egypt as much as 50%.

The use of herbicides is often only successful in cotton if the cotton plants are taller than the "weeds" to be controlled. But, even if this is the

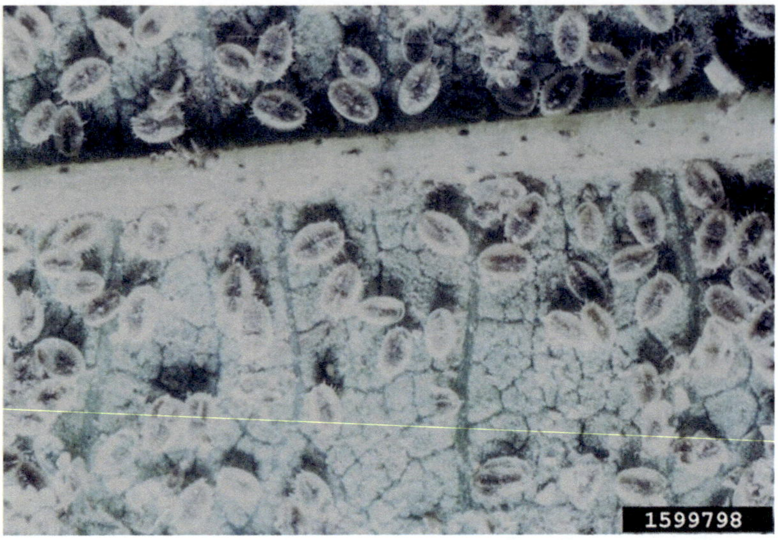

Fig. 5.121 White fly (French)

Fig. 5.122 Use of an aircraft for spraying pesticides

case, the use of herbicides can lead to later maturity of the cotton plants or to quality losses in terms of fiber properties.

Alternatively, therefore, pheromone and light traps are used or predators, for which the pests serve as food. Some studies show that the use of pheromone traps significantly reduces the cost of insect control. However, a pheromone trap appears to be effective only within a relatively small radius (a few rows of cotton plants)

(Tillman & Cottrell, 2019). Therefore, depending on the amount of pheromone delivered, multiple traps per hectare may be required.

Typical beneficial insects in cotton fields that feed on the pests are big-eyed bugs, ladybirds, damsel bugs and various spider species (Fig. 5.123). They also feed partly on the juices of the cotton plant, but the resulting damage is minor.

Other pest control options include:

- a regular crop rotation and
- the manual collection of eggs or larvae or infested branches.

Both methods are used in particular in the cultivation of "organic cotton".

5.12 Genetic Engineering ("Engineered Cotton")

Another way to reduce pest infestation is the targeted use of genetic engineering to change the genetic structure of the cotton plant. This has made it possible to make cotton plants resistant to certain insects, for example. But it, however, also involves risks.

5.12.1 General Objectives of Genetic Engineering

The general objectives of plant breeding are not fundamentally changed by the inclusion of genetic engineering methods. In most cases, the aim is to produce lines with greater resistance to biotic stress (fungi, viruses, bacteria, nematodes, insects) or abiotic stress (drought, cold, heat, etc.) or with properties that make breeding and cultivation of the crop more efficient (e.g. early flowering, herbicide resistance). These breeding objectives relate to the agronomic traits of plants and have long been pursued in conventional plant breeding. They are also referred to as "input traits". Agronomic traits are primarily of importance to breeders and farmers. The transgenic properties of most genetically modified varieties currently being commercialized relate to such input traits.

The second group of breeding objectives concerns the quality of plant products intended as food or feed, so-called "quality traits" or "output traits". Changes or improvements in the quality of seeds, fruits or other plant parts can be made by modifying ingredients. Since the development of genetic engineering methods has removed the biological restrictions on the exchange of genetic

Big eyed bug (Dykinga, 2005) Damsel bug (Schulz, 2015)

Fig. 5.123 Typical beneficial insects in cotton fields

information, the possibilities in the field of quality breeding have expanded considerably. However, many details of plant metabolism are still unknown and therefore the use of genetic engineering is associated with a certain risk.

5.12.2 Early Research

Already the first farmers practiced a form of genetic selection of plants and animals by selectively cultivating or breeding existing varieties with special characteristics. This resulted, for example, in cereal varieties with higher yields or stronger and larger livestock, such as new breeds of sheep. It was not until the beginning of the twentieth century that "hybrid varieties" were created through targeted genetic crossing on the basis of the results of G. Mendel's research. From 1950 onwards, mutations of existing varieties were obtained by irradiation with γ-rays, and since 1978, the genetic material of plants has been specifically modified by the transfer of individual genes.

5.12.3 Genetic Modification of Cotton

Cotton is genetically modified to make it resistant to herbicides and insects in particular. Furthermore, it has also been possible to dye cotton already in the capsule (blue, reddish) (cf. Sect. 5.2.3).

5.12.3.1 Herbicide Resistance

Most of the herbicides currently used are selective, i.e. they kill only certain plant species. Selective herbicides can therefore be used to kill certain weeds without damaging the crop. Non-selective herbicides (also known as broad-spectrum or total herbicides) kill all species by blocking a key metabolic step in higher plants. Non-selective herbicides can only be used where all vegetation is to be destroyed, e.g. before sowing and before harvesting (defoliation of cotton plants).

The production of herbicide resistant varieties is interesting for the following reasons:

- Weed control with selective herbicides is labor intensive and problematic in some rotations.
- The development of new herbicides is more costly than the breeding of new varieties, i.e. it is easier and cheaper to produce one new herbicide-resistant variety than several selective herbicides for various weeds.

The breeding objective of "herbicide resistance" already existed before the introduction of genetic engineering. As early as 1980, for example, the first rapeseed variety resistant to the pesticide atrazine was approved in Canada. In 1988, the first transgenic cotton with resistance to a herbicide (glyphosate) was bred and successfully used.

In addition to easier and more flexible weed control, there are other advantages to herbicide resistance:

- Species with lower economic importance, where the development of new herbicides is too expensive, can be equipped with this trait.
- The choice of the appropriate complementary herbicide can be limited to those compounds that cause less environmental impact.

At present, BXN cotton is of particular importance. A gene from the soil bacterium Klebsielle oazenae, which produces the enzyme nitrilase, has been inserted into it. It converts nitriles, the main components of the widely used herbicide Buctril, into harmless substances. This allows the use of this herbicide against numerous "weeds" without danger to the cotton plants.

5.12.3.2 Insect Resistance

Cotton plants are attacked by several hundred insect species. These either impair growth, e.g. by feeding on leaves, or damage the cotton boll directly. Large quantities of insecticides are consumed every year to protect the plants from the most important pests. For this reason, it seems very worthwhile to protect the cotton plant directly from insect feeding by genetic modification. The Bt toxin is a protein naturally produced by Bacillus thuringiensis soil bacteria. This protein destroys the intestinal wall of some feeding insects (Fig. 5.124).

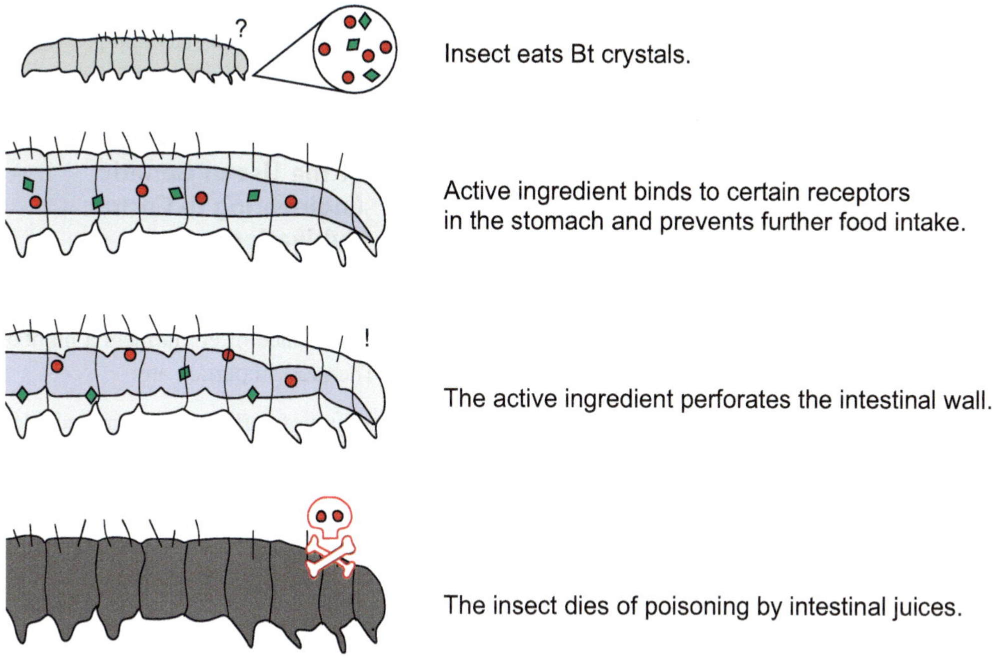

Fig. 5.124 Effect of Bt toxin

There are different Bt toxin variants that act specifically against certain groups of insects, such as beetles or butterflies. Normally, beneficial insects are little affected.

Genetic engineering has opened up the possibility of transferring the bacterial gene to plants. These now produce the Bt toxin and thus provide their own crop protection agent (Fig. 5.124).

Ideally, the use of Bt plants can eliminate the need for additional insecticide sprays. This is reported, for example, from the USA, where the use of insecticides against the boll weevil has in some cases has become completely obsolete on plantations with cotton containing Bt toxin. After the Bt gene could be "produced" economically in 1989, the first Bt cotton was grown in the USA in 1990. Currently, the main Bt toxin used for cotton is Bollgard®, developed by Monsanto in 1996. In 2003, the second generation of this Bt toxin was developed (Bollgard II®), which also makes the cotton resistant to herbicides. Since 2018, Bollgard 3 has been on the market, which is intended to provide even better protection, for example against the bollworm.

5.12.3.3 Development of Resistances

Especially in the case of large-scale cultivation of Bt plants, it was expected that insects that are resistant to Bt toxin could spread strongly. This would render both Bt plants and the classic Bt preparation ineffective. To slow down this process, resistance management strategies have been developed.

Depending on the size of the field, different methods are used and 5–20% of a cotton plantation is not planted with genetically modified cotton. This creates a habitat for Bt-susceptible beneficial insects, which should enable them to combat Bt-resistant insects. This strategy has so far proved very successful worldwide.

Nevertheless, transgenic cotton sometimes led to unexpected problems under commercial growing conditions. In 1997, for example, capsule deformations and early shedding occurred in some growing areas with a herbicide-resistant cotton. According to USDA findings, the insect-resistant variety has led to infestations by secondary pests in certain areas, which had previously hardly appeared and in turn led to increased pesticide use.

Possible resistance developments of the affected pests by insect-resistant plants play a major role in the context of risk assessment. Under field conditions, no extensive resistance developments have yet been detected. However, laboratory experiments give cause for concern. In particular, recent studies on the inheritance pattern of Bt resistance in Bt sensitive pests suggest that besides a recessive inheritance pathway also dominant inheritance occurs. Therefore, resistance management plans need to be developed. Figure 5.125 shows two different concepts (Tabashnik et al., 2020). Shown on the left is a refuge strategy to delay the evolution of BT-resistance. A refuge of non-Bt cotton planted next to a Bt cotton field leads to susceptible moths (blue) which mate with the Bt-resistant moths (red) producing Bt-susceptible offspring if the resistance to Bt is recessive. In the concept on the right, Bt-susceptible and sterile moths are released from an aircraft and mate with Bt-resistant moths producing Bt-susceptible offspring that dies when feeding on Bt cotton bolls as larvae.

The problems caused by resistant weeds through the cultivation of transgenic herbicide-resistant plants could far exceed the previous herbicide resistance problems because of the possibility of gene transfer and owing to the expected increase in the use of a few complementary herbicides (increased selection pressure).

5.12.4 Advantages and Disadvantages of the Use of Genetic Engineering in Cotton Cultivation

Currently grown transgenic cotton with herbicide and insect resistance offers farmers a number of benefits. The most important are:

- Reduced use of herbicides and insecticides.
- Lower groundwater contamination,
- Protection of beneficial insects and birds, and
- Higher yields and thus higher profits (approx. US$ 40 per ha).

However, the use of transgenic cotton also involves some risks:

- By reducing insecticide use, more pests survive that are not affected by the Bt toxin.
- Under certain circumstances, Bt-resistant pests may develop, against which there would then be no antidote (risk of large-scale crop failures).
- Higher costs due to the use of Bt toxin (Monsanto quasi monopolist).

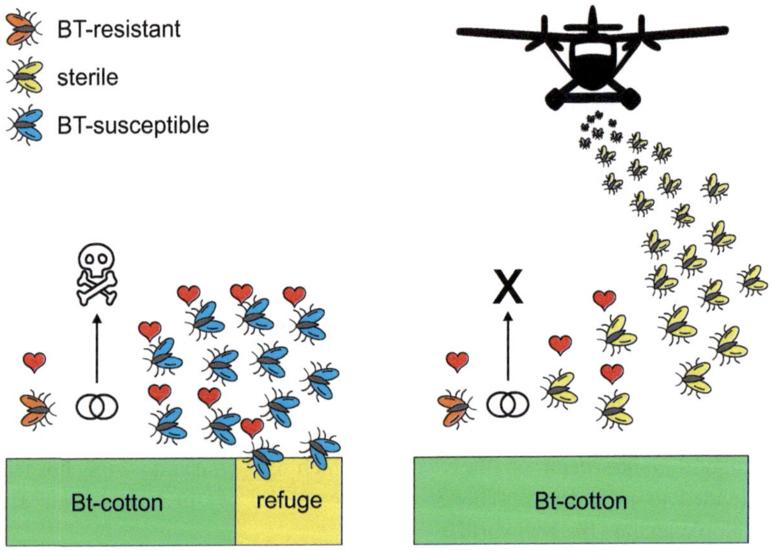

Fig. 5.125 Pest resistance management scheme acc. to (Tabashnik et al., 2020)

- The Bt toxin is only 90–95% effective, i.e. in years with a heavy insect plague, losses due to the boll weevil can still be very high.

Although transgenic varieties have been shown to reduce the use of certain chemical pesticides, they have often not resulted in noticeable yield effects. Yield increases of insect-resistant cotton in the US and China average less than 10%.

At present, the advantages seem to outweigh the possible disadvantages, so a further increase in the proportion of transgenic cotton is to be expected.

5.12.5 Market

According to estimates by the ISAAA (International Service for the Acquisition of Agribiotech Applications), transgenic cotton varieties were cultivated on 24.1 m ha worldwide in 2017. In relation to the total area, genetically modified cotton is thus planted on around 80% of the global cultivated area. Of these, 75% are insect-resistant varieties, 3% are herbicide-resistant cotton and 22% are insect- and herbicide-resistant varieties.

5.12.5.1 China

Improved harvests are reported from China through the cultivation of insect-resistant cotton plants originating from the USA. With average yields of 1050 kg per hectare, the transgenic plants outperformed conventional varieties by up to 30%. However, this figure should be treated with caution because the corresponding differences in US cotton yields were only 5–10%. It is possible that traditional varieties grown in China differ from conventional American cotton plants in terms of yield. Despite the higher seed price, farmers achieved an average increase in revenue of 4500 yuan per hectare, which, in addition to the increase in yield, was also due to the use of insecticides being reduced by half.

Since the introduction of genetically modified cotton, pesticide consumption has been reduced by up to 80% despite increasing productivity. This development has also benefited smallholders: apart from the reduced labor input, the cost of spraying has also fallen by around US$ 750/ha. The costs of producing cotton fell by around a quarter overall. Accordingly, the cultivation of GM cotton has rapidly increased from 2000 ha (1997) to about 3 m ha (2017), which is almost the entire cultivated area of China.

In addition to the positive aspects, there are also negative effects. For example, the boll weevil is being combated on a large scale with genetically modified Bt cotton. In two Chinese provinces, however, pests have apparently already developed that are resistant to the genetically modified cotton. The plants equipped with an insecticide alter the natural balance of the insect world. The likelihood of new infestations increases and the pests become more and more resistant, so that the protective effect is lost after eight to ten years. Thus, while the population of the bollworm is significantly reduced, the number of other pests such as aphids, spiders, fringe flies, grasshoppers and the like increases.

China has meanwhile developed its own Bt cotton plant and has also patented it. This competes with Monsanto's or Delta & Pineland's Bt cotton, but is not yet exported to other countries on a large scale.

5.12.5.2 India

The bollworm is the most feared cotton pest in India. Early studies have shown that bollworm related crop losses can amount to 50–60%. As a result, it causes an economic loss of about US$ 300 m every year. Pesticides worth US$ 350 m are therefore used annually to control it. Hence, there is a large market for the use of transgenic cotton in India.

According to official data, the share of transgenic cotton grown in India was about 90% in 2017. In 2002, the figure was only 0.5%. Illegally acquired varieties containing Bt toxin are grown mainly by small farmers in the Gujarat region. The seeds come from a small company that brought a new, supposedly unmanipulated variety onto the market. Apparently, it used GM seeds from a Monsanto field trial in the breeding process or its seeds were contaminated with pollen from GM cotton.

In India, the cultivation of genetically modified and thus pest-resistant cotton has let to significantly higher yields and considerably reduced pesticide use compared to conventional varieties. In one study, average yields for Bt cotton were found to be 80% higher compared to the non-modified same hybrid and 87% higher than the locally popular hybrid. In addition, Bt cotton fields were sprayed three times less against bollworm than the two comparison varieties. For sucking pests, such as aphids and scale insects, against which the genetic modification offers no protection, there were no significant differences in pesticide use when comparing the three cotton hybrids grown. The market power of a few suppliers of Bt cotton has left many Indian cotton farmers in debt and bankrupted when crops fail which led to a high suicide rate among cotton farmers.

The main difference between India and many other countries is that the region suffers from higher pest pressure. The use of effective chemical pesticides, on the other hand, is lower. Under such conditions, transgenic varieties have greater potential to increase yields than in other countries.

5.13 Organic Cotton

On the one hand, the proportion of genetically modified cotton on the world market is steadily increasing. On the other hand, the demand for organic cotton, which is produced without the use of pesticides and genetic engineering, increased significantly until 2010. Due to the difficult availability of suitable seeds and the generally low price of cotton due to the economic crisis, production dropped by 50% from its peak in 2010 to 2016. In recent years, it increased sharply again and is now over 240,000 tons (1.1 m bales), Fig. 5.126. This represents about 1% of the global market. In 2019, 220,000 farmers worldwide grew organic cotton on about 420,000 ha (about 1.5% of the world's cultivated area). The price of organic cotton is US$ 2–4/kg, which is significantly higher than the world market price for conventional cotton.

In 2019, organic cotton was grown in 19 countries (Organic Cotton, 2020), with 7 countries producing about 97% of the cotton. The largest producer was India (51%), ahead of China (17%), Turkey (10%), Kyrgyzstan (10%), Tajikistan (5%) and the USA (2%). There is also a strong trend to produce organic cotton in Africa (Tanzania, 2%), with 7 countries already growing organic cotton. However, the absolute production volumes there are still relatively low with 9500 t (cf. Sect. 5.9.9).

Sales of clothing made from organic cotton currently reach around US$ 30bn worldwide, with an upward trend. Europe is by far the largest market for organic cotton.

5.13.1 Egypt

In the early 1980s, the biodynamic Sekem Initiative was founded. Today it includes not only the "mother farm", but also several companies, social institutions and markets products from 200 organic farms. A major problem in the cultivation of "organic" cotton is the enormous water requirement, which is solved by artificial irrigation.

Increased manual use in pest control has increased yields by 15–20% compared to conventionally tilled fields. This had a direct impact on the production of other farms after some organic farming methods were transferred to them. According to the Egyptian Ministry of Agriculture, this saves about 30,000 tons of pesticides per year.

5.13.2 India

In the Indian state of Madhya, cultivation of the first trial fields began in 1992. Around 167,000 farmers are now involved and the cotton products are sold mainly in Europe. In 2018, the area under cultivation was around 300,000 ha and the harvest volume reached 123,000 tons. This corresponds to around 2% of the Indian cotton harvest (Fig. 5.127).

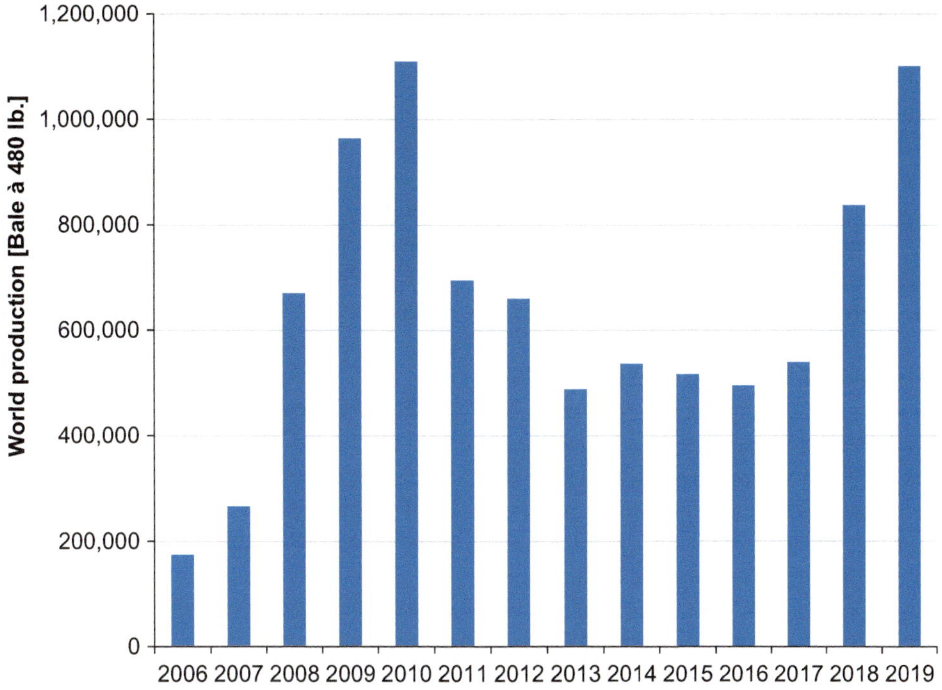

Fig. 5.126 Production volumes of "organic" cotton (data: Organic Cotton, 2020)

Fig. 5.127 Cotton collection point in India and bale warehouse (Bremen Cotton Exchange, 2021)

5.13.3 China

About 1300 cotton farmers produce 41,000 tons of organic cotton on 19,000 ha of land. This is more than three times the production of 5 years ago. The main cultivation area is the province of Xinjiang.

5.13.4 Turkey

A study conducted in the mid-1990s showed that farmers who grew cotton conventionally achieved a higher harvest than farmers who planted their cotton according to organic criteria. However, the latter achieved higher purchase prices at lower costs, so that the cultivation of organic cotton seemed worthwhile. The German trading company Otto, Hamburg, buys several hundred tons of organically grown cotton every year and was one of the first companies to promote such products in the mass market. Currently, 400 farmers produce about 23,000 tons of organic cotton on 11,000 hectares.

5.13.5 Kyrgyzstan

Only 220 cotton farmers produce around 24,000 t of organic cotton on 13,000 ha. Most of the production is exported to Turkey, only little is processed locally.

5.13.6 USA

Over 90% of the production of 5100 tons of organic cotton is produced in Texas. In total, there are only 66 farmers who cultivate 8000 ha of land.

5.13.7 Colored Cotton

Since the 1950s, there have been efforts, especially in the then Soviet Union, to grow colored cotton, especially shades of brown. However, these were not successful because the fibers could not be spun and hence this endeavor was discontinued. In the 1980s, Sally Fox in the USA succeeded in producing and spinning naturally colored cotton. This cotton grows in different shades of brown, red and green depending on the variety and is marketed by Vreseis, Ltd., under the name FoxFiber®. This could save some process steps in finishing, provided the consumer accepts the naturally grown colors (Fig. 5.128).

However, farmers fear that the colored cotton could possibly contaminate their white cotton.

Fig. 5.128 Field of colored cotton in the USA (Sally Fox, Vreseis, Ltd.) and yarn of colored cotton

Therefore, the distance to the nearest cotton fields used for seed production must be at least 5 km. Other government regulations prohibit the feeding of cottonseed because it could survive the digestion process and enter potential white cotton fields via fertilizer. In addition, all machinery used to process colored cotton must be inspected by government authorities before it can be used again for white cotton. As a result, this idea has not yet been commercially successful on a large scale, but this could change against the background of the demand for sustainable textile production.

5.14 Trade

Cotton is one of the ten “core commodities” of the UNCTAD (United Nations Conference on Trade and Development) commodity program. From the farmer (producer), the cotton usually first reaches the trader or trading houses. Cotton is then traded worldwide via national, partly state (e.g. Pakistan, China), private and international trading companies.

5.14.1 Forms of Cotton Trade

The cotton is initially sold by the farmers after the harvest to private ginning mills, trading companies or cooperatives. There are considerable differences depending on the producing country. While cotton is only ginned during the harvest season, cotton is traded all year round. This sometimes results in large price differences, which skillful traders can exploit to their advantage (Fig. 5.129).

Depending on the country, there are different systems for how cotton passes from the producer to the mill or to the exporter. In China, for example, all trading participants are state-owned. The farmers deliver to stately controlled ginning mills. From there, the cotton is either exported under state control or directly processed in state-owned enterprises. In India, there is a mixture of state-controlled market and free trade. However, there are regular export restrictions on raw cotton so that value addition takes place in India. In the USA, there is a completely free market.

5.14.2 Stock Exchanges

Members of the exchanges are the traders, forwarders, shipowners, banks and spinning agents. Purchases are made according to standards or type samples issued regularly by the producing countries. The traded cotton is compared with these samples. Price reductions can then be made depending on the quality. In addition, sworn experts also draw up official quality specifications, so-called quality arbitrage. In cases of dispute, the exchange also serves as a court of arbitration. Figure 5.130 shows the main cotton stock exchanges.

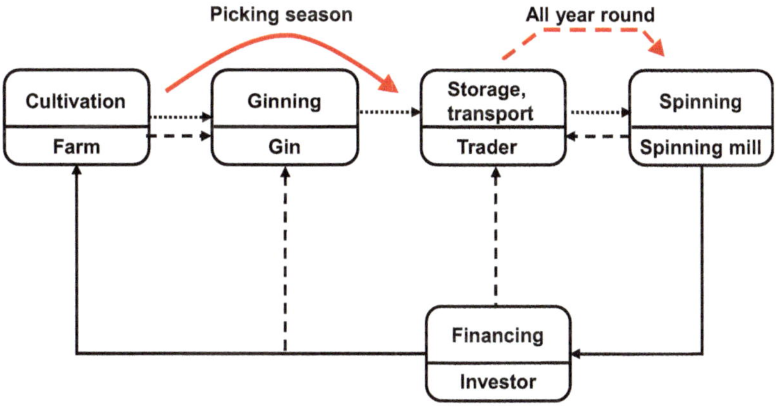

Fig. 5.129 Financial flows on the cotton market

Fig. 5.130 Important cotton exchanges

5.14.3 Futures and Options

As with other products traded on stock exchanges, the price of cotton is also subject to considerable price fluctuations in some cases. Even within a year, the price of cotton often changes abruptly. At the end of 2010, the price of cotton initially rose sharply within a short period of time as a result of speculators and reached its highest level since 1866. The price then collapsed again just as quickly and returned to its long-term average. In order to make such price fluctuations more manageable for buyers and sellers, "futures" and "options" of cotton are traded on many cotton exchanges:

Futures: A futures contract is a purchase contract for a certain quantity of a certain commodity with a certain quality, which is to be fulfilled at a certain future delivery date at a price agreed upon conclusion of the contract. It is important to note that delivery, acceptance and payment of the goods lie in the future. This hedges risks in the form of price fluctuations.

Options: An option is the right (not an obligation), extending over a specified period of time, to buy or sell an option item at a predetermined price. In an option transfer transaction, the option buyer pays a purchase price for acquiring the option right, called a premium. The prices of options generally follow the prices of futures.

In New York, for example, the Cotton No. 2 Future Contract is traded. It comprises 50,000 pounds of cotton with a grade of Strict Low Middling and a staple length of 1 1/16″. The minimum fiber strength is 25 cN/tex and the micronaire value must be between 3.5 and 4.9 (Cottonguide).

If a spinning mill buys the raw cotton at a low price, this can be a much greater financial advantage than improving its machinery. Hence, traders can decide a cotton mill's financial success.

5.14.4 Cotlook Indices

The world market price of cotton is published by the British company Cotlook, Ltd., in the so-called Cotlook-A-Index. This index is determined by averaging the 5 lowest selling prices of a trading day for the cotton quality "middling 1–3/32" from currently 19 countries (Fig. 5.131).

Since 2003, there has also been a corresponding index for the Far East; it is called the Cotlook FE Index and reflects the market situation in Asia.

Figure 5.132 shows the evolution of high and low prices for cotton since 1840. Clearly visible is the period of the American Civil War, which led to all-time high prices for cotton. In the 1980s and 1990s, the price of cotton was mainly influenced by developments in Uzbekistan and China. The speculative bubble in 2011 is also clearly visible.

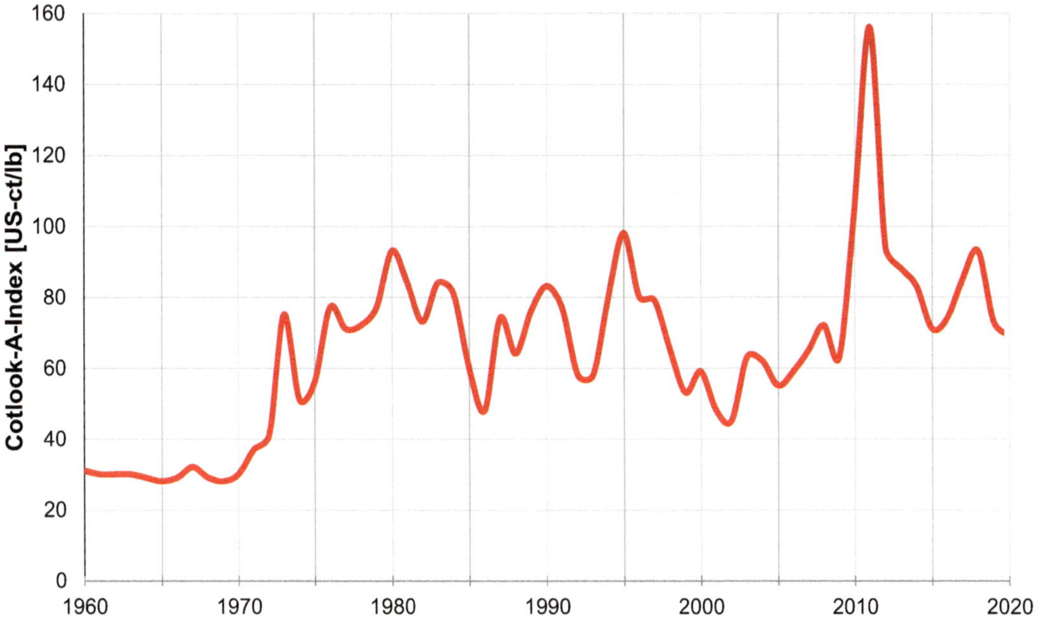

Fig. 5.131 Development of the annual mean values of the Cotlook A index (data: YCharts)

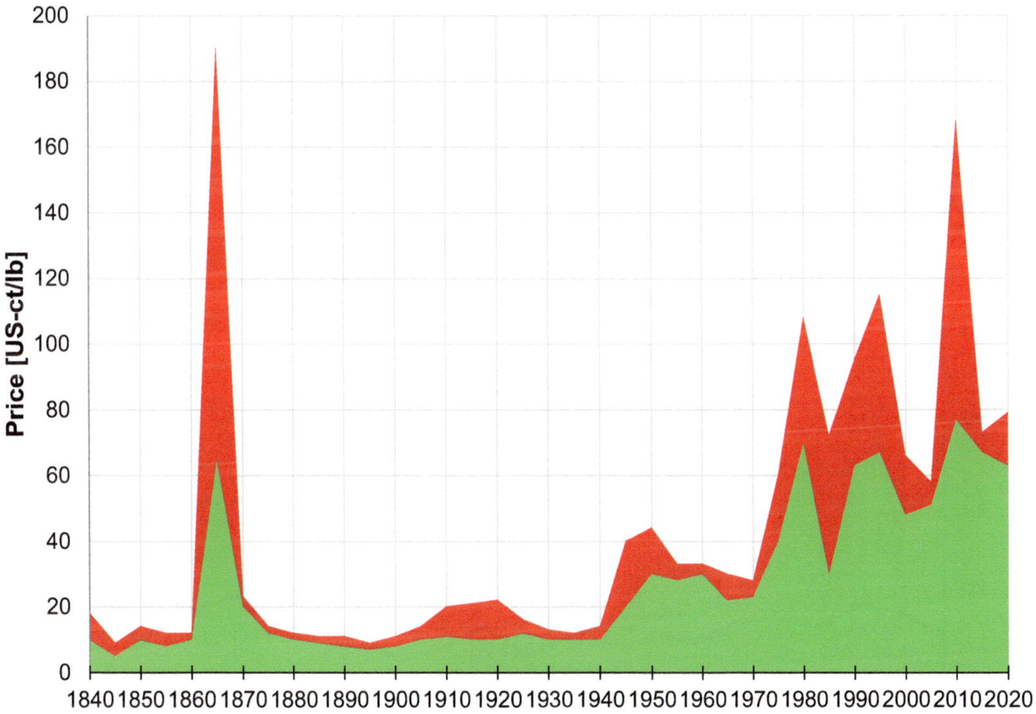

Fig. 5.132 Development of cotton prices on the New York Stock Exchange

5.14.5 Cotton Policy of Producing Countries

About half of the world's cotton production is subsidized by the producing countries (mainly USA, EU, China, African countries, Turkey). In some countries, money is paid directly to the farmers or insurance against possible income losses is financially backed (USA, India). In other regions, e.g. in Africa, seeds and fertilizer are financially supported. India recently expanded its infrastructure (roads, irrigation systems) and supported cotton farmers in marketing. China controls the import of cotton through a complex system of import restrictions (quotas, tariffs) and pays direct subsidies to its farmers. In addition, the Chinese government guarantees cotton farmers a minimum price ("floor price"). If the price falls below this, either the state buys the cotton or reimburses the farmers the difference to this minimum price. As a result, China has created a considerable overproduction of cotton, which is only slowly being reduced. Low-interest loans are also widespread. Such payments are usually not included in subsidies statistics.

It is true that such financial aid ensures an artificially high price. However, production costs in cotton-producing countries vary greatly, so that some compensation is necessary in order to be able to produce cotton economically even in high-wage countries. While the USA and China influence the market considerably through their policies, the EU's subsidies have virtually no influence on the world market because of the small quantities of cotton produced.

As Fig. 5.133 shows, subsidy payments are reduced when world prices rise and increased when prices fall.

Total subsidies flowing into the cotton sector worldwide amounted to about US$ 8bn (10 ct/lb) in 2020. Of this, China accounts for about US $4.7bn (37 ct/lb), the US for US$2bn (21 ct/lb) and Turkey for US$0.23bn (13 ct/lb). The EU subsidizes cotton farmers in Spain (46 ct/lb) and

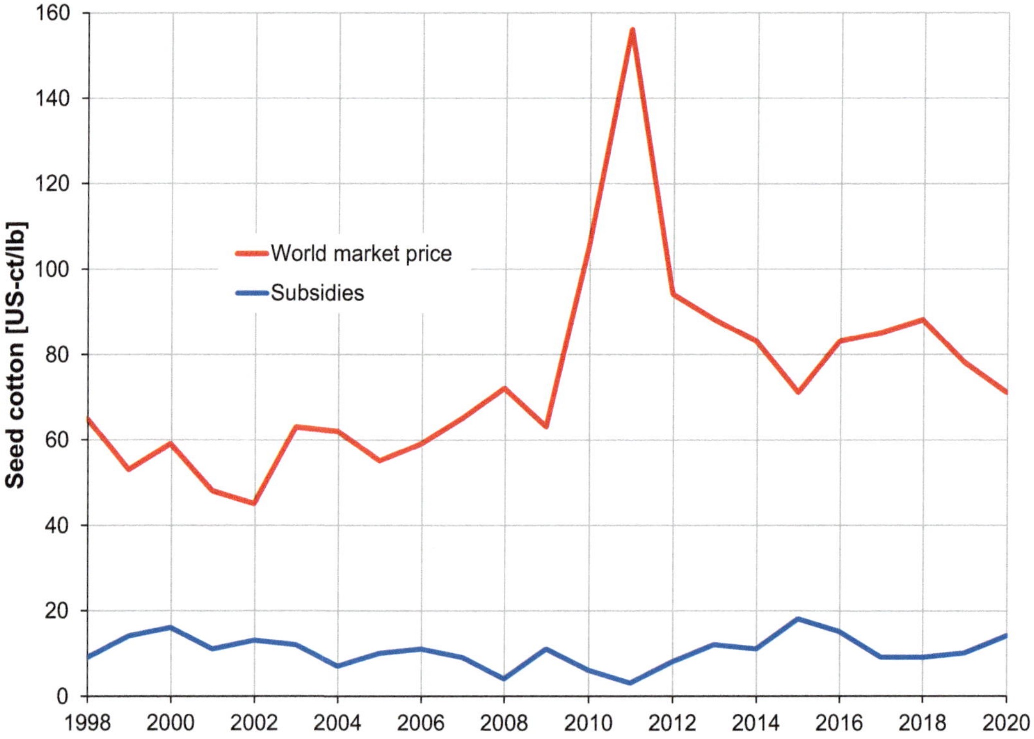

Fig. 5.133 Development of world market price and subsidies (data: ICAC, 2021)

Greece (33 ct/lb) to the amount of US$280 m, the highest of any cotton subsidy in the world. In contrast, African countries such as Burkina Faso and Mali pay subsidies of 6–8 ct/lb. to their farmers. However, production costs are also significantly lower there (Fig. 5.134).

The price-depressing subsidy policies of the major producing countries have a particularly strong impact on countries whose cotton exports are a major component of total exports, such as Benin, Togo and Burkina Faso. Because these and other African countries cannot produce at marketable prices, they are increasingly shifting to the production of organic cotton, for which a significantly higher price can be obtained.

Figure 5.135 shows the development of cotton subsidies for selected countries. Although Colombia produces very little cotton (2020: 17,000 tons), it paid extremely high subsidies to its farmers until a few years ago. China has greatly reduced its subsidy payments in recent years.

5.15 Areas of Application

Cotton is mainly used for clothing because of its good physiological properties, the comparatively low price and the large quantities available. In addition to underwear, T-shirts, shirts and blouses, this includes above all jeans. Cotton is also often used for hygiene products, e.g. pads, tampons, etc., because of its skin-friendliness and moisture absorbency (Fig. 5.136). Other typical products are bed linen, work clothes and shoes.

5.16 Trends

Cotton will remain the most important fiber for the production of textiles after polyester. The most important production countries will continue to be India, China, the USA, Brazil and Pakistan. Should Australia solve the problem of water scarcity, it will also play an important role. The size of cotton fields will continue to increase, which will be accompanied by increasing mechanization of harvesting, so that machine picking methods will continue to gain importance. Among the gin processes, saw gin is the most widely used. This will not change due to its economic viability.

Genetic engineering will become increasingly important, but it remains to be seen whether the largely positive results achieved to date will be sustained in the long term or whether negative effects will occur that call into question the use of genetically modified varieties. In some countries, almost 100% of cotton is already genetically modified (e.g. Australia, Mexico), while other

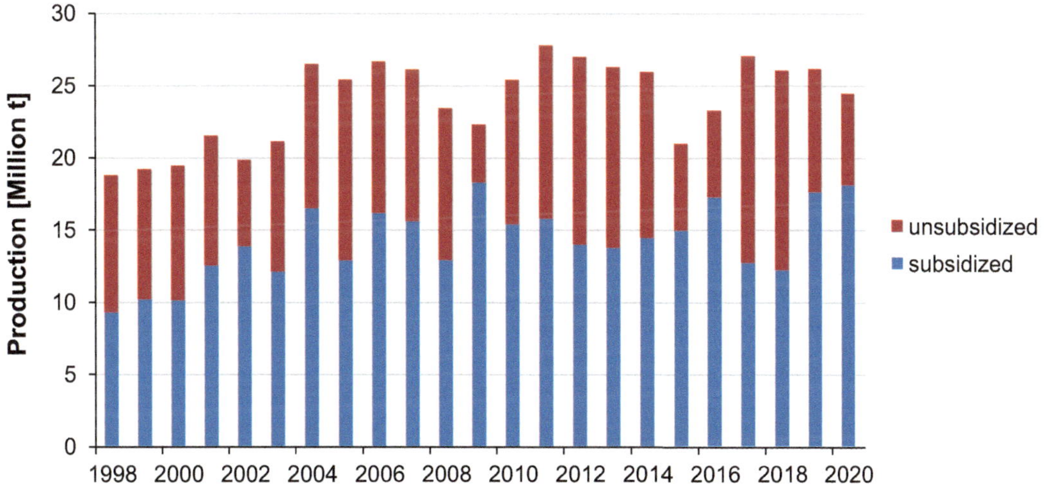

Fig. 5.134 Share of subsidized cotton in total production (data: ICAC, 2021)

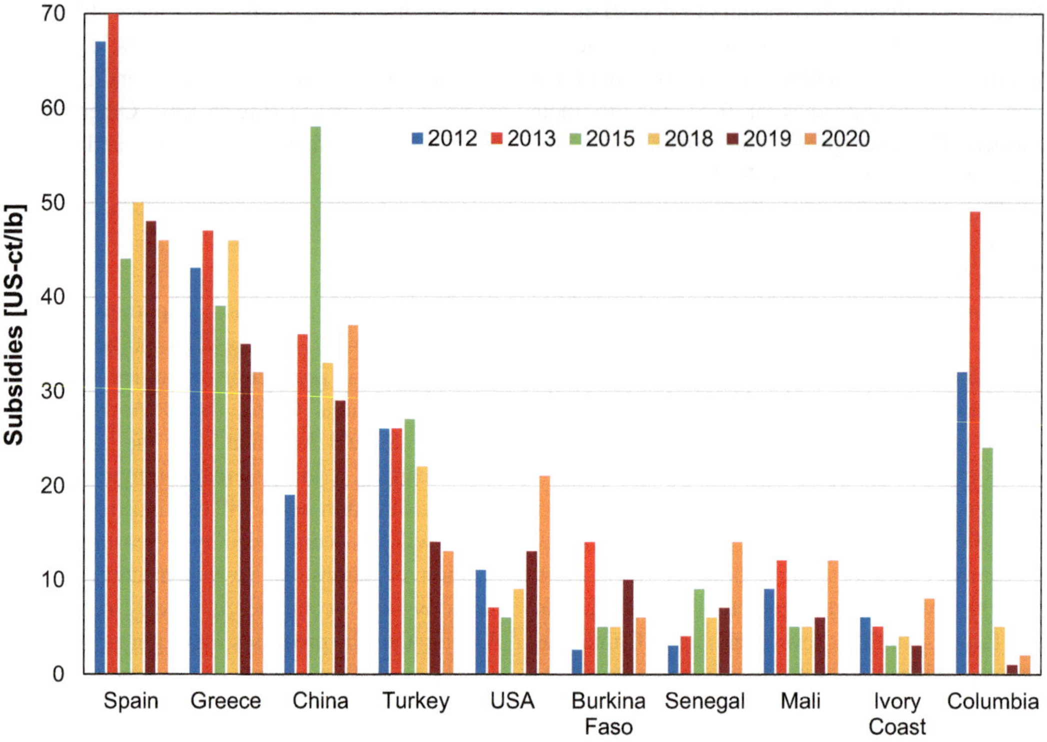

Fig. 5.135 Subsidies of selected countries for cotton farmers (data: ICAC, 2021)

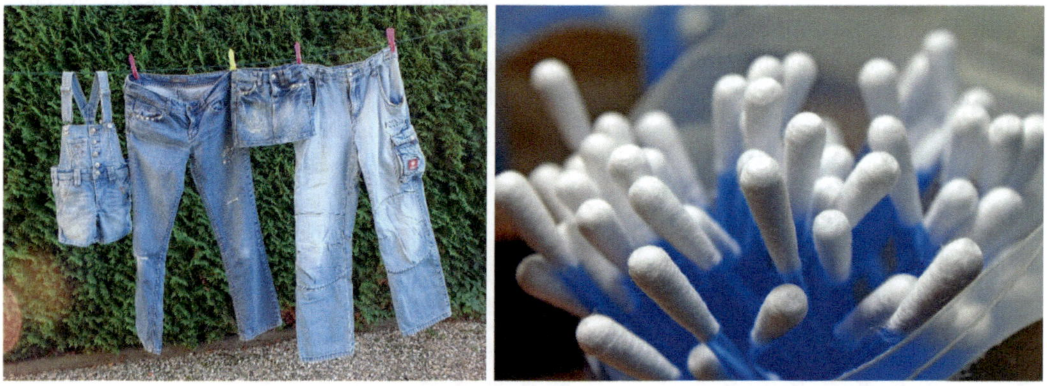

Fig. 5.136 Jeans and ear sticks

countries are acting more cautiously (e.g. Brazil with a share of 50%). It remains to be seen to what extent new spinning processes (e.g. air spinning) will lead to other fiber properties being desired by spinning mills (e.g. fiber length, fiber length distribution).

In general, a trend towards longer fiber lengths and higher strengths can be observed, because the spinning preparation and spinning processes are placing ever higher demands on the cotton fibers.

The wrapping of cotton bales will continue to consist mostly of PP or PET foils and to a lesser extent of jute fabrics. Knitted and woven fabrics made of cotton have not yet been able to establish themselves due to their high price, and a trend reversal is not in sight here. Therefore, the PP problem in further processing will continue to exist in the future.

References

Betten, F., & Kaufmann, A. (1919). *The modern world, from Charlemagne to the present time; with a preliminary survey of ancient times*. Allyn & Bacon.

Bremer Cotton Report. (2021). *Baumwollbörse*. Retrieved from https://baumwollboerse.de/

Chapagain, A. K., Hoekstra, A. Y., Savenije, H. H. G., & Gautam, R. (2006). The water footprint of cotton consumption: An assessment of the impact of worldwide consumption of cotton products on the water resources in the cotton producing countries. *Ecological Economics, 60*, 186–203.

CmiA. (2021). *Cotton made in Africa*. Retrieved from https://cottonmadeinafrica.org

Cotton Australia. (2021). Retrieved from https://cottonaustralia.com.au/

Cotton Grower. Retrieved from https://www.cottongrower.com/

Cotton Inc. (2021). Retrieved from http://www.cottoninc.com

Cottonguide. Retrieved from https://www.cottonguide.org

Dykinga, J. (2005). Retrieved from https://commons.wikimedia.org/wiki/File:Geocoris_punctipes.jpg

FAO. (2021). Retrieved from http://faostat.fao.org

French, J. C. University of Georgia. Retrieved from http://www.bugwood.org

Godot13. (2015). *Godot13*. Retrieved from https://commons.wikimedia.org/wiki/File:CSA-T13-$100-1861%E2%80%9362.jpg, CC BY-SA 4.0.

Gutenberg. (1919). Retrieved from https://commons.wikimedia.org/wiki/File:Eli_Whitney_sitting.jpg

Herodot. (2008). *Neun Bücher zur Geschichte* (3rd ed., p. 106). Übersetzung des marixverlags.

Hessler, L. E., Simpson, M. E., & Berkley, E. E. (1948). Degree of polymerization, spiral structure, and strength of cotton fiber. *Textile Research Journal, 18*(11), 679–683.

Hullwarren. (2012). Retrieved from https://commons.wikimedia.org/wiki/File:StGeorgeCottonIrrig.jpg, CC BY-SA 3.0.

ICA Bremen. (2019). *Cotton varieties by origin*. Retrieved from http://www.ica-bremen.org, CC BY-SA 3.0.

ICAC. (2001). *Cotton fiber development and processing – An illustrated overview*. International Textile Center, Texas Tech University.

ICAC. (2021). *Production and trade subsidies affecting the cotton industry*. Retrieved from http://www.icac.org

intracen. (2020). Retrieved from www.intracen.org

John Deere. (2021). Retrieved from https://www.deere.com

Kenpei. (2006). Retrieved from https://commons.wikimedia.org/wiki/File:Gossypium_arboreum1.jpg, CC BY-SA 3.0.

Kidoudeni. (2017). Retrieved from https://commons.wikimedia.org/wiki/File:Cotonculteurs_-_Ferk%C3%A9_-_C%C3%B4te_d%27Ivoire.jpg, CC BY-SA 4.0.

Kielmann, M. J. (2021). *Martin J. Kielmann für CmiA*. Retrieved from https://cottonmadeinafrica.org

Kling, W., & Mahl, H. (1952). Zur Morphologie der Baumwolle IV: Zusammenfassung der experimentellen Ergebnisse zu einem neuen, verbesserten Fasermodell. *Melliand Textilberichte, 33*, 829–832.

Kooyman, C. (2006). Retrieved from https://commons.wikimedia.org/wiki/File:Cotton_plant.jpg, CC BY-SA 3.0.

Lee, H. (1887). *The vegetable lamb of Tartary: A curious fable of the cotton plant to which is added a sketch of the history of cotton and the cotton trade*. S. Low, Marston, Searle & Rivington. Retrieved from https://commons.wikimedia.org/wiki/File:Vegetable_lamb_(Lee,_1887).jpg

Lummus Corp. Retrieved from https://www.lummus.com/

Mandeville, J. (14. Jh.) Retrieved from https://commons.wikimedia.org/wiki/File:Mandeville_cotton.jpg

Mapswire. Retrieved from https://mapswire.com

Matthews, G. A. (1962). Retrieved from https://commons.wikimedia.org/wiki/File:Boll_weevil.jpg

Mekonnen, M. M., & Hoekstra, A. Y. (2011). The green, blue and grey water footprint of crops and derived crop products. *Hydrological Earth Systems Science, 15*, 1577–1600.

N.N. (2001). *Cotton fiber development and processing*. Cotton Inc..

NCSU. Retrieved from https://cotton.ces.ncsu.edu/insect-scouting-guide/

Nichols, K. (2016). Retrieved from https://commons.wikimedia.org/wiki/File:COTTON_FIELD_(49168278973).jpg, CC0 1.0.

Organic Cotton Market Report. (2020). Retrieved from http://www.textileexchange.org

Ottens, R. University of Georgia. Retrieved from http://www.bugwood.org

Renault, C. (2005). Retrieved from https://commons.wikimedia.org/wiki/File:Cotton_picking_in_India.jpg, CC BY 2.0.

Sandlapper. (2015). Retrieved from https://commons.wikimedia.org/wiki/File:Elko_cotton_field_4.JPG, CC BY-SA 4.0.

Schmidt, M. (2004). Retrieved from https://commons.wikimedia.org/wiki/File:Gossypium_sp_MS4007.JPG, CC BY-SA 2.5.

Schulz, K. (2015). Retrieved from https://www.flickr.com/photos/treegrow/22209238916/, CC BY 2.0.

Sheppard, W. (1869). Retrieved from https://commons.wikimedia.org/wiki/File:William_L._Sheppard_-_First_use_of_the_Cotton_Gin,_Harper%27s_weekly,_18_Dec._1869,_p._813.png

Smith, R. (2009). Retrieved from https://commons.wikimedia.org/wiki/File:Aphis_gossypii01.jpg, CC BY 3.0 US.

Starr, K. (2008). Retrieved from https://commons.wikimedia.org/wiki/File:Starr_080716-9494_Gossypium_barbadense.jpg, CC BY 3.0.

Tabashnik, B. E., Liesner, L. R., Ellsworth, P. C., & Carrière, Y. (2020). Transgenic cotton and sterile insect releases synergize eradication of pink bollworm a century after it invaded the United States. *PNAS, 118* (1).

Textechno, Herbert Stein GmbH & Co. KG Textile Mess- und Prüftechnik, Mönchengladbach. Retrieved from https://www.textechno.com

Tillman, P. G., & Cottrell, T. E. (2019). Influence of pheromone-baited traps on stink bugs in cotton. *Journal of Insect Science, 19*(1), 1–7.

Trading Economics. Retrieved from https://tradingeconomics.com

USDA. (2020a). *Cotton report*. Retrieved from https://www.fas.usda.gov/data/cotton-world-markets-and-trade

USDA. (2020b). Retrieved from https://commons.wikimedia.org/wiki/File:Boll_weevil.jpg

Uster Technologies AG. *Uster (Schweiz)*. Retrieved from https://www.uster.com

Vardeman, K. (2009). Retrieved from https://commons.wikimedia.org/wiki/File:Cotton_field_kv06.jpg, CC BY 2.0.

Vreseis Ltd. Retrieved from https://www.vreseis.com

www01. Retrieved from https://commons.wikimedia.org/wiki/File:Cotton_Picking,_Milwaukee_Public_Museum_Miniature_Group,_Milwaukee,_Wisc_(NYPL_b12647398-74,637).tiff

www02. Retrieved from https://commons.wikimedia.org/wiki/File:Gandhi_spinning.jpg

www03. Retrieved from https://commons.wikimedia.org/wiki/File:Emblem_of_Uzbekistan.svg

www04. (n.d.). Retrieved from https://commons.wikimedia.org/wiki/File:Aral_Sea_1989-2008.jpg

YCharts. Retrieved from https://ycharts.com/indicators/cotlook_aindex_cotton_price

Zell, H. (2009). Retrieved from https://commons.wikimedia.org/wiki/File:Gossypium_herbaceum_001.JPG, CC BY-SA 1.2.

6 Bast Fibers

Bast fibers include flax, hemp, jute, ramie and kenaf. The fibers are all obtained from the stems of the corresponding plants. Their biological function is to stabilize the stalk against kinking. For this reason, the wet strength of bast fibers is always greater than 100%, since the risk of buckling is greater in wet weather. They consist mainly of cellulose fibrils, hemicellulose and varying percentages of lignin and pectin (Fig. 6.1).

Figure 6.2 shows the typical structure of a bast fiber plant and the position of the fiber bundles within the stalk.

World production is about 4–5 m tons, yields vary widely reaching 3600 kg/ha for flax, 2220 kg/ha for jute and 600 kg/ha for abaca. Figure 6.3 shows the production volumes of important bast fibers. Depending on weather conditions and demand, the quantity of bast fibers produced varies, sometimes considerably. This applies in particular to flax, which is used mainly for clothing and is therefore fashion-dependent. Jute is used almost exclusively for packaging, so the production volume is relatively constant and has even increased slightly (before the pandemic) in recent years. In 1985, there was a bumper crop of jute fibers, the quantity of which has never again been equaled. It is estimated that some 8 million families worldwide earn their income mainly from the production of bast and hard fibers.

6.1 Flax (Linen)

Flax fibers (abbreviation: LI) are obtained from the stems of Linum usitatissimum Linné. After harvesting, either straw flax with or without seeds (mature, air-dried flax plants) or roasted straw flax (straw flax without seeds, which has been prepared in a retting process and then dried) is obtained (Fig. 6.4).

A distinction is made between fiber flax and oil flax. Originally, flax was mainly used as a fiber plant. The seeds, from which oil can be obtained, were not processed, or only in small quantities. The reason for this was that the capsules of the flax burst open when ripe (spring flax, Linum crepitans) and eject the seeds. In order to be able to use the seeds as well, a so-called closing or threshing flax (var. bulgare) was bred, in which the capsules no longer open. This meant that the seeds could also be harvested and processed.

The most important cultivation areas are in Central and Western Europe, the countries of the former USSR, China, India, North America and Argentina.

Three different materials can be obtained from flax:

- Linseed (as a food and medicinal product),
- seed oil (for varnishes, paints, linoleum, edible oil, press residues as cattle feed)
- stalk fibers (as raw material for the textile industry).

D. Veit, *Fibers*, https://doi.org/10.1007/978-3-031-15309-9_6

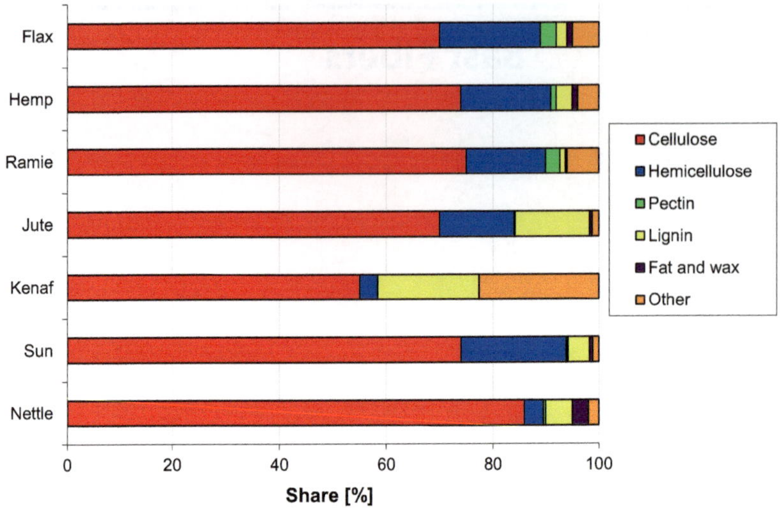

Fig. 6.1 Composition of selected bast fibers (without water content)

6.1.1 History

Like barley and wheat, flax is one of the oldest cultivated plants. It was already cultivated more than 6000 years ago by the Egyptians and the Sumerians in what is now Iraq and came to southern Central Europe in the later Stone Age (in the third millennium BC). The parent form is considered to be the narrow-leaved, perennial wild species Linum angustifolium with dehiscent capsules. It occurs in an area extending from the Mediterranean to southwest Asia.

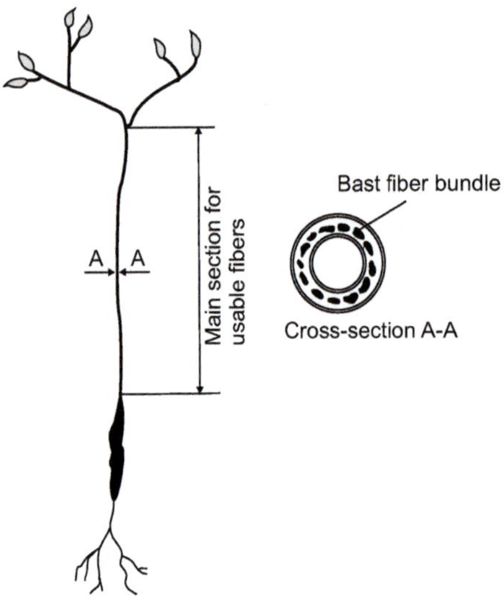

Fig. 6.2 Structure of bast fiber plants and cross-section through the stalk

The oldest finds of linseed come from Mesopotamia. Fragments of linseed and capsules were found in Iran (approx. eighth millennium BC). Finds of fabrics come from ancient Egyptian graves (e.g. mummy bandages) and from the lakeside settlements ("pile dwellings") in Switzerland from the time around 5000–3000 BC. In addition to the manufacture of clothing fabrics, belts, cords, strings, lamp wicks and sails were also made from flax.

As Fig. 6.5 shows, flax was harvested in Egypt by pulling the whole plant out of the ground. It was then placed in water ("water retting") to separate the fibers from the bast.

The goddess Isis was the patron goddess of flax processing. Linen was considered a symbol of light and purity and was therefore used especially for fine garments. Yarns with a fineness of up to Nm 200 (5 tex) could be produced from linen. Such fine fabrics could be manufactured that one could see through them ("woven wind"). Such yarn counts are no longer achievable even with modern spinning machines. Because fine cords, with the help of which straight lines could be drawn, were also made of flax fibers, the Latin word for flax, "linea", became synonymous with straight strokes and our modern word "line".

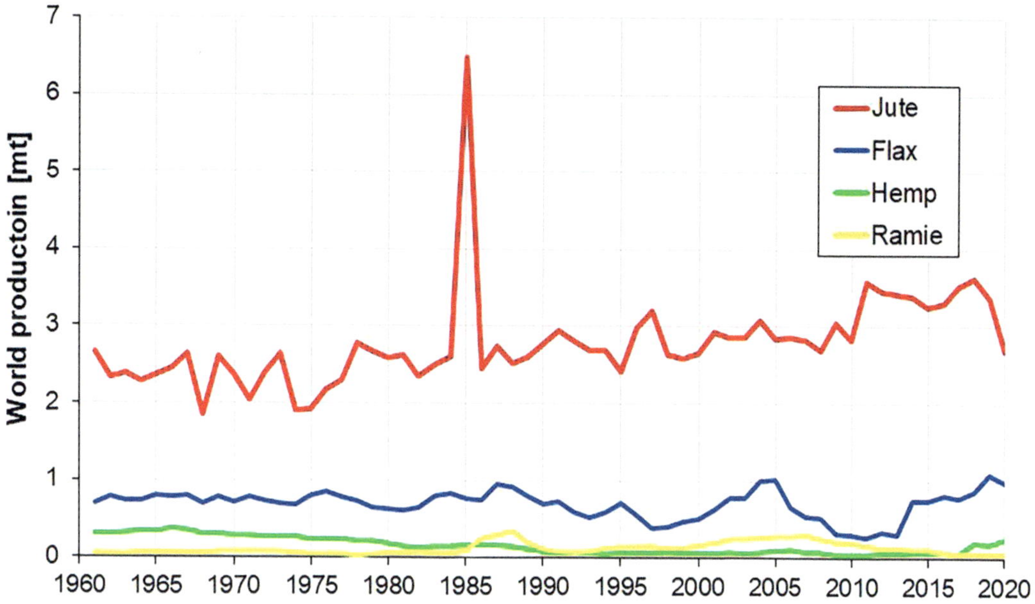

Fig. 6.3 Production of important bast fibers (data: faostat, 2022)

Flax remained dominant in Egypt even after it had already been displaced by wool as the most important fiber in other countries. From Egypt, flax came to Europe via Greece and Italy. The Romans introduced linen from the Orient. The historian Titus Livius (59 BC–17 AD) mentions the garments and armor of the Etruscans made of flax.

According to the historian Strabon, the worst linen of antiquity grew near Amporias in Spain and was almost unusable. It was considered a variety of reed (lat. "iuncus"). The end product was called junkaric linen and developed via a slang word of the sailors for damaged ropes and ropes to "junk" in the meaning of "rubbish".

The emperor Charlemagne considered the cultivation of flax to be so important that he ordered his subjects in several decrees to plant flax. From the thirteeth to the sixteenth century was the heyday of flax cultivation. Swabian, Flemish, Silesian and Upper Lusatian linen articles were sought-after trade goods. At this time, linen was the most important textile raw material in Europe. Especially the Fugger in Augsburg systematically

Fig. 6.4 Flowering flax field

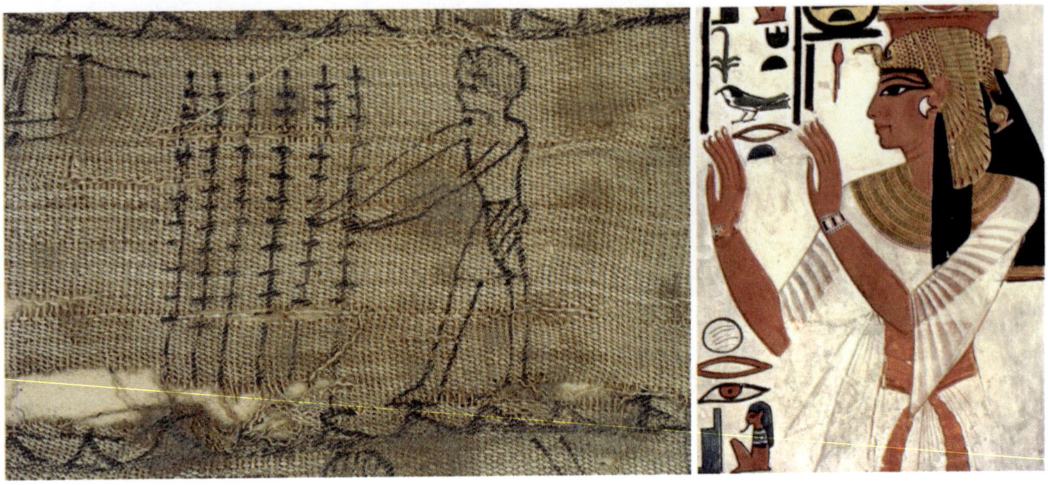

Fig. 6.5 Flax extraction in Egypt and flax dress of Nefertari (ca. 1200 BC)

traded in linen. Later, among others, Frederick the Great (Silesia and Westphalia) were eager promoters. Other important centers in Europe were Ireland, France, the area of present-day Belgium and the Netherlands. In Eastern Europe, Russia and the Baltic states became major producers and processors. For many regions in Europe, the cultivation of flax was so economically important that it is still commemorated today in the respective coats of arms. Figure 6.6 shows selected examples.

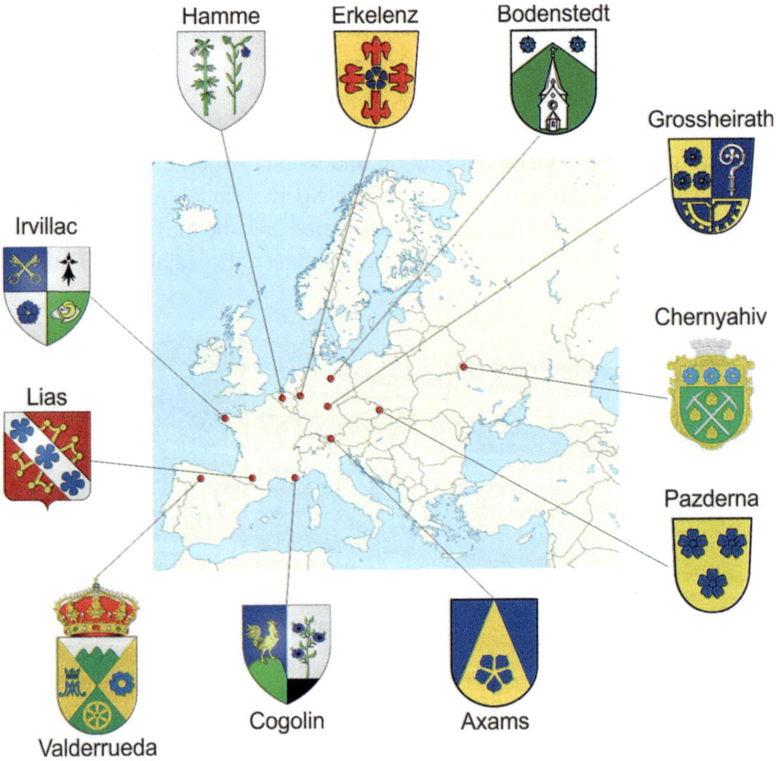

Fig. 6.6 Cities with flax symbols in their coat of arms

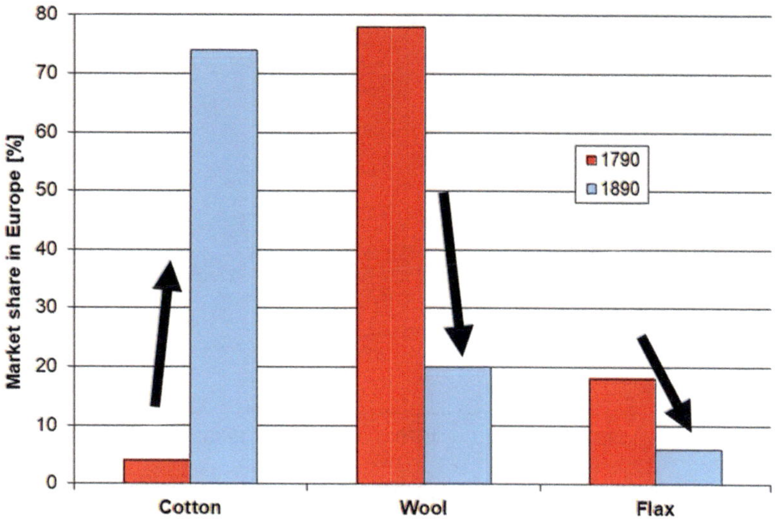

Fig. 6.7 Flax cultivation in the eighteenth and nineteenth century

The Thirty Years' War marked the beginning of the decline of flax cultivation, because many fields were devastated and it was difficult to cultivate them again. The rapidly developing textile industry at the beginning of the nineteenth century, the mechanization of processing and, above all, the possibility of spinning cotton to the same fineness as standard linen warp yarns then displaced flax not only in Europe but worldwide. Until the end of the eighteenth century, flax was the most important plant fiber in Europe, accounting for approx. 18% of fiber consumption, and the most important textile raw material alongside wool (approx. 78%). By the end of the nineteenth century, the share of cotton in Europe had risen to 74%, while the share of sheep's wool had fallen to 20% and that of flax to 6%. The American Civil War (1861–1865) led to another boom in flax growing, mainly in the Netherlands, and flax processing, mainly in Ireland (Belfast was called "Linenopolis"), because of the resulting cotton shortage. However, the decline of flax as a fiber crop could not be permanently halted (Fig. 6.7).

The reasons for this development are manifold:

- The production of flax is highly weather-dependent and costly,
- the areas of application for flax fibers are limited by their susceptibility to creasing, low thermal insulation, heaviness and susceptibility to rot,
- the raw material linen was not available in the required quantity.

After World War II, hardly any flax was grown on the territory of the Federal Republic of Germany (Fig. 6.8) and from 1980 onwards, flax no longer appears in official statistics. In 1951, the International Flax Association CELC ("Confédération Européenne du Lin et du Chanvre") was founded. Since then it has endeavored to improve the image and marketing of flax fibers.

Due to the discussion about ecologically compatible crop cultivation, flax briefly regained importance in the 1980s. In Germany, especially in Brandenburg, there was a renaissance of flax cultivation in the mid-1990s because it was subsidized. However, production has since fallen to quantities that are no longer recorded statistically. A similar development could be observed in other European countries.

In Western Europe, flax is now mainly grown in northern France and Belgium. In these countries, great efforts have been made in recent years to rationalize production and to increase the market share of flax fiber, especially in blends with other types of fiber, in the fashion sector as well as for home textiles. Flax also became fashionable again in collaboration with Italian fashion

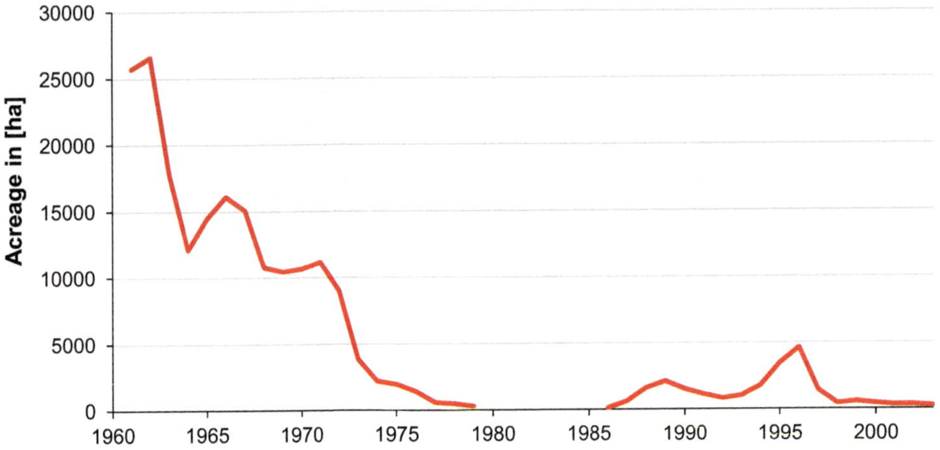

Fig. 6.8 Area under flax cultivation in Germany (data: statista, 2022; faostat, 2022)

designers. Its susceptibility to creasing, for example, was described as "noble crease" and was accepted as such by consumers. In 2019, the production volume of flax worldwide was approximately 1 m tons, of which 94% was produced in Europe.

6.1.2 Cultivation

Flax belongs to the Linaceae family and there to the genus Linum. The genus Linum includes many different species, which are mainly found in temperate and subtropical regions. They are wild growing, robust, small perennials. Only the annual flax plant, Linum usitatissimum ("the extremely useful flax") is cultivated.

6.1.2.1 Crop Growing

Fiber flax can be grown relatively extensively, but sufficient water must be available for optimum fiber formation. Sowing takes place at the end of March or September (southern hemisphere). In contrast to oil flax, the seed for fiber flax is sown very densely with approx. 300 plants per m^2. This avoids undesirable strong branching and seed formation. A lower seed density may lead to yield losses and poorer fiber quality (e.g. coarser fibers).

With a favorable germination temperature and uniform soil irrigation, the seed germinates after approx. 3 weeks. The flax plant has a vertically downward growing taproot with pronounced lateral roots. Branching strength and root length depend on soil type, water supply and nutrient content. Under normal cultivation conditions, the length of the root corresponds to that of the stem. From a growth height of about 8 cm, rapid longitudinal growth sets in.

Around 6 weeks after sowing, the flax flowers. The most common flower color is blue, but white, purple and pink flowers are also known (Fig. 6.9). Each plant bears 4–8 flowers. The flowering period is about 2 weeks. Maximum capsule and seed size is reached after about 8 weeks. Depending on the species, the capsule can contain up to 10 seeds. 3 to 4 weeks later the field is ready for harvest.

Location

Loamy sands, sandy loams and loess soils are suitable for growing flax. Heavy clay and light sandy or peaty soils are not suitable. From April–June (northern hemisphere), a rainfall of 120 mm is required to ensure good growth. The heat requirement is somewhat lower than that of summer cereals (heat sum 1600–1850 °C; it is determined by summing up the temperatures of all days of a defined period at a certain time when a certain minimum value is exceeded and a certain maximum value is not reached). Young flax plants are relatively cold-resistant, but

Fig. 6.9 Flax flower and flax ready for harvesting with seed pods

Fig. 6.10 (Cranshaw) and earth owl (Peairs)

temperatures should not fall below −3 °C for any length of time.

Crop Rotation
Fiber flax is not self-compatible ("flax fatigue"), so it cannot be grown on the same land in successive years without fertilization. The necessary cultivation interval is 6–7 years. Suitable preceding crops are cereals (e.g. oats), rape and maize. They leave a low residual nitrogen content in the soil and therefore provide good growing conditions for flax. Flax itself is a good preceding crop for cereals. Nowadays, fertilization usually shortens the cultivation period to 3–5 years.

Fertilization
During sowing and the growth phase, the flax must be fertilized with nitrogen (max. 40–100 kg N/ha). If the soil already contains more than 60 kg N/ha, additional fertilization is not necessary. In total, the flax plant absorbs approx. 50 g nitrogen, 30 g potassium, 20 g phosphorus and 8 g sulfur per kg flax during its growth. A large part of this (20 to 80%) can be removed again from the seed.

Plant Protection
Flax plants must be protected against weeds and fungal infestations. Furthermore, earth fleas, earth owls, lygus bugs, moths, aphids and the capsule worm (bollworm) are a problem in some areas. They feed on the buds, flowers, capsules and stems (Fig. 6.10). For weed control, mechanical measures are usually sufficient at the beginning of the growth phase, but, later, herbicides must be used because of the risk of plant damage.

6.1.2.2 Growth

During the vegetative growth of the plant, flax requires a steady supply of water. Alternations

Table 6.1 Maturity stages of fiber flax

Maturity stage	Amount of leaf loss on stem	Color		Seed color	Average number of days after blossoming
		Stem	Bud		
Green ripening	1/5	Light green	Green	Milky white	ca. 24
Yellow ripening	2/5	Light green, yellow at the base	Yellow	Light yellow	ca. 33
Fully ripe	3/5–4/5	Yellow	Yellow, brown	Light brown	ca. 43
Overripe	>4/5	Yellow to brown	Brown	Shiny brown	ca. 52

between dry and wet periods lead to a reduction in yield and quality. The ripening process begins about 12 weeks after sowing. The stages of maturity are defined by the coloring of the stalks, capsules and seeds. Stem color depends on the vegetative stage of the plant, growing conditions and diseases. It ranges from green to yellow-green and yellow to brown and black-brown. Maturity for fiber flax can be divided into four periods (Table 6.1).

6.1.2.3 Varieties

According to the EU aid regulation, 25 flax varieties are currently approved for cultivation. The best yields are achieved with the varieties Ariane, Argos, Viking, Elise, Escalina and Ilona. Breeding new varieties is a lengthy process and takes around 8–18 years.

In Fig. 6.11, different types of flax are depicted schematically:

1. Long ideal type, susceptible to storage, low seed yield
2. Fine-flat type, susceptible to storage, low seed yield
3. Seeded fiber flax with strong branching, low long-fiber yield
4. Sturdy fiber flax with medium fiber and seed yields
5. Early fiber flax, short
6. Primitive form, small-seeded bush form without fiber value
7. Oil flax, high seed yield, without fiber value
8. Oil flax, good seed yield, fibers coarse to medium fine
9. Oil fiber flax with medium seed and fiber yields
10. Similar to 9, but better fibers
11. Ideal form of an oil fiber flax, high fiber and seed output

Depending on the respective type of use, the breeding aims are either fiber flax or oil flax. The breeding objectives for fiber flax are high straw and fiber yields in good quality. For fiber production, types with long, thick, unbranched stems and a high fiber content are desired (Fig. 6.12).

In contrast, oil flax has a high seed yield. Oil flax types should be rather short-growing and have strong inflorescence branching as a prerequisite for a high number of capsules per plant. Seed size, oil content and oil composition of the

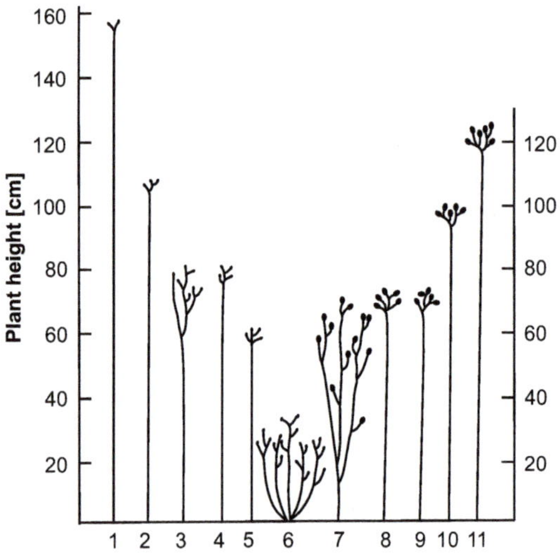

Fig. 6.11 Various forms of cultivated flax plants according to Kind et al. (1930)

Fig. 6.12 Types of flax (left: fiber flax; right: oil flax; Bjørtvedt, 2008)

Table 6.2 Important flax species for breeding purposes

	Fiber flax	Oil flax
Plant length [cm]	110–130	40–80
Fiber content [%]	20–25	gering
Fiber yield [kg/ha]	4000–6000	500–1200
Fiber quality	Premium	Medium
Oil content [%]	30–35%	40–45%
Branches	Few	Many
Seed size	Small to medium	Medium to large
Usage	Long and short fibers	Oil production
	Mainly textile industry	Chemical and food industry

seeds must meet the specific requirements of the processing industry (Table 6.2). The seeds of the oil flax are rich in the essential fatty acids linoleic and linolenic acid. It is converted into anti-inflammatory substances in the body. In 2018, around 3.2 m tons of linseed were produced, with an average yield of around 1000 kg/ha.

From these two extreme forms, intermediate forms such as oil flax were developed by breeding, but these did not become established.

6.1.2.4 Harvest

The vegetation period of fiber flax is 100–120 days, and harvesting takes place 20–30 days after flowering. The harvesting time for flax depends on the variety and the stage of maturity. For high fiber yield, maximum fiber content and good fiber quality, yellow ripening is considered to be the most favorable time.

For harvesting, the plant is pulled out of the ground by hand or with a root digging machine (Fig. 6.13).

After the pulling, the capsules are separated from the straw ("rippling"), Fig. 6.14.

The rippling can also be done during the plucking, linseed is a by-product.

6.1.3 Structure

A variety of layers are distinguished in the flax stalk: The flax stalk is surrounded on the outside

Fig. 6.13 Flax harvest (Sparshatt) and sheaves in the field (Lokilech, 2009)

Fig. 6.14 Rippling (left) and linseed (right)

by the cuticle, a wax-like layer. It protects the stalk from drying out. The smooth, hairless outer skin is formed by the epidermis. It contains stomata, which regulate transpiration and respiration of the stem. They also allows the penetration of bacteria during retting. The shape, arrangement and frequency of the stomata allow flax and hemp fibers to be distinguished, as both are very similar. The epidermis is followed by the cortical layer. It contains chlorophyll and therefore influences the color of the flax stalk. Approx. 10–50 separate bast fiber bundles are arranged in a ring in the cortical layer (Fig. 6.15).

The bast fiber bundles are separated from the cortical layer by destroying the non-woody middle lamella. The next very fine layer is the cambium, the growth zone. It forms the cortical cells on the outside and the woody cells (xylem) on the inside during the growth. The lignified cells take up the largest part of the cross-section. They serve as a strengthening and supporting organ for the flax stalk. If the woody part is not sufficiently developed, there is a risk that flax will buckle. The proportion of wood cells depends on the flax variety, the cultivation conditions such as spacing (distance between plants) and nutrition, as well as the position in the stem. The proportion of wood is greater in the lower part of the stalk than at the top. During retting, the wood cells become brittle and fall off as shives during further processing. The next layer is the pith. In young plants, it fills the entire inner part; in older plants, the pith recedes and a cavity, the lumen, is formed.

Fiber yield and quality are influenced by the formation of fiber bundles in the stalk. There are differences in the number of fiber bundles, the number of fiber cells they contain, the outline of

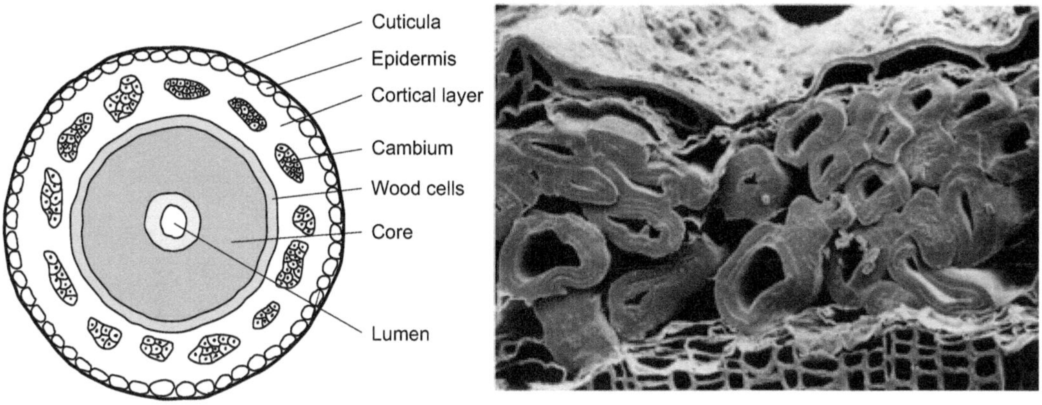

Fig. 6.15 Structure of the flax stalk (at half plant height) and arrangement of the elementary fibers in the fiber bundle

Fig. 6.16 Flax fiber model acc. to Haudek and Viti (1980)

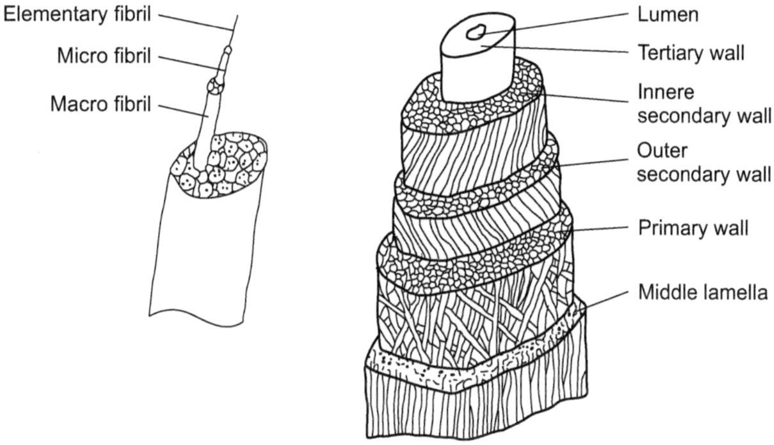

the fiber bundles, the structure of the fiber cells and their chemical composition.

Possible causes are:

- Flax variety,
- growing conditions (nutrition, water supply, stand spacing, diseases),
- position of fibrous elements in the stem (base, middle or tip).

The technical flax fiber consists of fiber bundles, which are made up of 10–25 individual fibers, the elementary fibers (Fig. 6.16).

The creation of the fiber occurs very early in the growing point at the top of the plant. Initially, the fiber grows downwards until it reaches mature tissue, while the upper end of the fiber still penetrates growing tissue. Once the cell has completed its length growth, thickness growth begins.

The elementary fibers have an average length of 20–40 mm, the average fiber diameter is 15–20 μm and the width for fibers from the lower stem area reaches values of 30 μm. These figures vary widely in the literature. Elementary fibers are similar in structure to cotton. However, they do not have a cuticle because, unlike seed fibers, they have no contact with the outside world and therefore require no protection. In general, the fiber cell has a relatively large wall thickness and a small lumen.

As a characteristic structure, the flax fiber exhibits nodular swellings in the longitudinal direction, which are also known as displacements (Fig. 6.17). They occur during growth and as a result of mechanical processing.

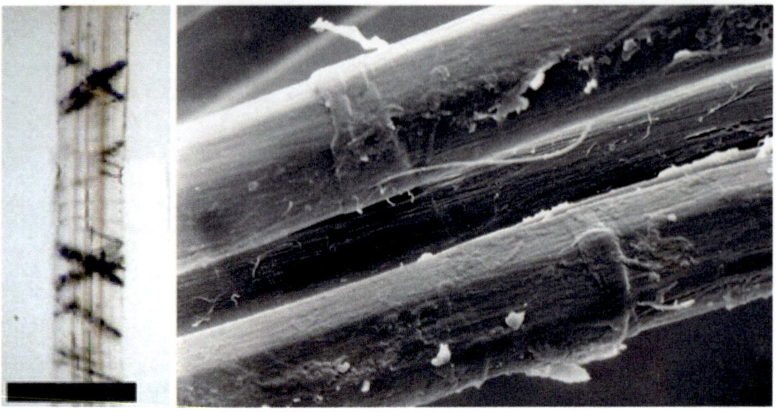

Fig. 6.17 Flax fibers with displacements

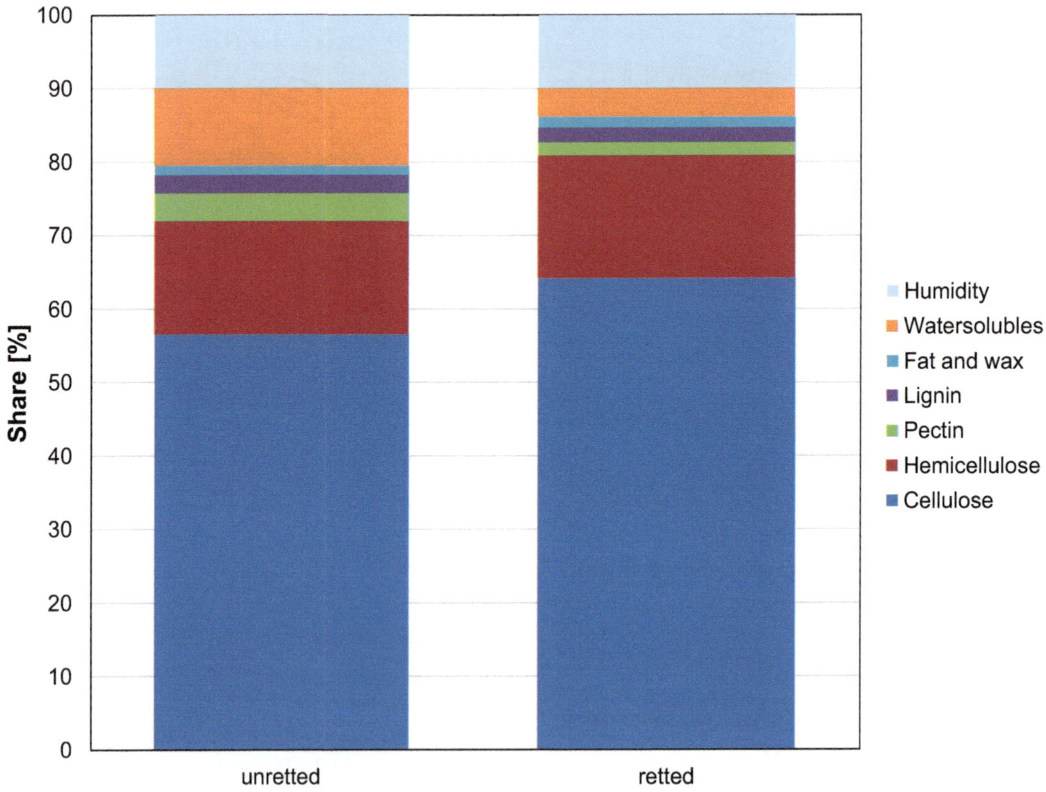

Fig. 6.18 Main components of flax fiber (data: Liebscher, 1983)

The main constituent of flax fiber is cellulose, although the content may vary. The better the retting and subsequent cleaning, the higher the cellulose content (approx. 64%). In addition, flax fibers contain other accompanying substances, of which hemicellulose is the most important in terms of quantity (16%), Fig. 6.18.

6.1.4 Fiber Extraction

There are three different processes for obtaining flax fibers:

- Biological digestion ("retting"),
- chemical pulping,
- mechanical solubilizing.

Fig. 6.19 Flax after harvest

6.1.4.1 Biological Solubilizing

In this process, microorganisms (bacteria or fungi) destroy the pectin-containing lamellae (plant glue) of the fabric layer surrounding the fibers. By dissolving the plant glue substance in the stem that holds the fiber bundles together, the separation of the fiber bundles can be done mechanically. During normal retting (Fig. 6.19), the retting organisms do not attack the cellulose fibers. If this does occur, it is called "over-retting".

The retting processes can be divided into dew and water retting. In the case of dew retting, the flax straw is spread out on the field for about 3–6 weeks, depending on the weather. Dewing promotes the development of microorganisms present on the flax stalk. These develop enzymes, which cause the breakdown of pectins and hemicellulose in particular. After this retting process, the flax color is gray-green or gray-yellow. The advantage of this process is that the resulting degradation products are immediately fed back into the material cycle on the field, which in turn helps to save fertilizer. In the event of prolonged rain, however, the retting can continue in an uncontrolled manner and thus the flax fibers can become damaged. In extreme cases, this can lead to the total loss of the crop. Repeated turning with special machines accelerates the retting process and it proceeds more evenly. Dew retting is the predominant method on around 75% of the world's cultivated area (Fig. 6.20).

Water retting takes place in open waters, basins or tanks. The retting time depends on the temperature and is 3–5 days in hot water (28–40 °C) and 2–3 weeks in cold water. The color of the flax after retting is light yellow. This process is still carried out in China, Eastern Europe and Egypt and pollutes the waters relatively heavily.

6.1.4.2 Chemical Solubilizing

In this process, fiber generation is carried out under heat and the addition of chemicals (e.g. sulfuric acid, chlorinated lime, caustic soda, potassium soap, soda ash). The process time is relatively short (30 min), the process costs are usually high and the fibers are heavily stressed by the chemicals. This leads to poor spinnability and insufficient yarn properties. However, the process can be easily controlled at any time.

6.1.4.3 Mechanical Solubilizing (Green Flax Solubilizing)

Mechanical solubilizing is performed directly after harvesting, thus eliminating the need for retting. Flax can be debarked green, but should contain 7–14% of moisture. Straw, which is too dry, may cause fiber loss and straw, which is too wet, cannot be separated from the wood. Decortication is carried out after breaking, scutching and hackling either on the sliver or on the roving spun from it. Chemical solubilizing is then used to extract the pectins from the fibers. Green decorticated flax can only be spun into coarse yarns.

6.1.4.4 Other Methods

In order to improve fiber quality on the one hand and to protect the environment on the other, new solubilizing processes have been developed. These include high-frequency treatment, in which pectins and lignin are removed by evaporating the water in the flax stalk. In the steam pressure process, the biomass is heated with saturated steam at high pressure. After a reaction time, the pressure is abruptly reduced to atmospheric pressure. This results in both chemical and mechanical solubilizing. This process is

Fig. 6.20 Flax in a field for dew retting in France (Mark Felske)

used for the production of cottonized flax fibers. Enzymes and ferments can also be used to dissolve the pectins. This can considerably shorten the retting time. In the surfactant solubilizing process, a reagent is used which removes pectin, hemicellulose and lignin by dispersing and emulsifying processes.

After retting, the flax is dried and either packed in bundles or pressed into round bales of about 100 kg and sent to the scotching mill for further processing.

6.1.4.5 Rippling

The dried retted flax is rippled, where the seeds are extracted. In the process, branches and leaves are cut off at the same time (Fig. 6.21). When using hot water retting, rippling occurs before the retting process.

6.1.4.6 Breaking

After the retting process, the flax straw is dried. This is done either by placing the bundles outdoors or by means of channel dryers. Water-retted flax has a high moisture content and is therefore crushed before drying.

To remove the stalk parts, the flax is bent or broken to shred the wooden layer of the stalk ("shives"). These shives are removed without damaging the flexible fiber bast. For this purpose, the flax straw is passed between pairs of toothed, intermeshing rollers. The result is flax and shives which are bent or crushed (Figs. 6.22 and 6.23).

6.1.4.7 Scutching

After buckling, the flax stalks are scutched. In this process, the broken flax straw is freed from the still adhering wood particles and short fibers, so that the bast fiber bundles are exposed. During scutching, the shives and short fiber material

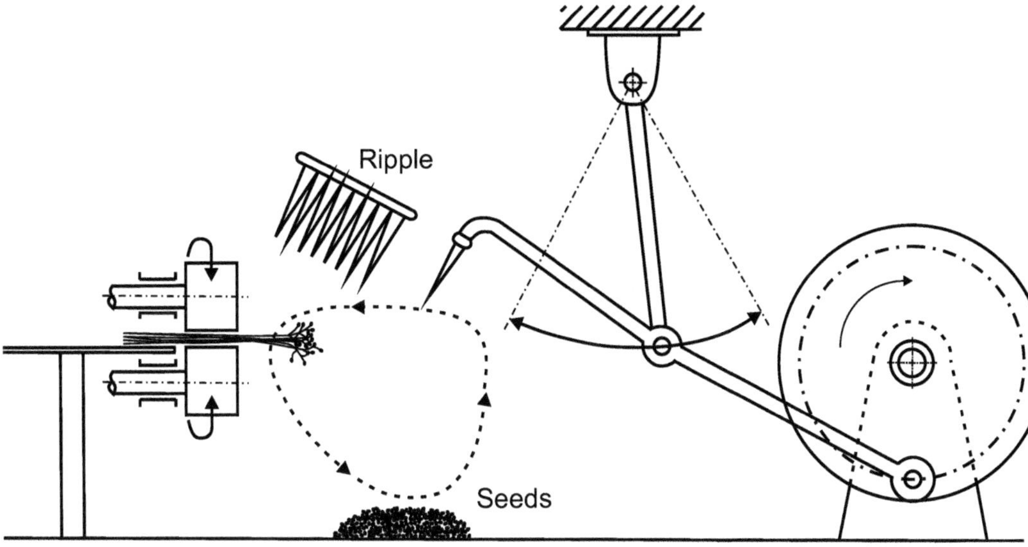

Fig. 6.21 Design of a rippling machine

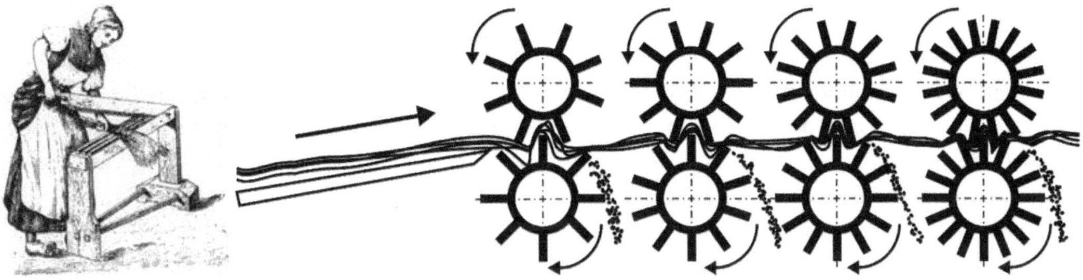

Fig. 6.22 Principle of breaking (left: historical; right: modern)

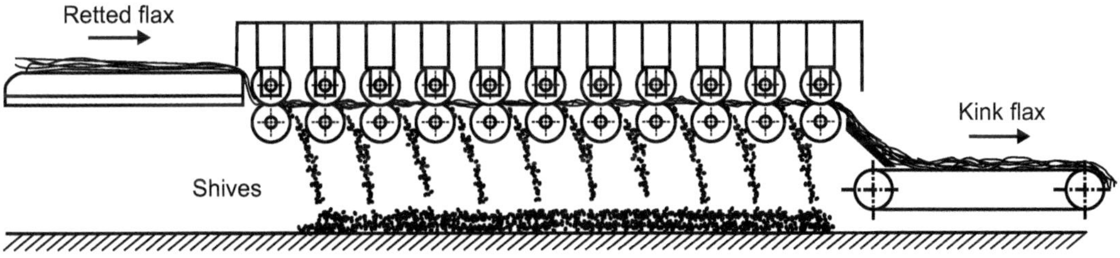

Fig. 6.23 Structure of a breaking machine

(scutching tow) are knocked out of the hanging material by rapidly rotating rails (beater blades) mounted on a shaft (Figs. 6.24 and 6.25).

This leads to

- Scutched flax: long fibers freed from wood, pith and bark particles
- Scutching tow: short fibers and shives

The operations of rippling, breaking and scutching can be combined. The shives are

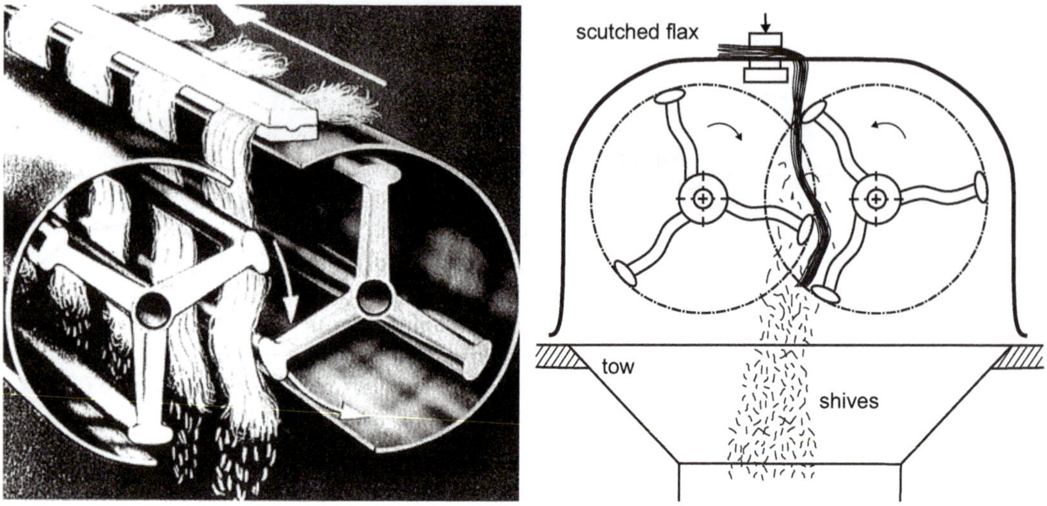

Fig. 6.24 Principle of scutching

removed from the scutching tow and processed in the tow spinning mill.

6.1.4.8 Hackling

Flax enters the spinning mill as scotched flax. Before spinning, the exposed fiber bundles are divided by hackling and freed from any wood particles and short fibers still adhering to them. Clamped oscillating fiber beards are moved up and down on the hackling machine and combed out by circulating rows of pointed needles (hackle bars). In this way, the short fibers (hackling tow) are removed and the bast fiber bundles are broken up. The tow remains caught in the hay bars and is brushed out (not shown in Fig. 6.26). From 1000 kg of flax stems, 65 kg of flax and 35 kg of tow are obtained (Fig. 6.27).

Hackled flax is available as a technical fiber in parallel layer and is processed in the long flax spinning mill. After sliver formation and homogenization of the sliver by several drawing passes, a roving is produced on the roving frame. Fine spinning is carried out on ring spinning machines using either the wet or the dry spinning process.

Figure 6.28 shows the most important steps in the extraction of flax fibers

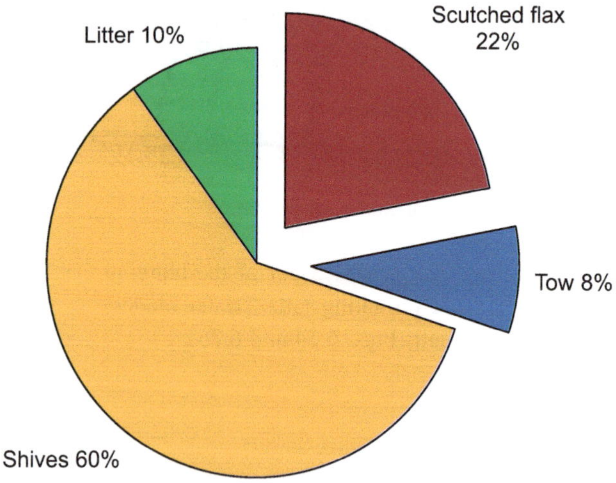

Fig. 6.25 Scutching yield

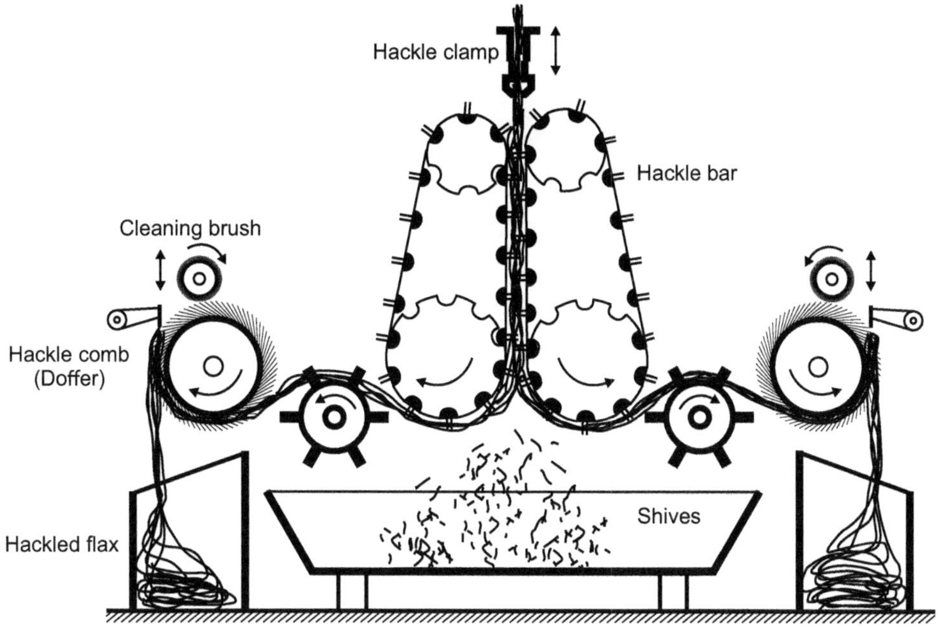

Fig. 6.26 Hackling machine in cross-section

6.1.4.9 Yarn Production

In the flax tow spinning mill, both scutching tow and hackled tow are processed into yarns, usually in a blend. The fibers in the tow are in an entangled state and heavily interspersed with shives and dust and therefore need to be cleaned. This is done on tow cleaning machines which is followed by parallelization and sliver formation on the card, levelling and refining on draw frames and roving formation on the roving frame. Fine spinning is carried out in the same way as in the

Fig. 6.27 Mass balance of the hackling process

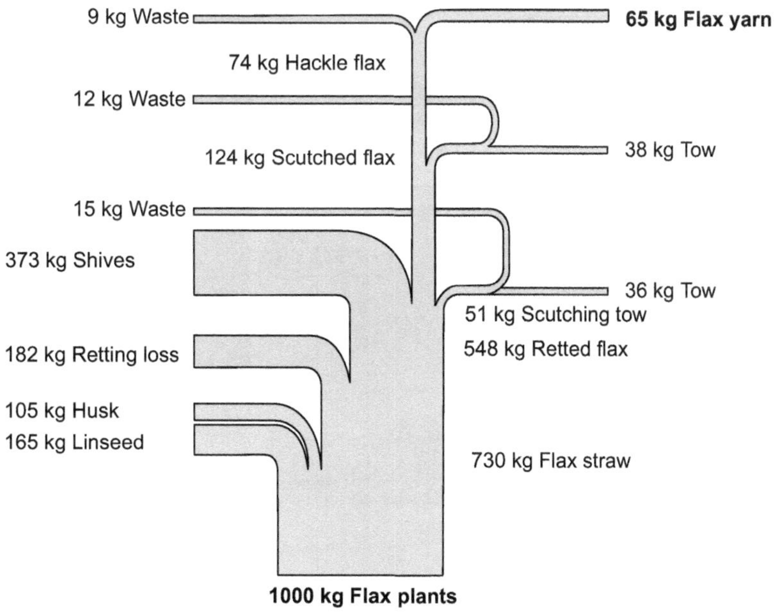

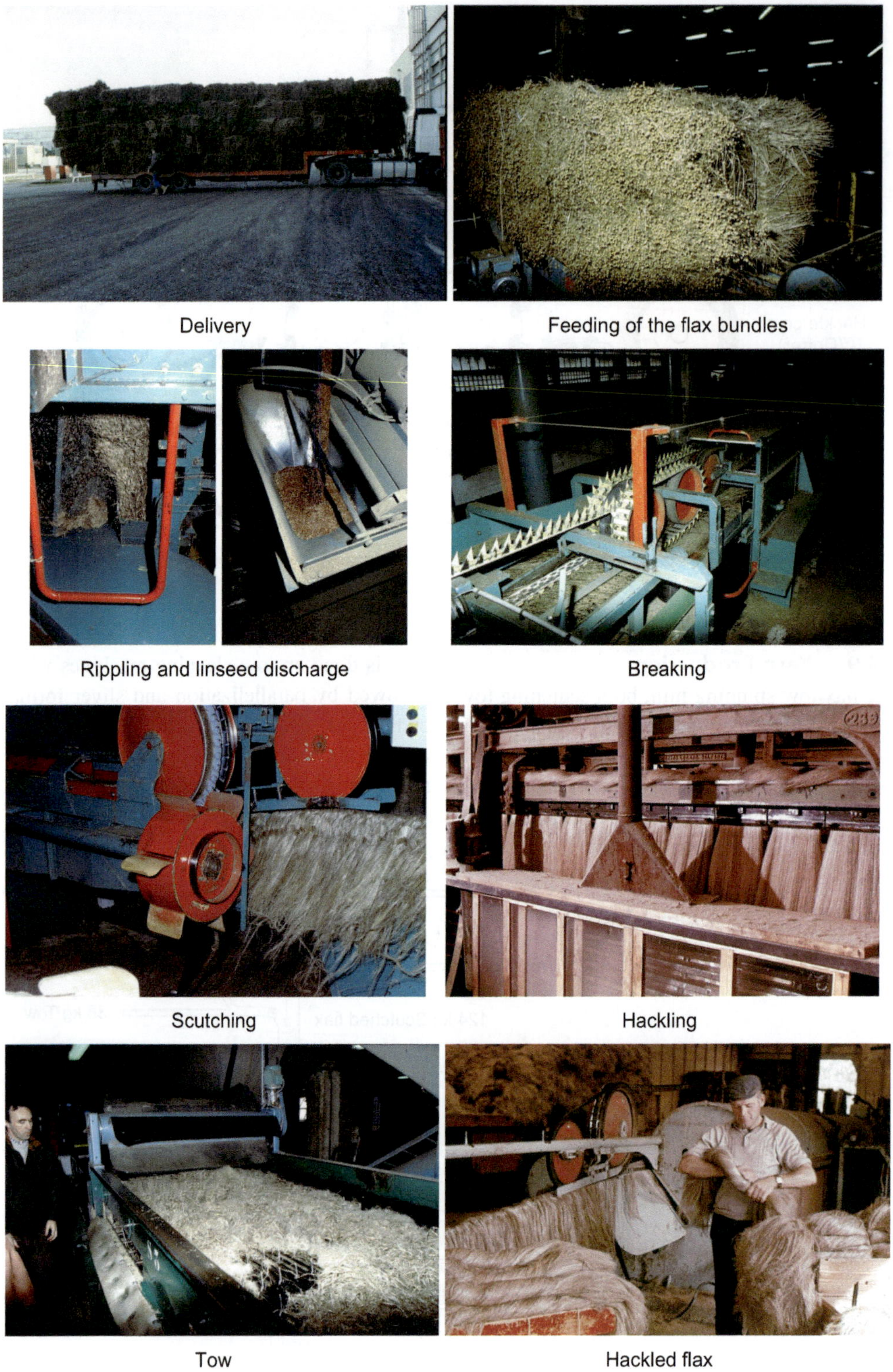

Fig. 6.28 The most important steps in flax fiber extraction

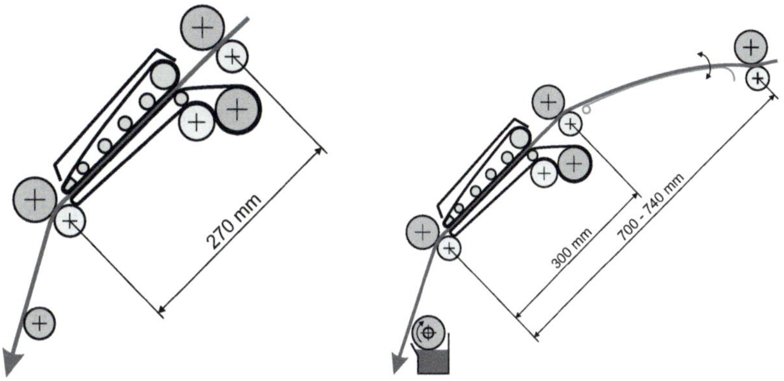

Fig. 6.29 Ring spinning drafting system for flax processing (left: dry; right: wet)

long fiber flax spinning mill using the wet or dry spinning process.

Wetting the flax fibers increases their strength (Fig. 6.29). This means that the required yarn count can be achieved with fewer fibers. This makes it possible to spin finer yarns.

The yarn is then spun on slightly modified ring spinning machines (Fig. 6.30).

Before industrialization, yarn production was home work and was carried out by women using spinning wheels. When the women were old, they would be called "spinster".

6.1.5 Properties

The most important characteristics of flax fibers are length, fineness and degree of maturity and color. In terms of fiber length, a distinction is made between technical fibers consisting of fiber bundles and single fibers. The length of the

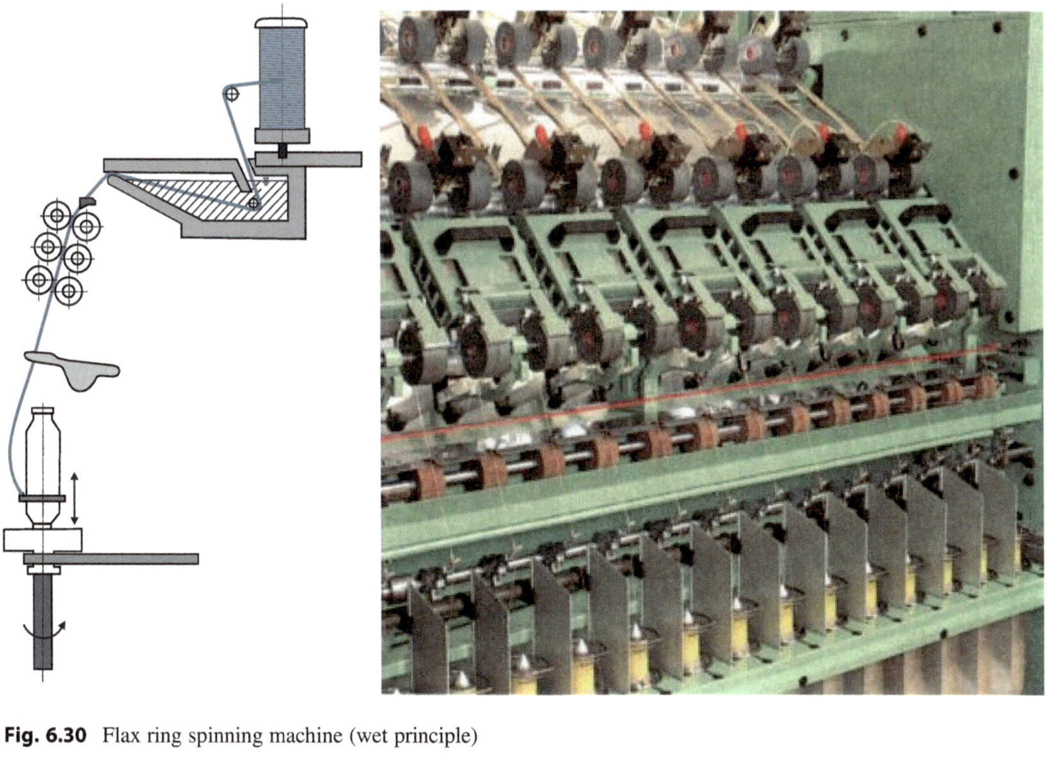

Fig. 6.30 Flax ring spinning machine (wet principle)

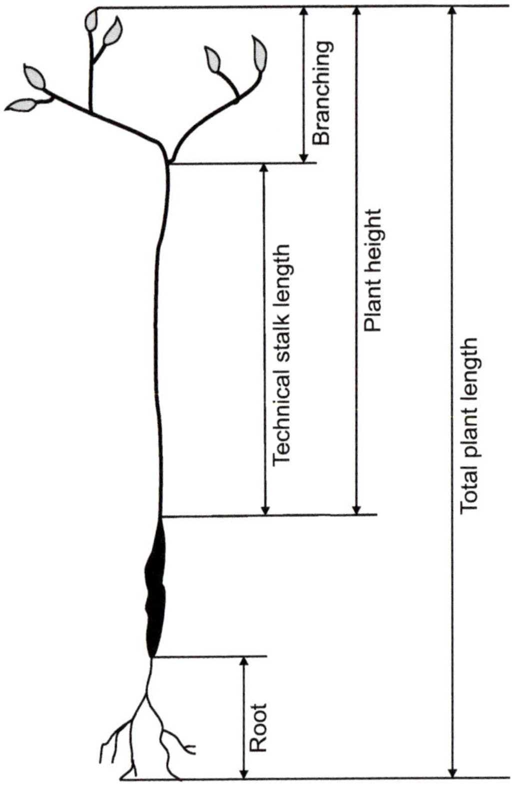

Fig. 6.31 Stalk sections of a flax plant

technical fiber is between 200 and 800 mm. Hackled flax with a fiber length of less than 400 mm is known as tow. The fiber length depends on the flax variety, the growing conditions and the time of harvest. The solubilizing or retting process and mechanical preparation also have an influence. The stalk zone from which the fibers are taken is decisive (Fig. 6.31 and Table 6.3).

The fiber fineness for the technical fiber normally is in a range of 10–40 dtex. It depends on the number of individual fibers. For the single fiber, the fineness is 1–7 dtex. It depends on the stalk section, the time of harvesting, the flax variety and the stalk thickness. Increasing stalk maturity causes the fiber to lignify and reduces the divisibility of the fiber bundles and thus the fineness. If harvesting takes place at the green stage, finer fibers of lower strength are obtained. The diameter of the individual fiber is 15–37 μm and averages 20–25 μm. The density is 1.43 g/cm^3 (raw) or 1.52 g/cm^3 (bleached), Table 6.4 and Fig. 6.32.

Flax fibers have a high tensile strength of around 20–65 cN/tex. The fineness wet strength is 105–120% of the dry value and is the highest of all natural fiber materials. The elongation in the dry state is 1.5–4.0% and is the lowest elongation of all major fiber materials. This is particularly disadvantageous during processing. The elongation in the wet state is 110–125% of the dry value. The degree of maturity depends on the time of harvest. A high fiber yield with good fiber quality is achieved at the stage of yellow ripeness. The most important influencing factors are the flax variety, the distance between the plants and the nutritional conditions. The color of the flax depends on the retting method used. In the case of water retting, it is yellowish (flax blond, straw yellow), in the case of dew retting silvery-grey to greenish. Chemically roasted flax is yellowish to whitish. The luster varies depending on the flax variety and retting method. Cold water and dew retted flax have a natural linen luster, since accompanying substances (plant glues, fats, waxes) are largely retained. Over-roasted flax, on the other hand, is dull. Luster may be increased by finishing (bleaching, calandering). Flax has a cool feel, which is caused by the plant glue which surrounds the individual fibers. It can be altered by bleaching. Flax is also characterized by a high tendency to crease, which is marketed in fashionable clothing as "noble crease".

Table 6.3 Properties of the elementary fiber as a function of stalk height

	Number of fibers in cross-section	Length [mm]	Width [μm]	dtex	Content [%]
Upper stem quarter	200–600	38–42	11–16	2.0–1.0	15–20
Second stem quarter	600–1200	33–38	17–19	2.6–2.4	22–30
Third stem quarter	800–1300	28–32	19–25	4.0–2.6	25–33
Lower stem quarter	100–800	20–26	30–33	6.3–4.2	12–22

Table 6.4 Properties of flax fibers

Length [mm]	Fineness [µm]	[dtex]	Tenacity [cN/tex]	Elongation [%]	Density [g/cm^3]
13–60	8–38	1–7	20–65	1.5–4	1.43–1.52

The decomposition temperature of flax is 260–320 °C, with a slight browning occurring from approx. 220 °C onwards. The plant glue, which results in a smooth surface of the fibers and strong cohesion of the elementary fibers within the fiber bundle (no air inclusions), results in very high thermal conductivity. This in turn results in a low heat retention capacity of linen fabrics, which in turn results in the characteristic "cool" feel. Even the Knights Templar knew to appreciate this. Thus the Rule of the Order, written in 1128 by Bernard of Clairvaux, states in Article 67: "The choir shirts shall be of wool, but because of the great heat in the eastern countries, anyone who wishes may wear a linen choir shirt, from Easter to All Saints' Day; the linen choir shirt, however, is permitted only "out of charity," not by right."

Flax is resistant to boiling and has a high swelling and absorption capacity. The moisture absorption in a normal climate is 8–10%, at 100% humidity approx. 17%. Bleached linen is more absorbent than unbleached linen because the plant glue is partially removed by the bleaching process.

6.1.6 Classification

Flax is not classified according to standards like cotton. Instead, it is traded according to samples. The quality of flax is determined subjectively by the buyer. Color, purity, luster, length, fineness and strength are the most important criteria. Because these quality characteristics depend heavily on the preparation process, there are no precisely defined standard types as there are for cotton (Fig. 6.33).

6.1.7 Production

Depending on the climate of the growing region, either fiber flax or oil flax is cultivated. The longer fiber flax is mainly found in colder regions, e.g. Russia and Belarus, but also in Western Europe. In areas with warmer climates, faster-maturing oil flax is grown. These include, for example, Egypt and China.

World production of flax fibers amounted to around 1 m tons in 2020. Production figures fluctuate greatly due to fashion trends and other factors (e.g. subsidies), Fig. 6.34. The total area under cultivation worldwide was about 285,000 ha in 2020. So the fiber yield was around 3.4 t/ha. This is about 4–5 times as much as can be obtained from cotton cultivation. However,

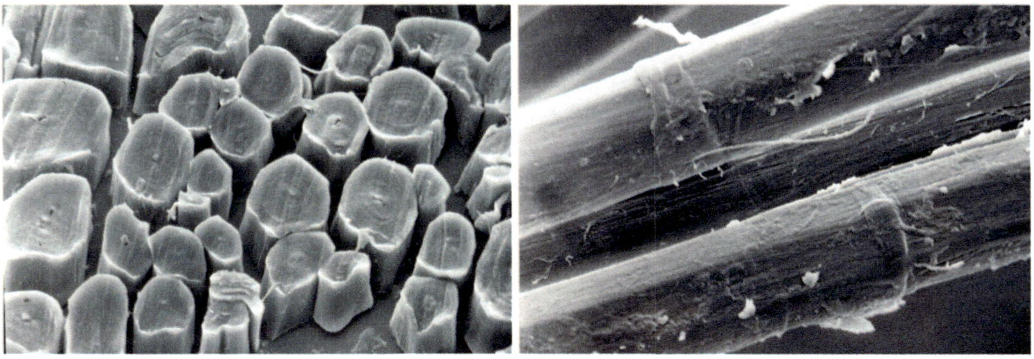

Fig. 6.32 Cross-sectional and longitudinal view of flax fibers

Fig. 6.33 Classified flax at the trader's premises

depending on the country, yields vary widely from 900 kg/ha (Belarus) to 7000 kg/ha (France).

Flax was grown in 32 countries in 1961, as Fig. 6.35 shows. Because production volumes are subject to large fluctuations due to climatic influences and changing fashions, flax is still produced in about a dozen countries today (Fig. 6.36).

The most important cultivation countries today are shown in Fig. 6.36. France (85%) dominates by far, ahead of Belgium (9%), Belarus and Russia, each with a 4–5% share of the world market. China, dominant until 2005 with a production of up to 700,000 t, no longer plays a major role, as do all the other flax-producing countries.

The main processor of flax is Italy with more than 250 companies, followed by France (more than 50 companies) and Germany (European Linen & Hemp, 2021).

6.1.8 Trade

Important trading centres for flax are Riga, Leningrad, Lille and Kortrijk. The trade names for flax are distinguished according to

- origin (e.g. Russian, French, Belgian flax with the addition of the growing districts, e.g. Kortrijk, Namur),
- method of production (green or retted flax),

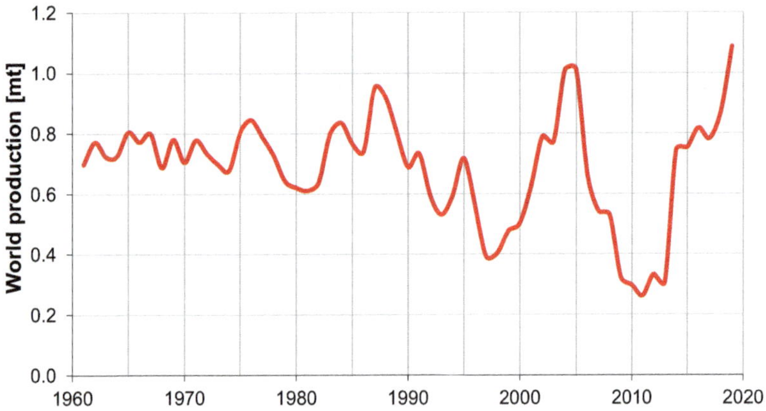

Fig. 6.34 World production of flax fibers (data: faostat, 2022)

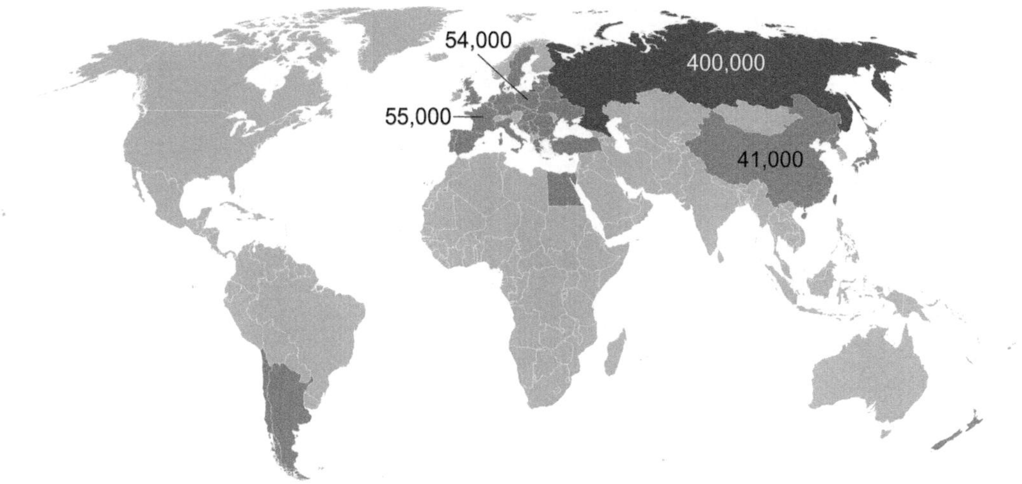

Fig. 6.35 Production of flax 1961 in [t] (data: faostat, 2022)

- retting processes (e.g. Slanetz—flax of Russian origin obtained by dew retting, Motschenetz—flax of Russian origin obtained by water retting)
- Degree of processing (buckling, swinging, stalking flax)

Typical prices are 3–6 €/kg depending on the quality of the fibers (Fig. 6.37).

6.1.9 Applications

Due to their special properties, flax fibers are mainly used for clothing as well as house and home textiles (85%) and so far only a small amount for technical textiles (15%). Flax is processed purely as long fibers in the traditional linen industry into linen yarns, linen damask, tableware and towels (low hairiness) as well as decorative and upholstery fabrics. Flax short fibers are spun in blends with cotton, wool,

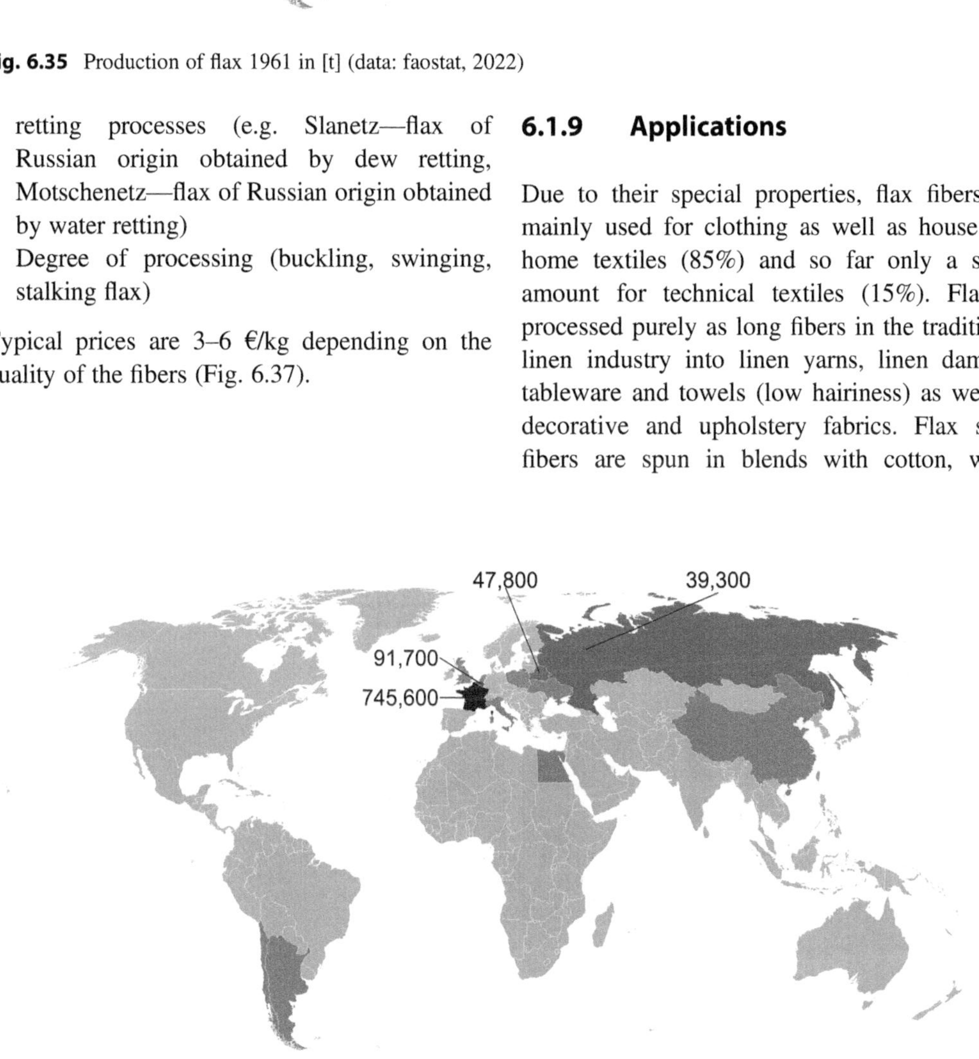

Fig. 6.36 Current flax-growing countries and production volumes in [t] (data: faostat, 2022)

Fig. 6.37 Hackled flax

viscose, polyester or polyacrylonitrile. The proportion of flax is often in the range of 5–50%.

The appearance of linen is often imitated, in particular the uneven yarn structure, the luster and the color. According to the Textile Labelling Act of 1986, only fibers from the stem of the flax plant "Linum usitatissimum" may bear the designation "flax" or "linen". To protect consumers and flax products in the trade, the German flax industry introduced a quality mark in 1927, the so-called "Schwurhand" (swearing hand). Today, products made from flax in Germany are marked by a newer version of this mark, the so-called "Leinensiegel" (linen seal), Fig. 6.38.

Pure linen is a protected designation for textiles made of 100% linen; in the case of half-linen, the warp must be made of pure cotton yarn and the weft of 100% linen yarn and the linen content must be at least 40% in total. Textiles with a linen content of at least 30% that are combined with other materials, e.g. silk, cashmere or wool, can bear the Linen Plus seal.

There are also European linen seals, e.g. the Masters of Linen, which confirms the European origin of flax products.

Due to fashion trends, flax fiber experiences a major upswing in the clothing sector at irregular intervals, sometimes lasting several years. Typical areas of application in the textile sector are:

- Fabrics for shirts, blouses, outerwear (also workwear),
- table and bed linen,

Fig. 6.38 Leinensiegel (Schwurhand-Zeichenverband e.V)

Fig. 6.39 Products made of flax (typical fabric made of flax with thick and thin places in the yarn, sailing rope)

- glass tissues, handkerchiefs,
- handicraft fabrics,
- decorative and upholstery fabrics,
- wallpaper and mattress ticking,
- money and mail bags,
- cabinet blinds and carpet backing (Fig. 6.39).

In the technical sector, flax is used as a substitute for asbestos, glass fibers and rock wool. Other possible applications include the reinforcement of cement-bound building materials, thermal and acoustic insulating materials, compression molded parts (composites), friction linings (brake and clutch linings). The decisive factor here will be that flax fibers can be produced and processed cost-effectively in the required quantities and with consistent quality. In terms of sustainability, linen products can be produced locally as flax grows all over the world.

6.2 Hemp

Botanically, hemp belongs to the Cannabinaceae family, of which mainly Cannabis sativa is cultivated in about 20 cultivated forms (Fig. 6.40). Its abbreviation is HE.

6.2.1 History

Hemp is already known around 1500–1000 BC in China and India. It is first used for the production of intoxicating drinks and medicines from the fruits and only later cultivated as a fiber plant.

Fig. 6.40 (Cannabis sativa; WIKI01) and hemp field (Aleks, 2000)

Fig. 6.41 Enlightened Chinese scientist (ca. 2700 BC) with hemp plant and advertisement for sails and ropes made of hemp (early twentieth century)

From India, hemp culture first reached southern Russia via the Scythians (1000–500 BC) and from there northwards to the Slavs and Germanic tribes. A southern route led hemp from the Black Sea countries via Asia Minor to the Greeks and Romans (mentioned by Pliny the Elder in 100 BC) and from there to the Gauls.

In the Middle Ages, the use of hemp was regarded as a sign of poverty, so it was hardly ever used as a textile raw material. It was not until the advent of sailing that hemp, due to its high wet strength (ropes, sails), became very important and widespread in the sixteenth century (Fig. 6.41). To equip a single sailing ship, 50–100 t of hemp were needed, which had to be replaced approx. every 2 years.

Hemp also reached Chile and North America via Spain, where it was cultivated in Kentucky from 1775. Hemp was cultivated on a large scale, especially in Russia, which thus achieved a dominant position on the European market. For many regions in Europe, hemp cultivation was so economically important that it is still commemorated in the respective coats of arms. Figure 6.42 shows selected examples. The city of Hamme in Belgium has both hemp and flax symbols in its coat of arms, which is unique in this form.

Due to the different growing conditions in the individual growing regions, numerous cultivated forms developed. These originated from the same parent plant, but differ in stem height (e.g. Russian type: 1.5–1.8 m; Italian type: 3–3.5 m).

Hemp was not only used for sails and ropes, but also to seal ship planks. This was already common practice among the Romans in antiquity.

In the nineteenth and twentieth century, hemp, like flax, gradually lost importance due to the advance of cotton and the import of cheaper jute and sisal fibers, even though the first jeans by Levi Strauss were still made of hemp. Before and during World War II, hemp experienced a renaissance as a raw material important to the war effort (Figs. 6.43 and 6.44). Thus, the world production amounted to about 500,000 t in 1938.

After 1945, the cultivation was forbidden in many countries, because, from some sorts, narcotics can be produced. In the USA, until 2019, hemp was not generally forbidden, but, in most states, it was punishable to possess the flowers, capsules or leaves of hemp plants, because hashish and marijuana can be produced from them. Since it is not possible to grow hemp without flowers, capsules and leaves, there was thus no hemp cultivation in those states. In 2019, 47 states in the USA changed their legislation so that today, hemp cultivation is legal.

6.2.2 Cultivation

The stiff, herbaceous stems of the annual, dioecious plant reach a height of 1–5 m depending on climatic and soil conditions. In dense stands, they branch at the top and then have a diameter of

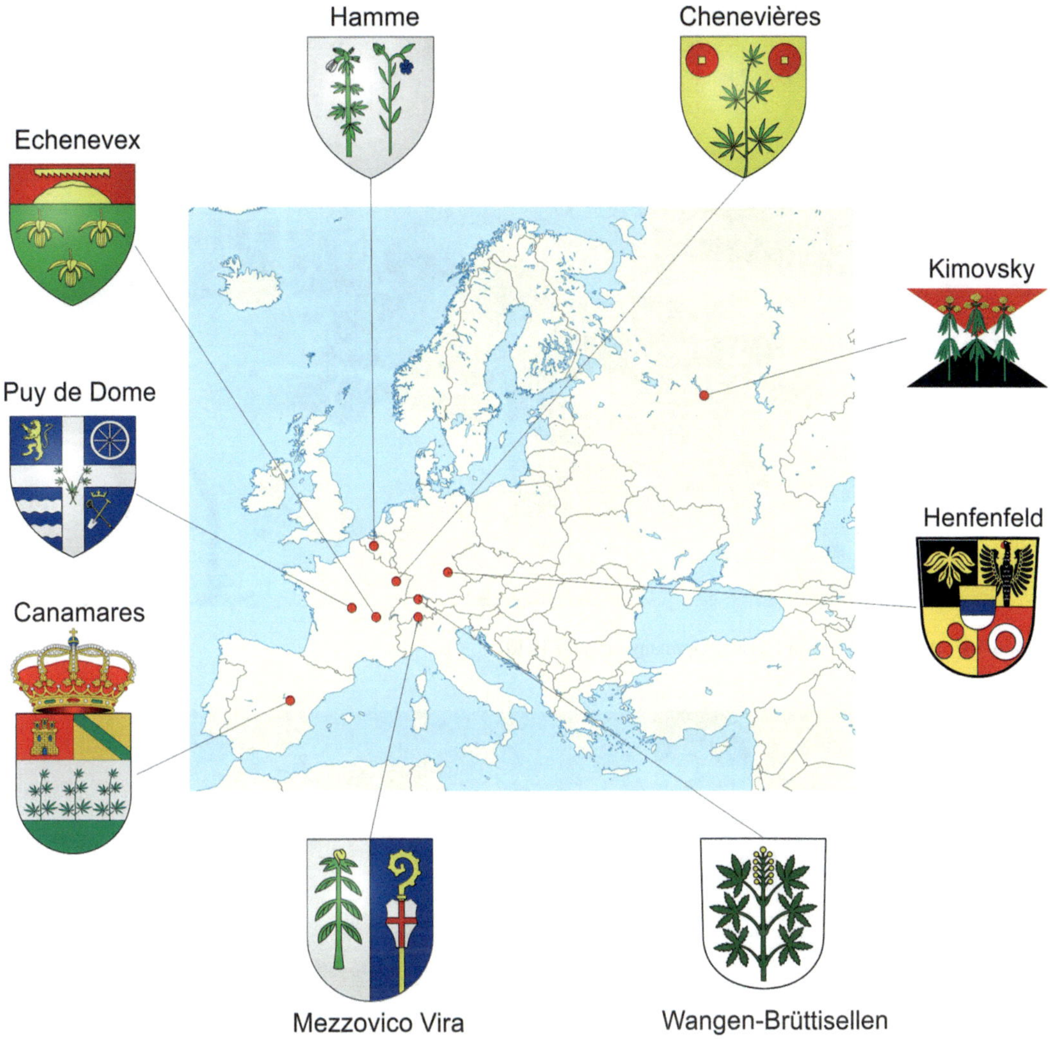

Fig. 6.42 Cities with hemp symbols in their coats of arms

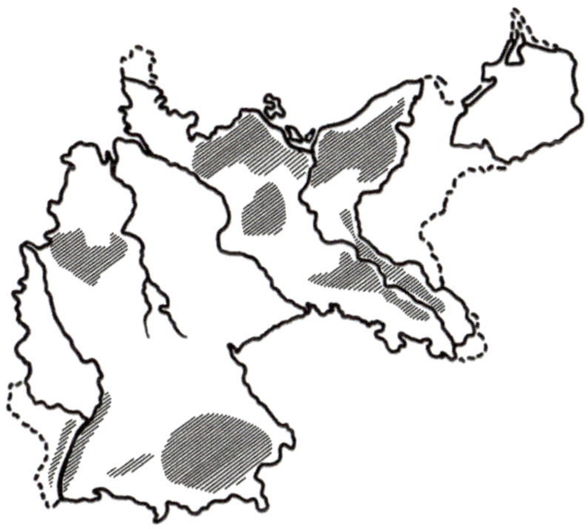

Fig. 6.43 Hemp cultivation areas in Germany 1938

6–20 mm. In nutrient-poor soils, the taproot can reach a length of up to 2 m. Approx. 35–50 kg of seeds are required per hectare.

Hemp plants thrive in almost all climates, but prefer mild, humid conditions. Well-fertilized mineral and moor soils are particularly suitable. Per hectare, approx. 100–120 kg of nitrogen fertilizer and around 200 kg of K_2O fertilizer are required.

6.2.2.1 Crop Growing

Hemp is dioecious, so there are half male and half female plants in a hemp field (Fig. 6.45). The male supplies finer fibers due to its faster growth

Fig. 6.44 Flax processing in Füssen, Germany, around 1940

Fig. 6.45 Female (left) and male (right) hemp plants (Tuason, 2017; Fendersson, 2006)

and matures 2–3 weeks earlier than the female plant. If the seed of the female plants is to be harvested and lignification of the fibers of the male plants is to be prevented, the male hemp is harvested earlier in small-scale cultivation. If, on the other hand, the entire crop is harvested when the seed of the female hemp is ripe, the male plants have already died and are thus lost for fiber production. In large-scale cultivation, therefore, seed collection is usually dispensed with and all plants are harvested in the green state as soon as the male plant has withered. However, this then leads to inconsistent fiber quality.

Fig. 6.46 Growth of hemp plants (Oregon State University; Aleks, 2000)

To avoid these disadvantages, a monoecious hemp variety was bred, in which both sexes are combined on one plant. Furthermore, the development of a simultaneously maturing hemp type succeeded, in which the male plants show the same growth and are just as long viable as the female plants. Compared to dioecious hemp, the simultaneously maturing, monoecious new breed delivers high fiber and seed yields in addition to a more uniform and consistent fiber quality, even on mineral soils.

6.2.2.2 Growth

Sowing takes place from April–May (northern hemisphere). It is carried out with drills at a sowing depth of 4–5 cm and a row spacing of 16–20 cm. Thus, approx. 80–90 kg of seeds per hectare are required. After 3–5 days, the seeds germinate, it then takes 3–4 months until the plant matures (Fig. 6.46).

6.2.2.3 Varieties

In the case of hemp, a distinction is made between fiber hemp (fiber production) and smoke hemp (production of drugs).

Smoking Hemp

It was created by selecting and breeding the Indian hemp (Cannabis indica). Its female inflorescences contain a psychoactive resin whose main component is THC (tetrahydrocannabinol). This is a diterpene with an alcohol function. It is formed by aging processes of tetrahydrocannabinol carboxylic acid. This occurs in the resin of hemp. THC is an essential component of hashish and marijuana (Fig. 6.47).

Fiber Hemp

Currently, only hemp varieties with a THC content of less than 0.2% are approved by the EU for fiber production. These include above all the varieties Fibrimon, Férimon, Fedora, Felina, Fedrina and Futura. They produce a straw yield of 10–14 t/ha, with a fiber content of 25–40%. This results in fiber yields of 2.5–4 t/ha. The plant height is approx. 2.5–4 m.

6.2.2.4 Harvest

The densely standing hemp stalks are harvested in September, when the male plants are ripe. The female plants yield equivalent fiber qualities, but the fiber yield is somewhat lower. Because of the heavily woody lower stem parts of the hemp plant, the fiber is obtained—in contrast to flax—by cutting (either by hand or mechanically, Fig. 6.48).

The drying of the hemp plants takes place in the field. A rippling takes place only if the seeds are to be extracted. This is then done with special threshing machines.

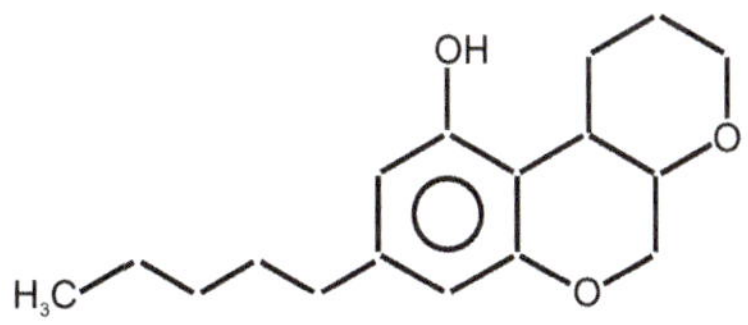

Fig. 6.47 THC formula

Fig. 6.48 Hemp harvest (left: manual; right: mechanical, Aleks (2007))

6.2.2.5 Pests

Among other things, hemp is attacked by aphids, spider mites, the white fly and the corn borer. However, to a much lesser extent than, for example, cotton. The use of special pesticides is therefore not necessary. In most cases, the pests are controlled by biological means.

6.2.3 Structure

A hemp stalk has a similar structure to a flax stalk. As with flax, the bast fibers form the technical framework of the stalk and are therefore arranged in a ring. There are two different bast bundle layers: on the one hand, the primary bast fibers of the outer layer, which are formed from the outset and usually united in groups, and on the other hand, the secondary bast fibers of the inner layer, which are formed only during further growth of the plant to reinforce the skeletal substance. They are arranged in individual bundles.

Hemp fibers are hollow fibers consisting of high-strength cellulose, in which many elementary fibers form bast fiber bundles. They are bonded together by natural adhesive substances (lignin, pectins, etc.). These bundles fray during preparation (Fig. 6.49).

The finer primary fibers already begin in the lower part of the stalk and extend to the tip. They are therefore more valuable, when long fibers are to be produced, than the short, coarse, more lignified and therefore less readily divisible secondary fibers. However, these are of particular importance for the extraction of hemp tow and for the production of cottonized bast fibers.

In the lower parts of the stem, often two or more bast fiber rings exist (Fig. 6.50).

The fiber content of the stalk is 16–24%, depending on the variety, soil conditions, stalk thickness, gender of the plants, etc.

The structural properties (fineness, length, width) of the individual hemp fiber cells are largely identical to those of flax single fibers. The divisibility of the bast bundles of hemp, which are similar in structure, is nevertheless considerably more difficult than in the case of flax. This is due above all to the greater lignification of the middle lamella (lignin content 3.3% compared with 2% in flax) and to the irregular boundary of the fiber cross-sections with recessed angles. Hemp is referred to as "rubble-like masonry" and flax as "brick masonry". For this reason, retting is difficult in the case of hemp and hemp fibers are suitable only for producing coarse yarns.

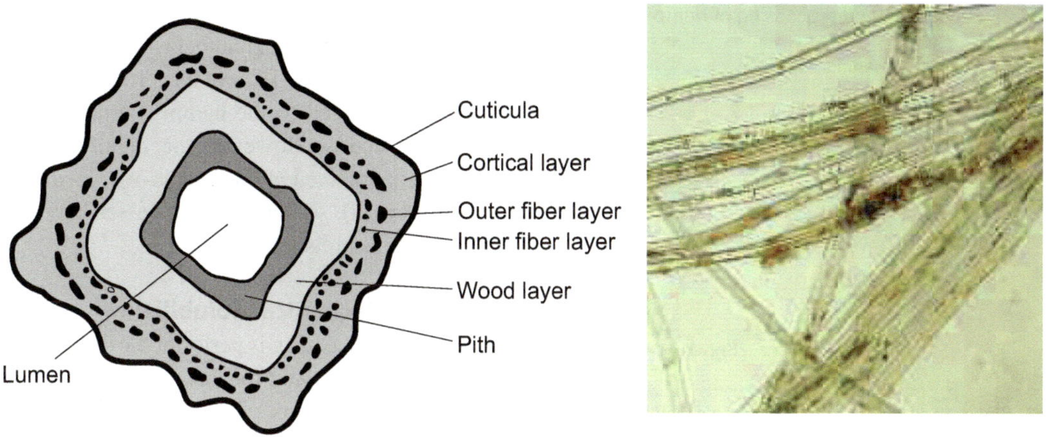

Fig. 6.49 Structure of hemp stem and longitudinal view of the fibers

Fig. 6.50 Cross-sectional view of a hemp stalk at different heights

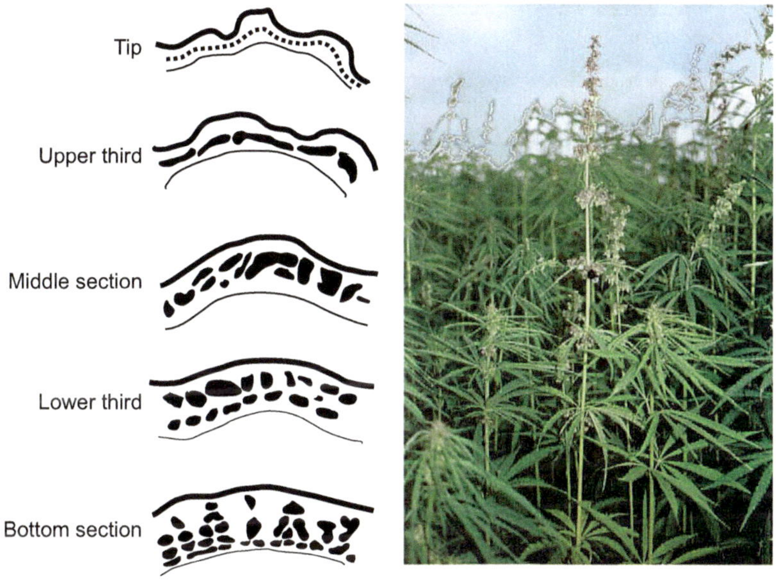

6.2.4 Fiber Extraction

Hemp fibers are processed in the same way as flax fibers, with minor differences due to fiber length and fiber structure (Fig. 6.51). 100 kg of hemp yields approx. 20–40 kg of fibers.

6.2.4.1 Retting

Dew retting takes 6–8 weeks and produces dark-colored, low-quality fibers ("black hemp"). Cold water retting takes only 3–4 weeks and yields good fiber quality. It is carried out in standing or flowing waters (Fig. 6.52). Hot water retting is used by large companies by using flowing water at a temperature of 28–30 °C. For this purpose, the hemp is submerged in water in retting boxes (capacity approx. 350 kg. Retting is carried out according to the countercurrent principle (run of the retting liquor in the opposite direction to the migration of the retting boxes), so that the fresh stalks immediately get into the water enriched with retting bacteria.

After retting, the hemp has a light yellow, bright color ("white hemp"), Fig. 6.53. The retted

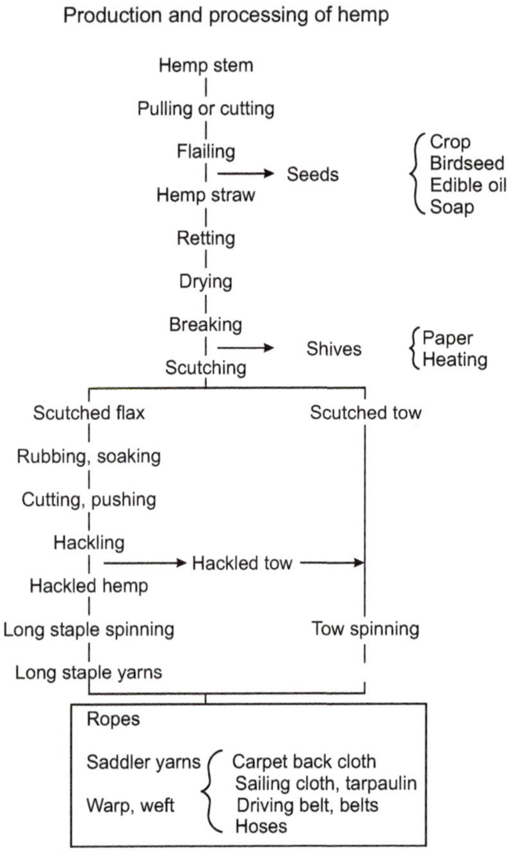

Fig. 6.51 Steps in hemp processing

hemp is then dried standing up in the open air or in special drying chambers.

In order to be able to produce hemp on a larger scale, enzymatic retting processes have been developed in recent years. These processes take place, for example, in an autoclave at 50 °C and then last only approx. 18 h. In this process, certain enzymes are added to the hemp in order to dissolve the pectins. However, the quality of the hemp fibers obtained in this way is only mediocre (color, strength), so that these processes have not yet become established.

6.2.4.2 Mechanical Solubilizing

Mechanical processing is performed using similar machinery to that used for flax (Fig. 6.54). After scutching, the result is scotched hemp (commercially known as raw hemp) with 2–3 m long bast bundles and scotching tow and shives.

Compared to flax, scutched hemp has longer fibers, a higher strength and is more glued than flax. For processing on the hackling machine, the scutched hemp must be cut to a length of 60–75 cm. The hackling process refines and parallels the bast fiber bundles and removes impurities. The result is hackled hemp (long fibers) and hemp tow (short fibers), Fig. 6.55.

For applications of hemp in construction, retting is not required. Instead, the green hemp is directly processed. However, it is difficult to remove the shives, so the material remains very coarse and has a high recoil force, its coloration being blond to ocher. The hemp bundles are dried standing upright in a channel dryer (2–5 h, temperature approx. 70–80 °C). If this has not already been done before drying, the seeds are then

Fig. 6.52 Water retting (Theodor von Hörmann) and dew retting (Čebotarev, 2009)

removed on a threshing drum with beater bars. The long-fibered, light-colored and tear-resistant “green hemp” thus obtained is processed both purely into hemp yarns and in a blend with jute tow into coarse yarns.

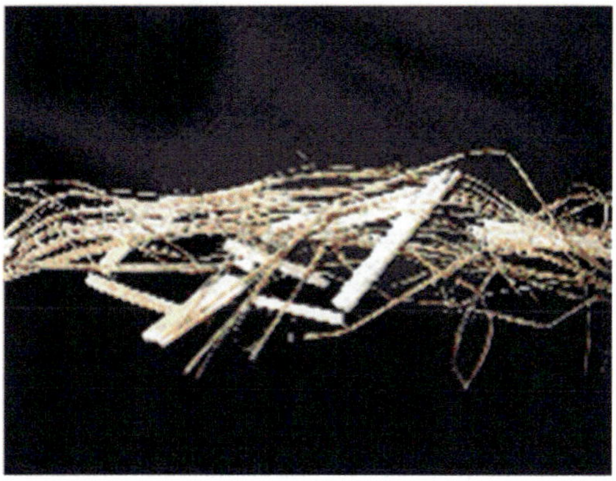

Fig. 6.53 Retted hemp

6.2.5 Properties

The length of the bast fiber bundles ranges from 1–3 m (technical length), depending on the variety. When cut, they average 600–750 mm. The elementary fibers are 10–30 mm long. The fineness of the elementary fibers is 2–6 dtex (15–50 μm), their number per fiber bundle is

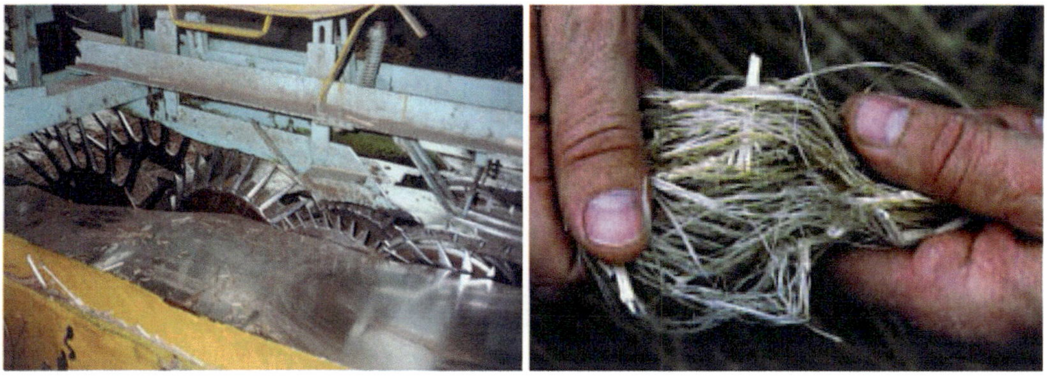

Fig. 6.54 Breaking machine (left) and shives (right)

Fig. 6.55 Hackled hemp

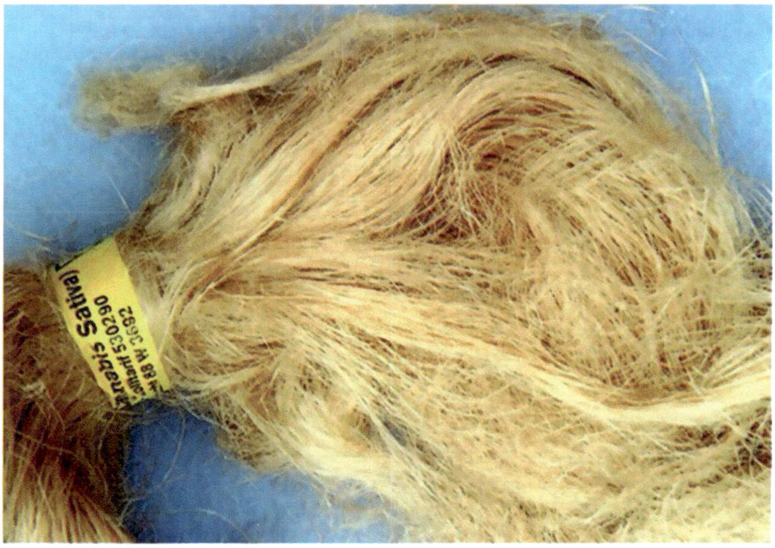

Table 6.5 Properties of hemp fibers

Length [mm]	Fineness [μm]	[dtex]	Tenacity [cN/tex]	Elongation [%]	Density [g/cm^3]
5–55	12–50	2–6	35–70	1–6	1.48–1.5

2–40. The density of hemp is 1.48–1.50 g/cm^3. Dry tenacity is 35–70 cN/tex and depends on variety, growth, retting and processing. The wet strength is 106–108% of the dry values. The tensile strength exceeds the values of flax by up to 20%, making hemp, together with ramie, one of the strongest natural fibers. The elongation in the dry state amounts to 1–6%, in the wet state 104–107% of the dry value are reached. The elasticity of hemp fibers is low, as with all bast fibers. Hemp has a high, mild, natural luster, which depends on the variety and the retting process. Its color ranges from white to yellowish and greenish to brownish, also depending on the retting process. Because impurities can be removed very well, hemp fibers are very pure (Table 6.5).

6.2.6 Production

The fiber hemp used today is mainly grown in China, North Korea, the Netherlands, Russia and North Korea. European hemp cultivation is rather insignificant compared to flax cultivation. Thus, until the 1990s, only France produced hemp on a significant scale in the EU. This is also due to the fact that in many countries of the EU, hemp cultivation was prohibited due to drug laws. With the relaxation of the legislation on the cultivation of hemp for fiber in recent years, Germany, the Netherlands, Lithuania and Austria have joined France in hemp cultivation, of which only France and the Netherlands still play a role. The area under cultivation in the EU is 38,838 ha (2020), of which France accounts for 16,880 ha. This means that the area under cultivation has doubled since 2010. However, part of the land is not used for fiber production, but for the production of medicinal cannabis products. The production volume of fiber in Europe is around 154,239 tons. France has the largest share with 102,580 tons, Poland with 14,290 tons and the Netherlands with 12,790 tons are also important. Worldwide, a total of 245,271 tons of fibers are produced on about 76,484 ha (Figs. 6.56 and 6.57). On average, the fiber yields are 3200 kg/ha, in France about 6000 kg/ha are achieved.

In the Federal Republic of Germany, the cultivation of commercial hemp was only permitted again in 1996. However, it does not appear in official statistics because of the very small quantities produced.

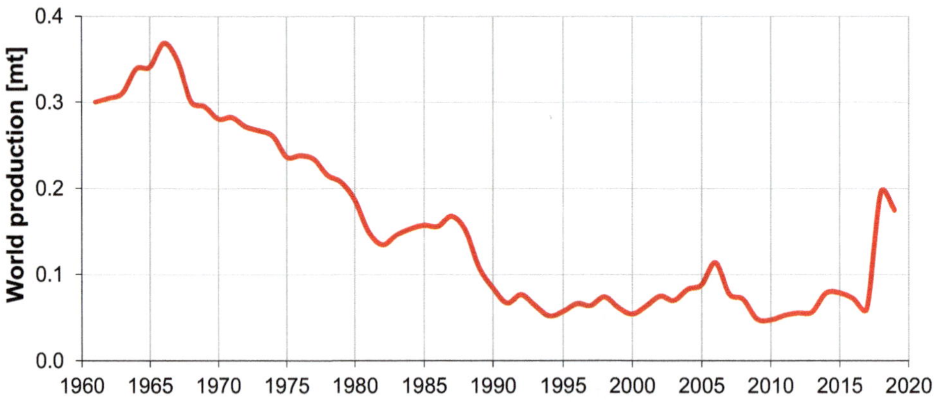

Fig. 6.56 Production of hemp (data: faostat, 2022)

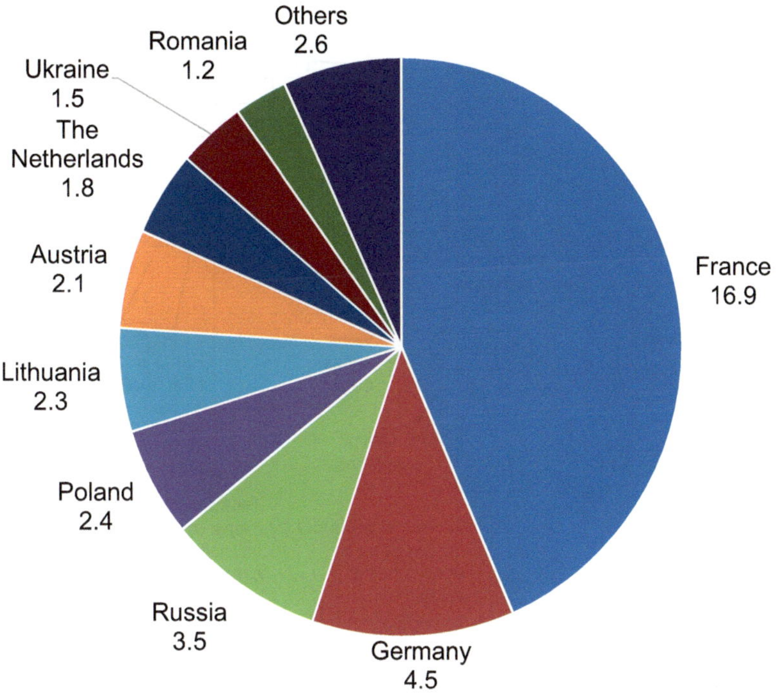

Fig. 6.57 Area under cultivation of hemp producers in Europe in [1000 ha] (data: faostat, 2022)

6.2.7 Trade

As with flax, there is no uniform classification system. Instead, hemp is traded on the basis of quality characteristics such as color, purity, luster, length, fineness and tensile strength. The main trading centers for hemp in Europe are Lille and Bologna. Typical prices range from € 5–9/kg.

6.2.8 Areas of Application

Hemp is a very versatile plant and, apart from the production of fibers, can also be used for the production of paper, oils, tea and animal feed (Fig. 6.58).

Hemp fibers may be processed pure or in blends (mostly with flax). Due to the poor divisibility of the bast fiber bundles, only relatively coarse yarns are usually produced (hemp: 500 dtex; flax: 85 dtex). Hemp is therefore rarely used for clothing (e.g. jeans, shirts) and mostly for technical applications. These include tents, tarpaulins, bags, nets, twine, ropes and sealing material. There are also suppliers of hemp underwear and other clothing items that are otherwise mostly made from cotton. In addition, hemp yarns can be used as a base warp for carpets as well as for the production of sewing threads for shoe manufacturing (Fig. 6.59).

6.3 Jute

The jute plant belongs to the lime family and there to the species Corchorus. Corchorus capsularis and Corchorus olitorius are of greatest importance for fiber production. They grow wild, but are also cultivated (Fig. 6.60). Jute is the most important bast fiber in terms of volume, with an annual production worldwide of around 3 m tons. Its abbreviation is JU.

6.3.1 History

Jute has been cultivated for several thousand years in India for fiber production, as a vegetable plant and for medicinal purposes. At the end of the eighteenth century, jute also became known in

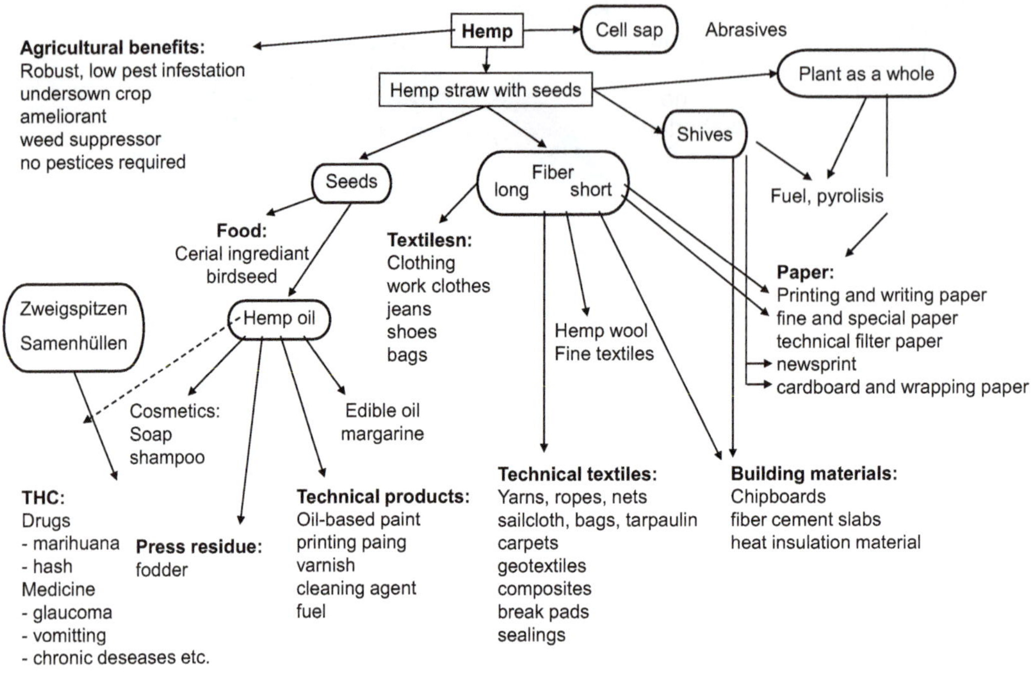

Fig. 6.58 Possible hemp uses acc. to Herer and Bröckers (2017)

Europe and from 1828 to 1832, experiments were carried out with the "new" fiber on mechanical spinning machines in Scotland. The Crimean War (1854–1856) and the American Civil War (1861–1865) led to shortages in the supply of cotton, flax and hemp for the textile industry. Therefore, jute was imported and processed, especially from India. For example, the first German jute spinning mill was established in Braunschweig in 1861.

After cotton, jute is the most important natural plant fiber used in the production of yarns. However, it often does not appear in statistics on textile raw materials because jute is mostly used

Fig. 6.59 Cord and rope made of hemp

Fig. 6.60 Jute plant and jute field before harvest (Manners, 2017)

for technical purposes, e.g. packaging material and carpet backing, and is not traded on stock exchanges.

6.3.2 Growth

Jute is an annual, herbaceous plant. It grows to a height of 2–5 m and thrives mainly in tropical and subtropical countries. It needs a humid and hot climate as well as a nutrient-rich and well-drained soil. One week after sowing (February to May) the seed emerges. When the plants are 30 cm high, they are separated. After 3–4 months, the flowering starts. The flowers develop into capsules (pods) and after another 4 weeks, the harvest begins, even before the capsules mature.

6.3.3 Harvest and Preparation

The 1.5–3 m long jute stalks are cut. This is followed by drying in the field, the leaves are removed by beating and then ginned or rippled like hemp (Fig. 6.61).

Water retting of jute is the dominant pulping process (Fig. 6.62). Thereafter, as in the case of flax and hemp, the fibers are obtained by mechanical preparation (swinging, panting).

The duration of water retting depends on the type of water, its temperature (ideally 26 to 30 °C) and the maturity of the stems. Normally, this process takes 10 to 25 days.

The water and energy consumption is very high with industrial water retting. Therefore, spray retting is also used as an alternative. In this process, the green stalks are first debarked and dried. They are then sprayed with water on retting stations. Exposure to air and water shortens the retting process. Brushing and beating of the dried bast opens the fiber bundles and smoothes the fibers, at the same time removing adhering impurities.

After retting, stripping (peeling) is carried out by hand. The bast is then loosened by breaking or tapping, and the fibers obtained are washed out in water and then dried. In order to make the jute fibers pliable, they are sprinkled with water, petroleum or oil, this is also called batting. They are then squeezed and dried (Fig. 6.63).

Jute is mostly grown in small fields. Processing is also often done on small farms and transport to market is often by very simple means of transport (Figs. 6.64, and 6.65).

Fig. 6.61 Jute bundles in a field in Bangladesh

6.3.4 Structure

The stem structure is similar to that of flax. The major difference is the position of the fiber bundles in the bark layer (Fig. 6.66).

The fiber strands consisting of the individual fiber bundles run in a star shape to the edge of the stem. During longitudinal growth, the primary fibers develop to reinforce the stem and then the secondary fibers grow, which lie between the primary fibers and the wood layer.

The chemical composition of the elementary fiber is similar to that of flax. The lignin content of jute is 12%, the highest of all bast fibers. It leads to strong lignification.

Water retting (Ganguly, 2017)

Drying (Auyon, 2011)

Fig. 6.62 Steps of jute processing

Fig. 6.63 Extraction and processing of jute

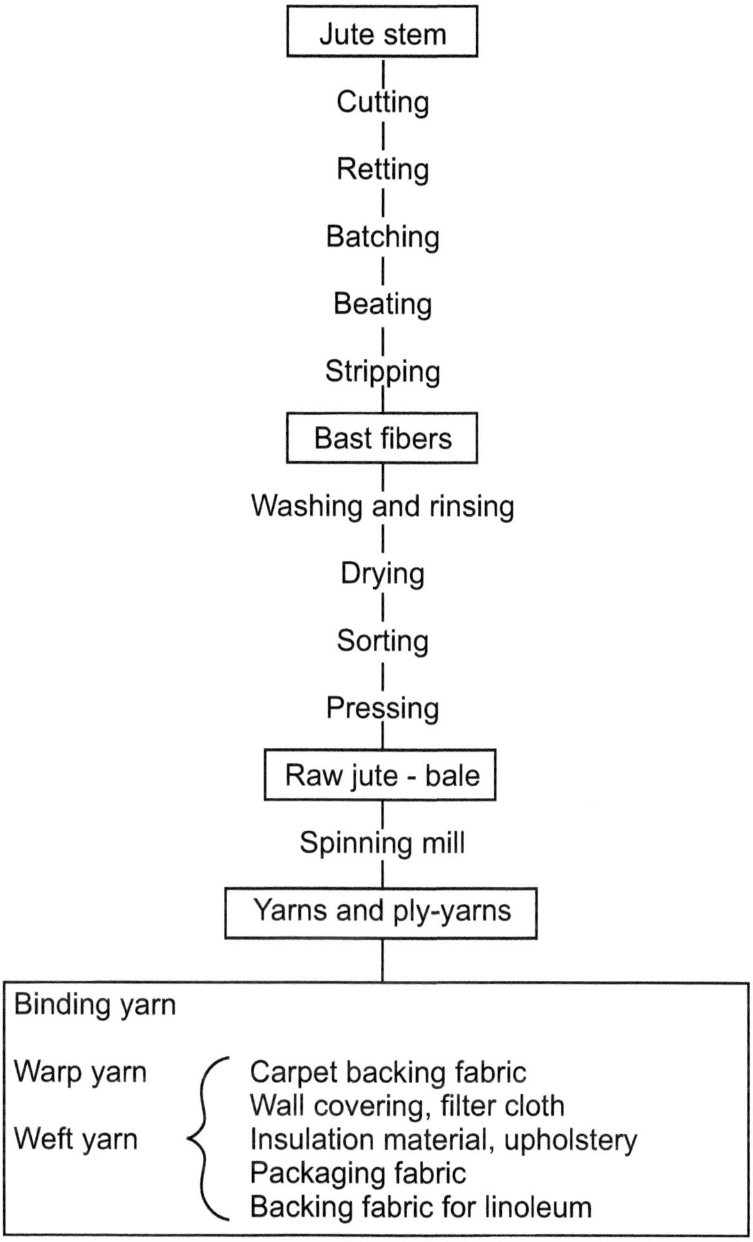

The morphological structure is similar to that of flax and hemp. However, jute has lignin not only in the middle lamella, but also in the layers of the elementary fibers (Fig. 6.67).

6.3.5 Properties

The length of the bast fiber bundles (technical length) is 1.5–3.0 m, while the length of the elementary fibers is only 1–5 mm. The elementary fibers have a diameter of 15–25 μm, which corresponds to 2–3 dtex. The number of elementary fibers per stalk is 6 to 20. Good qualities have a mild natural luster similar to that of linen. In the raw state, jute appears in various shades, usually white, yellow or reddish to light brown and darkens on exposure to light. Owing to its high lignin content, jute is difficult to bleach. Due to the preceding processing steps, jute is only

Fig. 6.64 Transport of the raw jute (Touhid, 2017)

Fig. 6.65 Transporting jute to market (Meraj, 2015)

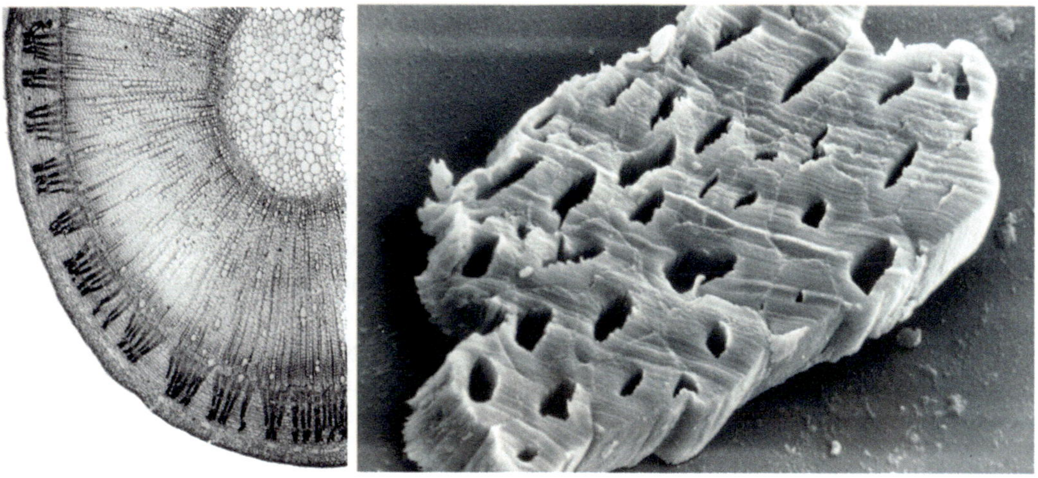

Fig. 6.66 Structure of the jute stalk (detail) and jute fiber bundle

slightly contaminated. Its density is 1.44 g/cm^3, the tensile strength is the lowest of all bast fibers (Table 6.6). This is due to the high lignin content and the short fiber length of the elementary fibers. The fineness strength in the dry state is 30–34 cN/tex, the elongation 0.8–2%. In the wet state, 99–104% of these values are achieved. Moisture absorption is 12.5% in standard climates and up to 35% in tropical climates (24 °C, 95% r. h.). The water retention capacity is 45%.

6.3.6 Production and Trade

A total of around 3 m tons of jute fiber are currently produced each year. 1985 was a record year, with a harvest of 4.4 m tons, after particularly large quantities of jute were grown to take advantage of the high prices at the time (Fig. 6.68). The main producing countries are India (1.8 m tons) and Bangladesh (0.8 m tons). Yields are around 1900–2300 kg/ha.

In Bangladesh, 3 jute leaves adorn the seal of the prime minister (Fig. 6.69, left), in Pakistan the national coat of arms shows jute as well as cotton, wheat and tea (Fig. 6.69, right). This shows the

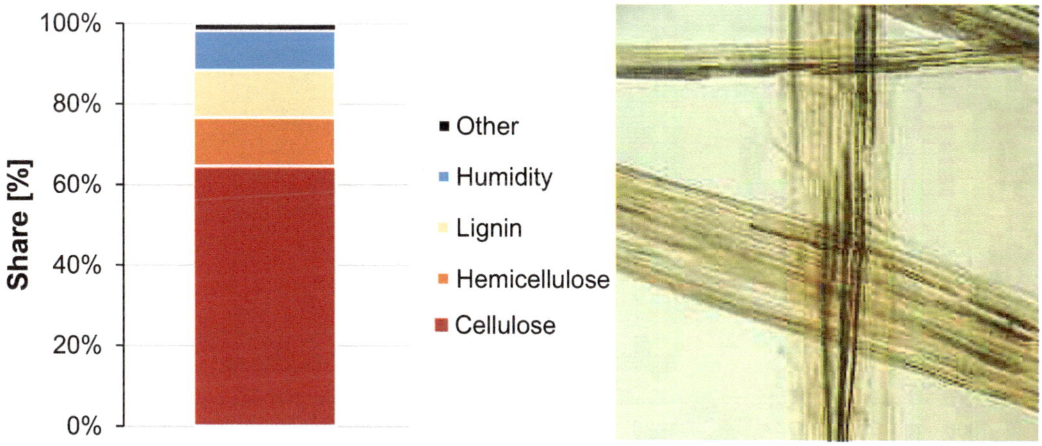

Fig. 6.67 Chemical composition and microscopic view of jute fibers

Table 6.6 Properties of jute fibers

Length [mm]	Fineness [µm]	[dtex]	Tenacity [cN/tex]	Elongation [%]	Density [g/cm^3]
1–6	10–25	2–6	3–4	0.8–2	1.44

great importance attached to jute in these countries.

Like other bast fibers, the jute traded is assessed according to the quality criteria of color, purity, luster, length, fineness and strength and also according to its degree of softness.

The price per ton has ranged from US$300 to US$800 in recent years, and is currently around US$750, depending on quality.

6.3.7 Areas of Application

Jute can be used to make coarse yarns and twisted yarns, which are usually woven (Figs. 6.70 and 6.71).

Due to its low price, jute is mostly used purely as

- packaging material such as sacks, rope goods,
- carpet base fabric,
- wall coverings and
- for technical articles (also in blends), Fig. 6.72.

6.4 Ramie

The ramie plant belongs to the family of non-stinging nettle plants (Urticaceae) and there to the genus Boehmeria. It is perennial and therefore only needs to be planted once every 15–20 years. It is found almost exclusively in Southeast Asia. Because of its lustre, ramie is regarded as the noblest of all bast fibers. Its abbreviation is RA.

6.4.1 History

Ramie has been used as a source of fiber for at least 4500 years. The oldest finds come from Egypt (mummy bandages), the first written mention from China (1797 BC, as perennial hemp). At the beginning of the eighteenth century, ramie reached Europe from India and was processed into yarns in England.

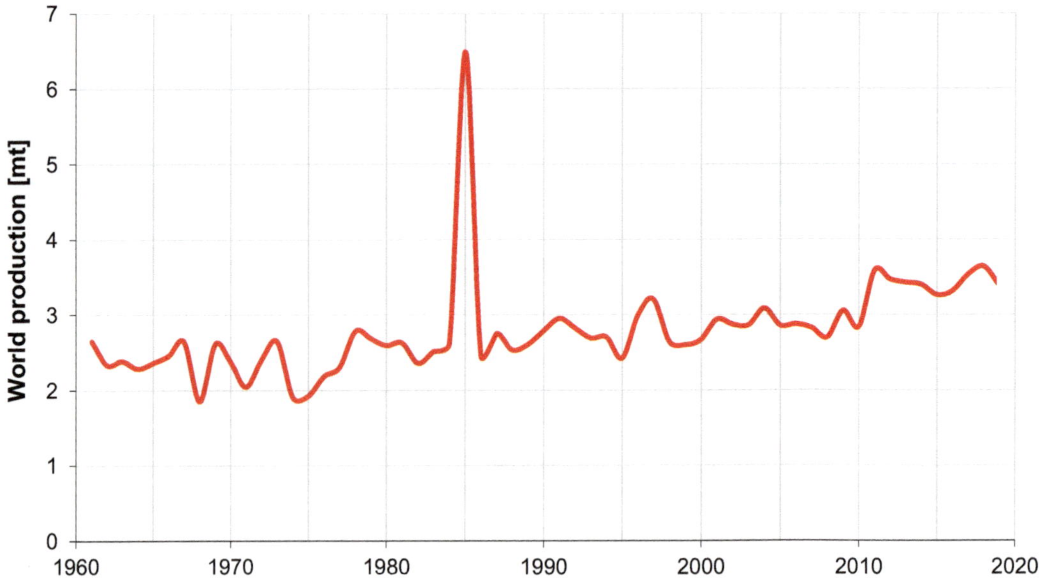

Fig. 6.68 Production of jute (data: faostat, 2022)

Fig. 6.69 National emblems of Bangladesh (left) and Pakistan (right) with jute leaves

6.4.2 Crop Growing

Ramie thrives in tropical, subtropical and temperate climates. It requires a warm, humid climate, nutrient-rich and water-retaining soil and a rainfall of 1000 to 1200 mm evenly distributed throughout the year.

The perennial plant has a taproot 1–2 m long and a rhizome that runs underground. This constantly forms new sprouts (rhizomes), whereby the individual ramie plant becomes more and more extensive and thus gradually occupies an area of 1 m^2. While the young plant has only about 8 to 10 stems, this number increases tenfold in the fifth year. Ramie plants can reach an age of up to 20 years.

The stem reaches a height of 2 to 3 m. It then sets yellow flowers from which a very fine, brown seed develops. The large, oval or heart-shaped leaves grow upwards with the stem, so that later, its lower third is leafless (Fig. 6.73).

Because in a well-designed planting the stems are very dense, the leaves take all the light from the ground and thus prevent the emergence of weeds.

When the stem has finished growing, it starts to turn brown at the base. Similar to hemp, the stalk is then cut off a hand's breadth above the ground. The cutting thus takes place at the stage of so-called green maturity, when the fiber is fully formed, but the structure of the stem has not yet hardened. After only a few days, the plant forms new shoots, so that the cut can be repeated several times a year. Yields are typically around 1600 kg/ha, resulting in approx. 1200 kg of ramie fiber after degumming (Table 6.7).

Fig. 6.70 Ringspinning frame (left) and cord winding machine (right)

Fig. 6.71 Jute spinning mill

Due to the multiple harvesting, the need for mineral fertilizer for ramie cultivation is very high. According to estimates, it is around 7 t/ha.

6.4.3 Varieties

The ramie comes in two varieties: Boehmeria nivea ("white ramie" because of the light underside of the leaves) and Boehmeria utilis tenacissima ("green ramie"). White ramie is also known as Chinese grass and grows preferentially in warm climates, especially in China (Fig. 6.74). There, ramie plays a major role in fabric and rope manufacture because flax is not known.

The variety Utilis, originating from Indonesia, is a plant of the tropical and subtropical zone and grows accordingly fast and long. Thus, 5 harvests per year are possible with this variety, whereas only 3 are possible with Chinese grass. It is cultivated in India, Mexico and the equatorial regions of South America. The fibers of this

Fig. 6.72 Typical products made of jute (carpet, sack, dress)

Fig. 6.73 Ramie plant (Kojian, 2010)

variety are easier to process than those of white ramie and yield about twice as much fiber (Table 6.8).

Approximately 1–3.5 kg of ramie fiber can be obtained from 100 kg of ramie stalks.

6.4.4 Structure

The structure of the stem is similar to that of the other bast fibers. However, the elementary fibers are not arranged in bundles within a bark layer, but individually or in groups (Figs. 6.75 and 6.76).

In cross-section, ramie fibers resemble cotton. Therefore, they are often processed together in blends.

The chemical composition is similar to that of flax. After cotton, ramie has the highest cellulose content of all plant fibers and the lowest lignin content (Fig. 6.77).

The elementary fibers of the white ramie plant are usually between 120 and 250 mm long, making them the longest of all bast fibers. Extreme values go even beyond this.

6.4.5 Fiber Extraction

Because ramie pectin is considerably harder than that of flax and hemp and also hardly forms any bacteria, ramie fibers cannot be extracted by a retting process. After harvesting, the stalks are therefore often placed in running water for several days so that the plant glue (mixture of pectins, hemicelluloses, vegetable protein, colorants and salts) is dissolved. Alternatively, the stalks are pulled through a crushing mill with the addition of copious amounts of water, which removes a large proportion of the glue. If ramie is not cut during the green ripening stage but during the yellow ripening stage, removal of the plant glue

Table 6.7 Yield per hectare of a ramie plantation

Year	Cuts	Green stems [kg]	No. of stems per plant
1	1	20,000	8–10
	2	30,000	15
2	1	40,000	20–40
	2	55,000	20–40
	3	60,000	20–40
3	5	per 80,000	60–80
≥4	5	per 100,000–120,000	80–100

Fig. 6.74 White ramie (Kao, 2010)

Table 6.8 Comparison of the two types of ramie

	White Ramie	Green ramie
Proliferation region	Warm climate	Tropical zone Subtropical zone
Planting	Medium dense Not too much shade	Very dense Lots of shade
Stem		
Length	1.5–2.2 m	2.5–3 m
Thickness	10–15 mm	20–25 mm
Wood body	A lot of voids	Woody Not many voids
Leaf		
Form	Oval	Heartshaped
Color	Light green	Dark green
Underside	Dense, white hairs	Almost no hairs
Seed		
No. of seeds	Many	Scarce
Germinability	Good	Bad
Bast		
Length	0.8–1.6 m	1.4–2.2 m
Color	Green–yellow	Pure yellow
Fiber content	68–72%	72–76%
Yield		
Cuts/year	3	5 (subtropics)

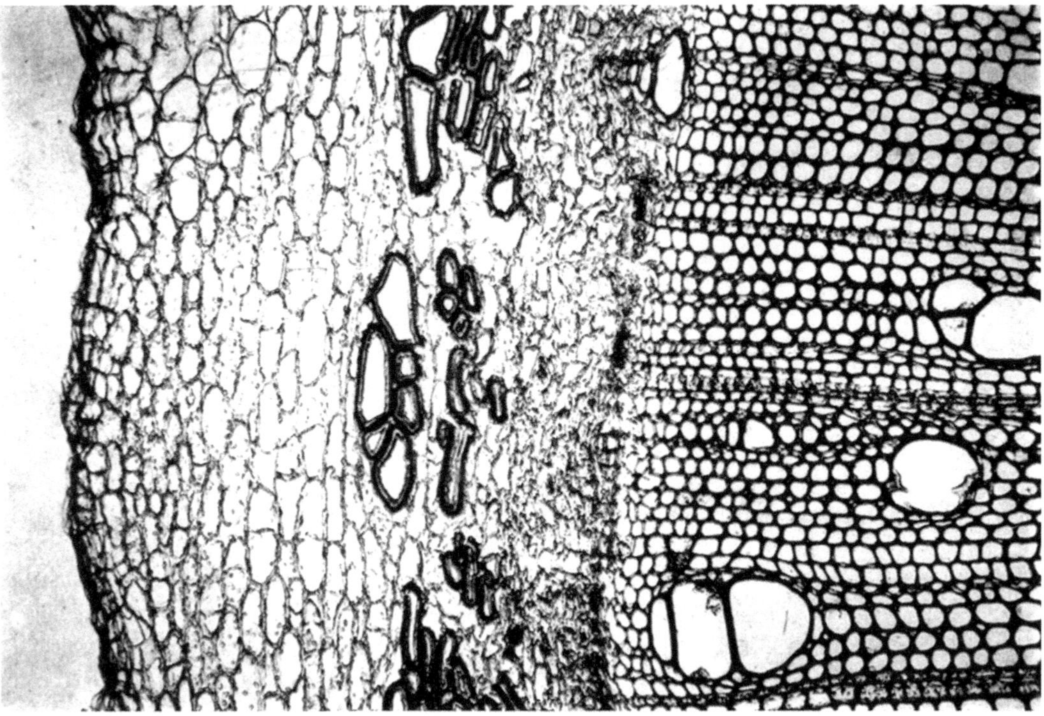

Fig. 6.75 Cross-sectional view of ramie stem with fiber bundles

is much more difficult because it has then hardened considerably.

During subsequent degumming, the residual plant glue in the bast is removed and the fibers are exposed. There are chemical, biological and physical processes. Chemical degumming exploits the fact that cellulose hydrolyzes more slowly than the accompanying substances to be removed (pectin, hemicellulose etc.). For example, the raw ramie is treated with 1% sodium hydroxide solution at 96 °C for 2 h. More modern processes also use Na_2CO_3. The fibers can then be bleached with H_2O_2. They are then neutralized, rinsed and dried. In biological degumming, the non-cellulosic components are decomposed by microorganisms, possibly with the addition of enzymes. This process may take several days, and cellulose decomposition may also occur. As a preliminary stage of chemical or biological degumming, the raw amylaceous

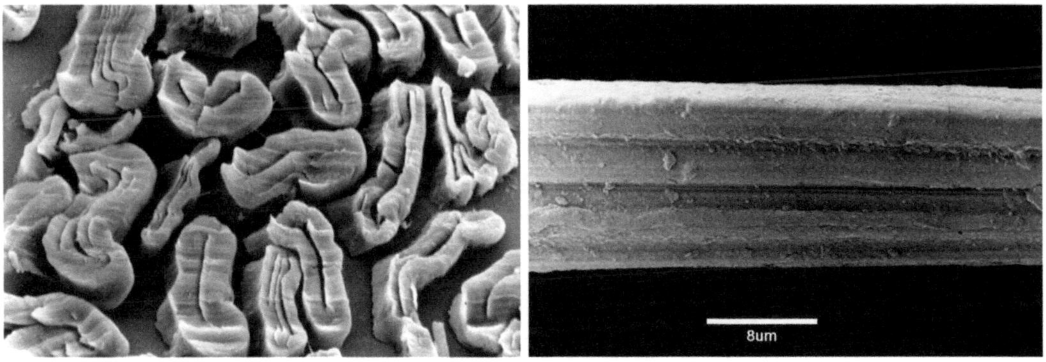

Fig. 6.76 Cross-sectional and longitudinal view of ramie fibers (right: CSIRO, 2008)

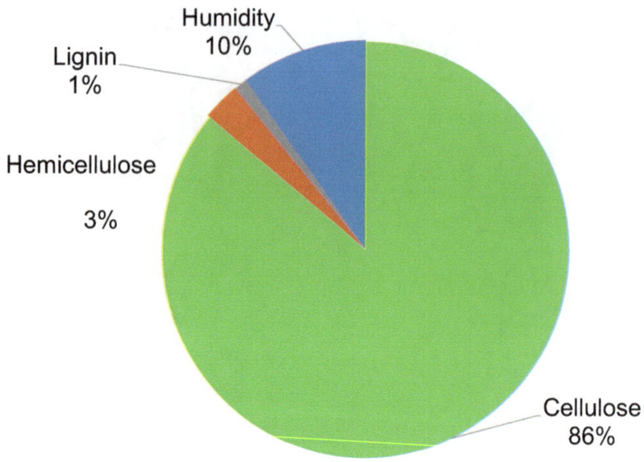

Fig. 6.77 Chemical composition of ramie (before degumming)

material can be treated with rollers, ultrasound, supercritical CO_2 or steam. In this process, the internal bonds between the cellulose chains and the accompanying substances are loosened and thus chemical reagents or microorganisms can attack more easily (Cheng et al., 2020).

The degumming process is of decisive importance for the quality of the ramie fibers and must therefore be carried out very carefully. During degumming, the plant glue is not dissolved immediately. Rather, it undergoes swelling under the influence of the lye. This loosens the molecular structure and the chemical attack takes place "from the inside", so to speak. The glue is then dissolved by the splitting of the molecules. In the process, dyes also go into solution, and the lye therefore turns a deep brown. The fibers also take on this color. Degumming is followed by an intensive washing process in which the chemicals must be completely removed. The water consumption is therefore very high. The washing process lightens the color of the ramie fibers (to brownish-grey). Actually, the fiber is white, so the fibers are last put into a chlorine bath to remove any remaining dyes. After that, the fiber is highly swollen and in this state it is easy to plasticize. Therefore, the water must be gently removed by pressing, pressure rolling or centrifuging. If the fibers were now dried, a brittle, poorly divisible material would result, which would be torn during the opening process and thus be unusable for spinning. Therefore, a lubricant must be incorporated between the individual fibers. Usually, soaps, softeners, glycerine, oils and fats, olein, paraffins and hydrocarbons are used in a suitable mixture for this purpose. This is sprayed onto the fibers before drying (Fig. 6.78).

Degumming at the producer could, on the one hand, make it easier to remove the glue (in the case of green waste it is soft) and, on the other hand, also save transport costs, because the glue

Fig. 6.78 Spinning preparation (Doshi Group)

accounts for approx. 20–30% of the bast weight. Nevertheless, degumming is not carried out until the spinning mill. This is mainly because degumming is a very difficult process to control and therefore cannot be carried out with satisfactory results by most producers. In addition, the degummed ramie fiber is very sensitive (kinking, soiling) and the risk of damage during transport is therefore very high.

Ramie fibers retain up to half their weight in water even after crushing. Furthermore, 300–500 individual fibers lie next to each other in close packing in the ramie stripes. With conventional dryers, the fibers would be destroyed, therefore either high frequency dryers are used (expensive) or dryers with low temperature and strong air circulation.

Ramie fibers are mainly classified according to their length. The best quality is fibers with a length of up to 1500 mm, while the worst quality has a fiber length of only 400 mm (technical fiber). Further classification criteria are the softness and the color of the fibers.

6.4.6 Properties

After preparation, the bast fiber bundles usually have a technical length of approx. 500 mm, the elementary fibers from 80 to 250 mm, in individual cases also over 400 mm. The fineness of the elementary fibers is 40–80 μm or 5–13 dtex. Typical degrees of polymerization are 600–3000. The tenacity is very high and, depending on the type, is approx. 40–70 cN/tex. When wet, 116–125% of the dry value is achieved. Ramie can absorb 12% of moisture. The elongation is 2–3% both dry and wet, the modulus of elasticity is in the range of 300–450 cN/tex. The density is 1.50 g/cm^3 and the color is yellowish to brownish in the raw state, bleached snow-white. Ramie fibers have a silk-like sheen, which is why they are often used in fiber blends, e.g. with cotton (Table 6.9).

6.4.7 Spinning

The dried ramie fibers can be spun into yarns by various processes:

- Schappe spinning,
- worsted spinning,
- cotton spinning,
- carded yarn process.

The Schappe process uses the longest fibers and produces the finest yarns (up to 20 tex). In the worsted process, 30–35 tex are achieved. In the cotton spinning process, the fibers are strongly shortened on the card, which enables economic further processing on conventional spinning machines. However, the minimum yarn count is then only 50 tex. With the carded yarn process, yarn counts of up to 80 tex are possible (Fig. 6.79).

Table 6.9 Properties of ramie fibers

Length [mm]	Fineness [μm]	[dtex]	Tenacity [cN/tex]	Elongation [%]	Density [g/cm^3]
40–250	16–125	5–13	40–70	2–3	1.5

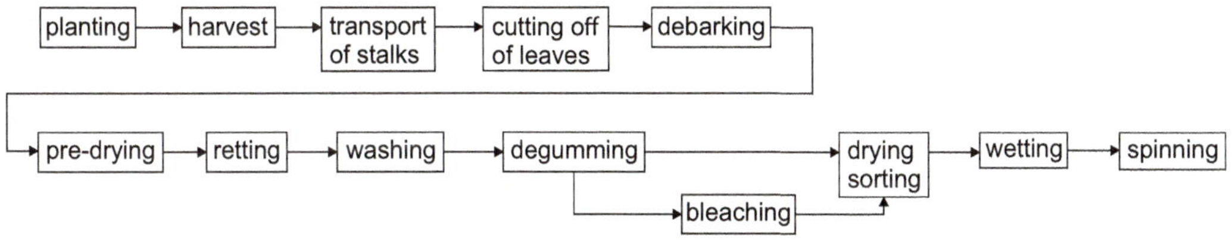

Fig. 6.79 Processing of ramie

6.4.8 Production

The most important producing country is China (58,000 t) ahead of Laos (2500 t), Fig. 6.80. Other countries do not play a role. Attempts have also been made to grow ramie in Europe (France, Italy, Spain) and in the USA. There, however, it did not attain any economic importance. The price is approx. 80 US-ct/kg.

6.4.9 Areas of Application

Typical products are clothing (sweaters, shirts, blouses and dresses, often blended with cotton), sewing threads and, in small quantities, technical articles. such as ribbons (Fig. 6.81).

6.5 Kenaf

Kenaf (Hibiscus cannabinus), like cotton, belongs to the mallow family and probably originates from Africa (Sudan). The species name "cannabinus" comes from the hemp-like leaf shape. It is an annual, jute-like fiber plant, cultivated mainly in India. Kenaf is also cultivated in Africa ("gambo hemp") and in Indonesia ("java jute"). After the partition of British India in 1949 and the resulting kenaf shortage, kenaf was also cultivated in South Africa, Florida, Cuba and the North Caucasus Its abbreviation is KE, Fig. 6.82.

The kenaf plant is not very demanding on the soil, but requires a warm climate (minimum temperature 13–16 °C). Sowing takes place in April and is done mechanically. With a planting width of 5 cm and a row distance of 20 to 30 cm, 3 to 5 m high, almost unbranched canes with a stem thickness of 2 to 3 cm are obtained.

Harvesting is carried out 3–5 months after sowing with mowing machines. In order to obtain a high fiber yield with low lignification at the same time, the stems must be cut shortly after flowering and before the fruit ripens (Fig. 6.83).

There are several possibilities for obtaining the fibers. On the one hand, the bast fibers are loosened by cold water retting followed by mechanical processing similar to that used for hemp. On the other hand, warm water retting, which lasts approx. 2 weeks at a water temperature of 24 °C. The fiber content reaches approx. 20%, resulting in a fiber yield per hectare of 1.8 to 2.5 t.

Kenaf fibers consist of 45–55% cellulose, 8–20% lignin, 20% pentosans and 2–3% ash.

The fiber bundles, which can be up to 300 cm long, are less lignified than jute and are therefore easier to divide. The elementary fibers have irregular wall thickenings and a length of approx. 2 mm (maximum up to 10 mm) with an average diameter of 20 μm. They have a high tenacity of 40–50 cN/tex and a good luster (Table 6.10).

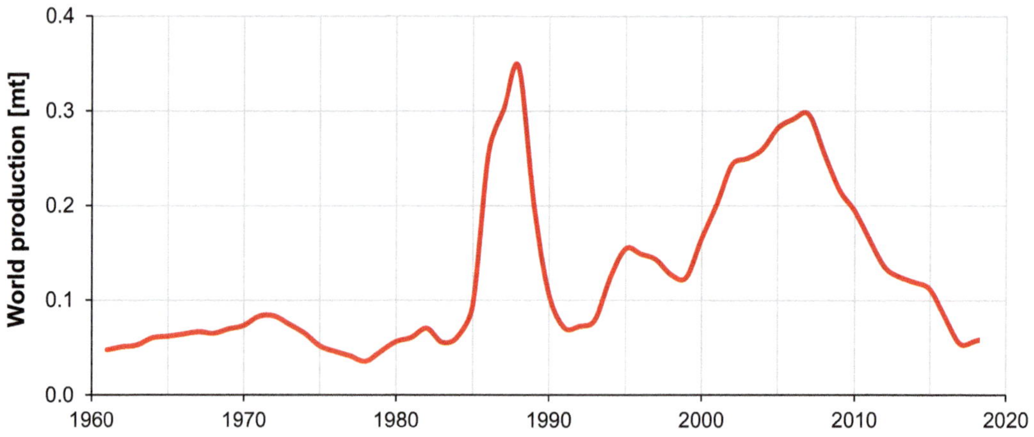

Fig. 6.80 Production of ramie fibers (data: faostat, 2022)

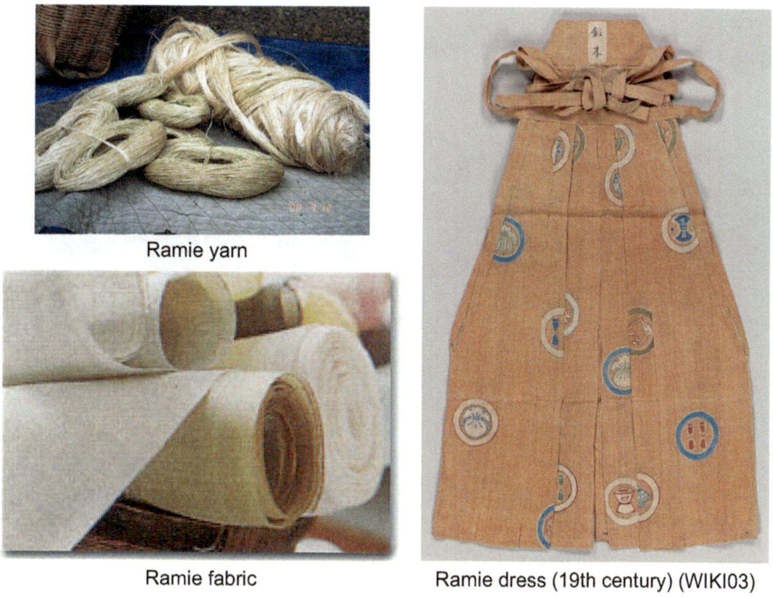

Fig. 6.81 Typical products made from ramie

Currently, around 340,000 t of kenaf fibers are produced worldwide. The most important producing countries are India (140,000 t), China (90,000 t), Thailand (36,000 t), Brazil (26,000 t), Vietnam and Cuba (10,000 t each). The price is 0.8–1.2 €/kg.

Kenaf fibers can be used to produce twine and sacks as well as special papers and packaging material. The shives and short fibers are used to produce gritting material for absorbing oil and cattle and cat litter (Fig. 6.84 Kenaf fibers (left: Bauer, 2009)).

Fig. 6.82 Kenaf plant and field (WIKI04)

Fig. 6.83 Kenaf field in Malaysia (WIKI05)

Table 6.10 Properties of kenaf fibers

Length [mm]	Fineness [µm]	[dtex]	Tenacity [cN/tex]	Elongation [%]	Density [g/cm^3]
1–12	10–40	6	40–50		1.5

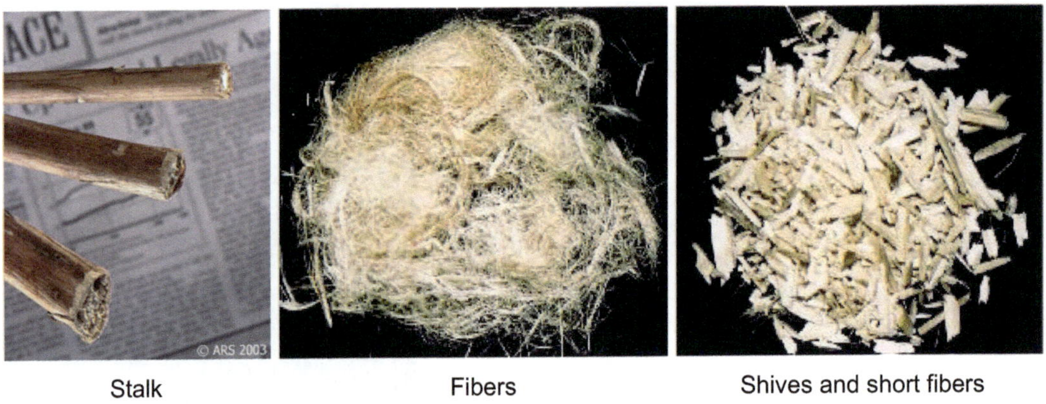

Fig. 6.84 Kenaf fibers (left: Bauer, 2009)

Kenaf is increasingly being used in recent years, especially in the automotive sector (interior lining, insulation). For this purpose, the fibers are cut and usually blended with polypropylene in a 50:50 ratio, carded and processed into nonwovens. For example, the interior lining of car doors is reinforced with around 2 kg of kenaf fibers. The parallelogram-shaped mat blanks consist of three layers: a middle layer of kenaf and PP as well as two outer layers of 70% PP and 30% PET each. These are thermally formed and simultaneously laminated with fabric. For this purpose, they are heated to over 200 °C together with the decorative material. In the mold,

Fig. 6.85 Typical products made from kenaf (top: coarse yarns and insulation material; bottom: interior lining of the BMW i3, Dräxlmaier Group, 2013)

the solidifying PP seals the kenaf fibers in the now three-dimensionally formed interior door lining. This protects the fibers from moisture and the car occupants from odors. When this very stiff yet lightweight component leaves the mold, it is still 2.5 mm thick and the fibers are between 40 and 60 mm long.

Especially in sound insulation, such products have very good properties compared to conventional materials made of purely synthetic materials. In impact protection, their mechanical properties are comparable to those of conventional materials. In Germany, kenaf fibers therefore account for approx. 20% of all natural fibers in automotive construction (flax 50%), Fig. 6.85.

Yarns made from kenaf are relatively coarse and very hairy, similar to jute yarns.

Kenaf also has one of the highest CO_2 adsorption rates of all plants. During growth, 1 t of kenaf can store approx. 1.5 t of CO_2.

6.6 Nettle Fiber

Nettle fibers are obtained from stinging nettles (mostly Urtica dioica), which were bred by Bredemann (1959) for a high fiber yield. In total, there are more than 500 different species of nettle. Fibrous nettle, called "great nettle" as a wild herb, is a perennial plant. The four-edged stem grows up to 2 m high. The stinging nettle is

Fig. 6.86 Nettle plant (Thomé, 1903) and advertising poster from the 1940s

dioecious, i.e. there are both male and female plants (Fig. 6.86).

The stinging nettle is widespread in temperate climates worldwide and is known as a cultivated wild herb and medicinal plant. In the Middle Ages, the long, firm fibers were processed into ropes, cords, fishing nets and nettle cloth (today, "nettle cloth" is made of 100% cotton). They also appear in fairy tales, for example in Hans-Christian Andersen's fairy tale "the wild swans". The use of cotton from the eighteenth century onwards displaced the stinging nettle in the manufacture of textiles, but it was rediscovered as a domestic raw material in Germany during the two world wars because it could be processed with the machines normally used for cotton. The fiber nettle is extraordinarily undemanding. Nevertheless, nutrient-rich sites with a good water supply are advantageous. It is mown at the end of flowering in July/August, yielding around 1000 kg/ha with a fiber content of 5–15%. As with almost all bast fibers, the fibers are obtained by dew and cold water retting. It is also possible to use "green" fibers, which are produced by a mechanical de-wooding process similar to that used for sisal. The fiber nettle can be extensively cultivated and harvested for up to ten years.

Typical fiber lengths are 5–50 mm, the fineness is 3–9 dtex (15–25 μm). The tenacity reaches typical values around 25–35 cN/tex, the density is approx. 1.5 g/cm^3, Table 6.11.

Blended yarns (up to 70% nettle fiber content) in combination with e.g. cotton or lyocell as well as nonwovens are produced from nettle fibers, so far only in small quantities. Nettle fibers are also suitable in principle for fiber composites. Fig. 6.87 shows SEM images of nettle fibers.

Table 6.11 Properties of nettle fibers

Length [mm]	Fineness [µm]	[dtex]	Tenacity [cN/tex]	Elongation [%]	Density [g/cm^3]
5–50	15–80	3–9	25–35	3–4	1.5

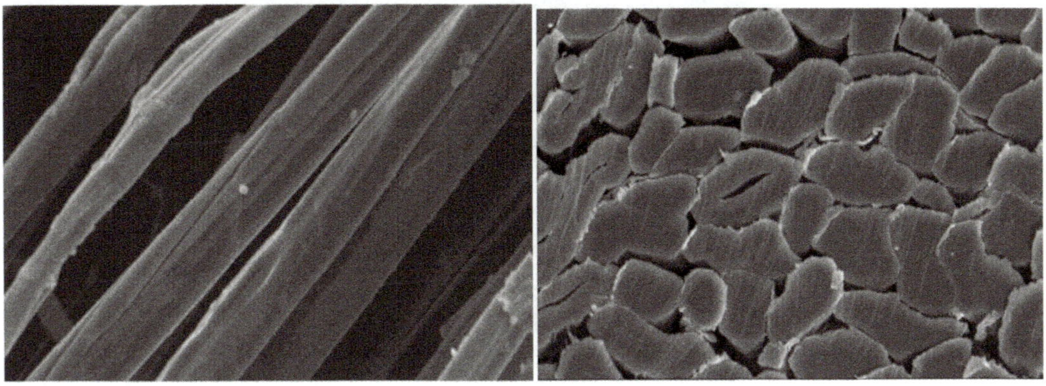

Fig. 6.87 Transversal and cross-sectional view of nettle fibers

From the root, together with alum, a yellow dye can be produced which is suitable for dyeing wool. A green dye can be obtained from the stems and leaves.

In addition to the European nettle Urtica dioica, the nettle Girardinia diversifolia, which grows in Nepal, is also used locally for fiber production. It thrives at altitudes of up to 3000 m above sea level (Fig. 6.88).

6.7 Bamboo

Bamboo is a subfamily of the sweet grass family (Gramineae). Bamboo fiber is a cellulosic fiber obtained from the stems of the bamboo plant. It is used both as a bast fiber and as a raw material for regenerated fibers (e.g. viscose).

6.7.1 History

As early as 552 AD, the first silkworm eggs were smuggled from China to Constantinople in bamboo canes. Bamboo did not come to Europe until the middle of the nineteenth century. It was silk importers who brought this exotic plant from China and Japan as a gift for their princes and patrons. These plants had little chance of survival because there was no knowledge of proper cultivation and the tropical species barely survived the cold winters. In 1879, Thomas Alva Edison carbonized a bamboo fiber into a filament for an electric lamp that withstood the test of time after many other materials had already burned out. Graham Bell recorded the first sound of his phonograph with a bamboo needle in 1876.

6.7.2 Cultivation Regions

Over 1000 different species of bamboo grow worldwide (Fig. 6.89). There are grass-like varieties and those that reach a height of up to 30 m with a trunk diameter of 30 cm. Bamboo can grow very fast, within 24 h up to 1 m. Only a few species are used by humans and cultivated in plantations.

The bamboo grows in tropical forests up to over 3000 m above sea level and from the 40th latitude degree south to the 40th latitude north of the equator (Fig. 6.90). In the Andes and in the Himalayas, bamboo grows up to the snow line at an altitude of approx. 4000 m a.s.l. Bamboo is mainly cultivated in Asia, in South America, South Africa and Australia, but its economic importance is very low.

Fig. 6.88 Girardinia diversifolia (Valke, 2012)

Fig. 6.89 Bamboo plants (Heyde, 2007)

Fig. 6.90 Geographical distribution of bamboo (map: mapswire)

Worldwide, about 20 m tons of bamboo are harvested annually. The largest producing countries are China, India and Burma with a cultivation area of about 20 m hectares.

6.7.3 Properties

In the cross-section of a bamboo cane, the bamboo bark lies on the outside, adjoined on the inside by the zones with the bamboo fibers. The zone with high fiber density is called "bamboo green", the zone with loose fiber organization is called "bamboo yellow".

The length of bamboo fibers is 10–80 mm, their diameter approx. 50 µm, the tenacity is around 20 cN/tex, the elongation 20%. Most bamboo fibers are significantly shorter than 20 mm and therefore cannot be spun in their pure form. They can only be processed into nonwovens. The lignin content in the culm is approx. 32%, the cellulose content approx. 61%. Bamboo contains a high proportion of silicic acid. It therefore shows very good resistance to chemicals and fire.

Bamboo fibers are hollow and have an irregular cross-section with a density of 0.8 g/cm^3 (Fig. 6.91). Therefore, the fiber is very breathable and has a very good moisture absorption, which is 4 times higher than that of cotton. In longitudinal view, the fiber is corrugated.

Bamboo fibers are often first dissolved and spun out as regenerated fibers, e.g. using the viscose process. In this way, their properties correspond to those of viscose fibers produced from wood. Textiles made from such regenerated "bamboo fibers" have a soft and velvety feel and a natural luster. They are crease-resistant and retain their shape.

A special feature of the bamboo fiber is that it has antibacterial properties because it contains the active ingredient "bamboo kun". Furthermore, it contains honey pectin, which is especially beneficial for sensitive skin. However, both properties are lost during the transformation of bamboo fibers into regenerated fibers and are therefore no longer present in such bamboo textiles.

6.7.4 Areas of Application

Today, bamboo fibers, mainly regenerated fibers made from bamboo cellulose by the viscose process, are used in various textiles, e.g. towels, bathrobes, socks and ladies' and men's outerwear. They are mostly used in blends (e.g. 40%

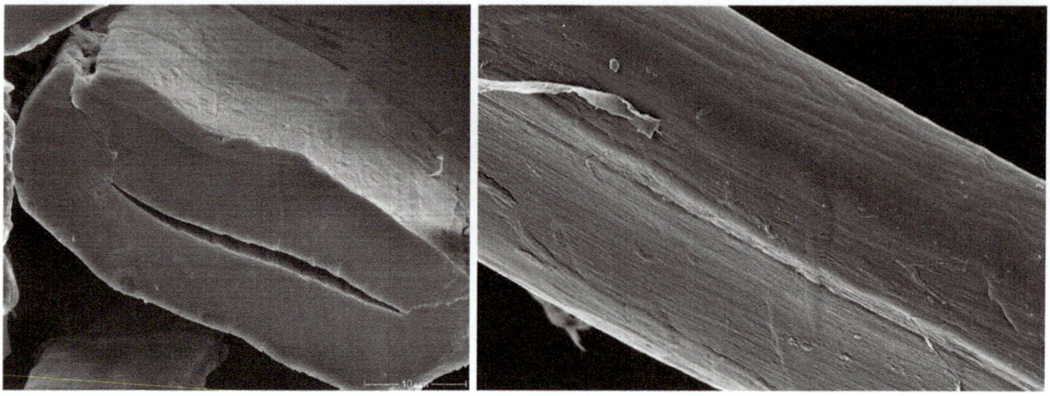

Fig. 6.91 Bamboo fibers (Lenzing AG)

Fig. 6.92 Card sliver and non-woven fabric made from bamboo fibers

cotton and 60% bamboo fibers or 72% bamboo fibers, 27% PA and 1% EL), Fig. 6.92.

Around 2010, Litrax developed a process in which the bamboo fibers are extracted from the stalk mechanically and with the help of enzymes. They can be processed into yarns with a cashmere-like feel using the ring spinning process. This process has not yet been commercialized for cost reasons.

6.8 Property Comparison

In Table 6.12, important properties of bast fibers are summarized. Flax competes primarily with hemp, while ramie is the coarsest bast fiber, with high strength, especially when wet (Table 6.12). The elementary fibers of jute are too short for yarn production, therefore only the technical fiber is processed into coarse yarns. The elongation of all bast fibers is low, so their tendency to crease is very high.

Table 6.12 Properties of the most important bast fibers

	Flax	Hemp	Jute	Ramie
Fineness	1–7 dtex	2–4 dtex	2–3 dtex	5–13 dtex
Length	25–40 mm	10–30 mm	1–5 mm	60–260 mm
Diameter	10–35 μm	15–25 μm	15–25 μm	40–80 μm
Tenacity	30–55 cN/tex	50–70 cN/tex	28–34 cN/tex	40–70 cN/tex
wet (% of dry value)	105–120%	106–108%	99–104%	116–125%
Breaking elongation	1.5–4%	0	0.8–1.3%	2–3%
wet (% of dry value)	110–125%	78%	99–104%	
Moisture adsorption	8–10%	8.5–10%	12–13%	7.50%
Water retention	50–55%	50–55%	50–55%	
Density	1.47 g/cm^3	1.48 g/cm^3	1.44 g/cm^3	1.5 g/cm^3
Acid resistance	o	–	–	–
Alkali resistance	+	–	–	–
Abrasion resistance	++	++	–	–
Heat retention	–	–	–	–
Air permeability	+	+	+	+
Electric charge	Low	Low	Low	Low
Creasing tendency	Very high	Very high	Very high	Very high
Shape retention	Low	Low	Low	Low

References

Aleks. (2000). Retrieved from https://commons.wikimedia.org/wiki/File:Industrialhemp.jpg, CC BY-SA 3.0.

Aleks. (2007). Retrieved from https://commons.wikimedia.org/wiki/File:R%C3%A9colte_chanvre1.jpg, CC BY-SA 3.0.

Auyon. (2011). Retrieved from https://commons.wikimedia.org/wiki/File:Jute_under_Sun.jpg, CC BY-SA 3.0.

Bauer, S. (2009). *USDA*. Retrieved from http://www.ars.usda.gov/is/graphics/photos/k7234-38.jpg

Bjørtvedt. (2008). Retrieved from https://commons.wikimedia.org/wiki/File:Linum_usitatissimum_01.jpg, CC BY-SA 3.0

Bredemann, G. (1959). *Die Große Brennessel, Urtica dioica L.* Akademie-Verlag.

Čebotarev, A. (2009). Retrieved from https://commons.wikimedia.org/wiki/File:USO-xx_and_Zolotoniski-xx_hemp_strains_019.jpg, CC BY-SA 3.0.

Cheng, L., Duan, S., Feng, X., Zheng, K., Yang, Q., Xu, H., Luo, W., & Peng, Y. (2020). Ramie-degumming methodologies: A short review. *Journal of Engineered Fibers and Fabrics, 15*, 1–7.

Cranshaw, W. Colorado State University. Retrieved from http://www.bugwood.org

CSIRO. (2008). Retrieved from https://commons.wikimedia.org/wiki/File:CSIRO_ScienceImage_8439_Ramie_fibre.jpg, CC BY 3.0.

Doshi Group. Mumbai, India. Retrieved from http://www.doshi-group.com

Dräxlmaier Group. (2013). Retrieved from https://commons.wikimedia.org/wiki/File:2013-09_BMW_i3_Tuer_Naturfaser_(9)_Dr%C3%A4xlmaier_Group.jpg, CC BY-SA 4.0.

European Linen & Hemp. (2021). Retrieved from http://news.europeanflax.com

faostat. (2022). Retrieved from http://faostat.fao.org

Fendersson, E. (2006). Retrieved from https://commons.wikimedia.org/wiki/File:Male_hemp_flowers.jpg

Ganguly, B. (2017). Retrieved from https://commons.wikimedia.org/wiki/File:Jute_Extraction_From_Retted_Jute_Stem_-_River_Jalangi_-_Nadia_2017-08-15_1827.JPG, CC BY-SA 1.2.

Haudek, H. W., & Viti, E. (1980). *Textilfasern*. Bondi & Sohn.

Herer, J., & Bröckers, M. (2017). *Hanf*. Nachtschatten Verlag.

Heyde, M. (2007). Retrieved from https://commons.wikimedia.org/wiki/File:Bamboo_Richelieu.jpg, CC BY-SA 1.2.

Kao, S.-S. (2010). Retrieved from https://commons.wikimedia.org/wiki/File:Boehmeria_nivea_leaves.jpg, CC BY-SA 2.0.

Kind, W., König, P., Müller, W., et al. (1930). *Der Flachs*. Verlag Julius Springer.

Kojian, R. (2010). Retrieved from https://commons.wikimedia.org/wiki/File:Gardenology.org-IMG_0086_rbgm10dec.jpg, CC BY-SA 3.0.

Liebscher, U. (1983). *Faserstoffe – Begriffe, Eigenschaften, Bezeichnungen*. Technische Textilien 26, Deutscher Fachverlag.

Lokilech. (2009). Retrieved from https://commons.wikimedia.org/wiki/File:Adinkerke_Garbe1.jpg, CC BY-SA 3.0.

Manners, M. (2017). Retrieved from https://commons.wikimedia.org/wiki/File:Jute_field.jpg, CC BY-SA 2.0.

Mapswire. Retrieved from https://mapswire.com

Meraj, I. H. (2015). Retrieved from https://commons.wikimedia.org/wiki/File:Transporting_Jute_In_Rural_Bangladeshi_Village_15.jpg, CC BY-SA 3.0.

Peairs, F. Colorado State University. Retrieved from http://www.bugwood.org

Schwurhand-Zeichenverband e.V. Bielefeld.

Sparshatt, J. Retrieved from geograph.org.uk/p/747095, CC BY-SA 2.0.

Statista. (2022). Retrieved from http://www.statista.de

Thomé, O. W. (1903). *Flora von Deutschland, Österreich und der Schweiz*. Gera.

Touhid. (2017). Retrieved from https://commons.wikimedia.org/wiki/File:Joining_two_boat_in_one.jpg?uselang=de, CC BY-SA 4.0.

Tuason, T. (2017). Retrieved from https://commons.wikimedia.org/wiki/File:Cannabis_sativa_2.jpg, CC BY-SA 4.0.

Valke, D. (2012). Retrieved from https://commons.wikimedia.org/wiki/File:Girardinia_diversifolia_(8,071,156,699).jpg, CC BY-SA 2.0.

WIKI01. Retrieved from https://commons.wikimedia.org/wiki/File:Marijuana.jpg

WIKI02. Retrieved from https://commons.wikimedia.org/wiki/File:Bambusoideae_World_map.png

WIKI03. Retrieved from https://commons.wikimedia.org/wiki/File:MET_DP158227.jpg, CC0 1.0.

WIKI04. Retrieved from https://commons.wikimedia.org/wiki/File:Hibiscus_cannabinus0.jpg

WIKI05. Retrieved from https://commons.wikimedia.org/wiki/File:Kebun_sawit_di_tanam_kenaf.jpg, CC BY-SA 4.0.

7 Hard Fibers

Hard fibers are obtained from the leaves of various tropical and subtropical plants. These are usually fleshy and require thick and stiff fiber bundles for support. The most important hard fibers are agave (sisal, henequen) and musa fibers (manila, abaca). Lily, grass, palm, pineapple and peat fibers are also of regional importance (Fig. 7.1).

7.1 Agave Fibers (Sisal, Henequen)

Agaves belong to the asparagus family. The word "agave" comes from Greek and means "noble, splendid". Which, however, refers more to the plant as a whole, less to the fibers. Agaves bloom only once and the development of the flower often takes decades. The inflorescence can then grow up to 12 m high. The abbreviation for sisal is SI, for henequen HE.

7.1.1 History

Agaves have already been cultivated in ancient Mexico and already the Aztecs made paper and sewing threads from their fibers. The thorn at the end of the leaf served as a needle. Furthermore, sisal was used to make cords, ropes, straps, hats, mats and shoes. After the conquest of Mexico by Spanish conquistadores, the first agaves reached Spain and Italy as ornamental plants in the sixteenth century. At the end of the nineteenth century, they came to Florida and East Africa and the production of fibers from the plant began. The main producing countries were Indonesia and the Philippines. In 1892, the German botanist Richard Hindorf introduced Agave plants to Tanzania. Only 62 plants formed the basis of todays East African sisal production (Fig. 7.2).

Cultivation began in Kenya in 1903 and simultaneously in Brazil. However, it was not until the 1930s that significant quantities of sisal fiber were produced in Brazil. From 1946, Brazil exported sisal and a few years later was the second largest producer. Since the 1970s, sisal has increasingly been replaced by man-made fibers as a basic material for ropes and cordage, but nevertheless remains one of the most important natural fibers in terms of volume.

7.1.2 Cultivation

Among the 250–300 agave species known today (60 of them in Mexico), only 3 are suitable for fiber production:

1. The Agave sisalana, whose name comes from the first export port "sisal" in Yucatan. It is most important as a supplier of fibers because its leaves have few marginal spines, which facilitates the work of the picker. The main areas of cultivation, apart from Africa, are Indonesia and Brazil.

D. Veit, *Fibers*, https://doi.org/10.1007/978-3-031-15309-9_7

Fig. 7.1 Main cultivation areas of important hard fibers (map: mapswire)

2. Agave cantala, which originates from Mexico, is cultivated in the Philippines, Indonesia and India.
3. The Agave fourcroydes is Mexico's most important fiber plant, providing "henequen fiber", the name of which comes from an Indian language of Haiti (Fig. 7.3).

The fleshy and elongated or heart-shaped leaves of the agaves are arranged in the form of a rosette. The sisal agave forms about 300 leaves until it dies after 7–12 years. Their length (50–150 cm) and weight (approx. 600 g) depend on the climate and the soil conditions. The young leaves, which provide a finer and softer fibrous

Fig. 7.2 Sisal production in Tanzania at the beginning of the twentieth century (Dobbertin, 1906)

Agave sisalana Agave cantala Agave fourcroydes

Fig. 7.3 Important agave plants for fiber production (Wiki, 2003; Valke, 2007; Michiel, 2012)

material than later cuts, are erect vertically and assume a more or less horizontal position as they grow. Only those leaves over 50 cm long and about 100 mm wide which form an angle of 45° against the stem are considered ready for cutting. The first pruning is done after 2–3 years with the sisal agave, with the hennequen agave only after 4–5 years (Fig. 7.4). Propagation is exclusively via seedlings in the form of root shoots (rhizomes) or bud bulbs from the inflorescence (bulbils) and not via seeds.

Below the leaf bark lie the bast strands, which form the mechanical framework of the leaf. The fiber content of agave leaves is only 4–5%, so that large quantities of plant flesh are produced during fiber extraction. The fiber yield per hectare varies greatly, depending on the plant density. Usually there are about 4000 plants on one hectare, from which 60,000–80,000 leaves are harvested annually with a fiber yield of 1–1.5 tons.

7.1.3 Harvesting and Extraction of the Fibers

The leaves are cut off close to the ground, trimmed for the defoliating machine and tied into bundles. The ground tissue is then scraped

Fig. 7.4 Sisal field after harvest (Navez, 2008)

Harvested sisal leaves (Talbot, 2007a)

Mechanical opening (Schuiling, 2011)

Opened fibers (Talbot, 2007b)

Drying scaffolds (Talbot, 2007c)

Fibers (Raschka, 2008)

Bale (Wiki, 2008)

Fig. 7.5 Processing steps during fiber extraction

off. This is done either manually or mechanically by squeezing off the flesh using rotating drums fitted with blunt blades. In modern machines, this is done with automatic treadmill feed and turning of the sheets in the air stream. The wet fibers are then soaked in water to dissolve the pectin and chlorophyll. They are then dried, either by hanging in the sun or in centrifuges and belt dryers. Finally, any leaf tissue residue that is still attached is removed with a brushing machine (Fig. 7.5). This treatment gives the fiber a finer and glossier appearance. The fiber content of the leaves is normally around 4–7%.

The fibers are processed into yarns, ropes and fabrics on special machines (similar to those used for bast fibers), Fig. 7.6.

Fig. 7.6 Loom for sisal mats and rope machine (Sánchez, 2017)

7.1.4 Production

The leaves of the agave are harvested for the first time at the age of 5 years. Until the eighth year, the fiber yield increases every year, then it decreases again (Fig. 7.7). At the age of about 12 years, agaves bloom for the first and only time, after that they die. After each harvest, the agave must still have at least 35 leaves so that it does not die.

World production of sisal has been around 300,000–500,000 t per year for the last 20 years. In 2012, production was halved compared to 2011 to only 220,000 t (Fig. 7.8). The main reason for this was an extreme drought in Brazil, the main producing country. As a result,

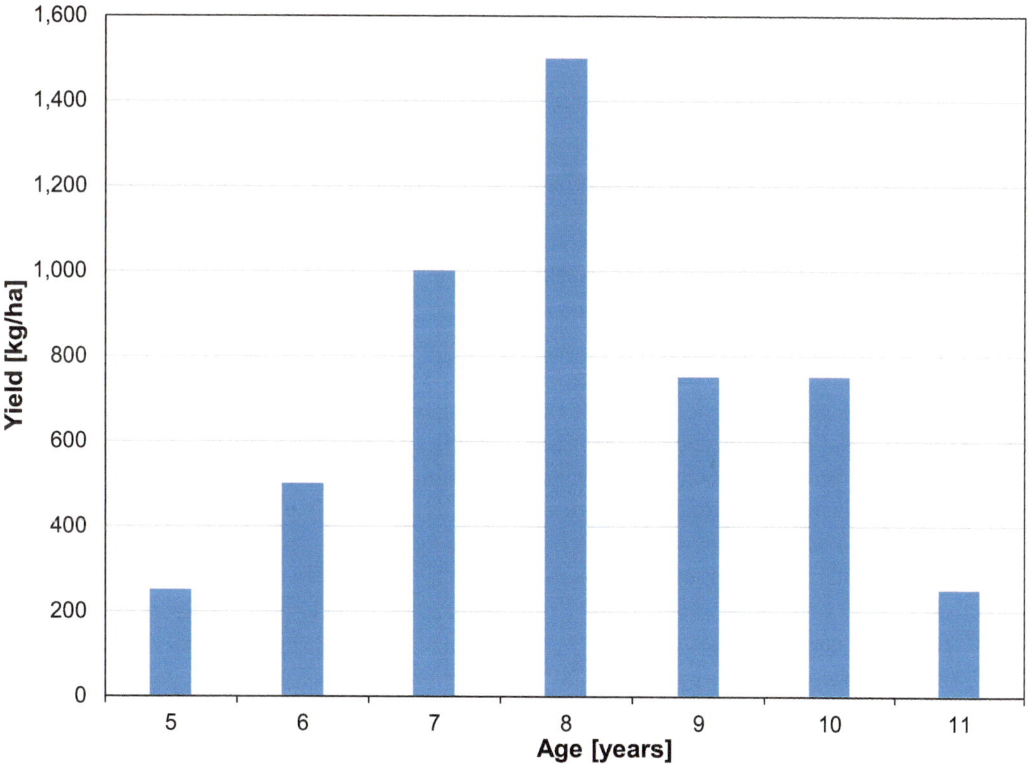

Fig. 7.7 Fiber yield of the agave

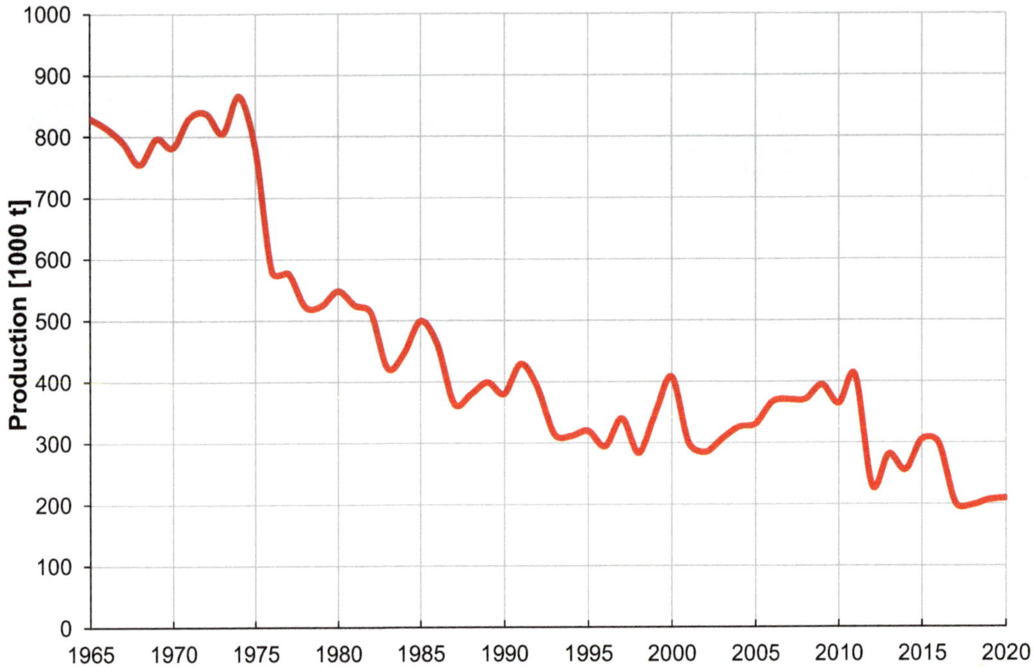

Fig. 7.8 Production of sisal fibers (data: faostat, 2022)

production there fell to just 86,000 t (2020). In contrast, production in all other countries remained almost constant and is currently around 209,000 t in total.

Brazil is nevertheless still by far the most important producer for sisal, with a market share of 42%. Around 2/3 of its production is exported to China. In Tanzania, the sisal plant was smuggled in by a German agricultural engineer in 1893 because this country was a German colony and he recognized the great value of the plant for rope and carpet production. In 1961, Tanzania was the main producer of sisal with a production of about 200,000 t. With the nationalization of the sisal industry in 1967, production quickly collapsed, partly due to increasing competition from man-made fibers. Today, Tanzania is the second largest producer of sisal fibers after Brazil, with a market share of 16% (33,000 t), Fig. 7.9.

The fibers are classified according to length and color. The price of sisal is currently around € 0.5–2/kg, depending on quality. The total value of the sisal fiber harvest is around €100 million (faostat, 2022).

7.1.5 Properties

The density of sisal is approx. 1.16 g/cm^3. The length of the elementary fiber strands is 500–1100 mm, the length of the elementary fibers is 1–5 mm. Therefore, the single fibers are not spinnable, but only fiber bundles. The thickness of the elementary fibers reaches 20–25 μm. The tenacity reaches 35–50 cN/tex when dry and approx. 85% of this value when wet. The loop strength is 7.5 cN/tex, which is a high value. The elongation at break in normal climate is 3% and in wet condition 135% of this value is reached (Table 7.1 and Fig. 7.10). Sisal fibers have a medium luster and are yellowish-white in color. Sisal is very resistant to moisture and can be dyed in many colors.

When sisal fibers are treated with caustic soda, hemicellulose and pectins dissolve and the fibers become finer and shorter. This is advantageous when they are used as reinforcing fibers for fiber composites, because the fiber-matrix bond is thus improved.

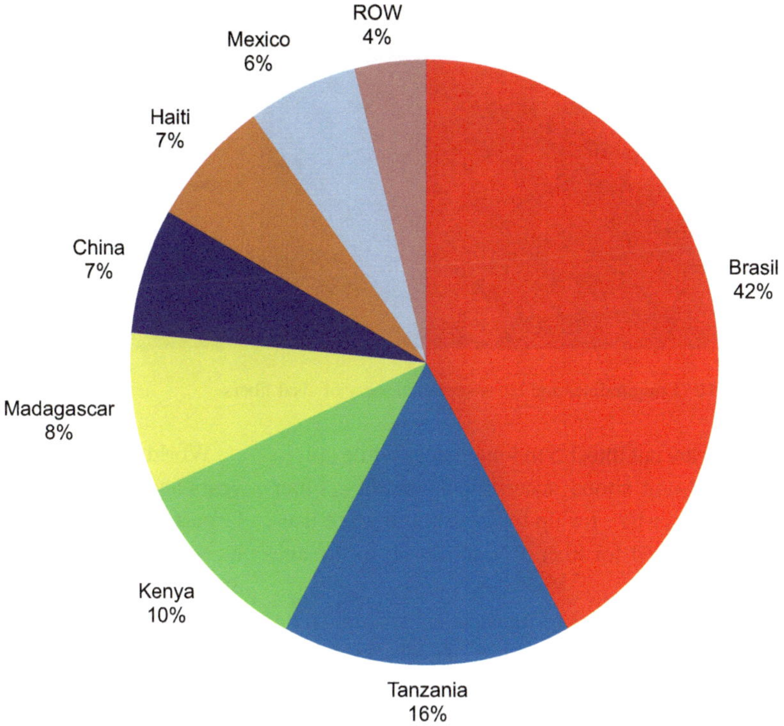

Fig. 7.9 The main sisal producing countries (data: faostat, 2022)

7.1.6 Areas of Application

Typical applications of sisal are ropes, mats (carpets, runners), packaging material and fillers of all kinds as well as wrapping paper (short fibers). A special application is the use of sisal in steel cables as a lubricant reservoir.

In the 1990s, sisal also became popular again in Germany, especially in the form of coarse carpets. Dartboards and cat scratching posts are also often made from sisal (Fig. 7.11). Other products made of sisal are polishing cloths, mattresses and furniture upholstery.

In order to make sisal extraction more economical, the fiber waste is refined by a special retting process to produce a high-quality sisal plant ("flume tow"). This is used for the production of sacks. The plant flesh is often used as fertilizer. Furthermore, tequila can be obtained from the heart of the blue agave.

7.2 Musa Textilis (Manila Hemp, Abacá)

There are more than 70 different species of banana, which botanically belong to the berries. The fibers of the fiber banana are obtained from the leaf sheaths of the Musa textilis, which thrives in tropical areas (Fig. 7.12). The country of origin and still the main producer is the Philippines. The fiber banana is also known as abacá and manila hemp. Its abbreviation is AB.

Fibrous banana plants are usually lower than edible banana plants and form numerous shoots from the rootstock which, as they grow, reach heights of 3–8 m. This consists of nested leaf sheaths, which unfold at the tip to form leaves 20–40 cm wide and 1.5–2.0 m long. After the long inflorescences have sprouted, only the 2 to 3 year old sheaths are cut. The harvested leaf

Table 7.1 Properties of sisal fibers

Length [mm]	Fineness [µm]	[dtex]	Tenacity [cN/tex]	Elongation [%]	Density [g/cm^3]
1–8	20–32	4–5	35–50	3	1.16–1.45

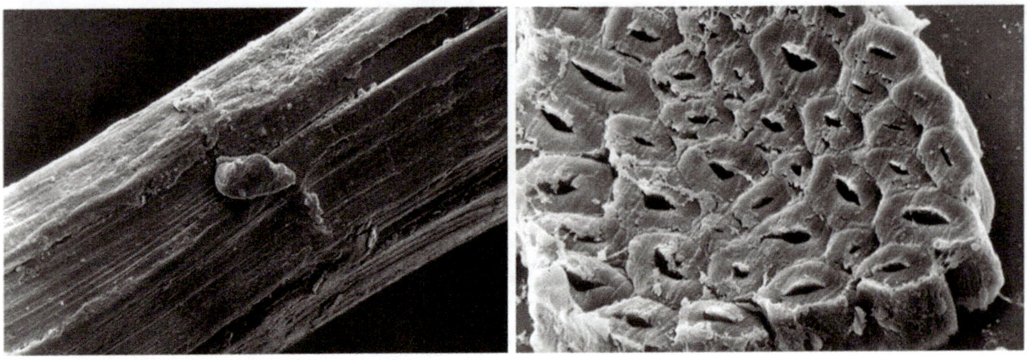

Fig. 7.10 Longitudinal and cross-sectional view of sisal fibers

sheaths are cut into 1.5 m long pieces while still in the field and sorted according to quality. Fiber bananas can live for up to 25 years, but are usually only used for fiber production up to the age of 15.

The fiber bundles of the abacá are arranged on the cross-section of the leaf sheath, as in the agave. Therefore, the extraction of the fiber is similar. The fibers are obtained without prior retting by separating the bast from the leaf flesh by hand or with special defibering machines. The daily output is approximately 9 kg of fiber per worker. The fiber bundles are then dried in the air (Fig. 7.13).

Depending on the degree of cleaning and the color of the fibers, different qualities are distinguished. The color ranges from white to yellowish to brown. The best qualities are obtained when the leaves are harvested immediately before flowering or before fruiting.

World production has been constant for many years and is currently around 109,000 t/year. The largest producers are the Philippines with a world market share of 61% followed by Ecuador with 36%. Of lesser importance are Costa Rica, Indonesia, Equatorial Guinea and Kenya (Fig. 7.14).

Abacá fibers are 2–12 mm long (elementary fiber) or 1000–2000 mm (technical fiber), they have a tenacity of 45 to 70 cN/tex, an elongation of 8% and their density is 1.35–1.50 g/cm^3. They are very resistant to salt water (Table 7.2).

Traditional main applications are therefore ship ropes and fishing nets. Furthermore, they are used for hammocks and transmission ropes. In Japan, they have also been used to make kimonos since the thirteenth century. The waste from fiber extraction is used for paper production ("manila paper") and also for tea bags.

Fig. 7.11 Typical products made of sisal (cat scratching post, dartboard, carpet)

Fig. 7.12 Musa textilis (Bikolano, 2015)

Fig. 7.13 Extraction of fibers (Tagra, 2012; Steger, 2008)

A new application is the use of banana fibers as reinforcement for fiber composite components in automotive construction. The spare wheel well of the Mercedes A-Class, for example, consists of 25% Abacá fibers (PP matrix) from the Philippine island of Leyte (Fig. 7.15). The price for 1 kg of Abacá fibers is approx. 1.2–2 €.

7.3 Pineapple

Pineapple plants belong to the bromeliad family, which thrive mainly in tropical and subtropical areas of South America and the Philippines. There are approx. 3000 species, of which Ananas comosus and Ananas lucidus (Carauá, Caroá) in particular are used for fiber production (Fig. 7.16). The fiber content of the fleshy leaves is 1.5–2.5%. With approx. 20,000 plants per hectare, a fiber yield of 1.5–3 t/ha is obtained.

Pineapple fibers consist of fiber bundles which can be separated from one another by a preparation process. They have an irregular, circular cross-section with longitudinal ridges and a silky luster. The length of the fiber bundles is 10–90 mm, that of the elementary fiber 3–8 mm, the fineness is 2–6 tex for the fiber bundle and 2–4 dtex for the elementary fiber. The tenacity is

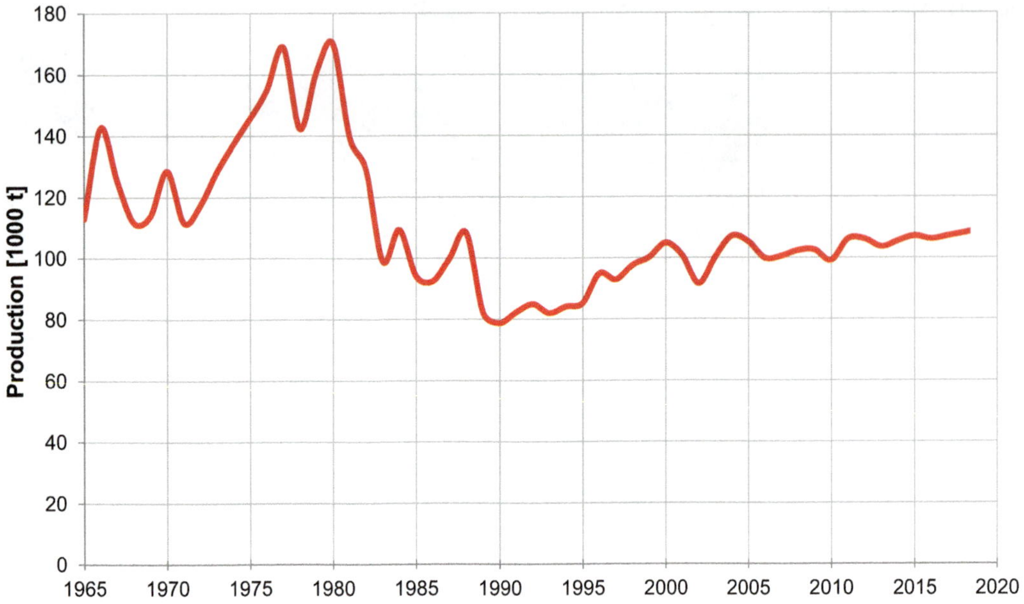

Fig. 7.14 Production of abacá fibers (data: faostat, 2022)

Table 7.2 Abacá fiber properties

Length [mm]	Fineness [µm]	[dtex]	Tenacity [cN/tex]	Elongation [%]	Density [g/cm^3]
2–12	10–50	1.5–35	45–70	8	1.35–1.50

Fig. 7.15 Ruiz (2015) and spare wheel cover (Mercedes A-Class) with reinforcement of abacá fibers

30–40 cN/tex and the elongation 2–5%. The modulus of elasticity is 570–700 cN/tex and the density is 1.44–1.54 g/cm^3 (Table 7.3).

The fibers are obtained from the fiber bundles either by a mechanical extraction process, similar to that used for sisal, or by a retting process in water. This takes approx. 1–2 days at a water temperature of 15–30 °C. Enzymes are used to produce particularly fine fibers. After degumming, the cellulose content is around 94%. Degumming must not be carried out completely, because the elementary fibers can

Fig. 7.16 Ananas comosus (Stang, 2007) and Ananas lucidus (Ziarnek, 2016)

Table 7.3 Pineapple fiber properties

Length [mm]	Fineness [μm]	[dtex]	Tenacity [cN/tex]	Elongation [%]	Density [g/cm^3]
3–8	7–18	2–4	30–40	2–5	1.44–1.54

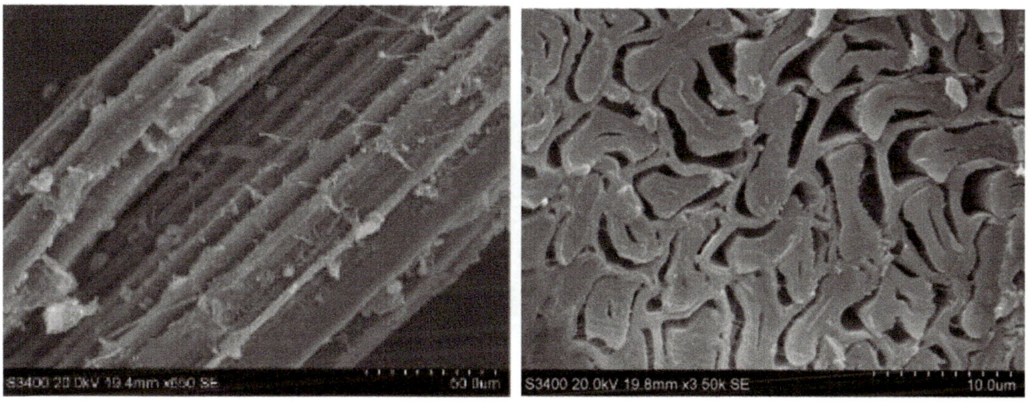

Fig. 7.17 Longitudinal and cross-sectional view of pineapple fibers

then no longer be spun due to their short length (Fig. 7.17).

Pineapple fibers are usually processed in blends with cotton, wool, polyester or silk. Their handle is harder than that of textiles made from flax and ramie, but softer than that of jute and kenaf. Typical products are dresses (pineapple batiste), cords and ropes (technical fibers), hammocks, and nonwovens. Pineapple fibers have also been used in combination with polyamide for fiber composites.

The main areas of cultivation are Brazil and the Philippines, where the pineapple fibers are also called Piňa.

7.4 Lily Fibers

These are fibers obtained from the leaves of the lily family (Liliaceae). The average length of the leaves is 50–200 cm. Lily fibers are used as a substitute for jute, agave and banana fibers. Only 2 species are of significance (Fig. 7.18):

- **Phormium** (New Zealand flax, flax lily) is derived from the genus Phormium tenax, which grows mainly in New Zealand. Compared with agave and banana fibers, the fiber is more lignified, rougher and less firm. The individual fibers are 2–15 mm long and have a diameter of 5–25 µm. Its abbreviation is NF.
- **Yucca fibers** are obtained from the palm lily (Yucca filamentosa), which is native to the southern part of North and Central America. The individual fibers are 1–6 mm long with a diameter of 4–26 µm. Their tenacity reaches 30–65 cN/tex.

7.5 Grass Fibers

Esparto fibers (abbreviation: AL) are obtained from the leaves of the halfa or alfa grass (Stipa tenacissima), which is native to Spain and North Africa, and are used primarily to obtain cellulose (paper) and to manufacture mats and ropes. The fiber yield is 40–50%.

Seagrass is produced from the dried leaves of the grass species Zostea marina, which is endemic to the North Sea and the Baltic Sea (Fig. 7.19). Seagrass is particularly used as an insulating material. Even the first refrigerator was equipped with seaweed as an insulating material. It is

Fig. 7.18 Phormium tenax (Kahuroa, 2006) and Yucca filamentosa (Rebou, 2008)

Fig. 7.19 Alfa grass (Niehaus, 2005) and sea grass (Faulkingham, 2009)

Posidonia oceanica (Gronk, 2007)

Neptune grass ball on the beach (Neptu GmbH)

Roof insulation - interior (Neptu GmbH)

Roof insulation - exterior (Neptu GmbH)

Fig. 7.20 Neptune grass under water, as a ball on the beach and as roof insulation

resistant to vermin and therefore also suitable for people allergic to house dust.

While the two grass fibers mentioned above are of only minor market importance, the fibers of Neptune grass (Posidonia oceanica) are already being used commercially as insulation material (Fig. 7.20). For this purpose, the fiber balls are collected on the beach (Mediterranean Sea), cleaned and processed and marketed under the trade name NeptuTherm®.

7.6 Palm Fibers

Palm fibers are only of local importance in the respective producing countries.

Dwarf palm fibers are obtained from the leaves of the wild-growing dwarf palm (Chamaerops humilis) which grows in Mexico.

Date palm fibers are produced from the dried leaves of the date palm (Phoenix dactylifera). These fibers compete with coconut fiber because the fiber yield is significantly higher at more than 50% and the roasting process is not required (Fig. 7.21).

Oil palm fibers have a cellulose content of about 60%, a tenacity of 10–20 cN/tex and an elongation of 10–14% (Sreekala et al., 1998).

Other fibers, such as peat fibers, are currently of no significant importance and are only used in times of austerity. These also include piassava (for brooms) and rotang fibers.

Fig. 7.21 Dwarf palm (Tato, 2006) and date palm (Jowaninpensans, 2019)

References

Bikolano, M. (2015). Retrieved from https://commons.wikimedia.org/wiki/File:Abaca_Tree.jpg, CC BY-SA 4.0.

Dobbertin, W. (1906). Retrieved from https://commons.wikimedia.org/wiki/File:Bundesarchiv_Bild_105-DOA0876,_Deutsch-Ostafrika,_Sisalfabrik.jpg, CC BY-SA 3.0 DE.

faostat. (2022). Retrieved from http://www.faostat.fao.org

Faulkingham, C. (2009). Retrieved from https://commons.wikimedia.org/wiki/File:Zostera.jpg

Gronk. (2007). Retrieved from https://commons.wikimedia.org/wiki/File:Posidonia.JPG?uselang=de, CC BY-SA 1.2.

Jowaninpensans. (2019). Retrieved from https://commons.wikimedia.org/wiki/File:Date_palm_in_Morrab_Gardens,_Penzance.jpg, CC BY-SA 4.0.

Kahuroa. (2006). Retrieved from https://commons.wikimedia.org/wiki/File:NZflaxPiha02.jpg

Mapswire. Retrieved from https://mapswire.com

Michiel. (2012). Retrieved from https://commons.wikimedia.org/wiki/File:Lanzarote_-_Agave_fourcroydes.jpg, CC BY-SA 3.0.

Navez, B. (2008). Retrieved from https://commons.wikimedia.org/wiki/File:EuropaIsland_Sisal_02.JPG, CC BY-SA 3.0.

NeptuTherm®. Retrieved from http://www.neptugmbh.de, Neptu GmbH, Karlsruhe.

Niehaus, C. (2005). Retrieved from https://commons.wikimedia.org/wiki/File:Stipa_tenacissima.jpg, CC BY-SA 1.2.

Raschka, A. (2008). Retrieved from https://commons.wikimedia.org/wiki/File:Sisal_fibres.JPG, CC BY-SA 4.0.

Rebou. (2008). Retrieved from https://commons.wikimedia.org/wiki/File:Yucca_filamentosa_fh_1182.6_SC_B.jpg, CC BY-SA 1.2.

Ruiz, L. (2015). Retrieved from https://commons.wikimedia.org/wiki/File:Abaca_weaving_in_Banton.JPG, CC BY-SA 4.0.

Sánchez, G. (2017). Retrieved from https://commons.wikimedia.org/wiki/File:Hacienda_Sotuta_de_Pe%C3%B3n_-_M%C3%A1quinas_8.jpg, CC BY-SA 2.0.

Schuiling, R. (2011). Retrieved from https://commons.wikimedia.org/wiki/File:2011.10-386-145ap_sisal,processing(decortication),factory_Athinai_Sisal_Estate@Mogotio(Rift_Valley_Prv.),KE_sat15oct2011-1534h.jpg, CC BY-SA 4.0.

Sreekala, M. S., Kumaran, M. G., & Thomas, S. (1998). Oil palm fibers: Morphology, chemical composition, surface modification, and mechanical properties. *Journal of Applied Polymer Science, 66*(5), 821–835.

Stang, D. (2007). Retrieved from https://commons.wikimedia.org/wiki/File:Ananas_comosus_17zz.jpg, CC BY-SA 4.0.

Steger, J. (2008). Retrieved from https://commons.wikimedia.org/wiki/File:Abaca_sachsenleinen_01.jpg, CC BY-SA 3.0.

Tagra, R. (2012). Retrieved from https://www.flickr.com/photos/kamsky/8223921119, CC BY 2.0.

Talbot, J. (2007a). https://commons.wikimedia.org/wiki/File:SisalLeavesDecortication.jpg, CC BY 2.0.

Talbot, J. (2007b). Retrieved from https://commons.wikimedia.org/wiki/File:SisalDecorticationMachine.jpg, CC BY 2.0.

Talbot, J. (2007c). Retrieved from https://commons.wikimedia.org/wiki/File:SisalFarmDrying.jpg, CC BY 2.0.

Tato, G. (2006). Retrieved from https://commons.wikimedia.org/wiki/File:Chamaerops_humilis_(Zingaro)029.jpg, CC BY-SA 2.5.

Valke, D. (2007). Retrieved from https://commons.wikimedia.org/wiki/File:Agave_cantala_(505,790,915).jpg, CC BY-SA 2.0.

Wiki. (2003). Retrieved from https://commons.wikimedia.org/wiki/File:Plantsisal.jpg, CC BY-SA 1.2.

Wiki. (2008). Retrieved from https://commons.wikimedia.org/wiki/File:Pacas_de_sisal.jpg

Ziarnek, K. (2016). Retrieved from https://commons.wikimedia.org/wiki/File:Ananas_lucidus_kz1.jpg, CC BY-SA 4.0.

8 Fruit Fibers

Fruit fibers either grow outside the fruit and protect it from mechanical damage (coir) or they grow inside the fruit body and help the seeds to disperse (kapok, poplar). Fruit fiber plants grow primarily in Southeast Asia (Fig. 8.1).

8.1 Coir Fibers

Coir fibers (abbreviation: CC) are the framework of the outer covering of the coir, which is the fruit of the coconut palm (Cocos nuciferia). They are therefore classified as fruit fibers. Their property profile is similar to that of hard fibers which is why they are often regarded as hard fibers. Coconut palms for fiber production thrive mainly on the southwest coast of India, the south coast of Sri Lanka, in Indonesia and Malaysia as well as in Vietnam and Thailand. The main producer is India with a market share of approx. 45% (as of 2018).

The coconut palm probably originated in the Philippines. About 3000 years ago, it was known in India and from about 300 BC it is documented in Sri Lanka. The coconut contains up to 60% oil, which is why it is also one of the most important oil fruits. Since the middle of the nineteenth century, coir fibers have been used industrially for the production of ropes and textiles.

Coir palms grow up to 25 m high and bear up to 200 fruits per year after 8 years (Fig. 8.2). They remain fertile for up to 100 years. The coconut consists of an outer, firm and tough shell, the underlying fibrous cushion about 8 cm thick, the hard inner shell and the inner white kernel containing the milk. The elastic and air-containing fibrous layer serves as mechanical protection for the core (Fig. 8.2, right) and as an insulator for the coconut milk. It accounts for 30–35% of the nut weight and is composed of a short-fibered intermediate filling and the vascular bundles, which can be up to 30 cm long. These consist of a central cell strand which is surrounded in a cylinder-like manner by a thick layer of bast cells. Each fiber pad yields approx. 60–100 g of staple fibers and 10–40 g of brush fibers. Thus, approx. 10–15 nuts are required to produce 1 kg of fibers (Wagner, 1981).

Approximately every 2 months, the fully grown coconuts are harvested and the fiber husks removed. The fibers are then subjected to water or mud retting for approx. 5–6 months, then rinsed and knocked out. The fiber bundles are then broken up, the fibers separated, parallelized and spun (Fig. 8.3). The average fiber yield is 1500 kg/ha.

Coir fibers have a technical length of 15–30 cm and a diameter of 100–450 μm. The elementary fiber has a length of 0.3–1 mm and is therefore naturally not spinnable (Fig. 8.4). The tensile strength of the technical fiber is 12–18 cN/tex when dry, its elongation is 15–40%. Its

D. Veit, *Fibers*, https://doi.org/10.1007/978-3-031-15309-9_8

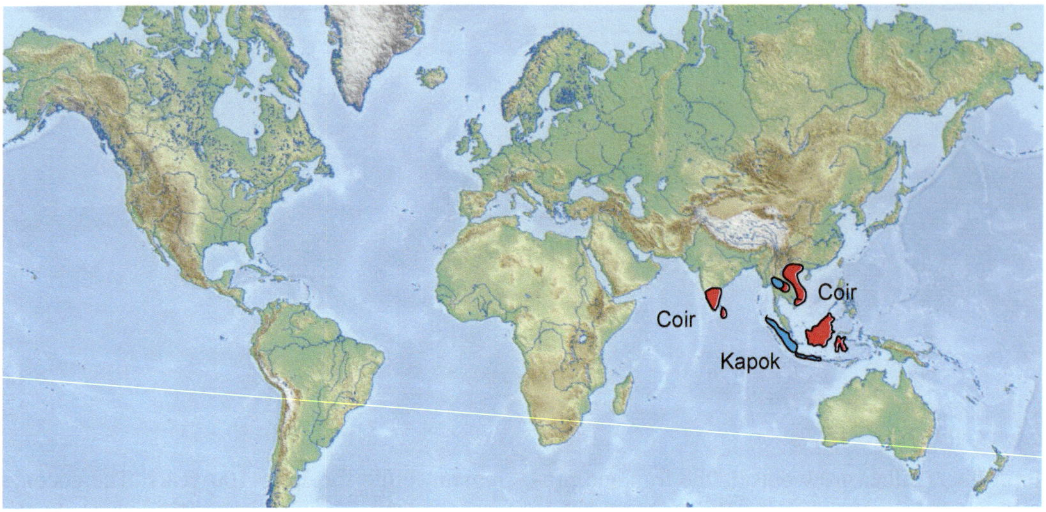

Fig. 8.1 Main cultivation areas of important fruit fibers (map: mapswire)

Fig. 8.2 Coconut palm and structure of the coconut (Grobe, 2008)

density is 1.4 g/cm^3. Coir fibers are particularly resistant to abrasion, moisture and rot as well as weathering and are difficult to soil.

Due to the high proportion of air inclusions (40–65% by volume), they are very light and float on water. Their color is reddish to brown (Fig. 8.5).

The main areas of application are ropes, mats and carpets as well as fillers. Coir fibers are also used in the automotive industry, e.g. in the back of seats (Fig. 8.6).

World production is currently around 1.2 m tons and has been relatively constant for around 10 years. Its value is around US$ 100 m. If all the coconuts harvested each year were used for fiber production (approx. 61.5 mt in 2020), this would result in a world production of approx. 19 mt (faostat, 2022, Fig. 8.7).

The production volumes of the main producers are shown in Table 8.1.

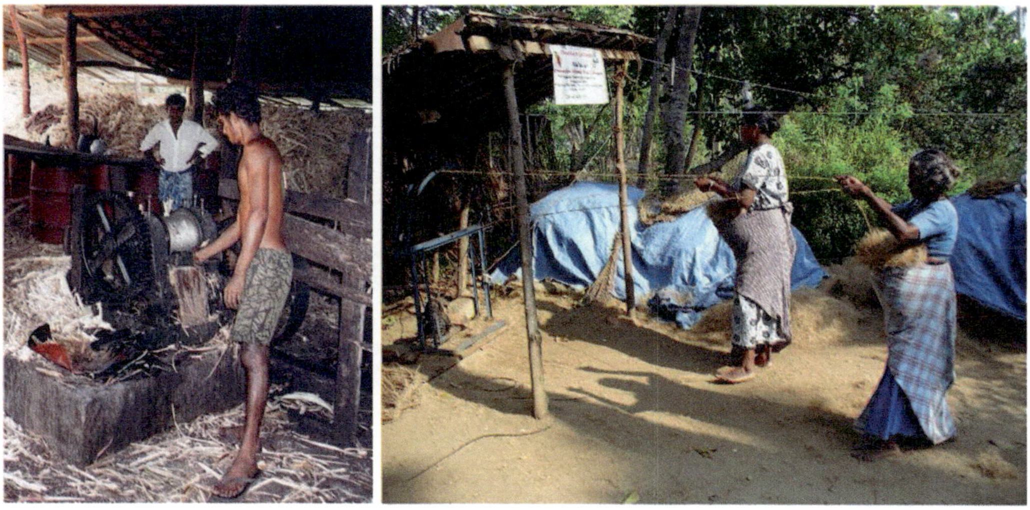

Fig. 8.3 Extraction of coir fibers (Krankman, 2006) and rope production (Bricaniwi, 2010)

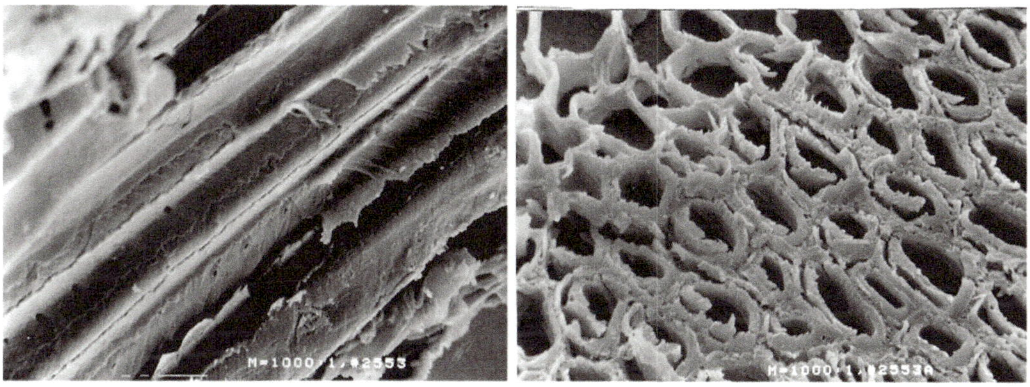

Fig. 8.4 Longitudinal and transverse view of coir fibers

Fig. 8.5 Coir fibers (Fotokannan, 2011) and coir rope (Maari, 2011)

Fig. 8.6 Car seat with coir fiber back and coir mats (Midori, 2011)

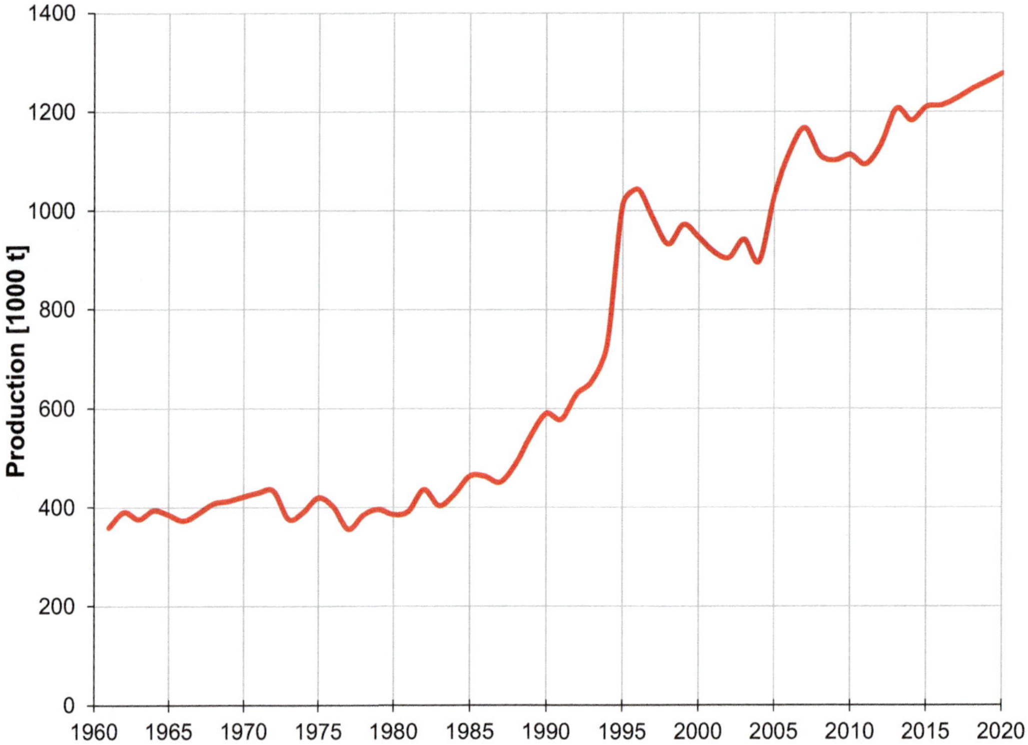

Fig. 8.7 Production of coir fibers (data: faostat, 2022)

Table 8.1 Main coir fiber producing countries (data: faostat, 2022)

	[t]	%
India	581,641	46.1
Vietnam	381,761	30.3
Sri Lanka	160,736	12.8
Thailand	62,833	5.0
Ghana	39,811	3.2
Malaysia	21,488	1.7
Bangladesh	9964	0.8
Ivory Coast	2186	0.2

8.2 Kapok (Wool Tree)

Kapok fibers (abbreviation: KP) are obtained from the fruit capsules of the woolly trees of the Bombaceae family that grow in tropical regions. Kapok trees grow mainly in the southern hemisphere and near the equator. To protect the seeds contained in the fruit capsules, unicellular, soft and shiny fibers develop on the inner fruit wall, which are also called plant down or plant wool. The two most important species of kapok are the African cotton tree (Eriodendron anfractuosum) and the Indian silk wool tree (Bombax ceiba, Ceiba pentandra). Kapok trees grow up to 70 m high and live up to several hundred years. The trees are fast-growing and have branches arranged in tiers. The flowers are white to pink (Fig. 8.8).

After pollination, the fruit develops in the capsule, which can be up to 15 cm long and from whose inner wall the fibers grow. Because they are attached to the fruit wall and not to the seed, they are not actually seed hairs (although classified there according to DIN 60001). The plant wool detaches from the wall of the fruit capsule as the seeds mature. Up to 20 kg of fruit are harvested per tree. 250 capsules yield approx. 1 kg of raw kapok (Fig. 8.9).

The fibers obtained from it are called kapok, bomba or ceiba wool. Kapok grows wild or semi-wild, particularly in East and West Africa, India, Java, Sumatra, Mexico and Brazil, and is not cultivated in plantations. Because kapok grew in the German colonies of the time, efforts were made to use kapok fibers as a textile raw material, but all failed (Ulbrich, 1913).

The yellowish fibers are 10–40 mm long, have a thickness of 20–30 μm, are smooth and have only low strength. Kapok is rarely spun pure, but mostly in blends. Because kapok fiber is a hollow fiber (70–80% by volume air inclusion), it has a very low specific weight, absorbs little moisture, dries quickly, is breathable and insulates well (Fig. 8.10). Kapok fibers contain natural bitter substances which largely prevent infestation by mites, moths and other animals.

Kapok fibers are hardly spinnable and are therefore used almost exclusively as fillers for cushions, pillows, beds and stuffed animals

Fig. 8.8 Kapok tree (Zezinho68, 2016) and kapok fruits (Donald, 2009)

Fig. 8.9 Kapok fruit on tree (Dupont, 2016), burst kapok fruit (Mead, 2013) and kapok fibers with exposed seeds

(especially for allergy affected), as mattress toppers (warm in winter, cool in summer) and as fillings for life jackets (load-bearing capacity up to 30 times its own weight) and lifebuoys. Since the 1950s, kapok has been largely replaced as a filling material by foams and therefore no longer plays a major role today (Fig. 8.11).

Otto GmbH & Co., KG, Dietenheim, markets a yarn made from a blend of 85% cotton with 15% under the trade name Piumafil®.

Indonesia (71%) and Thailand (29%) are the main producing countries. The total production volume in 2020 was around 90,000 t (Fig. 8.12). The main markets for kapok are in Japan, China, the EU and the USA. 1 kg of kapok costs around 2 €.

8.3 Poplar Fibers

For some years now, the Freiburg-based company Pappella has been processing the fibers of the poplar fruit into fillers for pillows and duvets under the brand name Pappelflaum®. The pea-sized fruits of the female trees are harvested either by cutting the branches using tethered balloons or by tree climbers. The still unopened capsules are dried, causing them to burst open and release the fibers. Approx. 1–1.5 kg of poplar fibers are obtained per tree, which is sufficient for 3 quilts.

Poplar fibers are very fine (diameter up to 8 μm), but do not have the necessary length (only approx. 5 mm) and strength to be spun. They are therefore processed into nonwovens blended with cotton, linen, tussah silk, wool or camel fluff (Fig. 8.13).

According to the world's only manufacturer (Pappella), the fibers keep you as warm as down and wick moisture as well as wool (Fig. 8.14).

The manufacturer lists the following advantages for its product:

- No pesticides or fertilizers are required for cultivation,
- low energy consumption during harvesting,
- regional product (short transport distances),
- no chemical pre- or post-treatment during cleaning.

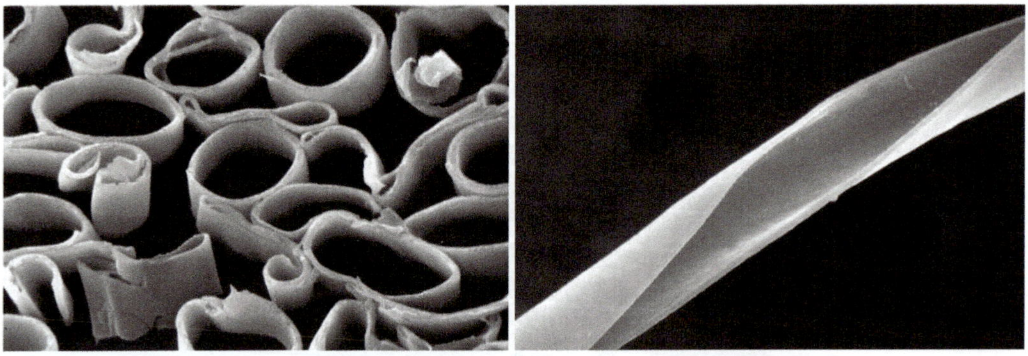

Fig. 8.10 Longitudinal and cross-sectional view of kapok fibers

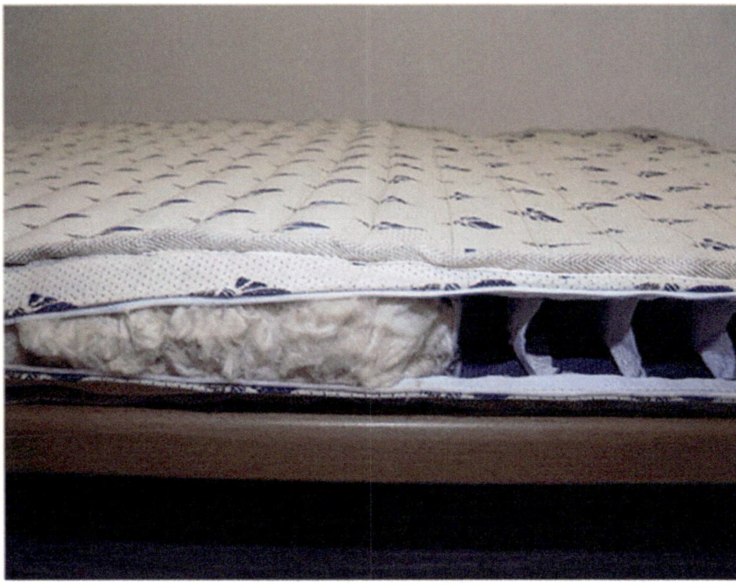

Fig. 8.11 Mattress with kapok filling

8.4 Betel Nut Fibers

The betel palm thrives mainly in Malaysia and Indonesia (Fig. 8.15). The nuts of the betel palm can be chewed and used as an intoxicant. The betel nuts are surrounded by a fibrous layer that cushions the fall, similar to coir. The fibers are obtained by a mechanical process. The fiber

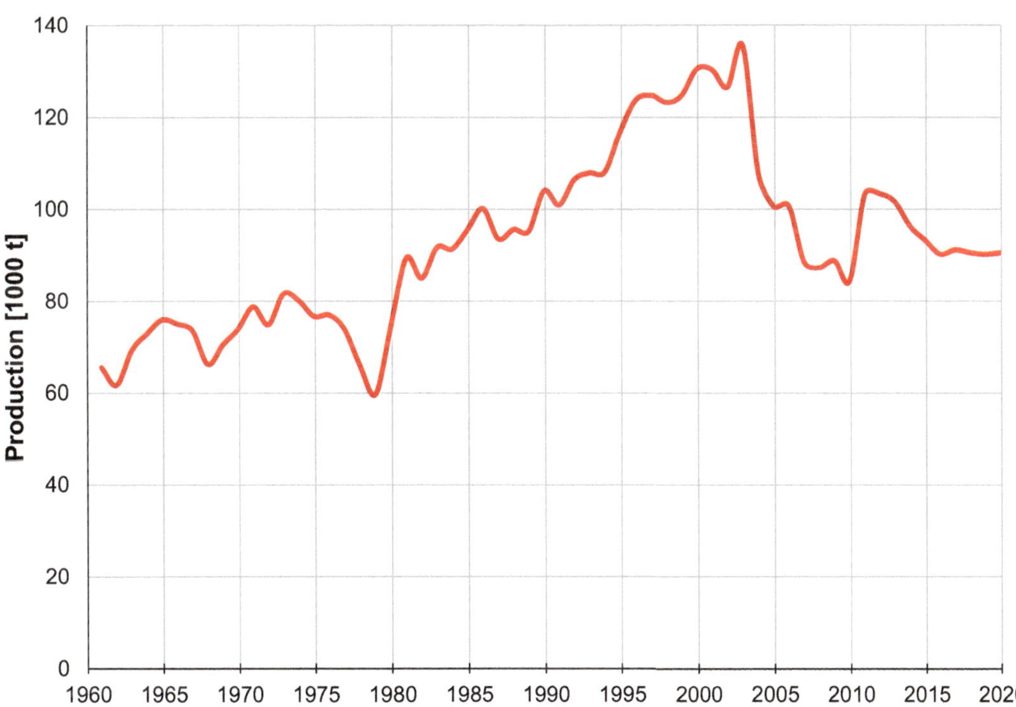

Fig. 8.12 Production of kapok fibers (data: faostat, 2022)

Fig. 8.13 Poplar fibers on the tree

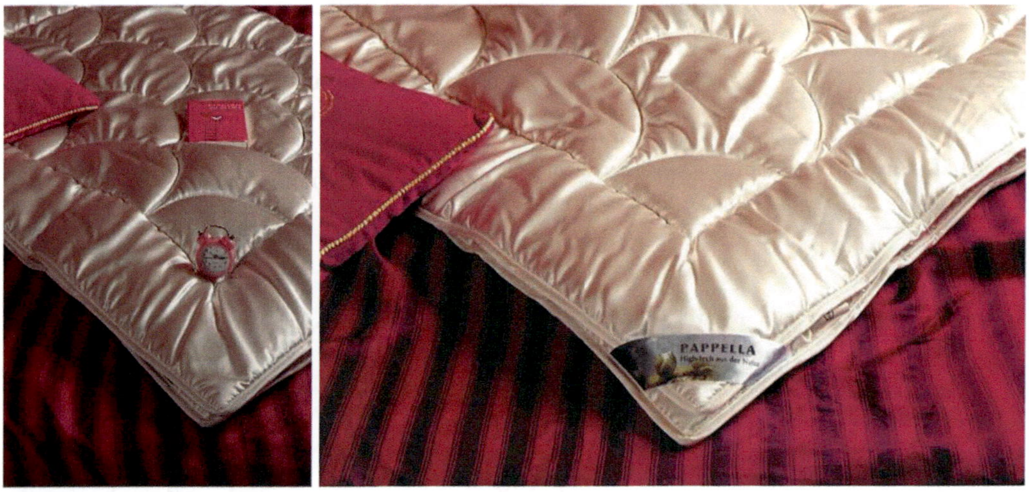

Fig. 8.14 Blanket with poplar fiber filling (Margrit Müller, Pappella)

diameter is 20–50 μm. The strength reaches similar values as coir fibers.

The fibers can be used to produce fiber-reinforced plastics (Yusriah et al., 2012). However, this has so far only been successful on an experimental scale and is economically questionable.

8.5 Cactus Fibers

Numerous cacti possess fibers on the outside of their shoot axils. Thus, strictly speaking, they are not fruit fibers, but still less do they belong to the bast or hard fibers. Cactus fibers have a very uniform structure with a diameter of 10–20 μm (Fig. 8.16). Their strength is 5–20 cN/tex. The special feature of cactus fibers are the characteristic spines (Fig. 8.16, right). This property predestines them for use in fiber composites, because the fiber-matrix adhesion is significantly improved when the fibers are embedded directionally.

This type of fiber should not be confused with the "cactus silk", which was already used by pre-Columbian peoples of Central and South America. These are hard fibers from agaves.

Fig. 8.15 Betel nut palm (Garg, 2007a) and ripe betel nuts before harvest (Garg, 2007b)

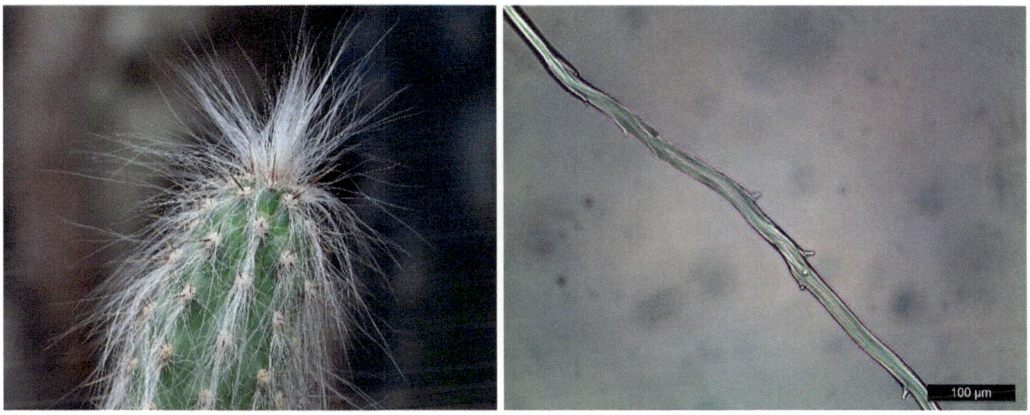

Fig. 8.16 Cactus with fibers and barbed cactus fibers

References

Bricaniwi. (2010). Retrieved from https://commons.wikimedia.org/wiki/File:Making_coir_rope_in_Kerala.JPG, CC BY-SA 4.0.

Donald, M. R. (2009). Retrieved from https://commons.wikimedia.org/wiki/File:Kapok3839674626_48e20e6274_o.jpg, CC BY-SA 4.0.

Dupont, B. (2016). Retrieved from https://commons.wikimedia.org/wiki/File:Kapok_Tree_(Pseudobombax_tomentosum)_(31754308746).jpg, CC BY-SA 2.0.

faostat. (2022). Retrieved from http://www.faostat.fao.org

Fotokannan. (2011). Retrieved from https://commons.wikimedia.org/wiki/File:Coir_fibery.jpg, CC BY-SA 3.0.

Garg, J. M. (2007a). Retrieved from https://commons.wikimedia.org/wiki/File:Palm_I_IMG_2081.jpg, CC BY-SA 1.2.

Garg, J. M. (2007b). Retrieved from https://commons.wikimedia.org/wiki/File:Palm_I_IMG_2108.jpg, CC BY-SA 1.2.

Grobe, H. (2008). Retrieved from https://commons.wikimedia.org/wiki/File:Cocco-nut_hg.jpg, CC BY 3.0.

Krankman. (2006). Retrieved from https://commons.wikimedia.org/wiki/File:Sri_Lankan_coir_fiber.jpg, CC BY-SA 2.5.

Maari, T. (2011). https://de.m.wikipedia.org/wiki/Datei:A_view_of_coir_pith_rope.JPG, CC BY-SA 3.0.

Mapswire. Retrieved from https://mapswire.com

Mead, D. (2013). Retrieved from https://commons.wikimedia.org/wiki/File:Kapok_(pods).jpg, CC0 1.0 Universal (CC0 1.0).

Midori. (2011). Retrieved from https://commons.wikimedia.org/wiki/File:Keset_sabut_kelapa.jpg#mw-jump-to-license, CC BY-SA 3.0.

Pappella, Freiburg. Retrieved from http://www.pappella.de

Ulbrich, E. (1913). Die Kapok liefernden Baumwollbäume der deutschen Kolonien im tropischen Afrika. *Notizblatt des Königl. botanischen Gartens und Museums zu Berlin, 6*(51), 1–34.

Wagner, E. (1981). *Die textilen Rohstoffe*. Deutscher Fachverlag.

Yusriah, L., Sapuan, S. M., Zainudin, E. S., & Mariatti, M. (2012). Exploring the potential of betel nut husk fiber as reinforcement in polymer composites: Effect of fiber maturity. *Procedia Chemistry, 4*, 87–94.

Zezinho68. (2016). Retrieved from https://commons.wikimedia.org/wiki/File:Kapok_Tree_Rio.jpg?uselang=de, CC BY-SA 4.0.

9 Wool

Wool belongs to the animal fibers together with the fine and coarse animal hairs and silk. Wool is understood to mean only the hair of the domestic sheep (Ovis aries). The abbreviation is WO.

The most important breeds of sheep for fiber production are Merino (Fig. 9.1), Crossbred and Comeback, the most important producing countries are Australia (fine fibers) and New Zealand (coarse wool, mainly for carpets).

9.1 Terms

Depending on the type of wool production, a distinction is made between:

- Pure new wool: obtained by shearing living sheep.
- Skin wool: obtained from the skin of slaughtered sheep.
- Grease wool: obtained by a thermal treatment of the fur.
- Lime wool: obtained by coating the flesh side of the hide with a hydrolyzing depilatory agent (e.g. lime) and a reducing agent (e.g. sulfide).
- Slipemaster wool: obtained from pieces of fur by mechanical means.
- Tanner's wool: obtained by immersing the whole hide in a liquor containing alkali and possibly a reducing agent. Mostly obtained during leather production in tanneries (severe damage to the fibers).
- Reclaimed wool: recovered from semi-finished or finished products by a tearing process (secondary raw material). The fiber length is shorter than that of pure new wool.

The age and sex of the sheep can also be used to characterize the wool:

- Lamb's wool: from young sheep shorn for the first time in their first year of life.
- Yearling wool: from young sheep which have already been shorn as lambs and not yet used for breeding.
- Ewe's wool: obtained from ewes.
- Mutton wool: obtained from castrated rams.
- Buck wool: obtained from rams.

Depending on the condition of the wool, a distinction is made between:

- Raw wool (grease wool): shorn from sheep, still contains all accompanying substances (wool grease, sweat, dirt).
- Back-scoured wool: surface-cleaned in fresh water on live sheep.
- Scoured wool: machine washed, largely degreased.
- Carbonized wool: treated with acid to destroy the vegetable components ("charring" of the vegetables) and then heated.
- Carded sliver wool: machine washed and formed into tape shape by carding.
- Wool top: machine-scoured and combed wool in sliver form (high degree of parallelization of the fibers).

D. Veit, *Fibers*, https://doi.org/10.1007/978-3-031-15309-9_9

Fig. 9.1 Merino sheep (Wiki, 2005) and wool fiber (DWI)

Wool may also be subdivided according to the parts of the sheep's body from which it was obtained:

- Fleece: the entire shorn coat of the sheep.
- Fleece wool: the skirted fleece, i.e. after adhering foreign parts have been removed.
- Belly and back wool: the belly or back part of the fleece.

More generally, a distinction can also be made simply according to the genus of sheep:

- Merino: from Merino sheep, average fiber diameter ≤ 25 µm.
- Crossbred: from crossbred sheep, mean fiber diameter > 25 µm
- Comeback: of sheep from backcrosses between Crossbred and Merino ewes.
- Coarse wool: average fiber diameter > 30 µm

9.2 History

It is now believed that the sheep was the third wild animal, after the dog and the goat, to be domesticated around 9000 BC in what is known as the Fertile Crescent (north-eastern Iraq and Iran, Anatolia, south-central Asia). The domestic sheep (Ovis aries) is a descendant of the wild sheep (Fig. 9.2). These are distinguished into the western wild sheep (mouflons) and the steppe wild sheep (Uriale). The former were widespread

Fig. 9.2 Wild sheep (Rufus, 2014; Stoger, 2017)

Fig. 9.3 Wool sheep on the standard of Ur (Geni, 2013)

from the Balkans to what is now eastern Turkey, while the latter lived in an area stretching from Iran to northern India. Wild sheep were initially bred primarily for meat, fat, milk and hides, and only later for wool.

Sheep's wool is probably the first textile raw material used by humans to make clothing. At first, felts were made from the plucked or shorn hairs. With the development of spinning and weaving technology, woven fabrics ("cloth") were also produced. With the keeping of sheep for wool production, the way of life of the people also changed, because wool can be transported over great distances, so that sheep can be kept far away from settlements in areas where no food can be grown. It is now thought, based on preserved bones, that the breeding of sheep for wool production was 'invented' several times at different times and in different areas (Schier & Pollock, 2018).

Cave paintings in Spain show that sheep were kept there as early as 3000 BC. Fine-wooled breeds of sheep were introduced to Spain by the Phoenicians, and later by the Greeks and Carthaginians. The fibers were mainly brown and black, had little crimp, but were considered "fine". The oldest written mention is on clay tablets from Babylonia ("land of wool") from the time around 2000 BC. There, besides grain and oil, wool is mentioned as another important product of the country. In addition to bread and beer. The servants at the royal court also received an annual allocation of wool (men 1.5–2.0 kg, women 1.0–1.5 kg and children 0.5–1.0 kg). This probably led to the invention of scales because wool is compressible and weight cannot be determined without scales (Büttner, 2019). Wool sheep also shows the peace side of the standard of Ur from the third millennium BC (Fig. 9.3), which underlines their economic importance.

The oldest finds of woollen fabrics come from Denmark, approx. 2300 BC. Asia Minor was known for particularly fine wools, which is why the predecessors of Merino sheep are assumed to have been domesticated there. While in Egypt rather coarse wools were produced (29–35 μm), in Italy, in the first century, there were wools with fiber diameters of 16–24 μm, which would also be considered fine today. In antiquity, sheep breeding was widespread, as the many references in the writings of Virgil, Varro and Columella from the first century show.

The Merino sheep, which is important for wool sheep breeding, probably originates from the areas around the Black Sea. From there, it came to Spain via North Africa. Breeding began around 1280 and was promoted by the Visigoths and the Moors as well as the kings of Castile and Aragon. The most important Merino sheep breeds

Fig. 9.4 Wool extraction and dyeing of wool cloth in the fifteenth century

in Spain were the Eskurial, the Paulars, the Infandatos and the Negretti, some of which are still bred today. A strict ban on the export of live sheep (punishable by death) ensured Spain's position as the sole producer of fine wool for centuries. In the sixteenth century, the English government banned wool exports to support the domestic textile industry. This led to competitors in Flanders and Italy switching to Spanish Merino wool, which rapidly increased their production output. Whereas in Spain large flocks were kept, especially by the nobility (up to 25,000 heads in the eighteenth century), in England there were many unorganized sheep farmers with correspondingly smaller flocks. Therefore, it was first possible in Spain to produce finer wool with a yield of 2–4 kg per sheep. Nevertheless, England dominated the market until the eighteenth century with its large production despite the coarser wool.

The sheep population in the world today is due to the intensive breeding of Merino sheep in Spain and the sheep breeds in England. The most important are Lincoln, Leicester, Romney-Marsh, Cheviot and Southdown.

In the Middle Ages, wool was the most important textile raw material after linen. Many countries and towns owed their prosperity to wool production and processing or the wool trade. These include the north of France, Flanders, Brabant and the area between the Meuse and the Rhine. Ypres, Ghent, Bruges, Maastricht, Aachen and Cologne were already respected cloth-making centers in the twelfth century. Brussels, Louvain and Mechelen followed in the fourteenth century, then Leiden and Amsterdam in the fifteenth century (Fig. 9.4). In the Mediterranean region, Genoa, Florence and Venice were of great importance. The mechanization of textile production increased the production of cloth at the expense of raw wool exports. Nearly half of all woolen goods produced in England were exported. In 1750, about 1.5 m people were employed in the English wool industry. At this time, half the wool processed in Europe came from England and half from Spain. Apart from the Rhineland (especially in Aachen), there was no commercial production of woollen goods in Germany until the eleventh and twelfth centuries. Citizens and peasants made their clothing from linen, only aristocratic gentlemen wore clothing made from woolen fabrics.

Flemish wool weavers settled in northern Germany as well as in Saxony and Thuringia. Because of their knowledge in mechanical fulling, finishing and dyeing of cloth, the wool trade developed there in a similar way as in England. Woolen goods were exported as far as Russia and Italy.

Except for the finishing of wool cloth in fulling mills (from the seventeenth century), wool processing (sorting, scouring, combing, spinning, weaving) was a manual process. Mechanization of the wool industry was slower than in the cotton industry because of the more difficult processing of the fibers. The first functional combing machines for removing the short fibers were

Fig. 9.5 Peltzer's wool scouring machine (Kerl & Stohmann, 1879)

built in 1843 by S. C. Lister and in 1853 by J. Noble (circular combers) and in 1845 by J. Heilmann (flat combers). The worsted carding machine was invented in 1820, the needle bar section in 1834 and the double needle bar section in 1860. The first wool scouring machine was built in 1855 by Peltzer (Fig. 9.5). The water enters from the left (E) and circulates around the center wall (B).

It was replaced in 1864 by J. G. Ry-Kaivers' Leviathan (Fig. 9.6), which dominated the next few decades and is still used in modern form today.

Merino sheep were introduced to Spain by the Arabs, where breeding began in 1280. The export of Spanish Merino sheep to other countries of the world began only in the eighteenth century (Fig. 9.7). Before that, export was strictly forbidden to secure Spain's monopoly on fine wool.

After the settlement of South America, South Africa and Australia by Europeans at the end of the eighteenth century, it was quickly recognized that the climate there was ideally suited to sheep farming. The first Merino sheep were brought to Australia via South Africa as early as 1797, and sheep farming began in South America in 1810 (Fig. 9.8).

The development of the different sheep breeds depends on climatic and breeding conditions and therefore proceeded differently. For example, Merino sheep could not survive in Sweden, England and New Zealand (too cold, too wet). In other countries, however, huge flocks developed (Australia, South Africa).

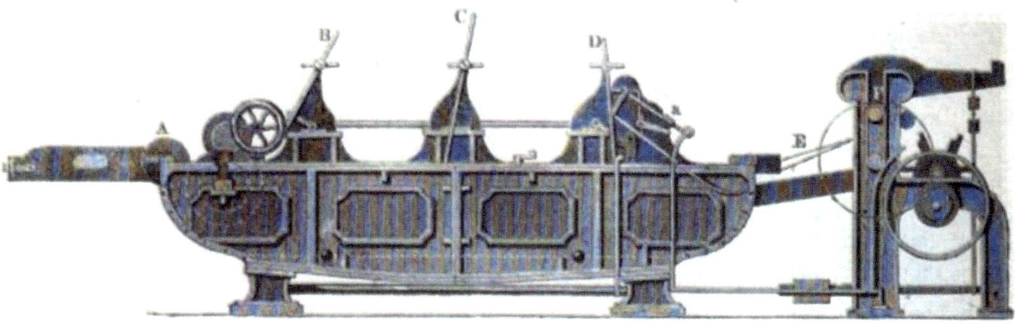

Fig. 9.6 Leviathan wool scouring machine (Kerl & Stohmann, 1879)

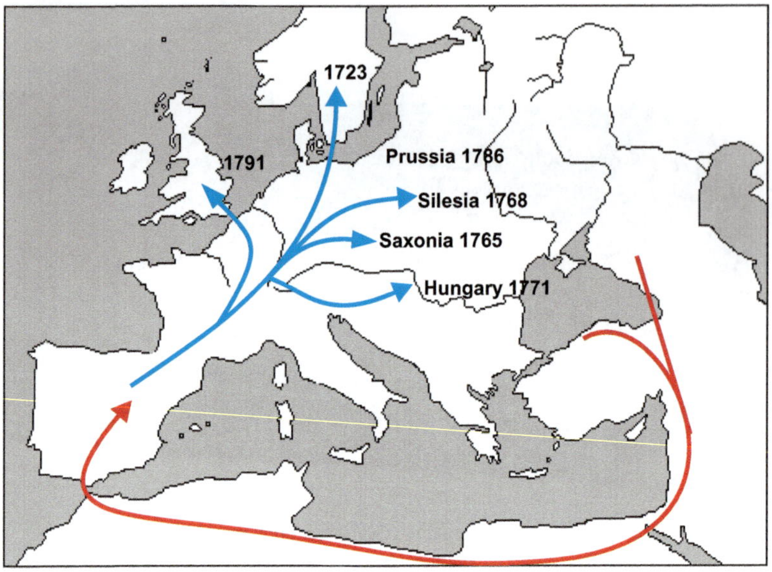

Fig. 9.7 Dispersal of Merino sheep in Central Europe in the eighteenth century

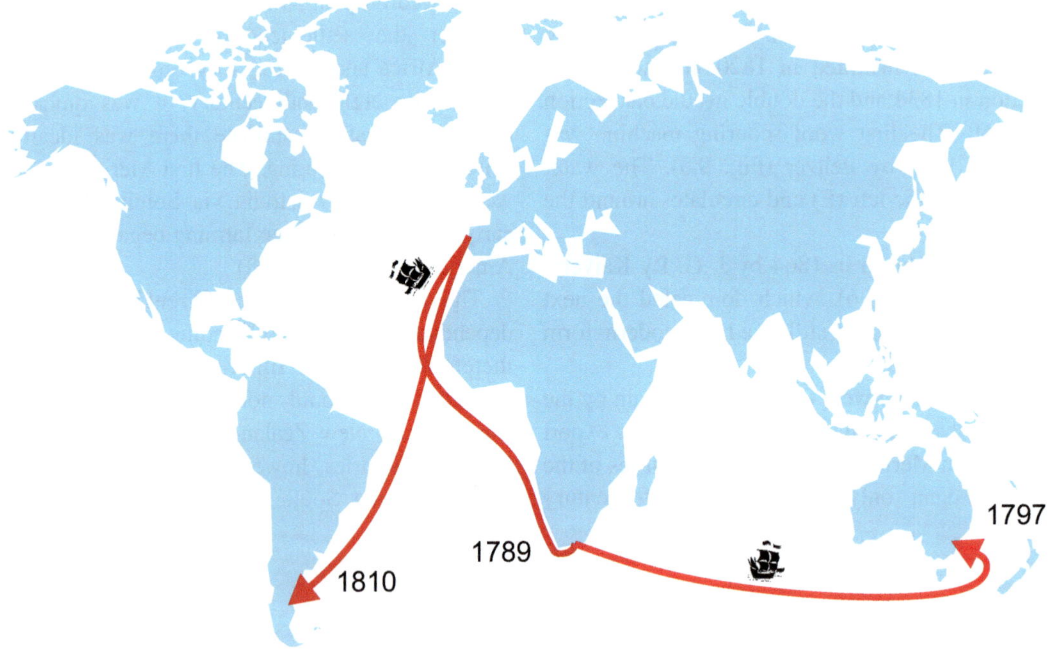

Fig. 9.8 Dispersal of Merino sheep to South America, South Africa and Australia

Before the export of Merino sheep overseas, the breeding of sheep with fine wool was the domain mainly of Germany, Spain and France. However, this changed quickly after the overseas herds were established at the end of the eighteenth century. Due to the large pastures, huge flocks were bred there and wool was soon exported to Europe.

In 1804, the first Australian wool was auctioned in London. Until about 1870, two thirds of wool imports from overseas were made via the London Wool Exchange. Starting in 1865, wool was also traded in Bremen. From 1872 onwards, wool combers were established in Leipzig, Hanover and in the Vogtland region. In 1883 the Bremer Wollkämmerei (BWK) was entered in the commercial register. It was in operation until 2009 and in the meantime was the largest company of its kind in Germany.

Around 1900, wool accounted for about 20% of the world fiber market (cotton 80%), and 1 m tons were produced annually. Although the production volume of greasy raw wool has doubled since then to approx. 2 m tons, the relative share of scoured wool in the total fiber production of around 120 m tons is now less than 1%.

9.3 Sheep Breeds and Breeding

Sheep thrive best in dry areas, e.g. on steppe and savannah pastures. Tropical areas, on the other hand, are unsuitable. The climatic factors of altitude, rainfall and air temperature are decisive for the growth of sheep and wool. Sheep acclimatize well, their bodies adapting quickly even to extreme conditions (e.g. poor feed supply). According to estimates, there are more than 1000 sheep breeds worldwide (Fig. 9.9).

The numerous breeds of sheep are usually classified according to the fineness of the corresponding fibers into those that produce fine, medium-fine and coarse wool. The finer the fibers, the shorter they usually are (Table 9.1).

Fig. 9.9 Important sheep breeds (selection)

Table 9.1 Selected sheep breeds and their fiber fineness and lengths

		Fiber fineness [µm]	Fiber length [mm]
Extra ultrafine wool	Merino, superfine wool	<16	70–75
Ultrafine wool	Merino, superfine wool	16–17.5	70–75
Superfine wool	Merino, superfine wool	17.6–18.5	70–75
Fine wool	Merino, fine wool	18.6–19.5	70–75
Medium fine wool	Merino, medium wool	20–22	85–115
	Merino, strong wool	23–25	100
	Polwarth	21–25	110–120
	Corriedale	24–31	90–150
Coarse wool	Perendale	28–35	125
	Suffolk	30–35	75–100
	South Suffolk	28–32	75–100
	Poll Dorset	27–32	60–100
	Ryeland	25–32	80–100
	Cheviot	28–33	75–100
	English Leicester	32–38	200–250
	Lincoln	33–41	300–400
	Romney	30–40	125–200

Systematic breeding of individual races began in the Middle Ages. From a genetic point of view, there are two breeding methods to choose from: crossbreeding and pure breeding. In crossbreeding, there are unequal hereditary characteristics, i.e. mating takes place between different breeds or lines. In pure breeding, there are equal hereditary characteristics, i.e. sheep of the same breed are paired with each other. The aim of breeding is to increase the performance of the parents (wool, meat, milk) in the offspring as much as possible. Selection (breeding choice) is very important for the breeding method. It can be done according to phenotype (external appearance) and genotype (genetic make-up). The latter method uses the knowledge and methods of population genetics. It works with mathematically determined values and has the better chances of success.

When selecting for wool performance, the factors of fiber fineness and its evenness ("fidelity") as well as the hair density of the sheep fleece are of decisive importance. In New Zealand, the Corriedale sheep was obtained by crossing English meat rams with Merino ewes. In Australia, it was bred from Merino rams and Lincoln ewes (Figs. 9.10 and 9.11).

These "crossbreds" became especially native to South America, apart from New Zealand. Besides relatively fine wool, they also provide plenty of meat. With the development of frozen meat technology in 1882, the Crossbreds gained great importance, because this meant that the meat could also be exported to Europe.

In addition to simple cross-breeding, "backbreeding" can also produce sheep with finer wool. For example, the combination of a Lincoln and a Merino sheep results in the "Half-bred". Crossing with a Merino sheep results in a "1/4-bred" and further crossing with a Merino sheep results in the "comeback" or the "7/8-Merino".

Due to climatic and economic conditions in the respective countries, sheep farming is either specialized in wool production (Australia, South Africa) or in wool and meat production (New Zealand, Argentina, Uruguay).

Some important breeds of sheep are briefly described below.

9.3.1 Merino

The Merino sheep is the economically most important sheep worldwide. Merinos are

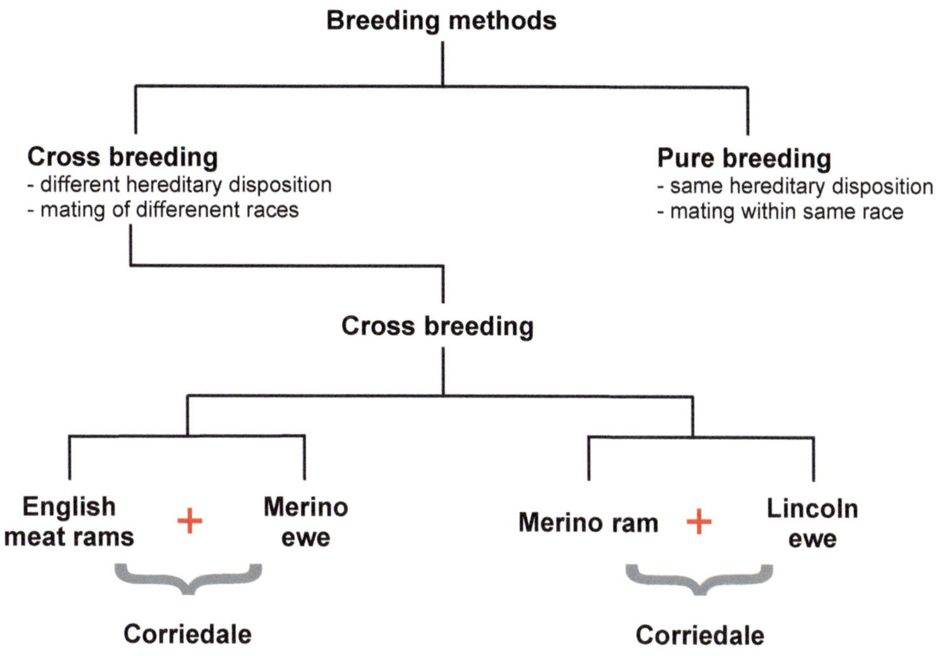

Fig. 9.10 Breeding of Corriedale from Merino and Lincoln

classified according to their wool quality as superfine, fine, medium and strong Merino (Table 9.1). Merino sheep originally come from North Africa and were brought to Spain by the Merinids, a Berber tribe. Today's Merinos are all descended from 3 sub-breeds, the Escurial, the Negretti and the Paula Merinos. From the fourteenth century onwards, Merinos were specifically bred for the production of fine wool. However, it took about 100 years until it was available in sufficient quality. Until the eighteenth century, the export of Merinos from Spain was forbidden. In 1765, the Spanish King Charles III gave his brother-in-law, Prince Xaver of Saxony, 210 Merino sheep as a gift, so that he could establish his own breeding. These sheep became known as "Electoral" because the Prince Regent, was allowed to elect the German King. This gave rise to the "Saxon Merino", which produced the finest wool. Frederick II of Prussia also acquired 300 Merino sheep from Spain (1768) and so after the conquest of Silesia and parts of Poland, Merino sheep were also bred there because of the good pastures. In 1808, the Spanish monopoly over Merino sheep was finally broken by Napoleon I's invasion and the Spanish Merino herds were largely destroyed or dispersed throughout Europe. Merino sheep, including the "Saxon Merino", reached Australia

Fig. 9.11 Merino, Lincoln and Corriedale

Fig. 9.12 Poll merino (Goodwin, 1998)

via South Africa, where they have since been bred for both fiber and meat. Around 40% of the wool produced worldwide comes from Merinos. A new breed is the Poll Merino, which has no horns (Fig. 9.12).

The Merino is therefore not a purebred sheep, but combines the genes of about 20 different breeds. Hence, it is particularly adaptable.

9.3.2 Lincoln

The Lincoln is considered the ancestor of all long-staple English sheep breeds. Its wool is coarse (33–41 μm) and exceptionally long (300–400 mm). Therefore, it was crossed with Merinos and Corriedale and Polwarth were created.

9.3.3 Corriedale

The Corriedale is the second most important sheep breed after the Merino (Fig. 9.13). It originated in Australia in 1874 from a cross between the Merino ram and the Lincoln sheep. Today it is raised especially in North and South America and in Asia. It is mainly used for meat production (lamb), but also provides coarse wool of 24–31 μm in diameter.

9.3.4 Polwarth

This breed is 75% Merino and 25% Lincoln. It produces medium-fine wool (21–25 μm) with a good yield. Apart from Australia, there are large populations for fiber and meat production, especially in South America.

9.3.5 Romney

Romney sheep originate from England and produce coarse wool (30–40 μm) with a large staple length (up to 200 mm), which is mainly used for carpets.

9.3.6 Cheviot

The cheviot originated in the fourteenth century along the English-Scottish border. It produces coarse wool of good fiber length.

9.3.7 Perendale

The Perendale is a cross between Romney and Cheviot and produces coarse wool (28–35 μm) with a good fiber length (up to 125 mm) for carpets and hand knitting yarns. It is also used for meat production.

9.3.8 Border Leicester and English Leicester

These sheep breeds produce coarse wool (32–38 μm) for carpet yarns. They are also used for meat production (Fig. 9.13).

9.4 Wool Extraction

90% of the wool fibers are obtained as pure new wool from live sheep. Shearing is divided into full shearing (after 12 months of wool growth, "full length" fleeces), half shearing (after 6 months,

Fig. 9.13 Border Leicester and Corriedale (The Woolmark Company)

"second shear"), second shearing and three-quarter shearing (after 7–8 months). Mostly full shearing is used, in New Zealand, half and three-quarter shearing dominate. The type of shearing determines the fiber length. The shearing yield per sheep depends on the breed, age and nutrition of the sheep as well as climatic conditions. It is about 5 kg per year for Merinos and up to 10 kg for coarse wool sheep. From 5 kg of wool, about 10 m of fabric can be made, which is enough to produce 6 sweaters and 3 suits, or upholstery fabric for a large sofa.

The wool of a sheep becomes coarser every year. After about 5 years, it is so coarse that the sheep is usually slaughtered.

9.4.1 Manual Shearing

For shearing, the animals are herded together, separated according to age and sex, sorted into lambs, yearlings, ewes, rams and mutton and driven into so-called 'paddocks'. This requires adherence to a precise schedule because the paddocks are small and provide little food for the sheep. Supplementary feeding would be unprofitable. A sudden rainfall can change the entire schedule, because wet wool can basically not be shorn. The shearers conclude a fixed contract with the farmer before shearing, in which the minimum number of sheep to be shorn and the price are specified. The shearer works on a piecework basis and shears a sheep within 1–4 min, which results in around 100 to 300 sheep in an 8 h working day (Fig. 9.14). The fleece is cut off in one piece from the belly side.

The fleece of a sheep is not of uniform quality. This depends rather on the region of the body on which the wool has grown (Fig. 9.15).

After shearing, the fleece is taken to the classification table where it is skirted (Fig. 9.16). In the process, bad corners, manure parts, markings and very bad wool parts are separated from the rest of the fleece.

The classifier assesses the wool on the basis of his experience according to fineness, staple length, fat content and staple strength ("sound"). Only after the basic qualities have been determined does is the fleece classified into different qualities, e.g. Super, AAA, AA. These qualities are stored in separate compartments and then packed into bales (Fig. 9.17). The wrapping is made of jute or polypropylene, similar to cotton. Each bale is numbered, marked with the farm name and quality number and then transported to the auction by trucks. Figure 9.18 shows typical proportions of fibers and other components of fleeces for Merino and Crossbred. Because Merino fibers are finer, they have more crimp and more dirt can be trapped. Their fat content is also higher.

Fig. 9.14 Sheep shearing and shorn sheep (Pot, 2007; Kidd, 2007)

9.4.2 Mulesing

The skin of Merino sheep is very wrinkled to produce more fibers on the thus increased skin surface. This also affects the rear of the sheep. On the one hand, the narrow folds make shearing difficult and, on the other hand, fly maggots settle in the folds of the skin, which can lead to inflammation (fly strike). For this reason, the skin folds on the sheep's hindquarters are cut off. This "mulesing" (according to J.W.H. Mules) is often done without anaesthesia and is very painful for the sheep. There are now efforts by animal rights activists to end this practice by boycotting large textile manufacturers. However, only about 10% of wool is obtained without mulesing, but painkillers are administered to as many as 45% of sheep. By breeding sheep with smooth skin, for example, attempts are being made to eliminate this practice completely. For example, the SRS (soft, rolling skin) Merino sheep, first bred in the 1990s by J.E. Watts, has a smaller surface area but a higher hair density. Furthermore, the hairs are said to grow faster, so that even 3 shearing cycles are possible within 2 years. The corresponding wool is marketed under the name New Merino® (NewMerino, 2021).

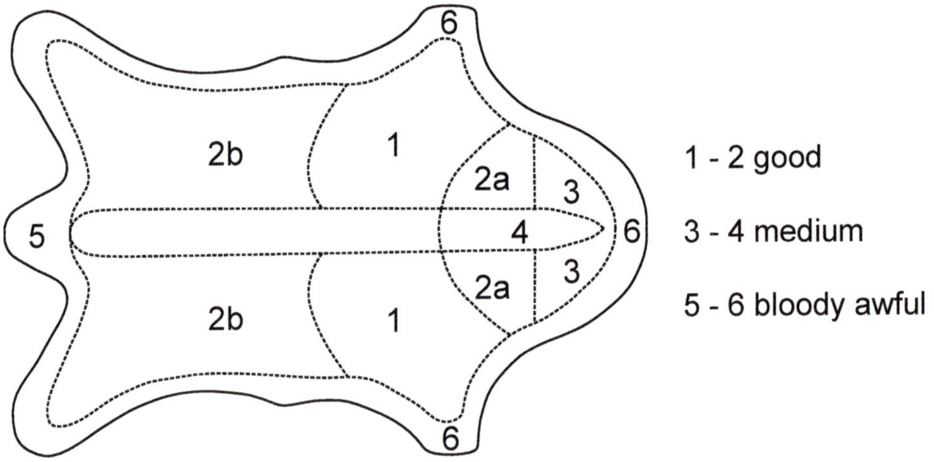

Fig. 9.15 Subdivision of the fleece parts

Fig. 9.16 Skirting

9.4.3 Mechanical Shearing

In the 1970s and 1980s, various attempts were made to develop a shearing robot (Fig. 9.19). E. g., the sheep is first clamped and then shorn with the help of a robotic arm. In early models, the electrical resistance between the shearing blade and the sheep's mouth was measured and the robotic arm moved accordingly. However, this was cumbersome and inaccurate. In the latest version, instead, the resistance of the shearing blade was determined during the cutting process. This increased the shearing speed up to 80 cm/s, but the shearing blade could not do this. In practical tests, 2 min were needed for shearing the belly, back, flanks and neck (approx. 90% of the wool), and another minute for the remaining 10%.

Machine shearing was not successful at the time, partly because of a crisis in wool production following the revolutions in Eastern Europe, and was later not pursued, partly because of the high investment costs of around US$ 35 m. However, a by-product was a sheep manipulator that can also be used by hand shearers. According to the manufacturer, it reduces the shearing time from 4 to 3.5 min per sheep.

9.4.4 Chemical Processes

As an alternative to mechanical shearing, chemical substances, e.g. cyclophosphamide, can also be administered to the sheep via the fodder. These cause the sheep to shed its wool. To avoid soiling the fibers, the sheep can be wrapped in plastic bags beforehand. In this way, only the plastic bags with the fibers contained therein need to be removed again afterwards. This method has so far only been tested on a pilot scale and has not gained acceptance. A detailed description of this and various other methods is given in Hudson (1987).

Fig. 9.17 Wool bales ready for transport to the stock exchange (Esson, 2006)

9.5 Structure of the Wool Fiber

Wool is round in cross-section, which is why its fineness is given in [μm]. Furthermore, it is the only fiber that has scales, which can be used for felting (Fig. 9.20).

9.5.1 Follicle

Wool is formed in follicles. These are infoldings of the epidermis into the skin, at the base of which are papillae of mesodermal cells. Follicles are dynamic organs. Therefore, several different processes occur in them simultaneously: The

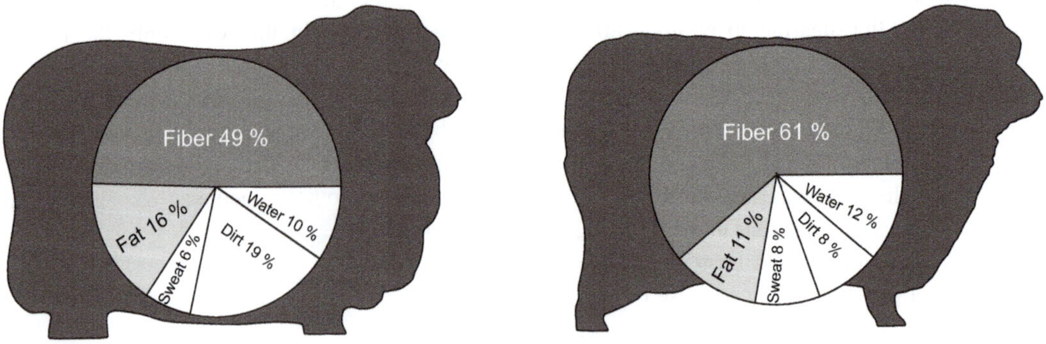

Fig. 9.18 Fleece components of Merino (left) and Crossbred (right)

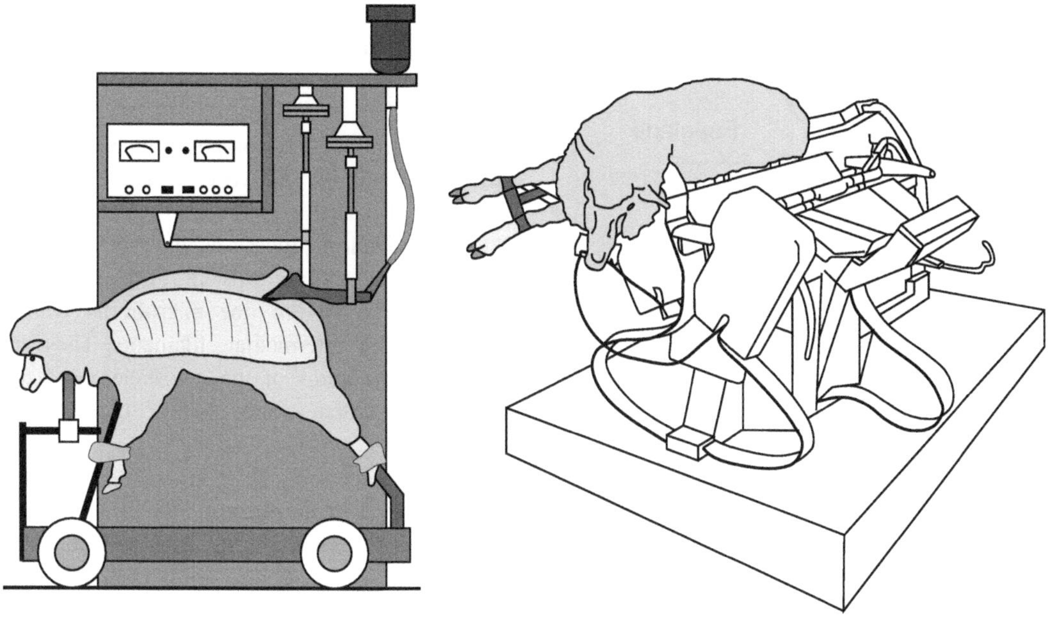

Fig. 9.19 Shearing robots 1974 (left) and 1983 (right) after (Trevelyan, 1999)

division, differentiation, and upward movement of fiber-forming cells (trichocytes), and the biosynthesis of monomeric wool proteins as gene products. This also includes the self-assembly of the monomers into supramolecular structures and their keratinization through the formation of cross-linking disulfide bridges, for example in the middle part of the follicle. In addition, cells of the root sheaths are degraded.

A distinction is made between primary and secondary follicles. The primary follicles develop earlier and consist, in addition to the hair canal, of a sweat gland and a fat gland (Fig. 9.21). Secondary follicles develop later and, in addition to the hair canal, have only a fat gland but no sweat gland. The ratio of the number of secondary to primary follicles varies depending on the sheep breed and is 20:1 for fine Merinos and only 2:1 for coarse wool.

After the completion of follicular development, the hair begins to grow. This does not occur continuously, but in cycles. Growth periods alternate with rest periods. The growth rate can be influenced by feeding. It should be noted that an oversupply of food leads to thicker fibers. On the other hand, if there is a lack of feed, the fibers form thin spots which can later lead to fiber breakage. Mother animals also form thin spots in their hair shortly before lambing, so ideally shearing should take place before that. Then the

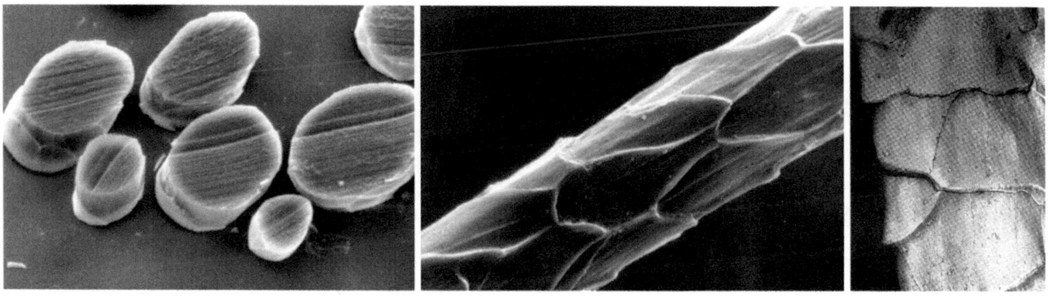

Fig. 9.20 Cross-sectional and longitudinal section of wool fibers and scale structure

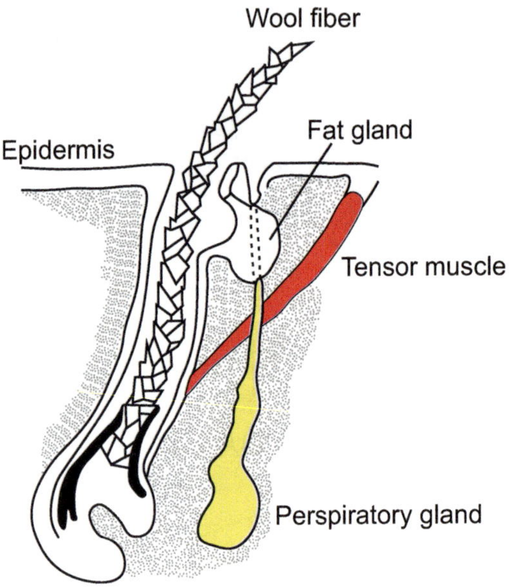

Fig. 9.21 Wool primary follicle

thin spots are in the tip of the hair, which often breaks off during further processing anyway.

9.5.2 Morphology

Wool fibers consist of two cell types, the cuticle (scale) and the cortex (fiber stem). In coarse fibers, a third type is added, the central medulla (pith), Fig. 9.22.

The cuticle consists of plate-like cells that overlap each other longitudinally and peripherally. Thereby, the approx. 1 μm thick scale edges are oriented towards the fiber tip. On the surface of the cuticle cells, the hydrophobic epicuticle forms a thin membrane of proteins and lipids and protects the fiber from environmental influences.

The cortex consists of densely packed, 80–100 μm long and 3–6 μm wide, spindle-shaped cells (Bradbury, 1973). Different types of cortex are distinguished: ortho-, para- and meso-cortex cells. In highly crimped Merino wool fibers, the softer orthocortex lies on the curved outer side of the crimping arc ("bilateral structure"), the harder paracortex on the inside. Coarser, less crimped wools are segmented radially into ortho- and paracortex (Figs. 9.23 and 9.24). Furthermore, the number of meso-cortex cells is greater in coarse wools than in fine wools.

Each cortex cell is composed of 50–200 nm thick macrofibrils, between which there are remnants of the cell components degraded during the keratinization process as intermacrofibrillar material. The macrofibrils in turn consist of approximately 10 nm fine microfibrils, which are also called intermediate filaments. They are important structures of the cytoskeleton of the cells of vertebrates and also of other living organisms.

The monomers of the intermediate filaments are embedded in the macrofibrils in sulfur-rich interfilament material. In the cross-section of an intermediate filament there are 8 protofilaments consisting of four-chain structural units. One unit contains 2 dimers in an antiparallel position. The dimers are pairs of one type I and one type II keratin in parallel arrangement. The medulla (pith) in coarser wool fibers consists of hollow cells with a skeleton of amorphous proteins and fine filaments. The natural coloration of brown wool comes from pigment granules of black-brown eumelanin and yellow-red pheomelanin.

Depending on the morphological structure of the cortical cells, the cuticle forms differently. Depending on the sheep breed, this results in very different scale structures, ranging from very coarse to very fine (Fig. 9.25).

9.5.3 Biological Composite Structure and Two-Phase Model

The complex morphological and molecular structure of wool reflects the design principle of all biological composite structures. Many artificial materials combine components with different properties into a material that is optimally adapted to its tasks (Fig. 9.22).

From the point of view of elastic properties, the decisive difference between the components is based on their microcrystalline and non-crystalline degree of order. If only the helical domains in the microfibrils are considered to be

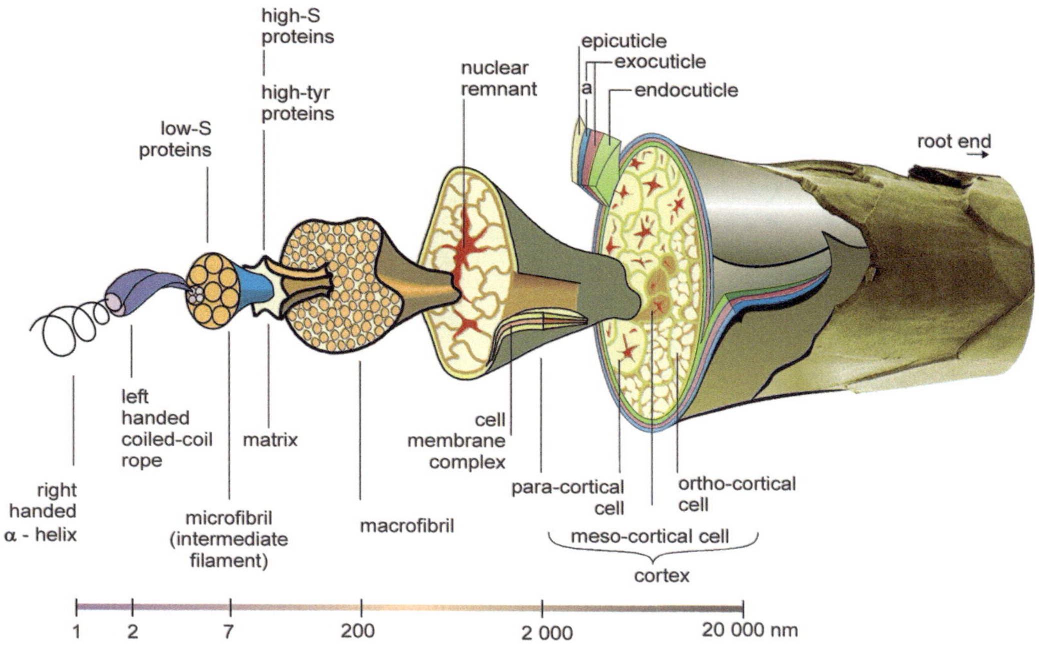

Fig. 9.22 Structure of a fine Merino wool fiber (CSIRO, 2008a)

microcrystalline, then the crystalline phase accounts for 30% of the fiber. These microcrystalline domains form 50% of the filaments (microfibrils). From the point of view of materials science, all other components are to be grouped together as the "matrix" phase, which is to be distinguished from the morphological matrix component, i.e. the interfilament material in which the microfibrils are embedded (Zahn et al., 1991).

9.5.4 Chemical Composition

Wool is a protein fiber and contains 17–24 different amino acid side groups (Fig. 9.26), with arginine, glutamic acid, cysteine and serine

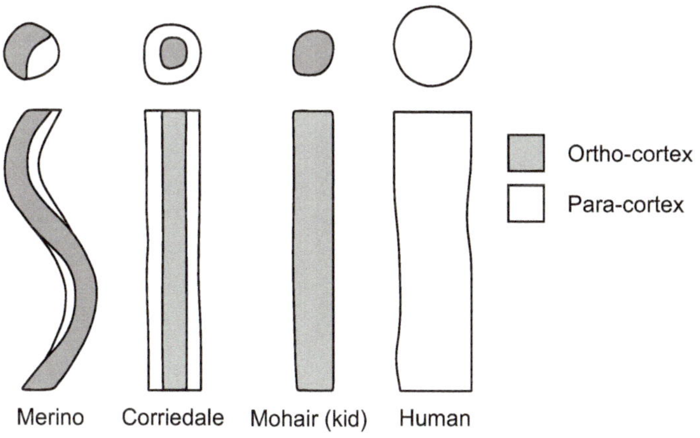

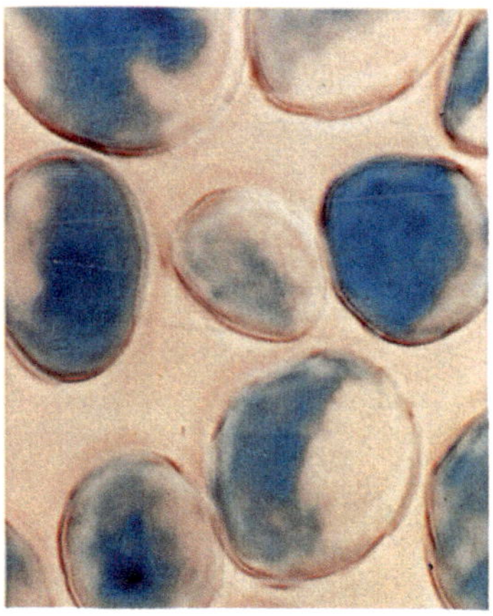

Fig. 9.23 Cortex types (right: orthocortex cells stained with methylene blue)

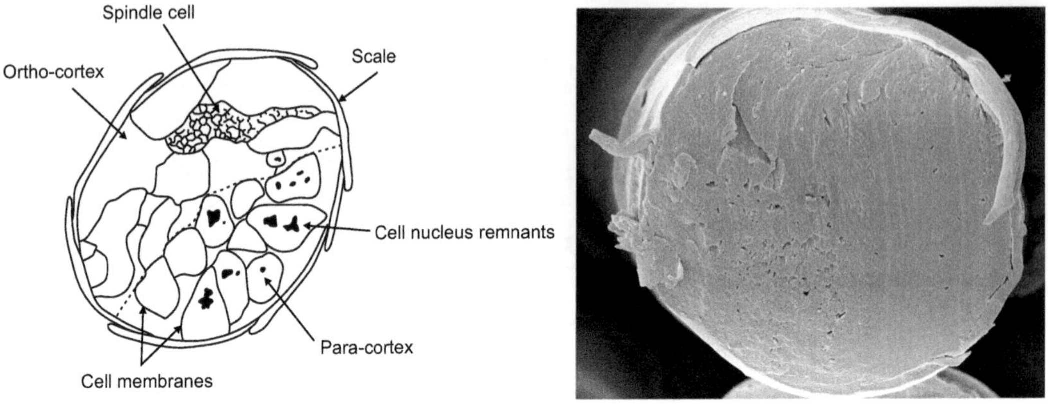

Fig. 9.24 Cross-section through a wool fiber acc. to (P. Kassenbeck) and (Lenzing AG)

accounting for about 44% and the others occurring in mostly similar proportions (Table 9.2). According to Zahn et al. (1980), wool consists of a total of approximately 170 different proteins.

Because proteins contain cationic and anionic groups, they are called poly-ampholytes. They can therefore neutralize both acids and alkalis to a certain extent. These polar groups also mean that wool can absorb and bind up to 30% of its own weight in water without feeling damp.

The composition of different wool qualities shows Table 9.2. "Amide" stands for amide residue and is not an amino acid, but is listed for the sake of completeness.

Wool consists of carbon, hydrogen, oxygen, nitrogen and sulfur (Fig. 9.27). This composition is—with the exception of the sulfur content—typical for all proteins. The high sulfur content is mainly due to the amino acid cysteine, which can form a disulfide bridge ("cystine bridge") with other cysteine groups via two sulfur atoms. The ash contains mainly potassium, sodium, calcium, aluminium, iron, silicon dioxide, sulfate, carbonate and traces of phosphorus pentoxide and chloride.

Similar to silk, according to Pauling et al. (1951), wool proteins consist of crystalline and non-crystalline domains. The former contain the short side chains without sulfur bridges and thus

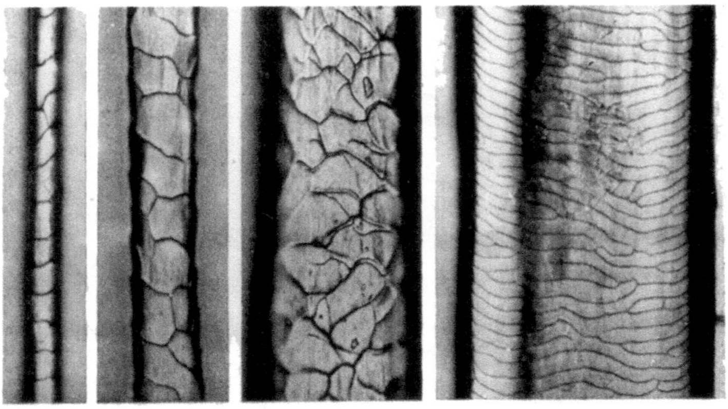

Number of scales per girth	1	3 - 4	6 - 10	10 - 14
Provenance	AAA	A	D/E	Angora rabbit

Fig. 9.25 Number of scales of different breeds of sheep as well as Angora rabbits

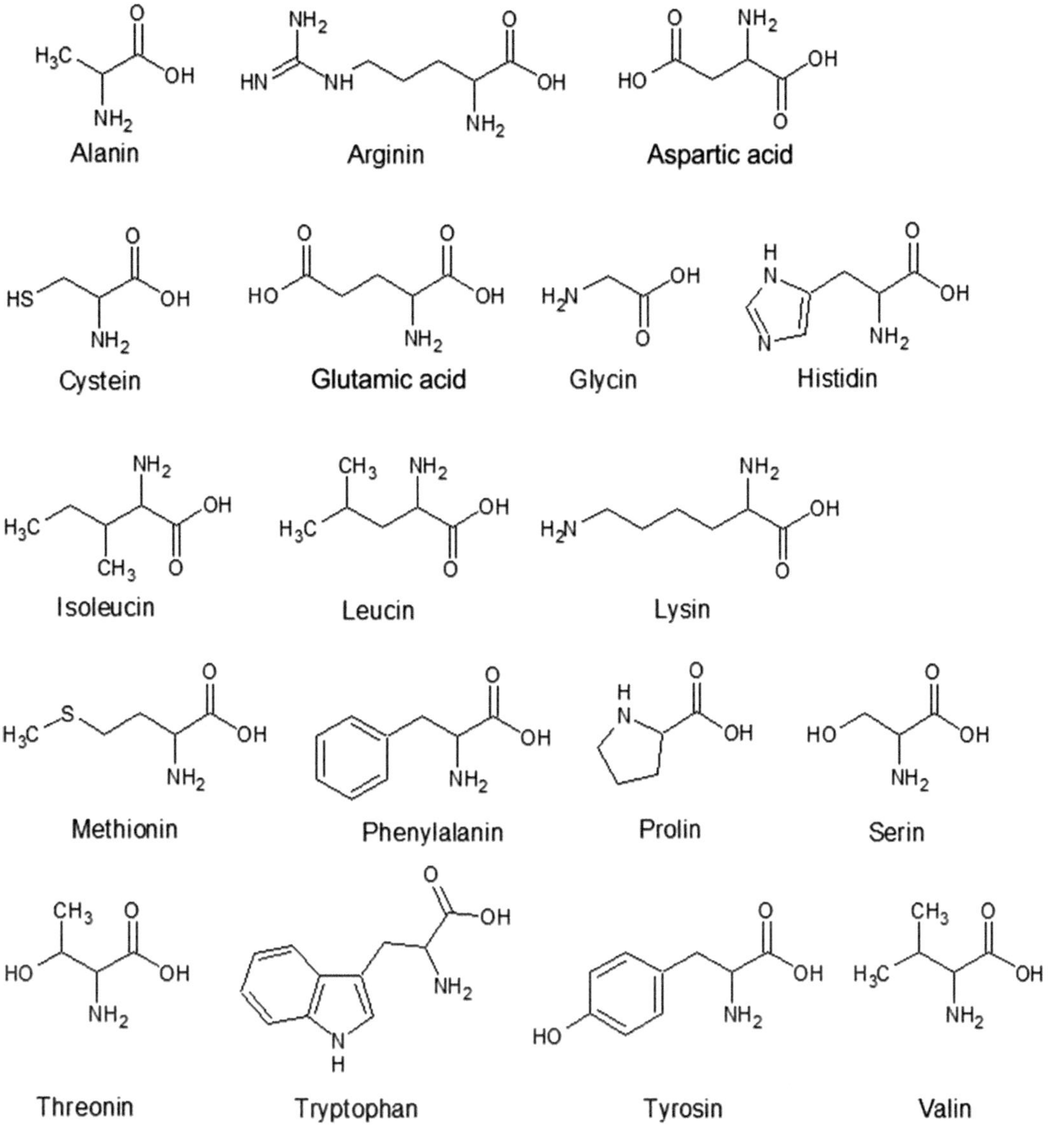

Fig. 9.26 Structural formulae of the amino acids of wool

can easily attach parallel to each other. The latter have sulfur bridges and, in particular, glycine and tyrosine side chains. These largely amorphous regions are tightly linked by numerous sulfur bridges. The covalent bonds between the sulfur atoms of the cysteine groups of adjacent chains or within the same chain provide the relatively high strength and elasticity of the wool (Fig. 9.28).

Attractive forces can also develop between other side groups, as Fig. 9.29 shows. For example, a covalent isopeptide bond forms between cysteine and lysine, a salt bridge between arginine and glutamic acid, a hydrogen bond between aspartic acid and serine, and an aromatic end group interaction between phenylalanine and tryptophan. The strength of the bond here decreases from top to bottom.

In raw wool, the peptide chains are present in the form of the α-helix (α-keratin) as described by Pauling et al. (1951). Here, two helices are

Table 9.2 Proportions of amino acids for different wool qualities according to Simmonds (1956)

Amino acid	Merino 64s	Merino 70s	Corriedale 56s
Alanin	3.5	3.5	4.4
Arginin	20.3	19.4	18.2
Aspartic acid	4.2	4.7	4.9
Amid	7.5	7.9	9.3
Cystein	7.9	6.5	6.8
Glutamic acid	8.6	8.5	9.7
Glycin	5.8	6.6	6.4
Histidin	1.5	1.5	1.6
Isoleucin	2.0	2.1	2.4
Leucin	4.9	5.4	5.5
Lysin	3.3	3.2	3.7
Methionin	0.3	0.4	0.4
Phenylalanin	1.8	2.3	2.4
Prolin	5.3	5.1	5.5
Serin	7.3	8.6	7.7
Threonin	4.6	4.1	4.8
Tryptophan	4.7	1.4	1.8
Tyrosin	3.0	3.1	3.1
Valin	3.6	3.6	4.5

connected to each other via hydrogen bonds. If wool becomes moist, water molecules attach themselves to these functional groups. This considerably reduces the intramolecular bonds between the helices. Thus, moist wool can be stretched to 125% of its original length. At the same time, the proportion of the α-helix structure decreases linearly with stretching and the proportion of the β-sheet structure increases linearly accordingly. This results in α-sheet structure with hydrogen bonds between adjacent protein chains. However, these are not stable and thus, after removal of the external force, the wool automatically reverts to the energetically more favorable α-helix structure (Fig. 9.30).

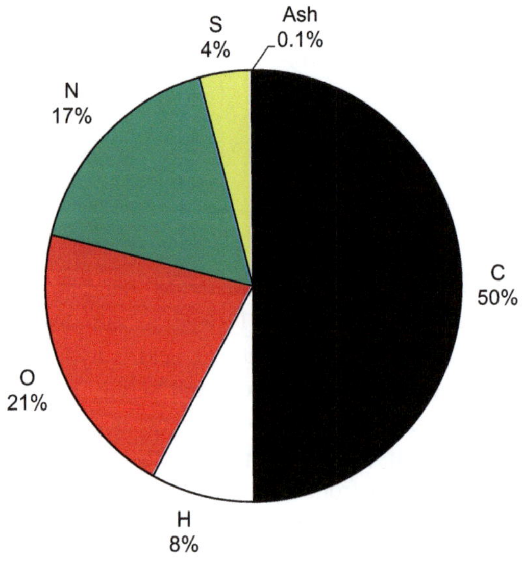

Fig. 9.27 Elementary composition of wool fibers according to Simmonds (1956)

A detailed description of the internal structure of wool can be found in Zahn et al. (1997). Various models describing the internal structure of wool keratin are explained in Wortmann (2009).

9.6 Features

9.6.1 Fineness, Length, Crimp, Density, Color, Shine

These properties are mainly influenced by the breed of sheep. Fineness, length and crimp are closely related: longer fibers are coarser and less crimped due to the higher area moment of inertia ($I \approx d^4$), shorter fibers are finer and more

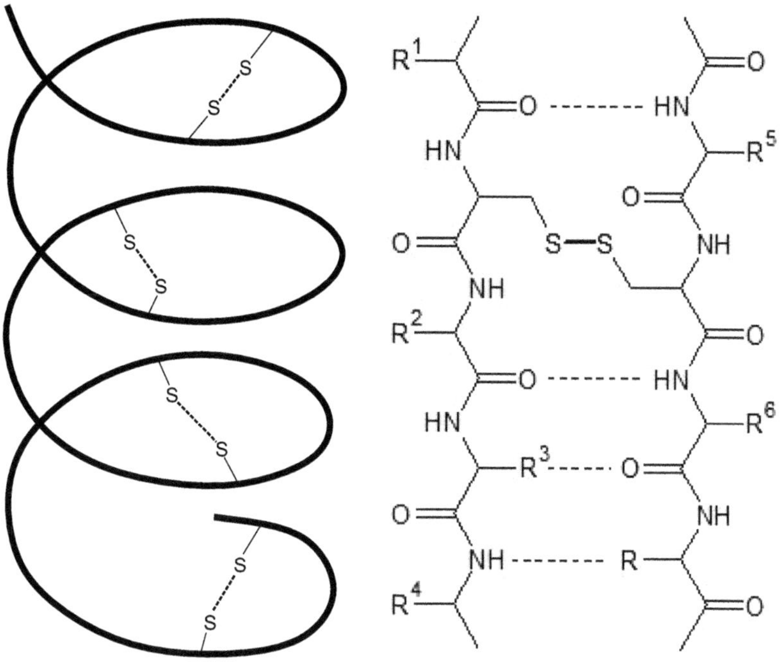

Fig. 9.28 Disulfide-bridges between cysteine groups (left: within α-helix; right: detail)

intensively crimped. To indicate the fineness, the fiber diameter is usually given. It lies between 16 and 60 μm (Fig. 9.31). The diameter of the wool fiber can vary greatly from root to tip ("unfaithful wool"). The more uniform ("faithful") the fineness of the wool fiber, the more uniform and tensile the yarn made from it. In the case of coarser fibers, the medullary canal in the middle is clearly visible in the microscopic image (medulla).

The length of wool fibers is 35–400 mm, depending on the breed of sheep. For wool fibers, a distinction is made between the length of the crimped fiber ('natural length') and the length of the stretched fiber ('true length'), Fig. 9.32.

In the case of a fiber length of more than 100 mm, the wool is generally processed further using the worsted spinning process; for fiber lengths of less than 100 mm, the carded spinning process is used.

For the characteristic values fiber diameter and fiber length, the coefficient of variation or the proportion of fibers below a certain value is often given.

The crimp is three-dimensional and ranges from flat to normal to high arched. The number of crimped arches per 10 mm is usually 2–12 (Fig. 9.33).

Table 9.3 shows examples of typical values for crimp as a function of fiber fineness for Australian Merinos.

The structure of the scales varies greatly depending on the breed of sheep and is much finer in Merino wool than in coarse wool, e.g. from Lincoln sheep (Fig. 9.34).

The density of wool is 1.31–1.32 g/cm^3. The color of wool ranges from cream to yellowish-white to ochre and brown. The French name for undyed fabrics made of sheep's wool, beige, became synonymous with this shade in the nineteenth century. The dark wool fibers (colored hairs) are of genetic origin and can hardly be removed by breeding. They are undesirable because they are clearly visible in light colors in the finished textile. Yellowing of wool is due to alkaline damage or to a photochemical process which may already occur on the sheep or during prolonged storage of the wool. Cystine bridges split by exposure to light oxidize with their SH groups to form sulfuric acid, which causes yellowing of the wool. Yellowing caused by high temperature during improper drying is

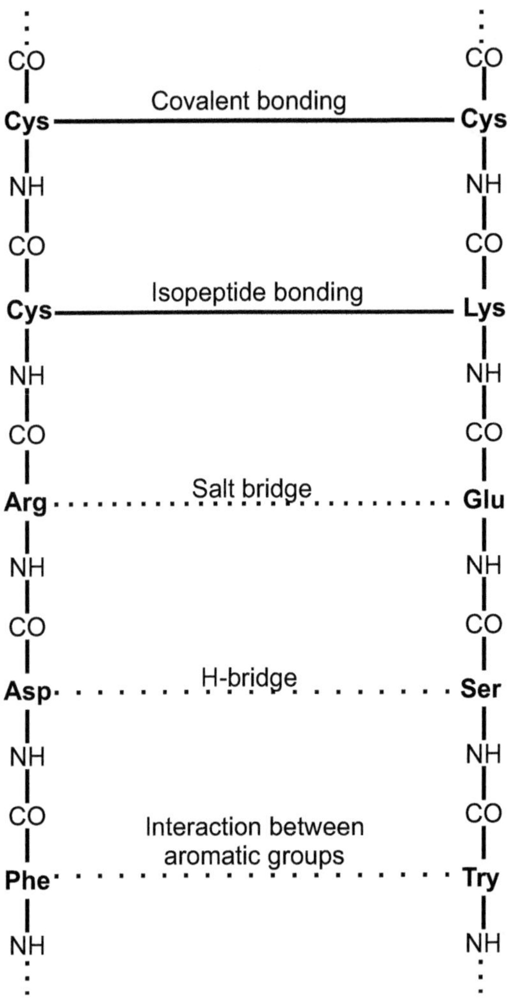

Fig. 9.29 Forces between side groups in the peptide chain

incipient charring. However, this can be remedied by a weakly alkaline hydrogen superoxide bleach. Dark wools cannot be bleached and can only be dyed with dark colors. The luster ranges from dull to very shiny.

9.6.2 Physical Properties

Wool is strongly hygroscopic and, at 65% relative humidity, may absorb around 15–17% of its own weight in water. At an r. h. of 95%, the value is as high as 25–30%. The water retention capacity is 40–45%. The absorbed water acts similarly to a plasticizer in thermoplastics and thus the mechanical properties are dependent on the humidity of the environment.

The electrostatic charge is low to medium.

9.6.3 Mechanical Properties

Wool has a relatively low fineness strength with a high elongation. When wet, wool loses considerable strength (Table 9.4).

Compared to other natural fibers, wool exhibits a relatively high elongation, while its tenacity is significantly lower (Fig. 9.35, left). These values change as a function of relative humidity (Fig. 9.35, right).

The high elastic elongation, which accounts for around 90% of the total elongation in wool, is an important quality-determining factor. It is mainly due to the elasticity of the protein structures and the crimping of the fiber. It is also reflected in the wool's tendency to return to its original position after ironing, stretching or pressing (low tendency to crease due to elastic recovery). In fabrics made of pure wool, pressed folds and creases therefore quickly even themselves out again. Although moistened wool fabrics are easily deformable under pressure, their elasticity is retained. These advantages make it possible to press, decatize and pleat wool fabrics. If wool fabrics are hung up in a damp environment, creases usually disappear overnight.

9.6.4 Chemical Behavior

Heating with water can lead to wool degradation depending on temperature, pH-value of the liquor and exposure time. Heating with water in an autoclave at 130 °C leads to the destruction of the crystalline areas in the microfibrils. Even at temperatures below 100 °C, chemical degradation of the wool occurs at pH values between 7 and 10. The cystine bridges incorporated in the wool proteins are decomposed, releasing hydrogen sulfide. This in turn attacks the wool cysteine and the decomposition is accelerated autocatalytically. Another decomposition product is ammonia,

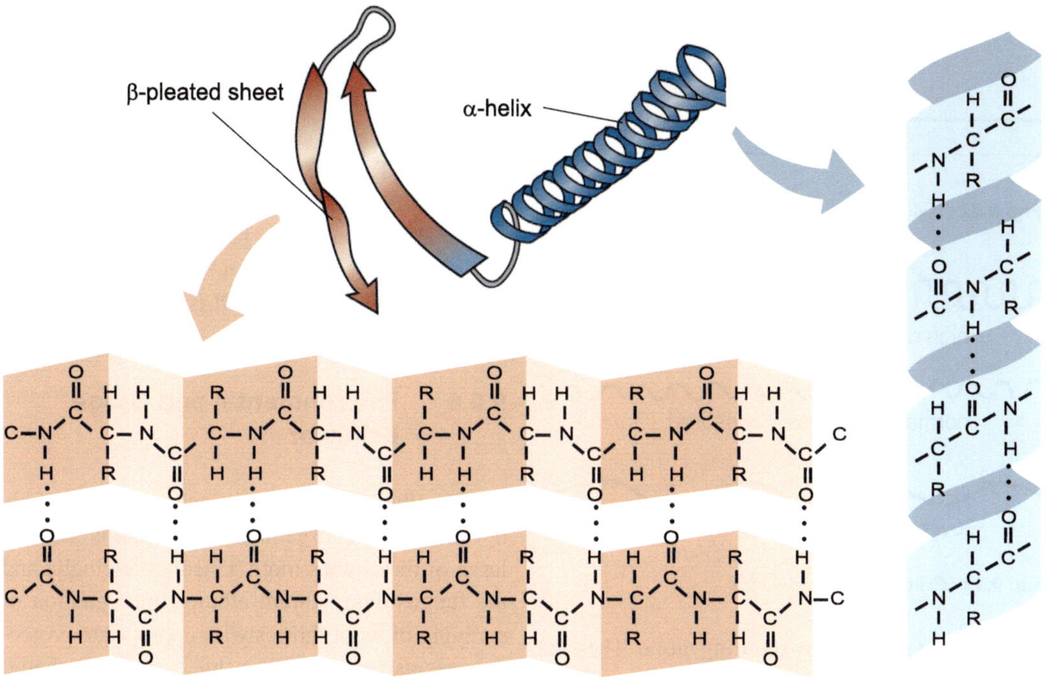

Fig. 9.30 Structure of the α-helix with hydrogen bonds

which is formed by hydrolysis of aspartic and glutamic acid groups. Therefore, wool textiles should not be put into the boil wash.

Hydrochloric, sulfuric and nitric acids can degrade wool proteins in the presence of salts and surfactants. Sensitive sites in wool proteins are the side chains of aspartic and glutamic acid groups and the peptide bonds, e.g. between serine and threonine, aspartic and glutamic acid groups. However, weakly concentrated acids can be used to remove vegetable constituents because they do not attack the wool (carbonization).

Wool is sensitive to alkalis. When wool is placed in alkali solutions, the alkali components bind to the wool proteins with swelling and simultaneous degradation. This gives the wool a yellowish coloration. The bond between alkali and wool fibers is so strong that the alkalis can

Fig. 9.31 Wool fiber (top) and human hair (bottom) (CSIRO, 2008b)

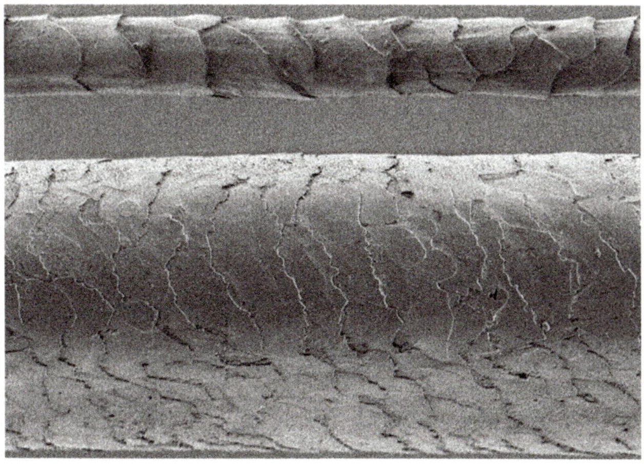

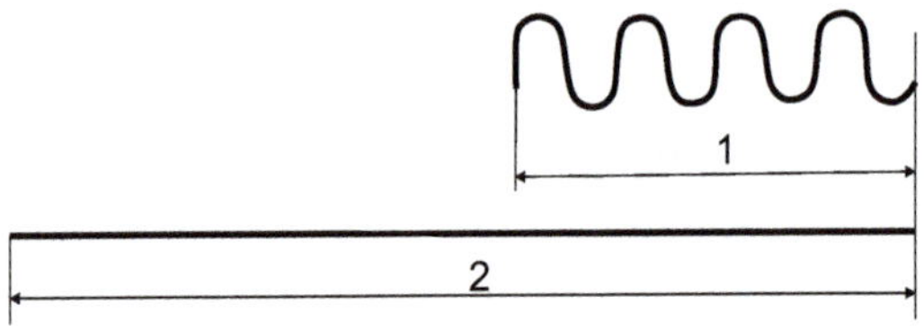

Fig. 9.32 Natural length (1) and true length (2)

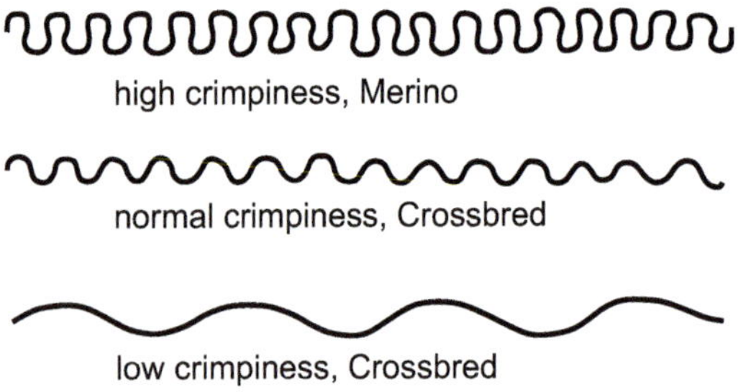

Fig. 9.33 Crimp of different wools

no longer be removed by scouring alone. They are first broken down by the addition of acids and can only then be removed.

9.6.5 Thermal Behavior

Dry and chemical-free wool can withstand a temperature of up to 150 °C for several hours. Short heating to 195 °C (30 s) leads to still tolerable fiber changes. The glass transition temperature is lower under wet conditions than dry because the absorbed water increases the mobility of the macromolecules by enveloping polar groups, thus reducing the intermolecular forces that bind the chain molecules together (Wortmann & Zahn, 1994). Temporary wrinkles can be created by heating the wool to a temperature above T_G, mechanical deformation and subsequent cooling below T_G (Table 9.5).

Wool fibers do not melt and continue to burn slowly after the flame is removed. This produces an odor like burnt horn. Carbonaceous ash remains as residue. The LOI is approx. 25.

9.6.6 Environmental and Usage Behavior

Unlike other proteins, keratins are not digestible by most living organisms. Exceptions are the larva of the clothes moth Tineola bisselliella and the fungus Tritirachium album. Impregnation or spraying of wool fabrics with appropriate protective agents (e.g. based on halogenated diphenyl ether) prevents moth feeding. Antiseptic chemicals such as zinc chloride, formaldehyde, phenol and salicylic acid prevent mold growth.

In use, wool clothing tends to wrinkle only slightly when dry and has a good recovery capacity. The heat retention capacity is good and the pilling tendency is low.

9.7 Classification

For the sale of greasy wool, the yield (often called rendement) is particularly important. It indicates

Table 9.3 Number of crimped sheets as a function of fiber fineness according to Duerden (1929)

Number	Microns	Crimps per inch
120s	14.0–14.7	28–30
110s	14.7–15.4	25–27
100s	15.4–16.2	22–24
90s	16.2–17.0	20–21
80s	17.0–17.9	18–19
70s	17.9–18.9	16–17
66s	18.9–20.0	14–15
64s	20.0–21.3	12–13
60s	21.3–23.0	10–11
58s	23.0–25.5	8–9
56s	25.5–29.0	6–7

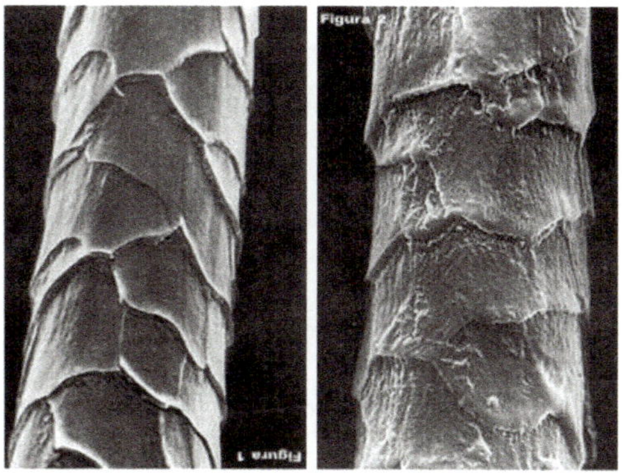

Fig. 9.34 Scale structure of Merino (left) and Lincoln (right)

the proportion of scoured wool in the raw wool and depends on the provenance and fineness (Fig. 9.36). The world average is 54%. Typical values for Australian wools are 60 to 70%, New Zealand wools reach 75%. Finer wool has a lower yield because more dirt is trapped in the more crimped fibers and finer wool has a larger surface area by mass and therefore contains more wool grease. If scoured wool is traded, the risk of estimating the yield is eliminated.

When buying wool for worsted spinning (worsted wool), the romaine plays a major role. It is the ratio of the comber content to the total fiber quantity. The romaine depends on the distance between the nipper and the pair of take-off cylinders on the combing machine (ecartement), the feed and the fineness, length and provenance of the wool fibers. The coarser and longer the wool, the lower the romaine.

During shearing, the wool accumulates as a coherent fleece. Immediately after shearing, the fleece is classified by viewing the entire fleece and simultaneously removing individual fleece pieces. The fleece parts are combined with equal parts of other fleeces to form a batch. The actual sorting of the raw wool batches according to fineness and length takes place in the processing plant (e.g. wool scouring mills) and is very labor-

Table 9.4 Mechanical properties of wool

Tenacity [cN/tex]	10–16
Cross-sectional tenacity [daN/mm^2]	13–21
Wet strength [%]	70–90
Knot tenacity [%]	85
Loop strength [%]	75–85
Elongation at maximum tenacity [%]	25–50
Wet elongation at maximum tenacity [%]	110–140
Modul at 100 % elongation [cN/tex]	150–300

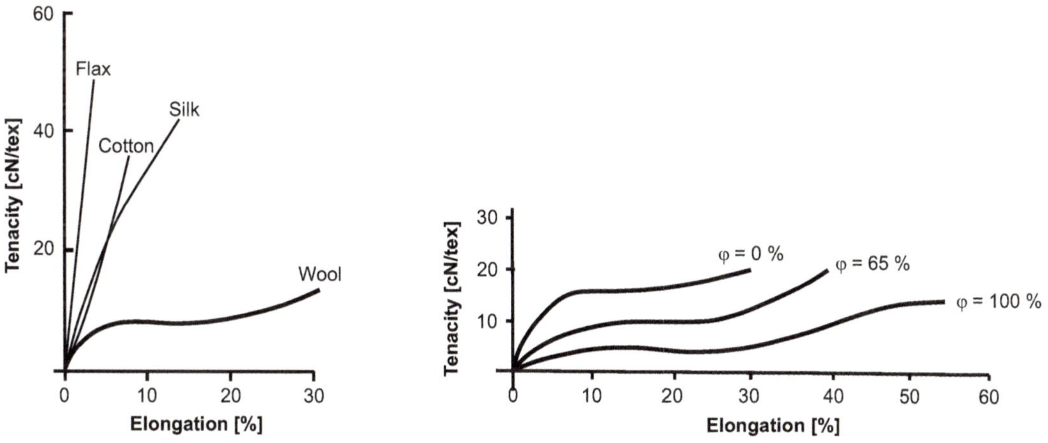

Fig. 9.35 Fineness-related tenacity-elongation curves of wool fibers

Table 9.5 Characteristic temperatures of wool

Temperature	[°C]
Glass	169 (wet)–174 (dry)
Color change	120
Ironing	160–170
Softening	240
Degradation	250
Self-ignition	590–600

intensive. Therefore, usually only coarsely contaminated fleece pieces are removed.

The climate in the producing country and the nutrition of the sheep influence the quality of the wool. Furthermore, the body region on which the wool grew is important (Fig. 9.37). The finest and most uniform wool in the pile is found on the shoulder areas. The parts at the core and the flanks provide average wool. Wool from neck, nape and back is long, tangled and coarse. Wool from belly and lower legs is heavily soiled and matted. Wool from the head is short and inferior. Wool from the thighs is coarse and clotty and wool from the blaze is coked.

The fineness of the wool (diameter in μm) is the most important classification criterion. The price is essentially based on this. This system has become established worldwide due to the round cross-section of the wool and the associated simple determination of the average diameter (Fig. 9.38 Wool fibers in cross-section).

In some countries, other systems are used to characterize the fineness of wool. In England, for example, the number of hanks of 560 yards

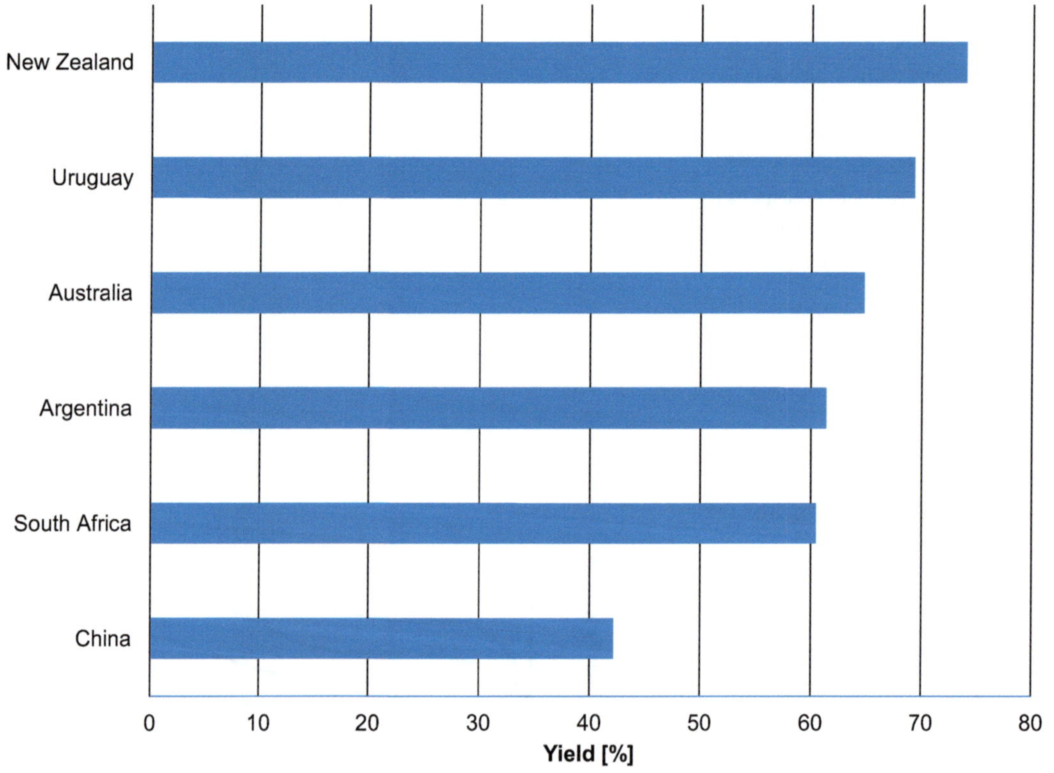

Fig. 9.36 Yield of selected producing countries (data: IWTO, 2019)

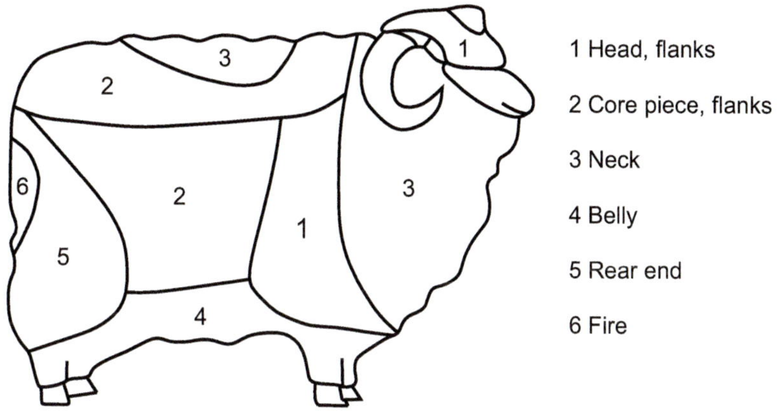

Fig. 9.37 Wool qualities on a sheep

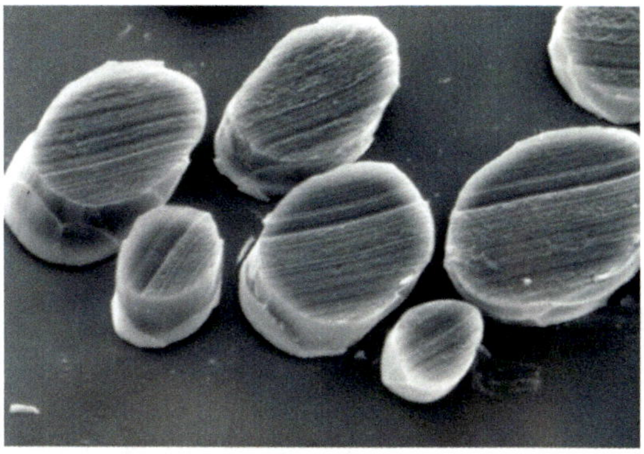

Fig. 9.38 Wool fibers in cross-section

(512 m) each that can be spun out of 1 pound (453.6 g) of wool is given ("spinning limit"). This results in values of around 100 for the finest wools and fineness of up to 36 for coarse fibers. A yarn with a fineness of approx. 14 tex can thus be spun from a wool with a fineness of 64. In Germany, the classification ranges from AAAA (very fine) to EE (very coarse), Table 9.6.

The distribution of world production among the wool qualities fine, medium and coarse shows Table 9.7. Since 1995, the percentage of fine wool has decreased from 43% to 35%, while the percentage of coarse wool has increased from 37% to 44% over the same period, with the percentage of medium-fine wool has remaining almost unchanged. This is mainly due to the fact that less wool fibers are used in the clothing sector today than 25 years ago.

However, this development does not affect all wool producers equally, as the example of Australia shows (Fig. 9.39). Whereas in 1990, only about 70% of the fibers were finer than 24 μm, this proportion had risen to 80% by 2000 and has since shifted further towards fine

Table 9.6 Wool fineness in different systems

Fiber diamter [μm]	Australia	England	USA	France	German#
16–18	Combing merinos (80s)	80s	Very Fine	130	AAAA
19–20	Combing merinos (70s)	70s/80s	Fine	120	AAA
21–22	Clothing merinos (64s)	64s	Fine Medium	110/115	AAA
23–24	–	60s/64s	High Half Blood	Prime croisé	A/AB
25–26	Low merinos (60s)	60s	Half Blood	Croisé I	AB/B
27–28	Crossed comeback (58s)	58s	3/8 Blood	Croisé II	B
29–30	Crossed comeback (58s)	56s	High 1/4 Blood	Croisé III	C1
31–33	Crossed Lincoln (50s)	48s/50s	1/4 Blood	Croisé III	C2/D
34–36	Romney Marsh (46s)	44s/46s	Low 1/4	Croisé IV	D1
37–39		44s	Common	Croisé V	D2
40–41		40s	Braid	Croisé V	E
42–44	Lincoln (36s)	36s		Croisé VI	EE

Table 9.7 Shares of wool qualities in world production (IWTO, 2019)

	Fineness [µm]	1000 t	%
Fine	<24.6 µm	406,407	35
Medium	24.6–32.5 µm	244,749	21
Coarse	>32.5 µm	504,588	44
		1,155,744	

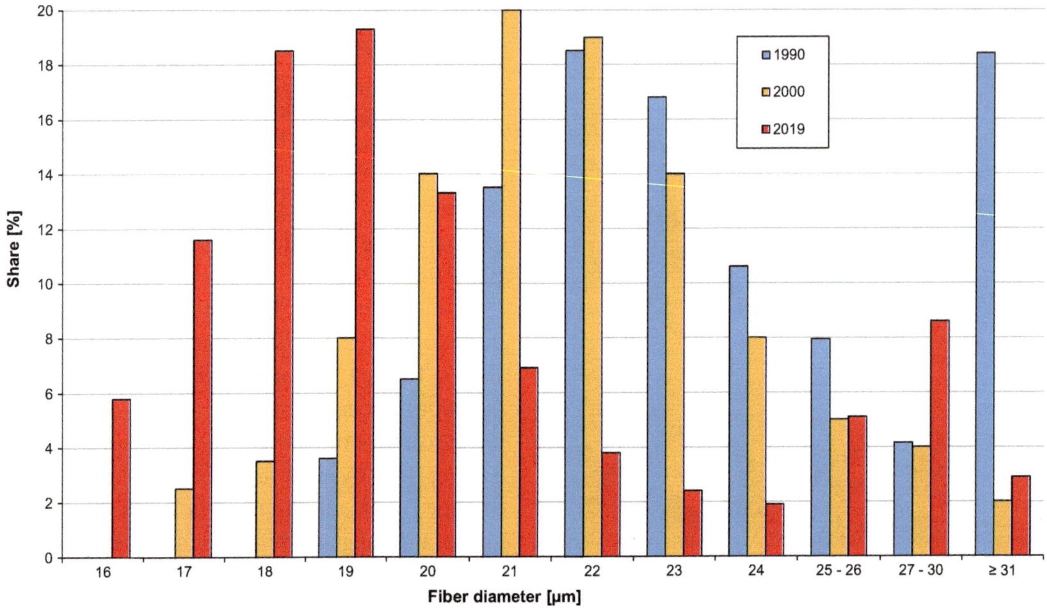

Fig. 9.39 Australian new wool production, breakdown by fiber diameter

wools (Fig. 9.39). This is due to the fact that since 2002 especially superfine (15–18.5 µm) wool has been produced ("superfine wool"). Even wool with a diameter of 11.5–15 µm ("ultrafine wool") is now produced. This has led to a rapid fall in the price of wool with a diameter of over 23 µm (Fig. 9.40).

Figure 9.40 shows that the wool price is strongly dependent on the fiber diameter. The price factor was calculated based on the average prices for Merino wool from 1991 to 2019 and the price for 23 µm fine wool was set equal to 1. The prices for fine wools obviously fluctuate significantly more in the long-term average than the prices for coarser fibers. The prices for wools in the range of 21 to 26 µm differ only slightly each year, as can be seen from the low dispersion.

At the beginning of the 1970s, it was started to test the wool arriving at the auction for yield, defects and fineness of the fibers already before the sale. Nowadays, all wool is examined and classified in this "pre-sale-testing". Appropriate certificates are attached to a representative sample of the lot, which can be manually examined by the potential buyer before the actual auction. Subsequently, all lots are auctioned.The typical lot size is 4 to 6 bales of approx. 120 kg each.

The aim is to completely dispense with the "inspection" of the bales or lots before the sale and instead to process all sales according to samples or certificates. This has not yet been achieved.

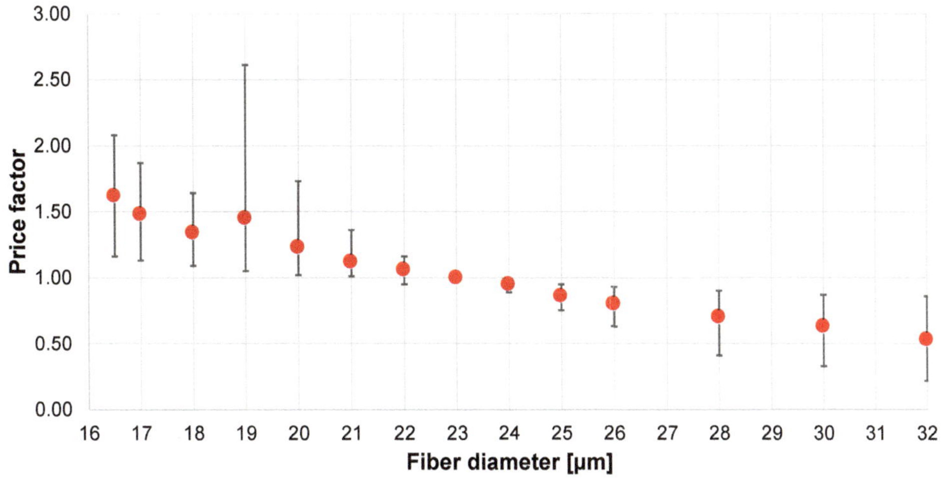

Fig. 9.40 Price factor as a function of the average fiber diameter (23 μm = 1)

9.8 Production

Wool production in the producing countries depends to a large extent on weather conditions (e.g. rainfall, droughts). Other animals (rabbits, wild dogs, kangaroos, etc.) and diseases (e.g. skin parasites, foot-and-mouth disease, brucellosis) also have an influence. The husbandry conditions naturally also play a role.

9.8.1 Sheep Flock

The world sheep population (Fig. 9.41) has been around 1.2bn animals for some years, of which around 360 m are used to produce wool for clothing and home textiles (Fig. 9.41).

9.8.2 World Wool Production

The annual wool production is about 2 m tons of greasy wool, which corresponds to about 1.1 m tons of scoured wool (Fig. 9.42). Of this, 0.5 m tons is used for clothing and 0.6 m tons for household and home textiles (e.g. carpets).

Merinos account for 40% of wool production and crossbreds for around 33%. About 80% of the Merino wool supply comes from Australia. Besides Australia and New Zealand, the most important wool exporting countries are the South Africa, Argentina and Uruguay. They have about 25% of the world's sheep and produce about 45% of the world's wool (scoured). Their export share is 70%. The developed countries import 80% of their wool from these countries. China is the largest wool producer, but consumes almost all of it in its own domestic market (Fig. 9.43).

Most wool is processed and consumed in Europe and China. Other regions, such as the producing countries Australia, New Zealand and South Africa, are of minor importance in absolute terms. However, taking into account the relatively small populations of these countries, wool plays a dominant role there with a per capita consumption of more than 1 kg/year. In comparison, worldwide consumption averages around 0.15 kg per capita annually.

9.8.3 Development of Wool Production

Australia has seen a 60% decline in wool production since 1990, while New Zealand's wool production has fallen by around a third over the same period. China, on the other hand, has nearly doubled its production since then and is now the

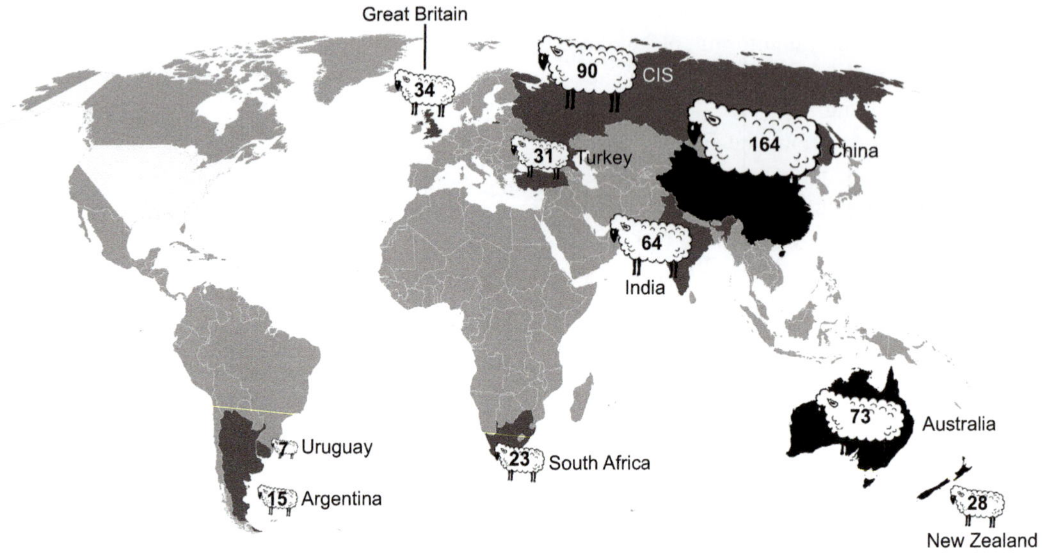

Fig. 9.41 Sheep population in millions (data: IWTO, 2019)

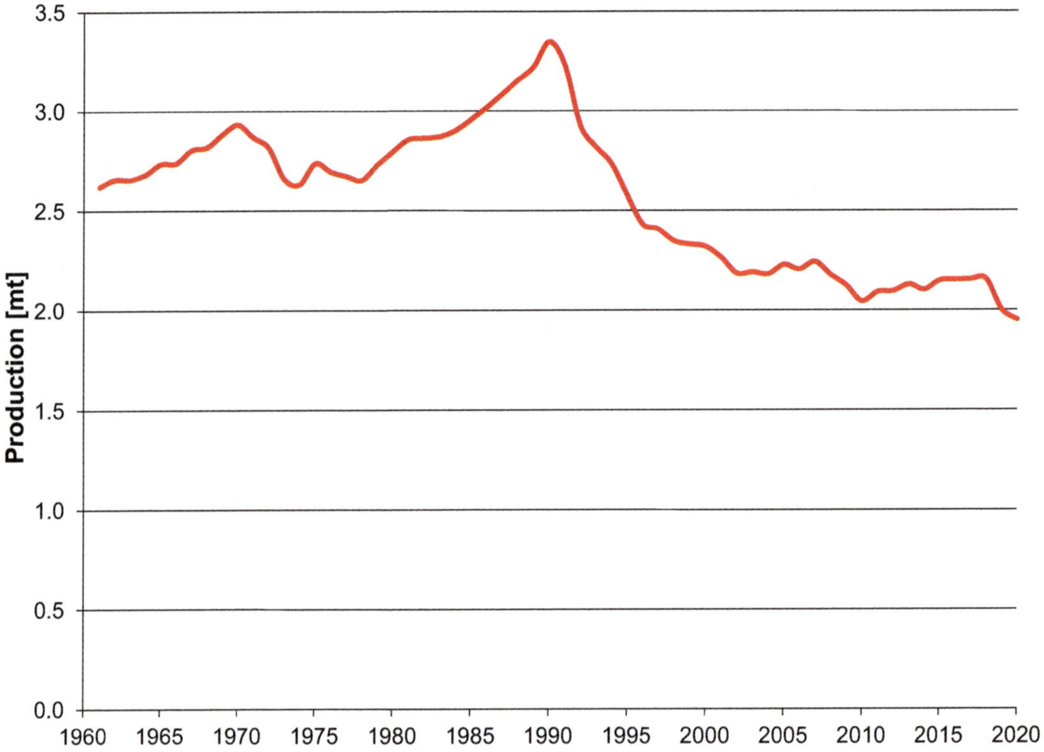

Fig. 9.42 Development of world production of greasy raw wool (data: faostat, 2022)

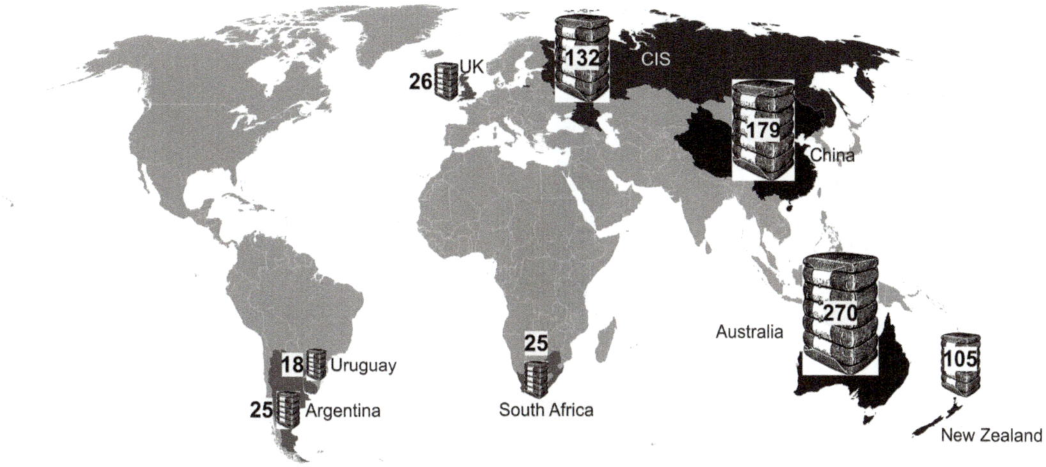

Fig. 9.43 Scoured wool selected countries in [1000 t] (Data: IWTO, 2019)

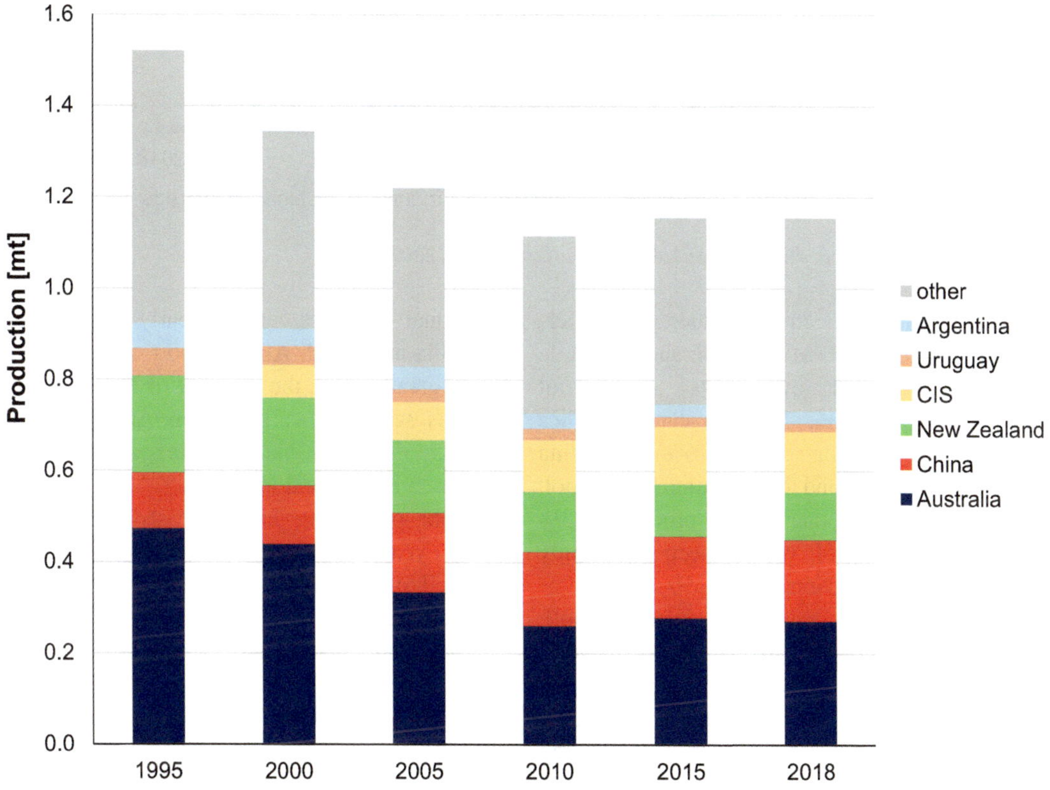

Fig. 9.44 Wool production by leading producers (data: IWTO, 2019)

world's second largest producer of scoured wool behind Australia (Fig. 9.44).

The amount of wool produced per sheep varies considerably between individual countries, as Fig. 9.45 shows. The trend reflects that more and more fine wool is being produced and the "yield" is therefore decreasing because the fibers are becoming thinner.

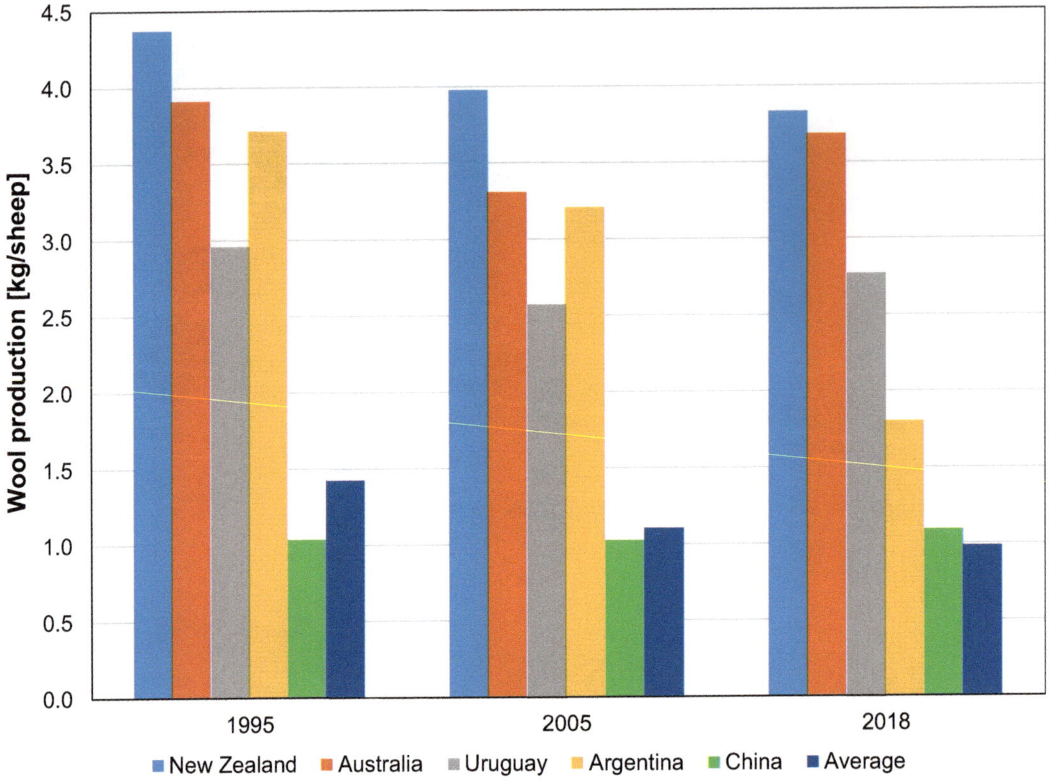

Fig. 9.45 Wool yield per sheep for selected countries (data: IWTO, 2019)

The individual countries produce very different qualities of wool, as Fig. 9.46 shows. While Australia produces over 90% fine wool (South Africa: 77%), New Zealand supplies almost exclusively coarse wool (76%). Argentina produces half fine and half medium-fine wool. Uruguay, on the other hand, produces 83% medium quality wool, where sheep are mainly kept for meat production.

Wool is sold as raw wool (greasy wool), scoured or carbonized wool, combed wool, combed wool, and as sheepskin. In New Zealand, so-called slips are also traded. This is wool obtained from the fur of slaughtered sheep using chemical or biological processes. Under the Textile Labelling Act (TKG), they are treated in the same way as virgin wool. The distribution of product groups in exports varies from country to country, as the Table 9.8 for selected products and countries shows. The different development of wool processing in individual countries can be clearly seen. While Australia and South Africa export almost exclusively greasy wool, the majority of New Zealand wool is scoured locally and in Argentina and Uruguay more than half of the exported wool is processed beforehand into tops.

Wool processing has increased steadily in recent years, particularly in India, but has declined or stagnated at a low level in most other countries, especially Italy.

The main imports into Germany are greasy wool (40%). Combed and scoured wool each account for about 25%, combed wool and waste for 4%.

9.8.4 Australia

In Australia, about 66 million sheep are raised on about 85 million ha by about 30,000 farmers, producing about 290,000 t of raw wool, from

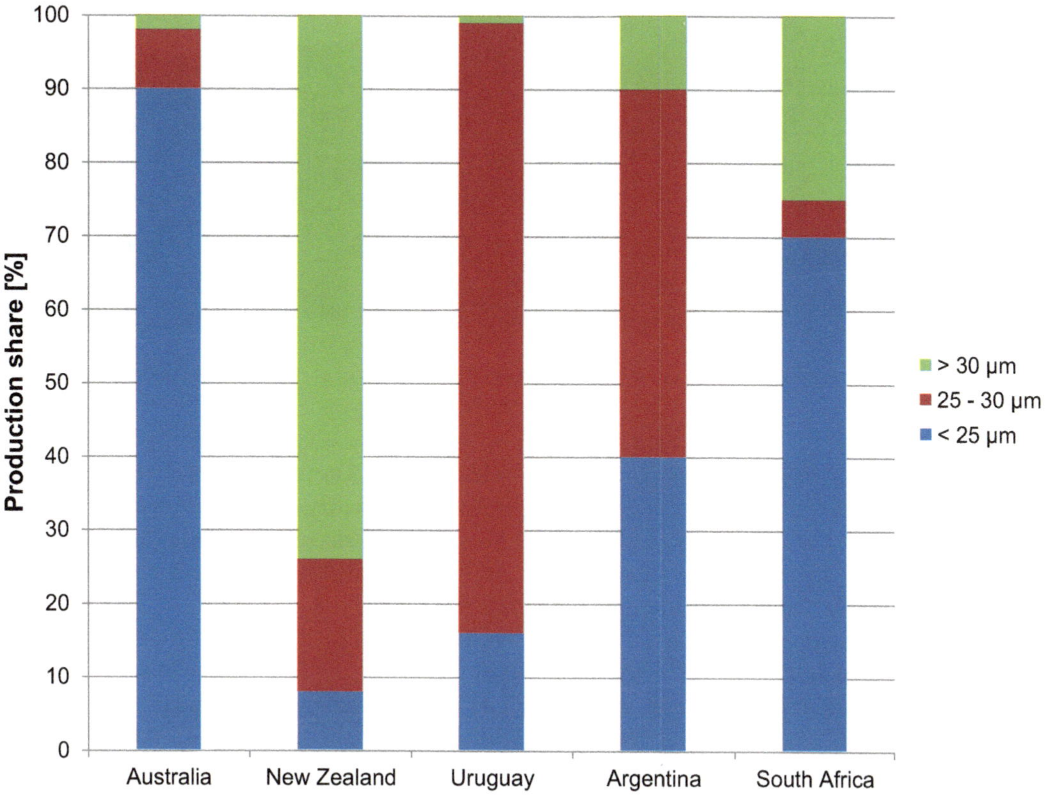

Fig. 9.46 Percentage of wool qualities for important producing countries

which 190,000 t of scoured wool is obtained (as of 2021). This corresponds to around 70% of the wool used for clothing worldwide.

Settlement of Australia by Europeans began in 1788, and even then some Indian and African sheep were introduced for meat supply. The first Merinos arrived in New South Wales in 1797 from Spanish stock via South Africa and formed the basis of today's Australian Merino farming. They were initially kept for meat to supply the fledgling colony, and only later were they used for wool. The most famous breeder was John MacArthur, whose farm was near Sydney (Fig. 9.47).

The fleece weights were about 2.5 kg. The herds of the time were small, the first wool export to London took place in 1804. Because the wool fetched a good price, the wool export increased rapidly. In 1813 the inland route was opened up from Sydney. As a result, sheep were also raised in Victoria and parts of South Australia by 1840. Queensland was not settled until between 1850

Table 9.8 Percentage shares of product groups in exports of selected countries (data: IWTO, 2019)

	Greasy wool		Scoured wool		Top	
	2004	2018	2004	2018	2004	2018
Australia	76%	97%	9%	3%	2%	
New Zealand	40%	39%	56%	61%		
South Africa	57%	96%	10%	2%	30%	2%
Argentina	36%	43%	44%	5%	16%	52%
Uruguay	10%	26%	10%	25%	80%	49%

Fig. 9.47 John McArthur on an Australian $2 note and Australian wool bales at the wool exhibition in Paris, 1867 (Rodolph, 2009)

and 1860. The wool was very soon successfully marketed in Europe.

After the importation of Spanish Merinos and the completion of settlement, breeding efforts were made to obtain the optimal Merino sheep for each region and climate. Saxon Merinos were therefore introduced as early as 1820. The superfine Merino wool from Tasmania and the Table Land of New South Wales go back to this.

After some time, however, it was found that this breed of sheep was not suitable for the hot and dry climate of Central Australia. Therefore, a more robust Merino sheep with somewhat coarser wool was bred. Thus, in 1861, the Peppin Merino was created, named after brothers George and Frederick Peppin, who had immigrated from England. Their French Rambouillet ram "Emperor" is considered the progenitor of the Australian Peppin Merino breed. About 70% of all Merinos in Australia today are said to come from this first breeding. The wool of these sheep has an extremely high ratio of secondary to primary follicles of about 21.

The crossing of Peppin Merinos with Saxon Merinos resulted in the so-called "Strong Wool Merino", which still dominates in South Australia, New South Wales and Queensland. A detailed account of the history of Merinos in Australia is given in Massy (2007).

Nowadays, Merinos with fine wool are mainly kept in New South Wales and in Tasmania, the Merinos with the somewhat coarser wool live in the so-called wheat belt in Western and South Australia as well as in New South Wales. In Victoria and Tasmania the Polwarth sheep is still bred, which is the result of a cross between Merino sheep and Lincoln rams. The first cross is again crossed with a Merino and then maintained by inbreeding.

The sheep density per km^2 is about 45 in the pastoral zones, in the regions with a lot of rainfall and correspondingly higher fodder supply up to 300 sheep per km^2 are kept (Fig. 9.48).

9.8.4.1 Pastoral Low Rainfall Zone

In the pastoral low rainfall zone, the annual rainfall is less than 250 mm. Drinking water for the sheep is therefore mostly taken from artesian wells. Common farm sizes range from 2000 to 20,000 ha. The flock sizes correlate not only with the farm size, but above all with the food supply and is usually between 3000 and 8000 head, with a maximum at 25,000.

The number of lambs per ewe covered per year is only 0.6–0.8, partly due to the low fertility of Merinos and partly due to the small food supply. With the exception of the summer months, lambing can take place at any time of the year, usually in spring in Queensland and in autumn in South Australia and New South Wales. Shearing is accordingly in autumn and winter in the south of the country and in spring in Queensland.

9.8.4.2 Wheat and Pastoral Zone

The farms in these areas grow wheat and raise sheep at the same time. The annual rainfall is 250–650 mm. Sheep are raised in this region both solely for wool and solely for meat. Crosses

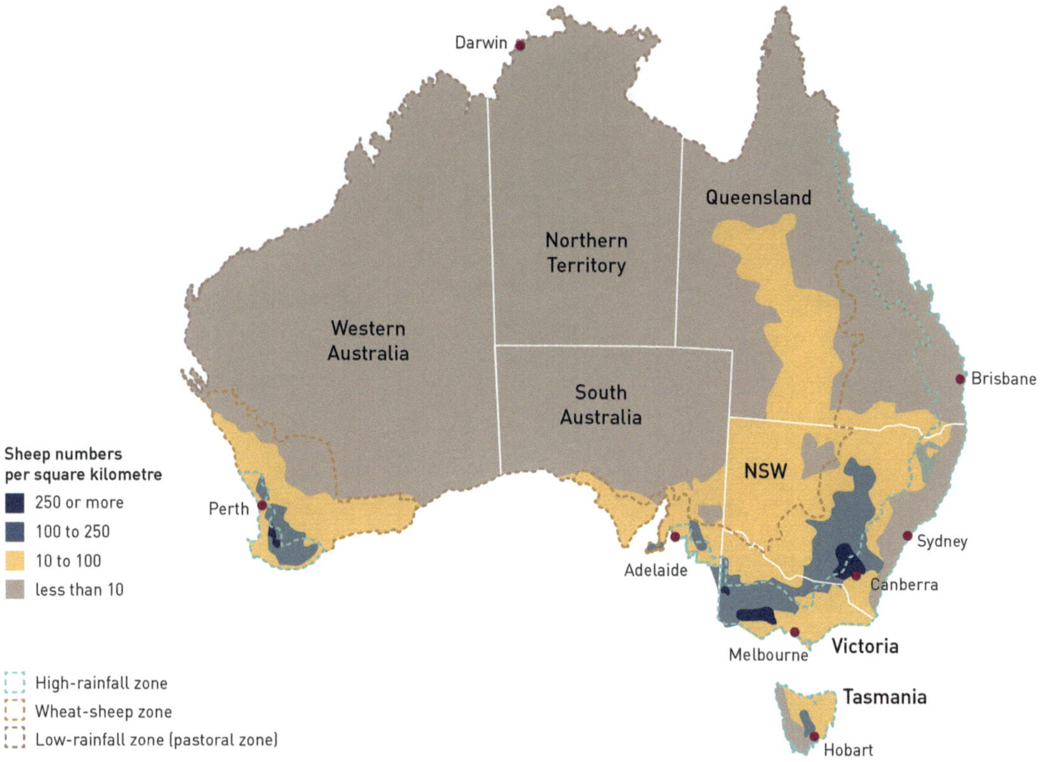

Fig. 9.48 Pastoral zones in Australia (The Woolmark Company)

of Merino ewes and Border Leicester rams are often kept for meat. The annual lambing rate is 0.8 to 1.0, with lambing and shearing taking place in spring.

9.8.4.3 Pastoral High Rainfall Zone

This zone is located in the southwestern part of Western Australia, the southeast of South Australia, southern and eastern Victoria, Tasmania and the Tableland of New South Wales. The annual rainfall is 500–1000 mm. The land is hilly and Merinos are kept exclusively for wool production. Corriedales and Romney Marsh sheep also occur occasionally. Lambing occurs in spring in Tasmania, and often in late autumn in Western and South Australia and Victoria due to winter rains (Fig. 9.49).

Australia produces 86% Merino wool and 14% cross-bred. Australian wool is therefore mostly fine to medium in diameter and is primarily used for clothing. Cross-bred wools are also finer than New Zealand's.

The number of sheep shorn is around 73 m (2019). The yield per head is about 5.6 kg, resulting in a raw wool production of about 414,000 tons (Fig. 9.50). It is almost completely exported to China.

The wool price is subject to strong fluctuations. This is related to changing fashions as well as different large production volumes due to climatic influences. After new record prices were achieved for fine and medium quality wool in 2018 and 2019, the wool price has now fallen almost to the long-term average, as Fig. 9.51 shows.

9.8.5 New Zealand

In New Zealand, 30 breeds of sheep are bred, of which, however, only 6 are of major importance. Sheep are kept for both wool and meat production (Fig. 9.52).

Fig. 9.49 Sheep in the high rainfall and low rainfall zone (The Woolmark Company)

Until the end of the nineteenth century, Merino sheep were dominant in New Zealand, but they have since been displaced by other breeds due to the unfavorable climatic conditions for Merinos. Halfbred and Corriedale are crosses of the Merino and are also important. Their wool is used for clothing.

The most important breed of sheep today is the Romney (66%). It was imported from England in 1853. Related to the Romney are Coopworth and

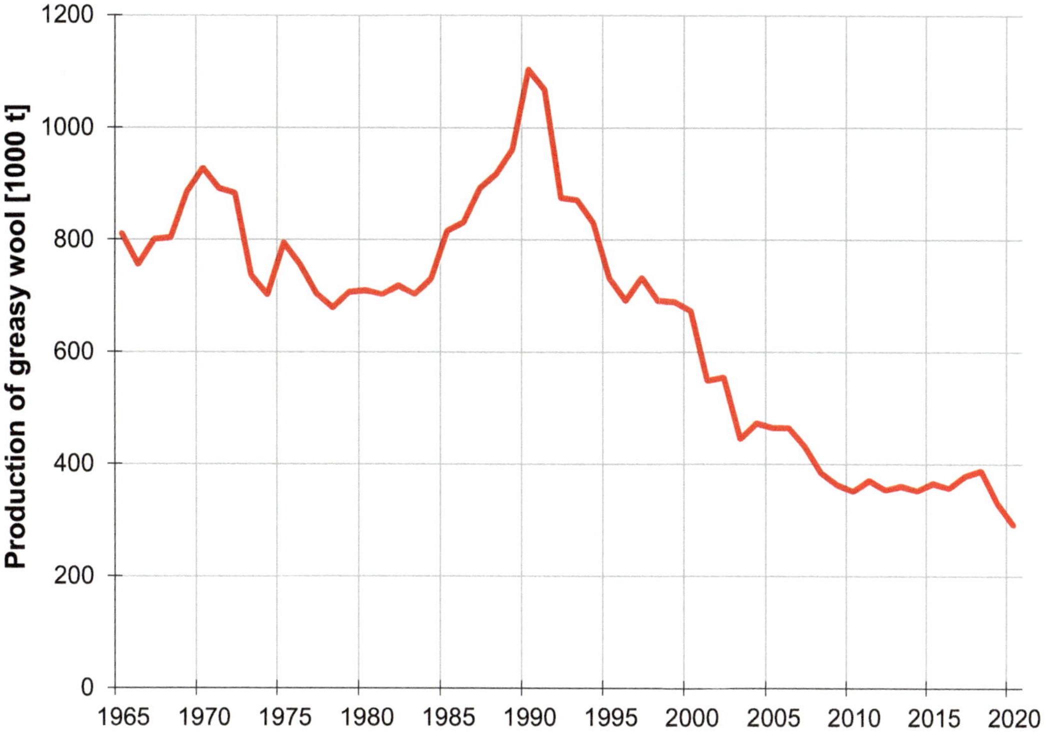

Fig. 9.50 Raw wool production in Australia (data: faostat, 2022)

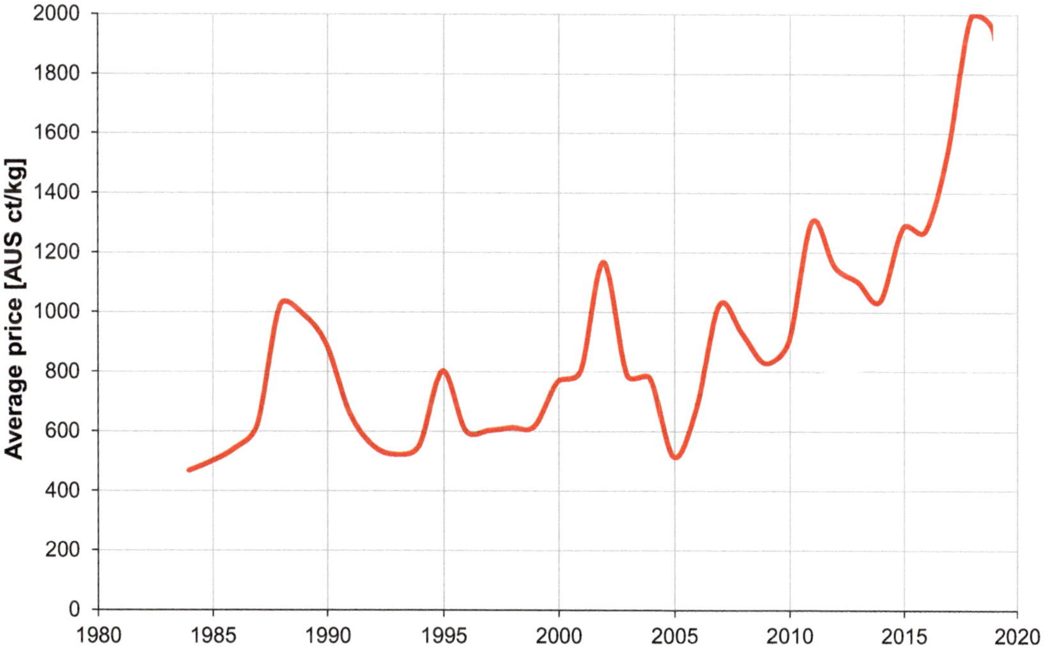

Fig. 9.51 Development of the wool price (data: Australian Wool Market Indicator)

Fig. 9.52 Sheep in New Zealand (Royan, 2020)

Perendale, which produce Crossbred wool used for home textiles and carpets (Fig. 9.53).

In total, the approximately 26.8 million sheep produce about 140,000 t of greasy wool, which results in 105,000 t when scoured (Fig. 9.54). Besides clothing, New Zealand wool is mainly used for high-quality carpets.

About 90% of the wool is exported, mainly to Asia (China) and the EU (Fig. 9.55). About 60% of this is scoured wool, 40% raw wool. Until 2008, the USA was one of the main export markets. After that, as a result of the housing crisis there, the demand for carpets dropped dramatically and so New Zealand wool production was directly affected. It has not recovered from this in the US market to this day.

The exported wool is worth around NZD 500 m (€ 280 m). Fine wool fetches around NZD 17/kg, medium qualities around NZD 7/kg and coarse wool (strong wool) around NZD 2.5/kg. With the current distribution of the wool clip, the average price is about NZD 5 to 6/kg. The export of 100,000 tons of lamb in recent years has yielded around NZD 3.3bn, many times the revenue from wool production. To this must be added NZD 700 m for the export of mutton.

9.8.6 South Africa

Besides gold, wool is one of South Africa's main export products. Merinos were imported from Spain as early as 1789. After the conquest of the Cape Province by the English in 1806 and the subsequent settlement by the Boers in 1834, large-scale sheep farming began. In 1846 already 3 million sheep were counted. At the beginning of the twentieth century Merinos were imported from Australia and crossed with the existing sheep (Merino, Rambouillet, Vermont). Therefore, today's South African Merinos are somewhat smaller than the Australian ones, their wool is finer and shorter. It has a very high felting capacity (Fig. 9.56).

Overgrazing caused major erosion damage in the mid-twentieth century. Today, about half of the sheep live in coastal regions and in the east of the country, the other half in the semi-deserts, which account for over 85% of the area.

Merino (Capper, 2020) Romney (Pam, 2020)

Corriedale (Nichols, 2020) Coopworth (McGrath, 2020).

Fig. 9.53 Important sheep breeds in New Zealand

In 2020, there were around 23 million sheep. About 70% of South African sheep are bred for wool production and 30% for meat production. Almost 2/3 of all sheep are Merinos, which produce about 43,000 tonnes of wool (Fig. 9.57). The wool fineness is usually between 18 and 22 μm with a fiber length of 40–80 mm. Accordingly, prices range from €9 (22 μm, 40 mm) to €12 (18 μm, 80 mm) per kg (Cape Wools SA, 2020). Due to the shorter fiber length, they are approximately 30–50% lower than the corresponding prices for Australian wools of the same fineness. Almost all South African wool is exported as raw wool.

9.8.7 Uruguay

Uruguay is one of the smallest countries in South America. The climate is warm, with rainfall up to 1250 mm/year. Farms are located in the north and center of the country. A total of 6.6 million sheep are currently kept. The number of sheep declined steadily until 2015, having reached its previous peak of almost 26 million in 1991. Since then, it has remained constant. Sheep in Uruguay, unlike most other countries, are kept in fenced pastures along with cattle (Fig. 9.58). Production is currently about 30,000 t (up from 94,000 t in 1991).

The dominant sheep breed in 2020 is the Corriedale (42%), followed by Merino (27%), Polwarth (9%), Merilin (4%) and Romney Marsh (3%). Sheep in Uruguay are often raised for "dual purpose", i.e. for both wool and meat production (lambs). Wool fineness is dominated by medium qualities (25–32 μm), although fine wool (<19 μm) has been increasingly produced in recent years. The average price for Uruguayan wool is 6–7 €/kg.

In Uruguay, a large wool processing industry has been built up in recent years. As a result, Uruguay now exports 52% of its wool as combed wool, 28% as scoured wool and only 20% as raw wool. Wool exports thus account for more than 1% of all Uruguayan exports. The most important customers are China, Germany, Italy, Turkey and India.

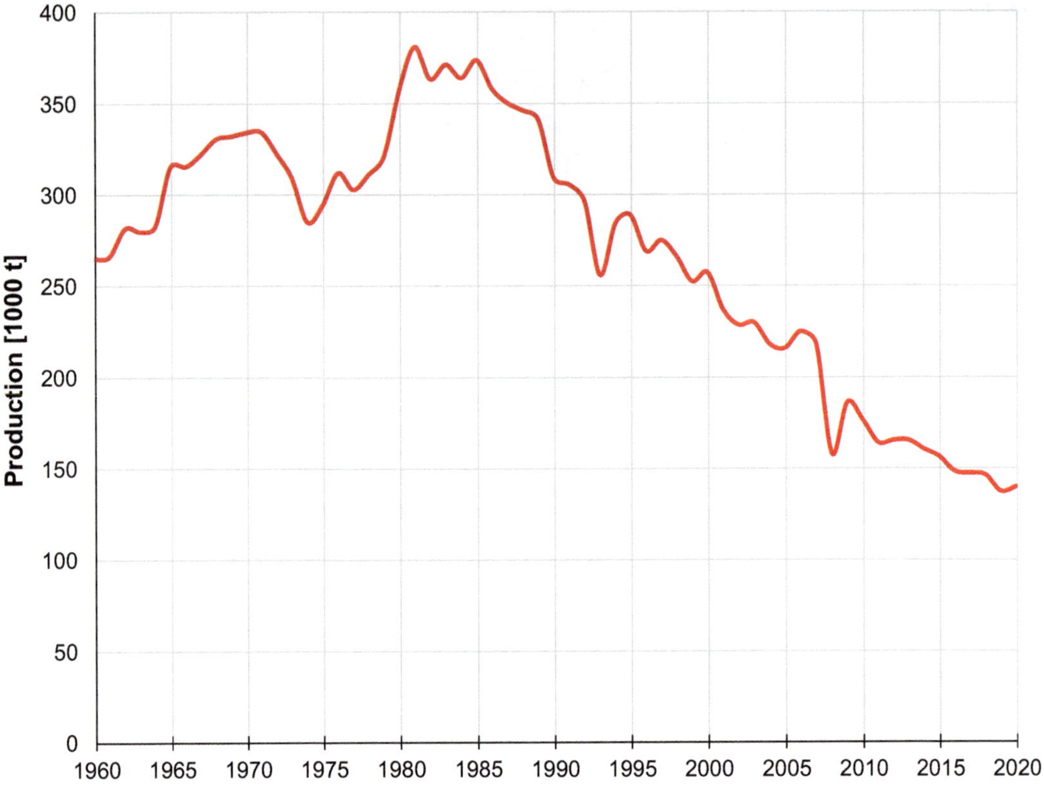

Fig. 9.54 Production of greasy wool in New Zealand (data: faostat, 2022)

Fig. 9.55 Export markets for wool from New Zealand (data: Beef + Lamb New Zealand, 2020)

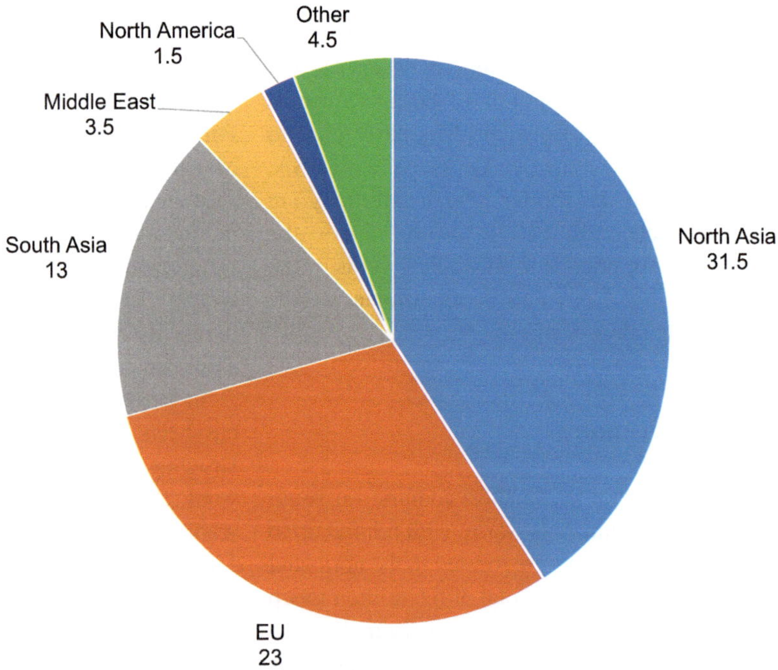

Fig. 9.56 Sheep flock in South Africa

The Uruguayan government has been particularly committed to animal welfare for many years and has issued guidelines for animal welfare, the application of which is regularly monitored.

9.8.8 Argentina

Around 1550, Juan Nuñez del Prado imported the first sheep to Argentina to meet local demand. When the ban on wool exports was lifted by Spain in 1810, the great era of the Argentine sheep farmers began. The improvement of the water supply and better fencing, combined with

Fig. 9.57 Wool Merino (left) and meat Merino (right) in South Africa

Fig. 9.58 Flock of sheep in Uruguay (Hoifodt, 2009)

the successful breeding of robust sheep, led to an upswing in wool production. Today, Romney, Corriedale, Merino (imported from Australia) and the typical Argentine Criolla sheep dominate the market.

Unusual in Argentine sheep is the large number of castrated rams, which are often kept until they are more than 6 years old. Because they are not exposed to the stress of pregnancy and lambing, they produce particularly fine wool.

Sheep are raised almost throughout the country, often separately from the cattle regions. Of particular importance are the northeast, the province around Buenos Aires and Patagonia (Fig. 9.59). Sheep are mostly raised where other forms of agriculture are not possible because of the dry soils. Sheep are kept for wool and for meat production, as in Uruguay.

The number of sheep was over 70 million at the end of the nineteenth century, was still at 40 million in the 1960s and is about 14.8 million today. The wool production is about 44,000 t per year, the prices per kg are 6–12 €. The main customer for the meat products is the EU. Argentine wool is exported about half as raw wool, and half as combed wool. The main customers for Argentine wool with a total value of € 120 million (2019) are Germany (43%), Peru (12.5%) and 8–10% each from China, Turkey, Mexico and Italy.

9.8.9 United Kingdom

Sheep farming in Britain began as early as 1900 BC. The Romans brought new breeds of sheep into the country (finer wool) and as a result of further invasions by Vikings and others, a wide range of different breeds of sheep developed. In the Middle Ages, Britain was one of the most important countries for wool production and wool was the basis for the prosperity of large regions (Fig. 9.60). Since the reign of Edward III in the fourteenth century, the Speaker of the House of Lords has therefore sat on a sack filled with wool at ceremonial occasions. With the decline of the English woollen cloth industry in the 19th and 20th centuries, the importance of wool steadily declined and that of cotton increased.

Fig. 9.59 Sheep in Argentina with Sheepboys (Proimos, 2020)

Fig. 9.60 Sheep flock in Great Britain (Halling, 2020)

Fig. 9.61 Black Face sheep (Simark, 2020) and Romney sheep (Sawyer, 2007)

Today there are still around 34.3 million sheep. About half of them live in England and a quarter each in Wales and Scotland. Northern Ireland is of little importance. English wool is relatively coarse (mostly >28 μm), the climate is too cold and damp for Merinos (Fig. 9.61). Prices are correspondingly low at 0.2 to 1 €/kg. Wool production is currently around 37,500 t/year, almost all of which is exported.

9.8.10 Spain

The number of sheep in Spain is currently around 16 million. Wool production has been constant over the last 10 years at 21,000 t (9500 t scoured wool). Of this, around 16,000 t are exported.

Typical of Spain is sheep farming in the form of 'transhumance' (Spanish transhumar: 'to lead to pasture'). The flocks of sheep move between widely separated areas according to the seasonal climatic rhythm (Fig. 9.62). Winter stabling is not necessary. The owners of the flocks usually do not migrate themselves, but have this work done by shepherds. In spring, the sheep are driven along the cañadas, the total length of which is 125,000 km, into the northern mountains, because fresh grass grows there and the summer heat is bearable. In autumn, the flocks migrate to the south, where rains again provide fresh grass. This type of sheep farming has existed since the eleventh century and was even legally protected in the thirteenth century.

9.8.11 Germany

In 1860, 28 million sheep were still kept in Germany. Due to industrialization and the associated decline in agriculture, the number of sheep fell to 6 million by 1912 (Wagner, 1981). Sheep farming is of little economic importance in Germany today. There are many different breeds depending on the region, but they are kept more for folkloric reasons than for economic considerations (Fig. 9.63). The wool is medium-fine and coarse and is only of limited use for the production of clothing. Therefore, sheep farming in Germany is mainly practiced from the point of view of meat and milk production.

The main breeds are:

- Merino-Land sheep (predominantly in southern Germany), approx. 31% of all sheep,
- black-headed meat sheep (northern and western Germany), approximately 11%,
- Merino longwool sheep (northern Germany, dykes), approx. 3%,
- crosses of these and other breeds.

The number of sheep kept has been constant for a number of years and was around 1.6 million in 2020. Sheep are bred by around 60,000 sheep farmers. Of these, about 40% of the total flock is kept in site-based sheep flocks and about 35% in transhumance flocks. Overall, only 4% of sheep farmers own about 57% of all the sheep.

Fig. 9.62 Sheep flock on transhumans (left: www.pikist.com; right: Oviso, 2003)

Currently, about 10,000 t of greasy wool are produced in Germany. However, this is not of textile importance because it is generally too coarse for clothing and the quantity is too small for the economic production of household and home textiles.

Similar to Spain and parts of South America, transhumance is also practiced in Germany in addition to conventional grazing. In this process, flocks of sheep move, for example, from the Swabian Alb and the Black Forest to lower-lying Upper Swabia and back again (Fig. 9.64). This is mainly of tourist interest (sheep runs).

9.9 Trade

In Australia and New Zealand, the farmer has the option of either offering his wool at auctions via brokers or selling it directly to the export trade or to the wool scouring mills. In the broker system, the farmer delivers the bales to the broker and leaves everything else to him (Fig. 9.65). The main task of a broker is therefore to channel the wool through the entire sales process. For this service, the broker receives a fixed amount per bale, about 10 ct/kg. He is not directly involved in the proceeds. Nevertheless, he tries to get the highest possible price because he wants to have satisfied farmers as customers. In South America,

Fig. 9.63 Typical German sheep breeds (left: Weller, 2020; right: Wiki, 2009)

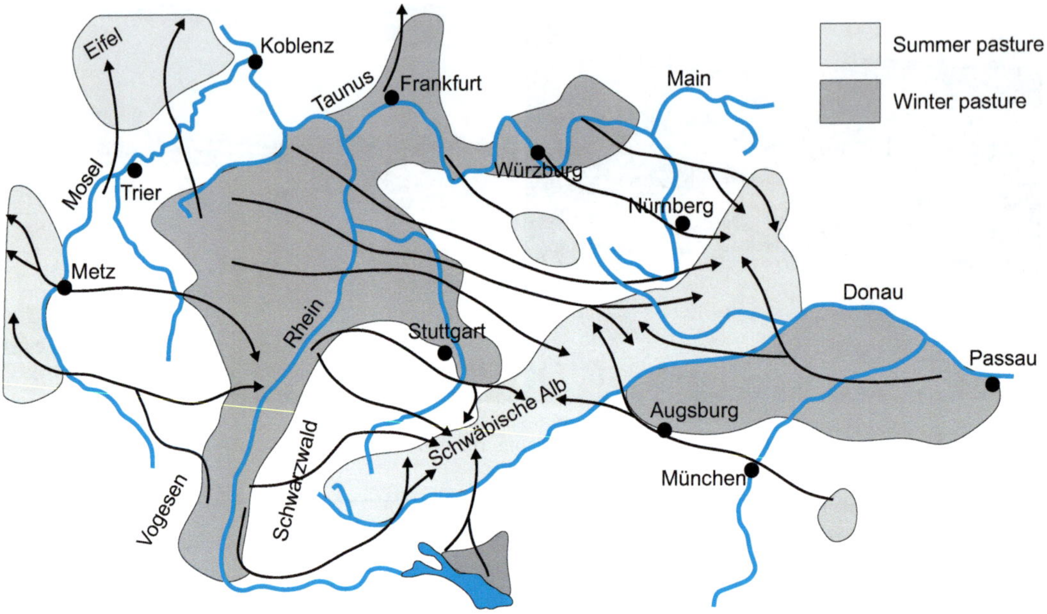

Fig. 9.64 Sheep transhumance in southern Germany

Fig. 9.65 Flow of the wool trade from the farmer to the spinner (example Australia)

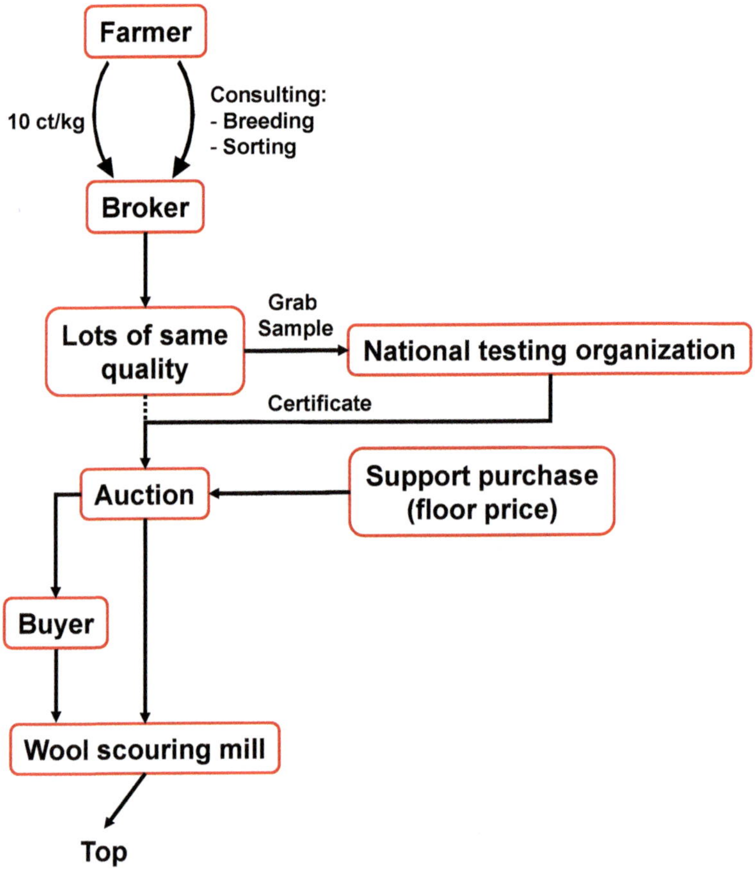

Fig. 9.66 Sampling from the bale (grab sample)

the wool is often purchased directly from the wool producer, without the detour via brokers and a stock exchange.

The bales delivered by the farmer are combined in lots by the broker, i.e. they are not sold individually. For each lot, the farmer has an inspection certificate drawn up by a state organization. Before the auction begins, a so-called grab sample is taken from each bale (Fig. 9.66).

The grab sample is placed in an open box, which is placed on the bale or by the respective lot in the warehouse (Fig. 9.67). In this way, every buyer can convince himself of the quality of the wool offered by inspecting and feeling it. The official test certificate with the most important fiber properties is also placed in this box.

The wool buyer has thus already decided before the auction which offers he wants to buy on the basis of the auction catalogue, the certificates with the preliminary test results and the Grab Sample.

The auction room resembles the lecture hall of a university. The auctioneer calls each item in quick succession using the "open cry" method and states the asking price. The highest bidder wins the lot. The auctioning of a lot thus takes on average 10–15 s. The buyers are either wool merchants who resell the wool to wool scouring mills, or the wool scourer themselves (Fig. 9.68).

The main export flows of raw wool of the main producing countries are shown in Fig. 9.69.

A total of around 835,000 tons of greasy wool are exported annually. Australia has a share of 42%, New Zealand 12% and South Africa 6%, Argentina and Uruguay together 3% (Table 9.9).

The most important importers of raw wool are China, India, Great Britain, the Czech Republic, Italy and Turkey (Fig. 9.70).

Fig. 9.67 Exhibition hall of a wool auction and contents of a sample box (Birgit Reiter)

Fig. 9.68 Auction room (Birgit Reiter)

The export and import quantities of scoured wool show Table 9.10.

A total of approx. 135,000 t of tops are traded, with China accounting for 23% of export (Table 9.11).

The wool price is subject to large fluctuations (Fig. 9.71). An initial record high in the late 1980s was followed by a prolonged period of low prices until an all-time high was reached in 2019. In the meantime, the mean price for raw wool for all qualities is approaching the long-term average again.

For example, for scoured Australian Merino wool, 18 μm, the price range over the last

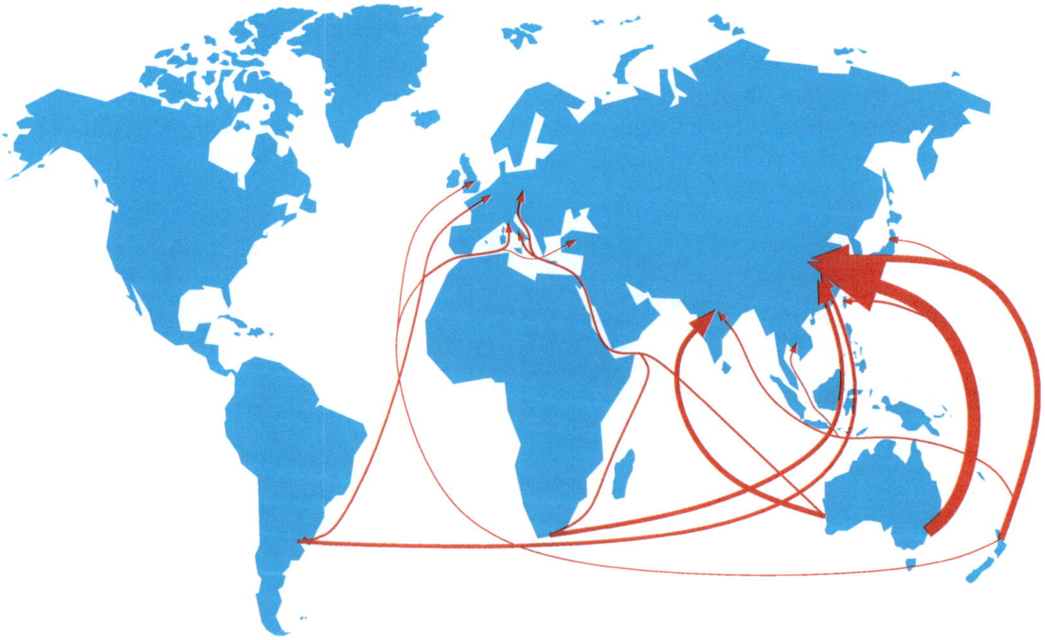

Fig. 9.69 Greasy wool exports by major producers (data: IWTO, 2019)

Table 9.9 Major producers of greasy wool (data: IWTO, 2019)

	China	India	Czech Rep.	Italy	Taiwan
Australia	264,000	20,000	18,000	13,000	2500
	China	UK	Italy	Germany	India
New Zealand	56,000	8600	8000	4500	3000
	China	Germany	Italy	Türkei	India
Uruguay	20,000	3600	2000	1400	800
	China	Germany	Italy	Czech Rep.	Turkey
Argentina	8400	5400	2400	2300	1900
	China	Czech Rep.	Italy	India	Bulgaria
South Africa	36,000	7600	2500	800	800

10 years was AUD 11–25/kg, for 21 μm it was AUD 9–22/kg and for 30 μm it was AUD 4–7.5/kg. The prices of finer fibers vary more than those for coarser wools (Fig. 9.72).

To protect farmers from the effects of such price fluctuations, for many years, a minimum price ("floor price") was paid by the Australian government. After this "Reserve Price Scheme" had functioned well for several decades, the organization and handling of the trade was transferred to the Wool Council of Australia (WCO) and the Australian Wool Corporation (AWC) in 1987. The farmers paid into a communal fund which, in addition to research and advertising, was used in particular to finance support purchases when the world market price fell below a certain level. As a result, a serious crisis occurred when the world market price for wool remained below the fixed minimum price for a prolonged period. As a result, the WCO and the AWC had to buy up huge quantities of wool that could no longer be sold on the world market. This led to the bankruptcy of the Fund within a very short period of time and subsequently to a complete collapse of this

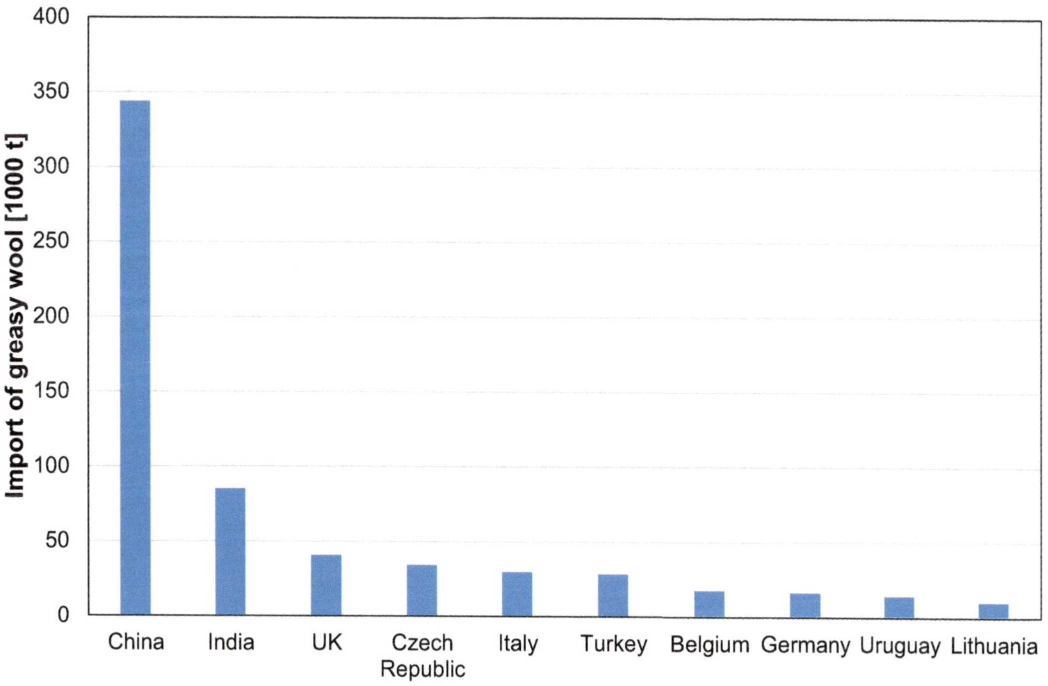

Fig. 9.70 Major importers of raw wool (data: IWTO, 2019)

Table 9.10 Major exporters and importers of scoured wool [1000 t] (data: IWTO, 2019)

Scoured wool-exports		Scoured wool-imports	
New Zealand	64,745	China	78,684
UK	24,221	India	49,809
Turkey	21,041	UK	23,617
Mongolia	16,159	Italy	10,489
Syria	9428	Lithuania	9564
Uruguay	7934	Belgium	8148
Russia	7814	Denmark	5593
Australia	6286	Russia	5276
China	5329	Germany	5048
Turkmenistan	4396	Thailand	4903

system in 1991, because the full stocks were unsaleable.

Other countries, such as New Zealand, South Africa and Uruguay, also support their wool farmers through subsidies. In the course of the liberalization of world trade, however, such systems are increasingly being abandoned.

Sheep meat is also an important export item for the wool-producing countries. The main markets are China, the USA, Europe and the Middle East. If sheep are transported alive and only slaughtered locally (e.g. in Islamic countries), sheep odysseys occur time and again if the sheep have a contagious disease so that they may not be unloaded at the destination ports (Fig. 9.73).

9.10 Wool Research

In line with its importance as a main producing country, the largest wool research institutes worldwide were located in Australia (CSIRO, Geelong) and New Zealand (AgResearch, Christchurch). The wool division of CSIRO was closed in 2021, but research at AgResearch continues. Furthermore, since 1952, there has been the German Wool Research Institute (DWI) in Aachen, which had a dominant position, especially for Europe, after corresponding institutes in England were closed (Ilkley). An international umbrella organization is the International Wool Secretariat (IWS).

9.11 Areas of Application

Wool is usually recycled because of the comparatively high price for the raw material. A distinction is therefore made between "pure new wool", i.e. wool obtained from living sheep, and "wool", which is reused.

Fine and medium-fine Merino and crossbred wools are mainly used in clothing and home textiles. In addition to underwear, these include in particular ladies' and men's outerwear. Medium to coarse wools are processed into

Table 9.11 Major exporters and importers of wool tops [1000 t] (data: IWTO, 2019)

Top-exports		Top-imports	
China	31,244	Italy	24,347
Czech Rep.	18,052	Germany	16,785
Uruguay	14,115	Romania	9662
Argentina	13,047	Turkey	9513
Italy	11,018	Bulgaria	8375
Germany	10,789	Poland	7274
		Czech Rep.	6412

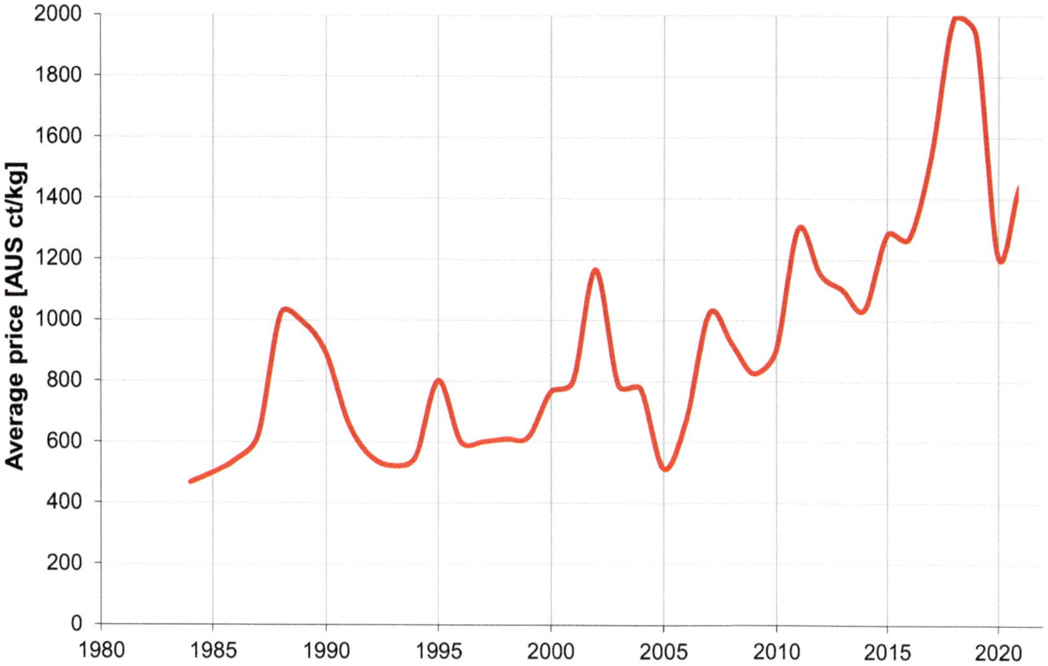

Fig. 9.71 Evolution of the average wool price in Australia (data: IWTO, 2019, www.wool.com)

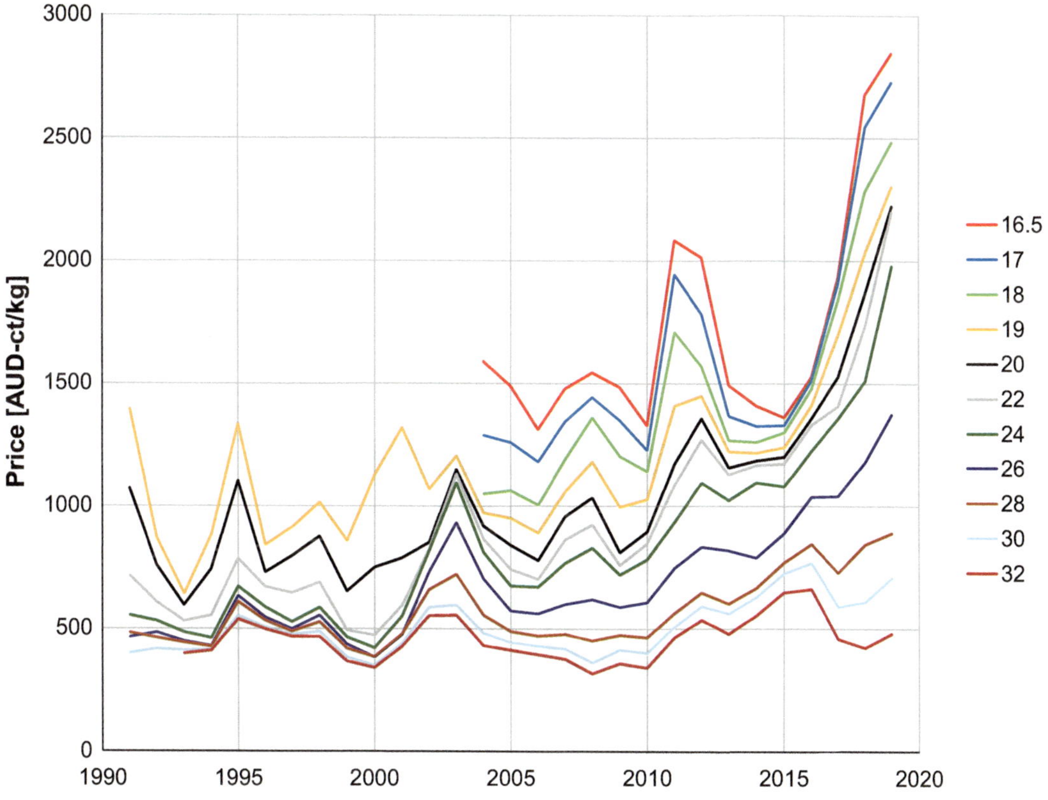

Fig. 9.72 Australian wool prices as a function of diameter [µm] (data: IWTO, 2019)

Fig. 9.73 Sheep on ship (Jervis, 2005a, b)

hand-knitting yarns, woolen blankets, blanket fillings and carpets (Figs. 9.74 and 9.75).

Because of its scaly structure, wool tends to felt. Therefore, it is also possible to produce dense, water-repellent fabrics (e.g. loden). However, wool can also be finished without felting and is then machine washable.

Another area of application is bulletproof vests, for which wool is combined with aramid. Due to the higher friction between wool and aramid fibers, energy absorption is improved,

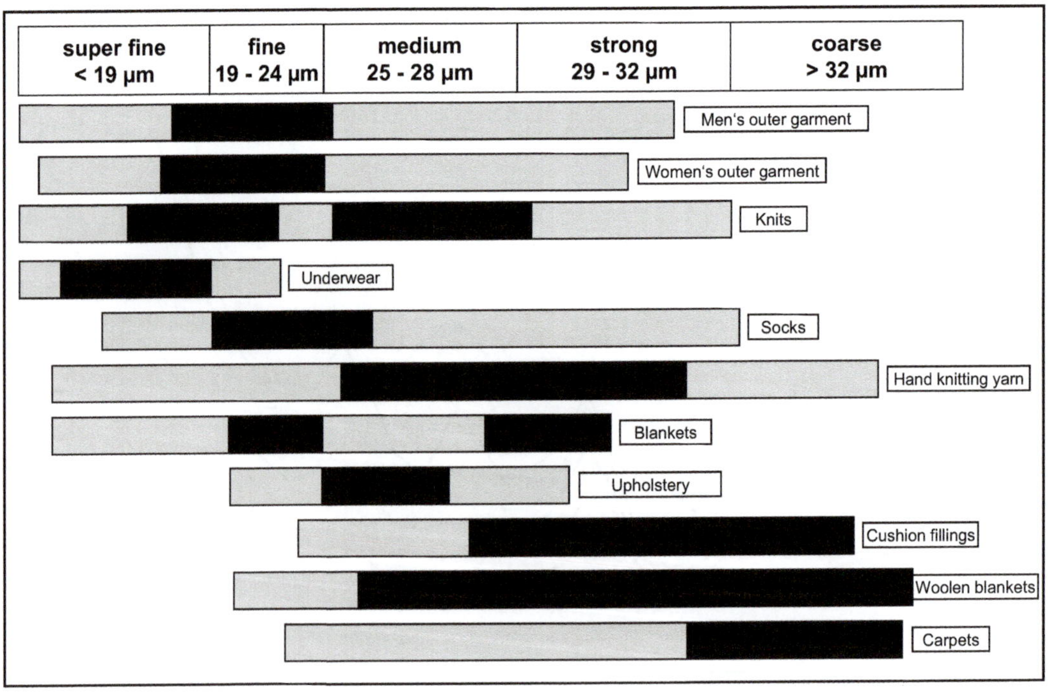

Fig. 9.74 Applications of wool depending on the fiber fineness

Fig. 9.75 Typical products made of wool (left: Carina Schauer, 2021)

Fig. 9.76 Wool seals (The Woolmark Company)

PURE NEW WOOL
WOOL RICH BLEND
WOOL BLEND PERFORMANCE

due to the swelling of the wool fibers, especially when wet (Sinnppoo et al., 2010).

9.12 Wool Seals

In 1964, the wool seal was introduced in Germany, Great Britain, Japan and the USA, and by 1966 in 93 other countries (Fig. 9.76, left). Textiles labelled with this seal must consist of 100% new wool. Small proportions of admixtures of wool obtained from the skins of slaughtered sheep are permitted. The license to award the label is granted by Australian Wool Innovation Limited (AWI) based in Sydney (Australia), a subsidiary of Woolmark Company Private Limited. It comes with conditions regarding compliance and monitoring of quality. In 1970, the term "pure new wool" was included in the Textile Labelling Act (TKG), and since 1972 it has applied throughout the EU. For products made of pure new wool with a small proportion of other fibers, e.g. man-made fibers, the Combi wool label was introduced in 1971. The minimum percentage of pure new wool is then 50 to 90% (Fig. 9.76, middle). In 1999, the Wool Blend label was introduced, which applies to products that consist of at least 30% virgin wool.

References

Beef + Lamb New Zealand. (2020). Retrieved from https://beeflambnz.com/data-tools/economic-reports

Bradbury, J. H. (1973). The structure and chemistry of keratin fibers. *Advances in Protein Chemistry, 27*, 111–211.

Büttner, J. (2019). Waage und Wandel – wie das Wiegen die Bronzezeit prägt. In *Innovationen der Antike*. Verlag Philipp von Zabern.

Cape Wools SA. (2020). Retrieved from https://www.capewools.co.za/documentlibrary/market-reports

Capper, P. (2020). Retrieved from https://commons.wikimedia.org/w/index.php?curid=3351067, CC BY 2.0.

Carina Schauer. (2021). *Aachen*. Retrieved from https://www.knitsbycari.de

CSIRO. (2008a). Retrieved from https://commons.wikimedia.org/wiki/File:CSIRO_ScienceImage_7663_The_structure_of_a_Merino_Wool_Fibre.jpg, CC BY 3.0.

CSIRO. (2008b). Retrieved from https://commons.wikimedia.org/wiki/File:CSIRO_ScienceImage_8115_Human_hair_and_Merino_wool_fibre.jpg, CC BY 3.0

Duerden, J. E. (1929). *Journal of the Textile Institute, 20*, 93–100.

DWI – Leibniz-Institut für Interaktive Materialien e.V. Retrieved from http://www.dwi.rwth-aachen.de

Esson, T. (2006). Retrieved from https://en.wikipedia.org/w/index.php?curid=7176435, CC BY-SA 1.2.

faostat. (2022). Retrieved from http://www.faostat.fao.org

Geni. (2013). Retrieved from https://commons.wikimedia.org/wiki/File:Standard_of_ur_peace_2013.JPG, CC BY-SA 4.0.

Goodwin, C. (1998). Retrieved from https://commons.wikimedia.org/wiki/File:Poll_Merino.jpg, CC BY-SA 3.0.

Halling, P. (2020). Retrieved from https://commons.wikimedia.org/wiki/File:Sheep_in_field_above_Lower_Cwmcoched_Farm_-_geograph.org.uk_-_570884.jpg, CC BY-SA 2.0.

Hoifodt, H. (2009). Retrieved from https://upload.wikimedia.org/wikipedia/commons/2/21/Estancia_El_Ceibo.JPG, CC BY-SA 3.0.

Hudson, P. R. W. (1987). High technology options for sheep shearing III: Chemical and biological defleecing. *Outlook on Agriculture, 16*(2), 68–76.

IWTO. (2019). *Market Information*. Edition 14.

Jervis, T. (2005a). Retrieved from https://commons.wikimedia.org/wiki/File:Sheep_Ship_2.jpg, CC BY 2.0.

Jervis, T. (2005b). Retrieved from https://commons.wikimedia.org/wiki/File:Sheep_Ship_3.jpg, CC BY 2.0.

Kerl, B., & Stohmann, F. (1879). *Muspratt's theoretische, praktische und analytische Chemie in Anwendung auf Zünfte und Gewerbe*. Schwetschke & Sohn.

Kidd, R. (2007). Retrieved from https://commons.wikimedia.org/wiki/File:Sheep_at_Llanddewi_Brefi,_in_Ceredigion,_Wales.jpg, CC BY 2.0.

Lenzing AG, Lenzing (Österreich). Retrieved from https://www.lenzing.com

Massy, C. (2007). *The Australian Merino*. Random House.

McCrath, M. (2020). Retrieved from https://commons.wikimedia.org/wiki/File:Coopworth_sheep.jpg, CC BY-SA 2.5.

New Zealand Wool. Retrieved from https://nzwool.co.nz

NewMerino. (2021). Retrieved from https://newmerino.com.au/

Nichols, B. (2020). Retrieved from https://commons.wikimedia.org/w/index.php?curid=11217471

Oviso. (2003). Retrieved from https://commons.wikimedia.org/wiki/File:Ovejas_merinas.JPG, CC BY-SA 3.0.

Pam. (2020). Retrieved from https://commons.wikimedia.org/w/index.php?curid=14782019, CC BY 2.0.

Pauling, L., Corey, R. B., & Branson, H. R. (1951). The structure of proteins: Two hydrogen-bonded helical configurations of the polypeptide chain. *Proceedings of the National Academy of Sciences, 37*.

Pot, M. (2007). Retrieved from https://commons.wikimedia.org/wiki/File:Sheep_shearing.jpg, CC BY 3.0.

Proimos, E. (2020). Retrieved from https://commons.wikimedia.org/wiki/File:Mustering_sheep_in_Patagonia.jpg, CC BY 2.0.

Rodolph. (2009). Retrieved from https://commons.wikimedia.org/wiki/File:Victorian_stand_in_1867.jpg

Royan, J. (2020). Retrieved from https://commons.wikimedia.org/w/index.php?curid=15897077, CC BY-SA 3.0.

Rufus. (2014). Retrieved from https://commons.wikimedia.org/wiki/File:Europ%C3%A4ischer_Mufflon_Ovis_orientalis_musimon_Wildpark_Poing-11.jpg, CC BY-SA 3.0.

Sawyer. (2007). Retrieved from https://commons.wikimedia.org/wiki/File:Colouredromneys.jpg

Schier, W., & Pollock, S. (2018). Die textile revolution. In *Innovationen der Antike*. Verlag Philipp von Zabern.

Simark, E. (2020). Retrieved from https://commons.wikimedia.org/wiki/File:A_flock_of_sheep_-_geograph.org.uk_-_615085.jpg, CC BY-SA 2.0.

Simmonds, D. H. (1956). The amino acid composition of keratins – III. The aminoc acid composition of different qualities of wool. *Australian Journal of Biological Sciences, 8*, 537–540.

Sinnppoo, K., Anrold, L. N., & Padhye, R. (2010). Application of wool in high-velocity ballistic protective fabrics. *Textilre Research Journal, 80*, 1083–1092.

Stoger, F. (2017). Retrieved from https://commons.wikimedia.org/wiki/File:20170309_2597_Ulley-Leh_Urial.jpg, CC BY-SA 4.0.

The Woolmark Company. Retrieved from https://www.woolmark.com

Trevelyan, J. (1999). *Redefining robotics for the new millenium*. Sage.

Wagner, E. (1981). *Die textilen Rohstoffe*. Deutscher Fachverlag.

Weller, K. (2020). Retrieved from https://commons.wikimedia.org/wiki/File:Flock_of_sheep.jpg.

Wiki. (2005). Retrieved from https://commons.wikimedia.org/wiki/File:Merino_sheep.png, CC BY-SA 1.2.

Wiki. (2009). Retrieved from https://commons.wikimedia.org/wiki/File:Aue_Deutsches_schwarzk%C3%B6pfiges_Fleischschaf.jpg, CC BY-SA 3.0.

Wools of Uruguay. (2020). Retrieved from https://www.wool.com.uy/production_systems.htm

Wortmann, F.-J. (2009). The structure and properties of wool and hair fibres. In *Handbook of textle fibre structure*. Woodhead Publishing.

Wortmann, F.-J., & Zahn, H. (1994). The stress/strain curve of alpha-keratin fibers and the structure of the intermediate filament. *Textile Research Journal, 64*, 737–743.

Zahn, H., Fohles, J., Nienhaus, M., & Schwan, A. (1980). Wool as a biological composite structure. *Industrial & Engineering Chemistry Product Research and Development, 19*, 496–501.

Zahn, H., Wulfhorst, B., & Külter, H. (1991). Faserstofftabellen nach P.-A. Koch: Wolle. *Chemical Fibres International, 41*(93), 521–553.

Zahn, H., Wortmann, F.-J., & Höcker, H. (1997). Chemie und Aufbau der Wolle. *Chemie in unserer Zeit, 31*(6), 280–290.

10 Fine Animal Hair

Animal hair is divided into fine and coarse animal hair. Only the fine animal hairs are important for clothing. This includes the hair of alpaca, llama, vicuña, guanaco, camel, angora rabbit, mohair and cashmere goats and yak. The fine underhairs are called "wool" and the coarser upper hairs are called "hair", each in connection with the respective animal name. The coarse animal hairs include cattle, horse and most goat hairs as well as the kemp (coarse hairs) of the camel (Fig. 10.1).

World production of fine animal hair (excluding superfine wool) is about 54,000 t/year, of which cashmere accounts for 30,000 t (Fig. 10.2).

Figure 10.3 shows the volume shares of all fine animal hair, based on a total production of 165,000 t, of which 109,000 t is superfine wool.

10.1 Camel

Camels are ruminant callipeds, very frugal and persistent and therefore ideally adapted to life in the desert. They are kept as beasts of burden and draught animals as well as in large herds for the extraction of their hair.

In addition to the dromedaries and the Bactrian camels, the camels also include the llamas and their corresponding subgroups. Figure 10.4 shows the distribution areas of llama and alpaca (1), one-humped (2) and two-humped (3) camels.

The pedigree of camels is shown in Fig. 10.5. About 10.4 million years ago, the alpacas evolved from the primitive llamas, about 6.4 million years ago, the vicuñas emerged and 1.4 million years ago the family tree split again and llamas and guanacos emerged.

10.1.1 Dromedary

The approximately 35 m one-humped camels (faostat, 2022) are mainly at home in North Africa (80%) and in the Near East as well as in India. They are therefore also called Arabian camels. In Australia there are large, mostly wild herds due to importation in the nineteenth century, but they are not used economically (about 300,000 animals). The wool of the dromedary is coarse and therefore only suitable for carpets. The very coarse mane hairs are also used for elastic interlinings and driving belts.

10.1.2 Bactrian Camel

This is a two-humped camel that lives in the high plains of western China and outer Mongolia (Fig. 10.6). It is also called the Bactrian camel and is kept as a caravan animal in Iran, Iraq and Afghanistan. It is estimated that there are around 1.4 m animals. The animal protects itself against the harsh climate and extreme cold with its very fine undercoat (18–26 μm) of relatively great length (30–120 mm), Fig. 10.7. Most of the camel wool used for textiles comes from Bactrian camels.

D. Veit, *Fibers*, https://doi.org/10.1007/978-3-031-15309-9_10

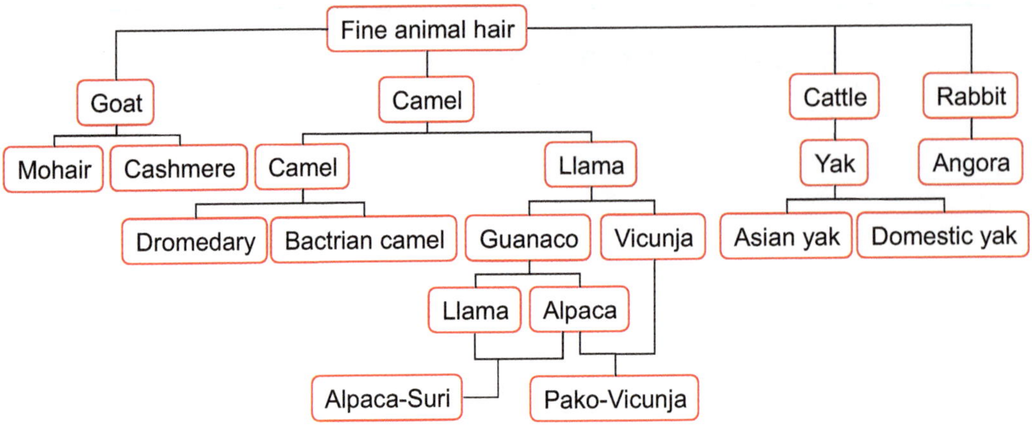

Fig. 10.1 Division of the most important fine animal hairs

Camels are generally not shorn as they usually shed their wool in tufts in the spring. This is collected and sorted according to fineness. The sale of Chinese camel wool takes place in May at a fair in Canton. An adult camel produces about 5 kg of wool per year. Young camels up to the age of 1 year are "natural blond". Their almost white, soft hair ("baby hair") is very valuable. The price is between 30 and 55 €/kg.

Camel hair cannot be completely separated from the coarse kemp by sorting. It is therefore unawned in a complicated process. The end product is very light and is highly prized for its silky, golden sheen and soft, warm feel. World production of camel hair has been steadily increasing

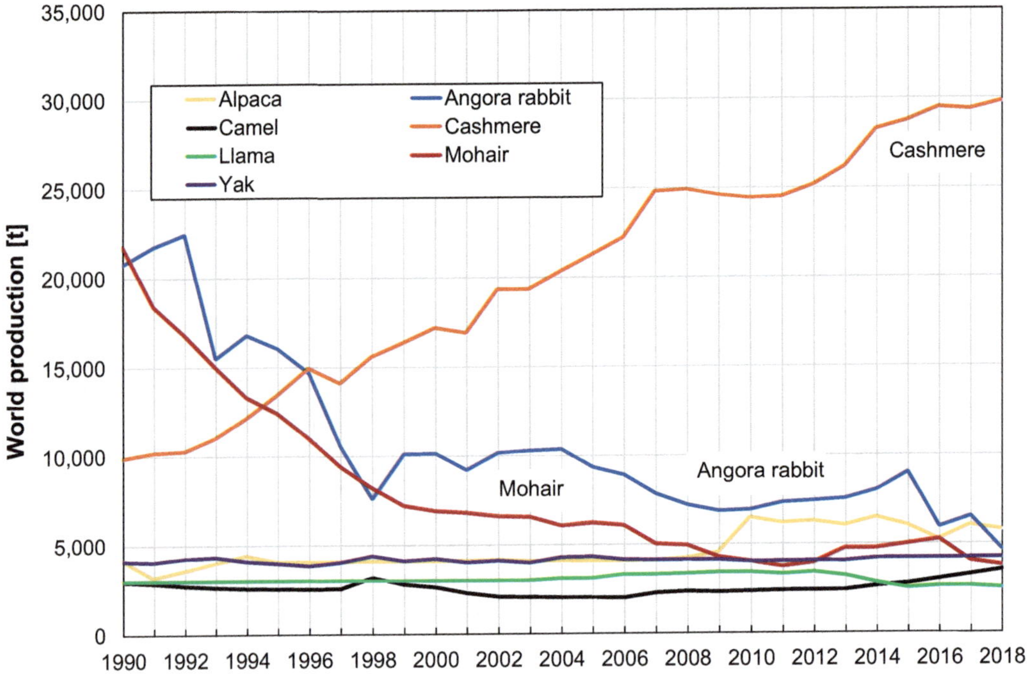

Fig. 10.2 Production of fine animal hair (data: IWTO, 2019)

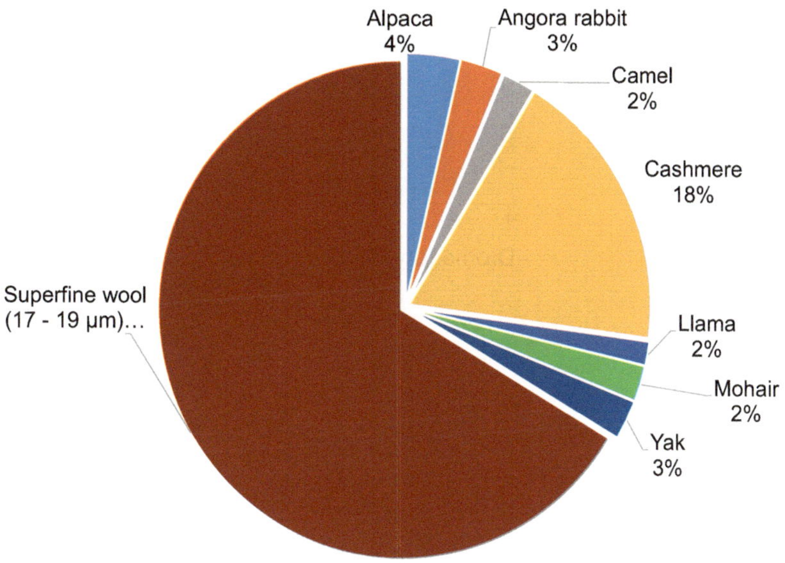

Fig. 10.3 Proportions of important fine animal hairs (data: IWTO, 2019)

since 2005 and currently stands at around 3500 t (Fig. 10.8).

10.1.3 Llama

Llamas are native to the high Andes of South America, especially Bolivia and Peru. They reach a shoulder height of up to 130 cm. Their moderately long-haired coat is yellow to black-brown, white or spotted. Llamas also serve as meat suppliers and male llamas are also used as pack animals (carrying capacity up to 50 kg). The first llamas are said to have been domesticated more than 4500 years ago. Llamas were the only large farm animals in the Americas before European animals were introduced. Llamas, alpacas, guanacos and vicuñas can be interbred, which has led to various hybrids, e.g. alpaca-suri and pako-vicuñja. The main aim is to increase the yield of fine fibers.

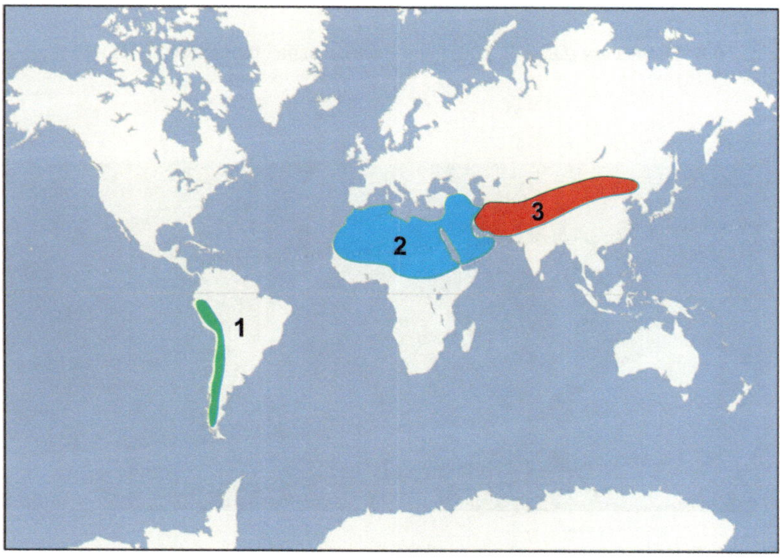

Fig. 10.4 Distribution areas of cameloids

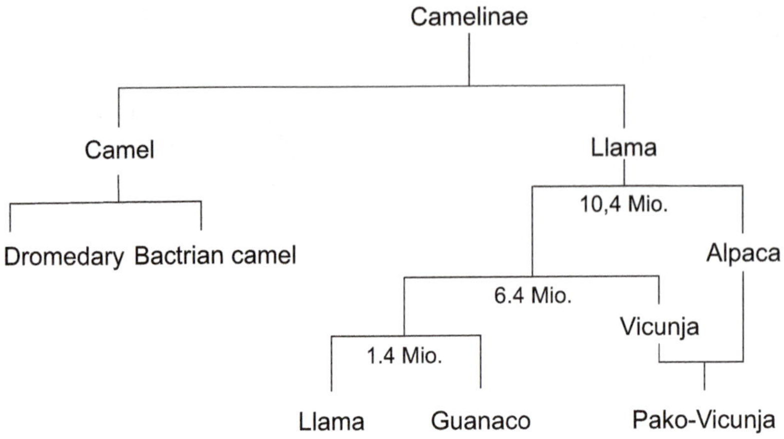

Fig. 10.5 Pedigree of camels according to Cui et al. (2007)

Fig. 10.6 Dromedary (O'Neill, 2007) and Bactrian camel (Giesenow, 2021)

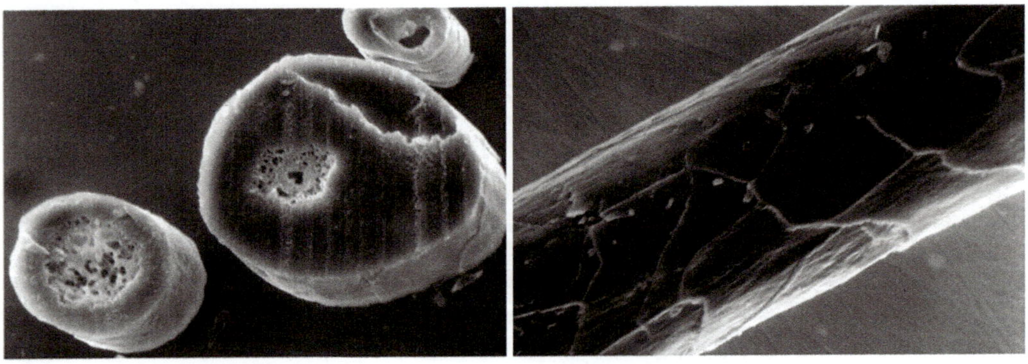

Fig. 10.7 Cross-sectional and longitudinal view of camel hair wool

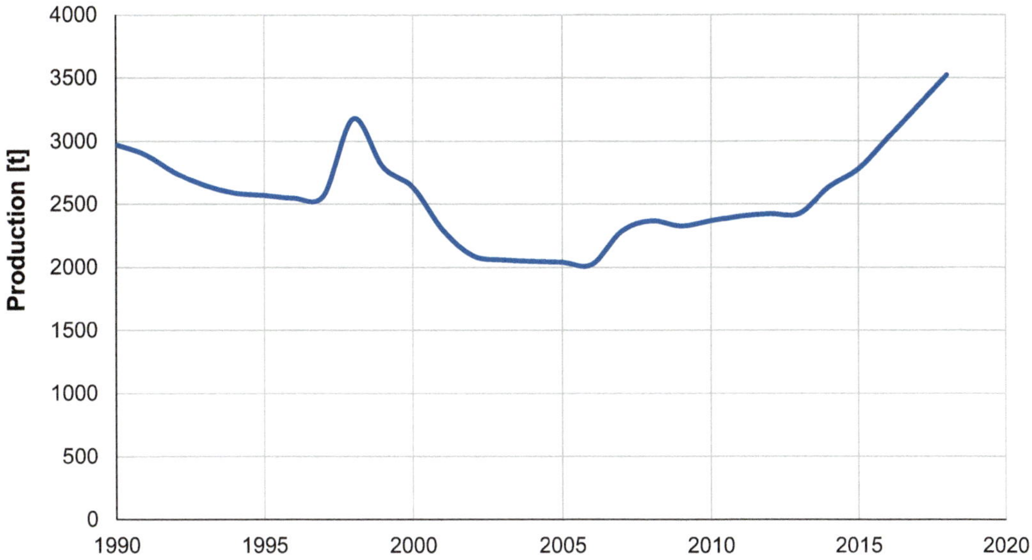

Fig. 10.8 World production of camel hair (data: IWTO, 2019)

The population of llamas is estimated at about 3.5 m heads. The herds graze mostly unguarded (semi-wild) the sparse grass of the mountain valleys (Fig. 10.9).

The wool of llamas is obtained about every 1–2 years. However, they are not shorn bald like sheep. Instead, a few centimetres of hair are left on the skin as protection against the cold. The fiber yield is 2–3 kg per shearing, the annual production is about 2500 t with a slight downward trend in recent years. Typical fiber diameters are 22–30 μm (Fig. 10.10).

10.1.4 Alpaca

There are about 4.2 m alpacas, of which about 3.7 m live in Peru. The alpaca is used exclusively for wool production and reaches a shoulder height of up to 120 cm. Characteristic for alpacas are the long neck and the narrow head. White, brown and black shades dominate the coat coloring (Fig. 10.11).

The alpaca suri was created from crosses between llamas and alpacas (Fig. 10.12). It is infertile and therefore not suitable for breeding.

Fig. 10.9 Llamas in Peru

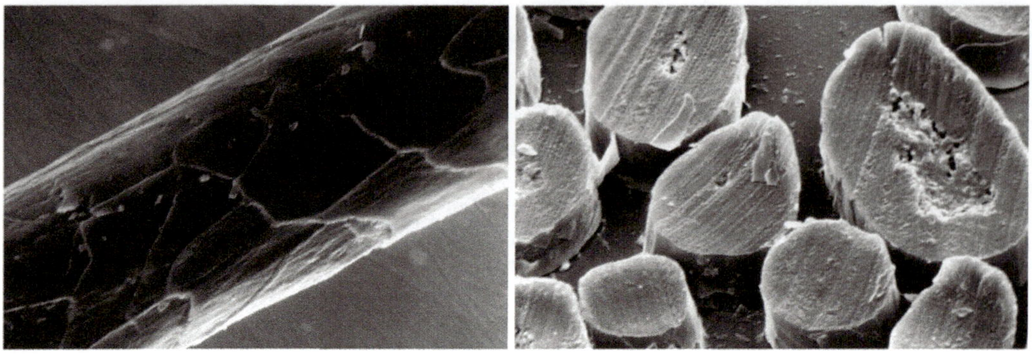

Fig. 10.10 Longitudinal and cross-sectional view of llama wool (DWI)

However, its hair is longer and shinier than that of the alpaca, has hardly any curl and forms straight strands. These animals grow to a height of approx. 130 cm, their fiber yield is approx. 3 kg per shearing.

About 80% of alpacas belong to the Huacayas, 12% to the Suris and 8% are hybrids. They differ in the structure of their wool: the huacaya alpaca has a fine, evenly curled fiber and only a few awn hairs. The fiber length of Suris is significantly greater, the fibers are softer and have a higher luster.

Alpacas are stubborn and unruly if not treated well (spitting). It can therefore be very tedious to catch and shear them (Fig. 10.13).

The wool of the alpaca is somewhat finer than that of the llama and has finer scales (Fig. 10.14). The fiber yield is approx. 2–4 kg per shearing, which takes place at intervals of 1–2 years. The yield is about 80% with a fat content on the fiber of only 3–4%. The fiber fineness is 10–15 μm (alpaca suri) or 22–25 μm (huacaya), the fiber length usually reaches 65–75 mm and the scale height is 0.25–0.3 μm. Alpaca fibers can absorb about 8% moisture at 20% relative humidity and up to 21% at 90% relative humidity. The LOI value is high and is 24–30, the self-ignition temperature is approx. 600 °C.

The wool of alpaca and llama is not monochrome, but consists of 3–4 or more different shades. These range from black, dark brown, chestnut, cinnamon, camel, beige and dark grey to silver grey, black and white, brown and white and pure white. A total of 22 shades are

Fig. 10.11 Alpacas in Peru (International Alpaca Association)

Fig. 10.12 Suri and Huacaya alpaca in Peru (International Alpaca Association)

distinguished. Pure white animals are albinos. Alpaca and llama wool must therefore be sorted by hand. Because white wool fetches the highest prices (up to 17 €/kg), livestock farmers strive to increase the stock of white animals. However, for genetic reasons, it is not possible to breed pure white animals (in contrast to pure white sheep).

Shearing begins with the rainy season in November and ends in April. The annual production has been around 5000–6000 t for some years (Fig. 10.15).

Around 90% of the alpaca fibers are processed in Arequipa, in the south of Peru. The fleeces are sorted by hand in the so-called "lavaderos" according to color and quality, pressed into bales and shipped. Sometimes the scouring process only takes place in the country of destination.

In Arequipa, around 20,000 people work in alpaca extraction and processing. Alpaca is thus an essential economic factor for this region. The value of the exported fibers, yarns, fabrics and knitted goods amounts to around US$ 150 m. The main markets for these are China and Italy. Made-up textiles worth around US$ 50 m are exported mainly to the USA, Germany and Japan (Blog Peru, 2019).

If a female alpaca is bred with a male llama, a barren huarizo results. The wool quality lies between that of llama and alpaca fibers (Fig. 10.16).

Fig. 10.13 Shearing of an alpaca (International Alpaca Association) and shorn alpaca

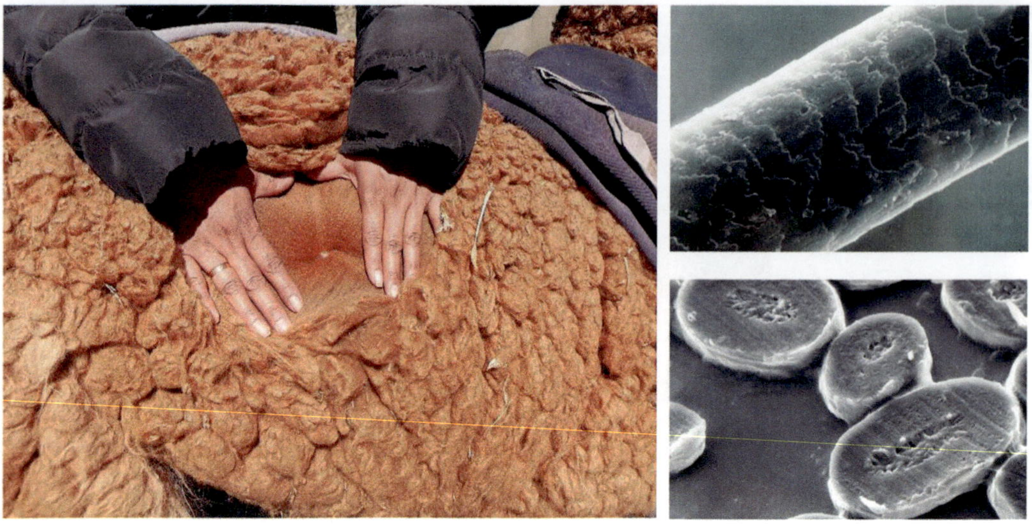

Fig. 10.14 Alpaca fleece (International Alpaca Association) and microscopic images

Depending on the fiber content, different seals are awarded to alpaca articles by the International Alpaca Association (IAA). A distinction is made between textiles made from 100% alpaca with a fiber fineness of no more than 23 μm (Fig. 10.17, left, seal on golden background), 100% alpaca with a maximum fiber fineness of 26.6 μm (seal identical, but on silver background) and articles with at least 30% alpaca content (Fig. 10.17, right, seal on white background).

10.1.5 Guanaco

Guanacos (llama guanicoe) are humpless wild camels that live primarily in arid areas of western

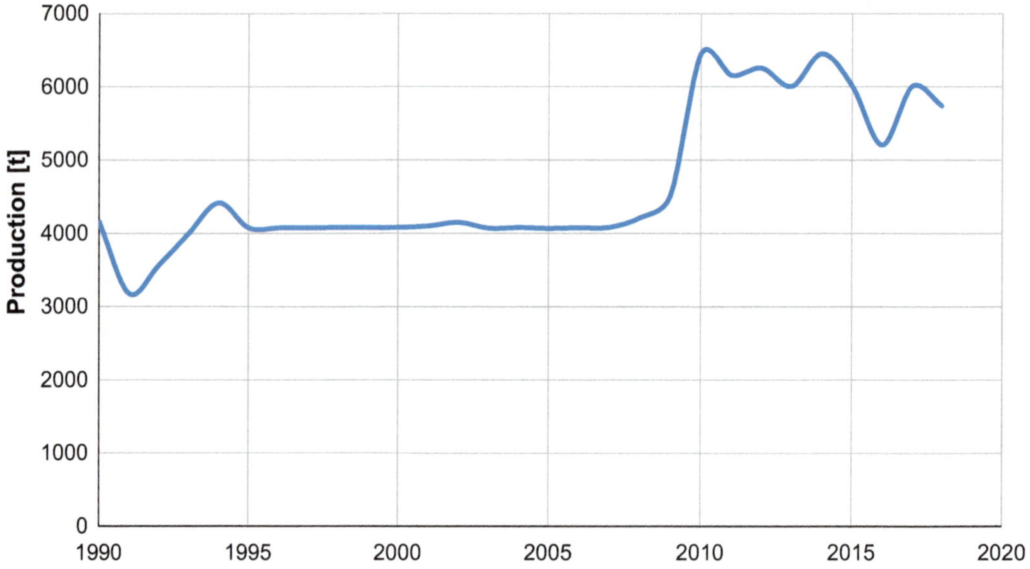

Fig. 10.15 Production of alpaca wool (data: IWTO, 2019)

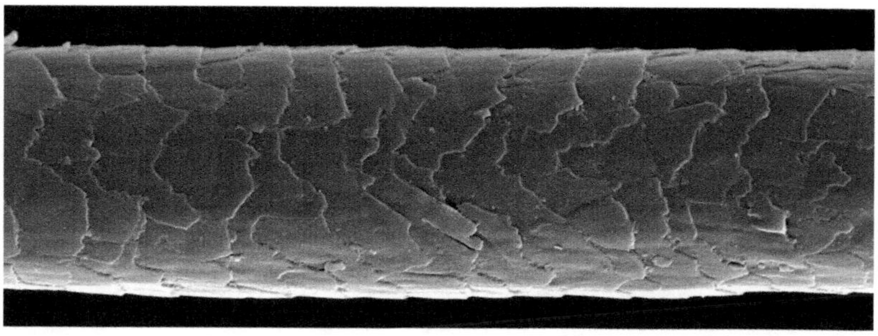

Fig. 10.16 Alpaca Huarizo fiber (DWI)

and southern South America. They are believed to be the ancestral form of llamas and alpacas and were probably domesticated 5000 years ago. Their shoulder height is about 90–120 cm, their weight 100–120 kg. In order to still be able to graze the barren and often very small areas with flat, extremely rough grasses, guanacos, just like llamas and alpacas, have teeth that continue to grow due to the great stress and wear. Guanacos live in small herds of about 20 animals, led by a male. In the Andes, they can be found at altitudes of over 4000 m above sea level.

Guanacos are very good swimmers and not only cross streams and lakes, but also swim through the sea to offshore islands. A special feature of the guanacos is that they deposit their excrement in certain places, which thus accumulates over the years to giant piles (Fig. 10.18).

The coat of guanacos consists of different colors, usually light brown on top and white underneath. There are long, wavy hairs that protect the animal from water and short, curly ones that protect it from the cold. The fibers are very light because they have air pockets inside. The scale height is low. The fineness of the fibers is 10–20 μm, their length is 30–50 mm (Fig. 10.19).

In order to gain the wool, guanacos were shot in the past and almost exterminated in the process. In the meantime, there are about 0.5 m animals again and they are protected. The world production is about 2 tons per year which makes it one of the scarcest commercially used fibers on earth.

10.1.6 Vicuña

Vicuñas are slightly smaller than guanacos and reach a body height of up to 110 cm. They live in small herds of about 20 animals each, which are led by a male as with the guanaco. They live

Fig. 10.17 Alpaca seals (International Alpaca Association)

Fig. 10.18 Guanaco (Howard, 2005) and giant dung heap (Fanny Habermeyer)

mainly at extreme altitudes (4000–5500 m above sea level) and are very shy. Vicuna wool was and still is highly sought after because of its high fineness (sometimes less than 10 μm, length 20–40 mm) and its golden-brown colour ("Gold of the Andes"). At the time of the Incas, it was only allowed to be worn by the kings. However, because the animals were shot in the twentieth century to obtain their wool, they were declared an endangered species and placed under protection in 1968, when it was estimated that there were only about 5000 animals left. Today, vicuña fiber is only processed by a few companies worldwide. Therefore, it is hardly possible to illegally process fibers obtained by shooting the animals into textiles without being noticed. Thus, the population recovered to about 350,000 animals today, half of which live in Peru (Fig. 10.20).

For shearing, the wild vicuñas are herded together during the so-called "Chaccu", then immediately shorn and then released again. About 200 g of fiber can be obtained per animal every 2 years. The annual world production currently amounts to about 7 t.

For one kilo of raw fibers the inhabitants of the Andean villages are paid about 500 US$ (the village, vicuñas do not belong to anyone), a running meter of finished vicuña fabric costs about 2500 € at a basis weight of 400 g/m^2, a finished suit up to 80,000 €. Sweaters are cheaper and are on sale at about 5000 €, scarves at 3000 € (Fig. 10.21).

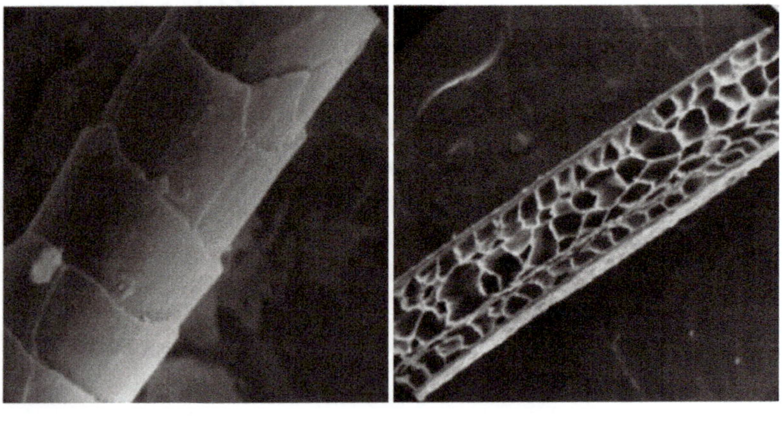

Fig. 10.19 Longitudinal view and longitudinal cut through a guanaco fibers

Fig. 10.20 Vicuñjas in Peru (Steffen Reppich) and fleece skirting

10.1.7 Pako vicuña

This is a breeding mix of vicuña and alpaca. For this, a young vicuña must be captured and wrapped in the fur of a slaughtered alpaca. This allows it to be adopted and raised by a mother alpaca. When it mates with an alpaca, a pako vicuña is created. Its wool is very fine and long, yielding 2–3 kg per shearing. Breeding is difficult, however, because most mares are infertile.

The high importance of the vicuña for Peru is shown by the fact that it adorns the Peruvian flag (Fig. 10.22).

Fig. 10.21 Typical vicuña products (Mahogany Cashmere)

Fig. 10.22 Pako vicuña and flag of Peru with vicuña (Wiki, 2019)

10.2 Goat

Goats were already domesticated around 9000 years ago. The wild ancestral forms of the domestic goat are the Bezoar goat (Capra aegagrus) from the region of the Euphrates and Tigris and the Markhor (Capra folconeri) from Inner Asia. The hair of most goats belongs to the coarse animal hair and is therefore not usable for clothing. Only the wool of the cashmere and mohair goats and of the musk ox (Qiviut) belong to the fine animal hair category (Fig. 10.23).

10.2.1 Cashmere

The habitat of the cashmere goat is the Mongolian steppes and the mountainous regions of western China, Afghanistan and Iran, up to 5000 m above sea level. Cashmere goats are kept in herds and

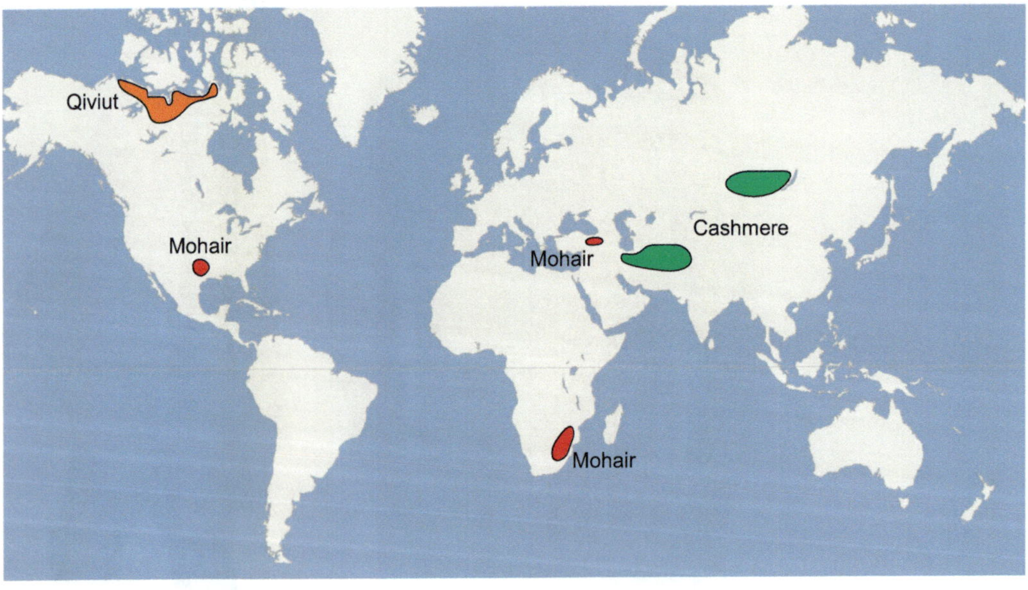

Fig. 10.23 Distribution area of mohair and cashmere goats and qiviut

Fig. 10.24 Cashmere goats

are not shorn. This prevents fine, valuable undercoat and coarse topcoat from mixing. Worldwide there are about 700,000 animals, in China as the largest cashmere wool producer about 120,000 (Fig. 10.24).

The goat sheds approx. 150–200 g of undercoat (raw cashmere) in spring during the change of coat, which is then combed out, plucked or collected from bushes and rocks. The cashmere wool obtained in this way still contains about 5–7% fat and about 30% coarse hairs and dirt. It is scoured, sorted according to color and subjected to an unawning process. Corresponding factories are located in China and outer Mongolia. The fiber diameter is 15–19 μm, the length 25–90 mm. The colour of cashmere ranges from white to dark brown. The scale height of 0.4 μm is only half that of sheep's wool (Fig. 10.25).

The annual production of raw cashmere is about 30,000 t with a slight upward trend, with about half coming from China and about 3000 t from Mongolia (Fig. 10.26). Depending on the quality, the price is up to 120 €/kg.

The extraordinary quality of this very noble fiber material, especially the great heat retention capacity in combination with its very low weight, was for a long time reserved only for privileged classes of the Asian population. In the seventeenth century, descriptions of this animal species are found in French literature. At about the same time, European merchants brought the first cashmere fabrics to Europe. Until today, articles made of cashmere have remained a luxury good. Cashmere is often blended with wool. In the past, textiles made of cashmere wool were called pashmina, but nowadays other fiber materials are often used for pashmina.

The demand for cashmere has been greater than the quantity produced for many years. Therefore, articles made of "100%" cashmere are often blended with other fibers, e.g. wool, yak and viscose or even contain no cashmere fibers at all. Such counterfeit products can only be detected by a few experts worldwide.

10.2.2 Mohair

The home of the mohair goat (formerly: Angora goat) is the Anatolian highlands (Fig. 10.27). There they are kept since the twelfth century. Their ancestors come probably from Tibet and

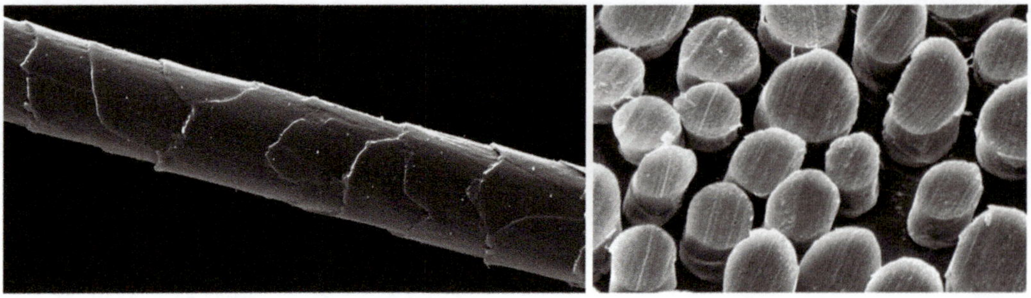

Fig. 10.25 Longitudinal and cross-sectional view of cashmere wool (DWI)

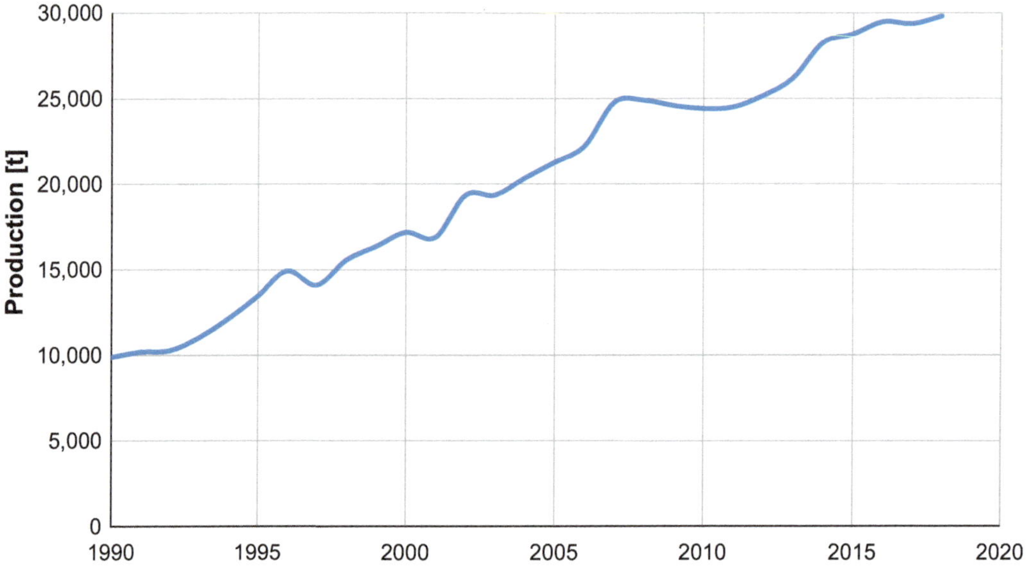

Fig. 10.26 Production of cashmere fibers (data: IWTO, 2019)

Fig. 10.27 Mohair goats (Angora goats)

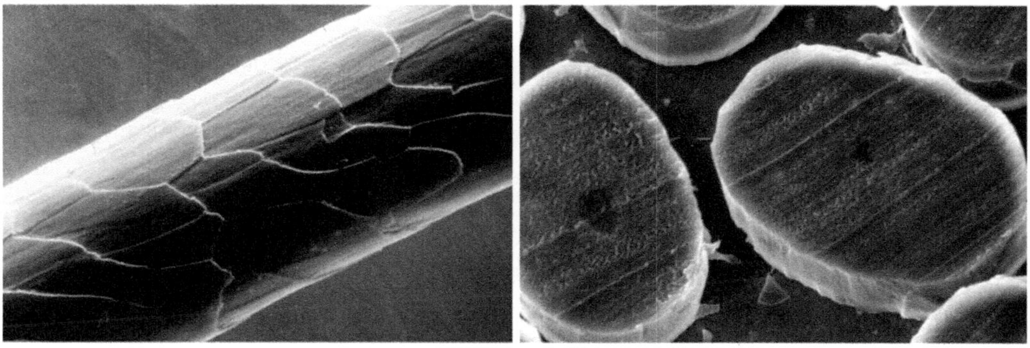

Fig. 10.28 Longitudinal and cross-sectional view of mohair wool

are related to the Markhor. After several attempts to settle them in Central Europe from the sixteenth century, they were exported to South Africa, Texas, Australia and New Zealand from the middle of the nineteenth century. These countries, along with Turkey, are the main producers of mohair today. The name "Angora" is derived from the name of Turkey's capital, Ankara. "Mohair" comes from the Arabic meaning "hair cloth".

The hair of the mohair goat grows up to 25 mm per month. Shearing therefore takes place from spring to autumn and the yield is 5–8 kg per year. The yield is high and lies at 70–90%. Decisive for the quality of the wool is the age of the animals. The fibers of 6–8 year old goats are too coarse for textile use. Mohair is graded according to fineness and length (75–350 mm) in a total of over 50 classes. Among others, the following designations are common:

- kids (lambs): 24–29 μm
- young goats: 30–33 μm
- adults (adult goats): 34–40 μm

Characteristic for mohair is its fine luster. For a long time, mohair wool was bred to be free of coarse and dark hairs. In the meantime, however, black and red Angora goats are also bred to produce naturally colored wool. The mechanical properties are similar to those of sheep's wool, but the felting capacity is much lower because the scale height is only 0.4 μm (sheep's wool: 0.8–1.0 μm). The crimp reaches 4 curls per 10 cm. Mohair wool has a high degree of thickness uniformity ("fidelity"), Fig. 10.28.

Due to scandals in the keeping of the animals, the production volume decreased by 80% from 1990 to 2010. Today, the world production is about 3800 t (greasy), Fig. 10.29. South Africa is the largest producer with a market share of 50%, Europe (Italy, UK) is the largest consumer (40%) followed by China (25%).

The price per kg is about 15–60 €, depending on the quality. Fine mohair fibers are often blended with wool or silk. Typical applications are scarves, shawls, jackets, sweaters. Rugs and home textiles are produced from coarse fibers.

10.2.3 Kashgora

By crossing female cashmere goats with male angora goats, the cashgora goat is obtained, which is mainly bred in Australia and New Zealand. The first cross produces the finest fibers. Further crossings with Angora goats produce coarser fibers. The corresponding wool is obtained by shearing. The fibers are similar to mohair in length (50–60 mm) and strength, while their fineness (19–22 μm) and luster are similar to cashmere. Prices range from 12 to 30 €/kg.

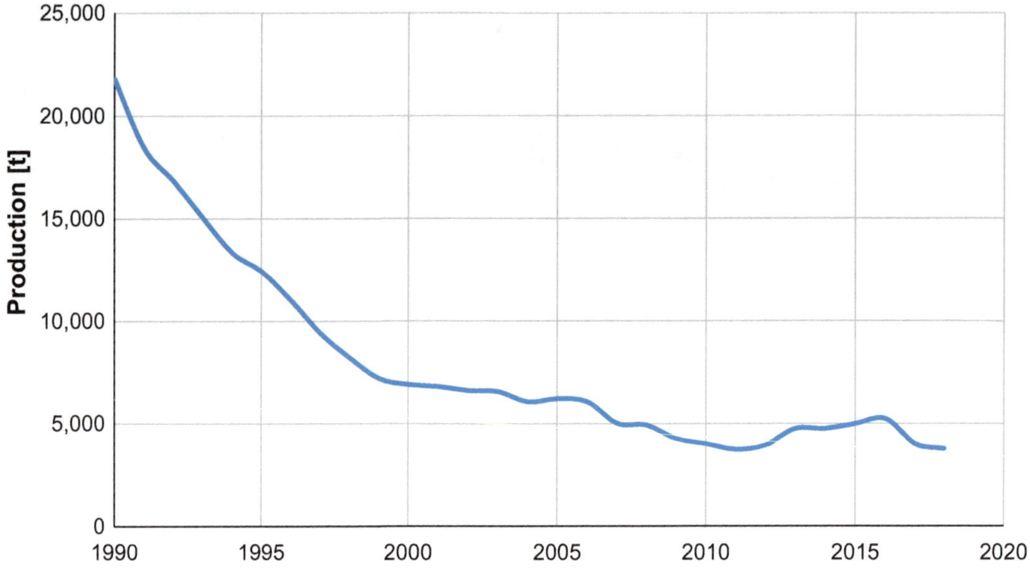

Fig. 10.29 Production of mohair fibers (Data: IWTO, 2019)

10.2.4 Qiviut

Despite their appearance, musk oxen belong to the goats (Fig. 10.30). Around 1930, there were only about 3000 animals left, but these have multiplied to about 160,000 specimens today. They live in small herds of 15–30 animals. The undercoat of the musk ox living in the Arctic and northern Canada is called qiviut or qiviuk and also "gold of the Arctic".

Musk oxen have the longest outer hairs of all mammals with a length of up to 180 cm. Their underhairs have a fineness of 11–22 μm (mostly 15–17 μm) and are thus among the finest fibers in the animal kingdom. Their length is 35–100 mm. The ratio of secondary to primary follicles is exceptionally high at about 37:1 (fine merino sheep: 20:1) and explains the very low proportion of fat and sweat in the greasy fibers at 7–10% (Rowell et al., 2001; von Bergen, 1931). The number of very fine scales, 6 per 100 μm, is similar to that of vicuña and cashmere fibers (Fig. 10.31). Thanks to these fibers, musk ox survives arctic storms with wind speeds up to 140 km/h and temperatures down to −60 °C. Qiviut fibers, unlike most other natural fibers, hardly shrink and do not develop an unpleasant odor like wool when they get wet.

Fig. 10.30 Musk ox with young (Rye, 2021) and greasy qiviut fibers

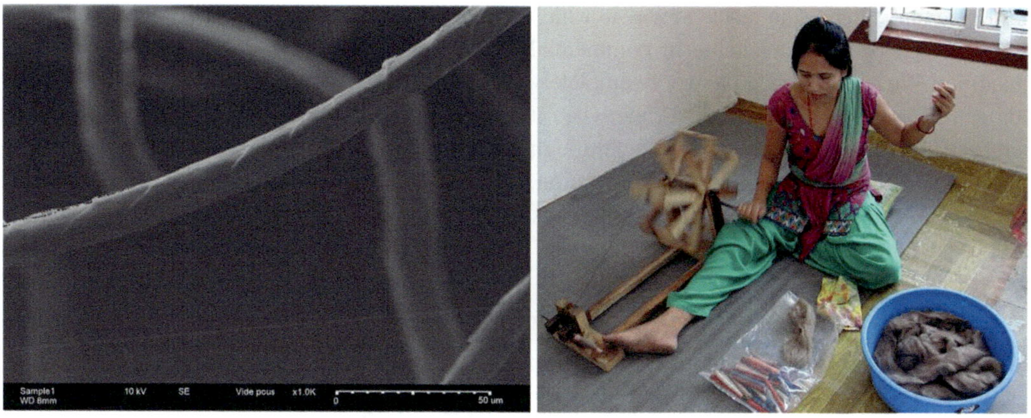

Fig. 10.31 Qiviut fiber and hand spinning of the fibers in India (Notter, 2021)

The fibers consist of about two-thirds of small air channels. This explains the effect of heat insulation comparable with hardly any other fiber. This is why for thousands of years the inhabitants of the Arctic have been lining their sealskin boots (mukluks) with qiviut.

Each animal combs out approx. 2–3 kg of fiber per year (shearing is not possible because the animals are only very cautiously happy about it). This results in an annual fiber production of approx. 2–6 t Qiviut. The yield is only 30–40%, because there are many plant parts in the raw wool. If these plant parts are excluded, the yield reaches 80%. The price for Qiviut fibers is 500–1000 €, depending on the quality.

Corresponding articles are therefore only produced by very few manufacturers for a wealthy clientele. Because warp yarns for woven fabrics can only be produced from qiviut fibers in a very laborious manner by hand, they are often blended with other materials, e.g. wool and silk and only rarely processed purely (Fig. 10.32).

Commercial use on a larger scale does not seem possible, as Prof. W.T Hornaday stated as early as 1920: “I do not believe that musk-ox wool ever can become a valuable commercial

Fig. 10.32 Jacket made of 100% qiviut fibers and scarf made of qiviut muga silk (Notter, 2021)

product. I think there is no practical way in which the wool can be made available in commercial quantities. We once combed the wool completely out of a two-year-old musk-ox at a time when it had been shed and was ready to work out through the long hair. It is a long and tedious operation, and about one month later that particular musk-ox died of pneumonia" (von Bergen, 1931).

10.3 Angora Rabbit

In reference to the long curly, soft hair of the mohair goat, originally called "Angora", long-haired rabbits are called "Angora rabbits" (Fig. 10.33).

Angora wool grows approx. 30 mm/month and is obtained by shearing, combing or plucking, the latter being associated with great stress and pain for the animals. The fiber yield is approx. 250–1000 g when shorn three to four times a year. Because Angora hair is very long and the wool is not scoured, the animals must be kept very clean.

The hair is very fine (11–15 µm) and long (25–50 mm). As a hollow fiber with an air content of 60–65%, it is very light (1.23 g/cm^3) and has a high heat retention capacity (Fig. 10.34).

Fig. 10.33 Angora rabbit (Little, 2006)

The main producers with a total annual production of approx. 4600 t are China, France and Chile (Fig. 10.35). Due to the very controversial method of fiber extraction, the quantity produced has decreased considerably in recent years. The price is around 25–35 €/kg.

The main applications of angora wool are health underwear, sweaters, scarves and socks.

10.4 Yak

Yaks live in the desert steppes of northern Tibet and western China at 5000–6000 m altitude. They belong to the cattle, whereby 2 types are distinguished: The wild yak (Bos mutus), whose peculiarity is its almost voiceless existence, and the domestic yak (Bos mutus grunniens), which is also called grunt ox because it grunts in case of imminent danger, or, when it wants to draw attention to itself. Despite its size and weight (about 200 cm, up to 700 kg), the wild yak is able to slide down icy slopes, swim across raging rivers and climb over rocks.

Both yaks are close relatives of the aurochs. They have massive skulls and curved horns. The coat is curly on the head, short on the back and mane-like on the shoulders and sides of the rump.

Wild yaks live in huge herds that can number up to several hundred animals. During the hair change in June, their fur comes off in large clumps and can be collected. Domestic yaks are somewhat smaller than wild yaks. Besides black and pure white animals, there are also brown, yellow, reddish, grey and spotted specimens. Today, the domestic yak is more common than the wild yak and is kept in alpine pastures. Besides wool, yaks also give a very fatty milk. They are also kept as riding and pack animals. Their dung is often the only source of fuel in these harsh areas (Fig. 10.36).

Shearing results in a maximum of 3 kg of hair per yak. However, only approx. 100 g of this can be spun into relatively fine yarns. The fineness of the yak hair is then 16–21 µm, its length 30–50 mm. Their feel is comparable to cashmere.

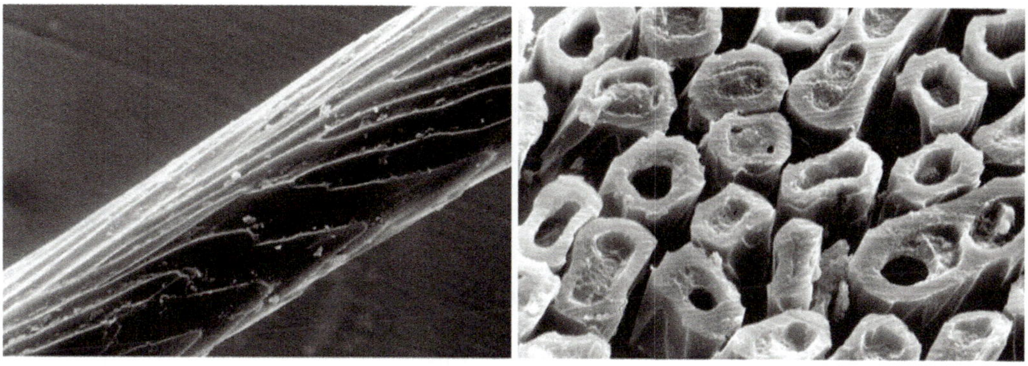

Fig. 10.34 Longitudinal and cross-sectional view of angora wool

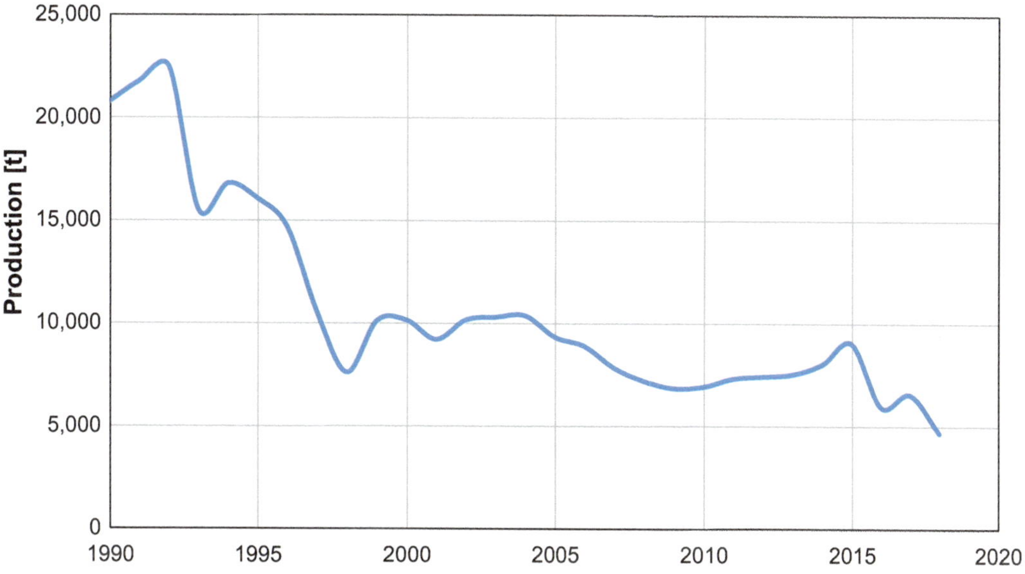

Fig. 10.35 Production of angora fibers (data: IWTO, 2019)

Coarser hairs (20–50 μm) are often processed into tent fabrics. Yak hair has extremely flat scales and therefore does not mat (Fig. 10.37).

World production of yak has been around 4000 kg per year for several decades, and the price per kg is around 35–65 €, depending on quality and demand.

Yak wool is processed pure or in blends, often with sheep's wool or other fine animal hair, e.g. into shirts, sweaters, caps and socks (Fig. 10.38).

10.5 Horsehair

Horsehair is obtained from the tail and mane hairs of many horse breeds. These awn hairs are smooth, stiff and have a high resilience. Therefore, they are often used for interlinings in suits, for upholstered furniture and as strings for the bows of stringed instruments. The strings are played against the scales in the upstroke and with the scales in the downstroke (Fig. 10.39).

Fig. 10.36 Mongolian yak (Giesenow, 2021)

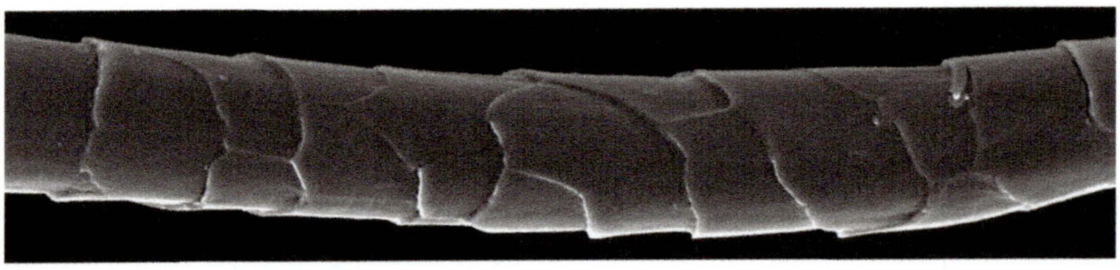

Fig. 10.37 Longitudinal view of yak hair (DWI)

Fig. 10.38 Socks made of yak wool (manufacturer: Happy Camel, Aachen)

In the past, they were also used to make wigs and, in combination with linen, to stiffen crinolines (French: crin for "horsehair"). Horsehair cannot be spun in its pure form, so it is combined with other fiber materials, e.g. cotton and flax. The diameter of horsehair is 15–200 μm for mane hair and 80–400 μm for tail hair. The length of tail hairs is about 400 mm, that of mane hairs up to 200 mm. Typically, the hairs are white, grey, brown or black (Fig. 10.40).

Fig. 10.39 Horsehair in longitudinal and transversal view

Fig. 10.40 Horsehair tails and jacket with horsehair interlining (Kürschner, 2009, 2017; Bennewitz, 2020)

10.6 Dog Hair

For some years now, the hair of Samoyed dogs, originally native to Siberia, has been used as a textile raw material. They are marketed under the trade name "Chiengora", mostly in blended yarns with silk or wool. Due to their almost pure white hairs, these fibers can be dyed well. Typical prices are about 600 €/kg. The hair of other dog breeds can also be spun in principle. These include, for example, Newfoundlands, Huskies, Golden Retrievers and Poodles (Fig. 10.41).

10.7 Fur Hair

The term "fur hairs" includes the fine hairs of mink, muskrat, opossum (American gopher), possum (Australian-Asian gopher) and raccoon. Furthermore, the coarser hairs of beaver, nutria, fur seal, fox et al. also belong to this group. Fur hairs are of local importance only.

Possum

The diameter of possum fibers is approx. 17–20 μm, their length is approx. 30–40 mm.

Fig. 10.41 Samoyed dogs (SheltieBoy, 2011)

Therefore, machines can be used to process wool. According to estimates, there are about 90 m possums living in New Zealand, which are not native there and are therefore considered a threat to the native flora and fauna. Therefore, they are hunted and their fur or hair is marketed as an ecological product to save the fauna and flora of New Zealand (Fig. 10.42). The hair is processed into knitting yarns and used to make sweaters, hats and socks for tourists (often blended with wool and man-made fibers).

Fig. 10.42 Possum (Harris, 2009) and opossum (Pope, 2007)

Table 10.1 Properties of selected fine animal hairs

	Fineness [μm]	Length [mm]	Price [€/kg]	World production [t]	Main producer
Camel	18–26	30–120	5–35	3500	China, Mongolia
Vicuňja/ Guanako	10–20	30–50	500	7	Peru, Chile
Alpaca/Llama	10–25	65–75	10–17	6000	Peru, Bolivia
Cashmere	15–19	25–90	40–120	30,000	China, Mongolia
Mohair	24–40	75–350	15–60	3800	South Africa, Texas, Turkey
Kaschgora	19–22	50–60	12–30		New Zealand, Australia
Angora	11–15	25–50	25–35	4600	China, Chile, France
Yak	16–21	30–50	35–65	4000	China, Mongolia
Qiviut	11–22	35–80	500–1000	6	Canada, USA

10.8 Comparison of the Most Important Fine Animal Hair

Fine animal hair almost always comes from dry areas of high altitude, especially in Asia and South America. The finest animal hair is produced by the vicuña, the qiviut and the angora rabbit. The longest hair comes from the mohair and cashmere goats and the camel. The highest prices are paid for qiviut and vicuña wool (Table 10.1).

References

AIA, International Alpaca Association. Retrieved from http://aia.org.pe/en/home-aia/

Bennewitz, O. (2020). Retrieved from https://commons.wikimedia.org/wiki/File:Otto_Bennewitz,_K%C3%BCrschner-Fachkunde-Ordner_ab_1961_-_Zutaten_die_in_der_K%C3%BCrschnerei_Verwendung_finden_(Rosshaar).jpg. CC0 1.0 Universell (CC0 1.0).

Blog Peru. (2019). Retrieved from https://peru.info/en-us/foreign-trade/news/7/32/peru-is-the-world-s-leading-producer-of-alpaca-fiber

Cui, P., Ji, R., Ding, F., Qi, D., Gao, H., Meng, H., Yu, J., Hu, S., & Zhang, H. (2007). A complete mitochondrial genome sequence of the wild two-humped camel (Camelus bactrianus ferus): An evolutionary history of Camelidae. *BMC Genomics, 8*(1), 241.

DWI – Leibniz-Institut für Interaktive Materialien e.V. Retrieved from http://www.dwi.rwth-aachen.de

faostat. (2022). Retrieved from http://faostat.fao.org

Giesenow, M. (2021). Retrieved from http://www.happy-camel.de

Harris, J. J. (2009). Retrieved from https://commons.wikimedia.org/wiki/File:Trichosurus_vulpecula_1.jpg, CC BY-SA 2.5.

Howard, M. F. (2005). Retrieved from https://commons.wikimedia.org/wiki/File:Guanaco_09.24.jpg

International Alpaca Association (IAA). Retrieved from http://aia.org.pe

IWTO. (2019). *Market information*. Edition 14.

Kürschner. (2009). Retrieved from https://commons.wikimedia.org/wiki/File:Horse_tails_(fur-skins).jpg

Kürschner. (2017). Retrieved from https://commons.wikimedia.org/wiki/File:Herrenjacket,_Half_-Canvas-Verarbeitung,_Peek_%26_Cloppenburg_(D%C3%BCsseldorf_2017).jpg

Little, R., (2006). Retrieved from https://commons.wikimedia.org/wiki/File:Fluffy_white_bunny_rabbit.jpg. CC BY-SA 2.0.

Mahogany Cashmere. *Angers (Frankreich)*. Retrieved from https://www.mahogany-cashmere.de

Notter, W. (2021). *Qiviut International – creation Walter Notter, Pontresina (Schweiz)*. Retrieved from http://www.qiviut.ch

O'Neill, J. (2007). Retrieved from https://commons.wikimedia.org/wiki/File:07._Camel_Profile,_near_Silverton,_NSW,_07.07.2007.jpg, CC BY-SA 1.2.

Pope, C. (2007). Retrieved from https://commons.wikimedia.org/wiki/File:Opossum_1.jpg, CC BY-SA 2.5.

Rowell, J. E., Lupton, C. J., Robertson, M. A., Pfeiffer, F. A., Nagy, J. A., & White, R. G. (2001). Fiber characteristics of qiviut and guard hair from wild muskoxen (Ovibos moschatus). *Journal of Animal Science, 79*, 1670–1674.

Rye, M. (2021). *Qiveut Designs, Matanuska-Susitna Valley*. Retrieved from http://www.qiveut.com

SheltieBoy. (2011). Retrieved from https://commons.wikimedia.org/wiki/File:AKC_Great_Falls_June_2011.jpg, CC BY 2.0.

Von Bergen, A. (1931). Musk-ox wool and its possibilities as a new textile fiber. *Melliand Monthly, III*, 472–846.

Wiki. (2019). Retrieved from https://commons.wikimedia.org/wiki/File:Escudo_nacional_del_Per%C3%BA.svg

11 Silk

11.1 Terms

Most of the silk used for textile production comes from the cocoons of the mulberry silkworm (Bombyx mori), Fig. 11.1. The silk of the tussah moth ("wild silk") is also important.

Silk is also produced by other animals, including mussels, which is of only local importance. Recently, attempts have been made to obtain spider silk from genetically modified plants and animals or to produce it synthetically.

The abbreviation for mulberry silk according to DIN 60001 is "SE", for tussah silk "ST".

Yarns made of silk are divided into:

- Reeled silk (grège, trame, organzin, grenadine),
- spun silk (Schappe silk, floss silk, Bourette silk).

The reeled silk is obtained by unwinding the cocoon as a continuous thread. The spun silk is produced from the waste from silk extraction or from damaged cocoons (short silk threads) using a short-fiber spinning process.

11.2 History

Silk worms very first cultivated in China. There, the mulberry silkworm has been bred for around 5000 years. Initially, only threads were drawn from the intestinal contents of the caterpillars and used for fishing lines or strings of musical instruments. According to legend, in 2697 BC the empress Xi Ling Shi had the idea of reeling a thread from the cocoons of the silkworm and then weaving it. This is considered the birth of silk production.

In 2255 BC silk fabrics are mentioned for the first time in Chinese literature, around 2200 BC silk dyeing is reported. In a "silk calendar", the life phases of the mulberry moth were recorded. The production and wearing of silk fabrics initially remained the prerogative of the imperial court and the nobility. For a long time, therefore, China had a monopoly on silkworm breeding and on silk extraction. For example, the export of eggs and cocoons was punishable by death (Fig. 11.2).

Admiral Nearchos, who explored the coast from the mouth of the Indus to the Persian Gulf on behalf of Alexander the Great in 326 BC, is the first European to report seeing Chinese silk. He refers to them as "skin of the serer". The term "ser" probably goes back to the Chinese name for silk, Si. Virgil, Pliny and Aristotle ("koian silk") also report on silk in their works, without having a more precise idea of how it is obtained.

Due to the artificial scarcity of silk on the part of the Chinese, its weight was weighed in gold, e.g. in Europe. Dyed with purple, silk fabrics were worn by the Roman emperors as a sign of their power. In the meantime, the import of enormous quantities of silk put the Roman state finances in such a bad way that Emperor Tiberius allegedly saw himself forced to forbid men to

D. Veit, *Fibers*, https://doi.org/10.1007/978-3-031-15309-9_11

Fig. 11.1 Mulberry tree and silk moth (top: male, bottom: female)

wear silk robes. However, probably no one cared about that.

Silk was also used as a material for technical fabrics. The sails of the ships of the Chinese admiral Zheng He, for example, with which he undertook a total of seven sea voyages to India and East Africa from 1405–1433, are said to have been made of red-dyed silk.

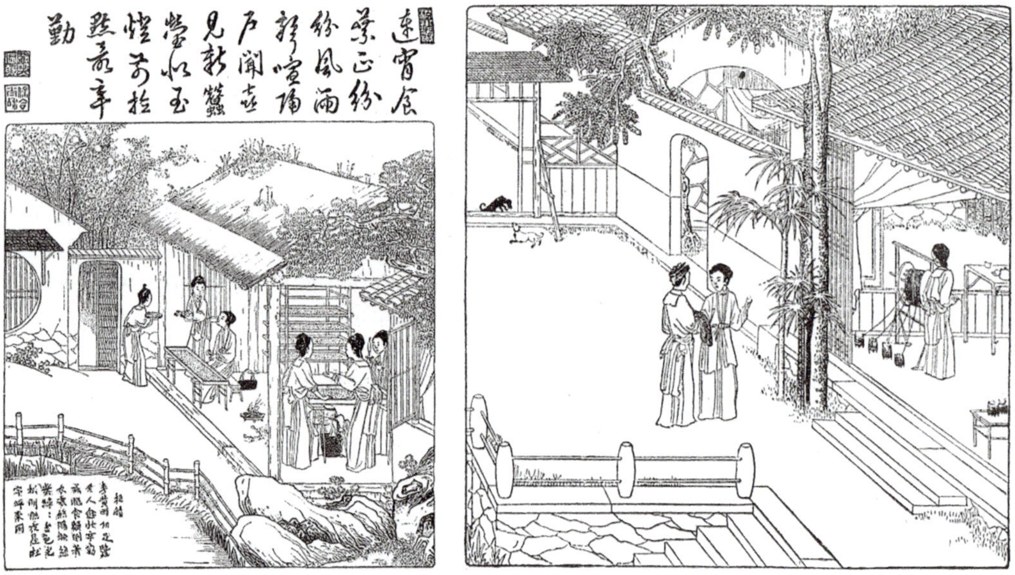

Fig. 11.2 Silk production in China (Tcheng-Ki-Tong, 1890)

Fig. 11.3 Silk roads in Antiquity (Furfur, 2019)

The trade in silk was conducted via the Silk Road(s). Chinese traders were forbidden to leave the Middle Kingdom. Therefore, the silk was transported via many middlemen from China to Europe in 2–3 years. The Silk Road is thus the oldest global trade route. It is still used today for the transport of goods. The "Silk Road" was first named so by the German explorer Ferdinand von Richthofen (1833–1905). From a Chinese point of view, it could also have been called the "Gold Road" or the "Glass Road", because these trade goods were transported from West to East in exchange for silk. Strictly speaking, there is not "one Silk Road", but a network of trade routes, by land and by sea, over which silk was transported from Asia to Europe (Fig. 11.3).

Despite the export ban on silkworm eggs and cocoons, the art of silkworm breeding was introduced to Korea by Chinese emigrants around 200 BC and later to Japan (around 300 AD), India and Persia. There were silkworms in the eastern parts of India and Persia even before that. By plucking off the cocoons after the butterfly had hatched, only silk of inferior quality could be obtained though. The technique of unwinding (reeling) the silk thread was not yet known.

It took until the sixth century for silkworm breeding to come from Central Asia to Europe. In 552, two Persian monks brought a large quantity of caterpillar eggs and seeds of the mulberry tree to Byzantium, hidden in bamboo sticks. As a result, silk production began there the same year under Emperor Justinian. Byzantium remained the most important and largest silk market for Europe from the seventh to eleventh centuries.

After the conquest of Persia by the Arabs in the seventh century, they also took over the knowledge of silkworm breeding and silk production. Through their further conquests, their knowledge reached Sicily and Spain in the ninth century (Fig. 11.4). In the tenth century, silk products were already exported from Spain. From Sicily, silk cultivation spread to northern Italy, where it was introduced to the Republic of Venice in 1204 under Doge Dandolo. In France, silkworm breeding began in the thirteenth century encouraged by Pope Gregory X, especially in the Avignon area. At about the same time, silk production also began in Switzerland, especially around Zurich. In 1450, Lyon was granted a monopoly on the silk trade throughout France by Charles VII, but it was not until the reign of his son, Louis XI, in the

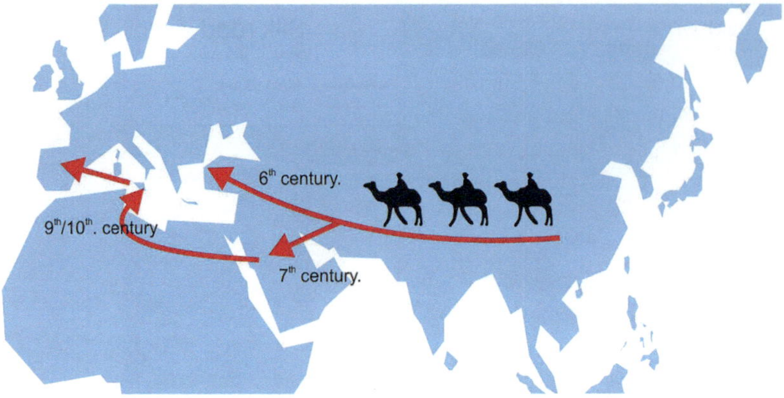

Fig. 11.4 Distribution of silk production in the early Middle Ages

second half of the fifteenth century, that silk culture gained greater importance.

In the fourteenth century, there were already silk weaving mills in Augsburg and Ulm. Therefore, attempts were made to introduce silkworm breeding in Germany (Krefeld, Brandenburg, Rothenburg o. d. Tauber, Stuttgart). As a result of the 30 years' War (1618–1648), however, further expansion was interrupted. Soon afterwards, with the revocation of the Edict of Nantes (1685) and the resulting emigration of Huguenots living in France to Germany, it was possible to further expand silkworm breeding through their know-how.

11.3 Breeding

Silkworm breeding is very labour-intensive. Today, it is particularly important in China, Japan and some non-highly industrialized countries, such as India and Brazil. Silkworm rearing is tied to the presence of mulberry trees (mainly Morus alba). In turn, certain climatic conditions are required for mulberry trees to thrive. These include moderate temperatures, cool and not too long winters, and sufficient rainfall (Fig. 11.5).

The mulberry moth, like other butterflies, passes through the stages in its morphological development: egg, caterpillar or larva, pupa and butterfly. After a 5-day development of the fertilized eggs, a diapause usually occurs (overwintering at about 4–5 °C). The eggs are stored cold and placed in incubators at about 27 °C for the start of growth in spring. At a certain temperature and humidity the caterpillars (blackling) hatch after 12 days and are then fed with fresh mulberry leaves. The caterpillars are 2–3 mm long and weigh about 0.5 mg at the beginning of their growth. The caterpillars grow in five life stages and within 3–4 weeks, they molt four times. During their growth, the caterpillars must receive fresh mulberry leaves regularly and constantly (even at night). When the caterpillars are ready to spin, they have a size of 80–90 mm and a weight of about 4 g (Figs. 11.6 and 11.7).

The rearing rooms must be evenly climatized (around 23 °C, 65–75% rel. humidity), dry and well ventilated (Fig. 11.8). The adult caterpillars stop feeding when they are ready to spin and look for a suitable place to make their cocoons within a few days. Thin, leafless twigs to which the cocoon is attached in a loose tangle of threads, are suitable for this purpose. Alternatively, elaborate constructions made of raffia braids are used, which are more easily accessible (Fig. 11.10, left). Today, rearing can be started at any time of the year.

The two silk glands occupy nearly ¾ of the body length of the silkworm. They are divided into an anterior, a middle and a posterior section. In the anterior section, fibroin is formed, and in the middle section, sericin is synthesized and accumulated in a layer around the fibroin nucleus. Fibroin and sericin move together into the anterior section, which narrows toward the exit. The two glands join to form a single duct to the

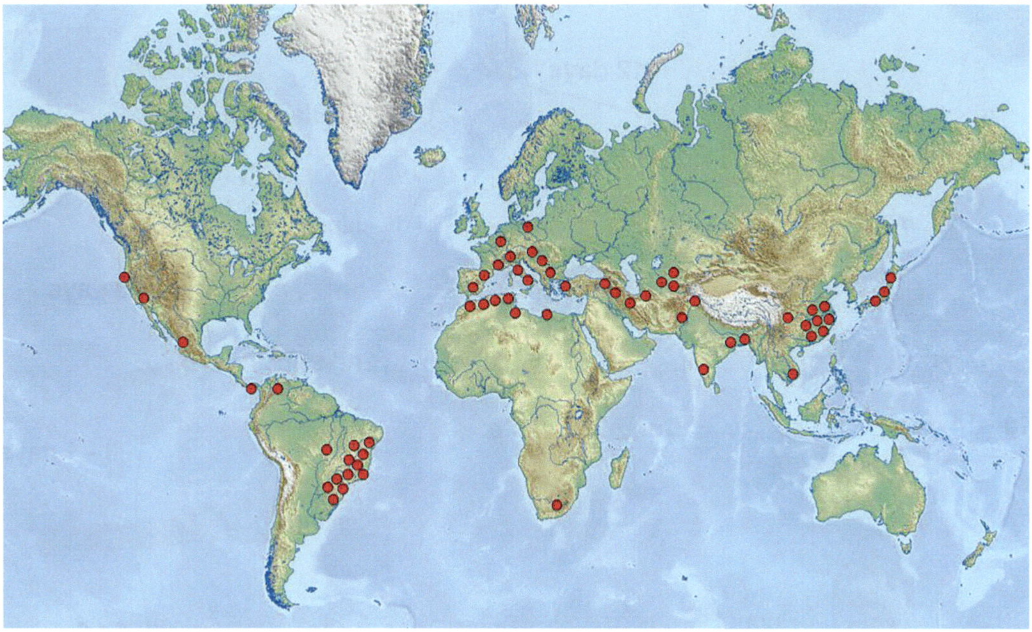

Fig. 11.5 Distribution area of the mulberry tree (Morus alba)

spinneret, which the caterpillar uses to spin the cocoon from the outside to the inside at about 60 cm/min. Figure 11.9 (right) shows the basic structure of a caterpillar (A—anterior gland section, M—middle gland section, P—posterior gland section and D—philippic glands).

The aqueous protein secretion solidifies in air to form a solid silk thread. This consists of two fibroin filaments, each of which is enclosed by sericin (silk bast or silk glue).

The transformation of the liquid fibroin to partially crystalline fibrillar fibroin (crystalline portion approx. 60%) takes place through the rearrangement and orientation of the folded fibroin molecules. This is due to the stretching and tension exerted by the larva on the filament during spinning with its head. First, the pupal bed (wadding silk) used for attachment is made by a continuously spun and loosely laid thread. Then the caterpillar forms the cocoon by constantly laying the endless thread over each other with figure-eight looping movements of the head. This process takes about 4–5 days and is not interrupted.

About 10 days after spinning in, the cocoons are collected for silking and sorted according to size, color and uniformity. For the rearing of new caterpillars, some butterflies are allowed to hatch. In this case, the transformation of the caterpillar into a pupa takes place in about 2 weeks in the cocoon. About 30 days after pupation, metamorphosis to the sexually mature butterfly is complete. By secreting a protein-degrading liquid, the cocoon softens and the butterfly works its way through the softened tangle of threads with the help of its head and legs. It is unable to fly and takes no food. Its short life span of only a few days is solely for reproduction. After mating, the female lays around 400–800 eggs in 24 h and the cycle begins again (Fig. 11.10).

The following characteristics are important for breeding:

- weight and number of eggs as well as their development time,
- the skin pigmentation of the caterpillars,
- the color of the haemolymph (has an influence on the cocoon color),
- their food requirements and resistance,
- the weight and shape of the cocoons as well as
- the fineness and mechanical properties of the silk thread.

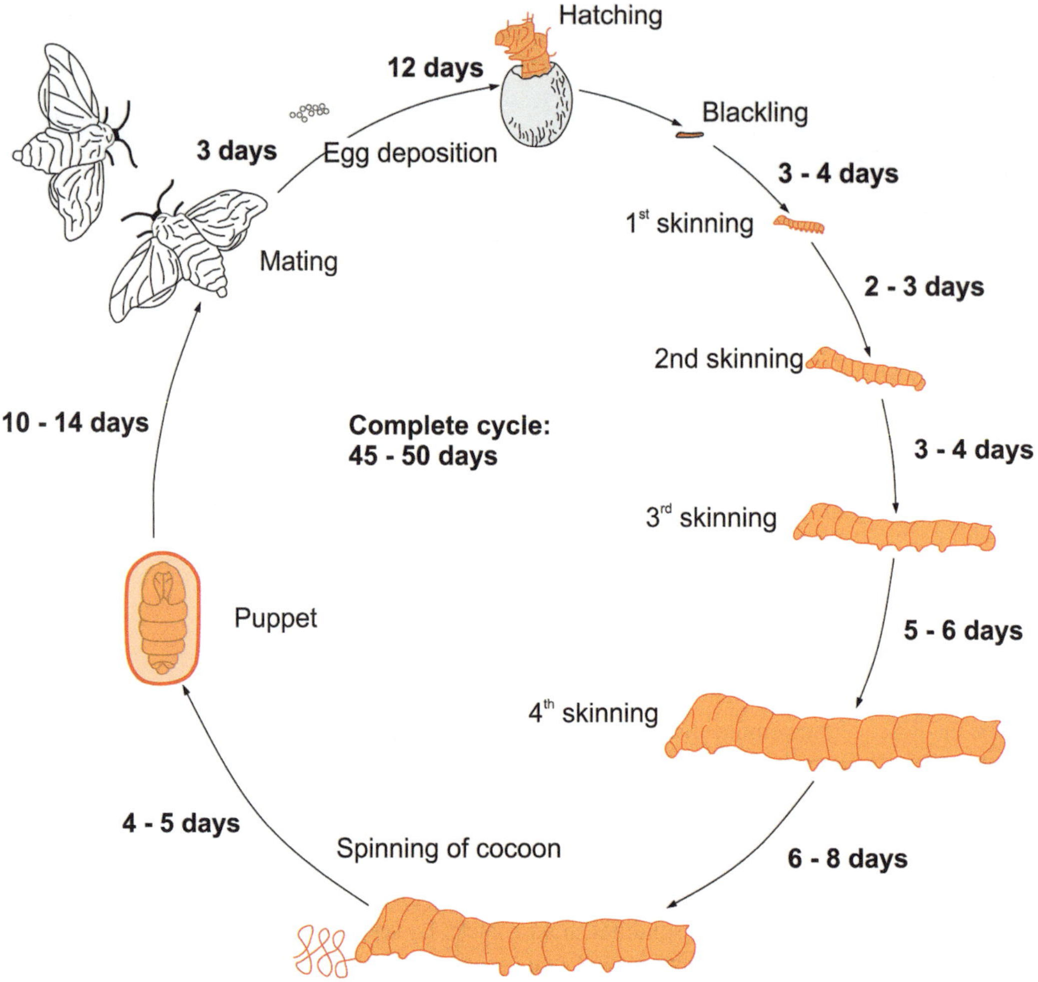

Fig. 11.6 Development cycle of the mulberry moth according to (Türk, 2014)

The weight of the cocoon depends on the breed and the rearing conditions. It is on average between 1.5 and 2.5 g. Its length is 18–25 mm and the width 8–30 mm. Until the butterfly hatches, the fresh cocoon loses about 23% of its original weight. The shape, on which the reeling

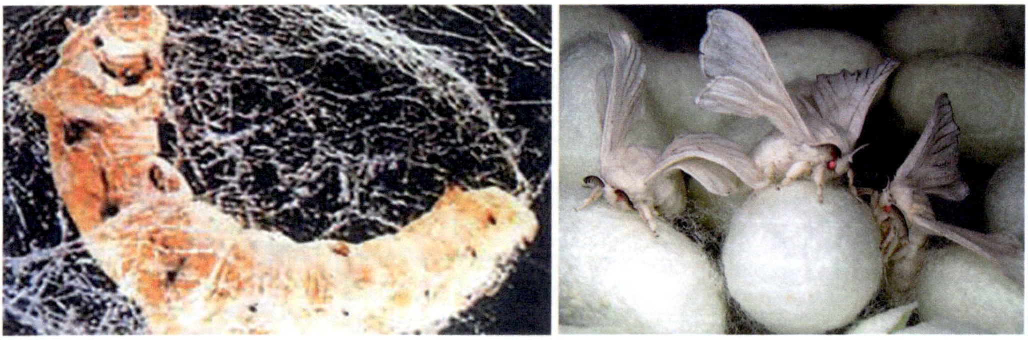

Fig. 11.7 Silkmoths building cocoons and hatched silkmoths (Kraig, 2016)

Fig. 11.8 Toughs with mulberry leaves and silkworm caterpillars (Tobias Löffler)

ability also depends, is usually oval. The color is white, yellow, orange, green, dark brown or pink, depending on the breed. The dye is found in the sericin. After removal of the sericin ("degumming"), the silk is therefore white.

11.4 Structure

The raw silk thread consists of two fibroin filaments, which are round to trilobal in cross-section and have an average fineness of

Fig. 11.9 Structure of a silk spinner and silk apparatus

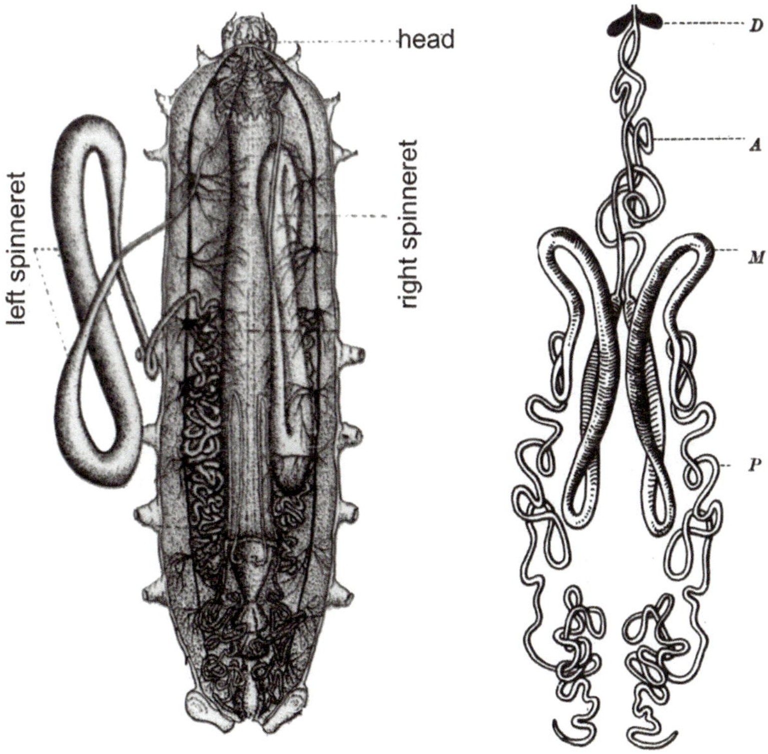

Fig. 11.10 Silk cocoons (Armstrong, 2006) and spinner in its cut open cocoon

10–14 µm. They are coated with sericin ("silk glue"). The silk thus has a core-mantle structure (Fig. 11.11).

The fibroin filaments consist of fibrils, these of microfibrils and these in turn are built up by protein chain molecules (Fig. 11.12).

11.4.1 Crystalline Areas

Silk is a protein fiber and consists of the elements carbon (35%), hydrogen (10%), oxygen (22%) and nitrogen (33%). Unlike wool and fine animal hair, it contains very little of the sulfur-containing amino acid cystine.

The structure of the peptide chain of fibroin is shown in Fig. 11.13. It consists essentially of a repeating sequence of 4 amino acids (hexapeptide: glycine-serine-glycine-alanine-glycine-alanine). Therefore, glycine, alanine and serine occur in a molar ratio of about 3: 2: 1.

There are several models for the spatial structure of the crystalline domains of silk fibroin. They all assume a folded sheet structure, comparable to the internal structure of wool (Fig. 11.14).

11.4.2 Fibroin

Silk fibroin consists of 18 different amino acids and accounts for approx. 75–83% of the weight of raw silk. The proportion of the four amino acids glycine, alanine, serine and tyrosine is more than 90 mol%. Glycine dominates in mulberry silk with a proportion of around 45%, whereas alanine dominates in tussah silk.

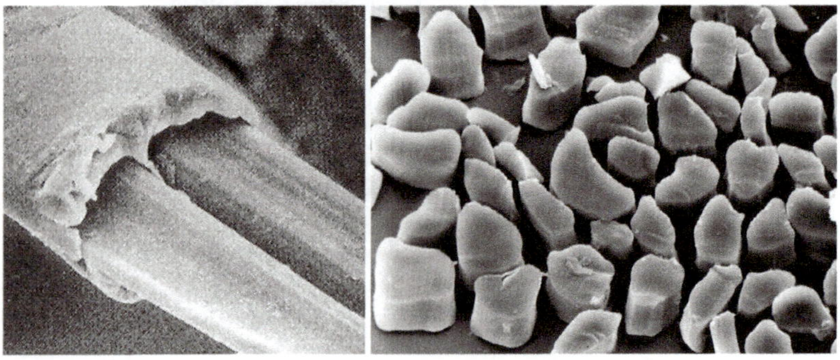

Fig. 11.11 Fibroin thread (left: longitudinal view with sericin sheath; right: cross-sections)

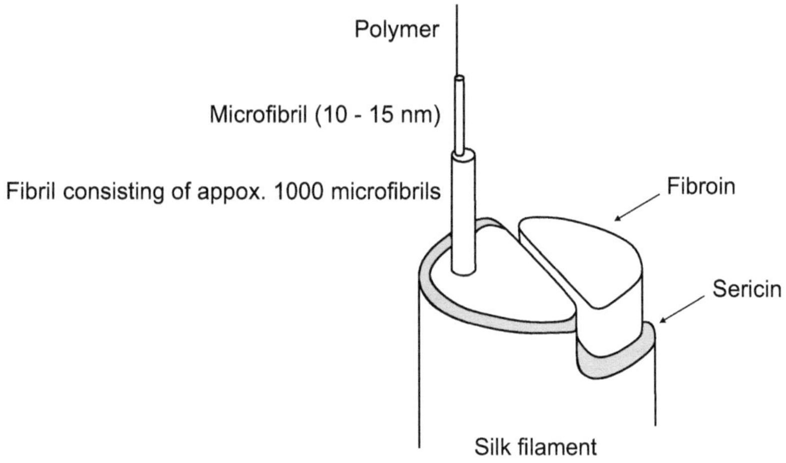

Fig. 11.12 Structure a silk thread from fibroin and sericin

The molecular mass of fibroin is about 370,000, which corresponds to about 1500 hexapeptide chains (Fig. 11.13).

11.4.3 Sericin

Sericin is a protein mixture with rubber-elastic properties. The sericin content of raw silk is 17–25% and depends on breed, origin and breeding conditions. The amino acid composition of sericin differs significantly from that of fibroin, the most abundant amino acid being serine (Fig. 11.15).

Waxes and fats (approx. 1.5%) as well as colorants and mineral components (1%) are found exclusively in sericin. Xanthophyll, carotene and flavones were detected as coloring substances.

11.5 Classification

In the classification of silk, a distinction is made between the classification of the cocoons and the classification of the actual raw silk thread.

11.5.1 Cocoons

The cocoons are classified according to appearance, size, color and uniformity (Fig. 11.16).

Silk accounts for about 15% of the total weight of the cocoon in the fresh cocoon and about 40% in the dry cocoon. The remainder consists of pupa and caterpillar skin. Classification is always carried out at the silk producer by visual inspection (Fig. 11.17).

11.5.2 Raw Silk

Raw silk, silk waste, silk yarns and fabrics as well as finished products are traded. Since the 1990s, the import of raw silk and silk fabrics in Europe has been steadily declining. In contrast, the import of finished products is steadily increasing.

The most important trading centers have accordingly shifted from Europe to the producing countries, especially China. Silk is now an important foreign exchange earner there. Besides Shanghai, the most important trading centers are

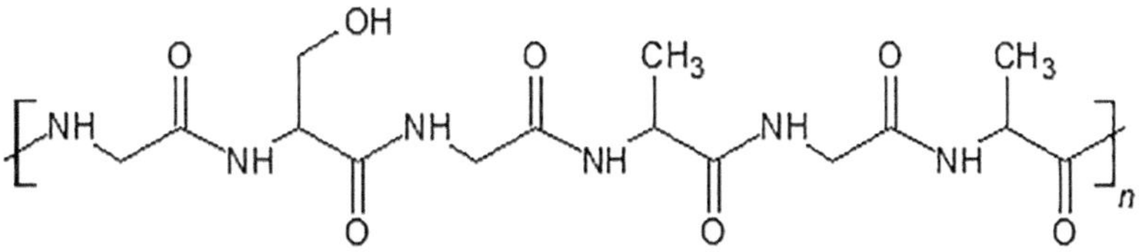

Fig. 11.13 Chemical structure of fibroin with repeating unit

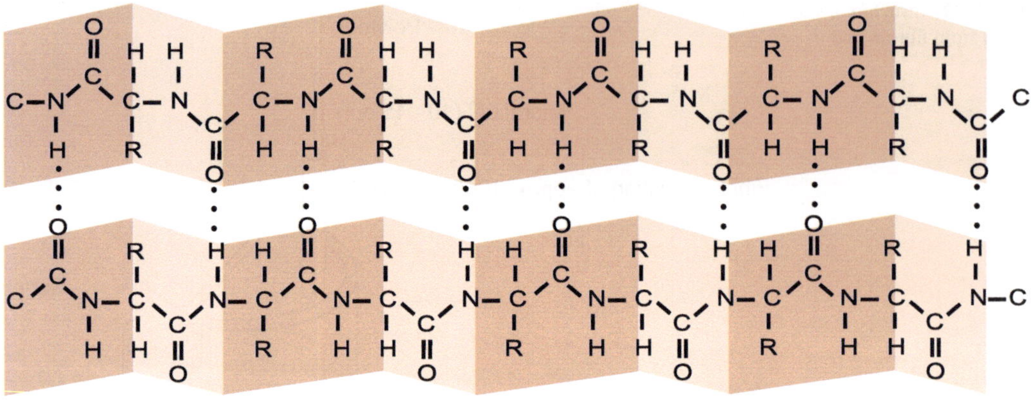

Fig. 11.14 Folded sheet structure of fibroin

Yokohama, Kobe, Milan, Como, Lyon and Zurich.

Raw silk is traded according to a standard classification issued by the Association Internationale de la Soie (AIS) in Lyon. It consists of 12 grades (6A, 5A, . . ., 2A, A, B, C, . . ., G), with 6A being the best and G the worst. The classification takes into account the uniformity of the thread, its purity, tensile strength and elongation at break, as well as the number of thread breaks during winding. Standard silk has a fineness of 20 den or 22 dtex and a grade of 3A.

The price of raw silk declined from 1990 to around 2000, falling to around 10 €/kg. Today it is around 30–45 €/kg for a fineness of 20 den or 22 dtex.

11.6 Reeling

To obtain the silk, the cocoon must be unwound. To do this, the silkworm is first sent to the happy spinning ground by heat in order to prevent the butterfly from hatching and thus destroying the cocoon. This is done either by steam and subsequent drying in heated rooms with warm air or—particularly gentle for the silk—in a vacuum at 65–75 °C. Too high temperatures can damage the silk.

The winding is done on so-called “reels”, normally in the producing country in factories or in small specialized businesses.

First, the outer, loose cocoon layer (wadding silk), is removed. It is less valuable and is used in

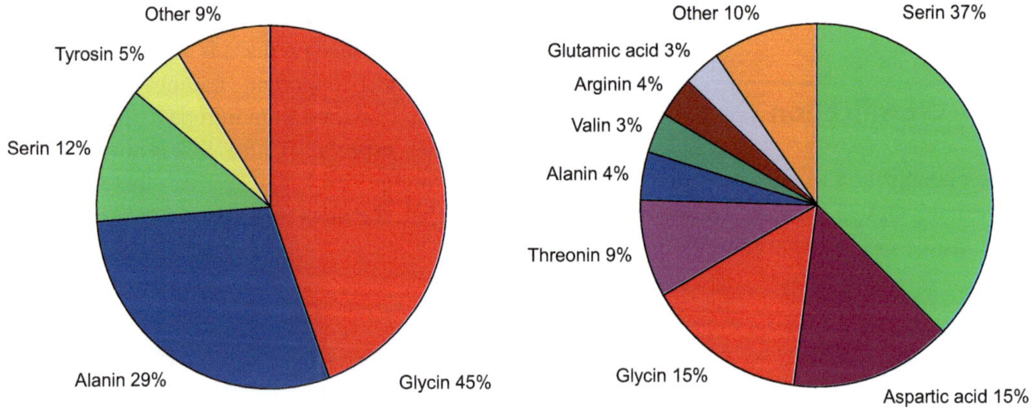

Fig. 11.15 Composition of fibroin (left) and sericin (right) in mol%

Fig. 11.16 Classification of cocoons by color and size (Rungminoz, 2016)

the Schappe spinning mill. Then the sericin is softened by treatment with hot water so that the individual thread windings no longer stick together. In small water basins the "thread searching" takes place. For this purpose, the cocoons (40–60) are treated with a rotating brush, whereby the thread beginnings get caught in the bristles of the brush (Fig. 11.18 and 11.19).

Because a cocoon thread does not have the same fineness over the entire length and because there can also be a changing color affinity over the length, several cocoon threads are always brought together. Depending on the desired fineness, the number of cocoons per silk thread is 5–10. In addition, it is necessary to replace the thread of the expired cocoon directly with a new one. This process is called thread piecing. It is done either manually or mechanically and, if not carried out properly, leads to considerable fineness fluctuations in the thread.

The smoothing and rounding of the cocoon threads brought together to obtain a perfect silk thread is carried out before winding by giving them a slight twist by means of a rotating funnel or disc, or by crossing the silk yarn with itself or with another silk yarn (Fig. 11.18, torta). This also removes some of the water. The threads are wound together into a skein and dried. Approximately 700–2000 m can be unreeled from one cocoon.

Fig. 11.17 Visual classification of cocoons (Partensky, 2009)

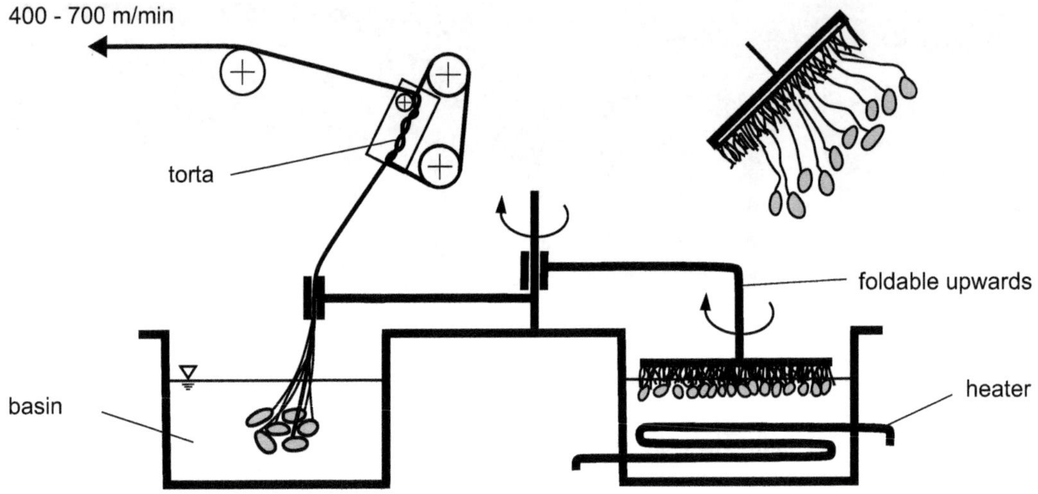

Fig. 11.18 Design of a reeling machine

11.7 Terms for Silk Products

Reel silks are produced by unwinding several cocoons at the same time and then twisting them. The product of the first stage of silk processing is called grège. It is formed by combining several cocoon threads and has no twist. Further designations for grège yarns are defined in DIN 60550. These include the double grège, an uneven, untwisted grège (poil) and a grège twisted up to 3000 T/m in the S or Z direction.

Grège is also the starting product for the production of other reel silks (Zahn et al., 1994):

- Trame: a twisted yarn produced from one or more untwisted, intertwined grège yarns. The twist direction is S, the number of twists is 100–150/m.
- Doppi-trame: a twisted yarn made of two double grège yarns. The twist direction is S, the number of twists is about 120/m.
- Special trame: a twisted yarn with weaker ($<$ 80 t/m) or stronger twist (150–200/m). The twist direction is S.
- Crepe: a twisted yarn with a particularly strong twist (2000–3500/m). The twist direction is S or Z.
- Organzine: a medium or high twist yarn made from one or more Z-twisted barge yarns twisted together in the S-direction.
- Grenadine: a twisted yarn made from two grège yarns twisted together in the Z direction

Fig. 11.19 Silk cocoons and reeling box on a spinning machine (Tobias Löffler)

and twisted in the S direction with a particularly strong twist.

Large differences in the quality of silk result from:

- Provenance (country of origin),
- Race,
- rearing conditions (upland, lowland),
- year of production and season of harvest,
- method of drying the cocoon,
- temperature and quality of the water during softening and reeling of the cocoon,
- type of yarn spreading.

11.8 Spinning Mill

The reeled, continuous silk threads are usually combined into several in order to increase the uniformity and strength of the yarn. These yarns are thus multifilament yarns (Fig. 11.20).

In the silk spinning mill, staple fiber yarns are produced from the non-reeling part of the cocoon, the silk waste. This is referred to as "spun" silk yarns and a distinction is made between Schappe and Bourette silk (Fig. 11.21).

11.8.1 Schappe Spinning

According to DIN 60001, Schappe silk (floret silk) is a yarn spun from the waste of cocoon processing (damaged cocoons, reel waste) and twisted yarns produced thereof. The spinning twist is mostly in the Z direction, the ply yarn-twist mostly in the S direction.

A distinction is made between two processes in Schappe spinning: the classic Schappe spinning process and the modified French worsted spinning process (Fig. 11.22). In the classic Schappe spinning process, an attempt is made to shorten the fiber length (120–200 mm) of the waste as little as possible. This makes the process very costly. In the modified French worsted process, the fibers are shortened, resulting in a fiber length of 30–120 mm. The spinning process is therefore simpler. The yarns spun by both methods can then be twisted. Schappe yarns are smooth and strong and have a regular appearance. Their fineness is 25–125 dtex, depending on the length of the combed yarn (Zahn et al., 1994).

11.8.2 Bourette Spinning

Bourette silks are spun yarns which are produced from the waste from card spinning (noils etc.) by

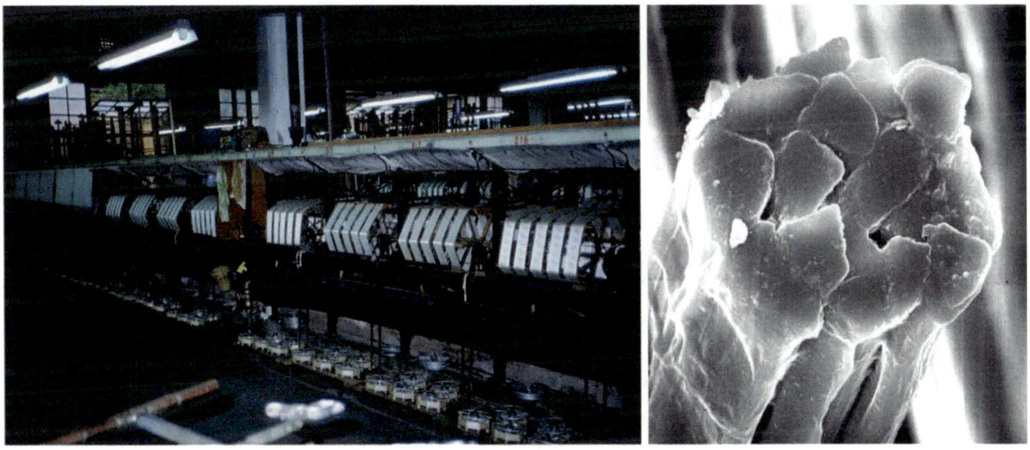

Fig. 11.20 Reeling machine and silk yarn from several single yarns

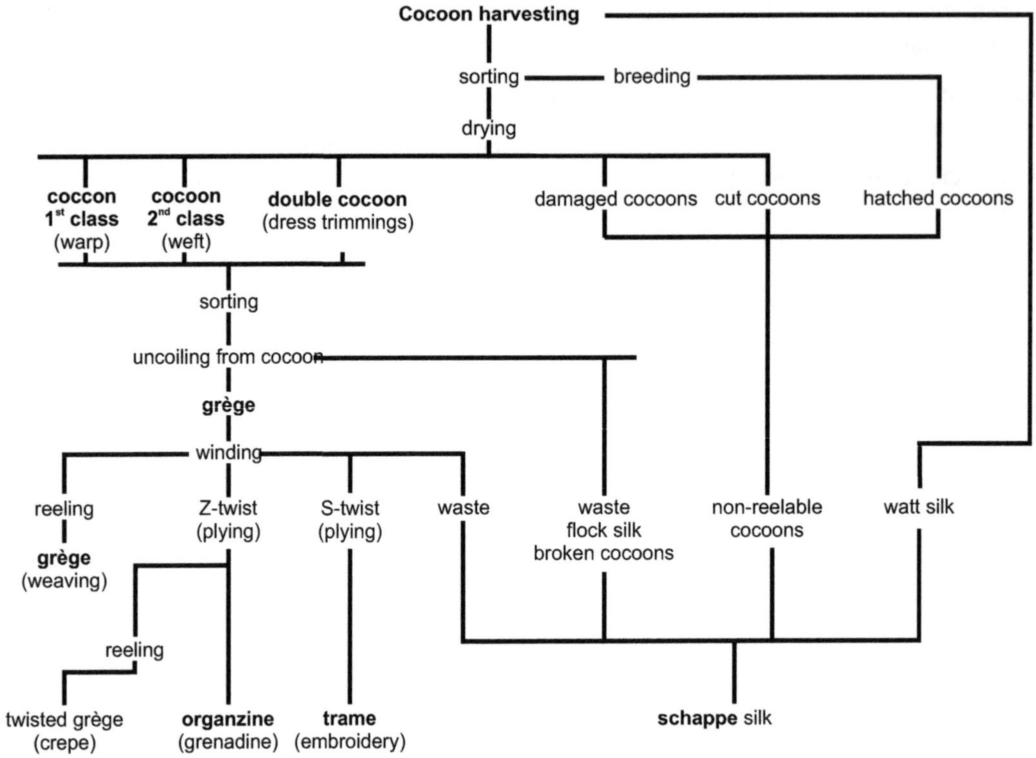

Fig. 11.21 Production of silk yarns according to (Ullrich, 1928)

the carded yarn process (Fig. 11.23). The fibers have an initial length of 10–50 mm. The yarns are medium-fine (approx. 167 dtex) and have an uneven nubby appearance. The spinning twist is mostly Z-direction. The waste from Bourette spinning is used, for example, as upholstery material (Zahn et al., 1994).

11.9 Production

The world production of raw silk in recent years was approx. 660,000 t/a. This results in approx. 200,000 t of reel silk. This means that silk accounts for around 0.2% of all fibers (Fig. 11.24).

Until the 1980s, Japan was the main producer and exporter of raw silk. Today, the main producing countries are China (66%) and India (27%). All other countries except Uzbekistan (3%) have market shares well below 1%. It is estimated that about 300,000 households worldwide make their living from silk production (Fig. 11.25).

The largest exporter of raw silk, silk yarns, fabrics and garments is Italy, ahead of India and Japan (Fig. 11.26). The total value of imported raw silk is currently US$ 1.8bn.

11.10 Properties

11.10.1 Fineness, Density, Color, Luster, Handle

The fineness of the cocoon thread (fibroin and sericin) and that of the fibroin thread depend on the breed and vary over the length of the thread (coarser at the beginning than at the end as the silk worm gradually improves its spinning process). The average fineness of the single yarn is 1.0–3.5 dtex. Commercially available yarn counts for reel silk are given in DIN 60550. The average diameter of the cocoon thread is 15–25 μm. The density

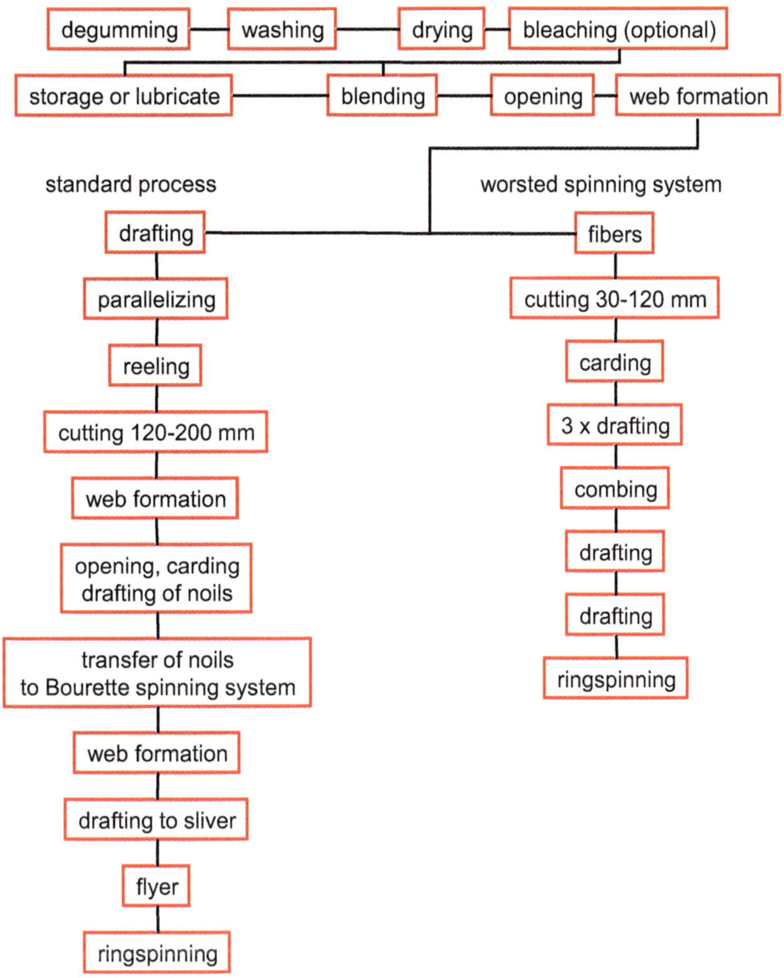

Fig. 11.22 Process steps in Schappe spinning

of the degummed silk is 1.37 g/cm^3. The color of the completely degummed cocoon thread is white. Incompletely degummed silk is yellowish to greenish in color. Degummed silk is shiny and has a supple handle.

11.10.2 Tenacity and Elongation

Silk has a great tenacity in combination with good elongation. When wet, the strength decreases and the stretch increases (Table 11.1).

The values for degummed silk are slightly lower than those given in the table above.

11.10.3 Mechanical Properties

Silk has a high moisture absorption and a high water retention capacity. Silk swells in contact with water, especially in the transverse direction (anisotropic behavior) (Table 11.2).

11.10.4 Chemical Properties

The resistance of silk to acids is somewhat lower and to alkalis somewhat better than that of wool. Silk is sensitive to water and very sensitive to perspiration. In concentrated sulfuric,

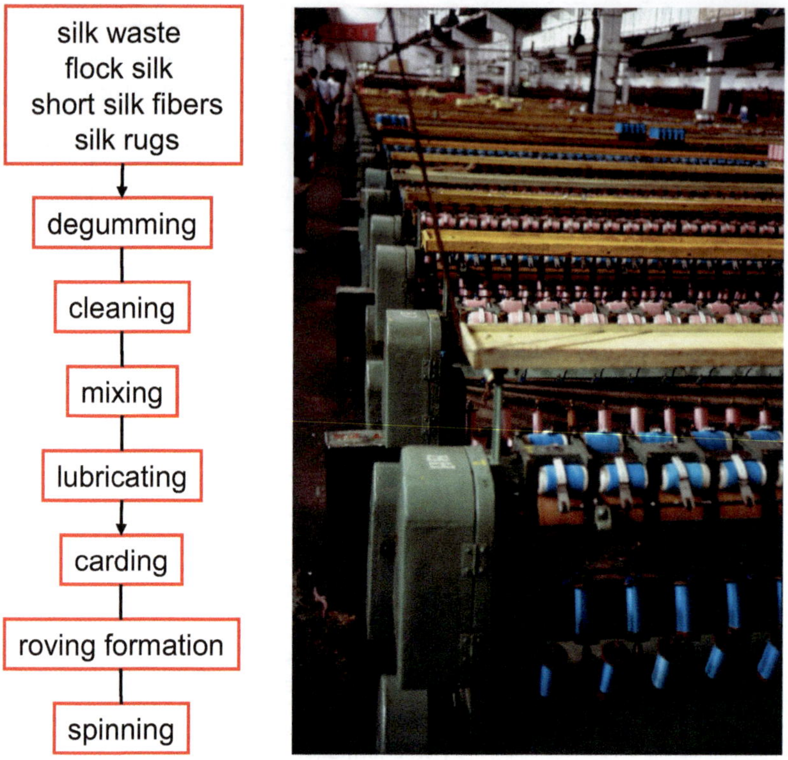

Fig. 11.23 Bourette spinning and spinning machine

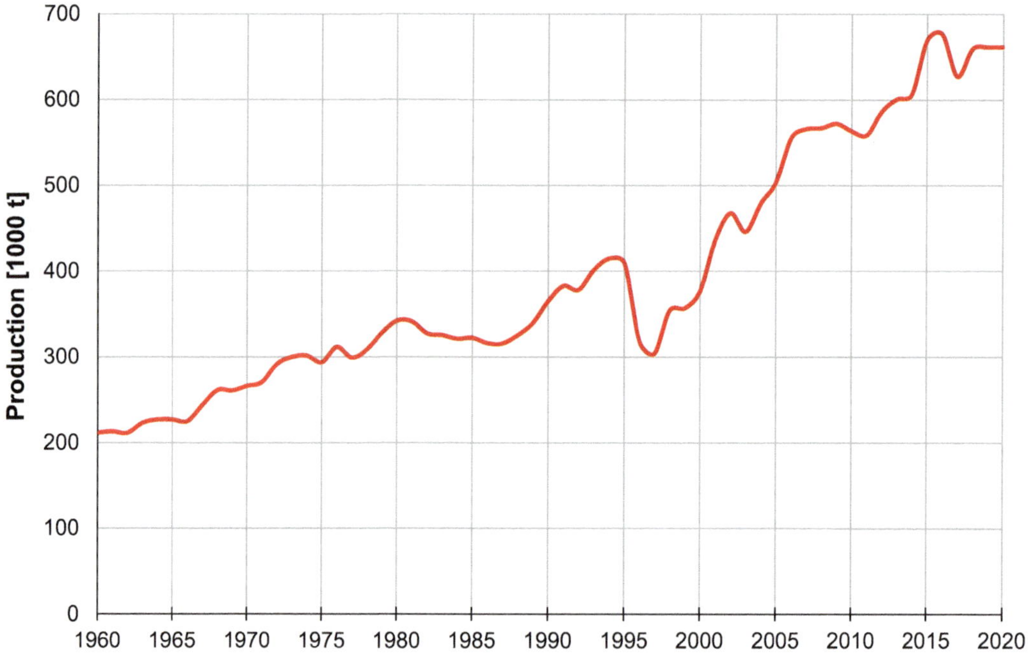

Fig. 11.24 Production of raw silk (data: faostat, 2022)

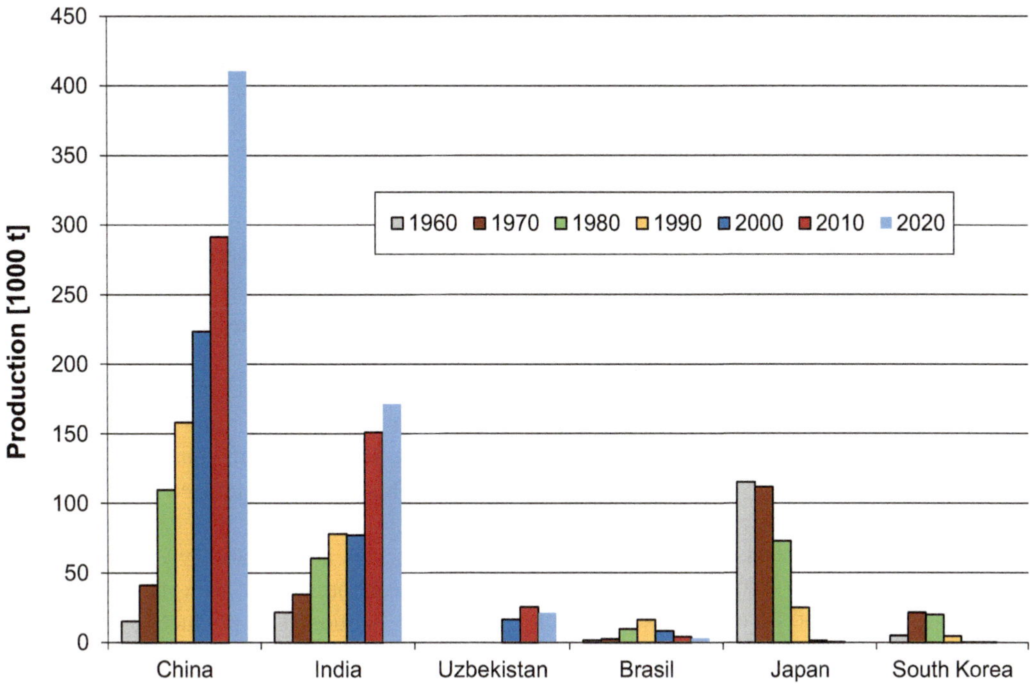

Fig. 11.25 Development of raw silk production in important countries (data: faostat, 2022)

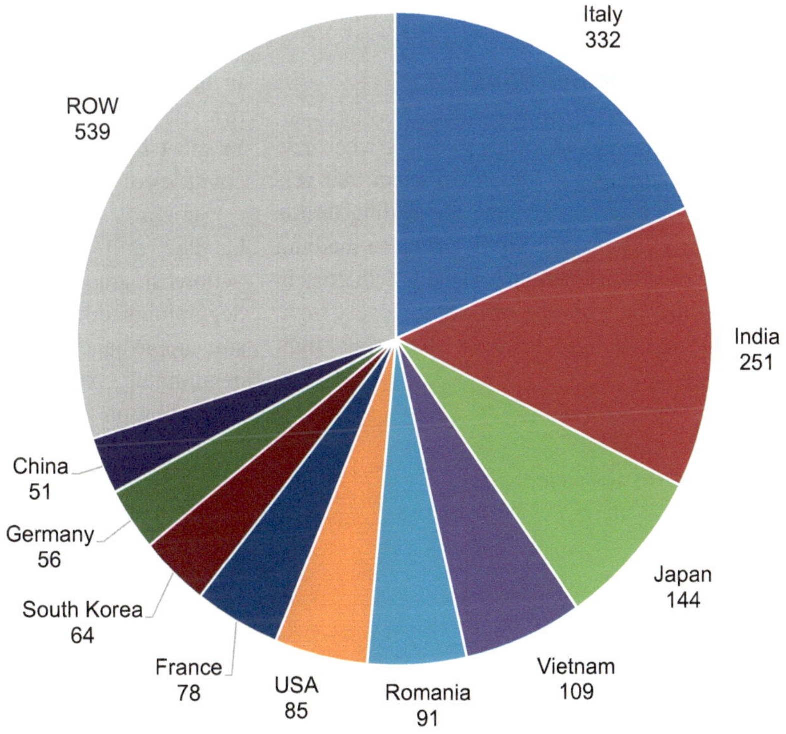

Fig. 11.26 Turnover of the most important importing countries with raw silk in [US$] (data: WRC, 2017)

Table 11.1 Mechanical properties of raw silk

Maximum tenacity [cN/tex]	25–50
Wet maximum tenacity [%]	75–95
Elongation at maximum tenacity [%]	10–30
Wet elongation at maximum tenacity [%]	120–200
E-modulus [N/mm^2]	8000–12,500

hydrochloric, phosphoric and formic acid, silk dissolves within 30 min. The isoionic point at which the ampholytic ions carry equal numbers of positive and negative charges is pH 5.0.

11.10.5 Thermal Properties

Silk can be washed at 30 °C. The ironing temperature is 140–160 °C, the discoloration temperature is 120 °C. At 170–180 °C, silk begins to decompose, at 300–400 °C charring sets in. Silk does not melt before ignition, but ignites in a pilot flame. When the pilot flame is removed, silk continues to burn slowly, producing an odor like burnt horn. Its LOI is 24, the thermal conductivity is 0.2–0.4 W/mK and the specific heat capacity 1.4–1.5 kJ/kgK.

11.10.6 Usage Behavior

Silk has a very high tendency to crease when dry. The shape retention is low to sufficient. Air permeability is high to very low, depending on the type of silk. The soilability is strong to medium and the soil release medium to heavy. Silk dries in a short time, the abrasion sensitivity is very high. Silk has a low light fastness and a low light resistance.

11.10.7 Other Properties

Silk is hygroscopic. The commercial moisture content is 11%. Silk is a polyampholyte with cationic and anionic side chains. Therefore, it can buffer strong alkalis and mineral acids to form salts. Silk can absorb considerable amounts of salts. This property is utilized when weighting the fiber.

Because silk has a relatively good alkali resistance, it can be treated with 16–18% sodium hydroxide solution. This produces a crèpe effect, which is used in particular for blended fabrics with a cotton content.

11.11 Finishing

11.11.1 Degumming

In this process, the sericin which coats the two raw silk threads is partly or completely removed. In most cases, it is already carried out in the countries of production.

Sericin is responsible for the brittleness, hard handle, dull appearance and yellowish to yellowish green color of raw silk. Only after degumming does the silk acquire its luster and soft supple handle. The weight loss resulting from degumming ranges between 2 and 30%, depending on the desired degree of degumming.

Table 11.2 Physical properties

Moisture absorption [%]	9–11 (21 °C/65% r.h.) 20–40 (24 °C/95% r.h.)
Water retention [%]	40–45 (degummed)
Electrostatic charge	Low–medium
Electrical conductivity	Good (loaded silk)
Electrical resistance	High
Swelling (degummed) [%]	Lengthwise: 1.65 Crosswise: 18.70

Fig. 11.27 Degumming unit

At the same time, the strength decreases by about 20%.

Marseille soap, an alkali-free olive oil soap, is mainly used as a degumming agent. In this classic degumming method, depending on the quality and type of fabric, the silk fabric is degummed over a period of 2–4 h in a soap bath at a temperature of 90–98 °C. The silk fabric is then repassed and rinsed with a weak ammoniacal solution (Fig. 11.27).

Other degumming processes use enzyme solutions, weak alkaline solutions, acidic solutions or boiling water under pressure. Due to the increasing importance of time, energy and water costs within the degumming process, new degumming agents have been developed in recent years. On the one hand, these are powder products consisting of sodium salts (e.g. phosphates, carbonates). On the other hand, liquid degumming agents are used which, in addition to sodium salts, also contain anion-active compounds (surfactants) as well as other organic substances (e.g. fiber protection agents). In contrast to soap, these degumming agents can also be used in continuous processes.

Depending on the degree of degumming, a distinction is made between:

- Cuite silk: completely degummed silk, weight loss approx. 18–30%.
- Souple silk: half degummed silk, weight loss approx. 6–12%.
- Ecrue silk: weakly degummed silk, weight loss approx. 1–4%.

For the degumming of tussah silk (see below), an alkaline decoction with soap or soda is sufficient due to the lower sericin content.

As yarn, silk can also be degummed in rope form or on perforated dye tubes. Woven fabrics are degummed on reels, jiggers or in continuous dyeing machines. Depending on the weight of the fabric, a longer degumming time may be required for piece goods.

In principle, uniform degumming is of decisive importance for the subsequent dyeing processes, because otherwise uniformity problems (uneven dye absorption) can occur.

11.11.2 Weighting (Loading)

The term "weighting" is used to describe the compensation of the weight loss caused by degumming by incorporating metal salts, vegetable substances or synthetic substances into the fibroin. At the same time, an improvement in the luster and handle of the fiber and an increase in the fiber volume are achieved. A distinction is made between "below par" (no complete weight compensation), "par" and "above par". In the latter case, values of up to 120% of the weight before degumming are possible.

The tin phosphate silicate weighting (mineral weighting), which was used for many decades, plays only a minor role today. In this method, the silk is treated successively with aqueous solutions of tin tetrachloride and disodium hydrogen phosphate. This produces basic tin phosphate in the fiber, which is converted into washed tin sodium silicate by treatment with a concentrated silicate bath. By varying the number of treatments with tin salt and phosphate, the degree of weighting can be varied, so that weight gains of over 100% are also possible.

Mineral-plant or purely plant-based weighting uses tanning agents or bluewood extract. This type of weighting is mainly used for material to be dyed black or is often even combined with black dyeing.

In recent years, both, silk yarns and fabrics, are loaded with synthetic resins such as methacrylamide. In this process, the material is almost exclusively dyed first and then loaded by graft polymerization with methacrylamide. This process achieves better fastness properties, although only dyes that are insensitive to the conditions of graft polymerization are suitable.

In general, the intended use determines the degree of weighting, although high mineral weightings may reduce quality to such an extent that the silk becomes brittle and decrepit. Sunlight and moisture can also damage the silk thread. Then, the affinity for dyestuffs and fastness to washing are reduced.

The loading is carried out in the same way as the debasing process, for the most part in the hank before dyeing in barks, on sticks or in special centrifuges, on hank washing machines or on continuous open-width washing machines. Today, loading is only used for certain end articles (e.g. ties, borders, embroidery, etc.) to improve luster, handle, fabric drape and fullness (Zahn et al., 1994).

11.11.3 Bleaching

The color pigments embedded in the sericin are usually already removed by the degumming process. Nevertheless, oxidative or reductive bleaching is required if clear whites and bright brilliant shades are to be achieved. Sulfur dioxide or hydrogen sulfite, for example, can be used as reductive bleaching agents. Sulfur dioxide bleaching is easy to carry out, gentle on fibers and inexpensive. Oxidative bleaching with hydrogen peroxide or $H_2 O_2$—emitting salts is of considerably greater importance. In contrast to sulfur dioxide bleaching, the resulting white is more durable and the fiber material is free from odor.

To achieve a full white tint, so-called optical brighteners are added to the bleaching bath. This step usually takes place after the oxidative pre-bleaching in the reduction bleach bath. The concentration and temperature of the bleaching bath are determined by the amount of color pigments remaining on the fiber. In the case of piece goods, degumming and bleaching often take place in the same bath.

Tussah silk must be bleached more intensively than mulberry silk due to the intensive inherent coloration of the fiber, which is still present even after complete degumming. If a full white is desired, this is very difficult to achieve, if at all, depending on the tussah silk quality, and is usually accompanied by fiber damage.

11.11.4 Weaving

Silk is mostly used to make fabrics. For this purpose, the silk is processed either raw white or bleached so that the fabrics can subsequently be printed. Alternatively, the silk can also be dyed in the yarn and then processed into colored fabrics using Jacquard weaving machines (Fig. 11.28).

11.11.5 Dyeing

Almost all dyes that are suitable for dyeing wool can also be used for dyeing silk. These include acid, basic, metal complex, reactive, vat and pigment dyes. Natural dyes such as bluewood or tannins can also be used. Acid dyes are common both for rope silk dyeing and for piece goods dyeing. With them, good to very good light fastness properties are achieved. At the same time, a wide range of colors is available. With basic dyes, one obtains glossy and clear dyeings, but the light fastness is not as good. Chrome complex dyes have a good levelling capacity and give high light and wet fastness properties. Furthermore, 1: 2 metal complex dyes are also used.

Vat dyes can be used to achieve washfast dyeings. However, considerable fiber damage occurs when exposed to light. Therefore, this type of dye is not applicable for decorative or curtain fabrics.

Good results are achieved when dyeing under alkaline conditions with reactive dyes (especially with tussah silk because of the even higher alkali resistance). In this process, covalent bonds are formed with the functional groups that are present in many cases, which leads to very good color fastness.

Fig. 11.28 Jacquard weaving machines with punch card control for silk

Silk has a much lower affinity for wool dyes than wool, especially for reactive dyes. In order to obtain the same depth of color as wool, twice the amount of dye is required, in the case of tussah silk almost three times the amount. Uniform degumming is a prerequisite for high color fastness, which is why silk is normally dyed only after degumming.

In general, all dyeing machines known for other textile fibers can be used for silk dyeing. However, due to their sensitivity with regard to tensile effects and mechanical stress, they are modified accordingly. Silk is dyed in strand form (yarn), on perforated dyeing tubes and in the piece (fabric). The dyeing units used for yarn dyeing are rope dyeing machines, cross-wound dyeing machines and spray-arm dyeing machines, and for piece dyeing mainly reel skids, jiggers, beam dyeing machines and jet dyeing machines.

11.11.6 Printing

The vast majority of silk fabrics are colored by printing (Fig. 11.29). This mainly includes dress fabrics (shawls, scarves, ties) as well as decorative and upholstery fabrics. Printing on silk also requires prior, uniform degumming.

As a rule, loaded silk fabrics are not printed. If this is nevertheless tried, the material can easily become brittle during steaming if the weighting is more than 15% above par. Furthermore, color

Fig. 11.29 Quality control of pure white silk fabrics

Fig. 11.30 Printing of silk fabrics

changes can occur during etching due to the formation of tin sulfide.

Both, direct printing and etching, are suitable printing methods. In most cases, silk prints can be produced with the same dye types as wool prints. To achieve better fastness, acid, metal complex and reactive dyes are preferred to other dye types. Vat dyes are also used for etching prints.

Silk printing is carried out mechanically as table or machine printing with flat printing and rotary printing machines (Fig. 11.30).

11.12 Areas of Application

The applications of silk range from sewing threads, high-quality ladies' clothing, scarves, shawls and ties to technical articles, e.g. medical wound dressings (Fig. 11.31). Silk powder is often added to lipsticks, skin creams and soaps. Shampoos may also contain fibroin or sericin.

Frequently used terms for special silk fabrics are chiffon (very fine fabric, strongly twisted silk threads), eolienne (made of half silk, warp thread made of silk), georgette (somewhat heavier than chiffon, warp and weft made of crèpe), velvet (fabric with pile, originally made of silk), satin (silk fabrics in atlas weave) and taffeta (dense fabric with horizontal stripes). Since the invention of man-made fibers, other materials, such as, e.g., viscose and polyester, are often used instead of silk for these articles.

Under the Textile Labelling Act (TKG), the designation 'silk'is permitted only for fibers

Fig. 11.31 Typical silk products

Fig. 11.32 International silk signet of the AIS

obtained exclusively from the cocoons of silk-spinning insects (mulberry or wild silk). The trademark for fabrics made of 100% “fine” or “wild” silk is awarded by the Association Internationale de la Soie (AIS) in Lyon, provided that certain conditions are met (Fig. 11.32).

11.13 Tussah Silk (Tussar or Tasar Silk)

Apart from the mulberry silk moth, it is mainly the tussah silk moth that is of economic importance for silk yarn production. The three most important species are:

- The Japanese tussah silk moth (Antheraea yamamai),
- the Chinese tussah silk moth (Antheraea pernyi),
- the Indian tussah silk moth (Antheraea mylitta).

The tussah silk moths live mostly in the open. Their webs, like those of all free-living silkworms, are called “wild silk”. The caterpillars of the tussah silk moth feed mainly on leaves of various oak species, sometimes also of other deciduous trees. The cocoons are larger and heavier (50 × 30 mm; 5–7 g) than those of the mulberry moth (Fig. 11.33). Since the tussah moth hatches, thus destroying the cocoon and the continuous filament, the tussah silk can only be processed by Schappe spinning. The silk filament has (theoretically) a length of up to 1400 m.

In cross-sectional view, two wedge-shaped individual fibers can be seen, which are glued together at the narrow sides. The single fibroin filament has a diameter of 25–30 μm. The longitudinal view shows distinct longitudinal stripes (Fig. 11.34).

The fibroin of tussah silk differs from the fibroin of mulberry silk in quantitative composition, especially in the ratio of the amino acids alanine and glycine, as regards the amino acids present in the hydrolysate. In addition, the ratio between fibroin and sericin is different from that of mulberry silk.

Tussah silk has a harder handle and is less lustrous than mulberry silk. It is more resistant to chemical treatment with acids and alkali. Because the fibroin of tussah silk also contains components of sericin, it must be degummed in stronger soap baths and for longer than mulberry silk. The inherent color of the fibroin is brown or yellow-green and cannot be completely removed even with intensive bleaching. For this reason, tussah silk is usually processed unbleached and often dyed with colors from the brown to green

Fig. 11.33 Tussah silk moth (Süpfle, 2012) and cocoons (both Antheraea yamamai)

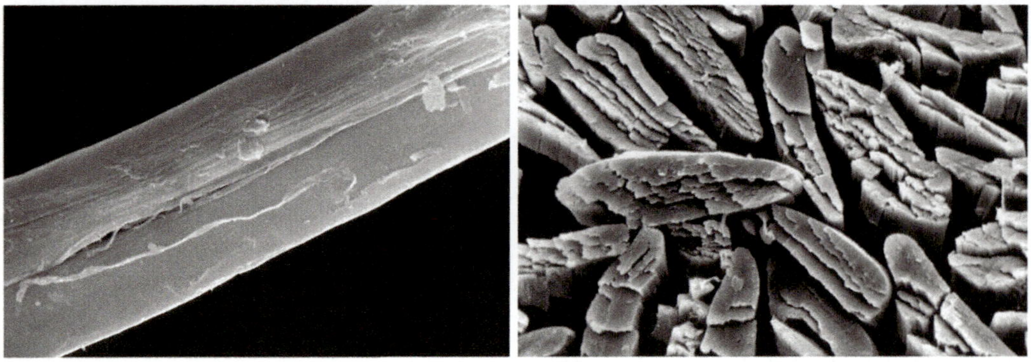

Fig. 11.34 Longitudinal and cross-sectional view of tussah silk

spectrum. The main uses of tussah silk are sewing threads, upholstery fabrics, imitation furs, dress and blouse fabrics and insulating material (Fig. 11.35). The share of wild silk in all silks is approx. 10–15%.

11.14 Other Silks

Of historical or local importance are silks from butterflies and shells. Spiders also produce a variety of different silk fibers.

11.14.1 Butterfly Silk

In addition to tussah spiders, other butterflies yield useful silks. These include the Eri silks from the largest butterflies living in East Asia Philosamia cynthia and Nester silk from the silk moths of the genus Anaphe found in Africa. Furthermore, there is the silk of the Muga silk moth Antheraea assamensis (Assamese Muga: golden-yellowish). It lives mainly in the Indian state of Assam and spins silk threads with a golden luster, which is increased by washing (Fig. 11.36). It absorbs up to 85% of UV radiation. This silk was once reserved for the maharajas and is still processed today on only a few hand looms in Assam. Because the Muga silk moth hatches, only staple fiber yarns are produced from this silk. Muga silk can also be combined with e.g. Qiviut fibers, resulting in interesting color effects and an extremely soft handle.

In South and Central Africa lives the Anaphe silk moth, whose cocoon is collected by the natives and processed into textiles. Also of only local importance is the silk of the Atlas moth (Attacus Atlas), which is native to Southeast Asia and is known as "Fagara silk"(Fig. 11.37).

Other silk producers are mussels and spiders. These include the noble penshell (Pinna nobilis)

Fig. 11.35 Typical products made of wild silk: raw silk (Distelfliege, 2013) and saris (Ganguly, 2016)

Fig. 11.36 Muga silk moth (Yakovlev, 2011) and cocoons of Muga silk (provided by Walter Notter, 2021)

living in the Mediterranean, from which the so-called byssus is obtained. Spider silk is produced among others by the giant spider (Nephila madagascariensis), which lives in East Africa and Madagascar. Their threads are up to 600 m long and have an average diameter of 7 μm.

11.14.2 Byssus

Byssus silk originates from the noble penshell (Pinna nobilis, Pinna rudis), which is widespread in the Mediterranean. They are up to 1 m long and have a gland on the underside from which a protein thread emerges and coagulates on contact with seawater. Several of these 5–20 cm long fibers form a beard ("byssus"), with the help of which the mussel clings to the rocky seabed (Fig. 11.38). In Italy, these fibers were known as "lana pesce" (fish wool).

The mussel silk is very difficult to obtain, but very fine (10–70 μm), soft, golden-glossy and non-flammable. In addition, it is very resistant to alcohol, dilute acids and alkalis.

In ancient times and until modern times, such fibers were processed into scarves and trimmings in Sicily, Sardinia and Corsica. For the production of 1 kg of silk, 1000–4000 shells had to be killed. Therefore, textiles made of byssus were

Fig. 11.37 Eri silk worm (Patniak, 2017) and Eri silk (Simtastic, 2017)

Fig. 11.38 Pinna nobilis (Hectonichus, 2007) and byssus harvesting in the nineteenth century

very valuable and reserved only for the highest nobility (Fig. 11.39).

Nowadays the pinna nobilis is protected and only used in Sardinia by a museum for the production of textiles. It is no longer killed, but only some silk threads are cut off and the pinna is then anchored in the ground again.

For science, the noble penshell is interesting because of its high life expectancy of up to 40 years. The higher the surrounding water temperature, the more oxygen isotopes it stores in its shell. Therefore, analyses of the isotope concentration allow conclusions to be drawn about the climate of past times.

11.14.3 Spider Silk

Unlike caterpillars, spiders do not produce silk threads with paired salivary glands, but with several different spinning glands in their abdomen. The thread is pulled out with the legs or the whole body from several spinnerets, the number of which is family-specific, after the spider has anchored one end to the substrate.

11.14.3.1 Types of Spider Silk

Almost all spiders produce silk threads that perform different tasks: For example, there are stable threads used as security threads and for the basic framework of the web. Furthermore, there are

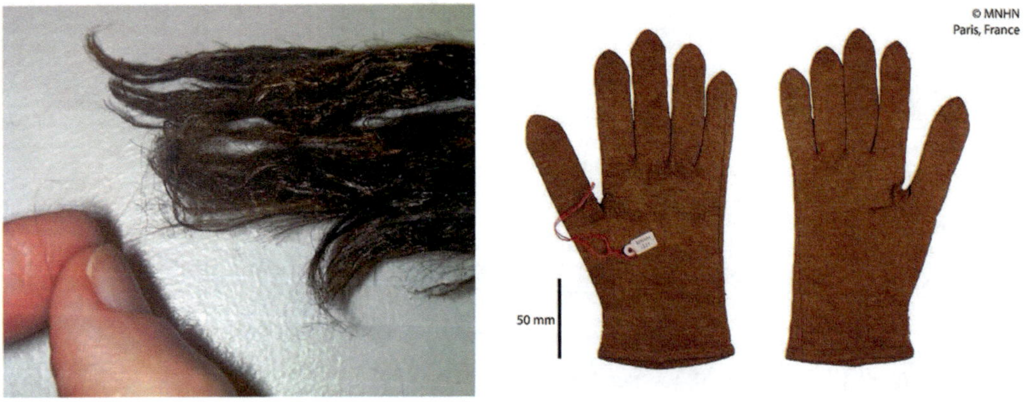

Fig. 11.39 Byssus fibers (Hill, 2008) and gloves made of byssus (Hennion, 2020)

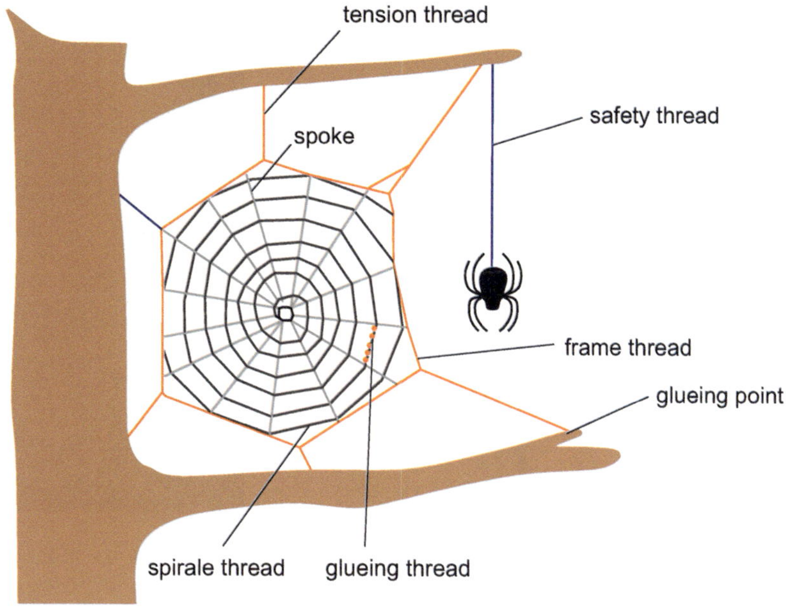

Fig. 11.40 Types of spider silk

threads for the non-adhesive spirals of a web as well as adhesive threads with which the spider can attach its web to a base. Other adhesive threads are used to catch prey, while other silk threads are used to spin in prey and build the egg cocoon. Furthermore, there are elastic threads for the axis threads of the catching spiral (Mukhopadhyay & Sakthivel, 2005), Fig. 11.40.

There are 23 tropical and subtropical species of silk spiders (Nephila) that weave particularly stable wheel webs. In Madagascar, for example, the wheel-shaped catching webs, holding threads and cocoon-like protective covers produced by the large Madagascar spider (Nephila madagascariensis) are used for textile and also for technical purposes. The threads, each 200–600 m long, are manually unreeled from a dozen spiders. Within its lifetime of 1 month, the Madagascar spider can produce this thread length 4–6 times. The tensile strength is 30 cN/tex, which is a typical value for protein fibers.

11.14.3.2 Properties

The density of spider silk is 1.3 g/cm^3. The diameter varies between 0.5 and 5 μm depending on the species. The main components of spider silk are the amino acids glycine (37%) and alanine (21%) (Kaplan & McGrath, 1997). Their arrangement in the protein chain has a significant influence on the tensile strength of the thread. In particular, longer sections of alanine side groups form crystalline structures, accounting for up to 30% of the internal structure. These are interconnected by amorphous region with longer segments of amino acid residues with large side groups (Xu & Lewis, 1990). The distribution and orientation of the crystallites largely determines the mechanical properties. The distinctive feature of spider silk is its high elongation combined with high strength, which no other fiber possesses in this combination. Some spider silk threads are so elastic and resilient that they can be stretched by around 40% of their total length. This gives them an extremely high energy absorption capacity, which is even many times greater than that of aramid, as the example in Fig. 11.41 shows. The mechanical data for spider silk apply to the holding thread of spiders of the species Nephila edulis and were taken from (Vollrath & Knight, 2001). In many publications, significantly higher strengths are given, but then nanofibers were always produced, which have almost no practical use. This phenomenon is observed in most fibrous materials and is due to the size effect.

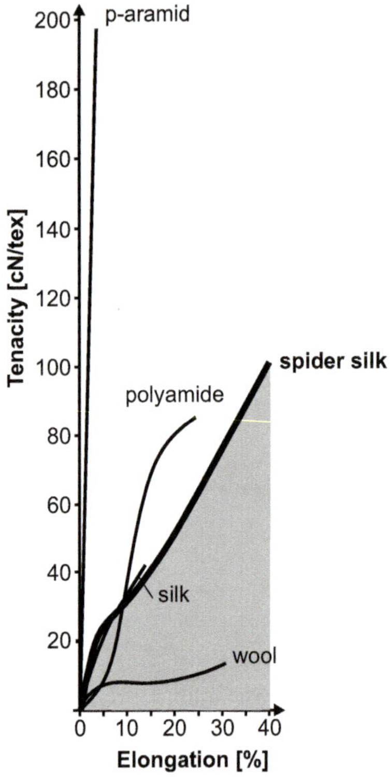

Fig. 11.41 Force-elongation curves of fibers with amide bond and spider silk

Fig. 11.42 Cloak of spider silk (Cmglee, 2012)

11.14.3.3 Applications

In earlier times, spider silk was used for fishing nets and bird traps or woven into small bags. Spider webs were used as hemostatic dressing material and even served painters as canvas when stretched and folded over. Spider silk, e.g. from cross spiders or the black widow, was also used as material for reticles for optical instruments, e.g. surveying instruments, microscopes and rifle scopes. The extremely small diameters of spider silk threads are now technically achievable with various polymers, such as polyamide. Therefore, spider silk threads have now been replaced by synthetic materials in optical applications.

In the nineteenth century, garments were still being woven from real spider silk. However, this involved an enormous amount of effort, as an experiment from 2007 to 2012 showed. 80 workers milked 3000 golden silk spiders daily in Madagascar and took 400 m of spider silk from each of them. 24 of these filaments were twisted and woven into a thread. This led to the cloak shown in Fig. 11.42, which is exhibited in the Victoria and Albert Museum in London.

Several factors have so far limited the potential applications of spider silk:

- Availability only in very small quantities (spiders kept in captivity produce less silk than wild ones and tend to eat each other).
- Maximum service temperature below 160 °C (decomposition).
- Poor reproducibility of mechanical properties, which depend not only on the spider‘s diet but also on environmental influences.
- Strong moisture dependence of the mechanical properties.
- Microorganisms decompose the silk threads.

Handling these spiders may be dangerous, because the bite of an adult specimen of a black widow can be fatal. Nevertheless, some brave contemporaries “harvest” the spinning thread of semi-domesticated specimens.

11.14.3.4 Artificial Spider Silk

Because it is difficult to collect sufficient quantities of silk from potentially lethal spiders for the reasons mentioned above, attempts have been made for some time to produce spider silk artificially. In particular, the safety and frame threads have been the focus of attention. For example, genes encoding the various silk proteins have been inserted into the genetic material of bacteria, which then produce these proteins. Corresponding genes were also introduced into dwarf goats so that they produced milk with spider silk proteins. However, the quantities produced were far too small and the extraction of spider silk was not economically feasible. Many other attempts have not yet led to viable technologies to produce spider silk economically in larger quantities.

The company AMSilk near Munich has been trying to develop a commercially successful process since 2008, but so far without resounding success. The production of artificial spider silk with the electrospinning process does not make sense because the stretching required for high strengths cannot be initiated (Lang, 2015). Wet or dry spinning appears to be a viable route. Spider silk fiber is marketed under the name Biosteel™ and is produced on a pilot scale. It has a tensile strength of about 135 cN/tex, which is above polyester (85 cN/tex) but below aramid (190 cN/tex) and well below high-strength PE (400 cN/tex). In 2017, the company reported that adidas AG would soon produce shoes made of spider silk, which would be about 15% lighter than conventional sports shoes. The spider silk used, which is produced by bacteria, is also biodegradable through the addition of an enzyme. However, none of these products are yet commercially available and there is no apparent technological sense in using spider silk for such articles.

The company Bolt Threads from the USA has been intensively marketing its outdoor clothing with an outer layer of spider silk for several years. So far, however, no products are known to be available for purchase. Kraig Biokraft Laboratories, USA, uses genetically modified silk moth caterpillars to produce spider silk proteins. The threads from this are marketed as Dragon Silk™ and Monster Silk™, so far without great success.

The problem with all these processes is the arrangement of the building blocks in a specific order, which is what gives spider silk its extraordinary mechanical properties in the first place. (Ayoub et al., 2007) estimate the number of amino acid building blocks of fibroin to be approximately 3500, although side chains of approximately 100 amino building blocks may be present. Furthermore, there is no such thing as "the" spider silk, but rather a multitude of more diverse filaments that differ considerably in their mechanical properties (Fig. 11.40). This problem has not yet been solved. Studies of silk protein before and after processing by the spider also show that the mechanical properties depend not only on the composition of the spider silk thread, but also quite significantly on the way in which it is spun out. For example, spider silk threads are constructed as an β-sheet structure, whereas spider proteins taken from the spider before the thread is formed are in the form of α-helix. This is probably one of the reasons why it has not yet been possible to technically reproduce spider silk.

The search for a product for which spider silk would be particularly suitable that could not also be produced from conventional man-made fibers, which are much easier to manufacture in large quantities, has so far been unsuccessful. It is true that spider silk has a very high elongation combined with high strength and thus a high energy absorption capacity. This combination of properties is useful in some technical applications, but spider silk is not available in sufficient quantities. Artificial production is far too expensive and therefore there are no commercially available products so far. A use in bullet-proof vests, which is often mentioned as a possible application, is nonsensical because the elongation of spider silk is much too high. One possible application where the high price of artificially produced spider silk would be justifiable is the regeneration of nerve cells. Because spider silk is extremely fine and at the same time biocompatible, cells grow along the spider silk fibers when they connect severed nerve tracts. However, there is still a long way to go before commercial application is possible.

References

Armstrong, I. (2006). https://commons.wikimedia.org/wiki/File:Silk_worm_cocoons.jpg, CC BY-SA 2.0.

Ayoub, N. A., Garb, J. E., Tinghitella, R. M., Collin, M. A., & Hayashi, C. Y. (2007). Blueprint for a high-performance biomaterial: Full-length spider dragline silk genes. *PLoS One, 2*(6), e514. https://doi.org/10.1371/journal.pone.0000514

Cmglee. (2012). https://commons.wikimedia.org/wiki/File:Spider_silk_cape.jpg, CC BY-SA 3.0.

Distelfliege. (2013). https://commons.wikimedia.org/wiki/File:Bfl_tussah_(10824027793).jpg?uselang=de, CC BY 2.0.

faostat. (2022). http://faostat.fao.org

Furfur. (2019). https://commons.wikimedia.org/wiki/File:Silk_Road_in_the_I_century_AD_-_de.svg, CC BY-SA 1.2.

Ganguly, B. (2016). https://commons.wikimedia.org/wiki/File:Tussore_Sarees_-_Phulia_2016-11-12_1891.JPG?uselang=de, CC BY 3.0.

Hectonichus. (2007). https://commons.wikimedia.org/wiki/File:Pinnidae_-_Pinna_nobilis.jpg

Hennion, M. (2020). https://commons.wikimedia.org/wiki/File:Pinna_nobilis_(MNHN-IM-2014-10867)_003.jpeg, CC BY-SA 3.0.

Hill, J. (2008). https://commons.wikimedia.org/wiki/File:Fine_sea_silk_threads.JPG, CC BY-SA 3.0.

Kaplan, D. L., & McGrath, K. (1997). *Protein-based materials*. Birkhäuser.

Kraig. (2016). Kraig Biocraft Laboratieries. https://commons.wikimedia.org/wiki/File:6_Monster_Silk_Moths.jpg

Lang, G. (2015). Herstellung und Charakterisierung von Fasern aus rekombinanten Spinnenseidenproteinen und deren potentielle Applikationen. Dissertation an der Universität Bayreuth.

Mukhopadhyay, S., & Sakthivel, J. C. (2005). Spider silk – Providing new insights in the field of high performance materials. *Journal of Industrial Textiles, 35*(2), 91–113. SAGE Publishing, Thousand Oaks.

Notter, W. (2021). Qiviut International – creation Walter Notter, Pontresina (Schweiz). http://www.qiviut.ch

Partensky, P. (2009). https://commons.wikimedia.org/wiki/File:Steamed_Silk_Worm_Cocoon_Sorting_(3968805970).jpg, CC BY-SA 2.0.

Patniak, S. (2017). https://commons.wikimedia.org/wiki/File:Eri_Silk_Worm.jpg?uselang=de, CC BY-SA 4.0.

Rungminoz. (2016). https://commons.wikimedia.org/wiki/File:Silkcocoon.th.jpg, CC BY-SA 4.0.

Simtasik. (2017). https://commons.wikimedia.org/wiki/File:Eri_silk_fiber.jpg?uselang=de, CC BY-SA 4.0.

Süpfle, H. (2012). https://commons.wikimedia.org/wiki/File:Antheraea_yamamai_-_female_1_(HS).jpg, CC BY-SA 3.0.

Tcheng-Ki-Tong. (1890). Chinese silk-lore. In *Popular science monthly* (Vol. 36). Bonnier Corporation.

Türk, O. (2014). *Stoffliche Nutzung nachwachsender Rohstoffe*. Springer Nature.

Ullrich, E. (1928). Unterscheidung der Naturseide-Garnsorten und deren gebräuchliche Titer. In *Melliand textilberichte* (Vol. 9, p. 305). Deutscher Fachverlag.

Vollrath, F., & Knight, D. P. (2001). Liquid crystalline spinning of spider silk. *Nature, 410*, 541.

WRC. (2017). http://www.worldsrichestcountries.com/top-silk-importers.html

Xu, M., & Lewis, R. V. (1990). Structure of a protein superfiber: Spider dragline silk. *Proceeding of the National Academy of Science, 87*, 7120–7124.

Yakovlev, A. (2011). https://commons.wikimedia.org/wiki/File:Antheraea_assamensis.JPG, CC BY 3.0.

Zahn, H., Wulfhorst, B., & Steffens, M. (1994). *Fiber table acc. to P.-A. Koch: Silk (Mulberry silk) – Tussah silk, Chemical fibers international (44/94) S* (pp. 40–59). Deutscher Fachverlag.

12 Asbestos

Asbestos is a silicate-based mineral fiber (Fig. 12.1). It has been used for the production of textiles for about 2500 years. A special property of this fiber is its very good temperature resistance. In addition, it is inexpensive and is therefore mined and processed in large quantities worldwide. In 2020, the world production was about 1.2 m t.

Fine asbestos dust is carcinogenic and asbestosis has been recognized as an occupational disease since 1936. The use of asbestos has therefore been banned in the EU since 2005. However, a number of alternative materials are available for most applications of asbestos (cf. Sect. 12.15).

12.1 Terms and Definitions

According to DIN 60001, T. 1, ISO 6938-1984, asbestos fibers are natural fibers and belong to the group of mineral fibers. They are spinnable fibers of crystalline nature. Asbestos occurs in serpentine (serpentine asbestos = chrysotile) and in hornblende (hornblende asbestos: crocidolite or blue asbestos; amosite). The abbreviation is AS.

Asbestos is the name given to natural inorganic crystalline silicates when they occur as fibers. Two groups are distinguished according to their mineralogical composition: the serpentine asbestos and the amphibole asbestos (Table 12.1).

Chrysotile is also called "white asbestos"because of its color. This designation can lead to confusion with the "white" amphibole asbestos. Hornblende is widely used as a companion to the asbestos-forming minerals or rocks. The names "Cape Blue" (Cape asbestos), "Transvaal Blue", "Australian Blue" and "Bolivian Blue" refer to the origin of certain crocidolites (blue asbestos), Fig. 12.2.

Varieties, especially of chrysotile, are also known as mountain leather, mountain flax, stone flax, mountain cork, or mountain wood. Quartzified crocidolite from Prieska (South Africa) is known as yellow-brown or blue tiger's eye and is a prized semi-precious stone.

12.2 History

The word "asbestos"comes from the Greek and has the meaning "imperishable" or "indestructible". This refers to the incombustibility of textiles made from asbestos.

The use of asbestos has been known for about 2500 years (Callimachus, fifth century BC; Plutarch and Pausanias, first and second centuries AD). Pliny the Elder writes in his Naturalis Historia 19 (19–20): "One has even discovered a flax that is not consumed by fire. We [...] saw tablecloths made of it, burning on the hearth, which, after the dirt had burned out, became cleaner by the fire than could have been done by water. [One makes from it] garments for the dead kings, to separate the ashes of their bodies from the rest of the ashes. [This kind of flax] grows in

D. Veit, *Fibers*, https://doi.org/10.1007/978-3-031-15309-9_12

Fig. 12.1 Asbestos fibers

the sun-scorched deserts of India [. . .]. It is difficult to weave because of its shortness."

In ancient times, the "Carpasian flax", "Lithos amiantos" or "Karystios lithos" of various smaller deposits was used in the Mediterranean region. These included, for example, the chrysotile of Cyprus and the tremolite asbestos of northern Italy. They were used to make special products, such as cremation cloths and mats and wicks for temple lamps. According to legend, Charlemagne impressed his guests by throwing a tablecloth made of asbestos into the fire and then pulling it out again undamaged. Marco Polo reported around 1300 about a fabric from the Tartar Empire that resisted the power of fire.

Around 1710, the extraction of chrysotile began in the Urals. Around 1850 the large chrysotile deposits were developed in the province of Quebec and Tremolite asbestos was discovered in 1860 in the Val d'Aosta. This was the real beginning of the asbestos industry. Initially, only longer fibers were processed, but soon short fibers were also used. This was followed by mining in Russia, where large companies quickly emerged. The mining of crocidolite in the Cape Province in South Africa began in 1893, and amosite, named after the Asbestos Mines of South Africa, was found in the Transvaal in 1907. In 1908, mining of chrysotile began in what is now Zimbabwe.

Asbestos was soon used as an insulating and filtering material and, in particular, to reinforce cement boards ("fiber cement", "Eternit"). This is still its most important field of application today. But asbestos was also carelessly used for everyday objects, as the examples in Fig. 12.3 shows.

Among the most important inventions for asbestos processing are (Külter & Wulfhorst, 1990):

- For asbestos-cement products: the wet process of L. Hatschek from 1899 ("Eternit"), the semi-dry process of Österheld from 1901 ("Fulgurit"), the Mazza process for pressure pipes from 1913,
- for sealings: the production of high-pressure gaskets made of asbestos, fillers and rubber by R. Klinger around 1900 ("Klingerit"),
- brake linings reinforced with asbestos fibers (high temperature resistance) 1896 in England,
- for yarns: first spinning of crocidolite in the Cape Province (1895), first spinning of amosite by S. Weingarten in South Africa (1909).

12.3 Origin, Geology

Both main mineral groups are transformation products of other minerals, which extend over earth-historical periods and depend on the local conditions. According to the two main groups, two transformation pathways can be formulated: serpentinization and amphibolization. In both paths, the starting materials are enriched with SiO_2 and water is either absorbed (serpentinization) or released (amphibolization).

Table 12.1 Types of the technically used asbestos

	Mineral	Asbestos type
Serpentine asbestos	Serpentine	Chrysotile
Amphibole asbestos	Riebeckite	Crocidolite, blue asbestos
	Grunerite	Amosite
	Anthophyllite	Anthophyllite-Asbestos
	Tremolite	Tremolite-Asbestos

Fig. 12.2 Important types of asbestos

Asbestos usually occurs in fibrous form parallel to the walls of fissures, crevices and cracks. This indicates crystallization from probably laterally secretive aqueous or gel-like substances. The so-called cross-fiber asbestos is formed. Due to oppositely directed parallel movements (slip) of the surrounding rocks, the initially formed cross fibers can become slanted. This process can lead to the fibers finally assuming an almost parallel position to the fissure walls (longitudinal fibers; slip fibers). Occasionally, asbestos also occurs in the form of nests of irregularly matted fiber bundles (mass fibers). None of the three fiber types is characteristic of any of the asbestos species, but crocidolite and chrysotile occur mainly as transverse fibers.

The proportion of raw asbestos in the rock and the proportion of staple fibers in the raw asbestos vary depending on the mining area. In Russia, for example, the proportion of raw asbestos in the rock is 5–10%, and the proportion of staple fibers in the raw asbestos is 20–25% (Külter & Wulfhorst, 1990).

The profitability limit for the raw asbestos content in the rock is about 3%, which many deposits do not achieve. Attempts have therefore been made to produce asbestos synthetically, e.g. by blowing a melt. For economic reasons, however, these processes have not been successful.

12.4 Deposits

Asbestos often emerges in the form of compact veins or fibrous bands, sandwiched between blocks or layers of fine-grained rock. The rock does not show any fiber formation in this case,

Fig. 12.3 Iron handle with asbestos insert (left) and toaster with asbestos insert (right)

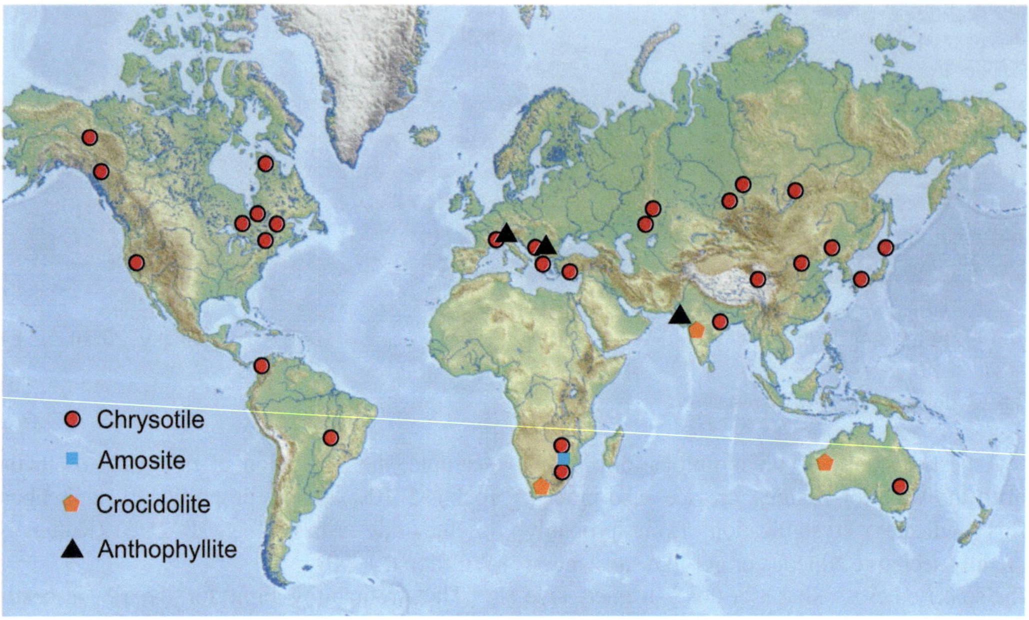

Fig. 12.4 Important deposits of asbestos according to (Virta & Mann, 1994)

although it often has approximately the same chemical composition as the intercalated asbestos. Asbestos deposits of chrysotile are found all over the world (Fig. 12.4). The main deposits are in Canada (Quebec, British Columbia and Yukon provinces), Russia (Bazhenov district in the Urals and Siberia), Zimbabwe and Eswatine (formerly Swaziland). Occurrences of amphibole asbestos are rare. Crocidolite was mainly extracted in the Republic of South Africa (Cape Province). Amosite was also mostly from South Africa (Transvaal Province). Anthophyllite and tremolite are rarely occurring asbestos species. According to estimates, the already known deposits could cover the demand until the end of the twenty-first century.

About 95% of the asbestos mined in the world is chrysotile, the rest being mainly amphibole asbestos. These are divided into:

- 60% crocidolite,
- 25% amosite,
- 15% anthophyllite asbestos.

12.5 Generation

The extraction of asbestos in Russia and Kazakhstan takes place almost exclusively in open cut mining (Fig. 12.5). The same applied to Canada and Australia. In Zimbabwe, South Africa and Eswatini, on the other hand, mining was carried out using underground methods.

Extraction in opencast mining is usually carried out in kilometer-long pits. For this purpose, 10–30 m high steps are cut into the ground on which the drills, excavators and dump trucks work. Initially, the rock is brought up via spirally laid ramps; at greater depths, inclined lifts are also used. When open cut mining becomes uneconomical, underground mining is used. This is done, for example, using the decline method. The undermined rock mass breaks apart and slides through shafts into a conveyor (Fig. 12.6).

The processing methods vary depending on the type of asbestos, the quality of the fibers and

Fig. 12.5 Surface mining in Canada (Hains, 2009)

the type of occurrence as transverse, longitudinal or bulk fibers. In general, wet and dry methods are possible. The rock often contains less than 10% raw asbestos, so the long fibers are extracted manually. In this process, the crude asbestos is freed from rock in the mine by hammering and sorted according to fiber length. This is how the crude asbestos is obtained (Fig. 12.7).

The bulk of the rock consists of lumps up to about 1 m in diameter. These lumps are crushed by squeezing and pressing and passed onto vibrating screens, where suction hoods collect the separated fibers. This is how the mill fiber asbestos is obtained. The fibers produced by each sieve pass are sorted according to fiber length. In the case of long fibers, dust may still need to be removed. In modern asbestos plants, the sorted and cleaned fibers are pressed under pressure into rectangular bags.

12.6 Classification

Classification is based on fiber length and degree of disintegration. The raw asbestos obtained is classified internationally for this purpose according to the Quebec Standard Test, whereby classes 1 and 2 can be processed into yarns and classes 3–9 are used in fiber form:

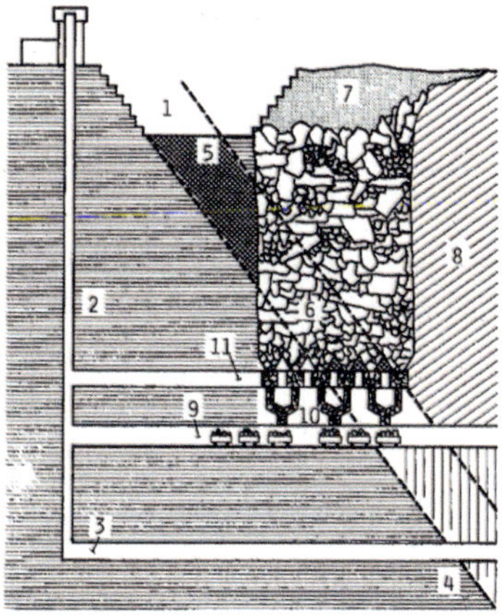

1 Open pit mine
2 Underground mining (fallback method)
3 Further development
4 Scraping
5 Mineral body containing asbestos
6 Broken ore
7 Refilling
8 Asbestos-free rock
9 Conveyor belt
10 conveyor chutes
11 Mine grate walkway with collection points

Fig. 12.6 Cross section through an asbestos mine (Lincoln, 1972)

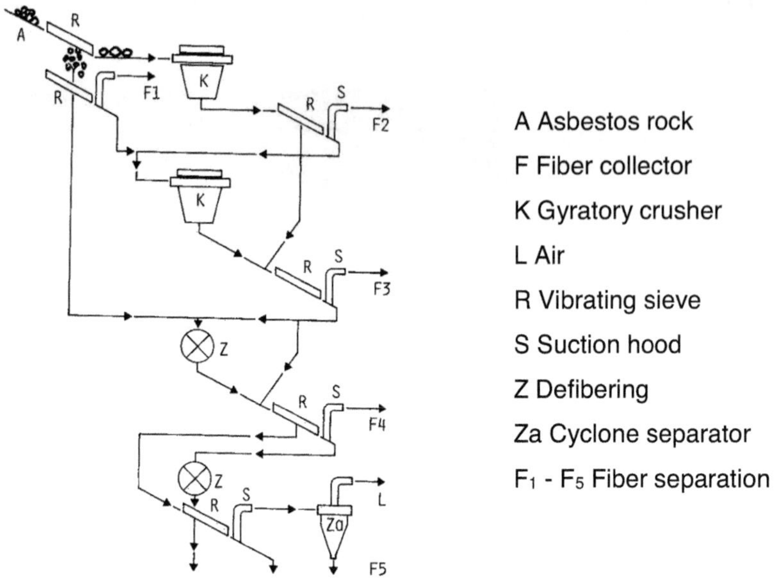

Fig. 12.7 Processing of asbestos rock (Lincoln, 1972)

Crude 1: Fiber length > 19 mm (0.75 in.)

Crude 2: Fiber length 9.5–19 mm (0.375–0.75 in.)

The main part of the asbestos production consists of non-spinnable short fibers. Their main area of use was and is predominantly the construction industry. Around 98% of all asbestos fibers are used for asbestos-cement products. Only about 1% is processed into fabrics and about 0.1% is used in blends with other materials for heat protection products.

12.7 Chemical Composition and Structure

Asbestos is a three-substance system consisting of silicon dioxide ($Si0_2$), various metal oxides and water. The most important data for practical use are water and metal oxide contents. The water content is characteristic for the classification into serpentine asbestos (chrysotile) and amphibole asbestos.

The chemical formulas of the most important asbestos are:

Chrysotile: $Mg_6\ [(OH)_8\ Si_4\ O_{10}]$

Crocidolite: $Na_2\ (Fe^{2+}\ Mg)_3\ Fe3^+\ [(OH)_2\ Si_8\ O_{22}]$

Amosite, anthophyllite: $(Fe^{2+})_2\ (Fe^{2+},\ Mg)_5\ [(OH)_2\ Si_8\ O_{22}]$

Tremolite, actinolite: $Ca_2\ Mg_5\ [(OH)_2\ Si_8\ O_{22}]$

The chemical composition depends on the group, within the amphibole asbestos on the type and, for all asbestos of the same type, on their origin (deposit location). Chrysotile consists of approx. 40% each of SiO_2 and MgO, while amosite is composed of 50% SiO_2 and 40% FeO. Crocidolite consists of more than 50% SiO_2 and 15–20% each of FeO and Fe_2O_3 (Fig. 12.8). Especially in amphibole asbestos, individual atoms (e.g. Si, Al, Ca, Mg) may be replaced by others (Fe, Mn, Ti, Ni, Cr, Li, Zn). This leads to a mixed structure of such asbestos (Table 12.2).

As with all silicates, the basic building block of asbestos minerals is the SiO_4-tetrahedron, as X-ray investigations showed as early as 1930. The oxygen can belong to two different tetrahedra at the same time. The arrangement and linkage of the SiO_4-tetrahedra in combination with various metal atoms characterizes the fundamentally different structural composition of chrysotile and amphibole asbestos.

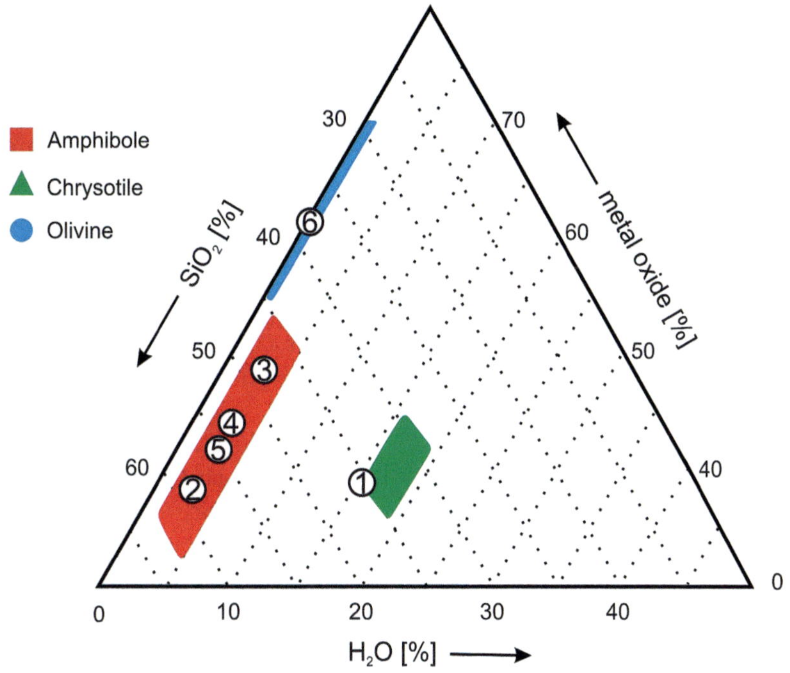

Fig. 12.8 Composition of asbestos acc. to (Berger, 1961)

12.7.1 Chrysotile

Asbestos fibers of this type usually grow perpendicular to the walls of rock crevices. Their fiber length can reach several centimeters, but is usually less than 10 mm. Due to displacements of the corresponding rock formations, the asbestos fibers often reach parallel positions relative to the rock wall ("slip fibers"). Chrysotile fibers can also accumulate in larger quantities of short fibers in a tangled position. The fiber content of such deposits can be over 50%, in which case they are also referred to as "mass-fiber deposits".

During the geothermal formation process, other minerals are also deposited. These include brucite, magnetite, calcite, dolomite, chlorite and tallow. Minerals containing iron can also be deposited. During extraction, these substances are often heavily crushed and thus reach the fiber material.

The diameter of chrysotile fibers can be as small as 25 nm. Usually, several fibers are clustered together, so that the diameter of such structures is 0.1–100 μm. The length ranges from a few millimeters to several centimeters.

Table 12.2 Chemical composition of different types of asbestos

	Chrysotile	Amosite	Crocidolite	Antophyllite asbestos	Temolite asbestos	Aktinolith asbestos
SiO_2	36–44	49–53	49–56	53–60	55–60	51–56
MgO	38–42	1–7	0–3	17–34	20–26	12–20
FeO	0–3	34–44	13–21	3–20	0–5	5–15
Fe_2O_3	0–5	0–5	13–20	0–5	0–5	0–5
Al_2O_3	0–2	0–1	0–1	0–3	0–3	0–3
CaO	0–2	0–2	0–3	0–3	10–15	10–13
K_2O	0–1	0–1	0–1	0–1	0–1	0–1
Na_2O	0–1	0–1	4–9	0–1	0–2	0–2
H_2O	12–14	2–5	2–5	1–6	1–3	1–3

The structure of chrysotile (also called serpentine asbestos) is characterized by a regular repetition of certain two-dimensional reticular planes. The base layer consists of SiO_4-tetrahedra linked in such a way that 3 oxygen atoms each form bridges to neighboring tetrahedra. The fourth oxygen atom forms the connection to the second layer, which consists of $Mg(OH, O)_6$-octahedra (brucite layer). The second layer is followed by the base layer (Fig. 12.9).

Chrysotile fibers have a hollow fiber structure. It is formed by curling or tubular growth of the monoclinic mineral lamellae built up from brucite and Si_2O_5 layers to form tube-like structures. The reason for the formation of the hollow fiber structure is the slight deviation of the lattice spacing between SiO_4—and brucite layers. The distance between the SiO_4-tetrahedral vertices is slightly smaller than the distance between the octahedral vertices. Therefore, the double layer curls in such a way that the SiO_4-tetrahedra are on the inside and the $Mg(OH,O)_6$-octahedra are on the outside. The deviations of the lattice spacing lead to a certain radius of curvature, so that limit values of the curvature arise both when the tubes grow over each other and when the spiral grows. A small section of the spiral (in Fig. 12.9 marked by arrows) shows the structure of the double layer and the single cell (Külter & Wulfhorst, 1990).

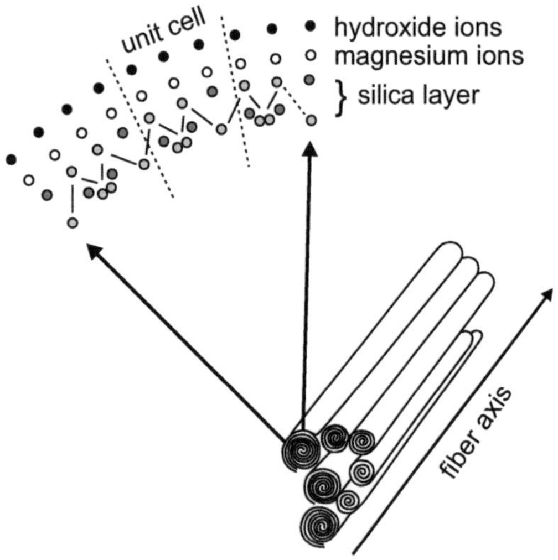

Fig. 12.9 Structure of chrysotile acc. to (Hodgson, 1979)

12.7.2 Amphibole

The minerals of this group can be found almost throughout the earth's curst. Their chemical composition varies widely and only a few form fibers. It is believed that asbestos fibers form as a result of a chemical reaction in which rocks containing iron ore solidify from a semi-liquid to a solid form. In addition, mechanical deformation is necessary for the formation of crocidolite, where mineral-bearing bands shear off within rock formations.

In the case of amphibole asbestos, one-dimensional infinite tetrahedral double chains (ribbons) are present, whereby two chains of Si_4 O_{11}-assemblies are connected with each other by cations and seven cations form the basic unit. The central cation is connected to two hydroxyl groups.

Amphibole asbestos fibers are solid and their surface is formed by the SiO_4 tetrahedra. This creates a preferential cleavage plane along which the fibers splice (Fig. 12.10). In contrast to chrysotile fibers, amphibole fibers usually occur singly and not in groups. They are therefore particularly easily respirable. The fact that the structure of amphibole fibers resembles coffins is a freak of nature.

12.8 Properties

12.8.1 Fineness, Length, Density, Color, Gloss, Handle

Asbestos fibers are very fine compared to other natural fibers and to man-made fibers. The diameter of the elementary fibers is between 0.02 and 0.04 μm for chrysotile and between 0.1 and 0.2 μm for amphibole asbestos. In practice, asbestos is usually used in the form of the technical fiber. It consists of bundles of many elementary fibers lying in parallel. The diameter of the technical fiber depends on its degree of disintegration and is between 0.75 and 1.5 μm (0.01–0.04 dtex) for chrysotile and between 1.5 and 4.0 μm (0.06–0.4 dtex) for amphibole asbestos. Due to

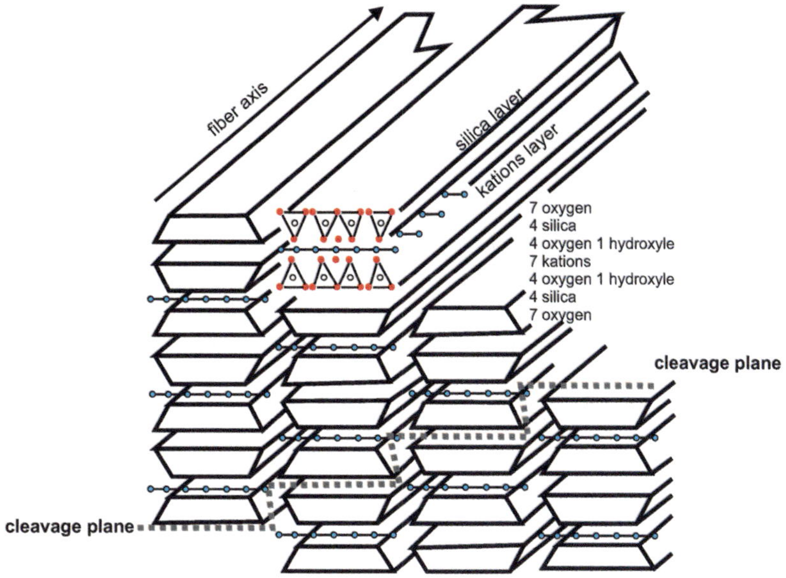

Fig. 12.10 Structure of amphibole acc. to (Hodgson, 1979)

their very high fiber fineness, asbestos fibers have a very large specific surface area of approx. 2–20 m^2/g, which is higher by a factor of 10–100 than, for example, for cotton.

The length of the fibers varies according to type and is usually 1–3 mm. Asbestos fibers also occur in lengths of 1–70 μm. Fibers of this length are respirable and can endanger and damage human health (cf. Sect. 12.13).

The density of chrysotile is 2.4–2.6 g/cm^3, for amphibole asbestos it is 3.1–3.6 g/cm^3.

The color of asbestos ranges from green to grey-white (chrysotile), blue to green (crocidolite) and yellow-brown to grey (amosite). The luster is described as silky to dull as well as glassy and pearly. The feel ranges from soft to rough to hard (Fig. 12.11).

12.8.2 Microscopic Image

The fiber cross-section is circular or elliptical.

12.8.3 Mechanical Properties

The tensile strengths listed in the literature vary greatly. These differences are due to different test

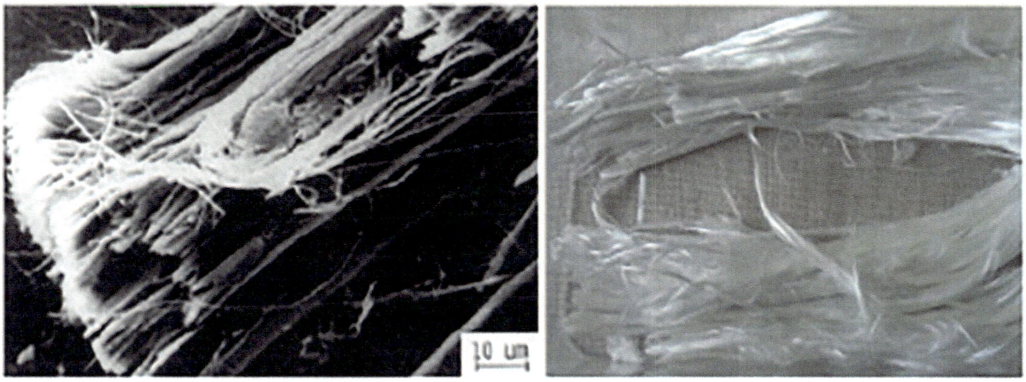

Fig. 12.11 Asbestos fibers (right: Dulyan, 2005)

Table 12.3 Important mechanical properties of asbestos fibers

Tenacity [cN/tex]	12–180
Cross-sectional tenacity [MPa]	300–6000
Wet tenacity ratio [%]	100
Loop tenacity ratio [%]	25
Elongation at maximum tenacity [%]	0.5–4
E-modulus [MPa]	50,000–275,000
Tenacity utilization fiber/yarn [%]	15–17

conditions and in particular to the different disintegration of the fibers in the bundle test. Typical cross-sections for technical fibers in the bundle test are 0.01–0.4 mm^2 (Table 12.3).

12.8.4 Physical Behavior

Moisture absorption [%]:	At 65% rel. h.: 1.6–1.9
	At 85% rel. h.: 1.8–2.2
Water retention [%]:	0
Electrical conductivity:	Low

12.8.5 Chemical Behavior

The resistance to acids and alkalis varies greatly for the different asbestos materials. Exposure to acid leads in all cases to a loss of weight, which is in some cases severe. It amounts to approx. 50% for chrysotile, 3–4% for crocidolite and 12% for amosite. The duration of exposure, the type of acid, the concentration and temperature of the acid and the degree of decomposition of the asbestos fibers all have an influence. Crocidolite is not resistant to concentrated alkalis.

12.8.6 Thermal Behavior

Chrysotile changes its chemical composition at a temperature above 500 °C, crocidolite and amosite already at temperatures above 200 °C. This change is accompanied by a sharp drop in strength. For all types of asbestos, the loss of weight and strength increases with higher temperature and longer exposure time. The maximum weight loss occurs for chrysotile and amosite at 600–900 °C, for crocidolite at 400–900 °C and for tremolite asbestos only at 950–1100 °C (Fig. 12.12).

The main characteristics for important applications are summarized in Table 12.4.

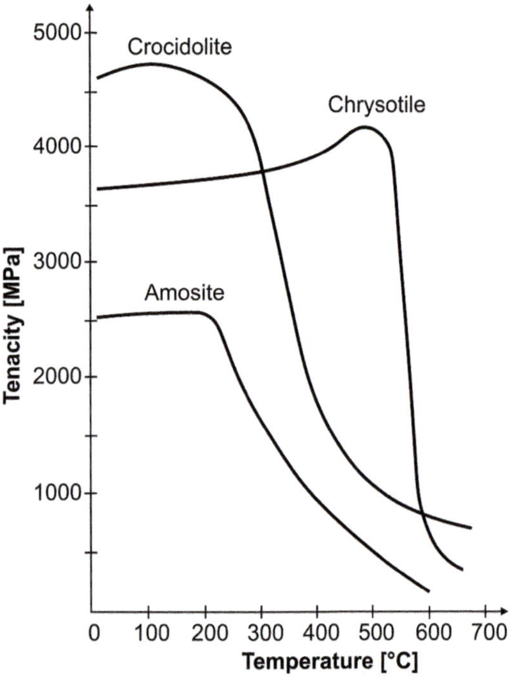

Fig. 12.12 Strength loss of asbestos as a function of temperature

Table 12.4 Properties of asbestos for special applications (Külter & Wulfhorst, 1990)

Applications	Mechanical properties		Chemical resistance		Thermal resistance	Insulation		Special friction properties	Special adhesive properties
	Tenacity	E-modulus	alkali	acids		thermal	electrical		
Asbestos in inorganic matrices									
Asbestos cement	+	+	+						
Fire protection slabs					+	+			
Spray coats					+	+			
Asbestos in organic matrices									
Friction pads	+				+	+			
Sealings			+	+	+	(+)	(+)		
Composites	+			(+)			(+)		
Asbestos products with low binder content									
Asbestos textiles					+	+	+		
Asbestos paper and cardboard					+	+	+		
Filter									+

12.9 Production and Consumption

Figure 12.13 shows the development of world production. According to this, it reached its peak value of 4.79 m tons in 1979 (Fig. 12.13). Since then, production has fallen to 1.2 m tons (2020).

The main producing countries of asbestos today are Russia, Kazakhstan, China and Brazil. China and Brazil have drastically reduced their production since 2015. However, Kazakhstan and Brazil have increased their production again in 2020 compared to 2019 (Fig. 12.14). Until the 1970s, Canada was the largest producer of asbestos, with a market share of around 50%. Only

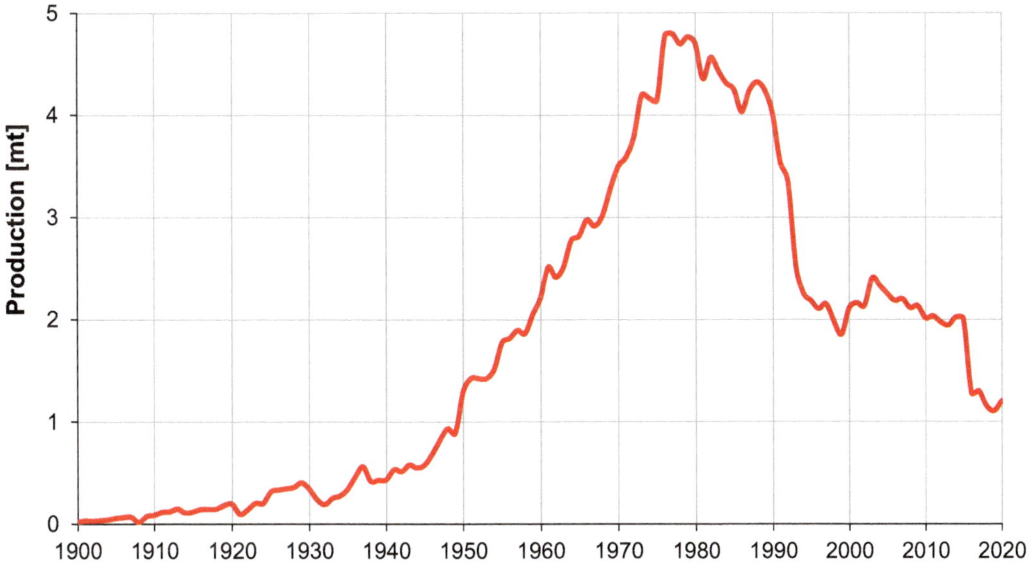

Fig. 12.13 World production of asbestos (USGS, 2021)

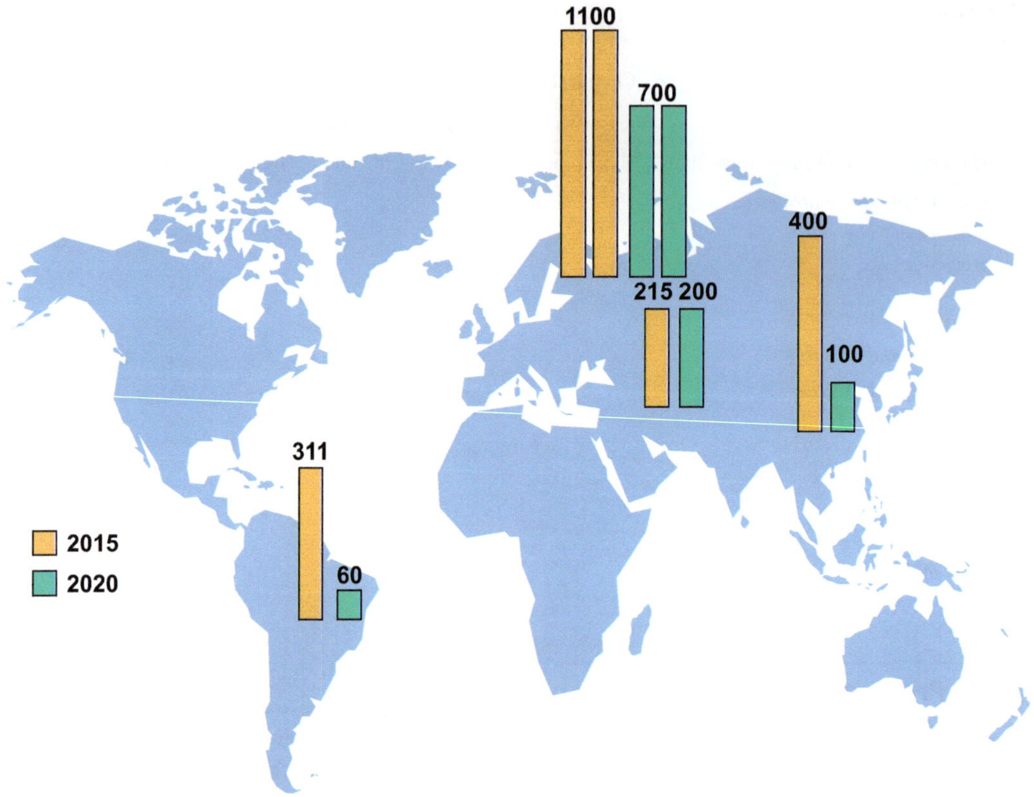

Fig. 12.14 Major producing countries of asbestos (2015 and 2020) in [1000 t] (USGS, 2021)

since 2012 has asbestos no longer been mined there, and in the USA since 2002 (Table 12.5).

Consumption in the world varies greatly depending on the region. Russia, China and India consume around two thirds of the asbestos produced worldwide, mainly for building products. In Europe, the use of asbestos is banned everywhere (Fig. 12.15).

Russia has the highest per capita consumption (4 kg), followed by Kazakhstan (2 kg).

The price of a ton of asbestos ready for processing rose sharply from 2000 to 2016, then fell and is currently around 2 US$/kg (Fig. 12.16).

The import of asbestos into the countries of the EU is prohibited. A typical example for the drop in consumption of asbestos is Germany. Even before the ban on the use of asbestos products there in 1989, imports were falling continuously. While it was still 400,000 t in 1980, it dropped to 200,000 t as early as 1981 and then fell to 50,000 t by 1988. Until 1985, Germany also exported products made of asbestos. The decline in imports, and thus in consumption, is mainly due to the increased use of substitutes since the 1980s, especially in the fiber cement industry. The switch to asbestos-free brake and clutch linings, floor coverings and road construction materials as well as the substitution of asbestos in the field of plastic reinforcement also played a major role.

12.10 Environmental Behavior

Lightfastness:

- behind glass very well,
- good when weathered, but loss of strength after weathering and burial possible.

By microorganisms and insects: no infestation.

Environmental pollution, health hazards and damage see Sect. 12.13.

Table 12.5 Production volume of the most important producing countries (USGS, 2021)

	1920	1930	1940	1950	1960	1970	1980	1990	2000	2010	2020
Brasil			454	844	3,538	16,329	1,69,173	2,05,000	1,70,000	2,70,000	60,000
Canada	1,47,000	1,99,255	2,84,406	7,94,100	10,14,647	15,07,420	13,23,053	7,25,000	3,20,000	1,00,000	
China	5	286	18,157		81,647	1,72,365	1,31,700	2,21,000	3,70,000	4,00,000	1,00,000
South Africa	5583	15,868	22,543	79,300	1,59,544	2,90,318	2,77,734	1,46,000	18,782		
USSR/Russia		49,063		2,17,725	5,98,743	10,65,943	20,70,000	24,00,000	9,83,200	10,00,000	7,90,000
Kazachstan									1,25,000	2,14,000	2,10,000
USA	1356	3491	16,509	38,496	41,026	1,13,683	80,079	10,000	5260		
Zimbabwe	15,491	31,080	46,093	64,888	1,21,529	79,832	2,50,949	1,61,000	1,45,000	25,000	8000
ROW	5296	7957	39,838	94,648	1,89,326	2,44,109	3,97,312	1,52,000	57,758	36,000	32,000

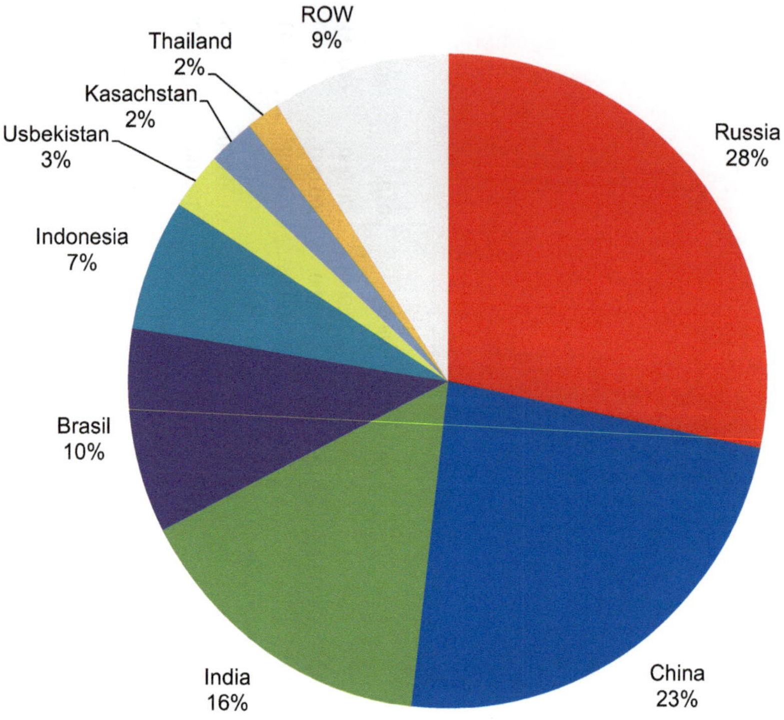

Fig. 12.15 Main consumers of asbestos in [1000 t] (2015)

12.11 Processing

Chrysotile fibers can be spun well, crocidolite fibers only partially due to their short fiber length. Short and brittle fibers are unsuitable for textile processing. Asbestos fibers can be spun either pure or with a small proportion of carrier fibers. Asbestos fibers supplied as crude or mill fibers are

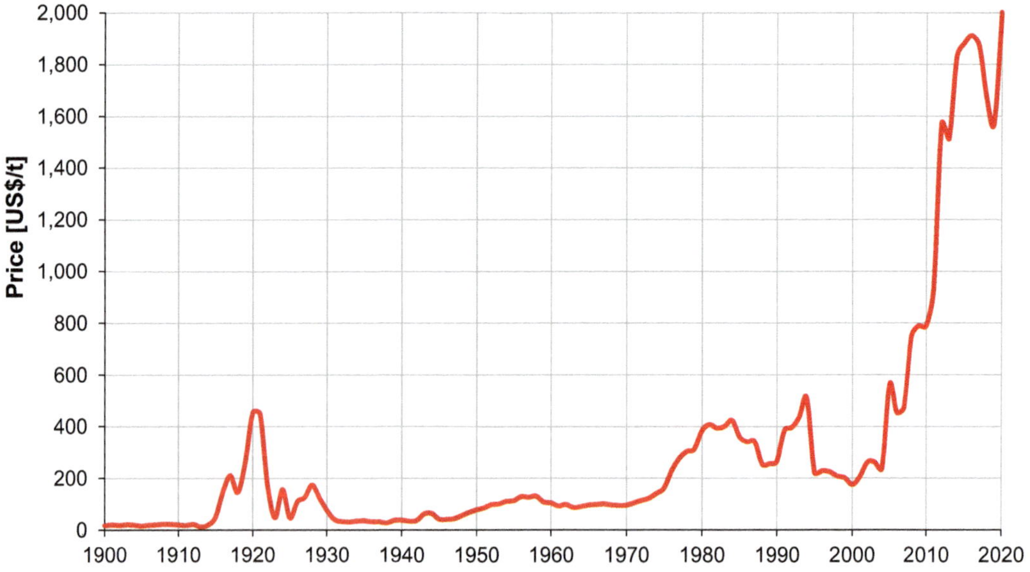

Fig. 12.16 Price development of asbestos on the world market (USGS, 2021)

opened in openers, carding or shredding mills, depending on the type and properties. The number of passes influences the degree of opening and thus the fineness as well as the length and strength. In general, soft fibers are less affected than hard fibers.

The blending of asbestos grades of different origin and fiber length, possibly also with cotton and/or with man-made fibers, depends on the intended use. Blending with glass fibers is also common. The roving is produced on card sets with sliver dividers. The yarn is produced without draft on flyer or ring spinning machines. There are also core yarns with cores made of various fibrous materials or wires.

Another method of producing slivers for the spinning process is the wet process. Its principle is similar to that of paper production. The mixing process produces a fiber dispersion, which is formed into a wet sliver and fed to the spinning machine. The wet process prevents the formation of dust. Additives such as pigments, graphite, lubricants, fillers and thermosetting strengtheners can be added to the dissolved fiber slurry. The dispersing agents or solvents are removed from the finished yarn by heat treatment in pressure vessels. Further textile processing, such as twisting, weaving, braiding and knitting, does not differ from the processing of other fibrous materials.

12.12 Areas of Application

It is estimated that there are over 3000 products made of asbestos in various fields of application. In Table 12.6, the nine most important areas of

Table 12.6 Most important applications of asbestos

Application	Product groups					
Work protection	Personal heat protection clothing	heat protection gloves	plane textiles	material for special work places		
Fire protection	Fire resistant boards and mats	Spray coats insulation plaster	Plastic masses, paint, sealant	Cardboard, string, nonwovens, inorganic foam	Textiles – Fire blanket – Fire curtain	Protective clothing for firefighters
Heat insulation	Boards and mats	Inorganic Spray coats	Materials to fill Seams and voids	Molded pulp Molded masses	Textiles Products	
Electrical insulation	Wire and cables	Insulation material	Molding mass	Household appliances		
Sealings	Static – flat seals	Dynamic – gasket	Cylinder-head gasket	Hot gas sealing	Compensators	
Filtration	Fluid filtration filter media	Gas filtration venting, process air, dedusting	Breathing filter respirator	Diaphragma separators		
Friction pads	Disc breaking pads	Drum breaking pads	Breaking pads soles	Breaking pads for industrial applications	Clutch pads	
Building products (asbestos cement)	Flat boards	Ripple	Pipes for underground work – Pressure pipes – Canal pipes	Rohre für Haus- und Grundstück – Abgas und – Lüftung	Garden applications	
Chemical products and others	Paint and spackle	Glues, sealing mass, sealant	Special products with bitumen or tar matrix	Molding mass with resin matrix	Molding mass with plastic matrix	

Fig. 12.17 Asbestos warning sign

application are listed, to which the main product groups are assigned.

12.13 Health Hazards and Damage

Asbestos fibers in respirable dimensions endanger and damage human health. For this reason, asbestos fine dust and asbestos-containing fine dust have been included in the Ordinance on Hazardous Substances as very dangerous carcinogenic substances for some years now (Fig. 12.17). However, it has not yet been sufficiently clarified which fiber sizes are effective. According to current knowledge, fibers with a length > 7 μm and a diameter of 0.1–0.2 μm in particular are considered to be particularly carcinogenic, as they persist in the body for many years. For fibers with a length < 5 μm and a diameter > 1 μm, only a low carcinogenic effect is assumed. However, this is uncertain due to the very long incubation period (in some cases several decades).

Most asbestos particles absorbed by the body are either coughed up or swallowed. However, those that are particularly long and thin often accumulate in the lungs, where they can cause cancerous ulcers. The risk of disease increases with the intensity and duration of contact with asbestos. After the start of industrial asbestos processing, it took decades to prove the long-suspected link between inhalation of fine asbestos dust and certain diseases.

In Fig. 12.18, the carcinogenic effect of asbestos fibers is shown schematically. Initially, the fibers are inhaled and thus gradually enter the narrower passages of the lungs, from the tracheae through the bronchi to the bronchioles. Fibers embedded along this pathway can produce lung cancer.

Fibers can also reach the honeycomb-like air sacs, the so-called alveoli. There they produce scarring, which can lead to respiratory problems if severe. This form of asbestos disease is called asbestosis. It is mainly caused by long-term

Fig. 12.18 Carcinogenic effect of asbestos fibers

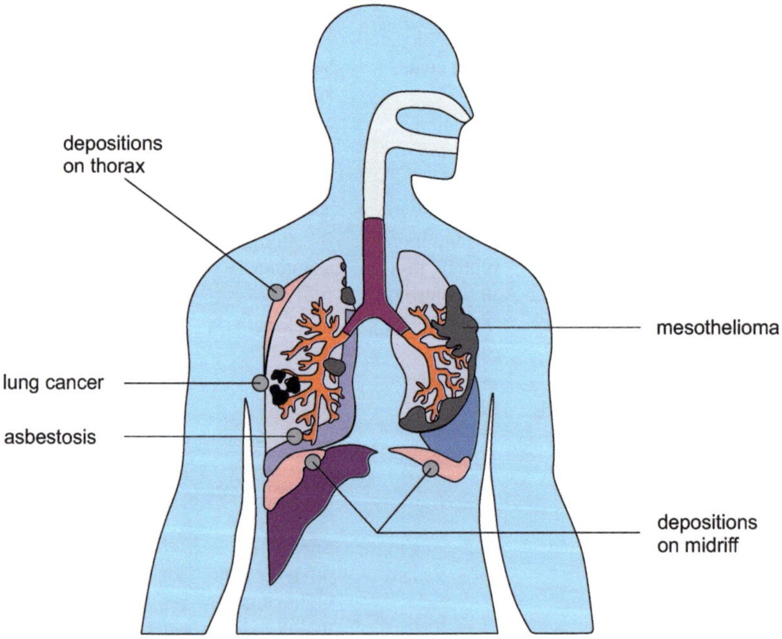

contact with asbestos and has been recognized as an occupational disease since 1937.

Fibers that reach beyond the alveoli to the pleura or peritoneum can damage the mesothelial cells there and thus produce malignant tumors. This rare cancer is called mesothelioma, usually occurs 25–40 years after the first contact with asbestos and has been recognized as an occupational disease since 1977. It is usually fatal to the affected patient within a year of discovery. Mesothelioma can also occur in stomach.

A study published in 2004 in the International Journal of Occupational and Environmental Health shows that the consumption of asbestos worldwide correlates with the number of cases of mesothelioma occurring about 25 years later. According to this study, an average of one case of mesothelioma can be expected for every 170 t of asbestos consumed 25 years later. For the example of Germany, this means that with an estimated consumption of approx. 10 m tons between 1950 and 1990, there will be around 40,000–60,000 cases of mesothelioma by approx. 2020.

One problem that has not been solved to date is determining the cause of such a disease. In many cases, it has not been possible to produce clear evidence that proves beyond doubt that the disease is caused by exposure to asbestos. This makes it extremely difficult to give precise figures on the number of people who have fallen ill as a result of handling asbestos.

Based on the results of animal experiments, the other inhalable fibrous substances are also included in the analysis in the same way as asbestos. Thus, since 1980, artificial mineral fibers with a diameter < 1 µm have also been identified as substances with a reasonable suspicion of carcinogenic potential. Products containing glass wool, rock wool and slag wool (non-textile glass fibers), ceramic fibers and gypsum fibers may also contain fibers in the respirable range and thus be harmful to health. The same applies to carbon nanotubes. Therefore, great care must be taken when handling such materials.

Fig. 12.19 Protective suit during the renovation of buildings containing asbestos (Serwis, 2008)

12.14 Refurbishment of Buildings Containing Asbestos

Despite the ban on the production and import of asbestos products in many countries, large quantities of asbestos are still installed in buildings. As long as these building parts are not renovated, asbestos is considered harmless (Fig. 12.19). The danger only arises when the fine asbestos fibers or also asbestos dust are released and can enter the respiratory tract.

The limit value at the workplace is one fiber per cubic centimeter of air for asbestos fibers with a length of more than 5 µm and a diameter of less than 3 µm and a length/diameter ratio of more than 3. This Technical Reference Concentration (TRC) applies to all asbestos fibers except crocidolite, the TRC value for which is 0.5 fibers/cm^3 air. A limit value of 0.25 fibers/cm^3 air applies to refurbishment work.

The health hazard posed by fine asbestos dust can be countered by extraction and continuous monitoring of workers handling asbestos. For example, the renovation of parts of buildings is carried out by sealing off the relevant sections and creating a slight negative pressure so that no asbestos dust can escape into the environment.

Table 12.7 Technical properties of possible asbestos substitutes

	Mechanical properties		Chemical properties		Spec. friction	Thermal resistance [°C]				Insulation	
Fiber material	Tenacity	E-modulus	Alkali	Acids	Properties	350	550	750	1250 and higher	Thermal	Electrical
Inorganic amorphous fibers											
textile glass fibers	+	(+)					+			+	
C-glass				+							
E-glass											+
Z-glass			+								
Silica fiber ($SiSO_2$)	+	(+)			(+)			+		+	
Non-textile glass fiber									+		
Ceramic wool				+			+				
Glass wool								+			
Rock and slag wool					(+)						
Inorganic crystalline fibers											
Carbon fibers (PAN-based)	+	+	+	+	+		+		(+)		
Carbon fibers (pitch-based)	+	+	+	+	+		+		(+)		
Carbon felt			+	+			+		(+)	+	
Steel fibers	+	+	(+)	(+)							
Kaliumtitanate									+	+	
Silicacarbide-whisker	+	+	+	+							
Organic fibers											
Oxidized PAN	+		+	+	+	(+)				+	+
PAN	+		(+)	+							
Polyester	+			+						+	+
Polyaramid	+	+	+	+	+	(+)				+	
Polytetrafluorethylene	+		+	+		(+)				+	+
Viscose	+				+					+	
Polypropylene	+		+	+						+	

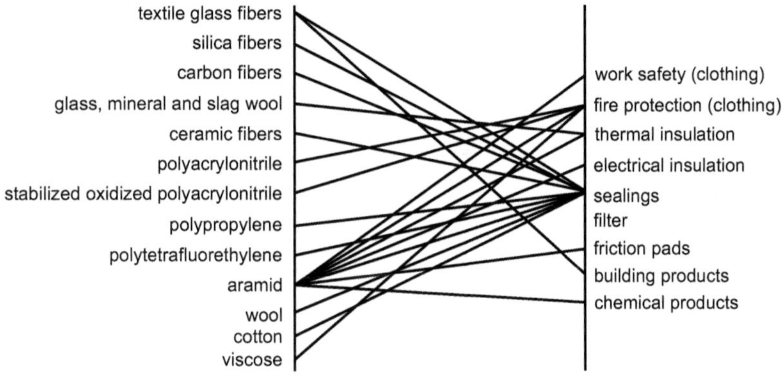

Fig. 12.20 Alternative fibrous materials as asbestos substitutes for typical applications (Külter & Wulfhorst, 1990)

12.15 Substitute Materials

Due to the severe health hazards caused by improper handling of asbestos, this fibrous material has been increasingly replaced by other materials worldwide in recent years. It has not yet been possible to replace asbestos with a single other fiber material. However, there are alternative materials for a large number of asbestos applications, which are described in Table 12.7 with their most important properties.

Even if it has thus not been possible to find a single substitute material for asbestos that combines all the essential properties of asbestos, there are many alternative materials that possess some of the essential properties of asbestos for a particular purpose. As the example in Fig. 12.20 shows, asbestos is being replaced as a sealing material by a variety of other—non-hazardous—

Table 12.8 Asbestos-replacement fibers for selected applications

	Work protection	Fire protection	Heat insulation	Electrical insulation	Sealings	Filtration	Friction pads	Building products	Chemical products
Textile glass fibers	+		+	+	+		+		+
Silica fibers	+	+	+	+	+				
Carbon fibers					+				+
Non-textile fibers									
Glass, rock and slag wool	+		+			+	+		+
Ceramic fibers	+	+	+	+	+		+		
Plaster fibers			+						+
Woolastonite	+		+	+			+	+	+
Attapulgite									+
Sepiolith									+
Polyacrylonitrile						+	+	+	+
Stabilized oxidizes PAN	+		+			+	+		+
Polyvinylalcohol								+	
Polypropylene					+				
Polytetrafluorethylene				+	+	+			
Aramid	+	+	+	+	+		+		+
Wool	+	+							
Cotton	+			+					+
Viscose					+	+	+	+	+
Flax and hemp								+	
Glimmer					+				+
Talc					+				+
Bentonite							+		
Diatomite							+		+

fibrous materials. At the same time, it is also clear that aramids, for example, can be used for a large number of typical asbestos applications, but they cannot replace asbestos nevertheless (Fig. 12.20).

The so-called asbestos substitute catalogue, which was compiled in 1985 by the Battelle Institute on behalf of the Federal Environment Agency, provides detailed information on all the important properties and areas of application of possible substitutes for asbestos. Selected examples are shown in the following table (Table 12.8).

References

Berger, H. (1961). *Asbest-Fibel. Aus: AWG-Reihe Gummi Asbest Kunststoffe* (Vol. 53). A.W. Gentner Verlag KG.

Dulyan, A. (2005). https://commons.wikimedia.org/wiki/File:Asbestos_fibres.jpg

Hains, M. (2009). https://commons.wikimedia.org/wiki/File:Asbestos_Jeffrey_pit_-_panoramio.jpg, CC BY-SA 3.0.

Hodgson, A. A. (1979). Chemistry and physics of asbestos. In *Asbestos, properties, applications and hazards* (Vol. 1). Wiley.

International Journal of Occupational and Environmental Health. (2004). Taylor & Francis Group.

Külter, H., Wulfhorst, B. (1990). Fiber table asbestos and alternative fibers. In *Chemiefasern/Textilindustrie* (40/92, pp. 62–82). Deutscher Fachverlag.

Lavinsky, R. (2010). https://commons.wikimedia.org/wiki/File:Tigers-Eye-Riebeckite-Hematite-214889.jpg, CC BY-SA 3.0.

Lincoln, B. (1972). *Asbest, Abbau und Aufbereitung*. Ciba-Geigy AG.

Serwis, D. (2008). https://commons.wikimedia.org/wiki/File:Warsaw_New_Town_Warsaw_Poland_-_panoramio_(5).jpg, CC BY-SA 3.0.

Siim, S. (2006). https://commons.wikimedia.org/wiki/File:00054_3_cm_crocidolite_riebeckite.jpg, CC BY-SA 3.0.

USGS. (2021). United States Geological Survey. https://pubs.usgs.gov/periodicals/mcs2021/mcs2021-asbestos.pdf

Virta, R. L., & Mann, E. L. (1994). Asbestos. In D. D. Carr (Ed.), *Industrial minerals and rocks* (6th ed., pp. 104–105). Society for Mining, Metallurgy, and Exploration, Inc.

Zimbres, E. (2006). https://commons.wikimedia.org/wiki/File:Chrysotile_1.jpg, CC BY-SA 2.5.

13 Other Natural Fibers

In addition to the fibers described so far, there are a large number of other plants and animals that can be used as fiber suppliers. They are all of only regional importance, because fiber extraction is usually laborious and the fiber properties are generally only mediocre. Therefore, they were and are mostly used only when nothing else is available. In some publications, such fibers are described as an ecological and sustainable alternative to today's dominant fiber materials. This is usually nonsensical because neither the necessary quantities of fiber material are available in sufficient quality nor are their textile properties competitive. Some are luxury products and should hence be regarded as such. Typical examples are described in more detail below.

13.1 Cellulosic Fibers

Many plants have fibers to strengthen the stem or leaves. Only a few form seed fibers.

13.1.1 Akon

Akon, which is also called asclepias wool, is one of them. The fibers are 10–55 mm long, 35–40 μm in diameter, hollow and slightly yellowish in color. Because of their silky sheen, they are also known as plant silk (silk cotton). They were already obtained at the beginning of the twentieth century from the seeds of various Asclepiadaceae species, including Calotropis gigantea, but were of no commercial importance because they cannot be spun. Due to their high air content of 70–80%, they can be used as filling material for pillows and life jackets, similar to kapok.

13.1.2 Floss Silk Tree

The floss silk tree (Ceiba speciosa) belongs to the woolly tree family (Bombacoideae). It is endemic to South America and grows up to 15 m high. Its fruits are up to 20 cm long and 5 cm thick. When ripe, they burst open to reveal the white fibers inside. As with the kapok tree, the fibers are not attached to the seed, so they do not have to be laboriously separated. These fibers give the tree its name, as "floss silk" is the name given to base silk in which the outer fibers of the cocoon are also processed (Fig. 13.1).

These fibers could be used as a filling material similar to kapok and poplar fibers.

13.1.3 Willow

Fibers on the seeds of the salt willow, called willow catkins, are 4–6 mm long and have a diameter of 4–15 μm (Markova, 2014). They are hollow fibers and their structure resembles poplar and kapok fibers. Textile use is not possible, the fibers are too short for this.

D. Veit, *Fibers*, https://doi.org/10.1007/978-3-031-15309-9_13

Fig. 13.1 Fruits of the floss silk tree with fibers (Philmarin, 2007)

13.1.4 Woolgrass

The cotton grasses (Eriophorum) belong to the salt grass family and therefore thrive particularly well in boggy areas. The cotton grass only appears after flowering, when it bears fruit (Fig. 13.2).

13.1.5 Typha

The cattail, also called Typha, belongs to the order of sweet grasses. It is an aquatic and marsh plant and therefore grows mainly in wetlands in Central Europe (Fig. 13.2).

This plant produces both seed hairs and bast fibers that support the stem. The seed hairs are brittle and short and have been processed into cotton wool. The yield of bast fibers is exceptionally high at about 33% of dry weight. Therefore, "In 1916, the Study Commission for Typha Research was founded in Berlin, and in early 1917, the German Typha Exploitation Society was founded in Charlottenburg. [...] The Sächsische Kunstweberei Claviez, A.-G. in Adorf i. V., processes Typha, which has been made particularly soft by a freezing process, as a wool stretching agent in the manufacture of ready-made fabrics and carpets" (Lueger, 1920).

According to Lueger (1920), the technical fiber reaches a length of up to 600 mm, the average fiber fineness is approx. 200 μm. Analogous to the other bast fibers, the elementary fibers are considerably shorter (0.3–1.8 mm) and finer (approx. 10 μm). The moisture absorption is 9%, the strength of 9 cN/tex is very low for a bast fiber. Therefore, textile use is hardly possible.

13.1.6 Urena-Ur, Aramina, Congo Jute

This plant is grown in the Congo, Brazil and Madagascar. It reaches heights of 0.5–2 m. Its fibers are white and shiny, with an elementary fiber length of 1.5–5 mm and a diameter of 9–12 μm. The fiber content of the stalk is approx. 5–7%. Thus, yields of 500–800 kg/ha can be achieved (Wagner, 1981). Preparation is similar to that for jute, and the areas of application are the same as for kenaf, mainly sacks and twine.

13.1.7 Rosella, Java Jute

This plant belongs to the mallow family. Its fibers are similar to those of kenaf, but less pliable. It is used as a source of fiber, vegetables, medicinal and oil. The length of the elementary fibers is 2–5 mm. The main cultivation countries are India, Java, Australia, Philippines, Malaysia, Sri Lanka and Vietnam. The main textile application is packaging.

13.1.8 Sun (Crotolaria juncea)

Also called Indian hemp and Madras hemp, it is native to India and West Africa (Fig. 13.3). It reaches heights of 0.3–3 m. Its fibers have a density of 1.07 g/cm^3, a strength of 30–40 cN/

Fig. 13.2 Bulrush (Blaine, 2011) and cotton grass fibers (Rice Salad, 2006)

tex and a modulus of elasticity of about 35 GPa. Yields reach 700 kg/ha.

13.1.9 Lotus Silk

These fibers are obtained from the stems of the Indian lotus flower (Nelumbium speciosum) (Fig. 13.4). They measure about 5 μm in diameter and have a large number of small holes inside. So far they are only produced in small quantities in Myanmar. The linen-like fabric produced from the hand-spun yarns has very little tendency to wrinkle, unlike many other bast fibers. Samatoa Lotus Textiles in Cambodia markets corresponding products.

13.1.10 Other Bast Fibers

Other fibers include turka or kendyr fiber, broom fiber, okra fiber, Chinese hemp (kanatnik), lupine fiber, willow fiber, hop fiber, lime bast fiber, straw and peat fiber, as well as asclepias fiber (for cellulose extraction) and Spanish broom (Spartium junceum Linné) for reinforcing composites. All of the above fibers are of only local and generally minor importance because it

Fig. 13.3 Java jute (Valke, 2011) and Madras hemp (Wiki, 2011)

Fig. 13.4 Extraction of lotus silk fibers (Wiki, 2012) and yarn production by twisting the fibers (Wiki, 2013)

has not yet been possible to extract them in economically viable quantities.

13.2 Animal Fibers

In principle, almost all animal hair can be used for textile purposes if they have the necessary length and strength. However, they are mostly brown, grey or black and can therefore only be poorly dyed in bright colors. Many animal hairs are stiff and therefore difficult to spin into textiles or offer poor wearing comfort. Because many animals are rather averse to domestication, they were often killed in the past to obtain the fibers.

13.2.1 Siberian Ibex (Yangir)

The Siberian ibex (Capra Sibirica) lives in southern Russia, northern India and western China in mountainous regions up to 6700 m a.s.l. (Fig. 13.6). The fiber diameter is 13–15 μm, the scales are very flat and enclose the entire hair. Because the animal refuses to be sheared, it used to be shot to get access to the fibers. This is now banned and small quantities of yangir are obtained in an animal-friendly way by combing. Yangir fibers are spun either by hand or by machine. The resulting textiles are extremely soft and very comfortable to wear (Fig. 13.5).

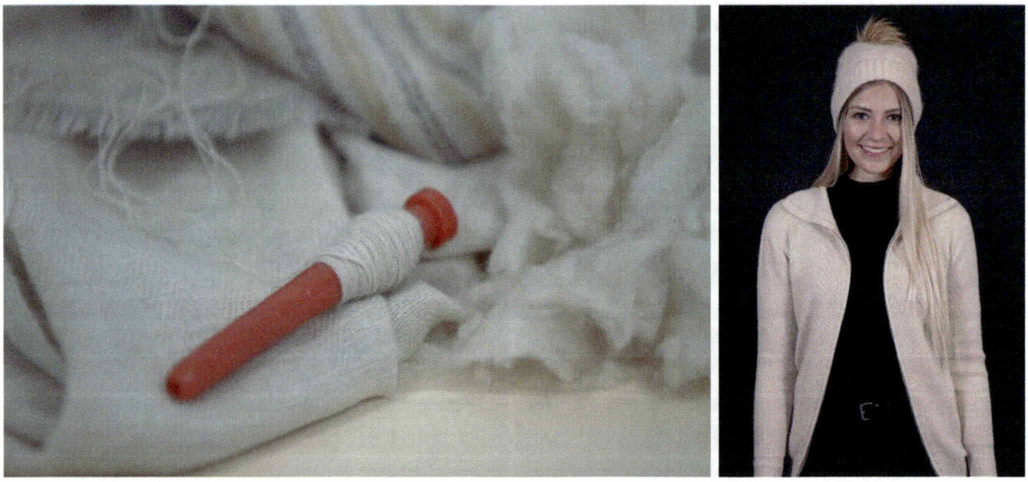

Fig. 13.5 Handspun yarn from yangir and jacket and cap from yangir (Notter, 2021)

Fig. 13.6 Siberian ibex (Ries, 2007) and Tibetan antelope (Acabashi, 2016)

13.2.2 Tibetan Antelope (Chiru)

Shatush (shahtoosh) is obtained from the hide of the Tibetan antelope (Pantholops hodgsonii) (Fig. 13.6). The diameter of the fibers is 10–13 μm, making them among the finest animal fibrous materials, which were made into shawls until the 1970s. Each antelope produces about 150 g of fiber. Because the animals cannot be shorn, they were killed, whereby 3–5 antelopes had to give their lives for one scarf. The possession of such scarves is therefore now forbidden and will be prosecuted.

13.2.3 Mangalica

The Mangalica pig (woolly pig), which originates from Hungary, has a hairy woolly undercoat alongside the coarse bristles typical of wild boar. Typical diameters of the fine underhairs are about 50 μm. The fiber strength reaches up to 25 cN/tex with an elongation of 38–44%. The fiber length is 60–80 mm (Figs. 13.7 and 13.8).

The woolly pig has so far only been used for gastronomic purposes and not for fiber production.

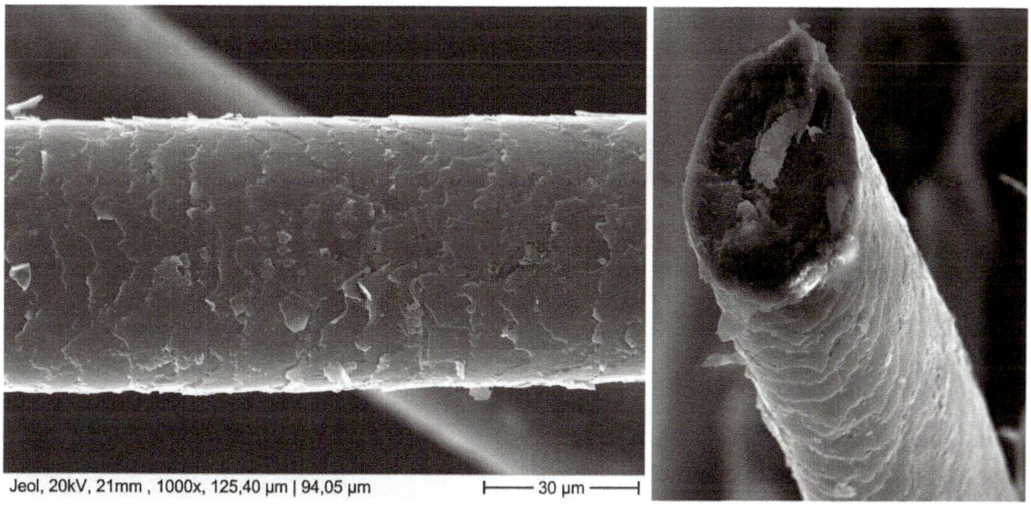

Fig. 13.7 Cross-sectional and transversal view of the fine undercoat of woolly pigs

Fig. 13.8 Mangalica pig (Nienetwiler, 2010)

13.2.4 Whale Whiskers

Until the beginning of the twentieth century, the baleen hairs of whales (e.g. blue, fin and humpback whales) were used, among other things, to make corsets for women. They (the corsets) consist of horny cells, i.e. cells filled with keratin. Because hunting whales for baleen hair has been banned since the 1980s, baleen hair is nowadays replaced by plastic or steel rods in corset construction (Fig. 13.9).

References

Acabashi. (2016). *CC BY-SA 4.0*. https://commons.wikimedia.org/wiki/File:Tibetan_antelope_or_Chiru_(Pantholops_hodgsonii)_displayed_at_the_Powell-Cotton_Museum,_Birchington_Kent_England_(cropped).jpg

Blaine, E. (2011). *CC BY 2.0*. https://commons.wikimedia.org/wiki/File:Bullrushes_(6002786093).jpg

Lueger, O. (1920). *Lexikon der gesamten Technik und ihrer Hilfswissenschaften* (Vol. 1, 2nd ed., p. 647). Typhafaser.

Markova, I. (2014). *Textile fiber microscopy*. Wiley.

Monniaux, D. (2007). *CC BY-SA 3.0*. https://commons.wikimedia.org/wiki/File:Baleen_P1180087.jpg?uselang=de

Nienetwiler. (2010). *CC BY 2.5 CH*. https://commons.wikimedia.org/wiki/File:101028_Wollschwein-Eber_Alphons_002.jpg

Notter, W. (2021). *Qiviut international – Creation*. Walter Notter. http://www.qiviut.ch

Philmarin. (2007). *CC BY-SA 3.0*. https://commons.wikimedia.org/wiki/File:C._speciosa-1.JPG

Rice Salad. (2006). *CC BY-SA 2.0 EN*. https://commons.wikimedia.org/wiki/File:Wollgras_I.jpg

Ries, G. (2007). *CC BY-SA 2.5*. https://commons.wikimedia.org/wiki/File:Steinbock1000943.jpg

Valke, D. (2011). *CC BY-SA 2.0*. https://commons.wikimedia.org/wiki/File:Pumdibija_(Kannada-_%E0%B2%AA%E0%B3%81%E0%B2%82%E0%B2%

Fig. 13.9 Baleen hair (Monniaux, 2007) and corset advertisement (1894)

A1%E0%B2%BF%E0%B2%AC%E0%B3%80%E0%B2%9C)_(6310158090).jpg?uselang=de
Wagner, E. (1981). *Die textilen Rohstoffe*. Deutscher Fachverlag.
Wiki. (2011). *CC BY-SA 3.0.* https://commons.wikimedia.org/wiki/File:Crotalaria_juncea_D9991.jpg
Wiki. (2012). *CC BY 2.0.* https://commons.wikimedia.org/wiki/File:Lotus_silk_yarn.jpg
Wiki. (2013). *CC BY 2.0.* https://commons.wikimedia.org/wiki/File:Silk_yarn_from_lotus_stalk_cropped.jpg

Part III

Chemical Fibers: Processes and Machines

14 Man-Made Fibers: Polymer Formation Processes

In this chapter, we look at the processes and aggregates used to produce polymers for synthetic fibers, as well as the actual polymer formation. Most synthetic polymers are produced from petroleum or natural gas. Polylactide fibers are obtained from lactic acid, cellulosic fibers from wood and from waste from other processes (e.g. straw).

14.1 Extraction of Raw Materials from Oil, Coal and Natural Gas

Crude oil consists of more than 500 different components. These include the hydrocarbons, e.g. alkanes, cycloalkanes and aromatics, the napthenic acids (Greek naphta: petroleum), e.g. cyclohexanoic acids, the phenols and many others. Figure 14.1 shows the process chain from crude oil to some important fiber polymers in a simplified way.

The separation or production of these substances takes place in large distillation towers, as exemplarily shown in Fig. 14.2. First, the crude oil is heated to 360 °C in a tubular furnace, whereby almost all components evaporate. The mixture is then fed into the distillation tower. This consists of several bubble trays to separate the individual components from the blend. The different boiling temperatures of the crude oil components are exploited and the high-boiling components are separated from the compound in the first stage and then the lower-boiling components in further stages using a targeted temperature control. The components rising in vapor form are brought into contact in countercurrent with their condensate collected in the respective bubble trays. This increases the proportion of each component that is separated from the mixture. This process step is called rectification.

At the initially very high temperatures around 350 °C, many hydrocarbons would decompose under atmospheric pressure. Therefore, the pressure is reduced and thus also the boiling temperature of the hydrocarbons (vacuum distillation).

In further process steps, the precursors of the monomers are produced from the components of the crude oil obtained in this way (cf. Fig. 14.1). This process requires approx. 4 t of crude oil per ton of polyester (PET).

Similar processes are used to extract the corresponding raw materials from coal and natural gas.

14.2 Raw Material Extraction from Cellulose

Cellulose occurs in all plants, but not in pure form. It is accompanied by other substances that interfere with further processing and must therefore be removed (Fig. 14.3).

Wood and, to a lesser extent, cotton linters and other plants are of major industrial importance for the production of cellulose, which is also called

D. Veit, *Fibers*, https://doi.org/10.1007/978-3-031-15309-9_14

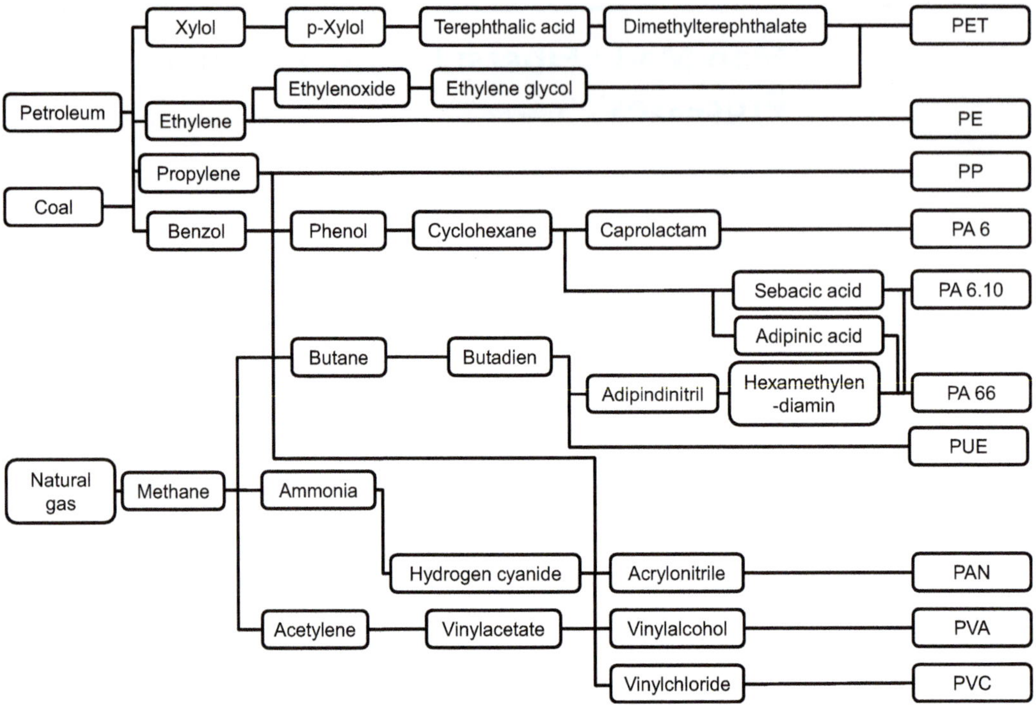

Fig. 14.1 Production paths of important basic materials for man-made fiber production

pulp. In wood, the soft lignin is reinforced by firm and stiff cellulose fibers, while the soft, non-crystalline hemicellulose (also composed of glucose groups) acts as an "adhesion promoter" between the lignin and the cellulose. The cellulose in plants has a fibrous shape and is 1–5 mm long.

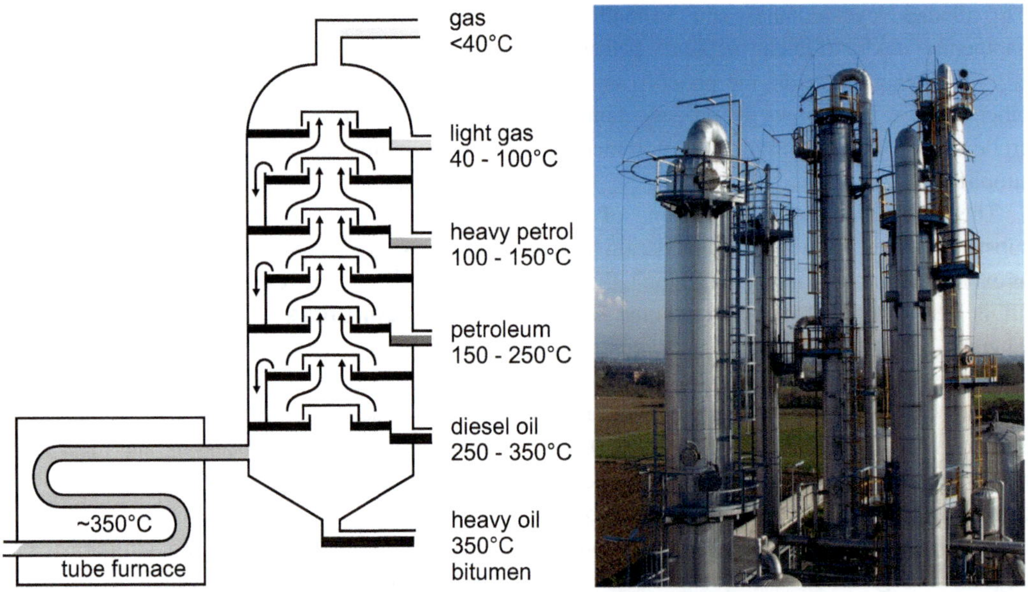

Fig. 14.2 Distillation tower acc. to Chiesa (2005)

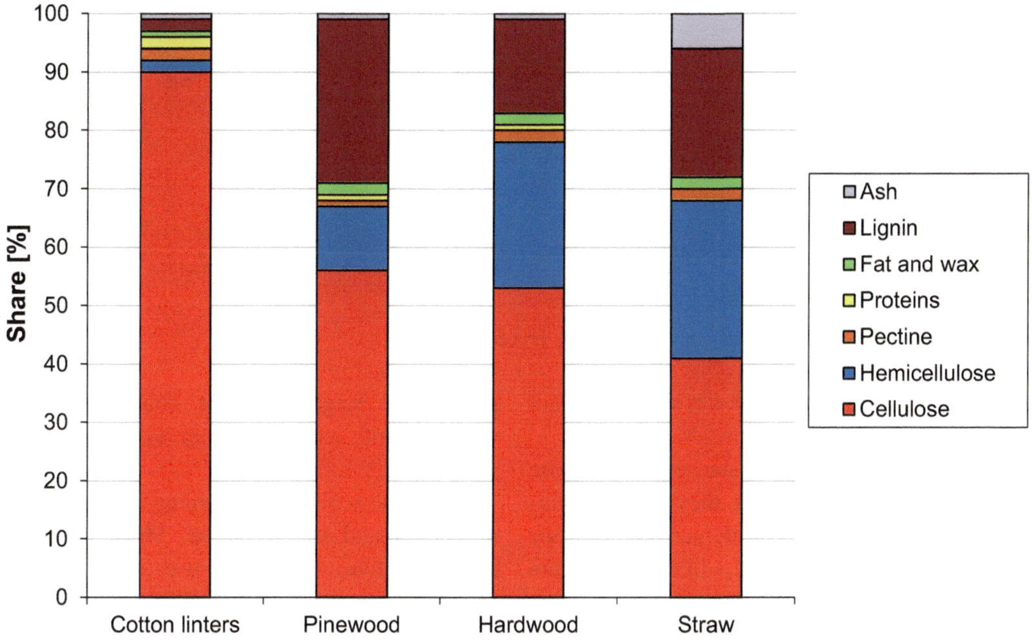

Fig. 14.3 Percentage composition of some plants acc. to Rogowin (1982)

The processes for obtaining pulp from wood are shown in Figs. 14.4 and 14.7. In this process, the logs are first sawn into smaller pieces, debarked, and then wood chips are produced. These can either be stored or fed directly into pulp production. Depending on the type, approx. 2 t of wood are required for 1 t of pulp.

Wood consists essentially of cellulose, hemicelluloses and lignin (Fig. 14.3). The aim of wood digestion is to break the bonds between the cellulose and lignin molecules and then remove the lignin. Lignin reacts readily with various chemicals and can therefore be oxidized, nitrated and chlorinated. The corresponding substances react with sulfurous acid or its salts and with alkalis and are therefore soluble in dilute alkali and even in water. Thus, they can be easily removed.

The wood chips are first boiled for several hours. The sulfate process uses caustic soda with sodium sulfides and sulfates. Lignin is first depolarized by hydroxy ions and then broken down into soluble components by the addition of sulfide ions. High cellulose yields are possible, but unpleasant odors develop (Fig. 14.5).

An alternative is the sulfite process. In this process, the lignin is dissolved with a calcium hydrogen sulfite solution at a temperature of

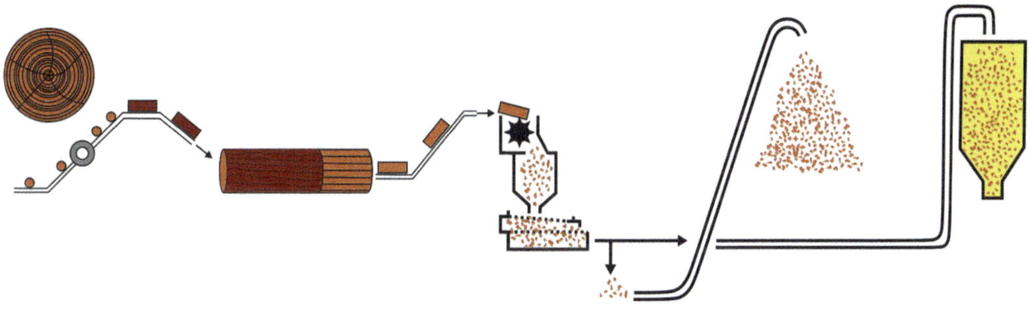

Fig. 14.4 Principle of the production of chips from wood acc. to Lenzing AG (n.d.)

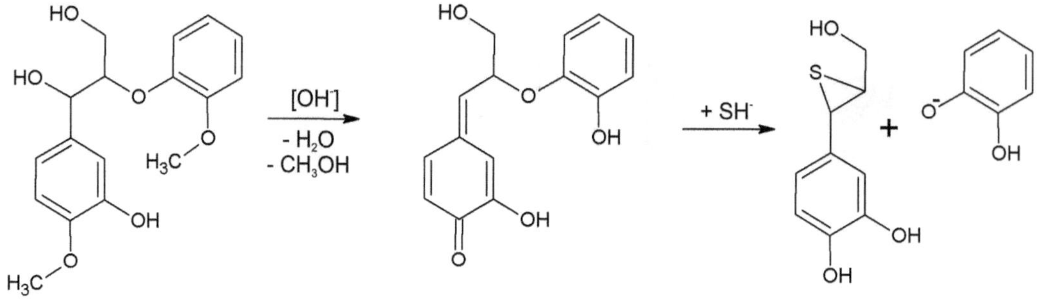

Fig. 14.5 Degradation reaction of lignin in the sulfate process

130–170 °C. The sulfite process is simple, but not suitable for resinous wood species.

Other accompanying substances of cellulose are similar to cellulose, but they have significantly lower degrees of polymerization. They are easily hydrolyzed and, like pectin, are degraded to sugars (Fig. 14.6).

After the cooking treatment, the pulp is extracted and separated from the water. This is followed by a bleaching and dewatering process. Finally, it can either be dried to produce pulp sheets or the pulp is crushed after dewatering and processed into flakes. Figure 14.7 shows the individual process steps from the wood chips to the finished pulp (Fig. 14.8).

Worldwide, about 600 m tons of cellulose are extracted from wood and other raw materials every year. It is mainly used for the production of paper and cardboard. Approximately 7 m tons are used to produce fibers.

The quality of the pulp is of great importance for the production and properties of man-made cellulosic fibers. For the production of cellulose fibers, it is further refined. In this process, accompanying substances and low-molecular hemicellulose (polysaccharides) are removed and the proportion of pure cellulose (α-cellulose) is increased. Depending on the number of purification stages and the conditions, pulps with a purity content of up to 99% can be produced.

After bleaching or refining, the pulp is dewatered and usually dried. Moist pulps are more reactive, but their use requires short distances between the pulp mill and fiber production. The pulp is usually supplied as sheets or flakes.

The pulp quality and the possible applications in man-made fiber production are determined by its chemical composition (α-cellulose and hemicellulose content as well as accompanying substances), its physico-chemical properties (degree of polymerization, absorbency, swellability) and its technological characteristics (degree of contamination, moisture, degree of bleaching). Because wood treatment processes pollute the environment, alternative routes have been sought for many years. An important result of these efforts is the development of the Lyocell process, which will be discussed in detail later.

Figure 14.9 shows the structure of a "wood fiber" and the orientation of the microfibrils in the individual layers.

Straw, grass, sugar cane, reeds and other cellulose carriers have so far been of no major significance for cellulose production. However,

Fig. 14.6 Structural formula of a pectin (poly-D-galacturonic acid)

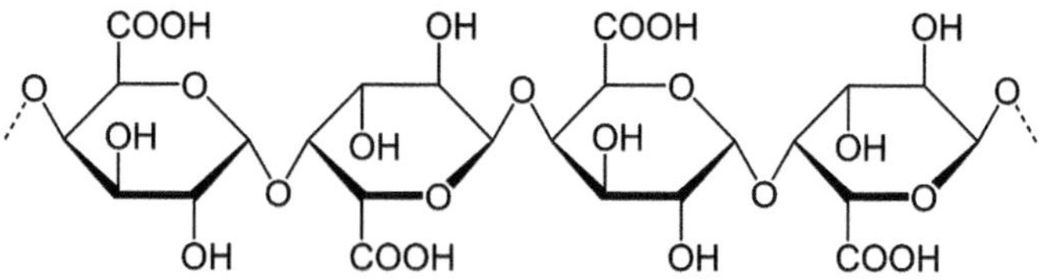

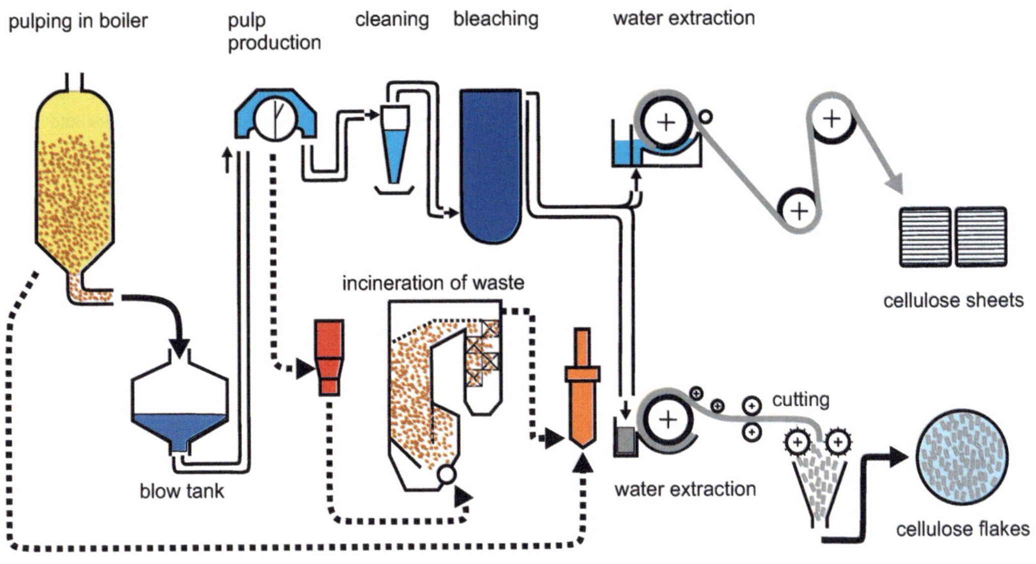

Fig. 14.7 Principle of pulp production from wood chips acc. to Lenzing AG (n.d.)

efforts are also being made to utilize these. Bamboo, for example, has been increasingly used in recent years and marketed as an environmentally friendly viscose fiber.

14.3 Raw Material Extraction from Protein

Protein-based raw materials can be obtained from casein, for example. In this process, the casein is precipitated from skimmed milk with the aid of acids or acid salts below the isoelectric point (pH 2.5–4.7). This yields casein in a concentration of approx. 25%. Alternatively, casein can also be precipitated by rennet ferments. This may result in hydrolytic degradation of the macromolecules; on the other hand, longer molecules should be present in the rennet casein than in the acid casein.

In both processes, the casein is washed with water, pressed and dried. To produce the spinning

Fig. 14.8 Pulp boards (Lenzing AG, n.d.)

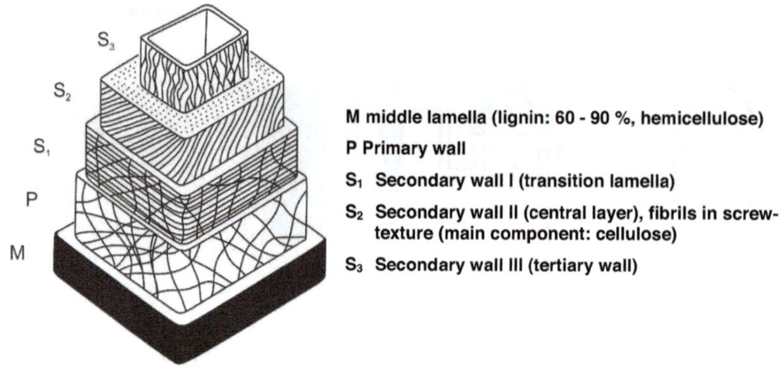

Fig. 14.9 Schematic structure of the cell wall of a wood fiber acc. to Götze (1967)

mass, the water-insoluble casein is converted in aqueous sodium hydroxide solution into a water-soluble compound with a casein content of 18–20%. Similar to the extraction of spinning solutions from cellulose, a maturing process takes place after filtration. The temperature control and the duration of the maturing process have a decisive influence on the resulting fiber properties. The viscosity of the spinning solution is 0.5–0.6 Pas.

Proteins can also be extracted from peanuts (ardein) and corn flour (zein) for fiber production. However, this has not yet obtained market relevance.

14.4 Polymer Formation

The formation of the macromolecules of which all fibrous materials are composed can take place acc. to three different principles:

- Polymerization,
- polycondensation
- polyaddition.

In this process, a varying number of individual building blocks (monomers) are combined to form long chains, the polymers, sometimes with the elimination of small molecules (e.g. water, hydrochloric acid). Depending on the polymer, this takes place either in a chain-growth polymerization or in a step-growth polymerization.

14.4.1 Polymerization

In polymerization, similar, unsaturated monomers always combine to form long chains in a chain growth reaction. It is initiated by the splitting of a double bond of a monomer with the help of radicals, anions or cations (Fig. 14.10).

Typical examples are the polyolefins polyethylene (PE), polypropylene (PP) and polyacrylonitrile (PAN).

14.4.1.1 Basics

After breaking a double bond, covalent bonds subsequently form between the monomers and a polymer is formed (Fig. 14.11).

No by-products are formed during polymerization, so reaction control is comparatively simple.

14.4.1.2 Radical Polymerization

To start the chain reaction, a radical R* is first generated. This has a free electron and can

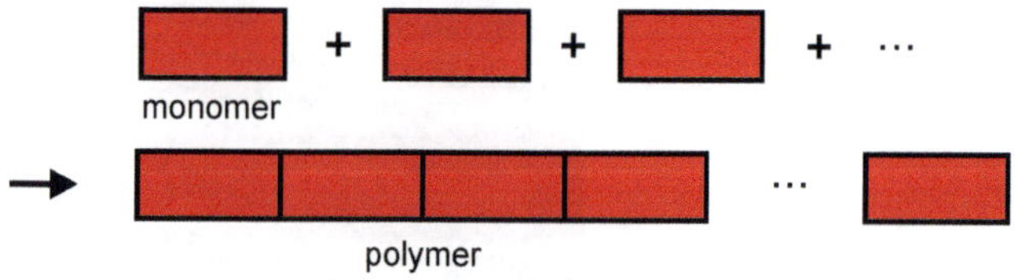

Fig. 14.10 Principle of polymerization

Fig. 14.11 Formation of polypropylene from prop(yl)-enes by polymerization

therefore form a covalent bond with a monomer that has a double bond. This in turn creates a radical, with the free electron now sitting on the monomer:

$$R^* + M \rightarrow R - M^* \qquad \text{(start reaction)}$$

The chain reaction is thus started and continues until the radical chain meets another radical (initiator radical or another radical chain) and both radicals are eliminated from the process by forming a covalent bond:

$$R_1\text{–}M_1 - M_2 - M_3 - \ldots - M_n{}^* + R_2{}^* \rightarrow R - M - M_1 - M_2 - M_3\text{–}M - \ldots - M_n - R_2 \qquad \text{(termination reaction)}$$

In addition to other factors (e.g. speed of the radical-forming initiation reaction), the average degree of polymerization depends very strongly on the viscosity of the polymer solution or melt. If the viscosity increases during chain formation (gel effect), the mobility of the individual macromolecule decreases and with it the probability that two radical chains meet. As a result, the average degree of polymerization can become very large. If this is to be avoided, diluted solutions are used or radical scavengers are added. Figure 14.12 shows examples of fiber-forming polymers produced in this way.

A chemical reaction can only take place if:

$$\Delta G = \Delta H - T \cdot \Delta S < 0,$$

where G is the free energy of the reaction, H the enthalpy of reaction and S the entropy at temperature T. Usually H < < 0 and S is also less than 0 because of the formation of ordered chains from the randomly distributed monomers. Therefore, a temperature must exist at which:

$$\Delta \mathrm{H} = \mathrm{T} \cdot \Delta \mathrm{S}$$

Then polymerization and depolymerization (decomposition of the macromolecules) are in equilibrium and the reaction comes to a standstill. This temperature T is called the ceiling temperature. Therefore, cooling is particularly important in this exothermic process in order to achieve high chain lengths.

For the synthesis of polyethylene, values of H = −92 kJ/mol, S = −143 J/(mol/K) and thus a ceiling temperature of approx. 372 °C apply.

During polymerization, only a small fraction of the monomers are involved in the reaction at any one time. Typically, long chains with high molar masses are formed (Fig. 14.13).

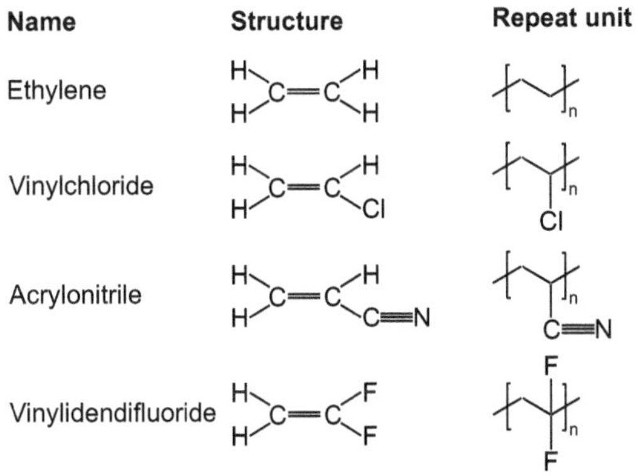

Fig. 14.12 Fiber-forming polymers produced by free-radical polymerization

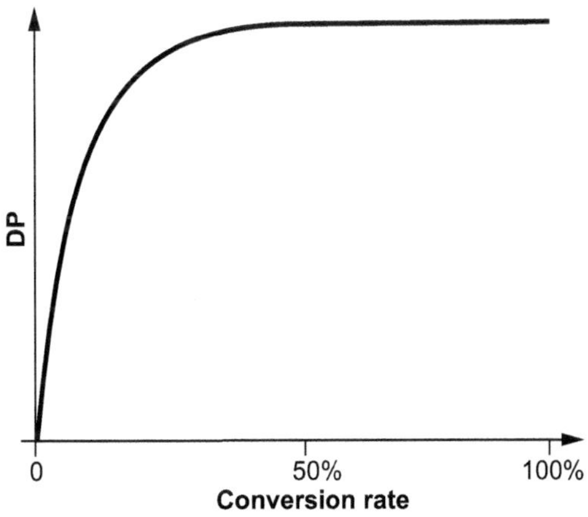

Fig. 14.13 Average degree of polymerization as a function of conversion in a chain growth reaction

14.4.1.3 Ionic Polymerization

In addition to radical polymerization, there are also anionic and cationic polymerization as special processes. Here, anions or cations are used as initiators. Cationic polymerization has only little technical significance.

14.4.1.4 Process Engineering

A polymerization reaction can be carried out by various methods. These include:

- Emulsion polymerization,
- suspension or dispersion polymerization,
- precipitation polymerization,
- solution polymerization,
- ionic polymerization and
- substance or bulk polymerization.

Emulsion Polymerization

The water-insoluble monomers are emulsified in water and polymerized using water-soluble initiators (e.g. potassium persulfate, redox systems). The advantages are the low process temperature and the low viscosity even at high molar masses of the polymers. Disadvantages are the often complex preparation and a relatively low degree of purity of the product.

Suspension or Dispersion Polymerization

These two possibilities of polymerization are not always clearly distinguished. In many cases, only suspension polymerization is mentioned, in which, in contrast to dispersion polymerization, the initiators and dispersants are insoluble in water. A redox system is usually used as the initiator. Polymerization in aqueous solution results in good heat dissipation of the often exothermic reaction and a high degree of purity of the polymer produced. The presence of oxygen often inhibits or prevents polymerization. Therefore, polymerization is usually carried out in an inert gas atmosphere.

Precipitation Polymerization

In this process, the resulting polymer is not soluble in the solvent of the monomer and therefore precipitates. This facilitates the extraction of the polymer.

Solution Polymerization

In this process, both the monomer and the polymer are soluble in the solvent. The advantages lie in the simple polymer production and filament formation. The viscosity increases only slightly during polymer formation. The polymer concentration is usually 20–30%. A disadvantage is the relatively low conversion rate of the monomer of 50–70%. Unreacted monomer is recovered from the polymer solution for re-polymerization by the thin film process. However, residues of the monomer may remain in the polymer solution, which can lead to problems in the spinning mill.

Ionic Polymerization

Ionic polymerization is not used industrially, among other reasons because of the high requirements regarding the purity of the solvent and the high price of the catalysts.

Substance or Mass Polymerization

If the polymerization of monomers takes place in the absence of solvents or diluents, it is referred to as substance or bulk polymerization. The reaction mixture contains only the pure monomers and the initiators or catalysts required to start the polymerization. Therefore, the polymers produced are particularly pure. If the resulting polymers are insoluble in the monomers and the substance polymerization is carried out below the melting point of the polymers, these will precipitate during the polymerization. An advantage is that an additional separation step for the non-reacting monomer is not necessary, nor is waste water produced as in other polymerization processes. A disadvantage is the exothermic reaction, which requires cooling of the spinning apparatus. Therefore, this process is not used commercially.

14.4.2 Polycondensation

Polycondensation belongs to the step growth reactions. Depending on the stability of the monomers, it can be carried out in the melt, in solution and at phase interfaces as well as in crystals and in the gas phase.

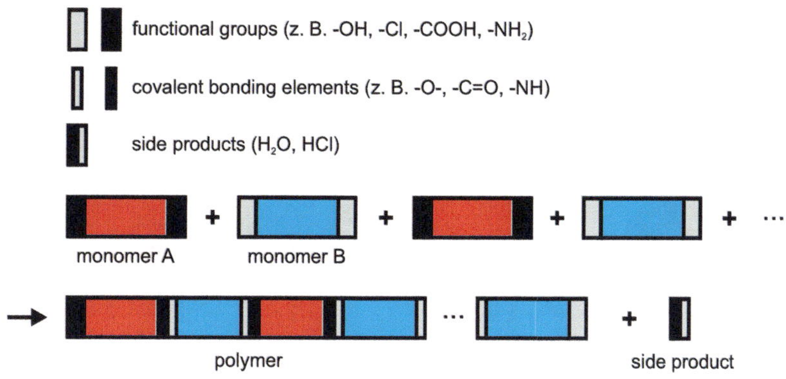

Fig. 14.14 Polycondensation of two different monomers with different end groups

14.4.2.1 Basics

Chain formation in a step growth reaction occurs via reactive end groups of the monomers reacting with each other to form covalent bonds with splitting off of a by-product.

In Fig. 14.14, there are two different types of monomers which each have identical end groups at both ends. These react with each other and in the process form a by-product.

A typical example is the formation of polyester from terephthalic acid and ethanediol ("PET"). Polyamide 66 from hexamethylenediamine and adipic acid is also produced in this way (Fig. 14.15).

Illustrated in Fig. 14.16, there is only one type of monomer, which has different end groups at each end. In this case, the monomers are often present in the form of a ring and the ring must first be opened. The ring opener consumed in the process, e.g. water, is formed again during the polycondensation reaction and thus no water is released to the outside. This is therefore also referred to as polymeric polycondensation.

A typical example of such a ring-opening reaction is the formation of polyamide 6 ("nylon 6") from ε-caprolactam (Fig. 14.17):

Only a few polycondensation reactions occur spontaneously. Most of them take place only after the addition of a catalyst (e. g. titanium or tin alcoholates). During polycondensations, by-products are formed which would influence the further course of the reaction, e.g. H_2O or HCl. Therefore, they must be removed during the reaction to shift the reaction equilibrium to the product side. This is the only way to achieve a large chain length of the macromolecules to be produced. The molecular weights are usually between 10,000 and 50,000 g/mol.

14.4.2.2 Reaction Rate and Turnover

The relationship between the conversion, i.e. the proportion of monomers that have already reacted

Fig. 14.15 Formation of polyethylene terephthalate (PET)

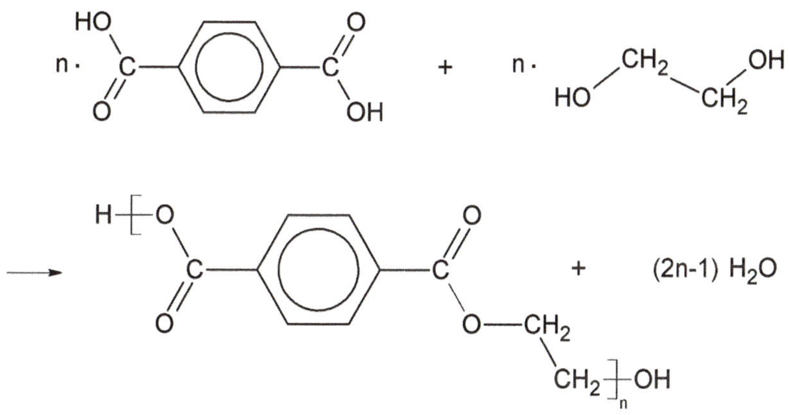

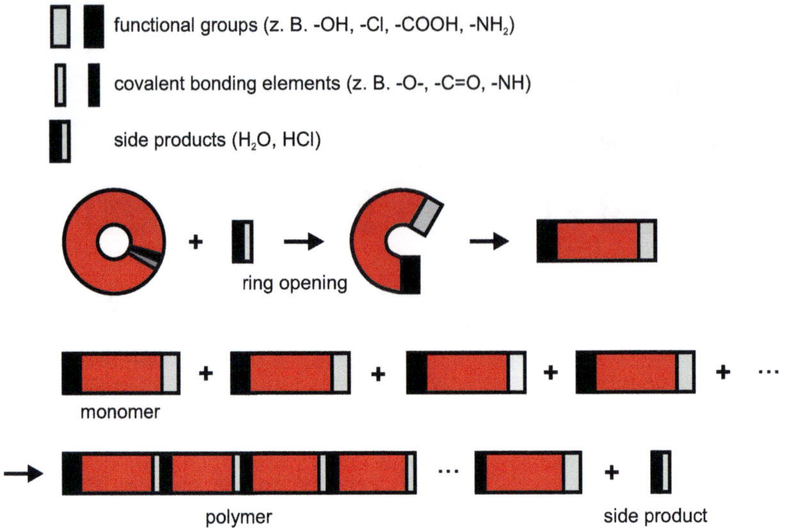

Fig. 14.16 Polycondensation of monomers with different end groups on the same monomer

with each other, and the average degree of polymerization DP is described by the Carothers equation for a stoichiometric molar ratio:

$$\frac{N_0}{N_t} = \frac{1}{1-p} = DP$$

Where N_0 is the number of monomers at the beginning of the reaction, N_t is the number of monomers that have not yet reacted at time t, and p is the conversion rate ($0 \leq p \leq 1$).

This relationship is illustrated in Fig. 14.19. It is obvious that in polycondensation, the average degree of polymerization (DP) is high enough to form fibers from the polymer only at a conversion of more than 99%. In polymerization, on the other hand, long macromolecules are formed even at low conversion rates.

In the ideal case, a single, very long macromolecule is thus formed during polycondensation and a conversion of 1. In practice, a larger number of very long molecules are normally formed, but this does not alter the need for a very high conversion to achieve high molecular weights. Figure 14.18 illustrates this with a simple example. Only when almost all monomers have been converted (right), 76% of the functional groups have reacted and formed bonds to neighboring molecules. But even here, despite the high conversion, mainly oligomers have formed and not yet polymers.

Fig. 14.17 Formation of polyamide 6 (perlon, nylon 6) from ε-caprolactam

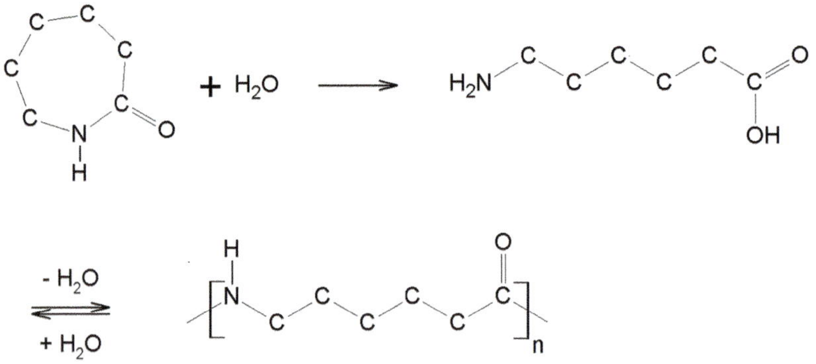

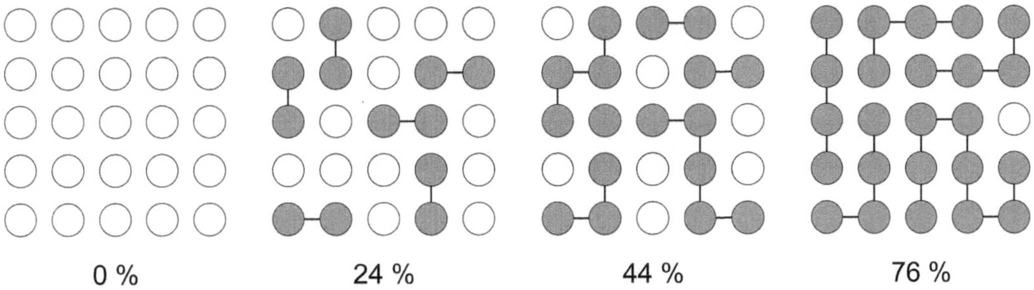

Fig. 14.18 Influence of the conversion on the chain length in the step growth reaction

In the case of a reaction of two different monomers (A-A and B-B), it is essential that their stoichiometric ratio r (molar ratio) is set exactly. If it is unequal, the Carothers equation changes to:

$$\frac{N_0}{N_t} = \frac{1+r}{1+r-2rp} = DP$$

Where r is the ratio of the number of molecules A-A to the number of molecules B-B, where $r < 1$ must always apply.

The importance of a stoichiometric ratio for a large chain length is shown in Fig. 14.19. Already at an imbalance of the two components by only 1%, a large chain length is no longer possible (on average, the max. DP is 198). At a deviation of 2%, the DP is only 98 and is thus too low for a chain length sufficient for fiber formation (e.g. polyester: min. 120).

14.4.2.3 Solid State Polycondensation (SSP)

To increase the molecular weight and hence the chain length, polymers that have already been formed can be post-condensed. The dried polymer chips are then heated in a controlled manner above the glass transition temperature under exclusion of water and air until the temperature is reached at which the crystallization rate becomes maximum (PET: 160–170 °C). There is a risk that the amorphous chips will stick together, which must be avoided at all costs. Usually either a fluidized bed reactor or a vacuum dryer is used. The polycondensation reaction is exothermic, so the container must be cooled to prevent the temperature from rising. After this process step, the previously amorphous polymer chains are crystallized and the post-polycondensation reaction begins.

Fig. 14.19 Relationship between molar ratio r, turnover p and DP

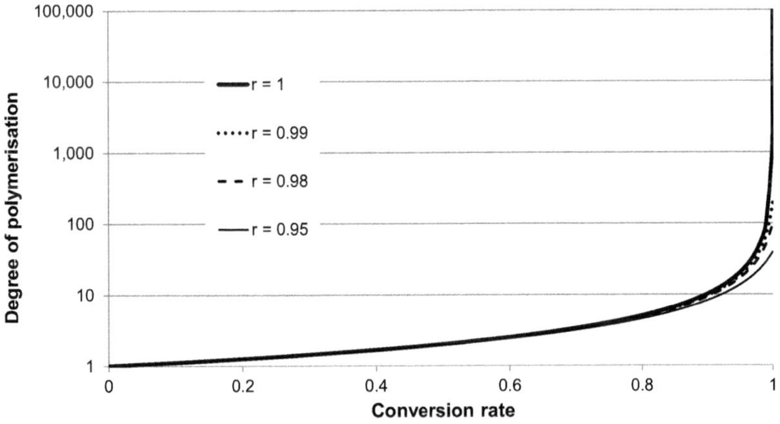

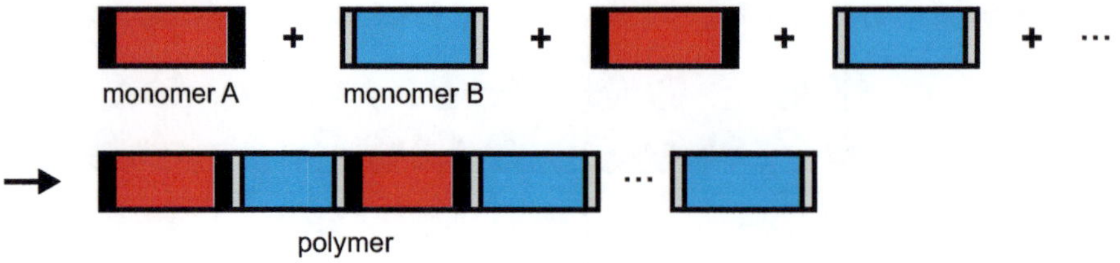

Fig. 14.20 Principle of polyaddition

In solid state polycondensation, the by-products formed only have to be removed from the reaction mixture over a short distance (1–2 mm) and they are immediately transported away by a gas stream when they reach the surface. In contrast, the by-products in the melt must travel much longer distances before they can be removed. Although solid state polycondensation is much slower than the polycondensation reaction in the melt (because of the lower mobility of the macromolecules), this disadvantage is somewhat reduced by the much faster removal of the by-products. The temperature of this reaction is below the softening temperature of the crystallized polymer chains, so there is little risk that the chips will now still "stick" together. In addition, the reverse reaction (degradation of the chains) is lower than with a melt due to the lower process temperature.

After completion of this process step, the polymer chains are in crystalline form, the average degree of polymerization has increased as well as the intrinsic viscosity (IV) of the polymer and the mechanical properties are better. SSP is therefore often used to produce fibers for tire cord and other applications requiring high strength and a high modulus of elasticity.

14.4.3 Polyaddition

In polyaddition, different types of monomers, each with at least two functional groups, are combined with each other in a step growth reaction. Hydrogen atoms are exchanged between the monomers and no by-products are formed. Polymers are formed via di-, tri- and oligomers. Only with almost complete conversion do sufficiently large chain lengths result for fiber formation, similar to polycondensation (Fig. 14.20).

Polyurethanes, which also include elastane, are produced acc. to this principle. The Fig. 14.21 shows this by the example of the reaction of 1,4-butanediol with hexane-1,6-diisocyanate.

14.4.4 Comparison of Chain Growth Reaction and Step Growth Reaction

The Table 14.1 summarizes the essential differences between the two reaction types.

14.4.5 Molecular Weight Distribution

Chain growth occurs randomly and leads to a distribution of chain lengths that can be calculated using statistical methods. For the case of polycondensation of monomers with two different functional groups (A-B), this is shown below for the example of polyamide 6 (cf. Fig. 14.22).

First, we are interested in the probability P (i) that exactly i monomers have reacted to form a chain after time t. This corresponds to an

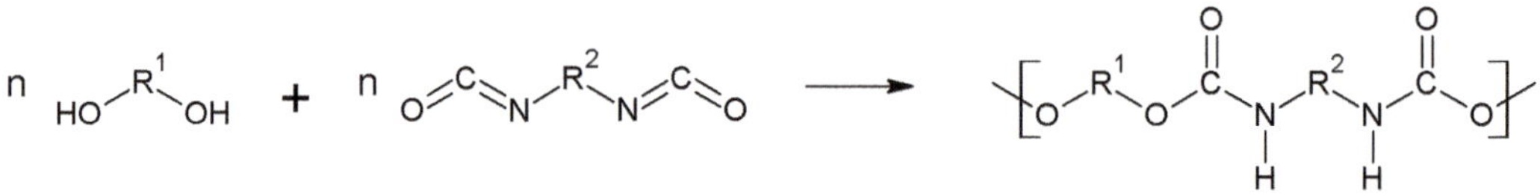

Fig. 14.21 Formation of polyurethane from butanediol and hexanediisocyanate

Table 14.1 Comparison of chain growth and step growth reaction

Chain growth reaction	Step growth reaction
The macromolecules that form grow very quickly.	The macromolecules that form grow slowly.
Even with low turnover, long chains are created.	Long chains are only formed when all monomers are almost completely converted.
The reaction starts only with an initiator.	All monomers can react directly.
The chain growth is terminated by a termination reaction.	The reaction is not complete until all monomers have been consumed.
Only a few monomers react with each other at a time t.	All monomers can react with each other at a time t.
There is usually only one type of monomer.	There is only one type of monomer with two different functional groups A and B per basic building block or two monomers with two functional groups A or B per basic building block.
	In the case of two different monomers, a stoichiometric ratio (1:1) is required.

average degree of polymerization DP of this chain of i. The chain thus consists of i monomer units, in the example therefore of (i−1) amide bonds.

The probability P(i) is then equal to the product of the probabilities p of the formation of the single amide group in this molecule. In this example, it is $p^{(i-1)}$. The resulting macromolecule then still has one unreacted NH_2 and COOH group at each end. The probability for such a group not to have reacted is (1−p). This gives the probability P(i) to:

$$P(i) = p^{(i-1)} \cdot (1-p)$$

The number of molecules N_i with i monomer units at time t can be calculated from the total number N_t of molecules at time t with:

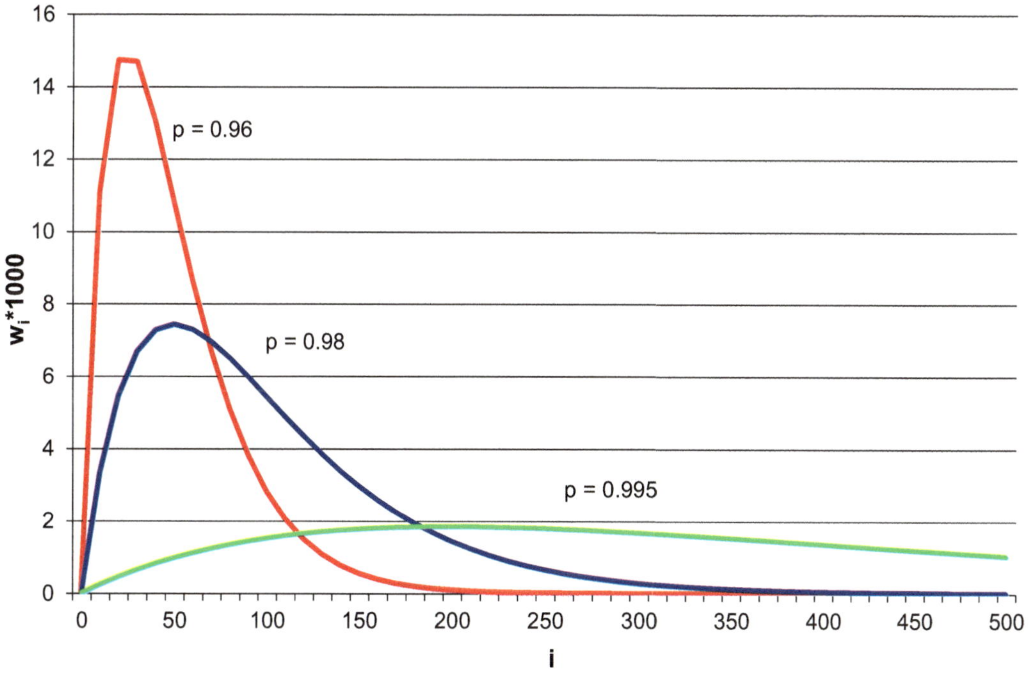

Fig. 14.22 Relationship between weight fraction w_i, chain length i and turnover p

$$N_i = P(i) \cdot N_t = N_t \cdot p^{(i-1)} \cdot (1-p)$$

From $N_t = N_0 \cdot (1 - p)$ then follows:

$$N_i = N_0 \cdot p^{(i-1)} \cdot (1-p)^2$$

The weight fraction w_i is defined as the quotient of the mass of the molecules of length i and the total mass of all molecules acc. to:

$$w_i = \frac{N_i \cdot (i \cdot M_0)}{N_0 \cdot M_0} = \frac{i \cdot N_i}{N_0}$$

where M_0 = molar weight of the monomers.

Thus it follows:

$$w_i = i \cdot p^{(i-1)} \cdot (1-p)^2$$

The Fig. 14.22 shows the distribution of weight fractions for different values of p as a function of i. The number of short molecules initially predominates ($p = 0.96$), but decreases with greater turnover in favor of the long-chain molecules ($p = 0.995$). This type of distribution is also referred to as the Schulz-Flory distribution.

The number average molecular weight is defined as

$$\overline{M}_n = \sum x_i \cdot M_i$$

With $x_i = \frac{N_i}{N}$ and $N = \sum N_i$ then applies

$$\overline{M}_n = \frac{\sum N_i M_i}{\sum N_i}$$

The weight fraction w_i defined above can also be calculated acc. to

$$w_i = \frac{N_i \cdot M_i}{\sum N_i \cdot M_i}$$

The weight average molecular weight is then given by

$$\overline{M}_w = \sum w_i \cdot M_i$$

which is equal to

$$\overline{M}_w = \frac{\sum N_i \cdot M_i^2}{\sum N_i \cdot M_i}$$

As a rule, the number average of the molecular weight is smaller than the weight average, as the example in Fig. 14.23 shows.

Example:

A polymer P consists of 100 molecules with the following distribution (Table 14.2):

This then results in

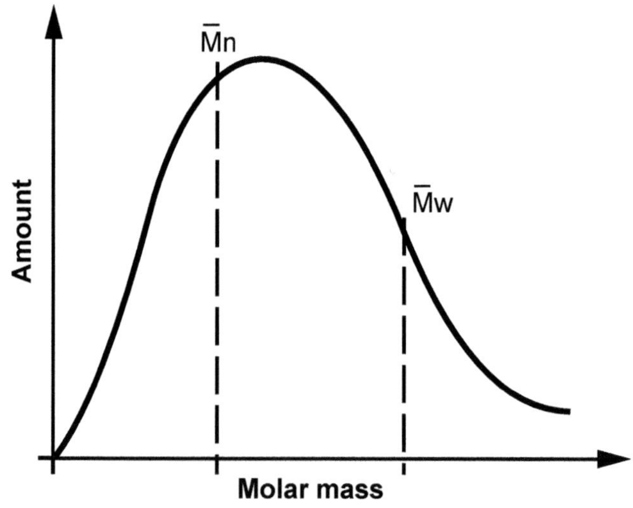

Fig. 14.23 Molar mass distribution of a polymer

Table 14.2 Chain length distribution of a polymer

Number of molecules	Chain length
5	100
30	120
40	140
20	150
5	160

$$M_n = \frac{(100\cdot 5)+(120\cdot 30)+(140\cdot 40)+(150\cdot 20)+(160\cdot 5)}{5+30+40+20+5} = 135g/mol$$

and

$$M_w = \frac{[5\cdot 100^2]+[30\cdot 120^2]+[40\cdot 140^2]+[20\cdot 150^2]+[5\cdot 160^2]]}{(5\cdot 100)+(30\cdot 120)+(40\cdot 140)+(20\cdot 150)+(5\cdot 160)} = 137g/mol$$

From these values, the average degree of polymerization DP can be determined using the molecular weight M_o of the repeating unit.

It applies to the number average:

$$DP_n = \frac{\overline{M}_n}{M_o}$$

and for the weight average, we get:

$$DP_w = \frac{\overline{M}_n}{M_o}$$

The most common and descriptive is the weight average. For most fiber polymers it is in the range of 80–150.

Typical molecular weights of macromolecules for fibers are thus usually between 20,000 and 50,000 g/mol (e.g. PET, PA 6, PA 66).

14.5 Formation of Polymers from Several Components: Copolymers

In addition to the homopolymers described above, which consist of only one or two types of monomer, there are also so-called copolymers. These consist of at least two components and, with the exception of alternating copolymers, have no regularly repeating monomer units. Strictly speaking, homopolymers of two monomers, such as polyamide 6.6 or polyester, are also copolymers. However, since the individual monomers cannot form chains, they are still considered homopolymers (Fig. 14.24).

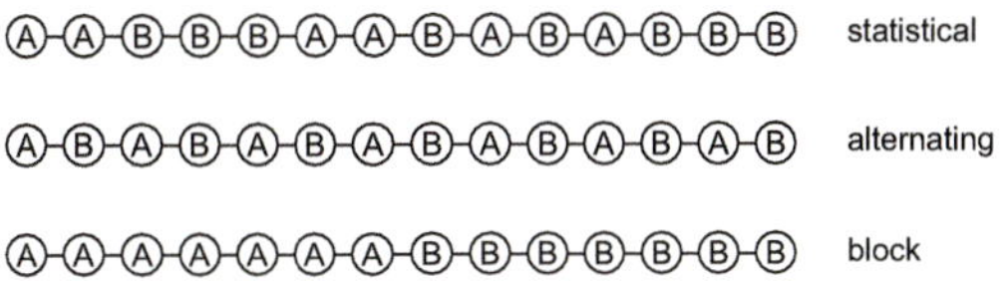

Fig. 14.24 Typical copolymer structures

When another type of monomer is linked to a homopolymer backbone, it is called a graft copolymer (Fig. 14.25).

If two monomers are combined, the melting temperature of the resulting copolymer depends on the mixing ratio. In some cases, it is thus possible to process copolymers consisting of infusible monomers in the melt spinning process if the melting temperature of the copolymer is below the decomposition temperature of the monomers (Fig. 14.26).

14.5.1 Grafting and Macromonomers

The production of such structures is usually done by grafting. In this process, an existing homopolymer of monomer A is exposed to gamma radiation, resulting in the formation of free radical places along the main chain. Radical chains consisting of monomer B can then dock onto these.

An alternative process uses so-called macromonomers. In this process, marginal groups (e.g. OH-groups) of a homopolymer consisting of monomer A react with functional groups (e.g. COOH-groups) of other chains consisting of monomer B (e.g. by esterification).

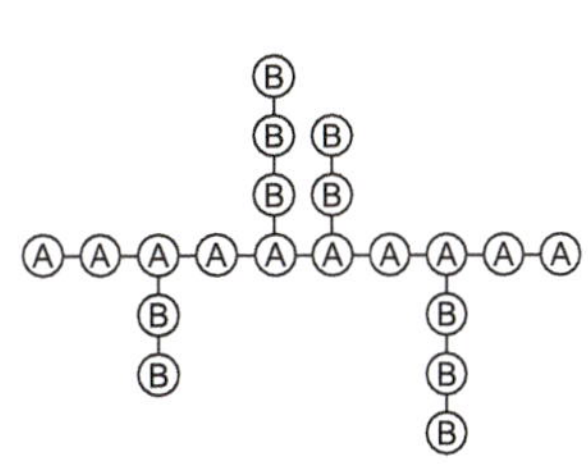

Fig. 14.25 Structure of a graft copolymer

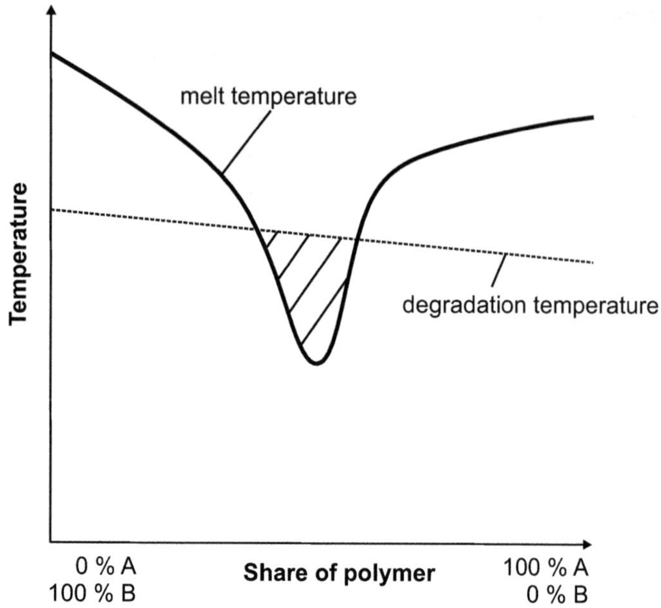

Fig. 14.26 Change in melting temperature during the formation of copolymers acc. to Sloan (2017)

14.5.2 Coextrusion

If two different components are already brought together during extrusion, this is referred to as co-extrusion. Bicomponent yarns are also produced acc. to this principle when two extruders are used.

14.5.3 Additivation

Homopolymers can be spun out with additives in order to change their properties or also the spinning process. This can be done both in the reactor and shortly before the actual filament formation, e.g. by a sidestream extruder (Fig. 14.27).

Typical examples are TiO_2 pigments for delustering, carbon black particles and color pigments for blackening, and the addition of carbon nanotubes (CNT) to change the mechanical properties and electrical conductivity. The hydrophilization of homopolymers by adding hydrophilic groups to the main chain also belongs to this category. This is done, for example, with 2-acrylamido-2-methylpropane sulfonic acid, which improves the dyeability of polyacrylic and polyester fibers.

14.6 Recycling

Waste from each process stage is often reused to a large extent in modern plants for economic

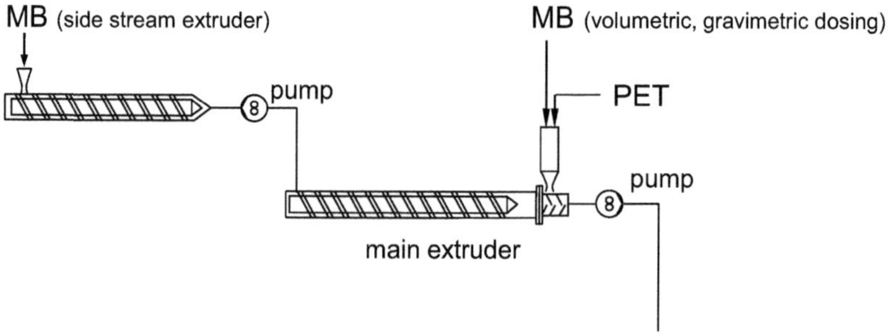

Fig. 14.27 Masterbatch (sidestream extruder, volumetric/gravimetric metering)

reasons and because of environmental protection, and is reintroduced into the process cycle after appropriate reprocessing. Non-recyclable waste is thus reduced to amounts below 1%, apart from unavoidable by-products.

References

Chiesa, L. (2005). *CC BY 3.0*. https://commons.wikimedia.org/w/index.php?curid=516874

Götze, K. (1967). *Chemiefasern nach dem Viskoseverfahren*. Springer.

Lenzing AG. (n.d.). Lenzing, Österreich. https://www.lenzing.com

Rogowin, Z. A. (1982). *Chemiefasern*. Georg Thieme Verlag.

Sloan, F. (2017). Liquid crystal aromatic polyester-arylate (LCP) fibers: Structure, properties, and applications. In *Structure and properties of high performance fibers* (pp. 113–140). Elsevier.

15 Aggregates for Polymer Formation

Polymers are synthesized in reactors. They are then either fed directly to the spinning machines via pipes or—in the case of melt-spinnable polymers—often processed into chips, stored and later remelted with the aid of extruders.

15.1 Reactor Types

In the reactor, the monomers and the polymer are mixed and the chemical reaction takes place. In most cases, heat is also added. Depending on the choice of the reactor speed and the residence time of the polymer in the container, its viscosity can be specifically adjusted.

The synthesis of the monomers to form the polymer can either be carried out discontinuously (stirred vessels), i.e. batchwise in small to medium-sized vessels, or continuously in large reactors (flow vessels), into which reactant is continuously fed and product removed. While stirred and flow-through vessels can be used for all types of polymer formation reactions, flow tubes can only be used for chain growth reactions (polymerizations). This is because active mixing cannot take place in the flow tube itself, which is mandatory for step growth reactions (e.g. polycondensation) in order to form long polymer chains.

Figure 15.1 shows the setup and concentration curves for a discontinuously operating reactor, the so-called batch stirred tank. At the beginning of the reaction, the components are fed into the reactor and the reaction starts. This constantly increases the amount of product (polymer) while decreasing the amount of reactant (monomer). In the ideal case, the concentrations of both, monomer and polymer, as well as the temperature, are the same throughout the reactor at any given time t. At the end of the reaction, the products are removed and the reactor is refilled with reactants.

In Fig. 15.2, the structure and the ideal concentration curves for a continuously operating flow-through vessel are shown. Reactants are continuously added and products removed. Because reactants are constantly being turned into product and thus "consumed", the concentration of reactants within the reactor is always lower than in the reactor feed. In the ideal case, it is therefore assumed that the concentration of the reactants decreases sharply immediately after the inflow ("concentration jump"). The concentration of reactant and product is then ideally the same at all locations and at all times (quasi steady equilibrium).

Frequently, several flow-through vessels are used in series, which are then referred to as "cascades". This improves mixing and makes the average chain length more uniform (Fig. 15.3).

The structure and the concentration ratios in a flow tube are shown in Fig. 15.4. Again, in the ideal case, it is assumed that the concentration of reactant and product at a given location is

D. Veit, *Fibers*, https://doi.org/10.1007/978-3-031-15309-9_15

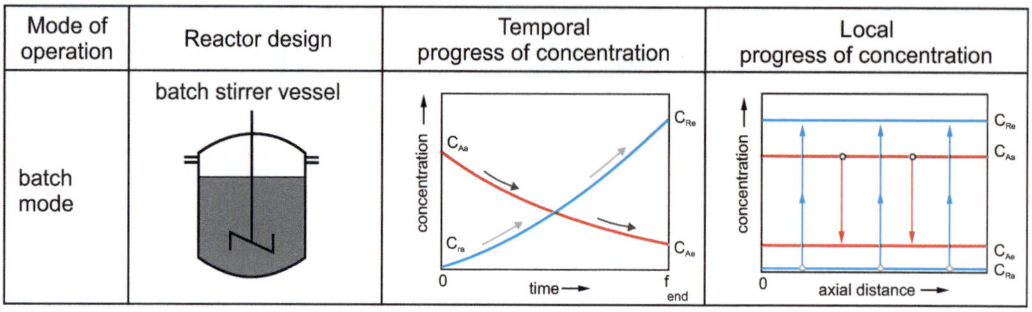

Fig. 15.1 Details of a batch stirred tank acc. to Ignatowitz (2015)

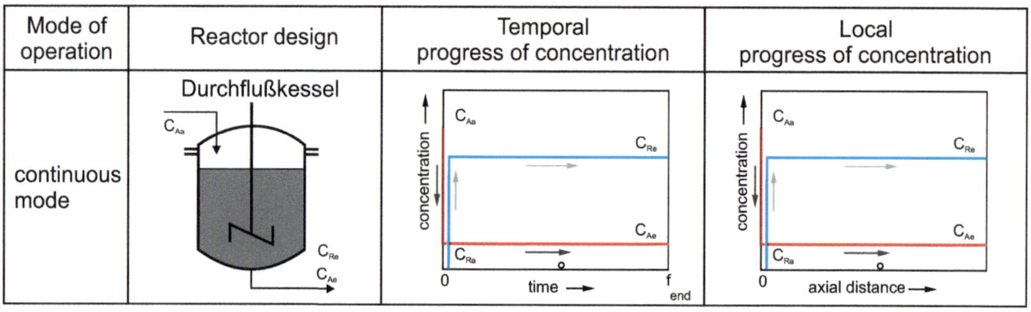

Fig. 15.2 Details of a flow-through vessel acc. to Ignatowitz (2015)

constant at any time t. If we look at the concentration curve at a certain point in time for different locations, the concentration of the reactants decreases continuously towards the outlet, while the concentration of the products increases accordingly.

Fig. 15.3 Flow through vessel with stirring elements (Raysper, 2005)

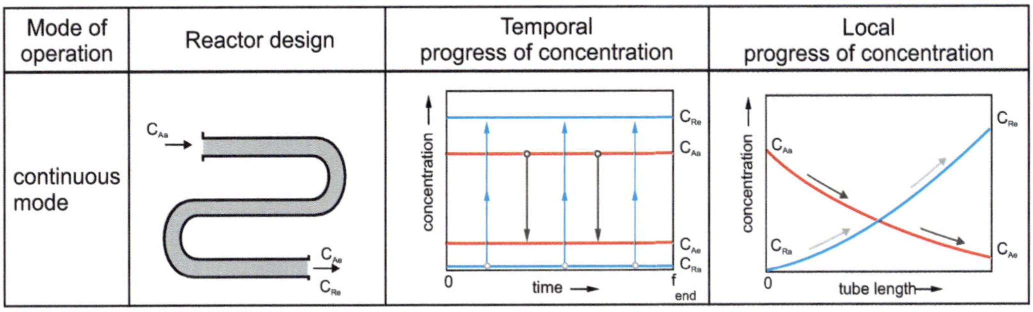

Fig. 15.4 Details of a flow tube acc. to Ignatowitz (2015)

15.2 Autoclave

The autoclave (stirred tank reactor) is a batch stirred tank in accordance with Fig. 15.1. The first functional autoclave in the form of a pressure cooker was patented by Denis Papin in 1681 (Papin, 1695). Autoclaves are pressure vessels that can be sealed gas-tight and are used for a variety of tasks, usually under positive pressure. In the textile sector, these include, for example, the production of polymers and the consolidation of fiber composites.

The Fig. 15.5 shows the one-step polymer formation with the aid of an autoclave using the example of polyamide 66. Here, the AH salt is first dissolved (1) and then mixed with water (3) under vacuum and in nitrogen (2). Via a drain valve (4) the mixture flows into the autoclave (5, pressure approx. 18 bar), which is heated with diphyl (6). Stabilizers and matting agents can then be added. The reaction time is about 3 h at a temperature of approx. 240 °C. The temperature is then increased to 270 °C and the reaction proceeds for a further 2 h. During this time, the pressure is reduced to atmospheric pressure. A pump (7) removes the finished polymer, which is passed over a cooling drum (8) and sprayed with water (9). Dried by air (10), it is then cut into chips (11). The typical capacity of such an autoclave is 20 t/day with a cycle time of approx. 7 h.

An alternative process control with two autoclaves shows Fig. 15.6. Here, the AH salt is conveyed from the storage tank (1) via a feed pump (2) and a weighing device (3) into the

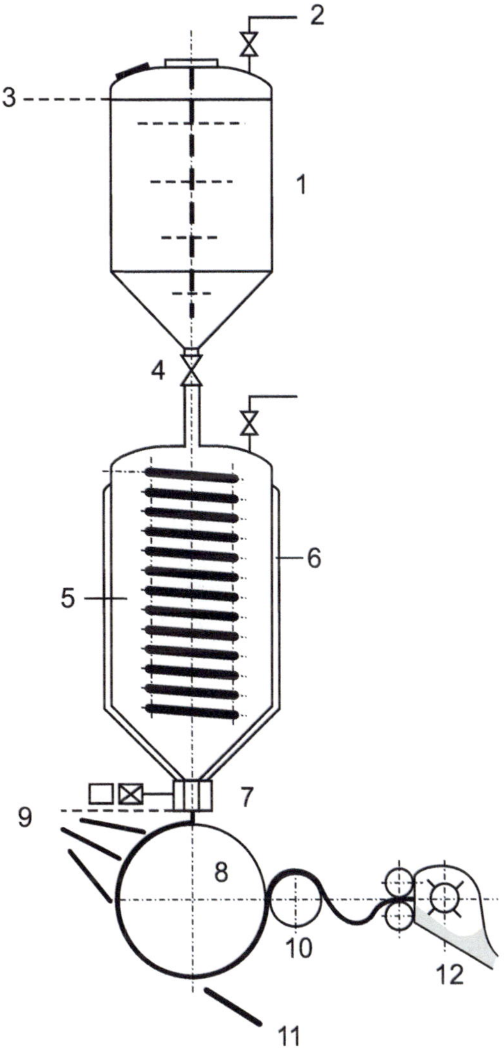

Fig. 15.5 Autoclave polycondensation of PA 66 acc. to Fourné (1995)

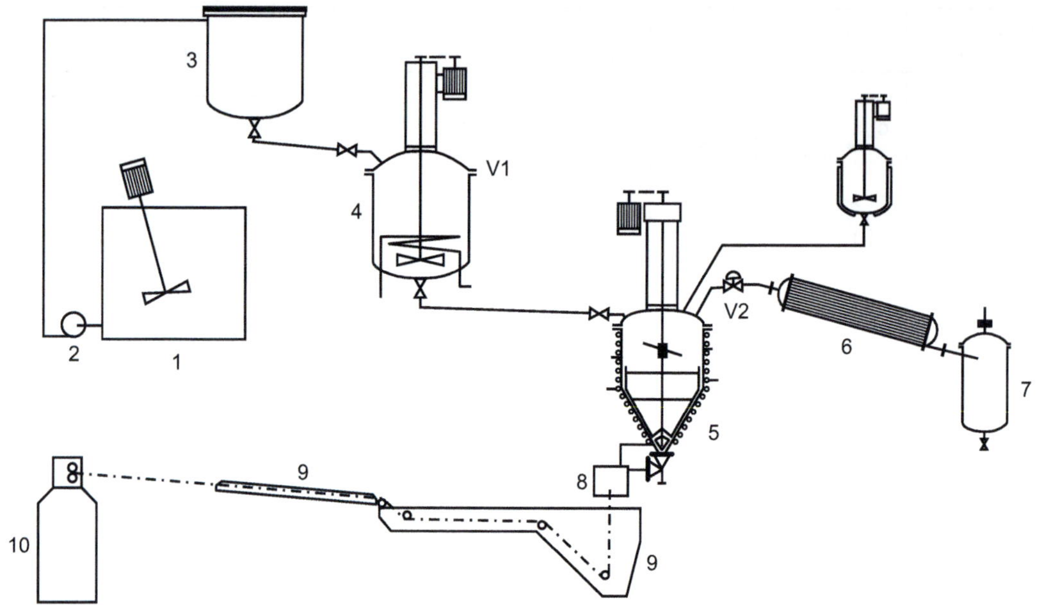

Fig. 15.6 Two-stage polycondensation in the autoclave of PA 66 acc. to Fourné (1995)

pre-condensation autoclave (4). The reaction temperature is around 260 °C and the pressure is 18–20 bar. After 1 h, the pressure is reduced to 1.5–2 bar and the polymer is conveyed to the finishing polycondensation autoclave (5) after another hour. Steam can be released via a valve (V1). This autoclave also has an expansion valve for steam (V2) as well as a steam condenser (6) and a water collection tank (7). The polymer then passes to the pelletizer (10) via a pouring head (8) and a cooling device (9). For an autoclave size of approx. 2000 l each, typical capacities of 8–9 t/day result.

15.3 Disk Ring Reactor

Disk ring reactors (DRR) operate acc. to the principle of the flow-through vessel. The cylindrical reactor is divided into individual chambers by disk elements. It is always about two thirds full and the flow is horizontal. The rotating discs simultaneously mix and convey reactant and product, which adhere to the discs as a film and are thus drawn up and then fall down again.

The daily production of such reactors is normally up to 500 t. For particularly large polymer quantities, several reactors can also be operated in parallel (Figs. 15.7 and 15.8).

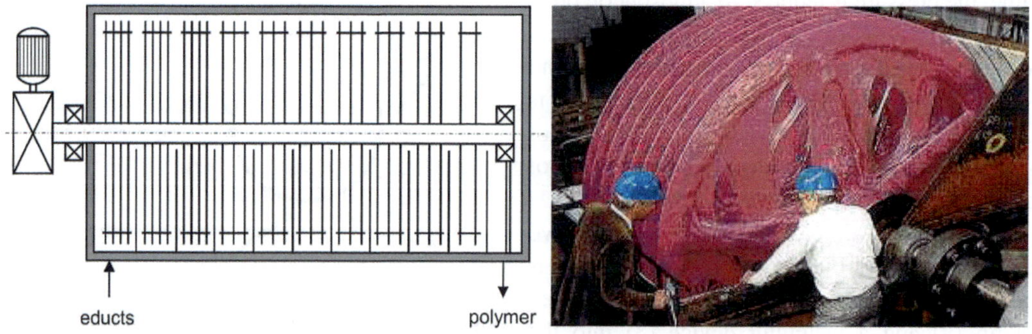

Fig. 15.7 Disk ring reactor for continuous polymer production (Zimmer AG, n.d.)

Fig. 15.8 Disk ring reactor (Zimmer AG, n.d.)

Figure 15.9 shows the process of continuous polymer processing as it is often used for polyester. Two reactors are connected in series. In the first reactor, pre-condensation takes place, forming polymers with medium chain length. In the second reactor, these chains are then increased to the desired length by further polycondensation (Fig. 15.10). Excess glycol and water are removed in the process.

Typical residence times are in the range of 2–20 h. The end product is a polymer melt that can be spun out directly.

For smaller polymer quantities, the discontinuous process is used (Fig. 15.11). Here, the polymer formed in the reactor is not spun directly, but first converted into chips with a granulator. These can be stored and transported for processing at a later stage.

15.4 VK Tube

Polymer formation with the aid of a plug-flow reactor is carried out, for example in the case of polyamide 6, with the aid of a simplified continuous tube ("VK tube"). In this process, the raw material (here: lactam) is first melted under inert gas and then pre-condensed at approx. 220 °C. The melt is then heated under atmospheric pressure. The melt is fed into the VK tube under atmospheric pressure and usually polymerized in three stages with increasing temperatures (e.g. 230 °C at the top, approx. 260–270 °C in the middle, approx. 260 °C at the bottom). The heating is usually carried out via diphyl. The inner diameter of such a pipe is limited to approx. 1.60 m for heat conduction reasons, the height to 15–20 m. Capacities of up to 70 t/day are possible. The throughput time is 15–40 h (Fig. 15.12).

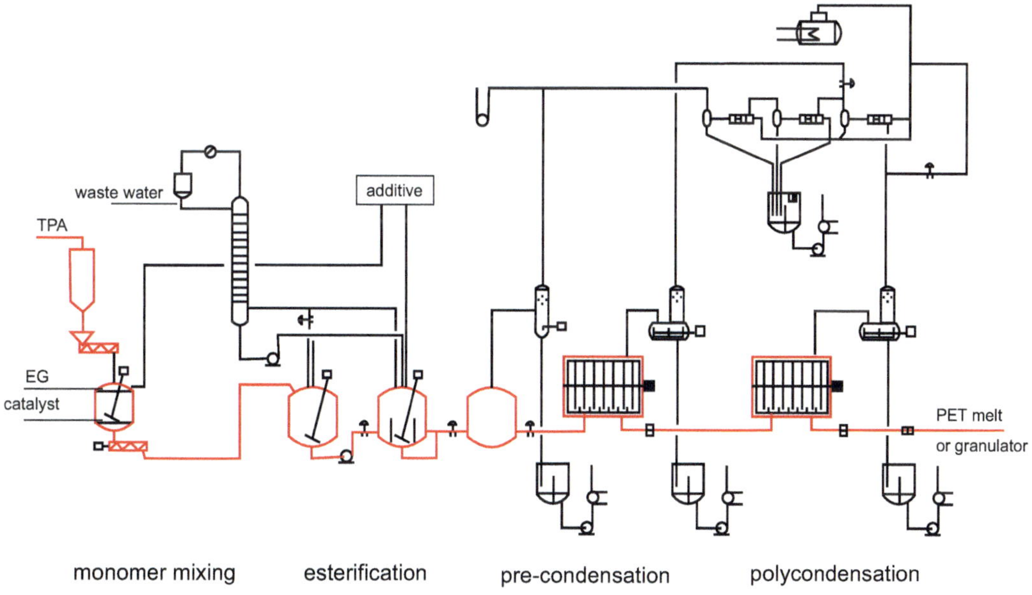

Fig. 15.9 Continuous process of polymer processing of PET (Zimmer AG, n.d.)

Fig. 15.10 Reactor (finisher) in a direct spinning plant (Oerlikon, n. d.)

15.5 Discharge and Booster Pumps

To ensure a uniform polymer flow in a spinning plant, special pumps are used at various points. The polymer flow is kept constant by a discharge pump. Its delivery volume is 100–3000 cm^3/revolution, the maximum pressure is approx. 700 bar. Besides, after the reactor, additives (e.g. TiO_2) can be added by special metering pumps. Such pumps can also be connected downstream of an extruder to pre-dose the polymer flow. There are heated and unheated versions, depending on the application (Fig. 15.13).

A booster pump compensates for pressure losses caused, for example, by static mixers. The spinning pump ensures precise polymer throughput in the spinnerets, and a spin finish pump is used to apply oil to the yarn. Figure 15.14 shows typical pumps in a spinning plant.

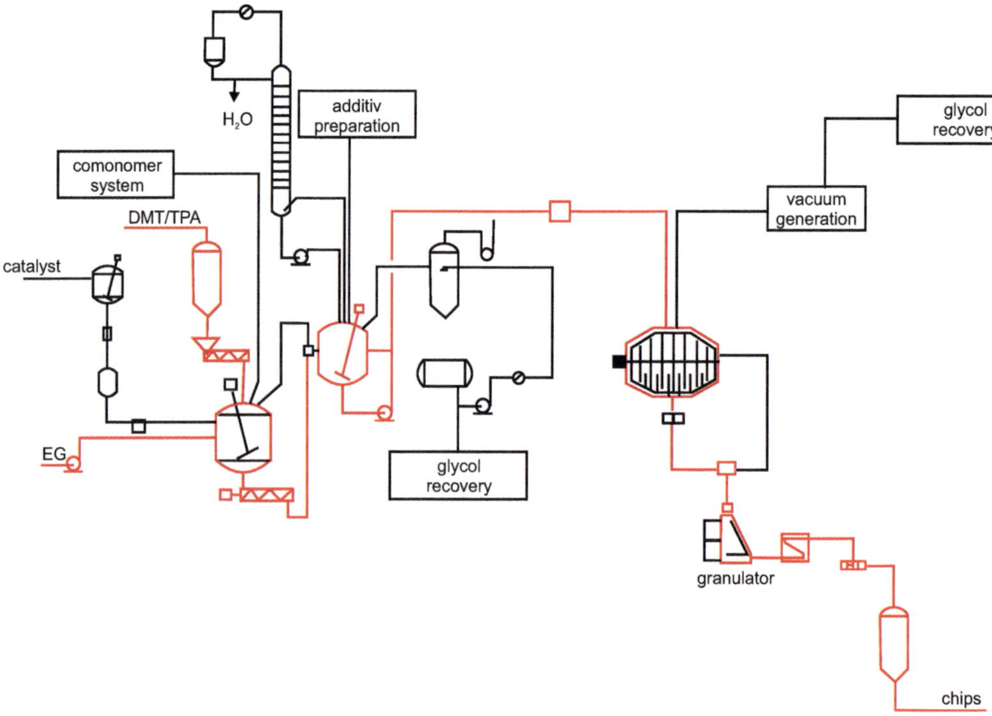

Fig. 15.11 Discontinuous process of polymer processing of PET (Zimmer AG, n.d.)

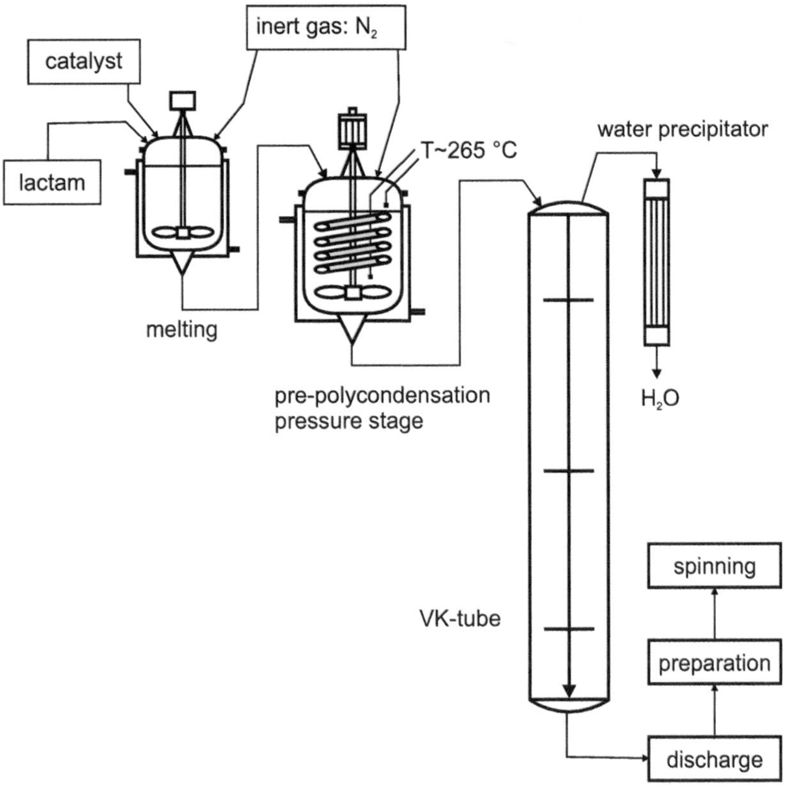

Fig. 15.12 Polymer formation with VK tube

15.6 Tube Flows

For the calculation of pipe flows, the law of Hagen-Poiseuille must be observed. This assumes a homogeneous, Newtonian fluid flowing laminarly through a pipe. This is the case when the Reynolds number Re < 2050:

$$\mathrm{Re} = \frac{\rho \cdot \upsilon \cdot d}{\eta}$$

with ρ density of the fluid, ν = flow velocity, d = characteristic length of the body and η viscosity of the fluid.

Figure 15.15 shows the increase in shear stress with increasing shear rate for selected fluid types.

In polymers, viscosity decreases under the influence of high shear forces (Fig. 15.16). They are therefore not Newtonian fluids, but are referred to as structurally viscous fluids or Bingham fluids.

This behavior can be explained by the fact that in polymer solutions and melts the macromolecules are entangled. With increasing shear force, the macromolecules are detached from each other, they become mobile and the viscosity therefore decreases.

The force on the cross-sectional area due to the pressure difference between the inlet and outlet of the pipe is equal to the force exerted on the liquid by the wall along the length l of the pipe. Thus, for an ideally round pipe, we have:

Fig. 15.13 Vacuum discharge and booster pump (Mahr, n.d.)

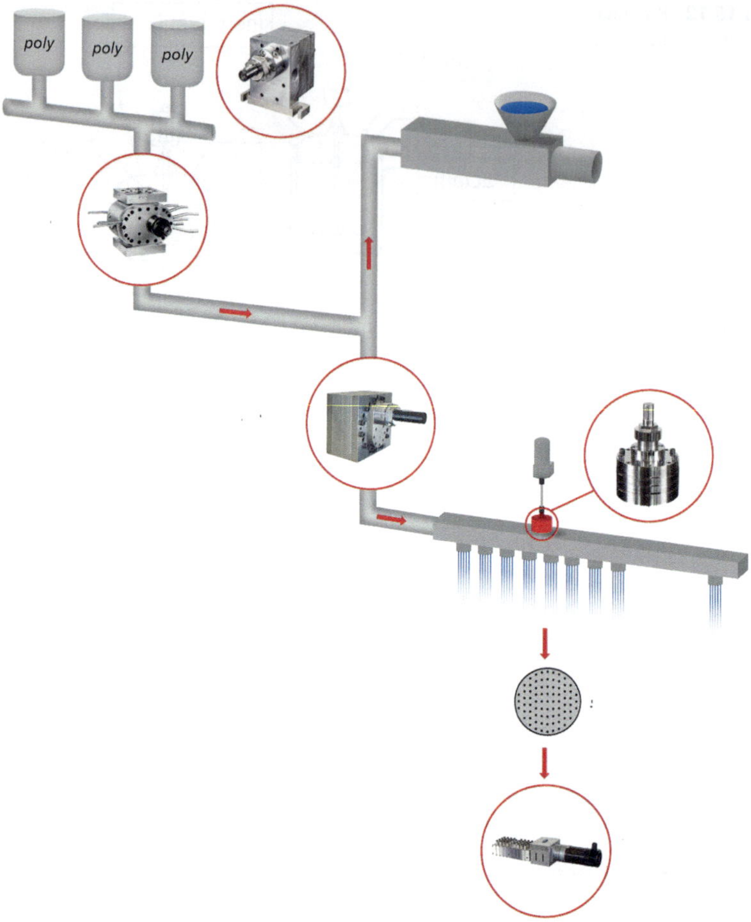

Fig. 15.14 Pumps in a spinning plant (Mahr, n.d.)

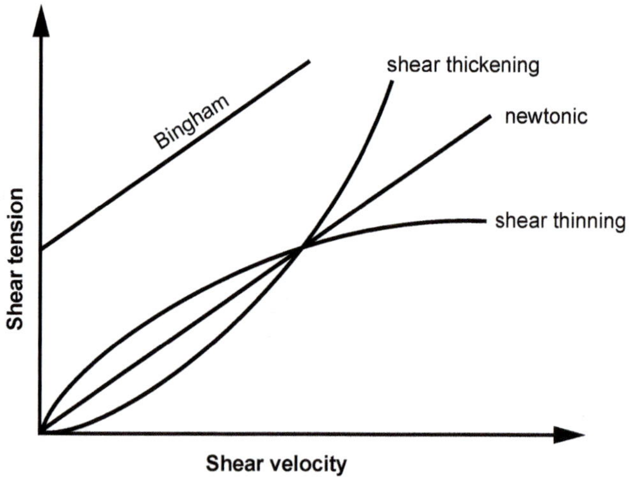

Fig. 15.15 Shear stress as a function of shear rate for fluids

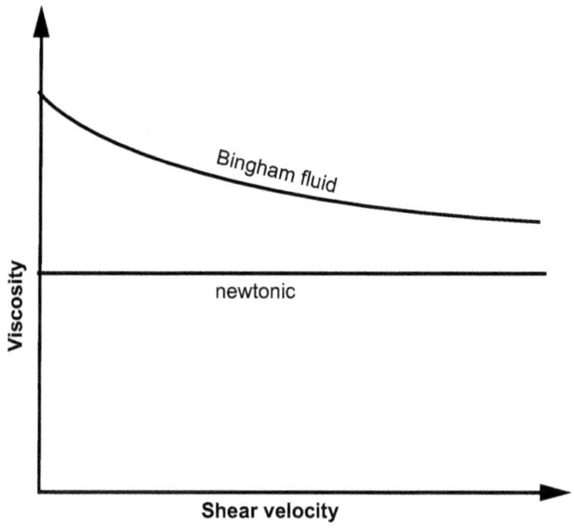

Fig. 15.16 Viscosity as a function of shear rate for different fluids

$$\sum F = 0 = (p_1 - p_2) \cdot \pi r^2 = \eta \cdot 2\pi r l \cdot \frac{dv}{dr}$$

At the pipe wall, the flow velocity is due to the adhesion condition: v = 0. Thus, after integration over the pipe radius, we obtain:

$$v(r) = \frac{\Delta p}{4\eta l} \cdot \left(\frac{d^2}{4} - r^2\right)$$

The following then applies to the volume flow:

$$\dot{V} = \int_{r=0}^{r=d/2} v(r) \cdot 2\pi r \cdot dr = \frac{\pi}{8\eta} \cdot \frac{\Delta p}{l} \cdot \left(\frac{d}{2}\right)^4$$

Doubling the pipe diameter therefore results in 16 times the volume flow.

The average flow velocity results from

$$v_m = \frac{\dot{V}}{A} = \frac{\Delta p \cdot d^2}{32\eta l}$$

and for the pressure loss along the pipe thus applies

$$\Delta p = \frac{v_m \cdot 32\eta l}{d^2}$$

With the Reynolds number

$$\mathrm{Re} = \frac{v \cdot d}{\upsilon},$$

we get

$$\Delta p = \left(\frac{64}{\mathrm{Re}}\right) \cdot \left(\frac{l}{d}\right) \cdot \left(\frac{\rho \upsilon^2}{2}\right)$$

Hagen-Poiseuille's law applies only to laminar flows (present here, Fig. 15.17, top) and strictly speaking only for Newtonian fluids. Thus, it does not actually apply to structurally viscous (visco-elastic) polymer flows, but nevertheless represents a good approximate solution when a quasi-steady state is considered.

Figure 15.18 shows the effect of changing the pipe diameter on the throughput (left axis). The right axis shows the pressure increase required to maintain the original throughput with a reduced pipe diameter. For example, a 10% reduction in pipe diameter results in a 34% reduction in throughput. The pressure would therefore have to be increased by 52% to achieve the original throughput. If the pipe diameter is halved, the throughput is reduced to 6% of the original value and the pressure would have to be increased by a factor of 15 to compensate.

Furthermore, when designing pipelines, it must be ensured that the distance from the reactor to each spinneret is the same for all spinnerets. The number of right and left bends of the pipes should also be the same. This is the only way to ensure that all macromolecules in the polymer are exposed to the same thermal conditions and that there are no quality problems due to different residence times in the product lines (Fig. 15.19). Possible differences usually only become apparent after dyeing.

15.7 Heat Exchanger

Heat exchangers are needed in polymer processing to keep the temperature of the melt constant at a certain value.

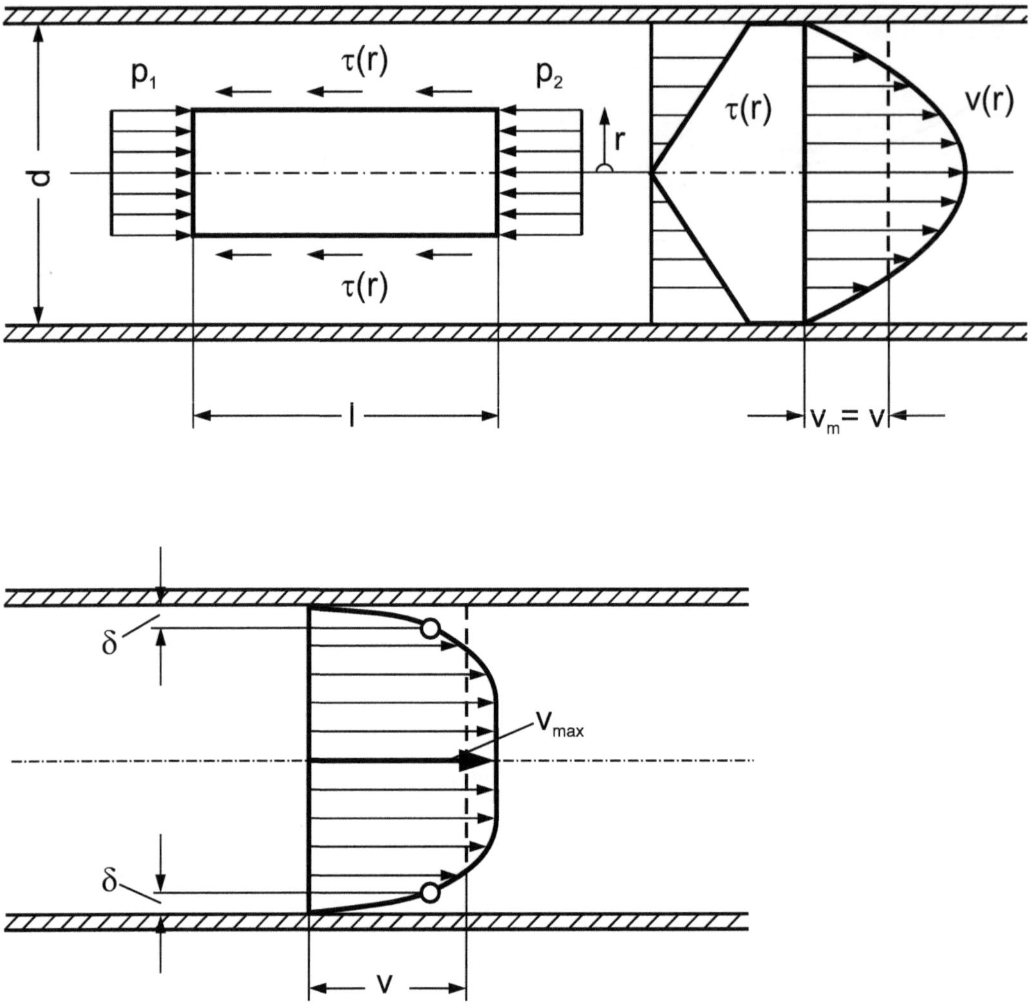

Fig. 15.17 Pipe flows (top: laminar, bottom: turbulent)

15.7.1 Principles of Heat Exchange

In heat exchange with liquid or gaseous media, a distinction is made between the principles of "co-current" and "counter-current" (Fig. 15.20).

The heat flow $\dot{Q}$ at a given time t is then calculated from

$$\dot{Q}(\dot{t}) = k \cdot A(t) \cdot \Delta T(t)$$

where k = heat transfer coefficient, A = cross-sectional area of the heat exchanger, T = temperature difference of the two media.

With the co-current principle (left), the medium to be cooled comes into direct contact with the cold cooling medium, which means that the temperature difference ΔT is initially very high and the heat exchange is therefore fast. However, the material of the tube may be very stressed. With the counterflow principle (right), the medium to be cooled initially comes into contact with an already heated cooling medium. This means that the temperature difference is initially significantly lower, but the heat exchange is gentler on the material and possibly also more uniform. The efficiency of the counterflow principle is somewhat higher than that of the co-current principle due to the higher mean temperature difference.

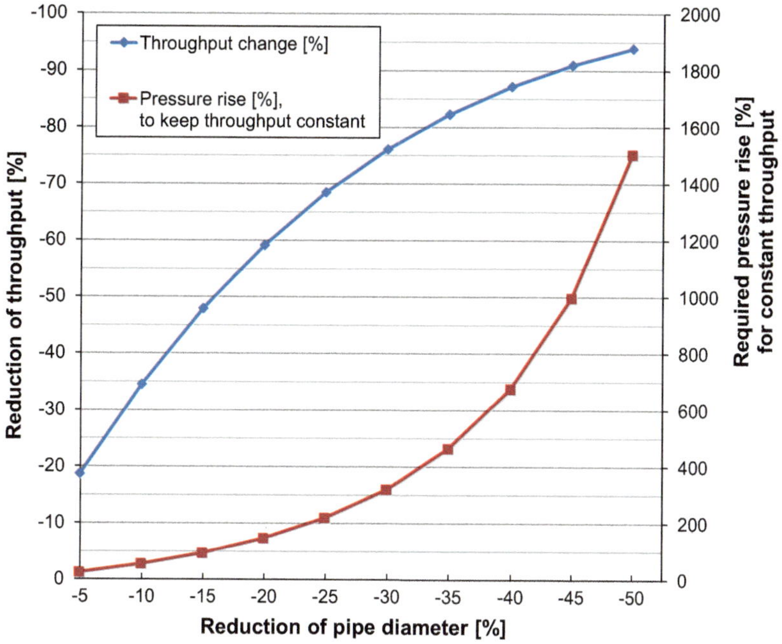

Fig. 15.18 Flow rate and pressure change depending on pipe diameter

There are also other principles of heat exchange (e.g. eddy current, cross current), but these are not used with polymers.

The total transmittable heat can be generally determined by

$$\dot{Q} = k \cdot A \cdot \Delta\theta,$$

where k is the heat transfer coefficient in [W/m^2 K], A is the transfer area and $\Delta\theta$ is the mean temperature difference in [K].

It is calculated from:

$$\Delta\theta = \frac{\Delta T_1 - \Delta T_2}{\ln\left({}^{\Delta T_1}/_{\Delta T_2}\right)}$$

The index "1" indicates the larger temperature difference and the index "2" the smaller one.

Heat exchangers used in polymer production are either electrical heating tapes for lab equipment (inexpensive, temperature can only be roughly adjusted) and heat exchangers that work

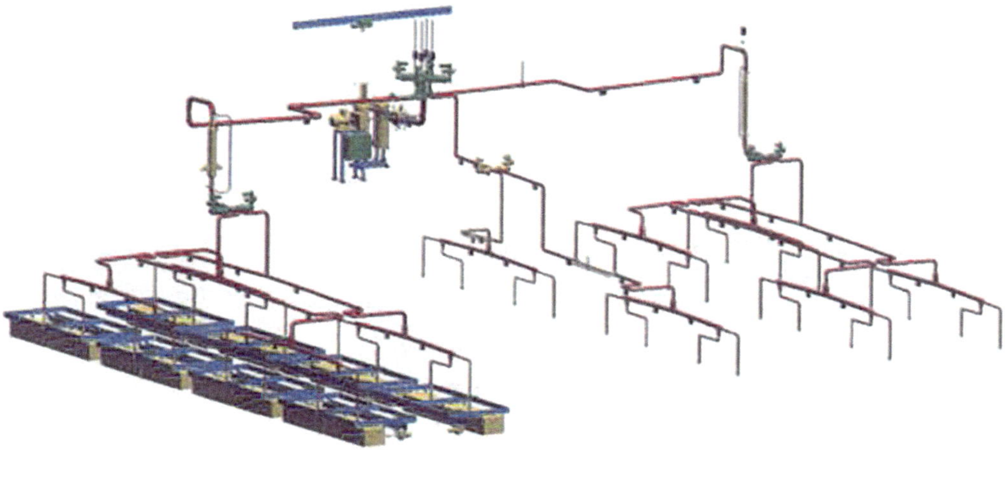

Fig. 15.19 Typical course of the piping of a melt spinning plant (Zimmer AG, n.d.)

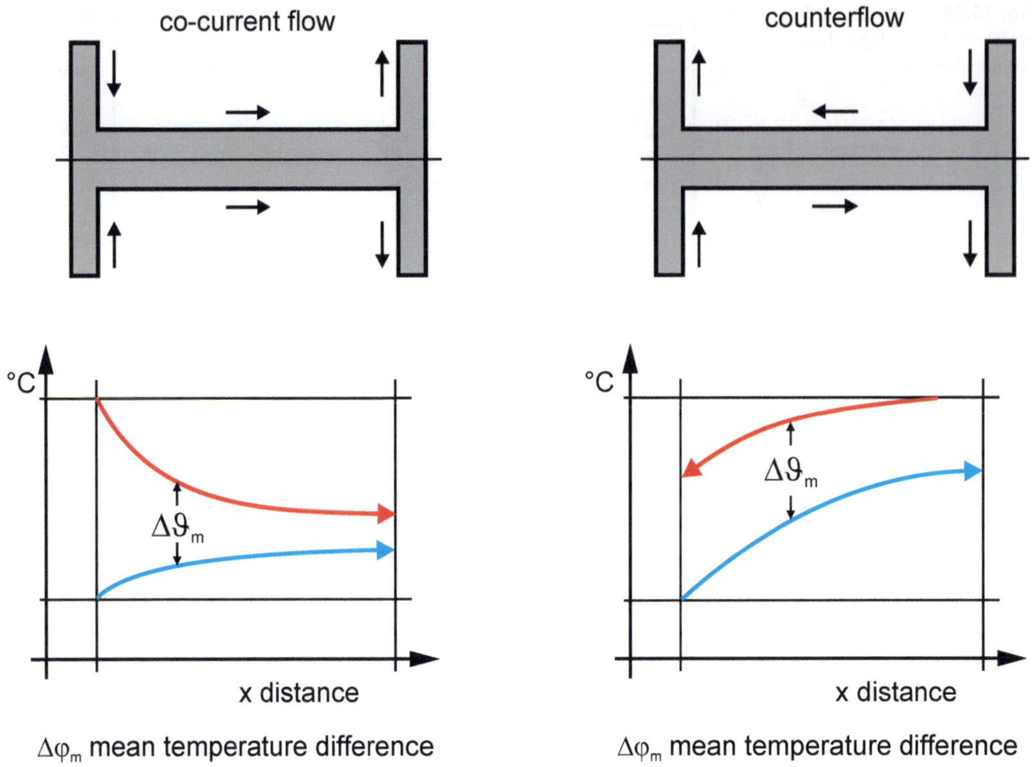

Fig. 15.20 Principles of heat exchange

on the principle of temperature adjustment as a function of the boiling pressure for large scale production units (expensive, temperature can be reasonably precisely adjusted).

15.7.2 Tube Bundle Heat Exchanger

The principle structure of a frequently used tube bundle heat exchanger is depicted in Fig. 15.21. It operates on the principle of boiling pressure. This allows the temperature of the heating or cooling medium to be adjusted very precisely. The principle of counterflow is shown here. However, it is also possible to exchange the heat in co-current flow.

15.7.3 Electric Heating Tapes

Electric heating tapes are particularly suitable for laboratory and pilot plants when a heating

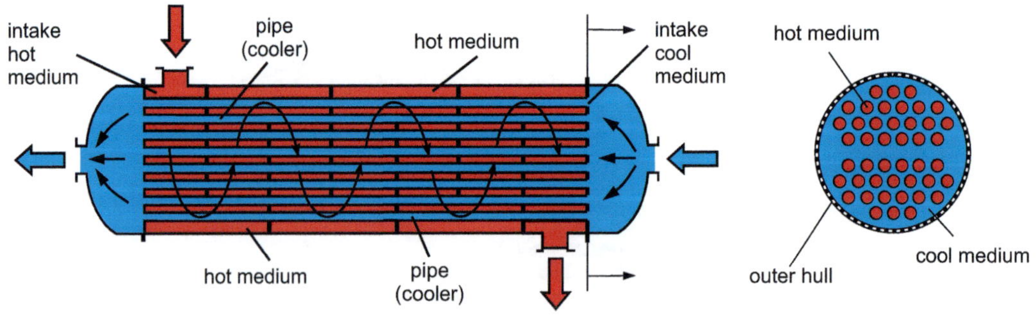

Fig. 15.21 Design of a tube bundle heat exchanger acc. to Ignatowitz (2015)

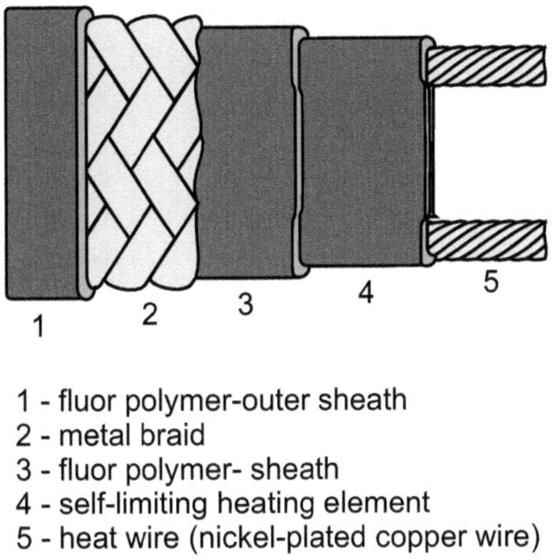

Fig. 15.22 Design of an electric heating tape acc. to Klöpper Therm (n.d.)

medium circuit is too costly and thus uneconomical because of the insufficient quantities of polymer. The pipe or vessel to be heated is wrapped with the heating tape (Fig. 15.22). This type is suitable for a temperature range of up to 120 °C and has a power consumption of 15–60 W.

15.7.4 Evaporator with Heat Transfer Oil

The most commonly used heat transfer oil is diphyl as its evaporation pressure is comparatively low (boiling temperature 258 °C at atmospheric pressure) and the material is cheap. It is evaporated at a defined pressure via electric heating pipes and fed to the appliance. The condensate produced is recycled so that a closed circuit is created (Fig. 15.23).

In the case of polymers, a maximum of 300 °C is usually required, which fits in very well with the boiling temperature of many heat transfer oils, which—depending on the manufacturer and product—is approx. 250–400 °C. A temperature of 200 °C is already reached at vapor pressures of well below 1 bar.

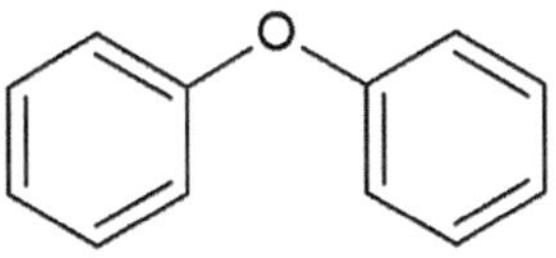

Fig. 15.23 Diphyl structural formula

15.7.5 Heat Exchanger Types and Their Application

In production plants, heat exchangers with liquid or vaporous heating media are mostly used due to their higher temperature accuracy. The Table 15.1 table gives an overview of typical heating media and heating methods for common temperature ranges.

15.8 Static Mixers

Static mixers are used to maintain a polymer stream at a uniform temperature and/or to mix it with other streams (e.g. masterbatch). The fluids to be mixed flow through rigid mixing elements built into the product pipe. Figure 15.24 shows the design of a typical static mixer.

Figure 15.25 illustrates the mixing effect along a static mixer with eight mixing elements. After only a few mixing elements, the different polymer streams are almost completely mixed.

More mixing elements therefore lead to better mixing. However, the pressure loss also increases strongly, depending on the mixer type. If the flow in the mixer changes to the turbulent case (Re > 100–200), the mixing is even better.

In order to assess the quality of the mixing, the coefficient of variation $\sigma/\overline{c}$ is often calculated from the standard deviation of the concentration of a component and its average concentration at the end of the mixer. Ideally, this coefficient is zero. It decreases with increasing length/diameter ratio (*L*/*D*). Thus, longer and thinner mixers improve mixing. However, the smaller the diameter of the mixer, the higher the pressure drop acc. to the Hagen-Poiseuille equation:

$$\Delta p = \zeta \cdot \mathrm{Re} \cdot \frac{\eta \cdot \upsilon_m \cdot L}{2 \cdot D^2}$$

where ζ = pressure loss coefficient, η = dynamic viscosity of the medium and υ_m = mean flow

Table 15.1 Values and applications of typical heating media and methods acc. to Fourné (1995)

Temperature range [°C]	Pressure range [bar]	Medium	Temperature accuracy [± °C]	Application
7–160	1–8	Water	1	Vacuum dryer
100–240		Water vapor	1	Dryer drums
140–240		Induction heating with water vapor coat	1	Godets
120–300	0.02–2.5	Liquid carrier oil (e.g. diphyl)	2–4	Heat dissipation
180–370	0.15–7.6	Diphyl vapor	1	Spinning beam
60–400		Induction heating with inductive cooling	3–7	Godets
60–400		Electric heating tape	3–7	Godets
400–700		Radiator	4	HT-godets

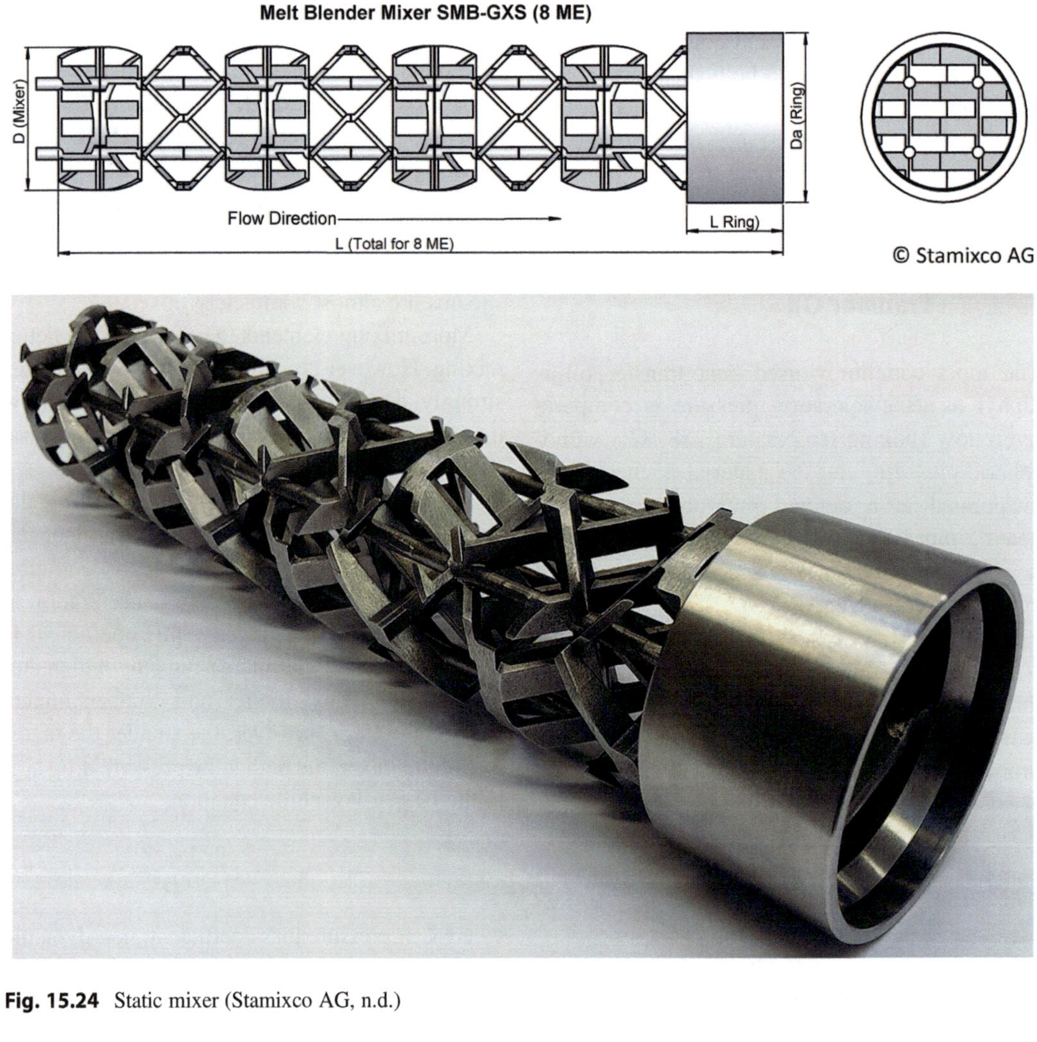

Fig. 15.24 Static mixer (Stamixco AG, n.d.)

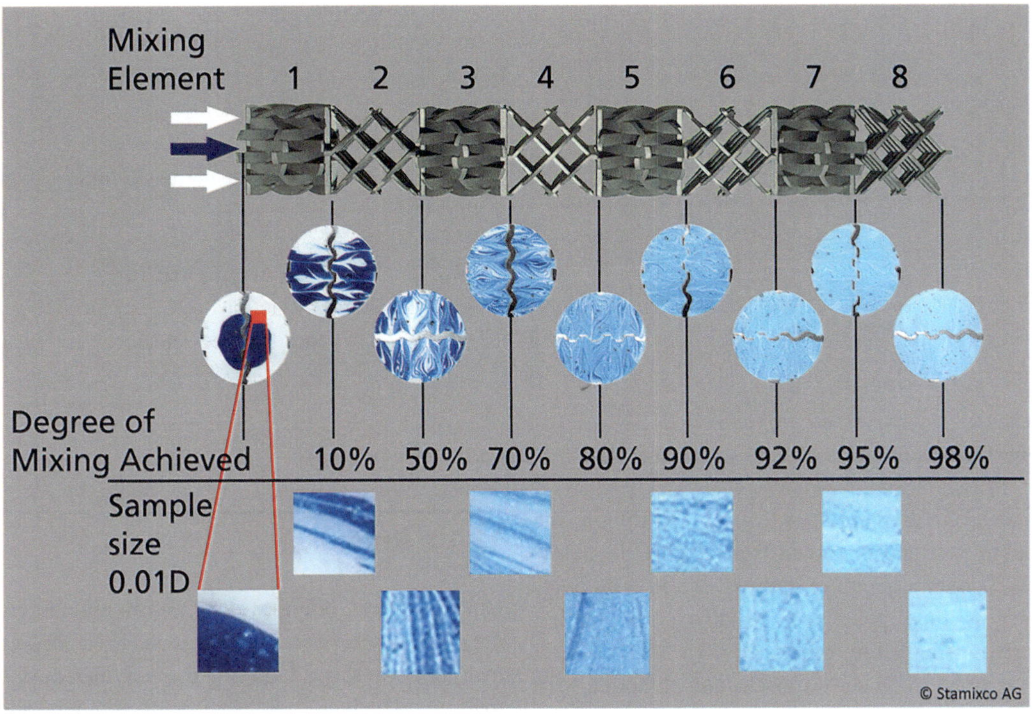

Fig. 15.25 Effect of a static mixer on two parallel polymer streams (Stamixco, n.d.)

velocity. Typical pressure loss coefficients range from 0.02 (empty pipe) to 12 for a cross-mixing element analogous to Fig. 15.25.

A halving of the pipe diameter thus leads to a quadrupling of the pressure loss. If the volume flow is to remain constant, the pressure needs to be increased 16 fold. Therefore, it must be pondered how many mixing elements are required and how large they should be in order to minimize the pressure loss for a sufficient mixing quality.

Furthermore, it must be avoided that polymer adheres to the mixing elements, because otherwise thermal degradation ("cracking") of the polymer can occur quickly. Regular cleaning of the static mixers (after approx. 2–4 spinneret changes) is therefore absolutely necessary.

15.9 Granulator

Granulators are used to bring the produced polymer into a form that can be stored for a longer period of time. On the one hand, this is useful for polymers that cannot be spun directly (e.g. because the residence time in the product lines is then too high or for economic reasons). On the other hand, it is used to process polymer surpluses in direct spinning plants when the capacity of the reactor and the spinning plant do not exactly match, which is almost never the case due to frequent product changes. The diagram in Fig. 15.26 shows the basic structure of an underwater pelletizing unit. From the extruder (1) the melt is transported to the filter by a gear pump (2). Via a start-up valve (4), the polymer reaches a perforated plate, where it is pressed through holes into the cutting chamber, where it is cut into pellets (5). Excess process water is removed in the dryer (6). An optical measuring system (7) analyses the granulate quality. The dry chips (pellets) are then packed into bags for storage.

Typical chip dimensions are in the range of 2–4 mm. The capacities of such granulators are 2–36,000 kg/h (Fig. 15.27).

An alternative process is the strand pelletizer. Here, the polymer melt is first cooled in a water

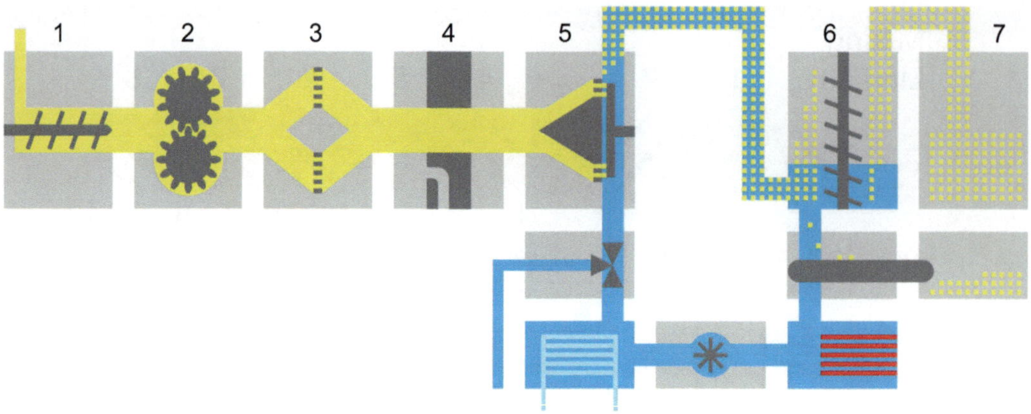

Fig. 15.26 Structure of an underwater granulator (MAAG, n.d.)

bath. At a drawing speed of e.g. 50 m/min, the polymer strand is cooled from 200 °C to below 50 °C over a length of approx. 2–5 m. The polymer is then pelletized, usually with rotating knives arranged on rollers (Fig. 15.28).

Figure 15.29 shows virgin and shredded PET bottles, which are marketed as rPET and can be added in small quantities for certain products.

After cutting, the process water must also be removed from the polymer chips by drying. Typical throughputs are up to 20,000 kg/h. Figure 15.30 shows a complete system.

15.10 Dryer

During storage, polymer chips absorb moisture in the ppm ("parts per million") range. This can lead to hydrolytic degradation of the polymer during melting, which in turn can lead to filament and thus yarn breakage. Therefore, polymer chips usually have to be dried before spinning.

The following drying principles are used for this purpose:

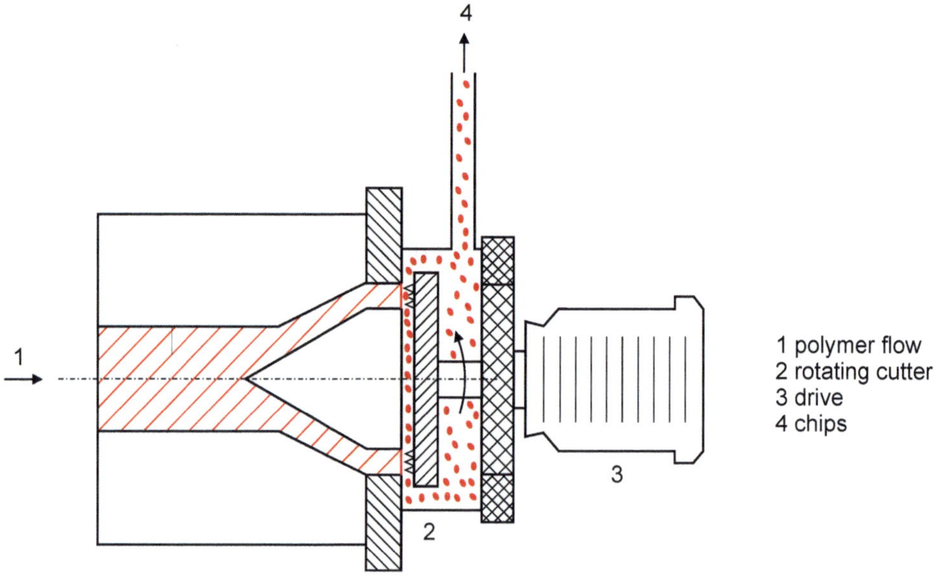

Fig. 15.27 Design of an underwater pelletizer cutting unit acc. to MAAG (n.d.)

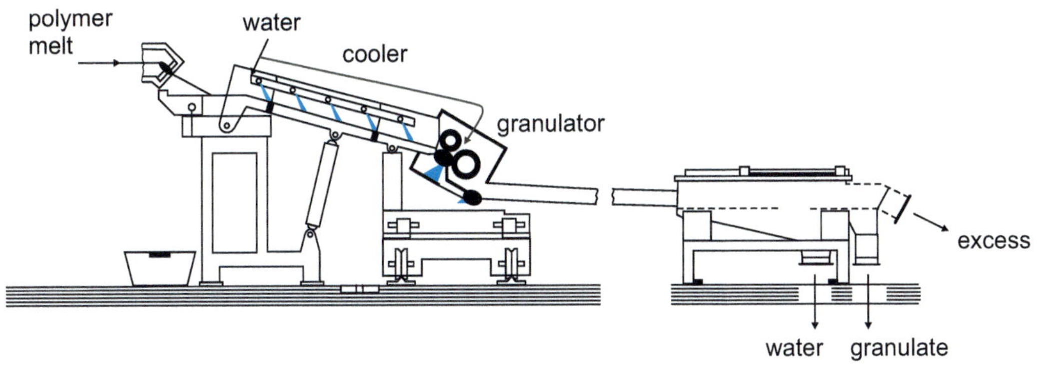

Fig. 15.28 Structure of a strand pelletizer

- Contact drying,
- convection drying,
- radiation drying (mostly together with convection),
- high frequency drying,
- ultrasonic drying (together with convection).

Of these drying processes, contact and convection drying dominate; high-frequency and ultrasonic drying have been used only to a limited extent to date.

In general, a distinction can be made between direct and indirect drying. In indirect drying, the drying medium (e.g. steam, diphyl) has no direct contact with the polymer to be dried. In direct drying, the drying medium (e.g. air, inert gas, highly heated steam) flows directly through the polymer to be dried. Although the latter is technically simpler, it can lead to contamination of the polymer.

The designs of dryers vary considerably, as do the throughputs. It is important that the residence time spectrum of the pellets is as narrow as possible. If the polymer, such as polyethyleneterephthalate, tends to stick during heating, a crystallizing device is often installed upstream.

15.10.1 Shaft Dryer (Flash Dryer)

Continuously operating shaft dryers are often used for the (pre-)drying of polymer chips, in which the granulate moves slowly with or against the air flow. Hot, possibly pre-dried air or a hot inert gas is usually used as the drying medium. An example of such a dryer is shown Fig. 15.31. In the system shown here, fine and coarse particles are simultaneously separated from each other in the shaft dryer by the air flow. The coarse particles are further dried in a cyclone and then

Fig. 15.29 New pellets (left) and pellets from recycled PET bottles (right)

Fig. 15.30 Underwater strand pelletizer (MAAG, n.d.)

discharged. The fine particles are separated from the air flow in downstream pocket filters and collected separately. This type of drying is adiabatic and therefore particularly suitable for temperature-sensitive polymers. Typical throughputs are in the range of a few kilograms to several tons per hour.

15.10.2 Tumble Dryer

For smaller polymer quantities, tumble dryers are often used (Fig. 15.32). They have volumes of up to 40 m^3, in which the granulate is constantly moved and operate discontinuously. A drying cycle takes up to 20 h. Hot air is often used as the drying medium (often under vacuum). The chips are constantly moved, firstly by the rotation of the drum, and secondly by the drying gas flowing through. This type of dryer is widely used, both in pilot plants and in production plants.

Polyamide is often treated in tumble dryers at temperatures of 70–80 °C, i.e. far below the glass transition temperature. This causes the moisture content to drop from typical values around 5–10% to 0.1% (1000 ppm) within 10–24 h. As polyamide is very hygroscopic, moisture is difficult to remove. Therefore, the specific air requirement is about 2.2 m^3/kg (compared to PET with 1.7 m^3/kg). If the moisture in the polymer is to be reduced further, fluid bed dryers and paddle dryers are suitable.

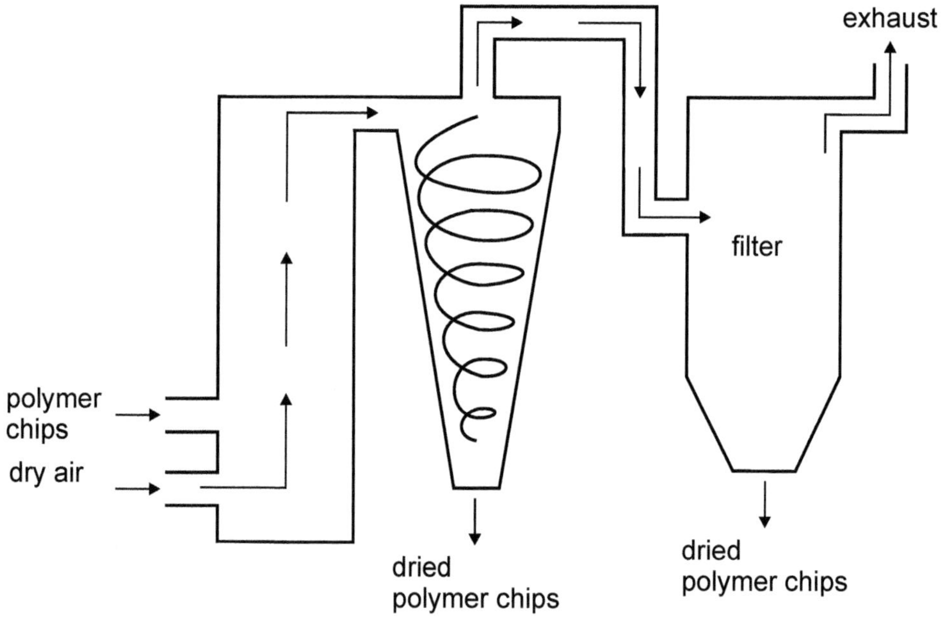

Fig. 15.31 Design of a shaft dryer with cyclone and filter unit

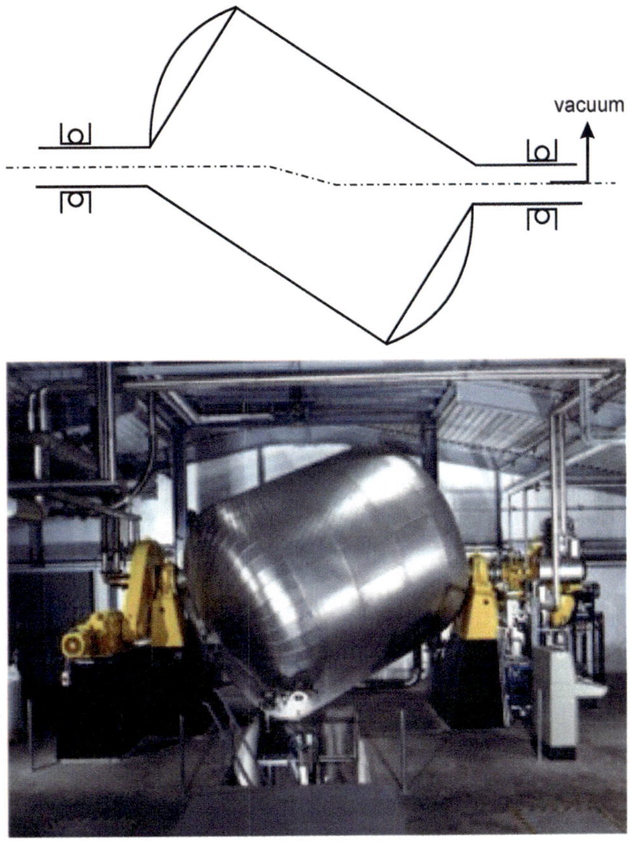

Fig. 15.32 Tumble dryer

15.10.3 Fluidized Bed Dryer

A process for continuous polymer drying developed only a few years ago is shown in Fig. 15.33. Air flows through a porous plate above a "bed" of polymer chips. In this principle, each polymer particle, surrounded by hot air or an inert gas, is kept in a state of suspension. Solids then behave much like a boiling liquid. This leads to a very good and uniform drying effect with a high energy efficiency. At the same time, the polymer chips are mixed and transported to the outlet. Usual drying times are below 1 h.

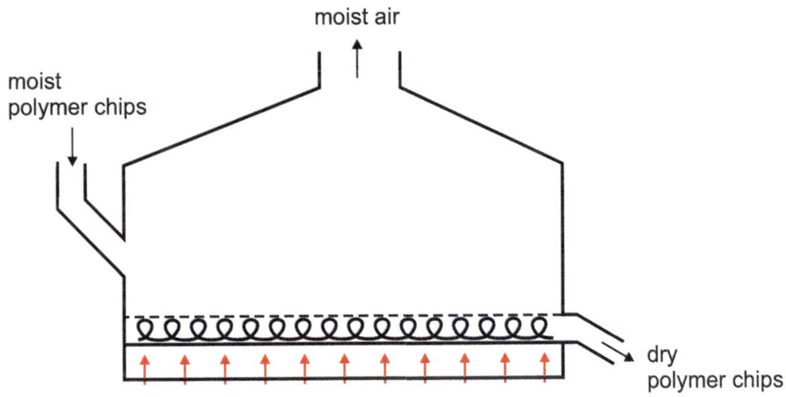

Fig. 15.33 Design of a fluidized bed dryer

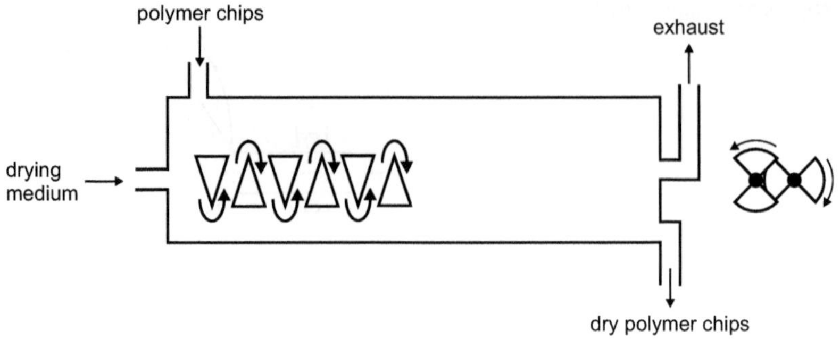

Fig. 15.34 Structure of a paddle dryer

The porous plate can also be vibrated (e.g. 5–25 Hz), which further improves the mixing and thus the uniformity of the drying. In addition to the flat design shown, this principle is also available as a cylinder dryer.

15.10.4 Paddle Dryer

Another principle that has rarely been used so far is the paddle dryer (Fig. 15.34). This uses hollow paddles through which the heating medium flows. The pellets to be dried are heated by the paddle surface and at the same time continuously transported in a kind of helical motion through the dryer to the outlet. These very compact dryers require approx. 50% less energy than conventional shaft dryers. They are used, for example, for polyamide and polypropylene.

15.10.5 Crystallization

When polymer chips tend to stick together, they are often crystallized before drying. A typical example of this is PET. Figure 15.35 shows the basic structure of a drying process for PET chips with upstream crystallization. In the crystallizer, the surface water evaporates and the PET macromolecules are converted from an amorphous to a crystalline state. This process is exothermic, so the temperature in the crystallizer rises above the glass transition temperature and there is thus a risk of agglomerates forming.

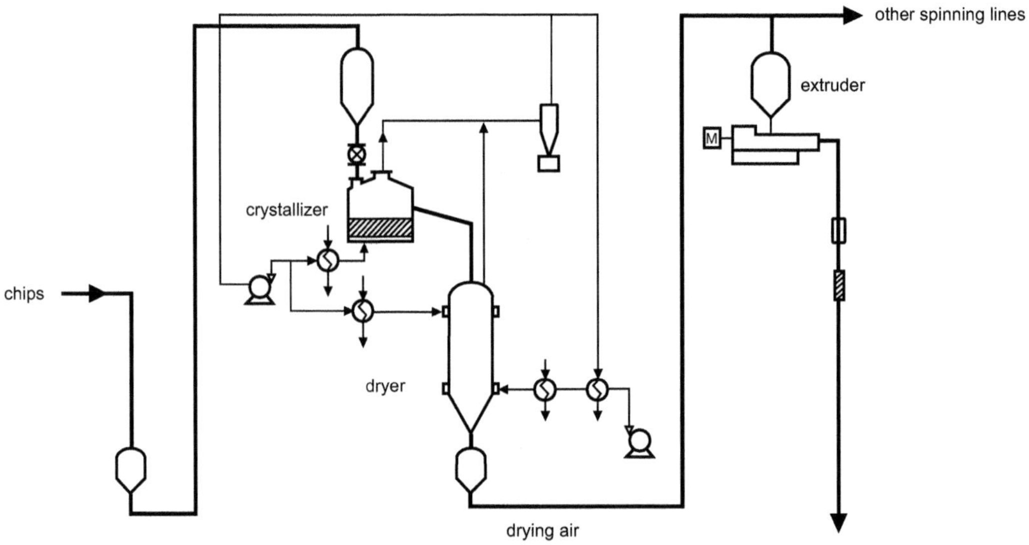

Fig. 15.35 Scheme of drying with upstream crystallization

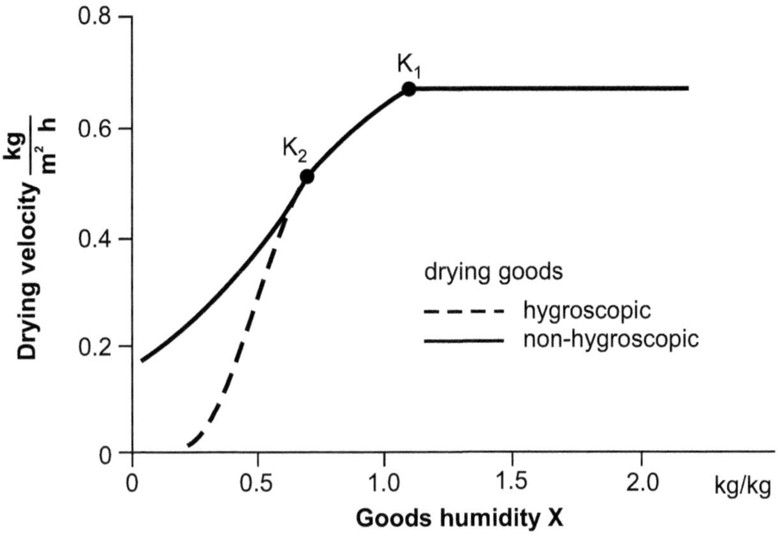

Fig. 15.36 Schematic representation of a drying process

Baffles constantly divert the polymer chips in the hot air flow so that such lumps cannot form. The combination of crystallization and drying reduces the moisture content of the PET chips from 0.5% before crystallization to 0.05–0.1% afterwards and through the subsequent drying to 50 ppm (standard filaments) or 20 ppm (microfilaments).

15.10.6 Drying Speed

From an energy point of view, it is advantageous to start the actual drying process with as little moisture as possible, because the heat of evaporation of water is very high.

The drying process of a capillary-porous granulate is shown in Fig. 15.36. During the first drying stage, the moisture evaporates on the surface and liquid is supplied from the inside by capillary forces. During this time, the surface temperature remains constant at the "adiabatic steady-state temperature" (evaporation temperature) and the drying rate is constant. From the point K_1, the drying rate decreases continuously with the decrease in fabric moisture content. The evaporation zone shifts from the surface to layers further inside. The moisture must thus reach the surface by vapor diffusion through the already dry layers in order to be dissipated to the ambient air. The transport resistance constantly increases and the drying rate decreases accordingly.

From the point K_2, the third section of the drying curve begins. Here, the moisture bound in microcapillaries is evaporated to a material-dependent residual moisture (equilibrium moisture).

15.10.7 Energy Balance

As an approximation, the minimum energy required for all drying processes acc. to the evaporation principle (heating above 100 °C) can be calculated from:

$$\dot{Q}_{overall} = \sum\left(\dot{Q}_{textile} + \dot{Q}_{water} + \dot{Q}_{evaporation}\right)$$

with:

Heating of the granulate:

$$\dot{Q}_{granulate} = \dot{m}_{granulate} \cdot c_{p,granulate} \cdot (u_2 - u_1)$$

Heating up the water:

$$\dot{Q}_{water} = \dot{m}_{water,1} \cdot c_{p,water} \cdot (u_2 - u_1)$$

or

$$\dot{Q}_{water} = \dot{m}_{textile} \cdot c_{p,water} \cdot (\varphi_2 - \varphi_1) \cdot (u_2 - u_1)$$

Evaporation of the water:

$$\dot{Q}_{evaporation} = \dot{m}_{water} \cdot (u_2 - u_1) \cdot \Phi$$

Whereby:

$\dot{Q}$: power (kW)

$\dot{m}_{granulate}$: mass flow of dry granulate (kg/s)

$\dot{m}_{water,1}$: mass flow rate of water at the inlet of the dryer (kg/s)

$c_{p,\ granulate}$: Heat capacity of the dry granulate (kJ/kgK)

$c_{p,water}$: Heat capacity of water (kJ/kgK)

u_2, u_1: Temperature at dryer outlet, temperature at dryer inlet (K)

φ_2, φ_1: Output humidity, input humidity (%)

Φ: Heat of evaporation (kJ/kgK)

References

Fourné, F. (1995). *Synthetic fibers*. Hanser.

Ignatowitz, E. (2015). *Chemietechnik*. Europa Lehrmittel.

Klöpper-Therm. (n.d.). Klöpper-Therm GmbH & Co. KG, Dortmund. http://www.kloepper-therm.de

MAAG. (n.d.). Maag Germany GmbH, Großostheim. https://maag.com/

Mahr. (n.d.). Mahr GmbH, Göttingen. https://metering.mahr.com/de/

Oerlikon. (n.d.). Oerlikon Barmag, Remscheid. https://www.oerlikon.com/en/brands/oerlikon-barmag/

Papin, D. (1695). *Recueil de diverses Pieces touchant quelques nouvelles Machines*. Estienne Huter.

Raysper, Y. (2005). https://commons.wikimedia.org/wiki/File:Chemical_reactor_CSTR_AISI_316.JPG

Stamixco. (n.d.). Stamixco AG, Wollerau, Schweiz. https://www.stamixco.com

Zimmer AG. (n.d.). Frankfurt a. M.

16 Processes for the Production of Man-Made Fibers

16.1 Process Steps

The most important process steps after polymer generation and before spinning out are shown in Fig. 16.1. After the polymer has been generated, it is transported to the spinnerets via pipes. In continuous processes, the liquid polymer is fed directly to the spinnerets and spun into fibers. Alternatively, the polymer can be processed into granules or powder, dried and stored. In the next step, depending on the material, it is either melted or dissolved and then spun.

The actual spinning process is divided into the sub-steps of spinning, solidifying, drawing and winding. Other process steps, such as washing, drying and texturing, may also be involved. In the production of staple fibers, a cutting or tearing process follows.

In general, a distinction is made between three basic spinning processes for fiber production:

- solution spinning: wet and dry,
- melt spinning

Figure 16.2 gives an overview of the process routes.

The spinning raw material is liquefied by melting or dissolving and pressed through a plate with small holes (spinneret), Fig. 16.3. The resulting polymer strands are called filaments. They are passed downstream of the spinneret through a gaseous or liquid medium, in which they are solidified by cooling, by removal of the solvent or by chemical reactions. The solidified filament bundle is then drawn off as yarn. This is usually done many times faster than the mass exits the spinneret holes, orienting the macromolecules, which increases the strength of the yarn.

In the spinning liquid in front of the spinneret, the chain molecules are statistically disordered or at best show a close order in small areas, but without preferential orientation. In the narrow capillary of the spinneret, a flow dominated by the shear viscosity is created. The bore is usually very short, so that an ideal tube flow cannot form. In addition, the spinning fluid is viscoelastic. The pressure drop at the lower opening of the capillary therefore obeys Hagen-Poiseuille's law all the less, the smaller the ratio of the residence time in the bore to the relaxation time of the spinning fluid.

16.2 Solution Spinning

Solution spinning processes are mainly used to produce cellulosic fibers, but polyacrylonitrile, elastane and others can also be produced in this way. The polymer is brought into solution, often by changing it chemically, and spun out. The solvent is then removed either in a precipitation bath or with warm air. If the fibers remain chemically altered, they are referred to as derivative fibers (e.g. acetate). If, on the other hand, the chemical change is reversed, regenerated fibers are produced (e.g. viscose). Typical winding speeds are between 100 and 1000 m/min. It

D. Veit, *Fibers*, https://doi.org/10.1007/978-3-031-15309-9_16

Fig. 16.1 Process steps in the production of polymers

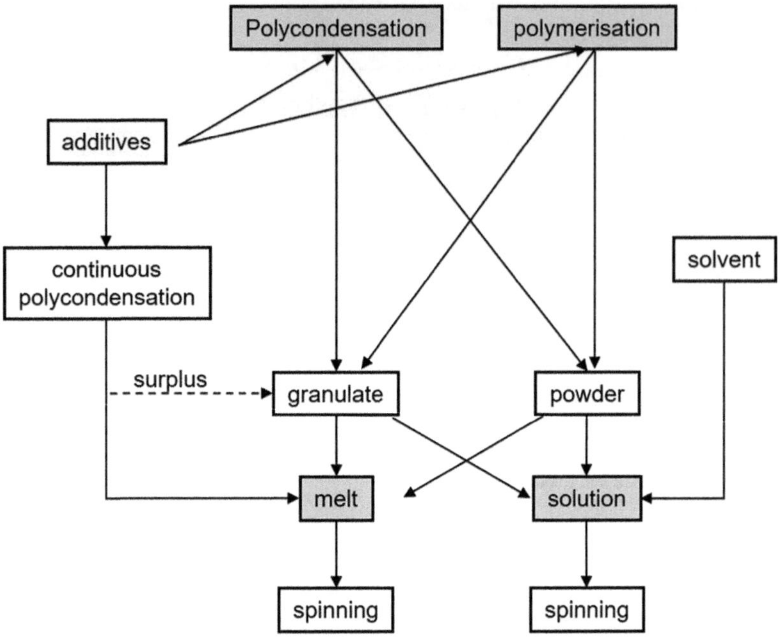

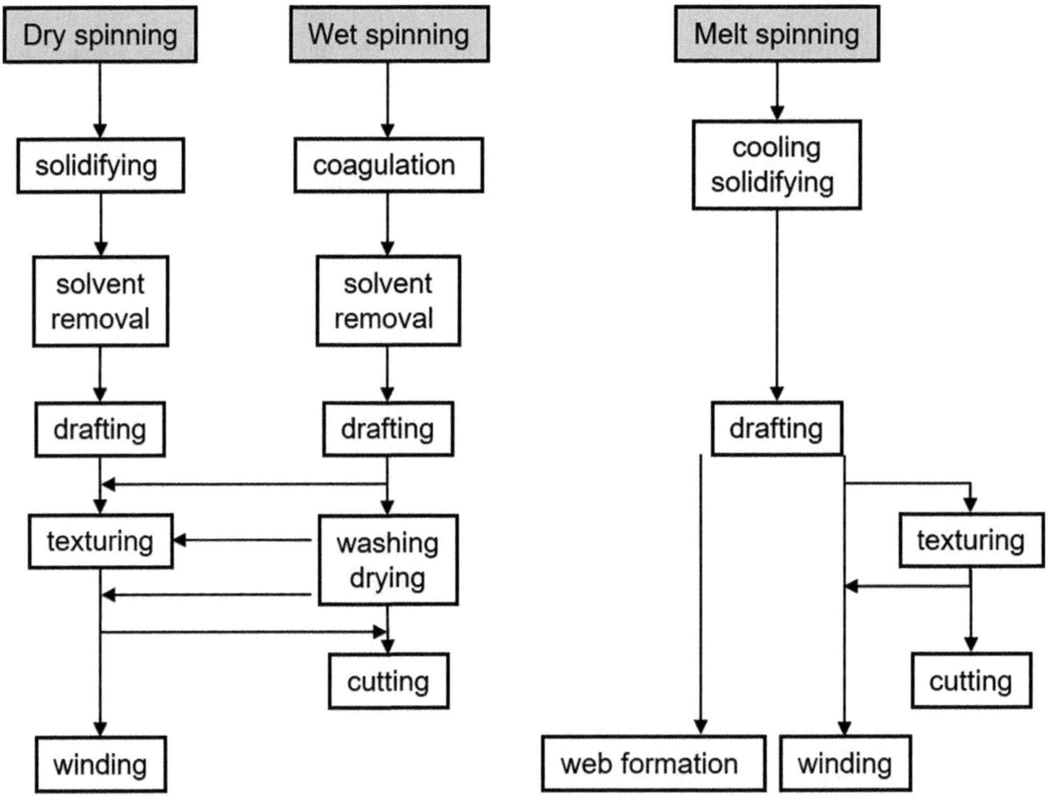

Fig. 16.2 Process steps in yarn or fiber production

Fig. 16.3 View from below of a spinneret during melt spinning (top left: round spinneret; top right: ring spinneret; bottom: rectangular spinneret, Oerlikon Barmag, n.d., Zimmer AG)

should be noted here that the higher the proportion of the solvent in the dope the lower the mass throughput. Typical polymer concentrations are 5–40%.

16.2.1 Dry Spinning

The dry spinning process is mainly used to produce polyacrylonitrile and acetate fibers. Low-boiling, ideally non-explosive and non-toxic chemicals that are easy to evaporate are used as solvents. These include, for example, acetone, ethers and alcohols. Decompression of the spinning solution causes some of the solvent to evaporate immediately after the spinning solution leaves the spinneret. Subsequently, the solvent diffuses from the forming filaments into the quench air. This causes the filaments to coagulate and a core-sheath structure is usually formed. To increase the process speed, the quench air is often heated to a temperature of 100–250 °C, depending on the solvent and the polymer. This increases its absorption capacity for the solvent. Typical polymer concentrations are 15–40%. Winding speeds are up to 1000 m/min as the extraction of the solvent from the solution is much quicker than with wet spinning owing to the lower viscosity of air compared to water. For reasons of economy and environmental protection, the solvent must be recovered as completely as possible, which requires additional machinery (Fig. 16.4).

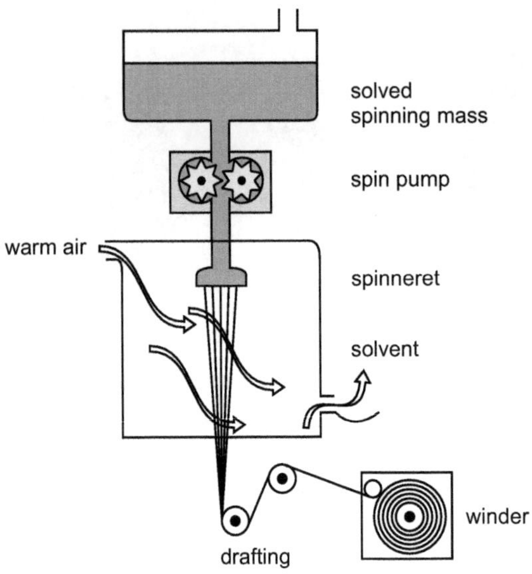

Fig. 16.4 Principle of the dry spinning process

The Table 16.1 gives an overview of the factors influencing the yarn properties during dry spinning.

16.2.2 Wet Spinning

The wet spinning process is mainly used to produce viscose, but modal, polyacrylonitrile, protein and alginate fibers are also produced in this way. The polymer is first dissolved and spun out through the spinneret into the precipitation bath. There the solvent reacts with the precipitant and the filaments coagulate. The sheath areas coagulate first and a thin skin is formed. Then, the solvent diffuses through the skin into the precipitating bath due to the osmotic pressure inside (Fig. 16.5). In the case of viscose, this results in the formation of small cavities, so that the fiber cross-section changes from a round shape to a wrinkled shape in the course of coagulation ("cauliflower structure", Fig. 16.23). Furthermore, a core-sheath structure is formed in the filament. The sheath consists of more and smaller crystals than the core, thus its strength and elongation are higher. Fibers with a high sheath content therefore have a higher fatigue resistance.

To produce viscose, the cellulose is usually solubilized in sodium hydroxide solution and by xanthogenation with CS_2. Common precipitants are mixtures of sulfuric acid and water (viscose) or nitric acid and water (PAN). Typical solvents for polyacrylonitrile are dimethylformamide, nitric acid, sulfuric acid and ethylene glycol carbonate, usually in combination with water.

If necessary, several precipitation baths are used in succession to remove the solvent completely. Often, the fibers then undergo several washing baths to remove solvent and precipitant residues.

The polymer concentrations reach 5–30% and are thus lower than in dry spinning. This is mainly due to the fact that the removal of the solvent by diffusion in an aqueous precipitation bath is significantly slower than the evaporation of the solvent in air during dry spinning. Typically, take-off speeds are 50–300 m/min, depending on the polymer.

The spinneret in the precipitation bath is usually directed upwards because the fibers formed are less dense than the precipitant and therefore rise to the top, where they can be easily drawn off. In addition, the spinning process is thus easier to observe.

16.2.3 Funnel Spinning

A variation of wet spinning is the funnel spinning process. It is used, for example, to produce cupro fibers from cellulose. In this process, the cellulose is first reacted with amine copper(II) hydroxide to form a complex that is soluble in ammonia. The less ammonia and the more electrolyte the spinning solution contains, the faster the thread coagulates, which in turn allows higher spinning speeds.

The spinning solution is pushed under pressure through the spinnerets and enters a precipitation bath filled with water. The thin jets of the highly viscous spinning solution are entrained by the water rushing downwards (Fig. 16.6). The speed of the water increases continuously due to the conical shape of the spinning funnel and the filaments that form are stretched. In the cross-

Table 16.1 Influencing variables on yarn properties during dry spinning acc. to Falkai (1981) and Mark et al. (1968)

Spin parameter		Viscosity of spinning solution	Spinnability	Residual solvent content in filaments	Drawability	Luster	Mechanical properties: Strength	Mechanical properties: Elongation
Solvent	Boiling point	×	⌒	↗	⌒	↗	↘	↗
	Density	↗	↘	—	↘	—	↘	↗
	Content of non-solvent	⌣	⌒	—	↘	↘	↘	↘
Concentration		↗	⌒	↘	↘	↗	↗	↘
Spin bath viscosity (0.2 ... 2 kg/m•s bei 40 °C)		×	⌒	—	⌒	↗	↗	↘
Degree of polymerisation		↗	⌒	—	⌒	—	↗	↘
Capillary	Number	×	↘	↗	↗	—	↘	—
	Diameter	×	⌒	—	↘	↗	↗	↘
	Length	×	↘	—	↘	↗	↗	↘
Temperature	Spinneret	↘	↘	↘	↘	↘	↘	—
	Quench duct	×	⌒	↘	↘	—	⌒	↘
	Quench air	×	⌒	↘	↘	↘	⌒	↘
Spinning duct	Length	×	⌒	↘	—	↗	↗	↗
	Diameter	×	⌒	—	—	—	↗	↘
Air velocity	Counter current	×	↘	↘	↘	↗	↗	↘
	Co-current	×	⌒	⌣	—	↘	↘	↗
Spinning speed		×	↘	↗	↘	↗	↗	↘
Draft (in spinning duct)		×	↘	—	↗	↗	↗	↘

↗ Property improvement
↘ Property deterioration
⌒ Property reaches maximum
⌣ Property reaches minimum
× No correlation
— No connection

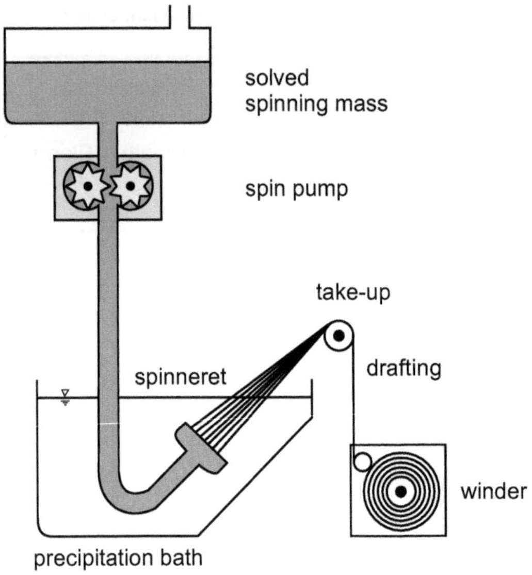

Fig. 16.5 Principle of wet spinning

section C–D (Fig. 16.6) the filament speed is approx. 10–15 times higher than in cross-section A–B. Stretching factors of 100–150 fold are achieved. This makes it possible to use capillary holes with—compared to conventional wet spinning—significantly larger diameters and still produce very fine fibers. In a second bath, the cellulose is regenerated from the copper oxide-ammonia complex by adding weakly concentrated sulfuric acid (1–2%).

The spinning speed is 50–150 m/min and is essentially limited by the fact that the water flow in the lower area must not become turbulent in order to produce uniform filaments. Filament formation is faster with increasing precipitation bath temperature. At the same time, coagulation should be slow so that smaller crystals form, which in turn can be more easily stretched. The precipitation bath ideally has a temperature of 35–40 °C.

Typical single filament counts for this process are just over 1 dtex for textile applications, and 6–10 dtex for carpet yarns.

16.2.4 Gel Spinning

Gel spinning is a combination of dry and wet spinning and is also known as air gap spinning. It is mainly used for polymers with extremely high molecular weight, e.g. polyethylene (UHMW-PE). Due to the large chain length, the macromolecules can hardly be stretched in pure form. Therefore they are dissolved, e.g. in paraffin oil, with a polymer concentration of 1–2%. In this way they remain mobile and can be oriented with low shear forces, resulting in very high strengths. The capillary holes in the spinneret are relatively large (0.5–1 mm) because of the very low polymer content in the spinning solution. After leaving the spinneret, part of the solvent is immediately evaporated and the polymer coagulates. Then, the coagulated filaments are immersed in the aqueous solution of the precipitation bath, where the polymer is regenerated. The gap between the spinneret and the spin bath is usually between 10 and 200 mm. Figure 16.7 shows a model of the spinning solution after the spinneret. Partially crystalline areas are already formed, but the macromolecules remain mobile due to the still relatively high proportion of solvent. Typical degrees of crystallinity are 50–80%.

In the next step, the macromolecules, which are only very slightly oriented, are stretched over heated rollers or a hot plate. Stretching of up to 100 times is achieved, usually in several stages. Optionally, drying can be performed downstream. In this way, degrees of crystallinity of up to 95% can be achieved in the end product (Fig. 16.8).

16.2.5 Liquid Crystal Spinning

Related to gel spinning is the liquid crystal spinning process. The main difference is the significantly higher polymer concentration of approx. 20%. This is possible because partially crystalline areas are already present in the solution phase, so-called mesogens (Fig. 16.10). They are obtained by the polycondensation of block building segments or copolymers before they react with each other to form the macromolecule. As the Fig. 16.9 shows, the viscosity of the spinning solution increases as expected with increasing polymer concentration. Until the maximum is reached, no liquid crystalline regions are formed yet, the material is isotropic. After

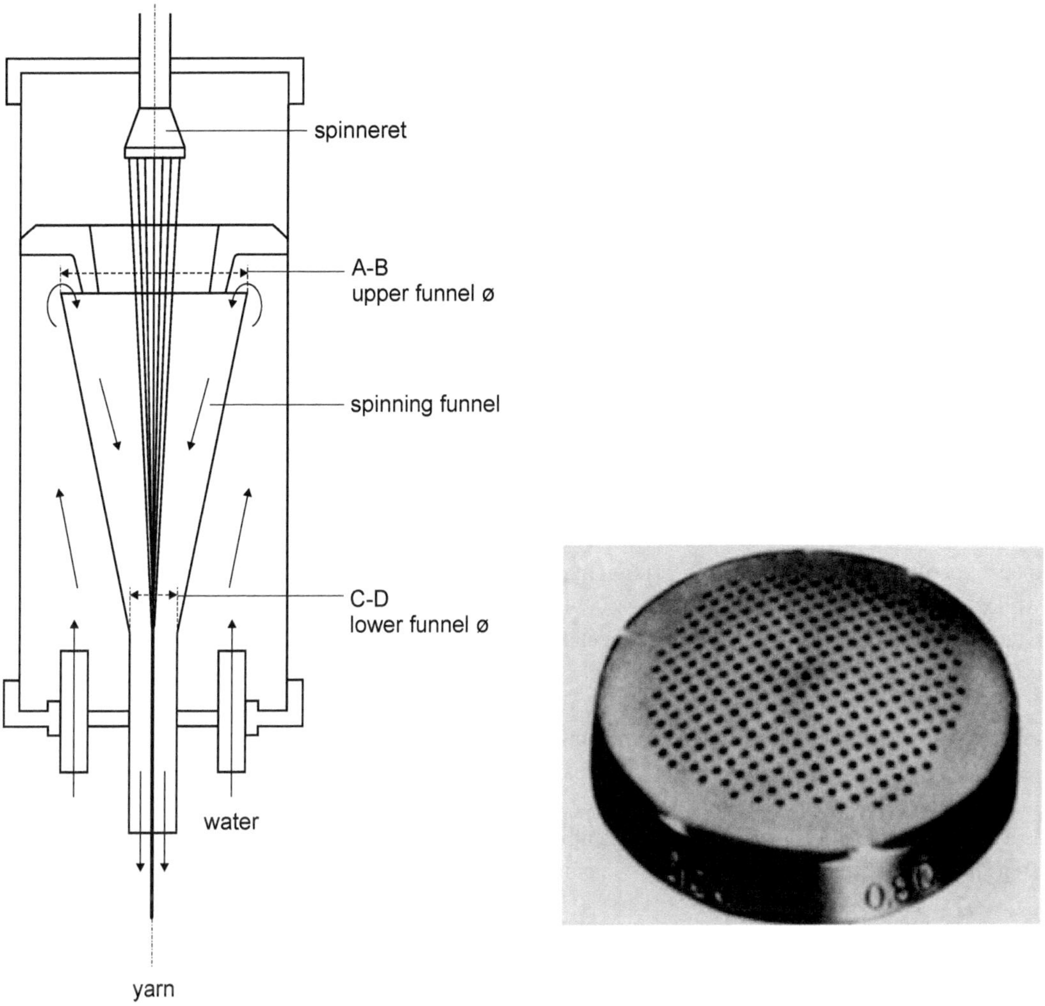

Fig. 16.6 Principle of funnel spinning with the Thiele spinning funnel acc. to Rogowin (1982) and spinneret (Götze, 1940)

exceeding a critical concentration, at which the viscosity is at a maximum, anisotropic crystalline regions form rapidly, which are more mobile than the previously unoriented macromolecules. As a result, the viscosity decreases rapidly. Thus, relatively high polymer concentrations are possible. Polymers that are formed by dissolving mesogens that have both hydrophilic and lipophilic properties are called lyotropic (Greek: lyo: dissolve; tropic: change).

In the liquid crystal spinning process, part of the solvent evaporates immediately below the spinneret before the polymer coagulates completely in the precipitation bath. The crystalline regions are already oriented somewhat in the direction of the fibers by the take-up tension and are compacted by the evaporation of the solvent (Fig. 16.10).

In fibers, the mesogens are present in the main chain. Accordingly, this results in the extremely high tensile strength of such materials. Depending on the arrangement of the macromolecules in the crystalline regions, one speaks of a nematic or smectic structure. The Fig. 16.11 shows this phenomenon by the example of p-aramid, which is produced using this process. The core area is

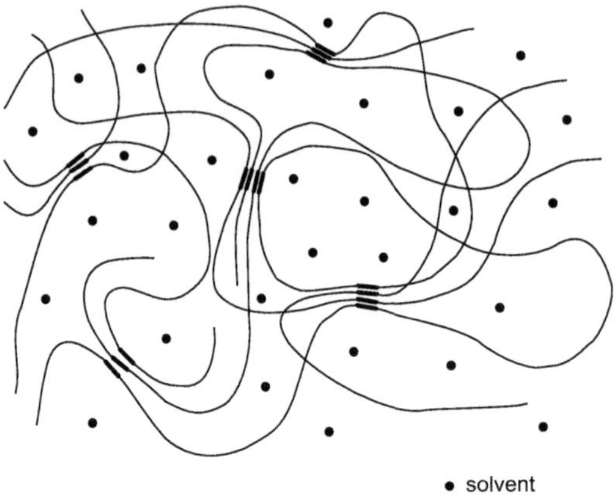

Fig. 16.7 Model representation of macromolecule assembly downstream of the spinneret during gel spinning acc. to Zhang (2014)

made up of crystals in a smectic arrangement, while the sheath crystals have a nematic structure. Cracks propagate through the gaps in between the crystals.

The situation with melt-spinnable liquid crystalline polymers is analogous to that with soluble liquid crystalline polymers. The viscosity increases with the concentration of polymer up to a critical temperature at which liquid crystalline areas form. After that, the viscosity drops again.

16.2.6 Dispersion Spinning

This process is used for polymers that cannot be dissolved and are also not melt-spinnable. Depending on the aggregate state of the polymer (liquid, solid), it is also referred to as emulsion or suspension spinning. In this process, the polymer is dispersed in a spinnable polymer, usually in an aqueous solution. The fiber-forming polymer is then spun out using dry or wet spinning. Thereafter, the fiber-forming polymer is dissolved and the non-spinnable polymer is sintered in its then present fiber form. A typical example is the production of PTFE fibers using polyvinyl alcohol, which can be easily dissolved.

16.2.7 Reaction Spinning

In this process, the basic substances are dissolved in different solvents. The solvents are not or only slightly miscible. At the interface between the solvents, the two components react with each other and may already form crystalline areas. Thus, they are extruded through the spinneret and the solvent is removed by dry, wet or air gap spinning. Alternatively, one component is injected in dissolved form through the spinneret directly into the solution of the other component, where both react immediately to form the polymer and the resulting filament is drawn off directly from the solution and wound up.

This process is used, for example, for the production of elastane fibers, in which long-chain (elongation) and short-chain (strength) monomers react with each other. Aramides and special

Fig. 16.8 Principle of gel spinning

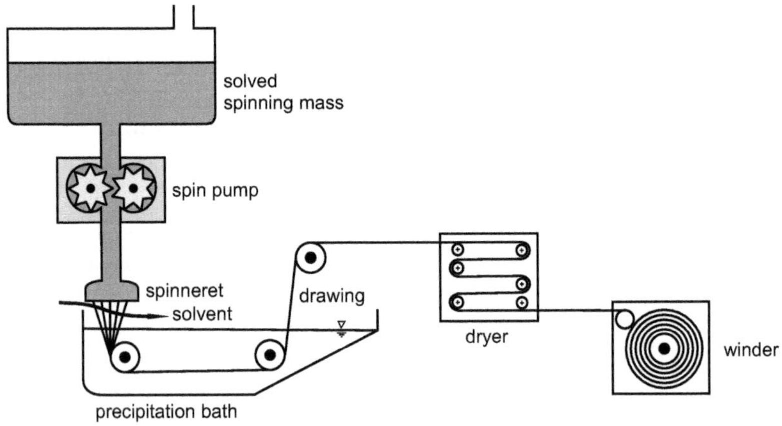

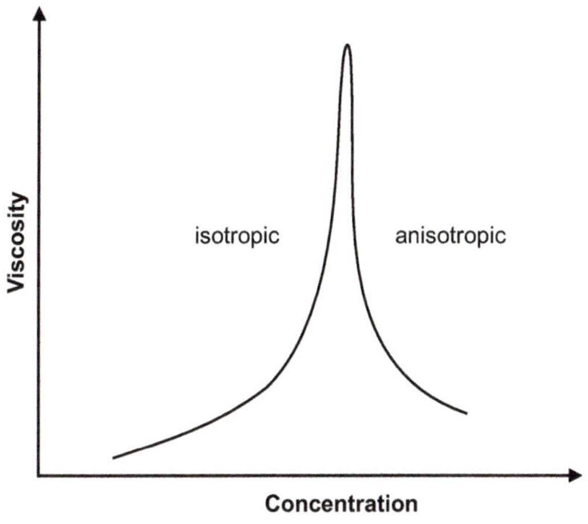

Fig. 16.9 Viscosity of lyotropic liquid crystalline polymers as a function of concentration acc. to Zhang (2014)

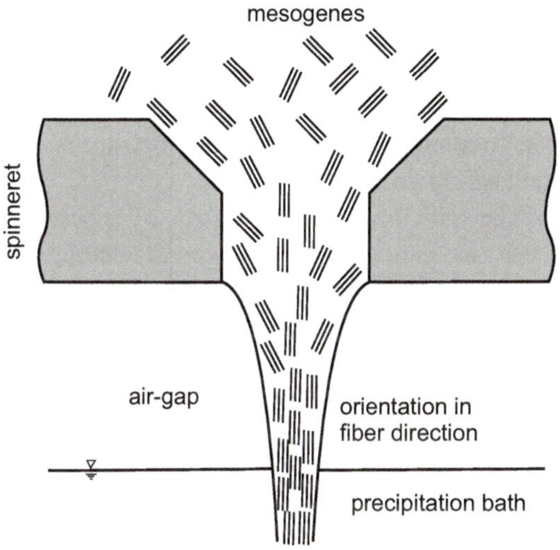

Fig. 16.10 Conditions at the spinneret during liquid crystal spinning

polyamides as well as polyimides can also be produced in this way.

16.3 Melt Spinning

Melt spinning is the most widely used spinning process because of its economy and variety of processable polymers. The Fig. 16.12 shows the basic principle. The molten polymer is first fed to the spinnerets via a metering (spinning) pump.

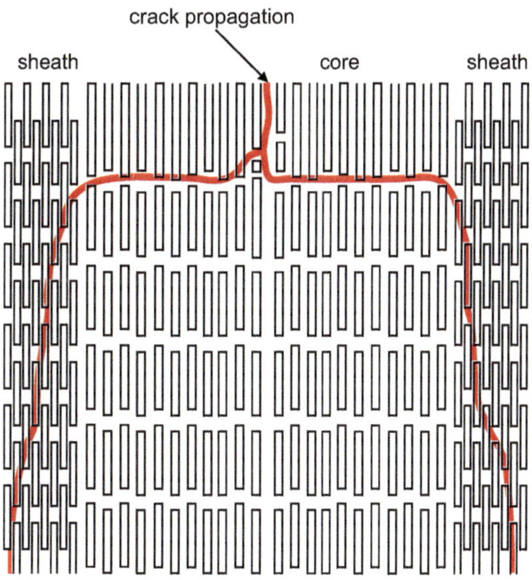

Fig. 16.11 Internal structure of an aramid with core-sheath structure and crack propagation

Upon leaving the spinneret, the polymer initially expands slightly due to relaxation caused by the pressure difference between the spinneret and the environment. It then solidifies as a result of the blowing of cool air along the spinning path. Before the first godet, the filament temperature should fall below glass transition temperature in order to avoid wrong draft and filament breaks. Depending on the process, the yarn is then stretched to a greater or lesser extent and wound up.

Depending on the yarn type, typical take-up speeds are 3000–5000 m/min, in special cases even up to 10,000 m/min. The mass throughput is thus orders of magnitude higher than with solution spinning due to the take-up speed alone. This is further enhanced by the fact that the entire polymer flow can be processed into fibers or yarn because no solvent is required. This process is mainly used for polyester, polyamide and polypropylene.

16.4 Spinnability

Of the large number of polymers, only a few can be processed into fibers under certain conditions. These include a minimum chain length,

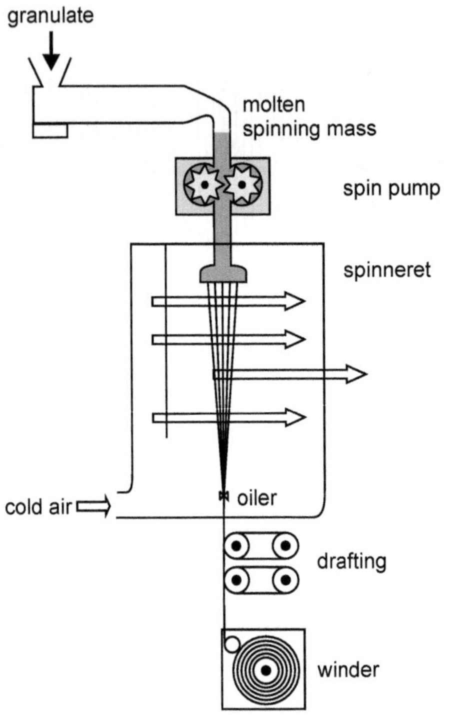

Fig. 16.12 Principle of melt spinning

sufficiently large intra- and intermolecular forces and sufficient viscosity.

When the spinning fluid exits the spinneret, the lateral restriction of the flow is omitted. In the case of viscoelastic spinning liquids, normal stresses are released from the liquid which have not yet been reduced during the short residence time in the spinneret hole. They cause the spinning jet to widen behind the spinneret opening. This phenomenon is also known as the Barus or memory effect. The greater the relaxation time and the shorter the residence time of the spinning mass in the capillary, the greater the expansion. In extreme cases, the polymer jet emerging from the spinneret can become rough and unevenly thick; this is referred to as melt fracture (Fig. 16.13).

Immediately after the spinneret, it also becomes apparent whether a spinning fluid can be spun at all. If its interfacial surface tension against the spinning medium is too high in relation to its viscosity, the polymer jet contracts into drops and the filament breaks (capillary oscillation breakage). This can be avoided by increasing the viscosity and/or the filament take-off speed.

If, on the other hand, the viscoelasticity and thus the relaxation time of the spinning fluid is too great and spinning is too fast, then the yarn tension exceeds the tensile strength of the filaments and yarn breakage occurs (cohesion defect). This can be avoided by reducing the viscosity and/or the take-off speed.

The spinning range of a polymer lies between these two extremes. If the necessary settings of the spinning parameters are mutually exclusive, the polymer cannot be spun.

The Fig. 16.14 on the left shows the strength as a function of the product of viscosity and spinning take-off speed v for a cellulose acetate

Fig. 16.13 Orientation of the macromolecules in the spinneret (left) and jet expansion behind the spinneret hole (right) acc. to Fourné (1995)

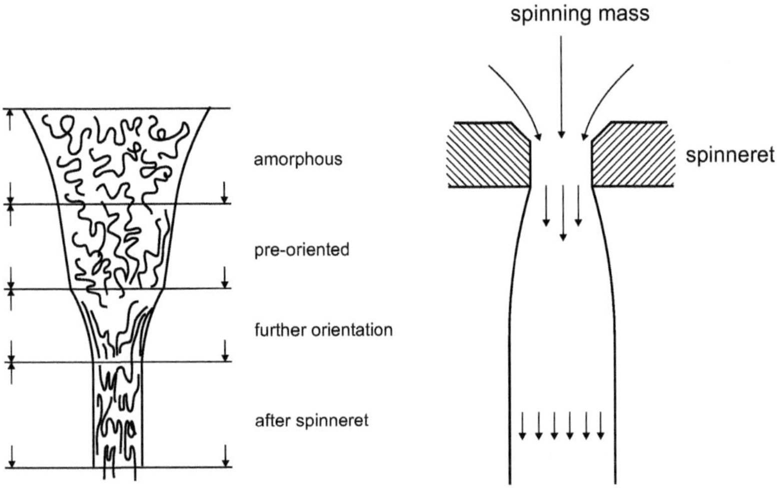

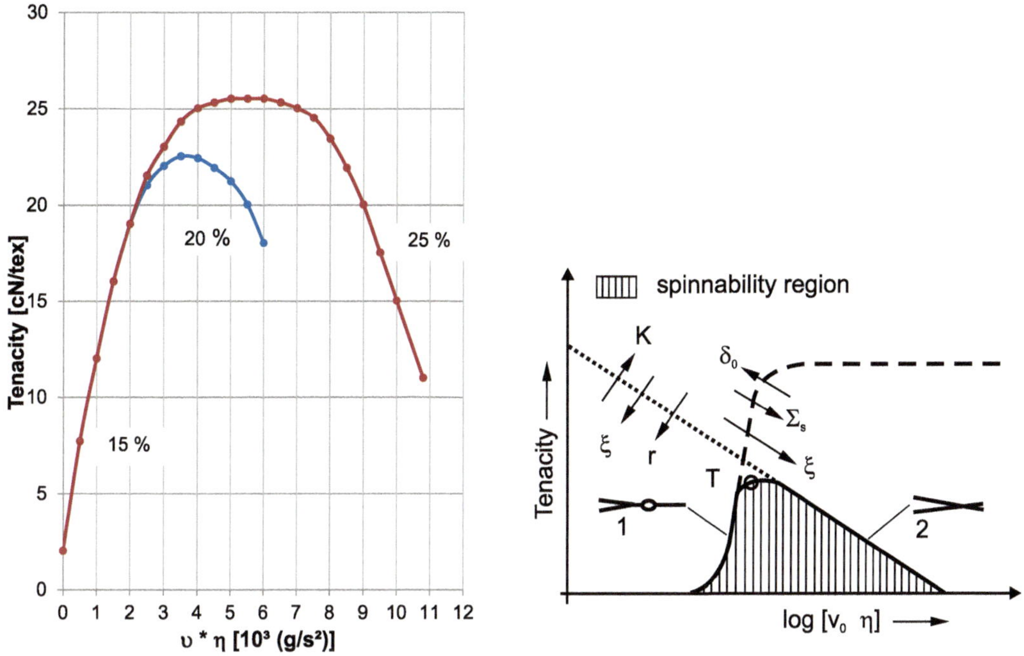

Fig. 16.14 Strength and range of spinnability acc. to Fourné (1995) and Ziabicki (1976)

solution of different concentration. For a stable process, a maximum of the strength should be achieved, resulting in corresponding combinations of values. In Fig. 16.14 on the right, the spinnability range of a typical polymer is shown. It is obtained by superimposing the theoretical curves for capillary vibration fracture (1, dashed line) and for cohesive failure (2, dotted line). Shear viscosity is a function of surface tension σ_s, cohesive energy K, relaxation time τ, and initial diameter d ($d = 2r$), as well as the axial deformation gradient $= \mathrm{dln}(\nu/\nu_0)/\mathrm{dx}$. The arrows indicate the direction in which the theoretical curves shift when the values of these parameters are increased.

Figure 16.15 shows the equilibrium of forces on the spinning thread. The following forces are opposed to each other: On the one hand, the surface tension force $F_{surface.}$, which gives the yarn a minimum surface area due to shrinkage, the viscosity force F_{rheo}, the friction force F_{fric} between the yarn and the spinning medium (e.g. air or water) and the mass inertia force F_{inert} of the yarn. On the other hand, there are the external take-off forces F_{ext}, as well as the gravity force F_{grav}, which acts on the yarn (with vertical take-off downwards).

Within the spinneret, the macromolecules become somewhat oriented. However, the time for this is very short, so this effect can be neglected. Furthermore, the polymer relaxes when leaving the spinneret (Fig. 16.15). As a result, this orientation is largely lost again directly.

The Fig. 16.16 gives a qualitative overview of the spinning process. Apart from the jet expansion just below the spinneret opening, the filaments become thinner and thinner due to the rapid take-up speed v_A from the spinneret, the tension in them increases. At the same time, they gradually solidify.

The temporal and local course of the cross-sectional change in the filament depends on the mass flow rate, the extensional viscosity as a function of the solidification state, the course of the solidification and the take-off speed.

For the tensile force in the thread at any point along the spinning path holds:

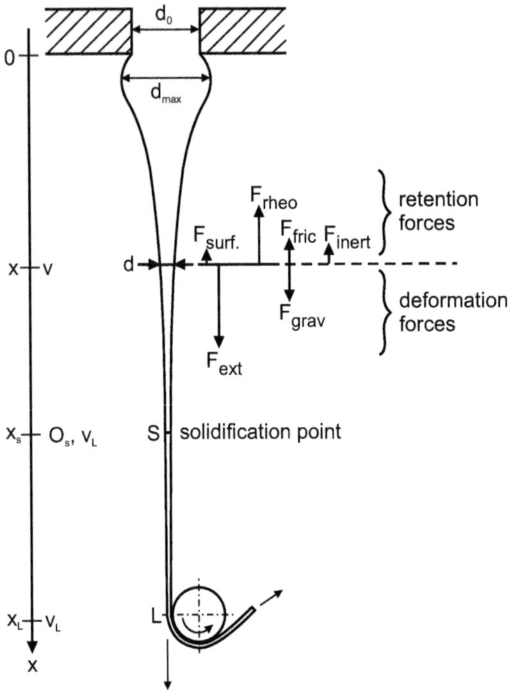

Fig. 16.15 Equilibrium of forces on the spinning thread acc. to Fourné (1995)

$$F_{drawing} = F_{deformation} + F_{acceleration} + F_{limit} + F_{friction} - F_G$$

F_{limit} denotes the forces against the interfacial tension and $F_{friction}$ denotes the frictional forces against the spinning medium. F_G stands for gravity and is negative when spinning downwards. In the viscose spinning process, where the draw-off is upwards, a positive sign results accordingly.

The forces depend on the respective spinning process and the spinning parameters. This balance also applies to the force at the end of the spinning path, the so-called spinning force.

Especially according to the constant diminution of the filaments, the tension in them gradually increases starting from the spinneret, from the solidification point downwards only as a result of friction against the spinning medium. Due to the increase in tension, a molecular orientation is constantly built up in the polymer. As long as the mass is still thin, the relaxation proceeds very quickly, and the orientation that is created immediately reverts. This also applies to the flow orientation created in the spinneret hole. While the filament gradually becomes more solid, the relaxation speed decreases. When it becomes comparable with the increase in tension and orientation, a measurable orientation is formed in the filaments, the formation of which predominates during further solidification. It remains in the spun yarn when it finally solidifies.

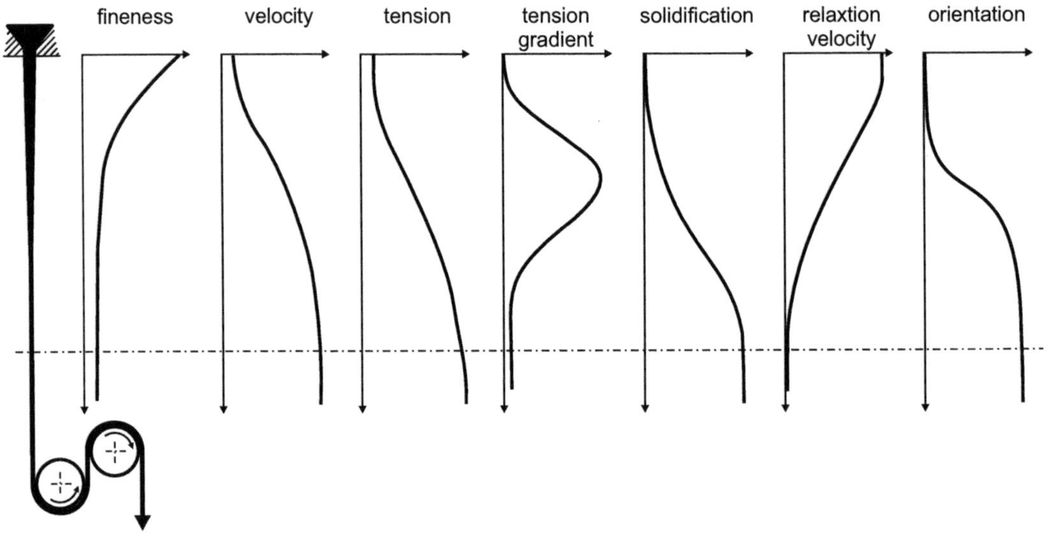

Fig. 16.16 Situation during spinning (spinning path shortened) acc. to Fourné (1995)

16.5 Orientation of Macromolecules and Mechanical Properties

Depending on the crystallization tendency of the polymer, a partially crystalline order forms early on. The crystallization rate, and thus the crystallinity in the spun material, grow with increasing orientation, provided that the polymer crystallizes at all. The close order that forms can now in turn influence the relaxation processes and the formation of the orientation, so that there is often reciprocal effect between polymer properties, spinning and solidification conditions, orientation and crystallization. This is particularly true in the case of high crystallinity. Crystallization can be favored by diffusing plasticizers. However, there are also orientation-enhancing measures that hinder crystallization. For example, rapid solidification of the filaments can mean that the polymer no longer has enough time on the spinning path to crystallize to the desired degree of orientation. This can be observed, for example, with very thin, rapidly solidifying spinning filaments (Fig. 16.17).

The formation of crystalline regions begins at a temperature just below the melting temperature. However, the rate of crystal formation is then still low because the macromolecules are very mobile. As the temperature increases, the diffusion rate increases, while the nucleation rate increases as the temperature decreases. At a polymer temperature just above the glass transition temperature, the rate of crystallization is again low because the mobility of the macromolecules is correspondingly low. In between, there is a temperature $T_{c,max}$, at which the crystallization rate is maximum.

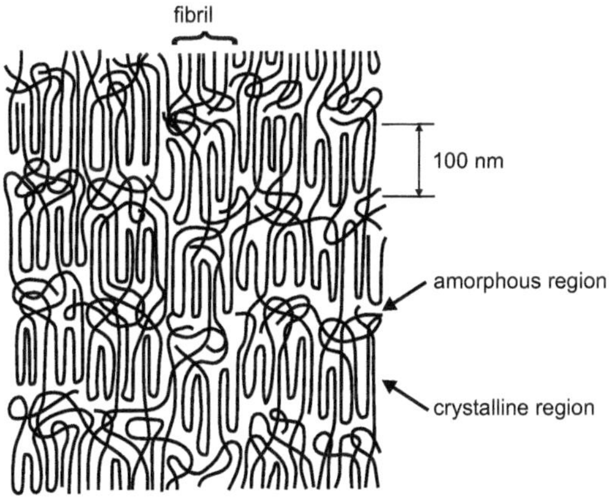

Fig. 16.17 Amorphous and crystalline regions in a polymer acc. to J. Hearle

As the molecular weight M_w increases, the mobility of the macromolecules decreases, so that the crystallization rate curve is correspondingly flatter. This results in the bell-shaped curve of the crystallization rate shown in the Fig. 16.18 as a function of temperature and molecular weight (where $M_{W2} > M_{W1}$).

It is obvious that a certain crystallization rate C can be achieved at two different temperatures (T_1, T_2). At the lower temperature, many small crystals are formed due to the high nucleation rate, and at the higher temperature correspondingly few, but large crystals appear. Rapid cooling can therefore have a positive influence on the subsequent drawability, because small crystalline areas can be displaced more easily than large ones.

The Fig. 16.19 shows for the most common polymers the temperature at which the crystallization rate becomes maximum. In most cases, it is approximately midway between the glass transition temperature and the melting temperature.

In melt spinning, the cooling rate is usually very high in the order of 100–1000 K/s. A higher cooling rate reduces the time the polymer has to crystallize at a given temperature. Therefore, the formation of crystalline regions usually occurs at temperatures well below the melting temperature. If the cooling rate is very high, it may even result in the polymer not crystallizing at all. In the Fig. 16.20 this is shown for different cooling rates (1 = low, 6 = high). The so-called C-curve represents the dependence of the onset of crystallization on the temperature and the time available. If the polymer is given less time, the onset of crystallization shifts to lower temperatures. As can be clearly seen, this can result in the cooling curve of the polymer not meeting the C-curve at all at a very high cooling rate (4, 5, 6). In these cases, the polymer therefore remains amorphous.

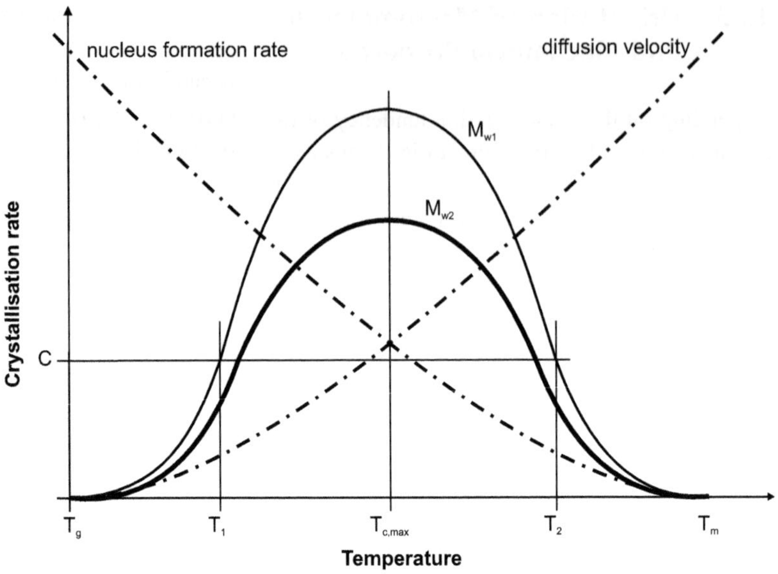

Fig. 16.18 Crystallization rate as a function of temperature (after Bonnet)

In the melt spinning process, the crystallization rate is increased by inducing a stress by pulling the filaments into the macromolecules. As a result, crystallization starts earlier and the C-curve shifts to the left (Fig. 16.21). Thus, crystal formation is initiated even in the polymers with a high cooling rate (4, 5). This is referred to as stress-induced crystallization. Only at very high cooling rates (6) does the polymer remain amorphous.

The Fig. 16.22 shows the course of the filament temperature and the degree of crystallinity as a function of the distance to the spinneret. As soon as crystallization begins (A), heat is released so that the filament cooling by the blowing air is partially or completely compensated. If crystallization is then largely complete (B), less heat is released and the filaments continue to cool rapidly.

The solidification of a filament always starts on the outside first and then continues towards the

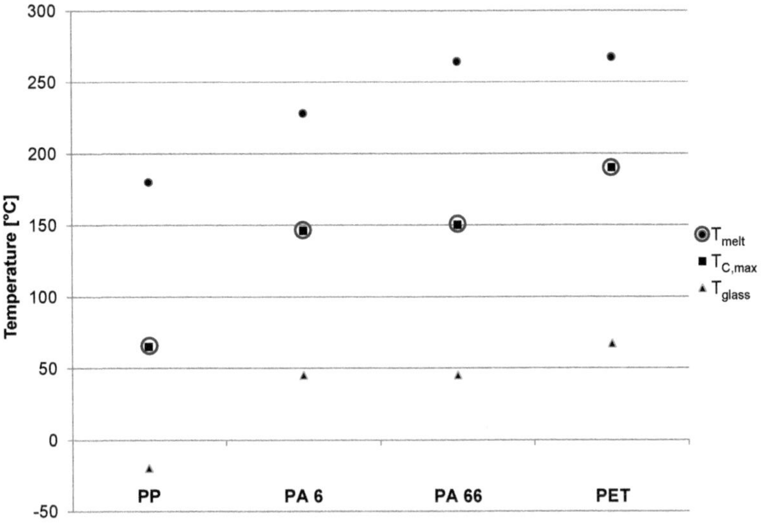

Fig. 16.19 Typical temperatures for the maximum crystallization rate of selected polymers acc. to Ziabicki (1976)

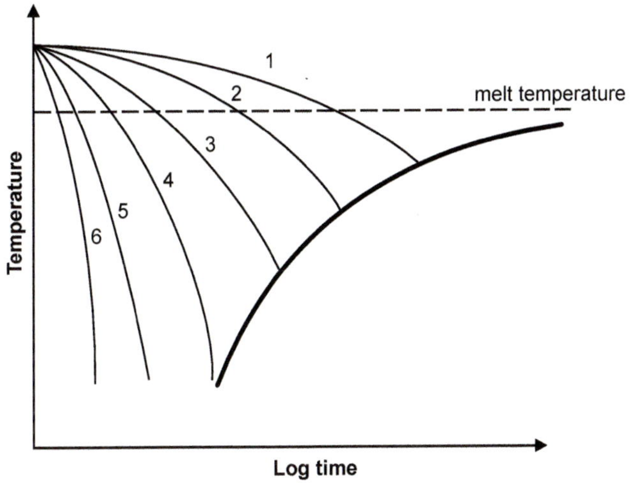

Fig. 16.20 Temperature-time diagram of crystallization of polymers (strainless) acc. to Zhang (2014)

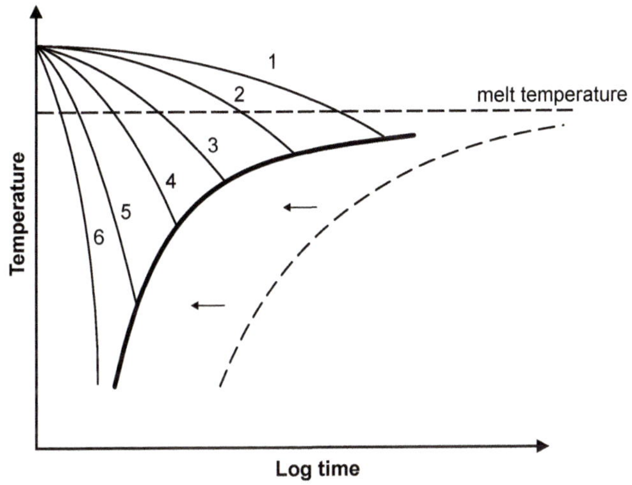

Fig. 16.21 Temperature-time diagram of crystallization of polymers (stress-induced) acc. to Zhang (2014)

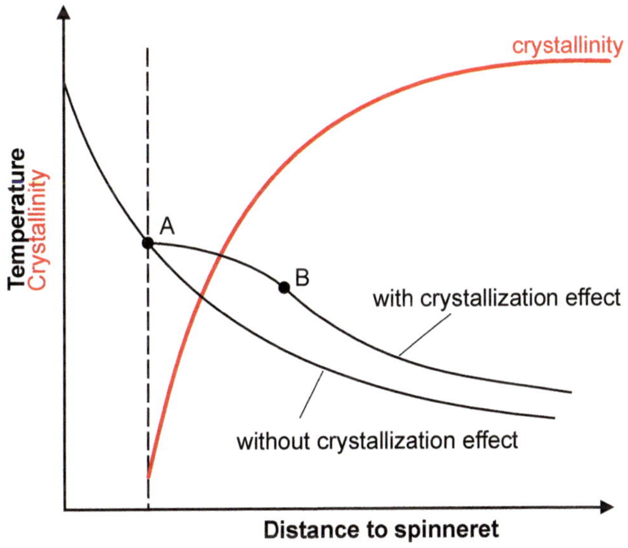

Fig. 16.22 Temperature of filaments as a function of crystallinity during the spinning process acc. to Zhang (2014)

Fig. 16.23 "Cauliflower effect" in viscose fibers in cross-section (wet spinning process)

inside due to heat or mass transfer. A filament is therefore initially more highly oriented on the outside and often more strongly crystallized than on the inside. This radial inhomogeneity is also called core-sheath effect. Especially in wet spinning, it has a considerable influence on the processes in the spinning path and the textile and dyeing properties of the end product. In viscose, for example, it leads to the typical "cauliflower" cross-section of the fibers (Fig. 16.23).

In melt spinning, this effect is usually smaller or even non-existent. However, here the cooling of the filaments takes place from the outside to the inside. This results in a temperature gradient, according to simulations in the order of 0.1–1 K/μm, transverse to the fiber axis, which can influence crystal formation. Often, therefore, a core-sheath structure is also formed, whereby the orientation is usually higher in the sheath than in the core and the crystallinity is correspondingly higher. In the case of cross-flow blowing (cooling from one side only), which is usually used, there is an additional temperature gradient in the direction of the air inflow velocity. Similarly, the cooling of the filaments varies depending on their position in the spinneret and their crystalline structure differs. In the case of very thin filaments, this can lead to differences in dyeing during further processing.

Within a certain framework given by the polymer, the crystallinity of the fibers can thus be influenced. A more crystalline fiber is less sensitive during its further handling, but on the other hand a crystalline structure must be broken up again with each subsequent drawing. Depending on the spinning process and further processing, this results in different orientations of the macromolecules, ranging from amorphous to semi-crystalline to highly crystalline (Fig. 16.24).

Thus, the time and location of the extent of orientation in the spinning filament depend significantly on the solidification conditions.

Although the relationships between the spinning conditions and the sequence of filament formation have been studied in detail in numerous research projects, it has not yet been possible to derive a closed theoretical model from them. Therefore, spinning trials are still necessary to determine the ideal settings of the spinning process for a specific polymer.

16.6 Theory of Drawing

The textile-mechanical properties of the filaments immediately after spinning are rarely sufficient for further processing: strength, modulus of elasticity and thermal stability are often still too low, while plastic ductility is frequently too high. This is mainly due to the fact that the orientation of the macromolecules in the crystalline and non-crystalline segments is usually too low. In addition, crystalline areas or correspondingly acting lateral bonds are often not or hardly present. If the spun fiber is now stretched beyond a certain extent, the orientation of the macromolecules becomes larger and the number of cross-links increases accordingly. The processes during stretching depend essentially on whether the filaments are in an amorphous or crystalline state.

Various ideas have been developed for the structural transitions, but these are as certain or uncertain as the model ideas of the structure of semi-crystalline oriented and non-oriented polymers. A generally accepted model shows Fig. 16.25. According to this model, partially crystalline regions are already present in the undrawn filament and are aligned in the axial direction of the fiber during drawing.

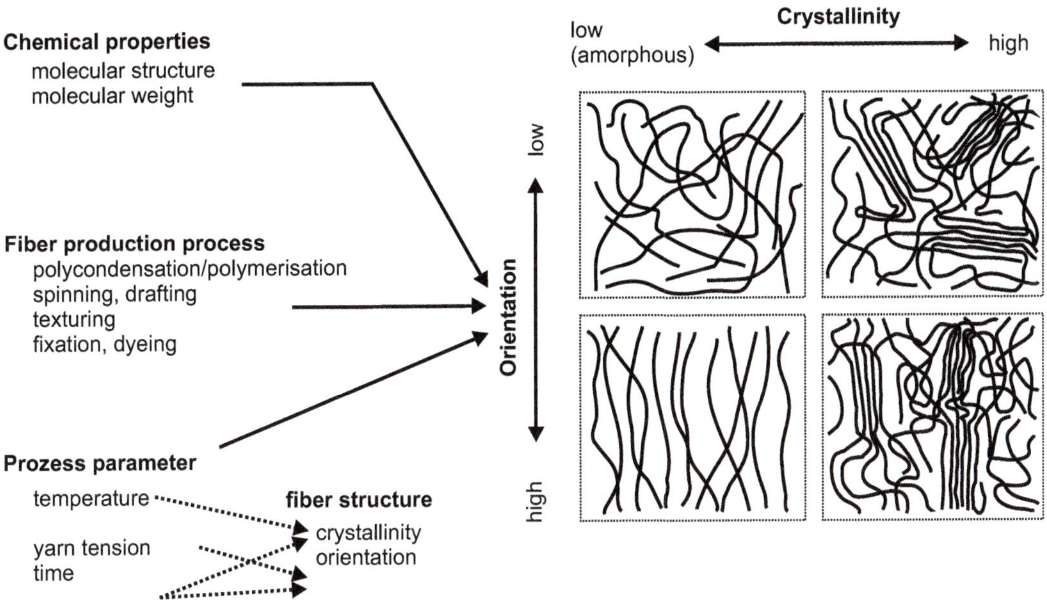

Fig. 16.24 Effect of fiber production, processing and finishing methods on fiber structure and fiber properties acc. to Fourné (1995)

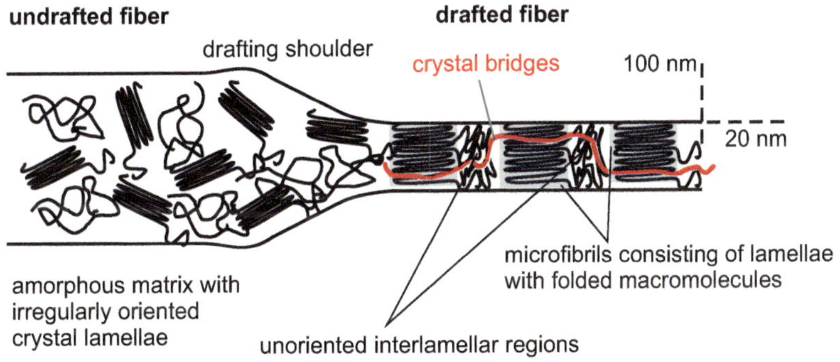

Fig. 16.25 Structural model acc. to Elias (2001)

16.6.1 Drafting of Amorphous Filaments

Some polymers require more time to form crystals than is available during the spinning process. Therefore, they exist in an amorphous state prior to drafting. Typical examples are polyethyleneterephthalate (PET) and polyacrylonitrile (PAN). If they are slowly drafted, the macromolecules orient themselves in the tensile direction, but no crystallization takes place yet. At a certain stretching ratio (Fig. 16.26A) stress-induced crystallization begins (cf. Fig. 16.21). Simultaneously with crystallization, the orientation of the macromolecules increases rapidly until a critical stretching ratio (B) is reached. Here, the first tie molecules are now present in stretched form, connecting the crystalline domains. This massively hinders the disentanglement and orientation of the macromolecules and the crystalline regions, and the curves for orientation and crystallinity flatten out.

The Fig. 16.27 shows the model representation in point B.

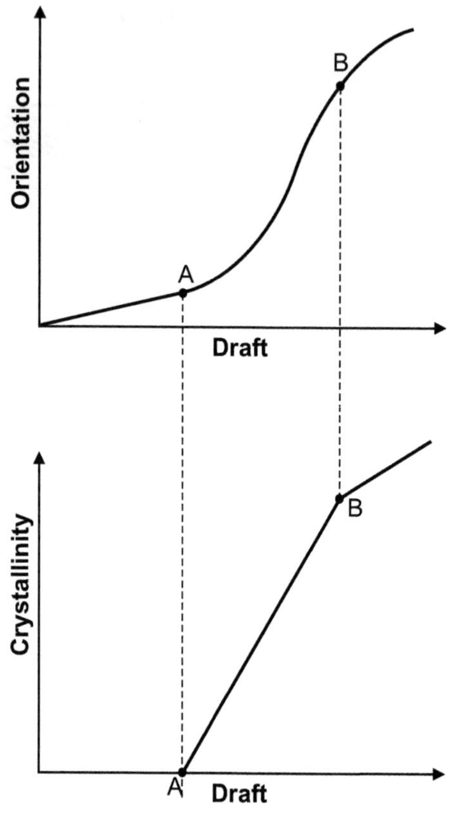

Fig. 16.26 Change in orientation and crystallinity during stretching of amorphous filaments after Zhang (2014)

16.6.2 Drafting of Crystalline Filaments

Some polymers are already in a partially crystalline state directly after spinning. These include polyamide (approx. 35% degree of crystallinity), polyethylene and polypropylene (50–80%). However, the crystalline regions are usually not oriented in the direction of the fibers, but are distributed randomly in the polymer. Therefore, the mechanical properties of such filaments are not yet sufficient for further textile processing. In Fig. 16.28, the course of the orientation of amorphous and crystalline areas as a function of the draw ratio is shown. Basically, at low stretching ratios, the amorphous areas are oriented first. With increasing tension due to drawing, the crystalline areas also begin to orient themselves in the direction of tension. In the process, large crystals can break into smaller crystals as a result of the stress, but they are still connected to each other. These then orient themselves more easily in the tensile direction than large crystalline areas.

With increasing stretching, a transformation from amorphous to crystalline regions may occur, as Fig. 16.29 shows for polyamide.

With some polymers, a transformation of the crystal lattice can also occur. This can usually be supported by the addition of heat (e.g. in the case of polyamide). In this case, the hexagonally densely packed crystals are transformed into more easily orientable, sheet-like crystals, which facilitates drafting (Fig. 16.30).

16.7 Force-Elongation Curve During Drafting

During drafting, depending on the polymer, temperature and stretching conditions, different curves of the drawing force occur (Fig. 16.31). It is measured by stretching the yarn between two clamps and then applying a tensile force. If the temperature is far below the glass transition point

Fig. 16.27 Model representation of stretched tie molecules between crystalline domains after Zhang (2014)

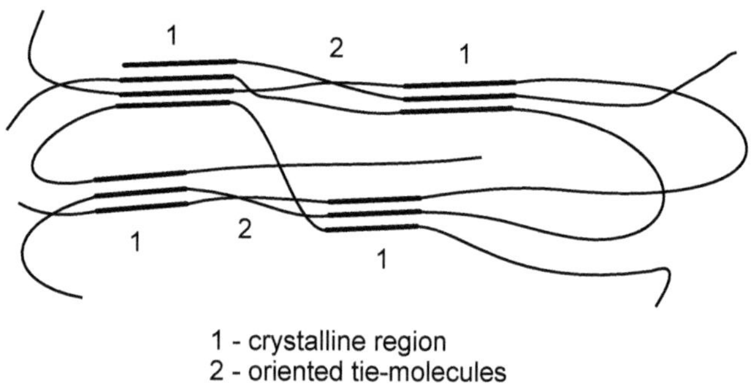

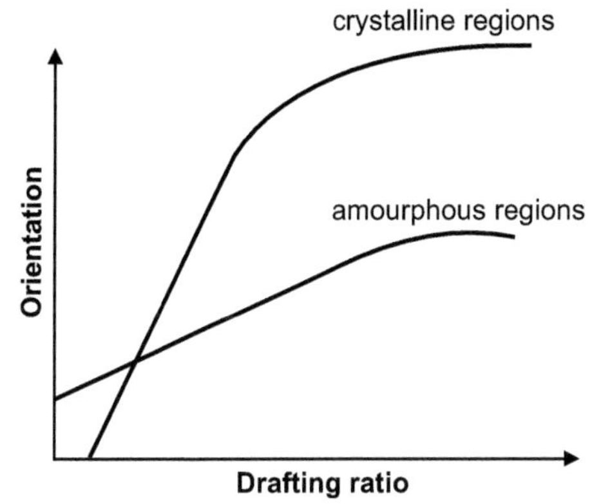

Fig. 16.28 Change in orientation during stretching of crystalline filaments after Zhang (2014)

or the yarn is already very solidified internally due to crystalline bonds, the force increases rapidly and the yarn breaks even at low elongations (Fig. 16.31a). If the temperature is significantly above the glass temperature or if the spun yarn is plastic due to plasticizers, for example due to residues of the solvent when spinning from solutions, then the yarn elongates considerably more, with the force increasing uniformly and only very slowly (c).

Between the two described curves lies the force-elongation diagram with stretch saddle (Fig. 16.31b), as occurs with many man-made fibers when the temperature of the yarn is approx. 50–60 °C above the glass temperature. With increasing elongation, an elastic stress first builds up until a stretching point forms at a random

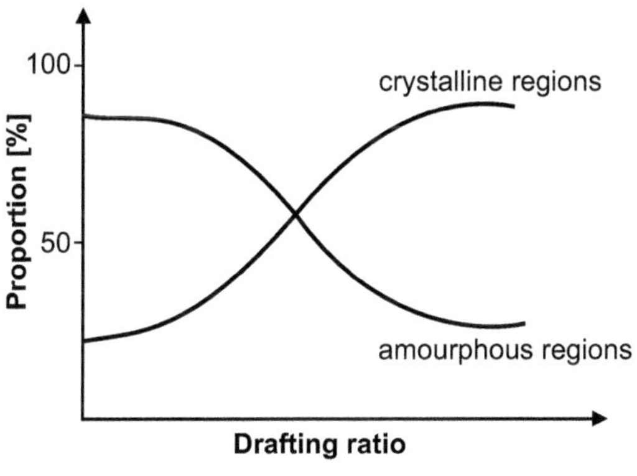

Fig. 16.29 Proportion of amorphous and crystalline regions during stretching acc. to Zhang (2014)

point, which moves along the clamped thread. During this process, the stretching force remains constant. It only increases again when the stretching shoulder has travelled along the entire thread between the clamps. The elongation at which this occurs is called the *natural stretch ratio*. This is also the difference between the fineness of the thread on either side of the stretching point (Fig. 16.32).

This phenomenon of cold stretching, which is also extremely important for industrial practice, was first observed by Carothers and Hill (1932a, b). There has been much and also controversial discussion about how it comes about. The viscoelastic behavior of the spun fiber and the heat of elongation concentrated at the point of elongation probably play a role.

The Fig. 16.33 shows the dependence of tensile strength and elongation-at-break as a function of filament temperature. It can be clearly seen that at a temperature above the glass transition temperature, the tensile strength decreases sharply and the elongation at break increases rapidly. The supply of heat during stretching can therefore be advantageous in terms of the forces required and thus the uniformity of the stretching. After drawing, internal stresses may nevertheless prevail in the polymer, which will have a negative effect in the subsequent processing steps (e.g. by shrinkage of the fabric). For this reason, drawn yarns are often thermally treated during further processing, which relieves internal stresses and simultaneously stabilizes the drafted state.

The cold draw test characterizes the spun fabric. The more strongly it is oriented, the smaller the natural draw ratio and the higher the yield strength. The yield stress, related to the linear density after natural stretching, is approximately independent of the spinning orientation and, with otherwise unchanged stretching conditions, depends essentially on the polymer and on the position of the stretching temperature in relation to the glass temperature. The product of orientational spinning elongation and natural draw ratio is approximately constant and is in the range of about 5–7 for all polymers.

The dependence of strength and elongation of the spun fiber on the stretching is exemplified by

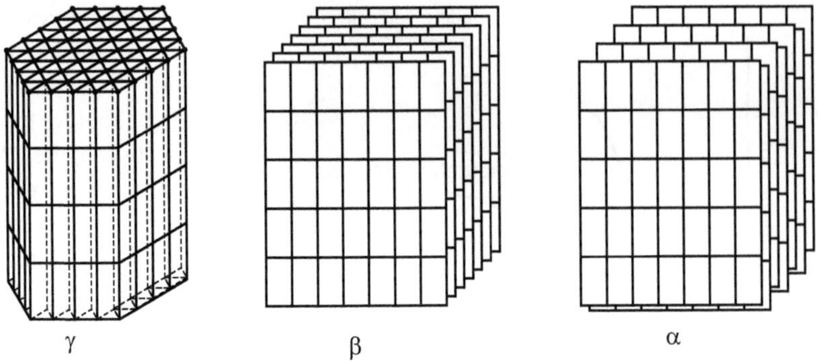

Fig. 16.30 Typical crystal lattice types for polymers

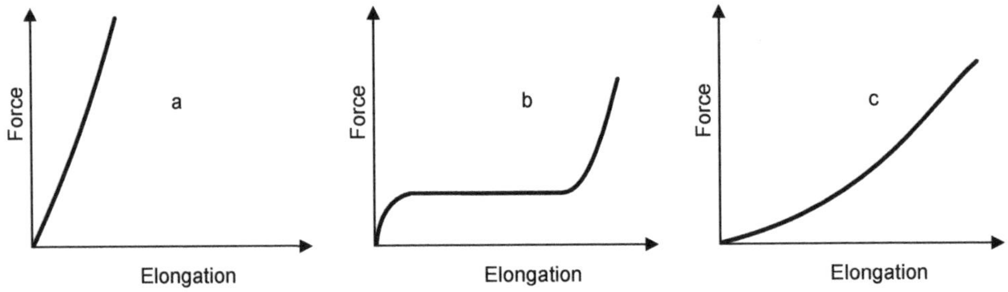

Fig. 16.31 Force-elongation diagrams of spun threads acc. to Fourné (1995)

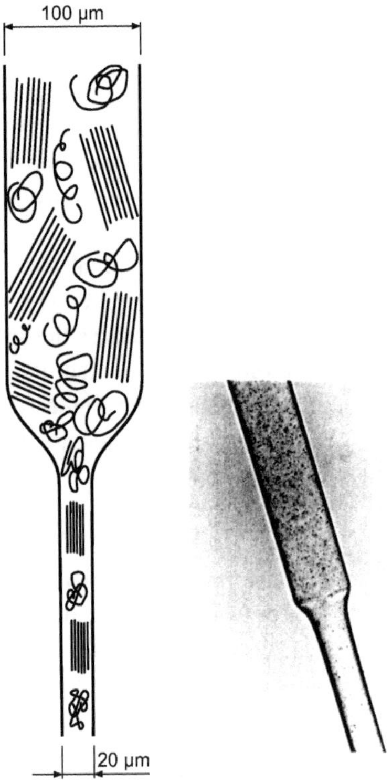

Fig. 16.32 Stretching point theoretically and in practice

Fig. 16.34. With increasing drafting, the strength increases and the elongation decreases accordingly.

16.8 Spinning Orientation and Take-Up Speed

The spinning orientation increases with increasing molar mass, decreasing spinning temperature, finer linear density and higher winding speed. The Fig. 16.35 shows the relationship between the spinning speed, i.e. the speed at which the filaments are withdrawn from the spinning duct, and the draw ratio required to obtain fully drawn fibers. This is shown for polyester and polyamide for textile filaments and staple fibers. For example, the draw factor for PET for a winding speed of 3200 m/min (POY) is approximately 1.8, meaning that the POY yarn can still be drawn to 1.8 times its length. A draw factor of 1:1 means that the yarn is fully drafted. The drafting factor is independent of the yarn fineness.

Fig. 16.33 Temperature-dependent force-elongation curve for melt-spinnable polymers acc. to Michaeli

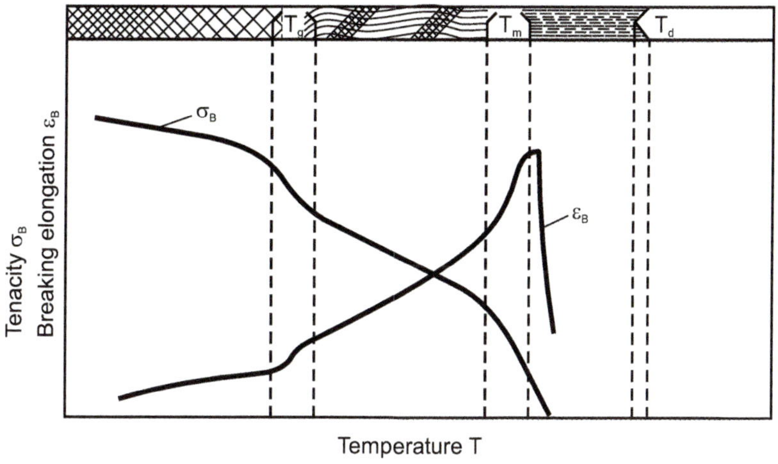

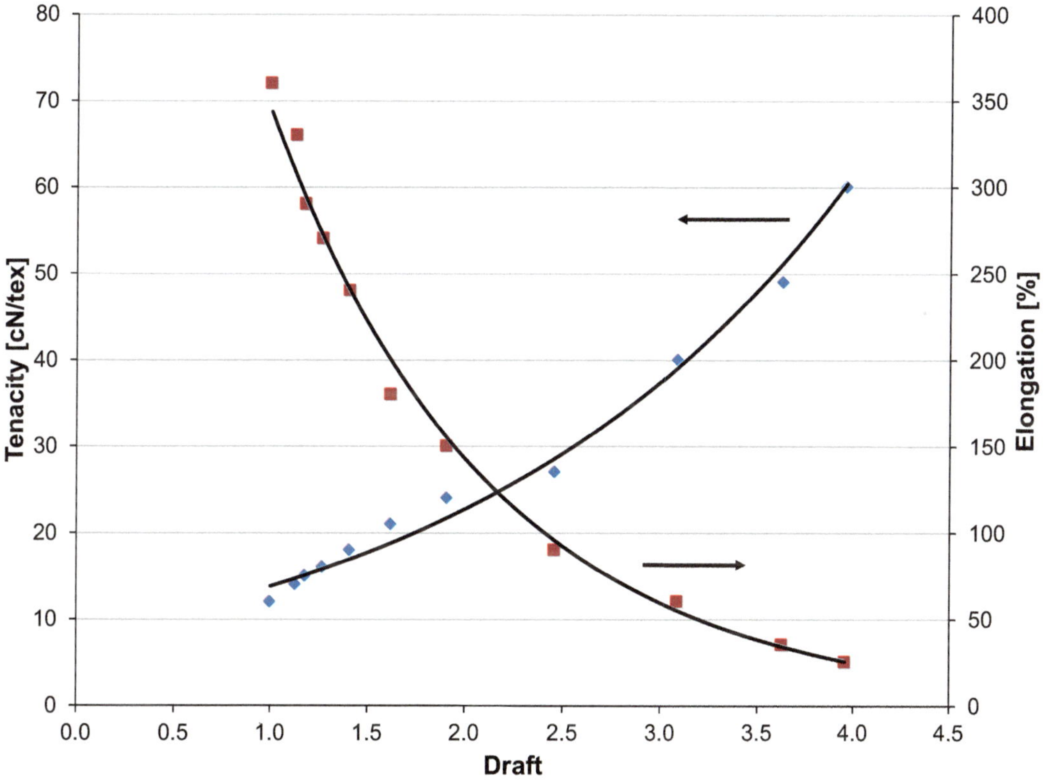

Fig. 16.34 Strength and elongation of PA 66 as a function of stretching acc. to Fourné (1995)

As already mentioned, the following spinning ranges and product designations are distinguished depending on the spinning speed: LOY (low oriented yarn), MOY (medium oriented yarn), POY (partially oriented yarn), HOY (highly oriented yarn), FOY (fully oriented yarn). Of those, mainly LOY, POY and FOY are actually used.

Of importance here are the classic spinning range for polyamide and polyester, which produces LOY at take-off speeds of up to

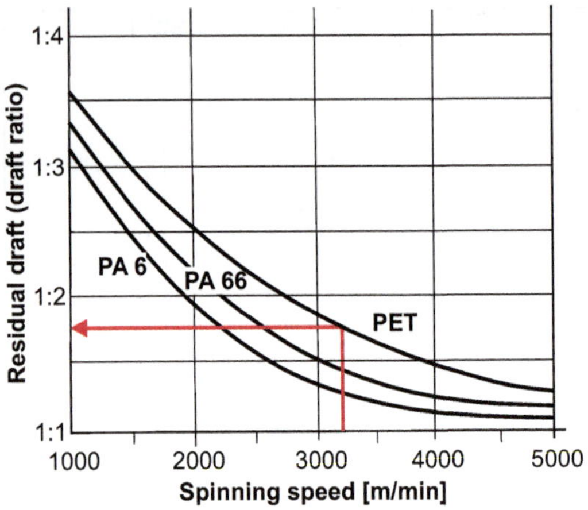

Fig. 16.35 Relationship between spinning speed and draw ratio for important polymers acc. to Lünenschloß (1973)

1500 m/min. Fast spinning produces POY at spinning speeds around 3000 m/min for polypropylene, 3200 m/min for polyester and 4200 m/min for polyamide. Fully drawn yarn, which is already fully drafted after take-up from the spinning zone, plays a role for polyester, textile filaments and spunbonded nonwovens made of polypropylene and polyester. The spinning speeds must be above 5000 m/min for polyester and above 3500 m/min for polypropylene. These numerical values are reference values, for fine filaments and highly viscous polymers these characteristic speeds are up to 20% lower.

The dependence of the draw ratio on the spinning speed is polymer-specific. It can only be changed by chemical or physical modification of the polymer. The aim is to achieve a higher take-off speed and thus a higher throughput at the same draw ratio. Corresponding systems have been built for textile filaments made of PET, but there has been no major commercial success, partly because the further processing of the filaments is altered by the addition of additives.

16.9 Spin Dyeing

Filaments can already be dyed as a liquid spinning mass. This so-called spin dyeing has considerable cost advantages over later dyeing of the finished filaments. In particular, time-consuming and thus expensive work in textile piece and strand dyeing and the associated environmental pollution are eliminated in the production of spin-dyed yarns. However, the main disadvantage of spin-dyeing is the limited choice of dyestuffs, because they have to be either temperature-resistant (melt spinning) or solvent-resistant (wet and dry spinning). Also, the production of small quantities of yarn is uneconomical. Changing the color is usually associated with time-consuming cleaning work or it is accepted that a certain amount of spun filaments cannot be used during a color change without stopping the process.

For the production of spun-dyed filament yarns and staple fibers, special color pigments are usually used which are added to the polymer as a masterbatch. This route is also suitable for feeding matting agents and additives. During melt spinning, the masterbatch can be added to the pellets in a specific ratio and, after mixing, melted together with the pellets. The dosage may be gravimetric or volumetric, the former being more accurate. However, the masterbatch can also be melted separately in a sidestream extruder and fed into the mixing zone of an extruder. Static mixing distributes the pigment evenly throughout the polymer stream and prevents agglomerates (Fig. 16.36).

Typical applications of such yarns are car seat covers and BCF yarns for carpets, because the dyestuffs also reach the inside of the fibers and thus the color impression is maintained even in the case of prolonged abrasion. Here, the small selection of possible dyes does not play such a large role.

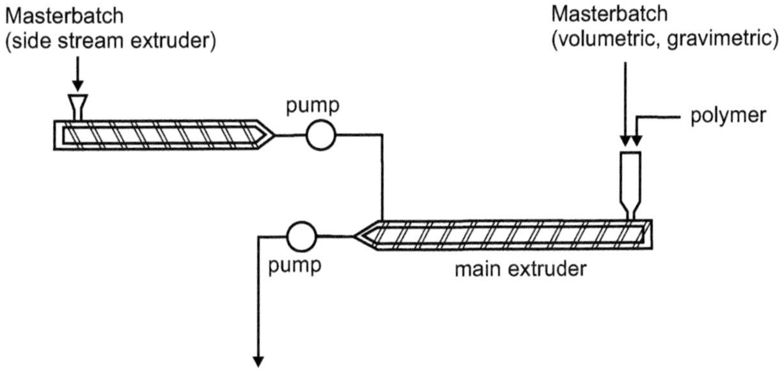

Fig. 16.36 Principle of masterbatch dosing

References

Carothers, W. H., & Hill, J. W. (1932a). Studies on polymerization and ring formation. XIII. Polyamides and mixed polyester-polyamides. *Journal of the American Chemical Society, 54*, 1566–1569.

Carothers, W. H., & Hill, J. W. (1932b). Studies of polymerization and ring formatlon XV. Artificial fibers from synthetic linear condensation superpolymers. *Journal of the American Chemical Society, 54*, 1579–1587.

Elias, H. G. (2001). *Makromoleküle*. Wiley.

Falkai, B. (1981). *Synthesefasern*. Wiley.

Fourné, F. (1995). *Synthetic fibers*. Hanser.

Götze, K. (1940). *Kunstseide und Zellwolle nach dem Viskose-Verfahren*. Springer.

Lünenschloß, J. (1973). Die Strecktexturierung,- Stand und Entwicklung. *Chemiefasern-Textilindustrie, 23*(11), 1067–1080. Deutscher Fachverlag, Frankfurt a. M.

Mark, H. F., Atlas, S. M., & Cernia, E. (1968). *Man-made fibers, science and technology*. Wiley.

Oerlikon Barmag. (n.d.). https://www.oerlikon.com/polymer-processing/de/

Rogowin, Z. A. (1982). *Chemiefasern*. Georg Thieme Verlag.

Zhang, X. (2014). *Fundamentals of fiber science*. DEStech Publications.

Ziabicki, A. (1976). *Fundamentals of fibre formation*. Wiley.

17 Aggregates for the Production of Synthetic Filaments

In this chapter, the aggregates required for the production of synthetic fibers, e.g. polyester and polyamide, are described.

17.1 Extruder

If the polymer has been processed into pellets (chips) after production, an extruder is used for melting after drying (and possibly crystallization). These have long been known in the plastics processing industry and have been used in fiber production since around 1960. Before that, the so-called grid spinning process dominated, in which the polymer was melted over a grid of heated spirals.

The Figs. 17.1 and 17.2 show the basic internal structure of a typical three-zone extruder, consisting of the feed hopper, the screw, three heating zones and the polymer outlet.

The polymer is metered gravimetrically or volumetrically and continuously fed to the feed zone via a hopper.

The central component of an extruder is the screw. With the help of the screw, pressure is built up and the polymer is liquefied via frictional heat between the wall and the screw as well as via external heating. As the screw speed increases, the proportion of frictional heat increases accordingly. Typical temperatures are 220–320 °C, typical pressure are around 80–150 bar. The rotation of the screw moves the polymer towards the outlet. At the same time, thermal and material uniformity of the polymer is achieved. Their geometry must be adapted to the properties of the polymer and the dimensions of the pellets. Typically, extruders are classified acc. to the length to diameter ratio (L/D) of the screw. Typical values for fiber polymers are from 24 to 36, meaning that the screws have a length that is 24–36 times their diameter. A typical screw designation is, for example, 75/30D. The screw then has a diameter of 75 mm and a length of 2250 mm. The flank depth of the screw is shallower in the compression zone than in the feed zone, so the polymer is further compacted and plasticized. In the metering zone, it is finally homogenized before being transported through the outlet towards the spinneret. The typical geometry of a three-zone screw shows Fig. 17.3.

For polyester, for example, the feed rate of an extruder ranges from about 2.5 kg/h for single spinning extruders (18 mm screw diameter, speed 150 rpm) to 6000 kg/h for very large extruders (360 mm screw diameter, speed 70 rpm), which can feed large or multiple spinning lines. Typical values are compiled in Table 17.1.

The end of the screw usually carries a cylindrical core with wart-shaped protrusions. This is the mixing head, which compensates for thermal irregularities in the melt. A considerable part of the heating power comes from the mechanical power of the screw drive. The extruder also has additional heaters in individual zones.

D. Veit, *Fibers*, https://doi.org/10.1007/978-3-031-15309-9_17

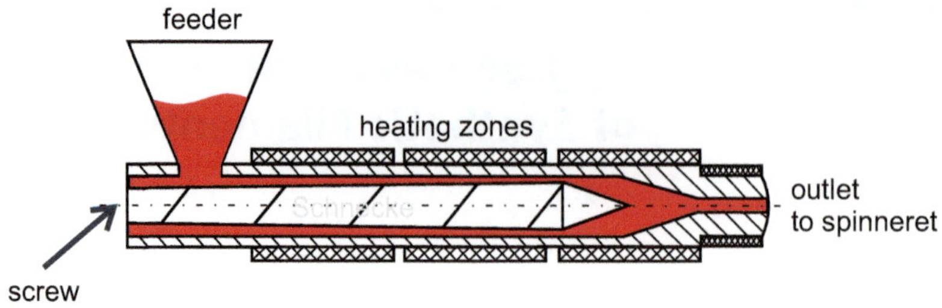

Fig. 17.1 Design of a single-screw extruder

In addition to the most commonly used single-screw extruders, twin-screw extruders are also common. They can be used for reactive extrusion (e.g. PA 6). Furthermore, there are so-called degassing extruders with up to five zones for the removal of low-volatile components (and water). Depending on the direction of rotation of the second screw, a distinction is made between co-rotating and counter-rotating twin-screw extruders. In the latter, the polymer is forced to be conveyed. Such extruders are only used in special cases, e.g. for polymers presented in powder form and for recycled polymers as well as for the production of masterbatches with additives. Especially for the processing of polyester recyclate, the extrusion zone can be extended and the screw geometry adapted.

Fig. 17.2 Typical design of a spinning extruder with chip silo (top) (Oerlikon Barmag, n.d.)

Extruders are heated either by electrical heating tapes (simple, temperature control inaccurate) or by a heat transfer medium (oil, steam), which is more complex but much more accurate.

From the extruder, the melt passes through a heated pipe to the spinning beams, where it is distributed to the individual spinning positions by a pipe system. The residence times of the melt in the distribution system must be as equal as possible for all spinning positions in order to avoid uneven degradation of the polymer.

With one extruder, several spinning positions are usually supplied with polymer, large throughputs are possible and the control effort is lower than with direct spinning systems.

17.2 Filter

Often, during the polymer's journey through the product line, impurities are created that can later lead to yarn breaks. These include agglomerates of cracked polymer due to thermal overstress and gels that form in the lines. If one takes into account that the diameter of a single filament is in the range of 10 μm (approx. 1 dtex), it becomes clear that even small interfering particles can lead to a breakage of the melt in the spinneret and thus to a filament or yarn breakage. Furthermore, it has been proven that even small impurities can lead to a strong increase of the filament tension in the filament and thus to a filament break. These impurities must therefore be removed before the spinneret.

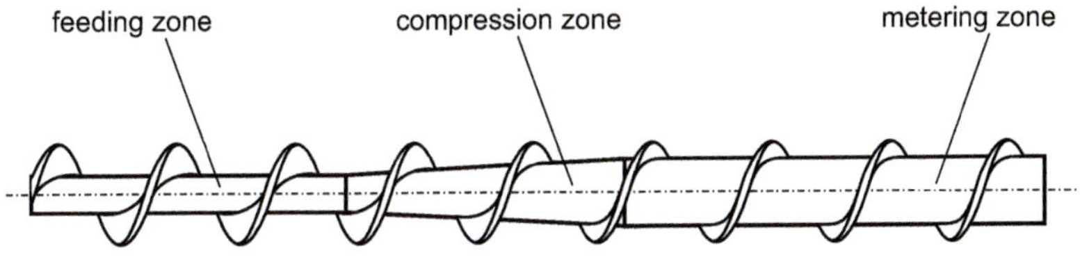

Fig. 17.3 Structure of a three-zone screw

A central melt filter is therefore often used downstream of the extruder to clean the polymer and relieve the filters upstream of each individual spinneret. These filters are constructed as alternating filters, i.e. two sets of filters are arranged in parallel and flow through them alternately while the other is cleaned. The filter media used are usually sand as well as nonwovens and woven fabrics made of metal.

17.2.1 Sand Filter

Sand filters consist of sands of different grain sizes, which are usually built up in layers (from coarse to fine), Fig. 17.4. Both silicates (Al_2O_3) and metal powders are used. The filtering effect depends essentially on the size and shape of the particles. Sand filters are mainly used to retain gels from the polymer stream.

17.2.2 Fabric Filter

Simple filters consist of one or more layers of metal mesh, usually in plain weave. Depending on the fineness of the metal wires, filter pore sizes of e.g. 5–25 μm are achieved, which is necessary for the production of microfibers. For the production of BCF fibers, pore sizes of 200 μm are sufficient. Depending on the wire thickness, this results in 20,000–80,000 pores per cm^2. This results in deposition rates of up to 98%. The fine wire cloths have to be supported, therefore a multilayer structure is often used. A typical example is a structure with 6000–40,000–6000–200 meshes/cm^2 (Fourné, 1995). Filter fabrics are enclosed in aluminum, which also serves as a seal against the spin pack wall up to 350 °C.

The Fig. 17.5 shows typical examples of metal filters. In addition to flat filters (a), candle filters (c, d) are frequently used. Pleating multiplies the filter area without significantly increasing the installation space (b). Disc filters (e) are also known, in which the melt flows from the outside to the inside and is discharged centrally. Compared to flat filters, cylindrical filters have a filter area that is about 20 times larger for the same installation space, but they are also considerably more expensive.

Figure 17.6 shows typical designs of pleated sintered metal meshes for round and for flat filters. Depending on the application, there are a large number of variations.

Metal filters are often combined with upstream sand filters (height: 20–30 mm) in order to

Table 17.1 Typical extruder throughputs (Oerlikon Barmag, n.d.)

		Screw diameter							
24D		45 mm	60 mm	70 mm	90 mm	100 mm	120 mm	150 mm	170 mm
PET	kg/h	70	130	200	300	400	520	820	1120
PA 6	kg/h	40	80	125	200	270	350	550	750
PA 66	kg/h	40	70	120	180	240	315		
30D		45 mm	60 mm	70 mm	90 mm	100 mm	120 mm	150 mm	170 mm
PET	kg/h	90	160	260	380	500	660	1000	1380
PA 6	kg/h	45	90	145	230	310	400	630	
PA 66	kg/h	45	90	140	200	280	380		

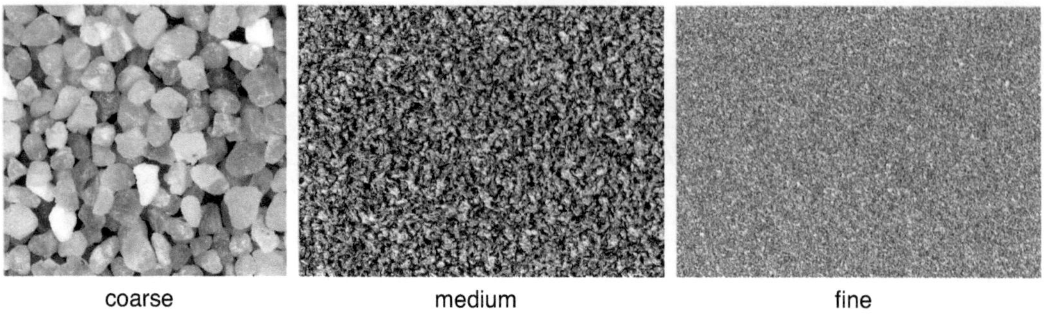

Fig. 17.4 Sand with different fineness (grain size)

increase their service life through this pre-filtering. This is usually in the range of 1–3 months, depending on the load and polymer.

During operation, spontaneous self-cleaning of the filter pack can occur. In this case, a certain pressure is exceeded due to the formation of a filter cake and a deformation of the embedded gels then occurs, which are subsequently pressed through the filter medium. This leads directly to a large number of filament breaks and at the same time to a significant reduction of the pressure in front of the filters. After some time, a stationary state results again and the filter effect resumes.

17.2.3 Nonwovens Filter

Metal fleeces are also used as an alternative to metal meshes (Fig. 17.7). Compared to fabric filters, they filter out more smaller particles with the same average pore size, because their pore size varies more and there are thus also smaller pores. For this reason, however, their filtering effect is not as adjustable as that of fabric filters, which are "precisely woven holes", as one filter manufacturer put it. Especially gels can be removed well from the polymer stream by non-woven filters. Sintered textile structures are also used for special polymers.

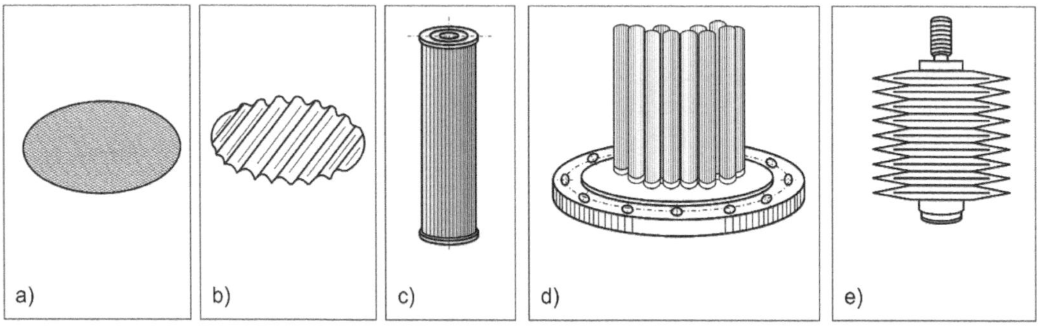

Filter type	Filter area (cm^2)	Effective filter area at 300 °C (cm^2)	P_{max} (kg/cm^2)
a Flat filter	33	33	3 - 5
b Pleated flat filter	150	90	5
c Pleated cylinder	970	590	210
d Many small cylinders	580	580	350
e Disc filters	450	450	300

Fig. 17.5 Structure of metal filters and typical data acc. to Fourné (1995)

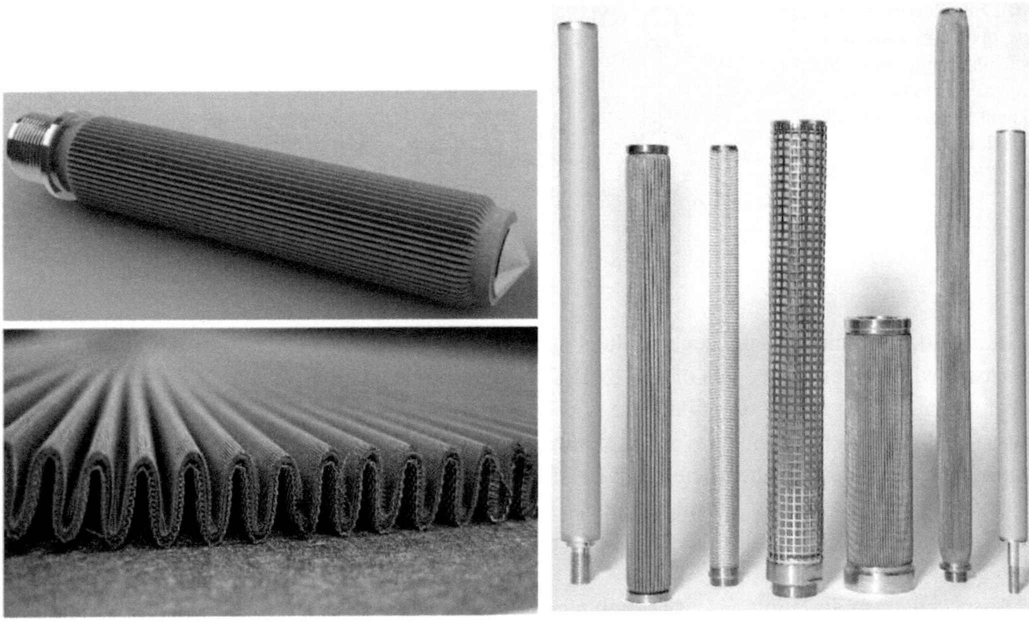

Fig. 17.6 Filter made of pleated, sintered metal mesh (Bekaert & Bekaert, n.d.)

Normally, filters are exchanged and cleaned from the polymer by pyrolysis at certain intervals. Sand filters are often also "disposed of" directly if cleaning is not economically viable.

17.2.4 Revolver Filter

Figures 17.8 and 17.9 show the design and operating principle of a revolving filter. The medium to be filtered flows through a filter segment until the pressure in front of the filter exceeds a limit value. The filter is then moved one segment further until the pressure in front of the filter again becomes too high, and so on. In this way, a relatively constant filter effect is achieved.

Filters of this type have a service life of 1–16 weeks. The filter change takes about half an hour.

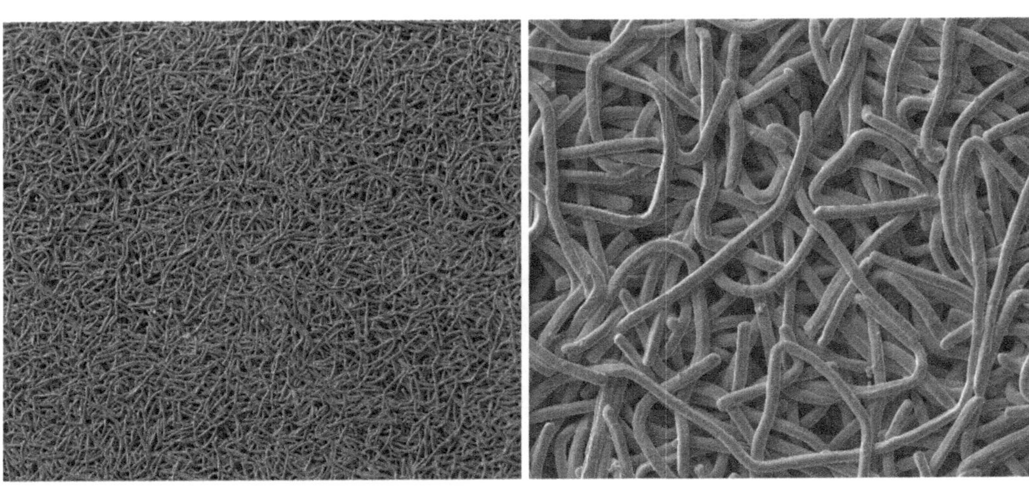

Fig. 17.7 Metal nonwovens for polymer filters (Bekaert & Bekaert, n.d.)

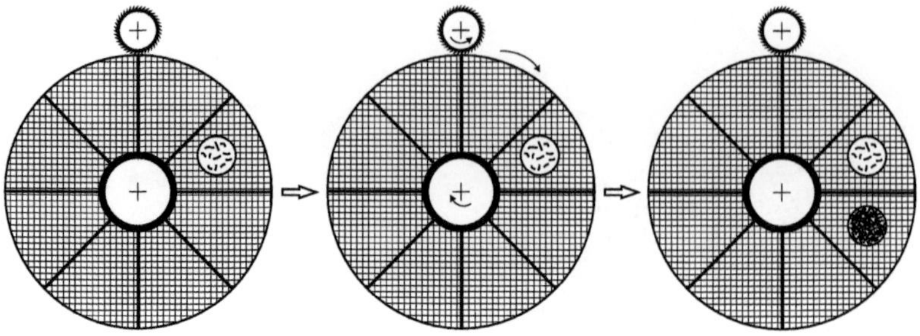

Fig. 17.8 Revolver filter (left: operation; middle: switching; right: after filter change) acc. to Gneuss (Gneuß Kunststofftechnik GmbH, n.d.)

In extruder spinning plants, central filters are also used, which offer a filter area of 0.4–8 m^2. In direct spinning mills, large-area filters are nowadays installed which supply several spinning beams. The filter area can then be e.g. 40 m^2. In the meantime, pleated filter cartridges are used exclusively. Typical flow rates are 150–300 kg/(h m^2).

A special form for high viscosity polymers shows the Fig. 17.10. Here, a pleated round filter is used, whose filter area is larger by a factor of 5–11 compared to a flat filter of the same diameter.

When designing a filter, a compromise must be found between the desired filter efficiency, the longest possible filter service life and the lowest possible pressure loss at a given volume flow of the melt.

17.3 Spin Pump

For all man-made fibers, especially for filament yarns, the linear density must be constant over time and also from spinning position to spinning position. Deviations of more than ±1.5% would, for example, have a negative influence on the fabric appearance of the finished textile. For this purpose, the spinning mass must be supplied to all spinnerets with the same accuracy. Depending on the polymer, pressures of up to 400 bar build up at the outlet due to the flow resistance of the capillary holes and upstream filters. It is therefore advantageous to supply each spinneret with polymer by means of a separate conveying device—a spin pump. For this purpose, gear pumps with a delivery volume per revolution of about 0.2–200 cm^3 are used. Several pumps are often combined in one pump block. In filament yarn processes, up to ten fold pumps are used in special cases, depending on the number of yarns per winder position.

Figure 17.11 shows the operation of a gear spinning pump. The melt enters the pump and fills the gaps between the wheels and the housing and is pressed out of the housing after one revolution.

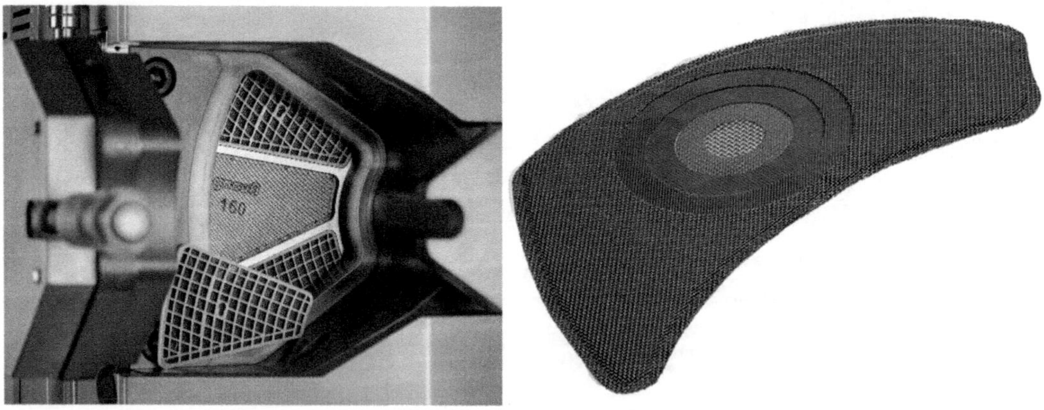

Fig. 17.9 Revolver filter and single segment (BB Engineering GmbH, n.d.)

Fig. 17.10 Pleated filter element in the spin pack for high-viscosity polymers (BB Engineering GmbH, n.d.)

Spin pumps are made of high-alloy steel (tungsten, chromium, vanadium, molybdenum), their operating temperature can be up to 450 °C. The differential pressure between inlet and outlet is usually between 300 and 400 bar.

The volume flow through a spinning pump in (cm^3/min) is calculated from:

Fig. 17.11 Design and mode of operation of a gear spin pump (Mahr AG, n.d.)

$$\dot{V} = \frac{T_t \cdot V_{take-up}}{10{,}000 \cdot \rho}$$

where Tt = spinning titre (dtex), $V_{take\text{-}up}$ = winding speed (m/min), ρ = density of melt (g/cm^3).

Typical values are in the range of 0.1 up to 30 cm^3/U. This results in flow rates of 7 g/min to 2.5 kg/min at speeds of 50–70 rpm for PET.

For fine fibers, the spinnerets are often smaller than the spinning pump. Therefore, one spinning pump then supplies several spinnerets and therefore needs several outputs. The Fig. 17.12 shows the structure of a spinning pump with eight outputs and two gear levels.

If more gear levels are required (for more outputs), the spinning pumps become very long. This makes it difficult to keep the temperature constant inside the pump, which directly affects the viscosity and thus the flow rate. Therefore, pumps with many outputs are often realized in planetary gear design (Fig. 17.13, left).

This type of pump is also used for metering the spin preparation. A typical example shows the Fig. 17.14. Since this metering takes place under room temperature, there is no temperature problem.

The choice of spinning system depends on the throughput per spinning position and the number of individual filaments. The spinning capacity of a melt spinneret results from the take-off speed v_A, the number of holes n of the spinneret and the individual fibre fineness $Tt_{sp,single}$ of the spinning filaments:

$$\dot{m}[g/min] = n \cdot V_{take-up}\left[\frac{m}{min}\right] \cdot Tt_{sp,single}[dtex] \cdot 10^4$$

In solution spinning, the equation must take into account the proportion of the solvent.

To calculate the spinning capacity $\dot{m}$ for a certain final fineness $Tt_{,end}$ the drafting ratio DR (drafting factor)

$$W = \frac{Tt_{sp}}{Tt_{end}}$$

be taken into account.

Thus:

Fig. 17.12 Sixteen-gear spinning pump with two gear levels (two fold eight-gear pump) (Mahr AG, n.d.)

$$\dot{m}\left(\frac{g}{\min}\right) = V_{take-up}\left(\frac{m}{min}\right)\cdot DR\cdot Tt_{end}(dtex)\cdot 10^{-4}$$

Thus, the extraction increases more slowly than the extraction velocity $v_{take\text{-}up}$, because as $v_{take\text{-}up}$ increases, VV decreases (Fig. 17.15).

17.4 Melt Spinneret

In the spinnerets, the polymers are formed into the actual fibers or threads and the thread shape (cross-section) is generated. Spinnerets for melt and solution spinning are fundamentally different, therefore they are described separately.

The polymer mass flow must be filtered through an appropriate filter combination during melt spinning in the spin pack as explained above. A minimum and a maximum area-specific melt throughput must be achieved as design criteria. This depends on the polymer and the process and is in the range of 0.5–2.5 g/cm^2 per minute. If the spinneret load is too low, uniform distribution of the melt over all capillaries is impeded and the residence time of the melt in the spinneret package may be too high (risk of thermal degradation). At higher filter loads, the filter service life decreases and the additional polymer heating due to the increased internal friction must be taken into account.

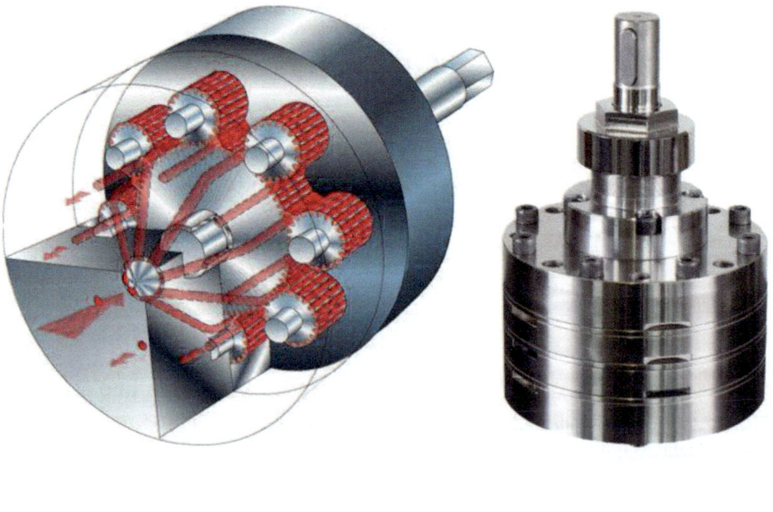

Fig. 17.13 Spinning pump designs (left: interior view of a planetary gear pump; right: assembled spinning pump (Mahr AG, n.d.))

Fig. 17.14 Spin preparation pump (Mahr AG, n.d.)

17.4.1 Hole Pattern

When designing a spinneret, a minimum spacing of the filaments is required so that the cooling air in the quench duct can penetrate the filament curtain and a sticking together of the filaments in the spinning area is avoided. The minimum filament spacing is between 6 and 8 mm, depending on the polymer and the cooling air. Furthermore, with regard to the number of capillaries, it must be taken into account that the cooling air, which is generally introduced from one side transverse to the spinning direction, can uniformly cool a maximum of 13–15 filament rows. Figure 17.16 shows typical circular and ring melt spinnerets.

17.4.2 Hole Cross Sections

The quality of a spinneret is determined by the uniformity of the dimensions of the fine bores, the quality of the bore walls, the sharpness of the outlet edge of the bores and the quality of its outer surface. For spun yarns with a non-round cross-sectional profile of the individual filaments, profiled spinneret bores are used, the cross-sectional profile of which is considerably sharper than that of the filaments to be spun from them (Fig. 17.17). Due to relaxation processes, the cross-sections of profiled fibers are always somewhat "rounder" than the corresponding jet bores.

The Fig. 17.18 shows trilobal fibers in transverse and longitudinal view. Such a cross-sectional shape causes a change in lustre and handle, and also an increase in the adhesive

Fig. 17.15 Throughput as a function of spinning speed using the example of a melt-spun yarn with final titre 167 dtex (150 den)

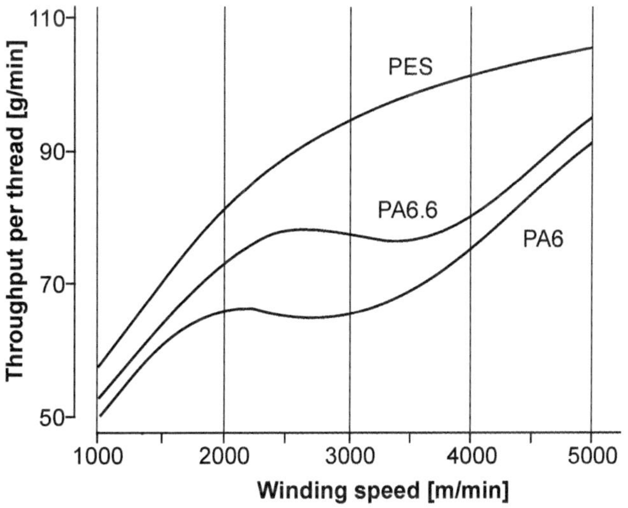

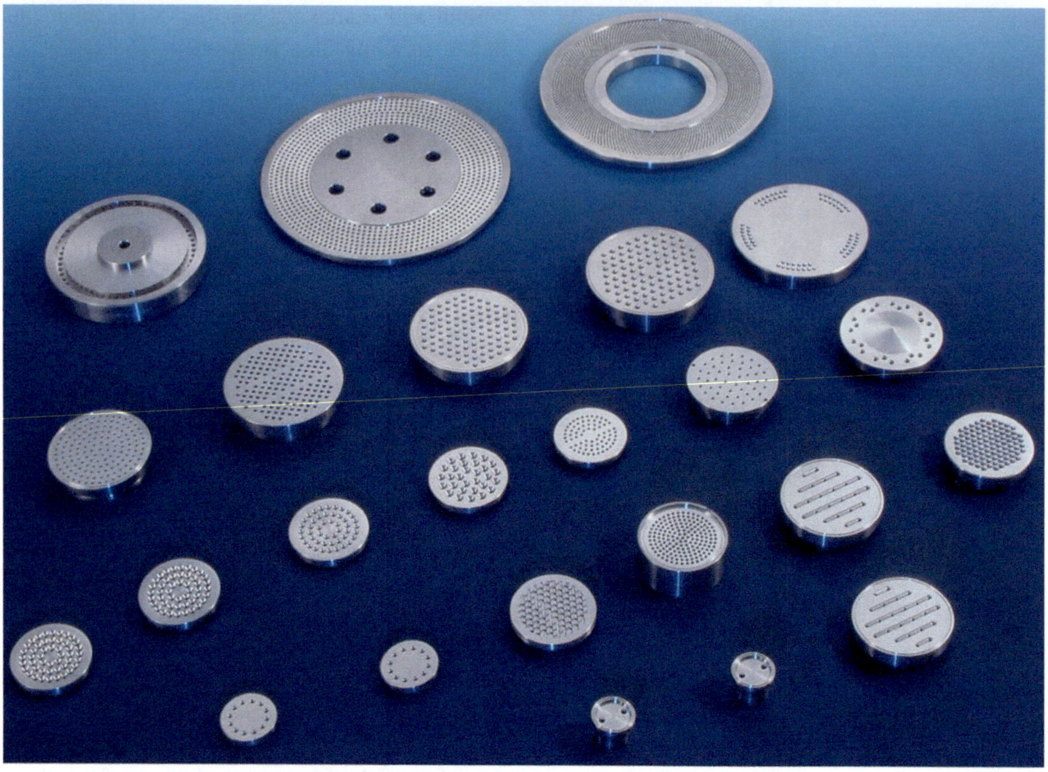

Fig. 17.16 Melt spinning spinnerets (circular and ring spinnerets) (Sossna GmbH, n.d.)

strength of the fibers in the yarn structure, which reduces the tendency to pilling.

Before the melt solidifies, there is a risk that the profiled cross-section will change into a circular one because of the surface tension. To

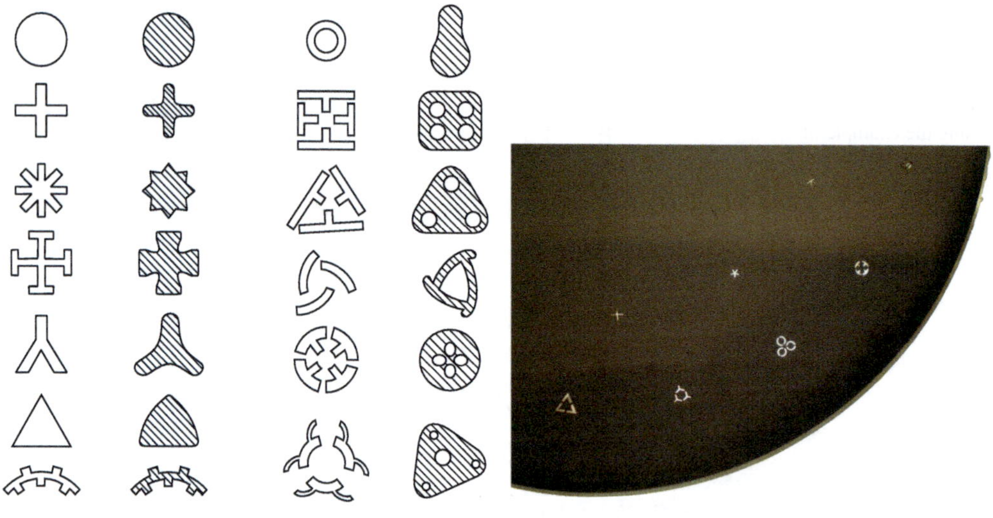

Fig. 17.17 Spinneret cross-sections and resulting fiber cross-sections acc. to Fourné (1995) and capillary holes in a spinneret (Sossna GmbH, n.d.)

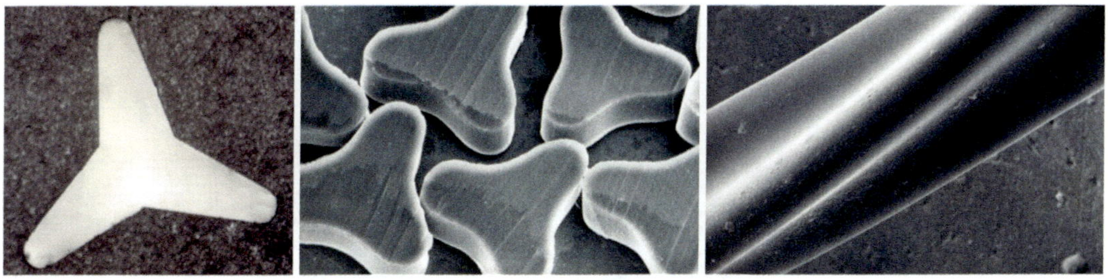

Fig. 17.18 Spinneret bore and trilobal PA fibers (transverse and longitudinal view)

prevent this, a sufficiently high melt viscosity and rapid cooling of the filaments must be ensured. The Fig. 17.19 shows this Barus effect, which occurs when the polymer relaxes too much as it leaves the hole. This results in a distorted fiber cross-section, which can be avoided by increasing the viscosity of the polymer.

Additives that reduce surface tension can also be used.

17.4.3 Spinneret Types

Spinnerets for filament yarns are usually round, spinnerets for staple fiber yarns usually rectangular. This means that their capacity is greater, even if the cooling of the filaments becomes more uneven. However, this does not play a major role with staple fibers because they are intermixed in later process stages.

The following limits apply to the selection of spinning systems:

Round spinning package: up to 300 filaments, up to 500 g/min/spinneret

Rectangular spinning package: up to 2500 filaments, up to 1500 g/min/spinneret

Ring spinning package: up to 7500 filaments, up to 4.5 kg/min/spinneret

These limit values are only indicative. Small round spinnerets are often combined into several spinning packages in a rectangular spinneret block (Fig. 17.20). If particular importance is given to a particularly high uniformity of dyeing, care must be taken to ensure that all spinning packages are heated evenly around the entire circumference. In the case of circular packages, this leads to individual round spinnerets which are separately inserted in a spinneret housing and heated all round with steam.

Fig. 17.19 Capillary bore (snowflake) and fiber cross-sections with Barus effect

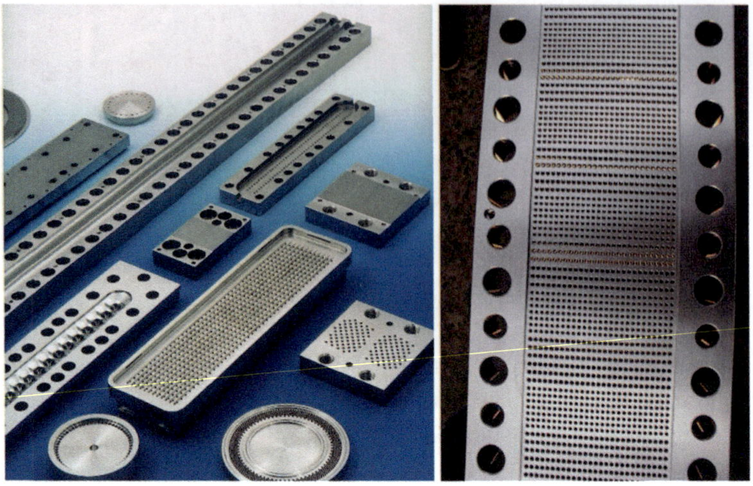

Fig. 17.20 Rectangular spinnerets for staple fibers and slot spinnerets for spunbonded nonwovens (Sossna GmbH, n.d.)

The Fig. 17.21 shows longitudinal sections through typical capillary holes of melt spinning spinnerets. The tapering of the capillary bore cross-section increases the polymer velocity, which leads to a pre-orientation of the macromolecules.

Round and rectangular spinning packages (Fig. 17.22) are usually cooled by means of cross-flow blowing. If round spinnerets are used for high-speed spinning of textile filaments with spinning speeds above 3000 m/min, cross-flow blowing may not be necessary. The filament draws itself and the necessary cooling air through a perforated tube. This self-suction principle is used in special cases, e.g. for spun-dyed yarns for car seat covers. Ring spinning packages have a ring-shaped spinneret arrangement and ring-shaped melt filtration. They are usually combined with radial blowing. Cooling air can be supplied either from the outside to the inside or from the inside to the outside (see Sect. 17.7).

Spinnerets for common polymer temperatures below 400 °C are mostly made of steel, e.g. 1.4571. Round spinnerets are used for filament yarns, rectangular spinnerets mostly for staple fibers.

Melt spinnerets have to withstand pressures of up to approx. 200 bar at temperatures of up to 300 °C due to the high dynamic viscosity of most melts (PA 6.6: 75 Pas; PET: 1000 Pas). Spinnerets are usually approx. 10–30 mm thick. Depending on the spinning program, the number of holes ranges from 1 to 2000 for filament yarn spinning, approx. 10,000 for spinning fiber cables for the two-stage process and up to 150,000 holes for single-stage compact spinning of staple fibers. The hole diameters are about 0.2–2 mm and the length of the hole in relation to the diameter ("L/D ratio") is usually 5:1 to 20:1.

Spinnerets—like metal filters—are cleaned by pyrolysis (Fig. 17.23) and can thus be used for many years.

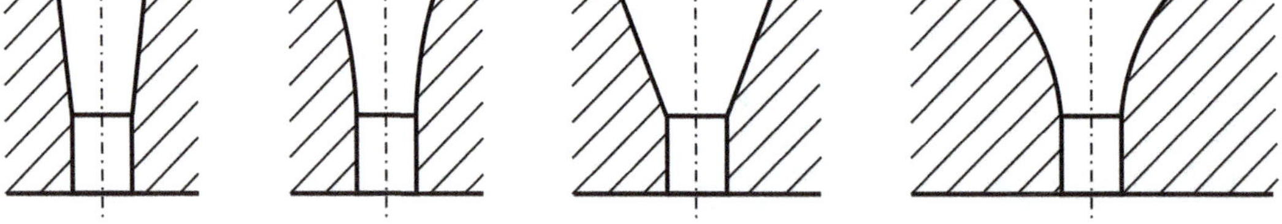

Fig. 17.21 Typical longitudinal geometries of capillary holes

Fig. 17.22 Ring spinneret (left: Sossna, n.d.; right: Zimmer AG)

17.4.4 Hollow Fibers

For spinning hollow fibers, mostly ring-shaped spinneret openings with a centrally pierced core are used for metered gas supply. In another method, spinneret openings in the form of an interrupted ring lead to hollow fibers which are formed by fusion only after spinning out. This is simpler from a process engineering point of view, but the cross-sections of the hollow fibers produced in this way are more uneven (Fig. 17.24).

Hollow fibers trap air and therefore have a lower density than solid profile fibers. They are softer and fuller and result in better thermal insulation. Optical effects can be achieved by light refraction and reflection at the inner interfaces, which are interesting for applications in the carpet sector. Other areas of application are mass transfer (reverse osmosis) and optical fibers.

17.5 Spinning Packages

The spinnerets are assembled together with various filters and the distributor plate to form so-called spinning packages (Fig. 17.25).

The spinning packages are designed in such a way that they can be easily removed and installed and also easily disassembled and reassembled. One prefers designs in which the package seals itself by the pressure in the spinning mass. Sometimes these packages also contain the spin pumps.

Fig. 17.23 Spinneret before (left) and after (right) pyrolysis

Fig. 17.24 Hollow fibers made of PET

Often, several spinning packages are combined next to each other to form a spinning beam with collective heating. Their distance is the so-called spacing, which is 30–120 cm depending on the size. In filament yarn processes, each spinning position is usually assigned to one take-up and drafting position as well as a winder position.

In the spinning filter, the melt usually first passes through a packing of quartz sand or similar sharp-edged inert material (steel sand, Al_2O_3). This is followed by a set of metal screens, possibly a metal fiber fleece, which is supported by a coarser support screen, which in turn rests on a distributor plate with multiple holes. This is followed at a short distance by the spinneret. The filter arrangement is not only intended to clean the melt, but also to mix it and make it uniform as well as to break up inhomogeneities.

There must also be no points of slowed or even stagnant flow in the spin pack. In the case of thermally susceptible polymers with a high molar mass, the melt is kept as cold as possible in the extruder and in the pipe system and its viscosity is reduced for a short time by additional heating of the spin pack (annealing) or by high internal friction. This facilitates the drawing process.

The Fig. 17.26 shows the design of a self-sealing spin package. The housing is screwed onto the mating thread in the spinning head. It is sealed at the top by the aluminum gasket. The flange of the spinneret seals downwards onto the spin pack housing in case of overpressure, thus the whole system is tight.

The polymer is supplied under high pressure, up to 600 bar in the case of high-pressure spinning. The pressure is then converted into a temperature increase of the melt via the friction in the filters and the other components. As a rule of thumb, we have:

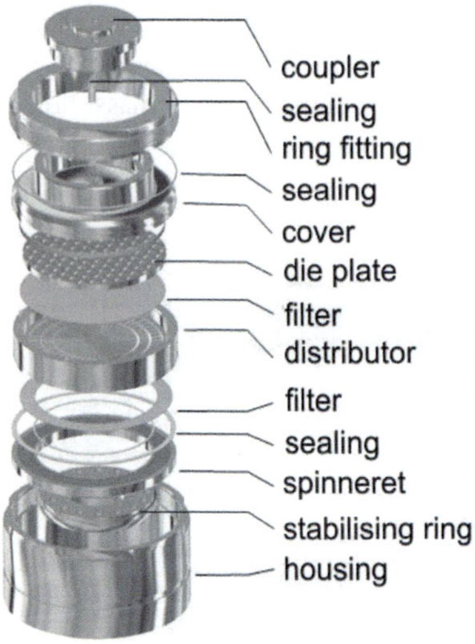

Fig. 17.25 Design of a spinning package and spinning package with distributor plates and spinneret (Sossna GmbH, n. d.)

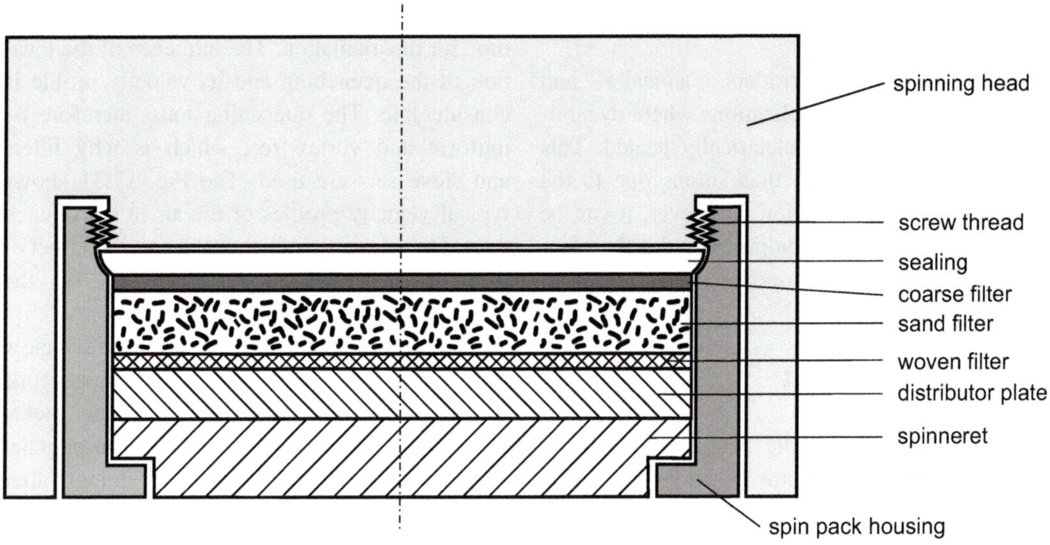

Fig. 17.26 Design of a self-sealing spin pack

$$\Delta\vartheta = (p_{in} - p_{out}) \cdot \left({}^{4K}\!/_{100\ bar}\right)$$

With an inlet pressure of the melt of 600 bar and a spinneret pressure of 30 bar, this results in a temperature increase of the melt of 22.8 K.

17.6 Spinning Beam

Spin packages are combined in groups of e.g. 2, 3, 4, 6, 8 or 10 in the so-called "spinning beam" (Fig. 17.27). For its heating, diphyl is usually used as the heat transfer medium. It consists of a mixture of diphenyloxide and biphenyl. This type of heat transfer medium is suitable within a temperature range of −30 to 400 °C. Liquid heating has the advantage that heat can be both supplied and removed (heating and cooling). However, due to the heat transfer, up to 10 °C temperature difference between the heat transfer media is necessary. For this reason, liquid heating is not suitable for the temperature-sensitive area of the spinning head. Here, an accuracy of the temperature of max. 0.5 °C is required in order to avoid dyeing differences in the textile end product. For this reason, spinning beams are almost exclusively heated with vaporous heat transfer media (HTM). The HTM steam condenses and thus releases its evaporation enthalpy to the heat consumer. Due to the boiling delay, the temperature is constant and can be adjusted very precisely via the boiling pressure. As a result, this type of heating guarantees the highest temperature consistency as long as the same pressure difference prevails at all spinning positions. The enthalpy of evaporation is approximately one power of ten greater than the specific heat capacity. Therefore, vapor heating systems require lower volume flow rates. However, because the density of the liquid

Fig. 17.27 Spin blocks for round spinnerets (Sossna GmbH, n.d.)

HTM is greater, the mass flow rates are lower for liquid heating systems.

Only spinning extruders, annealer and spinning systems for applications where dyeability is not an issue are electrically heated. This heating has temperature fluctuations due to the necessary control deviation. However, it can be implemented with less equipment and is therefore often used in pilot plant operations.

17.7 Quench Duct

In order to accelerate the heat transfer during cooling and to stabilize the position of the filament bundle, cooling is usually takes place underneath the spinneret perpendicular to the spinning direction. Depending on the linear density, the number of filaments and the spinning speed, the quench air usually flows at a speed of 0.2–2 m/s. The quenching takes place either by cross-flow quenching (Fig. 17.28, left), or radially from the inside to the outside (Fig. 17.28, middle) or radially from the outside to the inside (right). Cross-flow quenching is the most common because it is easy to implement and the yarn remains accessible. The quench air temperature should be as constant as possible over the entire width in order to cool the filaments evenly. The length of the quenching zone is generally 60–150 cm. Then the filaments are solidified and can be combined into a yarn or cable by take-off devices.

The Fig. 17.29 shows the structure of a quench duct for cross-flow blowing. Of particular importance is the door, whose profile significantly influences the air flow inside and thus the cooling effect.

The Fig. 17.30 shows the force conditions on a yarn in the blow shaft. The blowing air flowing in transversely causes the yarn to bulge (shown highly exaggerated). Care must be taken to ensure that the spinning tension does not exceed the melt strength. The take-up tension S_1 must be less than the sum of the spinning tension S_0, the weight G of the thread and the friction R between the thread and the quench air.

The finer the titer and the stronger the quenching, the faster the filaments solidify and the greater the spin orientation. There is then less time for disorientation. The influence of the location of the quenching and its velocity profile is considerable. The quenching must therefore be uniform and vortex-free, which is why filters and sieve sets are used. The Fig. 17.31 shows typical velocity profiles of the air in the quench duct. They directly influence the cooling effect of the air and thus the orientation of the macromolecules.

If the mobility of the macromolecules below the spinneret is to be maintained for a longer time for more uniform drawing, a heating jacket (annealing) is used. This increases the homogeneity of the filaments and the resulting strength after drawing is higher. If the lower part of the quench duct is blown with warm air or at reduced speed, the spinning orientation increases there, because the internal relaxation is only slight in this area.

Basically, the filament can transfer heat to the environment by convection (via the cooling air), contact (via machine elements) and radiation. The heat transfer by direct contact can be neglected and also the share of radiation heat in the total heat at polymer temperatures around 300 °C is only approx. 10–15% of the energy transferred by convection. It is therefore neglected in the following. However, the proportion of radiation is relevant in the case of glass and basalt, where the melt temperatures exceed 1000 °C.

The equilibrium between the heat flow of the cooling air in the blow shaft and the heat extracted from the polymer filament in the process is then given by:

$$\dot{m}_{air} \cdot c_{air} \cdot \left(T_{bottom} - T_{top}\right) = \dot{m}_{yarn} \cdot \left[c_{yarn} \cdot \left(T_{melt} - T_{yarn}\right) + S\right]$$

T_{bottom} and T_{top} stand for the temperature of the air at the top and bottom of the blowing chamber, c stands for the respective heat capacities and S is the enthalpy of fusion of the polymer.

The air flow rate Q_{air} required to cool the filament bundle can then be calculated following (Fourné, 1995) from:

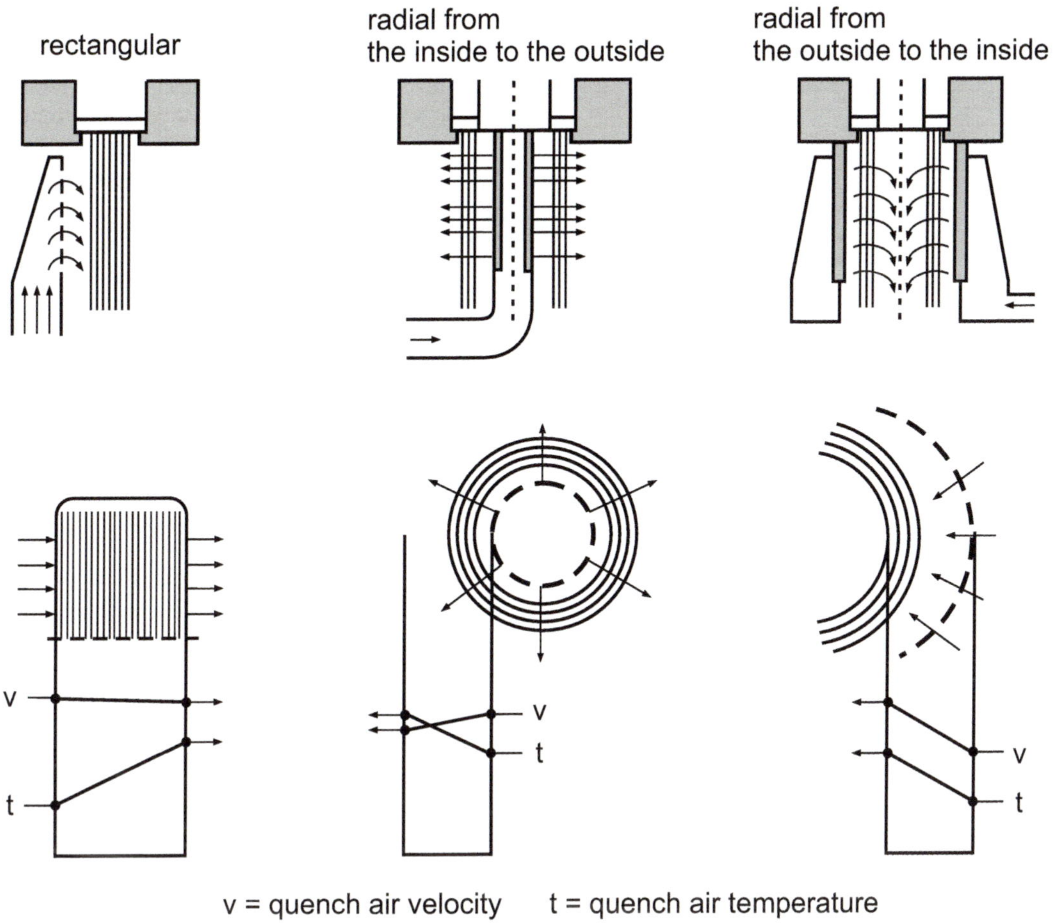

Fig. 17.28 Possibilities of quenching and important physical quantities acc. to Fourné (1995)

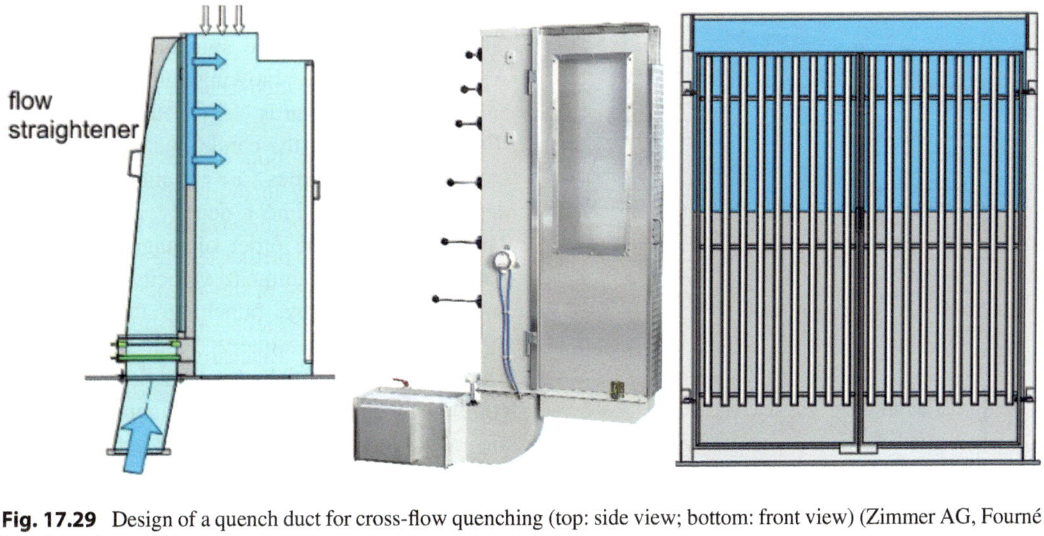

Fig. 17.29 Design of a quench duct for cross-flow quenching (top: side view; bottom: front view) (Zimmer AG, Fourné Maschinenbau, n.d.)

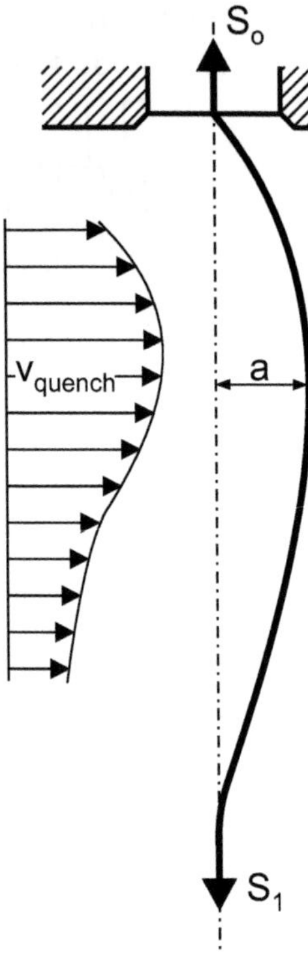

Fig. 17.30 Forces on the thread in the quench duct acc. to Fourné (1995)

$$Q_{air} = \frac{c_{yarn} \cdot \left(t_{melt} - t_{yarn}\right) + S}{c_{air} \cdot \left(t_{bottom} - t_{top}\right) \cdot \rho_{air}}$$

In Table 17.2, typical values for the cooling of a thread bundle of common polymers are compiled (Fourné, 1995). The heat capacity of air is assumed to be 1045 kJ/kgK (at 20 °C and ambient pressure 1 bar). The enthalpy of fusion of polymers depends mainly on their crystallinity. For a typical crystallinity of 27%, the value for PET, for example, is 130 kJ/kg (Auras et al., 2005).

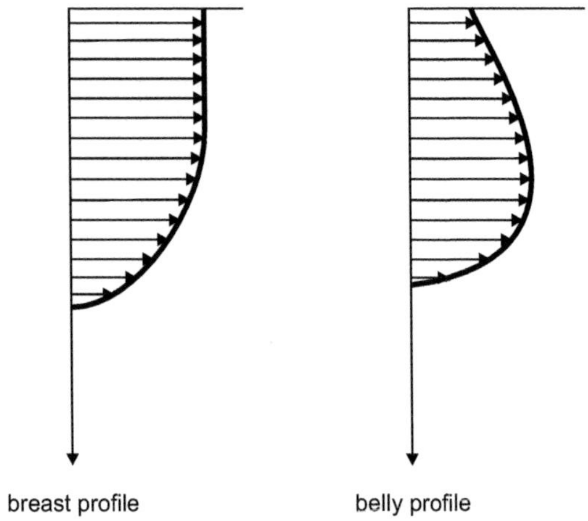

Fig. 17.31 Typical velocity profiles of the quench air

Thus, a first estimate for the required cooling air volume at a take-up speed of 5000 m/min is obtained, as shown in Table 17.3. It was assumed that the cooling air in the blow shaft heats up by 10 K due to the yarn cooling and that the yarn is cooled down to a temperature of approx. 20 K below the glass temperature in order to avoid yarn breaks.

Polymers that form crystalline regions even at low take-up speeds emanate heat in the process. These polymers include polyamides and olefins. In the case of polyester, this effect only becomes relevant at high take-up speeds. This should be taken into account.

Due to the cooling of the filaments, the blowing air heats up when flowing through the filament bundles across the spinneret bar. Therefore, the cooling air volume must be increased accordingly if numerous spinnerets are combined in 1 bar. This is the only way to ensure that the rear filament bundles are also sufficiently cooled.

The equation derived above is mainly valid for the upper region of the quench duct. During cooling, an air boundary layer forms around the individual filaments. After a certain distance (about 30 cm below the spinneret, Fig. 17.32, A), the transverse blowing air is therefore entrained downwards by the filaments (Fig. 17.32, B) and the conditions change. Typical vertical velocities are about 0.2–3 m/s, depending on the linear density and polymer. This is of the same order of magnitude as the horizontal air entrainment velocity. When the yarn is formed by bunching the filaments together, the air is squeezed out between the filaments (Fig. 17.32, C). This reduces the cooling effect of the transverse incoming quench air accordingly. That results in an actual air volume requirement that is up to ten times higher than calculated above. It is extremely difficult to determine the exact boundary layer thickness

Table 17.2 Typical thermal characteristics of common polymers

	Melt temperature [°C]	Glass temperature [°C]	Melt enthalpie [kJ/kg]	Heat transfer coefficient [kJ/kg K]
PET	260	75	130	2
PA 6	220	45	192	3.3
PA 66	260	45	197	2.8

Table 17.3 Typical values for the amount of blowing air required to cool a yarn bundle in the blow shaft (in Nm^3/h)

Cooling air flow [m^3/h]			
	75 den (84 dtex)	150 den (167 dtex)	300 den (332 dtex)
PET	105	210	420
PA 6	168	335	670
PA 66	180	360	720

because the model of the flat plate is not applicable due to the small radius of curvature of the filaments.

The Fig. 17.33 shows another difficulty in calculating the amount of cooling air required. The transverse incoming quench air passes through the filament bundles in the upper part and exits the quench duct through the door. The profile of the door must be designed in such a way that the air can flow through the door and there is no uncontrolled backflow. The latter could lead to turbulence in the quench duct, as evidenced by increased yarn unevenness. Due to the drag air in the lower part of the quench duct, ambient air is sucked in (Fig. 17.33). This should therefore also be air-conditioned or conditioned to ensure defined cooling conditions.

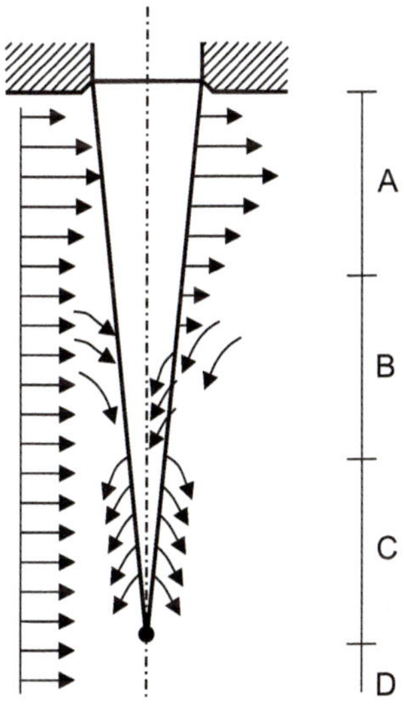

Fig. 17.32 Flow of cooling air around the filaments acc. to Fourné (1995)

The blowing process bulges the yarn path somewhat: the filaments rest on a cushion of air which softens disturbances that hit them from elsewhere. Disturbances to the yarn path which lead to variations in thickness and orientation must therefore be avoided as far as possible. Air flow in the quench duct in the opposite direction to the spinning direction should also be avoided (Fig. 17.34).

The uniformity of linear density and orientation of the spun fibers are decisively influenced by the quenching conditions and can be accordingly set by them.

On the spinning path, the flow orientation created in the filaments relaxes immediately below the spinneret holes. This is followed by a zone in which the filaments are still easily deformable and can be slightly stretched. However, a permanent spinning orientation cannot yet develop here; it relaxes immediately. As the filaments move further down into the solidification region near the glass transition temperature, they are still further distorted, but less and less so because their extensional viscosity increases very rapidly. The increase in orientation associated with stretching now does not relax completely because the relaxation times become longer and longer as the temperature decreases. At the glass transition temperature, the filaments have assumed their final thickness, and the spinning orientation built

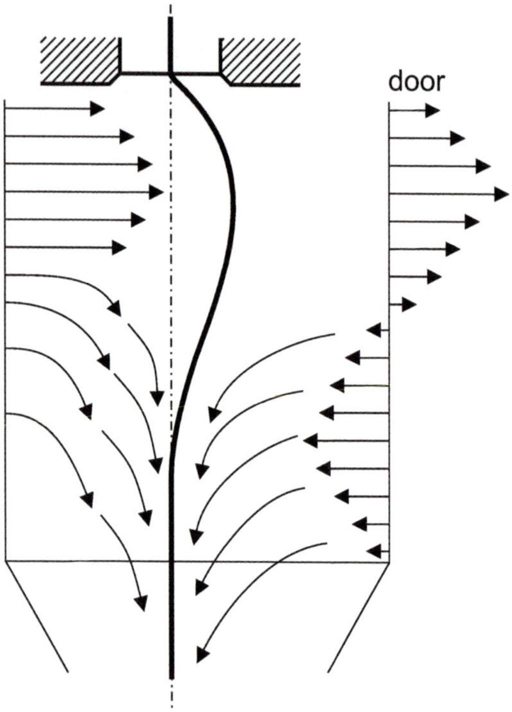

Fig. 17.33 Velocity profile of the quench air between converter and door acc. to Fourné (1995)

up until then freezes as spinning strain. This amounts to only a fraction of the total elongation of the filaments, i.e. the spinning distortion.

Frequently, a tubular quench duct, the length of which depends on the height of the spinning facility, is connected to the downward blowing. The free spinning length from the spinneret to the point where the filaments are combined to form the yarn is 1.5–4 mm, the total drop height to the take-off unit is about 2–5 m.

17.8 Spin Preparation

Single filaments spun fresh from a melt or dry from a solution show hardly any cohesion among each other ("thread compactness"). Friction on yarn guides causes them to become easily electrically charged and therefore to diverge. In addition, their surface may be rough, which can lead to undesired friction when in contact with yarn guides. For this reason, so-called spinning preparations are applied to the filament bundle, which is usually still open. These are often aqueous emulsions of surface-active agents, the combination, selection and quantity of which depend on the polymer in question and on the requirements of the further process steps. Spin preparations are also referred to as spin finish, spin oil and avivage. Typical values for the oil pick-up are in the range of 0.5–1.0%.

In LOY processes, the filament bundle often still runs over a rotating roller (roller oiler) that is partially immersed in the preparation liquid and by rotating transfers the preparation to the fibers (Fig. 17.35). The advantage here is that all filaments are oiled individually; the disadvantage

Fig. 17.34 Quench ducts in a filament spinning plant (Oerlikon Neumag)

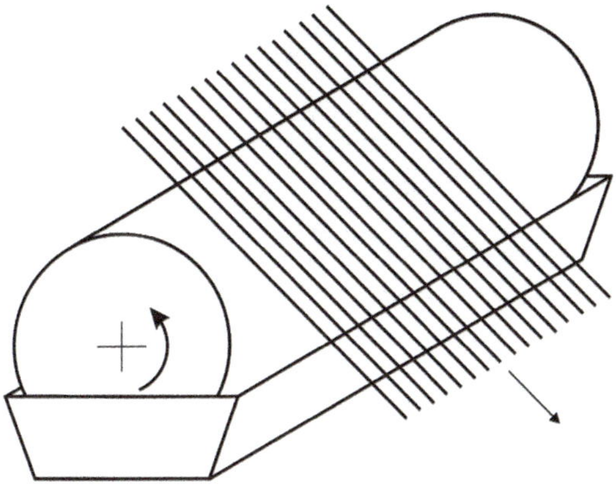

Fig. 17.35 Roller oiler with spread out filament bundle

is that when the relative speed between filaments and oiler is increased, the friction becomes too great and the filaments break.

For higher process speeds (POY, FOY), the preparation is therefore metered with a small pump and applied evenly to the passing filament bundle via a perforated filament oiler (slot oiler). For high capillary numbers, the lubricator is designed to be correspondingly wide and is then referred to as a lip or wide slot lubricator (Fig. 17.36). Although there is no danger of filament breakage because the filaments are grouped together in a bundle in the lubricator, only the outer filaments on one side of the thread are lubricated. For this reason, two lubricators are often used in series (twin lubricators), Fig. 17.36.

In order to bring the oil into contact with the inner filaments ("oil migration"), an air jet (Fig. 17.37) is often used, which at the same time slightly interlaces the filaments and thus

Fig. 17.36 Slot oiler (Debicki, 2020)

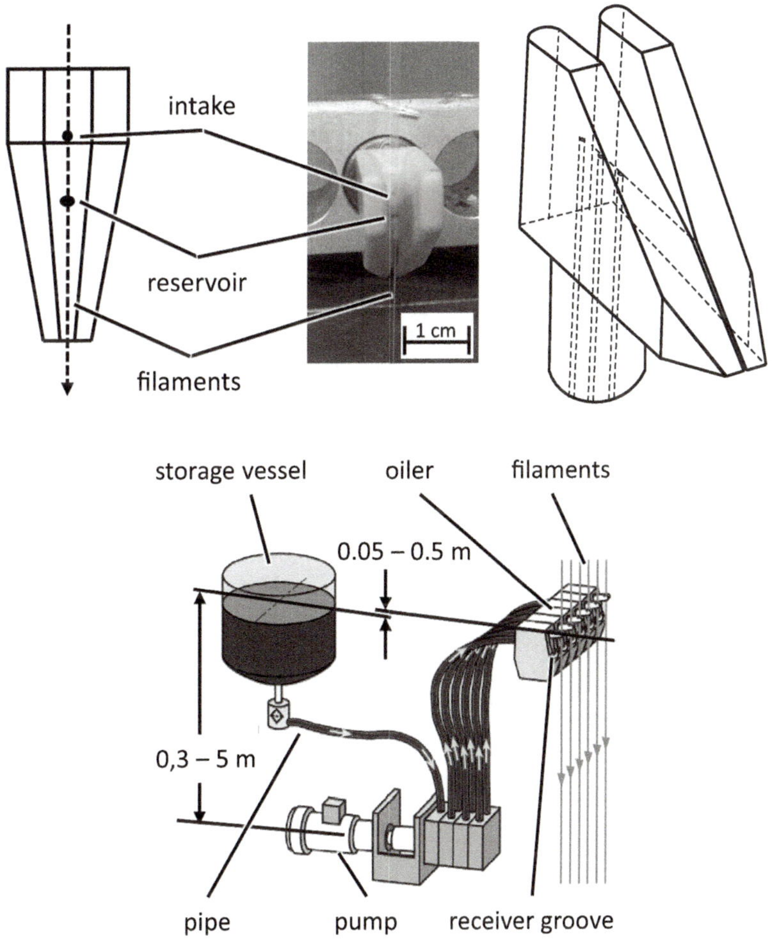

Fig. 17.37 Air jet for better distribution of the preparation within the yarn (Heberlein AG, n.d.)

improves the yarn compactness (entangling). Typical air pressures in the jet are 0.2–2 bar.

The preparation gives filaments all the surface properties required for winding and other subsequent processes. If required, a winding oil is applied to the filament at the very end. In the production of staple fibers, the spinning preparation is only required for the stable placement of the fiber cables. The surface properties required for further textile processing are then obtained by applying a spin finish at the end of the drawing and after-treatment processes to the cable or to the already cut fibers. The composition of the oil or the spin finish is usually confidential and has recently been increasingly influenced by environmental protection requirements.

17.9 Drafting

Synthetic filament yarns are guided over at least two cold or heated godets for drawing, which are arranged in pairs (DUO) or with a separate small roller (Fig. 17.38, left) and differ in their circumferential speed. Typical values are in the range of 400–5500 m/min, usual godet diameters are 250 mm. Stretching is more uniform with a higher number of godet DUOs. In the 4-DUO process, the last DUO runs somewhat slower than the third in order to allow the yarn to relax already in the process (low-shrinkage yarns).

For drawing filament yarns to high strengths, the first hot-godet drawing stage is followed by a second one—usually with double-godet (DUO) drawing. Godets are often heated to temperatures from 110 to over 200 °C, depending on the polymer (Fig. 17.39).

If the spun yarn tends to form a drawing point during drafting, this must be stabilized and fixed in the drafting zone. In the case of fusible polymers whose glass transition temperature is close to room temperature, a drafting pin of a few millimeters diameter around which the yarn is wrapped is sufficient for this purpose. This principle is frequently used for polyamides and polyesters. If the spun yarn has to be additionally heated for drafting, heated drafting pins or heated friction bodies with a larger diameter are used around which the yarn is wrapped. The drawing

Fig. 17.38 Configurations for drafting synthetic filament yarns

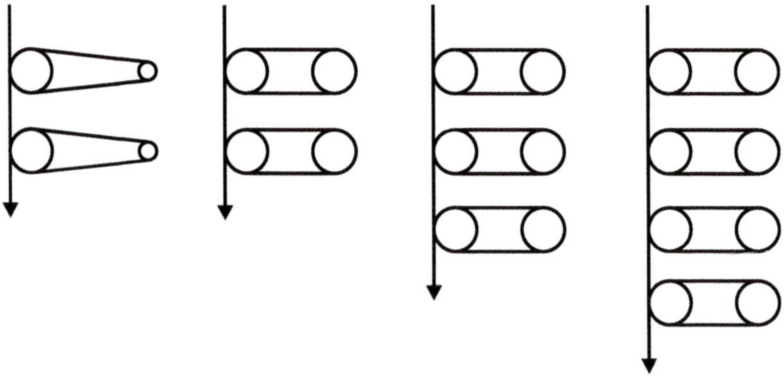

Fig. 17.39 Godets with separate small roller (left) and DUOs (right) (Oerlikon Barmag, n.d.; Oerlikon Neumag)

tension is built up on them by friction and the drawing point is also located on them. The actual stretching that can be measured on the finished yarn is usually somewhat lower than the stretching set on the machine, because a small part of the stretching elongation is elastic.

The filaments must have a minimum degree of plasticity so that the stretching forces do not become too great and cause them to break. Their temperature should therefore be at least a few degrees above the glass temperature. Alternatively, a sufficient amount of plasticizer may be sufficient, either as a solvent residue or as a swelling agent diffused in from a drawing bath. Softening of the spinning fabric shortens the relaxation time of the tensions in the filaments, above the glass transition temperature even considerably. Therefore, the introduction of an external stress not only leads to a stretching of the macromolecules, but also has a non-orienting flow component. This is very large directly below the spinneret and becomes smaller as the temperature of the filament decreases. The proportion of plasticizer and the intermolecular forces also play a role. The flow portion is greater at lower yarn speeds.

With increasing stretching, tensile strength and modulus of elasticity increase, while elongation-at-break decreases. If the value of the stretching exceeds the internal strength of the filaments, thread breakage ("ends down") occurs. To avoid this, the drawing is therefore often divided into several zones. In the first zone, a higher orientation occurs and the crystallinity often increases. Therefore, higher drawing can then take place in the second zone.

The drafting point or drafting zone should always be located at the same point within the drafting field, because otherwise fluctuations in the yarn fineness and yarn orientation can occur. For energetic reasons, the drafting point often moves against the running direction of the filaments. To avoid this, attempts are often made to hold the drafting point in the drafting zone by means of friction elements (drafting pin) or by sudden heat input which reduces the drafting tension. These measures are also often combined.

The stretching conditions not only influence the textile data of the stretched fabric via its orientation and semi-crystalline structure, but also strongly influence its dyeing behavior. The human eye is very sensitive to even small differences in color depth, both along a yarn and between individual yarns. Therefore, all process conditions must always be kept as constant as possible.

In general, the drawing tensions of solution-spun filament yarns are relatively low, while for yarns made of synthetic polymers they reach values of about 3–30 cN/tex.

For technical articles, the processes "spinning" and "drawing" are sometimes separated. The latter is then referred to as "draw winding". Drawing can also be combined with warping and twisting.

17.10 Entangling

Yarns are often interlaced before or during drawing or immediately before winding. This improves the cohesion of the filaments ("yarn compactness"). Oil application can also be integrated into the interlacing jet, which then allows the oil to reach the center of the yarn better.

The structure of a typical entangling jet and the air flows as well as the tangle knots are shown in Fig. 17.40. The interlacing of the yarns takes place in a component made of sintered ceramic. The typical interlacing knots ("tangle knots") are formed by the interaction of the lower and upper air flow.

The quality of the interlacing knots is determined by their stability and even distribution. It is directly influenced by the air jet pressure (usually 2–6 bar), the yarn overfeed, the yarn guidance before and after the air jet and thus directly via the yarn tension in the air jet.

The Fig. 17.41 shows various designs for multi-thread operation.

17.11 Winder

Yarns made from man-made fibers are wound using the same winding principles as yarns made from staple fibers (random winding, precision winding, etc.). Today, cylindrical and conical packages dominate.

For draw winding and draw texturing, winding units are used in which the bobbin is driven at constant speed by friction wheels on the circumference. The yarn is laid down on the bobbin with a yarn guide. The yarn guide movement is generated by a reverse thread shaft or by a gearbox. When using a gear transfer, conical cross-wound packages are also possible. The use of a stepper motor-driven yarn guide laying system enables a particularly flexible package design (Fig. 17.42).

Tension-controlled individual winders are used for wet spinning, yarn bundle drawing, slit film and monofilament production. The yarn tension is detected by a dancer arm or a yarn tension sensor with little deflection and thus the winding speed is controlled. The thread guide is usually moved by a helical thread shaft. The bobbin is

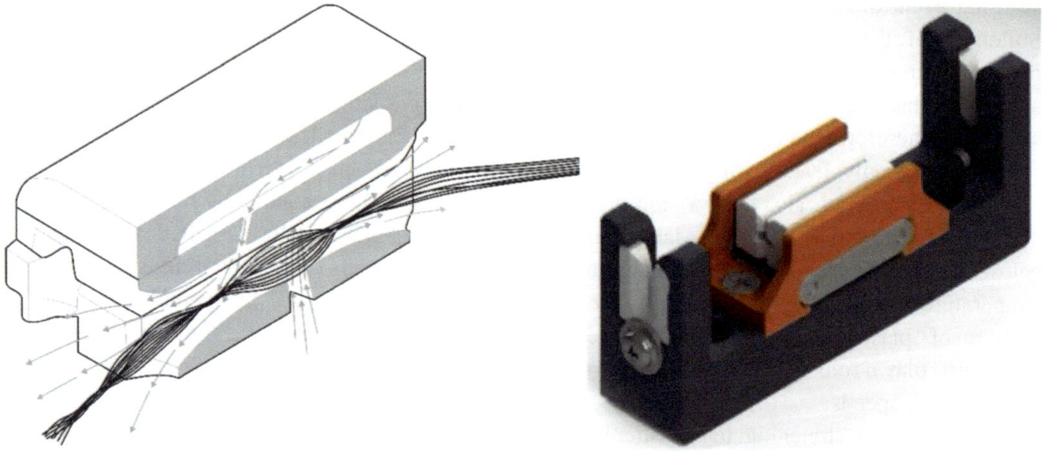

Fig. 17.40 Structure and design of an entangling jet (Heberlein AG, n.d.)

Fig. 17.41 Multi-thread interlacing jets (left: Saurer Technologies, n.d., right: Heberlein AG, n.d.)

driven by the mandrel. Stiff monofilaments are often wound with parallel winding.

When spinning textile and technical filaments, several yarns are always wound up simultaneously. At take-up speeds between 3200 and 5500 m/min, yarn transfer with a reversing thread shaft would be associated with high wear and very high forces at the reversing points of the yarn guide. For this reason, the flyer or birotor transfer was developed, which today dominates in all filament melt spinning processes.

The birotor winding system consists of two rotors rotating in opposite directions under a yarn guide ruler. Each rotor usually has three blades with which the yarn is moved back and forth on the guide ruler by the bobbin stroke. The yarn is transferred from one rotor blade to the other at the bobbin edges (Fig. 17.43).

As no transverse movement is required and due to the relatively low masses to be moved, this transfer method is also suitable for very high winding speeds. It is the standard winding method for the production of partially and fully oriented filament yarns (POY, FDY), Fig. 17.44.

The Fig. 17.45 shows the typical construction of such a winder. Typical dimensions are for the height 1.4 m, the width 0.6 m and the length of the winding mandrel (1.2–1.6 m). Usually between

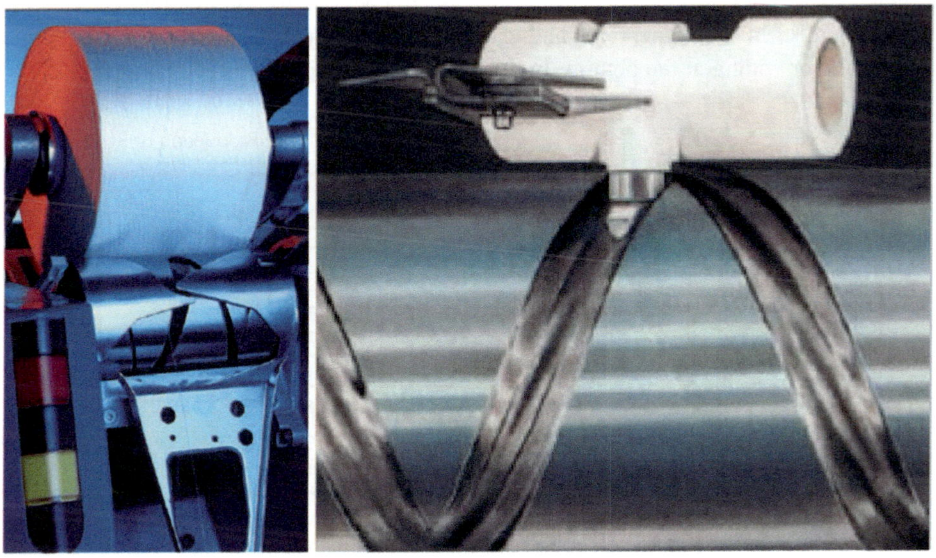

Fig. 17.42 Grooved roll (left) and sweep thread shaft (right) (Saurer Spinning Solutions, n.d.)

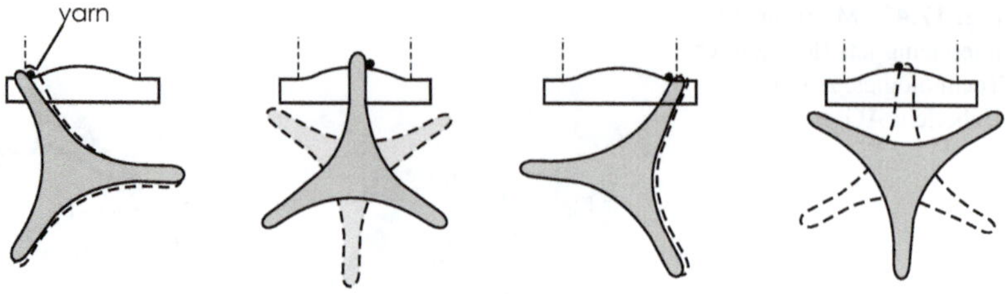

Fig. 17.43 Phases in bi-rotor yarn transfer acc. to Oerlikon Barmag (n.d.)

Fig. 17.44 Detail of the birotor transfer mechanism with birotors and guide bar (Oerlikon Barmag, n.d.)

6 and 16 bobbins are wound simultaneously per mandrel, the bobbin diameter is approx. 440 mm. The bobbin weights are then around 10–25 kg.

The Fig. 17.46 shows an FDY machine with the enclosed godets in the upper part and the winder with birotor widing at the bottom. In this example, a "double winder" is used, i.e. two winders on a shared machine frame.

For coarse yarns such most technical and carpet filament yarns (BCF), only 2, 3 or 4 filament yarns per winder position are wound at the same time; for textile filaments, there are six or eight

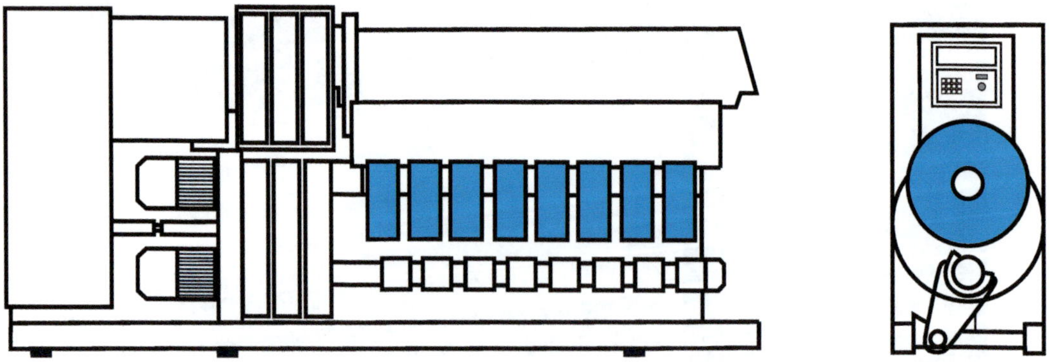

Fig. 17.45 Typical structure of a filament yarn winder acc. to Oerlikon Barmag (n.d.)

Fig. 17.46 FDY machine (Oerlikon Barmag, n.d.)

yarns per winder if large bobbin formats and thus long bobbin running times are required in downstream processing. Also, up to 14 positions per winder are available, which produce smaller bobbins.

For reasons of automation, each winder usually has two winding mandrels, which are located on a turret and one chuck of which is in winding position. When the bobbins of this chuck have reached their target weight, they swivel into the parking position and the second chuck, on which new empty tubes are located, swivels into the working position and starts winding on the threads without interruption (Fig. 17.47).

The full bobbins are picked up ("harvested") by an operator or a robot. For this purpose, the full cores are automatically pushed from the winding mandrel onto the mandrel of the robot by means of a disengaging unit and transported away to the bobbin storage or to the automatic packaging unit (Fig. 17.48).

In the warehouse, the coils are automatically packed and usually assembled into larger units (Fig. 17.49).

The Fig. 17.50 shows the schematic structure of a direct spinning plant from the reactor to the take-up. Such plants are about the size of one to four football grounds.

Fig. 17.47 Winder level of a POY plant (Oerlikon Barmag, n.d.)

Fig. 17.48 Automatic doffing of full bobbins and bobbin bearings (Oerlikon Barmag, n.d.)

Fig. 17.49 Packaging the spools in cartons (Oerlikon Barmag, n.d.)

Fig. 17.50 Schematic of a direct melt spinning plant from reactor to take-up (Oerlikon Barmag, n.d.)

References

Auras, R. A., Singh, S. P., & Singh, J. J. (2005). Evaluation of oriented poly(lactide) polymers vs. existing PET and oriented PS for fresh food service containers. *Packaging Technology and Science, 18*, 207–216. Wiley.

BB Engineering GmbH. (n.d.). Remscheid. https://www.bbeng.de

Bekaert, N. V., & Bekaert, S. A. (n.d.). https://www.bekaert.com

Debicki, L. (2020). *Industrie 4.0-Technologieansätze zur Effizienzsteigerung im Schmelzspinnprozess*. Dissertation am Institut für Textiltechnik der RWTH Aachen, Aachen.

Fourné, F. (1995). *Synthetic fibers*. Hanser.

Fourné Maschinenbau GmbH, Alfter-Impekoven. https://fourne-maschinenbau.de

Gneuß Kunststofftechnik GmbH. (n.d.). Bad Oeynhausen. https://www.gneuss.com

Heberlein, A. G. Wattwil. (n.d.). https://www.heberlein.com

Mahr GmbH. (n.d.). https://metering.mahr.com/

Oerlikon Barmag. (n.d.). Remscheid. https://www.oerlikon.com/polymer-processing/de/

Saurer Spinning Solutions GmbH & Co. (n.d.). GK, Übach-Palenberg. https://saurer.com

Saurer Technologies GmbH & Co. (n.d.). KG Engineered, Hammelburg. https://saurer.com/de/komponenten/temco

Sossna GmbH. (n.d.). Dorsten. https://www.sossna.de

18 Aggregates for the Production of Solution Spun Man-Made Fibers

Cellulose and its derivatives are not meltable. Therefore, man-made cellulosic fibers are always produced by the solution spinning process. The aggregates required to produce cellulosic fibers, e.g. viscose, acetate and lyocell, are described here.

18.1 Reactor

When producing a spinning solution, several reactors are usually connected in series. Either the polymers are first produced by polymerization or polycondensation in a solution (e.g. PAN) or the polymers already present (e.g. cellulose) are dissolved in the solvent, possibly first chemically modified.

The solution is then prepared in the first reactor, whereby the polymer is first solved. During this process, the molecules or ions of the solvent are deposited around and between the polymer chains. Polar groups, van der Waals forces and possibly the charge of the ions cause interactions between the polymer and the solvent. If the lattice energy of the polymer is greater than the solvation energy, the solution cools. If it is smaller than the solvation energy, it heats up.

The dissolution of the polymer only occurs if the binding forces between the macromolecules are sufficiently small. They can be reduced by a lower degree of polymerization. Some polymers dissolve better at higher temperatures, others at lower temperatures (e.g. cellulose xanthate in sodium hydroxide solution at 5–10 °C). The use of copolymers can also contribute to an increase in the mobility of the macromolecules by disturbing the internal structure.

Table 18.1 gives an overview of typical values for the polymer concentration, the solvent and the kinematic viscosity of frequently used spinning solutions. The kinematic viscosity is obtained by dividing the dynamic viscosity (ratio of shear stress to velocity gradient perpendicular to the flow direction) by the density of the solution. For melts, the kinematic viscosity is in the range of.

In the second reactor, the unreacted monomers are removed if necessary (e.g. by evaporation) or the solution is deaerated. In this process, the spinning solution flows over thin plates or along the inside of a vessel. This shortens the path of the gas bubbles to the surface and thus also the deaeration time. In the third reactor, the solution is homogenized (e.g. by stirring).

18.2 Filter

In the filtration of solutions, filters similar in principle to those used in melt spinning are used. The most commonly used is quartz sand, whose surface area is up to ten times larger than that of textile filters. Filter throughputs of 500–1000 l/m^2 h are thus achieved.

D. Veit, *Fibers*, https://doi.org/10.1007/978-3-031-15309-9_18

Table 18.1 Typical polymer concentrations, solvents and viscosities of spinning solutions acc. to Rogowin (1982)

Polymer	Solvent	Polymer concentration [%]	Kinematic viscosity [m^2/s]
Viscose	Caustic data	6–10	0.003–0.01
Cupro	Cuoxam	6–12	0.06–0.1
Triacetate	Methylenchloride-alcohol	20–22	0.04–0.08
PVC	Carbon disulfide-acetone	25–30	0.5–0.8
PAN	Dimethylenformamide	15–20	0.015–0.03
PVA	Water	12–15	0.01–0.02

18.3 Spinning Pump

The spinning pumps used in solution spinning are basically similar to those used in melt spinning. However, they differ in throughput (1–500 cm^3/min) and pressure range (15–30 bar).

18.4 Spinneret

The dynamic viscosity of the polymer solution in solution spinning is in the range of 10–50 Pas (in melt spinning approx. 50–100 Pas). Accordingly, temperatures of only up to 150 °C are reached and the spinneret pressures are usually in the range of 3–15 bar. The spinnerets for solution spinning are therefore significantly thinner and lighter than those for melt spinning, because little or no heat has to be applied (and maintained). The Fig. 18.1 shows typical spinnerets for solution spinning.

Spinnerets made of stainless steel or tantalum are often used for solution spinning. The shape of the holes shows Fig. 18.2. The fine capillary hole is usually about one to three times as long (L) as wide (D). The most favorable L/D ratio in each case depends on the spinning process and the spinning parameters.

Spinnerets for the wet spinning process are generally loaded with pressures below 10 bar and are therefore relatively thin (0.2–0.8 mm). They are mostly cup-shaped. Depending on the intended use, they have about 20–2000 holes for filament yarn spinning and 2000–200,000 holes for fiber cable spinning, arranged in a wide variety of patterns. The hole diameters are about 0.04–0.1 mm, the L/D ratio (length to thickness) is usually 0.7–2.

The Fig. 18.3 shows the process of thread formation during solution spinning. Initially, a thin sheath layer ("skin") is formed by withdrawal of the solvent from the edge areas, in which the polymer already coagulates and solidifies. In the

Fig. 18.1 Spinnerets for solution spinning (Sossna GmbH, n.d.)

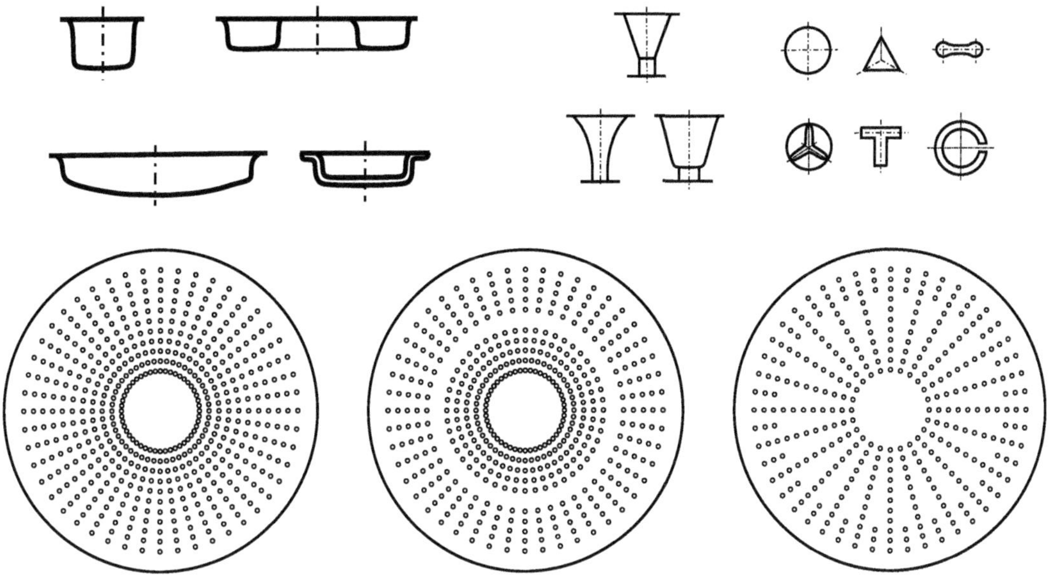

Fig. 18.2 Typical spinneret cross-sections and hole patterns of solution spinnerets based on Fourné (1995)

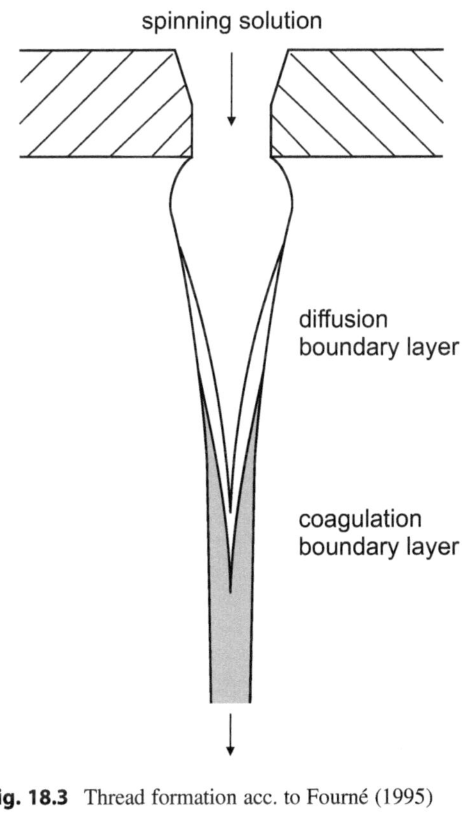

Fig. 18.3 Thread formation acc. to Fourné (1995)

interior, a diffusion boundary layer is thus formed through which the solvent must pass from the core to the surface, which is facilitated by higher temperatures.

In wet spinning, the concentration of the solvent in the precipitation bath essentially determines the diffusion rate. Reducing the concentration accelerates the process. If the precipitant penetrates the interior of the fiber more slowly than the solvent diffuses outwards, the vacuum in the fiber core creates cavities and the sheath can collapse. In the case of viscose, for example, this leads to a lapped cross-section ("cauliflower cross-section"), Fig. 18.4.

However, by skillful selection of the coagulation conditions, circular profiled cross-sections can also be achieved (e.g. with Lyocell fibers).

18.5 Winder

Yarns made of cellulosic man-made fibers are wound according to the same principles as yarns made of staple fibers (random winding, precision

Fig. 18.4 Transversal and longitudinal view of viscose

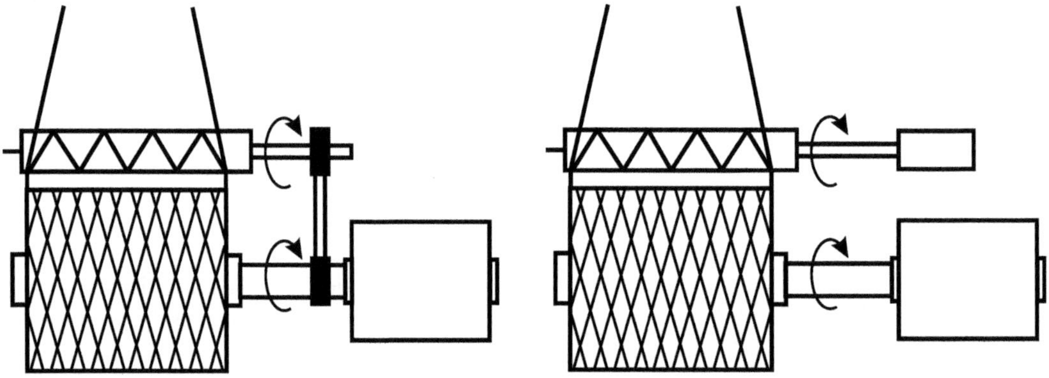

Fig. 18.5 Drive of the coil via coupling with laying unit (left) or individually (right)

winding, etc.). Today, cylindrical or conical packages dominate. Typical winding speeds are 100–1000 m/min.

Winding units are often used in which the bobbin is driven at a constant speed by friction wheels on the circumference. The thread is laid on the bobbin with a thread guide. The yarn guide movement is generated by a helical screw shaft or by a gearbox. When using a gear transfer, conical cross-wound packages are also possible. Tension-controlled single winding heads are frequently used. The yarn tension is detected by a dancer arm or a yarn tension sensor with little deflection. This signal is used to control the winding speed (Fig. 18.5, right). A particularly flexible package design is possible with a stepper-motor-driven yarn guide laying system.

The Fig. 18.6 shows the two most frequently used yarn laying methods in solution spinning.

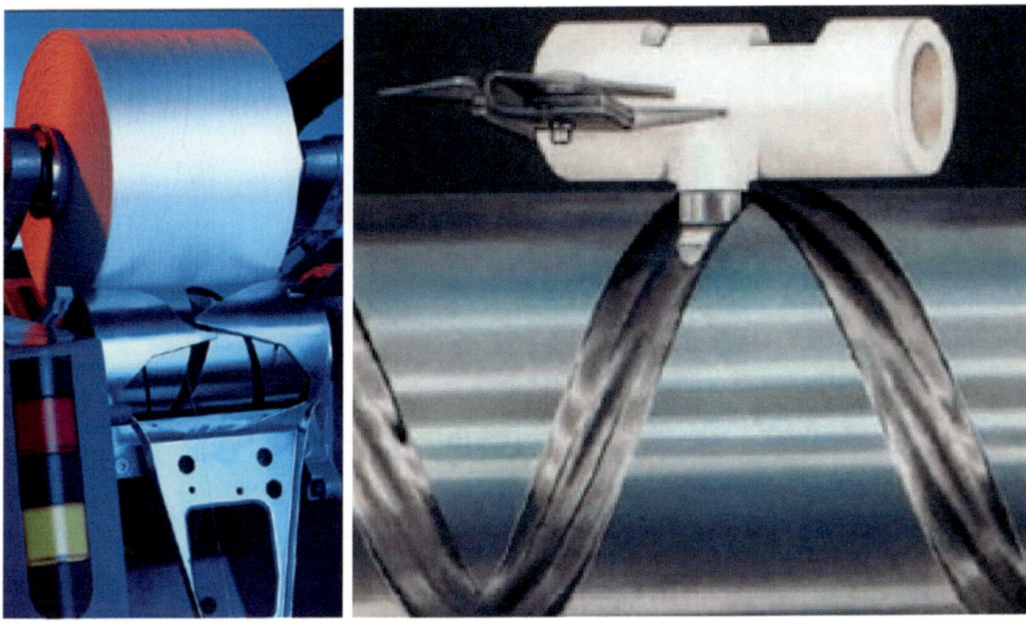

Fig. 18.6 Grooved roller (left) and reverse thread shaft (right) (Saurer Spinning Solutions, n.d.)

References

Fourné, F. (1995). *Synthetic fibers*. Hanser.

Rogowin, Z. A. (1982). *Chemiefasern*. Georg Thieme Verlag.

Saurer Spinning Solutions GmbH & Co. (n.d.). KG, Übach-Palenberg. https://saurer.com/de

Sossna GmbH. (n.d.). Dorsten. https://sossna.de/

19 Processes for the Production of Textile Filament Yarns

Man-made fibers can be produced in very different forms. The Fig. 19.1 shows the main types according to Thomas Gries.

In general, a distinction is made between technically endless filaments and so-called staple fibers of finite length (e.g. cotton type). The filaments are in turn subdivided into monofilaments (one fiber per yarn), multifilaments (many fibers per yarn, usually 7–400) and cables (very many fibers per yarn, e.g. 10,000). In the case of multifilaments, a further distinction is made between textile filaments of medium strength, technical filaments of high strength, high-modulus yarns of maximum strength and carpet yarns.

The manufacturing processes are basically similar, the main differences are the subject of this chapter.

19.1 POY Process

Originally, textile filaments were spun at LOY speeds (PET: approx. 1400 m/min) and then drawn and optionally textured in at least one further process stage. This was mainly due to the fact that the winding speed with the technology available at that time was too low for the production of fully-drawn yarns in a single process. In the meantime, the take-up speeds of modern winders reach 6000 m/min and more. Therefore, the lower limit of spinning speed for partially oriented yarns (POY) is now significantly higher, e.g. 3200 m/min for polyester.

During the first attempts to increase the performance of spinning plants by means of higher take-off speeds, it was soon realized that this would not only considerably increase orientation and crystallinity, but that the spun yarns would then also become less sensitive and have a longer storage life, and that they would need less subsequent drawing. This makes it possible to combine fully drafting with the most common further processing process for textile filament yarns, texturing. Fast spun and drawn products have a looser structural composition, because less additional crystallinity is generated by the reduced post-drawing. This facilitates dye uptake. Due to the higher spinning tension, the filaments stand more steady in the spinning shaft and it is basically easier to produce uniform yarns even without draw-off godets. However, if the spinning tension is considerably higher than the optimum winding tension, cold or heated take-off godets are used. For these reasons, POY spinning with take-off speeds of 3200–5500 m/min has almost completely replaced the LOY spinning process for filament yarns.

Spinning beams or winders usually have 6–10 positions if large bobbin formats and long bobbin running times are required in downstream processing. If the highest possible productivity is required with almost identical equipment, the number of spinning positions can also be 10, 12 or even 24.

D. Veit, *Fibers*, https://doi.org/10.1007/978-3-031-15309-9_19

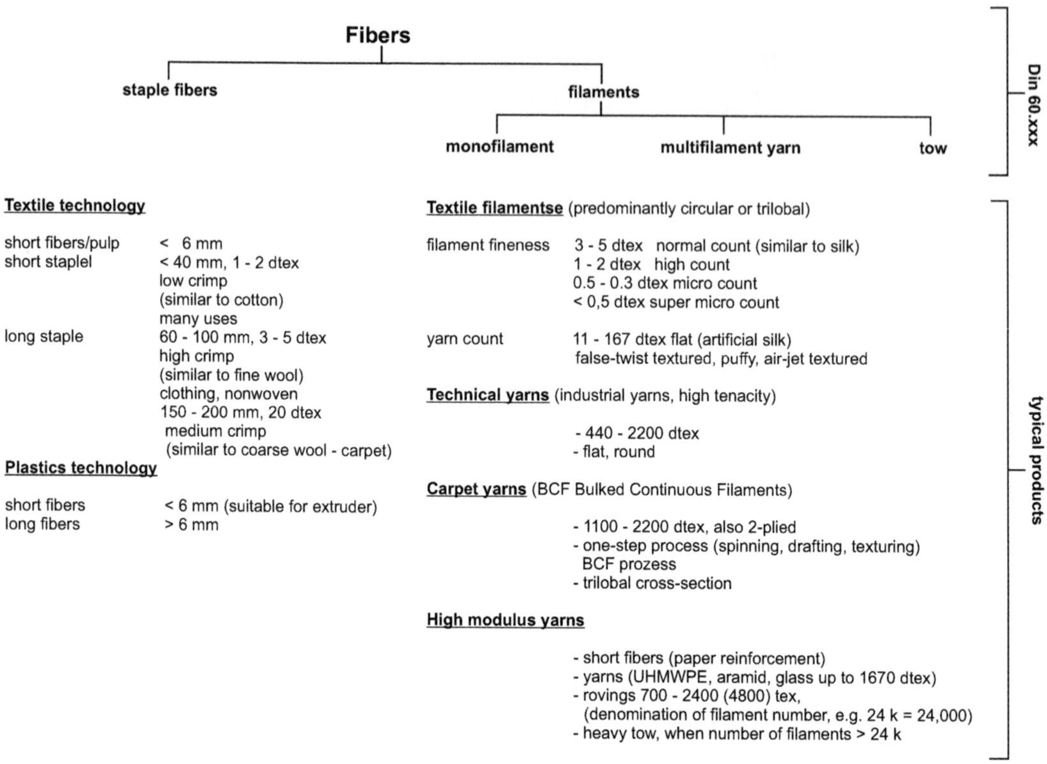

Fig. 19.1 Fiber types acc. to Gries

The Fig. 19.2 shows the spinning processes commonly used today to produce POY yarns. In the case of the godetless process, drafting is generated solely by the tension induced as a result of winding. In the process with S-godets, the yarn path is smoother and the yarn is usually interlaced between the godets (higher yarn compactness). In the process with cold godet DUO, the drafting takes place in two stages.

Usually, in order to save space, the spinning beam and winder are arranged at an angle of 90° to each other. Therefore, the filament bundle is also rotated accordingly on its way from the spinneret to the winder. More rarely, the spinning beam is arranged parallel to the winder. Parallel spinning systems are lower, but wider. The vast majority of textile filaments today are produced with the POY process.

19.2 FDY Process

Fully drawn, flat textile filament yarns (FDY) are nowadays mostly produced in a single-stage spun-draw process. In modern plants, the yarn runs onto the first godet at POY spinning speed. The arrangement of spinning beam and winder corresponds to that in the POY process. The Fig. 19.3 shows an overview of the most important processes. Drawing usually takes place with the addition of heat in several stages. This is usually done by heated godets. Alternatives are a steam chamber (polyamide) or a heated channel (thermochannel). This makes the amorphous and crystalline areas within the filaments more mobile, the stretching requires less force and is therefore more uniform.

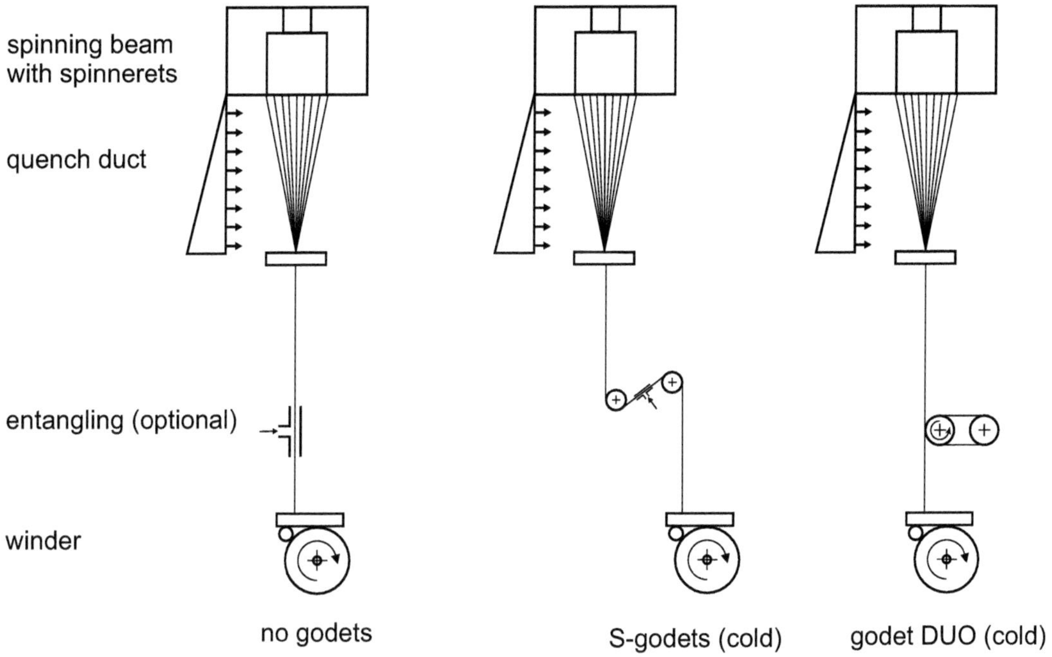

Fig. 19.2 POY spinning processes for textile filaments acc. to Gries

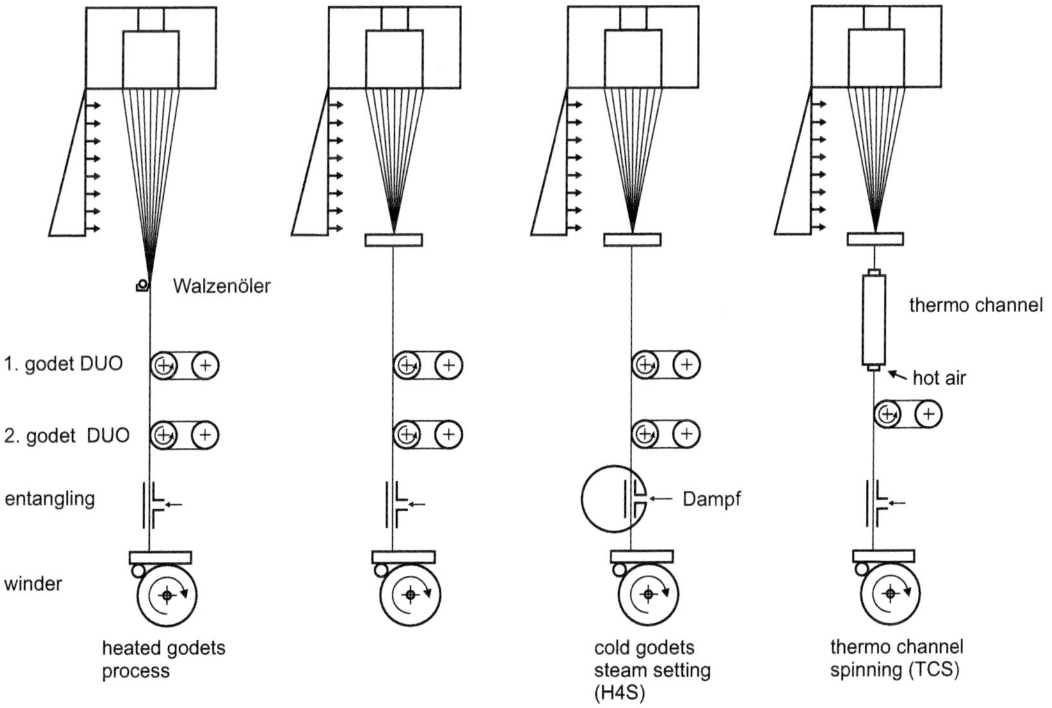

Fig. 19.3 FDY spinning processes for fully drawn, textile filament yarns acc. to Gries

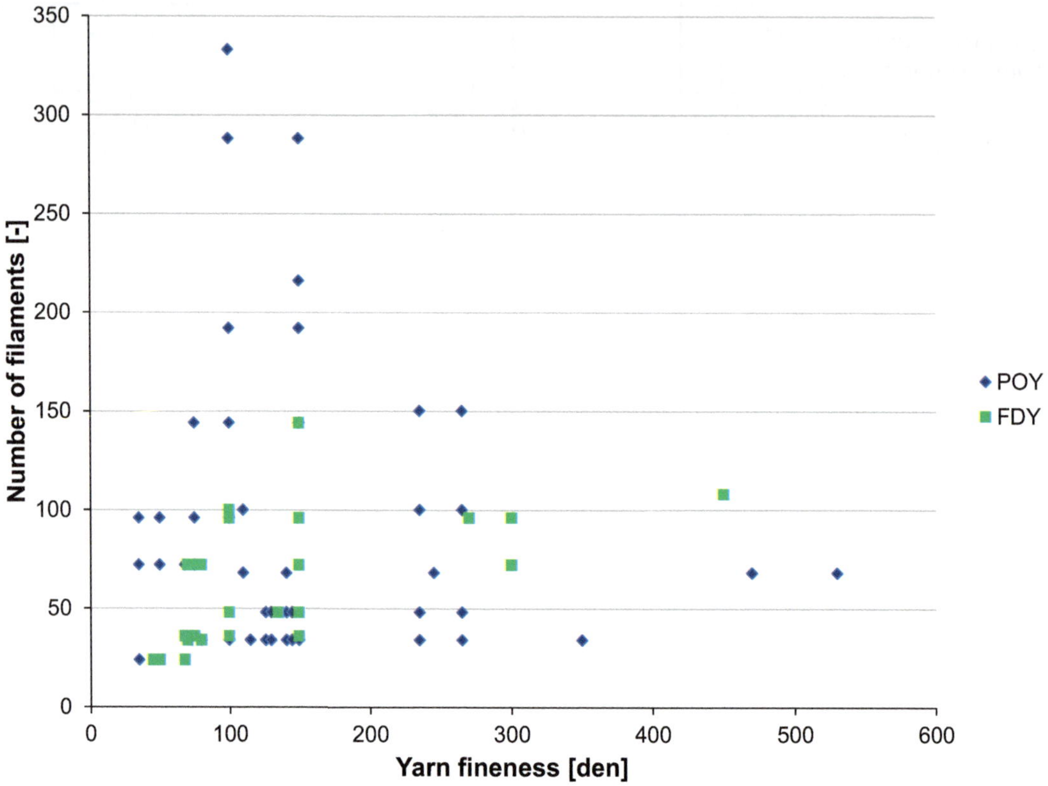

Fig. 19.4 Typical yarn counts and number of filaments

Typical yarn and filament counts of different manufacturers are shown in Fig. 19.4. The wide range of yarn counts is immediately apparent. However, a few filament counts dominate. There are country-specific differences here, e.g. filament counts of 36, 72 and 144 are often spun in Turkey, while in India the corresponding figures are 34, 68 and 136. Filament counts of 48 and 96, on the other hand, are produced in both countries. As is usual with man-made fibers, the fineness is given in Fig. 19.4 in [den]. This results in numbers that are easier to remember (e.g. 150 den instead of 167 dtex).

19.3 FOY Process

Fully drawn, flat textile filament yarn can also be produced by very high take-off speeds from the spinning zone without subsequent drawing (Fig. 19.5). This so-called FOY (also called HOY- or Super High Speed Spinning SHSS-FOY) was developed only for polyester. The necessary take-off speeds are in the range of 6000–7000 m/min. This clear cost advantage is offset by the disadvantage of other yarn properties. These include, due to the distorted crystal structure, a high elongation compared to FDY, a lower modulus of elasticity, a worse dyeability and a higher boiling shrinkage. Furthermore, only a limited linear density range is possible, because filaments with a fineness of less than 1 dtex cannot be produced with it. For this reason, these processes have only been implemented in a few plants.

19.4 Performance Increase in POY Spinning of Polyester

With the desired final fineness, the throughput of a spinning position is $\dot{m}$ of a spinning position is

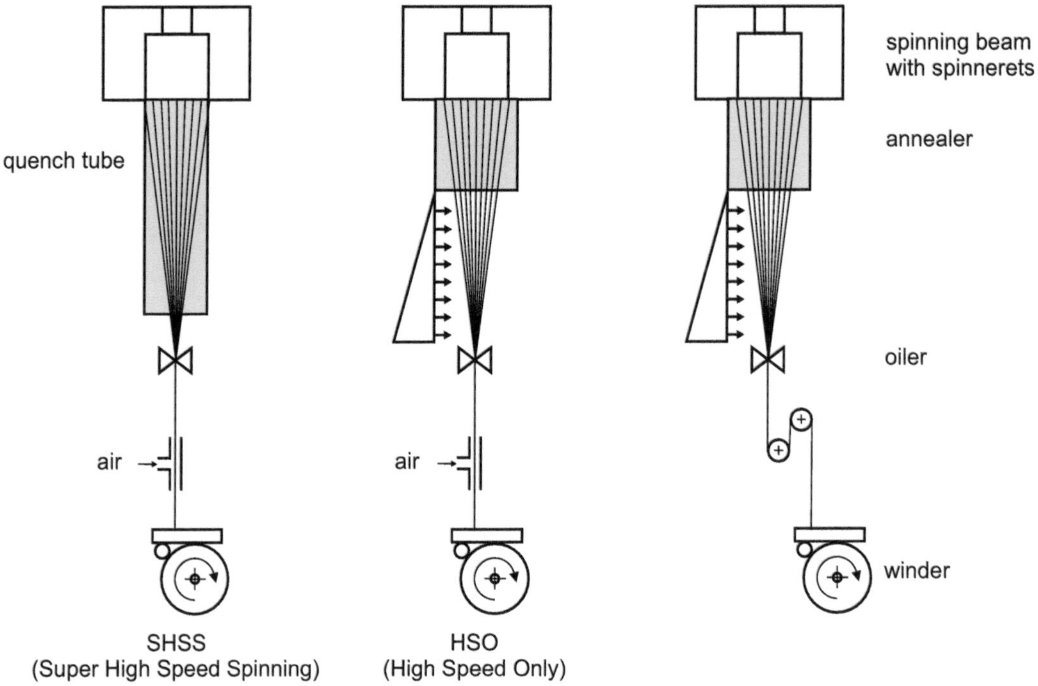

Fig. 19.5 Spinning processes for fully drawn, smooth, textile filament yarns acc. to Gries

determined by the spinning speed and thus limited. The final yarn count results from the product of the spinning titre and the drafting ratio i, which in turn is determined by the spinning speed.

$$\dot{m} = Tt_{tex} \cdot i \cdot V_{spin}$$

For example, if the spinning speed of polyester is increased from 3200 m/min (POY) to 5000 m/min, the induced stress increases the orientation of the macromolecules, thus the draw ratio decreases from i = 1.8 to i = 1 and the throughput therefore increases by only about 10%. In addition, stress-induced crystallization occurs in the polymer, which makes subsequent stretching more difficult or even prevents it. The Fig. 19.6 shows the throughput and draw ratios as a function of spinning speed for yarns with a fineness of 150 den (167 dtex) for polyester, polyamide 66 and polyamide 6. The “shoulder” in the function of throughput for polyamide is due to the recrystallization of the polyamides at the corresponding spinning speeds. Marked with an arrow is the determination of the residual draft for PET for a spinning speed of 3200 m/min. It results in i = 1.8, irrespective of the yarn count. A POY with a length of 1 m can therefore be stretched to a length of 1.8 m before it breaks.

If the throughput of a spinning position in POY spinning is to be increased while keeping a residual elongation of i = 1.8, either the tension or crystallization must be reduced or the latter even completely prevented. Three main approaches are known to achieve this. These have been developed to the point of a commercial application because of the great market importance of polyester.

19.4.1 Modification of the Spinning Process

The spinning stress results essentially from the acceleration forces in the drawing zone in the spinning duct due to the speed differences between extrusion and take-off speed. By means of targeted air and temperature control in this area, the spinning stress and thus crystallization can be suppressed accordingly. Such a process

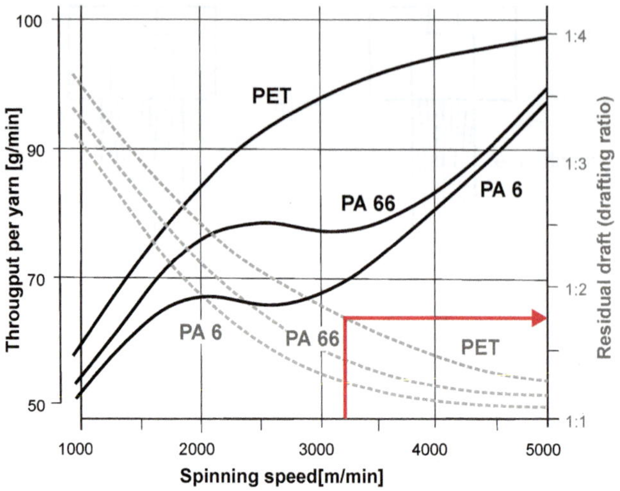

Fig. 19.6 Relationship between spinning speed, draw ratio and melt throughput for common polymers

was patented by DuPont and developed to commercial maturity by Oerlikon Barmag (n.d.). It allows PET-POY spinning speeds of up to 4200 m/min.

19.4.2 Chemical Modification

The formation of crystals is hindered and thus slowed down by chemical modification of the polymer. Chain branching agents (trimellitic acid, penthaerythritol, dipentaerythritol) are mainly used here, and chain kinkers (isophthalic acid and other comonomers) are used less frequently. The increase in performance is approx. 10%, because higher comonomer contents have a negative influence on the drafting behavior despite higher spinning thread elongation and lead to greatly changed end product properties.

19.4.3 Physical Modification

In this process, a non-soluble, highly viscous polymer with different crystallization behavior (glass transition temperature higher than that of PET) is finely dispersed and added to the polymer stream (Fig. 19.7).

This approach goes back to ICI and AKZO. A first commercial solution based on polymethylmethycrylate (PMMA) was developed and patented by Zimmer AG (n.d.). The additive, which is added at less than 1%, solidifies earlier than PET and, due to its high viscosity, captures most of the spinning stress during stretching. Modifier microfibrils are formed in the process. Thus, PET-POY spinning speeds of up to 5000 m/min can be achieved. In Klein and Wandel (2002), the process as well as the other and earlier patents are described.

Figure 19.8 shows the shift of the residual draft (drawing ratio) towards higher take-up speeds.

An important aspect in the use of modifiers is the consistency of the essential yarn properties compared to the unmodified polymer. This includes, in particular, color and surface and, more generally, unaltered further processing with, as far as possible, the same machine settings that are used for processing conventional POY. For this purpose, it is imperative that the modifier is homogeneously distributed in the polymer and does not agglomerate or change its structure by deposition on the surface.

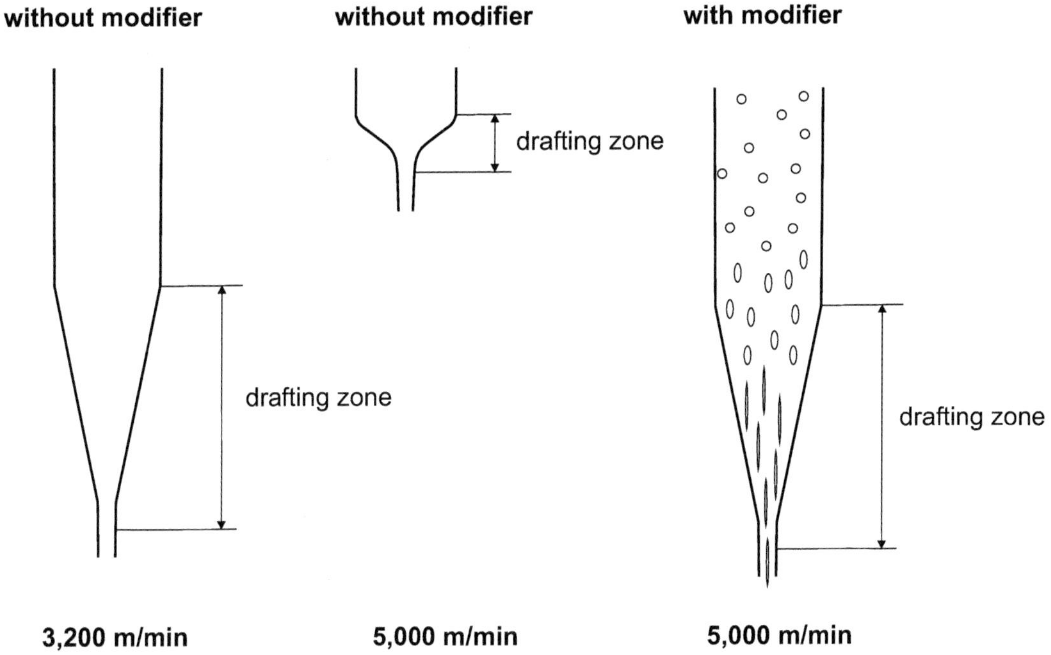

Fig. 19.7 Polyester POY spinning process with and without modifier acc. to Zimmer AG (n.d.)

Fig. 19.8 Influence of the modifier on residual draft and winding speed for PET-POY acc. to Zimmer AG (n.d.)

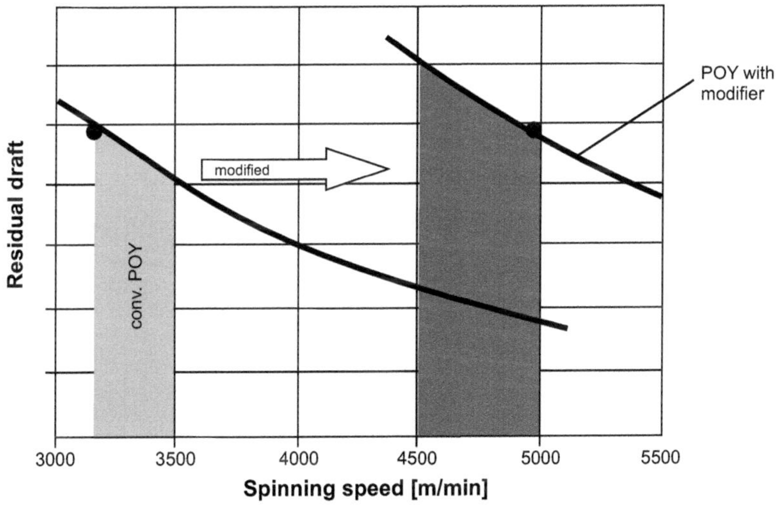

References

Klein, A., & Wandel, D. (2002). "Verfahren zum Spinnen und Aufspulen von PET-Filamenten", Patentschrift, WO 02/50348 A1, Deutsches Patentamt, München.

Oerlikon Barmag. (n.d.). Remscheid. https://www.oerlikon.com/polymer-processing/

Zimmer AG. (n.d.). Frankfurt a. M.

20 Processes for the Production of Industrial Yarns

Compared to textile filament yarns, industrial yarns have a significantly higher strength, a lower elongation and thus a comparatively high modulus of elasticity as well as mostly a low shrinkage.

20.1 Reactor

Depending on the required mechanical properties of the yarns to be produced, one or two reactors are used for polymer formation (Fig. 20.1). In the first reactor, a polymer of normal viscosity and usual chain length is produced. In the second reactor, the chain length is then further increased, as is the viscosity. This is not possible in a single reactor.

20.2 FDY Process

Conventional technical filament yarns made of polypropylene, polyamide and polyester are spun at LOY speeds on the first godet of approx. 800–1000 m/min (Fig. 20.2, left). Today, they are mainly processed and wound directly into fully drawn technical filament yarns in the single-stage spun-draw process. Round spinnerets are used for fine technical yarns, otherwise rectangular spinnerets with annealer and cross-flow quenching. The maximum capacity per spinning line is usually limited by the extruder and the melt distribution due to the high melt viscosity. The spinning preparation is applied before the godets.

Since about 1995, PET technical filament yarns have been spun at POY speeds on the first godet of 2500–2800 m/min for high-viscosity PET (Fig. 20.2, right). Rectangular spinnerets are used for normal counts and ring spinnerets for coarse fibers. The single filament fineness is usually finer compared to LOY-spun yarns. In most cases, an annealer is used to facilitate drawing and thus make it more uniform. The spinning preparation is applied in the quench duct, usually with slit oilers. Due to the increased orientation and crystallinity of the spun yarn, technical yarns are produced which are characterized by a high modulus of elasticity combined with low hot air shrinkage. Such a combination of properties is therefore also called high-modulus low-shrinkage (HMLS) or dimensionally stable (DST).

In the case of technical filaments, a separate heating system is sometimes used for the melt line, spinning pump and spinning head for precise temperature control that is gentle on the polymer. An annealer is also common for post-treatment of the freshly extruded filaments. The winders mostly process 6–8 threads, in extreme cases also up to 12 (Fig. 20.3).

20.3 Draw Twisting

To increase the strength of threads, they are often twisted. If this process is combined with drawing,

D. Veit, *Fibers*, https://doi.org/10.1007/978-3-031-15309-9_20

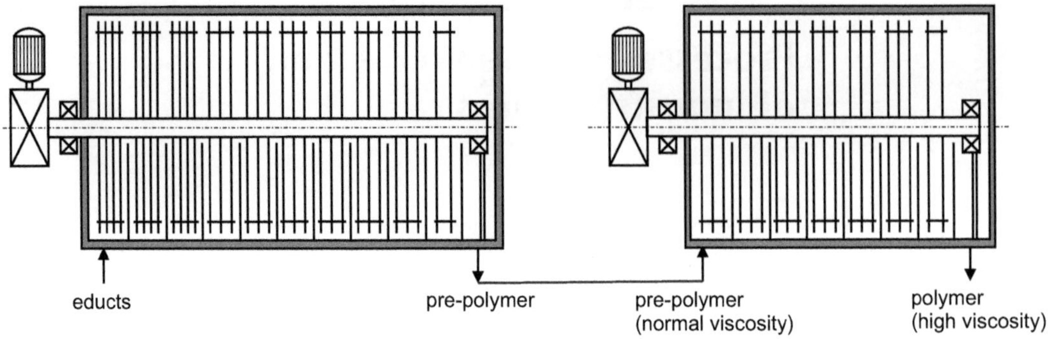

Fig. 20.1 Reactors connected in series to increase viscosity

it is referred to as "draw twisting". In this process, a POY yarn is first drawn, usually in several stages, sometimes via heated godets. Then, similar to ring spinning, the drawn yarn is wound onto a very fast rotating bobbin via the ring traveller running freely on the twisting ring. The usual twist height is 5–20 turns per meter, the winding speed up to 2000 m/min. Typical counts are in a range of 10–1000 dtex.

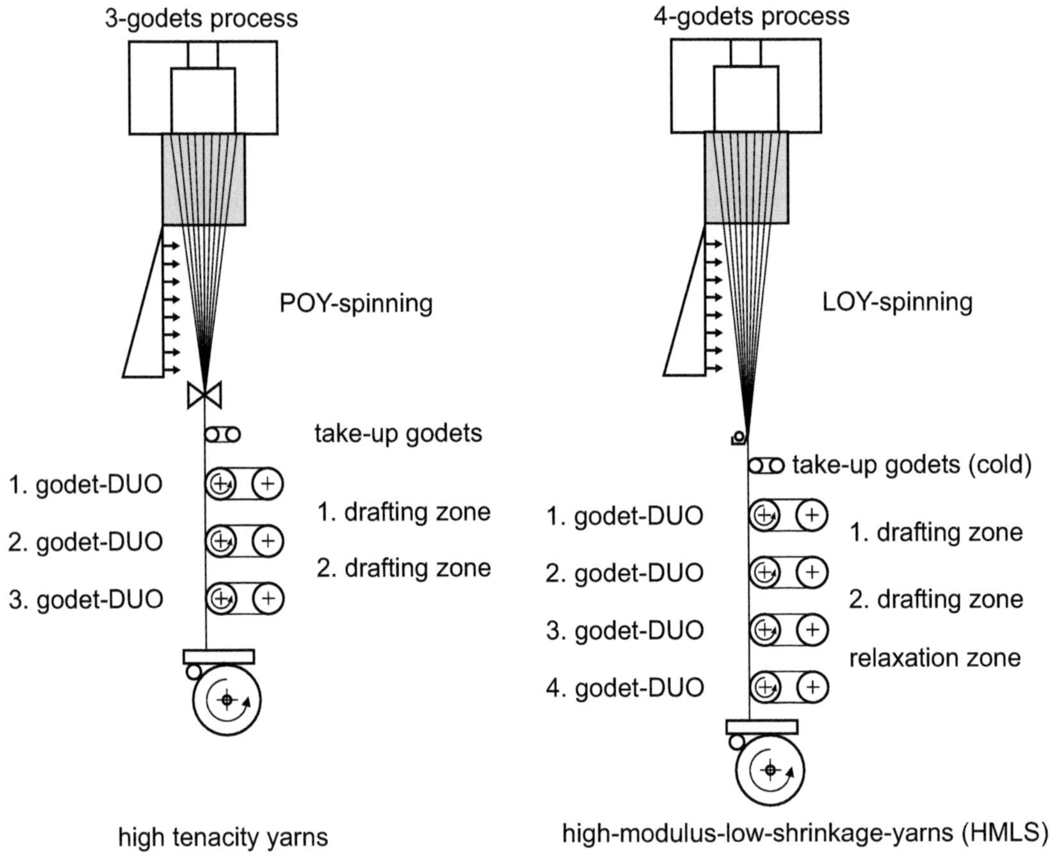

Fig. 20.2 Processes for the production of fast-spun technical PET filament yarns acc. to Gries

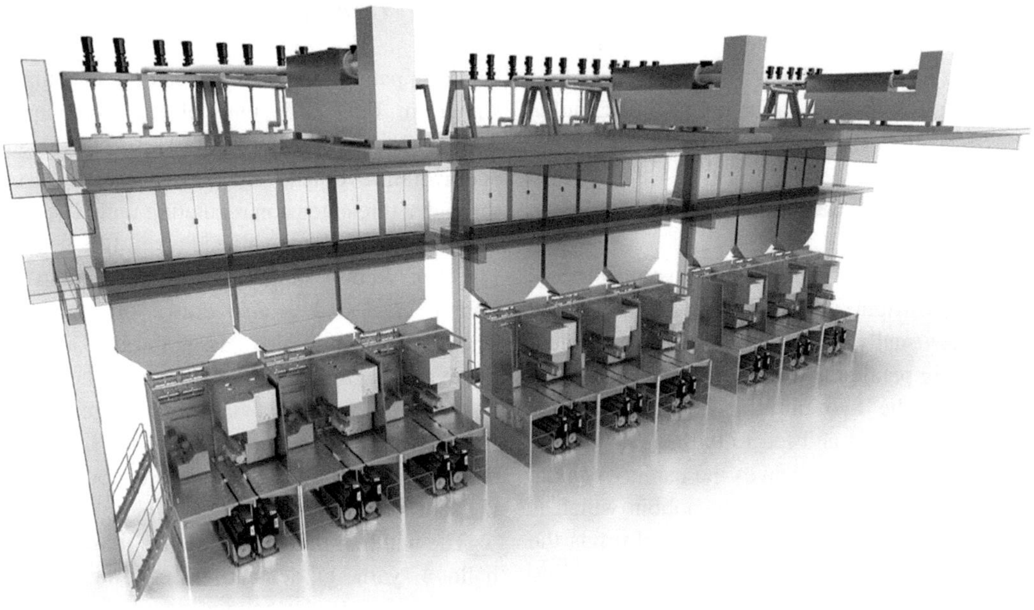

Fig. 20.3 Machine for technical filament yarns (Oerlikon Barmag, n.d.)

20.4 Cabling

Cabling has long been used for the production of carpet yarns. The term "cabling" originally referred to a two-stage twisting process for multiple yarns. Basically, two methods are distinguished: two-step process and direct cabling (Fig. 20.4).

20.4.1 Two-Step Process

In the first, the so-called pre-twisting stage, a twist is given to each of the two single yarns. Subsequently, the high twisted single yarns are plied and twisted back together in the second process step (Fig. 20.4, left). This process is complex, as a

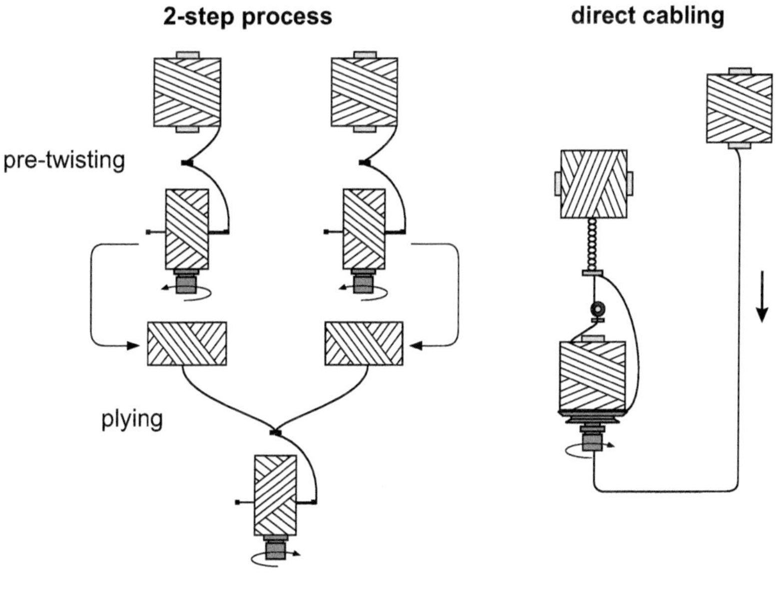

Fig. 20.4 Cabling methods acc. to Saurer Volkmann (n.d.)

total of three spindles (and two different machines) are required. There are also combination machines in which several pre-twisting spindles are connected to the untwisting spindle.

20.4.2 Direct Cabling

Direct cabling is a single-step twisting process (Fig. 20.4, right). Two untwisted continuous yarns are usually processed, with the individual yarns lying parallel to each other.

One of the two untwisted yarn feeders is inserted in a creel and fed to the hollow spindle from below. The yarn leaves the cabling spindle at the spindle rotor (below the bobbin which is plugged onto the hollow spindle) and orbits the second yarn feed set in the pot of the cabling spindle as a yarn balloon. The bobbin, which is plugged onto the hollow spindle, is only pulled off at the top. Thread brakes ensure that both yarns are under the same tension. As a result, the yarns are wrapped around each other evenly. In direct cabling, the individual yarns do not receive their own twist during the process.

20.4.3 Yarn Path

The yarn path through a direct cabling machine is shown in Fig. 20.5. One of the two supply bobbins is in the bobbin pot. This "pot thread" is drawn off overhead without rotation and passes through a thread tension regulating unit. The rotation imparted by the overhead take-off is negligible. The second supply bobbin is fitted in the bobbin creel and is also drawn off overhead. The "creel thread" also passes through a thread brake and is guided from below through the hollow shaft of the spindle via the thread guide tube. It leaves the spindle at the rotor, wraps around the storage disc and rotates as a yarn balloon around the stationary bobbin pot.

The width of the yarn balloon is limited by the balloon yarn guide and, depending on the machine design, the balloon diameter is limited by a balloon limiter. In the balloon yarn guide, the two yarns are wrapped around each other and fed together via the deflection rollers of the yarn path monitoring system of the leading roller. Finally, the traversing yarn guide transfers the yarn to the twisted yarn package.

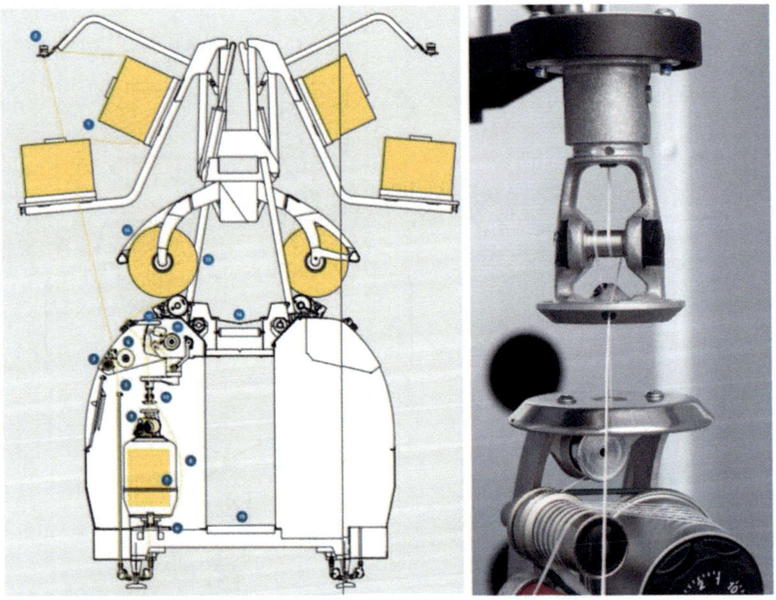

Fig. 20.5 Construction of a two-sided cabling machine (left) and yarn tension regulating unit (right) (Saurer Volkmann, n.d.)

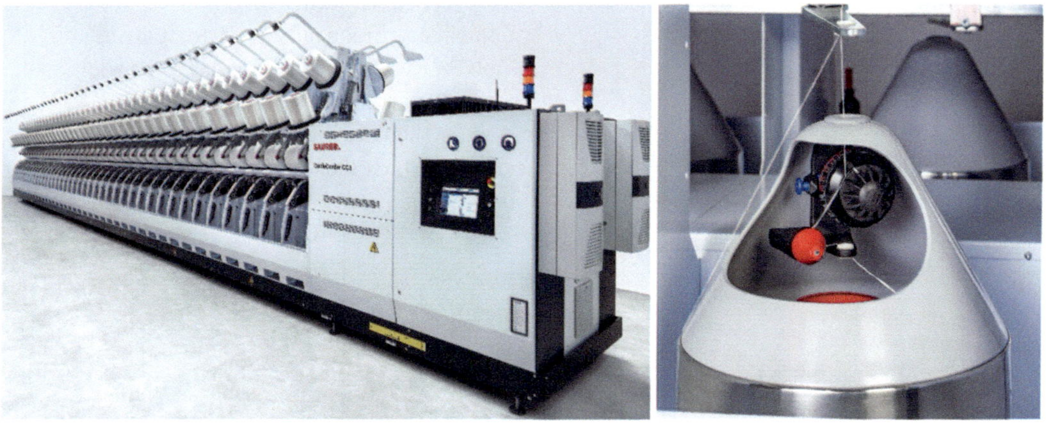

Fig. 20.6 Cabling machine (left) and view of the bobbin pot (right) (Saurer Volkmann, n.d.)

20.4.4 Machine

The maximum number of spindles of such a machine is 200, the spindle speeds are up to 12,000 rpm. The twist level is adjustable from 25 to 1500/m, the maximum delivery speed is 150 m/min (Fig. 20.6).

Cabled yarns are particularly suitable as pile yarns for carpets because they support each other and thus ensure high pile stability. Tire cord is also often cabled.

References

Oerlikon Barmag. (n.d.). Remscheid. https://www.oerlikon.com/polymer-processing

Saurer Volkmann. (n.d.). Krefeld. http://www.saurer.com

Process for the Production of Monofilaments

21

Monofilaments consist of a single, technically endless fiber and are normally produced by melt spinning. When they are cut to a specific length, they are known as 'bristle'.

21.1 Production

Monofilaments with fiber diameters of more than 50 µm (approx. 30 dtex) are generally spun from the melt into water or into a higher boiling mixture for economic reasons, in order to dissipate the heat quickly and evenly. For this purpose, the spinneret is placed a few centimeters above the cooling bath into which the monofilaments enter perpendicular to its surface. In the bath they are deflected and taken-up at an angle. The take-off speed is usually only slightly greater than the extrusion speed, and there is hardly any spinning orientation.

From the draw-off, the filament usually passes directly through the drawing unit. This often consists of several rollers between which the monofilaments are stretched in one to three stages with the aid of hot air. Alternatively, a heated drawing bath can be used. Typical winding speeds are 100–300 m/min (Fig. 21.1).

A line for drawing and winding round monofilaments of high strength consists of several drawing and heating units arranged one after the other.

The Fig. 21.2 shows the drawing unit of a monofilament plant for artificial turf yarns.

21.2 Properties and Typical Applications

The decisive quality parameters of monofilaments are their linear density uniformity between them and along each individual filament as well as, if spun round, a cross-section that is as circular as possible. This requires a number of measures during spinning and in the cooling bath, relating to the geometry of the machine, the deflection in the cooling bath, its temperature and the length of the cooling section.

The number of filaments per spinneret and per filament bundle in the cooling bath and during the subsequent drawing depends on the extruder capacity and the filament thickness, typical numbers are up to 30. Wires of usually about 0.05–1 mm diameter (30–15,000 dtex) are spun. A comparison of the mechanical properties of monofilaments made from different polymers is given by Guo et al. (2020).

Typical applications of such monofilaments are paper machine clothing (polyester, polyamide), geotextiles (polypropylene), fishing rods, brush bristles, doll's hair and, more recently, tea bags (e.g. made of PLA), Fig. 21.3.

D. Veit, *Fibers*, https://doi.org/10.1007/978-3-031-15309-9_21

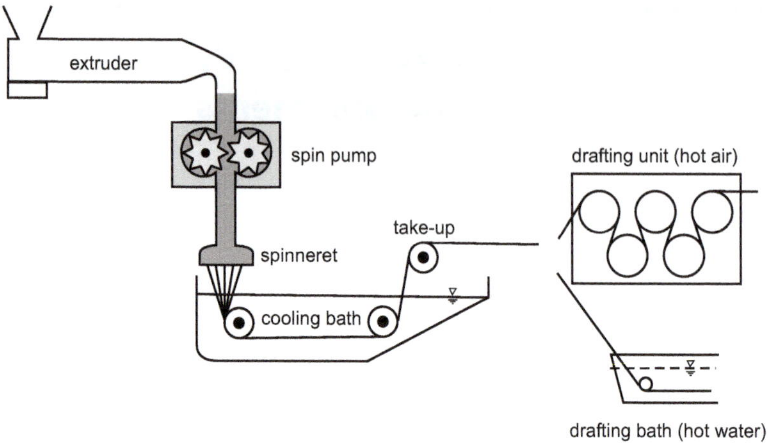

Fig. 21.1 Basic structure of a monofilament spinning machine

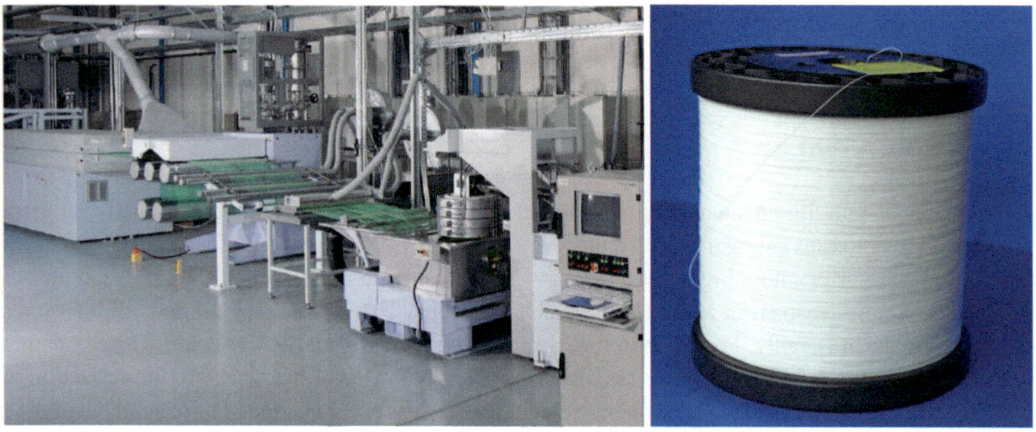

Fig. 21.2 Drawing unit of a monofilament line (Oerlikon Barmag, n.d.) and spool with monofilament

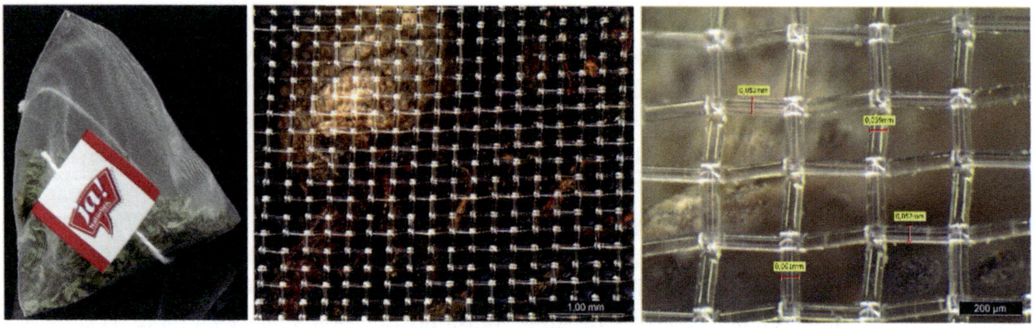

Fig. 21.3 Tea bag made of monofilament fabric (left: Wetzig, 2009; middle and right: detail photos)

Monofilaments are also used as polymer optical fibers and as sensors.

References

Guo, H., Dafaalah, A. H., Liu, Y., Zhang, Y., & Qiu, G. (2020). A comparative study on the mechanical properties of polymeric monofilaments. *Journal of Engineered Fibers and Fabrics, 15*, 1–11. SAGE Publishing.

Oerlikon Barmag. (n.d.). https://www.oerlikon.com/de/marken/oerlikon-barmag/

Wetzig, E. (2009). *CC BY-SA 3.0.* https://commons.wikimedia.org/wiki/File:Teebeutel_Polylactid_2009.jpg

22 Process for the Production of Carpet Yarns

Carpet filament yarns are usually produced from polyamide and polypropylene in a melt spinning process with extruder. These fibers are relatively coarse and are textured directly after spinning. They have a particularly good crease recovery. Carpet filament yarns are also referred to as BCF yarns (Bulk Continuous Filament).

22.1 Spinneret

BCF yarns are usually spun with rectangular spinnerets. They typically consist of up to 700 filaments with a single fiber fineness of up to 2.5 dtex (Fig. 22.1).

22.2 Process

Cooling takes place by cross-flow quenching. The filaments are taken-up from the spinning zone with cold godets and immediately drawn, textured, thermofixed and wound up in one process. The capacity-limiting factor here is the maximum possible speed of the air-jet texturing chamber (Fig. 22.2).

The essential components of a BCF machine that distinguish it from other spinning machines are the air texturing unit and the cooling drum. Here, the texturing effect is usually achieved in a combination of air-jet texturing and stuffer box crimping. The yarn is first preheated on hot godets and pressed through an air-jet into a pitot tube with hot air or steam. In the pitot tube, a slowly moving plug of compressed and three-dimensionally crimped filament yarn is formed. If the yarn is heated to just below the melting point at the moment of crimping, the crimping effect is significantly enhanced.

The structure of a typical air nozzle in the BCF process is shown in Fig. 22.3. Common air pressures are up to 10 bar.

Figure 22.4 shows the middle part of a BCF machine with the drafting godets, the air-jet and the cooling drum (bottom). The godets are always wrapped several times because of the better power and heat transfer.

Fig. 22.1 Rectangular spinneret in the BCF process (Oerlikon Neumag, n.d.)

D. Veit, *Fibers*, https://doi.org/10.1007/978-3-031-15309-9_22

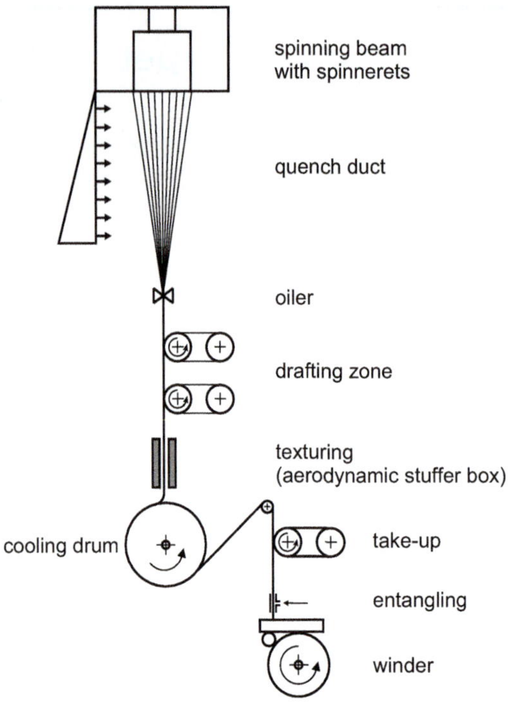

Fig. 22.2 Design of a machine for the production of BCF yarns

After texturing, the yarn is cooled in a tension-free state by passing the plug emerging from the crimping device over a cooling drum. On this drum it is pulled apart to form the crimped yarn. After the yarn has cooled, it is interlaced in an air-jet to give it greater cohesion and then wound up. Usually BCF machines are equipped with three-ply winders (Fig. 22.5).

The winding speeds are 1800–3700 m/min. Typical yarn counts are 900–3300 dtex, depending on the polymer, and the usual single filament counts are 2.5–40 dtex. In most cases, medium fineness yarns are produced with three-fold and coarser yarns on twofold winders. Fine BCF yarns are rarely spun in fourfold winders. The production of multicolored yarns (up to three colors) is also possible. They are used, for example, for melange yarns.

The Fig. 22.6 shows a BCF machine from the extruder to the winder.

The typical, crimped structure of a BCF yarn is shown in Fig. 22.7. The main application of such products are tufted carpets.

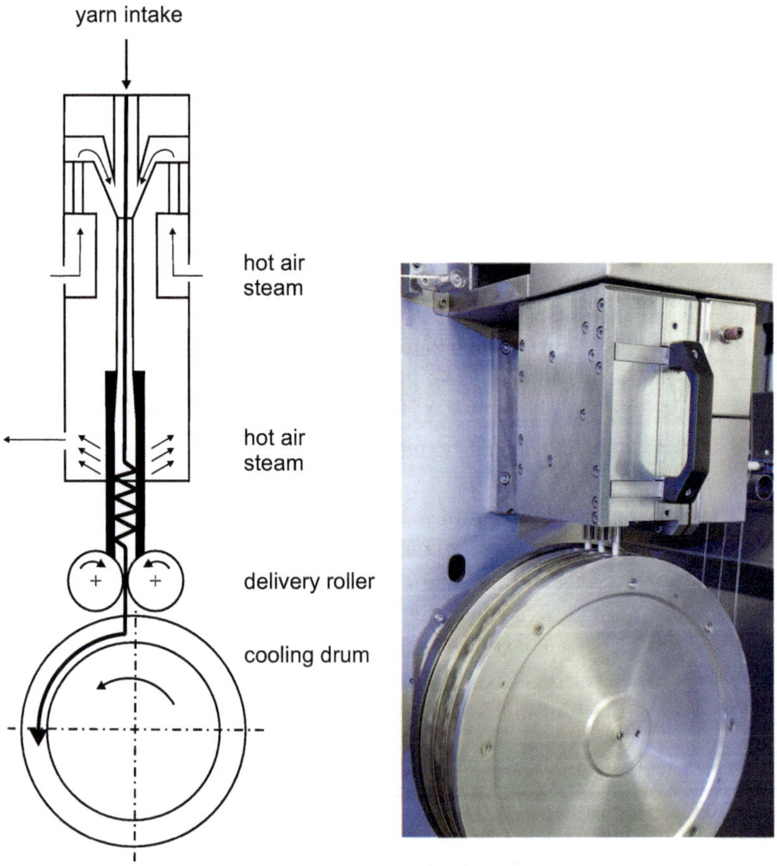

Fig. 22.3 Air nozzle and cooling drum in the BCF process (Oerlikon Neumag, n.d.)

Fig. 22.4 Drawing godets and detail of the yarn path (Oerlikon Neumag, n.d.)

Fig. 22.5 View of the winder level of a BCF plant (Oerlikon Neumag, n.d.)

Fig. 22.6 Machine for the production of BCF yarns (Oerlikon Neumag, n.d.)

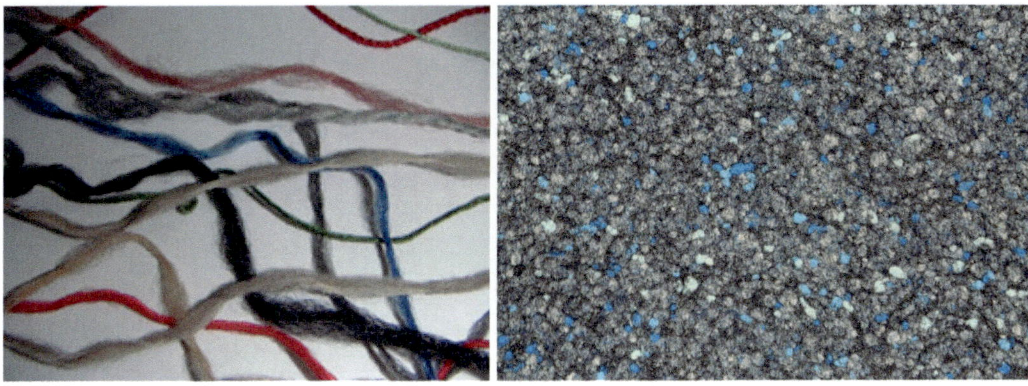

Fig. 22.7 BCF yarn (Oerlikon Neumag, n.d.) and tufted carpet

Reference

Oerlikon Neumag. (n.d.). Neumünster. https://www.oerlikon.com/polymer-processing/de/

23 Process for the Production of Film Yarns

From well orientable and crystallizable polymers, such as the polyolefins (e.g. polypropylene), it is very easy to produce endless tapes by splitting a film extruded from a slit die in the longitudinal direction.

23.1 Basics

The Fig. 23.1 shows the sequence of the process steps.

First, a film is extruded, cooled in a water bath and cut into narrow strips (ribbons). This is particularly easy with polyolefins due to their low secondary valence forces perpendicular to the direction of orientation. The slitting unit is located between the stretching and fixing units. The cut tapes are usually wound or twisted directly or, if necessary, crimped and then wound.

The Fig. 23.2 shows a complete line from extrusion to rewinding.

23.2 Technical Data

Typical technical data of such plants is given in the Table 23.1.

23.3 Aggregates

The Fig. 23.3 shows how the film is cut into ribbons and these are separated.

Typical stretching or stretching units (calendars) as well as the rewinding are shown in Figs. 23.4 and 23.5.

23.4 Products

The common fineness of film yarns with a width of 0.8–6 mm ranges from 500 to 5000 dtex. The film thickness is usually 25–50 μm.

The film threads take on even more of the character of textile yarns if their splitting ability is further exploited. By means of brush or needle rollers, knife mills, intensive jet streams or by twisting, they can be further split into thread-like fibrillar networks or largely separated individual fibers of relatively uniform fineness.

Film yarns are used in particular for ropes, harvest binder yarns, packaging fabrics and carpet base fabrics. In recent years, the natural coarse fibers jute, hemp and sisal have thus been displaced in many areas.

D. Veit, *Fibers*, https://doi.org/10.1007/978-3-031-15309-9_23

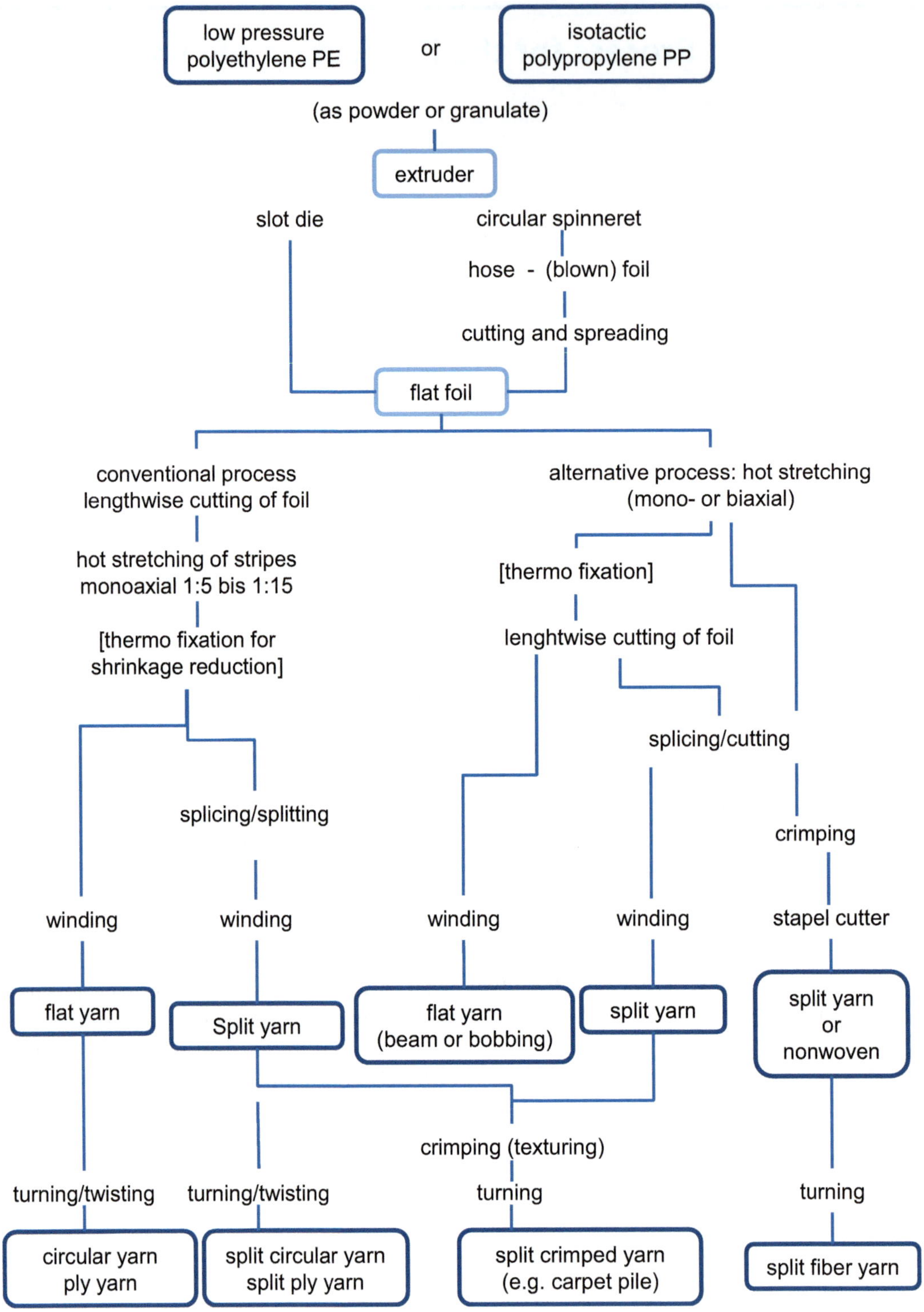

Fig. 23.1 Process stages for the production of film yarns from polypropylene (PP)

Fig. 23.2 Plant for the production of film yarns (Starlinger GmbH, n.d.)

Table 23.1 Typical technical data for film yarn systems according to Starlinger GmbH (n.d.)

Spinneret width [mm]	930	1650	1800
Throughput [kg/h]	330	600	720
Winding speed [m/min]	420	420	550
Fineness of foil yarns, 4 mm wide [dtex]	560	576	480

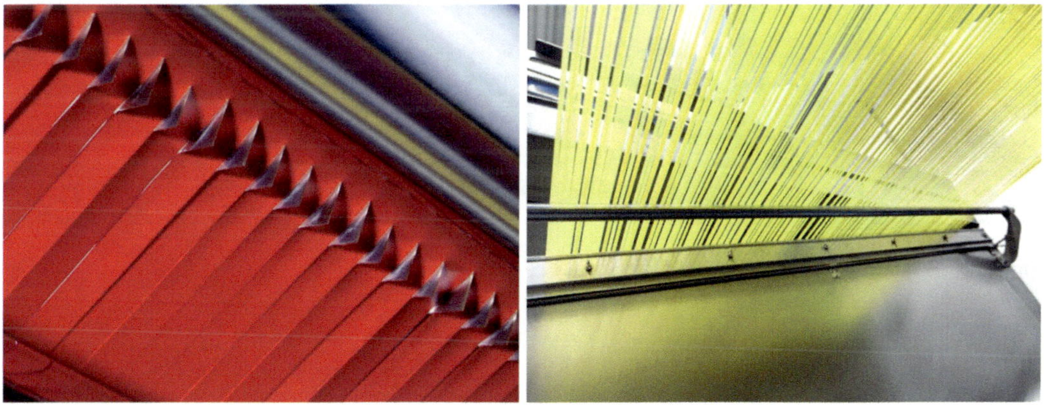

Fig. 23.3 Cutting unit and cut film yarns (Starlinger GmbH, n.d.)

Fig. 23.4 Unit for stretching the film yarns (Starlinger GmbH, n.d.)

Fig. 23.5 Winding unit (Starlinger GmbH, n.d.)

Fig. 23.6 Artificial turf

One of the best known applications of fabrics made from PP film tapes is artificial turf in stadiums and for other house and home products (Fig. 23.6). The film yarn count here is 1500–20,000 dtex.

Reference

Starlinger GmbH. (n.d.). Vienna. http://www.starlinger.com

24 Processes for the Production of Staple Fibers

To produce staple fibers, filaments are first produced and then cut or, less frequently, torn. Staple fibers can be processed in pure form or in blends with other staple fibers (e.g. cotton, wool, synthetic fibers). Their properties (mechanical, length, fineness) must be adapted accordingly.

24.1 Market

In 2020, 59 mt of staple fibers were produced. This corresponds to around 50% of total fiber production. The Fig. 24.1 shows the market shares of the materials for staple fibers. Natural fibers, especially cotton, dominate with about 32 mt. The cellulosic man-made fibers have an increasing market share of meanwhile about 7 mt since 2008 (growth since 2008: 90%), while the share of synthetic man-made fibers in the staple fibers with 20 mt recently grew by 1% annually.

The development of the production of cellulosic and of synthetic staple fibers shows Fig. 24.2. The quantity of synthetic staple fibers has been increasing almost linearly by 400,000 t per year since 1970. Cellulosic fibers have been experiencing a renaissance since about 2000. Their annual growth was around 6–10% until 2014 and has since slowed down to 2–3%.

24.2 Production

There are various processes for the production of staple fibers, which differ in terms of the spinning systems (Fig. 24.3).

24.2.1 Two-Step Process

The dominant technique is the two-stage process. Here, the filament bundles are spun out with rectangular or ring nozzles (Fig. 24.4).

The spinning systems are operated at maximum throughput. The possible throughput depends above all on the cooling effect of the blowing. The plant capacities common today, especially for polyester (PET: 300 t, PA: 45 t, PP: 80 t) in a spinning line predestine ring spinnerets so that the plants remain compact and operable (Fig. 24.5).

The spinning speeds rarely exceed 2500 m/min. The spinning cables of many spinning positions arranged next to each other (mostly 20–40) are combined into one spinning cable, drawn off together and deposited. Below the spinning ducts, the cables are prepared, mostly with roller oilers. They are then combined into one spinning cable and drawn off by rollers. Afterwards, the cables are slightly crimped by

D. Veit, *Fibers*, https://doi.org/10.1007/978-3-031-15309-9_24

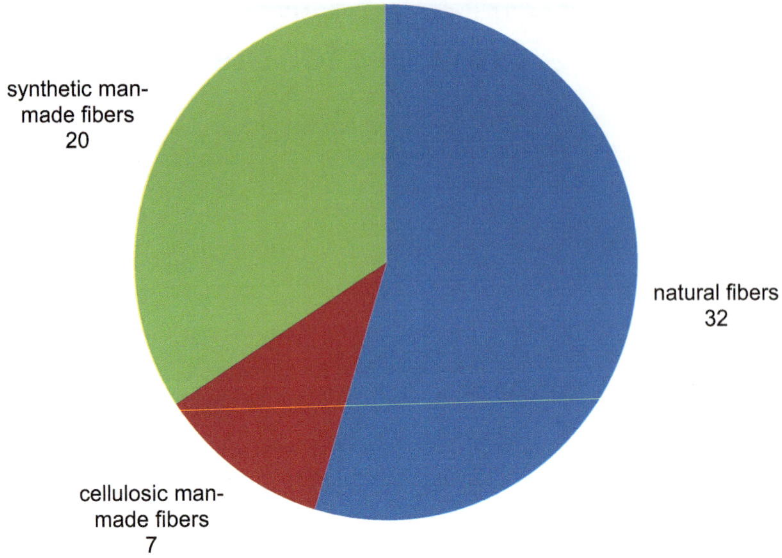

Fig. 24.1 Market shares [mt] of the materials in all staple fibers (The Fiber Year, 2021)

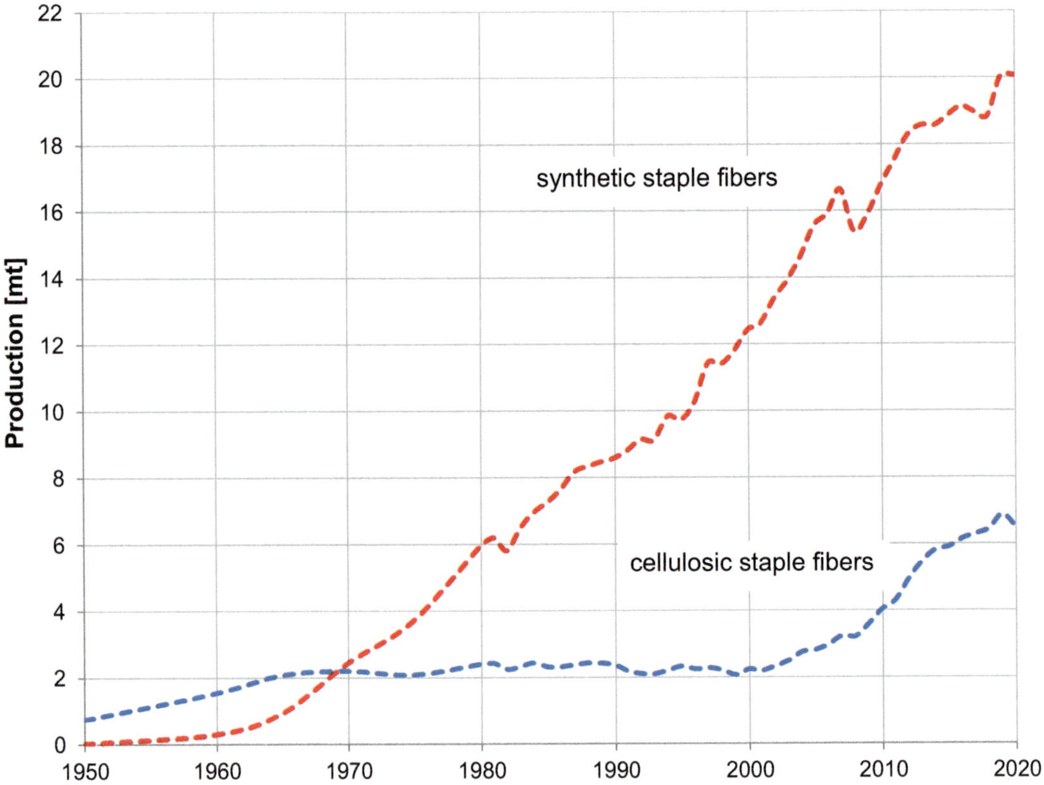

Fig. 24.2 Development of cellulose and synthetic staple fiber production (data: The Fiber Year, 2021)

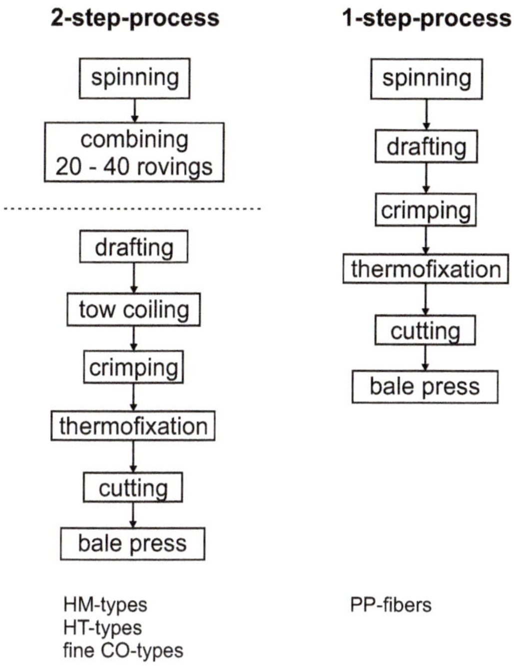

Fig. 24.3 Process lines for the production of staple fibers

guiding them between two so-called sunflower wheels, thus slightly kinking the fibers.

Figures 24.6 and 24.7 shows the first stage of staple fiber production from spinning to collection of the fiber cables in large cans. The capacity of typical spinning cans is up to 10 m^3 (2.5 × 2.5 × 2.5 m).

A 3D view of a single-stage spinning plant from fiber production to deposit in large cans is depicted in Fig. 24.8.

Fig. 24.4 Rectangular spinneret (Oerlikon Neumag, n.d.)

In the second process stage, the cables are drawn off from the cans, stretched, crimped, cut and pressed into bales. As the Fig. 24.8 shows, the cutting unit is usually located on the upper level above the baler for reasons of space. The size of the forklift at the bottom right of the picture illustrates the dimensions of the plant.

Each spinning line usually supplies a tape line on which the spinning cables are moistened, drafted in several stages (Fig. 24.9) and then crimped.

They then pass through a steam chamber for heat-setting and then through a calendar dryer. Finally, they are crimped in the crimper. In this process, the cable is pressed by two rollers into a chamber whose walls, which are parallel to the plane of the cable, move against each other and exert an adjustable pressure on the fiber material in the chamber. As it is pressed in over the feed rollers, the cable is compressed in the chamber, folded and thus crimped. The cable is pushed out of the chamber as a crimped package due to the constant re-feeding. The corresponding crimping unit shows Figs. 24.10 and 24.11.

The crimp of the staple fibers is relatively coarse (Fig. 24.12). After crimping (texturing), the fiber cables are cut.

The Fig. 24.13 shows the schematic layout of typical plants for the production of polyester staple fibers with different properties.

Typical plant capacities are 10–300 t/day. Typical production speeds in the spinning mill are 600–2500 m/min, depending on the fiber material, and 100–300 m/min in the drawing zone.

24.2.2 Single-Stage Process

Staple fibers can also be first spun in a single process and then further processed on a tape drawing unit. Behind the tape storage unit, the fibers are first heated with steam, then crimped and then fed via a conveyor belt to a dryer. From there they are transported to the cutting unit and to the baler (Fig. 24.14).

This process is also called compact spinning. Due to the speed limitation in the tape section, it

Fig. 24.5 Fiber formation with ring spinnerets (left: Oerlikon Neumag, n.d., right: Sossna GmbH)

is necessary to spin very slowly. To compensate for this capacity disadvantage, rectangular packages with up to 50,000 holes and ring spinnerets with up to 150,000 holes are used. The quenching takes place either in cross-flow or radially. This allows the capillary spacing to be reduced to 2–3 mm without too much negative influence on the cooling effect. However, the strong quenching leads to poorer stretching behavior. As a result, this process was not established for high modulus, high tenacity and fine cotton type staple fibers. However, this process is of market relevance for spinning polypropylene staple fibers.

In addition to polypropylene, polyester and polyamide fibers can also be produced with this process. Typical production speeds are 50–200 m/min. Typical plant capacities are 4–70 t/day. Typical yarn counts are 1–90 dtex, strengths 25–60 cN/tex and elongations 25–300%.

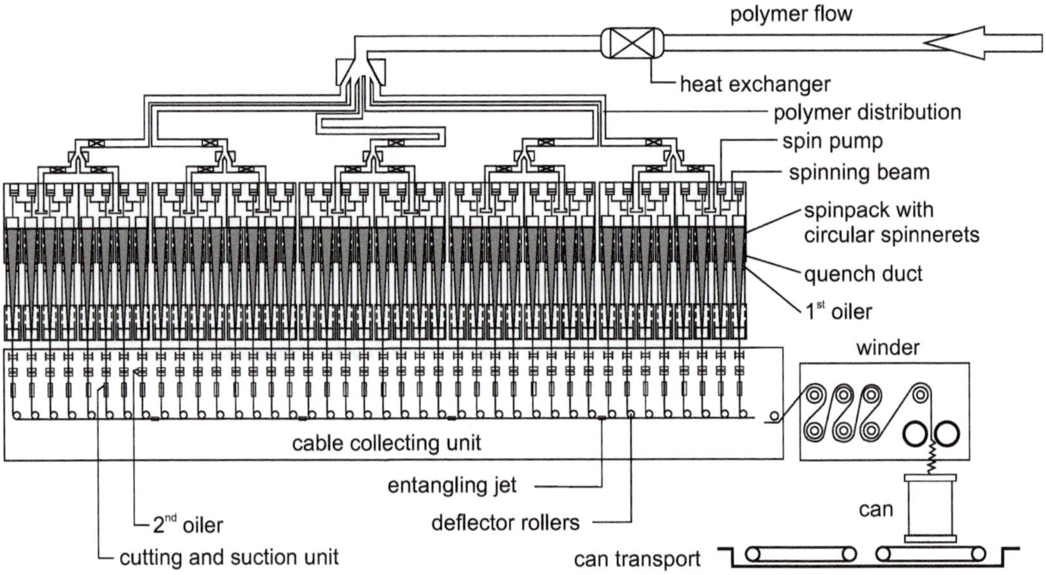

Fig. 24.6 Plant for the production of staple fiber cables acc. to Zimmer AG

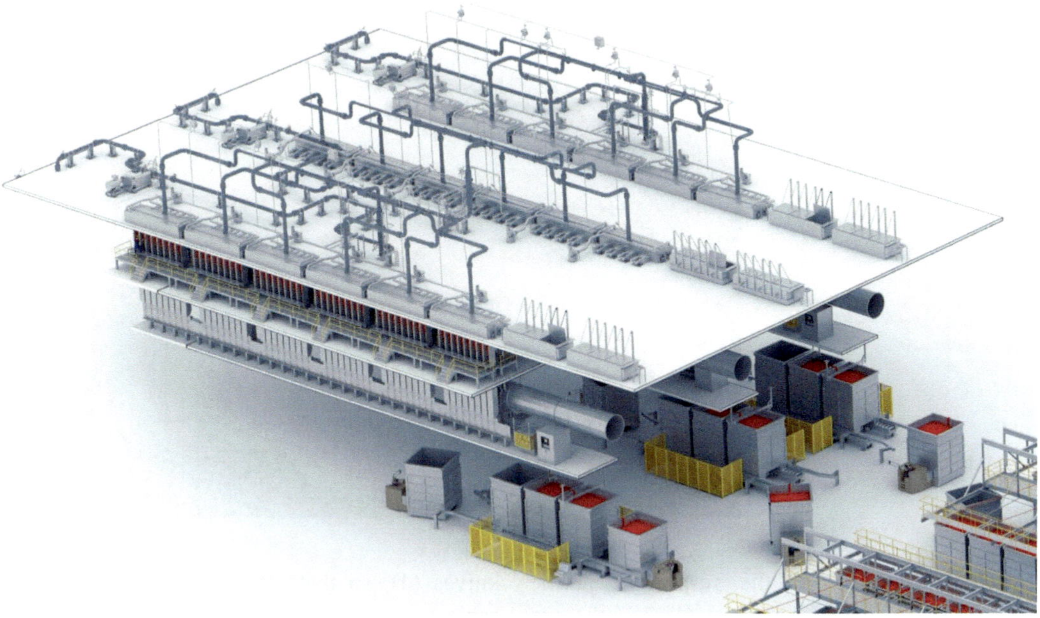

Fig. 24.7 First process stage of a two-stage plant (Oerlikon Neumag, n.d.)

24.2.3 Spin-Draw Texturing Process

The third possibility of producing staple fibers is based on the spin-draw texturing process. Following the texturing process, the spun cable is cut and, if necessary, heat-set. This process is only suitable for small fiber quantities.

24.3 Conversion of Fiber Cables

At the end of the process, stretched and crimped endless fiber cables must be cut ("converted") into staple fibers. Their length, the so-called staple length, should be as uniform as possible and is

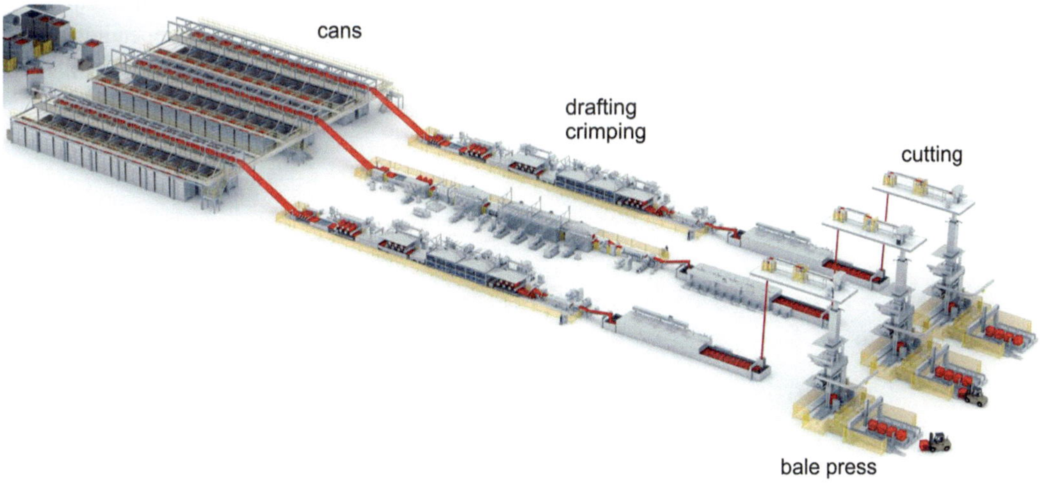

Fig. 24.8 Second process stage of a two-stage plant (Oerlikon Neumag, n.d.)

Fig. 24.9 Unit for drafting the fiber cables with calendar rolls (Oerlikon Neumag, n.d.)

selected according to the respective textile processing method. Typical staple lengths are 25–45 mm for cotton types and 60–120 mm for wool types. In the past, beaking conversion dominated, but nowadays almost all man-made fibers are cut.

24.3.1 Cutting Conversion

Today, a cutting machine has essentially become established in which the cable is wound onto a cutter wheel with peripherally arranged knives pointing outwards (Fig. 24.15, 2). A pressure roller (3) presses the fiber cable (1) against the knives (4), so that after a few revolutions the cable is pressed so firmly onto the knives that the inner layers are cut. The cut fibers fall through to the inside.

If the blades are too blunt to cut the cable, individual fibers may not be cut, resulting in fibers that are double or triple the desired cut length ("crackers").

This machine, also known as the Lummus cutter, is suitable for cable titers of up to 6,000,000 dtex at production speeds of up to 350 m/min. This corresponds to a cutting capacity of approx. 12 t/h. The cutting length is adjustable and ranges from 1.5 to 200 mm depending on the cable count and cutting speed (Fig. 24.16).

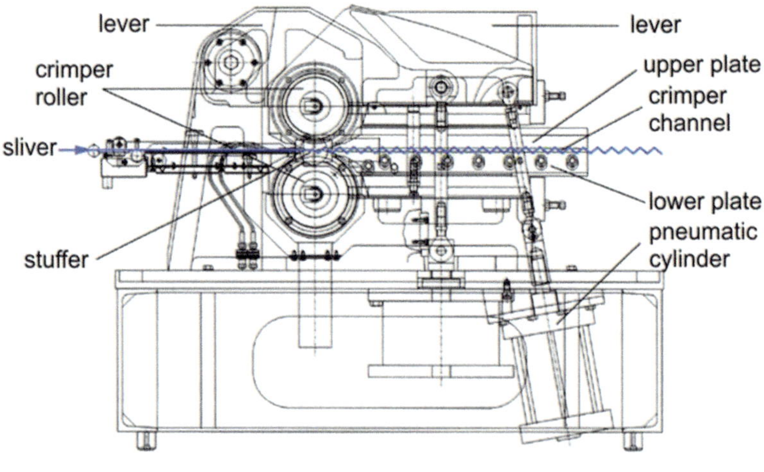

Fig. 24.10 Design of a crimper unit (Oerlikon Neumag, n.d.)

Fig. 24.11 Crimper (Oerlikon Neumag, n.d.)

Another principle is shown in Fig. 24.17. Here, the cable is drawn in and guided between two rubberized slotted wheels. Rotating blades reach into the gaps between the slots at an angle of 90° to the cable and cut the fibers.

The drive technology is extremely complex and changing the cutting length is costly. The knife service life is often only a few hours, so a spare machine must always be kept ready for continuous operation. This process is suitable for cables with up to 1,100,000 tex and operates at approx. 300 m/min.

It is actually nonsensical to cut parallel filaments in the fiber cable and deliver them unordered to the processor, who then, similar to natural fibers, parallelizes them again in complicated operations and spins them into a fiber yarn. It was therefore early on that the final fiber product was delivered as uncut crimped cable. The processor cuts these cables into fibers without affecting their parallel position. To do this, they are either cut under tension by helical knives or torn between increasingly fast-running rollers.

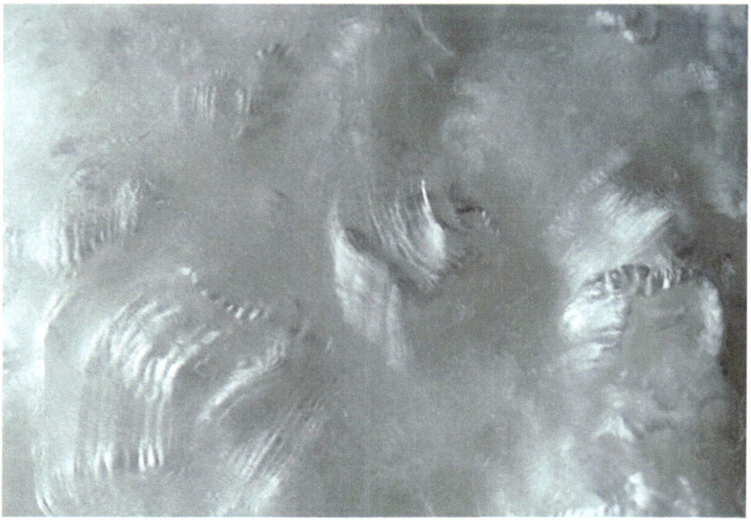

Fig. 24.12 Crimped staple fibers

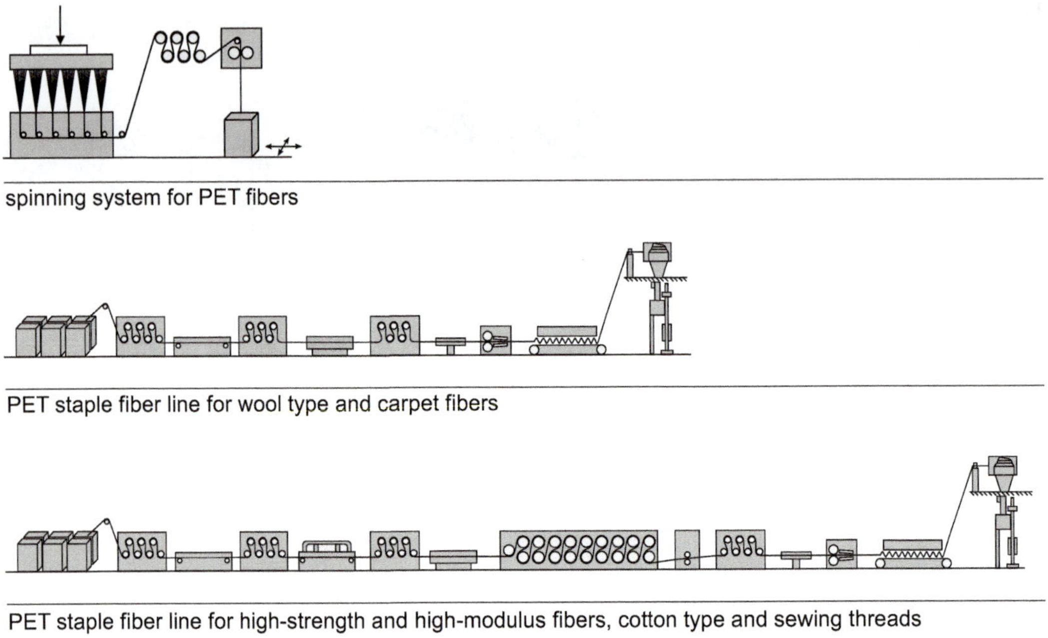

Fig. 24.13 Plant diagrams for the production of staple fibers of different polyester types acc. to Oerlikon Neumag (n.d.)

24.3.2 Breaking Converter

An alternative to cutting is the tearing of chemical fiber cables. In this process, the cable is first loosened and pre-drafted. After a pre-tearing process, the actual tearing usually occurs in two zones, between which compaction and damping take place. Subsequently, the torn sliver is crimped (Fig. 24.18).

Typical cable counts are 1,000,000–2,000,000 dtex, production speeds are 150–190 m/min. Today, break conversion is only a niche process.

Depending on the conversion process, a different length distribution of the cut fibers results

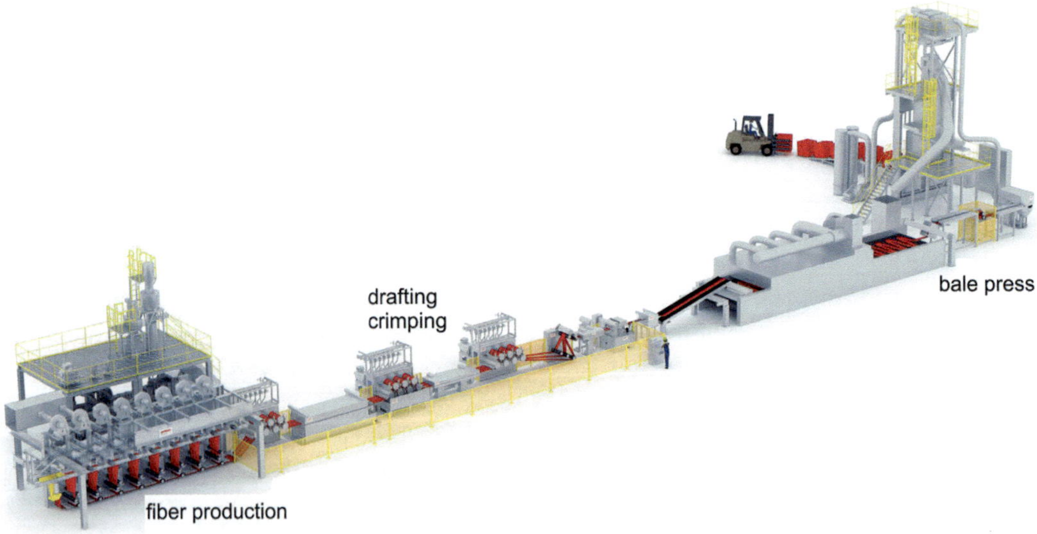

Fig. 24.14 Plant diagram for the production of staple fibers (Oerlikon Neumag, n.d.)

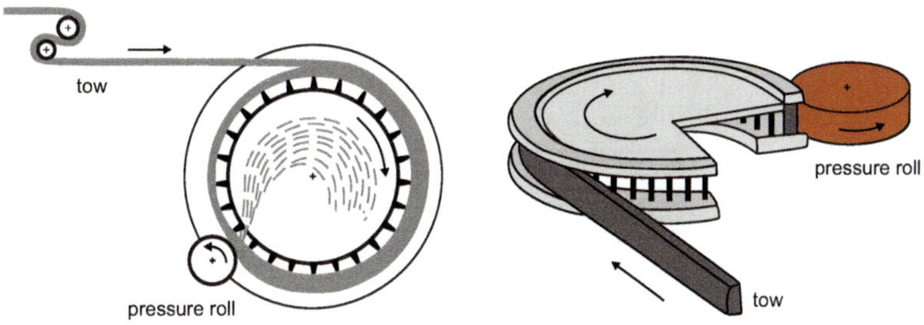

Fig. 24.15 Principle of a Lummus cutting unit

(Fig. 24.19). In the case of break conversion, this corresponds more to the natural staple length distribution, e.g. of cotton and wool, but this process is slower and therefore less economical.

Cut man-made fibers can be blended with natural fibers before the first processing stage. This increases the uniformity of the blending, but man-made fibers are then also "cleaned" together with the natural fibers, which can lead to fiber breakage.

24.3.3 Flock Fibers

If the fiber cables are cut to lengths of 0.2–5 mm, they are referred to as flock fibers. These can be electrostatically charged and applied to a surface coated with adhesive, e.g. with the aid of compressed air. In order to achieve a uniform flock fiber density, the fibers are conveyed from a reservoir via a sieve into the electrostatic field, where they align themselves along the field lines and thus become stuck vertically in the adhesive (Fig. 24.20). This results in a velvety surface.

Flocked articles are used for decoration (Fig. 24.21, left), for automotive interior design (e.g. glove compartment, headliner) and for clothing. In addition to a soft surface, flocked structures also provide heat and sound insulation. Lettering, e.g. of T-shirts, via screen printing is also possible. Flocked yarns can also be produced from the flock fibers (Fig. 24.21, right).

Typical fiber materials for the flocking technique are polyester and polyamide (kink resistance) as well as viscose. If the fibers are cut precisely, the result is a uniform fabric appearance. If the cut length varies, the fabric

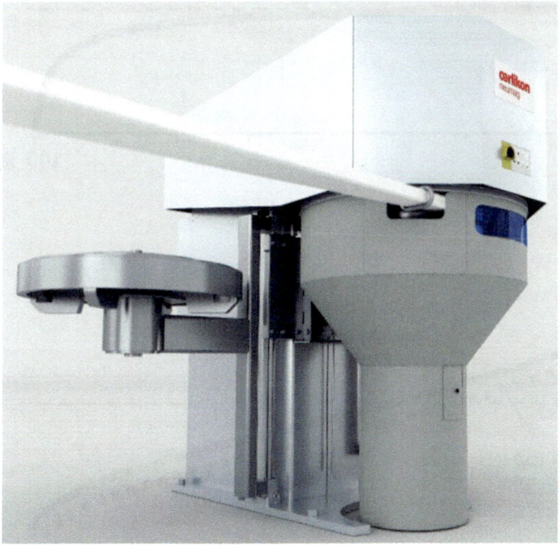

Fig. 24.16 Staple cutter (Oerlikon Neumag, n.d.)

Fig. 24.17 Principle of the Gru-Gru cutter

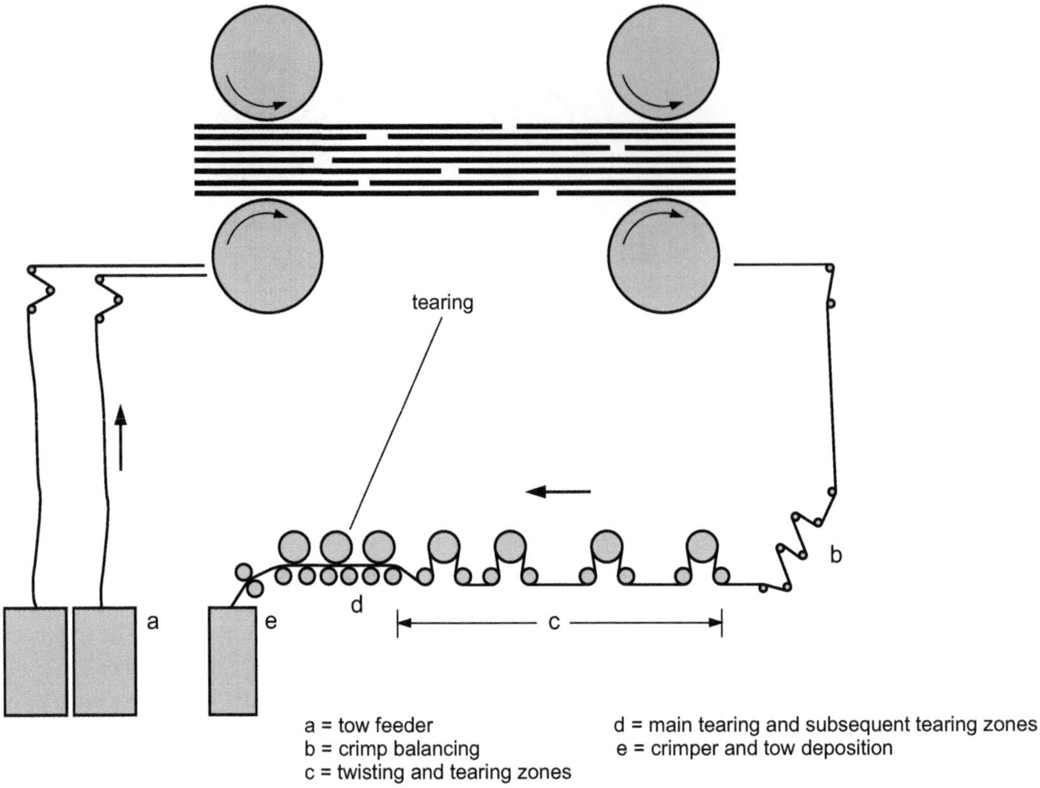

Fig. 24.18 Principle of break conversion

Fig. 24.19 Staple diagrams for converted fibers

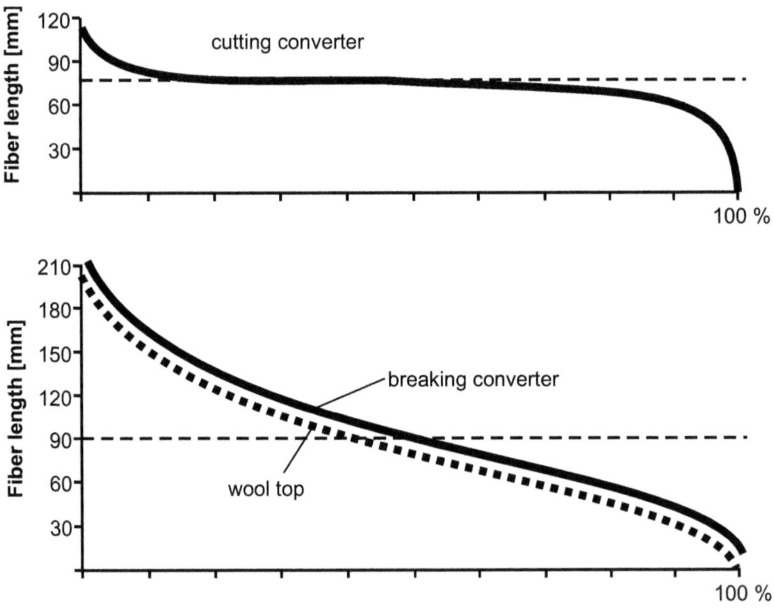

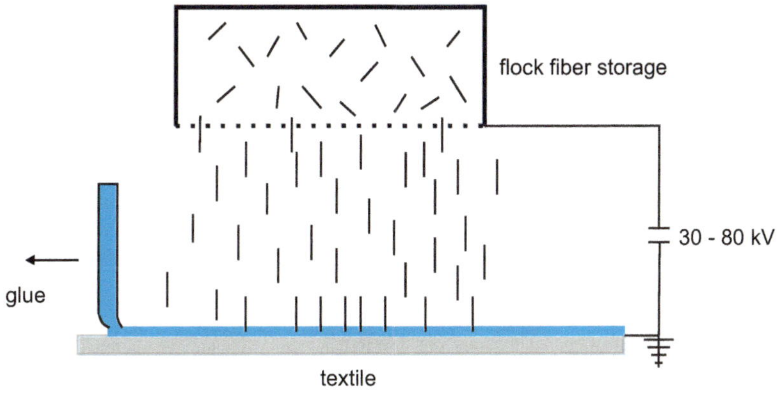

Fig. 24.20 Principle of flocking acc. to ITM of TU Dresden (n.d.)

appearance is more uneven, which leads to interesting design effects. Typical fiber prices are 2–5 €/kg.

Already several thousand years ago in China, natural fibers were embedded in resin to create a soft surface on decorative objects. In the Middle Ages and until early modern times, wallpapers were also produced in this way. Since the 1930s, the alignment of the fibers on the textile coated with adhesive has been improved with the help of electrostatics.

The name "flock fibers" is said to originate from the fact that, due to inadequate cleaning of non-adhesive fibers, they detached from an automobile roof lining, for example, looking like snowflakes trickling down.

24.3.4 Baling

The staple fibers fall from the converting machine via an intermediate container into a baler or are pneumatically conveyed there. The baler compresses the fiber material to approx. 0.5 g/cm^3. While still in the baler, the bales weighing up to 500 kg are wrapped in film and further compacted by steel belts (Fig. 24.22).

If the fibers have not been cut, but are packaged as fiber cables, they are usually placed in cardboard boxes.

24.4 Features

Synthetic staple fibers can be processed into a variety of different products. Typical examples

Fig. 24.21 Flocked ceramic sheep (Marion Veit) and flock yarn

Fig. 24.22 Baler (left) and bale storage (right) (Lenzing AG)

Table 24.1 Typical properties of synthetic staple fibers acc. to Oerlikon Neumag (n.d.)

	Polyester	Polyamide	Polypropylene
Fineness [dtex]	1–18	0.67–45	0.89–70
Tenacity [cN/tex]	36–60	32–60	12–100
Elongation [%]	18–55	35–100	18–350
Staple length [mm]	1.5–300	1.5–300	1.5–300

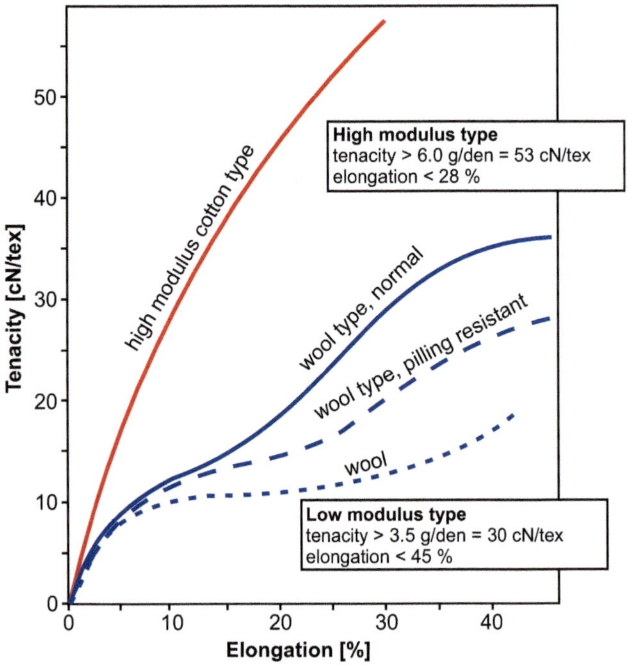

Fig. 24.23 Typical tenacity-elongation curves for staple fibers

are underwear, socks, sportswear, linings, sewing threads, curtains, filters, geotextiles and carpets.

The properties of the staple fibers must therefore be varied over a wide range. Typical values for selected polymers are listed in Table 24.1.

Figure 24.23 shows typical tenacity-elongation curves for staple fibers whose properties were modelled on natural fibers.

An overview of typical fineness values of important natural fibers is depicted in Fig. 24.24.

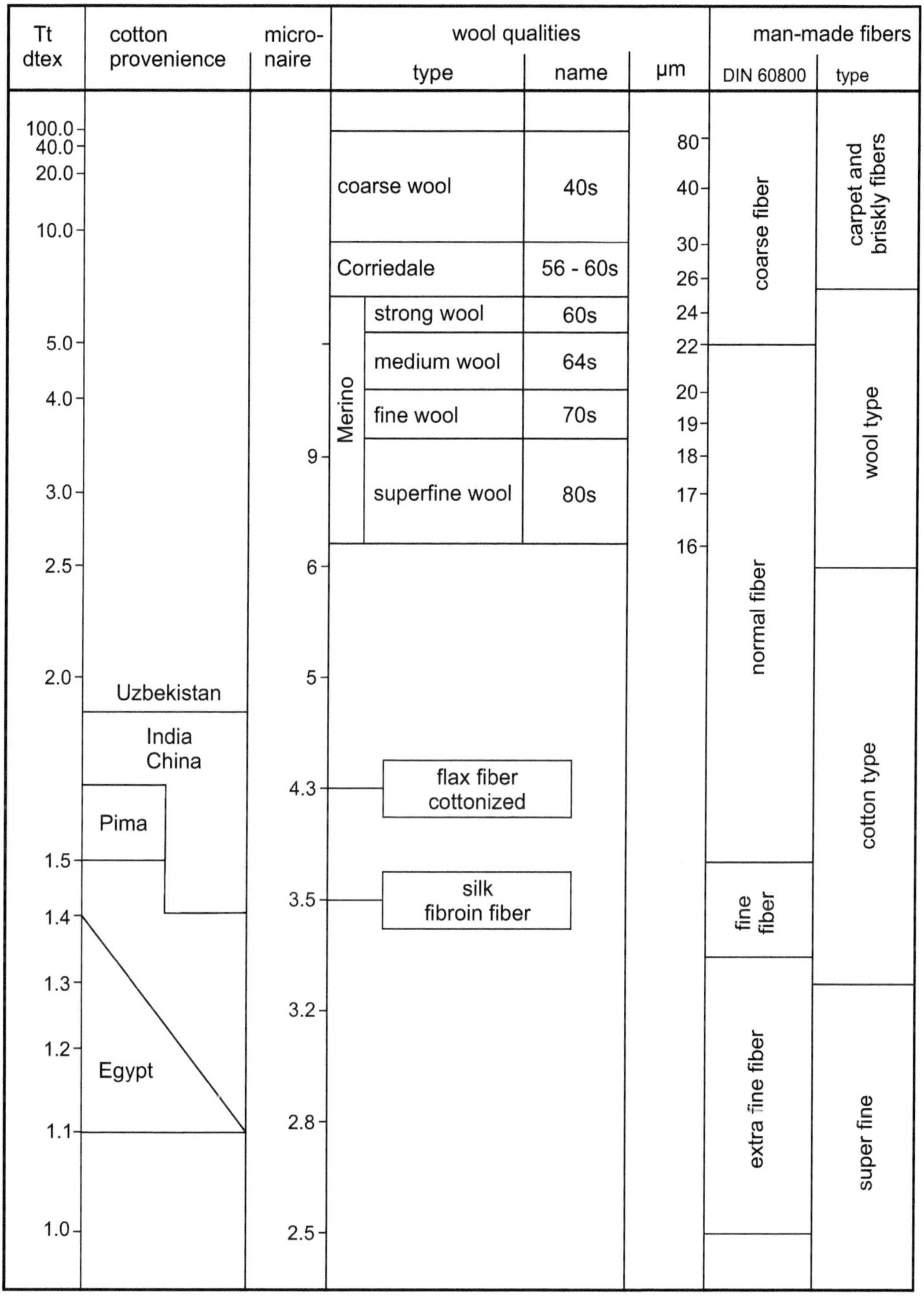

Fig. 24.24 Typical fineness ranges of natural and man-made staple fibers

If staple fibers made from synthetic and cellulosic polymers are to compete with natural fibers, the fineness must achieve similar values in addition to the mechanical properties. This applies in particular to blends.

References

ITM of the TU Dresden. (n.d.). https://tu-dresden.de/ing/maschinenwesen/itm

Oerlikon Neumag. (n.d.). Neumünster. https://www.oerlikon.com/de/marken/oerlikon-neumag

The Fiber Year. (2021). The Fiber Year GmbH, Roggwil. https://www.thefiberyear.com

25 Processes for the Production of Nonwovens

25.1 Spunbond

Spunbonds (meltblown) are textile fabrics produced directly from filaments without the detour via yarn production as in the case of woven and knitted fabrics. They are usually produced in a single-stage melt spinning process.

In this process, a large number of filaments are first formed with very wide spinnerets and deposited on a wire belt. To increase the basis weight, several spinning systems are often used in parallel. The filaments are cooled with air and thus consolidated. Both isotropic and anisotropic nonwovens can be produced by the arrangement of the spinning beams and the air flow during quenching. The web formation is followed by the web bonding, mostly via heated calendar rolls, after which finishing can still take place before the spunbonded web is wound onto rolls.

Typical fiber materials are polypropylene (80% market share) and polyester, rarely polyamide. World production is around 9.2 mt (2020).

25.1.1 Desing of a Spunbond Line

Figures 25.1 and 25.2 show the setup of a typical spunbond line, here for polyester roofing membranes.

25.1.2 Web Formation

Segmented or continuous rectangular spinnerets are used as spinning systems (Fig. 25.3). The delivery width can be up to 5 m.

Drawing speeds of 3500 m/min for polypropylene and 5500 m/min for polyester are necessary for spinning fully drawn filament yarns (Fig. 25.4). Filament strengths of over 30 cN/tex are thus achieved. The spinning systems always operate at the maximum cooling capacity. The melt throughput is constant and independent of the single filament diameter. Only for very fine filaments, the capacity has to be reduced.

The stretching and consolidation of the filaments in the spinning duct is usually achieved by a combination of pressure and suction flow. The compressed air in the quenching zone cools and stretches the filaments, while the suction air deposits the filaments evenly on the wire belt. A higher filament speed leads to greater strength and lower elongation in the case of PET, whereas no such effect can be detected for PP.

Due to the force-fitting stretching, the air velocity is usually higher than the filament velocity. This can lead to a sharp increase in air consumption to achieve high strengths, which increases costs. For this reason, attempts are made to work with as little excess air as possible

D. Veit, *Fibers*, https://doi.org/10.1007/978-3-031-15309-9_25

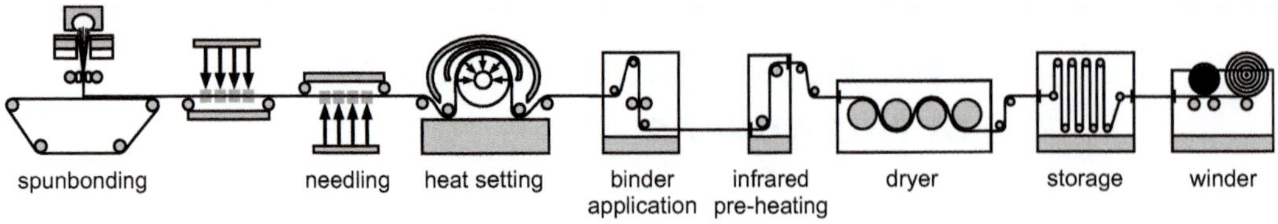

Fig. 25.1 Design of a spunbond line for roofing membranes acc. to Oerlikon Neumag (n.d.)

in closed systems. A typical spinning system of this type shows Fig. 25.5. The drafting and deposit zones are arranged within a closed system, which significantly reduces air consumption.

With regard to aerodynamic extraction, there are three basic differences in the process. Firstly, a round or slit-shaped injector jet, which operates with more or less high overpressure, can be used to draw off very quickly. Only this injector process is capable of producing highly oriented, low-shrinkage polyester filaments. Several injector jets are arranged over the entire delivery width. Since the filament speeds are several orders of magnitude higher than the speed of the base (20–1200 m/min), the filaments are deposited in a largely random manner.

To increase performance, the individual jet segments or the entire spinning beam can be set at an angle. In this way, either isotropic or—by means of special air guidance—anisotropic web structures are achieved (Fig. 25.6).

Fig. 25.2 Spunbond line (Oerlikon Neumag, n.d.)

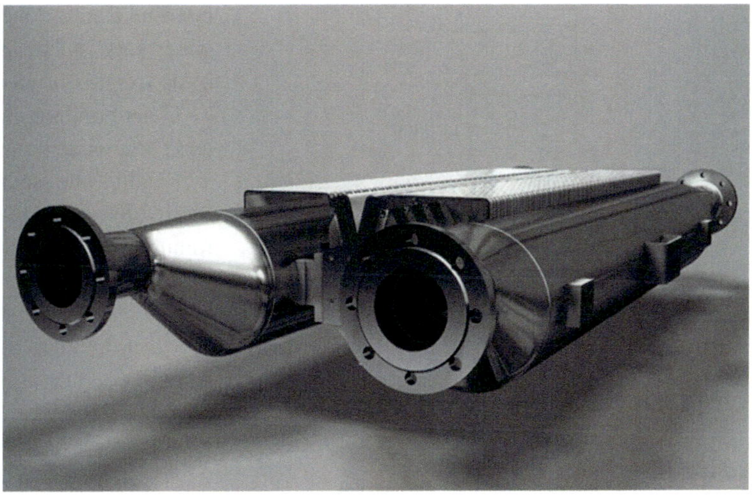

Fig. 25.3 Rectangular spinneret and drawing jet (Oerlikon Neumag, n.d.)

The running wire belt is usually designed as an endless, seamless sieve and is subjected to a vacuum.

25.1.3 Nonwoven Bonding

The still unbonded nonwoven is then fed to the nonwoven bonding unit. Nonwovens with a basis weight of up to 120 g/m^2 are continuously compacted and thermally bonded between two heated, smooth or engraved rollers. This results in adhesive dots between the individual filaments. In the case of engraved rolls, 5–20% of the nonwoven surface is bonded, resulting in relatively soft nonwovens (e.g. hygiene products). The use of hot melt adhesives is also possible. If a film-like character is to be achieved, smooth calendar rolls are used. Then up to 100% of the nonwoven surface can be bonded.

Figure 25.7 shows an unconsolidated spunbond and adhesive dots in a close-up.

Basis weights from 80 g/m^2 are often mechanically bonded by “needling”. In this process, individual fibers are partially pulled out of the nonwoven plane, thus bonding the nonwoven perpendicular to the nonwoven surface. Typical products produced in this way are geotextiles and filters.

Fig. 25.4 View into the spinning duct from below (left) and filament deposit on the wire belt (right) (Oerlikon Neumag, n.d.)

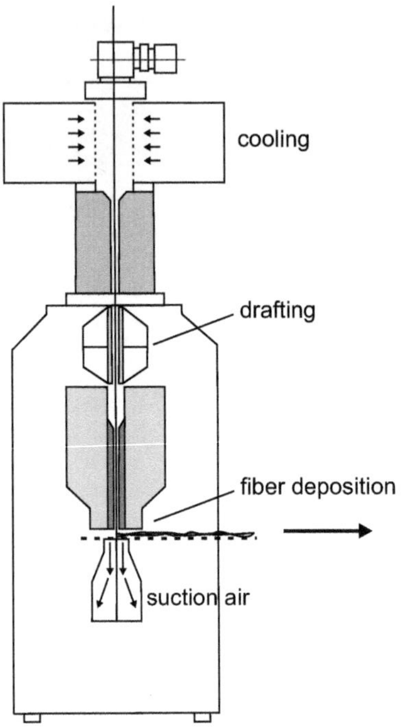

Fig. 25.5 Structure of a suction-compressed air spinning system (Reicofil system)

The bonding can also be done chemically with the help of binders in a padder or by hot air, or by using hot melt adhesives. This leads to nonwovens with a large volume. Alternatively, bonding can be done by continuous high-pressure water jets. Such spunbonded nonwovens have basis weights of 40–1000 g/m^2 (Table 25.1).

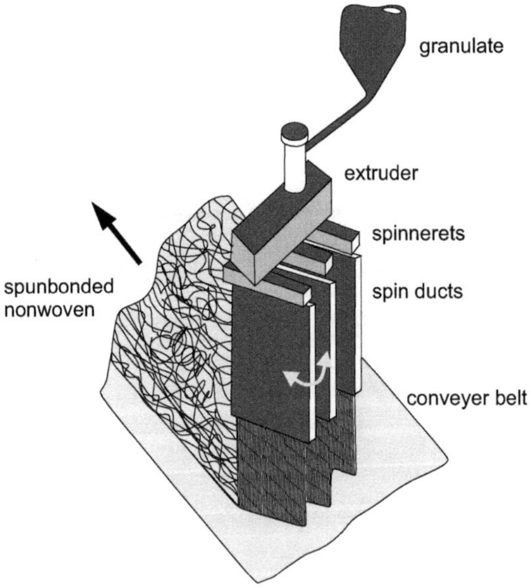

Fig. 25.6 Inclined spinnerets to increase the basis weight

Other bonding methods include stitch-bonding and spun lace.

Finally, the spunbond nonwoven is wound onto rolls at up to several hundred meters per minute (Fig. 25.8).

The weight per unit area G_F (g/m^2) of a spunbonded fabric can be calculated from the ratio between the mass throughput $\dot{m}$ per hour and the spinning width as well as the speed V of the deposit on the wire belt:

$$G_f = \frac{\dot{m}_{width}}{V_{deposit}}$$

Bicomponent fibers can also be used to produce spunbonded nonwovens. This is done either in core-sheath form, with the sheath acting as a melt-adhesive bonding component. On the other hand, fibers of the orange-pie type are also common, whereby the components are separated from each other by hydroentanglement, resulting in a fine fiber nonwoven (e.g. Evolon®, Fig. 25.9).

Typical applications of spunbonded nonwovens are hygiene articles (diapers, wipes), medical products (mouthguards, surgical drapes), furniture components (mattresses, disposable bedding, upholstered backings), technical textiles (geotextiles, early harvest nonwovens, sound insulation) and packaging articles (tea bags).

25.2 Meltblown Process

The meltblown process was developed in the 1950s at the Naval Research Laboratories (NRL) in the USA. It is used to produce nonwovens consisting of particularly fine fibers (Fig. 25.10). In this process, a low-viscosity, melt-spinnable polymer (usually polypropylene or polyester) is extruded from a single spinneret that extends across the entire width of the nonwoven. As the fibers leave the spinneret, hot air flows past the filaments at very high speed (6000–30,000 m/min) in spinning direction (Fig. 25.11). The friction between the air and the still slightly

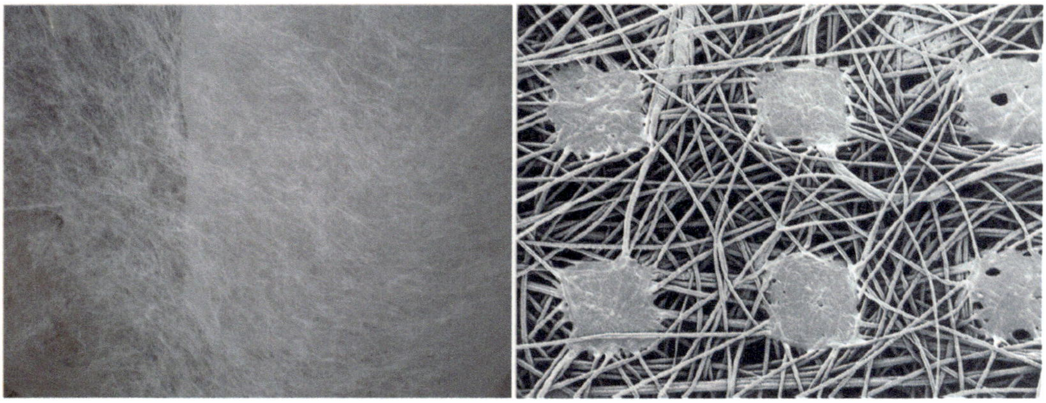

Fig. 25.7 Spunbonded fabric (left: unbonded; right: adhesive dots)

deformable filaments causes them to be drawn into very fine fibers. Typical individual linear densities are 0.01–0.1 dtex (approx. 1–3 μm).

The meltblown spun fibers are sprayed directly onto a wire belt, a wire drum or a spunbond.

Figure 25.12 shows a meltblown spinneret block in side view (left) as well as a view from below of the slit-shaped spinneret with the air blow channels on both sides (right).

Figure 25.13 shows the difference between meltblown and spunbond nonwovens (same scale). Both products are often combined, e.g. for diapers and protective masks. In diapers, an outer layer of printable meltblown nonwoven ensures impermeability; in protective masks, the meltblown layer prevents the penetration of viruses or fine dust particles. The spunbond layer ensures moisture absorption (diaper) or increases wearing comfort (protective mask).

Spunbond and meltblown processes can also be combined in one process. This is particularly interesting for diapers and filter products. In Fig. 25.14, as a typical example, a process designed by Reifenhäuser is shown with two spunbond beams, followed by three meltblown beams and finally another spunbond beam.

25.3 Products

Spunbonded nonwovens are used for filtration applications and wherever low-cost textiles with adjustable porosity are needed. The Fig. 25.15 shows typical products (coffee pads, tea bags, diapers, air filters).

A typical example of a combination of spunbond and meltblown nonwoven is the corona protective mask in Fig. 25.16. Here, two antiviral and self-cleaning cover layers of knitted fabrics (1, 4) have been combined with spunbonded and meltblown nonwovens. The small pore size reliably retains viruses, while at the same time the very high number of small pores makes breathing easier.

Table 25.1 Consolidation methods depending on the desired weight per unit area acc. to Fuchs and Albrecht (2012)

Principle	Process	Area weight [g/m^2]
Thermal	Calendar	5–100
	Thermofusion	80–1000
Mechanical	Needling	50–1000
	Stitch-knitting	60–1000
	Water-jet bonding	20–200
Chemical	Padding	20–300

Fig. 25.8 Winder of a spunbond line (Freudenberg Performance Materials KG, n.d.)

In Table 25.2, the characteristics of typical spunbonded and meltblown products are compiled.

25.4 Market

Figure 25.17 shows the rapid development in the production of spunbonded nonwovens since 1990. Annual growth over the last 10 years has been around 400,000 t and there is no end in sight to this development. The most important producer today is China with a market share of about 55% (2005: 18%). Due to the huge demand for protective masks in the wake of the Covid 19 pandemic, the demand for spunbond and meltblown nonwovens rose sharply once again. The growth from 2019 to 2020 was therefore over 30%, a value never achieved before.

Acc. to a study by The Fiber Year (2020), around 120 new spunbond lines have been installed worldwide since 2011, mainly in China, India, the US and Egypt (Fig. 25.18).

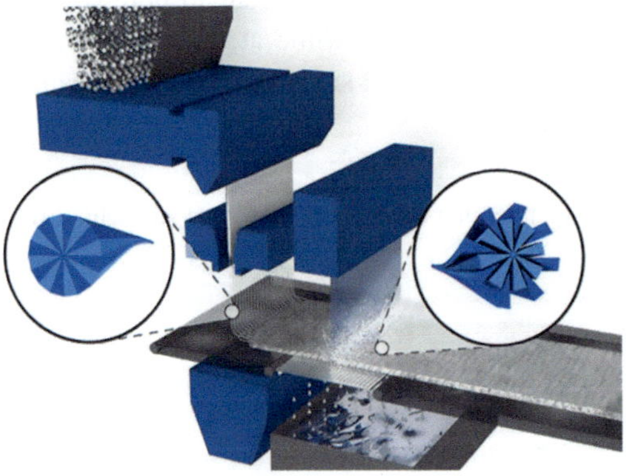

Fig. 25.9 Principle of the Evolon® process (Freudenberg Performance Materials KG, n.d.)

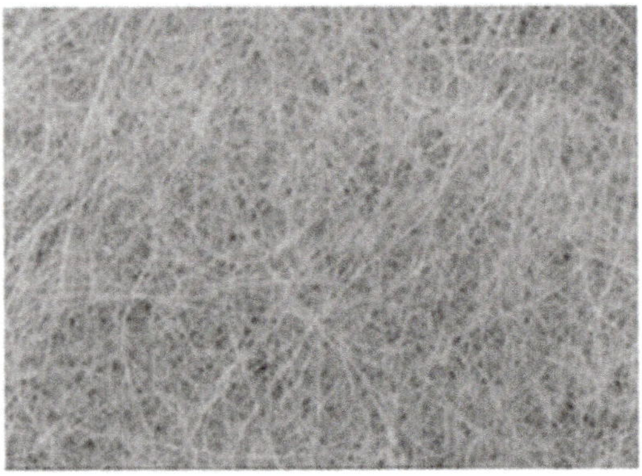

Fig. 25.10 Meltblown nonwoven

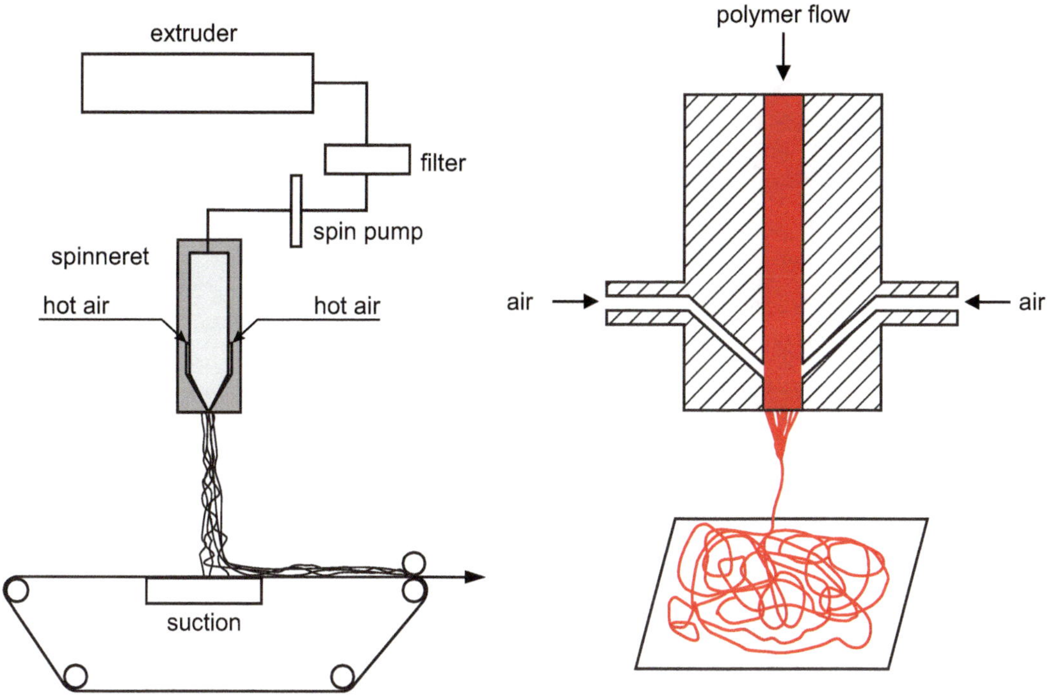

Fig. 25.11 Principle of the meltblown process and a meltblown spinneret with quenching

Fig. 25.12 Meltblown spinneret with blowing channels (Freudenberg Performance Materials KG, n.d.)

Fig. 25.13 Comparison of meltblown nonwoven (left) and spunbonded nonwoven (right)

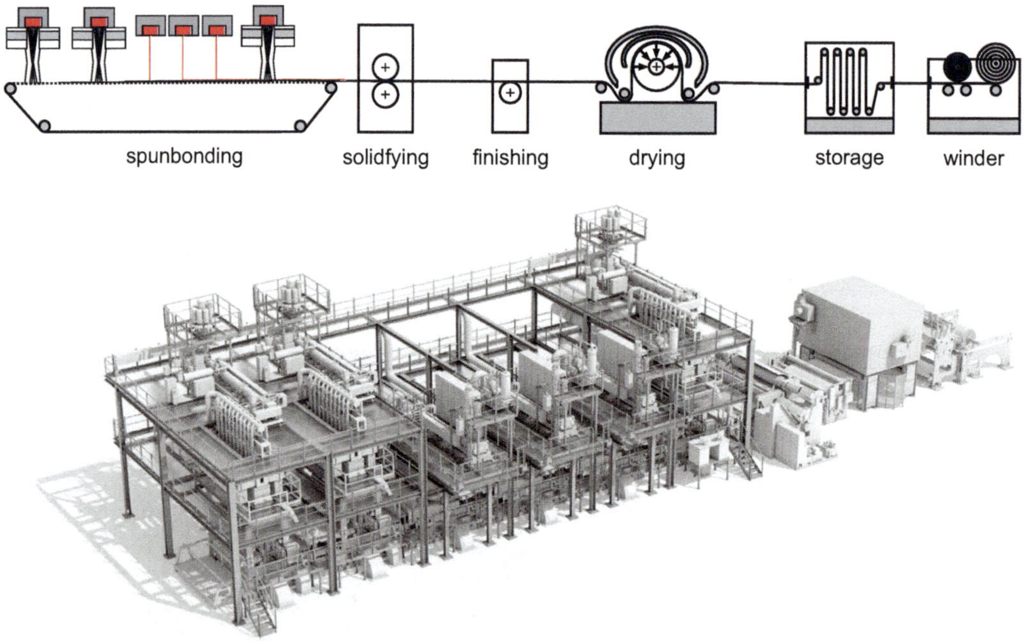

Fig. 25.14 Combination of meltblown and spunbond line (Reifenhäuser GmbH & Co. KG Maschinenfabrik, n.d.)

Fig. 25.15 Typical products made from nonwovens (bottom left: Coop41, 2007; bottom right: Evitomotua, 2007)

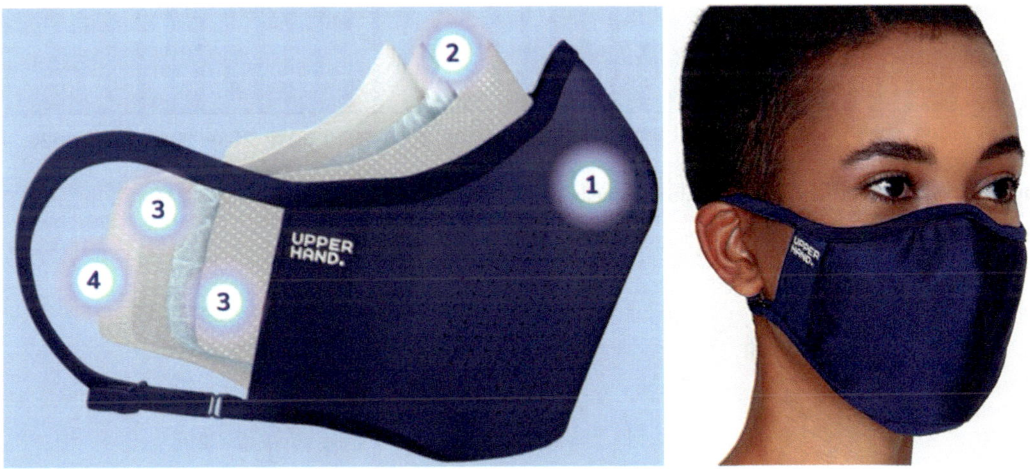

Fig. 25.16 Mouth-nose mask against viruses (Impact Products GmbH, n.d.)

Table 25.2 Typical product characteristics (Oerlikon Neumag, n.d.)

	Geotextiles	Roof sheet support	Roof sheet backing	Hygiene nonwovens
Polymer	PP, PET	PET	PP, PE, PET	PP, PET
Area weight [g/cm^2]	80–1000	100–300	50–120	12–35
Filament fineness [dtex]	4–10	2–6	4–7	2–5 (spunbond) 0.04–0.1 (meltblown)

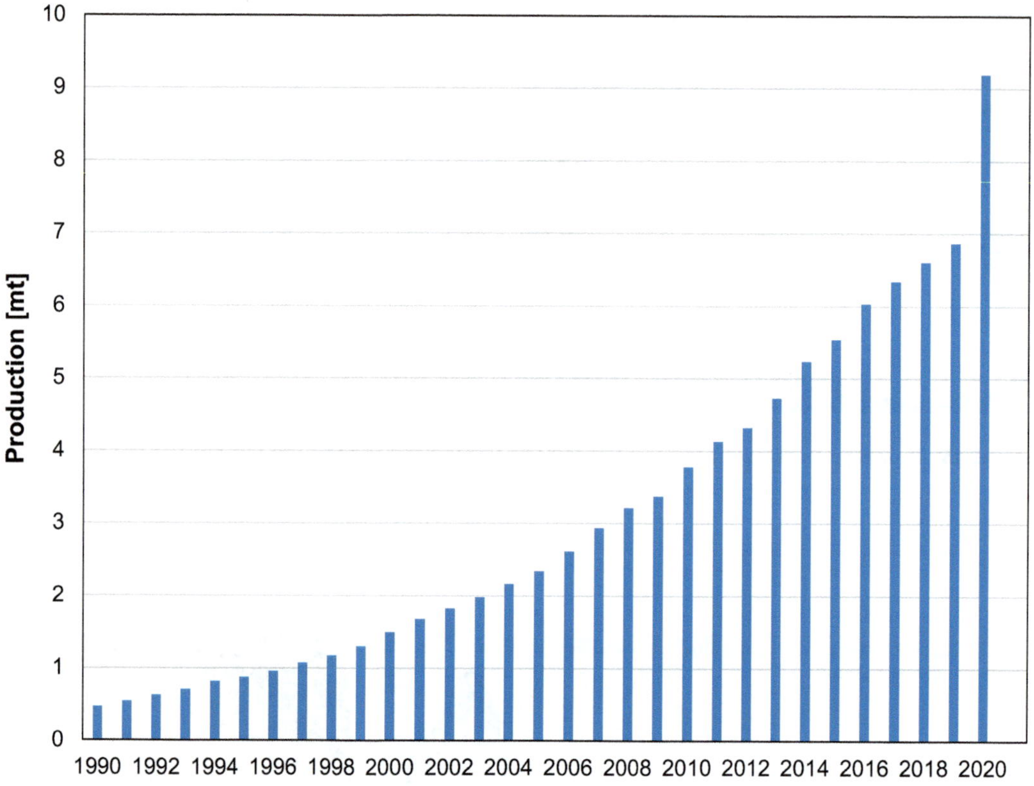

Fig. 25.17 Development of the production of spunbonded nonwovens (data: The Fiber Year, 2021)

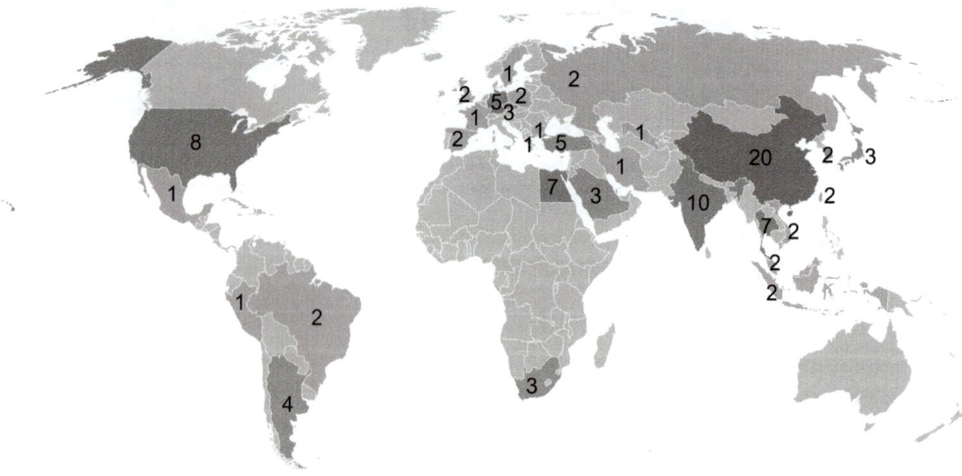

Fig. 25.18 Number of spunbond lines built from 2011 to 2020 (data: The Fiber Year, 2020)

References

Coop41. (2007). https://commons.wikimedia.org/wiki/File:Disposablediaper.JPG

Evitomotua. (2007). *CC BY-SA 2.5*. https://commons.wikimedia.org/wiki/File:Kombifiltermedium_Kabinenluftfilter.jpg

Fuchs, H., & Albrecht, W. (2012). *Vliesstoffe. Rohstoffe, Herstellung, Anwendung, Eigenschaften, Prüfung*. Wiley.

Freudenberg Performance Materials KG. (n.d.). Weinheim. https://www.freudenberg-pm.com/Materials/Nonwoven

Impact Products GmbH. (n.d.). Berlin. https://upperhand.de

Oerlikon Neumag. (n.d.). Zweigniederlassung der Oerlikon Textile GmbH & Co. KG, Neumünster. https://www.oerlikon.com/de/marken/oerlikon-neumag

Reifenhäuser GmbH & Co. KG Maschinenfabrik. (n.d.). Troisdorf. https://reifenhauser.com

The Fiber Year. (2020, 2021). The Fiber Year GmbH, Roggwil. https://www.thefiberyear.com

References

26 Bicomponent Fibers

When different polymers are combined in a single fiber, the term bicomponent fibers is used. By combining the often very different properties of the polymers, completely new types of fibers can be produced. It is also possible to produce ultra-fine fibers if the two polymers can be easily separated after spinning. Mostly synthetic polymers are used, but metals, carbon and cellulose can also be processed in this way.

For the production of bicomponent and microfibers, the melt spinning process, the solution spinning process (wet and dry) and, for special products, the electrospinning process are used.

Typical applications of bicomponent fibers are clothing, filter technology, upholstery fabrics, cleaning cloths and medical textiles.

26.1 Terms

Bicomponent fibers are man-made fibers produced from two firmly, but separably, bonded materials of different chemical and/or physical make-up (Houis et al., 2008). For this purpose, only one spinneret is used and the two components do not form a new polymer. Depending on the design of the spinneret, a variety of different bicomponent fibers can be produced (Fig. 26.1).

In general, three different types are distinguished: the side-by-side type (S/S type), the core/sheath type (C/S type) and the matrix/fibril type (M/F type) (Fig. 26.2).

26.1.1 Side-by-Side Type

This type of fiber consists of two polymers arranged side by side. The arrangement can be concentric or eccentric. If both polymers have a different degree of recovery, the result is a fiber crimp, similar to wool.

26.1.2 Core/Sheath Type

The sheath polymer completely encases the core polymer. These fibers are very versatile, e.g. as a combination of high-strength core and flame-retardant sheath or as a combination of biocompatible sheath and any core material for medical implants.

26.1.3 Matrix/Fibril Type

This type is used to produce microfibers. Either one component can be dissolved after spinning (islands-in-the-sea) or the two components are split, e.g. by a water jet (segmented-pie, multi-layer type). Extremely fine fibers can be produced in this way. Actually, the multi-layer type belongs to the S/S type, but is used exclusively for

D. Veit, *Fibers*, https://doi.org/10.1007/978-3-031-15309-9_26

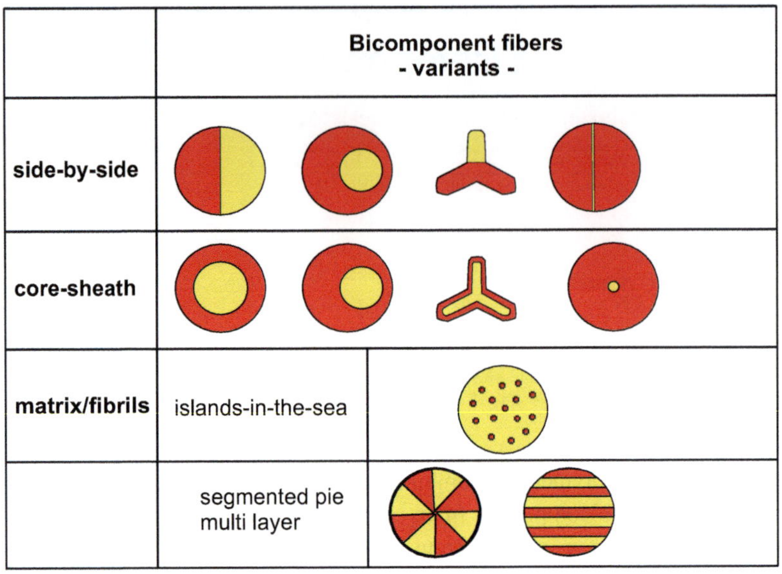

Fig. 26.1 Types of bicomponent fibers acc. to (Houis et al., 2008)

microfiber production. A typical application is artificial leather (Alcantara®).

26.2 History

The oldest bicomponent fiber is the curly angel hair, which was already produced in the eighteenth century on the basis of glass and used as a Christmas tree ornament. In 1886, J. de Brunfaut used a combination of a narrow strip of table glass and a round rod made of lead crystal. When the glass rods, which had been melted with the aid of a glassblowing lamp, were pulled off, the two combined to form a thread with a fineness of 6–12 µm.

In the 1930s, the first actual bicomponent fibers of the S/S type were produced at the Wolfen plant of the then I.G. Farben AG using specially designed double-bottom nozzles. This also resulted in the first patent for the production of bicomponent fibers of the S/S type in 1943. In the 1940s, various conjugated bicomponent fibers of the C/S type made of viscose and rayon were described in American patents by W. Sisson (Sisson, 1948a, 1948b). In 1954, (Breen, 1954) patented a crimped composite fiber of two components in S/S configuration. In 1959, the first S/S type synthetic bicomponent fiber made of acrylic was marketed under the name Orlon Sayelle (or Orlon 21) by DuPont, and in 1963, also by DuPont, a polyamide-based S/S type bicomponent fiber was marketed under the name Cantrece. Bosley (1966) obtained a patent for a bicomponent fiber based on polyester. Various fibers with many different functions such as

Fig. 26.2 Bicomponent- and microfiber types (Hills, Inc. n.d.)

crimping, thermal bonding, electrically conductive, antistatic, light conductive, absorbent, and splittable fibers are subsequently developed mainly in Japan. In 1965, the first multilayer fiber was produced and subsequently the Kanebo company succeeded in splitting these fibers to produce superfine fibers. This was the birth of microfibers. Based on these multilayer fibers, the companies Okamoto and Toray worked on a multi-core and multi-island fiber, respectively, and finally Toray launched the first imitation suede, which became known as Alcantara (Ecsaine). This was followed in 1972 by the multi-radial bicomponent fiber BelimaX and in 1977 by Bellseime for imitation suede from the Kanebo company. Also in the 1970s, DuPont developed S/S type electrically conductive bicomponent fibers (Antron and Ultron) to eliminate the hazards caused by static electricity in fiber products. Then, in the 1980s, numerous new variants of the previously developed bicomponent fibers and their products came on the market, for example, super high-density fabric (Kanebo, 1981) and high-performance cleaning cloths (Kanebo, 1985) made of microfibers and fourth generation electrically conductive fibers (Shimadzu, 1983). Toyobo (Japan) developed a highly water-absorbent bicomponent type C/S fiber in the 1990s. It consists of a highly water-absorbent polymer (core) and an acrylic polymer (sheath). The fiber absorbs water up to 150 times its own weight. At about the same time, Freudenberg Nonwovens developed the Evolon split fiber. It is split by water jet needling on the one hand and the nonwoven is bonded directly on the other. The main field of application is cleaning cloths. A detailed overview of the history of bicomponent fibers can be found in (Houis et al., 2008).

To date, bicomponent fibers are being developed further and further and they meet more and more technical requirements for a wide range of applications. For cost reasons, however, these fibers do not compete too much with monocomponent fibers. Better properties can be achieved in some cases, but the costs for bicomponent spinning lines are about 20% to 40% higher than for monocomponent spinning lines. In addition, the process control is more difficult.

26.3 Source Materials

There are a variety of polymers that can be combined in bicomponent fibers (Table 26.1). In order to create a bicomponent fiber from different polymers, various requirements must be met. On the one hand, both the melting temperatures and the viscosities of the materials should be in similar ranges. On the other hand, bonding forces of the interfaces to each other are advantageous. Because they meet these requirements well, they are inexpensive and available in large quantities, and they are easy to process, PET, PA and PP are often combined.

Figure 26.3 shows an overview of selected bicomponent fiber types with regard to typical materials, fineness and proportion of the two components.

26.4 Manufacturing Technologies

Bicomponent fibers can be produced with almost all common spinning processes. Figure 26.4 shows the process steps required in each case. The melt spinning process dominates worldwide.

Special spinnerets are required for spinning bicomponent fibers, in which the two different polymers are brought together in solution or as a melt from separate spaces. This can be done either before the polymers exit the spinneret or afterwards. When the filament leaves the capillary hole, it consists of two non-mixed components that are virtually welded together at their junction line (Fourné, 1995). This connection must not loosen even when the filaments are subsequently stretched. An exception is the Evolon® process, in which the two components are separated from each other by water jets directly after leaving the spinneret in order to produce microfibers (Fig. 26.5).

Due to the considerably lower viscosities (melt spinning: 80–1000 Pas; dry spinning: 20–200 Pas), the spinneret holes in the solution spinning

Table 26.1 Frequently used materials for bicomponent fibers acc. to (Houis et al., 2008)

	Repeat unit	Density [g/cm^3]	Melting temp. [° C]	Features
PA	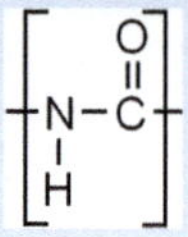	1.13	230 °C (PA 6) 270 °C (PA.6.6)	Good chem. Resistance, (except halogens and acids)
PET	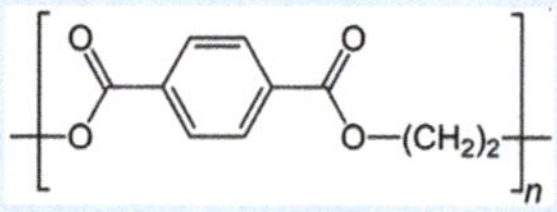	1.34	245 °C	Good chem. Resistance, (except halogens and alkalis)
PP	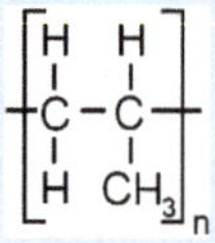	0.9	168–175 °C	Resistant to alcohol, organic solvents and greases, not resistant to petrol, benzene and hydrocarbons.
PAN	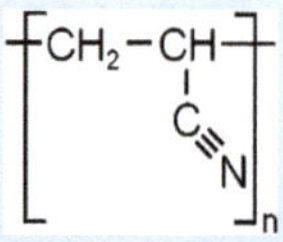	1.17	130–150 °C	Heat sensitive, chemical and solvent resistant
PUR	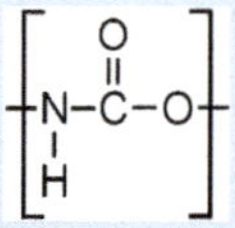	1.15	40–150 °C	High elasticity
PS	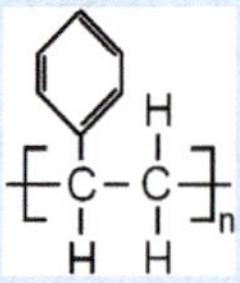	1.06	240–270 °C	UV-sensitive, resistant to alkalis and unstable to petrol, ketone and aldehydes
PTFE	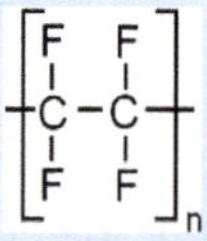	2.2	325 °C	Very good electrical insulation properties, extremely low friction, resistant to acids, alkalis and solvents
PVDF	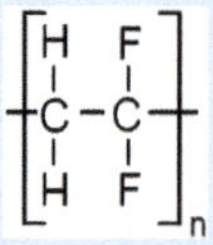	1.77	155–170 °C	Very good electrical insulation properties, low service temperature. Generally good chemical stability (exception: sulphurous acids).

process are narrower than in the melt spinning process (melt spinning: 0.2–2 mm, dry spinning: 0.15–0.25 mm) (Fourné, 1995).

In most cases, all components up to the spinneret are similarly constructed, irrespective of the type of fiber produced. The components are metered by spinning pumps to the spin package, where they pass through the spinneret. The filaments are then cooled or coagulated for bonding, drawn and wound up (Figs. 26.6 and 26.7).

In bicomponent electrospinning, very thin fibers (50 nm << 2 m) are produced from a polymer solution or melt by treatment in an electric field and deposited in the form of nonwovens.

Fig. 26.3 Typical bicomponent fibers (material, fineness, component ratios) acc. to (among others, Oerlikon Barmag n.d.)

			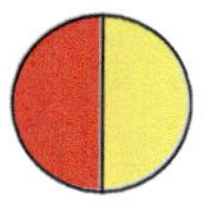	
material	PET/CoPET PET/PE PP/PE	PET/CoPET PP/PE	PET/CoPET PET/PE PP/PE	PET/PA PET/CoPET
titer [dtex]	1.7 - 20	1.7 - 20	2 - 20	0.1 - 0.2
ratio of components	30/70 - 80/20	30/70 - 70/30	50/50	50/50 - 80/20

		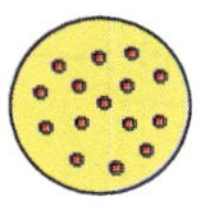		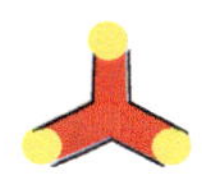
material	PET/PA	PA/CoPET PET/CoPET	PET/PTT PET/PA PP/PA	PET/PTT PET/PA PP/PA
titer [dtex]	0.1 - 0.2	< 0.05	15 - 20	15 - 20
ratio of components	50/50 - 70/30	50/50 - 80/20	30/70 - 50/50	30/70 - 50/50

26.4.1 Side-by-Side (S/S)

To produce an S/S bicomponent fiber, two polymer streams are fed to the spinneret orifice through narrow, separate channels. The viscosities of the two polymers must largely match during melt spinning.

(Bromley & Yu, 1992) developed a second variant of the melt spinning process in which the two polymers are fed through two separate orifices and are only combined below the spinneret. The speed of the polymer streams is defined by the size of the spinneret orifices and by the pump speed. Incompatible polymers can also be spun with this process (Yu, 1998).

In Fig. 26.8, the structure of typical spinnerets is depicted. On the left, the polymers meet only at the spinneret exit, while on the right they meet somewhat earlier within the nozzle. Therefore, the arrangement on the left results in very uniform bicomponent fibers, while irregular bicomponten fibers are produced on the right, in which the distribution of the components in the cross-section varies along the fiber length.

26.4.2 Core/Sheath (C/S)

In the production of core/sheath fibers, the polymers are fed to the spinneret through two separate channels (Fig. 26.9). As with S/S fibers, the meeting of the two polymer streams can occur upstream or close to the spinning orifice.

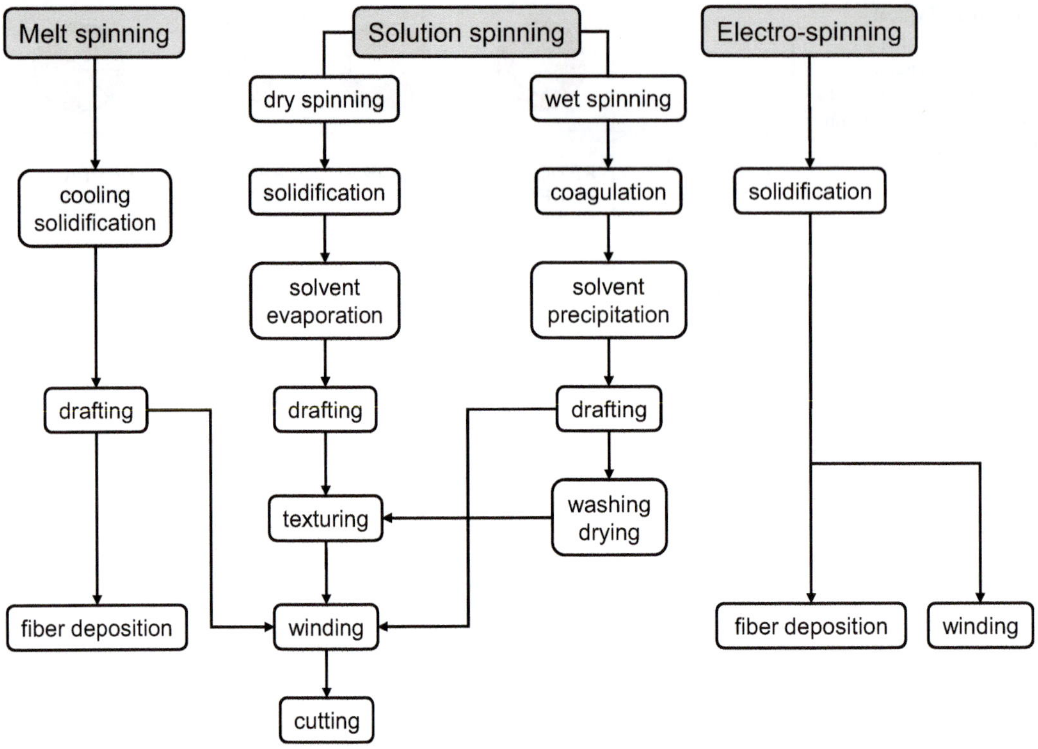

Fig. 26.4 Process stages of the bicomponent spinning processes

In the case of three-component fibers, the two sheath polymer streams either meet shortly before the spinneret opening in an annular channel, in the middle of which the core polymer flows, or they only join to form a stream outside the spinneret. Depending on the positioning of the core polymer opening, concentrically or eccentrically arranged components are obtained.

Fig. 26.5 Bicomponent spinneret and detail (segmented pie) (Sossna GmbH n.d.)

Fig. 26.6 Typical set-up of a bicomponent melt spinning machine

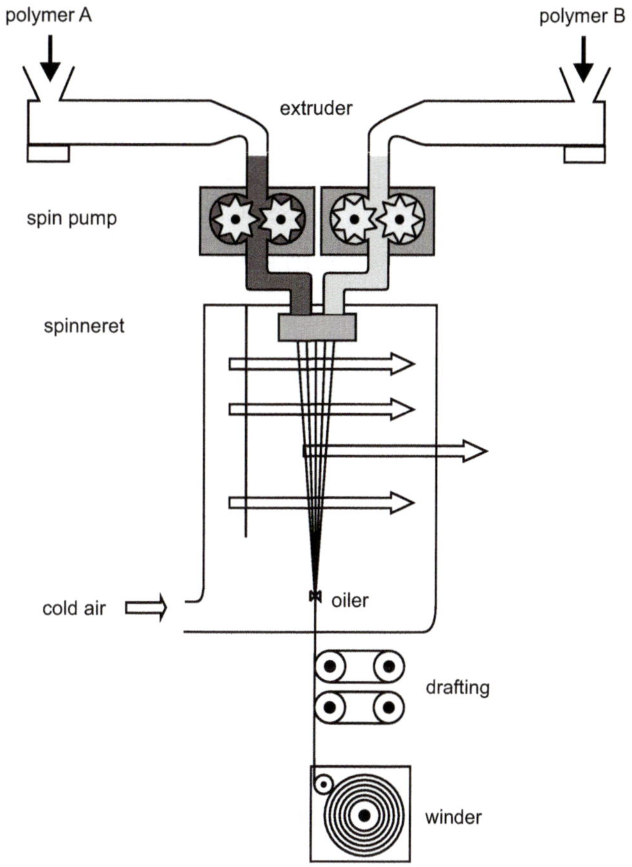

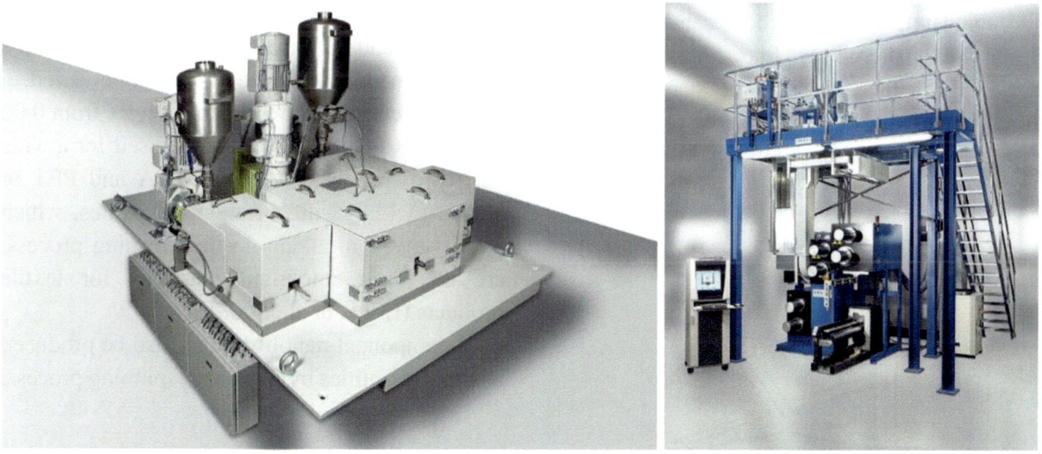

Fig. 26.7 Bicomponent melt spinning machine (left: extruder level; right: general view) (Fourné Maschinenbau GmbH n.d.)

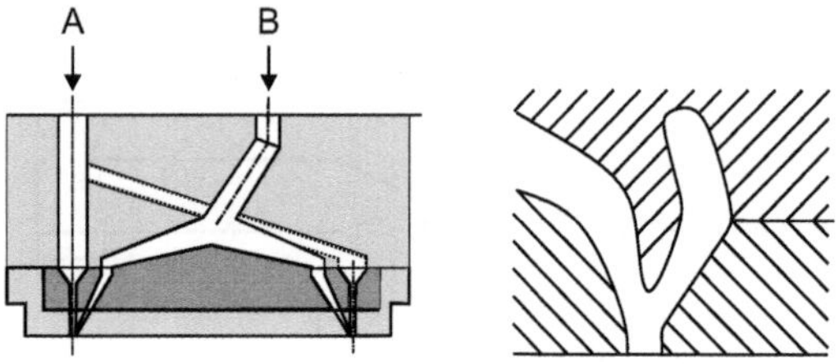

Fig. 26.8 Schematic structure of S/S spinnerets acc. to (Houis et al., 2008)

26.4.3 Matrix/Fibril Fibers or Islands-in-the-Sea

There are various methods of producing islands-in-the-sea fibers. In all methods, the two polymers are brought together within the spinneret, compacted and then spun out. The Fig. 26.10 shows the basic design of such a spinning apparatus.

In solution spinning, the matrix component can be dissolved after spinning and nanofibers are formed from the islands (Fig. 26.11).

26.4.3.1 Segmented Pie (Matrix/Fibril Type)

Segmented pie fibers (also known as orange pie fibers) are produced by combining two polymers that adhere poorly to each other, e.g. PET and PA. In cross-section, the fibers resemble a cake divided into pieces. Depending on the design of the spinneret, ultra-fine fibers can be produced from one material or fine fibers from both materials. In the first case, the star-shaped core of the second polymer is removed after spinning. Up to 16 segments are common. With a filament fineness of 2 dtex, fibers with a width of 0.5–2 μm and a height of 6 μm are then obtained. With more segments, even finer fibers are possible (Fig. 26.12).

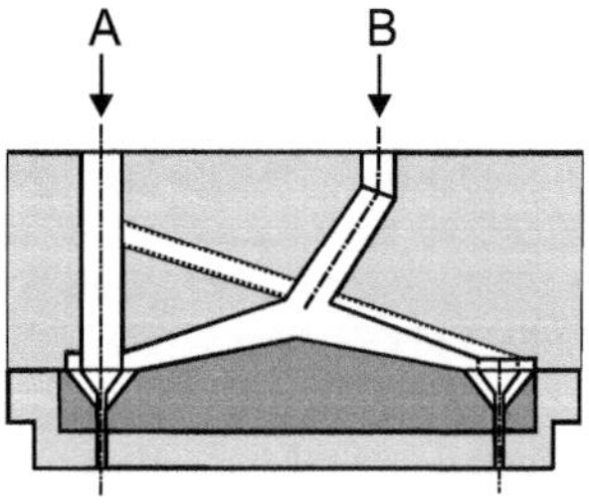

Fig. 26.9 Schematic structure of C/S spinnerets acc. to (Houis et al., 2008)

26.5 Fine Fibers

In the case of islands-in-the-sea fibers, the sea portion is usually dissolved either by an organic solvent or by alkaline compounds. Due to a high islands content, fine microfibers in the fineness range from 0.1 to 0.01 dtex can be produced easily and economically.

In order to achieve an islands polymer content of more than 90% and at the same time a high fineness of the spun fibers, the polymer viscosity and the good solubility of the sea component are decisive (Rupp & Yonenaga, 2000).

When the islands polymer stream is extruded through a sand filter, extremely fine island fibers (up to 0.0001 dtex, approx. 100 nm) can be produced. This makes them roughly equivalent to the fineness of collagen fibers in natural leather.

In the split and multilayer type, two non-compatible polymers are combined. The advantage of this process lies in the simple separation process, which eliminates the need for special organic solvents and allows the entire extruded polymer to be utilized. By choosing the type of polymer, the cross-section and the proportion of the two components, the appearance, feel and physical properties of the textile products made from these fibers can be varied.

The fineness of the segments ranges from 0.06 to 0.5 dtex and the fibers can be used for a wide variety of products. Blends of PA and PET or blends of PET with different viscosities, which are dissolved in alkalis in the finishing process, are the most important materials for textile products (Houis et al., 2008).

Bicomponent nanofibers can also be produced in small quantities by the electrospinning process.

26.6 Properties

The properties of bicomponent fibers depend on the respective components of the fiber and their

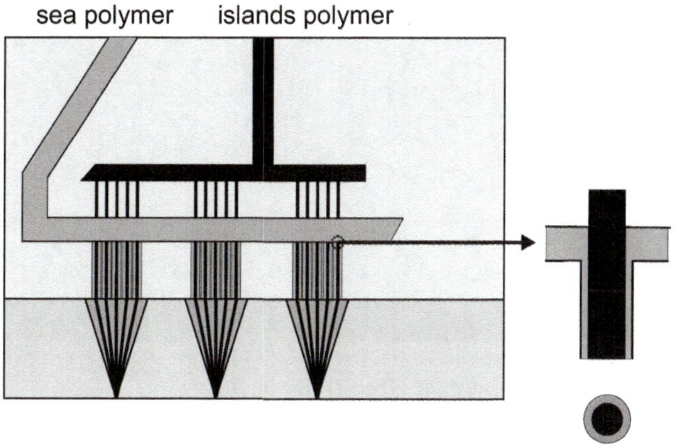

Fig. 26.10 Structure of a spinneret for islands-in-the-sea fibers acc. to (Houis et al., 2008)

structure. By combining different materials and the cross-sectional structure of the bicomponent fiber, very different products can therefore be manufactured.

26.6.1 Side-by-Side (S/S) Fibers

With bicomponent fibers in S/S configuration, among other things, a durable, reversible crimp can be achieved (Figs. 26.13 and 26.14). This occurs, for example, when the polymers absorb different amounts of moisture. After a drying process or through swelling, the crimp structure then develops. The more asymmetrically the components are arranged in the cross-section, the stronger it is.

The strength of the crimp can also be adjusted by the proportion of the two polymers in the fiber. Corresponding theories on this were developed early by (Abhiraman, 1974) and (Gupta & George, 1974).

26.6.2 Core/Sheath Fibers (C/S)

Fibers in C/S configuration always have the same structure. They consist of a core surrounded by a sheath. The ratio of sheath to core can be varied. The core can also be arranged asymmetrically (Fig. 26.15). An eccentric arrangement can lead to a crimping of the fiber.

The proportions of the core-sheath components vary from 1/99 to 50/50, depending on the function of the different components and the area of application of the C/S fiber (Fig. 26.16).

Fig. 26.11 Islands-in-the-sea fibers (longitudinal and transversal views) (Hills, Inc.)

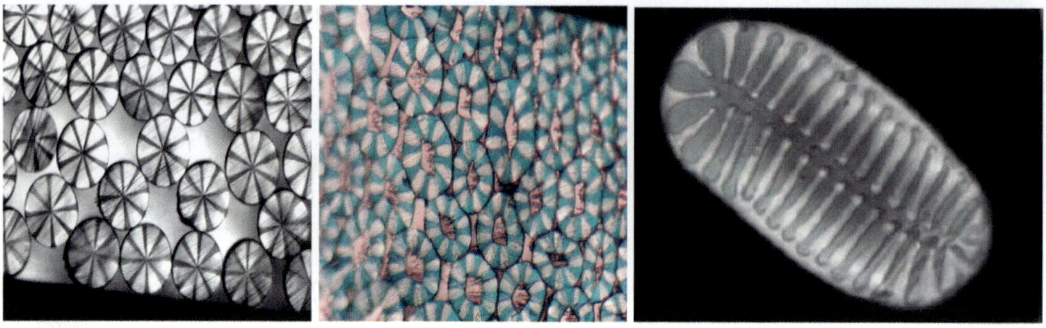

Fig. 26.12 Segmented Pie Fibers (Hills, Inc.)

26.6.3 Matrix/Fibril Fibers (M/F)

M/F fibers are used almost exclusively for the production of microfibers. Therefore, the number of fibrils in the fiber is decisive for the economic utilization of the fiber cross-section.

For islands-in-the-sea fibers, the number of islands can vary from a few to 1200 islands. For fibers with 600 islands, the individual fiber diameter is 300 nm (hills). For fibers with a segmented pie structure, the number of segments varies from 4 to 36. The minimum average pie cross-section width depends on the total diameter of the fiber and is a minimum of 0.5 μm (Fig. 26.17).

Multilayer fibers are also used to produce microfibers. The microfibers produced by splitting are characterized by properties such as softness, pleasant handle, high draping capacity, shine and fullness (Fig. 26.18).

26.6.4 Tri-component Fibers

Fibers can also be made from three components. The Fig. 26.19 shows the cross-section through a corresponding spinneret and the structure of the resulting fiber. It is of crucial importance that the viscosities of all three polymers are very similar,

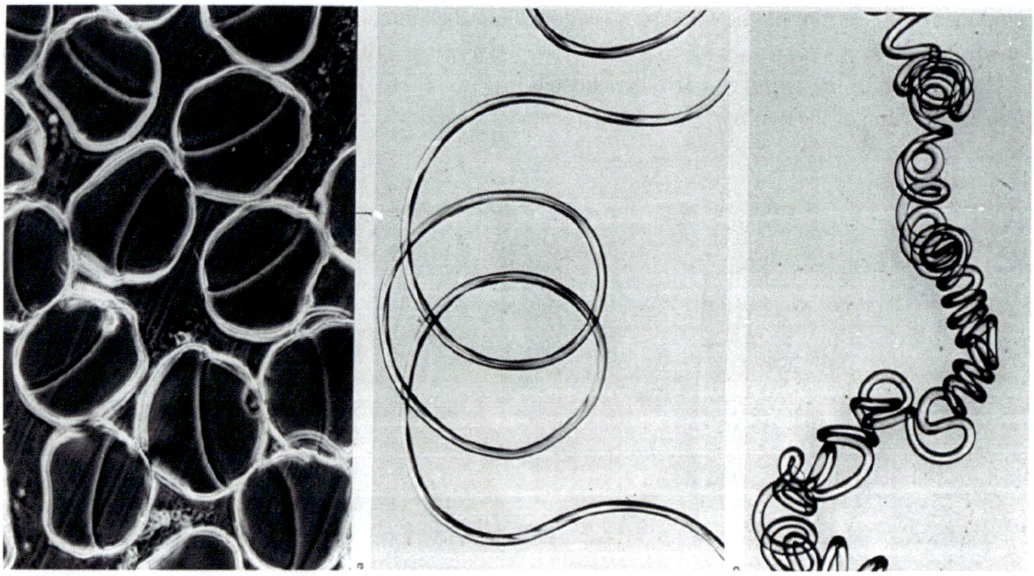

Fig. 26.13 Side-by-side fibers and resulting crimp after thermal treatment

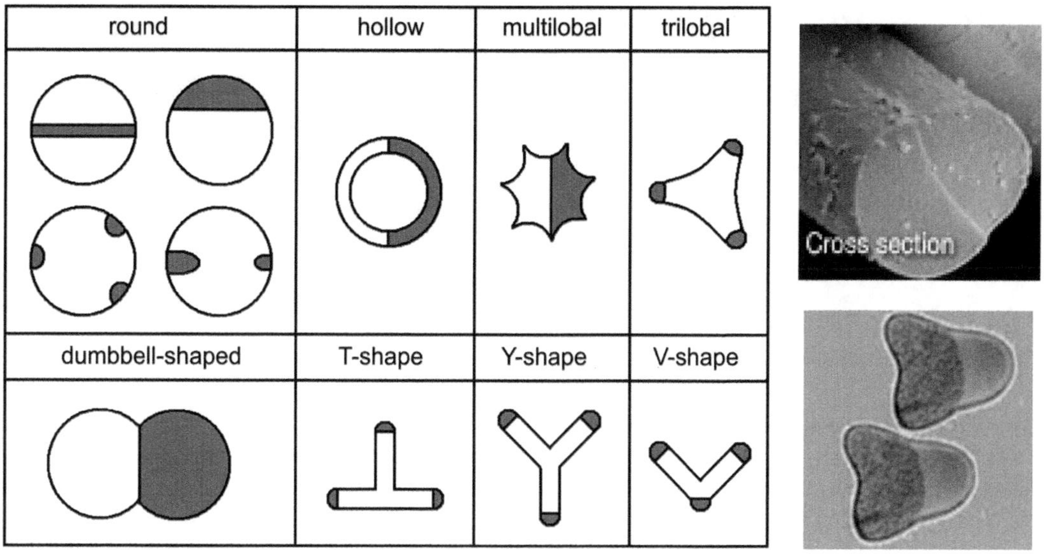

Fig. 26.14 Typical cross-sections of S/S fibers (Houis et al., 2008)

because this is crucial to achieve good bonding of the components.

26.6.5 Calligraphic Fibers

With special spinnerets, it is also possible to "label" fibers and thus make them forgery-proof (Fig. 26.20).

26.6.6 Physical and Textile-Technological Characteristic Values

The physical and textile-technological properties of bicomponent fibers depend on the respective components of the fiber, its structure and the respective proportion. (Houis et al., 2008) provide a good overview of the textile characteristics of bicomponent fibers from many manufacturers. Current data can be found on the manufacturers'websites (Table 26.2).

26.7 Areas of Application

Bicomponent fibers are used in a wide range of products. They undergo the same further processing as conventional textile fibers.

26.7.1 Side-by-Side Fibers (S/S)

S/S fibers are used mainly in carpets and garments because of their self-crimping property, as well as in resilient upholstery for furniture and as filling materials in mattresses and pillows. Other typical products are ladies' stockings, knitted fabrics of all kinds and material for thermal insulation. If

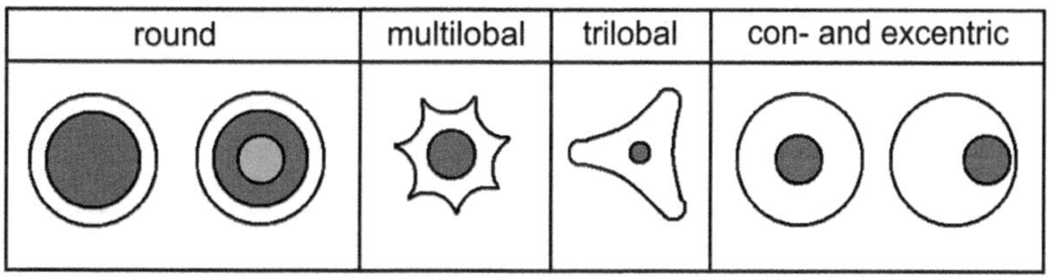

Fig. 26.15 Typical cross-sections of C/S fibers (Houis et al., 2008)

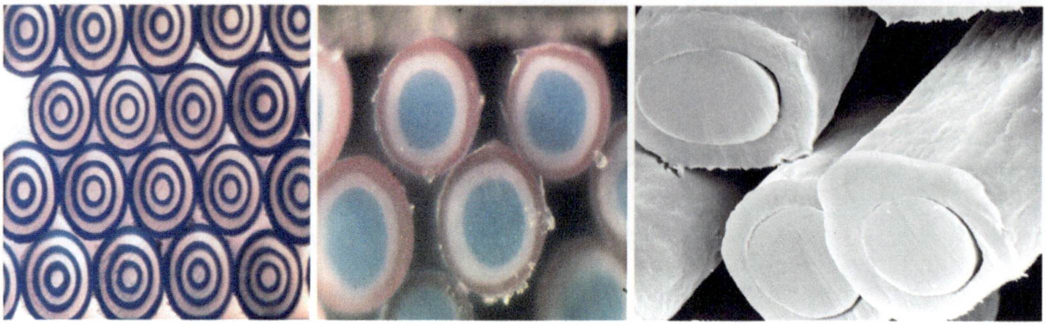

Fig. 26.16 Typical cross-sections of C/S fibers (Hills, Inc.)

one of the two components has a low melting point, S/S fibers can also be used as bonding fibers in nonwovens.

26.7.2 Core/Sheath Fibers (C/S)

If the melting point of the sheath component is lower than that of the sheath, C/S fibers can be used as bonding fibers in nonwovens. A typical material combination is PA 6/PA 66. During heat treatment by calendaring or with hot air, one component melts and so the fibers stick together. This also increases the dimensional stability of the nonwoven.

A combination of a solid core material with a flame retardant sheath is also possible. Likewise, C/S fibers are used to save costs by supplementing the expensive sheath material with a cheaper core. Additives can also be added to the core or sheath to selectively release drugs in textile implants. There are also cost advantages in dyeing if only the sheath is dyed. (Hu et al., 2018) developed antistatic polyester fibers by adding carbon black.

Analogous to S/S fibers, C/S fibers with an eccentric core are also used as crimping yarns.

C/S fibers are used in carpets, upholstered furniture, nonwovens, bedding and hygiene products, as a filling product in mattresses and pillows, and in garments. One of the most commonly used core/sheath fiber products is Colback™ from Colbond bv, Arnhem, The Netherlands. Colback is used in carpets and is a thermally bonded melt spunbond. The core is made of polyester and provides good mechanical stability. The polyamide (PA 6) sheath acts as a binder and has a good affinity for various coatings.

26.7.3 Matrix/Fibril Fibers (M/F)

M/F fibers are mainly used for the production of microfibers. As a result, a special softness, a pleasant handle, a high draping capacity, shine and fullness can be achieved. Typical applications include wipes, filters, towels, sports and all-weather textiles, and men's and women's underwear. (Duo et al., 2021) produced artificial

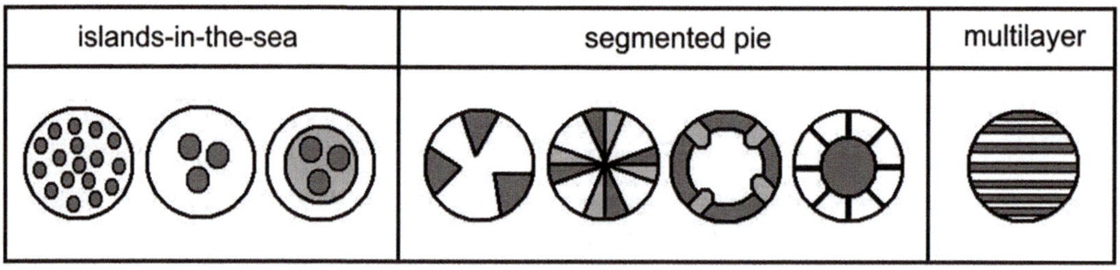

Fig. 26.17 Typical cross-sections of M/F fibers acc. to (Houis et al., 2008)

Fig. 26.18 Cross section of M/F fibers

leather from a combination of PET segmented pie and electrospun PAN nanofibers. Another approach is presented by the same authors in (Zhao et al., 2018), where they produced artificial leather from a microfiber nonwoven combined with a PU coating.

In the medical field, these fibers are used for diapers, gowns, face masks, surgical drapes, and tissue engineering.

Islands-in-the-sea fibers are also used in the production of plastic explosives. TheiIslands polymer, which is still present after the explosion, is used to provide precise evidence of the origin of the explosive.

One of the best-known matrix/fibril fiber products is Evolon™ from Freudenberg Performance Materials KG. This is a segmented-pie structure made of polyester and polyamide that is laid down as a spunbond and split and consolidated by water jets. It is mainly used to produce cleaning cloths and towels as well as bed linen for allergy sufferers.

Fig. 26.20 Calligraphic yarns (Hills, Inc.)

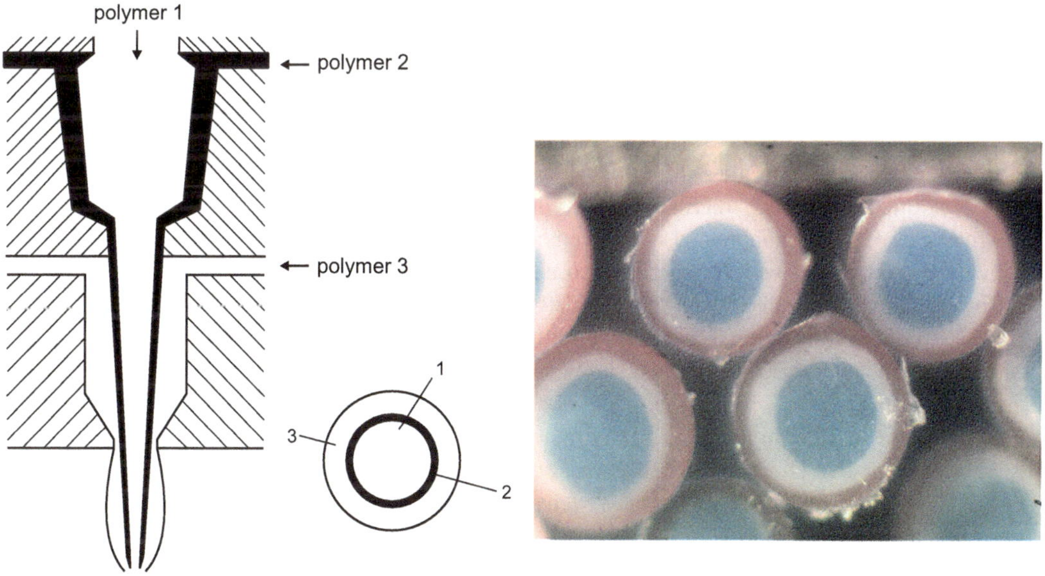

Fig. 26.19 Spinneret geometry for a tri-component fiber and fiber cross sections (Hills, Inc.)

Table 26.2 Selected manufacturers of bicomponent fibers

Producer	Website	Country
Dupont de Nemours, Inc.	http://www.dupont.com	USA
Far Eastern New Century Corp.	http://www.fenc.com	China
Fiber Innovation Technology, Inc.	https://fitfibers.com	USA
FiberVisions Corporation	https://www.fibervisions.com	USA
Hubei Botao Synthetic Fiber Co., Ltd.	http://www.bt-fiber.com	China
Huvis	http://www.huvis.com	Korea
Hyosung	http://www.hyosung.com	Korea
Jiangnan High Polymer Fiber Co., Ltd	http://www.jngx.cn	China
Kolon	http://www.kolonindustries.com	Korea
Mitsubishi Chemical	http://mccfc.com	Japan
Nan Ya Plastics	https://www.npc.com.tw	Taiwan
Ningbo Dafa Chemical Fiber Co., Ltd.	http://www.nbdafa.com	China
RadiciGroup	https://www.radicigroup.com	Italy
Toray Chemical Korea, Inc.	http://www.toray.com	Korea
XiangLu Chemical Fibers	http://www.xmxl.com	China
Xinghui Chemical Fiber	http://www.chinaxinghui.com	China
Yangzhou Tinfulong Co., Ltd.	http://www.tinfulong.com	China

26.8 Market

Bicomponent fibers as special fibers account for a small but steadily growing share. It is difficult to determine exact market figures because the manufacturers'data are not centrally recorded. According to estimates, sales of bicomponent fibers amount to approximately €1.8 million (2018) and this market is expected to grow by approximately 6% annually. The most important segment in terms of volume is core-sheath fibers. Here, polymers with a high melting point (core) and low melting point (sheath) are mainly combined to produce binding fibers, for example. They are often used for thermally bonded nonwovens (especially spunbonded nonwovens) for hygiene products. New areas such as household products, medical and agricultural products and geotextiles are still of minor importance.

Most bicomponent fibers are produced in Asia, especially in Japan and China. The USA and Europe are of lesser importance. Important manufacturers include Huvis Corp. (20% market share), Toray Industries, Teijin Ltd., Owens Corning, DuPont de Nemours, Eastman Chemical and JNC Corp.

References

Abhiraman, A. S. (1974). Mechanics of crimp formation in a bicomponent fiber model. *Textile Research Journal, 44*(4), 319–323.

Bosley, D. E. (1966). Bicomponent polyester textile fiber, US Patent US 3,454,460.

Breen, A. L. (1954). Verfahren zur Herstellung gekräuselter Verbundfäden. DE1202932B, Deutsches Patentamt, München.

Bromley, Y. E., & Yu, J. P. (1992). Deep dyeing conjugate yarn processes. US Patent 5,093,061.

Duo, Y., Qian, X., Zhao, B., Qian, Y., Xu, P., & Qi, J. (2021). Micro/nano microfiber synthetic leather base with different nanofiber diameters. *Journal of Industrial Textiles, 50*(7), 1127–1142.

Fourné, F. (1995). *Synthetische Fasern*. Hanser.

Fourné Maschinenbau GmbH, Alfter-Impekoven. https://fourne-maschinenbau.de

Gupta, B. S., & George, W. (1974). A theory of self-crimping bicomponent filaments. *Textile Research Journal, 45*(4), 338–349.

Hills Inc., West Melbourne (FL), USA. http://www.hillsinc.net

Houis, S., Schreiber, F., & Gries, T. (2008). Bikomponentenfasern. Faserstofftabelle nach P.-A. Koch, Institut für Textiltechnik der RWTH Aachen, Aachen.

Hu, C. C., Chang, S. S., & Liang, N. Y. (2018). Fabrication of antistatic fibers with core/sheath and segmented-pie configurations. *Journal of Industrial Textiles, 47*(5), 569–586.

Oerlikon. http://www.oerlikon.com

Rupp, J., & Yonenaga, A. (2000). Microfasern – Das neue Image der Chemiefaser. *ITB International Textile Bulletin, 46*(4), 12–24.

Sisson, W. (1948a). Composite thermoplastic fibers, US Patent 2,439,815.

Sisson, W. (1948b). Method of manufacturing artificial filaments, US Patent 2,443,711.

Sossna GmbH, Dorsten. http://www.sossna.de

Swicofil AG Textile Services. http://www.swicofil.com/belltron.html

Yu, J. P. (1998). Producing bicomponent fibers: Spinnerette design is crucial. *International Fiber Journal, 13*(3), 49–53.

Zhao, B., Qian, Y., Qian, X., Fan, J., Liu, F., & Duo, Y. (2018). Preparation and properties of split microfiber synthetic leather. *Journal of Engineered Fibers and Fabrics, 13*(2), 15–21.

27 Fine Fibers

Figure 27.1 shows the classification of fibers as a function of the fiber diameter or the cross-sectional area. The fineness of fibers produced by conventional methods ends at approx. 10 μm, which corresponds roughly to 1 dtex.

Ultrafine fibers are often also referred to as nanofibers. The Table 27.1 shows an overview of typical fiber diameters.

Special processes are required to produce very fine fibers. An overview gives Fig. 27.2. The ranges given are typical, but not limiting. Thus, a fineness of 10 nm to 5 μm is also achievable for electrospun fibers. Carbon nanotubes are discussed in another chapter of this book.

The most economically important processes are:

- Meltblown process (for the production of nonwovens),
- electrospinning,
- bicomponent spinning.

Flash spinning and centrifugal spinning are of minor importance.

27.1 Electrospinning

Electrospinning can be used to produce particularly fine fibers with diameters of a few nanometers. Common polymers used to generate electrospun fibers include PE, PP, PCL, PLA, PLLA, PVDF, PA6, and PET. Ceramic fibers can also be electrospun (Chapman et al., 2019).

Electrospinning from solution is used by almost all research groups worldwide, because it is relatively simple to implement without complex plant technology and many polymers can be spun with it. Melt electrospinning (Sect. 27.2) is much more complex, but allows much higher throughputs. However, it can only be used for thermoplastic polymers, which limits the choice of materials.

27.1.1 History

The electrostatic attraction of a fluid was first described by William Gilbert, court physician to Queen Elizabeth I in 1600 (Gilbert, 1600). Georg Matthias Bose (1710–1761) described the formation of aerosols on the surface of electrically charged droplets in 1745. John Strutt (1842–1919) determined the charge required to deform droplets in 1885, and John F. Cooley and William J. Morton received a patent for an apparatus for electrospinning (Cooley, 1900; Morton, 1902). Subsequently, patents were granted for the production of cellulosic fibers (Hagiwaba et al., 1929) and for synthetic man-made fibers (Formhals, 1934). In 1938 Nathalie D. Rozenblum and Igor V. Petryanov-Sokolov produced electrospun fibers from which they made filters. In the 1960s, Geoffrey I. Taylor developed the theoretical basis of electrospinning (Taylor, 1964, 1966, 1969). The Taylor cone is also named after him. The fundamental process

D. Veit, *Fibers*, https://doi.org/10.1007/978-3-031-15309-9_27

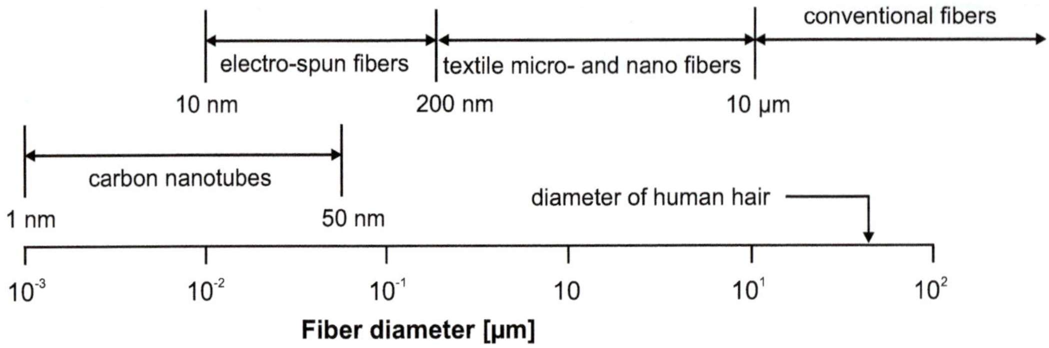

Fig. 27.1 Classification of fibers acc. to their fineness

description by Doshi and Reneker (1995) marked the beginning of a renaissance of electrospinning in the context of the general hype about nanotechnology. They showed that many organic polymers could be solution spun into nanofibers using electrospinning. Since then, the number of publications has literally exploded to several thousand per year (Fig. 27.3).

In particular Tucker et al. (2012), Teo and Ramakrishna (2006) and Amna et al. (2020) provide a very good overview on the state-of the-art and historical development of electrospinning. Hutmacher and Dalton (2011) report in detail on the state-of-the-art of melt electrospinning of various polymers.

27.1.2 Sense and Purpose

For many applications, e.g. for biomedical uses, fibers with a cross-section in the range of 10–100 nm are required (Fig. 27.4). This is the only way to ensure that, for example in medical technology, cells grow on and between them in a controlled manner. Electrospun fibers in nonwoven form are also used for filter applications (air, liquids) and for battery separators. To increase the throughput, several spinnerets can be arranged in parallel. However, due to the relatively large charge, the polymer streams may influence each other, resulting in uneven fibers (Theron et al., 2005).

In principle, yarns can also be produced from nanofibers. However, this makes little sense because the yarn diameters are in the range of "normal" fibers and the essential advantage of electrospun fibers, their small diameter, is lost. Therefore, no application has been successfully marketed using electrospun yarns. Also, productivity is very poor due to the low throughput and low process speed (up to 30 m/min), which inevitably increases costs considerably.

Table 27.1 Typical sizes of fibers based on (Hills, 2021)

Carbon atom	0.16 nm
Nanofiber	50–500 nm
Melt blown fiber	2000–5000 nm
1 dtex fiber (PET)	10,000 nm
Human hair	20,000–80,000 nm

27.1.3 Principle

If a conductive liquid is electrically charged, an electric field emanates from it which points to a point with a lower electric potential. The field normal to the surface of the liquid exerts an electrostatic force on the liquid which is opposite to its surface tension. As soon as the induced charge on the surface of the liquid is greater than the surface tension, the liquid is finely atomized. The droplets thus produced, continuing to carry a certain charge, tend towards a pole whose electrical potential is lower. This phenomenon was described as early as 1745 by G.M. Bose (Keithley, 1999). The charge that a liquid can sustain before the surface tension is overcome was determined by Lord Rayleigh in 1882. This

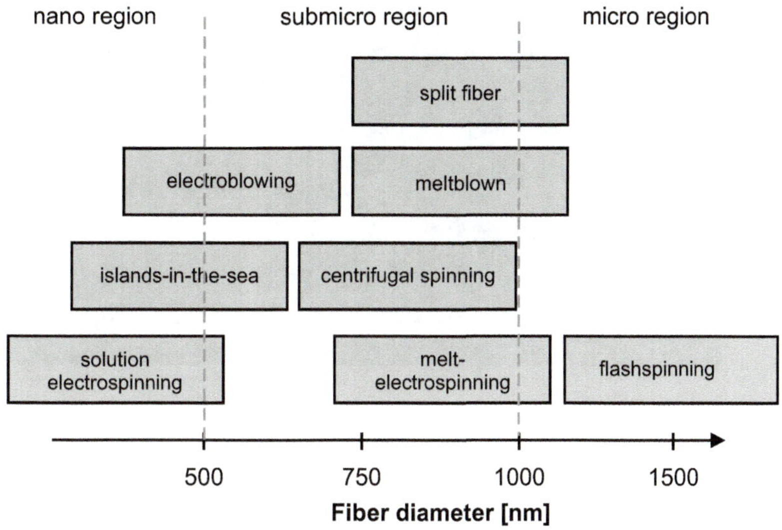

Fig. 27.2 Classification of the processes for the production of ultra-fine fibers acc. to (Hacker, 2014)

process of creating individual droplets in an electric field is commonly known as electro spraying and is used in various applications such as powder coating of metallic surfaces. When polymer solutions or melts are electrically charged instead of liquids, a coherent jet is formed instead

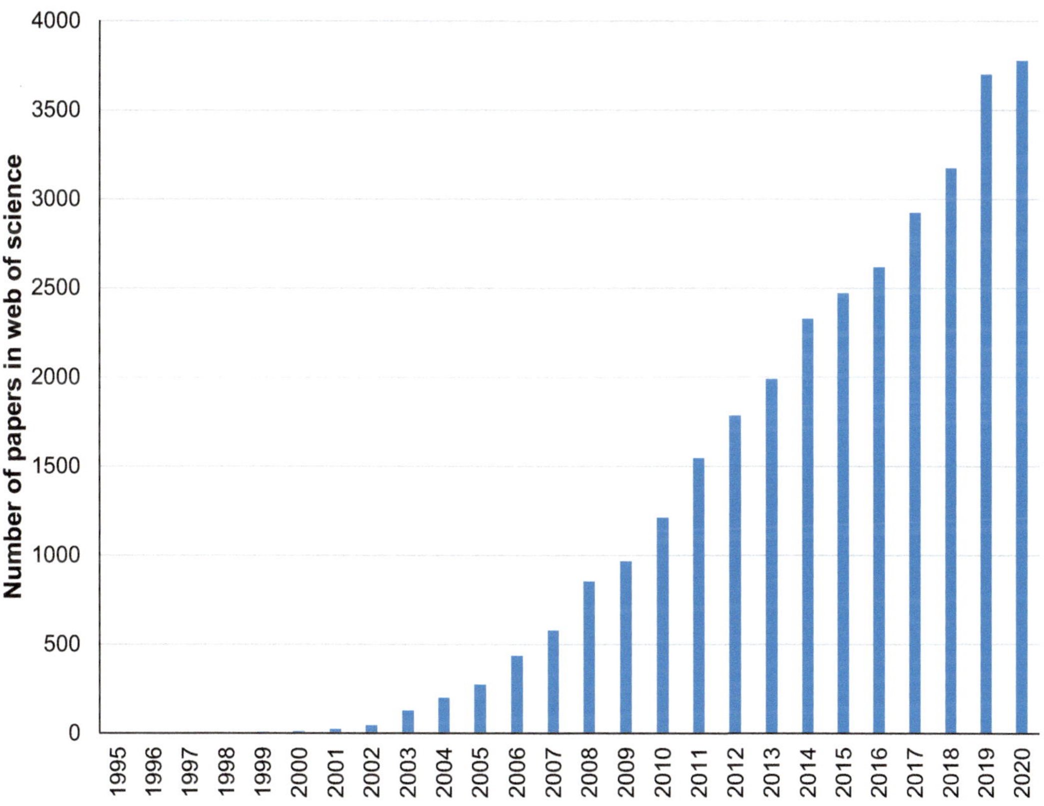

Fig. 27.3 Number of publications on electrospinning by Web of Science

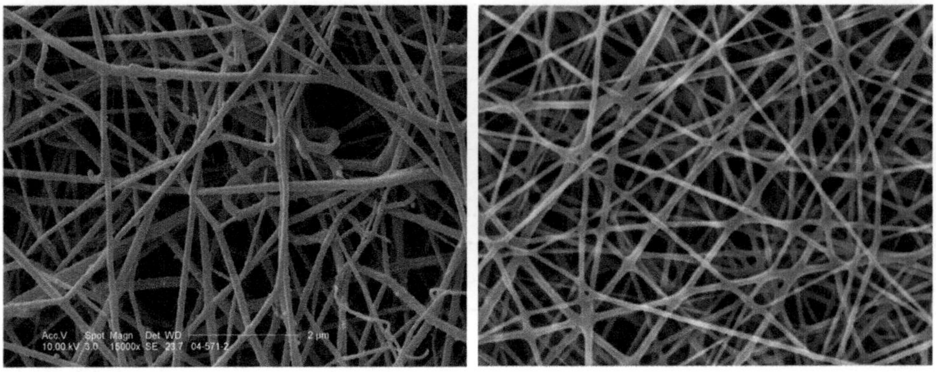

Fig. 27.4 Nanofiber nonwoven (left: extra-cellular matrix in humans; right: electrospun nonwoven; image width: 7 µm)

of individual drops due to the long-chain polymer molecules. This process is referred to as electrospinning.

In Fig. 27.5, the electrospinning process based on the solution spinning method is shown schematically. A polymer solution is delivered (mostly from a syringe) at a low delivery rate so that a drop forms at the needle's tip. The metal needle is connected to a DC high voltage source. It is located over a grounded or negatively charged collector, called the target. Once the high voltage is applied to the needle, the droplet becomes electrically charged and the charge spreads evenly over the surface. Because equal charges repel each other and Coulomb forces are exerted on the charges by the electric field, the droplet deforms and forms the so-called Taylor cone at its tip. As long as the electrostatic forces on the solution droplet are smaller than its surface tension, the Taylor cone forms an angle of 49.3°, measured between the cone center line and flank. As soon as a certain threshold voltage is overcome, which can vary depending on the polymer solution and process parameters, a jet emerges from the tip of the cone. This jet is accelerated towards the opposite pole due to the attraction of

Fig. 27.5 Principle of electrospinning

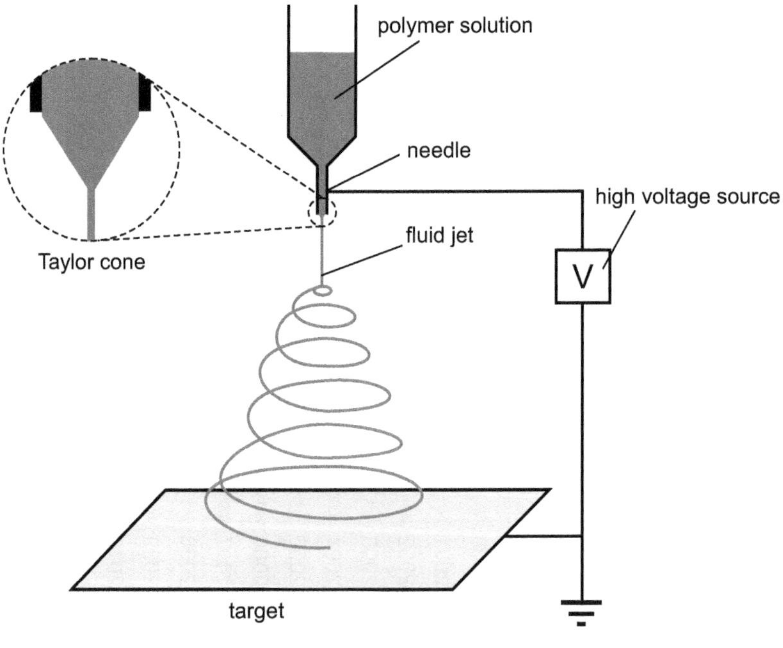

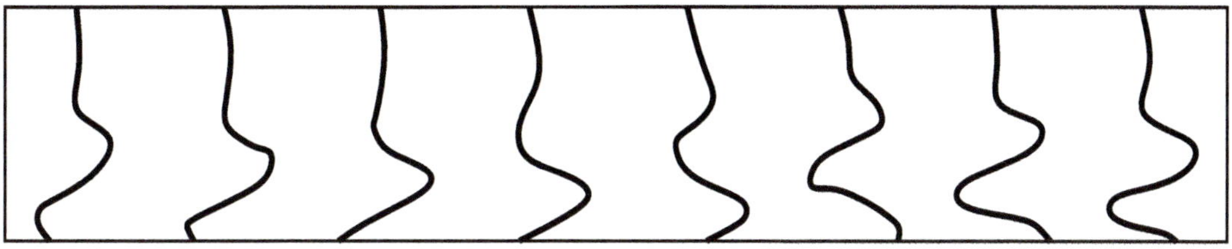

Fig. 27.6 Onset of a first order "bending instability" acc. to (Koombhongse et al., 2001)

different polarities, where it is deposited in a disordered form. This results in the formation of a tangled web. The target is usually a solid, but can also consist of a spinning bath, which also serves as a precipitation bath when solution electrospinning.

The jet of liquid emerging from the cone is accelerated for a few centimeters on a straight-line path, after which it apparently changes to a cone shape. The cone is perceived because the jet is moving so fast that it can no longer be detected as a single jet by the naked eye.

Because of this cone shape, it was assumed in early studies of electrospinning that the beam would branch into many thin beams and in this way the final fiber fineness would be achieved. However, high-speed imaging shows that the cone shape is not caused by rapidly bouncing fibers and their branching, but represents the enveloping curve of a single rapidly moving fiber.

Due to the mutual repulsion of the charges on the beam and the attraction from the target, the beam is accelerated and stretched. The turns of the beam become larger and larger as the beam is stretched. Due to conservation of mass, the diameter of the beam decreases as the length increases. The Fig. 27.6 shows the redrawing of high-speed images, from left to right, of the chronological sequence of the formation of the spiral. The area in which the beam is deflected from the straight line is called bending instability or whipping instability.

Bending instabilities are mainly caused by the interaction of the external electric field with the charged fiber. In addition to the external influences, the mutual repulsion of the electric charges on the beam also plays a non-negligible role. Instabilities occur in several orders, both axisymmetric and non-axisymmetric with respect to the beam. The type of instability shown in Fig. 27.6, where the straight beam is diverted into a spiral shape, is of the first order and non-axisymmetric. The second-order non-axisymmetric instability puts the ray into further rotation about the trajectory, as shown in Fig. 27.7 indicated. High-speed imaging can be used to detect this second-order twist.

Axisymmetric instabilities are mainly due to charge differences within the solution beam, which interact with each other due to Coulomb forces (Fig. 27.8). This causes the beam to oscillate. The influence of the external electric field plays an important role in the charge arrangement

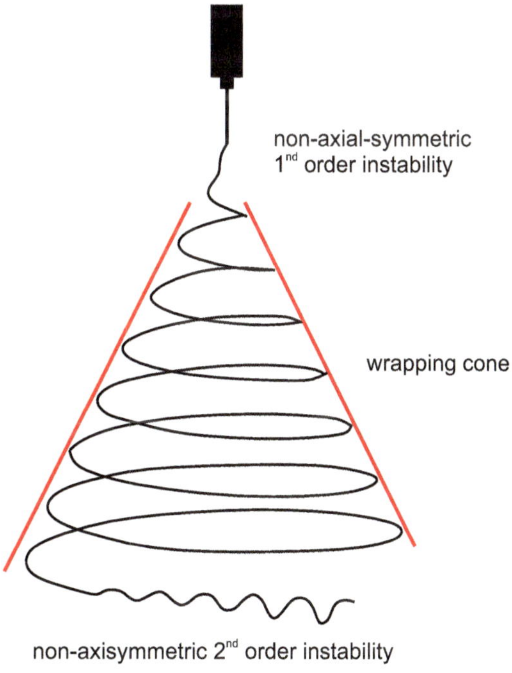

Fig. 27.7 Non-axisymmetric bending instabilities of first and second order

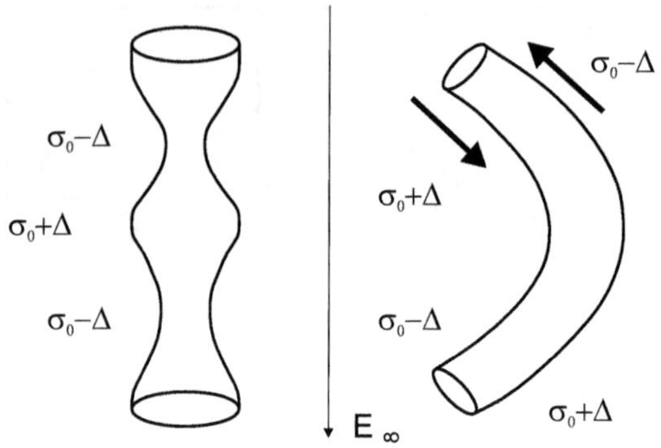

Fig. 27.8 Axisymmetric (left) and non-axisymmetric (right) instabilities (σ_0 stands for the average charge density on the surface) acc. to (Shin et al., 2001)

in the beam. The following figure shows the formation of the axisymmetric and the non-axisymmetric instabilities.

There are different approaches to calculate and simulate the instabilities. One way is to consider the fluid jet as a string of many individual viscoelastic elements with a charged droplet at each end. The viscoelastic properties are modeled by a spring-damper system. Material properties and environmental influences are also considered in the calculation. With these calculations, the path that the jet travels can be calculated (Yarin et al., 2001).

Another approach is to consider the beam as a long, thin object that exhibits the various properties of the material. Due to the mutual repulsion of the surface charges with each other, the beam is stretched. In particular, in the regions of bending instabilities, where the beam leaves its initial path and is accelerated in spiral paths, it undergoes a very strong stretching. Thus, the beam diameter can be tapered by a factor of approximately 10^6 on the path from the syringe needle to the target. The choice of polymer and solvent also has a direct effect on the acceleration and hence elongation of the beam. For example, for solutions containing polyethylene (PEG) and polyacrylonitrile (PAN), the main charge carrier is the polymer itself and the charges remain in the fiber until after deposition on the target. In contrast, for solutions with polycaprolactone (PCL), the solvent (e.g. chloroform) is the charge carrier. It evaporates during the spinning process. Therefore, the remaining charges in the beam are lower and it elongates less. By adding methanol, the charge distribution can be improved in PCL. Shin et al. (2001) have developed a model to calculate the final diameter of the fiber. In this calculation model, some properties of the polymer solution are assumed to be negligible. For example, the liquid is considered as a Newtonian fluid and the elastic properties resulting from the drying of the jet are ignored. Moreover, the evaporation of the solvent, which causes an additional reduction of the volume and therefore of the final diameter, is neglected. The achievable final diameter can then be calculated as a good approximation acc. to:

$$h_t = \left(\gamma \cdot \overline{\varepsilon} \cdot \frac{Q^2}{I^2} \cdot \frac{2}{\pi \cdot (2 \cdot \text{In}\chi\text{-}3)} \right)^{1/3}$$

Where γ is the surface tension, ε the average dielectric constant, χ is the dimensionless wavelength of the non-axisymmetric instability, Q is the volumetric and I is the electric current.

27.1.4 Needleless Spinning

In the needle electrospinning process, clogging of the orifice may occur and consequently the polymer flow may break off. To avoid this, a needleless spinning process was developed by Yarin and Zussman (2004). First, a magnetic fluid based on silicone oil is generated and an electrode is immersed in the spinning bath. A toothed grounded sheet is placed above the spinning solution. After adding the polymer solution to the spinning bath, a magnetic field is generated and numerous polymer jets are formed. These are attracted to the toothed sheet and fibers are formed which are deposited on the sheet (Fig. 27.9).

An alternative discontinuous process was developed by Sun et al. (2006). Other processes using a spiral, cylinder and disc to produce PAN fibers were published by Niu et al. (2012). They achieved a throughput of up to 23 g/h using this method. Ali et al. (2017) produced up to 6 g of PVA fibers per hour using a spoked wheel.

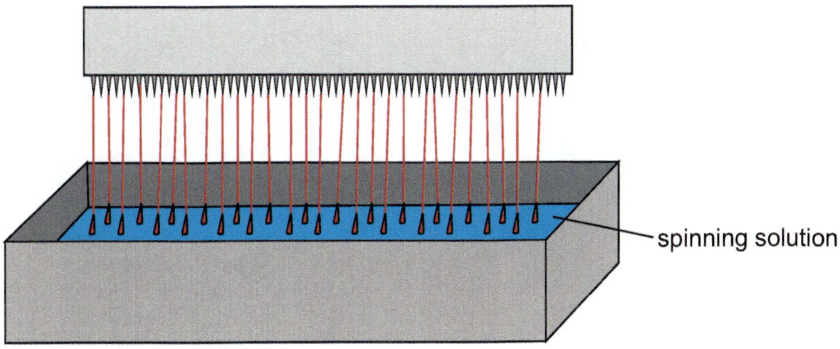

Fig. 27.9 Principle of needleless electrospinning acc. to (Teo & Ramakrishna, 2006)

The Czech company ELMARCO offers a needleless electrospinning machine with up to four spinning systems (Fig. 27.10).

According to the manufacturer, up to 38 m^2/min (20 million m^2/year) of material can be coated with PA 6 fine fibers (Ø 150 nm) over a width of up to 1.6 m per spinning position. The basis weight is then 0.03 g/m^2 (Fig. 27.11).

27.1.5 Parameters Influencing the Spinning Result

The most important factors influencing the properties of the spun fibers in solution electrospinning are the molecular weight of the polymer, the concentration of the polymer in the spinning solution, the solvent itself, the electric field and the ambient conditions (especially humidity).

27.1.5.1 Polymer Molecular Weight

The viscosity and thus the molecular weight of the polymer used have a great influence on the spinnability of the solution. If the viscosity is too low, so-called beads form in the fibers. In this case, a string of droplets forms on the target and not a straight, coherent fiber. This phenomenon is caused by axial instabilities of the polymer solution jet. This can be counteracted by increasing the polymer concentration in the solution or by increasing the molecular weight of the polymer. With a higher molecular weight, the viscosity of the solution increases. This is particularly important for the formation of fibers, as the fibers achieve their cohesion by the molecular chains

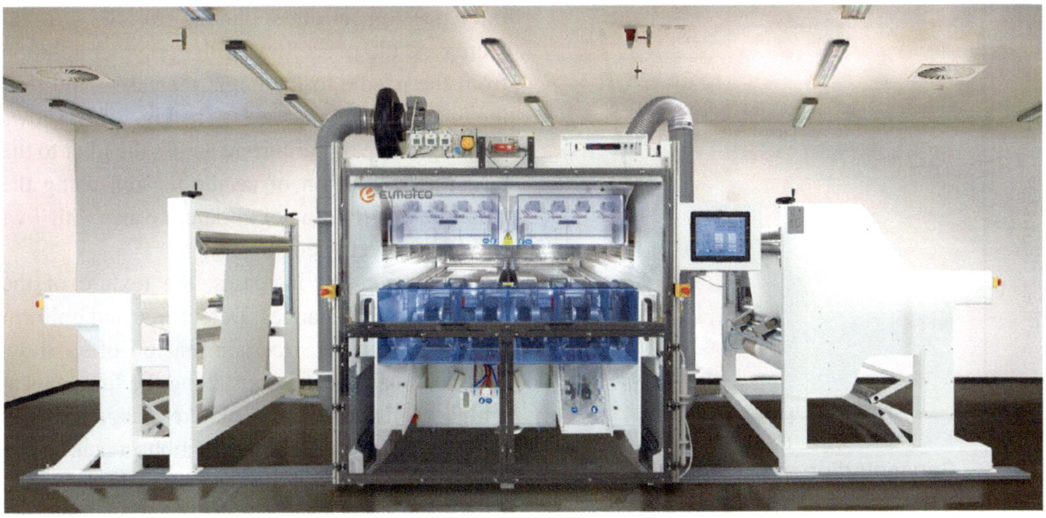

Fig. 27.10 Nanospider™ NS 8S1600U electrospinning machine (ELMARCO, n.d.)

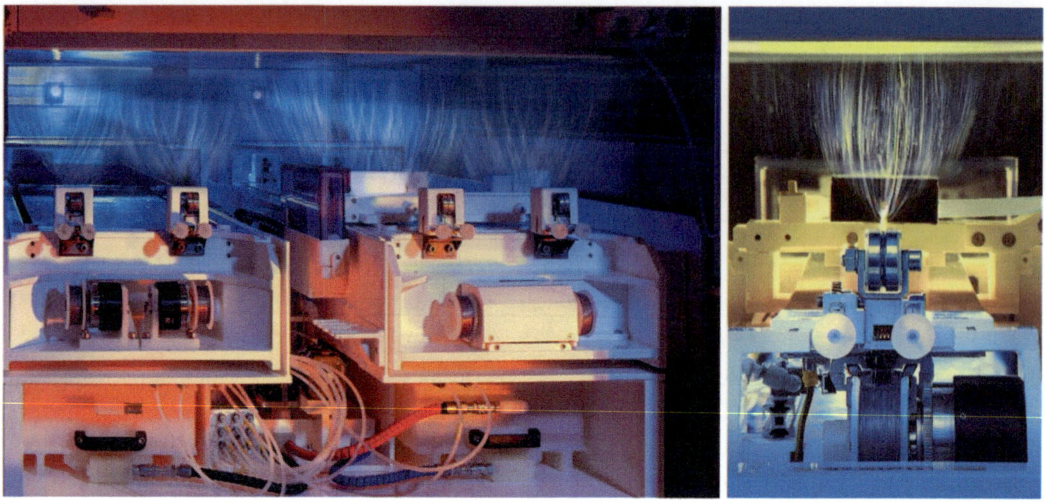

Fig. 27.11 Details of the spinning process (ELMARCO)

entwining with each other during acceleration in the electric field and thus adhering to each other.

27.1.5.2 Concentration of the Spinning Solution

Figure 27.12 shows the dependency of the fiber fineness on the concentration of the spinning solution using the example of polyacrylonitrile with dimethylformamide (DMF) as solvent. Below a certain minimum concentration, spinning of fibers is not possible. Above a certain value, the fibers become coarse on the one hand and very uneven in diameter on the other. This behavior is observed for all polymers, only the corresponding limit values of the process window differ.

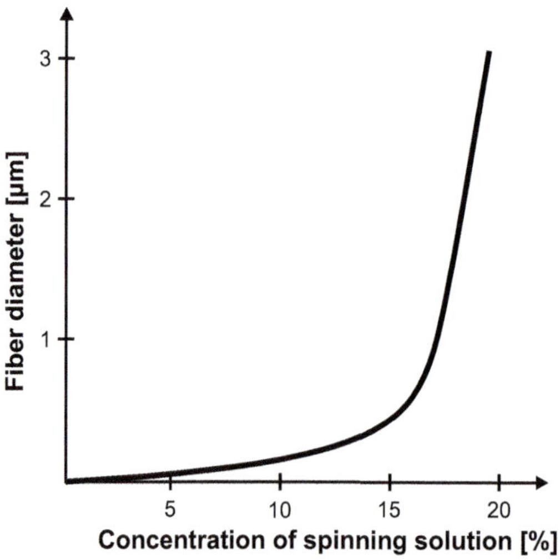

Fig. 27.12 Dependency of the fiber fineness on the concentration of the polymer solution

27.1.5.3 Electric Field

Even though it is very small, an electric current still flows from the needle through the fiber to the target. The greater the applied voltage, the greater this current. The current measured here is on the order of a few nanoamperes. Although it is so small, the current strength has a non-negligible influence on electrospinning. As the current increases, the mass flow of the polymer solution from the needle to the target increases, and thus the fiber diameter increases. The Fig. 27.13 shows the magnitude of the current across the fiber to the target as a function of needle tension using the example of a solution of ultrapure water with 7 wt % PEO (polyethylene oxide).

As the tension increases, the texture of the spun fibers also changes. Defect-free and uniform fibers produced at a certain tension often show spherical mass fluctuations along the fiber axis at a higher tension, which look like strings of beads ("bead effect"). These fluctuations occur more frequently when the change in the slope of the curve in Fig. 27.13 is approximately zero. Since more material is spun at higher current, the

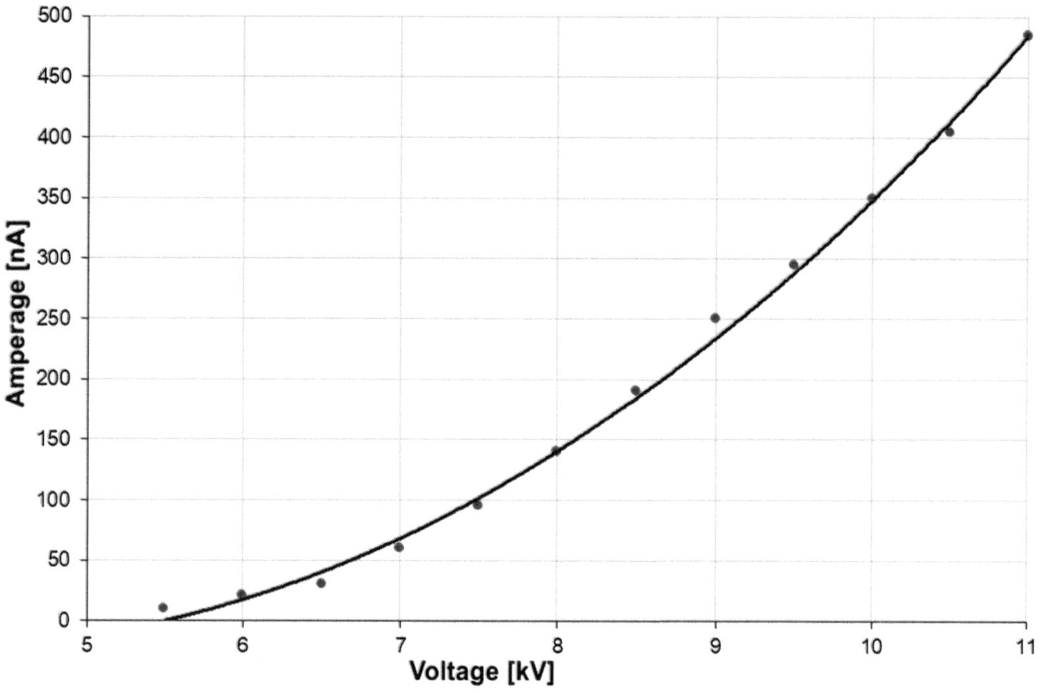

Fig. 27.13 Relationship between voltage applied to the needle and the current flowing across the beam to the target acc. to (Deitzel et al., 2001)

formation of beads can be counteracted by increasing the feed rate so that a uniform mass flow from the needle to the target flows.

In general, the speed of the polymer fiber in the electric field increases at higher voltages because the higher charge difference between the polymer beam and the target exerts a greater attractive force on the fiber. However, since more polymer solution is carried by the higher voltage, the applied voltage must be distributed over more material and the voltage density decreases. This in turn has a negative effect on acceleration and the fibers become thicker. In order to produce the thinnest possible fibers, the conveying rate must be matched to the tension so that the accelerating effect of the tension increase is not cancelled out by the influence of a lower tension density in the jet.

27.1.5.4 Humidity

Humidity is one of the decisive parameters in electrospinning. Some polymers (e.g. silk fibroin) can only be spun out in a specific humidity range ($< 30\%$). Other polymers, e.g. PCL, react with pore formation to a change in humidity. In extreme cases, these interconnect to form a coherent network.

Humidity also plays a major role in the alignment of fibers by electrostatic forces. Only electrically charged fibers experience an attraction by the target and can be aligned. Humidity in the air can cause the fibers to discharge. In a high relative humidity, the fibers therefore align less well than in a dry environment.

27.1.6 Influence on Fiber Deposition

In solution electrospinning on a flat target, the fibers are deposited as a disordered tangled web on the target. However, for many applications in which electrospinning could be used, for example the production of nanofiber-reinforced components made of plastics, glass or concrete, directed, defined fibers are required.

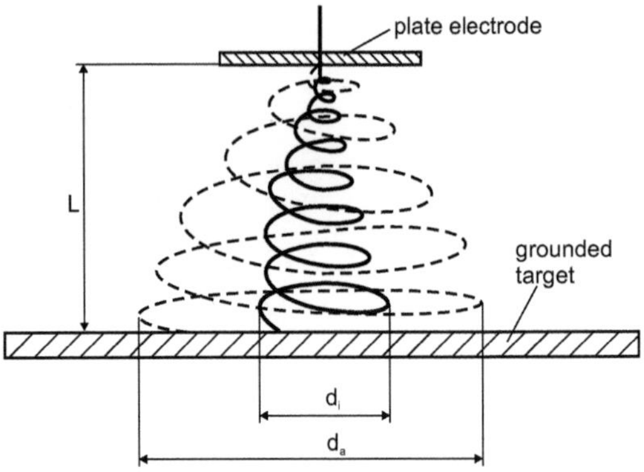

Fig. 27.14 Reduction of the propagation area on the target acc. to (Bunyan et al., 2006)

If spun directly onto a flat target, the fibers are deposited isotropically in a relatively large area on the target. To limit the propagation area on the target, a round metal plate attached to the needle exit can be used. Between the electrode and the target, the electric field propagates in a straight line. The dashed line in Fig. 27.14 shows the path that the beam would travel without the plate electrode.

The path that the beam describes with the device is then more narrowly constricted than without a plate electrode. Acc.ly, the deposition area (d_i) on the target is also smaller than without an electrode (d_a). The larger the plate diameter becomes, the smaller d_i. Doubling the electrode diameter from, for example, 0.01 L to 0.02 L (L is the distance between target and electrode) leads to a target diameter d_i that is four times smaller.

Another way of narrowing the fiber path by influencing the external electric field is to influence the path of the electric field lines by placing rings of the same polarity as that of the fiber, but with lower voltage, around the path of the beam. This type of cage, through which the beam passes, has the same effect as a plate electrode. The construction of this multiple-field electrospinning is illustrated in Fig. 27.15.

27.1.7 Fiber Orientation

The fiber orientation is of crucial importance for the subsequent use of the electrospun fibers. On the one hand, it influences the mechanical properties, but, on the other hand, it can also determine cell growth when the fibers are used in implants.

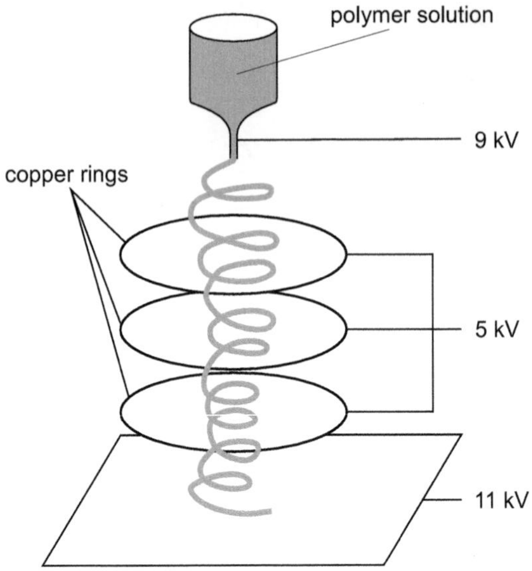

Fig. 27.15 Reduction of the propagation area on the target with rings of the same polarity around the path of the solution beam acc. to (Wienecke, 2007)

27.1.7.1 Rotating Target

The diameter of the beam emerging from the Taylor cone can be reduced to one millionth of its initial diameter from the tip of the cone to the deposition on the target. Due to the volume conservation of the material, up to one kilometer of fiber can theoretically be spun from one millimeter of initial beam length. To do this, the fiber must be accelerated very strongly. To prevent these fibers from being deposited in a disordered manner on the target, they must be accelerated at a take-off speed v_f of at least.

$$v_f = 3.5–10^4 - v_0$$

where v_0 represents the initial velocity at the cone. These pull-off velocities are achievable, if a sufficiently large rotating collector is used (e.g. with a diameter of 20 cm).

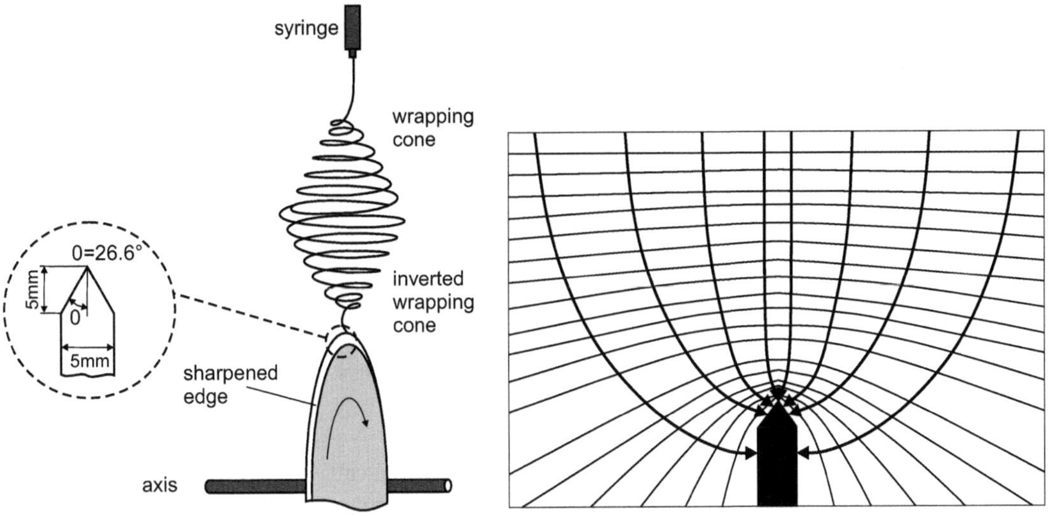

Fig. 27.16 Schematic structure of the rotating disk collector and field lines acc. to (Theron et al., 2001)

27.1.7.2 Rotating Sharpened Disc

Another approach is to use a rotating target as a disk collector, mounted vertically under the needle. The edge of this rotating disc is sharpened (Fig. 27.16).

The sharpening of the edge of the disc is based on the same principle as a lightning conductor. The electric field lines converge at the tapered tip and are closer together than with a non-sharpened disc edge. The closer the field lines are to each other, the higher the electric field force and correspondingly the attraction to the fibers.

27.1.7.3 Rotating Disc with Directional Polymer Application

Another way to use a rotating target is to inject the solution from the side (Fig. 27.17). The modified scanning tip allows a very small distance between the tip and the target. In the previously discussed variants of electrospinning, the solution was always delivered continuously from a syringe and spun out at the tip of the needle. Here, the spinning tip is formed from silicone and is immersed in the polymer solution. When it is pulled out, a droplet remains attached to the tip. This drop can now be spun out in the electric field as soon as a voltage is applied. Since the drop adheres relatively firmly to the tip, the tip can be brought to within a few centimeters of the rotating target before a beam is formed. This method has been used very little to date and is discontinuous due to its principle (Wienecke, 2007).

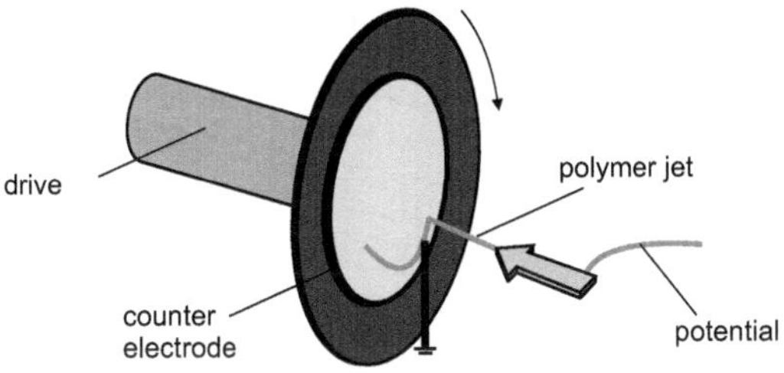

Fig. 27.17 Setup of the electrospinning experiment with special tip shape acc. to (Kameoka et al., 2003)

27.1.7.4 Rotating Collector

A grounded rotating roller as a target provides a relatively directional pull-off of the fibers. However, since the entire roller is grounded, there are many fibers in this target shape that do not have a preferred direction. The idea that a sharp edge leads to converging field lines and therefore a higher field force, as explained above, suggests that a needle tip should exert the greatest attractive force on the fiber. In order for the strong attraction effect of the needle to result in a directed fiber, a rotating collector was placed between the negatively charged needle and the spinning tip in experiments. The collector can be quickly moved axially by means of a second

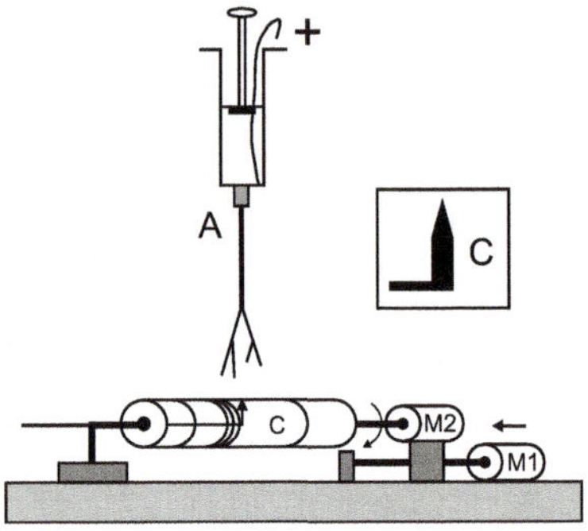

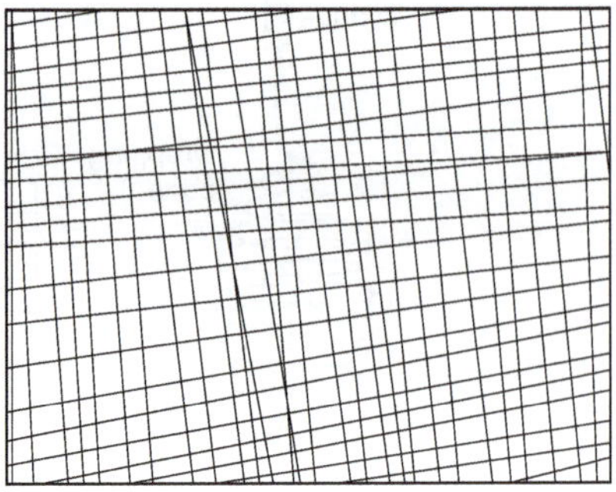

Fig. 27.18 Electrospinning plant with rotating drum collector (left) acc. to (Sundaray et al., 2004) and schematic of the fiber deposit

motor, thus enabling the fibers to be aligned perpendicularly to each other. The Fig. 27.18 diagram on the left shows the construction of the electrospinning system with rotating drum collector. Motor M2 rotates the drum, M1 moves it axially. The needle (C), to which negative voltage is applied, is positioned directly below the spinning tip (A). The splitting of the polymer beam is clearly visible. On the right is an electron micrograph of the fiber array produced on the collector. The vertically running fibers were created by a rapid axial displacement of the drum, the horizontal ones by the rotation (Wienecke, 2007).

27.1.7.5 Rotating Cable Drum

In another approach, using a rotating drum, the drum is grounded directly. The drum consists of two non-conductive discs between which, spaced at regular intervals around the circumference, copper cables are stretched. The fibers stretch between the cables running parallel along the axis of the drum. The side view schematically shows the copper cables arranged around the circumference (Fig. 27.19).

On the one hand, the forces exerted by the external electric field accelerate the fiber from the syringe outlet to the target. There, the fiber attaches itself to the copper cable closest to the tip of the needle. Due to rotation, the fiber extends from one cable to the next. Second, an electrostatic force is generated by charges that have remained on the fibers just spun and have not yet dissipated. As a result, fibers are attracted to the grounded cables and they align themselves in such a way that they are stretched across the distance between the cables in the shortest possible way (Wienecke, 2007).

27.1.7.6 Fiber Alignment Via an Interstitial Space

The fiber alignment of the drum collector is also achieved with a split target. The fibers spun in this process must not be thinner than ≈ 150 nm, otherwise they cannot support their own weight and will break. The Fig. 27.20 shows the structure of the spinning device and the magnetic field between the target halves in principle.

The alignment of the fibers is achieved by at least two different electrostatic forces. On the one hand, the charged fibers are attracted to the target. The split collector causes a change in the electric field so that the fiber is deposited on either one or

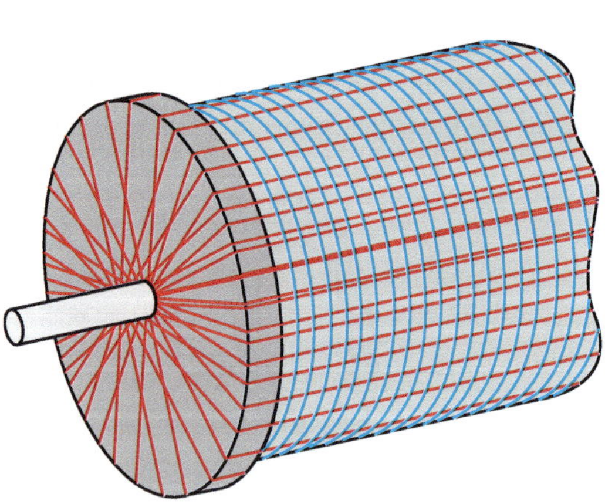

Fig. 27.19 Rotating drum acc. to (Katta et al., 2004)

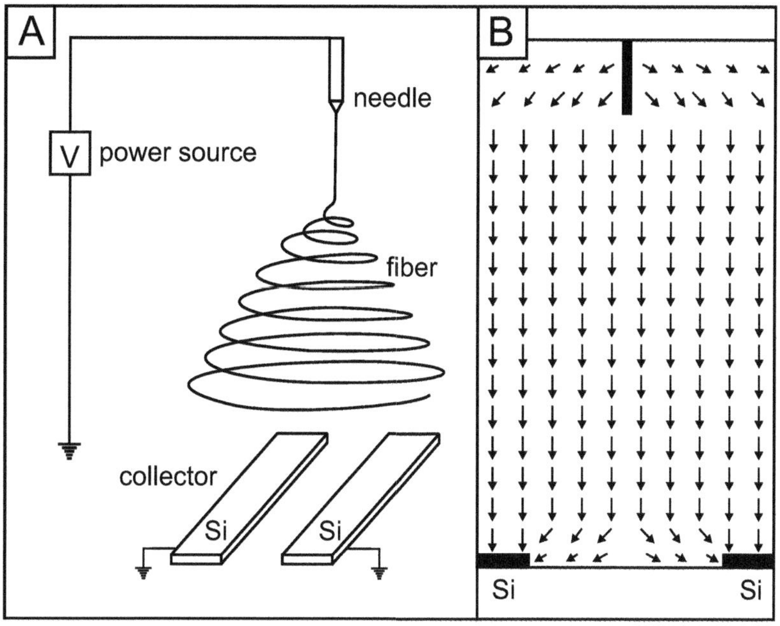

Fig. 27.20 (**a**) Schematic structure with split target. (**b**) Simulated electric field between positively charged needle tip and grounded target parts acc. to (Li et al., 2003)

the other half of the target. Due to the bending instabilities, the fiber orientation across the gap is random. Secondly, as soon as the fibers come into contact with the grounded opposite pole, negative charges, called image-charges, are created, which exert an attraction on the positively charged fibers. The Fig. 27.21 diagram on the left shows the forces on the fibers, where F_1 is the force of the electric field and F_2 is the attraction of the image charges on the target pieces to the positive charge carriers. On the right is an SEM image of the fiber orientation between the target halves.

With this method it is possible to spin several fiber layers with different orientations on top of each other. If several parallel target pairs are arranged at different angles to each other, more than just two orientations of the fibers can also be achieved. During spinning, the parallel pairs are alternately grounded and the fibers align between each grounded target pair for that period of time. Figure 27.22 (b) shows the fiber arrangement centered on the target when the electrode pairs 1/3 and 2/4 in (a) are alternately grounded. (d) shows the fiber orientation between the

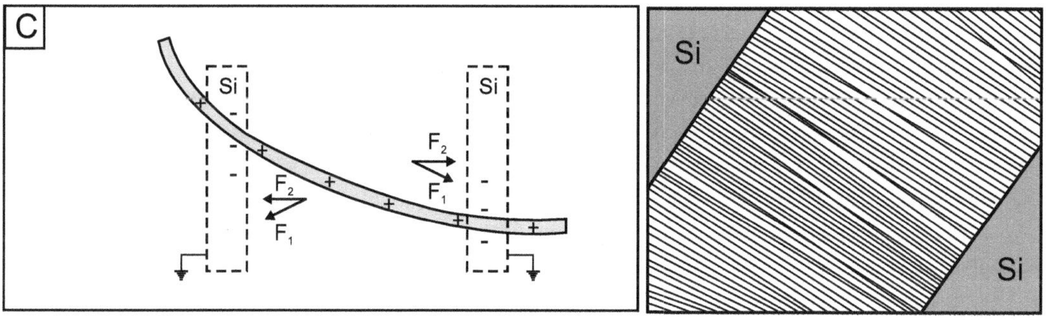

Fig. 27.21 Electrostatic forces on the charged fiber. F1 is the resulting after (Li et al., 2004) and fiber deposition (schematic)

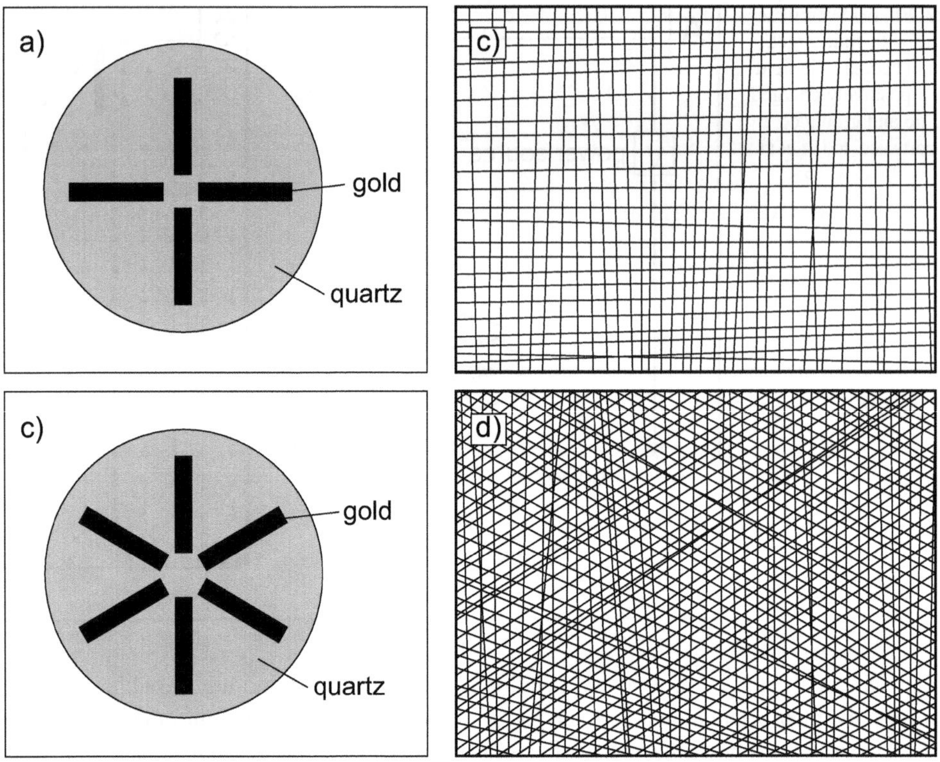

Fig. 27.22 Alignment of the fibers between the poles acc. to (Li et al., 2004)

electrodes when electrode pairs 1/4, 2/5, and 3/6 in (c) are alternately grounded (Wienecke, 2007).

If all poles are grounded at the same time, the fibers also tend to align mainly between the parallel poles, but there are more misalignments.

27.1.7.7 Influence of the Distance Between the Target Halves

The density of the fibers stretched between the target halves and their orientation are influenced by the distance between the target halves. When the charged beam is accelerated towards the target halves, their voltage difference causes the beam to be attracted. If the halves are close together, the beam is more likely to skip from one target half to the other, thus laying a fiber across the gap, than if the halves are far apart. If the distance between the halves is large, the beam is initially attracted to one half and thus the path of the beam is deflected a little. The beam therefore settles there. Only when the spiral beam path gets closer to the other target half, the fiber is spun to there and so on. This happens less and less frequently as the distance increases, and the density of the fibers extending across the gap decreases. Because the dead weight of the free-hanging fibers becomes greater as the space between the target halves increases, the fibers must become thicker, and thus more tensile, to avoid tearing. The greater the distance, the higher the tension of the polymer solution must be selected in order to produce optimum spinning results (Wienecke, 2007).

27.1.7.8 Influence of a Negative Voltage Applied to the Target

To increase the voltage difference between the needle and the target, and thus achieve higher attractive forces on the fiber, a negative voltage can be applied to the Taget. Experiments confirm that fiber alignment is better at higher amounts of negative voltage than would be the case with only grounded targets. However, if the amount of voltage applied during electro-airgap spinning is

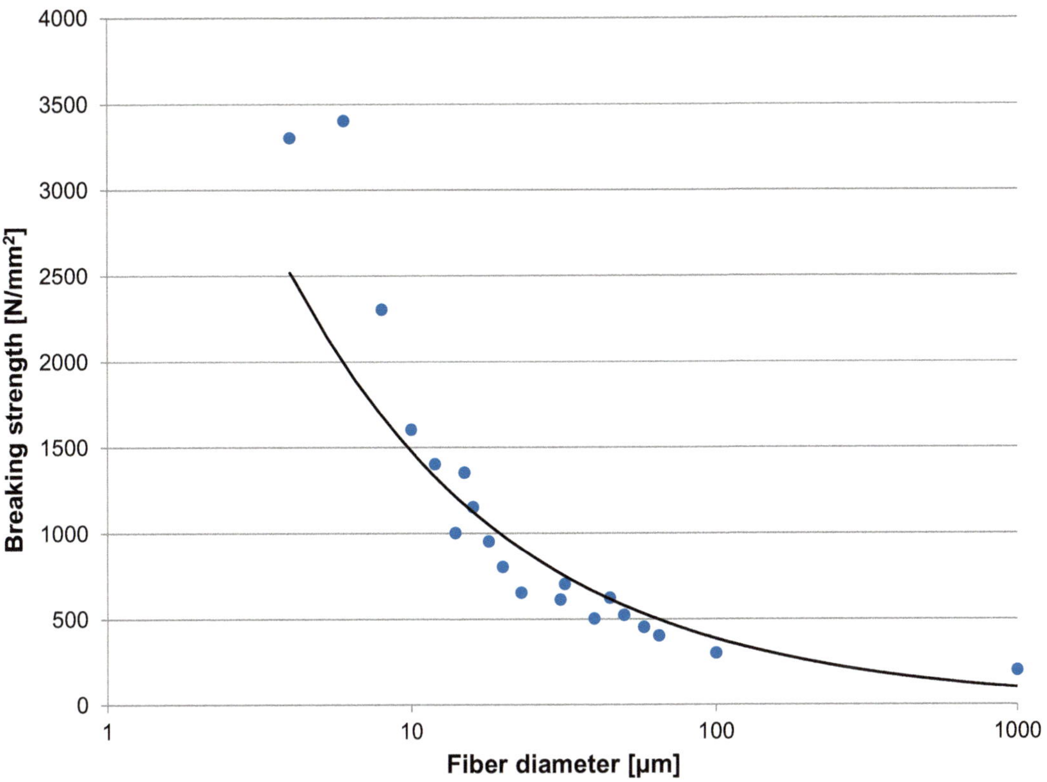

Fig. 27.23 Dependence of fiber strength on fiber fineness (Griffith effect)

increased above a threshold value, which depends, among other things, on the distance between the target halves, fiber alignment no longer occurs. Instead, the fibers are repelled. One possible explanation for this phenomenon is the ionization of the air due to the high voltages between the halves. The positively charged fibers attract the negative ions and the fibers experience a negative charge. Thus, the fibers are repelled by the negative targets (Wienecke, 2007).

27.1.8 Fiber Strength

The fiber strength depends directly on the fiber fineness, as the Fig. 27.23 shows. The coarser the fiber, the lower its specific strength. Acc. to Alan A. Griffith, this is because the strength of a fiber is essentially determined by microscopic defects in its internal structure. The coarser a fiber is, the greater the likelihood of internal defects and therefore the lower its specific strength. This phenomenon is also known as Griffith's criterion and was first published as early as 1920. It is also typical for many other fiber materials.

27.2 Melt Electrospinning

Most of the fiber materials, although solution electrospun to date, are thermoplastic. Therefore, they can also be spun out of the melt - analogous to solution electrospinning. For this purpose, the polymer is heated to melting temperature in the spinneret. This is done either with hot air (simple), hot fluid (complex) or via three laser beams crossing each other at the exit of the polymer from the nozzle, which ensures high temperature accuracy. Although the throughput can be increased many times over by melt electrospinning, it is still - depending on the spinning apparatus - a few μl/min or even less. Hacker et al. (2009) already

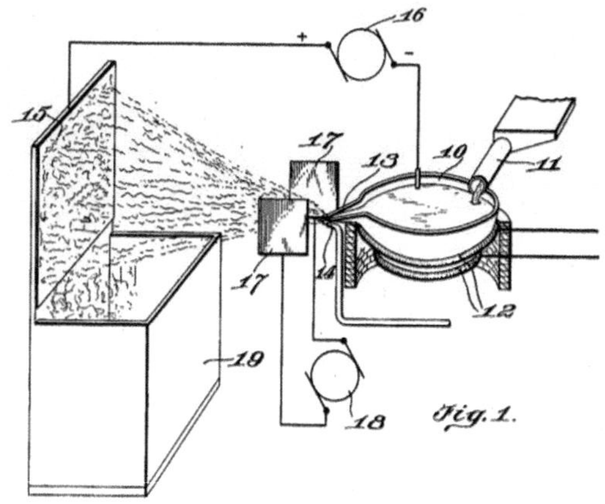

Fig. 27.24 Melt electrospinning process (Norton, 1936)

produced appreciable amounts of fibers with a multi-hole spinneret and simulated the melt flow through the capillaries of the spinneret. The diameters of the fibers produced are typically 1–60 µm, much larger than solution spun fibers. With special equipment, fiber diameters of about 200 nm have also been achieved with PCL and PEG copolymers (Dalton et al., 2007). The fibers are usually much more uniform than solution-spun ones, partly due to the larger diameter. After leaving the spinneret, the polymer needs to be cooled in a controlled manner, whereas in solution spinning only the solvent needs to be evaporated. If the cooling distance is too short, the fibers stick together on the target. The equipment required is considerably higher than for solution electrospinning and requires more spinning know-how. Therefore, this method is only successfully used by a few research groups.

Charles L Norton received a patent for a corresponding spinning process as early as 1936 (Norton, 1936), Fig. 27.24. Larrondo and Manley (1981a, 1981b, 1981c) laid the theoretical foundations of melt electrospinning. Reneker et al. (2001) produced melt electrospun fibers from PE, PP, PET and PEN. Since then, the number of publications has been steadily increasing, reaching a new high in 2020 with 135 registered in the Web of Science.

Hutmacher and Dalton (2011) combined melt electrospinning with the principle of additive manufacturing to create scaffolds for tissue engineering. Due to the significantly higher viscosity of the polymer in melt electrospinning, the path of the resulting fiber is not helical as in solution electrospinning, but straight from the nozzle to the target. This is also due to the lower charging of the polymer. This means that fibers can be positioned precisely. Furthermore, since they do not repel each other (lower charging of the polymer), a layered structure can be produced if the collector is moved in a controlled manner. This process is described in detail in (Dalton et al., 2013) and is referred to as the "electrowriting" process due to the high accuracy of the fiber deposition. The main parameter influencing the fiber fineness is the molecular weight. The production speed is typically a few meters per minute (Brown et al., 2011), which is sufficient for the production of scaffolds in tissue engineering. This process can thus be used to produce geometrically complex structures, but the mechanical properties of the fibers (strength, elongation) cannot yet be adjusted (Fig. 27.25).

27.3 Flash Spinning

This process, also known as the evaporation spunbond process, is used to produce very fine fibers from polyethylene (HDPE), which are processed into nonwovens (spunbond) directly after spinning out. The spinning mass is injected into the spinning bath under high pressure (20–30 bar) and at a temperature above the boiling point of the solvent (160–180 °C), where it is suddenly cooled down. When leaving the spinneret, the solvent evaporates abruptly due to the pressure drop to atmospheric pressure. This results in the formation of very fine, net-like fibrils, which can be skimmed off and then dried. Typical proportions of the polymer in the solution are around 20%. Hexane and pentane, for example, are used as solvents.

The fiber finenesses are 0.2–10 µm, depending on the polymer and equipment technology. Typical fiber densities of the resulting nonwovens are 10–18 mg/cm^3. The nonwovens are compacted with calenders. The materials produced in this way are resistant to most acids, alkalis and salts.

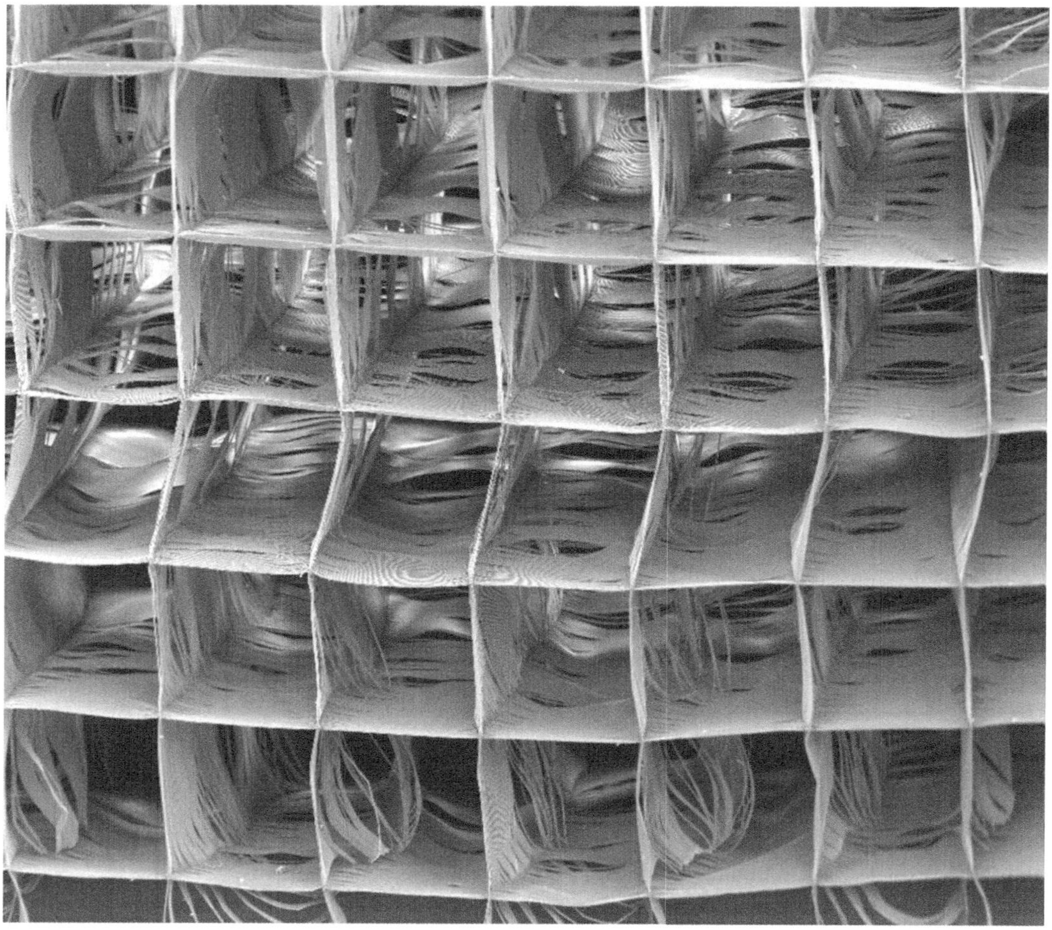

Fig. 27.25 Honeycomb structure produced by "electrowriting" (Dietmar Hutmacher; Brown et al., 2011)

The nonwovens can be processed similarly to paper (hardly any hairiness, printable) and are mostly used for protective clothing and packaging (weatherproof). The temperature resistance ranges from −73 to 100 °C, the self-ignition temperature is around 400 °C. Dupont is the only manufacturer to use this process commercially and has been producing Tyvek® since 1967.

27.4 Centrifugal Spinning

This process is used to produce extremely fine fibers with a fineness well below 1 μm. Figure 27.26 depicts the basic structure of a spinning unit. The spinneret has two or more nozzles from which the spinning liquid emerges when the centrifugal force on the spinning solution due to the speed of rotation is greater than the surface tension of the spinning mass.

The jet is decelerated by air friction and at the same time the diameter is reduced under the action of inertial forces in the direction of flight. The generated fibers are collected on a collector. This can be round (Fig. 27.26) for batch production or flat for a continuous process where mostly substrates are directly covered with fibers, e.g. for filter applications.

Up to now, almost exclusively spinning solutions have been used as spinning mass. Although melt centrifuge spinning is possible, it is technically complex due to the required heating unit and is therefore hardly ever used.

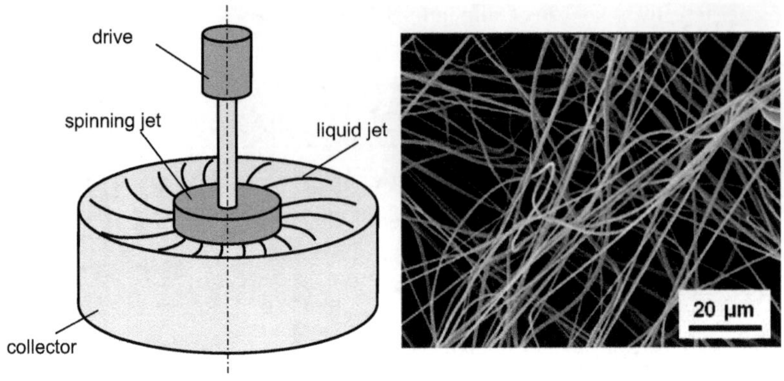

Fig. 27.26 Basic principle of centrifugal spinning and SEM image of an ultrafine fiber web acc. to (Zhang, 2014)

Figure 27.27 shows the design of a spinning unit with which a yarn can be produced directly. The principle essentially corresponds to the wet spinning process.

The main factors influencing the fiber properties are the viscosity and the surface tension of the spinning mass. If the viscosity increases, the risk of bead formation is reduced, and at the same time the fiber diameter increases. The surface tension should be as low as possible so that no droplets form. These parameters can be directly adjusted in spinning solutions via the concentration and chain length of the polymer and the type of solvent, among other things. To reduce the fiber diameter, the rotation speed of the nozzle can be increased. This increases the centrifugal force and, in combination with air friction, elongates the jet. However, a higher rotational speed reduces the time it takes for the fibers to reach the collector. The jet may then have had too little time to reduce to the desired extent. This can be compensated for by increasing the distance between the nozzle and the collector.

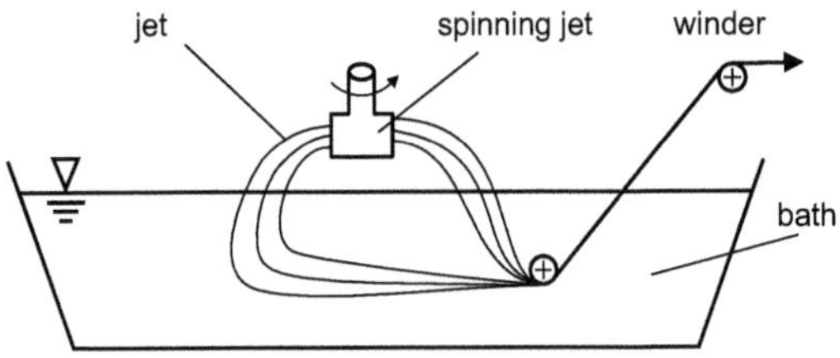

Fig. 27.27 Centrifuge spinning process with subsequent yarn formation acc. to (Zhang, 2014)

27.5 Areas of Application

For many years, the two main applications of electrospun fibers have been membranes and filters (Lee & Obendorf, 2006; Filatov et al., 2007). Filters equipped with nanofibrous webs are very breathable and at the same time have an excellent protective effect, even against minute particles and aerosols, due to the very small pore size. In particular, van der Waals forces bind even the smallest particles to the fibers and thus filter them out. In tissue engineering, electrospun fibers are mainly used as scaffold structures (Pham et al., 2006; Dalton et al., 2013). Kruse et al. (2017, 2018) developed yarns from electrospun PLA/PGA fibers for tissue engineering and a scaffold structure for artificial corneas from electrospun fibers. Schneiders et al. (2021) developed ultrafine fibers of chitosan for tissue engineering. Kopp et al. (2020) developed an additive-free spinning process for fibroin-based nonwovens. Electrospun fibers can also be used to reinforce composite structures, provided the fiber material has sufficient strength (Molnar et al., 2011). This largely excludes the polymer-based materials most commonly used to date. Another application would be the use of electrospun ultrafine fibers as sensors. Florczak et al. (2019) produced fibers from PVDF with a fineness of 17–55 µm, where a piezoelectric effect was demonstrated. Blachowicz and Ehrmann (2020) describe a method for generating magnetic electrospun fibers by doping with metal oxide nanoparticles.

References

Ali, U., Abbass, A., Khurshid, F., Aslam, S., & Waqar, A. (2017). Needleless electrospinning using a flat wheel spinneret. *Journal of Engineered Fibers and Fabric, 12*(3), 81–89.

Amna, R., Ali, K., Malik, M. I., & Shamsah, S. I. (2020). A brief review of electrospinning of polymer nanofibers: History and main applications. *Journal of New Materials for Electrochemical Systems, 23*(3), 151–163.

Blachowicz, T., & Ehrmann, A. (2020). Most recent developments in electrospun magnetic nanofibers: A review. *Journal of Engineered Fibers and Fabrics* (15),S., 1–14.

Brown, T. D., Dalton, P. D., & Hutmacher, D. W. (2011). Direct writing by way of melt electrospinning. *Advanced Materials, 23*, 5651–5657.

Bunyan, N., Chen, I., Chen, J., & Farboodmanesh, S. (2006). Electrostatic effects on electrospun fiber deposition and alignment. In Reneker, D. H., & Fong, H. (Eds.), Polymeric nanofibers, ACS Symposium Series 918. ACS Publications, Washington.

Chapman, B. S., Mishra, S. R., & Tracy, J. B. (2019). Direct electrospinning of titania nanofibers with ethanol. *Dalton Transactions, 48*, 12822–12827.

Cooley, J. F. (1900). Apparatus for electrically dispersing fluids. US Patent 692,631.

Dalton, P. D., Grafahrend, D., Klinkhammer, K., Klee, D., & Möller, M. (2007). Electrospinning of polymer melts: Phenomenological observations. *Polymer, 48*(23), 6823–6833.

Dalton, P. D., Vaquette, C., Farrugia, B. L., Dargaville, T. R., Brown, T. D., & Hutmacher, D. W. (2013). Electrospinning and additive manufacturing: Converging technologies. *Biomaterials Science, 1*, 171–185.

Deitzel, J. M., Kleinmeyer, J., Harris, D., & Beck Tan, N. C. (2001). The effect of processing variables on the morphology of electrospun nanofibers and textiles. *Polymer, 42*, 261–272.

Doshi, J., & Reneker, D. H. (1995). Electrospinning process and applications of electrospun fibers. *Journal of Electrostatics, 35*(2–3), 151–160.

ELMARCO s.r.o., Liberec, Tschechien. https://www.elmarco.com

Filatov, Y., Budyka, A., & Kirichenko, V. (2007). *Electrospinning of micro- and nanofibers: Fundamentals and applications in separation and filtration processes*. Begell House.

Florczak, S., Lorson, T., Zheng, T., Mrlik, M., Hutmacher, D. W., Higgins, M. J., Luxenhofer, R., & Dalton, P. D. (2019). Melt electrowriting of electroactive poly(vinylidene difluoride) fibers. *Polymer International, 68*(4), 735–745.

Formhals, A. (1934). US Patent. 1,975,504.

Gilbert, W. (1600). *De Magnete Magnetcisque Corporibus, et de Magno Magnete Tellure (On the Magnet, Magnetick Bodies also, and on the Great Magnet the Earth; a new Physiology, demonstrated by many arguments & Experiments)*. The Chiswick Press, London.

Hacker, C. (2014). *Anlagenentwicklung für das Elektroschmelzspinnen von Feinfaservliesstoffen für die Abwasseraufbereitung. Dissertation am Institut für Textiltechnik der RWTH Aachen*. Shaker Verlag.

Hacker, C., Jungbecker, P., Gries, T., & Thomas, H. (2009). Mehrdüsen-Elektrospinnen aus der Polymerschmelze - Die Entwicklung zum Upscaling. In *Proceedings of the Chemnitzer Textiltechnik-Tagung* (12), Chemnitz.

Hagiwaba, K., Oji-Machi, O., & Ku, K. (1929). Japan Patent. 1,699,615.

Hills Inc. (2021). http://hillsinc.net/assets/pdfs/polymeric-nanofibers.pdf

Hutmacher, D. W., & Dalton, P. D. (2011). Melt electrospinning. *Chemistry – An Asian Journal, 6*, 44–56.

Kameoka, J., Orth, R., Yang, Y., Czaplewski, D., Mathers, R., Coates, G., & Craighead, H. (2003). A Scanning tip electrospinning source for deposition of oriented nanofibers. *Nanotechnology, 14*, 1124–1129.

Katta, P., Alessandro, M., Ramsier, R. D., & Chase, G. G. (2004). Continuous electrospinning of aligned polymer nanofibers onto a wire drum collector. *Nano Letters, 4*, 2215–2218.

Keithley, J. F. (1999). *The story of electrical and magnetic measurements*. Wiley.

Koombhongse, S., Liu, W., & Reneker, D. H. (2001). Flat polymer ribbons and other shapes by electrospinning. *Journal of Polymer Science, 39*, 216–223.

Kopp, A., Smeets, R., Gosau, M., Kröger, N., Fuesta, S., Köpf, M., Kruse, M., Krieger, J., Rutkowski, R., Henningsen, A., & Burg, S. (2020). Effect of process parameters on additive-free electrospinning of regenerated silk fibroin nonwovens. *Bioactive Materials, 5*, 241–252.

Kruse, M., Greuel, M., Kreimendahl, F., Schneider, T., Bauer, B., Gries, T., & Jockenhövel, S. (2018). Electro-spun PLA-PEG-yarns for tissue engineering applications. *Biomedical Engineering/Biomedizinische Technik, 63*(3), 231–243.

Kruse, M., Walter, P., Bauer, B., Rütten, S., Schaefer, K., Plange, N., Gries, T., Jockenhövel, S., & Füst, M. (2017). Electro-spun membranes as scaffolds for human corneal endothelial cells. *Current Eye Research, 43*(1), 1–11.

Larrondo, L., & Manley, R. S. J. (1981a). Electrostatic fiber spinning from polymer melts. I. Experimental observations on fiber formation and properties. *Journal of Polymer Science Part B-Polymer Physics, 19*, 909–920.

Larrondo, L., & Manley, R. S. J. (1981b). Electrostatic fiber spinning from polymer melts. II. Examination of the flow field in an electrically driven jet. *Journal of Polymer Science Part B-Polymer Physics, 19*, 921–932.

Larrondo, L., & Manley, R. S. J. (1981c). Electrostatic fiber spinning from polymer melts. III. Electrostatic deformation of a pendant drop of polymer melt. *Journal of Polymer Science Part B-Polymer Physics, 19*, 933–940.

Lee, S., & Obendorf, K. (2006). Developing protective textile materials as barriers to liquid penetration using melt-electrospinning. *Journal of Applied Polymer Science, 102*, 3430–3437.

Li, D., Wang, Y., & Xia, Y. (2003). Electrospinning of polymeric and ceramic nanofibers as uniaxially aligned arrays. *Nano Letters, 3*, 1167–1171.

Li, D., Wang, Y., & Xia, Y. (2004). Electrospinning nanofibers as uniaxially aligned arrays and layer-by-layer stacked films. *Advanced Materials, 16*, 361–366.

Molnar, K., Vas, L. M., & Czigany, T. (2011). Determination of tensile strength of electrospun single nanofibers through modeling tensile behavior of the nanofibrous mat. *Composites Part B: Engineering, 43*(1), 15–21.

Morton, W. J. (1902). *Method of dispersing fluids*. US Patent 705,691.

Niu, H., Wang, X., & Lin, T. (2012). Upward needleless electrospinning of nanofibers. *Journal of Engineered Fibers and Fabrics*, Special Issue – Fibers, 17–22.

Norton, C. L. (1936). Method of and apparatus for producing fibrous or filamentary material. US Patent 2,048,651.

Pham, Q. P., Sharma, U., & Mikos, A. G. (2006). Electrospinning of polymeric nanofibers for tissue engineering applications: A review. *Tissue Engineering, 12*(5), 1197–1211.

Reneker, D. H., Yarin, A., Evans, E. A., Kataphinan, W., Rangkupan, R., Liu, W., Koombhongse, S., & Xu, H. (2001). Electrospinning and nanofibers. In *New frontiers in fiber science, proceedings of the the fiber society meeting*, Raleigh, NC, USA.

Schneiders, T., Vogel, L., & Gries, T. (2021). *Crosslinking electrospun chitosan nanfibers with genipin for medical applications and tissue engineering*. Nanofiber Applications and Related Technologies Conference (NART), Istanbul.

Shin, Y. M., Homan, M. M., Brenner, M. P., & Rutledge, G. C. (2001). Experimental characterisation of electrospinning: The electrically forced jet and instabilities. *Polymer, 42*, 9955–9967.

Sun, D., Chang, C., Li, S., & Lin, L. (2006). Near-field electrospinning. *Nano Letters, 6*(4), 839–842.

Sundaray, B., Subramanian, V., & Natarajan, T. S. (2004). Electrospinning of continuous aligned polymer fibers. *Applied Physics Letters, 84*, 1222–1224.

Taylor, G. I. (1964). Disintegration of water drops in an electric field. *Proceedings of the Royal Society A, 280*(1382), 383–397.

Taylor, G. I. (1966). The force exerted by an electric field on a long cylindrical conductor. *Proceedings of the Royal Society A., 291*(1425), 145–158.

Taylor, G. I. (1969). Electrically driven jets. *Proceedings of the Royal Society A, 313*(1515), 453–475.

Teo, W. E., & Ramakrishna, S. (2006). A review on electrospinning design and nanofibre assemblies. *Nanotechnology, 17*, R89–R106.

Theron, S. A., Yarin, A. L., Zussman, E., & Kroll, E. (2005). Multiple jets in electrospinning: Experiment and modeling. *Polymer, 46*, 2889–2899.

Theron, A., Zussman, E., & Yarin, A. L. (2001). Electrostatic field-assisted alignment of electrospun nanofibers. *Nanotechnology, 12*, 384–390.

Tucker, N., Stanger, J. J., Staiger, M. P., Razzaq, H., & Hofman, K. (2012). The history of the science and technology of electrospinning from 1600 to 1995. *Journal of Engineered Fibers and Fabrics*, Special Issue – Fibers, 63–73.

Wienecke, S. (2007). *Herstellung komplexer, dreidimensionaler Fasermatten mit definierter Faserorientierung durch Elektrospinning*. Diplomarbeit am Institut für Textiltechnik der RWTH Aachen und am Institut für Mehrphasenprozesse und Zentrum für Biomedizintechnik der Universität Hannover.

Yarin, A. L., Koombhongse, S., & Reneker, D. H. (2001). Bending instability in electrospinning of nanofibers. *Journal of Applied Physics, 89*(5), 3018–3026.

Yarin, A. L., & Zussman, E. (2004). Upward needleless electrospinning of multiple nanofibers. *Polymer, 45*, 2977–2980.

Zhang, X. (2014). *Fundamentals of fiber science*. DEStech Publications.

28 Aerogels

Aerogels are highly porous, rigid materials consisting of up to 99.98% pores. They therefore have an extremely low density of 0.0016 g/cm^3 (aerographs) to about 0.5 g/cm^3 (for comparison: air: 0.00125 g/cm^3). This is associated with an extremely large surface area of 100–2000 m^2/g, depending on the fiber material. Aerogels are usually made of silicate, but there are also aerogels made of metal oxides, cellulose and other materials.

28.1 Manufacturing Principle

The first aerogels were developed in 1931 at Stanford University, USA, by Kistler (1931). He also describes the principle of their production:

"The gel forms in a suitable liquid in dilute form. The liquid is then replaced by another liquid which does not decompose the structure and has a sufficiently low critical temperature. Alcohol is well suited for inorganic gels, ether has advantages for light-reduced materials, and propane has been used for all organic gels. The substitution of liquids requires that the two liquids can be completely mixed together. For example, water can be replaced by alcohol and the latter by ether. Pure evaporation [of the liquid] would result in shrinkage. The gel with an excess of liquid is heated in an autoclave above the critical temperature while the pressure is kept constant and above the vapor pressure, so that no evaporation occurs and the gel does not shrink due to resulting capillary forces. Above the critical temperature, the liquid is directly converted to a gas without a transition state. The gel "does not know" that the liquid in its pores has become gaseous. Therefore, the gas only needs to be removed and a coherent aerogel remains." This basic principle has not changed significantly to this day.

Figure 28.1 shows the transition of a liquid into the state of a supercritical fluid at constant pressure as well as the subsequent expansion by pressure reduction, whereby it passes directly into the gaseous state. This is also referred to as "supercritical drying".

Kistler thus produced silicate aerogels with a density of up to 0.02 g/cm^3. He had also already experimented with aluminum, various metal oxides (including tin and tungsten oxide), gelatine, agar and cellulose and noted that this list could be continued indefinitely. The new physical properties were "unusually interesting".

28.2 Aerogel Fibers

Three main processes are known to produce aerogels. The most widely used is the sol-gel process, which is also used to manufacture ceramic fibers. In addition, there is the template synthesis and the nanostructure assembly.

D. Veit, *Fibers*, https://doi.org/10.1007/978-3-031-15309-9_28

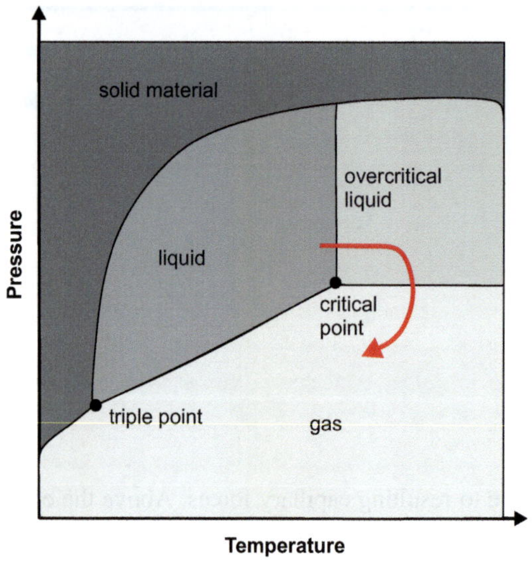

Fig. 28.1 State diagram for supercritical drying (arrow) acc. to (Saperaud, 2016)

28.2.1 Sol-Gel Process

In this process, the subsequent aerogel material is brought into solution and then combined into long chains in a polymer formation reaction (aggregation). With increasing chain length and decreasing solvent concentration, networks of macromolecules form, the so-called gels. The viscosity decreases in the process. Fibers can then be produced from this solution, e.g. using the dry or wet spinning process. After spinning, the solvent is removed by supercritical drying and aerogel fibers are produced (Fig. 28.2).

Depending on the solvent, a distinction is made between aqua and hydrogels (water), alcohols (e.g. methanol, ethanol) and organogels (organic solvent, e.g. acetone). Liquid CO_2 or nitric oxide can also be used as a solvent.

Pico et al. (2017) produced aerogel fibers from silicate for the first time using the air gap spinning process. Starting from tetraethylorthosilicate (TEOS), it is partially hydrolyzed in a mixture of ethanol, water and hydrochloric acid. This is followed by stepwise polymer formation by polycondensation, during which the solution transforms into a gel. The mixing ratios that allow a spinning process are shown in Fig. 28.3.

After washing of the wound filament yarn, supercritical drying takes place by extraction of CO_2. The fibers produced have a cross-section of 50 μm, the inner surface area is 755 m^2/g with an average pore size of 6 nm (Fig. 28.4).

28.2.2 Electrospinning

For example, the electrospinning process can be used to generate polyacrylonitrile fibers and then convert them into aerogel fibers (Lai et al., 2020).

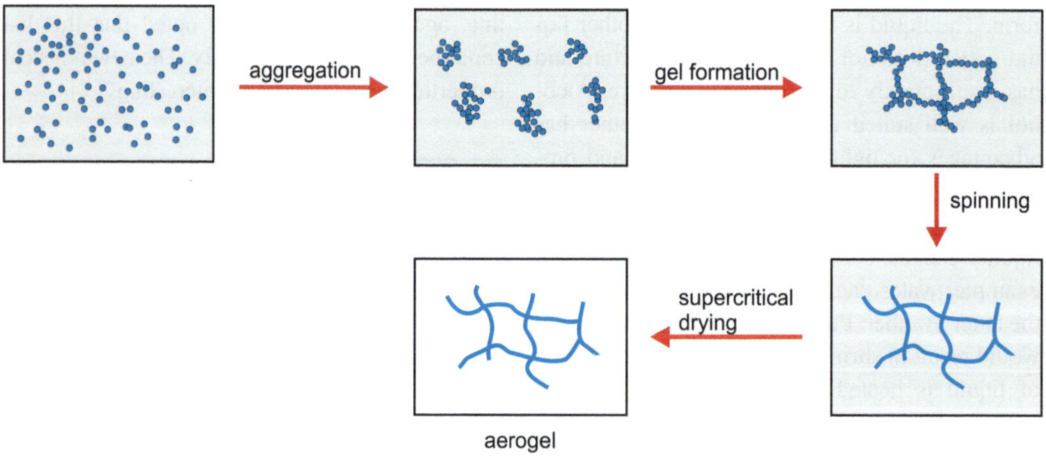

Fig. 28.2 Sequence of the sol-gel spinning process

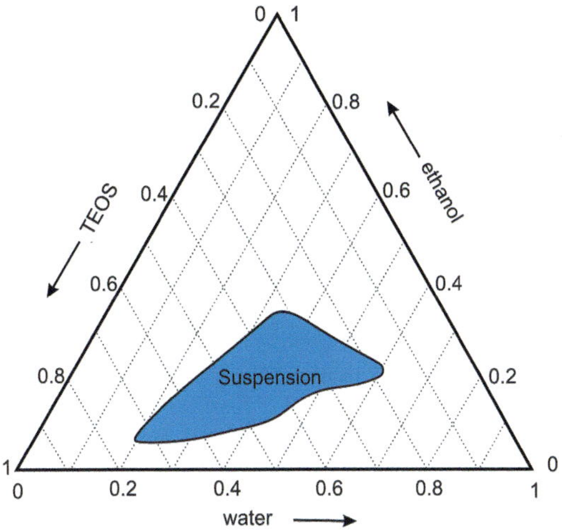

Fig. 28.3 Three-phase diagram for the production of aerogel fibers acc. to (Pico et al., 2017)

28.3 Features

Aerogels can be produced from inorganic and organic materials. Cellulose and SiO_2 are particularly interesting for fibers.

The pores of aerogels are divided into micropores (2 nm), mesopores (2–50 nm) and macropores (> 50 nm). By selectively adjusting the pore size, applications can be developed for insulation and insulating materials, for adsorption and filtration.

Aerogels have a very low thermal conductivity of 0.01–0.02 W/mK compared to other insulating materials, e.g. glass wool (0.03–0.04 W/mK). This is due to the fact that

- the thermal conductivity of the material is very low due to the high proportion of gas-filled pores and
- convection is extremely hindered by the microstructure of the pores (Knudsen effect).

By clever selection of the material, heat conduction can be further reduced so that aerogels, e.g. made of silicate, are very well suited as insulating material. However, heat radiation can easily penetrate aerogels because very little material stops the radiation. However, it usually only contributes a small part to the thermal conduction of a system.

The production of aerogel fibers from cellulose was successful (Schulz, 2016) using the wet-spinning process. Depending on the process route, the fineness of the filaments was 1.5–100 dtex, the specific surface area was 100–500 m^2/g, the strength reached 2–15 cN/tex and the elongation-at-break was 4–40%. The nonwovens produced from these fibers could be used, for example, for insulating material and highly porous electrodes.

28.4 Applications

In the last 10 years, intensive research has been carried out on aerogel fibers and different

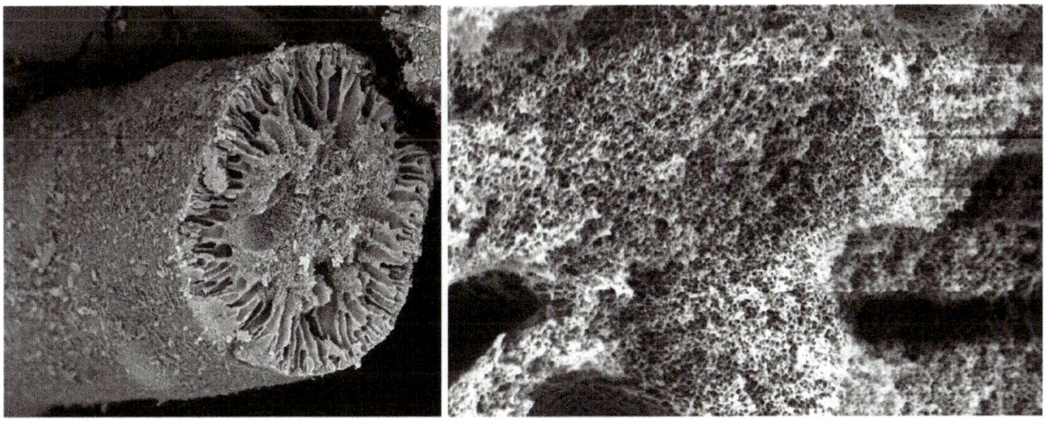

Fig. 28.4 Aerogel fiber (right: detail)

materials have been tested. In addition to the silicates and mullites that have been used for a long time, organic materials (cellulose, chitosan) and conventional polymers have also been investigated. Some of the latest research results are briefly presented below.

Batista et al. (2020) prepared aerogels from a mixture of chitosan and alginate for medical wound dressings. Chitosan is antimicrobial, while alginate serves as a moisture retainer. Both polymers were processed into a hydrogel using the emulsion gelation process. Subsequently, the water was first replaced by alcohol and this then by CO_2 and supercritically dried. Thus, a free surface of 162–302 m^2/g could be obtained.

Electrospun PAN aerogel fibers functionalized with glycerol developed (Lai et al., 2020). These fibers are able to bind both aqueous and oily substances to themselves and could be used, for example, to purify wastewater.

A different approach was taken by Meng et al. (2021). They produced silicate-based wet-spun aerogel fibers and functionalized them with silanes. Such fibers are to be used for the purification of dyeing wastewater.

A fundamental problem of aerogels is their brittleness. Although they withstand compressive loads very well, they fail abruptly in the event of damage. For this reason, fiber-reinforced aerogels are being developed that combine the good thermal insulation properties of aerogels with the good strength properties of reinforcing fibers. Aluminum-based aerogels reinforced with aluminum-silicate ceramic fibers resulted in a strong reduction of thermal conduction in the composite structure while maintaining good strength values (Peng et al., 2021). Similar results were obtained by Qin et al. (2021) by reinforcing cellulose aerogels with carbon fibers and graphene and many other researchers with different combinations of materials.

At present, there are no marketable products made from aerogel fibers. However, this is expected to change as soon as process costs fall and products are developed for which customers are prepared to pay a relatively high price, e.g. in medical technology and aerospace.

References

Batista, M. P., Gonçalves, V. S. S., Gaspara, F. B., Nogueira, I. D., Matias, A. A., & Gurikove, P. (2020). Novel alginate-chitosan aerogel fibres for potential wound healing applications. *International Journal of Biological Macromolecules, 156*, 773–782.

Kistler, S. S. (1931). Coherent expanded aerogels and jellies. *Nature, 127*, 741.

Lai, H., Wang, Y., Wang, Y., Liu, W., Bao, X., Liu, F., Li, X., Lei, Z., Jiao, H., & Fan, Z. (2020). Macroscale amphiphilic aerogel fibers made from nonwoven nanofibers for large active mass loading. *Journal of Power Sources, 474*, 228612.

Meng, S., Yu, S., Tang, F., Hu, X., Lu, J., Fei, X., & Zhu, M. (2021). Fiber engineering of silica-based aerogels with surface specificity and regenerability for continuous removal of dye pollutants from wastewaters. *Microporous and Mesoporous Materials, 314*, 110874.

Peng, F., Jiang, Y., Feng, J., Cai, H., Feng, J., & Li, L. (2021). Thermally insulating, fiber-reinforced alumina–silica aerogel composites with ultra-low shrinkage up to 1500 °C. *Chemical Engineering Journal, 411*, 128402.

Pico, D., Meyer, E., Lüking, A., Milow, B., & Gries, T. (2017). SILICA-AERO. Production of lightweight silica aerogel fibers for excellent heat insulating application. *Chemical Engineering Transactions, 60*, 91–96.

Qin, W., Zhu, W., Ma, J., Yang, Y., & Tang, B. (2021). Carbon fibers assisted 3D N-doped graphene aerogel on excellent adsorption capacity and mechanical property. *Colloids and Surfaces A: Physicochemical and Engineering Aspects, 608*, 125602.

Saperaud. (2016). https://commons.wikimedia.org/wiki/File:Phasendiagramm_Verdampfungen.svg, CC-BY-SA 4.0.

Schulz, B. (2016). *Cellulose Aerogelfasern*. Dissertation an der RWTH Aachen, Aachen.

29 Texturing

Texturing makes flat filament yarns crimped and thus more similar to natural fibers in terms of their mechanical properties and feel (Fig. 29.1). They can then be blended with them or processed pure or in blends on the textile machines originally developed for natural fibers. Depending on the application and yarn count, there are different texturing processes.

The main processes are:

- false twist texturing (fine filament yarns, mostly for clothing),
- air-jet texturing (fine filament yarns, mostly for clothing),
- mechanical compression chamber texturing (production of staple fibers),
- gas-dynamic compression chamber texturing (carpet yarns, BCF).

In addition, yarns can be textured using the twist/untwist texturing process and the knit fixation process, both of which are hardly used today. In principle, bicomponent yarns can also be crimped when using differently shrinking polymers. This principle corresponds to wool.

29.1 False Twist Texturing

Fiber production and texturing were initially separate processes. It was not until the development of draw texturing that the texturing process and filament yarn production were combined in one process. In this process, the yarn is both, fully drafted and crimped, hence the name draw textured yarn (DTY), Fig. 29.2.

This process principle has become so widely accepted that today, the majority of all melt-spun filament yarns for use in apparel textiles is textured in this way.

29.1.1 Principle

The filament yarn is drawn off the feed package (POY) and enters a fixing zone, behind which, after a cooling section, there is a twist-generating element. This if followed by an optional second setting zone. The yarn is then wound on a bobbin (Fig. 29.3).

29.1.2 Twist Insertion

The twist is inserted into the yarn with the so-called false twist aggregate. In the process, twists are created in opposite directions in front of and behind the twister. The latter are quickly transported away at the beginning of the process and so a twist is only created in the yarn in front of the twister, which is permanently fixed before and dissolved behind the twister.

The following figure shows this schematically. If, in a thought experiment (Fig. 29.4, from the top), the yarn is clamped between two rollers and the twister not yet in operation, the yarn is of course not twisted. Upon turning on the twister

D. Veit, *Fibers*, https://doi.org/10.1007/978-3-031-15309-9_29

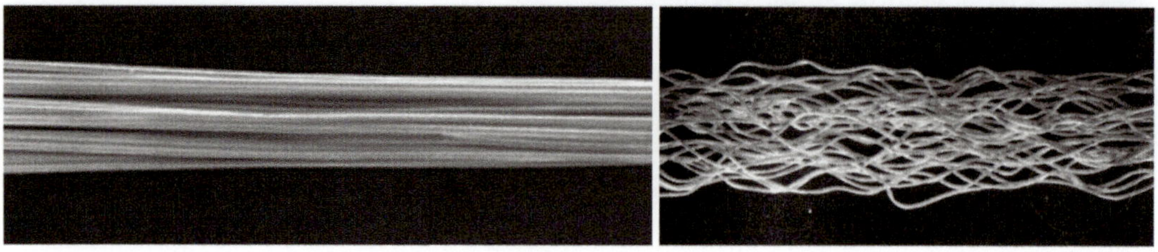

Fig. 29.1 Flat filament yarn (left) and textured yarn (right)

(the rollers stand still and the yarn is still clamped), a Z-twist is created left of the twister and a S-twist to the right of it. When the delivery rollers start turning, and the yarn now passes through the twister, a constant level of Z-twist builds up to the left of the twister and a S-twist is created on the right. Thereby, in the time Δt, a new Z-twist is created on the left hand side while another Z-twist slips through the twist aggregate and is compensated on the right by a just created S-twist. Hence, after the initially generated S-twists are transported away, on the right hand side of the twister, the yarn reaches a steady state and is untwisted. On the bobbin, the yarn is therefore wound up with no twist. Hence the name "false twist" for this process.

Initially, only rotating hollow spindles were used as twist generators, with a pin or bar in the bore around which the running yarn wraps. These spindles with magnetic bearing reached maximum speeds of up to 1000,000 rpm, which limited yarn take-up speeds to a few hundred meters per minute. An alternative method employed a pair of aprons. Today, friction disc texturing has become the commonly used method, in which the yarn is twisted by rotating discs (Fig. 29.5). On modern twist aggregates, more than 3,000,000 twists per minute can be inserted into a yarn allowing winding speed of more than 1200 m/min.

The twist in the yarn results from the rotations transferred from the discs on the yarn and the

Fig. 29.2 False twist texturing machine (Oerlikon Barmag n.d.)

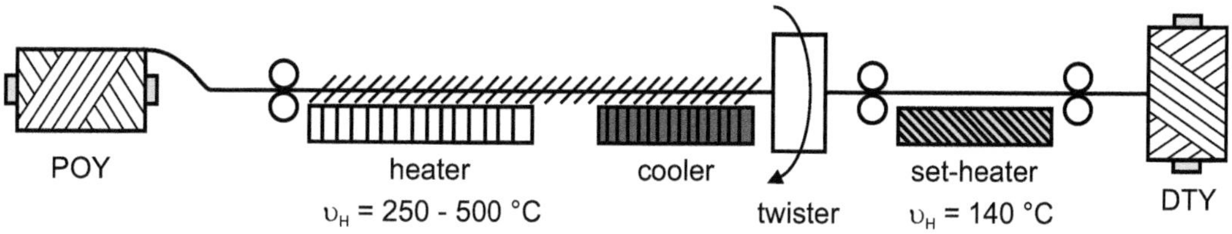

Fig. 29.3 Principle of false twist texturing

yarn's axial speed. In the case of magnetic spindles, the twist corresponds directly to their rotational speed; in the case of friction discs, it results from the rotational speed of the friction disc and from the ratio of its circumference speed to that of the yarn, with a friction-dependent slip of the order of up to 50% (Fig. 29.6). The crimp properties of the final product depend in a complicated way on the twist, the yarn speed, the fixing temperature in the heater and the residence time of the yarn in heater and cooler, and of course also on the properties of the POY yarn (Fig. 29.16, Schmenk, 2003). This also applies to the most important quality parameter, the dyeability of the finished yarn.

Today, texturing with discs made of ceramic or polyurethane is the prevailing technique, depending on the material and yarn count. Due to the higher friction coefficients with polyurethane, fewer discs can be used than with ceramic. However, the service life of PU discs is considerably shorter (approx. 2 to 6 weeks) than that of ceramic discs (several years). With ceramics, so-called "glazing" can occur under unfavorable conditions. In this case, the ceramic disc becomes smooth ("glazed") due to abrasion of the avivage and the friction is thus drastically reduced. Furthermore, the formation of abrasion in the aggregate is a problem with ceramic discs. This so-called "snow" consists of oligomers knocked out of the yarn by the rough ceramic surface. Mixed with avivage, they produce a greasy film in the entire texturing aggregate hence increasing slippage and thus impeding twist insertion.

The disc diameter is usually approx. 52 mm, the thickness approx. 9 mm. Special inlet and outlet discs (so-called knife discs) are often used

Fig. 29.4 Formation of the false twist in the yarn acc. to (Bruske, 1990)

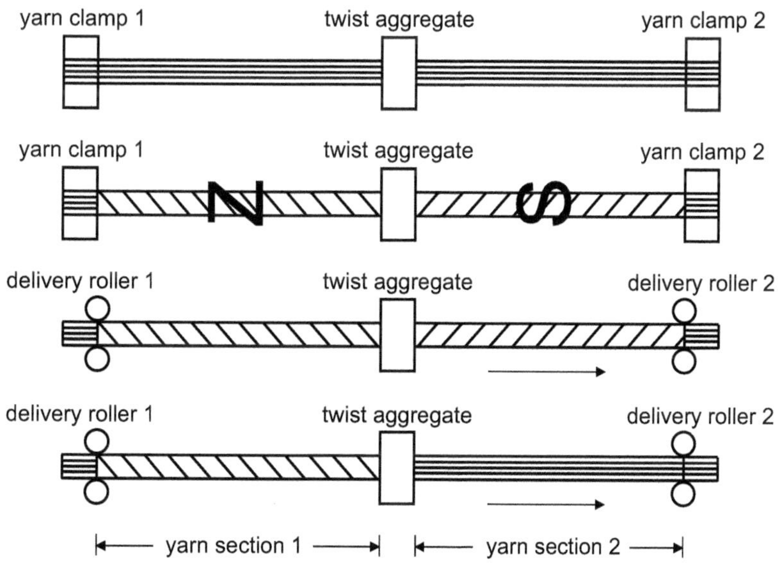

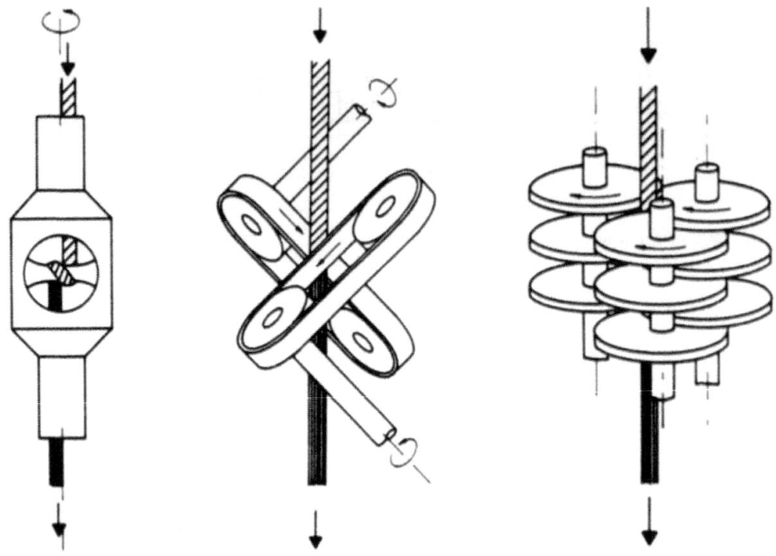

Fig. 29.5 Twisting aggregates (left: magnetic spindle, middle: apron; right: friction discs) acc. to (Lünenschloß et al., 1985)

to calm the yarn course in the false twister. Usually 3 to 5 working discs are used for PU and 5 to 8 for ceramic (Fig. 29.7). Typical configurations are referred to as 1-5-1 or 1-8-1.

Figure 29.8 shows the structure of a false twister (yarn path from top to bottom) and the spiral path of the yarn through the unit.

In addition to single false twist units, twin units are also available which are driven together, either by belts or by individual motors.

The ratio of disc circumferential speed to yarn speed is called the d/y ratio (disc-to-yarn ratio). Depending on the slip between discs and yarn, it is between 1.6 and 2.4.

The yarn path in the texturing machine is strongly influenced by the twist distribution. In extreme cases, a strong oscillation of the yarn tension occurs, which is then referred to as surging (Fig. 29.9). This is a mostly periodic fluctuation of the yarn tension, which can lead to differences in the coloring of the textured yarn. Surging can be reduced or even eliminated by specifically influencing the yarn tension, e.g. by a different disc geometry or a change in the disc material. An increase in the d/y ratio reduces surging but directly affects the crimp.

Various systems have been developed among others by Oerlikon Barmag and Saurer Technologies n.d., to enable online monitoring of the yarn tension behind the twisting unit as the most important quality parameter. Surging can thus be detected at an early stage and action can be taken (Fig. 29.10).

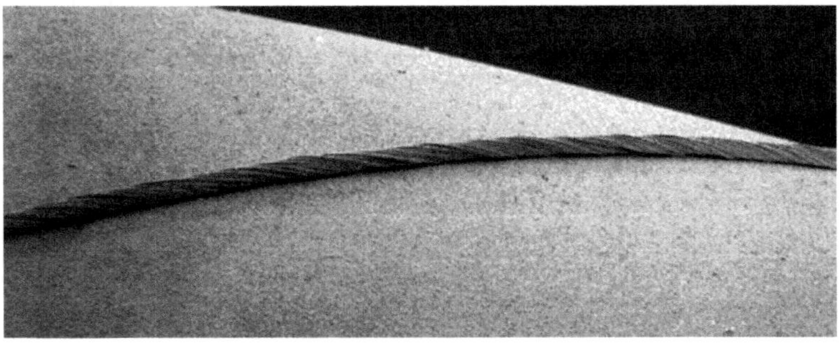

Fig. 29.6 Twisted yarn on a texturing disc

Fig. 29.7 Friction disc false twist units and drafting rollers (Oerlikon Barmag, Saurer Technologies)

29.1.3 Heater and Cooling Zone

The twist is fixed in the yarn by a combination of heater and cooling unit. As the following figure shows, the molecular chains in the undeformed state form secondary valence bonds, so-called hydrogen bonds. If the yarn is deformed by twisting, internal stresses are created in the yarn which counteract the twisting. Heating the yarn in the heater dissolves these tensions and new secondary valences can form in the subsequent cooling zone, thus fixing the twisted state of the yarn in front of the twister. After the twist aggregate, the formerly Z-twisted yarn meeting S-twists is turned back to an untwisted state. Therefore, it is now under high internal tension. This manifests itself in a pronounced crimp when the yarn is no longer subject to external tension, e.g. unwound

Fig. 29.8 False twister (Oerlikon Barmag), yarn path in false twister (Bröll n.d.) and yarn in a magnetic spindle twister

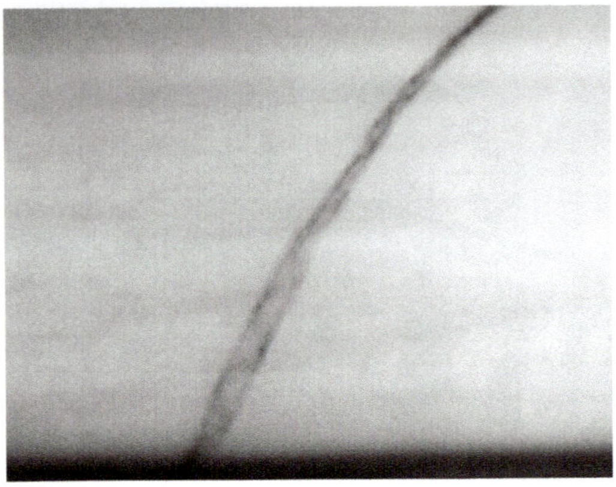

Fig. 29.9 Surging effect

from the bobbin. Because the twist is not visible after the twist generator as the yarn is still under tension, this is referred to as false twist (Fig. 29.11).

The heaters are built in two different versions. In the conventional heater, which has a length of approx. 2–3 m, the heat is transferred from the heater to the yarn by direct contact. The heater temperature is therefore below the melting temperature and for PET for example approx. in the range of 200–220 °C.

Alternatively, a high-temperature heater was developed in the 1990s. The heat is transferred by convection at a temperature of up to 600 °C. The main difficulty is therefore that the yarn does not come into contact with the inner wall of the heater, which is achieved by ceramic yarn guides that hold the yarn in the center of the heater channel. This type of heater is considerably shorter (approx. 1 m) than conventional heaters. Due to the significantly higher operating temperature, the heat transfer from the heater to the yarn is much faster, which in turn allows higher yarn speeds. This led to a doubling of the texturing speed. Such heaters can lead to dyeing irregularities with very fine titers in the subsequent process stages owing to uneven heat transfer into the yarn. On the other hand, due to its shorter length, the process is somewhat stabilized.

29.1.4 Yarn Structure

The result of false twist texturing is a highly elastic crimped yarn, also known as HE yarn. Part of the crimp or elasticity can be removed by subsequent heat-setting under tension, which reduces shrinkage and the tendency to curl. This results in a set yarn. The usual yarn counts are between 30 and 600 dtex (Fig. 29.12).

Figure 29.13 gives an overview of typical yarn counts and filament numbers of different manufacturers. It is obvious that, as with POY

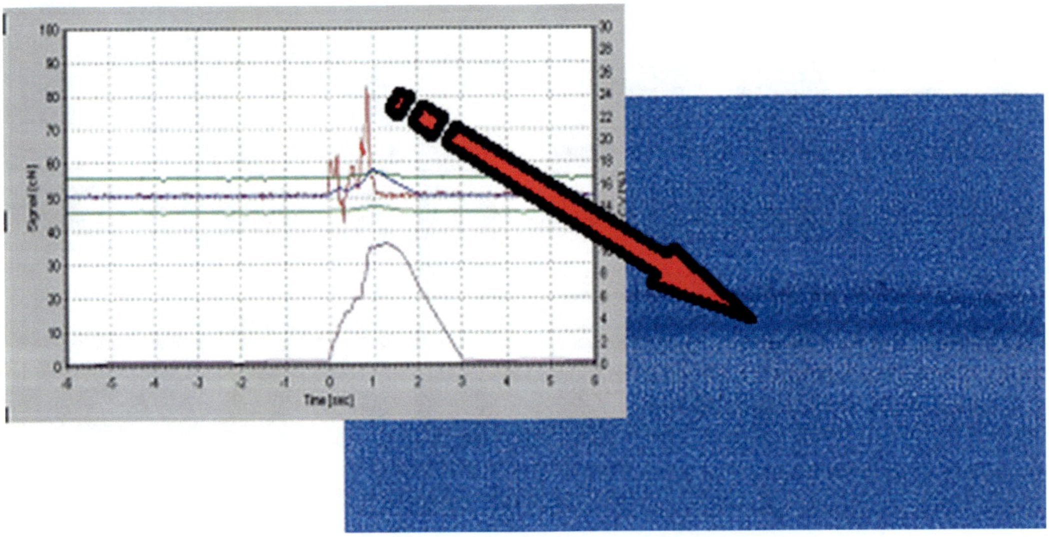

Fig. 29.10 Variation of yarn tension and dyeing defects in the textile (Oerlikon Barmag)

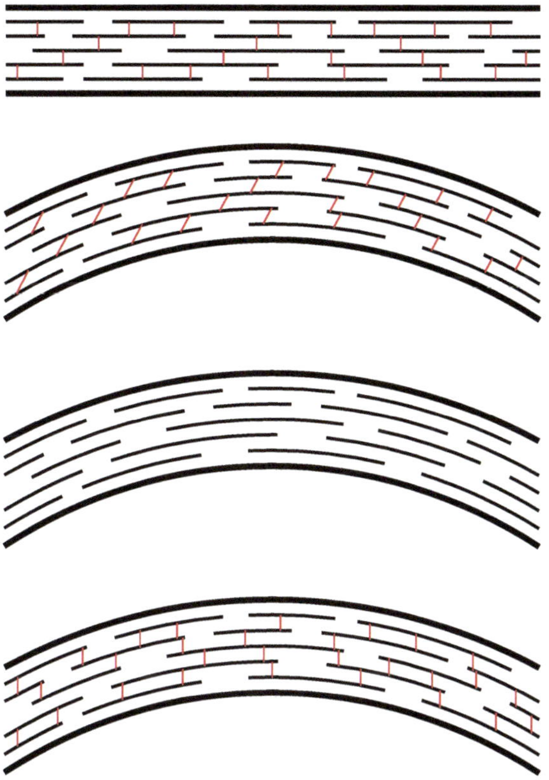

Fig. 29.11 Fixing the twist (crimp) in the yarn

yarn, certain combinations of yarn counts (35, 50, 68, 75, 100, 150, 300 den) and filament numbers (24, 34, 36, 72, 96, 144, 192, 288) are preferred. The filament numbers often differ slightly from country to country.

29.1.5 Structure of the Machine

Up to 576 texturing positions in 12 sections are arranged next to each other in a modern false twist texturing machine, which are driven either collectively by belts or by individual motors (Fig. 29.14). Typical texturing speeds are a maximum of 350 m/min for magnetic spindles, which have to rotate at 1000,000 rpm. With the dominant false twist texturing, yarn winding speeds of 700 to 1200 m/min are achieved, depending on the yarn count and individual fiber fineness. With the common yarn count of 167 dtex (150 den), for example, this corresponds to a yarn throughput of 0.7 to 1.2 kg/h per texturing position.

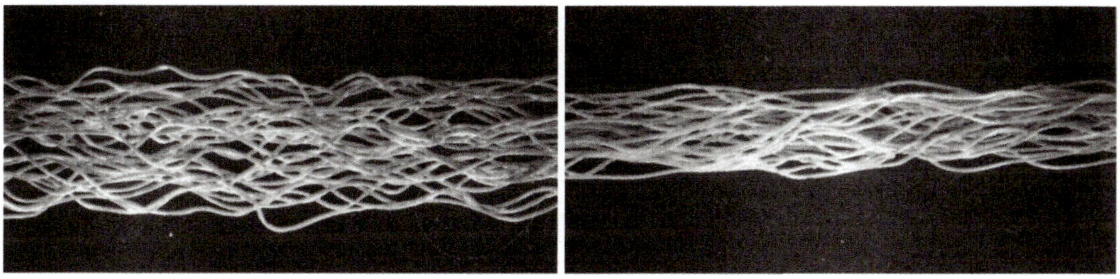

Fig. 29.12 False twist textured yarns (left: HE yarn, right: set yarn)

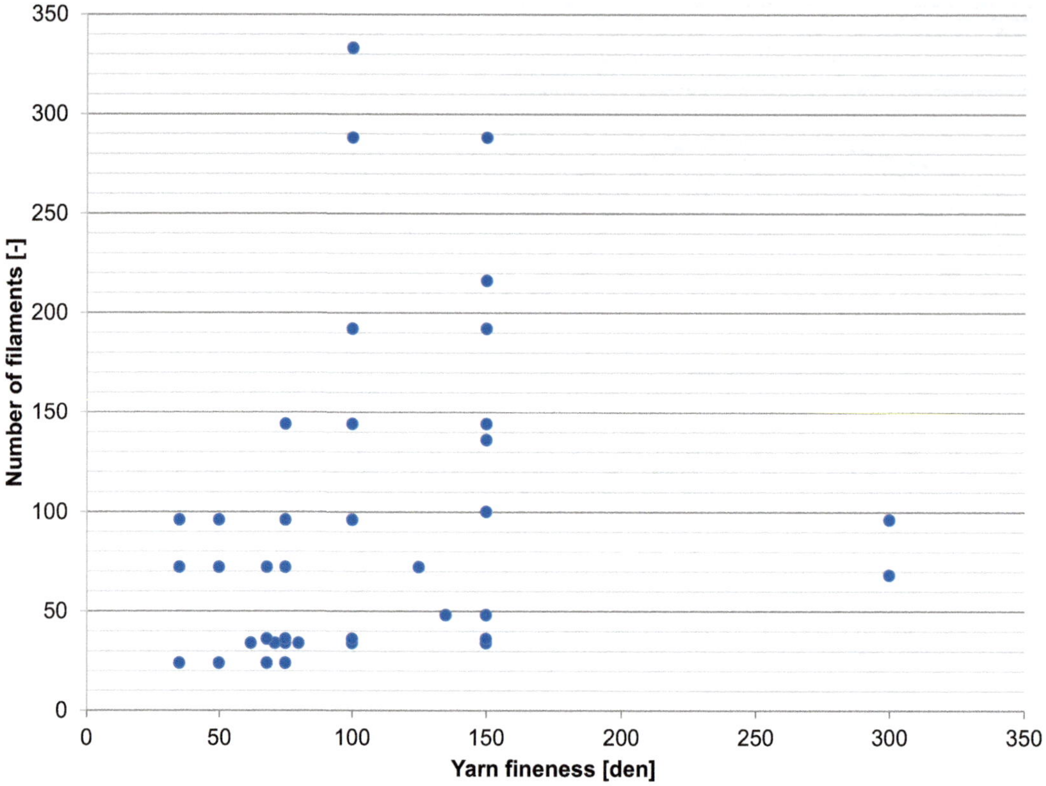

Fig. 29.13 Typical fineness and filament numbers of false twist textured yarns

False twist texturing machines usually have two machine sides, which can be operated with different yarn materials. Depending on the fiber fineness and the available installation space, there are different configurations of the aggregates, as Fig. 29.15 shows.

The set-ups 1 and 2 are referred to as the V-configuration because a V-shaped yarn path

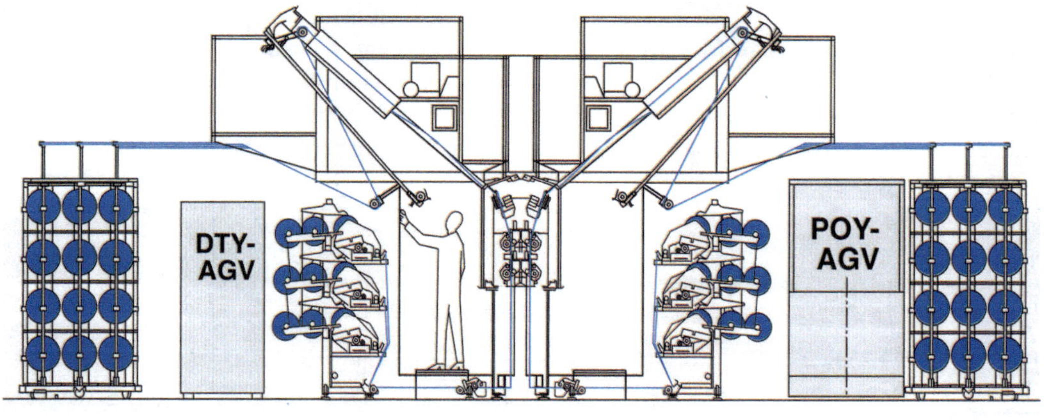

Fig. 29.14 Cross-section through a false twist texturing machine in V-configuration (Oerlikon Barmag), corresponding to set-up 1 and 2 of Fig. 29.15

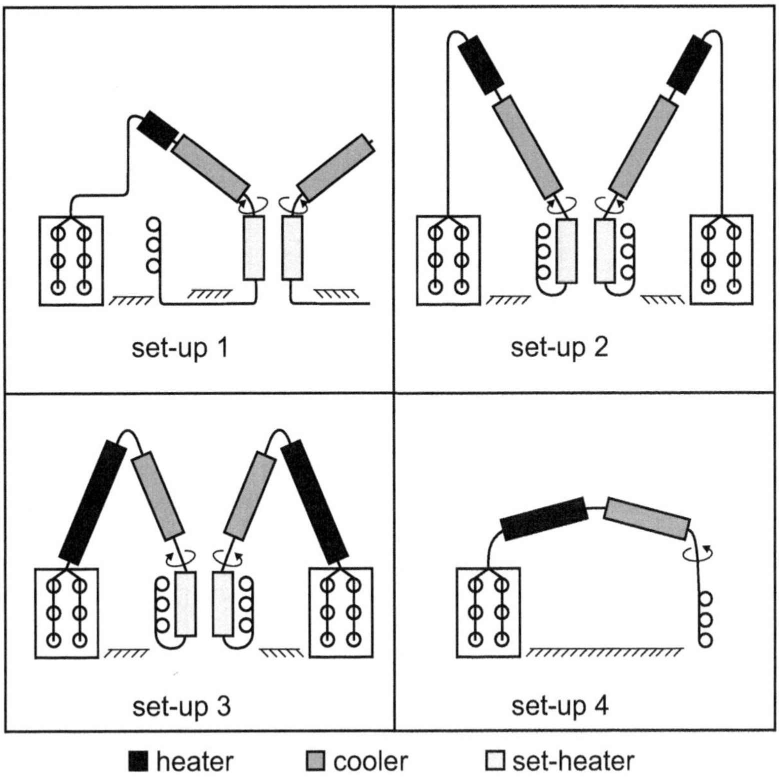

Fig. 29.15 Typical configurations of false twist texturing machines acc. to Gries

results when both sides of the machine are considered. They are mainly used for fine yarns due to the relatively straight yarn path. Set-up 1 has a short high temperature heater, whereas set-up 2 is equipped with a conventional heater. Set-up 3 is called the M-configuration and is used for medium to coarse yarns. Its overall height is considerably lower than that of set-up 1 and 2. The set-up 4 is a special design with a low overall height and relatively straight yarn path. However, its operation is somewhat more difficult and requires short operating personnel.

29.1.6 Influencing Variables and Their Effect

The Fig. 29.16 shows the complex relationships between the adjustable machine parameters and the resulting properties of the textured yarns. Many of these interrelationships cannot be quantified exactly. Therefore, the experience of the machine operator during texturing is of decisive importance for the product quality.

The most important factors influencing the quality of the false twist textured yarn are the yarn temperature, the d/y ratio, the ratio of the yarn tensile force in front of and behind the twister and the draft.

29.2 Air-Jet Texturing

This process is the most important after the false twist process. The resulting yarn structure resembles a ring yarn.

29.2.1 Taslan Process

The oldest process for crimping continuous fiber material through a flowing gaseous medium is the Taslan process (Fig. 29.17). It is still used today

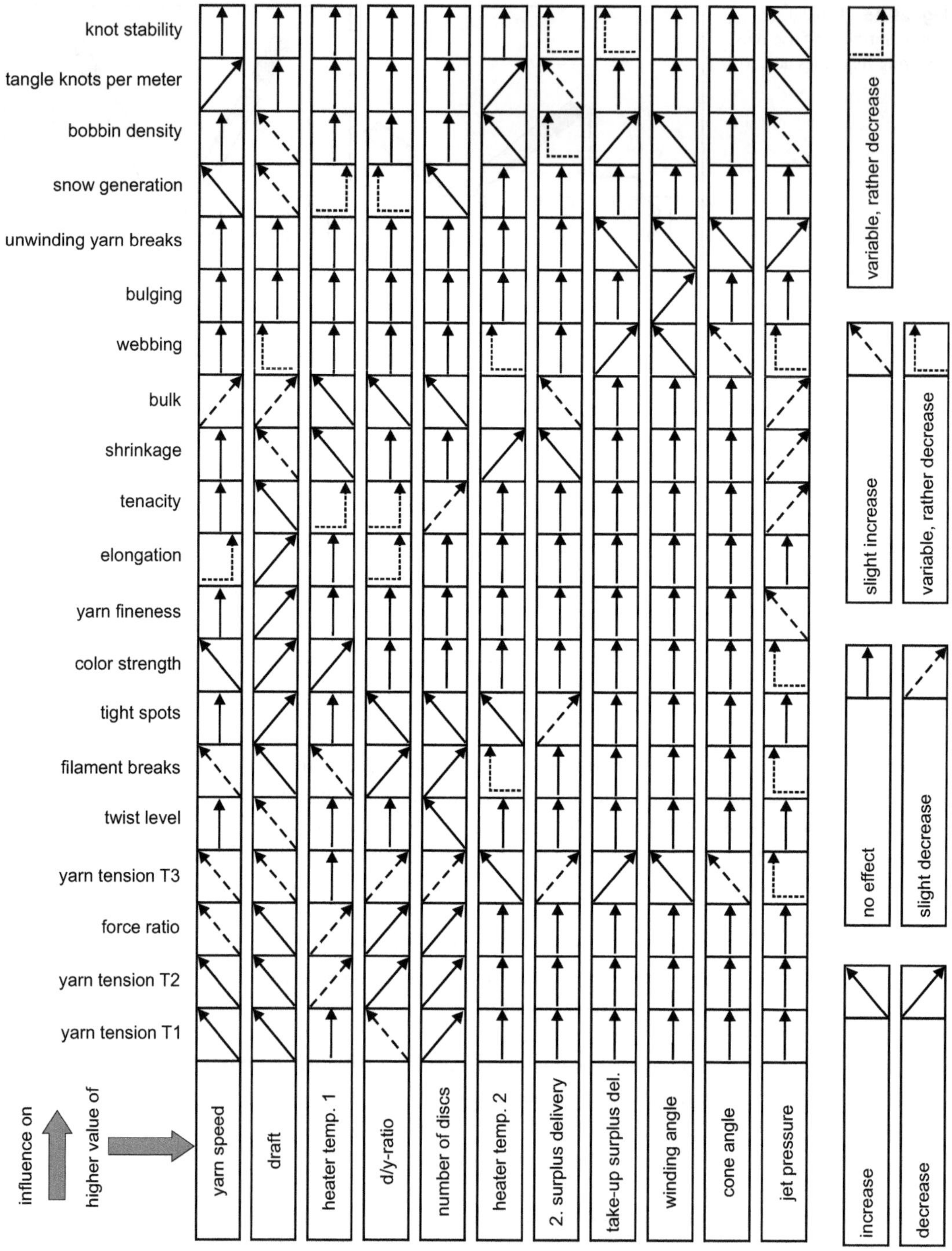

Fig. 29.16 Influencing variables and their effect during false twist texturing acc. to (Hearle et al. 2001)

because the yarn character is clearly different from that of false-twist textured yarns resulting especially in a different hand.

In this process, the smooth filament yarn enters the air-jet with an overfeed of 30–80%, depending on the yarn count (Fig. 29.18). Air

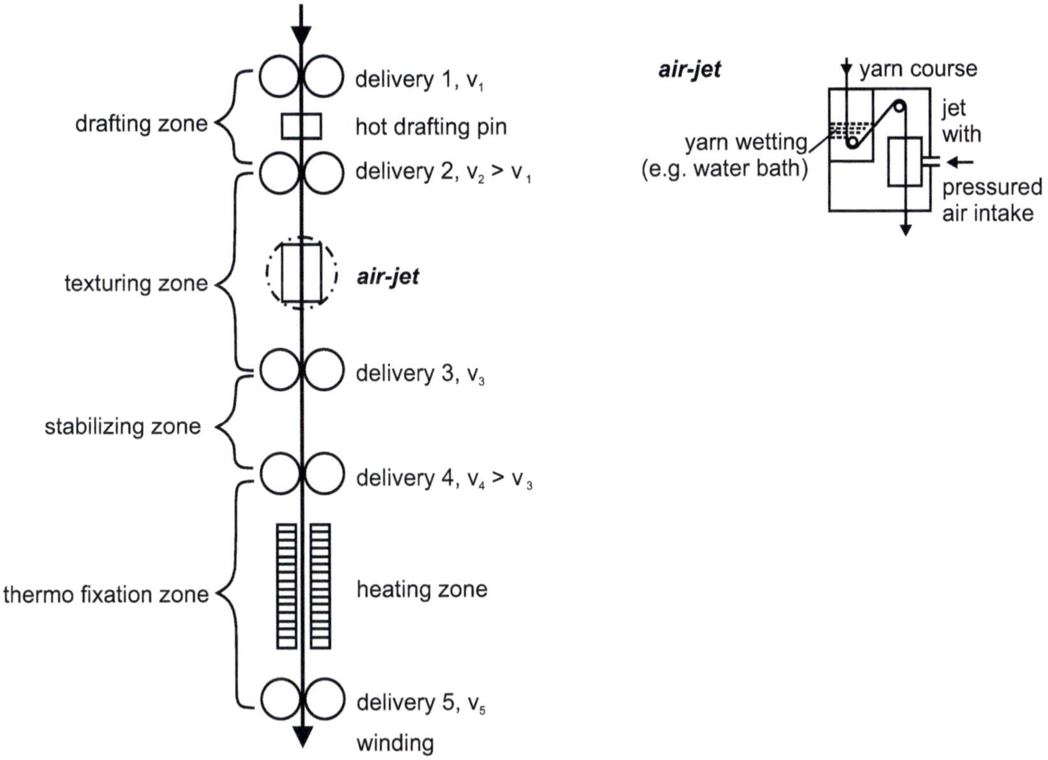

Fig. 29.17 Air-jet texturing process acc. to (Lünenschloß et al., 1985)

flows at very high speed through the air-jet. When the air flow behind the jet widens again, it splits the individual filaments, slightly crimping them and partially pulling them out of the yarn structure forming loops. The effect is further enhanced when the yarn is delivered from the jet onto an impact ball. The following illustration shows a typical air-jet.

Fig. 29.18 Air texturing jet (Heberlein AG n.d.)

The yarn crimp is then continuously fixed and the yarn is wound up. Speeds of up to 1000 m/min and yarn counts of over 1000 dtex are possible. Typical nozzle pressures are 7–14 bar. The air consumption V_{air} of an air texturing nozzle in [m^3/h] can be estimated by

$$V_{air} = 1{,}2 \cdot (p_{air} + 1)$$

Typical values for the setting of an air texturing jet are compiled in Table 29.1.

Air textured filament yarns for apparel textiles are similar in character to staple fiber yarns, especially in terms of cover factor and hand (Fig. 29.19).

Figure 29.20 shows a typical air-jet texturing machine module with 5 positions and a take-up

Table 29.1 Typical texturing settings for different yarns acc. to (Heberlein AG)

Titer [dtex]	Single fiber titer [dtex]	Max. overdelivery [%]	Max. yarn speed [m/min]
Voluminous yarns			
330–1100	0.8–5.5	160	600
600–2500	4.0–12.0	160	600
Compact, stable yarns			
10–50	0.5–15	60	400
80–250	0.8–3.5	60	1000
150–480	0.8–3.5	60	900
600–1100	0.8–5.5	60	800
1800–3500	4.0–12.0	40	600
Fancy yarns			
80–250	0.8–3.5	60	1000
150–480	0.8–3.5	60	900
600–1100	0.8–5.5	60	800

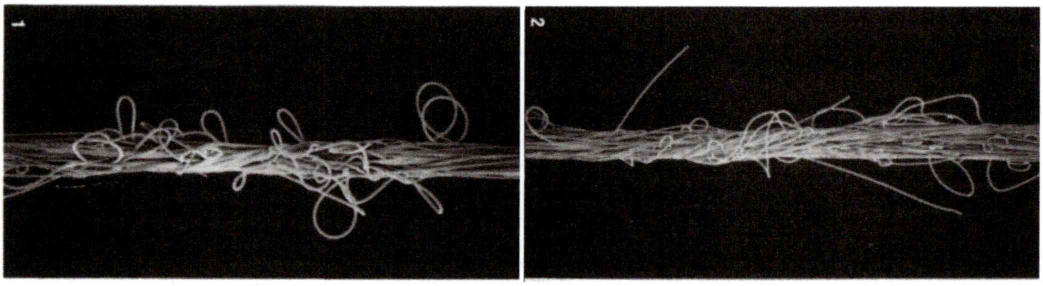

Fig. 29.19 Structure of air-jet textured yarns

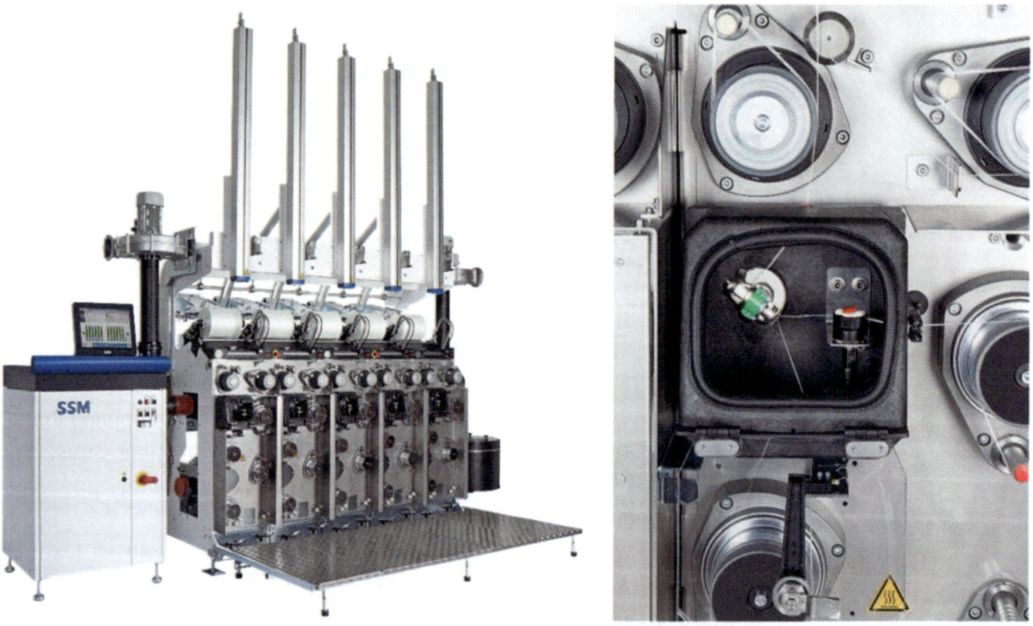

Fig. 29.20 Air-jet texturing machine DP5-TF/TC and air-jet in sound box (SSM n.d.)

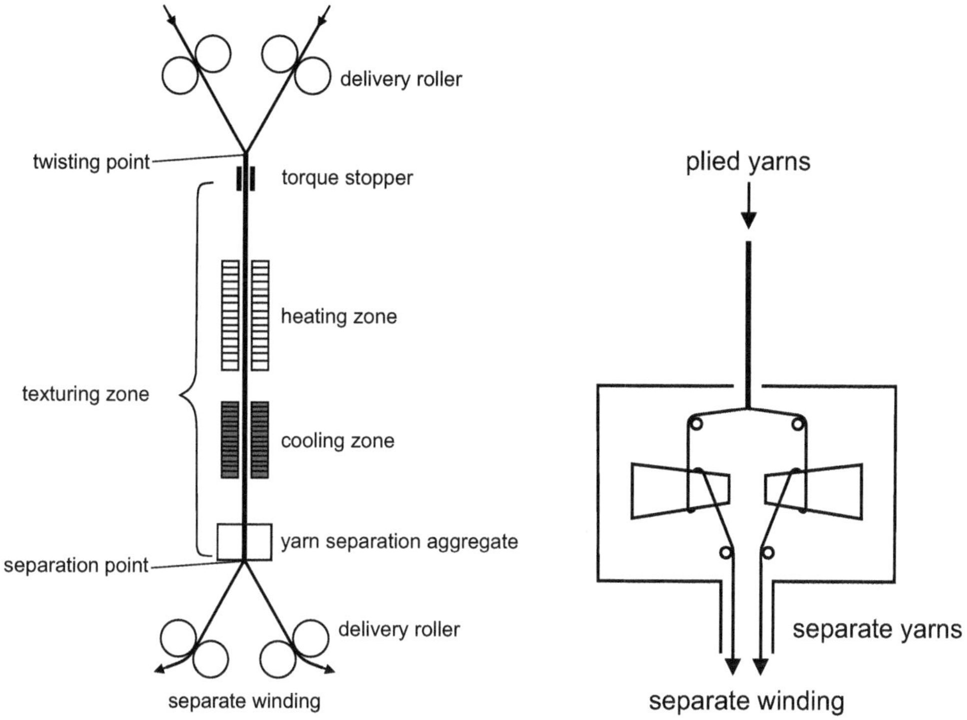

Fig. 29.21 Twist/untwist process (left: principle; right: twist/untwist unit) acc. to (Lünenschloß et al., 1985)

speed of up to 1500 m/min. From this type, up to 8 modules can be combined in a single machine.

29.2.2 BCF Process

Carpet filament yarns are textured by a different type of air-jet texturing. Here, the texturing effect is substantially enhanced in a combination of air texturing with the above-described stuffer box by heating the filament yarn to almost the melting point at the moment of crimping. The details are explained in the resp. chapter.

29.3 Other Processes

29.3.1 Twist/Untwist Texturing

In this process, smooth yarns made of fixable polymers are twisted together and this state is heat-set with a heating and a cooling zone. The twisted yarn is then untwisted and the yarns are wound up separately (Fig. 29.21).

The resulting crimp structure is much coarser than that obtained by false twist texturing, as Fig. 29.22 shows. The process has only marginal importance today.

29.3.2 Knit Fixation Process

In this process, the yarn is knitted on a circular knitting machine, followed by heat setting and

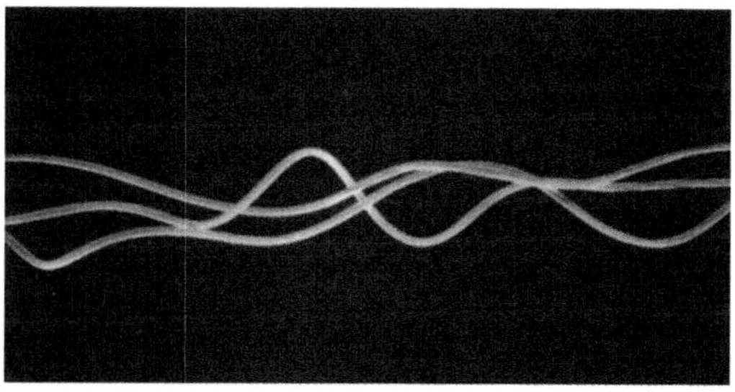

Fig. 29.22 Twist/untwist textured yarn

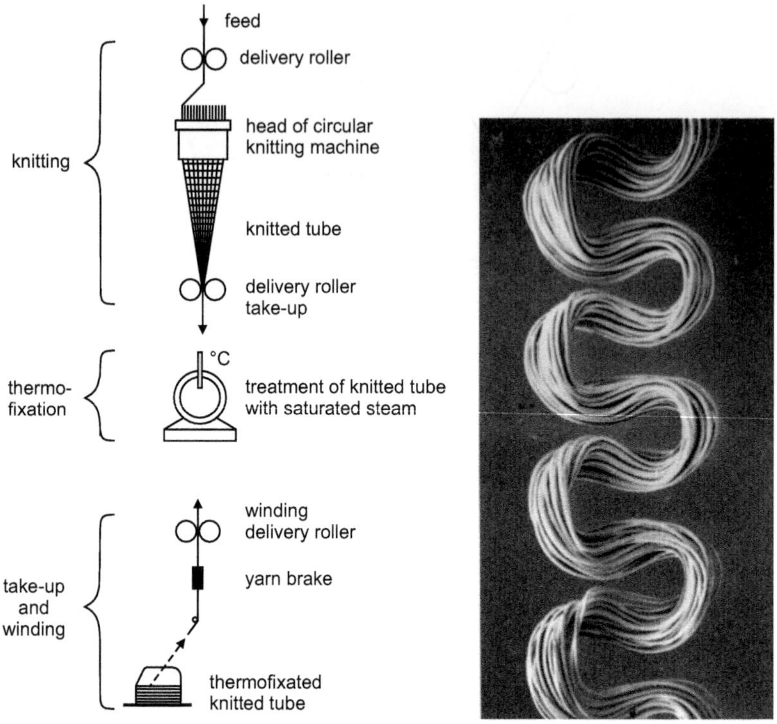

Fig. 29.23 Knit fixation process (left: principle; right: yarn) acc. to (Lünenschloß et al., 1985)

then the knitted fabric is unraveled. Figure 29.23 shows the principle and the corresponding yarn. This process is today only used on a laboratory scale for yarn characterization.

References

Bröll GmbH & Co. KG, Dornbirn, Österreich. https://www.broell.com

Bruske, J. (1990). *Beitrag zur Steigerung von Leistung und Qualität der FD-Texturierung*. Dissertation an der RWTH Aachen.

Hearle, J. W. S., Hollick, L., & Wilson, D. K. (2001). *Yarn Texturing Technology*. Woodhead Publishing.

Heberlein AG, Wattwil, http://www.heberlein.com

Lünenschloß, H.-J., Farber, K., & Steinmetz, J. (1985). *"Texturier-Verfahren", textil praxis international*. Konradin-Verlag Robert Kohlhammer GmbH.

Oerlikon Barmag, Remscheid. https://www.oerlikon.com/de/marken/oerlikon-barmag/

Saurer Technologies GmbH & Co. KG Engineered, Hammelburg. https://saurer.com/de/komponenten/temco

Schmenk, B. (2003). *Sicherung von Prozessstabilität und Garnqualität in der Falschdraht-Texturierung*. Dissertation at RWTH Aachen University. Shaker-Verlag, Aachen.

SSM, Schärer Schweiter Mettler AG, Wädenswil. https://www.ssm.ch

Part IV

Synthetic Chemical Fibers: Production and Properties

30 Polyester

30.1 Definition

Polyester fibers are by far the most important textile raw material in terms of quantity. They include all synthetic fibers which are made up of linear macromolecules and consist of at least 85% by mass of the ester of a diol and terephthalic acid.

In accordance with ISO standard 1043, the man-made fiber manufacturers united in BISFA (Bureau International pour la Standardisation de la Rayonne et des Fibers Synthétiques) have agreed on a uniform designation. Acc. to this agreement, the designation for polyester is PES. The most important polyester for fiber production is polyethylene terephthalate and is abbreviated as PET (Fig. 30.1).

30.2 History

The preparation of linear, filament-forming polyesters by reaction of aliphatic dicarboxylic acids and glycols had already been studied by W. H. Carothers as part of his work on high polymers (and on polycondensation reactions in particular) beginning in 1928 (Carothers & Arvin, 1929). However, because of the low melting points of these polyesters and their solubility in many organic solvents, which largely precluded textile applications, this work was not pursued. Nevertheless, the first synthetic man-made fiber presented by Carothers and Hill in 1931 was a polyester thread made from sebacic acid and ethylene glycol (Carothers & Hill, 1932; Schlack & Kunz, 1953).

Building on Carothers' research, Schlack and Ludewig realized in 1939 that terephthalic acid increases the melting point of the polymer and also the stiffness and elasticity of the thread (Ludewig, 1955). However, they initially used it for polyamide production. In 1942, Schlack condensed terephthalic acid dihalogenides and terephthalic acid esters with butylene glycol to form a primary polyester and further condensed this with diisocyanates to form nitrogen-containing polyesters. These had relatively high melting points and could be hot-drawn to three to four times their initial length while orienting the macromolecules (Schlack, 1952).

30.2.1 Polyethylene Terephthalate (PET)

Independently of this, in 1939, the English researcher John Rex Whinfield, together with his colleague James Tennant Dickson, discovered that when the aliphatic dicarboxylic acid is replaced by a dicarboxylic acid of the aromatic series, the solubility of the polyester in organic solvents is considerably reduced and its melting point is also substantially increased (Whinfield, 1941). The polyester they synthesized from terephthalic acid and ethylene glycol was given the name "terylene". Whinfield and Dickson first

D. Veit, *Fibers*, https://doi.org/10.1007/978-3-031-15309-9_30

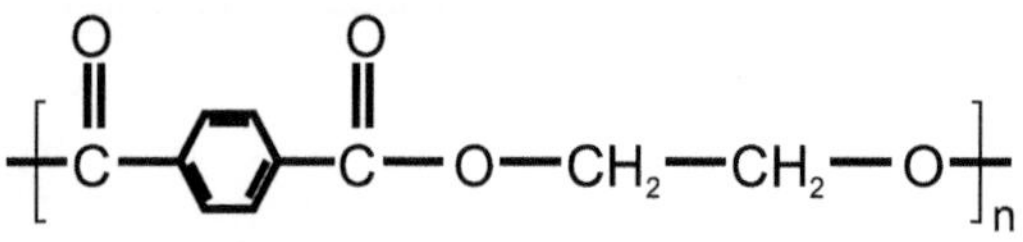

Fig. 30.1 Structural formula of the repeat unit of polyethylene terephthalate (PET)

produced usable filaments from this polyester in 1941 (Cook, 1946). In 1944, the first laboratory-scale production of terylene yarn was made by Imperial Chemical Industries, Ltd. (ICI) (Whinfield, 1952) and ICI acquired the patent rights for terylene outside the USA. They then took over the further development of yarn production (including the large-scale synthesis for terephthalic acid required for this purpose). Experimental production of terylene yarn began in 1947, with large-scale production starting in 1953.

The patent rights for the USA were acquired by DuPont de Nemours in 1946. Beginning in 1950, DuPont de Nemours produced trial quantities of filament yarn and staple fibers under the name "Fiber V." In 1953, it began production under the trade name "Dacron" at its Kingston, N. C., plant. In Germany, polyester production began in the mid-1950s under the trade names "Trevira" (Hoechst) and "Diolen" (Akzo). ICI, for its part, had granted production licenses for polyester fibers using its proprietary process to fiber manufacturers in various countries. From 1954 onwards, these began production under various trade names. The basic patents for the production of polyester fibers expired in 1966.

30.2.2 Polytrimethylene Terephthalate (PTT) and Polybutylene Terephthalate (PBT)

Since the 1990s, polyesters of terephthalic acid have been produced in combination with other diols as ethylene glycol. Their production costs are about 20% higher than those of PET due to the more expensive raw materials, but they are

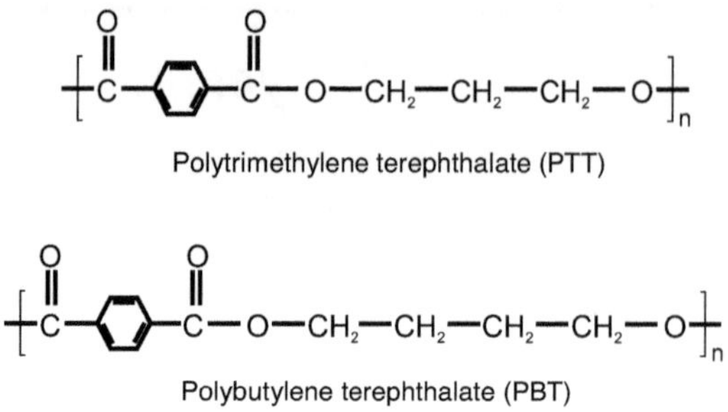

Fig. 30.2 Structural formulae of polyester terephthalates from other diols

usually only about 90% of the production costs of polyamide 6 and polyamide 66. PTT in particular competes with these fibers due to its high elasticity, its good dyeability already below the boiling temperature of water and its high recovery capacity, which is of great importance especially for carpet yarns. PBT has so far remained a niche product (Fig. 30.2).

30.2.3 Other Polyester

As early as 1958, the Eastman Kodak Company produced a polyester whose polycondensation used an aromatic diol instead of ethylene glycol (1,4-dimethylolcyclohexane), Fig. 30.3. The polyester fibers produced from this ('Kodel' or 'Vestan') have similar mechanical properties and, compared with PET, a slightly higher melting temperature of 285–295 °C (Eastman, 1958).

In order to improve certain properties of polyester fibers - in particular their only moderately good dyeability and the susceptibility of textiles made of staple fiber yarns to pilling - chemical or physical modifications were made which significantly altered the properties of these 'second and third generation polyester fibers' compared to those of the standard types. In the case of chemically modified polyester fibers, a small proportion

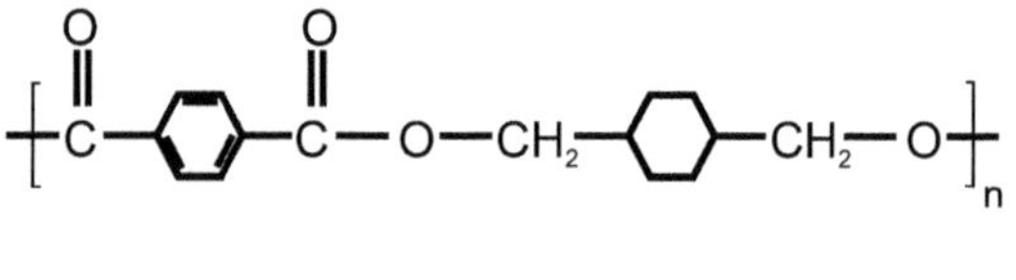

Fig. 30.3 Structural formula of Kodel™

of a third or fourth component, e.g. isophthalic acid or p-hydroxybenzoic acid, is incorporated into the polyethylene glycol terephthalate chain in addition to the terephthalic acid. Physical modifications are now made by cross-sectional profiling or spinning of bicomponent fibers. Other new developments led to polyester ether fiber. Recently, the production of polyester microfibers from PET has gained greatly in importance.

30.3 Raw Materials

Polyesters are formed by polycondensation of dicarboxylic acids with diols. Both substances have reactive end groups for polymer formation. Linear polyesters suitable for filament formation are produced from bifunctional starting materials. An alternative route is the self-condensation of x-hydroxycarboxylic acids or the reaction of corresponding tri- or polyfunctional compounds. However, it has not become established in the market.

30.3.1 PET

The most important raw materials for the production of polyethylene terephthalate (PET) are (Fig. 30.4):

Terephthalic acid in para position (1,4):

Until about 1976, the main material used in the production of polyethylene terephthalate was dimethyl terephthalate (DMT), which is much easier to dissolve and purify than the terephthalic acid that is predominantly used today. However, the use of terephthalic acid has proven to be more cost-effective with increasing capacities, combined with further advantages in process control. This applies in particular to the continuous manufacturing process up to direct spinning.

30.3.2 PTT

For the production of polytrimethylene terephthalate (PTT), the diol used is (1,3)-propanediol.

$$HO—CH_2—CH_2—CH_2—OH$$

The acid used for PET is either terephthalic acid or dimethyl terephthalic acid. PTT has a significantly higher elasticity than PET, can be dyed below the boiling temperature of water, unlike PET, and has a good recovery capacity. The diol used to produce PTT fibers is often produced from glucose by fermentation, resulting in PTT being sold as a "biopolymer".

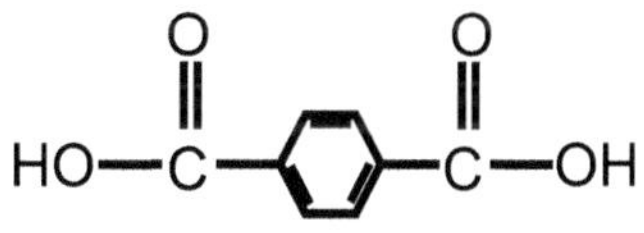

or (rarely)

dimethylterephthalic acid (DMT) in para position (1,4):

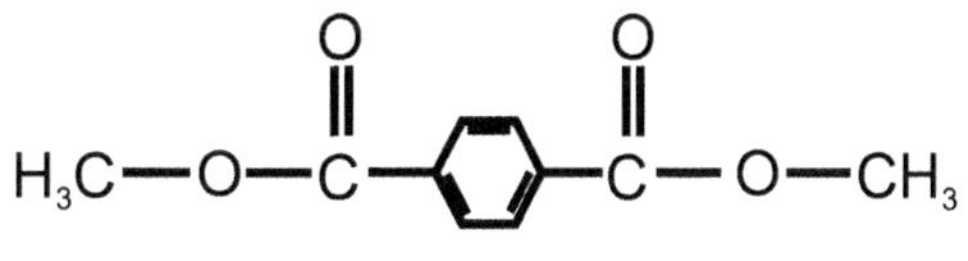

and ethylene glycol (EC)

$$HO—CH_2—CH_2—OH$$

Fig. 30.4 Structural formulae of the basic materials for the production of PET

30.3.3 PBT

For the production of polybutylene terephthalate (PBT), (1,4)-butanediol is used.

$$HO{-}CH_2{-}CH_2{-}CH_2{-}CH_2{-}OH$$

As with PET, the acid is either terephthalic acid or dimethylterephthalic acid.

30.4 Production of Polymers

In practice, the polymers are formed by transesterification processes based on DMT (continuously or discontinuously in the so-called batch process) or by esterification processes with terephthalic acid (continuously or discontinuously). In principle, all processes have two steps:

Step one

Transesterification of dimethyl terephthalate (DMT) to diglycol terephthalate (DGT) yields diglycol terephthalate + methanol (Fig. 30.5):

At the same time, pre-condensation takes place.

Step two

Polycondensation to PET takes place in a vacuum with a catalyst (Fig. 30.6):

When terephthalic acid is used instead of DMT, it is esterified directly with glycol in the first stage. This is referred to as "polyesterification" (Fig. 30.7).

The use of terephthalic acid (TA) in the production of polyethylene terephthalate is more advantageous than that of dimethyl terephthalate (DMT) in both economic and process engineering terms. For example, only 870 g TA are required to produce 1 kg of polyethylene terephthalate, as opposed to 1010 g DMT. From a process engineering point of view, the use of terephthalic acid results in a higher polycondensation rate and a lower COOH end group content. Furthermore, neither methanol preparation nor a transesterification catalyst are required.

30.4.1 Discontinuous Process

Small batches (< 20 t/day) and special types are produced discontinuously due to the higher flexibility. The corresponding reactors have a capacity of 50–100 t/d, which corresponds to around 20,000–40,000 t/year.

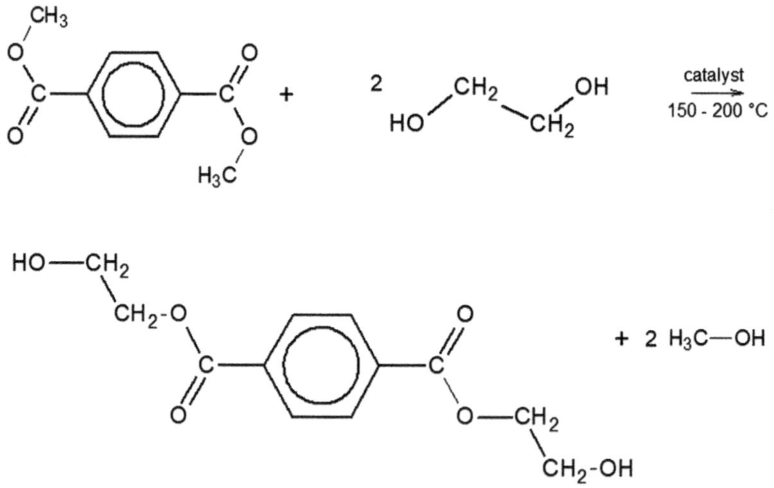

Fig. 30.5 Transesterification reaction

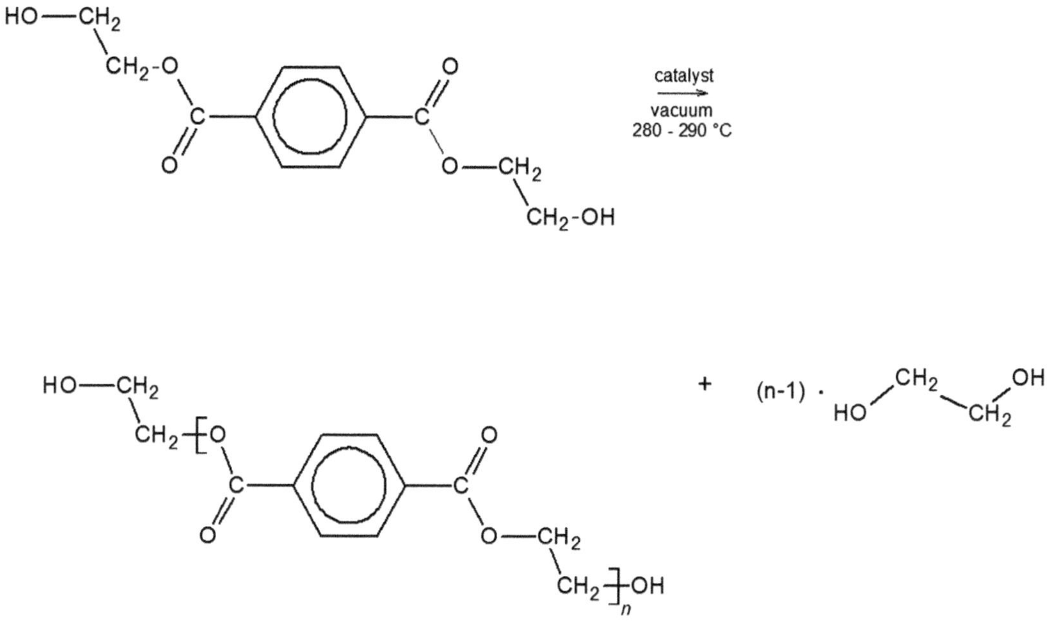

Fig. 30.6 Polycondensation reaction

In the discontinuous process (Fig. 30.8), the transesterification takes place between the DMT melted at 150°–200 °C and the ethylene glycol in a molar ratio of 1:2.5 to 1:4.5. The reaction process is positively influenced by an excess of glycol. As catalyst for the transesterification reaction metal salts, mainly soluble Mn-, Zn-, Mg-, Co-, Ti-compounds in amounts of 0.03–0.1 m/o are used. For dull polyester fibers, 0.1–0.5% titanium dioxide (TiO_2) is added to the glycol.

After transesterification, the temperature is increased to 250 °C in the pre-condensation stage and the excess ethylene glycol is removed. Subsequently, in the polycondensation stage, the polyester is further condensed to the desired viscosity at a temperature up to 285 °C and under vacuum up to 0.1 mbar and with the addition of special polycondensation catalysts (e.g. Sb, Ge and Ti). After polycondensation, the polyester is cooled in water and crushed with underwater or belt granulators (chips).

Due to the draining time of 20–60 min, thermal degradation of the polyester and thus inhomogeneity of the granulate may occur. This can be partially compensated by mixing, but the

Fig. 30.7 Formation of PET by polyesterification reaction

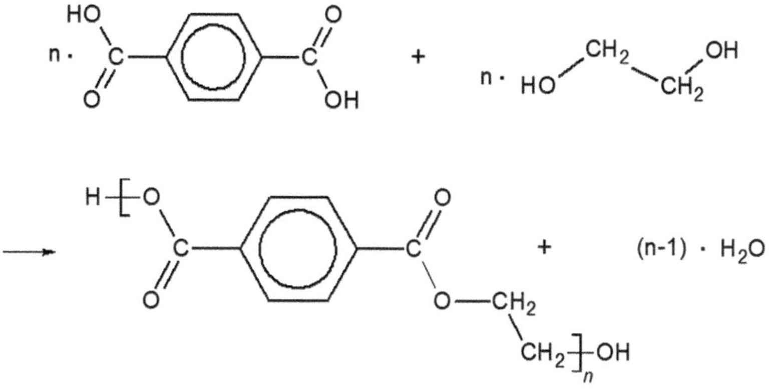

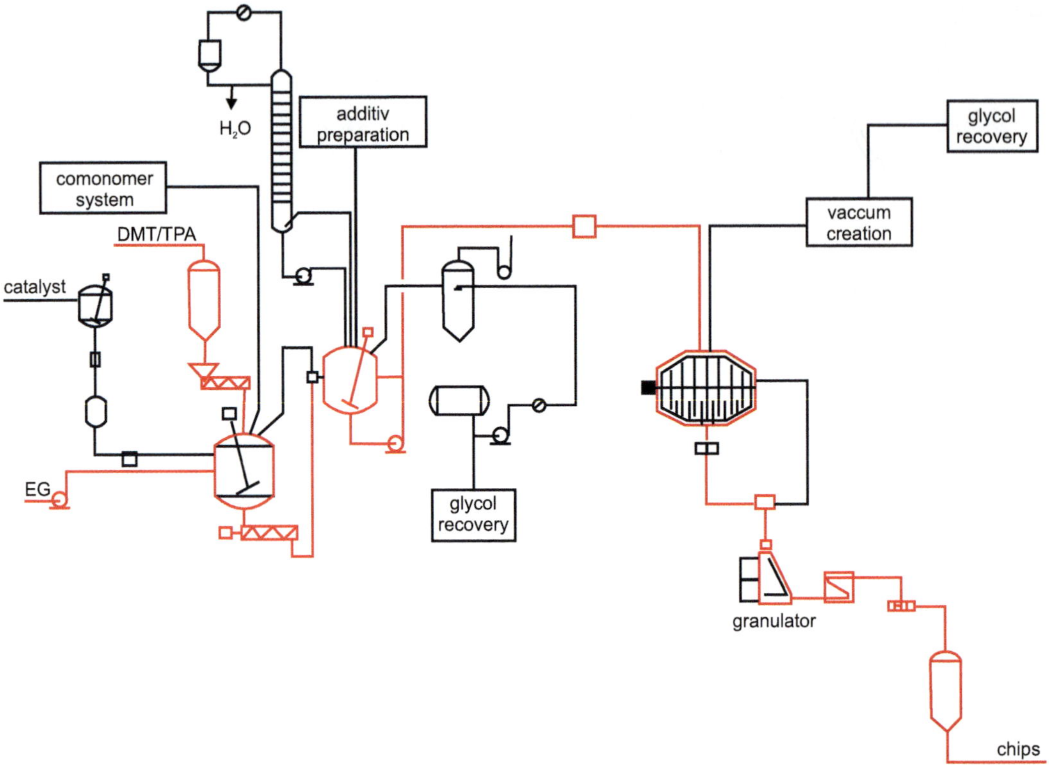

Fig. 30.8 Scheme of the discontinuous production of polyester from DMT/TPA

resulting PET is less uniform than continuously produced PET.

30.4.2 Continuous Process

The continuous process is mainly used for the production of standard types and for large throughputs and was first implemented industrially in 1965. The cascade reactors used in the continuous process based on DMT allow a narrow residence time spectrum within the process steps transesterification, precondensation and polycondensation. In the continuous process, the required viscosity is adjusted by varying either the flow rate or the vacuum. The polyester melt continuously discharged from the final reactor is fed directly into the spinning system via a finisher (Fig. 30.9).

Continuous polyester production from terephthalic acid has further advantages. In addition to the cheaper raw materials, there are lower specific investment costs and lower specific energy consumption compared to a DMT plant. Reactors with capacities of 500–1500 t/d are usually used (Fig. 30.10).

Since practically no esterification catalyst has to be used in the TPA process, the thermal degradation of the polyester caused by a catalyst is lower. Through appropriate reaction control in the cascade reactor, the risk of diglycol formation is significantly lower. The same catalysts are used in polycondensation as in the DMT process. This process has been used from the beginning for the production of polyester tire yarns (tire cord) because of the resulting higher strength.

Polyester is spun out into filament yarns or staple fibers in accordance with the processes described in the relevant chapters.

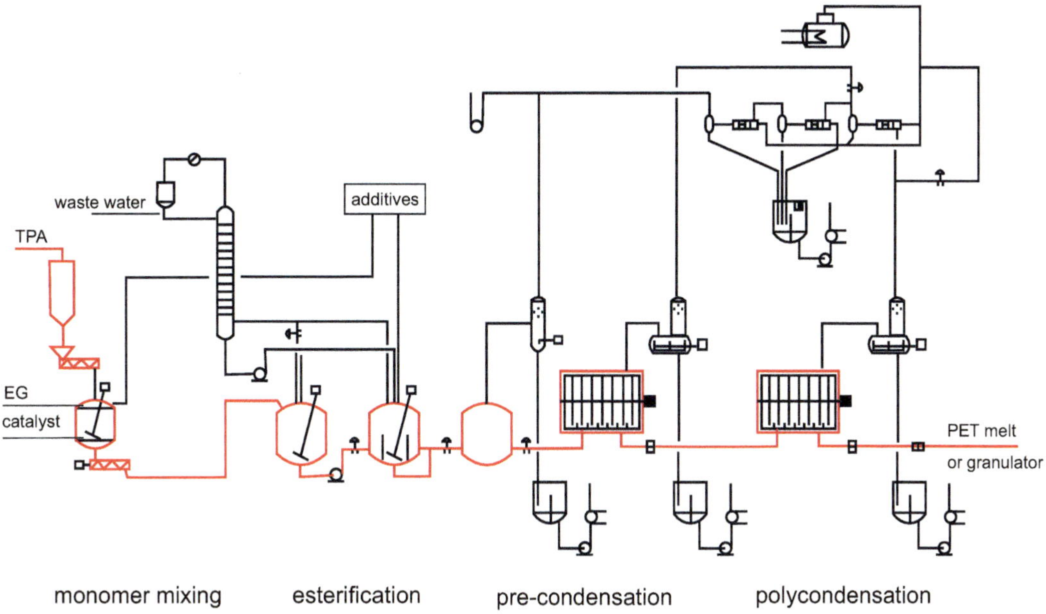

Fig. 30.9 Scheme of the continuous production of polyester from TPA

30.5 Fiber Modification

In general, fiber modifications are possible by producing copolyesters during polycondensation or by mixing different granules to form polymer blends.

Additional equipment can be used to add additives (master batches MB, matting agents TiO_2, antistatic agents, color pigments, etc.) either downstream of the final reactor of the polycondensation process or to the granulate (Fig. 30.11).

Fig. 30.10 Reactor for the continuous production of PET (Oerlikon Barmag n.d.)

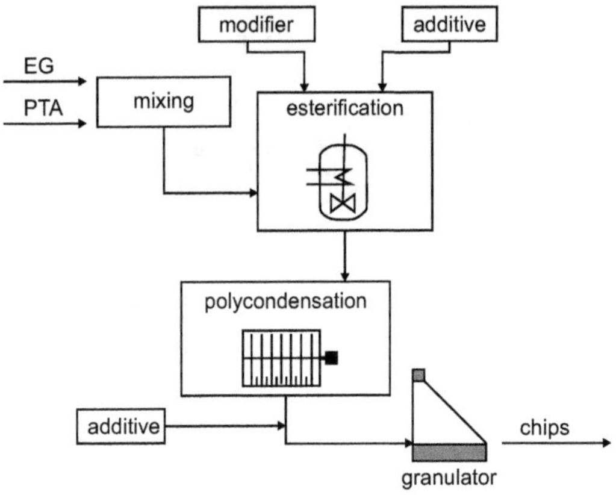

Fig. 30.11 Possibilities of dosing additives in the reactor

Alternatively, additives for small batches can also be fed to the mixing zone of the main extruder with a sidestream extruder via a metering pump (Fig. 30.12).

The use of spinnerets with a profiled cross-section influences the luster and hand of the filaments. For filter applications, for example, the filter effect is improved by the larger surface of the filaments.

PET can also be used to produce bicomponent fibers, e.g. with polyamide, as well as microfibers and microfilaments.

30.6 Chemical Structure

The structural formulae of the repeat units of the major filament-forming polyesters are for (Fig. 30.13):

The average degrees of polymerization (n) are 80–150.

The Fig. 30.14 shows the structural composition of PET, PTT and PBT. It is apparent that PTT chains have considerably more kinks along the molecular axis and an orientation in the fiber axial direction is therefore hindered. Therefore, the elongation of PTT fibers is considerably larger than that of PET and PBT fibers. PTT molecules form tightly coiled helical structures, therefore their elastic behavior resembles a spring and they possess the greatest elastic recovery of these three polyesters. PBT macromolecules are more oriented in the fiber direction than PTT molecules, so the strength of PBT is slightly higher than that of PTT. However, because PET has significantly more polar groups per unit length, the strength of PET is the highest among all 3 polymers (Table 30.1, Fig. 30.20).

Polyethylene terephthalate crystallizes in a triclinic unit cell (Fig. 30.15):

The molecular weight of polyethylene terephthalate is 15,000–27,000 for textile grades and up to 54,000 for high strength polyester grades. This corresponds to average degrees of polymerization of 80–140 or 280.

30.7 Properties of Polyester Fibers

The properties of polyester fibers can be adjusted within wide limits, both by the spinning conditions and by physical or chemical modification.

Fig. 30.12 Possibilities of dosing additives during extruder spinning

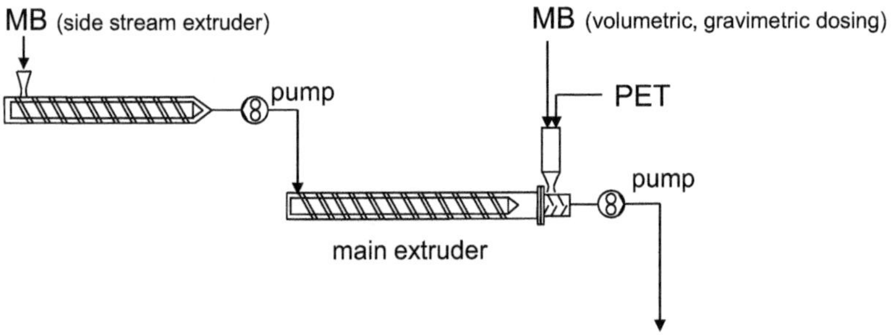

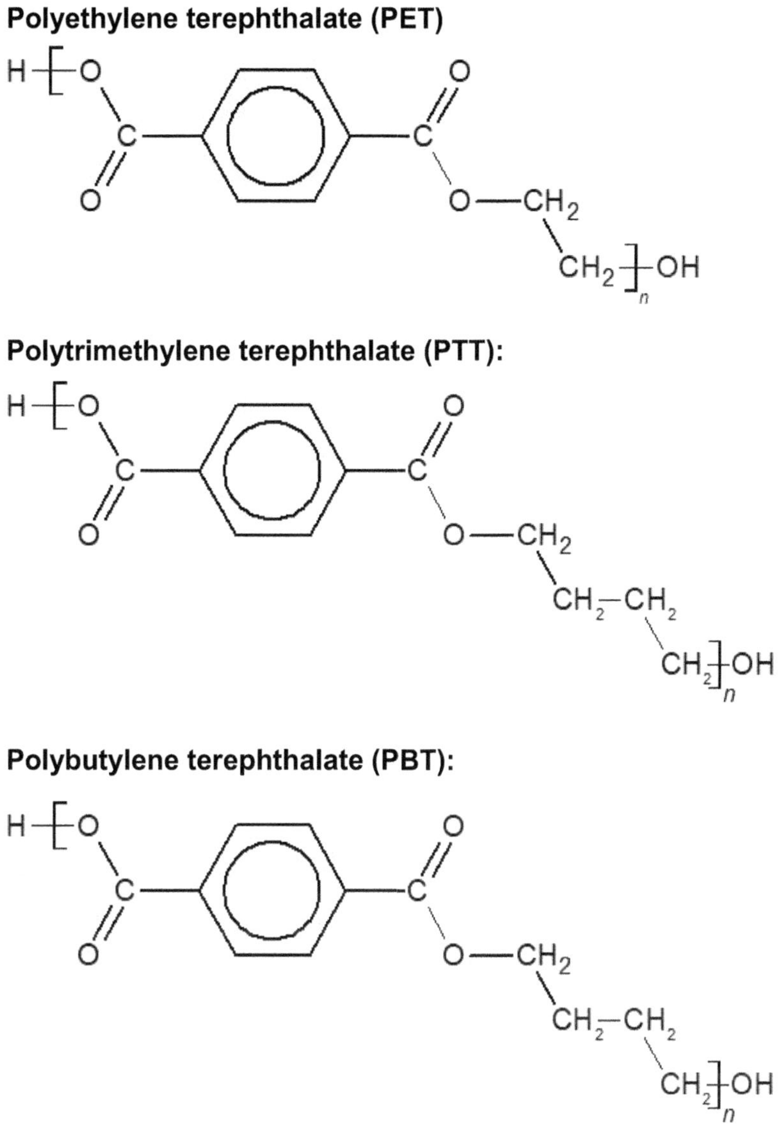

Fig. 30.13 Structural formulae of the repeat units of PET, PTT and PBT

30.7.1 Cross Section Shapes and Types

In longitudinal view, polyester fibers are mostly smooth and structureless, with the exception of embedded pigment particles of TiO_2 in the case of semi-dull or dull fibers. Profiled fibers have a corresponding longitudinal grooving.

The cross-sectional shape is usually round and smoothly bounded. In addition, there are fibers with deliberately profiled cross-sections, e.g. trilobal or pentalobal, as well as various hollow fiber types (Fig. 30.16).

In Fig. 30.17, common types of PET filament yarns are compiled. Yarns made of trilobal fibers are marked with triangles, all other yarns consist of round filaments.

In Fig. 30.18, common types of PET staple fibers are compiled. Fibers with a fineness of 5–10 dtex predominate, which corresponds to a fiber diameter of 18–25 μm.

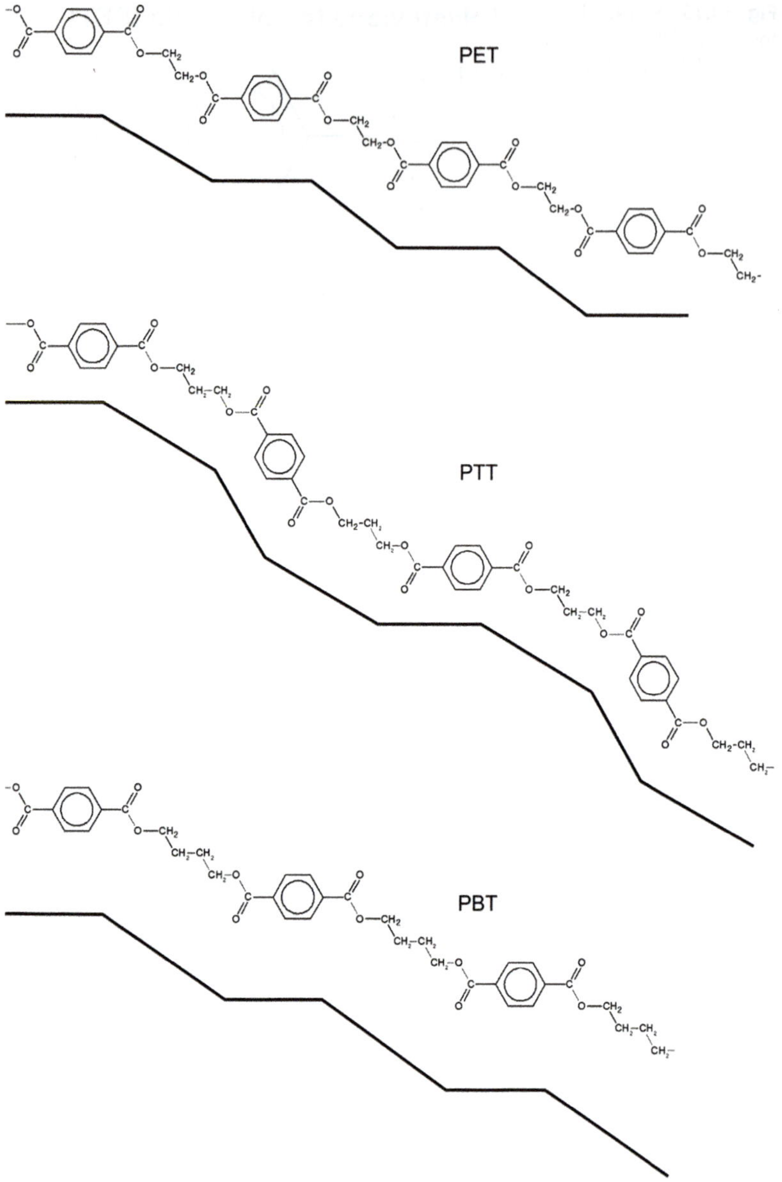

Fig. 30.14 Comparison of the crystal structures of PET, PTT and PBT

30.7.2 Mechanical Properties

The mechanical properties of polyester fibers, in particular the dynamometric characteristic values of maximum tensile strength and maximum tensile elongation, can be varied within wide limits depending on the intended use (e.g. cotton or wool type for staple fibers) by changing the process conditions. The degree of stretching is of particular importance. Depending on the type and

Table 30.1 Typical mechanical properties of textile PET, PTT and PBT fibers

	PET	PTT	PBT
Tenacity [cN/tex]	80	45	50
Breaking elongation [%]	6	12	8

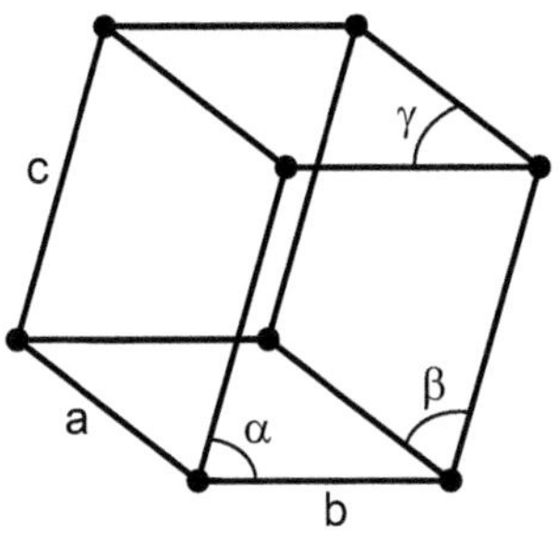

Fig. 30.15 Triclinic unit cell of PET

level of stretching, the same polymer can be used to produce both, a high-strength filament yarn with low maximum tensile elongation and a normal type with medium maximum tensile strength and medium maximum tensile elongation (Fig. 30.19).

Likewise, the degree of drawing influences the amount of break twist, which, together with the transverse brittleness angle, forms a criterion for the structural properties of the fibers. Due to the different processing requirements and service properties, the maximum tensile strength is generally lower for staple fibers, while the maximum tensile elongation is higher than for filaments. Accordingly, the transverse brittleness angle is also lower. In the case of a polyester yarn with profiled filaments, the transverse brittleness angle is lower than in the case of a yarn made from circular filaments due to the cross-sectional shape (Tetzlaff et al., 1993), Table 30.2.

In the force-elongation curve, PET fibers show a considerably higher initial modulus than corresponding polyamide fibers. This is favorable for processing because the low stresses that often occur in the process do not cause any additional stretching or elongation of the polyester yarns. The relatively high modulus of elasticity makes polyester particularly suitable for less stressed articles and those that change little in shape, such as underwear or outerwear, as well as for technical textiles that must exhibit high dimensional stability.

New types of polyester have been developed for some years, especially PBT and PTT. They have higher elongation values than PET, but also lower strengths. Their stress-strain characteristic up to approx. 5% elongation is similar to that of PA 6. PTT has the highest elastic elongation of all linear polyesters (Fig. 30.20).

The abrasion resistance of PET fibers is very high. Although it does not reach the extremely high values of polyamide fibers, it is considerably higher than that of most other chemical and natural fibers.

Figure 30.21 shows typical force-elongation curves of PET staple fibers. Their properties are often modelled on those of cotton and wool, especially for blended yarns.

30.7.3 Special Types

The large number of applications for polyester fibers has led to the development of a wide

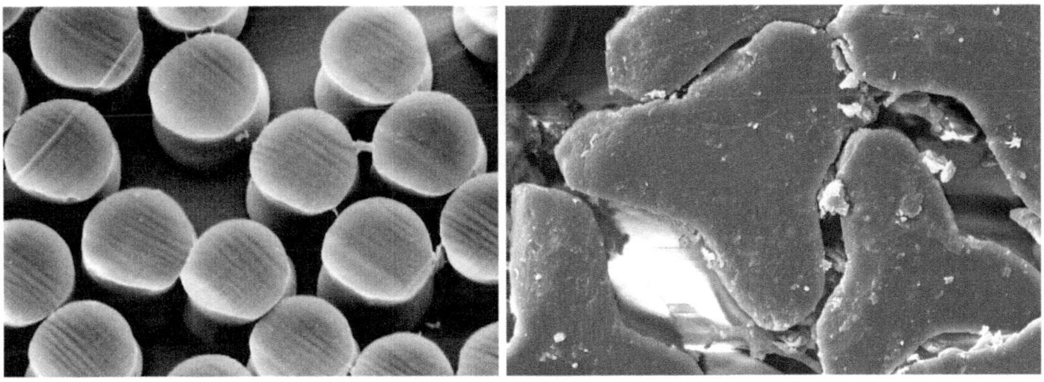

Fig. 30.16 Cross sections of round and trilobal PET fibers

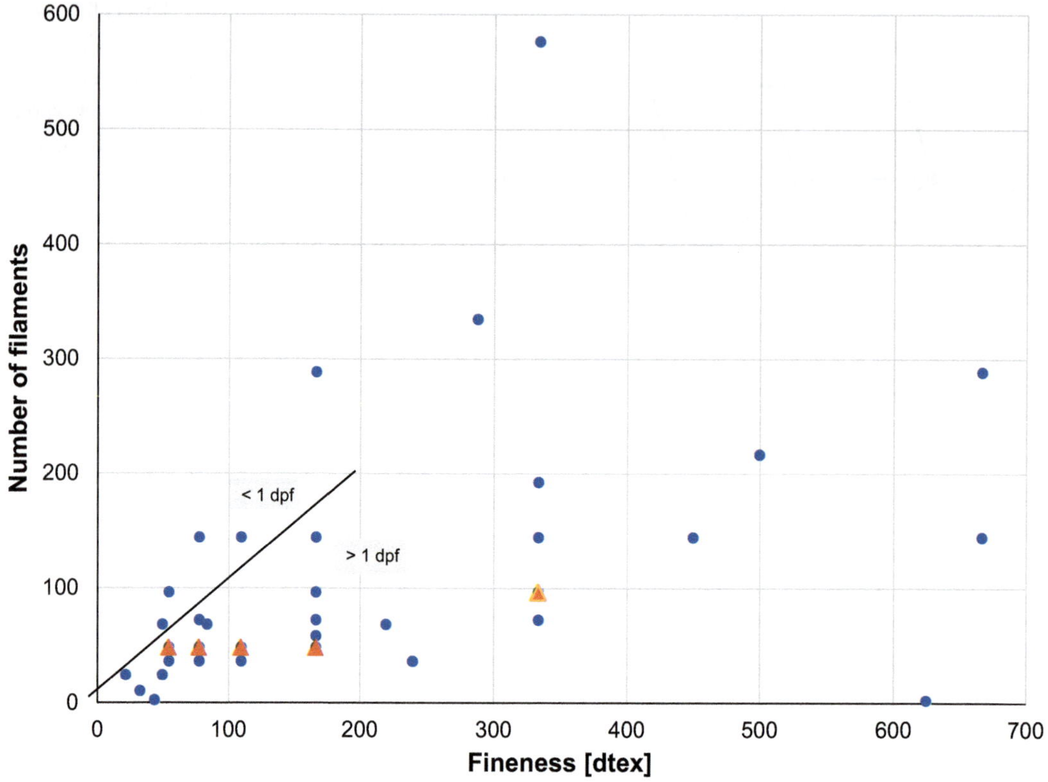

Fig. 30.17 Typical filament yarn types (fineness, number of filaments; circle: round; triangle: trilobal)

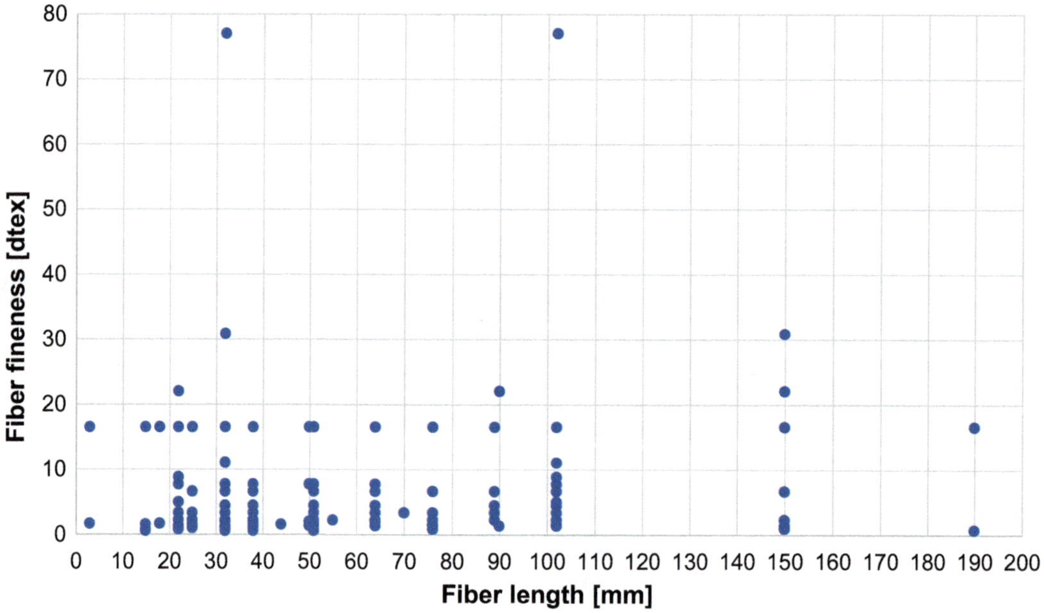

Fig. 30.18 Typical staple fiber types (fiber length, fiber fineness)

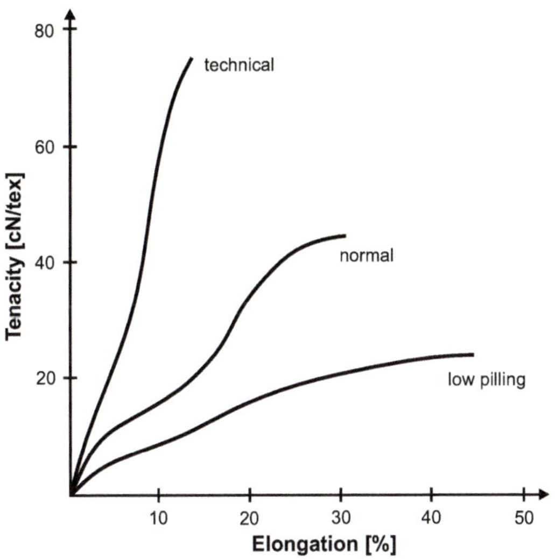

Fig. 30.19 Typical stress-strain curves for PET yarns

range of chemically or physically modified special grades.

30.7.3.1 High Tenacity Filament Yarns for Technical Applications

These are particularly highly stretched filament yarns. The high strength is achieved by a high orientation of the macromolecules along the longitudinal axis of the filament. High-strength filament yarns are used in the tire cord industry and for other technical purposes, e.g. for use in airdomes, for safety belts and conveyor belts.

30.7.3.2 Low-Pill Staple Fibers for Use in Apparel Textiles

By reducing the average molecular weight, the buckling strength is reduced to such an extent that pills already break off due to the mechanical stresses that occur during their formation.

30.7.3.3 High-shrink Fibers

In the production process, the boiling shrinkage is adjusted between 30 and 70%. This results in polyester high bulk yarns which are suitable for articles with a voluminous handle and as filling fibers.

30.7.3.4 Easier to Dye (Deep Dyeing or Carrier-Free Dyeing) Types

The incorporation of block polyethers in the production of polyethylene terephthalate by the transesterification process lowers the glass transition temperature of the polyester to such an extent that carrier-free dyeing at low temperatures is possible. In addition, the mobility of the molecular chains can be increased by incorporating additional methylene groups $(–CH_2)_n$. This makes it easier for the dye to go onto the fiber.

30.7.3.5 Types Dyeable with Cationic Dyes (Anionically Modified)

Bicolor effects can be achieved when blending these fiber types with homopolymer polyester fibers.

Table 30.2 Important mechanical properties of polyester fibers, e.g. acc. to (von Falkai, 1981; Fourné, 1995)

Type	Density [g/cm^3]	Tenacity [cN/tex]	Wet tenacity [%]	Breaking elongation [%]	Wet elongation [%]	E-modulus [cN/tex]	Transverse brittleness [°]
Staple fibers – pr	1.36–1.41	30–55	95–100	25–50	100–105	250–400	30–49
Staple fibers – pr		25–40		30–55			48
Filament yarns – hs		40–60		20–30		700–1200	41–48
Filament yarns – hs		60–95		8–20			47–48
Monofilaments		25–60		15–45			

pr pilling resistent, *hs* high strength

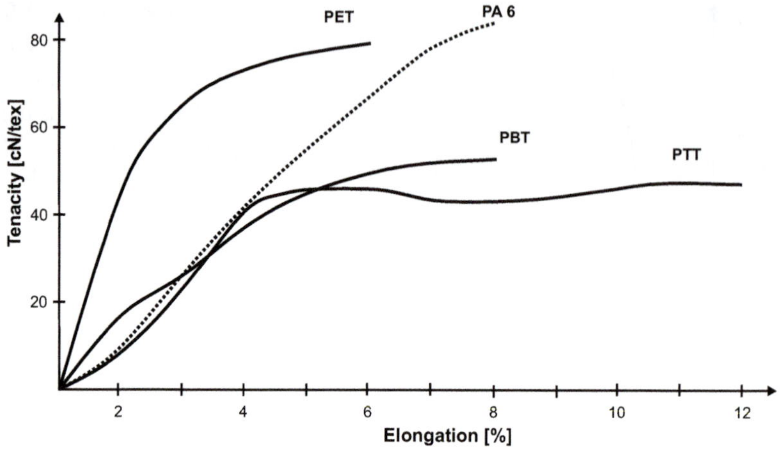

Fig. 30.20 Typical stress-strain curves of PET, PBT and PTT filament yarns compared to that of PA 6 filament yarns

30.7.3.6 Profile Fibers

These are fibers with specifically profiled filament cross-sections, e.g. triangular or pentagonal stars. The cross-section is profiled in a wide variety of ways in order to change the luster and feel on the one hand and to increase the opacity and adhesion of the finished goods on the other.

30.7.3.7 Bicomponent Fibers of the Type S/S (Side by Side)

The bicomponent fibers of this type are characterised by a stable three-dimensional crimp. They are used texturized for stockings and finished goods with a wool-like, puffy character and as filling fibers.

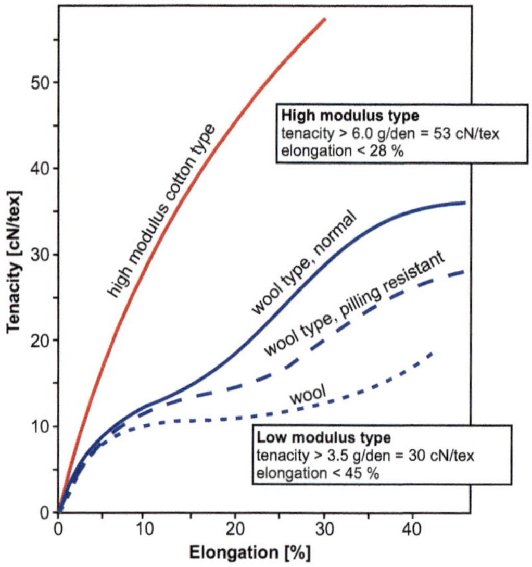

Fig. 30.21 KD diagram for different PET types (staple fibers)

30.7.3.8 Bicomponent Fibers of the M/F Type (Matrix Fibrils)

These are, for example, fibers made of polyamide and polyester which are used for tire cord yarns because they serve to improve the "flatspotting" (so-called "morning sickness") which is disadvantageous in polyamide tire yarns. Spectacle cleaning cloths made from PET/PA split fibers are also a typical application.

30.7.3.9 Microfibers

These are staple fibers and filaments with a fineness mostly well below 1.0 dtex (1 dpf), which are mainly used in apparel textiles. They have a particularly low bending moment, which leads, among other things, to a soft handle.

30.7.4 Physical Properties

The good physical properties, especially the relatively high glass transition temperature and melting temperature, are advantageous for further processing into a wide range of products. The moisture absorption is very low compared to other man-made fibers, the hot water shrinkage can be adjusted variably (Table 30.3).

Table 30.3 Properties and water absorption capacity of PET fibers

Type	Hot water shrinkage [%]	Specif. warmth [kJ/kgK]	Glass temperature [° C]	Melt temperature [° C]	Moisture take-up [%]	Water retention [%]
StF	0.5–1.0	1.1–1.4	70–100	250–260	0.2–1.0	3–5
pr StF						
Fil	5–10				0.2–0.5	
hs Fil	7–8					

StF Staple fiber, *pr* pilling resistent, *Fil* Filament yarns, *hs* high strength

The thermal properties of PBT and PTT differ greatly from PET. For example, their melting point is significantly lower and their glass transition temperature is also significantly lower than that of PET. The latter in particular has so far prevented their use in clothing textiles (ironing temperature). At 14%, the shrinkage of PTT is also far higher than that of PET, which can also have an unfavorable effect on further processing and textile use. Crystallization of PTT is approx. 15 times faster than that of PET, 3 times faster than that of PA 6 and approx. 20% faster than that of PA 6.6. The intrinsic viscosity of PET is approx. 0.4–0.7 dl/g, that of PTT is 0.8–1.2 dl/g (Table 30.4).

The molecular structure of polyester fibers is essentially dependent on the spinning speed. At low take-up speeds (< 3200 m/min), polyester fibers have a predominantly amorphous structure. With increasing spinning speed, the macromolecules crystallize and a higher orientation of the macromolecules along the fiber axis results. This reduces the so-called "residual elongation"(residual stretching), i.e. the plastic portion of the deformability of the PET fibers (Fig. 30.22).

In addition to stretching, heating processes, e.g. during thermofixation, also have an influence on the orientation and crystallinity of the macromolecules. With increasing fixation temperature, the crystallinity increases. Whether thermofixation is a physical reorientation process or a chemical transesterification process, in which chain molecules break apart in amorphous zones and pass into a more energetically stable state, is still unclear (Hearle, 1993).

30.7.5 Chemical Behavior

30.7.5.1 Flammability

During the burning test, polyester fibers melt together to form a brownish globule on the flame. This globule may ignite with soot, giving off an aromatic odor. After removal of the flame, the polyester fiber extinguishes and does not smoulder. However, the molten material shows a tendency to drip. In mixtures with other combustible fibers, polyester fibers continue to burn.

Table 30.4 Thermal properties of important polyester fibers (typical values)

	PET	PTT	PBT
Melting temp. [°C]	265	228	225
Glass temp. [°C]	80	45–65	20–40
Shrinkage [%][a]	8	14	18
Crimp [%]			
– HE	20	42	
– SET	42	50	
Density [g/cm^3]	1.4	1.33	1.35

[a]Hot water

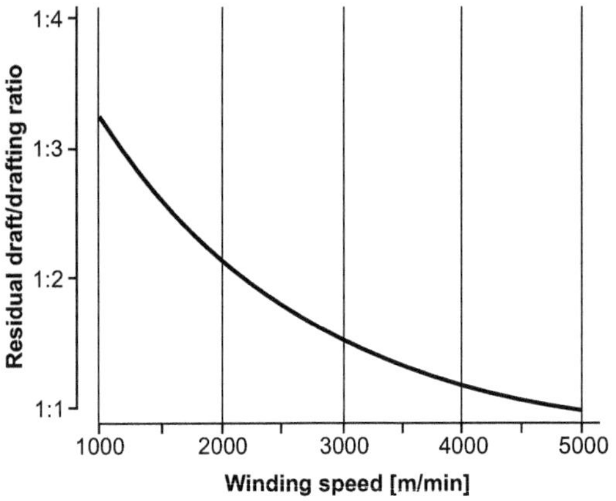

Fig. 30.22 Dependence of the residual elongation on the winding speed

The auto-ignition temperature for polyester fibers is 508 °C, which corresponds to that of polyamide and polyacrylic and lies between cotton (400 °C) and wool (590 °C).

The max. Flame temperature of burning polyester fabric is 697 °C. For cotton, on the other hand, this temperature is 860 °C, for wool 941 °C, for polyamide fibers 875 °C, and for polyacrylic fibers 855 °C (Tetzlaff et al., 1993).

30.7.5.2 Chemical Resistance

In general, the resistance to chemicals is very good. Modified polyester fibers are somewhat more sensitive in some cases. Polyester fibers have very good resistance to dilute mineral acids and organic acids at room temperature and still good resistance when boiled. Resistance to concentrated mineral acids, including phosphoric and hydrofluoric acids, with the exception of sulfuric acid, is good at room temperature. Boiling mineral acids decompose the fiber. Concentrated sulfuric acid rapidly dissolves polyester fibers even at room temperature.

The resistance to alkalis is sufficiently good. Concentrated or hot, diluted solutions attack the fiber, boiling concentrated alkalis dissolve it. Hydrolytic degradation by stronger alkalis first occurs at the fiber surface ("peeling"). This results in a loss of weight of the fiber (finer fiber) corresponding to the duration of treatment. Initially, the relative strength of the residual fiber remains approximately the same. With longer exposure, the structure of the residual fiber is also damaged.

Polyester fibers also exhibit excellent resistance to oxidizing and reducing agents. Even in bleaching processes with higher concentrations, no serious effects are observed. Modified polyester fibers are also slightly more sensitive in this respect.

Polyester fibers show excellent resistance to the usual organic solvents used in dry cleaning and stain removal at room temperature. In boiling water or in steam, a steady hydrolytic degradation takes place after a longer boiling time (von der Weyden & Becker, 1970).

30.7.6 Other Properties

The resistance to heat and cold is excellent. After dry heating to 150 °C for 250 h, the strength is still approx. 65% of the initial value, after 1000 h still approx. 50%. In comparison, polyamide and polyacrylic as well as cellulosic man-made fibers, cotton and silk have already completely lost their strength after 200–300 h of heating.

Short-term heating to 200 °C is tolerated without browning or pronounced degradation, the continuous heat resistance is about 120 °C. With moist heat, a stronger reduction in strength occurs due to the hydrolyzing effect of the water vapor above 100 °C.

In the cold range, the strength increases with a decrease in elongation (e.g. at - 40 °C: strength + 6%, elongation - 30%). The modulus of elasticity increases considerably.

After ½ year of weathering, the number of double bends to breakage is 99% of the unweathered yarn, after 1 year it is still 40–47%.

30.7.7 Dye Affinity

The stretching of polyester filaments has a marked effect on their affinity for dyes. Undrawn polyester filaments can be easily dyed with disperse dyes. After stretching, the affinity is significantly lower because the high crystalline content impedes the diffusion of the dyes. In addition, polyester fibers swell only slightly and are rather hydrophobic due to the absence of reactive side groups.

The dyeing of polyester fibers takes place:

- with suitable disperse dyes after pretreatment or after treatment with appropriate dyeing accelerators ("carriers"); these loosen the molecular bond and accelerate diffusion as a result of increased porosity of the fiber,
- with suitable disperse dyes without dyeing accelerators in closed dyeing apparatus under pressure at temperatures of 120–130 °C ("HT process"),
- with selected bases and naphthols using a modified diazotization technique,
- as a continuous padding process with the thermosol process.

The dyeing problems of polyester fibers have been solved for some years by the development of deep-dyeing, carrier-free dyeing and cationic dyeing grades.

PTT and PBT fibers can be dyed at lower temperatures than PET fibers and dye uptake is considerably faster. This is essentially due to the more open internal structure, which facilitates the penetration and attachment of the dye molecules.

30.8 Market and Trade Names

The price of crude oil is a significant factor in the manufacturing costs of synthetic man-made fibers, which will be illustrated by the example of PET. Approx. 4 kg of crude oil are required to produce 1 kg of PET fibers. At a price of 100 US $/barrel this corresponds to raw material costs of approx. 2.5 US$/kg (Fig. 30.23). It must be taken into account here that revenues are also generated from the 3 kg of residual materials, so that the actual raw material cost price is significantly lower. The world production of crude oil in 2020 was 4.2bn tons, of which about 8% was required for synthetic fiber production (about 80% of this for PET).

Terephthalic acid (TPA) costs about 90 US-ct/kg, ethylene glycol (EG) about 70 US-ct/kg (as of 2019). To produce 1 kg of PET, 0.72 kg of TPA and 0.28 kg of EG are required. This results in a

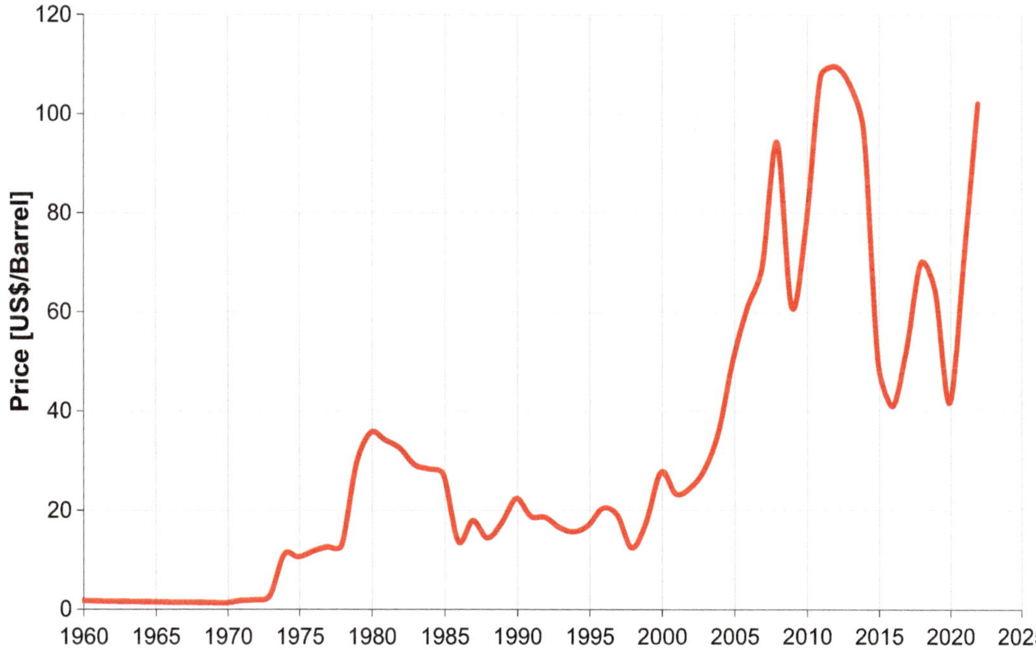

Fig. 30.23 Development of the oil price (WTI quality) (data: statista, 2022)

raw material price of approx. 85 US-ct/kg for 1 kg PET fibers. Thus, with the rising cost of crude oil, the price of PET gets in strong competition to cotton (1.50 US-$/kg) but is still significantly lower than that of viscose fibers (2–3 US-$/kg).

The production of PET filament yarns is currently around 46mt, which corresponds to a market share of 86% (Fig. 30.24) of all filament yarns. Furthermore, about 2.5mt of PET technical filament yarns are produced per year (market share 65%). In 2020, approx. 16.6mt t of staple fibers were also produced (Fig. 30.25), which corresponds to 28% of all staple fibers (incl. Natural fibers).

The increase in polyester production since 1995 is mainly due to numerous new plants in China. In 1995, China's market share for filament yarns and staple fibers was only around 16%; in 2010 it was already 70% (filaments) and 60% (staple fibers). Today, more than 80% of all PET filaments are produced in China and more than 61% of all PET staple fibers (Fig. 30.26).

Figure 30.27 shows the shares of fiber products in the production of PET for selected countries. While staple fibers are mainly produced in Europe and the USA, textile filament yarns dominate in China and India. The picture is more balanced in Korea, Taiwan and Indonesia.

China has the largest share in all market segments (The Fiber Year, 2021). For textile filament yarns it is 83% (35mt), for industrial filament yarns it is 72% (1.8mt) and for staple fibers it is 61% (10mt). Figure 30.28 shows the main producing countries for PET industrial yarns.

However, this rapid increase in production capacity in China has led to an oversupply of polyester, which will have to be reduced over the next few years before any significant new plants can be built. The 10 largest PET producers are all based in China (Fig. 30.29).

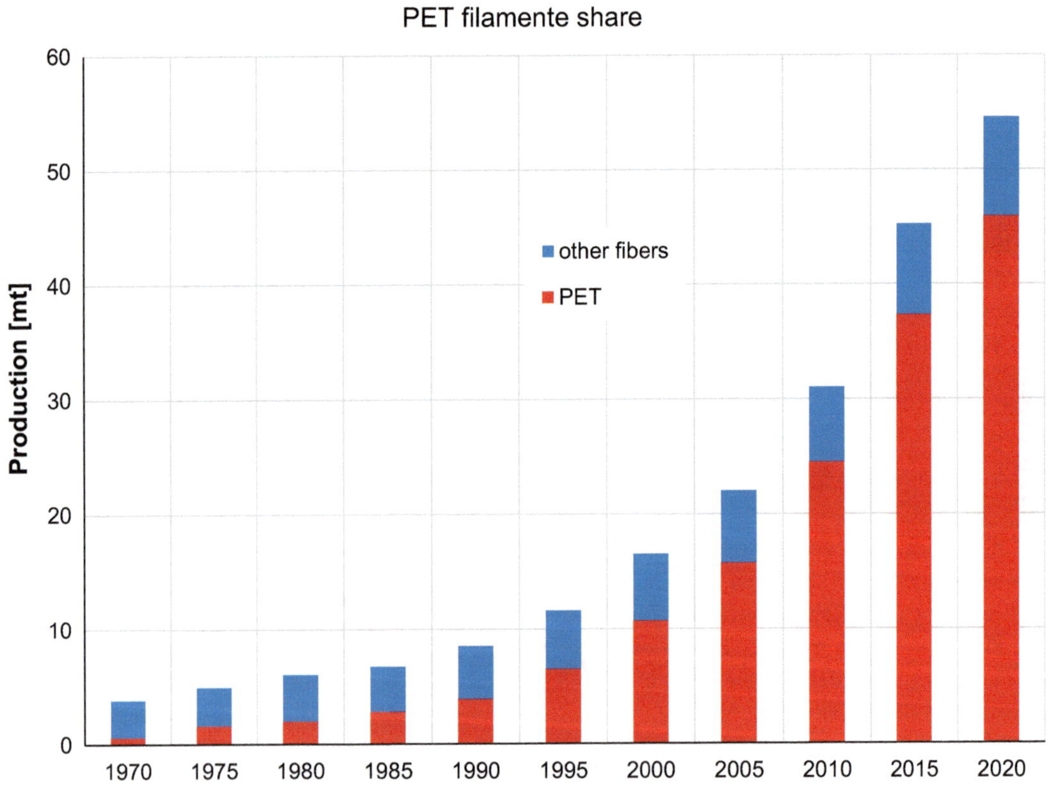

Fig. 30.24 Share of PET filaments in total filament fiber production (The Fiber Year, 2021)

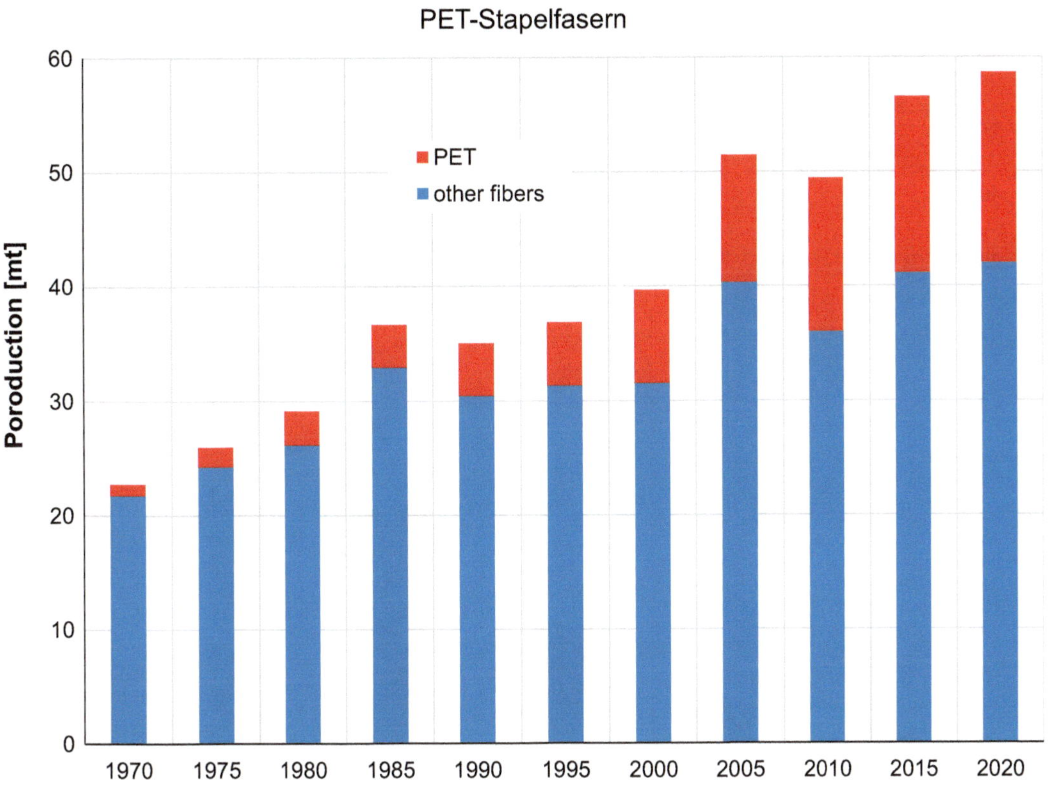

Fig. 30.25 Share of PET staple fibers in total staple fiber production (incl. Natural fibers) (data: The Fiber Year, 2021)

Important trade names of polyester (PET) are Trevira™, Dacron™, Terylene™, Diolen™ and Gualon™. PTT fibers are sold under the brand name Sorona™ and are often referred to as "elastomultiester". Blends of PET with PTT are also so called.

Selected manufacturers, on whose websites fiber data are mentioned, shows Table 30.5.

30.9 Processing Properties and Areas of Application

Compared to other types of man-made fibers, polyester filaments are characterized by a balanced combination of advantageous physical and chemical properties. In addition, the good processing properties and low production costs (approx. 1 €/kg for textile filament yarns) have contributed to polyester being used in a wide range of applications for clothing, home textiles and technical textiles.

The particular advantages of polyester fibers or textiles made from them are (Tetzlaff et al., 1993):

- High tensile strength and high modulus of elasticity; very good elastic properties with rapid recovery. Finished goods are therefore very resilient and largely crease-resistant and exhibit exceptionally good dimensional stability in dry and wet conditions and at high temperatures.
- Lower elongation than polyamide at low tensile stresses; this favors its use for sewing threads and tire cord.
- High abrasion resistance and durability.
- Excellent thermal resistance; due to the high softening range (> 230 °C) high ironing temperatures are possible.

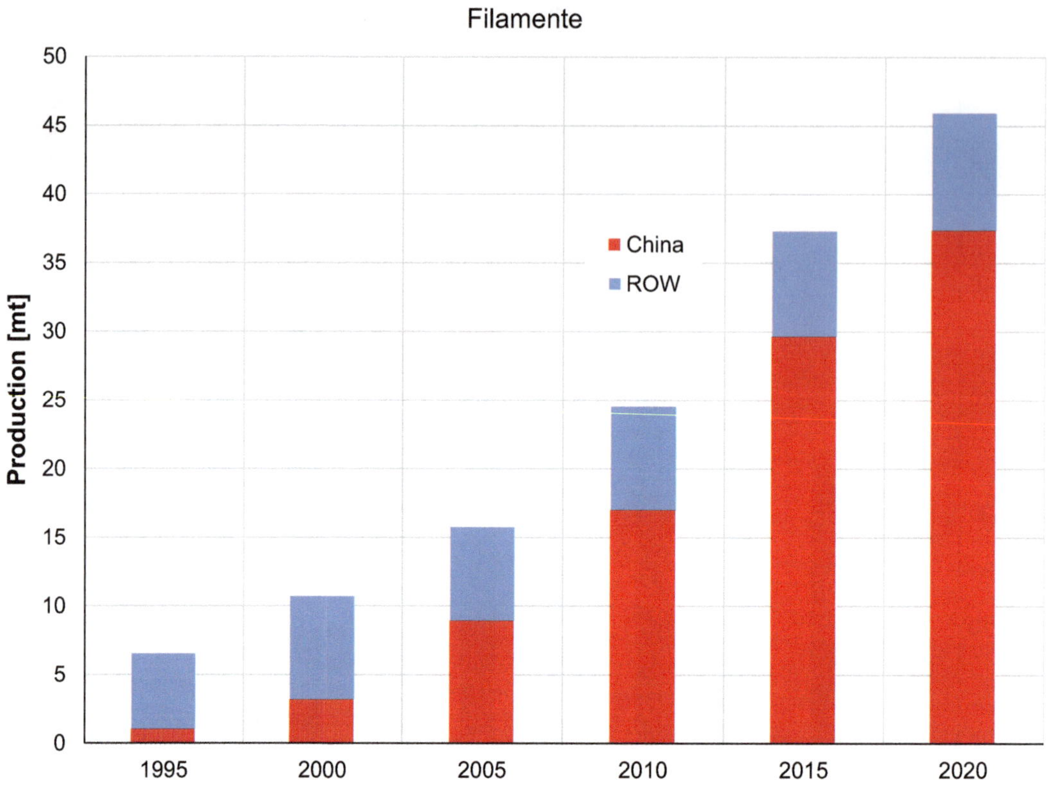

Fig. 30.26 China's share of PET filament yarn production (data: The Fiber Year, 2021)

- High resistance of embossed effects, pleats and creases even after repeated soaking and drying.
- Low moisture absorption, easy washability and quick drying.
- Very good light and weather resistance as well as permanent resistance to temperatures up to about 150 °C. These properties make polyester fibers particularly suitable for use in automotive textiles.
- Good electrical insulating capacity. It is advantageous in combination with thermal resistance and low moisture absorption.
- High resistance to a large number of aggressive chemicals, especially acids.
- Very good resistance to biological decay influences.
- In finished goods little "synthetic", rather "warm", pleasant feel.

The specific disadvantages of polyester fibers or finished articles made from them are:

- Dirt, grease and oil affinity.
- Strong electrostatic charge, which promotes rapid soiling.
- Difficulty in staining without specific modifications.
- From a clothing physiological point of view, poor water and perspiration absorption capacity. Through suitable article construction and the use of microfibers, moisture is transported to the surface of the textile by means of the capillary effect and thus a pleasant wearing comfort is achieved. Chemical modifications also allow an increase in water absorption.
- High tendency to pill in goods made from staple fiber yarns. This disadvantage is countered by the use of special pillarmer fiber types (important for knitted goods), which exhibit reduced transverse, tear and kinking abrasion resistance.

Polyester is used in a wide range of end products in adapted presentations:

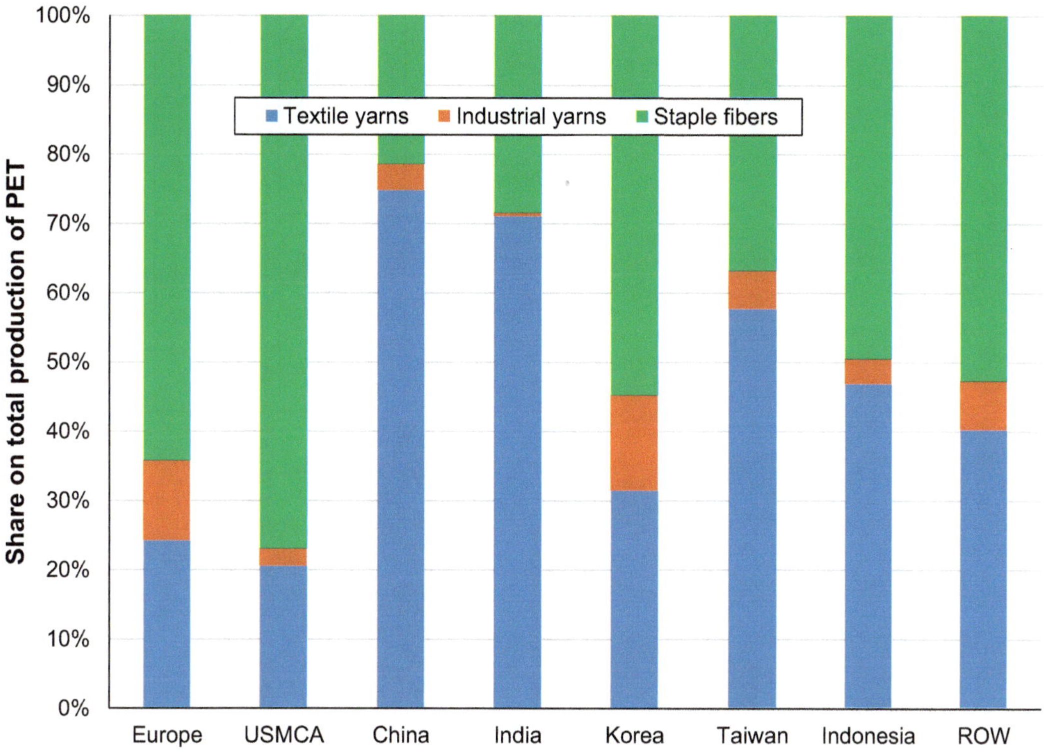

Fig. 30.27 Shares of fiber products in PET production of selected countries (data: The Fiber Year, 2021)

- Textured polyester filament yarn is used in knitting and weaving. It is used for blouse fabrics, ties, linings, sports and casual wear, umbrella fabrics, curtains, awnings and car upholstery covers.
- High-strength polyester filament yarn is used for technical purposes, e.g. for sewing and tire cord yarns, conveyor and transport belts, coating fabrics, large tents and air domes, driving belts, straps, hoses (high-pressure fire hoses),

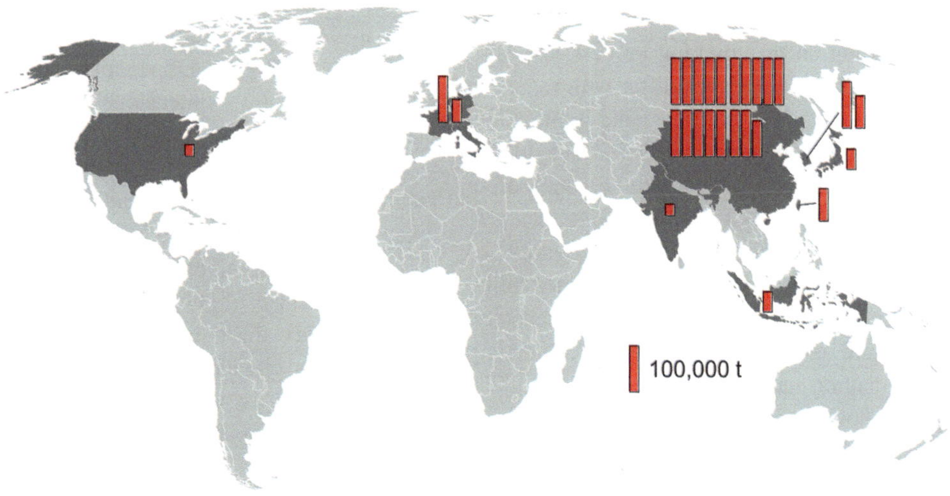

Fig. 30.28 Main producing countries of technical PET yarns (2019)

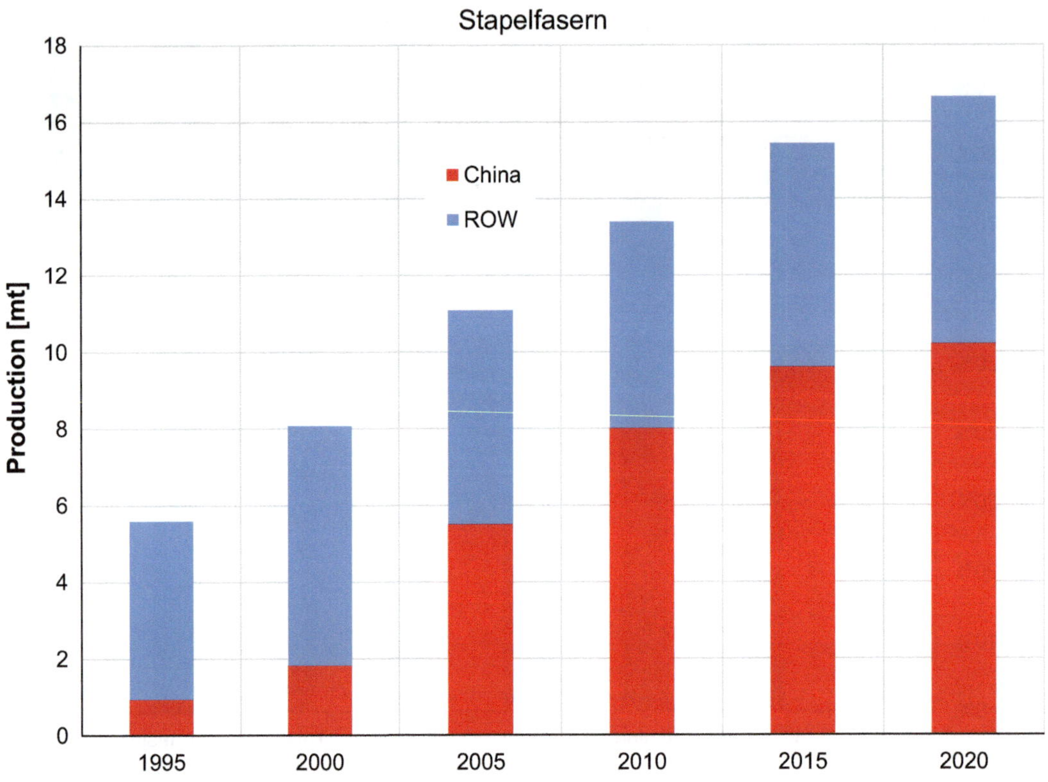

Fig. 30.29 China's share of PET staple fiber production (data: The Fiber Year, 2021)

ropes, sailcloth as well as for finest screens and filter fabrics made of monofilaments (e.g. for paper production).

- Polyester staple fiber yarns are mostly produced as blended yarns with cotton or wool. However, good spinnability with all common spinning processes is also given in pure form. Depending on the spinning process, the optimum fiber characteristics differ with regard to fiber length, fiber fineness and lubrication. Fields of application are shirt and blouse fabrics, trouser, suit, costume and dress fabrics, coats, workwear, underwear, scarves, bed linen, upholstery fabrics, carpet yarns.
- Polyester filling fibers serve as a bulky filling material with excellent recovery capacity. They are used in quilts, pillows, sleeping bags, batting and padding. Bicomponent fibers and hollow fibers favor the properties of building elasticity and heat retention, which are particularly desirable in filling material.
- Polyester is also processed into nonwovens and needle felts.
- Polyester fine fibers are used, among other things, to make artificial leather (e.g. Alcantara™).

Fabrics made of polyester filament yarn are heat-set at around 220 °C before being dyed or otherwise treated at higher temperatures. Heat-setting serves to ensure dimensional stability and prevents unwanted shrinkage.

The treatment in hot caustic soda, which results in an even removal on the filament surface, creates particularly fine fabrics.

Interesting fashion effects can be achieved by special coatings of the fiber surface. Fancy examples of this are temperature-sensitive pigments that change their color depending on the temperature as well as particles that change the color effect depending on the incidence of light.

Table 30.5 Selected manufacturers of PET and PBT fibers

Producer	Polymer	Filaments	Staple fibers	Monofilaments	Website	Country
DT Group	PET		x		https://dtcompany.com	Canada
Fujian JinLun Fiber Shareholding Co, Ltd	PET		x		http://www.fjsjlgx.com	China
HEDA Chemical Fiber	PET		x		https://www.cnfibers.com	
Huvis	PET		x		http://www.huvis.com	Korea
JB Ecotex LLB	PET		x		https://www.jbecotex.com	China
Märkische Faser GmbH	PET		x		https://maerkische-faser.com	Germany
NAN YA PLASTICS CORPORATION	PET	x	x		https://www.npc.com.tw	Taiwan
PPH Legs Sp. z o.o.	PET		x		https://legs.com.pl	Poland
Protex New Advanced Textiles GmbH	PET	x			http://www.protexonline.com	Germany
Recron (Reliance Industries, Ltd.)	PET		x		http://www.recron.com	India
Swicofil AG	PET	x		x	https://www.swicofil.com	Switzerland
Taşdelen Group	PET	x			http://www.tasdelengroup.com	Turkey
Thai Polyester	PET		x		http://www.thaipolyester.com	Thailand
Trevira GmbH (Indorama Ventures)	PET	x	x			Germany
TWD Fibres GmbH	PET		x		http://www.twd-fibres.de	Germany
WUXI LONGSHINE TEXTILE CO., LTD	PET		x		http://www.wxlongshine.com	China
DuPont de Nemours	PTT	x	x		http://www.sorona.com	USA
HaiXin Kaisheng Technology Co.,Ltd	PTT		x		http://www.haixingtec.com	China
Jiangsu Yinyu Chemical Fiber Co.,Ltd.	PBT	x			http://www.pbtyarn.cn	China
Minifibers, Inc.	PBT	x			https://www.minifibers.com	USA
Taşdelen Group	PBT	x			http://www.tasdelengroup.com	Turkey
Yinyuda	PBT	x			http://www.yinyuda.com	China

Polyester fibers are used as staple fibers and filament yarns for clothing, home textiles and technical textiles. Other applications are nonwovens, both staple fiber and spunbonded. Monofilaments are also used for technical articles (Fig. 30.30).

Fig. 30.30 Typical polyester products (tents, swimwear, air mattress, outdoor clothing, umbrellas, backpacks, sofa covers, soft toys, shoes)

References

Carothers, W. H., & Arvin, J. A. (1929). Studies on polymerization and ring formation, II. Polyesters. *Journal of the American Chemical Society, 51*, 2560–2570.

Carothers, W. H., & Hill, J. W. (1932). Studies of polymerization and ring formation. XV. Artificial fibers from synthetic linear condensation superpolymers. *Journal of the American Chemical Society, 54*, 1579–1587.

Cook, J. G. (1946). Terylene – new all British fibre – makes important advance in the synthetic field. *Silk J. & Rayon World, 23*(270), 28–30.

Eastman. (1958). Eastman adds a polyester. *Chemical & Engineering News Archive, 36*(38), 30–31.

Hearle, J. (1993). Structural mechanics from fibre to fabric. In Brunnschweiler, D., & Hearle, J. (Hrsg.), *Polyester. 50 Years of Achievement* (pp. 170–173). The Textile Institute, Manchester.

Ludewig, H. (1955). Mischpolyamidfasern der Trelon-Gruppe im Vergleich zu Polyamid- (Perlon-, Nylon-) und Polyesterfasern. *Faserforschung und Textiltechnik, 6*, 277–286.

Oerlikon Barmag, Zweigniederlassung der Oerlikon Textile GmbH & Co. KG, Remscheid. https://www.oerlikon.com/polymer-processing/de/

Schlack, P. (1952). Verfahren zur Herstellung von hochschmelzenden Polykondensationsprodukten. Schutzrecht DE 922 255.

Schlack, P., & Kunz, K. (1953). Historische Entwicklung der Polyamidkunstfasern. In: Pummerer, R. (Ed.), *Chemische Textilfasern, Filme und Folien* (pp. 631). F. Enke-Verlag, Stuttgart.

Statista. (2022). https://www.statista.de

Tetzlaff, G., Dahmen, M., & Wulfhorst, B. (1993). Polyester fibers. Fiber table acc. to P.A. Koch des Instituts für Textiltechnik of RWTH Aachen University, Aachen.

The Fiber Year. (2021). The Fiber Year GmbH, Roggwil. Retrieved from https://www.thefiberyear.com

Weyden, H. von der, & Becker, O. (1970). Kraft-Dehnungseigenschaften verschiedener Natur- und Chemiefasern unter Einwirkung von heißer Luft oder heißem Wasser. *Textilindustrie* 72, pp. 568–573 u. 631, 634.

von Falkai, B. (1981). *Synthesefasern*. Wiley-VCH Verlag.

Whinfield, J. R. (1941). Improvements relating to the manufacturing of highly polymeric substances. Britisches Patent Nr. 578097.

Whinfield, J. R. (1952). Fibres from aromatic polyesters. *Endeavour*, *11*, S., 29–32.

31 Polyamide

Polyamide is the second most important synthetic fiber material in terms of volume after polyester. A large number of different polyamides have been developed for technical use.

31.1 Definition

In polyamide fibers, at least 85% of the aliphatic chain components are linked by amide groups to form linear macromolecules. Characteristic of these are the recurring functional acid-amide groups [CO-NH] in the main chain.

Based on their chemical structure, polyamides can be divided into two groups:

- Aminocarboxylic acid types (AS) of one kind of monomer and
- diamine dicarboxylic acid types (AA-SS) from two different monomers,

where A denotes the amino groups and S the carboxy groups. The number of C atoms in the individual unbranched monomers serves as the code number. For the type AA-SS, the number of C atoms of the diamine is written first, then that of the dicarboxylic acid. For copolyamides, the components are indicated separated by slashes.

In English usage, the word nylon is often used as a synonym or generic term for polyamides. In the USA, other polyamides, e.g. PA 11, may also be referred to as nylon acc. to the definition of the term valid there. Originally, Nylon™ was the trade name chosen by DuPont for PA 6.6, whereas Perlon™ is a trademark protected for some producers of PA 6 fibers and may not be used to refer generally to caprolactam-based polyamide fibers. Perlon™ is the original trade name for PA 6 of IG Farben.

Polyamide fibers are produced either by polycondensation of diamines with dicarboxylic acids or by polycondensation of ω-aminocarboxylic acids containing at least 6 carbon atoms or by hydrolytic polymerisation of at least 7-membered ring systems containing a carbonamide group, e.g. caprolactam (PA 6).

The polyamides also include the aramides. They have aromatic chain links and are described in a separate chapter.

31.2 History

31.2.1 Polyamide 6.6 and Polyamide 6

The history of polyamide fibers begins at the beginning of the twentieth century. In 1899, 1902 and 1904, S. Gabriel, Th. M. Maass, A. Manasse and J. von Braun dealt analyzed the ring formation of ω-aminocaproic acid, 20–30% of which could be converted into caprolactam. However, the importance for industrial use was not recognized at the time (Gabriel & Maass, 1899; Manasse, 1902; von Braun, 1907).

In 1932, Carothers and Hill discovered that molecular distillation of molten polyesters produces so-called superpolyesters from which

D. Veit, *Fibers*, https://doi.org/10.1007/978-3-031-15309-9_31

threads can be drawn that solidify in air. Thus the principle for the production of synthetic fibrous materials was found (Carothers, 1929; Carothers & Hill, 1932a, 1932b; Carothers & Berchet, 1930).

In 1934, Carothers produced a polyamide from 9-aminononanoic acid in an experiment, an interesting starting material from a technical point of view. In 1935 he succeeded in synthesizing the polyamide from hexamethylenediamine and adipic acid. Due to the practically unlimited raw material base (benzene and phenol), this polyamide was selected for further development on an industrial scale and a patent application was filed in 1937. It became known as PA 6.6 or PA 66. DuPont first published the results of its research in 1938, and production began in the same year.

Unaware of Carothers' work, Paul Schlack had been working since 1928 in the scientific laboratory of the "Aceta" plant in Berlin, which belonged to I.G. Farben AG, on the production of a synthetic fiber and, among other things, on the polymerization of caprolactam. In a paper published in 1930, Carothers had ruled out the polymerization of ε-caprolactam with or without a catalyst. His experiments had resulted only in a polyamide with a low degree of polymerization, from which no filaments could be spun. Unaware of this work, Schlack continued to work and in 1938 he succeeded in splitting the lactam ring using aminocaproic acid hydrochloride as a catalyst, thus producing a linear polymer. This was the birth of polyamide 6 (Fig. 31.1).

In July 1938, a delegation from IG Farben set out for the United States to demonstrate the new fiber to Dupont and negotiate licenses for it. Already on the high seas, the delegation learned that another DuPont delegation was on its way to Germany to visit IG Farben and negotiate licenses for the production of a new synthetic fiber. To their dismay, they discovered that IG Farben was not interested in the new synthetic fiber polyamide 6.6, but had its own, which was based on a different monomer and had not been patented by DuPont.

An amicable agreement was then reached, the polyamide patents were exchanged and the respective global sales markets were determined. Eastern Europe and Asia went to IG Farben, while Western Europe and America went to DuPont. This division of the world market can still be seen today: In Eastern Europe and Asia, mainly PA 6 and comparatively little PA 6.6 is produced, while in the USA, Western Europe and South America, mainly PA 6.6 is spun.

In 1939, DuPont began the industrial production of nylon, as the fiber was now called. The first nylon stockings came onto the market in the USA in 1940. However, production soon concentrated exclusively on military applications, such as parachute fabrics and corduroy for aircraft tires, and amounted to approx. 5400 t in 1940. It was not until after 1945 that it was expanded for civilian and industrial applications. By the end of the war, nylon production was already 15,000 t/year, reaching 62,000 t in 1950. The sought-after nylon stockings quickly became a means of payment for American soldiers in Germany and thus soon spread to Europe as well (Fig. 31.2).

The production of PA 6 was started by IG Farben in 1939, and towards the end of the year the first monofilaments came onto the market under the name Perluran, later designated Perlon L. Large-scale production of perlon filament yarn began in 1943 at the Landsberg a.d. Warthe plant. The plant under construction in Premnitz (Westhavelland) was unable to start its planned large-scale production of perlon filament yarn because of the war. In 1946, production of PA 6 filament yarn was resumed at the VEB Thüringisches Kunstfaserwerk "Wilhelm Pieck" plant in Schwarza near Rudolstadt, after smaller quantities had already been produced there in 1945. In 1949, the Bobingen artificial silk factory near Augsburg followed, and in 1950, the Oberbruch glossy fabric factory with perlon filament yarns.

By 1953, the West German man-made fiber industry had the trade name "Perlon" protected, so the East German polyamide manufacturers were forced to create a new name: "Dederon" (after the German name for the GDR: "DDR"). This product, based on PA 6, was successful in East Germany until reunification in 1990.

From the 1950s onwards, outerwear was also made from polyamide instead of cotton. The

Fig. 31.1 W. H. Carothers (left), P. Schlack (center) and the first PA 6 fibers produced by Schlack (right)

so-called "Nyltest" shirt became particularly famous. It was made of 100% PA 6.6 and was the first easy-care outer shirt. The water absorption of PA is similar to that of cotton, but polyamide fibers store the absorbed moisture. This resulted in the wearer of such a shirt sweating very fast and very strongly and giving off bad smells after some time. Therefore, this article was not a lasting success (Fig. 31.3).

31.2.2 Development of Further Polyamide Types

Another method of producing a polyamide was found by W. Zeltner and M. Genas. This polyamide 11 is produced on the basis of castor oil, from which the 11-aminoundecanoic acid is synthesized. The patent for this was filed in 1944. The first trials were carried out in France in the mid-1940s, and large-scale production began in 1955. This fiber is still on the market today under the name Rilsan™ and is also known as biomaterial because of the biological starting material (Aelion, 1956; Genas, 1962).

Polyamide 7 is produced in two different ways: In the USA, 7-aminoheptanoic acid was obtained from caprolactone with subsequent conversion to methyl 7-aminoheptanoate. In the former USSR, in 1951–1955, A. A. Strepicheev and R. CH. Freidlina used a different method. It was based on telomerization, similar to polyamide 11. In this process, the starting material for polyamide 7 is

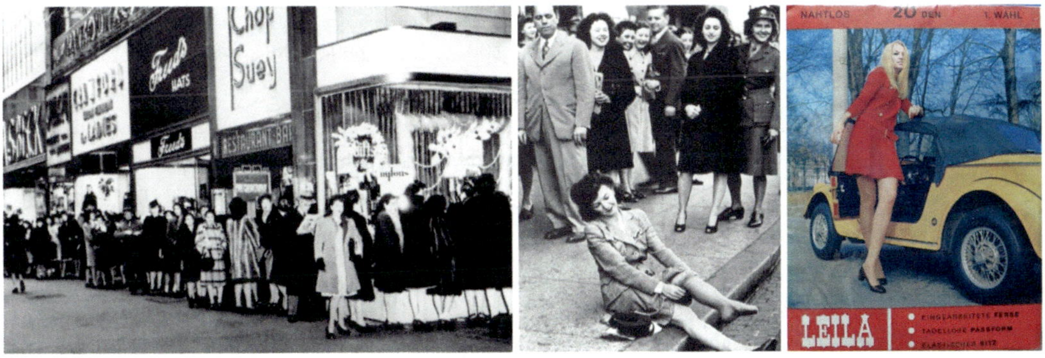

Fig. 31.2 Queue in front of a shop for nylon stockings in the USA, happy buyer and pantyhose Leila

Fig. 31.3 Hosiery production on Cotton knitting machines (left) and German advertising for Nyltest shirt (right)

ω-aminoenanthic acid, synthesized from ethylene and carbon tetrachloride. However, the process was discontinued, despite better properties of PA 7 compared to PA 6, due to inefficiency and lack of raw materials (Zaremba et al., 1997).

Polyamide 12, which is already mentioned in the basic patents of Carothers and Schlack, consists of 12-aminododecanoic acid or λ-laurinlactam. The process for the polymerization of laurinlactam originated from W. Griehl and S. Schaaf and was first patented in Switzerland in 1966. Polyamide 12 was produced as a monofilament under the name Grilon™. The monomer is also used for adhesive fibers in copolyamides.

Despite some favourable fiber properties, attempts to produce fibers from polyamides 3, 4, 8 and 9 did not lead to industrial success. In the former USSR, a fiber called Perlagon, a PA 9 grade, was temporarily on the market, but production was later discontinued.

31.3 Chemical Constitution and Raw Materials

31.3.1 Formation of Polyamides

Polyamides are chemically similar to proteins. Their formation can occur in three different ways (Fig. 31.4):

The polyamides formed by methods 2 and 3 are identical.

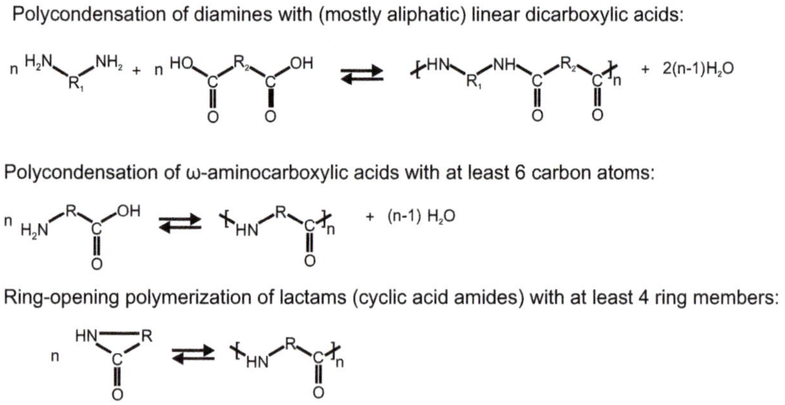

Fig. 31.4 Ways of forming polyamide

31.3.2 Polyamides from Diamines and Dicarboxylic Acids

The best known and economically most important representative of this group for fiber production is polyamide 6.6, which is also known as polyamide 66, nylon 6.6 or nylon 66 or simply as nylon.

31.3.2.1 Polyamide 6.6

The starting materials for PA 6.6 are hexamethylenediamine (1,6-hexanediamine) and adipic acid (hexanedioic acid). Hexamethylenediamine can be produced from 1,6-hexanediol. The reduction process of adipic acid dinitrile is used on a large scale today. Adipic acid is technically produced from cyclohexane, which is first oxidized to a mixture of cyclohexanol and cyclohexanone, the KA oil, and then further cleaved to adipic acid. The cyclohexane is obtained by hydrogenating benzene or directly from certain petroleum oils.

PA 6.6 can only be polycondensed to a high molar mass if the two starting components are present in an equimolar ratio. To ensure this, a detour via the so-called AH salt (adipic acid hexamethylene diamine or hexamethylene ammonium adipate) is preferred. It is produced by reacting both components in suitable solvents, e.g. methanol. The precipitated AH salt is a white, crystalline, water-soluble powder (melting point under water splitting 202–205 °C) with the following structural formula (Fig. 31.5):

Having succeeded in improving the purity of hexamethylenediamine and adipic acid, purifying crystallization can be dispensed with, and an aqueous 50–60% AH salt solution can be prepared directly (Fig. 31.6).

The polycondensation of AH salt is subject to the following special conditions:

- Due to the volatility of hexamethylene diamine, losses occur both in the preparation of the AH salt solution and in the evaporation of the water in the polycondensation process. These must be kept as low as possible or compensated by excess diamine.
- Phase separation within the batch must be avoided. For this reason, polycondensation must be carried out under positive pressure except for the final phase, and care must be taken to coordinate temperature and pressure control and water evaporation.
- Polyamides are sensitive to thermal stress and oxygen, so that the residence time must be as short and constant as possible. The temperature range between solidification and thermal damage is relatively narrow, which is particularly true for polyamide 6.6.

Polycondensation can be either discontinuous or continuous. The discontinuous process is currently the most widely used process. For this purpose, the aqueous approximately 45–60% AH salt solution is concentrated to 70–80% in an evaporator under exclusion of air at temperatures above 130 °C. After transferring the solution to an autoclave, it is heated further to 210–220 °C. After transfer to an autoclave, the solution is heated further to 210–220 °C, with the pressure rising to 16–20 bar depending on the water content. This is followed by a pressure holding period of 1–1.5 h duration, during which the pre-condensation takes place. During this period, the temperature can be raised further while steam escapes. During the subsequent expansion period, the overpressure is reduced within 1.5–2 h and the temperature of the reaction mass is adjusted to 270–280 °C. As a result, almost the entire amount of steam escapes. As a result, almost the entire amount of solvent and reaction water escapes, allowing the polycondensation to proceed. An antifoaming agent can be added to prevent foaming during expansion. The polymeric melt is then extruded in the form of strands into a cooling water bath and cut into granules. This process should not take longer than 15 min due to the low thermal stability of polyamide in this state.

The degree of polycondensation of the final product can be controlled by the addition of chain length regulators and the residual water content. This seemingly complicated process has two advantages. It provides sufficient water so that the reaction mass remains liquid until its melting point is reached. It also keeps the loss of diamine to a minimum. This process can be used to produce a polyamide 6.6 that is suitable for

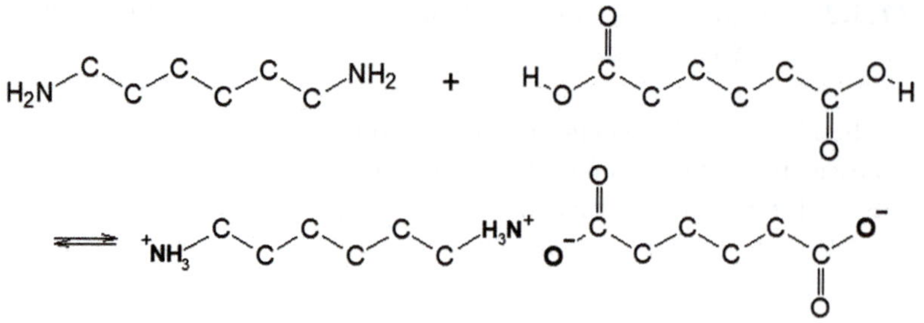

Fig. 31.5 Synthesis of AH salt from hexamethylenediamine and adipic acid

spinning into textile fibers. Higher molecular weights, as required for technical yarns, for example, can be produced by expansion at negative pressure. A second autoclave is often used for this purpose.

While the post-polycondensation takes place in the second autoclave, the first is available for a new batch. A completely different option, which is much used today, is the solid-phase post-polycondensation of the granulate at 170–200 °C, which includes drying.

The continuous production of the polymer was mainly developed by DuPont in the 1940s and modified by other companies in the meantime. Similar to the autoclave process, such a plant consists of an evaporator, pressure reactor, flash tank (separator) and final reactor. Continuous production requires considerable equipment and process engineering experience if good spinning quality is to be achieved. The main problem with polyamide 6.6 is the risk of thermal damage. In this respect, polyamide 6 behaves more favorably. If the reaction mass is almost free of water, there is a risk of decarboxylation (splitting off CO_2 from the carboxyl group) of the acidic end groups. The residence time between the final reactor and the melt extrusion should therefore be as short as possible. If the process is controlled, a more homogeneous product can be produced than with discontinuous production (Zaremba et al., 1997). A further advantage is the possibility of direct spinning, which, however, has so far only been used on a large scale in one single plant worldwide.

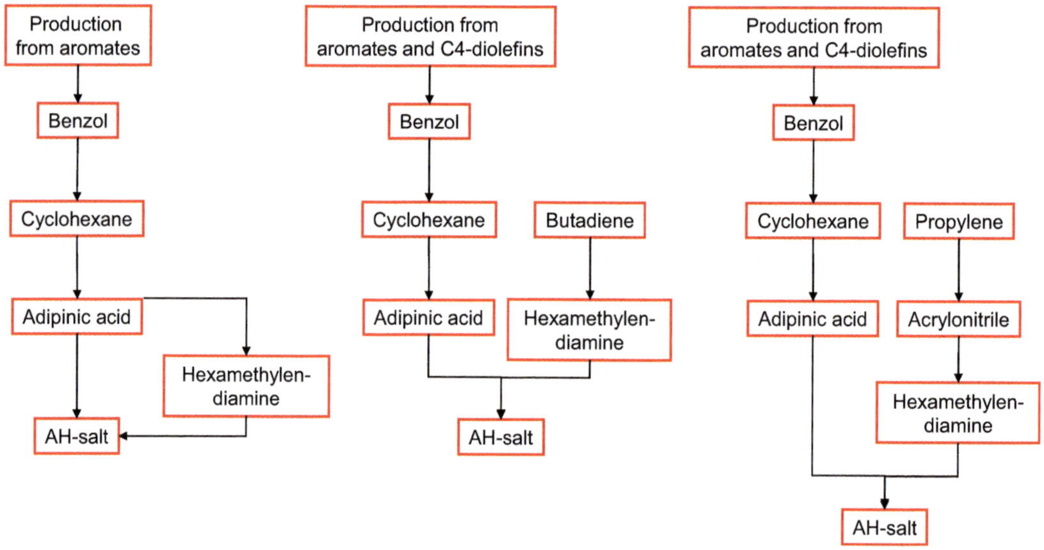

Fig. 31.6 Process for the preparation of AH salt acc. to (Koslowski, 2008)

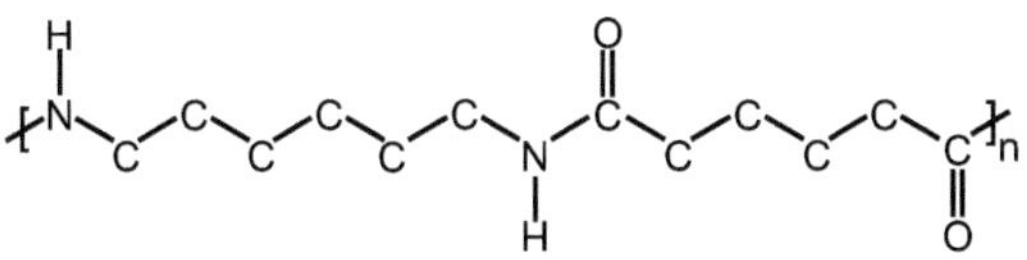

Fig. 31.7 Structural formula of the repeat unit of PA 6.6

After polycondensation, the polyhexamethylene adipamide with the following structural formula is obtained (Fig. 31.7):

31.3.2.2 Polyamide 4.6

Starting materials are tetramethylene diamine (1,4-butanediamine) and adipic acid (hexanedioic acid). The raw material base for 1,4-butanediamine is acrylonitrile and hydrogen cyanide. A salt is formed from the butanediamine and the adipic acid in a similar way to the production of PA 6.6. The salt is pre-polycondensed in aqueous solution at 225 °C and post-condensed in the solid phase. Due to the higher number of polar groups because of the shorter aliphatic chain segments compared to PA 6 and PA 6.6, the melting point is higher and the crystallinity is greater (Fig. 31.8).

31.3.2.3 Polyamide 6.10

Starting materials are hexamethylene diamine (1,6-hexanediamine) and sebacic acid (decanedioic acid). In the most commonly used process, sebacic acid is obtained by heating castor oil or ricinoleic acid with NaOH. Other methods are the KOLBE synthesis from adipic acid monomethyl ester or the oxidation of cyclodecanol. From both starting materials, which are present in equimolar ratio, a salt is formed, the melting point of which is 170 °C. Polycondensation in a discontinuous process takes place at 245–260 °C, first under pressure and then under pressure relief. Its course corresponds to the production of PA 6.6. Due to its low water absorption and high elasticity, this type of polyamide is suitable for the production of wires and bristles. Toothbrushes are an important area of application.

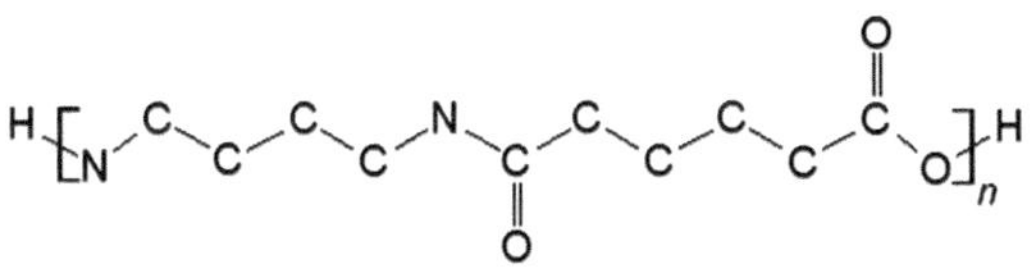

Fig. 31.8 Structural formula of PA 4.6

31.3.2.4 Other

Table 31.1 shows an overview of possible combinations of diamines and dicarboxylic acids, the names of the resulting polyamides and the ranges of melting temperatures. With an increasing number of C atoms, the melting temperature decreases because the number of polar groups per monomer decreases and thus the attractive forces between the macromolecules are reduced.

To ensure high quality polyamides, the following conditions must apply:

- The monomers must be in pure form,
- any impurities must be removed by filtration before polymerisation,
- before the polyamides are spun, they are filtered again,
- constant average molar mass of the polymer, uniform water content and constant content of amino end groups.

A constant temperature control, uniform residence time, the exact dosing of additives and the maintenance of a constant product flow are fundamental requirements, especially in continuous processes. Furthermore, the exclusion of oxygen must be ensured during the entire polymerization process as well as the subsequent extraction (polyamide 6) and drying of the granulate as well as during melting, because polyamides are sensitive to oxidation at high temperature.

The polymerization process can be discontinuous or continuous. In all three ways, with a few exceptions (ionic polymerization), water is added to the starting materials. In the polymerization of lactams, it is used for ring opening. In the polycondensation of diamines with dicarboxylic acids, it helps to avoid problems that would occur in the molten state. Without water, for example, it is more difficult to mix the two components sufficiently, to control the necessary equimolar ratio and, in the case of adipic acid, to limit the tendency to decarboxylation.

Table 31.1 Raw materials and melting temperatures of aliphatic polyamides acc. to (Ullmann, 1976)

Diamine	Chem. structure	Dicarboxylic acid	Chem. structure HOOC-(CH_2)n-COOH	Polyamide	Melt temperature [°C]
1,4-Butandiamine	H_2N-$(CH_2)_4$-NH_2	Hexanedioic acid (adipinic acid)	4	4.6	278–295
		Heptanedioic acid (pimelic acid)	5	4.7	233
		Octanedioic acid (suberic acid)	6	4.8	250
		Nonanedioic acid (azelaic acid)	7	4.9	223
		Decanedioic acid (sebacic acid)	8	4.10	239
1,5-Pentandiamine	H_2N-$(CH_2)_5$-NH_2	Pentanedioic acid (glutaric acid)	3	5.5	198
		Hexanedioic acid (adipinic acid)	4	5.6	223
		Heptanedioic acid (pimelic acid)	5	5.7	183
		Octanedioic acid (suberic acid)	6	5.8	202
		Nonanedioic acid (azelaic acid)	7	5.9	179
		Decanedioic acid (sebacic acid)	8	5.10	186–195
1,6-Hexandiamine	H_2N-$(CH_2)_6$-NH_2	Hexanedioic acid (adipinic acid)	4	6.6	250–265
		Heptanedioic acid (pimelic acid)	5	6.7	202–228
		Octanedioic acid (suberic acid)	6	6.8	220–232
		Nonanedioic acid (azelaic acid)	7	6.9	185–226
		Decanedioic acid (sebacic acid)	8	6.10	209–223
1,7-Heptandiamine	H_2N-$(CH_2)_7$-NH_2	Hexanedioic acid (adipinic acid)	4	7.6	226–250
		Heptanedioic acid (pimelic acid)	5	7.7	196–214
		Decanedioic acid (sebacic acid)	8	7.10	187–210
1,8-Octandiamine	H_2N-$(CH_2)_8$-NH_2	Hexanedioic acid (adipinic acid)	4	8.6	235–250
		Octanedioic acid (suberic acid)	6	8.8	205–225
		Decanedioic acid (sebacic acid)	8	8.10	197–210
1,9-Nonandiamine	H_2N-$(CH_2)_9$-NH_2	Carbonic diamine (urea)	H_2N-CO-NH_2	9.1	215
		Hexanedioic acid (adipinic acid)	4	9.6	205
		Nonanedioic acid (azelaic acid)	7	9.9	165

(continued)

Table 31.1 (continued)

Diamine	Chem. structure	Dicarboxylic acid	Chem. structure HOOC-(CH_2)n-COOH	Polyamide	Melt temperature [°C]
		Decanedioic acid (sebacic acid)	8	9.10	179
1,10-Decandiamine	H_2N-$(CH_2)_{10}$-NH_2	Hexanedioic acid (adipinic acid)	4	10.6	230–236
		Octanedioic acid (suberic acid)	6	10.8	208–217
		Decanedioic acid (sebacic acid)	8	10.10	194–203

In the course of the chain formation reaction, the water must be removed except for a small residue in order to achieve a higher degree of polycondensation in the reaction equilibrium. Chain length regulators, usually monofunctional acids (e.g. acetic acid, propionic acid or benzoic acid), which are added to the monomer, also serve to control this (Klare et al., 1963). On the one hand, they help to avoid strong shifts in the degree of polycondensation, e.g. when melting dried granules. On the other hand, they can be used to control the number of amino end groups, which allow dyeing with acid dyes. In the polycondensation of diamines and dicarboxylic acids, a slight excess of one of the two components also acts as a chain length regulator.

31.3.3 Polyamides from ω-Aminocarboxylic Acids

Among this group of polyamides, only polyamide 11 (Rilsan™) is relevant. The starting material is 11-aminoundecanoic acid. This is obtained from castor oil, 80–85% of which consists of the glyceride of ricinoleic acid. Polycondensation can be carried out discontinuously or continuously in three stages at 265 °C. Phosphoric acid is used as a catalyst. Phosphoric acid serves as a catalyst. In contrast to PA 6, the content of ring amides is low, so that no aqueous extraction is necessary. Spun products made of polyamide 11 absorb only little water and therefore show good dimensional stability under fluctuating humidity. Applications are technical articles such as wires, bristles and zippers. For the textile sector, the fibers are too expensive for standard articles.

31.3.4 Polyamides from Lactams

Lactams are molecules in organic chemistry that contain an amide bond between an amino and a carboxyl group of the same molecule. The name is preceded by Greek letters that indicate how many carbon atoms are present in the ring in addition to the carbonyl carbon. So the ε-caprolactam has 5 more C-atoms besides the carbonyl group (ε is the fifth letter in the Greek alphabet).

The most important representative of this class is polyamide 6, whose starting material is ε-caprolactam (6-aminohexanoic acid lactam), Figs. 31.9 and 31.10.

The production of ε-caprolactam takes place in 3 steps:

- Production of cyclohexanone by:cyclohexane oxidation orphenol core hydrogenation and cyclohexanol dehydrogenation or cyclohexylamine dehydrogenation and cyclohexanimine hydrolysis,
- oximation of cyclohexanone by hydroxylamine sulfate,

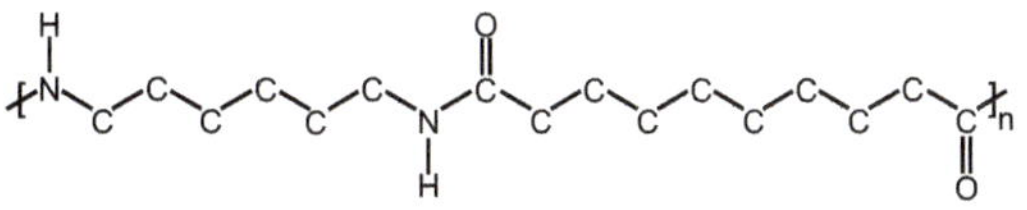

Fig. 31.9 Structural formula of PA 6.10

Fig. 31.10 Ring opening reaction and polycondensation to PA 6

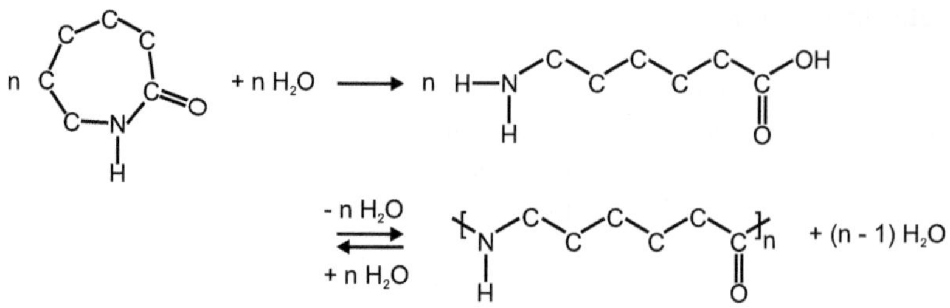

- formation of ε-caprolactam from cyclohexanone oxime by the Beckmann rearrangement (rearrangement of oximes under the influence of acid).

Alternatively, there are other processes without the use of cyclohexanone, such as the photonitrosation of cyclohexane (PNC process), the partial hydrogenation to oxime (DuPont) or the Snia cyclohexane carboxylic acid process. However, production using cyclohexanone is predominant worldwide.

31.3.4.1 Polyamide 7

The starting material is oenantholactam (7-aminoheptanoic acid). There are three processes for the production of polyamide 7, which were developed in the former USSR and in the USA:

- In the USSR, 7-aminoheptanoic acid was produced from ethylene and carbon tetrachloride by telomerization on the one hand, and from furfurol via tetrahydropyran and the dinitrile of pimelic acid on the other.
- In the USA, benzene, phenol or cyclohexane serve as starting materials from which cyclohexanone is produced. The reaction then takes place first to -caprolactone and then via further reaction steps to methyl 7-aminoheptanoate (Baeyer-Villiger reaction).

Despite favorable properties in textile and technical terms, polyamide 7 is no longer of economic importance today because no cost-effective raw material synthesis is available. Hydrolytic polymerization can be carried out both continuously and discontinuously at 250–260 °C in the presence of adipic acid or acetic acid. In contrast to polyamide 6, an almost 99% yield of linear polyamide 7 is obtained, which is why extraction is not necessary. The melting point of PA 7 is 223 ° C (Zaremba et al., 1997).

31.3.4.2 Polyamide 12

The starting material is λ-laurin lactam. The hydrolytic polymerization can be carried out continuously and, due to the good thermal stability, at temperatures of more than 300 °C. The catalysts are mainly acids. Acids are mainly used as catalysts. The properties of PA 11, apart from the melting point, are similar to those of PA 12. Spun products made from PA 12 and PA 11 and their areas of application are also similar (Zaremba et al., 1997).

31.3.4.3 Melting Temperatures of Polyamides from Lactams

The melting point of polyamides from lactams is a function of the number of C atoms in the aliphatic chain. As the number of C atoms increases, the melting temperature decreases because the distance between the polar groups and thus the cohesion of the macromolecules via hydrogen bonds decreases (Table 31.2).

This is less the case for even-numbered C-chains because the macromolecules can align themselves in such a way that the polar groups of adjacent chains can attract each other. For odd-numbered C chains, this type of alignment is less possible and therefore their melting point is lower (Fig. 31.11).

31.3.5 Copolyamide

For copolyamides, both a qualitative and a quantitative indication of the composition is common.

Table 31.2 Starting materials and melting temperature of polyamides from lactams (Ullmann, 1976; Müller, 1963)

Lactam		HN-(CH_2)n-CO [n]	PAm [m]	Melt temperature [°C]
β-Propiolactam		2	3	330
(2-Pyrrolidon)	γ-Butyrolactam	3	4	265
(2-Piperidon)	δ-Valerolactam	4	5	271
	ε-Caprolactam	5	6	223
	Oenanthlactam	6	7	233
	η-Capryllactam	7	8	200
	Pelargonlactam	8	9	209
	Caprinlactam	9	10	188
	λ-Laurinlactam	11	12	179

For example, the indication PA 6.6/6 (80:20) means that this copolymer consists of 1,6-hexanediamine, adipic acid and ε-caprolactam in the molar ratio 80:80:20.

31.3.5.1 PACM-12 (Qiana)

Starting materials are 4,4′-diaminocyclohexyl methane and dodecanedioic acid. The PACM-12 is an alicyclic polyamide whose diamine occurs in three stereoisomers, the trans-trans, the cis-trans and the cis-cis isomers. The diamine is prepared by catalytic hydrogenation of 4,4-′-diaminodiphenylmethane obtained from aniline and formaldehyde. The mixture of isomers is composed of 52% trans-trans, 9% cis-cis and 39% cis-trans. To achieve high tensile strength and low heat shrinkage of the fiber, the trans-trans content must be above 70%. For this purpose, the proportion must be enriched by fractional dissolution of the salts from the two starting materials. The polycondensation, which is carried out continuously, must initially take place under pressure. The melting range is between 270 °C and 280 °C when using a 70% trans-trans fraction. To achieve the desired degree of polycondensation, the temperature rises to 300–320 °C in the final phase of the reaction (Fig. 31.12).

Comonomers, such as isophthalic acid, are used to influence the crimping behavior. The Qiana™ fiber was marketed by DuPont as nylon types 472 and 473; production began in 1965. Fabrics made of Qiana™ have a pleasant, silk-like feel, are crease-resistant due to the high glass transition temperature (135 °C), and are permanently pleatable. The strength is approx. 30 cN/tex, the elongation 20–30%, the density

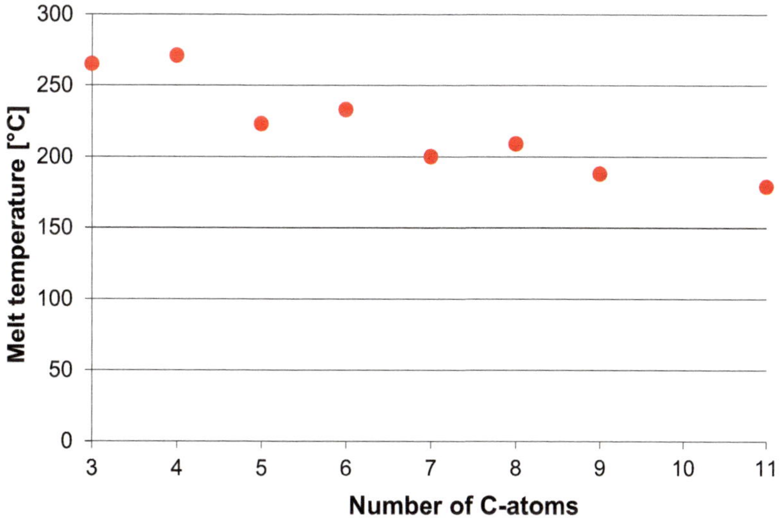

Fig. 31.11 Melting temperature of polyamides from lactams as a function of the number of C atoms acc. to (Ullmann, 1976; Müller, 1963)

Fig. 31.12 Synthesis of copolyamide PACM-12

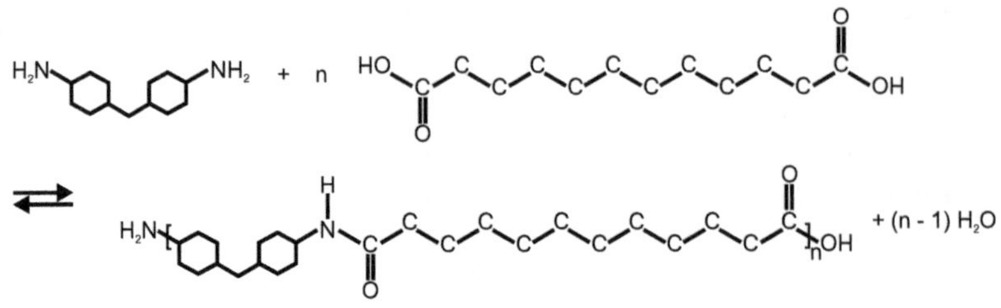

1.03 g/cm^3 and the melting point approx. 275 °C. Because the process was too expensive, the manufacturing of this fiber was soon abandoned.

31.3.5.2 Other Copolyamides

Copolyamides, which are composed of the known starting materials of the common polyamides, are characterized by a lower melting point, lower density, lower modulus of elasticity and better solubility. Certain products for the manufacture of hot-melt adhesive fibers and release yarns have become important. Their melting range is 110–150 °C. With appropriately composed binary and ternary copolyamides, these fibers have good adhesive strength and resistance to washing and dry cleaning.

31.4 Process for the Preparation of Polyamides

Polyamides can be produced either by polymerization or by polycondensation, depending on the starting materials used.

31.4.1 Polycondensation of Diamines with Dicarboxylic Acids and Polycondensation of Aminocarboxylic Acids, Respectively

In this process, the monomers, which are solid at room temperature, are heated to about 200–300 °C in aqueous solution. The two components or the two end groups must be present in a molar ratio of 1:1. Even a slight deviation leads to a too low degree of polycondensation. The reaction time is one to several hours. The reversible polycondensation process runs from start to finish in liquid phase. The water that is released must be removed to a large extent in order to achieve high degrees of polycondensation or molar masses. The final product of the diamine/dicarboxylic acid type contains only small amounts of cyclic oligomers, so separation is not necessary (Fig. 31.13).

31.4.2 Polymerization of Lactams

The polymerizability of lactams depends on the number of ring members ($n > 3$). Lactams can be polymerized in two ways:

- Hydrolytic polymerization: It takes place in the presence of water or water-releasing compounds. This process is only possible with 7 or more ring members.
- Cationic or anionic polymerization: This type of polymerization proceeds completely water-free by means of catalysts.

In hydrolytic polymerization, three reactions occur in parallel, the first being the starting reaction:

- Hydrolysis of the lactam with ring opening to the aminocarboxylic acid,
- polyaddition of the lactam to terminal amino end groups,
- polycondensation of amino and carboxyl end groups.

The position of the equilibrium of the polyaddition reaction determines how much non-convertible lactam remains. The polycondensation equilibrium determines, depending on the content of free water, the end group concentration and thus the chain length. Furthermore, side

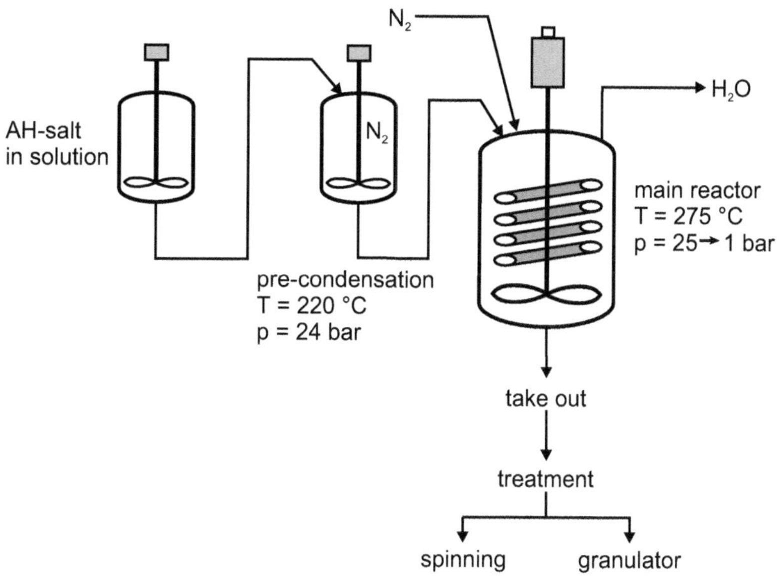

Fig. 31.13 Principle of polycondensation of dicarboxylic acids with diamines for the preparation of PA 6.6 from AH salt

reactions leading to cyclic oligomers have to be taken into account. All reactions are reversible and catalyzed by acidic end groups. Acids can be added as chain length regulators.

31.4.2.1 Single-Stage Process in the VK Tube

This process is used primarily to produce polyamide 6. The starting material is ε-caprolactam, a white, hygroscopic powder which melts at 69 °C, has a boiling point of 268 °C and is soluble in water and in organic solvents. PA 6 is produced exclusively by hydrolytic polymerization. The VK tube (German: vereinfacht kontinuierlich meaning simplified continuous), a heated flow tube developed by H. Ludewig, is mainly used for this purpose. Figure 31.14 shows the basic principle of this process.

In the classic version, which is still built today, the reaction takes place at atmospheric pressure. Molten lactam with 0.3–5% water is fed to the top of the VK tube with the addition of a chain length regulator and, if necessary, a matting agent, and the polymer melt is extruded at the bottom. As a rule, the VK tube has 2–3 heat exchangers with which the temperature profile can be specified along the tube. The upper one has the task of heating the reaction mass to over 230 °C in combination with the heat released during the reaction. This initiates first the ring opening and then the exothermic polyaddition and polycondensation. At the same time, the water evaporates down to a small residual amount. After passing through the heat exchanger, the temperature rises further to 270–280 °C due to the heat of reaction. With the middle heat exchanger, the temperature can be corrected if necessary and can be evened out over the tube cross-section. The lower heat exchanger is often used to reduce the temperature down to 250 °C. This allows the temperature to be reduced for a given total heat transfer rate. This allows a higher degree of polycondensation to form at equilibrium for a given total water content, consisting of free (dissolved) water and water bound in the end groups. At the same time, a lower equilibrium value for the unreacted monomer fraction is established. Baffles installed between the heat exchangers slow down the flow velocity in the area of the tube axis and thus help to equalize the residence time over the tube cross-section. The average residence time in the tube is 16–24 h. With this single-stage process, relative solution viscosities of 2.4–2.8 can be achieved (solvent: 96% sulfuric acid,

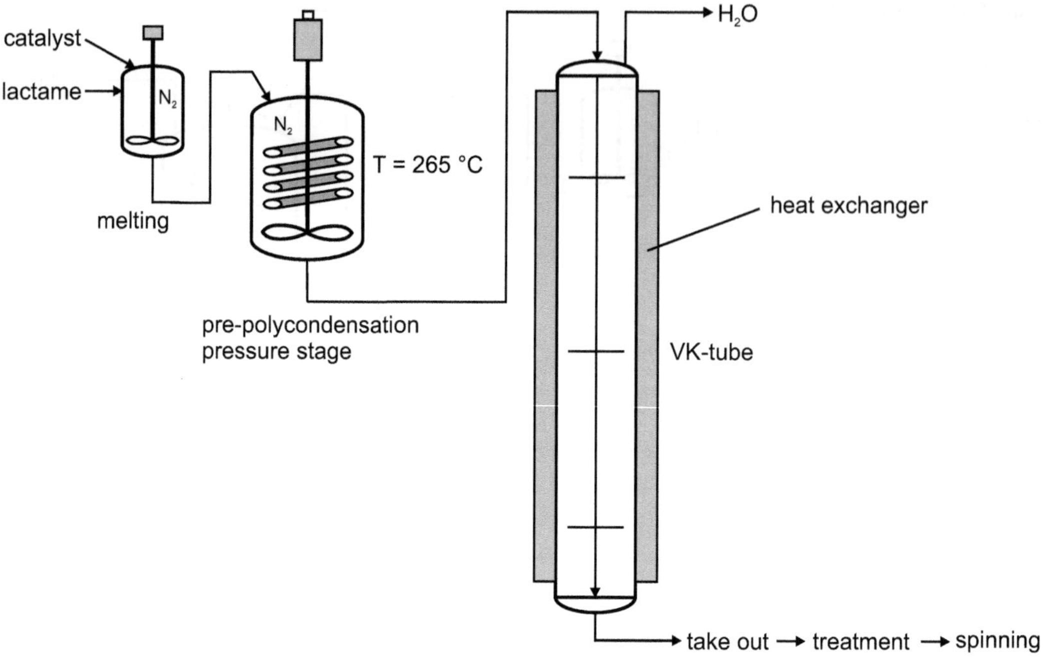

Fig. 31.14 Principle of polymerization of lactams in the VK tube

concentration: 1 g/100 ml, temperature: 25 °C), as required for the production of textile yarn and carpet yarn. The maximum throughput is about 100 t/d.

31.4.2.2 Two-Stage Process with VK Tube

The ring opening reaction makes a significant contribution to the long residence time. It can be accelerated by a higher water concentration in the reaction mass, which, however, requires an overpressure. This can be achieved with a 2-step process (Fig. 31.15).

Ring opening and pre-condensation take place in an upstream pressure stage, while post-polycondensation takes place after transfer and expansion in the VK tube. For higher degrees of polycondensation, e.g. for tire cord, the second stage is operated with negative pressure. In this way, relative solution viscosities of up to about 3.5 Pas can be achieved. The 2-step process is particularly suitable for high throughput rates and enables production of up to 200 t/d.

31.4.2.3 Discontinuous in Autoclave

In addition to the above, a discontinuous process in an autoclave is also possible, for the production of small batches or special grades. The polyamide 6 melt contains a water-soluble part which has a low molecular weight portion of 7–11%, consisting of monomer and oligomers. It must be removed to a large extent in order to avoid problems during further processing. This is usually done in the granulate.

31.4.2.4 Continuous with Tube

The particularly simple continuous process is often used for aqueous extraction of the granulate. Here, the granulate moving downwards in a vertical pipe is flowed through by upward-flowing water at about 100 °C. The water accumulates 8 to 14% extract, while the granulate extruded at the bottom only contains residual extractable water. In this process, the water accumulates 8 to 14% extract, while the granulate extruded at the bottom only has a residual extractable water

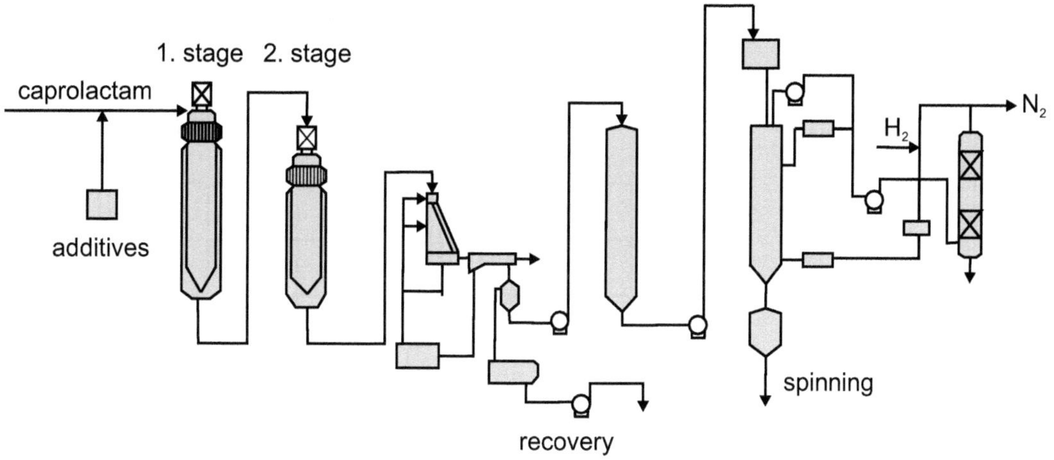

Fig. 31.15 Two-stage process

content of 0.6%. Internals prevent circulation flows due to density differences. Another option is vacuum de-monomerization of the polyamide 6 melt using thin film evaporators. Although this process allows direct spinning, it has not been able to gain acceptance. One of the reasons is that the oligomers can hardly be removed due to their low vapor pressure. In order to achieve higher molar masses, solid phase post polycondensation at 150–190 °C is also used. It can be carried out continuously or discontinuously.

31.4.2.5 Other Types of Polymerization

Cationic polymerization takes place using strong protonic acids (e.g. hydrogen chloride) as catalyst or amine salts of strong acids (e.g. butylamine hydrochloride). This polymerization proceeds at a similar rate to hydrolytic polymerization and produces similar oligomer contents at equilibrium. However, this process has not yet succeeded in producing useful spinning raw materials on a technical scale.

In anionic polymerization, strong bases such as alkali carbonates, alkali hydrides and alkali alcoholates are used as catalysts. In the case of molten, anhydrous caprolactam, for example, the reaction starts at 130 °C and lasts about 20 min. In the process, the compound reaches a temperature of 190 °C under adiabatic conditions as a result of the heat of reaction and crystallization released. The process is 50–100 times faster than hydrolytic polymerization. However, a technical application is so far only known in reaction injection moulding.

31.5 Process for the Production of Polyamide Fibers

Further processing of the polyamides into fibers is carried out using the melt spinning process. This can be done either in the direct spinning process by transporting the polymer from the reactor to the spinnerets (risk of thermal degradation) or via pellets as an intermediate product. For process-engineering reasons, the latter route predominates for the polyamides 6 and 6.6, which are important today. Also, the batch sizes for continuous production with reactors are usually not large enough.

31.5.1 Granulate Drying

Before melting, the granules must be uniformly dried to a certain residual water content in the range of 0.05–0.08%. Too high a value can reverse polycondensation after melting and lead to lower molecular weights. There is also the possibility that water vapor bubbles will form during spinning out, causing filament breakage. On the other hand, overdrying initiates an

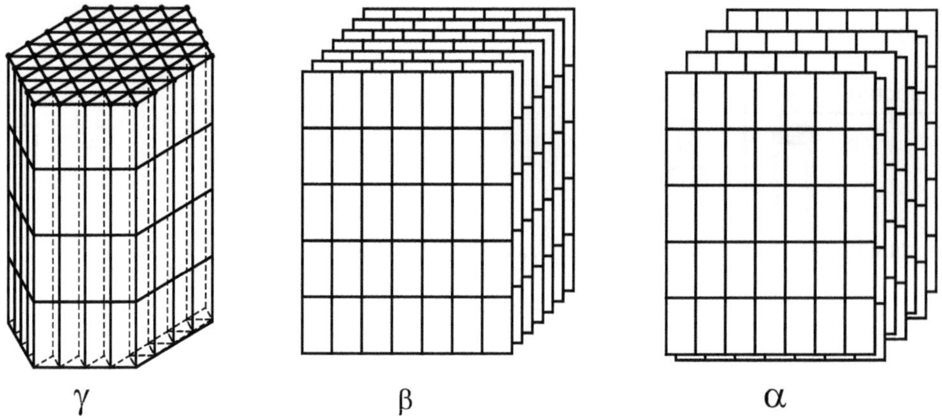

Fig. 31.16 γ-, β- and α-crystals (from left), (cf. Chap. 4)

undesirable, residence time-dependent post-polycondensation.

Granulate drying is usually carried out continuously in a shaft dryer with hot, pure nitrogen in counterflow, but can also be carried out discontinuously, e.g. in a tumble dryer under vacuum. In the continuous process, a duration of 16–24 h at 100–120 °C can be expected.

The subsequent pneumatic transport in pipelines to the storage tank and later to the spinning device takes place under pure nitrogen.

31.5.2 Melt Spinning

For the production of fibers from polyamide 6 and 6.6 granules, the grid spinning process was used until the 1980s. Due to its low throughput, it has since been completely replaced by extrusion spinning.

31.5.3 Drawing

For drawing, the yarn is guided over two godets whose circumferential speeds are set according to the desired drawing ratio. If heat is applied to the yarn from the outside, this is referred to as hot drawing, otherwise as cold drawing. Heated godets and rods are used to heat the yarn, while heated rollers, hot water baths and steam-heated channels are commonly used for cables. Higher stretching can be achieved by a two step process. The draw ratio depends on the pre-orientation of the spun yarn, but also depends on the desired yarn properties. For example, technical yarns requiring high tensile strength, low elongation-at-break and a high modulus of elasticity are sometimes stretched to the limit. For LOY made of polyamide 6 and 6.6, the draw ratio is 1:3 to 1: 5.

During the production of polyamide 6.6, different crystal structures are formed in the yarn, the so-called α- and γ-crystals. The hexagonal γ-crystals are formed immediately below the spinneret at relatively low temperatures (100–150 °C). They are particularly produced by external stresses ("stress-induced growth"). At higher temperatures, the thermodynamically more stable monoclinic α-crystals are formed. Their formation is much slower. The β-crystals represent a transitional form (Fig. 31.16).

The number and thickness of the α- and γ-crystals increase with increasing take-up speed. A minimum in the thickness of the γ-crystals can be observed at medium take-up speeds (Fig. 31.17).

If the yarn is wound onto the first godet at this speed and then heated there in a controlled manner, the γ- turn into α-crystals ("time-induced growth"). The α-crystals are much more sensitive to shearing. Therefore, the drawing of PA with a

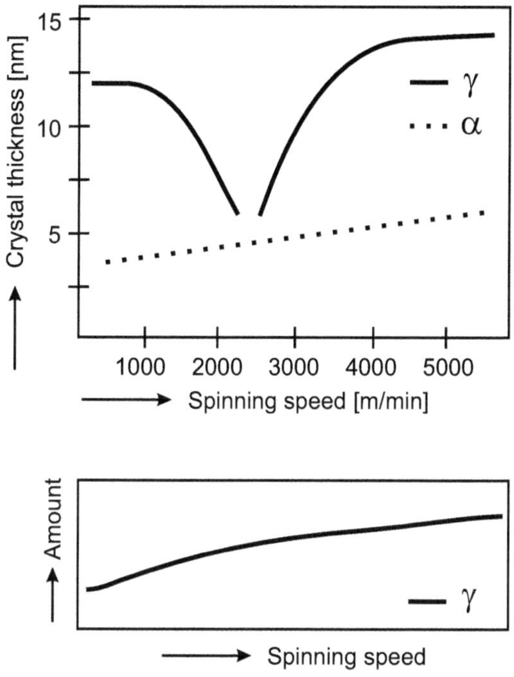

Fig. 31.17 Formation of α- and γ-crystals as a function of the winding speed (cf. Chap. 4)

high number of α-crystals is much easier and more uniform on the following godet duos (Fig. 31.18).

If POY filament yarns are first produced in a two-stage process and then drafted on draw winders, the yarn should already be relaxed before winding. In this way, an excessively hard or possibly deformed yarn package on the bobbin can be avoided. For this purpose, a heated drawing godet for heat setting and another godet for cooling are used in most draw-winding machines. The shrinkage can be adjusted between the two. In most cases, the yarn is interlaced for stabilization before winding. The drawing process can be followed by a twisting, warping or texturing process.

For economic reasons, stretching is usually combined with a subsequent production stage. Starting from POY, stretch texturing, warp stretching and stretch sizing are possible. In warp drawing and draw sizing, the yarns are drawn as a sheet with the aid of wide godets.

31.5.4 Production of Yarns from PA 6 and PA 6.6

A distinction is made between textile and technical filament yarns, the latter having, among other things, higher strengths.

31.5.4.1 Textile Filament Yarns

There are two different processes for the production of textile filament yarns, especially for clothing:

In the older process, a 2-step process is assumed. It consists of the spinning process at conventional take-off speeds and separate drafting of the LOY. The FDY obtained can be textured, twisted, warped or otherwise further processed.

Some experience is required in spinning to produce suitable package windings in terms of shape, hardness and uniform stretchability. Due to the hydrophilic nature of polyamides, a change in the length of the yarn may occur in moist air, resulting in slippage of yarn layers. Therefore, the yarn should absorb moisture even before winding. This is done by the conditioned quenching air and especially afterwards with the application of an aqueous preparation solution. The setting of an optimum water and oil layer is facilitated if there are two successive preparation devices, e.g. in the form of roller oilers, the first of which ensures wetting with water. Another important influencing factor is the room climate during winding. Finally, care must be taken to ensure a suitable winding tension of the yarn, which can be adjusted by matching the speeds of godets and bobbins.

Experience shows that, depending on the thread count and polyamide type, it is only possible to produce usable packages up to a certain winding speed. In the case of PA 6.6, this limit can be shifted upwards to about 1200 m/min by injecting steam into the spinning shaft, thus increasing the throughput. Undrawn polyamide yarns already start to crystallize on the spinning bobbin due to the absorbed water. To control this, the bobbins are usually conditioned at room temperature for several hours before drawing.

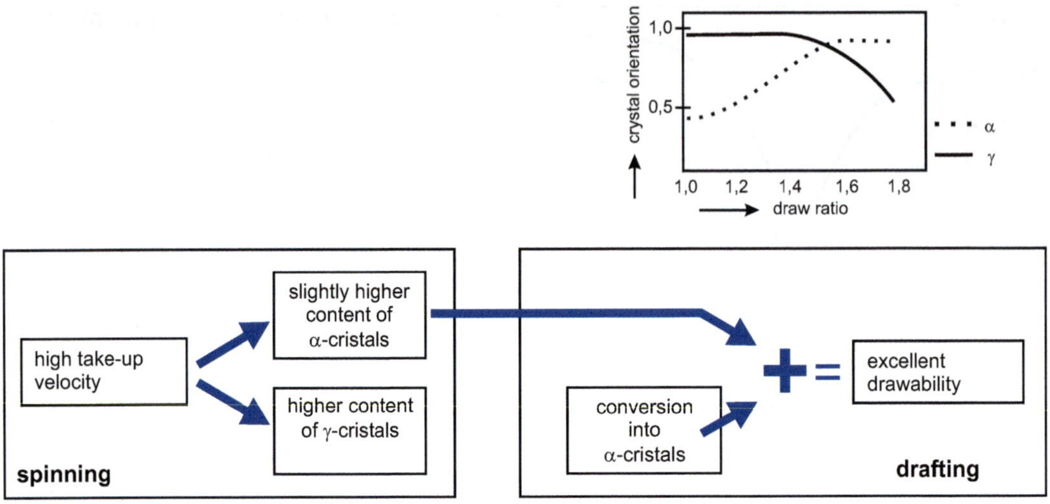

Fig. 31.18 Optimum spinning conditions for uniform drawing of PA (cf. Chap. 4)

Draw-twisting and draw-winding machines are used for drawing. The lowered glass transition temperature due to water absorption also makes cold drawing possible. A ceramic drawing pin wrapped by the yarn can be used to fix the drawing point. For fine smooth yarns and yarn specialities, the 2-step process can still be the most economical solution. However, the single-stage spun-draw process now dominates, of which there are various versions (Fig. 31.19).

In most cases, a high yarn take-off speed on the first godet initially creates a POY that is easy to draw. The main drafting then takes place between two godets with laying rollers or a godet DUO. The infeed godet can be unheated. After drafting, however, heat treatment is required to prevent excessive shrinkage of the yarn on the package. This can be done with a heated outlet godet or a downstream chamber operated with steam (Lückert et al., 1992). The winding speed is 4500–5500 m/min. The process has proved successful for the production of flat yarns. For textured yarns, on the other hand, draw-texturing is the most economical solution, starting from POY.

POY, which is mainly used for draw texturing, is spun at speeds from 2500–3200 m/min. Partially crystallized filaments are produced in the process. This makes the spinning bobbins easily transportable and storable for up to 3 months. During high-speed spinning, the filament tension increases considerably due to the air resistance. If it is in the range of the usual take-up tension, the upstream godets can be omitted. This can be achieved, for example, by bundling the filaments with a preparation pin located in the lower part of the quench duct. The typical draw in draw-texturing is below 1:1.4.

31.5.4.2 Technical Filament Yarns

As with textile yarn, production today also takes place with a spun-draw process. Spinning is carried out with both round and rectangular jet plates. A large hole spacing ensures fast and uniform cooling of the filaments. Ring-shaped jet plates can also be used. A heated tube, the so-called annealer is usually installed between the spinning beam and the quench duct. It delays the start of yarn cooling so that the quenching air hits a yarn that is already considerably drafted. The advantages are improved mechanical properties, in particular higher breaking strength. Low-water or water-free agents are used for preparation. Further processing of the yarn is normally carried out with four pairs of heated godets (Fig. 31.20).

Pre-drafting takes place between the first and second pair of godets and post-drafting between the second and third pair. When the third pair is circulated several times, heat setting takes place at

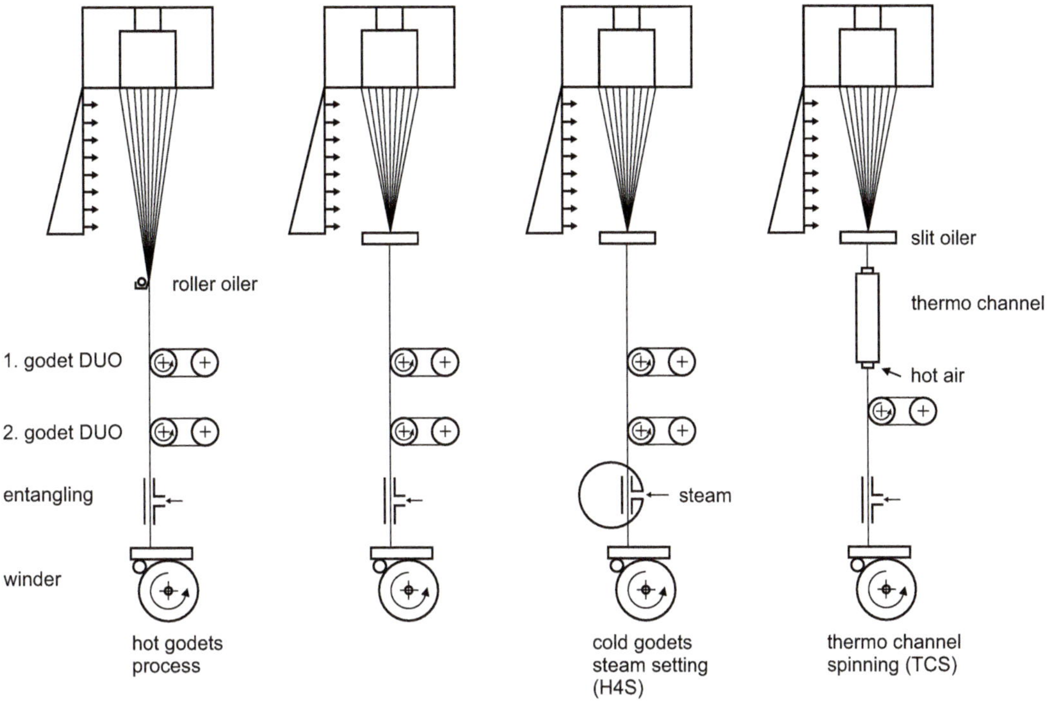

Fig. 31.19 Spin-draw processes for textile PA filament yarns acc. to Gries

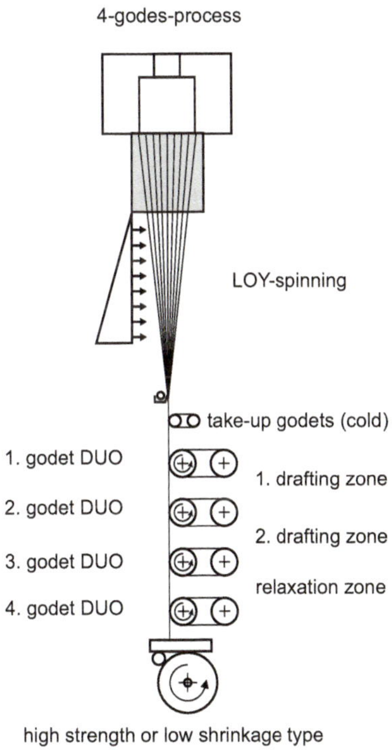

Fig. 31.20 Draw winding process for technical filament yarns (left) and draw winding unit (right)

a higher temperature, which is followed by adjustable shrinking with the aid of the fourth pair. Another task of the fourth pair is to cool down the yarn before winding. Depending on the fineness, a winding speed of 4500–5500 m/min can be achieved. An upstream, unheated roller increases the yarn tension in the infeed, which subsequently results in a smoother yarn run on the godets. With the various setting options, breaking strength, elongation-at-break, modulus of elasticity and heat shrinkage can be adjusted over a wide range.

31.5.5 BCF Yarn (Bulk Continuous Filament)

BCF yarn for tufted carpets is produced in the spun draw texturing process in a single operation. Usually 3 yarns are wound on each winder at the same time. The winding speed is 2500–3500 m/min, depending on the titer (600–3400 dtex). In the BCF process, spun-dyed fibers are often produced in order to improve the color fastness of the carpets made from them (abrasion).

31.5.6 Staple Fibers

In the production of staple fibers acc. to the 2-step process, the partial cables drawn off from cans are united and drafted with rollers. This is followed by compression crimping, drying, cutting into staple fibers and packaging in bales. In combination with a compact spinning plant, a 1-step process is also possible. Its advantage is the small space requirement and the good clarity of the plant.

By far the largest proportion comes from the 2-step process. The development here is going in the direction of ever larger numbers of holes per spinneret. Thus, several thousand filaments can be spun from one ring spinneret, which are cooled with radial quenching.

The second production stage is the fiber line. Its capacity is up to 300 t/d at drawing speeds of up to 300 m/min. In the case of direct spinning of PA 6, hot water extraction of the tow is required, for which a position located between crimping and drying is suitable. The treatment simultaneously leads to a low-shrinkage fiber type. A similar effect is achieved by a treatment with saturated steam. At the same time, the crimp is stabilized. It should be noted that for polyamide 6 and 6.6, the effect of dry heat setting at the same temperature is much lower. The production capacity of a plant in the 1-stage process with spinning speeds of 35–60 m/min is up to 80 t/d.

31.5.7 Spun Dyed Fibers

Polyamide fibers for automotive seat cover fabrics and floor coverings (BCF yarns) are mostly spun-dyed. The dyes are distributed over the entire fiber cross-section and thus the color impression is maintained even if the fibers are subjected to prolonged abrasion resulting in surface damage. In addition, large quantities of fibers of the same color are produced for these applications, so that the complex plant technology is economical and the lower flexibility is not disadvantageous compared to other dyeing processes.

31.5.8 Microfibers

Microfibers made of PA with a fineness of less than 1 dtex are mainly used in functional clothing, where water- and wind-repellent but water vapor permeable fabric properties are required. The extreme fiber fineness also opens up new areas of application in technical textiles, e.g. as filter fabrics or in combination with PET as split fibers for wiping cloths, e.g. for spectacles.

31.5.9 Profile and Hollow Fibers

To achieve special optical effects, e.g. for carpets, profiled fibers are produced from polyamide. This improves the adhesion of the fibers in the yarn, which reduces the tendency to pilling. An increase in the covering capacity as well as an improvement in the stability of the pile cover have a positive effect, particularly in the case of plush and carpets (BCF yarns). As a result, about 25% material can be saved.

Various brands of profile fibers include Antron™, Enkalure™, Grilon CS™ (multilobal and hollow profile), and Perlon NF™.

31.5.10 Differential Dyeing Types

Differential dyeing is a process in which fibers of the same basic type are dyed in different color depths and with different dye classes to achieve special color effects. For this purpose, the polyamide fibers are modified in such a way that dyeing with the desired dye class (acid, metal complex or disperse dyes) is possible. For example, dyeing with acid dyes is made possible by the amino end groups, the content of which is influenced both by the average molecular weight and by chain length regulators. A distinction is made between four types depending on the dyeing depth:

L = light type—barely dyeable,
R = regular type—normal dyeable,
D = deep type—darker than normal dyeable,
U = ultra deep type—can be dyed to a particularly deep shade.

By mixing different differential dyeing types, multicolor effects can be achieved, e.g. in carpets, by simple piece dyeing in a single bath process with acidic, basic and, if necessary, dispersed dyes (Zaremba et al., 1997).

31.5.11 Antistatic Polyamide Fibers

They are produced by modifying the molecular structure or by adding hydrophilic additives that are insoluble in the polyamide. The aim is to improve the conductivity of the fiber surface. Metallizing with nickel, grafting on a copper sulfide layer, adding superfine steel fibers to carpet yarns or adding carbon are further ways of reducing electrostatic charging. Antistatic polyamide fibers include Belltron™ from Comazo and Ultron™ from Monsanto (PA 6.6).

31.5.12 Flame Retardant Fibers

Added additives (compounds containing halogens, phosphorus and nitrogen) intervene physically or chemically in the various stages of the combustion process. A disadvantage is that higher concentrations can negatively affect the spinning behavior and the mechanical properties of the fibers.

31.5.13 Light Stabilized Fibers

Polyamide fibers show only moderate light resistance. In order not to increase this disadvantage by the usual fiber matting, special TiO_2 types (anatase modification) were developed for polyamides. This means that, as a rule, additional stabilization is not required for textile applications. For unmatted fibers, the addition of divalent manganese in the range of 10–50 ppm in the form of a salt is common, whereby neither color nor other fiber properties are negatively affected. For outdoor applications, the effect can be further improved by adding organic light stabilizers.

31.5.14 Heat Stabilized Fibers

The thermo-oxidative resistance of polyamide fibers, which is sufficient for most textile applications, is not sufficient for some technical applications, especially for tire cord. It can be considerably improved by the addition of suitable stabilizers. The stabilizing effect of copper (I) iodide has proven effective. The addition of antioxidants in concentrations of up to 1% is also possible.

31.6 Features

31.6.1 Chemical Proof Reactions

Differentiation of the various polyamide types can be made by their reaction to solvents or by

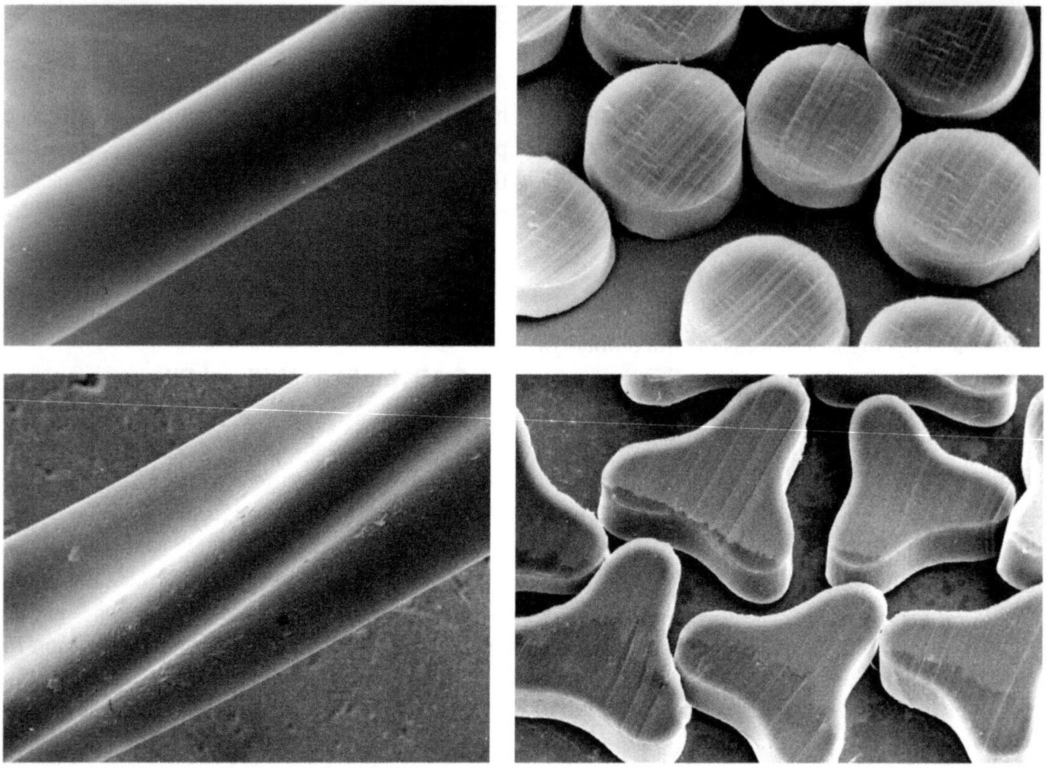

Fig. 31.21 Longitudinal and transversal view of round and trilobal polyamide fibers

dyeing. Another way to distinguish between different types of polyamide is to determine the melting point and the density of the fibers can also be used to differentiate. PA 11 and PA 12 have a density of approx. 1.04 g/cm^3 and thus float on a water surface, whereas all other polyamides sink (Fig. 31.21).

31.6.2 Physical and Textile Properties

Fibers made of polyamide 6 and polyamide 6.6 have very similar properties. Polyamide 11 and polyamide 12 deviate from this in terms of physical and chemical properties due to the significantly lower number of –CO–NH functional groups, as does polyamide 6.10.

Table 31.3 table contains the mechanical properties of the most important polyamide fibers.

The force-elongation curve of polyamide fibers shows a lower initial modulus than corresponding polyester fibers. At an elongation of 5–10% the modulus reaches a maximum (Fig. 31.22). At least up to this point, the elongation is almost reversible. Before reaching the breaking stress, the plastic strain increases strongly, resulting in a high work capacity and a long elongation section with low modulus. This course is less pronounced in the case of highly stretched fibers.

The abrasion resistance of polyamides is extremely high and exceeds that of almost all other textile fibers. The favorable course of the force-elongation curve, the excellent fatigue resistance, e.g. under bending stress, and the high durability are reasons for the widespread use of polyamide fibers in technical textiles.

The great similarity in the mechanical properties of PA 6 and PA 6.6 is due to the similarity in the structure of the macromolecules (Fig. 31.23).

Table 31.3 Mechanical properties of polyamide fibers acc. to manufacturer's data

		Density [g/cm^3]	Maximum tenacity [cN/tex]		E-modulus [cN/tex]	Maximum elongation [%]		Transverse brittleness angle [°]	DP
			Trocken	Nass	Trocken	Trocken	Nass		
PA 6	Fil	1.14	36–68	80–90	175–310	25–46	109–120	35–43	110–180
	Fil hat		68–85		350–730	14–22	107–150		
	SpF		32–63		97	40–70	107–113	27–41	
PA 66	Fil		36–63		265	25–46	104–122	35–41	50–80
	Fil-t		68–85		440–730	15–20	107–125		
	SpF		32–63		97	30–60	103–125	34	
PA 11		1.04	45–68	96–98	395	25–40	100	36–43	

Fil—Filament, textile
Fil-t—Filament, technical
SpF—Staple fiber

31.6.3 Thermal Properties

The internal structure of polyamide fibers consists of crystalline and amorphous regions. On the one hand, crystallization is favored by increasing spinning speed and increasing aspect ratio due to a higher molecular orientation to the fiber axis. On the other hand, crystallization is favored by water absorption due to a relatively low glass transition temperature (40–60 °C). Later heat treatment of the fiber leads to more stable crystalline regions and increases their proportion. For polyamide 6 and 6.6, the degrees of crystallization are 40–50%.

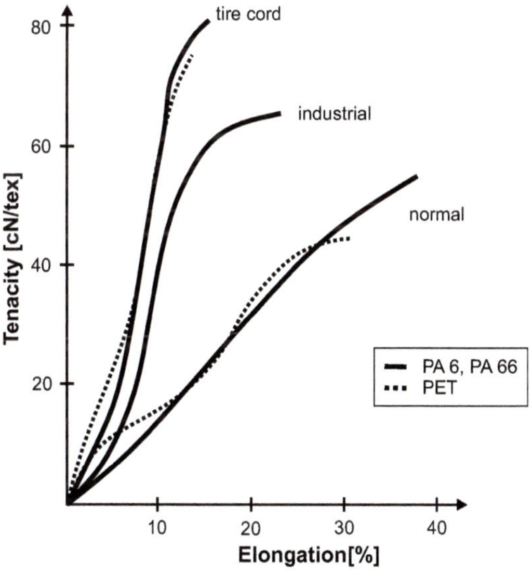

Fig. 31.22 Typical KD curves for PA 6 and PA 6.6 and comparison with PET

The Fig. 31.24 shows the differences in the thermal properties of PA 6 and PA 6.6. Depending on the manufacturer, these values differ slightly.

An overview of the most important thermal properties is shown in Table 31.4.

For car tires, polyamide 6.6 with its higher softening point is preferred because of the high vulcanization temperature used to save time. Where, on the other hand, longer vulcanization times are required, such as for truck tires, the use of polyamide 6 is also common.

31.6.4 Chemical Behavior

Polyamide fibers are classified as flame retardant. Finish-free, undyed polyamide fibers melt into a globule or drip off in the flame. They participate in combustion but do not continue to burn after

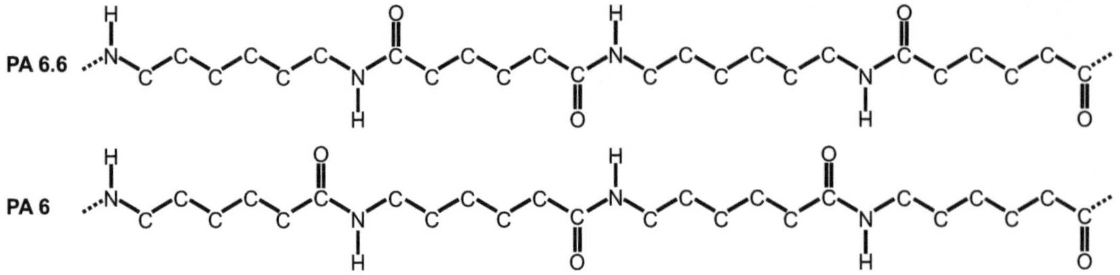

Fig. 31.23 Macromolecular structure of PA 6.6 and PA 6

the flame is removed. Finished or dyed polyamide fibers, on the other hand, can burn well, since dyeings with metal complex dyestuffs in particular, as well as textile auxiliaries, promote flame spread. The LOI value is around 20–21.

The self-ignition temperature of polyamide fibers is 510 °C, i.e. in the range of PET and PAN. The calorific value of polyamide fibers is around 20 MJ/kg, which corresponds to brown coal.

The resistance of the different polyamide fiber types to chemical and thermal stress correlates with the number of functional groups relative to the monomer length and decreases in the order PA 11 - PA 6.6 - PA 6. Polyamide 6 and polyamide 6.6 are more hydrophilic than polyester. During storage, the crystallinity can be changed by water absorption, which leads to a change in the coloring behavior. For the same reason, PA chips must be dried before spinning.

The sensitivity to weak mineral acids is more pronounced than for polyester. Concentrated mineral acids cause fiber damage or lead to hydrolytic degradation, depending on the type of acid, exposure time and temperature. Typical of PA 6 and PA 6.6 is their solubility in cold, concentrated

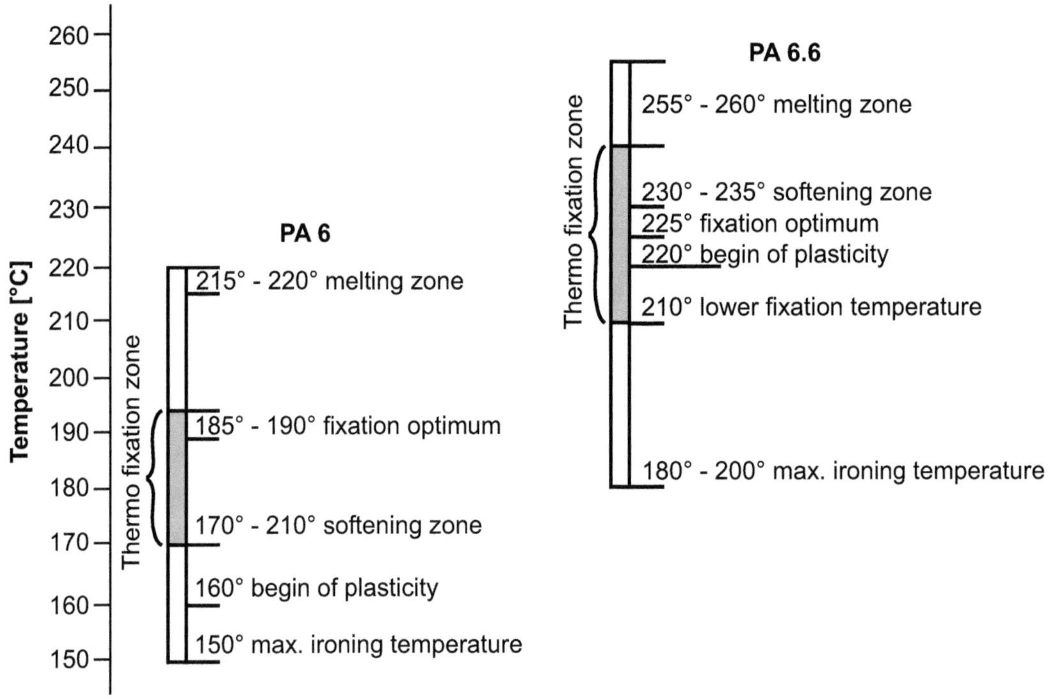

Fig. 31.24 Thermal properties of PA 6 and PA 6.6 (typical values)

Table 31.4 Thermal properties of important polyamides

Polymer	Fiber type[a]	Spec. heat [kJ/kg K]	Thermal conductivity [W/m K]	Melt temp. [°C]	Disintegration temp. [°C]	Softening temp. [°C]
Polyamide 6	Fill	1.5–1.9	0.24	215–220	310–380	170–210
	Fil hat					
	SpF					
Polyamide 66	Fill	1.5–1.9	0.24	255–260	310–380	220–236
	Fil hat					
	SpF					
Polyamide 11			0.29	190		165–175

Polymer	Fiber type[a]	Glass temp. [°C]	Moisture uptake [%]	Water retention [%]	Hot water shrinkage at 95 °C, set (non-set) [%]
Polyamide 6	Fil	40–60	3.5–4.5	10–15	1–5 (8–15)
	Fil hat				≈1 (≈10)
	SpF				0.5–1.5
Polyamide 66	Fil	45–65	3.5–4.5	10–15	1–5 (8–15)
	Fil hat				≈1 (≈10)
	SpF				0.5–1.5
Polyamide 11		46–56	0.9–1.3	3–7	

[a]Fil = filament yarn, ht. = high tenacity, SpF = staple fiber

formic and sulfuric acid; PA 6 is soluble in 14% hydrochloric acid, PA 6.6 in 30% hydrochloric acid. PA 11, on the other hand, is insoluble even in concentrated hydrochloric acid (cold).

At room temperatures the resistance to alkalis is very good. At high concentrations or high temperatures, destruction of the fibers occurs. Chlorine-containing bleaching agents and peroxides lead to fiber damage. Hydrogen peroxide bleaching requires the use of stabilizers to prevent loss of strength. Bleaching with sodium chlorite does not result in fiber damage.

At room temperature, there is excellent resistance to organic solvents, e.g. in dry cleaning. Polyamide fibers dissolve at room temperature in phenol, m-cresol and their chlorine substitution products, in chloral hydrate and in lactams. Furthermore, PA 11 and PA 12 dissolve in boiling o-dichlorobenzene and aniline, PA 6 in dimethylformamide.

31.6.5 Dye Affinity

The stretching of polyamide yarns influences their affinity for dyes. At a high degree of stretching, dye uptake is hindered due to the formation of crystalline areas. Therefore, spun fibers are easier to dye than highly drawn filament yarns. In general, they are easy to dye compared to other man-made fibers because of the large number of polar groups. There are hardly any differences in dyeing behavior between fibers made of PA 6 and PA 6.6. In contrast, polyamide 11 and 12 require higher temperatures and longer dyeing times due to their less hydrophilic character. The most important dye classes for PA 6 and PA 6.6 fibers are acid, metal complex and disperse dyes.

The dyeing properties can be well influenced by adding additives and chain length stabilizers during spinning. If monocarboxylic or dicarboxylic acids are used as stabilizers, polyamide fibers with low dyeability for acid dyes are obtained. The addition of primary aliphatic amines or diamines results in a polyamide with increased dyeability for acid dyes. An incorporation of tertiary amino groups into the polyamide chain molecules also increases the affinity to bind acid dyes.

Metal complex dyes are mostly based on Cu^{2+}, Co^{2+} and Fe^{3+}. If polyamides are to be dyed with

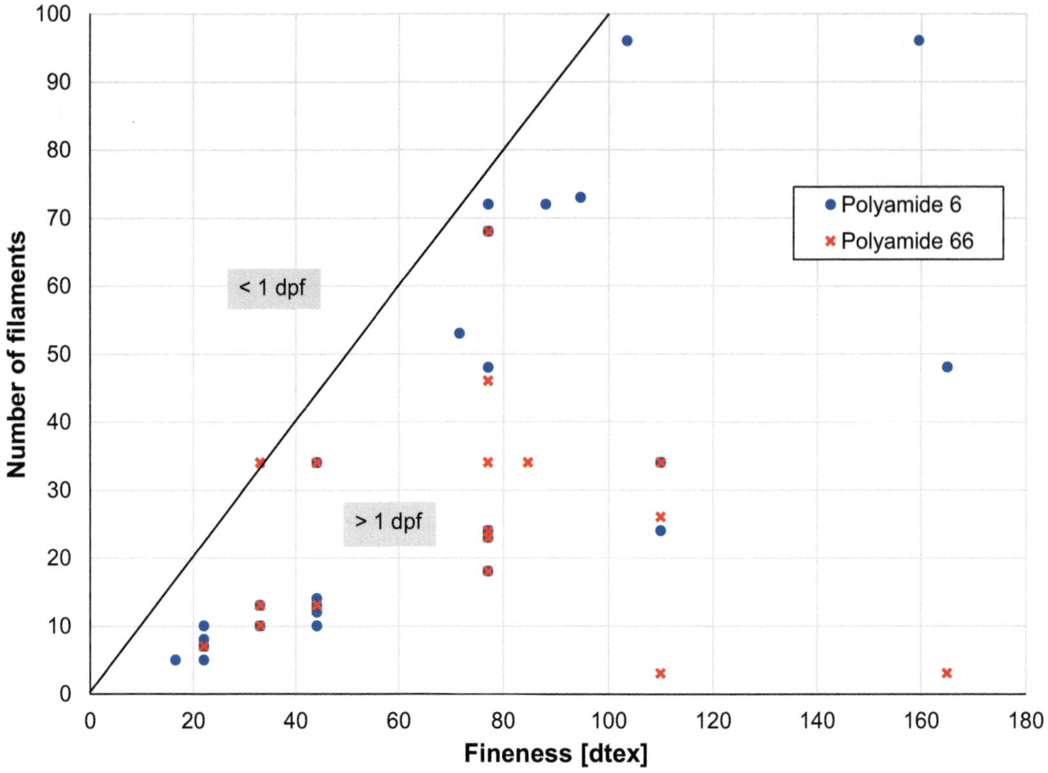

Fig. 31.25 Typical filament yarn types

basic dyes, this is possible by incorporating sulfo compounds, e.g. 5-sulfo-isophthalic acid, in equimolar ratio with 1,6-hexanediamine while simultaneously blocking the amino end groups. Such fibers are often marked with "B" (basic dyeable).

31.6.6 Other Properties

The resistance of the polyamide fibers to the effects of light is unsatisfactory without the addition of a stabilizer. The weather resistance is sufficient.

Permanent resistance to heat without loss of strength is given up to 90 °C, with the addition of suitable stabilizers and/or antioxidants up to 130 °C. Higher temperatures additionally lead to yellowing. Low temperatures (−40 °C) lead to a strong decrease of the ultimate tensile elongation and an increase of the ultimate tensile strength. The property changes at low temperatures are reversible. The resistance to cold is very good.

Flat fabrics made of polyamide fibers that are heat set (180–190 °C for PA 6 and 210–225 °C for PA 6.6) allow washing in warm water without becoming crumpled. They also exhibit better wrinkle resistance. Resistance to microorganisms and insects is very good. Polyamide fibers are mold resistant and cannot rot or decay. They are avoided by insects such as moths, ants, fur beetles and the like.

31.7 Yarn Types

The Fig. 31.25 shows typical textile filament yarn types of PA 6 (circle) and PA 66 (cross). In contrast to PET filaments, polyamides are rarely spun with a single filament fineness of less than 1 dpf. Yarn fineness of more than 200 tex are available for technical yarns.

In the case of staple fibers, there is a wide range of fineness from 1–140 dtex and a correspondingly large number of fiber lengths.

31.8 Market and Manufacturers

In 2020, global fiber production totaled around 120mt. The share of synthetic fibers was 67%, corresponding to around 81mt. Of this, polyamides accounted for about 7%, or about 5.2mt. In 1970, world production of PA was 1.9mt, corresponding to a market share of about 40%. By 1980, the market share dropped to 30%, in 1990 to 24%, in 2000 to 14% and in 2010 it was 9%. Since then, production increased by 25%, mainly due to greater demand for filament yarns (Figs. 31.26 and 31.27).

Polyamide 6 and 6.6 are used to produce textile and technical filaments as well as carpet yarns and, to a lesser extent, staple fibers. The textile filaments are mainly used in the hosiery industry, while around 75% of the technical filaments are used in the automotive industry, primarily for the production of airbags and tire cord. Carpet yarns are used primarily for carpeting. Staple fibers are often processed in blends with other fibers. In recent years, the production volume of staple fibers has decreased considerably due to the higher price of polyamides compared to other staple fibers (e.g. PET). In the area of carpet yarns, PA is also under strong price pressure from polypropylene.

Figure 31.28 shows the development of the production of PA filament yarns (textile, technical, BCF) since 1990. While the USA was the world's largest producer until 2005, China now dominates with a market share of over 60%.

The Fig. 31.29 shows the current situation on the world market for filaments (textile, technical, BCF).

There are major differences depending on the product group. China, for example, dominates the market for textile filaments with 2.5mt (78%) and also for industrial yarns with 725,000 tons (54%), while the USA is the leader for BCF yarns with 444,000 tons (67%).

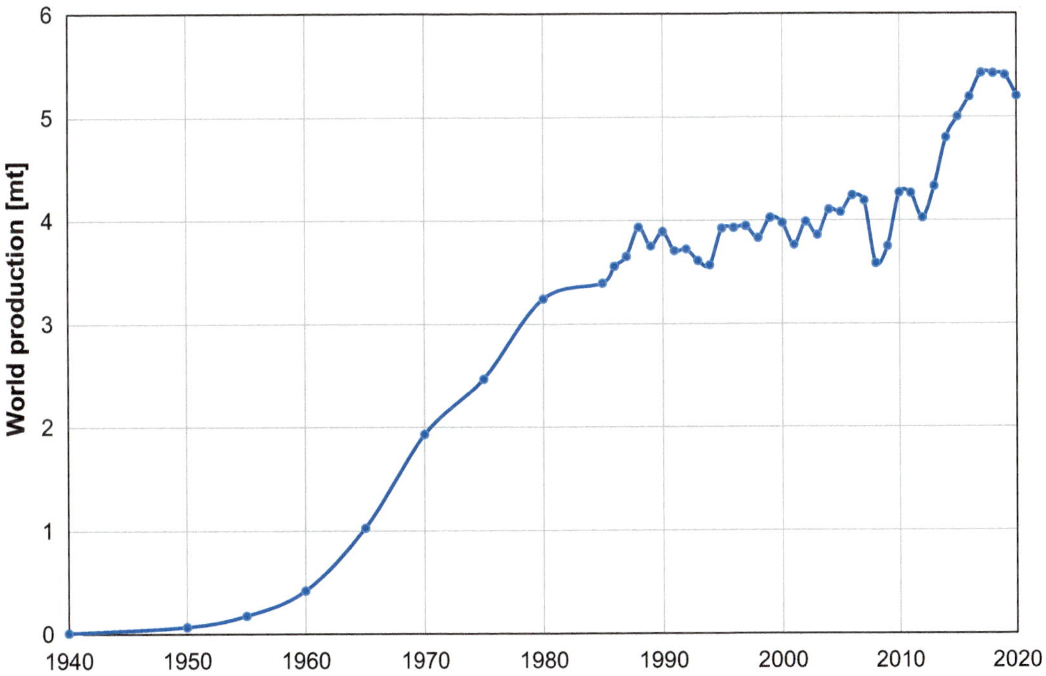

Fig. 31.26 World production of polyamide fibers (data: The Fiber Year, 2021; Falkai, 1981)

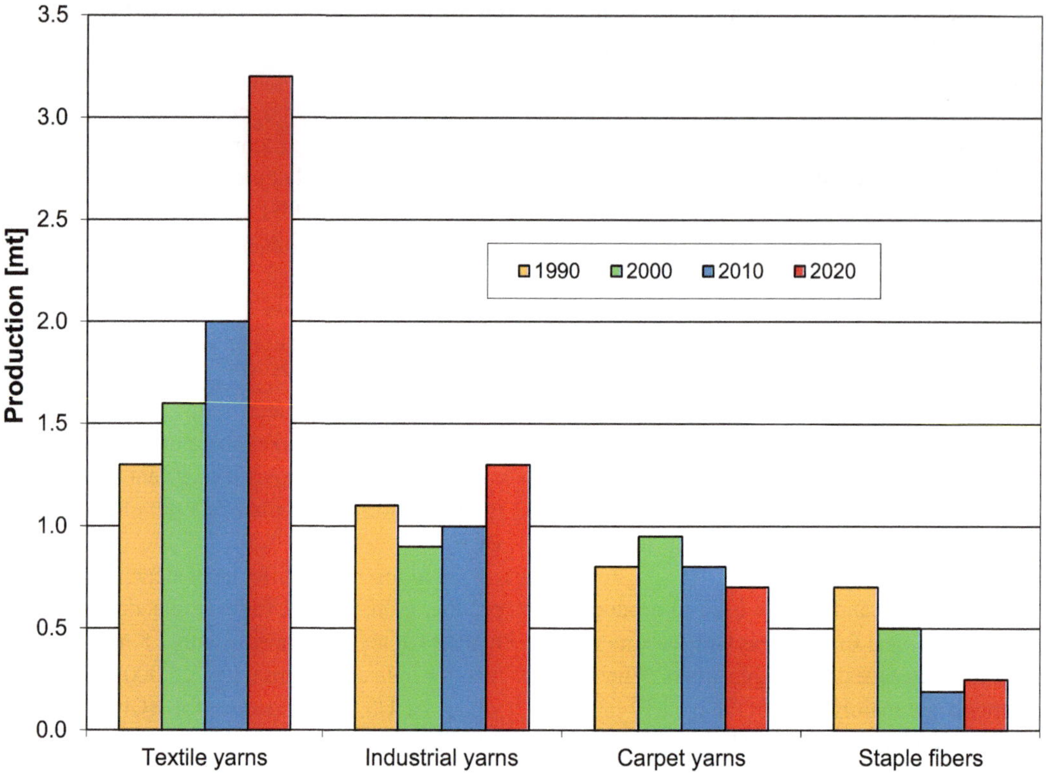

Fig. 31.27 Development of polyamide yarn production since 1990 (data: The Fiber Year, 2021)

Polyamide staple fibers are currently produced in China (187,000 t; almost exclusively PA 6), in the USA (33,000 t; almost exclusively PA 6.6) and in Europe (28,000 t; 67% PA 6.6, 33% PA 6). Other countries no longer play a major role.

Caprolactam, the basic material for PA 6, is produced by only a few manufacturers, half of them in China (2005: 8%) and 25% in Europe. Due to heavy investments in new plants in China in recent years, prices are not expected to rise soon. Adipic acid, the basic material for PA 6.6, is produced in several countries and with smaller plants. The price is therefore higher and is expected to remain so in the near future. Therefore, PA 6.6 is always slightly more expensive than PA 6.

PA 6 is historically represented mainly in Asia, Latin America and Europe, whereas PA 6.6 is represented mainly in North America and Western Europe (Sect. 31.2). The overall market share is 65% for PA 6 and 35% for PA 6.6. If we look at the polyamide market in China, the USA and Europe by area of application, we see different weightings here too. In Asia, textile filaments dominate (approx. 70% of the PA market), while in Western Europe textile, technical and carpet yarns each account for around 30%. In the USA, carpet yarns dominate with 75% (Fig. 31.30).

Widely used trade names for polyamide fibers are Nylon™, Perlon™, Grilon™, Nexylon™, Timbrelle™, Dorix™, Enka-Nylon™, Enkalon™, Zefran™, Stanylenka™. In the Table 31.5, selected manufacturers are compiled, on whose websites fiber data are published.

31.9 Processing Properties and Areas of Application

Compared to other man-made fibers, polyamide fibers are characterized by excellent physical and technological properties. Despite the higher raw

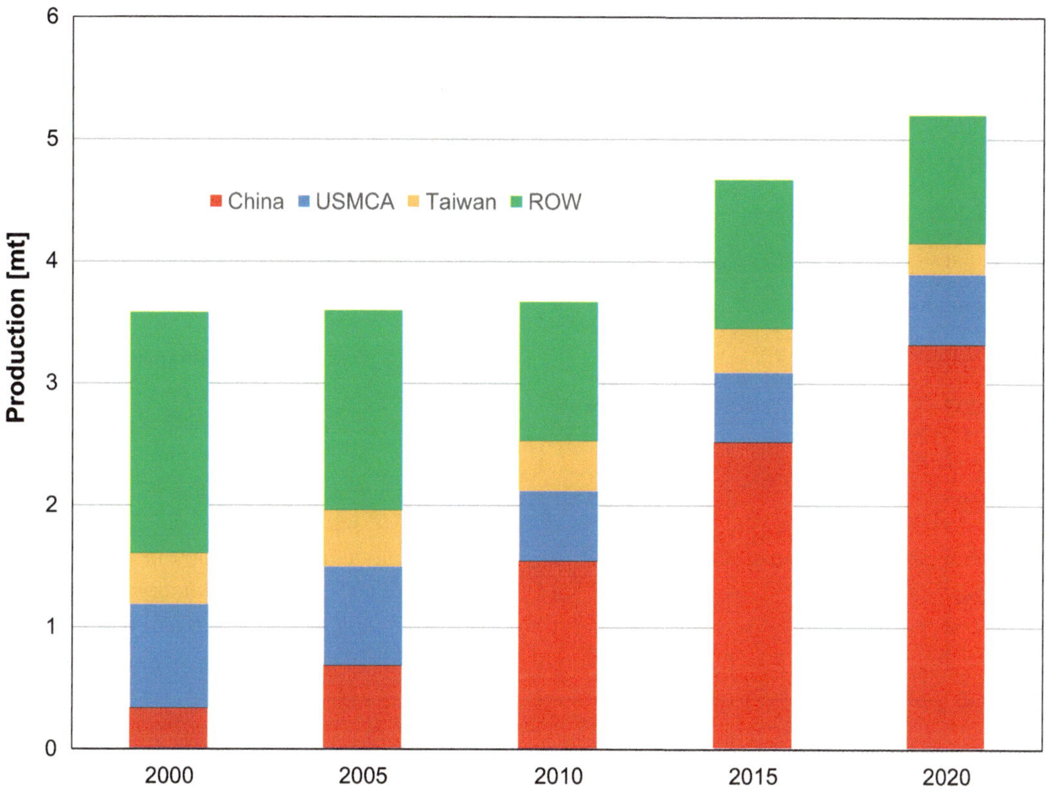

Fig. 31.28 Major producers of PA filament yarns (data: The Fiber Year, 2021)

material and production costs compared to polyester, the good processing and usage properties of polyamide fibers lead to a diverse range of applications for clothing, home textiles and technical textiles. The particular advantages of polyamide fibers or the finished goods made from them are (Zaremba et al., 1997):

- Excellent tear resistance,
- extremely high resistance to abrasion,
- high elasticity,
- very high fatigue bending strength,
- easy dyeing,
- easy washability and quick drying of the finished goods,
- good resistance to a large number of aggressive chemicals,
- very high resistance to biological decay influences.

Specific disadvantages of polyamide fibers or finished goods made from them are:

- Strong tendency to electrostatic charge, which favors rapid soiling,
- dirt, grease and oil affinity,
- high pilling tendency in goods made of staple fiber yarns,
- moderate stability to light; can be increased by light stabilizers,
- moderate resistance to higher temperatures (can be improved with heat stabilizers),
- from a clothing physiological point of view, poor moisture and perspiration absorption capacity; however, through suitable article construction and the use of microfibers, moisture is transported to the surface of the textile through the capillary effect and thus a pleasant wearing comfort is achieved,
- high elongation at small tensile stresses and slow elastic recovery, requiring careful processing.

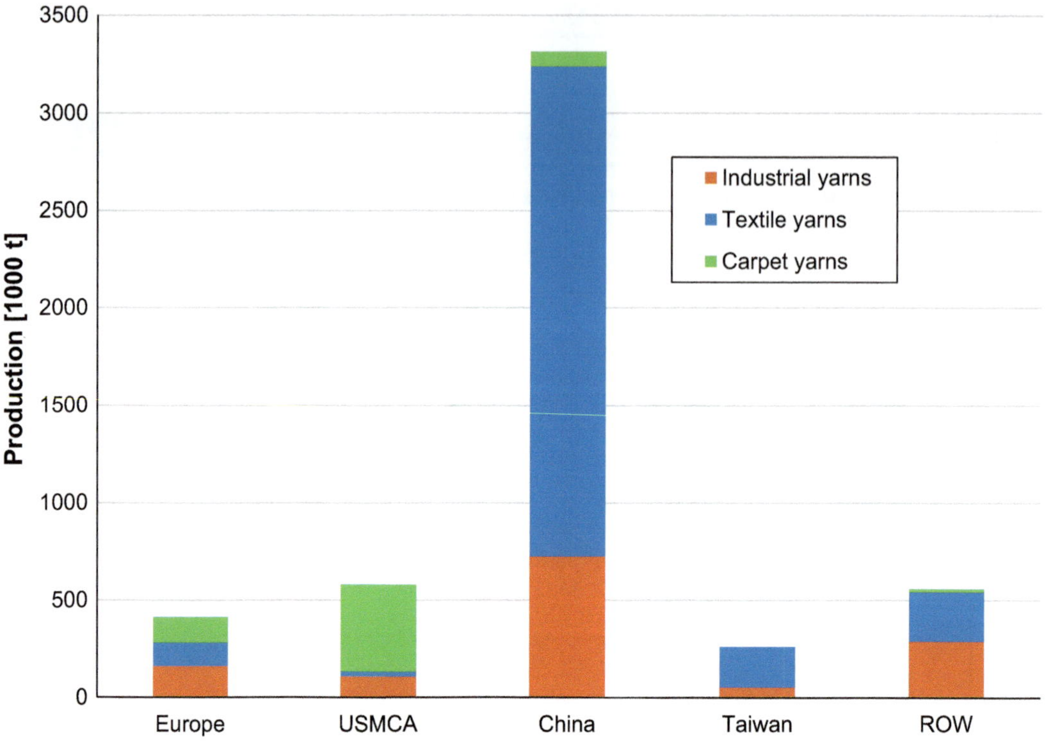

Fig. 31.29 PA filament yarn manufacturers 2020 (data: The Fiber Year, 2021)

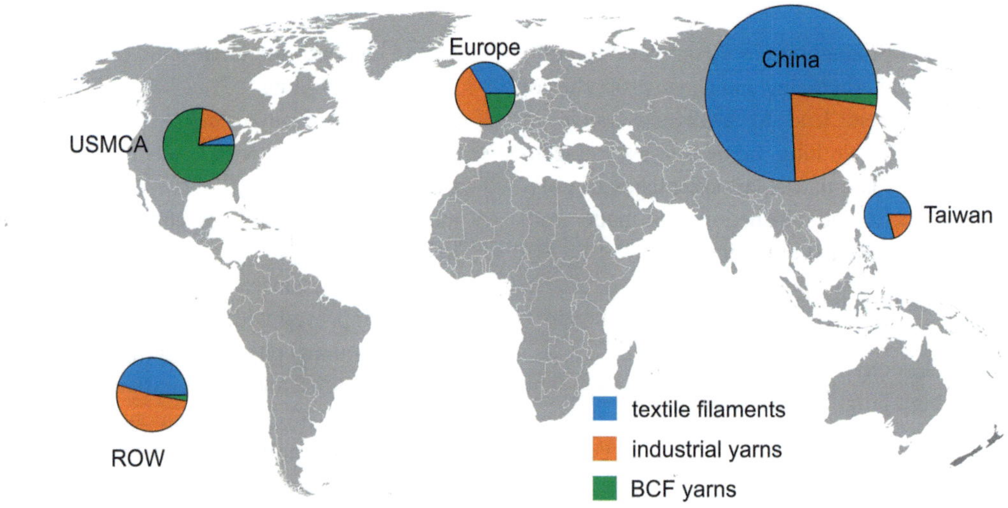

Fig. 31.30 Polyamide Filament Products and Producers (data: The Fiber Year, 2021)

Polyamide fibers are used in a wide range of end products:

- Fine polyamide filament yarns with a total fineness of up to approx. 200 dtex are processed to a large extent into stockings, pantyhose, ladies' underwear and corsetry. Important products made from textured polyamide yarns are swimwear and sportswear,

Table 31.5 Selected manufacturers of PA 6 and PA 66

Producer	PA 6	PA 66	Filaments	Staple fibers	Monofilaments	Website	
beti d.d.	x		x			https://www.beti.si	Slovenia
		x	x				
EMS-GRILTECH AG	x		x		x	http://www.emsgriltech.com	Switzerland
		x	x				
Franke Chemiefasern GmbH & Co. KG	x		x	x		https://www.franke-fibers.de	Germany
		x	x	x			
Hangzhou Dikai Industrial Fabrics Co.,Ltd	x		x			http://www.dikai-fabrics.com	China
		x	x				
Jiangyin Huacai Textile Co., Ltd.	x			x		https://www.huacaifiber.com	China
Ladylon Kft.	x		x			http://www.ladylon.hu	Hungary
		x	x				
Minifibers, Inc.		x	x	x		https://www.minifibers.com	USA
Neshin Spinning	x		x			http://www.neshin.com.tw	Taiwan
		x	x				
PHP Fibers GmbH (Indorama Ventures)		x	x			https://www.indoramaventures.com	Germany
PROTEX NEW ADVANCED TEXTILES GMBH		x	x			http://www.protexonline.com	Germany
Thread India, Ltd.	x		x			http://www.threadsindia.com	India
		x	x				
Toray Ind., Inc.		x	x			http://www.toray.us	USA
TWD Fibres GmbH		x	x			http://www.twd-fibres.de	Germany

women's dress, blouse and shirt fabrics, elastic outerwear, jacket and coat fabrics, umbrella, decorative and upholstery fabrics. In the technical sector, fine, high-strength polyamide filament yarns are used to produce fabrics for airbags and parachutes.

- Coarse polyamide filament yarns are mainly used in carpets and other floor coverings. Other applications include high-strength tire cord, nets, fishing tackle, as reinforcing fibers in driving belts and conveyor belts, as well as surgical sewing threads, luggage and shoe uppers (so-called Oxford yarns), all types of belts (e.g. safety belts in cars), tapes and ropes.
- Polyamide staple fiber yarns are mainly processed in blends with cotton, wool and other man-made fibers. Blending with polyamide improves the strength and wear properties of a fabric without changing its general character. Typical products include men's socks, leotards, dress fabrics, including those for work suits and children's clothing, linings and knitting yarns. Polyamide staple fibers are used in carpets and other floor coverings, as an admixture for needle felts, especially in the running layer, for the manufacture of nonwovens, e.g. for shoe inserts and dishwashing sponges, and as short cut fibers for electrostatic flocking. Polyamide staple fiber yarns are also processed into technical fabrics, filter fabrics, lacing belts and interlinings for the rubber industry.
- Wires and bristles made of polyamide, spun as monofilaments, have many uses, e.g. as tennis strings, fishing nets, fishing lines, tow ropes, climbing ropes and safety ropes, jumping

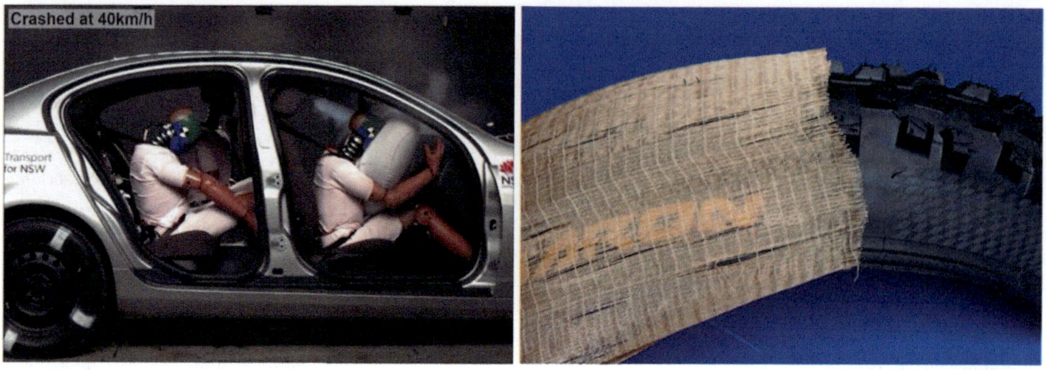

Fig. 31.31 Typical polyamide products (left: Transport for NSW, 2019)

cloths, screen cloths and other technical fabrics, brushes, toothbrushes (often made of PA 11, PA 12, PA 6.10) as well as paintbrushes.

The Fig. 31.31 shows typical technical products made of polyamide (airbag, tire cord fabric). Polyamide is otherwise used for almost all articles in which PET is also used.

References

Aelion, R. (1956). Rilsan – Polyamide 11. A newcomer in the field of synthetic fibres. *Fibres (Engineering and Chemistry), 17*, 96.

von Braun, J. (1907). Synthese von Amidosäuren aus cyclischen Iminen. *Berichte der Deutschen Chemischen Gesellschaft, 40*, 1834–1846.

Carothers, W. H. (1929). Studies on polymerization and ring formation. I. An introduction to the general theory of condensation polymers. *Journal of the American Chemical Society, 51*, 2548–2559.

Carothers, W. H., & Berchet, G. J. (1930). Studies on polymerization and ring fromation. VIII. Amides from epsilon-aminocaproic acid. *Journal of the American Chemical Society, 52*, 5289–5291.

Carothers, W. H., & Hill, J. W. (1932a). Studies on polymerization and ring formation. XIII. Polyamides and mixed polyester-polyamides. *Journal of the American Chemical Society, 54*, 1566–1569.

Carothers, W. H., & Hill, J. W. (1932b). Studies on polymerization and ring formation. XV. Artificial fibers from synthetic linear condensation superpolyamides. *Journal of the American Chemical Society, 54*, 1579–1587.

Falkai, B. V. (1981). *Synthesefasern*. Verlag Chemie.

Gabriel, S., & Maass, T. A. (1899). Über ε-Amidocapronsäure. *Berichte der Deutschen Chemischen Gesellschaft, 32*, 1266–1272.

Genas, M. (1962). Rilsan (Polyamid 11). *Synthese und Eigenschaften, Angewandte Chemie, 74*, 535–540.

Klare, H., Fritzsche, E., & Gröbe, V. (1963). *Synthetische Fasern aus Polyamiden*. Akademie-Verlag.

Koslowski, H.-J. (2008). *Chemiefaser-Lexikon*. Begriffe, Zahlen, Handelsnamen. Deutscher Fachverlag, Frankfurt a. M.

Lückert, H., Stibal, W., & Kemp, U. (1992). Stand und Trends der Technologien zur Herstellung schmelzgesponnener Synthesefasern. *Chemiefasern/Textilindustrie, 42*, 476–482.

Manasse, A. (1902). Synthese der ζ-Amoinoheptylsäure. *Berichte der Deutschen Chemischen Gesellschaft, 35*, 1367–1372.

Müller, E. (1963). Lactame – Grundstoffe für Fasern und Kunststoffe. *Melliand Textilberichte, 44*, pp. 484–487, 596–599.

The Fiber Year. (2021). The Fiber Year GmbH, Freidorf.

Transport for NSW. (2019). https://commons.wikimedia.org/wiki/File:Crash-test-with-airbag-and-safty-belt.jpg, CC BY-SA 4.0.

Ullmann. (1976). Fasern. In *Ullmanns Encyklopädie der technischen Chemie*, 4. Auflage, Bd. 11, pp. 193–411.

Zaremba, S., Steffens, M., Wulfhorst, B., Hirt, P., & Riggert, K. H. (1997). *Polyamide fibers. Fiber table acc. to P.-A. Koch*. Institut für Textiltechnik of RWTH Aachen University.

32 Aramid

Aramides belong to the polyamides, but are not melt spun but solution spun and have a completely different property profile than aliphatic polyamides such as PA 6.6 and PA 6. The name is the abbreviation for *A*romatic *A*mides and the abbreviation of this abbreviation is AR (Fig. 32.1).

32.1 History

After aliphatic polyamides had been developed in the 1930s and reached the market by 1950, research began into alternatives to this type of polyamide. In 1964, H. Mark showed that the substitution of the aliphatic chain segments by aromatic rings would increase the molecular stiffness of the chain molecules analogous to PET (Mark, 1964). This is accompanied by a lower melting entropy, a higher melting point and an increase in Young's modulus as well as reduced moisture absorption. The first aramids were developed in 1955–1965 (DuPont, Monsanto). In 1962, DuPont introduced its m-aramid fiber under the trade name HT-1, in 1966, industrial production started and since 1967, this fiber is called "Nomex". Chemstrand, a joint venture between Monsanto and Viscose America, also developed an m-aramid fiber under the name X-101. From 1970, Monsanto produced an improved version under the name X-400 and later Durette®. Commercially, however, this fiber was not a success, mainly because of its low lightfastness and poor dyeability.

J. Preston and W.B. Black experimented with oxadiazoles at Monsanto from 1966 onwards and produced the first p-aramid fiber, which they called X-500. Although it did not have the heat resistance of m-aramids, it did have extremely high strength (170 cN/tex) combined with a very high Young's modulus (750 cN/tex), properties that no previously known fiber achieved. Unfortunately for the researchers, the Monsanto marketing department recognized the uniqueness of the fiber, but did not want to use this for advertising purposes because there was no market for this new combination of properties (Preston & Smith, 1966; Preston et al., 1969). Development continued at Monsanto until the early 1970s, but the fiber never reached market maturity, partly because of the difficult process, partly because of external circumstances (economic crisis), and was finally abandoned. In parallel, Stephanie Kwolek, among others, worked at DuPont on the development of an aromatic polyamide fiber (Morgan & Kwolek, 1959). She discovered the liquid crystalline properties of such polymers in 1965 and produced a p-aramide for the first time in 1966, which was manufactured in a pilot plant with a capacity of 200 t/a from 1971. The first field of application was the reinforcement of tires. The corresponding spinning process, a combination of dry and wet spinning (air gap spinning) was developed by Herbert Blades. The new fiber is marketed under the name Kevlar® since 1972.

D. Veit, *Fibers*, https://doi.org/10.1007/978-3-031-15309-9_32

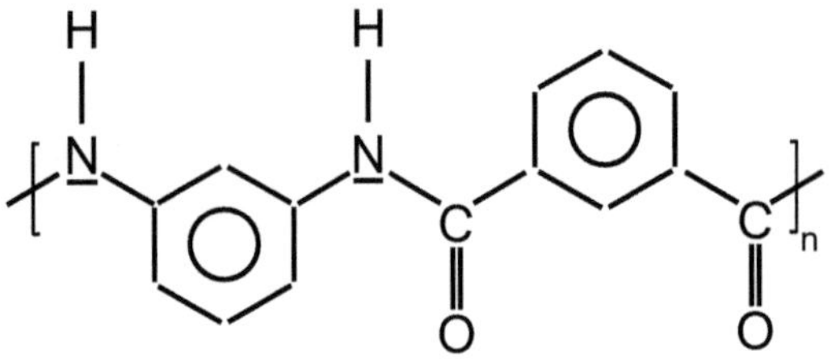

Fig. 32.1 Repeat unit of aramid (meta-form)

In 1974, large-scale production was started in Richmond with a capacity of 2700 t/a. Efforts were also made in the Soviet Union from the 1960s onwards to produce aramid fibers (Fenilon, Sulfon T, Terlon). However, none of these fibers could be produced economically on a large scale due to process engineering difficulties and the limited availability of raw materials and technical equipment. Bayer AG in Dormagen was also involved in the development of aramids in the 1960s and 1970s. The company had easy access to aromatic diamines from dye production, and so an m-aramide was developed under the name ATF2000®. It was heat-resistant and at the same time had a high moisture absorption capacity, which was advantageous for clothing. Bayer lacked access to manufacturers of technical textiles and the patent landscape was already largely covered in this area by DuPont, so they wanted to launch the new fiber as an alternative to polyester. However, because the raw material and process costs were too high compared to polyester and even Nomex, development was discontinued in 1973. The development of a p-aramid also remained stuck at an early stage of development for cost reasons. ICI in the UK also attempted to produce aramids (a4, a5). However, ICI lacked the know-how of solution spinning on a large scale because ICI had until then produced only melt-spinnable fibers. The production technology was largely covered by DuPont patents and DuPont showed no interest in licensing, so ICI ceased its activities in 1976. AKZO built its first pilot plant for the production of a p-aramid under the name Arenka® in 1976. The spinning process was very similar to that of DuPont, but different solvents were used. There was a fierce patent dispute between AKZO and DuPont until the 1980s, which was eventually settled. AKZO's aramid was given the name Twaron®. Teijin, Japan, produced an m-aramid fiber using the wet-spinning process starting in 1969, which was marketed under the name Conex® and later Teijinconex®. About 200 t/a were produced from 1974. A p-aramid copolymer was also developed under the name HM 50. Its process costs were lower than those of Kevlar, but the starting materials were more expensive. This fiber is now marketed under the brand name Technora®. In 2000, Teijin acquired the fiber business for Aramide from AKZO. Rhone-Poulenc began producing a polyamide-imide fiber under the name Kermel® in 1972. It belongs to the m-aramide class and is produced in relatively small quantities by Kermel. Its density of 1.34 g/cm^3 is the lowest of all common m-aramides. A Russian development marketed under the name Armos™ did not get beyond the experimental stage.

32.2 Raw Materials, Formation of the Polymer

The chemical structure of aramides corresponds to the known structure of polyamides. The basic building blocks are terephthalic acid (dicarboxylic acid) or the corresponding dichloride and poly-m-phenylene isophthalamide (meta-type) or poly-p-phenylene isophthalamide (para-type).

This results in the aramid types meta and para through a polycondensation reaction with the release of water (H_2O) or hydrochloric acid (HCl). Meta-aramides are "zig-zag" shaped, while para-aramides form elongated chains, resulting in significantly higher strength (Fig. 32.2).

Aramides consist of rigid macromolecules in which amide groups alternate with phenyl groups. This gives rise to high attractive forces between the chains, both through the amide groups and through the mutual attraction of the phenyl rings. The latter forces are of the order of the van der Waals forces in aliphatic polymers. Although the aramide types meta and para differ only in the position of the functional groups relative to each other, this leads to very different crystal structures.

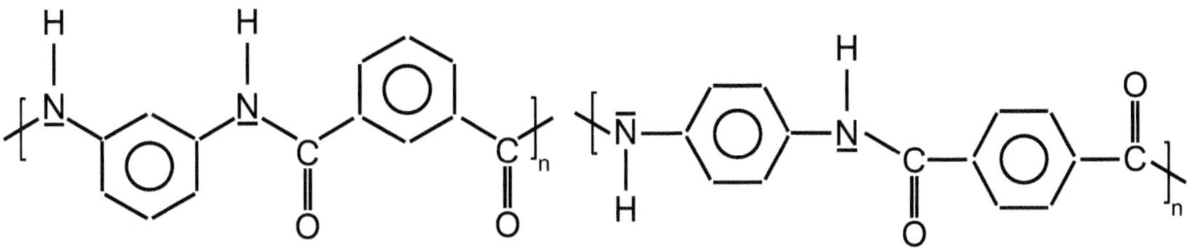

Fig. 32.2 Structural formula of the repeat units of m- and p-aramide

p-aramides are almost exclusively composed of -crystals and the amide groups point alternately in different directions. This allows extremely high degrees of stretching and orientation in the axial direction of the fibers. This results in the high strength and the high modulus of elasticity of p-aramids (Fig. 32.3).

p-aramide shows the superstructure shown in Fig. 32.4. Here, the crystallites form radial, leaflet-like structures. This results in high strength not only in the axial direction of the fibers, but also transversely, which can only be observed to a limited extent in PA 6, PA 66 and PET.

In m-aramide, the NH and CO functional groups all point in the same direction (Fig. 32.5). This massively hinders the formation of crystals in the solution and there are no liquid crystals as with p-aramid (see below). After spinning out, therefore, both partially crystalline and amorphous areas are formed, similar to polyamide and polyester. Accordingly, the mechanical properties, in particular the modulus of elasticity, are significantly lower than those of p-aramid.

32.3 Production

Aramides decompose already below their melting point. Therefore, they can only be spun in

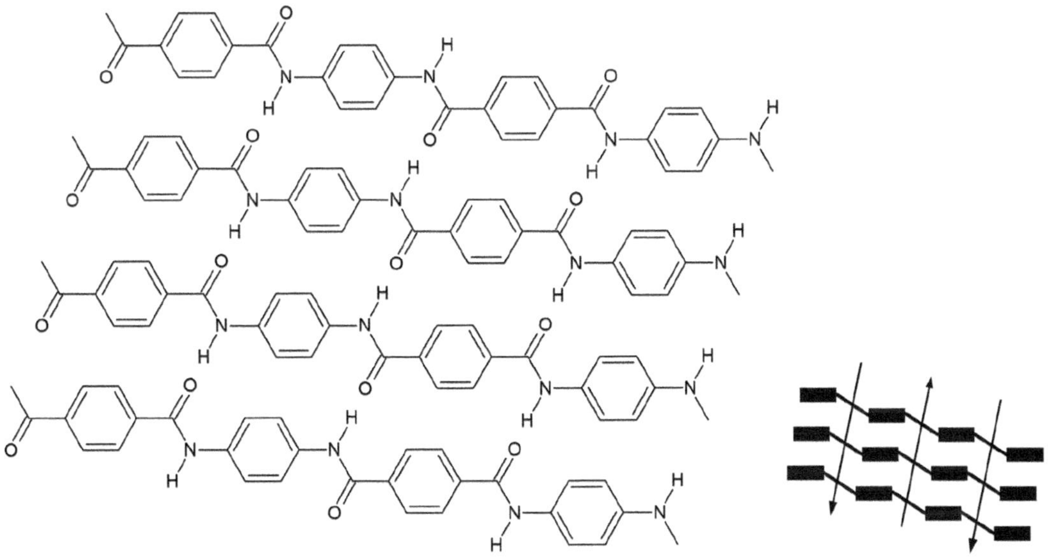

Fig. 32.3 Crystal structure of para-aramide (p-aramide)

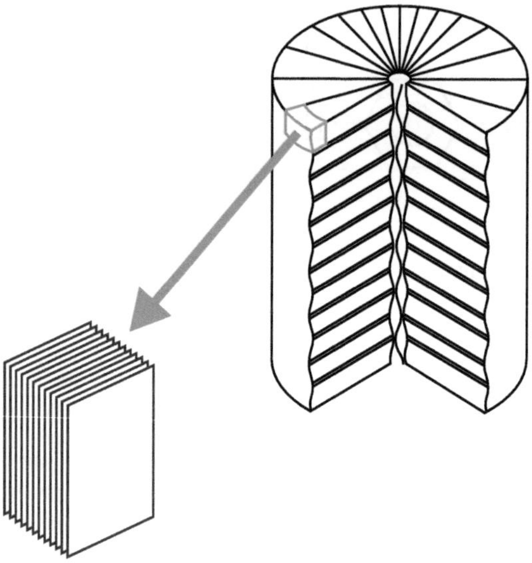

Fig. 32.4 Radial leaflet structure of para-aramid acc. to (Morton & Hearle, 2008)

solution. Concentrated sulfuric acid is usually used as a solvent.

32.3.1 Air Gap Wet Spinning Process

p-aramid fibers are produced using the air-gap spinning process. In this process, the polymer leaves the spinneret a few millimeters above the precipitation bath. The filaments formed are coagulated in an aqueous precipitation bath consisting of diluted sulfuric acid and water. This is followed by various drawing and washing processes before the filament yarn is wound up (Fig. 32.6).

Already in the spinning solution, small crystalline areas, so-called liquid crystals, are formed due to the strong mutual attraction of the macromolecules during the production of p-aramide. During coagulation of the fibers in the air gap and in the subsequent precipitation bath, these crystals become oriented and large crystalline zones with high packing density are formed. In between there are amorphous areas which can also be oriented under tensile stress and heat supply (Fig. 32.7).

The temperature of the precipitation bath for p-aramide is 0–20 °C, for m-aramide it is significantly higher. The filaments are withdrawn from the spinneret at a speed of up to 300 m/min. The polymer is present in high concentration in the spinning solution. This leads to a liquid crystalline, optically anisotropic solution. The non-oriented crystalline areas of the spinning solution are already pre-oriented within the spinneret by shear forces. Below the spinneret, they first lose their orientation again, before they are completely oriented in the following air gap. The orientation structure is then frozen in the coagulation bath.

During the preparation of m-aramides, the inorganic salts (e.g. LiCl, $CaCl_2$) required during solution preparation diffuse from the fiber into the solution during spinning. Therefore, they do not have to be removed subsequently. This is an advantage over the dry spinning process, where the inorganic salts remain in the fiber and may have an unfavorable effect on the properties (Fig. 32.8).

Various washing and drying processes follow to remove the solvent. After fixing and lubrication, the material is wound onto bobbins. The winding speed in the wet spinning process for aramide is approx. 300 m/min for m-aramide and up to 700 m/min for p-aramide. The fibers are stretched by about 400% during this process.

32.3.2 Dry Spinning

This process is currently only used to produce DuPont m-aramides. After heating, the solution is filtered and spun out. From a nozzle, it passes through a duct in which the nitrogen present removes the vapors of the solvent at approx. 160 °C. The spinning speed is approx. 200 m/min. The spinning speed is approx. 200 m/min, and the drafting rate is around 400%.

32.3.3 After-treatment

Both wet-spun and dry-spun yarns are afterstretched. m-aramids are first drawn by 300% in boiling water, then washed in hot water and subsequently dried on rollers. p-aramids are

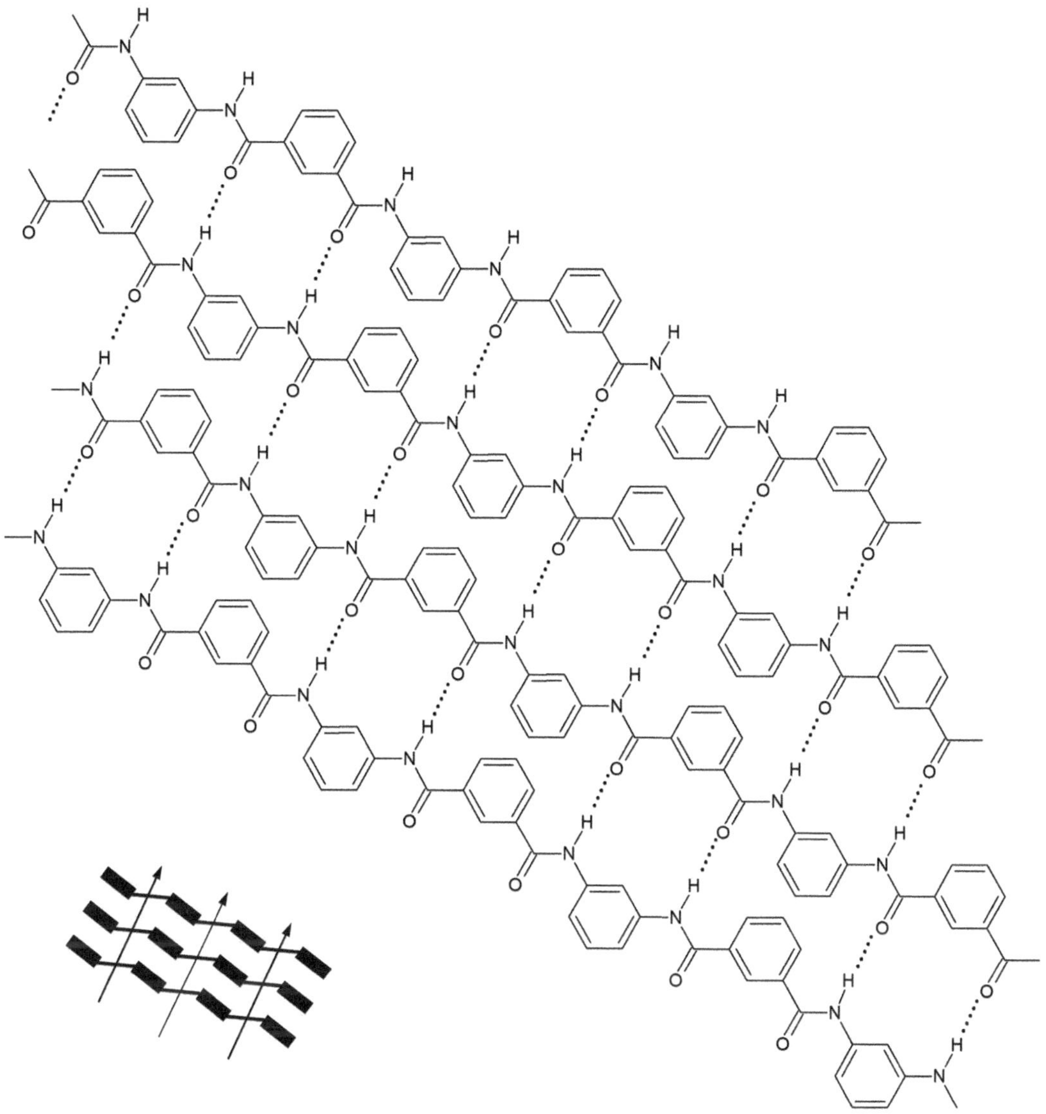

Fig. 32.5 Crystal structure of meta-aramide

heated and drawn over hot rollers (300–600 °C). This produces a fiber with high crystallinity of 50–75% (Uppal et al., 2012) and a high Young's modulus.

32.4 Properties

Aramides are characterized by particularly good temperature resistance in continuous use (m-aramide), good chemical resistance (Fig. 32.9), very high strength and a very high modulus of elasticity (p-aramide). Aramid fibers are only slightly UV-resistant because they absorb light, especially in the 300–450 nm range. For outdoor applications, Kevlar must therefore be protected from this radiation. Furthermore, aramid fibers are hygroscopic. Because of these properties, their suitability for aerospace applications in their pure form is limited.

The modulus of elasticity can be further increased by specific thermal after-treatment.

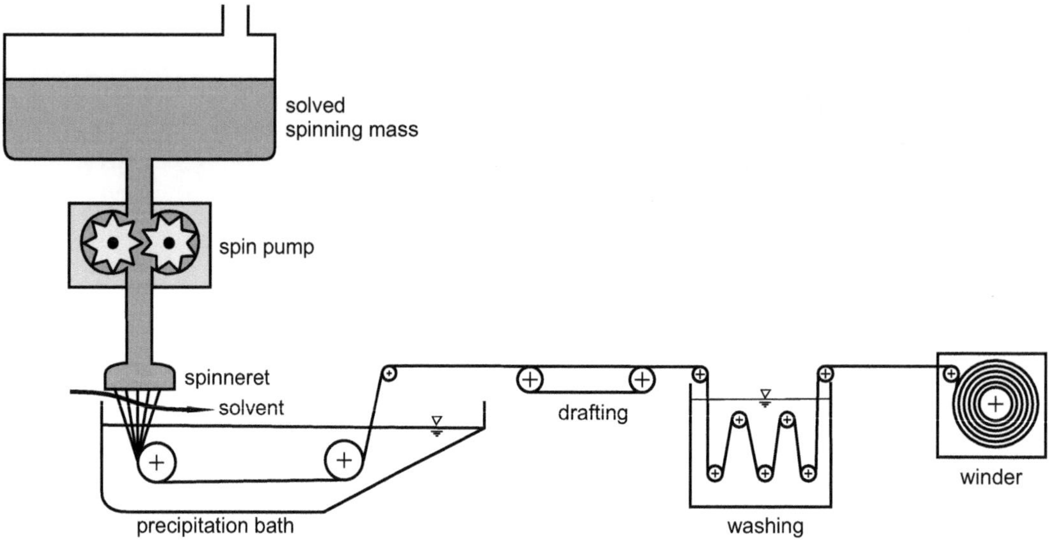

Fig. 32.6 Principle of the air gap spinning process for aramid

Such fibers are called high modulus fibers (Kevlar 49, Twaron HM).

Aramides lie between glass and carbon in the KD tensile test, as Fig. 32.10 shows by way of example. They break brittly.

Aramids have a core-sheath structure as shown in the Fig. 32.11 is shown. Here, the core regions are oriented in two dimensions (smectic arrangement), and the macromolecules do not overlap. The mantle regions are oriented in one dimension (nematic arrangement), with macromolecules overlapping. Crack propagation therefore takes place in the core region transverse to the fiber direction and in the mantle along the overlapping regions.

The Fig. 32.12 shows typical stress-strain curves for different p-aramids compared to carbon, E-glass and steel fibers.

Typical mechanical properties of m- and p-aramides as well as the copolymer Technora® (Fig. 32.13) and the polyamide imide Kermel® (Fig. 32.14) are given in the following tables. Typical yarn counts are 1200–1700 dtex with

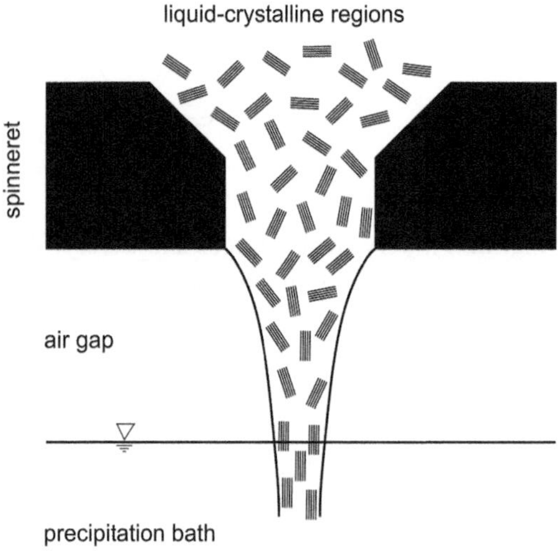

Fig. 32.7 Details of the processes in the spinneret in the air gap spinning process

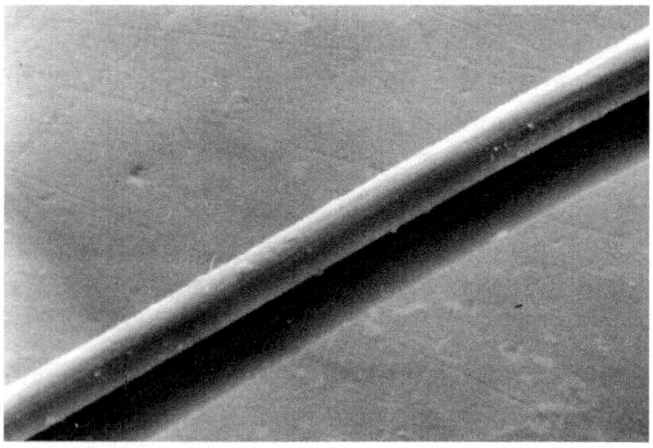

Fig. 32.8 Aramid fiber

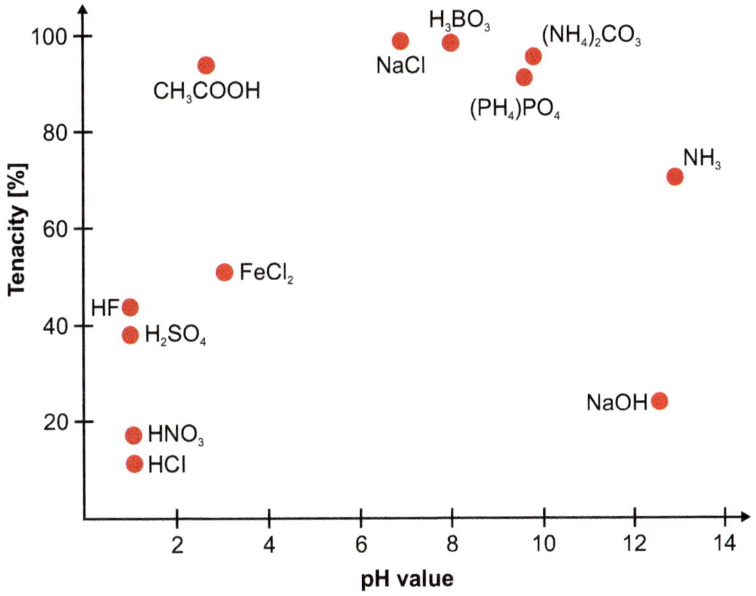

Fig. 32.9 Effect of chemicals on the residual strength of p-aramid after 3 months of storage in aqueous solution at 20 °C acc. to (Teijin, 2020)

filament counts of 500–1000 (Tables 32.1 and 32.2).

Another copolyamide of terephthalamide and benzimidazole is Armos™, which has a very high strength of about 340 cN/tex (Fig. 32.15).

The Table 32.3 shows typical values of the most important physical parameters of aramids.

Aramides do not melt, but char in the pilot flame. If it is removed, aramides extinguish by themselves. They are resistant to organic solvents, whereas hot, concentrated acids and alkalis (e.g. 60% sulfuric acid, 50% sodium hydroxide solution) destroy the fibers. When exposed to strong, inorganic acids, the tensile strength is reduced at concentrations above 35% or exposure times of more than 10 h or temperatures above 60 °C. Aramids are resistant to strong organic acids—with the exception of benzosulfonic acid. Strong alkalis of high concentration (> 40%) under simultaneous heating above 60 °C or a long exposure time reduce the tear resistance. Bleaching agents have only a medium influence when heated (e.g. sodium chlorite, 60 °C) and organic chemicals have no or only a weak influence on the tear strength.

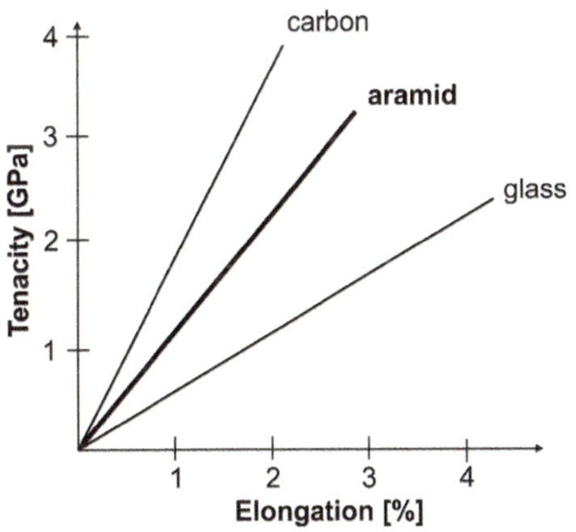

Fig. 32.10 Mechanical values for aramid compared to glass and carbon fibers acc. to manufacturer's data (Teijin, 2020; Dupont, 2020a)

When exposed to light, aramids are damaged by UV radiation. With direct exposure to sunlight, the tensile strength of m-aramids falls to half the original value after approx. 1 year. Even p-aramides lose around 65–80% of their tensile strength after approx. 4 months.

Aramid fibers are thermally stable up to 160 °C. When exposed to hot air of approx. 220 °C over a period of 1000 h, m-aramid fibers lose considerable strength. For short periods, m-aramides can withstand temperatures of up to 420 °C without any appreciable loss of strength, but at higher temperatures the fibers decompose very quickly. In the case of p-aramids, the

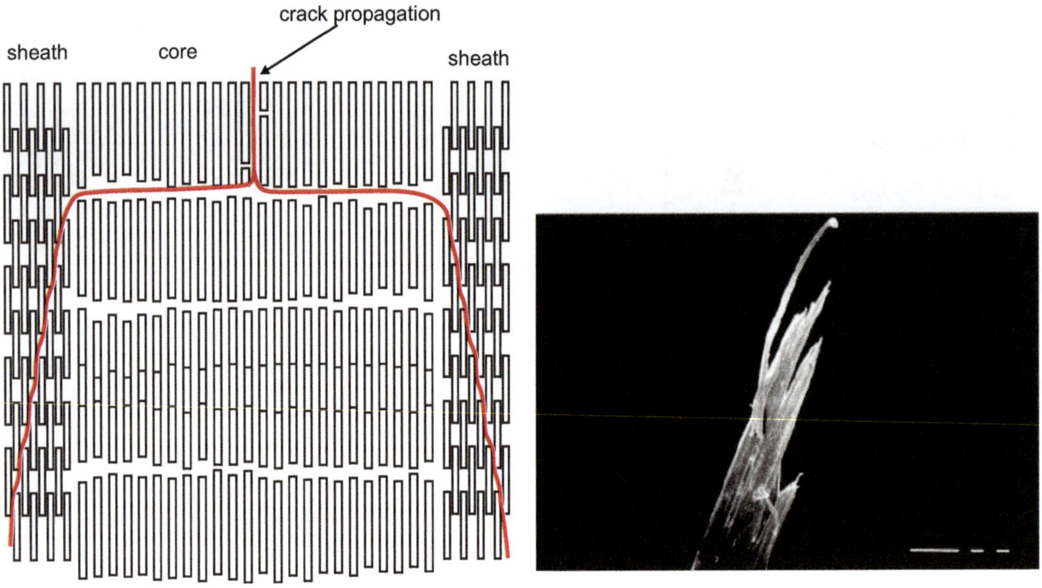

Fig. 32.11 Crack propagation in aramids and fracture surface of an aramid filament

Fig. 32.12 Typical stress-strain curves of p-aramids compared to other fibers (HM—high modulus, HT—high strength, N—standard, HE—high elongation) acc. to (Teijin, 2020)

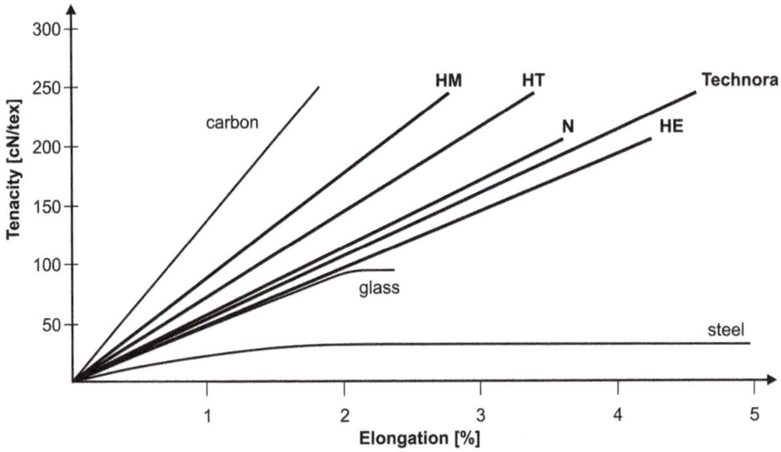

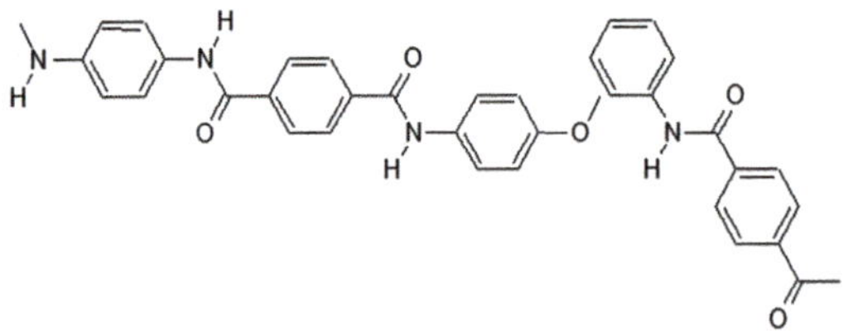

Fig. 32.13 Structural formula of the Technora™ repeat unit

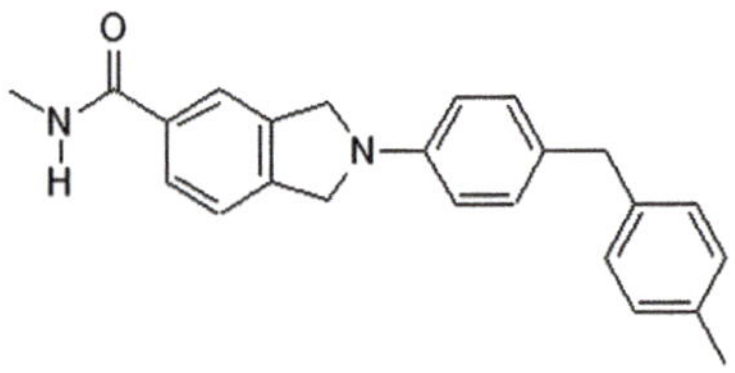

Fig. 32.14 Structural formula of the Kermel™ repeat unit

Table 32.1 Mechanical properties of aramid filaments (typical values) acc. to (Teijin, 2020; DuPont, 2020a, 2020b)

		m-Aramid	p-Aramid				Copolymer
			N	HM	HT	HE	Technora®
Tenacity	[MPa]	600–730	2760	2760	3385	2100	2450
	[cN/tex]	43–53	200	200	235	145	175
E-modulus	[GPa]	10–20	58	120	105	50	53
	[cN/tex]	750–1450	4500	8500	7100	3400	3800
Breaking elongation	[%]	15–30	3.3	2.4	3.3	4.3	4.6
rel. Loop strength	[%]	65–95	60	60	–		–
Knot strength	[%]	–	35	35	–		–

Table 32.2 Mechanical properties of aramid staple fibers (typical values) acc. to (Teijin, 2020; DuPont, 2020a, 2020b; Kermel)

			m-Aramid			Copolymer	Polyamidimid
		p-Aramid	N	MT	HT	Technora®	Kermel®
Tenacity	[MPa]	2700–3600	515	680	860	3200–3500	510–615
	[cN/tex]	190–250	37	49	62	230–250	38–46
E-modulus	[GPa]		1.1	1.7	3	75–82	3,2–5,9
	[cN/tex]		82	120	220	5500–6000	240–440
Breaking elongation	[%]	2.3–4.5	45	40	28	3.9–4.5	17–21

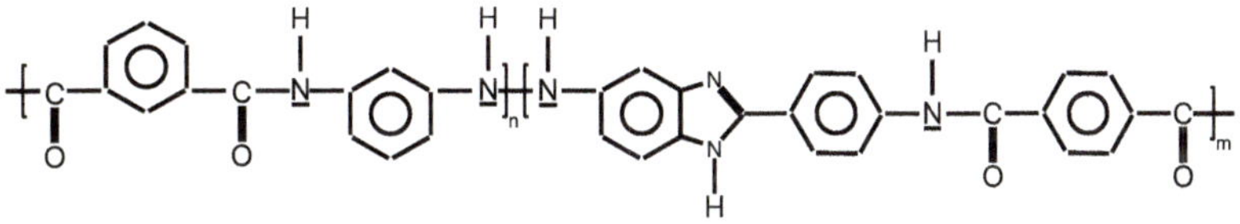

Fig. 32.15 Structural formular of the repeat unit of Armos™

Table 32.3 Physical properties of aramid acc. to (Teijin, 2020; DuPont, 2020a, 2020b)

			p-Aramid		
		m-Aramid	N	HM	Technora
Density	[g/cm^3]	1.38	1.44	1.44	1.39
Spec. heat	[l/(gK)]	1.2	1.4	1.4	–
Heat transfer capacity	[(W/(mK)]	0.13–0.25	0.05	0.05	–
Thermal expansion coefficient	[10^{-5}/K]	–	−3.5	−3.5	–
Disintegration temperature	[°C]	370	500	500	–
Glass transition temperature	[°C]	280–290	300	300	–
Moisture uptake	[%]				
(at 20 °C, 65% rel. humidity)		5	7	3.5	3
Flammability (LOI-index)		27–38	29	29	25
Hot air shrinkage	[%]				
(15 min, 190 °C)		0.5	0.1	0.1	–
Hot steam shrinkage	[%]				
(95 °C in water)		1.5	0.1	0.1	–
Water retention	[%]	12–17	7	7	2

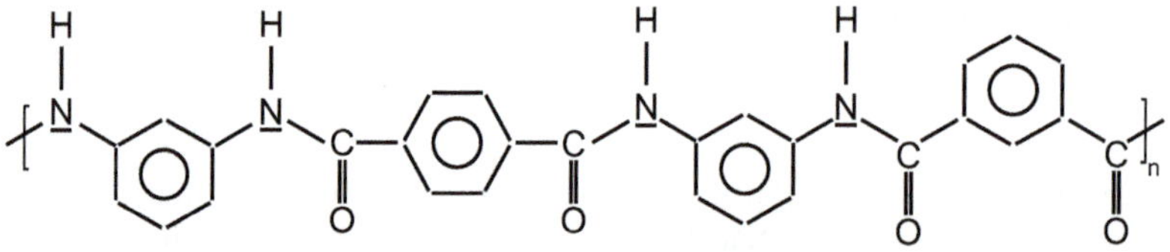

Fig. 32.16 Structural formula of the repeat unit of Togilen™

Fig. 32.17 Staple fibers and aramid-glass hybrid textile for composites (Cjp24, 2009)

strength is only reduced when exposed to temperatures above 200–300 °C, and decomposition begins above 500 °C. Aramid fibers only show signs of embrittlement at temperatures below −150 °C in cold conditions.

Up to a temperature of 400 °C, almost no CO, CO_2 and HCN gases are formed. At 800 °C, CO_2 (1850 mg), CO (50 mg), C_2H_2 (1 mg), N_2O (10 mg), hydrocyanic acid (HCN, 14 mg) and NH_3 (0.3 mg) are formed per gram of fiber material.

At the All-Russia Scientific and Research Institute of Polymer Fibers, Tverchimvolokno, and at the Scientific and Research Institute of Chemical Fibers and Composites, St. Petersburg, further aramid fibers were developed. These include Terlon™, representing p-aramids, Fenilon™, an m-aramid, and Tverlana™ and Togilen™, which are copolymers of p- and m-aramid building blocks (Fig. 32.16). The latter have temperature resistance up to 300 °C, LOI value of 35–44, combined with elongation of 12–30% and tensile strength of 30–50 cN/tex and moisture absorption of 11–15%. All of these fibers, developed in Russia, have not been commercialized on a large scale to date (Hearle, 2001).

32.5 Products and Applications

Aramid fibers are mainly produced as filament yarns, but there are also staple fibers (fiber fineness p-aramid: 0.9–2.5 dtex, m-aramid: 1.7–2.2 dtex (max. 14 dtex); strength p-aramid: 110–275 cN/tex), which can be processed into yarns and nonwovens using the same processes as other staple fibers (Fig. 32.17).

The most important applications of aramid fibers are limited due to the high price (approx. 45 €/kg):

Fig. 32.18 Typical applications for aramid fibers (fire protection clothing: Jens Eberhardt, Kermel; bulletproof vest: Simme, 2018)

Table 32.4 Brand names, manufacturers and market shares

Producer	Trade mark	p-aramid	m-aramid	copolymer	Website	
DuPont	Kevlar™	x			http://www.dupont.com	USA
	Nomex™		x			
Huvis Corp.	METAONE™		x		http://www.huvis.com	Korea
Kolon Ind., Inc.	Heracron™	x			http://www.kolonindustries.com	Korea
Teijin Aramid B.V.	Twaron™	x			http://www.teijinaramid.com	TheNetherlands
	Conexteijin™		x			
	Technora™			x		
Yantai Tayho Adv. Mat. Co.,Ltd	Taparan™	x			http://en.tayho.com.cn	China
	NewStar™		x			
	Market share	2/3	1/3			

- Fire suits,
- bullet-proof vests (Fig. 32.18),
- sails,
- ropes, cords,
- tire cord,
- composites (racing yachts, trains, aircraft wing components)

32.6 Market

Acc. to estimates, a total of around 100,000 t of aramids are currently produced worldwide, which corresponds to almost a doubling of production in the last 15 years. Of this, p-aramides account for two-thirds and m-aramides for one-third. The most important producers are Dupont and Teijin. China now also produces significant quantities of aramid fibers (e.g. by China Bluestar Chengrand Chemical, Yantai Tayho); exact production figures are not known, Table 32.4.

References

Cjp24. (2009). https://commons.wikimedia.org/wiki/File:Aramid_fiber.jpg, CC BY-SA 3.0.

DuPont. (2020a). https://www.dupont.com/content/dam/dupont/amer/us/en/safety/public/documents/en/Kevlar_Technical_Guide_0319.pdf

DuPont. (2020b). https://www.dupont.com/content/dam/dupont/amer/us/en/personal-protection/public/documents/en/Nomex(R)%20Fiber%20Technical%20Guide.pdf http://www.kermel.com/en/our-high-tech-fiber/

Hearle, J. (2001). *High-performance fibers*. Woodhead Publishing.

Kermel. http://www.kermel.com, photo by Jens Eberhardt: www.je-sign.de

Mark, H. (1964). Principles for influencing the essential fiber properties. *Lenzinger Berichte, 17*, 7–14.

Morgan, P. W., & Kwolek, S. L. (1959). Interfacial polycondensation 2. Fundamentals of polymer formation at liquid interfaces. *Journal of Polymer Science, 40*(137), 299–327.

Preston, J., & Smith, R. W. (1966). Novel aromatic copolyamides. *Polymer Letters, 4*, 1033–1038.

Preston, J., Smith, R. W., Black, W. B., & Tolbert, T. L. (1969). *Journal of Polymer Science, Part C/22*, 855–865.

Simme, D. (2018). https://commons.wikimedia.org/wiki/File:Bullet_and_Stab_Proof_Vest.jpg, CC BY-SA 4.0.

Teijin. (2020). https://www.teijinaramid.com/wp-content/uploads/2018/10/Product_brochure_Twaron.pdf

Uppal, R., Ramaswamy, G. N., & Loughin, T. (2012). A novel method to assess degree of crystallinity of aramid filament yarns. *Journal of Industrial Textiles, 43*(1), 3–19.

Polyolefins

33

Polyolefins consist of aliphatic carbon chains with more or less regularly repeating side groups. The best known representatives are polyethylene (PE), polypropylene (PP) and polyacrylonitrile (PAN). Polytetrafluoroethylene (PTFE) and polyvinyl alcohol (PVA) also belong to the olefins, but have completely different properties. There are also many different spinning processes for polyolefin fibers. Some can be melt spun, others only solution spun and some are produced by special processes. In this chapter, PE and PP are mainly discussed. Most of the special polyolefin-based fibers are explained in other chapters to which they are more appropriate, e.g. because of their special properties.

33.1 Polyethylene (PE)

Polyethylene is an aliphatic hydrocarbon. Extremely long macromolecules with an average degree of polymerisation of 100,000–200,000 are used to produce fibers. For this reason, PE for fibers is also usually referred to as UHMWPE (Ultra High Molecular Weight) in contrast to HDPE (High Density) and LDPE (Low Density), which are mainly used for films.

33.1.1 History

Polyethylene was discovered in 1898 by the chemist Hans von Pechmann (Fig. 33.1) and first produced industrially in 1933 in the ICI laboratories in England under a pressure of approx. 1400 bar and at a temperature of 170 °C. It was not until 1940 that an economical production process was developed. Ethene can also be polymerized at normal pressure with the aid of Ziegler-Natta catalysts (see PP). Alternatively, metallocene catalysts can be used as with PP. The use of metallocene catalysts allows the production of polyethylene with narrower distributions of molar mass and more uniform comonomer incorporation than is the case when Ziegler-Natta catalysts are used.

33.1.2 Structure

Polyethylene is produced by polymerisation of gaseous ethene. This process produces either long chains without or with branches (Fig. 33.2).

UHMWPE was discovered by Albert Pennings in 1968 and patented in 1979. It has been produced on an industrial scale since 1990. This fiber is characterized by an extremely high chain length combined with a very large orientation of the macromolecules in the longitudinal direction of the fiber (Fig. 33.3). The Van der Waals forces between the atoms of the chains are very weak, but the extremely high number of atoms per macromolecule results in a very high strength combined with a very low elongation (high modulus of elasticity). The molecular chains of UHMWPE fibers have a degree of

D. Veit, *Fibers*, https://doi.org/10.1007/978-3-031-15309-9_33

Fig. 33.1 Hans von Pechmann (WIKI, 1899)

orientation of 95%, the crystallinity is 85%. Both values are not achieved by any other polymer.

33.1.3 Production

UHMWPE fibers are produced either by the gel spinning process or by the solid state extrusion process.

33.1.3.1 Gel Spinning

In this process, the PE is dissolved in a very low concentration (<2%) in the solvent (usually paraffin oil) and then spun out using the air gap spinning process (gel spinning), Fig. 33.4. At concentrations of 2% PE in the solvent paraffin oil, dynamic viscosities of up to 20,000 Pas are already achieved due to the extremely large chain length (honey: 10,000 mPas). Therefore, higher PE concentrations are not possible.

The solvent and stabilizer are first mixed and heated in an autoclave. In a high shear mixer, the PE powder is added and then homogenized in the twin screw mixer. This gel is then extruded by the spinning pump through the spinning head into an air gap. The yarn bundle then enters the precipitation or coagulation bath where the filaments are formed. Drawing takes place between godets in a hot channel. After drying in hot air, the yarn is either wound directly or continuously hot-drawn (above 110 °C) and then wound up. Staple fibers can also be produced from the filaments by cutting.

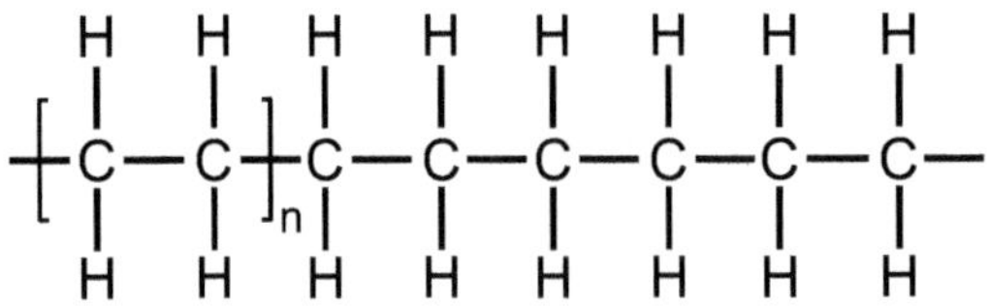

Fig. 33.2 Structural formula of polyethylene

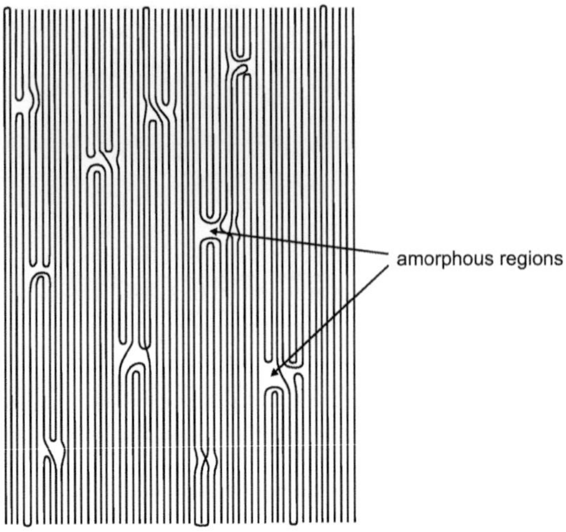

Fig. 33.3 Orientation and internal structure of UHMWPE fibers

The high strength is achieved the better, the more statistically unoriented the carbon chains are in the solution (or melt) and the less shear forces act during extrusion. Then, the total orientation is only achieved by the drawing process. Depending on the disorientation and entanglement of the molecular chains, stretching to 100 times the original length is easily achieved, even a 1000fold draft is possible.

This almost perfect alignment is only possible because the concentration of the PE polymer in the spinning solution is very low. If it exceeds a certain limit, the macromolecules "hook" together and full stretching can no longer be achieved. If, on the other hand, the PE concentration falls below a certain value, the stretching forces can no longer be absorbed and melt breakage occurs. The process window between these two limit values for the PE concentration in the spinning

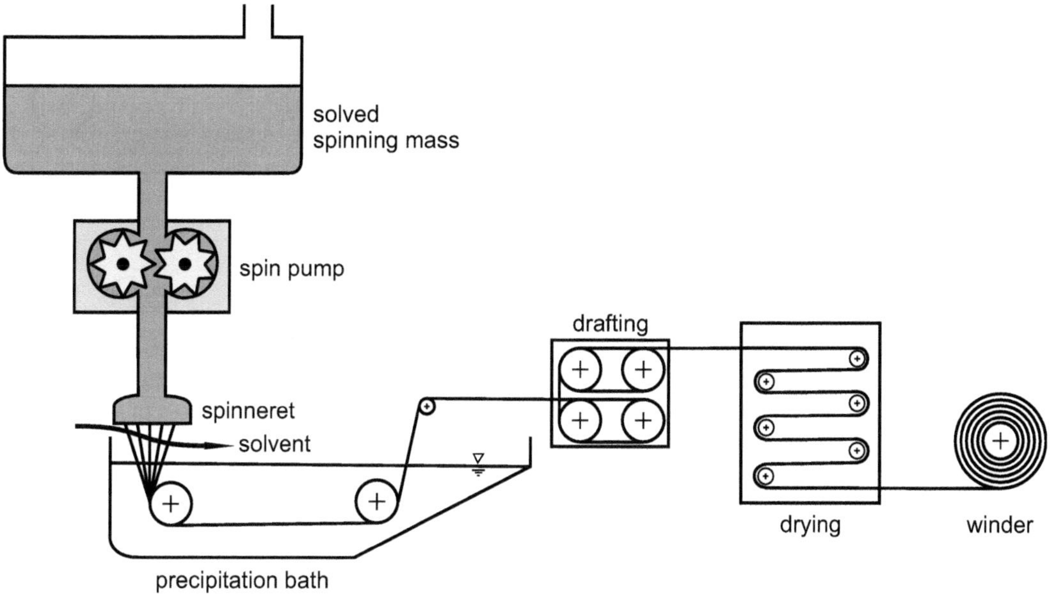

Fig. 33.4 Principle of the production of UHMWPE fibers in the gel spinning process

solution is very small and therefore difficult to achieve.

33.1.3.2 Solid State Extrusion Process

Alternatively, PE can also be produced using the solid-state extrusion process. In this process, PE powder is compressed without solvent, the resulting film is further compressed via pairs of rollers and then pre-stretched (Fig. 33.5).

Because the macromolecules are not as strongly oriented as in gel spinning, a so-called shish-kebab structure is formed (Fig. 33.6). The shish-kebab consists of chain molecules (shish) oriented in the direction of the fibers and the folded lamellae (kebab) in between. The lamellae can be stretched by a factor of approx. 10, so that almost all macromolecules are then oriented exclusively in the longitudinal direction. For this purpose, the film is cut into fine strips and passed over hot plates.

33.1.4 Properties

The strength of UHMWPE is up to 430 cN/tex, which corresponds to a tear length of about 430 km. UHMWPE is thus the strongest fiber of all. Standard grades achieve strength values of 250–380 cN/tex (Fig. 33.7).

The strength depends on the fineness of the individual fibers. The finer the individual filament, the greater the strength of the yarn. The elongation is approx. 3% and the modulus of elasticity reaches correspondingly very high values of 7750–16,000 cN/tex (73–152 GPa). UHMWPE has a density of 0.95–0.98 g/cm^3 and therefore floats on water. The LOI value is 17–18. The number of filaments is between 40 and 1560 per yarn and the yarn counts typically range from 83 dtex to 6600 dtex (Fig. 33.8).

The fiber is very durable and has a high resistance to abrasion, moisture, UV rays and chemicals. A disadvantage is the relatively low melting temperature of 144–152 °C. The service temperature is therefore only a maximum of 80–100 °C. Below −150 °C the fiber becomes brittle.

The molecular weight of UWHMPE is 2.8–5.6 m, which corresponds to approx. 100,000–200,000 repeat units per macromolecule and is thus a factor of 1000 higher than the average degree of polymerization of polyester and polyamide.

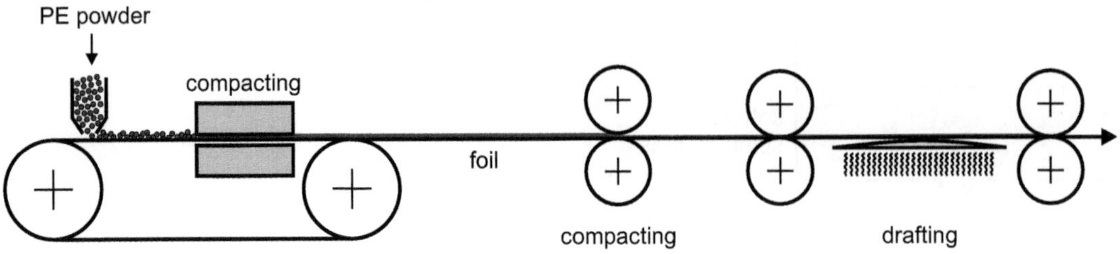

Fig. 33.5 Solid state extrusion process acc. to (Mather & Wardman, 2015)

33.1.5 Areas of Application

UHMWPE is used, among other things, for the manufacture of fishing lines, fishing nets, webbing slings, protective vests, motorcycle helmets, cut-resistant gloves, snowboards and various climbing aids. UHMWPE is frequently used as a line material in water sports in particular—since it has no stretch and is insensitive to salt water—as well as for paragliders and kites and for the winch launch of gliders. It is often coated with polyester (PET) to increase abrasion resistance. Ocean-going yachts are also often equipped with sails made of UHMWPE.

The application of UHMWPE in the fiber composite sector is particularly advantageous where high impact resistance at low weight is required (monoquoques in vehicles, underbody protection, boat hulls). The fiber is often only used on the surface as a wear layer because its wear is lower by a factor of 10 than that of PTFE under comparable conditions. The load-bearing laminate usually consists of glass or carbon fibers. Conversely, namely embedded between two layers of a high-strength fiber such as glass or carbon, UHMWPE acts as a lightweight, pressure-resistant core material. The high tensile strength combined with low weight is used in the manufacture of pressure vessels. Because of its permeability to X-rays, UHMWPE can be used as a reinforcing material for X-ray tables in hospitals. Another area is ballistic protection and the manufacture of ropes and cables.

UHMWPE fibers have an extraordinarily high energy absorption, especially against high-speed impacts. The fiber performs about 30% better than aramid. Another area of application is therefore the reinforcement of cockpit doors in aircraft construction, making them bulletproof. Bulletproof vests are another area of application. Since 2006, ropes made of UHMWPE have also been used by mountain rescue teams. With a length of 400 m in one piece and due to their low weight, these ropes are replacing steel ropes in many areas. UHMWPE is now also used in jeans to increase abrasion resistance (Fig. 33.9).

In addition to the white UHMWPE fibers, which continue to dominate, there are also spun-dyed black fibers.

33.1.6 Market

UHWMPE is mainly sold under the trade names Dyneema™ by DSM N. V. (Netherlands) and Spectra™ by Honeywell (USA). Toyobo Co. Ltd. (Japan) has been producing UHWMPE fiber, formerly produced jointly with DSM, under the name IZANAS™ since 2016. A new development is the fiber from Toyobo Co., Ltd., marketed under the trade name Tsunooga™. In Korea, PE fibers are produced by Huvis Corp. and marketed under the brand name Duraron™.

Exact production figures are not published. According to estimates, world production is a few 10,000 t per year. According to one study, annual growth of 6% is expected until 2025 (Globe Newswire, 2020).

33.2 Polypropylene (PP)

Polypropylene is the third most important synthetic chemical fiber in terms of volume after polyester and polyamide, with an annual

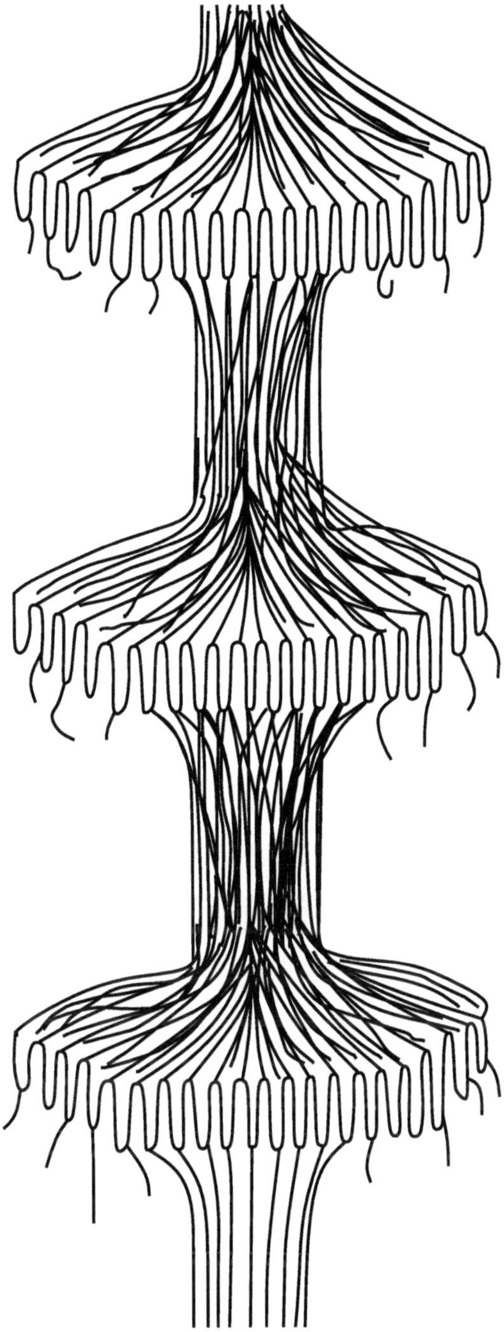

Fig. 33.6 Model of the PE-Shish-Kebab structure

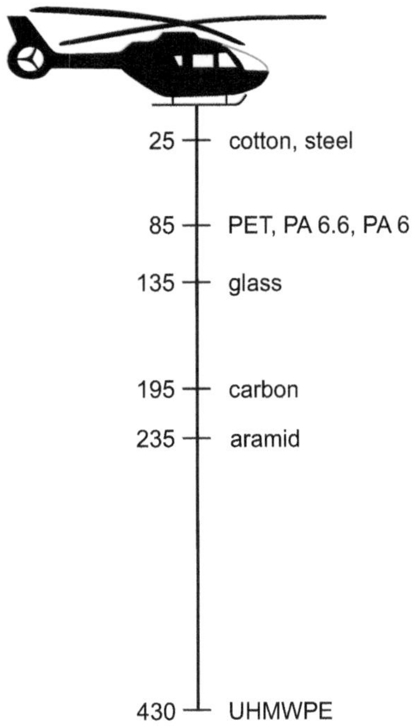

Fig. 33.7 Comparison of the strength of selected fibers in [tear kilometers or cN/tex]

production of 3.3mt (2020). It is produced by polymerization from propylene and is used primarily for technical textiles (e.g. filters) and packaging. The abbreviation for polypropylene is PP (Fig. 33.10).

33.2.1 History

In 1953, Karl Ziegler (1898–1973) developed a method for polymerizing ethylene at low temperatures and under normal pressure using transition metal catalysts (Ziegler et al., 1955). The resulting polyethylene is highly crystalline, similar to polyamide, in contrast to the radically polymerized, highly branched high-pressure polyethylene (HD-PE). This discovery laid the foundation for the polymerization of polypropylene. Giulio Natta (1903–1979), head of the Institute of Industrial Chemistry at the Polytechnic of Milan, succeeded in polymerizing α-olefins and styrene using the so-called Ziegler catalysts. In early 1954, this led to the production of isotactic polypropylene on a laboratory scale. Natta was able to elucidate the crystal structure and steric configuration of crystalline polypropylene in a short time (Natta, 1961). He also introduced the terms isotactic, atactic and syndiotactic for the different conformations of polypropylene. He succeeded in showing that the insoluble crystalline fraction obtained by solvent extraction is

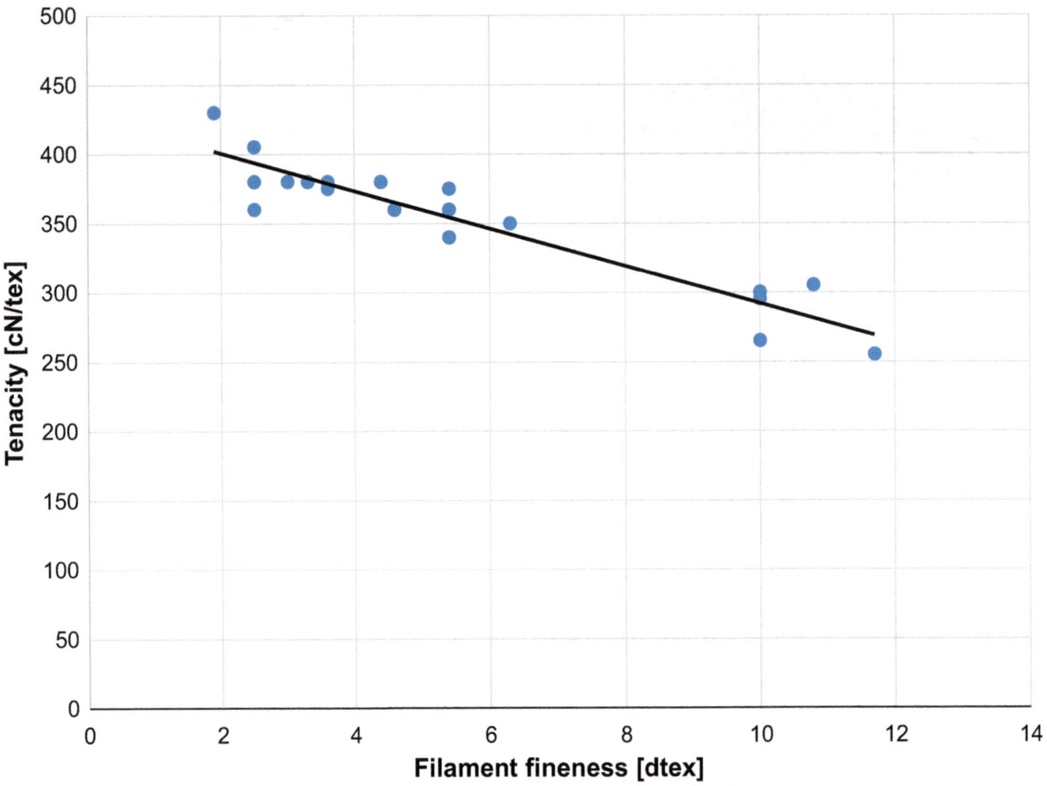

Fig. 33.8 Dependence of the strength on the single filament fineness acc. to (Spectra, 2020)

largely isotactic and suitable for the preparation of high-strength filaments. This isotactic structure is largely responsible for the good crystallization capacity and thus for the good mechanical properties of polypropylene. The monofilaments spun out by extrusion and subsequently drawn had an exceptionally high strength of about 75 cN/tex. Natta and Ziegler were awarded the Nobel Prize for their work in 1963 (Fig. 33.11).

Fig. 33.9 Typical applications of UHMWPE fibers (climbing ropes, cut protection trousers: Mertens, 2006, bullet-proof vest and PE-reinforced jeans)

Fig. 33.10 Transversal and longitudinal view of PP fibers (Lenzing AG)

Polypropylene fibers were first produced industrially by the Italian company Chimiche S.p.A. under the trade name Meraklon™, which still exists today. Shortly after its market launch, this new fiber established itself alongside the man-made fibers (polyester, polyamide and polyacrylonitrile) that had been produced industrially until then. An even faster upswing of the new fiber material was prevented on the one hand by the poor dyeing properties and on the other hand by the patent situation, because the Montecatini group owned almost all basic patents in Europe for melt spinning. Until the early 1970s, therefore, there was hardly any possibility for other companies to produce polypropylene fibers on an industrial scale. It was only after the first patents expired that other manufacturers started production. Due to the low manufacturing costs, the market for these fibers expanded very quickly (Wulfhorst & Meier, 1989).

Fig. 33.11 Karl Ziegler (left) and Giulio Natta (right)

33.2.2 Production of the Polymer, Chemical Structure

33.2.2.1 Monomer

The starting material for the production of polypropylene is mostly propene $CH_2{=}CH\text{-}CH_3$, an easily liquefiable gas with a boiling point of 47.7 °C. Most of the propene is a by-product of thermal or catalytic cracking of various petroleum fractions. Dehydrogenation of propane is also a possible process route. An overview is given in the Fig. 33.12.

In order to achieve the high degree of purity required for polymerization, the crude propene must be purified. In particular, polar and highly unsaturated compounds are removed down to very low residual contents.

33.2.2.2 Tacticity

In isotactic polypropylene all methyl side groups are aligned in the same way, in syndiotactic PP a regular change takes place and in atactic PP the sequence of alignments is random (Fig. 33.13).

Only iso- and, to a limited extent, syndiotactic polypropylene are suitable for fiber production, as only they form partially crystalline structures. Atactic PP is amorphous and is hence not spinnable.

33.2.2.3 Polymerization with Ziegler-Natta and Metallocene Catalysts

Isotactic polypropylene (iPP) suitable for the production of fibers was first prepared by

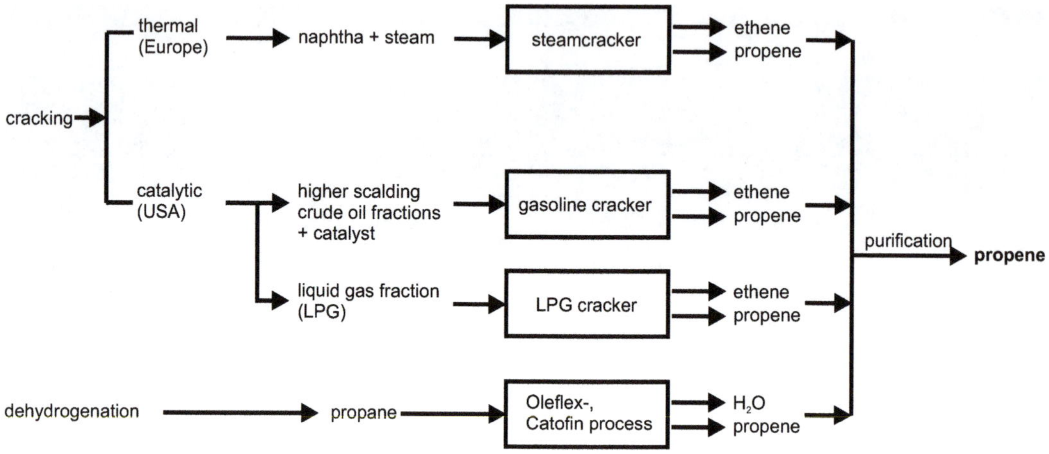

Fig. 33.12 Process routes for the production of propene

stereoselective polymerization with Ziegler-Natta catalysts. ZN catalysts consist of an organometallic main group compound of groups I, II or III (e. g. triethylaluminum) and a transition metal compound, mainly of groups IV to VI (e. g. titanium tetrachloride).

Metallocenes are sandwich compounds consisting of a central metal atom surrounded by

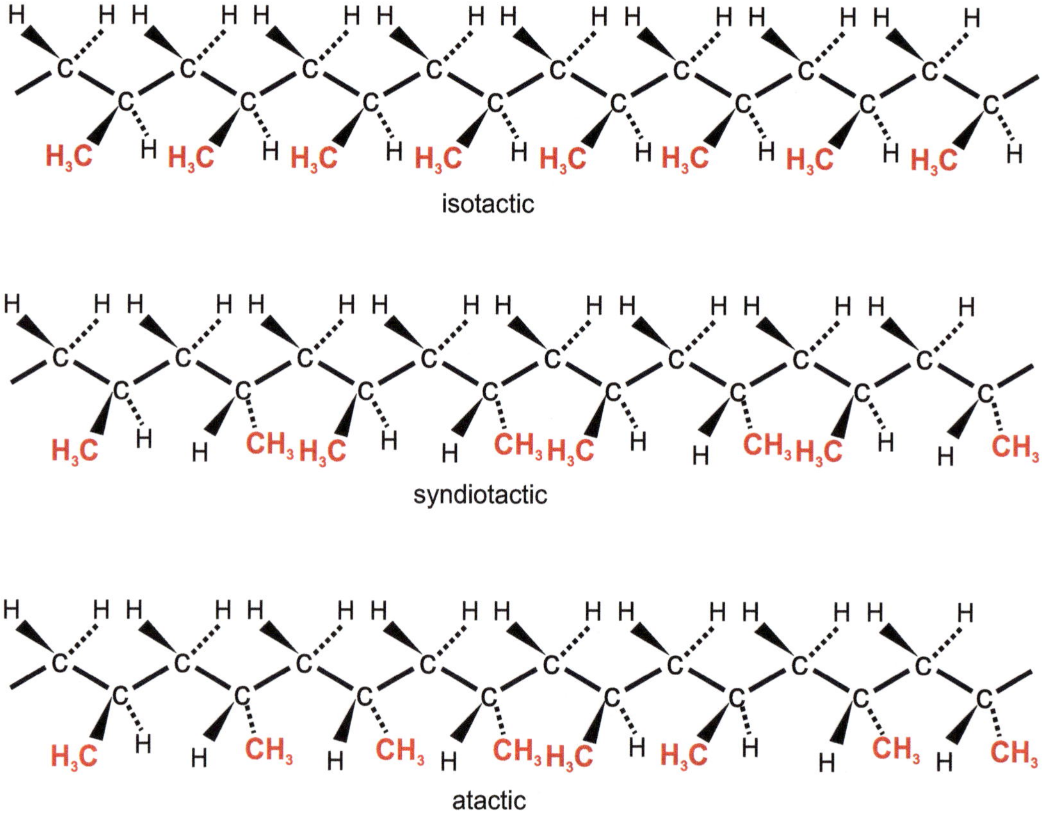

Fig. 33.13 Tacticity of polypropylene

two cyclopentadienyl radicals. Metallocene catalysts can also be used to produce syndiotactic PP (Fig. 33.14).

In the reaction known as coordination or insertion polymerization, the monomer is first coordinatively bonded to the transition metal M in the catalytic center (Ti in conventional ZN catalysts, Zr in most metallocene catalysts) and then intercalated (inserted) between the transition metal and the growing polymer chain (P):

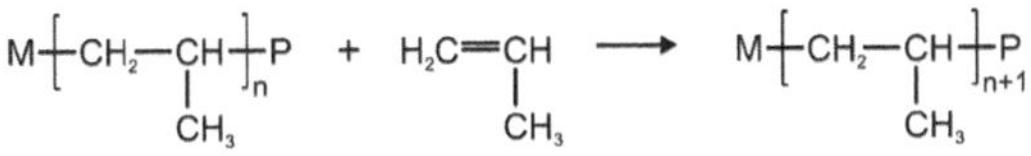

Insertion polymerization is largely regiospecific (primary or 1,2-insertion) and results in regular head-tail linkage.

In the case of PP produced with metallocene catalysts (mPP), minor regioirregularities (up to 1%) are detectable in the polymer, which leads to a somewhat lower melting point. The insertion is largely stereospecific, i.e. in the case of isotactic PP, the monomer is incorporated with the same steric configuration of the new asymmetric carbon atom in each case. In the case of syndiotactic PP, which only became technically accessible with the metallocene catalysts, a regular change of configuration takes place with successive tertiary C atoms.

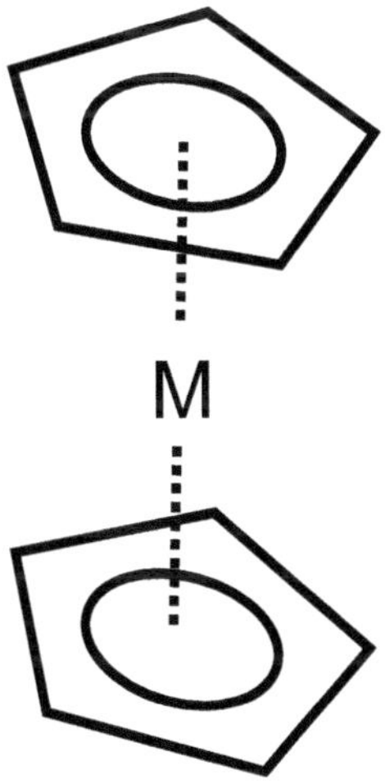

Fig. 33.14 Structure of a metallocene

The polymer chains are terminated mainly by transfer reactions, either with hydrogen added in small amounts to regulate the molecular weight:

or with the monomer:

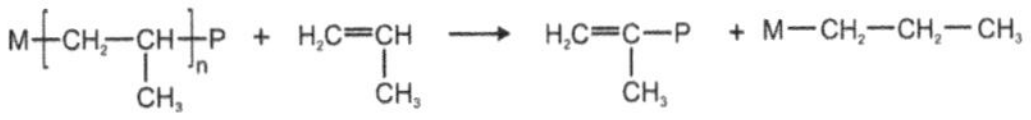

In the case of metallocene catalysts, in addition to transfer with hydrogen, –hydride elimination is also important for the termination of the polymer chains:

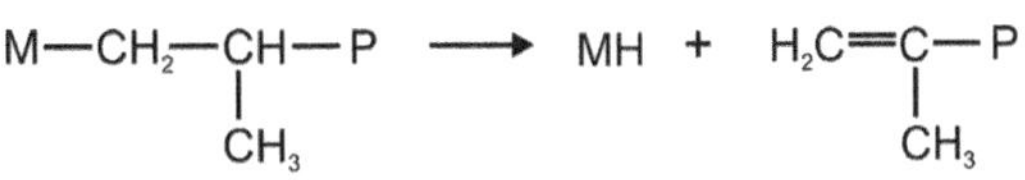

In the development of heterogeneous ZN catalysts, four generations are commonly distinguished to date, with significant improvements in stereoselectivity (isotaxy index), catalyst activity (kg PP/g cat) and control of polymer particle morphology.

Such a catalyst consists of two components, the actual catalyst and a co-catalyst, which together form a complex to which monomers can attach. The catalyst is a transition metal compound, usually titanium and vanadium, and the co-catalyst is usually an aluminum-organic compound, i.e. a compound in which organic radicals are bonded to an aluminum atom (Fig. 33.15).

Titanium can bind a total of 6 complex ligands. First, a complex of catalyst and co-catalyst is formed, in which, however, one binding site remains free: Here, a monomer can attach (for the other catalyst system, the complex formation works acc. to the same principle), Fig. 33.16.

The chain growth takes place through a so-called "insertion reaction": The monomer, which is bound to the titanium atom as a ligand, pushes itself between the titanium atom and the carbon atom, which is also bound to the titanium

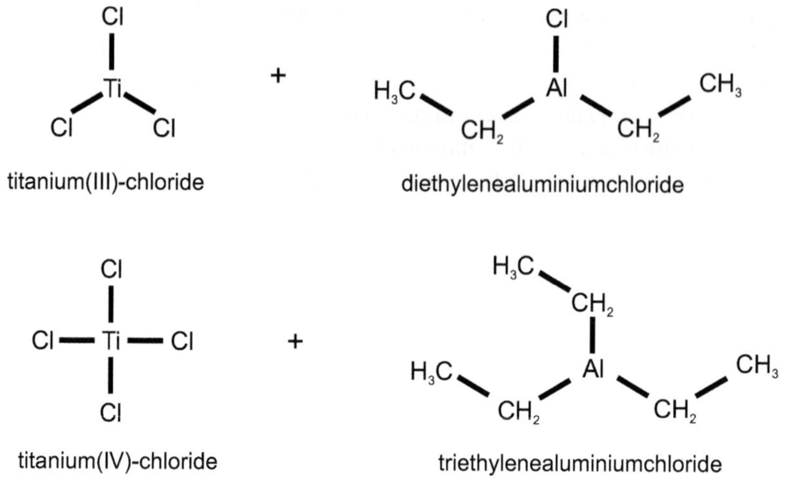

Fig. 33.15 Typical Ziegler-Natta catalyst systems

atom. In the complex there is now again a free bonding site to which another monomer can attach, which also pushes itself between the growing molecular chain and the titanium atom, and so on.

In Fig. 33.17, the first insertion reaction is shown. The co-catalyst with aluminum atom etc. is not shown to make the whole thing a bit clearer.

The tacticity of the PP is influenced by the choice of catalyst system. For the preparation of isotactic polypropylene, the system $TiCl_3$ and Al $(C_2\ H_5)_2$ Cl is used (Fig. 33.15), whereas syndiotactic polypropylene is obtained by using as catalyst system VCl_4 and $Al(C_2\ H_5)_2$ Cl, i.e. replacing titanium by vanadium, another transition metal. Then, during the insertion reaction, the monomers are rotated in such a way that the methyl groups on the chain at the end point alternately to the left and to the right.

Soluble catalysts were known since the early days of ZN polymerization and were studied as model systems, but for a long time remained uninteresting for technical application because the activities were low and stereoselective polymerization did not seem possible. The discovery by Kaminsky and Sinn (1980) that the

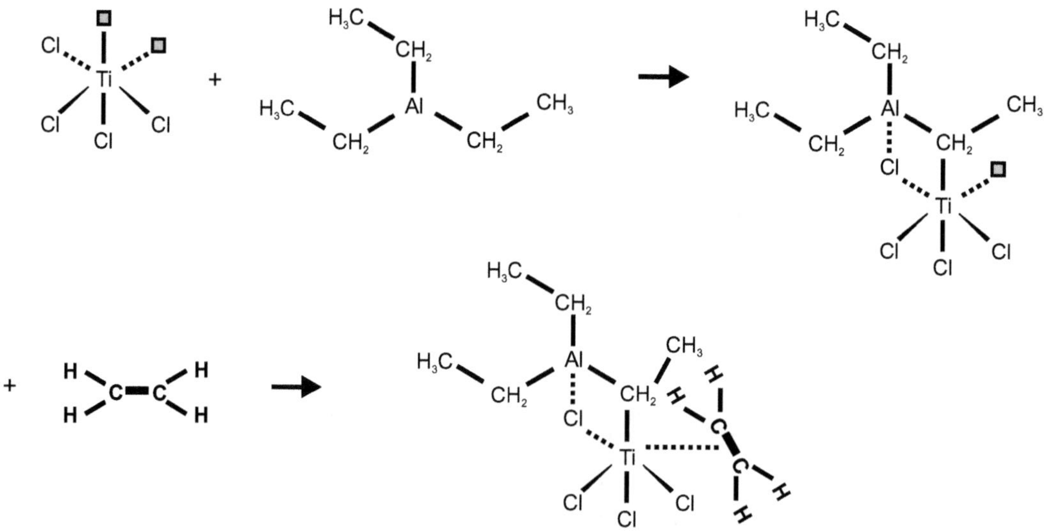

Fig. 33.16 Synthesis of PP with Ziegler-Natta catalyst

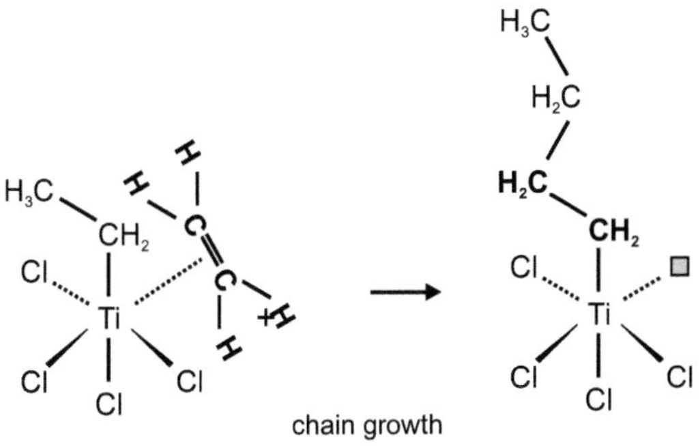

Fig. 33.17 Chain growth reaction

metallocene complex bis(cyclopentadienyl) zirconium dichloride with methylalumoxane (MAO) as cocatalyst exhibited extremely high activity in ethene polymerization rekindled interest in these homogeneous catalysts. A large number of these systems (zirconocenes, titanocenes, hafnocenes, variously bridged and exploiting the multiple substitution possibilities of the different ring structures) have since been studied in detail. Unprecedented control of molecular weight distribution (MGV) as well as tacticity and constitutive uniformity in copolymers thus seems possible.

Various polymers, such as syndiotactic PP (sPP), high molecular weight atactic PP or syndiotactic polystyrene could be produced industrially for the first time in 1995 with the aid of metallocene catalysts. These mPP grades are characterized by narrow MGV (M_w/M_n 2.5) as well as extremely low contents of oligomer and atactic PP. The melting point is about 15 °C lower than that of homopolymers produced with conventional ZN catalysts due to a low regioirregularity. However, the heat resistance is at the same level.

The use of metallocene catalysts results in several advantages in the production of mPP fibers:

- Higher processing speed and strength in spunbond and high-speed spinning,
- higher strength combined with higher elongation during compact spinning of staple fibers and
- improved thermobonding properties.

33.2.2.4 Polymerization Processes

The development of polymerization processes for PP is closely linked to advances in catalyst technology. Acc. to the polymerization medium, a distinction is made between suspension polymerization (slurry process) in an inert hydrocarbon, bulk polymerization in liquid monomer and gas phase polymerization. The newer processes are all continuous processes. Suspension polymerization is carried out in tank reactors, bulk polymerization in stirred tanks or loop reactors (polymerization mixture circulates in tube loops) and gas phase polymerization in stirred tanks or fluidized bed reactors.

33.2.2.5 Suspension Polymerization

With first generation catalysts, the slurry process was mostly used. Both catalyst and atactic PP (aPP) had to be removed. The catalyst residues were converted with alcohol into compounds soluble in the suspension medium and separated with water (ash removal). The iPP insoluble in the diluent was separated by filtration or centrifugation and dried. The aPP remained in solution and was recovered by evaporation of the solvent. Complex cycles were required to recover the alcohol and diluent. Typical polymerization temperatures for the slurry process are 60–80 ° C, the required pressure depends strongly on the diluent.

33.2.2.6 Mass and Gas Phase Polymerization

Bulk polymerization is also a suspension polymerization because the iPP is insoluble in the liquid monomer. At a polymerization temperature of 70 °C, the pressure is approx. 4 MPa, and the polymerization rate is about three times higher than in the slurry process. The third generation supported catalysts brought a significant process simplification by eliminating the ash removal step and the complex alcohol recovery. Finally, fourth generation catalysts require neither removal of the catalyst nor of the small atactic fraction. Deactivation of the catalyst is usually done with steam. Newly constructed plants usually have a capacity of 100–200 kt/a and operate either acc. to a gas phase or a combined liquid/gas phase process. Examples of modern processes are Montell's Spheripol process (combination of a loop reactor for homopolymerisation with a downstream gas phase reactor for impact copolymers) and Mitsui's Hypol process (liquid phase in a stirred tank) as well as the gas phase processes of BASF (Novolen, stirred tank), of Unipol-Shell (fluidized bed) and Amoco-Chisso (horizontal stirred tank).

Usually, the PP is finally processed into pellets by means of large extruders, whereby the thermostabilizers required for further processing are also added. In the Valtec process, pelletizing can be dispensed with due to the excellent morphology control of the polymer particles. The production of PP with narrow MGV ($M_w/M_n < 4$) directly in the reactor is not possible with conventional ZN catalysts. The products designated as CR (controlled rheology) grades are prepared by reactive extrusion with peroxides, in which statistical chain scission preferentially cleaves the long chains, resulting in a narrower molecular weight distribution.

33.2.3 Molecular Weight, Crystal Structure and Morphology

33.2.3.1 Molecular Weight and Molecular Weight Distribution

The weight-average molecular weight values M_w for PP used for fiber production, taking into account the extreme values for monofilaments and meltblown nonwovens, are in the range of 150,000–600,000, and between 200,000–350,000 for the majority of fiber applications. The molecular weight distribution of PP produced with heterogeneous ZN catalysts is rather broad with polydispersities M_w/M_n of 4–8, chemically degraded CR grades reach minimum values of about 3, typical values for mPP are about 2.5, theoretically even $M_w/M_n = 2$ is possible for single-site catalysts.

To characterize PP, either the melt viscosity in the form of MFI (melt flow index) in g/10 min at 2.16 kp/cm^2 and 230 °C or MVR (melt volume rate) in cm^3/10 min is usually considered. Melt viscosity is a good measure of chain length and is proportional to the 3.4th power of M_w at low shear rate. The proportionality constant is dependent on the MGV, so it is not possible to make a direct statement about the molecular weight.

33.2.3.2 Crystal Structure and Morphology

As with other crystallizable polymers with long chain length, crystallinity upon crystallization from the melt is also limited for iPP (and sPP) for kinetic reasons and rarely exceeds 70% even in highly isotactic samples, which is why the resulting polymer is referred to as semicrystalline.

There are 3 different crystalline forms and one form known as paracrystalline or smectic. The most stable under usual conditions is the monoclinic α-form. Crystallization in the hexagonal β-form can be achieved with special nucleating agents, and the ortho-rhombic γ-form is formed upon crystallization under higher pressure. It dominates iPP with low isotactic chain length (< 50) and is of interest in the context of mPP, where shorter isotactic chain lengths occur due to low regioirregularities. The mesomorphic smectic form occurs at high melt cooling rates and is the preferred initial form for high strength fibers. In all crystalline forms, the chains of iPP are present in a 3_1 helix. In the α- and β-form, parallel layers of helices build up lamellae, which in turn form radially symmetric spherulites, with the axes of the helices perpendicular to the lamellar surfaces. Since the length of the polymer chains is a multiple of the lamellar thickness, there is a multiple refolding of the chains and the connection of different lamellae by so-called binder molecules (tie molecules). In the γ-form, the helices lie parallel in layers, with the axes in successive layers forming an angle of 80°.

During the stretching of fibers, the microspherulitic structure is transformed into a fibril structure, whereby the fibrils oriented in the fiber axis are in turn composed of numerous microfibrils. In this process, the original lamellae are partially destroyed, resulting in the formation of numerous additional binder molecules that connect the individual microcrystallites across the amorphous interstices, thus having great significance for the fiber strength.

33.2.4 Production Method

The melt spinning processes used to produce polypropylene fibers for the production of multi- and monofilament yarns, staple fibers, spunbonded nonwovens and tapes are essentially the same as those used for other synthetic polymers. The specific properties of the material polypropylene have to be taken into account accordingly. Essential parameters of the melt spinning process, and thus important for the property profile of the fibers to be produced, are the molecular weight or melt flow index (MFI) and the molecular weight distribution of the raw material. It should be noted that PP has a relatively high molecular weight (150,000–600,000). With the same molecular weight distribution, the tensile strength of polypropylene filaments increases with increasing molecular weight. For textile applications, mainly polymers in the medium molecular weight range (MFI 230/2.16: approx. 10–25 g/10 min) are used. In the technical field for high-strength cables, polymers with higher viscosity (MFI 230/2.16: approx. 3–5 g/10 min) are selected.

The PP granulate is melted in a single or twin screw extruder. In the process, the polymer is homogenized and uniformly tempered. The melt must be free of impurities such as air bubbles and unmelted portions and fed to the spinnerets at a constant throughput rate. Important parameters of the extrusion process are temperature and viscosity of the polymer, number, diameter and length of the die holes, throughput per die hole, quenching air conditions, draw ratio, draw temperature and fixing conditions.

Depending on the application, various additives are often added to the melt in order to achieve certain fiber properties. These include in particular various stabilizers. In order to introduce additives, stabilizers or color pigments into the melt, various processes are used depending on the desired concentration and mode of action:

Masterbatch

The required additives, pigments or stabilizers are mixed with the starting polymer in relatively high concentrations (10–70%) to form granules. This masterbatch is then mixed with the actual polymer during the extrusion process. For color pigments in particular, predispersion with the polymer in the form of a masterbatch is necessary to ensure homogeneous distribution of the pigments in the melt and thus in the fiber cross-section.

Compound

If very high concentrations of additives are to be achieved in the actual spinning melt, a compound is fed directly into the extrusion process without adding PP granules.

Some additives can also be added in the form of copolymers. Examples are polymer-bound stabilizers of PP with polar monomers as adhesion promoters between PP matrix and natural or glass fibers in composites made of hybrid nonwovens.

There are two methods for mixing in a copolymer or masterbatch:

- Continuous mixing in the extruder feed: Metering is either volumetric or gravimetric. Volumetric addition has the disadvantage that the pigment distribution is usually not completely homogeneous. With gravimetric mixing, the addition is electronically controlled. The color uniformity is very good, since the weight decrease of the addition is electronically coupled with a throughput measurement of the spinning extruder.
- Continuous feeding from a side extruder via a gear metering pump with variable speed drive: The color pigments are added as late as possible. Therefore, this process is very flexible and allows reduced color transition times of the system. The disadvantage is the associated high investment costs.

Polypropylene can be processed into monofilaments, multifilament yarns, staple fibers and spunbonded nonwovens. BCF yarns are produced for carpets and film tapes for packaging.

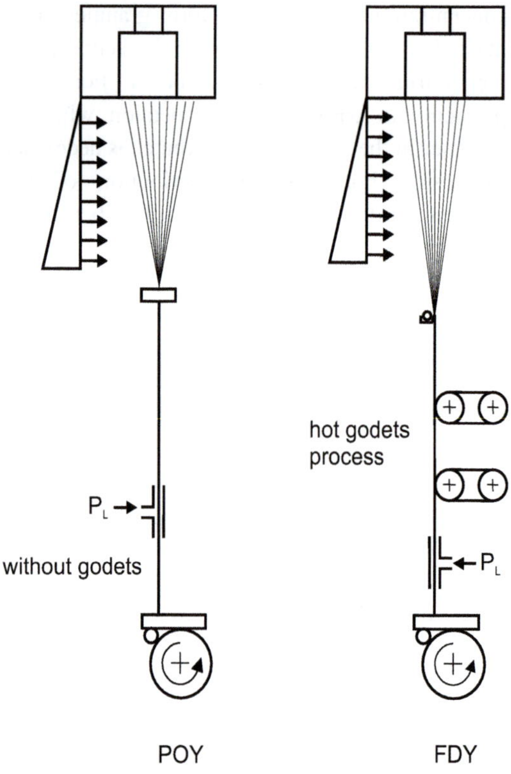

Fig. 33.18 PP filament spinning processes acc. to Gries

33.2.4.1 Filament Yarns

Multifilament yarns (POY, FOY) are produced on modified conventional spinning lines, the basic structure of which can be seen in the Fig. 33.18 can be seen.

Typical spinning speeds and product designations of the yarns thus produced are compiled in the Table 33.1 compiled.

The POY process for producing fully oriented yarns is a two-stage process. Because the macromolecules are not fully oriented at first, they are fully drawn in the second stage. The stretching process, during which the mechanical properties (strength, elongation and shrinkage) are formed, can be either one- or two-stage, depending on the desired strength. This process requires a polymeric raw material with a narrow molecular weight distribution due to the high drawdown rates.

PP filament yarns are often textured to improve the textile properties of the fibers. Common texturing processes for PP yarns are compression crimping, air texturing and disc freeze texturing, depending on the desired end product.

33.2.4.2 BCF Yarns (Bulk Continuous Filament)

These yarns are mainly used for carpets and home textiles. The BCF process is a continuous spinning, drawing, texturing and winding process with production speeds of 1500–3800 m/min, in which yarns with individual fiber counts of 8–30 dtex and strengths of 1.7–2.4 cN/tex can be produced. In particular, this process also offers the possibility of producing multicolored yarns.

33.2.4.3 Monofilaments

These coarse fibers are extruded in water and stretched to the maximum in one to three stages. The stretching can take place partly in a hot water bath and partly in hot air channels or completely in hot air channels. For filament diameters above 1 mm, there is a risk of vacuole formation.

33.2.4.4 Staple Fibers

Staple fibers can be produced by the conventional 2-step process or by the single-step compact spinning process. Typical finenesses are 1–30 dtex.

Table 33.1 Characteristic values of the spinning processes for PP filament yarns

	POY	FDY	
Winding speed [m/min]	1500–3000	Textile Filamente Technische Filamente	3800–4500 2500–3000
Single fiber fineness [dtex]	2.0–5.0 (standard) 0.35–1.0 (micro filament yarns)	0.8–2.0 (standard) 5–10 (technical filaments)	
Tenacity [cN/tex]	15–30	30–50 (standard) 50–73 (high tenacity yarns)	

The 2-step process is used for very high-quality and fine staple fibers as well as for production plants with a large throughput (more than 20,000 t/a). The spinning speed is higher by a factor of 10–20 than with the compact spinning process. The disadvantage is the large space requirement for the spinning plant, the can deposit and the can creel. The spinning chute has a length of approx. 2 m compared to only approx. 0.2 m space requirement for the blast air cooling in the compact spinning plant. In addition, care must be taken that the storage time of the undrawn sliver between the two process stages does not exceed 3 days, otherwise secondary crystallization will occur, which reduces the drawability and can cause difficulties in the second process stage. The advantages of the 2-step process are that each step can be run at optimum speed and parameter selection.

The single-stage compact spinning process is of increasing economic importance. It is characterised by a very low overall height of the plant and a high number of holes in the spinnerets. The limitation of the fineness is due to the cooling of the spun filaments and the maximum possible distortion during spinning. The cooling and the directly connected fiber line limit the production speed of compact spinning. It is mainly used in the production of spun-dyed PP fibers, whereas two-stage plants have proved their worth in the production of high-strength fibers. The economically still reasonable minimum batch sizes of a compact spinning plant are about a factor of 10 smaller than those of a two-stage plant.

33.2.4.5 Nonwoven Production

PP nonwovens are produced either in a single-stage process directly from the granulate to the nonwoven (spunbond) or in a multi-stage mechanical process via the spunbond production route.

Spunbonded nonwovens are produced with basis weights ranging from 5 g/m^2 to several kg/m^2, while meltblown nonwovens are usually 5–100 g/m^2.

In the meantime, combination lines for the formation of multi-layer nonwovens (SMS: Spunbonded-Meltblown-Spunbonded) are also used: While the outer layer as a spunbonded nonwoven provides the necessary mechanical strength, an inner meltblown nonwoven is responsible for good filter properties or for good and uniform coverage. In this way, the specific positive properties of both types of nonwoven are utilized, which supports a further expansion of the application range of PP nonwovens.

Depending on the application and the weight per unit area, PP fiber nonwovens are consolidated or mechanically stabilised by thermal or mechanical means. Thin nonwovens (especially for hygienic applications) are bonded by means of heated calender rolls with point-like structures, the elevations of which melt the nonwoven together at specific points. Thick nonwovens are consolidated by needle punching.

33.2.4.6 Split Film

There are two manufacturing methods for this. In the first, more common method, a film is extruded and cooled either on cooling rolls or in a water bath and monoaxially stretched up to 10 times. During monoaxial stretching, the macromolecules are very strongly oriented in the direction of loading. Since the PP macromolecules are only linked to each other by low secondary valence forces, the transverse strength of the film is low. Splicing into threads is therefore very easy. The film is either cut into tapes or fibrillated over needle rollers and torn into longitudinal stripes. An important property of film tapes is their high tensile strength.

In the second method, each ribbon is extruded separately through an individual slit opening. This method is very complex and therefore expensive.

Film tapes are almost exclusively processed into fabrics. End products are carpet backing, sacks, bags and technical textiles as well as artificial turf.

33.2.5 Properties

Compared with other synthetic fibers, PP fibers are characterized primarily by a relatively large

Table 33.2 Mechanical properties of PP

		Fibers	Filaments (normal)	Filaments (high tenacity)
Tenacity	[cN/tex]	13–60	15–60	70–95
Maximum elongation	[%]	30–350	20–150	8–10
Fiber fineness	[dtex]	0.8–400		
Heat resistance	Under long term heat >120 °C possible drastic tenacity loss			
UV resistance	Unstabilised low, by suitable stabilisators a very high UV resistance can be achieved			
Knot strength	[%]	70–90		
Loop strength	[%]	85–95		
Abrasion resistance		Good		
Wet tenacity	[%]	100		

elongation, a wide range of strength values and a relatively low melting temperature. The properties can be varied within wide limits by the spinning conditions and suitable additives.

33.2.5.1 Technological Values

The technological values, above all maximum tensile strength and maximum tensile elongation, can be varied over a wide range via the degree of draw ratio and the heat applied in the process. Thus, in addition to the normal strength types (13–25 cN/tex), there are also yarns with high strength (>50 cN/tex) for technical use. In most cases, higher strengths go hand in hand with higher elongation (Tables 33.2 and 33.3).

33.2.5.2 Chemical Behavior

- Burn test: Polypropylene burns like polyethylene
- Burning behavior of the fiber:
  - Before inflammation: shrinking, melting,
  - in pilot light: igniting,
  - when the pilot light is removed: continues to burn slowly,
  - odor: paraffin-like odor when extinguished,
  - residue: dark sphere.
- LOI value [%]: 18–19
- Resistance to chemicals: The resistance to all organic solvents, acids and alkalis at room temperature is excellent. Only strongly oxidizing substances, e.g. concentrated nitric acid or bleaching lye, attack PP at elevated temperatures, as do organic solvents such as hydrocarbons and chlorinated hydrocarbons at temperatures >100 °C (swelling and subsequent dissolution).
- Resistance to microorganisms: very good

33.2.6 Modification of PP Fibers

The fiber properties depend strongly on the production process used and on the polymer structure. The polymer properties (molecular weight, molecular weight distribution, tacticity) are determined by the polymerization catalyst (Sect. 33.2.2.3) and the process technology.

33.2.6.1 Dyeing Behavior

Polypropylene is usually dyed in the spinning mass with organic or inorganic pigments. This process offers in particular the advantage of very

Table 33.3 Physical properties of PP

Density	[g/cm^3]	0.90–0.91
Moisture take-up	[%]	0
Softening temperature	[°C]	150–160
Melting temperature	[°C]	160–175
Specific heat	[kJ/kgK]	1.6–2.0
Thermal conductivity	[W/mK]	0.1–0.3
Shrinkage	[%]	< 3

good color fastness, but a color change is costly. Therefore, this principle is only profitable for large batch sizes. The main applications of PP fibers, hygiene, carpet and geotextiles meet this criterion.

If unmodified PP is dyed in the bath, this leads to poor fixation and poor evenness of the dyeing. This is due to the high crystallinity of the drawn PP fibers (50–65%) as well as the low affinity between fiber and dye, caused by the lack of strong intermolecular forces between PP and functional groups of the dyes.

There are various approaches to make PP fibers also accessible for bath dyeing:

- Enhancement of the hydrophobic properties of the dyes by long-chain alkyl substituents,
- dyeing of chlorinated PP,
- mixture with PET dyeable with disperse dyes,
- formation of copolymers, one component of which binds the dye.

33.2.6.2 Behavior Towards Environmental Influences

UV and Light Resistance

PP is very susceptible to photo-oxidation, with this process being accelerated by the additional influence of heat. Light or UV stabilizers protect the PP from photo-oxidative degradation and fading. So-called HALS stabilizers (Hindered Amine Light Stabilizers) are often used. This protection is particularly important for applications of PP fibers in the outdoor sector, for carpets and for automotive interiors. Otherwise, PP loses its strength very quickly and decomposes.

Thermostabilizers

PP is susceptible to thermal oxidation. This is due to the tertiary carbon atoms carrying the methyl groups, which are preferentially attacked by free radicals. Therefore, thermostabilizers are necessary in polymer pelletizing as well as in fiber production (ensuring process stability) and processing. They are also important for the long-term service properties of textiles. Thermostabilizers in the form of primary antioxidants (e.g. sterically hindered phenols) serve as radical scavengers. Secondary antioxidants (mostly phosphites) cause a degradation of the hydroperoxides formed in the course of thermal degradation. Phenol-free stabilizers surpass the systems previously known as antigas fading stabilization in terms of resistance to discoloration by nitric oxide.

Antimicrobial Behavior

Antibacterial agents are intended to restrict the growth of microorganisms (bacteria and fungi) on textiles made of PP in the long term. This is particularly important for sportswear and floor coverings in hospitals or schools. Recent developments aim at the permanent embedding of antibacterial agents in the polymer matrix.

Flame Retardant

FR additives (FR = flame retardancy) are flame retardant agents that play a major role due to strict safety regulations depending on the area of application. This property is important, for example, for floor coverings in public buildings and for automotive and aircraft interiors. Conventional systems often contain halogen compounds as flame retardant components.

Antistatic Behavior

Antistatic additives often contain hydrophilic groups that act on the fiber surface by binding water to them. This reduces the surface resistance and thus also the static charge of the fibers. When using PP fibers for carpets as pile fibers, a finish is usually not necessary, since the triboelectric charge with rubber (shoe sole) as a friction partner is relatively low.

33.2.6.3 Modification of PP Fibers with Regard to Their Mechanical Behavior

Creep is the plastic deformation of the material under prolonged loading. It can be regarded as an extension of the stretching process. The reason for this behavior lies in the semicrystalline structure of long-chain polymers. This structure allows the macromolecules many possibilities of movement. For the majority of current applications of

PP, however, the creep behavior is not problematic.

Polypropylene with alternating isotactic and atactic areas has similar properties to thermoplastic elastomers. If this special type of fiber were developed to production maturity, the range of applications for polypropylene fibers could expand significantly.

33.2.7 Applications

Polypropylene can be used for a wide range of different applications. The market is dominated by carpet yarns and packaging material, mainly because of its low price compared to other polymers. The poor dyeability of polypropylene does not play a decisive role for such products. In addition to staple fibers and filament yarns as well as monofilaments, large quantities of spunbonded nonwovens are also produced from PP (Table 33.4).

33.2.7.1 House and Home Textiles and Clothing

Metallocene catalysis, modern processing technologies, new additives and dyeing methods now make PP interesting for textile applications. Particularly noteworthy here is the area of carpets and carpet base fabrics. Modern texturing processes and improvements in spin dyeing make PP-BCF yarns well suited as carpet pile material. Certain home textiles require properties that can be created using modern PP fiber chemistry. For example, mattress ticking made from PP can be finished with long-term substances that create a climate that is unfriendly to dust mites, bacteria and mould. Other home textiles made of PP are: Wipes, duvets and bed linen, inlets, bath mats, curtains.

PP is less interesting for widespread use in the clothing sector due to its poor dyeability in the dye bath. The first garments made of PP were thermal underwear. Today, mainly functional sportswear, socks and stockings are made from PP.

33.2.7.2 Hygiene and Medicine

Hygiene

The most important areas of application for PP fibers include hygiene products such as baby diapers, incontinence pads and sanitary towels. These products are now relatively complex structures in which PP nonwovens are often used as shell nonwovens in which another structure containing superabsorbents, mostly based on polyacrylamide, is embedded. Worldwide sales of

Table 33.4 Typical applications, properties and manufacturing processes

Application	Desired properties	Production process
Home textiles		
Pile yarns for carpets Carpet backing Carpet backing Upholstery	Voluminous, light High tenacity Cheap, high tenacity High abrasion resistance, dirt repellant	BCF FOY Slit film POY, staple fiber yarns
Packaging material	Cheap, high tenacity	Spun bond, staple fiber nonwovens
Medical products		
Diaper nonwovens Filter, one-way products Hygiene products and one-way products	Hydrophobic, skin-friendly Fine, homogenous structure Combination of high tenacity and fineness	Spun bond, staple fiber nonwovens Meltblown nonwovens Sandwich laminate of spunbonded and meltblown nonwovens
Technical textiles		
Geotextiles, drainage mats, Belts, ropes, cords	High tenacityHigh tenacity, resistent to chemicals, microorganis Hydrophobic	Spun bond, staple fiber nonwovens FDY, monofilaments
Cloting		
Sports clothing, socks	Light, no water take-up	POY, staple fiber yarns

diapers amount to approximately €50bn, of which adult diapers account for about 30%. This makes it the most important market for polypropylene nonwovens. It is estimated that around half of the global production of polypropylene is processed into a total of around 250bn diapers.

Medicine

The growing need for protection in medicine is due to the increasing number of known diseases transmitted through body contact. PP is well suited for the manufacture of protective medical products such as surgical masks, gowns or drapes due to various properties. Controversial, although very practical in use, are disposable absorbent products for medical or personal use (e.g. diapers, incontinence pads, hospital bed pads). They usually consist of different layers made of different materials and therefore complicate waste disposal.

33.2.7.3 Technical Textiles

PP is used here in the form of tapes and above all in the form of nonwovens.

Automotive Division

Currently, an average of 25 kg of synthetic fibers are used per automobile. With approx. 70 million new cars per year worldwide (as of 2019), this corresponds to approx. 1.5 million t of fibers, of which polypropylene accounts for a significant share. Thus, it is estimated that around a quarter of the PP fibers produced worldwide, i.e. approx. 0.9mt, are used in the automotive industry. The requirements for trim parts in the car interior (parcel shelf, floor assembly, side trim) are very high. They must be durable and non-fading, have good acoustic and thermal insulation properties and combine high dimensional accuracy, low weight and economical production. These cladding parts are often laminated with other textiles. PP is used both as a carrier material and as a surface material. This single-grade composite is comparatively easy to recycle and can serve as a raw material for new carrier parts. Material recycling is therefore relatively simple.

Geotextiles

Geotextiles made of PP are used in particular in garden and road construction: Woven fabrics up to 8.50 m wide, and beyond that also nonwovens, knitted fabrics and nets. They fulfil two functions: firstly, the mechanical function of reinforcement, i.e. the absorption of forces, made possible by the high tensile strength and elasticity of the geotextile. These properties are important in the reinforcement of roads and for the fastening of embankments. On the other hand, geotextiles are used for filtration and drainage. Here, the good water permeability of the textile is particularly important. Textile filters or drainages are used instead of mineral filters and are used to keep drainage packs fully functional over a long period of time.

Shipping; Ropes and Cables for Shipping

Particularly favorable for this application is the low density (PP ropes float on water), the good lightfastness of the spun dyeing and the resistance to mold and other microorganisms. Monofilaments are often used as the basic structure.

Building Textiles

PP fibers and fibrillated film tapes are used to prevent shrinkage cracks during curing and also for reinforcement (in combination with other fibrous materials). Concrete is an extremely cost-effective construction material whose great advantage lies in its high capacity to absorb compressive stresses. However, its disadvantages are its limited tensile strength, brittleness, tendency to shrinkage and poor flexural strength. To compensate for these weaknesses, concrete is also reinforced with PP fibers. PP has good physical, mechanical and chemical properties for these purposes.

33.2.8 Market

World production of PP fibers has increased from 2.8mt to 3.3mt over the last 20 years (Fig. 33.19). This is around 4.5% of synthetic fiber production.

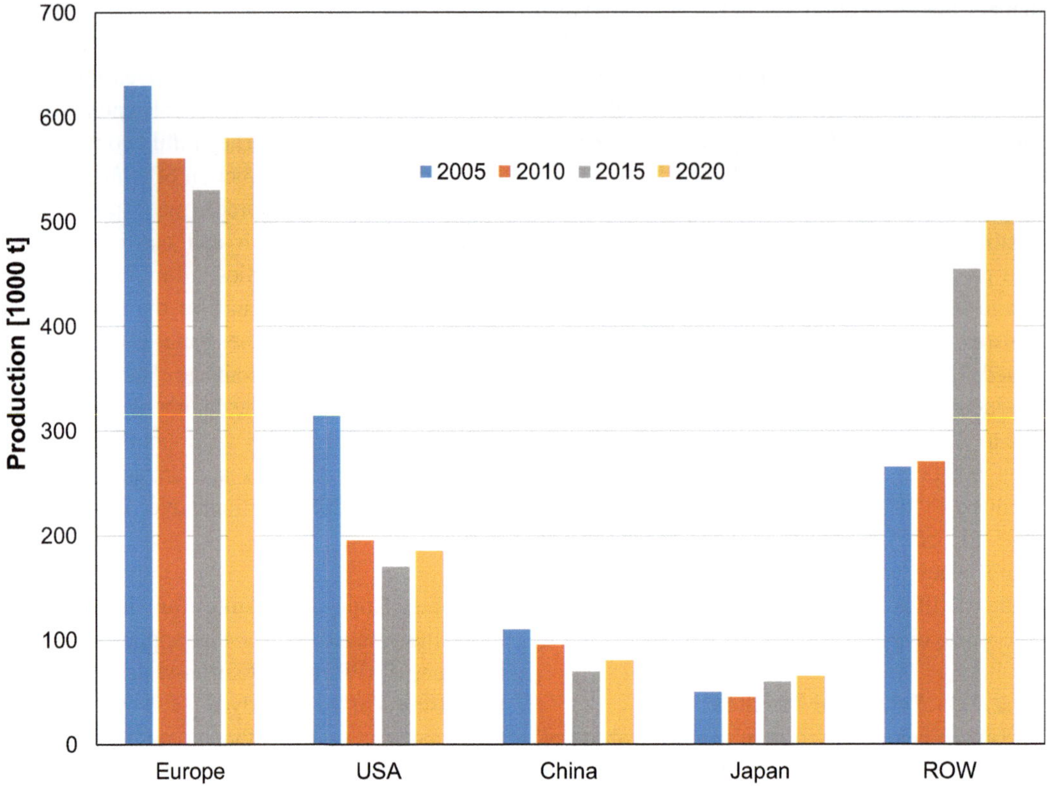

Fig. 33.19 World production of PP staple fibers (data: The Fiber Year, 2020)

Currently, about 1.4mt of staple fibers and 1.9mt of filament yarns are produced from PP annually (The Fiber Year, 2021). Because polypropylene is processed to a large extent into nonwovens and these are produced regionally, China is by far not as dominant in the PP market as it is for polyester. For staple fibers in particular, Europe is the most important, accounting for more than 40% of production (Fig. 33.20). Typical prices for PP fibers are 1–1.5 €/kg for standard grades and up to 2.5 €/kg for special grades.

The USA dominates the market for filament yarns with a share of around 30% (Figs. 33.21 and 33.22).

The production of PP film tapes for agro- and geotextiles and for packaging is not usually included in fiber statistics and is therefore not included in these figures.

33.2.9 Recycling and Environmental Aspects

PP is suitable for material recycling due to its chemical-physical properties. Material recycling means that new PP products can be manufactured from used PP products or from waste products resulting from PP processing. However, re-extrusion reduces the molecular weight and narrows the molecular weight distribution. This is helpful for certain applications or processing methods that require a high MFI (Melt Flow Index). Disadvantages may include: partial loss of strength, stability, color fastness and an increase in shrinkage. Various additives are also mixed in during re-extrusion to enhance the properties. An investigation of the re-extrudability and spinnability of

Fig. 33.20 PP staple fiber production by region (data: The Fiber Year, 2021)

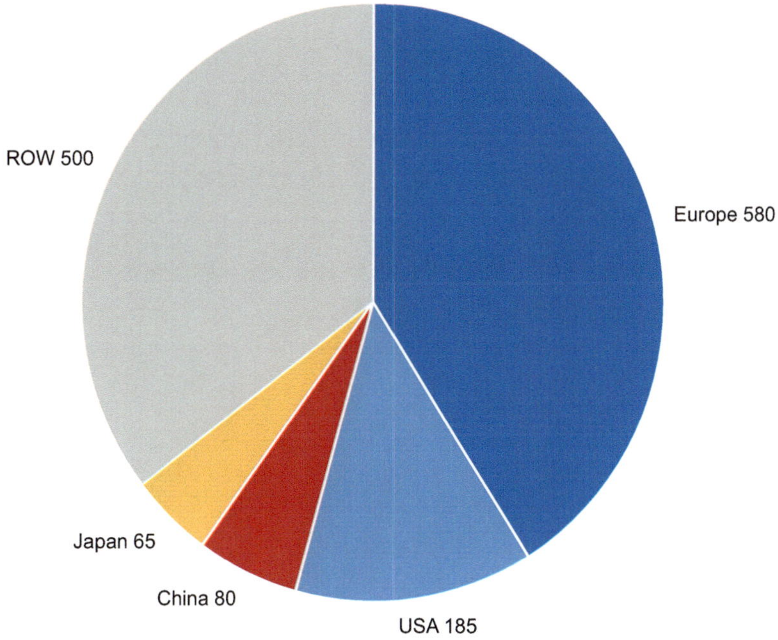

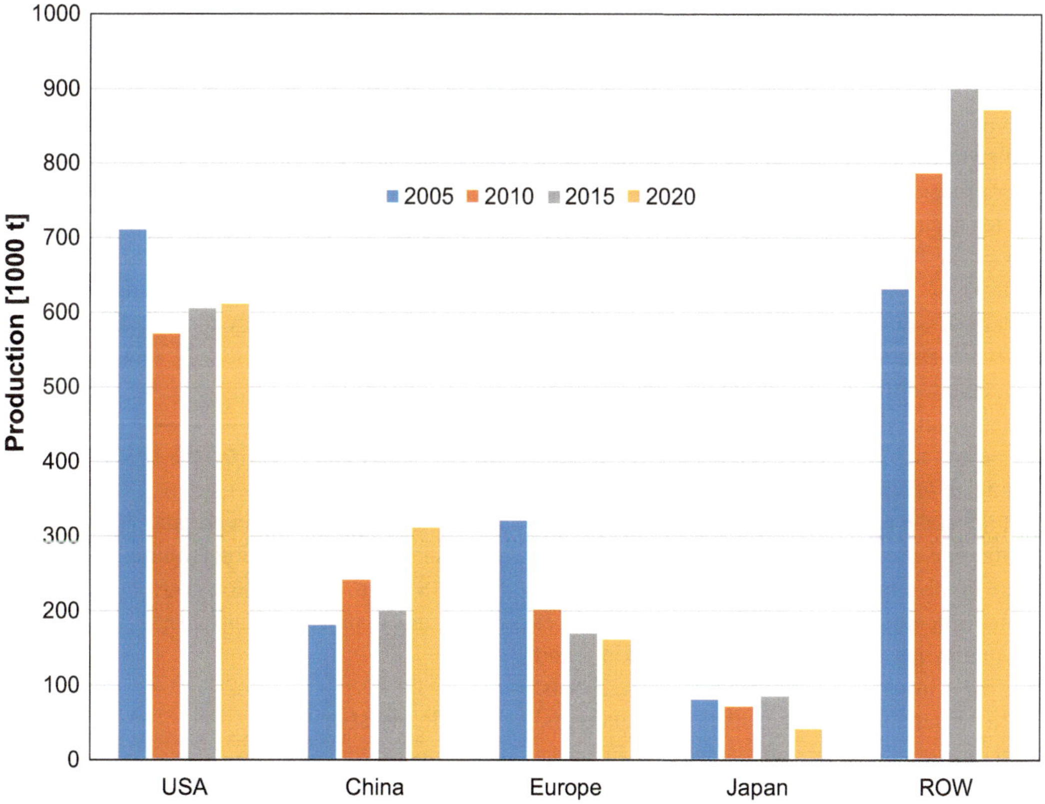

Fig. 33.21 World production of PP filaments (data: The Fiber Year, 2021)

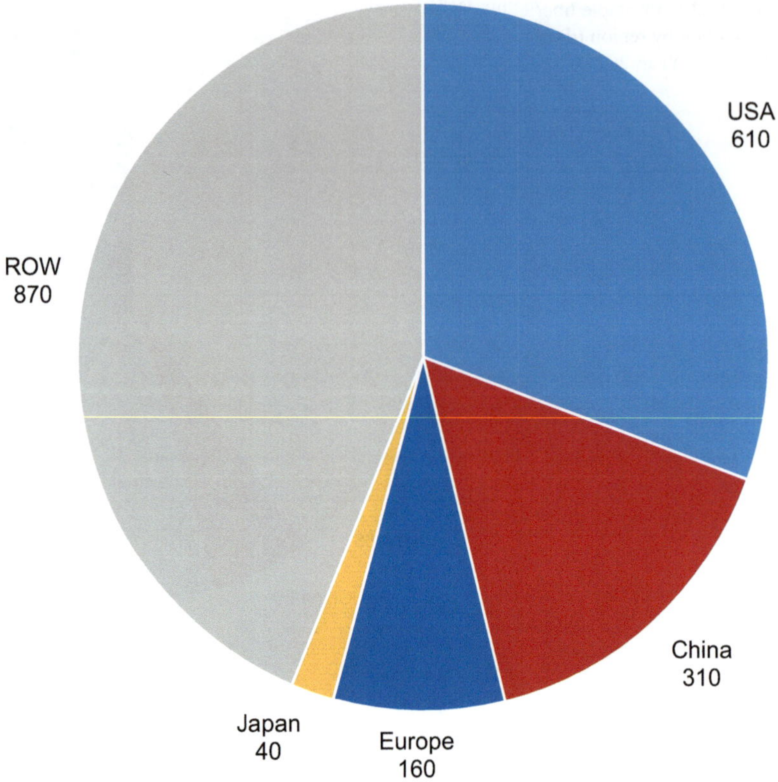

Fig. 33.22 PP filament production by region (data: The Fiber Year, 2020)

polypropylene fibers has shown that only after the sixth recycling cycle are there clearly deteriorated fiber properties. In principle, however, it is possible to manufacture products from pure, carefully processed PP recyclate that are comparable in quality to products made from primary raw materials.

The logistics for single-variety collection are the greatest difficulty, because PP is often already mixed with other materials during processing. These "foreign" substances can cause disruptions during re-granulation and secondary processing that affect the economic efficiency and properties of the end product. Different processing stages often involve material mixing that makes separation difficult, if not impossible. In carpet manufacturing, for example, the backing and pile are in many cases made of different materials.

In the automotive sector, interior trim parts and their textile covers are often made of PP. This single-variety composite should not be disposed of as waste later on, but serve as a raw material for new applications in the car interior, which would close the recycling loop.

The incineration of PP waste with the aim of generating energy can also be worthwhile. The combustion of PP is low in pollutants, its calorific value is the highest of all man-made fibers and corresponds approximately to that of heating oil (Fig. 33.23).

33.3 Polyvinyl Chloride (PVC)

PVC is one of the most important plastics of all, with an annual production volume of approx. 45mt (Statista, 2022). However, only a small part of this is processed into fibers.

33.3.1 History

PVC fibers were the first fully synthetic fibers. In 1913, they were produced for the first time by

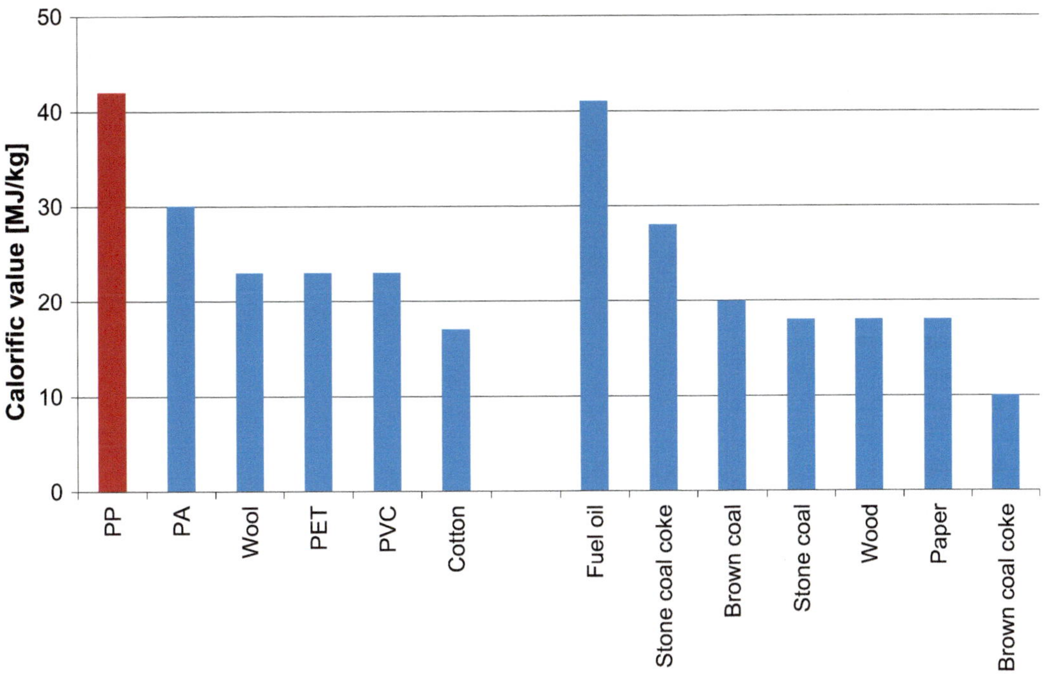

Fig. 33.23 Heating value of PP compared with other fibers and fuels

F. Klatte at Chemische Fabrik Griesheim. He received the first patent in the field of fully synthetic fibers (Klatte, 1913). Subsequently, however, not fibers but solid materials were produced from PVC and marketed under the name Igelit®. In 1931, E. Hubert at IG Farben (Agfa Filmfabrik Wolfen) discovered that PVC can be dissolved in cyclohexane and spun out wet in a precipitation bath of acetic acid. At about the same time, IG Farben's Wolfen plant was post-chlorinating PVC into what was known as PeCe, which could be dissolved in acetone and spun out in water. This was much simpler in terms of process technology and so large-scale production began in 1934 and PeCe fiber was the first synthetic man-made fiber to be produced on a larger scale. With a density of 1.44 g/cm^3, it is resistant to acids, alkalis and water and is therefore used mainly for technical articles, e.g. filters and strainers and parachutes, and later also upholstery fabrics and rheumatic underwear. Its water absorption is very low and it charges electrostatically, therefore it was unsuitable for clothing textiles.

33.3.2 Production

PVC is produced from vinyl chloride either by emulsion, suspension or bulk polymerization. Mixed polymers are also known, e.g. Saran™ (cf. Sect. 33.4). By lowering the polymerization temperature, larger proportions of syndiotactic PVC are formed. This increases the degree of crystallinity and thus the glass transition temperature and the melting temperature to values of approx. 110 °C and 275 °C, respectively (Fig. 33.24).

PVC fibers are produced using the dry spinning process, whereby the equipment and aggregates are similar to those used in PAN production. The main difference is the double filtration of the polymer before spinning. A mixture of

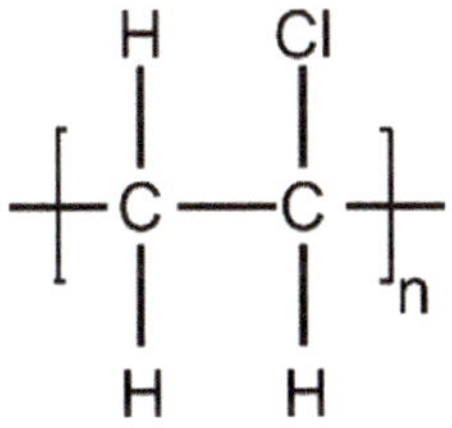

Fig. 33.24 Structural formula of the repeat unit of polyvinyl chloride

Table 33.5 Typical properties of PVC fibers acc. to manufacturer's specifications

	PVC
Typical fiber fineness [dtex]	1.5–60
Tenacity [cN/tex]	10–35
Breaking elongation [%]	10–40
E-modulus [cN/tex]	200–400
Wet tenacity [%] of dry value	100
Moisture absorption [%]	0–0.2
Hot water shrinkage [%]	30–95
Glass temperature [°C]	70–100
Melting temperature [°C]	160–200
Heat conductivity [W/mK]	0.2
Heat capacity [kJ/kg K]	0.8–1.3
LOI	37–45
Density [g/cm^3]	1.35–1.45

equal parts acetone and carbon disulfide or acetone and benzene is used as solvent. The polymer concentration is then about 30% and the spinning temperature is 70–125 °C, the take-off speed up to 400 m/min (Fourné, 1995). Wet spinning is also possible but rarely used. The DP of PVC is approx. 1000–2500.

The typical properties of PVC fibers are listed in Table 33.5.

PVC fibers have good light, weather and chemical resistance. Only concentrated hot nitric acid attacks PVC fibers. Furthermore, they show a high electrostatic charge, a low abrasion resistance, are flame retardant and poorly dyeable.

For special applications, e.g. coarse fibers for wig hair, PVC can also be melt spun. For this purpose, PVC is dry-mixed with heat stabilizers and other additives and processed into pellets. The pellets are then extruded and the resulting fibers are stretched and coiled. Because the decomposition temperature of PVC is only slightly above the melting temperature at around 210 °C, temperature control is of crucial importance.

33.3.3 Areas of Application

Monofilaments, multifilament yarns and staple fibers can be produced from PVC fibers. They are used for thermal underwear, upholstery and decorative fabrics, curtain fabrics, tarpaulins, nets and ropes as well as nonwovens, synthetic hair and imitation fur.

33.3.4 Market and Trade Names

The amount of PVC fibers produced is not statistically recorded. Important trade names are or were Rhovyl™, Fibravyl™, Thermovyl™, Clevyl™, Crinovyl™, Envilon™, Ternel™ and Teviron™. Some of these are now used for PVC plastic solids.

33.4 Polyvinyl Dichloride (PVDC)

PVDC fibers have been produced since 1939 and are a copolymer of 15% vinyl chloride and 85% dichloroethene (Fig. 33.25). They are melt-spinnable at 175 °C. There are monofilaments, multifilament yarns and staple fibers.

Typical properties are listed in the Table 33.6.

PVDC fibers have very good resistance to acids and alkalis (except conc. NH_3), good

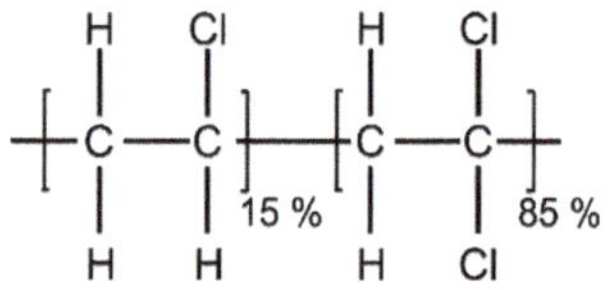

Fig. 33.25 Structural formula of the repeat unit of PVDC

Table 33.6 Typical properties of PVDC fibers acc. to manufacturer's specifications

Typical fiber fineness [dtex]	
Monofilament: Multifilament: Staple faibers	560–3400 400f3, 560f10, 1200f9 17–4400
Tenacity [cN/tex]	13–20
Breaking elongation [%]	15–40
E-modulus [cN/tex]	70–105
Wet tenacity [%] of dry value	100
Moisture absorption [%]	< 0.1
Glass temperature [°C]	115–130
Melting temperature [°C]	160–175
Heat conductivity [W/mK]	0.13
LOI	60
Density [g/cm^3]	1.68–175

resistance to most organic solvents (except ether). Brilliant, washfast dyeings can be achieved by spin dyeing. PVDC is not attacked by microorganisms and animals and does not decompose in humid environments. It can be made flame retardant.

Typical applications are doll's hair, filter screens, carrier material for biofilms in wastewater treatment, sun protection textiles and upholstery fabrics. An important manufacturer is Asahi Kasei Corp., Japan, which markets PVDC fibers under the brand name Saran™.

33.5 Polyvinyl Difluoride (PVDF)

PVDF is produced from the reaction of hydrogen fluoride with trichloroethane via the intermediate stage difluoroethane by polymerization:

$$HF + CCl_3CH_3 \rightarrow CF_2CH_3 \rightarrow CF_2CH_2$$

In macromolecules made of PVDF, there are two fluorine atoms on every second carbon atom. These are strongly electronegative and therefore attract the electrons of the covalent bond to the carbon. This creates a weak dipole perpendicular to the molecular axis. This can be used to create a small piezoelectric effect (Fig. 33.26).

PVDF is polymorphic, i.e. it has different crystal conformations. After spinning, PVDF is normally monoclinic in the -conformation, i.e. in the form TGTG (T = trans; G = gauche). The dipoles of the polymer chains neutralize and the material behaves like a normal dielectric with an electric field constant of aboutε_0 = 10 C/Vm, which corresponds to the values of glass and ceramics.

If a tensile force is applied, the -conformation transforms into the -conformation. This is orthorombic and the chain building blocks are almost completely in the trans form (TT). This results in the formation of a dipole (Fig. 33.27).

The transformation of the -shape into the -shape occurs more strongly with increasing deformation speed (pull-off speed) and a larger stretching ratio (Lund & Hagström, 2010). It is believed that the -crystals break into small pieces by shearing, which then transform into -crystals. Defects in the crystal structure promote this

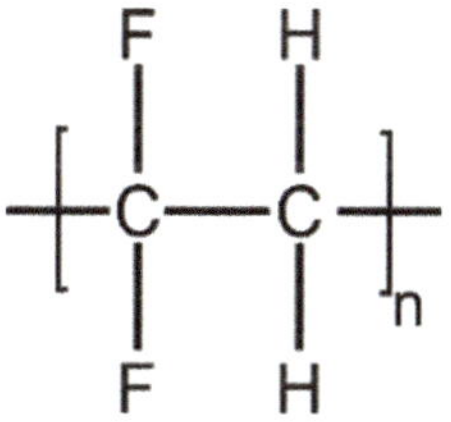

Fig. 33.26 Structural formula of the repeat unit of PVDF

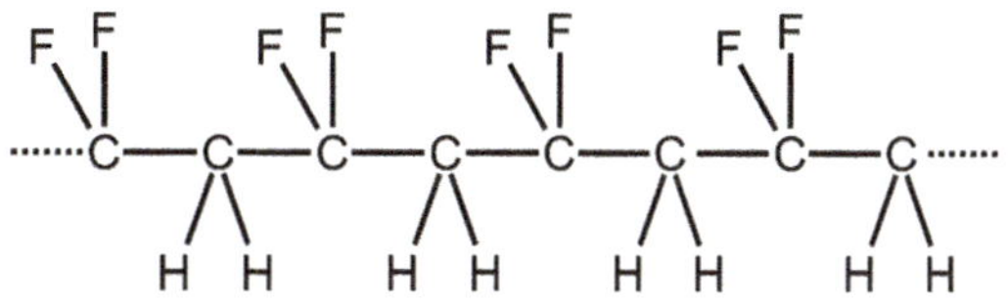

Fig. 33.27 Dipole structure of PVDF under external load resp. in -conformation

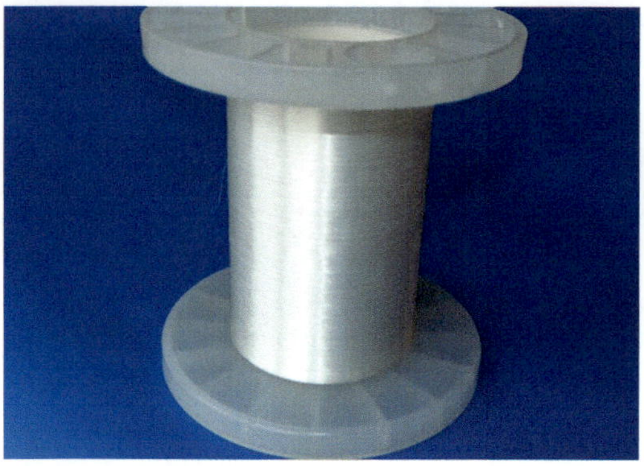

Fig. 33.28 Bobbin with PVDF multifilament yarn

process (Wu et al., 2000). If an external stress is now applied, the material deforms elastically. Conversely, elastic deformation of PVDF leads to an electrical voltage.

Semi-crystalline PVDF fibers can be wet-spun and melt-spun. In melt spinning, winding speeds of 2500 m/min are achieved on a pilot scale (Walter, 2012). The electrospinning process is also used for medical applications (Fig. 33.28).

Typical strengths of PVDF fibers are 40–45 cN/tex, elongation at break is 10–50%. The density of PVDF fibers is 1.77–1.79 g/cm^3, the moisture absorption is less than 0.1%, the glass transition temperature is −30 °C, the crystallization temperature is 140 °C, the melting range is 165–172 °C, and the decomposition temperature is 350–375 °C. PVDF fibers can withstand 150 ° C for a prolonged period and are virtually incombustible with an LOI value of 70. They do not react with strong acids, bases and oxidizing agents as well as organic solvents. They are hydro- and oleophobic and have a low coefficient of friction. PVDF fibers are chemically inert and therefore biocompatible.

The piezoelectric effect is used, for example, to integrate pressure, force and acceleration sensors in fiber form into textiles. The induced voltage can also be used to power small components or a light source. Other typical products are fishing lines, strings for plucked instruments and medical textiles (implants). Known trade names for PVDF are Kynar™, Solef™, Hylar™, Trofil™ and Sygef™.

33.6 Polytetrafluoroethylene (PTFE)

This fiber is discussed in the chapter on high performance fibers, because its properties have much in common with the fibers discussed there.

33.7 Polyacrylic (PAN)

This fiber is dealt with separately in Chap. 5 because of its great importance.

33.8 Polyvinyl Alcohol (PVA)

Polyvinyl alcohol (PVA or PVAL) is one of the few thermoplastic polymers that is water soluble (Fig. 33.29). PVA fibers are often combined with biopolymers. Their generic name is vinal fibers, but it is rarely used. They were first introduced in 1924 by W. O. Hermann and W. Haehnel from polyvinyl esters and spun into filaments. Nowadays, they are formed by hydrolysis (saponification) of polyvinyl acetate (Fig. 33.29, n is not the DP). Direct synthesis from vinyl alcohol monomers is not possible because of their tautomerism. The DP is 500–2500.

In the PVA chain, 98–99% of the monomers are in head-to-tail arrangement and only 1–2% in head-to-head arrangement. By changing these proportions during hydrolysis, the water solubility of PVA can be altered.

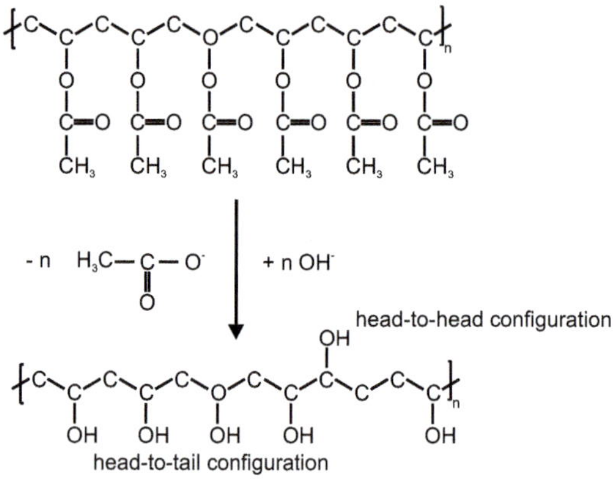

Fig. 33.29 PVA synthesis by saponification of polyvinyl acetate

Table 33.7 Selected manufacturers of PVA fibers and trade names

Producer	Trade mark	Website	Country
Briture Co., Ltd.		https://briture.en.made-in-china.com/	China
Changzhou Tian Yi Engineering Fiber Co., Ltd.		https://tianyifiber.en.made-in-china.com	China
Kuraray Europa GmbH	Kuralon™	http://www.kuraray.eu	Japan/ Germany
Monofil-Technik GmbH	Trofil PVA™	http://www.monofil-technik.com	Germany
Nantong Wanhe Chemical Co., Ltd.		https://63ed33d443bd3dd7.en.made-in-china.com	China
Nitivy Co., Ltd.	Solvron	https://www.nitivy.co.jp	Japan
Shenzhen Xiangu High-Tech. Co., Ltd.		https://xian-gu.en.made-in-china.com/	China

The fibers are produced using the wet-spinning process. The spinneret is made of gold or a gold-platinum alloy and has 2000–60,000 capillary holes with a diameter of approx. 0.1 mm. The take-off speed from the spinneret is approx. 5 m/min. The precipitation bath contains 15–16% sodium sulfate ($Na_2\ SO_4$), the bath temperature is 40–50 °C. Often 2 or 3 coagulation baths are used in succession. The PVA filament is then hot-stretched 4–15 times and thermally post-treated at 220–230 °C for 40–180 s. The heat is usually applied by means of a heated bath. The heat is usually introduced via heated godets or hot air. During this process, the filaments crystallize and are then wound up at a speed of approx. 50 m/min (Fourné, 1995). PVA fibers can also be made insoluble in water by cross-linking the hydoxyl groups. In rare cases, PVA fibers are also dry-spun.

The density of PVA fibers is 1.26–1.31 g/cm^3. Typical fiber finenesses are 1.5–10 dtex, strength 20–65 cN/tex (normal type) or 25–80 cN/tex (water insoluble type). The wet strength is 65–85%, the maximum tensile elongation is 10–30% dry and 12–42% wet, the modulus of elasticity is 300–600 cN/tex. The glass transition temperature is in the range of 75–130 °C, the melting temperature is 240–260 °C (water insoluble grade). For the water insoluble grade the hot water shrinkage is 2–3%. The LOI is 20. PVA fibers (normal type) are readily soluble in water but insoluble in almost all anhydrous organic solvents.

The main textile applications are plastic and cement reinforcement, binder fibers and ropes. The price for 1 kg of PVA fibers is 1.5–3 €, depending on the quality.

Kuraray, Japan, markets PVA fibers under the brand name Kuralon™. Other manufacturers are listed in Table 33.7.

References

DSM. https://www.dsm.com/dyneema/en_GB/home.html

Fourné, F. (1995). *Synthetic fibers*. Hanser.

Globe Newswire. (2020). https://www.globenewswire.com/

Kaminsky, W., & Sinn, H. (1980). Living polymers on polymerization with extremely productive ziegler catalysts. *Angewandte Chemie*, *19*(5).

Klatte. (1913). Patent specification DP 281,877.

Lund, A., & Hagström, B. (2010). *Melt spinning of poly (vinylidene fluoride) fibers and the influence of spinning parameters on b-phase crystallinity*. Wiley InterScience.

Mather, R. R., & Wardman, R. H. (2015). *The chemistry of textile fibers*. Royal Society of Chemistry.

Mertens, M. (2006). https://commons.wikimedia.org/wiki/File:Stihl_kettensaege.jpg, CC BY-SA 2.0 EN.

Natta, G. (1961). Polypropylene fibers. *Osterreichische Chemiker-Zeitung*, *7*, 207–215.

Spectra. (2020). http://www.spectra-fiber.com

Statista. (2022). http://www.statista.de

The Fiber Year. (2021). *The Fiber Year GmbH, Roggwil*. https://www.thefiberyear.com

Toyobo. https://www.toyobo-global.com/seihin/dn/izanas/index.html

Walter, S. (2012). *Entwicklung piezoelektrisch wirksamer Sensorfasern auf Basis von PVDF*. PhD thesis at RWTH Aachen University, Shaker Verlag, Aachen.

WIKI. (1899). https://commons.wikimedia.org/wiki/File:Hans_v_Pechmann_ca1899.jpg

Wu, J., Schultz, J. M., Yeh, F., Hsiao, B. S., & Chu, B. (2000). In-situ simultaneous synchrotron small- and wide-angle A-ray scattering measurement of poly (vinylidene fluoride) fibers under deformation. *Macromolecules, 33*, 1765–1777.

Wulfhorst, B., & Meier, K. (1989). *Polypropylene fibers*. Fiber table acc. to P.-A. Koch of the Institute for Textile Technology of RWTH Aachen, Aachen.

Ziegler, K., Holzkamp, E., Breil, H., & Martin, H. (1955). Polymerization of ethylene and other olefins. *Angewandte Chemie, 67*(16), 426.

34 Polyacrylonitrile

34.1 Definition

Polyacrylonitrile is a vinyl compound and consists primarily of an aliphatic chain with nitrile side groups on every other C atom (Fig. 34.1). For this purpose, other monomers are incorporated into the chain to form a copolymer. Acc. to DIN ISO 2076, "fibers made of linear macromolecules with at least 85% by mass" are referred to as polyacrylonitrile. Modacrylic fibers, on the other hand, are "fibers made of linear macromolecules containing at least 50% and less than 85% by mass of acylnitrile in the chain". In the USA, on the other hand, modacrylic fibers are defined as having at least 35% and less than 85% acrylonitrile by mass in the chain. The abbreviation for polyacrylonitrile fibers is PAN, and MAC for modacrylic fibers.

34.2 Inventors, Development

In 1893, the French chemist M.C. Moureau produced monomeric acrylonitrile, the starting material of polyacrylonitrile. A year later, he succeeded for the first time in synthesizing acrylonitrile by distilling acrylic acid amide and ethylene cyanohydrin over phosphorus pentoxide. In 1913, the German chemist Fritz Klatte proposed the use of synthetic raw materials for the production of fibers. His work on the polymerization of vinyl compounds resulted in a patent specification in which he mentioned, among other things, the processability of the resulting "horn-like mass" into threads (Klatte, 1913).

For a long time, acrylonitrile was considered technically insignificant. It cannot be melted because its decomposition point is below the melting point, and it was not soluble in the solvents known at the time. It was assumed that acrylonitrile was unsuitable for fiber production because it was totally cross-linked. It was insoluble in the solvents known at the time, except in concentrated sulfuric acid. In 1930 PAN was synthesized for the first time by H. Fikentscher and C. Heuck for the first time (Fikentscher & Heuck, 1937). They proposed to use it for fiber production. The first corresponding patents also date from this time. Since about 1931, the German chemist Herbert Rein had been searching for a suitable solvent for acrylonitrile. In 1934 he discovered the possibility of dissolving polyacrylonitrile in concentrated solutions of some inorganic salts—zinc chloride, sodium and calcium rhodanide—and in quaternary ammonium bases (Rein, 1934, 1936). However, these solvents were too expensive. Otto Bayer and Peter Kurtz developed the direct synthesis of acrylonitrile from hydrocyanic acid and acetylene in 1939, making large-scale production possible (Fig. 34.2).

In 1939, IG Farben made the first attempts to produce filaments from polyacrylonitrile under the name "PAN fibers". In 1941, Rein and the American Ryde C. Houtz independently discovered almost simultaneously dimethylformamide (DMF), the most economically suitable solvent

D. Veit, *Fibers*, https://doi.org/10.1007/978-3-031-15309-9_34

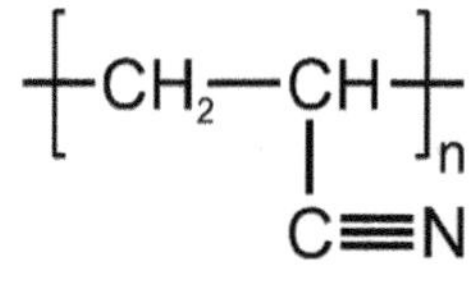

Fig. 34.1 Repeat unit of polyacrylonitrile

besides α-pyrrolidone and tetramethylene sulfone. In the USA, George H. Latham described a production process from polyacrylonitrile spinning solutions, which led to the trial production of polyacrylic filaments under the name "ANP" or "Fiber A" at DuPont. In 1950, DuPont launched a dry-spun polyacrylonitrile fiber in the USA under the name Orlon™. In Germany, the first industrially produced polyacrylonitrile fibers were launched in 1952 under the names "PAN"and "Redon™". The industrial production of Modacryl- began in 1949 at the Union Carbide Corporation. The main sites for the production of polyacrylonitrile in Germany were Dormagen, where Dralon™ was produced at Bayer-Werke after the 1955 takeover of the polyacrylic fiber production of Casella-Werke Mainkur, Frankfurt a.M., since 1956 in Wolfen, where Wolcrylon™ and later Wolpryla™ went into production, and from 1960 in Premnitz. The company Phrix, Hamburg, produced Redon™ and Süddeutsche Chemiefaser AG, Kelheim, Dolan™.

The production of polyacrylonitrile in Japan was different: In 1941, the Japanese Y. Mamiya, S. Matsui and S. Kambara developed a technical process for spinning polyacrylonitrile solutions in potassium or sodium rhodanide in a precipitation bath of ethylene or propyl alcohol. Later they developed other processes, such as dissolving the polyacrylonitrile in 72% sulfuric acid, hydrolysis at 40 °, then spinning in 60 to 80 °C hot water or dilute sulfuric acid, and then drawing the filaments by up to 300% at 200 °C. The fibers thus produced were called Sinsen™ (Tomonari, 1953).

The production of polyacrylonitrile fibers boomed in the 1950s and 1960s. A first fiber crisis in the mid-1970s, caused by the shortage of crude oil, and a second one at the beginning of the 1980s due to the change in consumer buying behavior, which turned to natural fibers and away from man-made fibers, had a serious impact on the polyacrylonitrile market. Polyacrylonitrile fiber production in Europe and North America declined, while production in Asia increased. Currently, stagnation and overproduction characterize the market. The last significant market development in the polyacrylonitrile sector dates back to 1977, when Bayer AG, Dormagen, entered the market with the absorbent fiber "Dunova". Today, PAN is almost exclusively used to produce staple fibers, 10% of which are still produced in Germany. PAN filaments are also an important starting material for the production of carbon fibers.

34.3 Raw Material, Formation of the Polymer

34.3.1 Synthesis of Acrylonitrile

Acrylonitrile (acrylic acid nitrile) H_2 C=CH-CN is the starting material for polyacrylonitrile. It is a colorless, pungent-smelling liquid with a density

Fig. 34.2 F. Klatte (1880–1934), O. Bayer (1902–1982) and H. Rein (1899–1955, Erkner, 2021)

of 0.81 g/cm^3. Its melting point is −82 °C and its boiling point is 77 °C. The liquid is explosive and its flash point is −5 °C. Acrylonitrile is highly carcinogenic and probably mutagenic and teratogenic. It is used industrially for fiber production and as a raw material in the manufacture of plastics. Dihydrodimerisation of acrylonitrile leads to the formation of adipic acid dinitrile.

$$2\,CH_2=CH-CN+2H^+ + 2e^- \rightarrow NC-(CH_2)_4-CN$$

This acid is used for the synthesis of hexanediamine, a starting product for the production of polyamide. Acrylonitrile is also used to produce insecticide. The production of acrylonitrile is about six million t/a, of which 25% is processed into PAN fibers.

The acrylonitrile can be synthesized in different ways. By the direct addition of hydrocyanic acid to acetylene or from ethylene oxide, which is reacted with hydrocyanic acid with elimination of water to form ethylene cyanohydrin and then dehydrated in the presence of metal oxides or alkali metal salts to form acrylonitrile. In the SOHIO process developed by Standard Oil Company Ohio (BP Chemicals since 1987), acrylonitrile is produced by reacting propene with air and water vapor with the participation of ammonia. This process is the most economical, which is why 95% of acrylonitrile is produced in this way today.

34.3.2 Polymerization

The polymerization of acrylonitrile to polyacrylonitrile takes place acc. to different processes. A distinction is made between

- Emulsion polymerization,
- suspension or dispersion polymerization,
- precipitation polymerization,
- solution polymerization.

34.3.2.1 Emulsion Polymerization

Dynel®, a flame-retardant modacrylic fiber based on vinyl chloride as a copolymer, was produced by Union Carbide Corp. using this process and was used by Pierre Cardin for a designer dress in 1968. Today, this fiber is used primarily for fiber-reinforced plastics (Fig. 34.3).

Fig. 34.3 Pierre Cardin designer dress from Dynel® (Cardin, 1968)

34.3.2.2 Suspension or Dispersion Polymerization

These two possibilities of polymerization are not always clearly distinguished. In many cases, only suspension polymerization is mentioned, in which, in contrast to dispersion polymerization, the initiators and dispersants are insoluble in water. A redox system is usually used as the initiator, consisting of, for example, ammonium, sodium or potassium persulphate (catalyst), sulphur dioxide, sodium bisulphite or metabisulphite and iron(II) compounds. The redox system is adjusted to a pH value between 2 and 3.5. The great advantage of this process is that it allows acrylonitrile to be dissolved in water to a limited extent (approx. 7% polymer in solution at 20 °C). In addition to good heat dissipation, polymerization in aqueous solution results in a high degree of purity of the polymer. The presence of oxygen inhibits or prevents polymerization. Therefore,

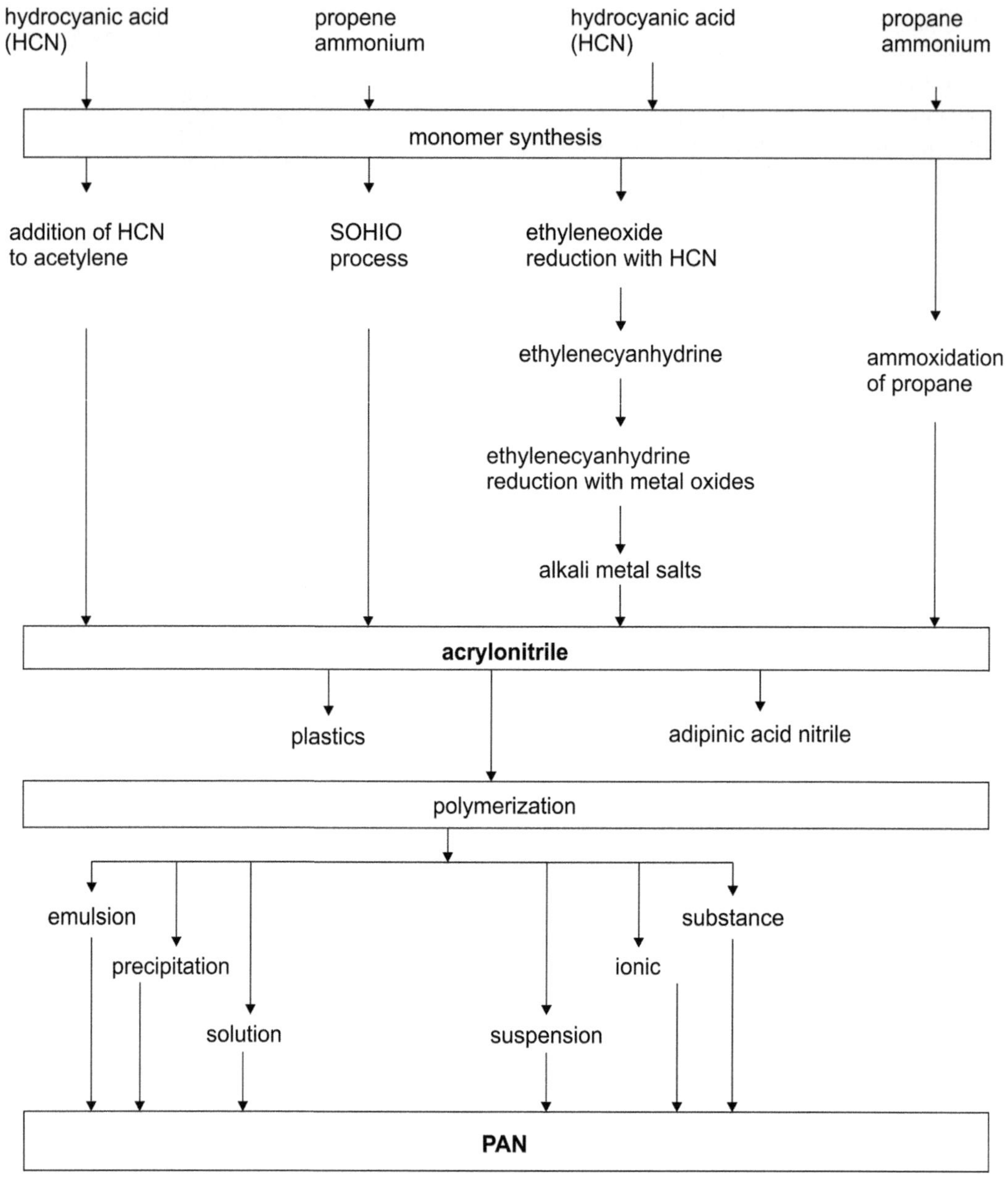

Fig. 34.4 Process routes for monomer synthesis and polymerization of PAN

polymerization is always carried out in an inert gas atmosphere.

34.3.2.3 Solution Polymerization

The advantages lie in the simple polymer production and subsequent filament formation The viscosity increases only slightly during polymer formation. The polymer concentration is usually 20–30%. A disadvantage is the relatively low conversion of the monomer of 50–70%. Unreacted monomer is recovered from the polymer solution for a new polymerization with the thin film process. However, approx. 0.2% of the monomer still remains in the polymer solution, which can lead to problems in the spinning mill.

Figure 34.4 shows the different process routes for the production of PAN polymer.

34.3.3 Copolymerization

This refers to a polymerization in which two or more structurally non-uniform monomers are incorporated into a polymer chain as comonomers. The arrangement of the comonomers is either statistically distributed, alternating or block or graft copolymers are formed. The polymer properties can be specifically adjusted by the type of arrangement and the proportions of the monomers. For the production of PAN copolymers, either methyl acrylate or vinyl acetate is usually used (Fig. 34.5).

34.4 Production Method

Polyacrylonitrile is usually present as a fine-grained powder and is not thermoplastic. Because its decomposition range is approx. 250–300 °C and thus below the melting point, the fibers are produced either by dry or wet spinning.

In special cases, PAN is also produced by the air gap spinning process, and the melt spinning process has also been used on a trial basis. The comonomers acrylic ester as well as methacrylic and unsaturated olefins are used.

34.4.1 Spinning Solution

In the dry spinning process, only dimethylformaid (DMF) is used as a solvent; in the wet spinning process, dimethylacetamide (DMAC), dimethylsulfoxide (DMSO) and aqueous solutions of inorganic salts and acids, e.g. NaSCN (sodium thiocyanate), HNO_3 (nitric acid), $ZnCl_2$ (aqueous zinc chloride solution) and 1,3-dioxolan-2-one are also used. Typical polymer concentrations are 20–30% for DMF and DMAC, and 10–15% for the other solvents.

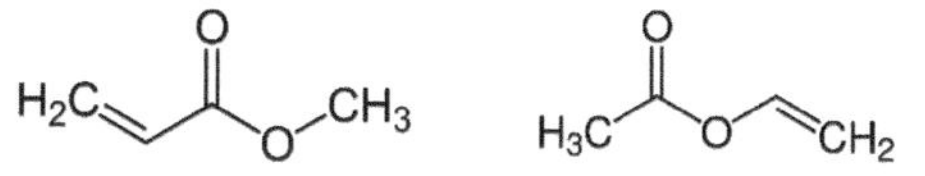

Fig. 34.5 Monomers of methyl acrylate (left) and vinyl acetate (right)

Because some of these solvents are toxic, appropriate caution should be used. Some also attack the equipment, which is therefore often made of special steel.

34.4.2 Spinning Process

The spinning process to be used determines the composition and properties of the spinning solution, i.e. polymer concentration and viscosity. Wet spinning requires solutions with low viscosities, whereas dry spinning requires higher viscosities. Currently, the wet spinning process dominates with a market share of about 80%.

34.4.2.1 Dry Spinning

In dry spinning, the spinning solution is heated to 100–150 °C immediately before spinning. Via a heated gear metering pump, the spinning mass is pressed through the spinneret with 300–2000 holes to produce staple fibers. Typical capillary diameters are 0.1 to 0.5 mm. For the production of polyacrylic filaments, spinnerets with max. 200 holes are used. The spinning material is led through a vertical spinning shaft of 6–10 m height. Temperatures of 200–250 °C prevail in the upper part of the spinning shaft. The spinning material is blown with inert gas at 300–400 °C, alternatively air or saturated steam can be used. Care must be taken to ensure an extremely low content of oxygen and water, so that on the one hand the spun yarns do not discolor and on the other hand the DMF does not decompose. In the further course, the temperature in the spinning shaft drops to 80–110 °C. With a solvent concentration of 30–40 g/m^3 and thread formation in a nitrogenous atmosphere, the risk of explosion is reduced. The geometry of the spinning system and the amounts of heat supplied to the spinning threads via the inert gas and the walls of the spinning shaft have a great influence on the spinning result. At a speed of 200–600 m/min the dry, consolidated spinning threads are wound up. A single spinning position produces approx. 8–15 kg per hour, outputs of 30 kg/h are possible with appropriately designed spinning conditions. The spinning threads of 10–70

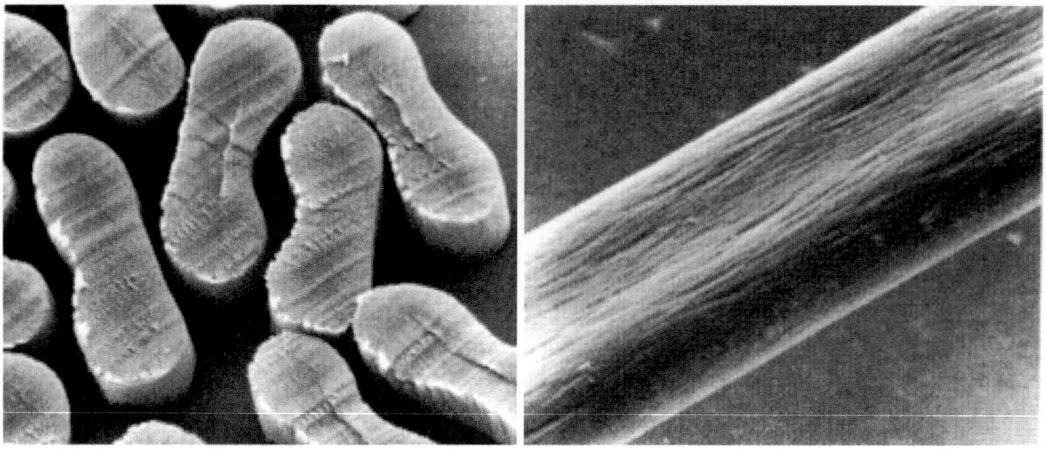

Fig. 34.6 Transversal and longitudinal view of dry-spun PAN fibers

spinning shafts are combined. On heated godets (155–170 °C), the threads are stretched to 8 to 12 times their original length, resulting in a tensile strength of 40–60 cN/tex and an elongation of between 20 and 40% (Gries et al., 2002).

Dry-spun PAN fibers can be recognized by their dumbbell-shaped fiber cross-section (Fig. 34.6).

The dry spinning process is more productive than the wet spinning process and is particularly suitable for the production of higher fineness and bicomponent fibers with stable three-dimensional crimp. The dry spinning process is used for the production of PAN filament yarns as well as PAN spun fibers.

34.4.2.2 Wet Spinning

The wet spinning process is used in particular for coarse titres (> 8 dtex) because coagulation is slower and the solvent therefore has sufficient time to diffuse out of the fiber interior. The spinning solution, preheated to 30–100 °C, is spun out through nozzles into the precipitation bath. The concentration of the polymer in the precipitation bath varies between 10–15% when using rhodanides and between 70–80% when using DMF. The greater the proportion of solvent, the more uniform the structure and elasticity of the fibers obtained. In aqueous precipitation baths as well as when using rhodanide solutions, the temperature is 10–12 °C, and only 0–5 °C when using nitric acid. Therefore, cooling is absolutely necessary to avoid hydrolysis of the polymer. Coagulation results in a highly swollen, barely oriented gel filament. The cross-sectional shape and porosity depend on the coagulation conditions. The same applies to the core-sheath structure of the spun yarns in the production of bicomponent fibers. The cross-section of wet-spun PAN fibers looks bean-shaped to round. Drawing off takes place at low draw-off speeds of mostly 5–20 m/min.

Greater bulkiness, more intense lustre, pleasant textile handle and better recovery distinguish the dry-spun PAN fibers from the wet-spun ones. The latter, in turn, are characterized by a softer handle, lower pilling tendency and better dyeability.

The Fig. 34.7 shows the main process steps in dry and wet spinning.

34.4.2.3 Gel Spinning

At the end of the 1970s, the gel or air-gap spinning process was developed to produce high-modulus fibers with high strength from flexible molecular chains. Gel spinning is suitable for polymers with ultra-high molecular weights, e.g. > 1,000,000 g/mol, and spinning solutions of low concentration. The low concentration of the spinning solution allows high stretching, which increases the Young's modulus. Gel-spun fibers have a high degree of crystallization and are

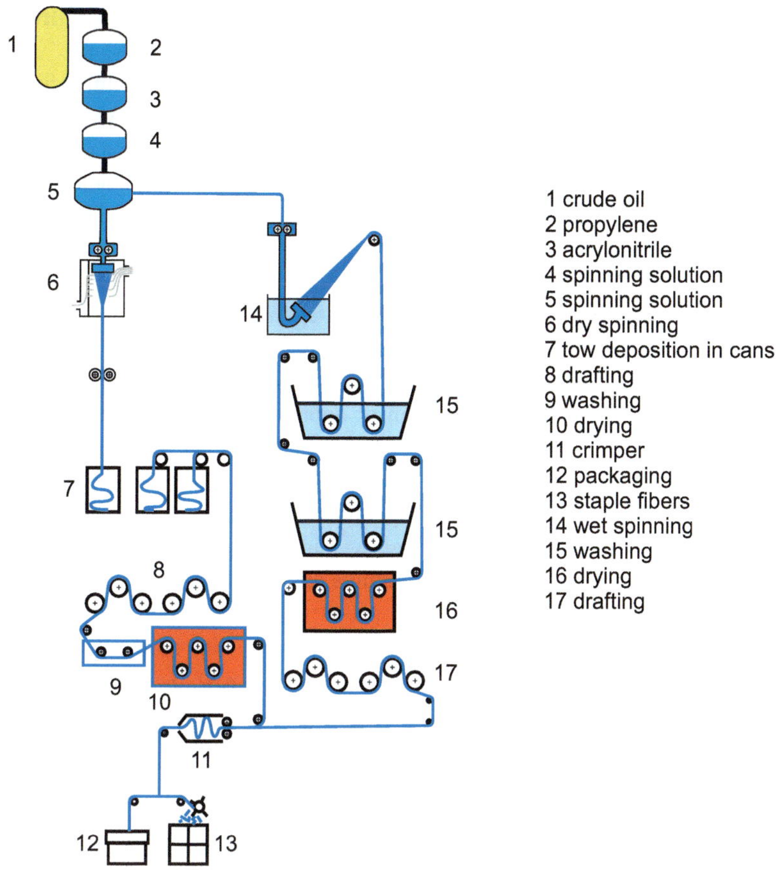

Fig. 34.7 Process lines in dry and wet spinning acc. to (IVC)

exceptionally highly oriented. This spinning process is particularly suitable for the production of high-strength and high-modulus PAN fibers for further processing into carbon fibers, because the fiber properties of the PAN produced in this way directly influence those of the precursor.

34.4.2.4 Melt Spinning

PAN microcrystals melt at approx. 320 °C, but the decomposition process already starts at temperatures around 250 °C. In order to obtain PAN fibers by melt spinning, the process temperature would have to be below 250 °C. To bring the melting point of PAN there, the intermolecular attraction of the CN dipoles must be weakened. This can be achieved chemically by copolymerizing acrylonitrile with other comonomers, because this increases the distance between the CN groups. Suitable comonomers are acrylic esters or methacrylics. Surrounding the CN groups with other dipoles is also possible, e.g. by H_2O molecules forming hydrogen bonds to nitrogen, or glycol and glycerol. In the early 1950s there were attempts to lower the melting point of the polymer by adding acrylic esters or methacrylic esters as comonomers. Polyacrylonitrile hydrates, e.g., a system of polyacrylonitrile, water, and carboxylmethyl cellulose (CMC), were described as early as 1948. Although these developments were further advanced in the 1970s and 1990s, they have not yet resulted in a marketable process.

34.5 After-treatment

Post-treatment in the production of staple fibers immediately follows the spinning process in the case of wet spinning; in the case of dry spinning, the spun yarns are bundled into slivers of 40–100 ktex and further processed on a post-treatment line. The post-treatment steps are hiding, washing, lubricating, drying, crimping/crimping, cutting and packaging.

34.5.1 Drawing

Polyacrylonitrile fibers can be stretched by a factor of 2–10, typical values are 6–8. Wet spun fibers can be stretched higher than dry spun fibers.

Single-stage stretching is carried out in hot water (90–100 °C) with the addition of approx. 5–8% solvent (e.g. DMF). If two-stage stretching is used, the first step is dry stretching at 50–100 ° C, followed by stretching at temperatures of 160–180 °C. Higher temperatures allow higher degrees of stretching. Higher temperatures allow higher degrees of stretching.

The shrinkage can be adjusted by a subsequent thermal treatment. Thus, it can either be wound onto coils under tension at 150 °C, which also leads to a higher fiber strength, or without tension. In this case, thermorelaxation leads to an increase in strength and elongation. Thermally treated fibers only shrink from 125 °C, untreated fibers already from 75 °C.

34.5.2 Washing

If washing takes place before drawing, the washing effect is better because of the lower process speed. However, it can be advantageous for the uniformity of the stretching if there are still residues of solvent in the filament. Washing is usually carried out in water at 50–80 °C, often under tension. Because the solvent diffuses only very slowly from the filaments into the washing medium, the countercurrent principle is often used, which significantly improves the washing effect. The tow is passed over one or more perforated drums and the solvent is effectively washed out by the pressure difference. Finally, the tow is passed through clear water and squeezed off.

34.5.3 Finishing

The tow is then finished by rolling, dipping or spraying. Typical finishes consist of ethylated fatty acids or salts of ethylated sulfonic acid and phosphoric acid esters of long-chain aliphatic alcohols. The finish content is usually 0.2–0.5%.

34.5.4 Drying

A drying process reduces the residual moisture content in the fibers to 1–2%. Drying takes place either by radiation, contact or convection. Drying without tension also triggers shrinkage, which can be assisted by the action of steam. If drying under tension is required, a subsequent relaxation step is necessary.

Screen drum dryers are frequently used, often with more than 20 screen drums arranged in a row. Typical drying temperatures are 120–170 ° C, the process speed is approx. 100–150 m/min. Higher temperatures lead to better heat setting, which, however, makes subsequent crimping more difficult.

34.5.5 Ripple

The compression chamber process is usually used for crimping. The tow is heated to 60–70 °C beforehand and then textured at approx. 150 m/ min.

34.5.6 Converting

PAN fibers are either torn or—more often—cut. Depending on the application, the cut length varies from 0.1 mm (flock yarn) to 180 mm (carpet yarns).

34.5.7 Dyeing

PAN fibers can be dyed in various ways. Dyes can already be added to the spinning solution,

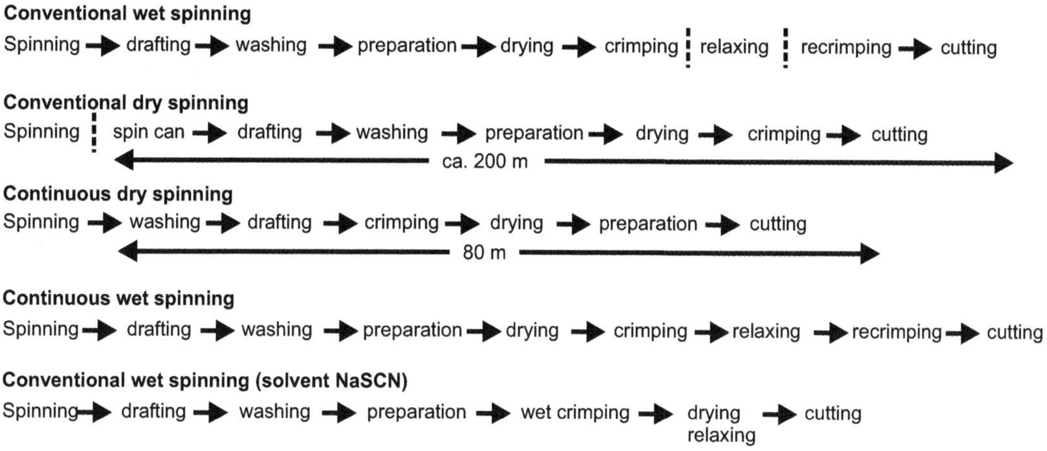

Fig. 34.8 Plant concepts for wet and dry spinning processes for PAN fibers acc. to (Gries et al., 2002)

which leads to a good dyeing of the fibers. Alternatively, dyeing can also be carried out in the precipitation bath, because the coagulated fibers have a high porosity and therefore absorb dyes well (gel dyeing).

Mostly cationic dyes are used, which leads to a high color brilliance. Disperse dyes are also possible, but the dyeing results are then poorer. Pigment dyes result in high color fastness. TiO_2 can be added for matting.

Spinning and gel dyeing are only useful for large batch sizes, otherwise dyeing is first carried out in the yarn or textile. PAN fibers are then dyed, e.g. in the cable, on the bobbin or in the warp beam, acc. to the exhaust or the continuous process with modified basic dyestuffs at pH values of 4–5 at approx. 100 °C.

34.5.8 Plant concepts

Various plant concepts exist for the production of PAN fibers. These include continuous wet and dry spinning plants as well as discontinuous dry spinning processes (Fig. 34.8).

In the discontinuous dry spinning process, the spinning process and the after-treatment are separated from each other. This is advantageous because the high spinning speeds of approx. 600 m/min can be utilized without being restricted by the after-treatment speed of only approx. 30 m/min. The length of such plants is about 200 m.

Bayer AG has developed a dry spinning process consisting of the process steps of initial spinning, washing, drawing, crimping/crimping, heat-setting and drying, preparation and cutting. Such plants are considerably shorter at 80 m in length.

Continuous processes are generally more favorable than discontinuous ones, even for small batches or special types. However, the aggregates must be well coordinated and, for example, insensitive to solvent residues in the spun material.

The discontinuous dry spinning process offers on the one hand the advantage of achieving higher capacities in after-treatment and on the other hand the possibility of controlling the quality of the spun product after the spinning process before after-treatment. However, disadvantages are higher personnel costs, overall lower throughputs and less consistent qualities because the production flow is interrupted (Gries et al., 2002), Fig. 34.8.

The main differences between wet-spun and dry-spun PAN fibers are their cross-sectional shape and their surface structure. Dry-spun fibers

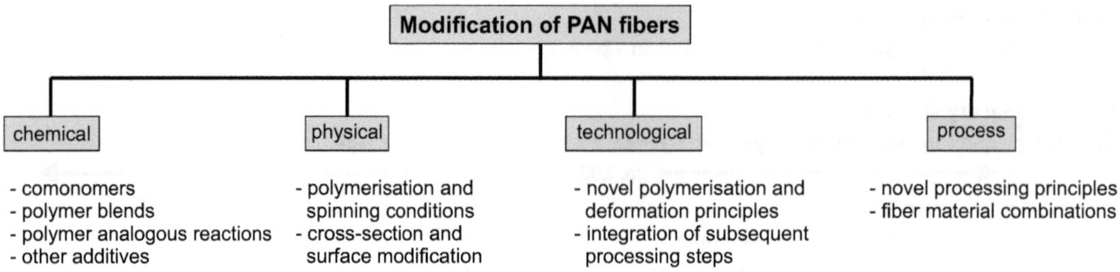

Fig. 34.9 Possibilities of modifying PAN fibers acc. to (von Falkai, 1981)

are smooth or only slightly structured, whereas wet-spun fibers are rougher, with a more pronounced structure. This has a direct influence on the amount of lubricant required, which is only 0.2–0.4% for dry-spun fibers, but double that for wet-spun fibers. Dry-spun PAN fibers have a higher luster. Due to their round cross-section, wet-spun fibers are more resistant to bending and their handle is therefore more "pithy".

34.6 Recovery of the Solvent

The recovery of the solvent is a technically complex, multi-stage process, but it is imperative from an economic and environmental point of view. To produce 1 kg of PAN fibers, approx. 2 kg of DMF are required in the dry spinning process, of which up to 99% can be recovered. In wet spinning, the polymer content in the spinning bath is 10–15%, and the required quantities of solvents are correspondingly large.

The solvent can be distilled from the aqueous solution by evaporating the water in cascades. When sodium thiocyanate (rhodanide) is used, recovery is more complicated because corrosion of the spinning apparatus causes iron atoms to enter the precipitation bath, which must be removed via ion exchangers.

34.7 Special Fiber Types

The possibilities of modifying PAN fibers to produce special types with particular properties are manifold (Fig. 34.9).

34.7.1 Modacrylic Fibers

The copolymerization of acrylonitrile with vinyl chloride, vinylidene chloride and vinyl bromide produces the so-called modacrylic fibers. They are the economically most important modification of PAN. Acc. to DIN 60.001 they consist of at least 50% and less than 85% by mass of polyacrylonitrile. Halogen contents of 25–35% are common.

The LOI value of modacrylic fibers is 26–31, which is significantly higher than that of PAN, and unlike polyacrylonitrile fibers, modacrylic fibers extinguish when the flame is removed and they do not drip. However, toxic chlorinated hydrocarbons are released in the event of fire. Modacrylic fibers are usually spun wet.

The technological characteristics of modacrylic fibers vary due to the different chemical composition. The values of strength and elongation are in the range of those of standard polyacrylonitrile fibers (Tables 34.1 and 34.2).

34.7.2 Bicomponent Fibers

A PAN bicomponent fiber was first introduced to the market in 1958 by DuPont under the name Type 21™. Of importance is the side-by-side (S/S) type, in which both components are arranged eccentrically in the cross-section. Due to the different content of comonomers in the two components, different shrinkage is induced when exposed to heat and the fiber curls similar to wool. If the components used differ in their dye uptake, they can be dyed with different dyes and thus

Table 34.1 Typical properties of PAN fibers acc. to manufacturer's specifications and (von Falkai, 1981)

	Homopolymer		Copolymer		
	Spun fiber	Spun fiber (wool type)	Spun fiber (cotton type)	Spun fiber (high modulus)	Modacryl
Density [g/cm^3]	1.14–1.19	1.17	1.17	1.18	1.3–1.42
Maximum strength [cN/tex] dry	20–50	17–45	28–38	35–55	10–25
Wet tensile strength ratio [%]	80–100	75–95	80–85		80–100
Maximum tensile strength [N/mm^2] trocken	230–590	200–530	330–440	400–650	130–400
Modulus [kN/mm^2]	10–12	3–7		8–12	
High traction work [Nm/cm^3]			47		
Maximum elongation [%] dry	15–70	20–60	28–33	15–40	25–50
Maximum elongation [%] wet	25–120	22–65	30–38		100–110
Maximum knot tensile strength ratio [%]	70	75–80			80
Schlingenhöchstzugkraftverhältnis [%]	30–70	70–80			50–70 (all)
Moisture absorption [%] at 65% rel. Humidity, 20 °C	1–2.5				0.4–4
Water retention [%]	4–30	5–12			10–20
Crystallinity [%]			40		
Hot water shrinkage [%] at 95 °C, non-set	14–16	16–40			16
Hot water shrinkage [%] at 95 °C, set	1	0.5–5			0.2–5
Hot air shrinkage [%] 3 min., at 130 °C					5–30
Mean refractive index Spec. Birefringence			−0.005 1.52		

mélange effects can be achieved with a single dye bath.

34.7.3 Absorbant Fibers

Fibers in core-sheath structure, whose core is more porous than the solid sheath, can absorb more moisture. Such fibers can absorb large amounts of moisture without feeling wet. Because their diameter does not change during moisture absorption, the clothing-physiological properties with regard to moisture transport through the textile are retained (Fig. 34.10).

If, on the other hand, the fibers are more swellable than standard types, they can also absorb more water. However, the associated increase in diameter reduces the water vapor permeability of textiles made from them, which impairs wear comfort.

Both types are mainly used for sports and leisure wear and for hygiene articles.

34.7.4 Asbestos Replacement

PAN fibers can also be used as asbestos substitutes, e.g. for filters, friction linings and fiber-reinforced concrete components. In order to achieve the high strength and high modulus of elasticity required for this purpose, the molecular weight is increased to 100,000–500,000 g/mol and the fibers are stretched more strongly. In this way, strength values of 50–80 cN/tex and

Table 34.2 Thermal properties of PAN and modacrylic fibers acc. to (von Falkai, 1981)

		Spun fiber (homopolymer)	Modacryl
Temperatur[°C]	Melting	320	130–170
	Disintegration	250–320	170
	Ironing	120–180	
	Washing	40–50	
	Discoloration	140–235	120
	Softening	200–330	130–200
	Glass	30–104	85–95
	Self ignition	510–560	650–700
	Charring		250–350
Spec. heat [J*g^{-1}*K^{-1}]		1.2–1.5	
Heat conductivity [W*m*K]		0.2	
LOI		17–18	26–31
Burning behavior	–	Softening and quickly flammable	
	–	Burns with yellow, sooty flame	
	–	Acrid, sweet odour	
	–	Dark, brittle residue	

E-modulus values of 800 to 1600 cN/tex can be achieved. The elongation at break is then 9–17%, the typical fiber fineness is 1.5–8 dtex.

34.7.5 Other Special Types

For antibacterial PAN fibers, titanium silicate particles (< 1 μm) are added to the spinning solution. The antibacterial effect is permanent. Underwear, bathroom and toilet articles (e.g. towels) and sportswear are made from these fibers.

Outlast™ was developed for thermoregulation. Paraffin-based microcapsules, so-called PCMs (phase change materials), are embedded in the fiber. These microcapsules melt or solidify within a certain temperature range (approx. 30–38 °C) and can thus absorb, store and release heat. This effect is reversible and corresponding fibers are used, for example, in outdoor and sports clothing. Outlast™ is also used in other fiber materials, e.g. viscose and polyester.

34.7.6 Low Pilling Types

The generally high tendency of PAN fibers to fibrillate leads to a high pilling tendency of the

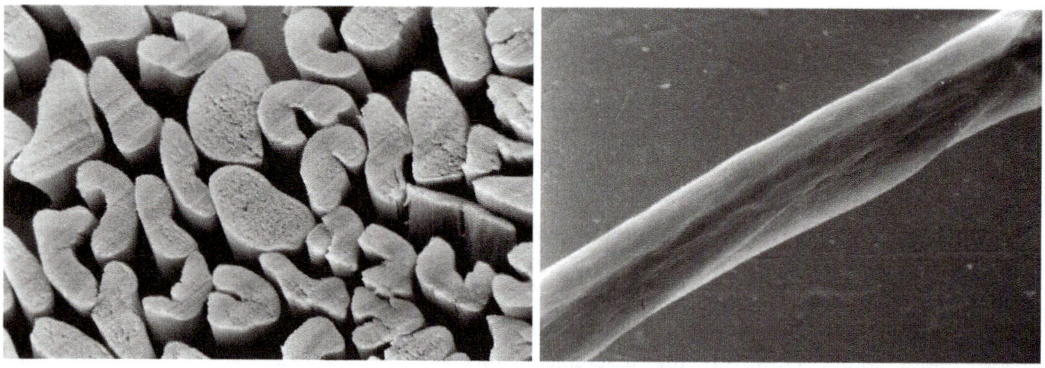

Fig. 34.10 Longitudinal and transverse view of absorbant PAN fibers

textiles made from them. There are several processes for producing low-pilling fibers:

- A round fiber cross-section which allows a more even arrangement of the fibers in the yarn structure, making it more difficult to pull out individual fibers.
- The setting of a higher drafting rate in conjunction with a low coagulation rate during the spinning process.

34.8 Properties

34.8.1 Fiber Types

Polyacrylonitrile is almost exclusively used to produce staple fibers with a fineness of 0.6–20 dtex. As with all staple fibers, a distinction is made between cotton types (28–35 mm), wool types (75–120 mm) and carpet types (150–200 mm). They are glossy, dull, semi-dull, raw white or spun-dyed. Filaments are used almost exclusively as precursor materials for the production of carbon fibers.

34.8.2 Physical properties

Polyacrylonitrile fibers are semi-crystalline (40%), have a wool-like handle and a high bulk and elasticity due to the often dumbbell-shaped fiber cross-section. The high elasticity leads to a low tendency to crease. In combination with their low density, textiles with good recovery and heat retention properties can be produced. The fibers prevent felting in appropriate admixture with articles made of wool. Since the fibers are absolutely rot-resistant and also have a high UV light resistance, they are often used for house and home textiles as well as for the outdoor sector (Gries et al., 2002), Table 34.1.

The Fig. 34.11 shows the typical stress-strain curve of PAN staple fibers for apparel textiles. Higher strength values are also possible for special types.

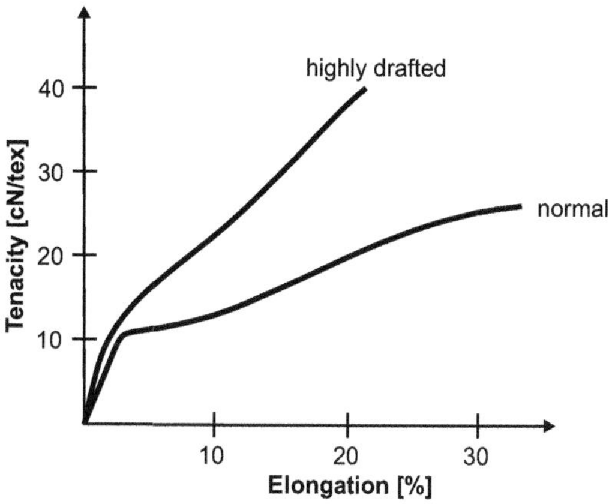

Fig. 34.11 Typical stress-strain diagram of PAN fibers

34.8.3 Thermal Properties

The typical thermal properties of PAN and modacrylic fibers are summarised in compiled in Table 34.2.

34.8.4 Microscopic Image

The cross-section of the polyacrylonitrile fiber depends on various parameters. These include,

- the composition of the polymer,
- the spinning process and
- the spinning conditions.

In wet-spun fibers, the ratio of precipitant to solvent influences the cross-sectional shape. If the spinning mass contains more precipitant than solvent, the fiber swells and a round cross-section is produced. If there is more solvent than precipitant in the spinning mass, the fiber shrinks. If the sheath zone is easily deformable, a round cross-section is formed, otherwise the initially formed outer skin collapses and a dumbbell-shaped cross-section is formed.

In dry-spun fibers, the ratio between the diffusion and evaporation rates of the solvent determines the cross-sectional shape. The outer layer of the forming filaments solidifies first, therefore a core-sheath structure is formed. As

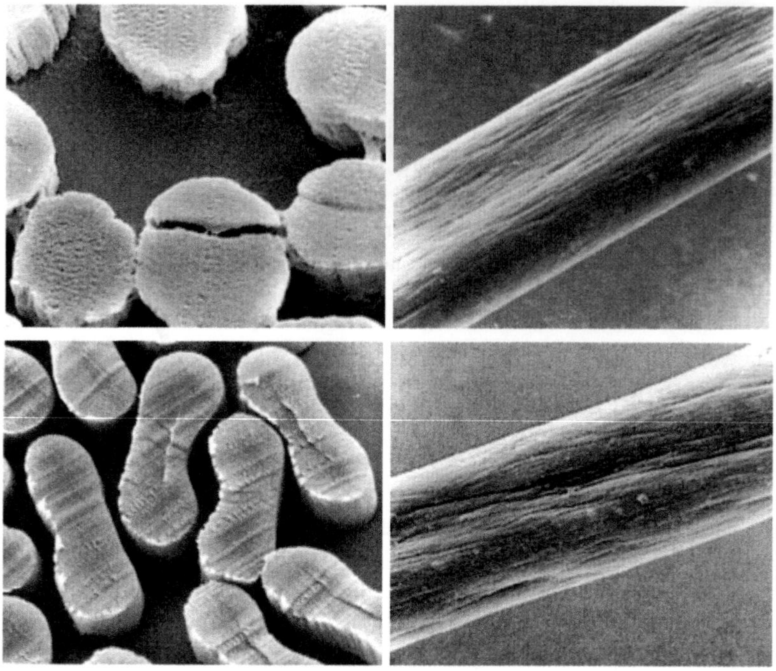

Fig. 34.12 Transversal and longitudinal views of PAN (top) and modacrylic (bottom) fibers

the solvent gradually evaporates, the mass of the core is reduced and the fiber collapses. This results in a dumbbell or kidney-shaped cross-section.

The cross-sectional shape affects the gloss, sorption capacity and mechanical properties of the fiber. For example, the bending stiffness of fibers with a round cross-section is greater than that of fibers with a dumbbell-shaped cross-section of the same area.

The surface structure in the longitudinal view varies greatly depending on the manufacturing process. Wet-spun fibers, for example, have a rougher and clearly structured surface. Dry-spun fibers, on the other hand, have a soft, slightly structured surface (Fig. 34.12).

34.9 Applications and Fields of Use

The tendency to pilling and the strong electrostatic charge of PAN fibers can be reduced by blending them with other materials, e.g. cotton, wool, polyester and viscose. At the same time, e.g. in PAN/wool blended textiles, the abrasion resistance is increased and the tendency to felting is reduced.

PAN fibers absorb very little moisture and therefore dry quickly. Therefore, in pure form or in blends, they are well suited for sportswear and socks. PAN is also often used for artificial fur, shoe linings and toy plush.

Because of their good dyeability and brilliant colors, PAN fibers are also processed into bedding and terry fabrics, pillowcases, etc. Due to their wear resistance, they are also often used in blends with other fibers as pile material for carpets and for upholstery fabrics.

Upholstery fabrics and blankets made of PAN are also frequently used in aircraft due to the material's flame retardancy.

For awning fabrics, sun sails, deck chair coverings, interior and exterior shades for conservatories as well as textile covers for garden and leisure furniture and other outdoor applications, the good light and weather resistance is an advantage. Because of its good resistance to rotting, PAN is also used for fishing nets and ship ropes. High molecular weight PAN is currently used as a precursor for around 90% of all carbon fibers.

Fig. 34.13 Typical products from PAN (Schmitz-Werke, 2015) and MAC (Greg2600)

34.9.1 Modacrylic Staple Fibers (MAC)

MAC fibers are used, among other things, for protective clothing, children's and nightwear and as imitation fur, pure and in blends. Wig hair can also be made from MAC.

Modacrylic fibers are particularly suitable for home textiles in public buildings due to their flame retardant properties (Fig. 34.13).

34.10 Market and Trade Names

Polyacrylonitrile is the fourth most important synthetic fiber material in terms of volume after PET, PA and PP. PAN is mainly used to produce staple fibers, the filaments are mostly processed into carbon fibers. With market prices for the monomer acrylonitrile of about 1.5 €/kg and prices for PAN staple fibers around 2.1 €/kg, the margins are acceptable. However, the price is much higher than for PET fibers and therefore PAN filaments are not price competitive with PET in the mass market. This is different in the staple fiber segment because PAN fibers have specific properties that justify a higher price for staple fiber applications, especially apparel.

In 1970, the volume of PAN produced was 1 mt, exactly the same as the volume of PET fibers produced at that time. The market share of all staple fibers (including natural fibers) was 6%, and in relation to man-made staple fibers it was 25%. The production volume increased to 2.6mt in 2005, but has been falling steadily since then. Today, PAN fibers account for 3% of all staple fibers and 6% of chemical staple fibers (Fig. 34.14).

The main producers are spread around the world, with still large production facilities in Europe, surpassed only by China. Since 2000, the market has been concentrated in a few countries and a large number of producers have withdrawn from PAN, as shown by the data for 'other' (Fig. 34.15).

Important manufacturers and trade names for PAN fibers are listed in the Table 34.3. The world's largest producer, with a market share of approx. 13%, is Dow Aksa AS; Dralon GmbH is also one of the largest producers of PAN fibers.

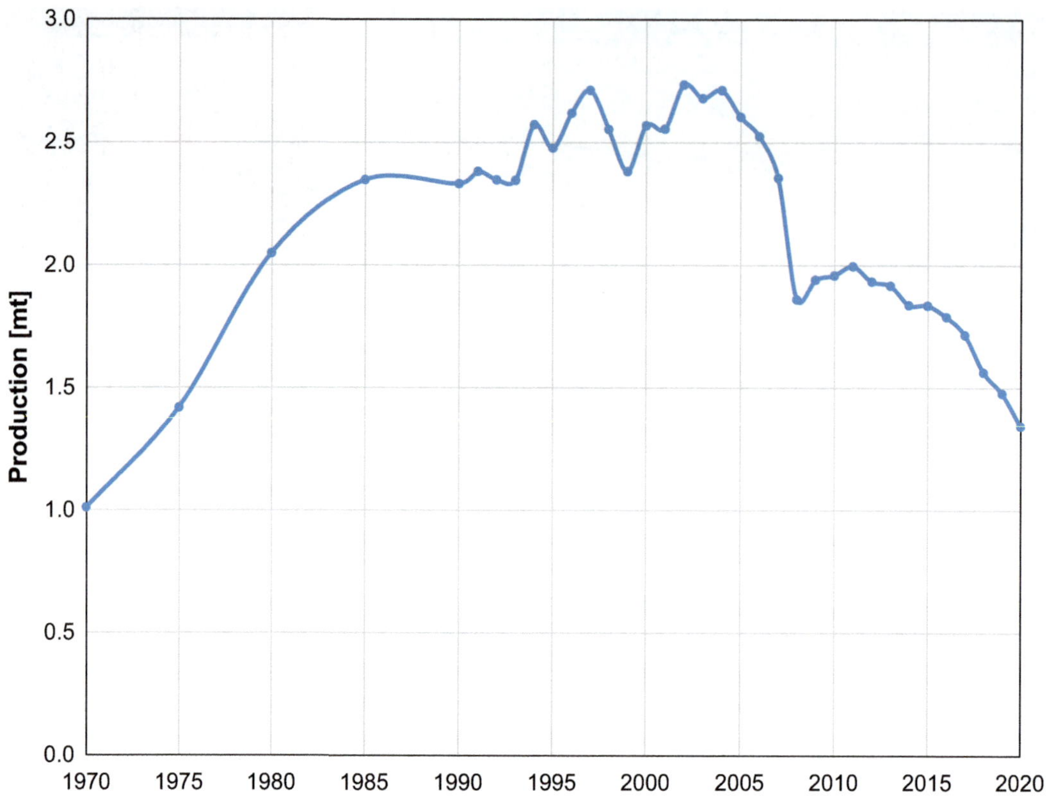

Fig. 34.14 Development of PAN fiber production (data: The Fiber Year, 2021)

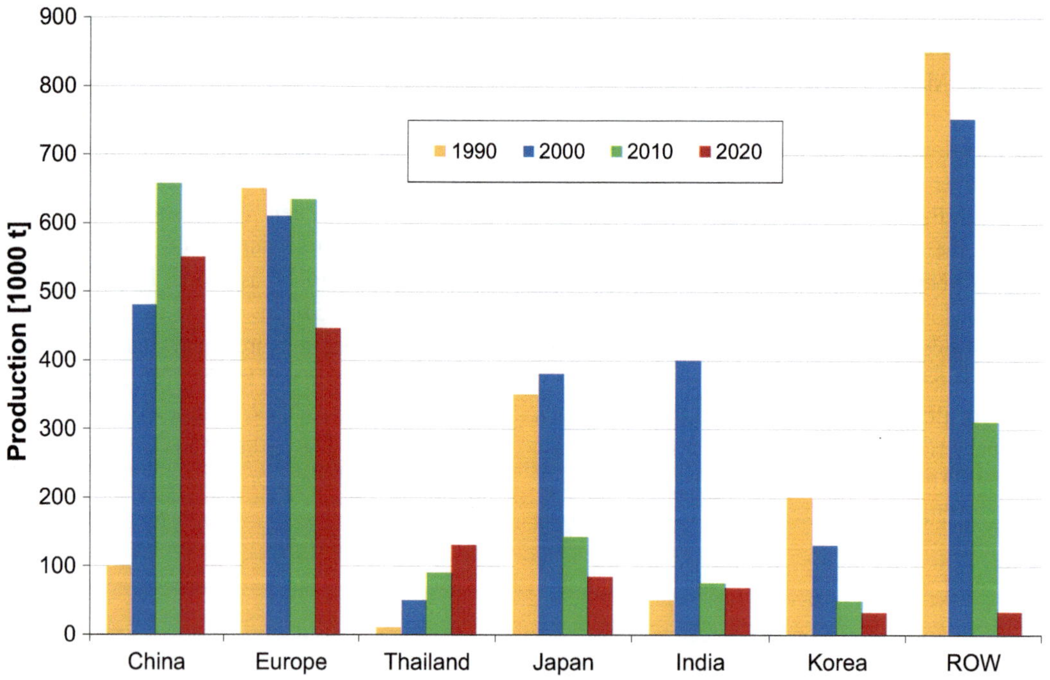

Fig. 34.15 Most important producers of PAN fibers (data: The Fiber Year, 2021)

Table 34.3 Selected manufacturers and trade names of PAN and modacrylic fibers

Producer	Trade mark	Website	Country
Companhia Sudamericana de Fibras S.A.	Drytex™	http://sdef.com/	Peru
	Drymix™		
	pimacryl™		
Dewan Textile Mills Limited		https://yousufdewan.com	Pakistan
Dolan GmbH (Lenzing AG)	Dolan™	http://www.dolan-gmbh.de	Germany
	Dolanit™		
Dow Aksa AS	Acryluna™	http://www.aksa.com	Turkey
	Acrysole™		
	Acrylusion™		
	Acryterna™		
Dralon GmbH	Dralon™	http://www.dralon.com	Germany
Drytex technical yarns	Drytex™	http://drytex.es	Spain
Formosa Plastics Corp.	Tairylan™	http://www.fpc.com.tw	Taiwan
Indian Acrylics, Ltd.		https://www.indianacrylics.com	India
Jiangsu Zhongxin Resources Group Co., Ltd.		http://www.china-acrylicyarn.com	China
Jilin Qifeng Chemical Fiber		http://www.jlhxjt.com	China
Kaltex	Kaltex™	http://kaltex.com	Mexico
Pasupati Acrylon		https://pasupatiacrylon.com	India
Polyacryl Iran Corp.		http://www.polyacryl.ir	Iran
Polymir	Nitron D™	https://polymir.by	Belarus
Sterling Fibers, Inc.	Conductrol™	http://www.sterlingfibers.com	USA
Thai Acrylic Fiber Co., Ltd.	Amicor™	http://www.amicorpure.co.uk	Thailand
	Birlacril™		
The Taekwang & Daehan Company	Acepora™	http://www.acepora-fiber.com	South Korea
Tong Hwa Synthetic Fiber Co.		www.tonghwa.com.tw	Taiwan
Toray Ind., Inc.	Toraylon™		Japan
TOYOBO CO., LTD	Exlan™	https://www.toyobo-global.com	Japan
Vardhman Acrylics, Ltd.		https://www.vardhmanacrylics.com	India
Modacrylic fibers			
Aksa AS	Armora™	http://www.aksa.com	Turkey
Kaneka Corp.	Kanecaron™ Protex™	https://www.kanecaron.com https://www.modacrylic.com	Japan

References

Cardin, P. (1968). https://commons.wikimedia.org/wiki/File:1968_Pierre_Cardin_dress,_pink_heat_moulded_Dynel.jpg, CC BY-SA 2.0.

Erkner. (2021). http://www.chemieforum-erkner.de/chemie-geschichte/personen/rein_h.htm

Fikentscher, H., & Heuck, C. (1937). *Verfahren zur Herstellung von Polymerisationprodukten*. Patent DE654989, Deutsches Reichspatentamt, Berlin.

Greg2600. https://commons.wikimedia.org/wiki/File:Miley_Cyrus_(23773086672).jpg, CC BY-SA 2.0.

Gries, T., Rixe, C., Steffens, M., & Cremer, C. (2002). *Polyacrylnitrilfasern. Faserstofftabelle nach P.-A. Koch des Instituts für Textiltechnik der RWTH Aachen*. Deutscher Fachverlag.

IVC, Industrievereinigung Chemiefaser e.V., Frankfurt a.M. https://www.ivc-ev.de

Klatte, F (1913). *Verfahren zur Herstellung einer auf Hornersatz, Filme, Kunstfäden, Lacke und dergleichen verarbeitbaren Masse*, DRP 281 877, Deutsches Reichspatentamt, Berlin.

Rein, H. (1934). *Verfahren zur Herstellung von Polyacrylsäurenitril-Lösungen*, DRP 631 527, Deutsches Reichspatentamt, Berlin.

Rein, H. (1936). *Verfahren zur Lösung von polymerem Acrylsäurenitril*, Patent DE631756, Deutsches Reichspatentamt, Berlin.

Schmitz-Werke. (2015). https://commons.wikimedia.org/wiki/File:Markilux_planet.jpg, CC BY-SA 4.0.

The Fiber Year. (2021). *The Fiber Year GmbH, Roggwil.* https://www.thefiberyear.com

Tomonari, T. (1953). New fiber developments in Japan. In McFarlane, S. B. (Hrsg.), *Technology of synthetic fibers* (pp. 234–250). Fairchild, New York.

von Falkai, B. (1981). *Synthesefasern*. Verlag Chemie.

35 Polylactic Acid

35.1 Definitions

Polylactic acid (PLA) belongs to the group of aliphatic polyesters and, despite its name, is not an acid (Figs. 35.1 and 35.2). PLA is produced from renewable raw materials (sugar, starch) for the manufacture of fibers and is therefore classified as a biopolymer. In this group of biopolymers, PLA is one of the materials whose raw materials can be produced by microorganisms (fermentation) and whose macromolecules are biodegradable. Depending on the molecular structure of the polymer, PLA decomposes slowly to quickly in water and decomposes into carbon dioxide and biomass.

Another advantage is that the production of PLA is climate neutral or even consumes CO_2 when considering the production chain from the raw material (corn) to the final product (Bogaert & Coszach, 2000).

Depending on the manufacturing process, the polymers are referred to differently. If the polymer is produced directly from lactic acid, it is called polylactic acid. However, if the polymer is produced indirectly via dilactide (lactide), it is referred to as polylactide.

35.2 Inventors, History of Development

The polymer PLA is based on lactic acid ($C_3H_6O_3$), which is also known as 2-hydroxypropanoic acid. In 1780, lactic acid was discovered by C.W. Scheele and initially produced from rice water and beetroot. In the nineteenth century, recipes for dilactide production were published by Gay-Lussac and Pelouze (1833) and Bischoff and Walden (1889), respectively. Further recipes for the production of dilactide were patented by the "Chemische Werke" in France (1913) and Grüter and Pohl in the USA (1914).

T.-J. Pelouze observed in 1845 that a polymer is formed when lactic acid is heated and water is removed at the same time. Carothers succeeded in 1932 in producing a polymer from lactic acid (Carothers et al., 1932). By heating the lactic acid under vacuum, he was able to produce low molecular weight polylactide. DuPont, USA and Ethicon Inc., USA produced PLA for medical applications such as sutures and implants from 1954. Shimadzu Corp. and Mitsui Toatsu Chemicals Inc., both Japan, produced small quantities of consumer articles from PLA (Linnemann et al., 2004).

In 1992, PLA is developed as a binder for cellulosic nonwovens because it can be hydrolytically decomposed. In the same year, Shimadzu, in collaboration with Kanebo Gohsen Ltd., Japan, succeeds in producing PLA fibers on a laboratory scale by melt spinning (Matsui & Kondo, 1996). In 1994, PLA fibers are produced at the University of Chemical Technology, Czech Republic, using the dry-spinning method at 250 m/min and hot-drawn (Horacek & Kalisek, 1994). In

D. Veit, *Fibers*, https://doi.org/10.1007/978-3-031-15309-9_35

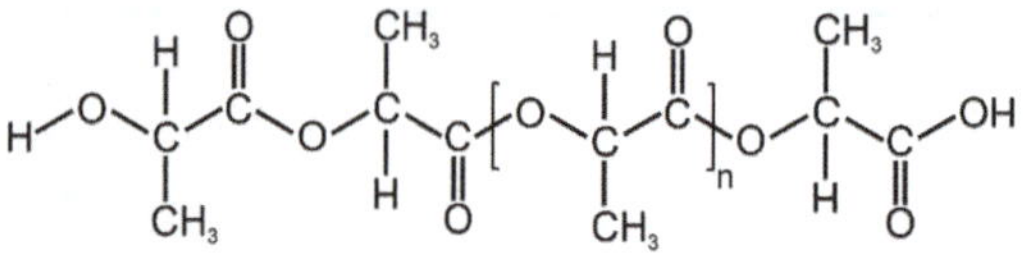

Fig. 35.1 Structural formula of PLA and repeat unit

the same year, Kanebo Gohsen starts industrial production of PLA fibers under the brand name Lactron™ and also spunbond nonwovens with an annual capacity of 3000 t/a. The nonwovens are used for agricultural textiles. In 1997, Fiberweb France S.A. launched spunbonded and meltblown nonwovens under the Deposa™ brand. From 1998, Cargill Inc., USA, and Purac Biochem B. V., Netherlands, jointly produce PLA with a capacity of 34,000 t/a. Kanebo Gohsen produces Lactron® for apparel textiles from 1998. From 2001, fine fibers are produced by separable bicomponent fibers, using PET/PLA bicomponents among others. Dow Cargill, USA, begins production of PLA (NatureWorks™) in 2002 with a 140,000 t/a plant. In early 2003, Cargill Dow introduced the PLA fiber Ingeo™, which is spun from the NatureWorks™ polymer. Meanwhile, NatureWorks B.V. is a joint venture between Cargill Inc. and PTT Public Co., Ltd.

35.3 Raw Materials and Synthesis

Lactic acid is a product of the fermentation of sugar or starch. Starch is obtained from corn, wheat, rye, potatoes or rice, sugar from sugar cane, sugar beet or molasses. The Fig. 35.3 shows the basic principle.

In principle, lactic acid could also be produced synthetically. However, this is significantly more expensive than fermentation by lactobacteria (Jamshidian et al., 2010). Furthermore, the stereochemical purity of lactic acid is not sufficient for further processing into PLA fibers.

The production of polylactide from lactic acid is carried out by two different methods. Either PLA is produced directly by polycondensation from lactic acid or indirectly via the dimer of lactic acid (dilactide) by ring-opening polymerization (Fig. 35.4).

35.3.1 Polycondensation

Polycondensation is carried out under high vacuum and temperature. A solvent is used to remove the water produced during the condensation reaction. The PLA produced in this way has a molecular weight of between 10,000 and 20,000, which corresponds to a DP of 140–280. The removal of water during the process shifts the reaction equilibrium towards the reactant and the average chain length increases. To produce long macromolecule, polycondensation can also be carried out with the addition of aromatic solvents or via azeotropic distillation. Methylene diphenyl diisocyanate can also be used to produce chain elongation and thus improve polymer strength. Polycondensation proceeds more slowly than ring-opening polymerization. However, due to

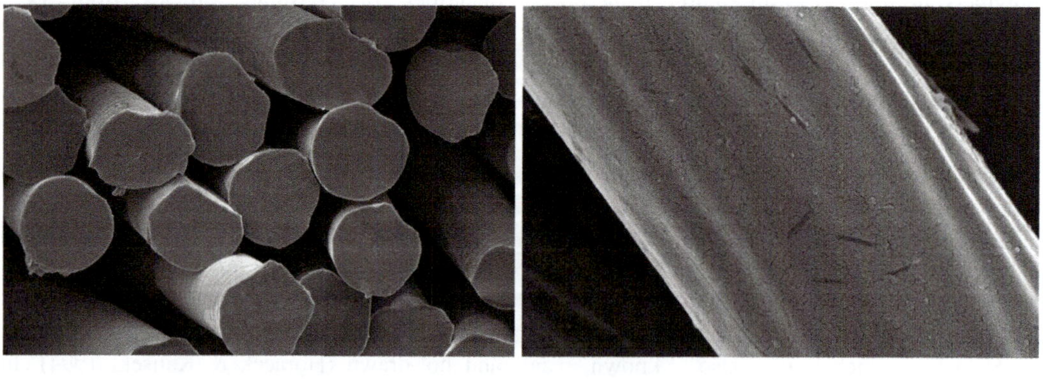

Fig. 35.2 Transverseal and longitudinal view of PLA fibers (Lenzing AG)

Fig. 35.3 Process stages in the extraction of lactic acid from starch

the mandatory continuous removal of the water formed, fewer monomers and catalysts remain in the reaction mixture.

35.3.2 Ring Opening Polymerization

In ring-opening polymerization, the water is removed without solvent to form lactic acid oligomers. Then the cyclic dimer of lactic acid is formed by depolymerization of the oligomers. Then, with the help of a catalyst, usually tin carboxylate, the ring of the dimer is opened. This process can be used to produce PLA with a molecular mass of at least 60,000 (DP = 770). For technical applications, DP up to 1300 can also be achieved. The extremely high DP compared to PET is due to the aliphatic structure of PLA (low stiffness of the chain), the low attractive forces between the macromolecules and the relatively short repeat units (442 nm compared to 1322 nm for PET). The ring-opening process is characterized by a relatively high polymerization rate, a high purity of the polymer achieved by

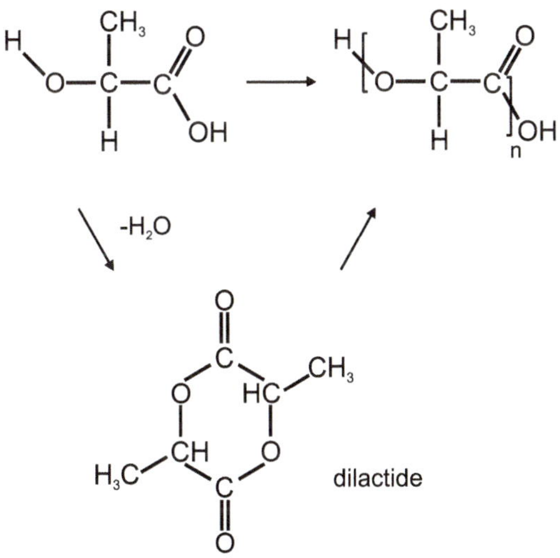

Fig. 35.4 Synthesis of PLA from lactic acid and dilactide respectively

distillation of the (D) and meso moieties, and a high relative molecular mass (Gruber et al., 1992). The Fig. 35.5 shows the details of the process steps for the production of PLA from starch (Linnemann et al., 2004).

35.4 Chemical Composition

Polylactide is an aliphatic polyester and is composed of ester and methyl side groups. Two optical isomers of lactic acid exist (Fig. 35.6). Both isomers have identical physical properties except for their effect on linearly polarized light. L(+) lactic acid rotates polarized light to the right and D(−) lactic acid rotates it to the left. This is the origin of the common names dextrorotatory and anticlockwise lactic acid.

Three different forms of dilactide can be formed from the different isomers. The dimer usually consists only of L(+)-lactic acid, but there are also dilactides composed exclusively of D(−)-lactic acid and those composed of both L (+)- and D(−)-lactic acid. The latter is also called meso-dilactide (Fig. 35.7).

A mixture of the isomers (L+) and (D−) in the molar ratio 1:1 is called racemic. Both the meso-dilactide and the racemic dilactide are optically inactive.

35.5 Production Method

Because of the small quantities, direct spinning is not economically viable. Therefore PLA is always spun in the extruder spinning process. In order to avoid depolymerization by hydrolysis during melting, the moisture content must not exceed 50 ppm. Therefore, the pellets must be dried before melting. Usually a single screw extruder with L/D ratio of 24:1 to 30:1 is used. Typical melt flow index values at 210 °C are 5–15 for high-strength monofilaments and 15–30 for

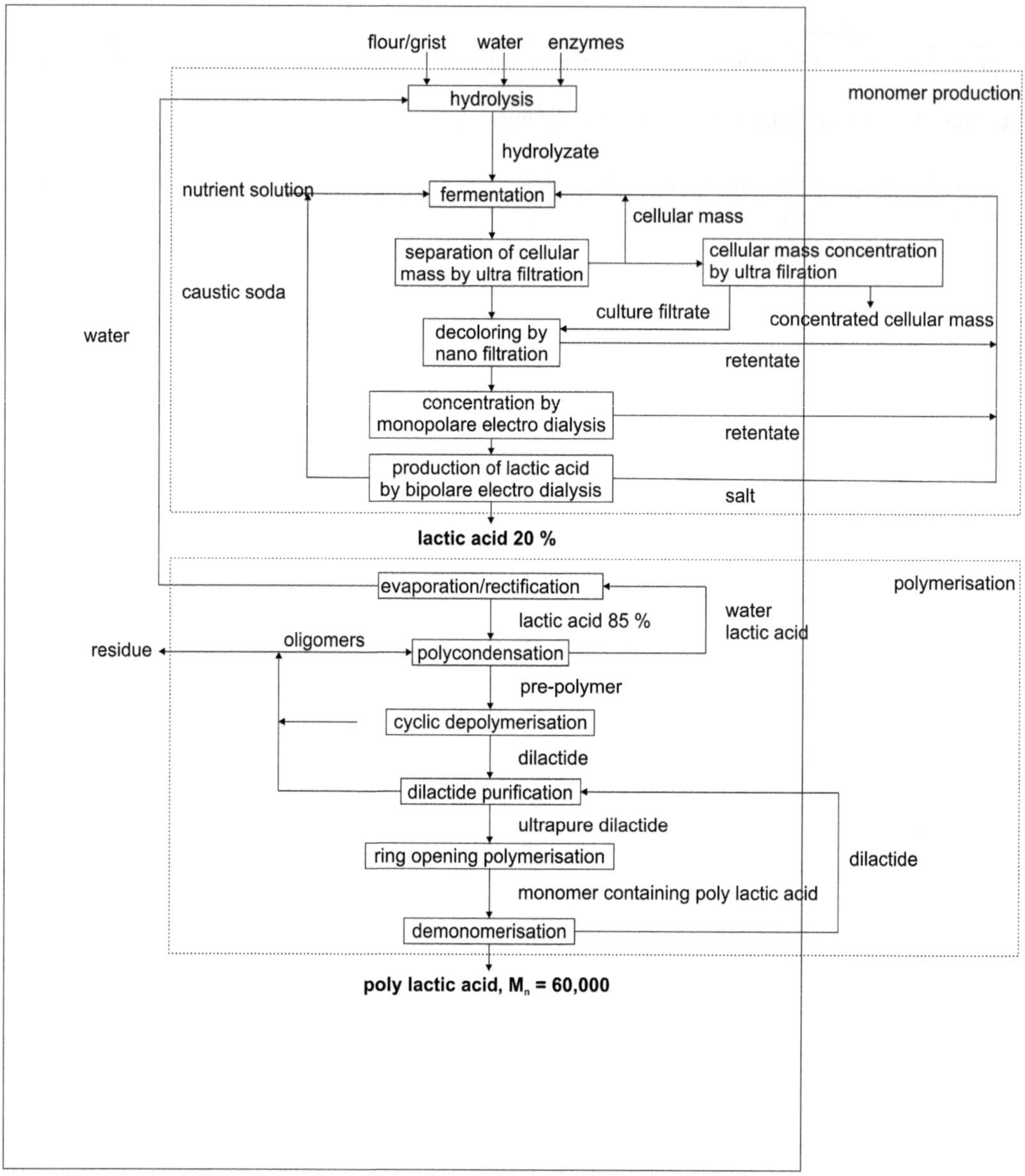

Fig. 35.5 Block diagram of ring opening polymerization acc. to (Hagen, 2001)

Fig. 35.6 Lactic acid isomers and repeating unit of PLA

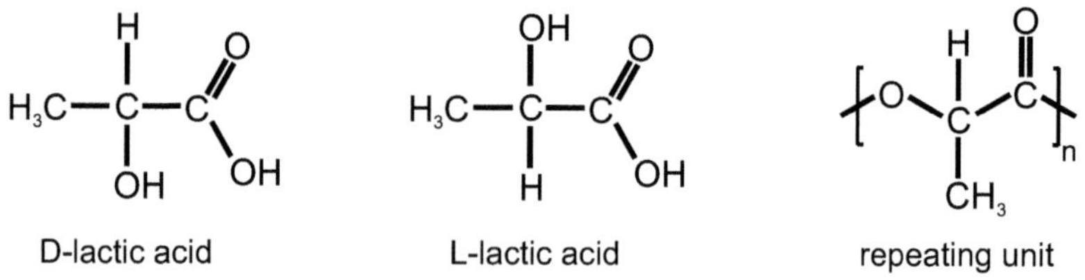

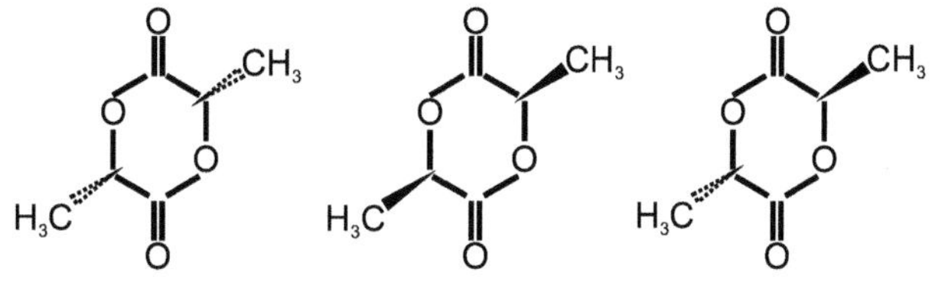

Fig. 35.7 Different forms of the dilactide (from left: L + L+, D–D–, L + D–)

other yarns. In the meltblown process the process temperature is usually 180–210 °C.

Textile filaments are spun at temperatures of 220–240 °C as LOY or POY, followed by drawing. The first godet duo should have a temperature above the glass temperature (45–65 °C). Thermal fixation on the second godet duo takes place in the crystallization range at 100–130 °C. Usual spinning speeds are around 3000 m/min, for fully drawn yarns with low shrinkage up to 5000 m/min can be achieved.

In addition to filament yarns, bico yarns, often in combination with other polyesters, e.g. PBT (Kawahara et al., 2020), and spunbonded nonwovens are also produced from PLA.

35.6 Properties

35.6.1 Polymer

Regardless of the manufacturing process (polycondensation or ring-opening polymerization), PLA fibers have almost identical mechanical properties. They can be specifically adjusted via the (D-)/L(+) ratio of the lactic acid monomers. A high proportion of L(+) lactic acid leads to a higher degree of crystallization. If the D(−) content exceeds 15%, the polymer is largely amorphous. The D(−) content also influences the melting temperature of the polymer (Fig. 35.8).

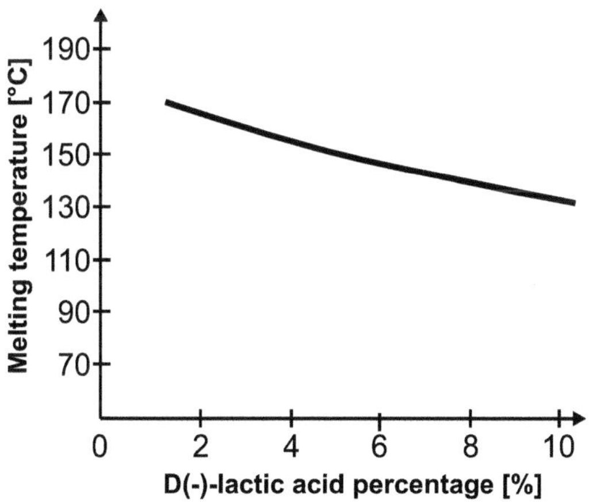

Fig. 35.8 Dependence of the melting point of PLA on the D(−) fraction acc. to (Lunt and Shafer 2000)

In Fig. 35.9, five copolymers in different isomer combinations are shown. The polymers composed solely of L(+)-lactic acid are referred to as P(L)LA. Polymers composed of L(+) and D(−) lactic acid monomers are referred to as P(LD)LA. As a rule, the ratio of the isomers is indicated (e.g. 60/40).

The P(L)LA crystal has a spiral structure, which leads to a high melting temperature (178 °), a high glass transition temperature (58 °C) and a high degree of crystallization compared to other aliphatic polymers. The crystallization rate is particularly high at a temperature of 110–118 °C (Di Lorenzo, 2005). After spinning, PLA initially exists as an orthorombic α-crystal. Above a temperature of 113 °C, trigonal β-crystals are formed, which are also formed by the action of external forces during stretching (Yasuniwa et al., 2006).

P(L)LA is semi-crystalline, relatively hard, brittle and optically transparent. P(LD)LA, on the other hand, is amorphous, flexible and softens already at 50 °C. PLA produced from meso-lactide is an amorphous and non-crystallizing

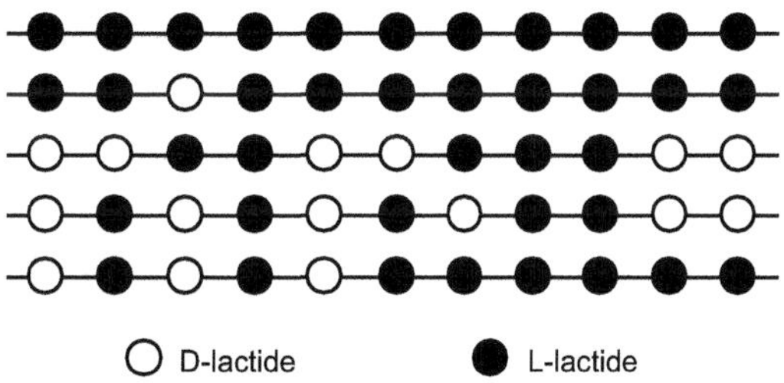

Fig. 35.9 Formation of different polymers by combining the lactic acid isomers acc. to (Kirschbaum, 2000)

Table 35.1 Physical properties of P(LL)A fibers acc. to manufacturer's data and (Linnemann et al., 2004)

Tenacity [cN/tex]	20–55
Elongation [%]	20–40
E-modulus [cN/tex]	60–70
Shrinkage [%]	8–15
Density [g/cm^3]	1.25–1.30
Moisture absorption [%]	0.4–0.6
Melt temperature [°C]	160–175
Glass temperature [°C]	57–58
Crystallisation temperature [°C]	110–118
Degree of crystallisation [%]	70–83

polymer with a glass transition temperature of approx. 30 °C. It is therefore unsuitable for fiber production.

35.6.2 Physical and Chemical Properties

Products made of PLA fibers hardly absorb any moisture and easily pass on water vapor. This improves the breathability of clothing.

PLA fibers have good strength and very high elongation, acc.ly the Young's modulus is very small compared to other polymers. The boiling shrinkage is relatively high, but can be reduced by thermal treatment under tension in the spinning process.

Typical properties of PLA fibers are summarized in Table 35.1.

The Fig. 35.10 shows the stress-strain curve of typical PLA fibers compared to other materials. It initially resembles that of wool, but continues because of the higher extensibility.

Dye uptake is much faster with PLA than PET, but is significantly lower at about 16% of the value for PET. The LOI is relatively low at 26 and PLA fibers continue to burn after removal of the flame in the burn test, although for a shorter time than PET fibers. UV resistance is good because PLA hardly absorbs light in the visible spectrum. However, it has little resistance to hydrolysis due to its aliphatic structure (Table 35.2).

Single filament counts of 1 dtex are achievable with multifilament yarns made from PLA. Because the melting temperature can be adjusted, PLA fibers are also suitable as components for bico fibers, e.g. in combination with PET and PA 6. PLLA fibers also possess piezoelectric properties.

35.6.3 Environmental Compatibility

Figure 35.11 shows the process routes for the production of PLA from the various starting materials. For the 100,000 t of PLA fibers currently produced, a minimum of 16,000 ha (sugar cane) and a maximum of 100,000 ha (wheat) of agricultural land are required. This is a comparatively small area, because in the EU alone the agricultural area is 174 million ha. However, the

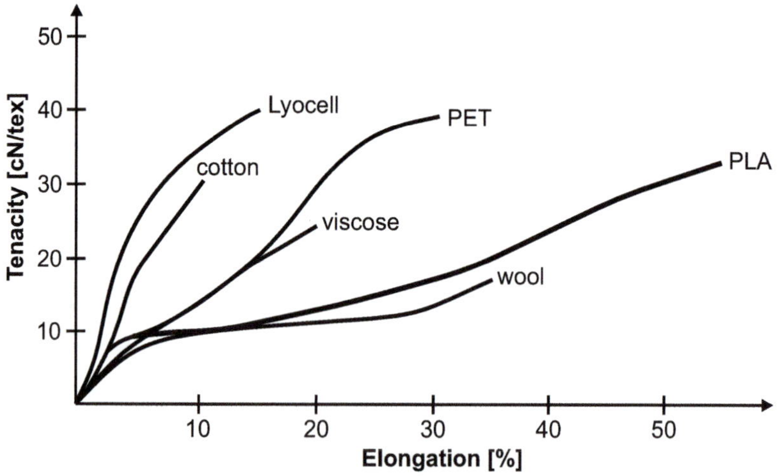

Fig. 35.10 Stress-strain diagram of PLA fibers compared to other fibers acc. to (Farrington et al., 2005)

Table 35.2 Chemical properties of P(LL)A fibers acc. to (Linnemann et al., 2004)

Molecular weight	50,000–120,000		
Chemical resistance	Solution resistance	Stabil	CH_3OH, isopropanole, hexane, petrolether and water
		Swollen	Acetone, methylbenzene, acetonitrile and methylethylenketone
		Solved	m-kresol, dioxane, chloroform and methylchloride
	Stability in alkaline solution	Very susceptible, quick degradation	
	Stability in acid solution	Rather stable	
Flammability	Low smoke development, hardly flammable after flame removal: burns for another 2 min.		
LOI	26%		
Dyeing	with disperse dyes at a temperature of approx. 100 °C		
Dyeing behavior	Deeper dye take-up compared to PET owing to lower degree of reflexion (PET: 1.58; PLA: 1.4)		

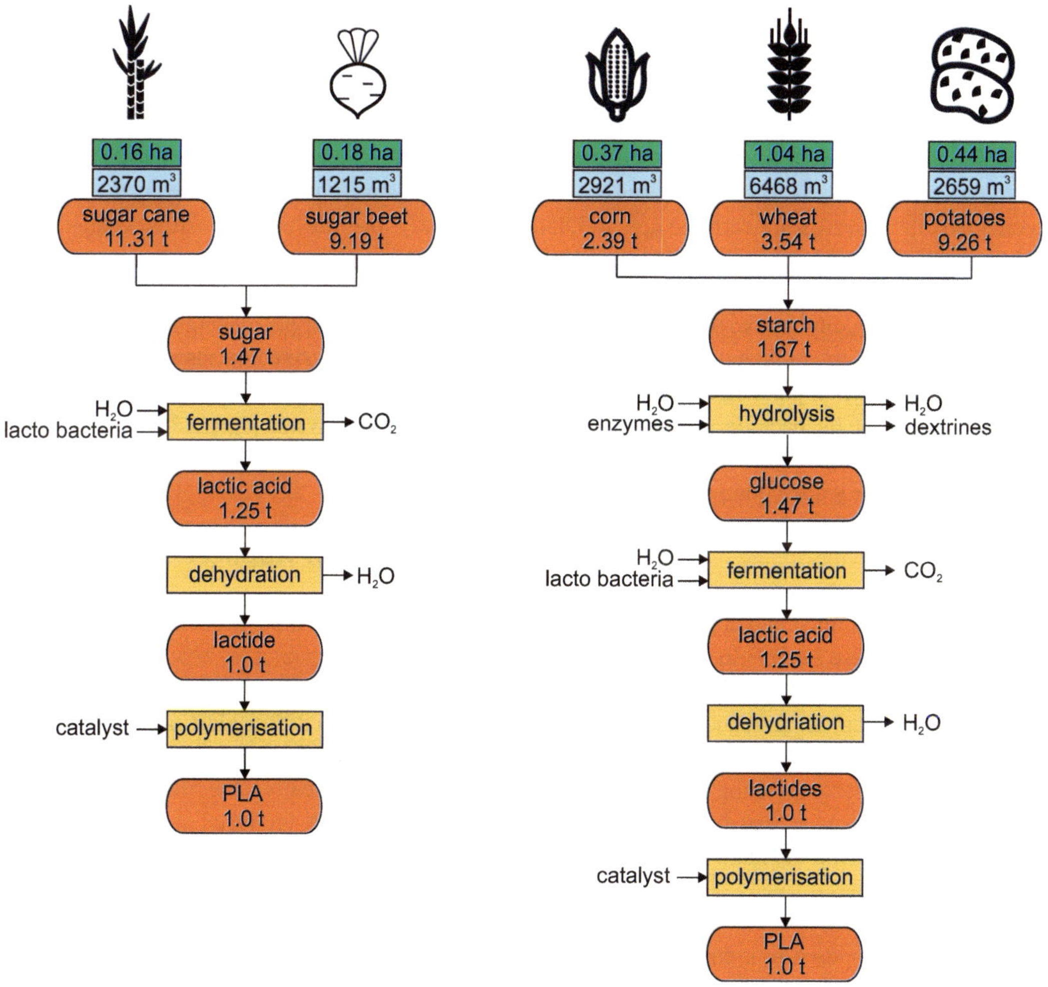

Fig. 35.11 Production of PLA from renewable raw materials acc. to (IFG, 2021)

Table 35.3 Various methods of PLA degradation and results acc. to (Mochizuki, 1998)

Type	Material	Duration	Result
Burial	PLA multifilament, 55dtex/f24	2 years	50% loss of strength 50% loss of relative viscosity
Weather test (sun, 0 m)	PLA Fibers PET fibers (for comparison)	500 h	5% loss of strength 40% loss of strength
Aerobic composting	PLA nonwovens (25 g/m^2) cellulose powder (for comparison)	45 days	79% biodegradation 90% biodegradation
Anaerobic decomposition	PLA nonwovens (25 g/m^2) cellulose powder	45 days	92% biodegradation approx. 99% biodegradation

water consumption is in any case considerably greater than in the production of synthetic polymers. Only about 4 t of water are required to produce 1 t of PET. Therefore, the environmental balance of PLA is mixed if artificial irrigation is required.

PLA is biodegradable. Depending on the molecular structure of the polymer, slow to rapid decomposition into water, carbon dioxide and biomass takes place. Pure PLLA takes years to decompose, whereas P(LD)LA is dissolved within months due to its lower crystallinity. The biodegradable materials are certified acc. to DIN V54900, DIN EN 13432 and ASTM D 6400. PLA composts well at higher temperatures (60 ° C) and higher humidity (above 98%). The molecular chains of PLA are broken down in a first step by hydrolysis under high temperature, followed by decomposition by bacteria. Because of this composting temperature of 60 °C is so far not suitable for private garden composting. However, there are efforts to make home composting possible by adding enzymes.

Various tests were carried out for degradation in the environment. First, when embedded in soil, the PLA fibers were buried. In the weather test, the material sample was exposed to the sun and in aerobic composting, the influence of composting conditions on PLA nonwovens was investigated. In the anaerobic decomposition test, biodegradation and decay of a material sample were evaluated acc. to the standard ISO/DIS 15985 (Table 35.3).

Other studies show that PLA is almost completely decomposed in industrial composters at 60 °C and 95% relative humidity within 2 weeks. In "home compost", however, this process takes several years due to the different climatic conditions (Farrington, 2005).

At first glance, rapid degradation of polymers seems advantageous. However, it is unpleasant for the consumer if his textile in the wardrobe degrades in summer. Therefore, this property of PLA is not necessarily an advantage. For applications of PLA in the human body, on the other hand, biodegradability, ideally adjustable, may well be desirable.

Material recycling of polylactide is also possible. Even after repeated melting in the extruder (7 passes), the polymer properties change only slightly. A quantification of the change cannot be found in the literature. Hardly changed polymer properties after multiple melting are significant for thermoforming. PLA could also be decomposed back into lactic acid by means of hydrolysis, in order to then be available as a starting monomer for polymerization into PLA (Kirschbaum, 2000).

35.7 Areas of Application

PLA is used in the form of filaments and nonwovens. Because it is biodegradable, it is suitable as a carrier material for absorbable medical textiles. Due to its low glass transition temperature, it is not interesting for many textile applications. Currently, acc. to manufacturers, about 70% of PLA produced worldwide is used for films, mainly packaging, and 25% is used to produce fibers, which is equivalent to about 100,000 tons. 3D printing of products from PLA has also been gaining in importance recently.

Typical examples of products made from PLA fibers are:

Medical Technology
Due to its biocompatibility, PLA is used as an absorbable suture material, for absorbable textile structures, also made of nanofibers, in tissue engineering, for artificial implants (vascular prosthesis, osteosynthesis, ligament replacement, etc.), as a carrier for drug-release implants and other medical applications such as bandages, plasters, etc.

Copolymers of lactide and glycolide (PLGA) are used for medical applications due to their good biocompatibility (Azimi et al., 2014).

Clothing Industry
Sportswear, as well as shirts and suit fabrics can in principle be made from PLA.

House and Home Textiles
The good resistance to UV light, flame retardancy and low smoke emission are properties that are important in the field of house and home textiles. Examples of possible products are curtains, seat covers, protective covers and bedspreads.

Technical Textiles
PLA fibers can also be used for industrial filters and geotextiles that are to be biodegraded over a period of a few years.

35.8 Market and Trade Names

Currently, PLA production volume worldwide is around 400,000 t (IFG, 2021), of which 68% is produced in Asia and 31% in North America and 1% in Europe (The Fiber Year, 2021). About a quarter is used for fiber production (Skoczinski et al., 2021). The largest producers are NatureWorks LLC (USA) and Toray Ind., Inc., (Japan). Total Corbion (Thailand, France) is the largest producer of PLA plastic granules but does not produce fibers. Although well-known manufacturers have stopped the production of PLA, the production volume continued to grow moderately in recent years. Major trade names of PLA fibers are Ingeo™, Lacea™, Lacty™ and Ecodear™. The market price is between 1.5 and 4.5 €/kg, depending on the quality.

Due to the high raw material and process costs compared to other polymers, PLA fibers are currently not competitive in the mass market from a purely economic point of view and this will not change in the foreseeable future. Therefore, PLA will only be able to conquer a larger market share in the textile sector if, above all, environmental considerations become more important to the buyer, i.e. biodegradability and the possibility of producing PLA from renewable raw materials. This is most likely to be the case in Europe, so it is expected that production capacities in Europe will increase to about 100,000 t/a by 2025, which would correspond to a market share of about 20% (Skoczinski et al., 2021).

However, it should be noted that the arable land required for the bio-based production of PLA, e.g. for sugar beet or maize, competes with that for food production (Bogaert & Coszach, 2000).

References

Azimi, B., Nourpanah, P., Rabiee, M., & Arbab, S. (2014). Poly (lactide-co-glycolide) fiber: An overview. *Journal of Engineered Fibers and Fabrics, 9*(1), 47–66.

Bischoff, C. A., & Walden, P. (1889). II. Über Derivate der Milchsäure. *Liebigs Annalen der Chemie, 279*, 71–99.

Bogaert, J. C., & Coszach, P. (2000). Polylactic acids: New polymers for novel application. *Nonwovens World*, 83–91.

Carothers, W. H., Dorough, G. L., & van Natta, F. J. (1932). Studies of polymersation and ring formation. X. The reversible polymerisation of six-membered cyclic ester. *Journal of the American Chemical Society, 54*, 761–772.

Di Lorenzo, M. L. (2005). Crystallization behavior of poly (l-lactic acid). *European Polymer Journal, 31*, 569–575.

Farrington, D. W., Lunt, J., Davies, S, & Blackburn, R. S. (2005). Poly(lactid acid) fibers. In *Biodegradable and sustainable fibres*. Woodhead Publishing, Cambridge.

Gay-Lussac, J. L., & Pelouze, T.-J. (1833). Über die Milchsäure. *Annalen der Pharmacie, 7*, 40–47.

Gruber, P. R., Hall, E. S., Kolstad, J. J., Iwen, M. L., Benson, R. D., & Borchardt, R. L. (1992). *Continuous process for manufacture of lactide polymers with controlled optical purity*. US Patent 142023A.

Hagen, R. (2001). *Ein neues Verfahren zur Herstellung von Polymilchsäure. Gülzower Fachgespräche, Workshop Moderne Polymere: Kohlenhydrate und*

Pflanzenöle als innovative Rohstoffe., 27./28.Nov. 2000, Kassel, S.153–158.

Horacek, I., & Kalisek, V. (1994). Polylactide. I. Continuous dry spinning-hot drawing preparation of fibers. *Journal of Applied Science, 65*, 1751–1757.

IFG. (2021). https://www.ifbb-hannover.de/files/IfBB/downloads/faltblaetter_broschueren/f+s/Biopolymers-Facts-Statistics-2020.pdf

Jamshidian, M., Tehrany, E. A., Imran, M., Jacquot, M., & Desobry, S. (2010). Poly-lactic acid: Production, applications, nanocomposites, and release studies. *Comprehensive Reviews in Food Science and Food Safety*, 9 (5).

Kawahara, Y., Takarada, W., Yamamoto, M., Kondo, Y., Tashiro, K., & Kikutani, T. (2020). Fiber structure, tensile behavior and antibacterial activity of polylactide/poly(butylene terephthalate) bicomponent fibers produced by high-speed melt-spinning. *Journal of Macromolecular Science, Part B: Physics, 59*(7), 440–456.

Kirschbaum, D. L. (2000). Polylactide (PLA) polymers. Where performance comes naturally. *Fiber Producer/Textile Industry Conference*, Clemson Univ., Palmetto Expo Center, Greenville, USA, Oct 24–25, pp. 1–11.

Linnemann, B., Sri Harwoko, M., & Gries, T. (2004). *Polylactidfasern. Faserstofftabelle nach P.-A. Koch.* Institut für Textiltechnik der RWTH Aachen, Aachen.

Lunt, J., & Shafer, A. (2000). Polylactic acid polymers from corn. Applications in Textiles. *Journal of Industrial Textiles, 29*(3), 191–205.

Matsui, M., & Kondo, Y. (1996). A biodegradable fiber made of poly-lactic acid. *Chemical Fibers International, 46*, 318–319.

Mochizuki, M. (1998). Polylactic acid fibers and nonwovens. *The Textile Machinery Society of Japan Journal, 1*, 139–142.

Natureworks. https://www.natureworksllc.com/

Skoczinski, P., de Guzman D., Carus, M., Käb, H., Chinthapalli, R., Ravenstijn, J., Baluts, W, & Raschka, A. (2021). *Bio-based building blocks and polymers – Global Capacities, Production and Trends 2020–2025*, nova-Institut GmbH, Hürth.

The Fiber Year. (2021). *The Fiber Year GmbH, Roggwil.* https://www.thefiberyear.com

Total Corbion. https://www.total-corbion.com/

Yasuniwa, M., Tsubakihara, S., Yoshinori, K., Iura, K., Ono, Y., Dan, Y., & Takahashi, K. (2006). Crystallization behavior of poly(l-lactic acid). *Polymer, 47*(21), 7554–7563.

36 Elastane

Elastane fibers consist of at least 85% by mass of segmented polyurethane. In the USA, they are also referred to as spandex. The abbreviation is EL (Fig. 36.1).

Elastane fibers have the outstanding property of an exceptionally high elongation at break of 200–800% and almost completely return to their original length after external forces have been removed. Elastane yarns usually consist of four or more filaments (Fig. 36.2).

36.1 History

The basis of today's elastane fibers is the diisocyanate polyaddition process developed in 1937 by O. Bayer and H. Rinke, among others (Bayer, 1963). In 1939, P. Schlack extended linear polyesters with diisocyanates and thus produced highly extensible, elastic fibers. From 1941 onwards, the diisocyanate polyaddition process was used for the specific construction of high-quality elastomers (Vulkollan principle). Reaction spinning was first investigated by H.A. Pohl in 1942; it was later used for linear and crosslinked polyurethanes. In 1949, E. Windemuth developed a process by which the high molecular weight polyurethane could first be synthesized and then spun directly (Oertel, 1965). Solution spinning for the production of polyurethane threads was first used by W. Brenschede in 1951. At DuPont, it was further developed by M.D. Snyder and J.C Shivers, among others, and thus large-scale production of the first elastane fibers began in 1959, which were marketed under the name Lycra™ from 1962 (Charch & Shivers, 1959). Around the same time, the US Robber Co. launched a polyester urethane under the brand name Vyrene™ and in 1964 Bayer AG followed with a similar product under the name Dorlastan™.

There have been several efforts by various companies to produce elastane fibers also by melt spinning. These include, for example, Mobilon™ by Nisshinbo Industries (1967) and Spantel™ by Kuraray Co. Ltd. (1991). None of these processes has yet been able to establish itself on the market (Fabricius et al., 1995).

36.2 Starting Materials, Formation of the Polymer

Polyurethanes are a class of polymers of very different compositions, the characteristic structural element being the urethane group. The diverse properties allow the production of materials ranging from those with high hardness and strength to those with high elasticity and low strength.

36.2.1 Polymer Formation

The basic principle of the diisocyanate polyaddition process is the reaction of

D. Veit, *Fibers*, https://doi.org/10.1007/978-3-031-15309-9_36

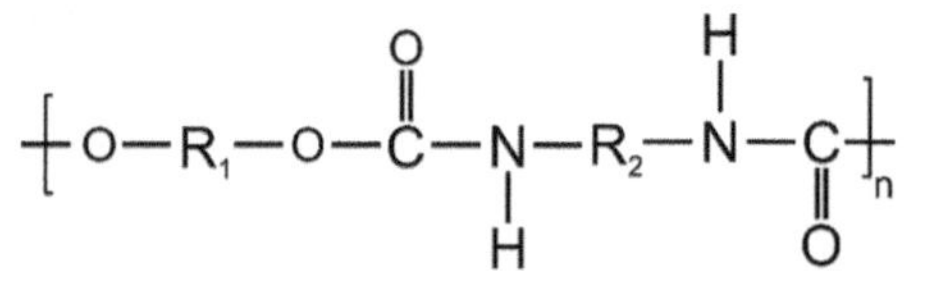

Fig. 36.1 Structural formula of the repeat unit of polyurethane

diisocyanates with bifunctional compounds containing reactive hydrogen atoms, e.g. dialcohols (R_1) and diamines (R_2), to form high molecular weight polyurethane (Fig. 36.3).

In segmented polyurethanes, units of different constitution alternate, and are therefore referred to as copolymers The blocks are referred to as "soft" or "hard" segments. The crystalline "hard" structure of the polyurethane is loosened up by the amorphous "soft" segments of the higher molecular weight dihodroxyl compound, whose macromolecular chains exhibit a high degree of internal mobility, to such an extent that segmented polyurethanes can be stretched by several hundred percent (Hicks, 1963).

36.2.2 Polymer Formation in 2 Steps

A polyurethane elastomer system is usually built up in two stages: First, long-chain dihydroxyl compounds, the so-called "macrodiols", are reacted with diisocyanates to form the macrodiisocyanate (NCO prepolymer). This then reacts with low-molecular diamines or diols, the "chain extenders", to form the segmented polyurethane. The diisocyanates link the macrodiol soft segments and act as a short, rigid urethane group, turning them into the hard segment.

Macrodiols are long-chain dialcohols with terminal hydroxyl groups, low melting point (below 50 °C) and low molecular weight (1000–3000). Typical macrodiols are polyether glycol, e.g. polytetrahydrofuran $H\text{-}[CH_2\text{-}CH\text{-}CH_{22}\text{-}CH_2]_n\text{-}OH$, which is the most widely used, and polyester glycol, e.g. polyethylene adipinate glycol $HO\text{-}[(CH_2)_2\text{-}O\text{-}CO\text{-}(CH_2)_4\text{-}CO\text{-}O]_n\text{-}(CH)_{22}\text{-}OH$.

Aromatic compounds are often used as diisocyanates, e.g. diphenylmethane-4,4-diisocyanate (Figs. 36.4 and 36.5).

Alternatively, isomeric toluylene diisocyanates can also be used.

First stage

NCO prepolymer formation is carried out analogously to the PU polyaddition process by reacting macrodiols ("soft segments") with excess diisocyanate. The Fig. 36.6 illustrates the reaction of a polyester diol (soft segment) with diphenylmethane-4,4-diisocyanate.

This moderately fast reaction succeeds both in solution and solvent-free under certain conditions.

Second stage

The NCO prepolymer is dissolved in a highly polar organic solvent in the second stage and reacts with a low molecular weight diamine or diol (especially in melt spinning) as a chain extender. Dimethylformamide or dimethylene acetamide are usually used as solvents.

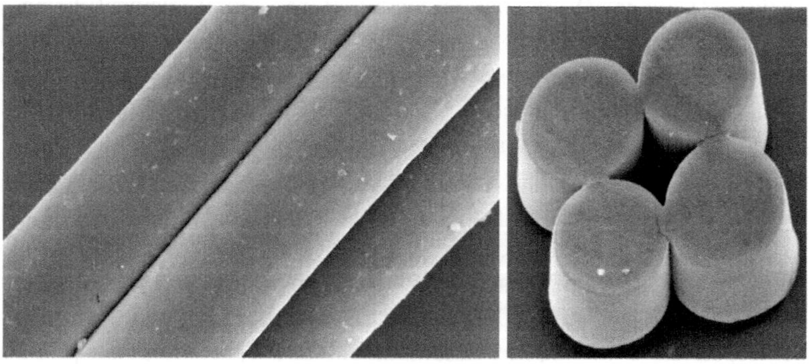

Fig. 36.2 Longitudinal and transversal view of an elastane yarn

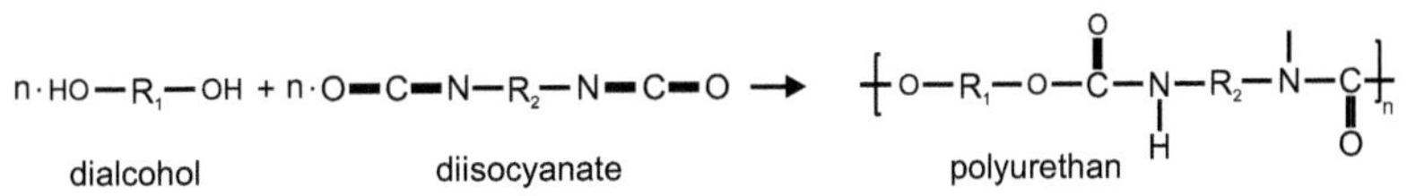

Fig. 36.3 Synthesis of polyurethane

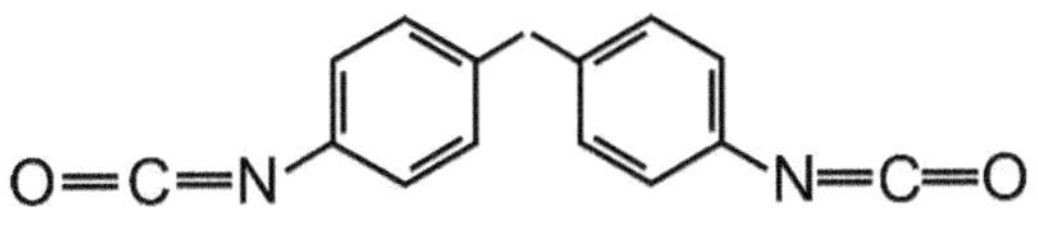

Fig. 36.4 Structural formula of diphenylmethane-4,4-diisocyanate

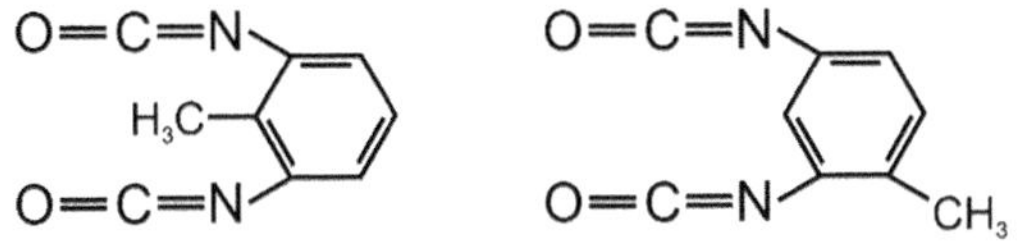

Fig. 36.5 Structural formula of the isomeric toluylene diisocyanates

Depending on the desired properties, 30–50 chain segment extenders form a soft segment (Fig. 36.7).

The diamines used are ethylenediamine, 1,2-propylenediamine, 2,3-diaminocyclohexane and 2-methyl-1,5-pentanediamine. The diols mostly consist of butylene glycol.

The reaction product in solution can form hydrogen bonds due to its urea or urethane groups. In polyurethane elastomers, the hard segments are thus not fixed by covalent bonds as in rubber, but by intermolecular hydrogen bonds between adjacent polyurethane chains. These secondary valence bonds can be dissolved at higher temperatures by mechanical stress or solvent action.

The textile-mechanical properties of elastane fibers are improved when the hard segments are stiffened by chain extenders with double bonds and subsequent cross-linking.

36.2.3 Internal Structure of Elastomers

The finished, elastomeric, segmented polyurethane consists of "soft" and "hard" segments. The soft segments consist of long, disordered and flexible chains. Under an external force, they form crystalline areas that dissolve again when the load is removed. This gives the fiber its high stretchability.

The soft segments are thus flexible and easily deformable at service temperature, while the hard segments are solidified like glass or are partially crystalline. Below the solidification temperature of the soft segments, the rubber-elastic extensibility disappears, so that the elastane filaments become hard and brittle. As the molar mass of the soft segments increases while the proportion of hard segments remains constant, the elongation force decreases for the same elongation. This is

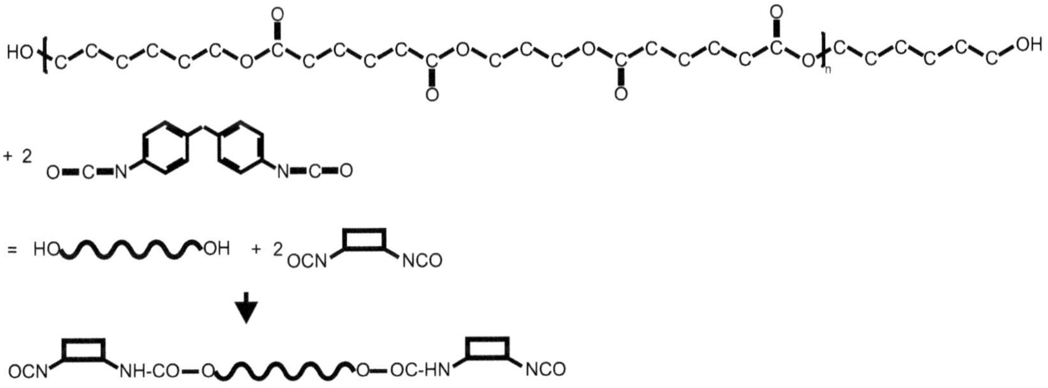

Fig. 36.6 NCO prepolymer formation (macrodiisocyanate)

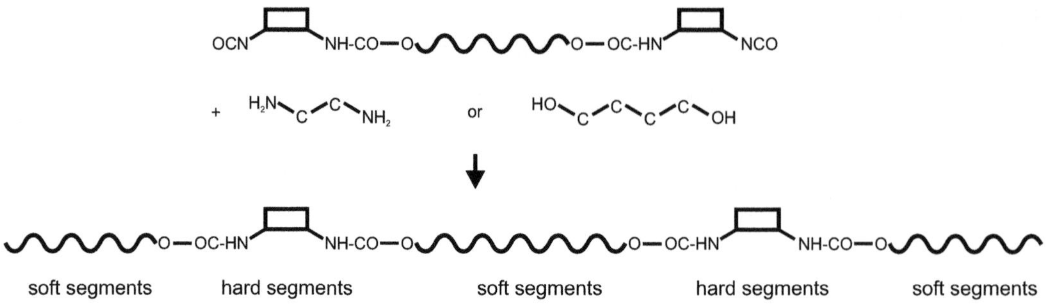

Fig. 36.7 Chain extension by ethylenediamine or butylene glycol

because the size of the segments and their distribution in the macromolecule change. Due to the longer soft segments, the hard segments are further apart.

The hard segments are short, crystalline areas. They provide the network linkage via hydrogen bonds. They cause the fiber to spring back after deformation when the external load is removed. Unsymmetrical hard segments result in elastanes with higher elongation at break and a flatter force-elongation curve than elastanes with even-numbered diamines because of the different hydrogen bonds. The more regular the structure of the hard segments and the higher their proportion in the elastane, the less sensitive the elastane is to thermal stress. The melting point therefore increases with the stability of the hard segments. As the proportion of hard segments increases, the force-elongation curve becomes steeper and the permanent elongation increases (Fig. 36.8).

36.3 Production

Elastane fibers can be produced by dry, wet, reactive and melt spinning. Around 80% of elastane fibers are spun dry.

36.3.1 Dry Spinning

The Fig. 36.9 shows the principle of a dry spinning machine for the production of elastane fibers. The highly viscous solution (50–1450 Pas) with a polymer concentration of 20–25% in dimethylformamide (DMF) or dimethylacetamide (DMAC) is spun vertically through a multi-hole spinneret into a spinning duct through which hot air flows. After evaporation of the solvent, a residual solvent content of less than 1% remains in the polymer. The individual filaments (0.5–2 dtex) are brought together in the spinning shaft and combined into a multifilament yarn by applying a false twist. The comparatively high oil application (paraffin or silicone oils mixed with calcium or magnesium stearate) of 2–6% is necessary to reduce the adhesion of the filaments. The filament cross-sections are round, oval or dumbbell-shaped. The winding speed is 200–1000 m/min (Fabricius et al., 1995).

36.3.2 Wet Spinning

In the wet spinning process, the polymer solution is spun out at a concentration of 20–25% into aqueous precipitation baths, e.g. with dimethylformamide solution. After passing through several washing baths, the filament yarn is wound at a speed of approx. 100 m/min, whereby the filaments stick together. The desired properties are adjusted by thermal post-treatment in hot water or with hot air (Fig. 36.10).

36.3.3 Reactive Spinning Process

In this process, polymer formation by chain extension of the NCO polymer and filament formation are combined in one process. The liquid NCO prepolymer is spun through a multi-hole spinneret into a bath with diamines. The reaction

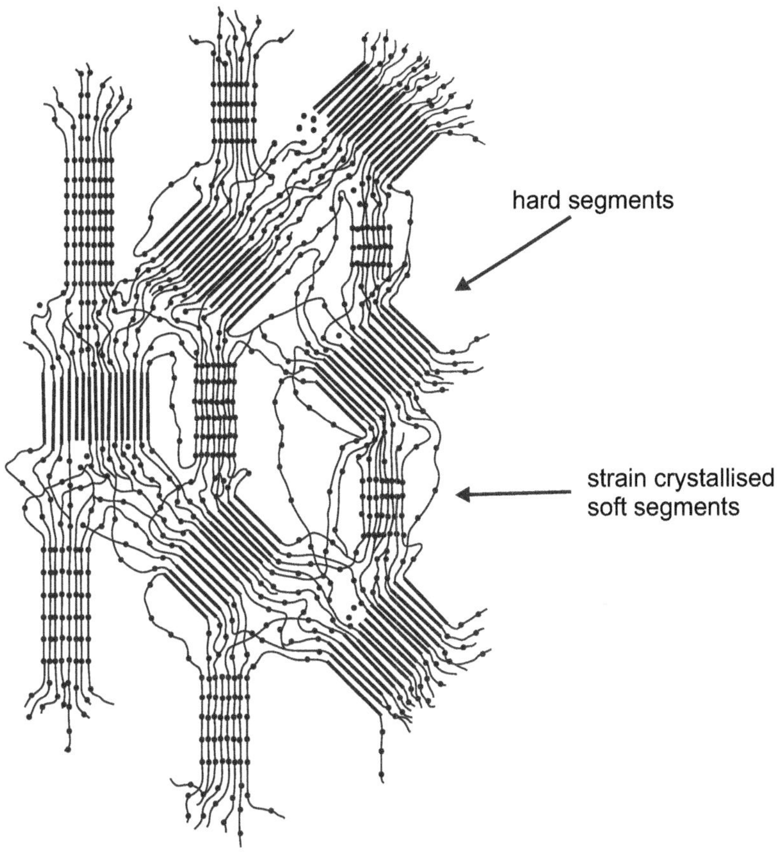

Fig. 36.8 Inner structure of a polyurethane elastomer acc. to (Rinke, 1968)

of the diisocyanate with the diamine forms a stable skin of cross-linked polyurea at the thread surface, which allows the thread to be pulled off. The interior of the yarn is then solidified in hot water under pressure or in diamine/alcohol or toluene solutions. The branched prepolymer structure thus produces a covalently crosslinked, segmented elastane filament yarn that is insoluble in solvents.

36.3.4 Melt Spinning

Diamine-extended polyurethane ureas cannot be melted without decomposing the urea groups. In contrast, thermoplastic polyurethanes (TPU) in which diols are used as chain extenders can be spun from the melt into elastic filaments. The melt is extruded through single- or multi-hole dies and the resulting filaments are cooled in the blow shaft. The desired filament properties can be adjusted by thermal post-treatment. Basically, the spinning equipment corresponds to that used for the production of polyester and other melt-spinnable polymers.

36.4 Features

Elastane fibers are blended with varying amounts of TiO_2 resulting in shiny, semi-matte and matte fibers. Typical cross-sections are shown in Fig. 36.11. Dry-spun elastane filaments are usually round, while wet-spun ones are usually lobed and irregular. Melt-spun fibers are often monofilaments and usually round. Filaments are often bonded together in multifilament yarns, especially in wet spinning.

While the elongation of elastane fibers is extremely high, the strength naturally reaches

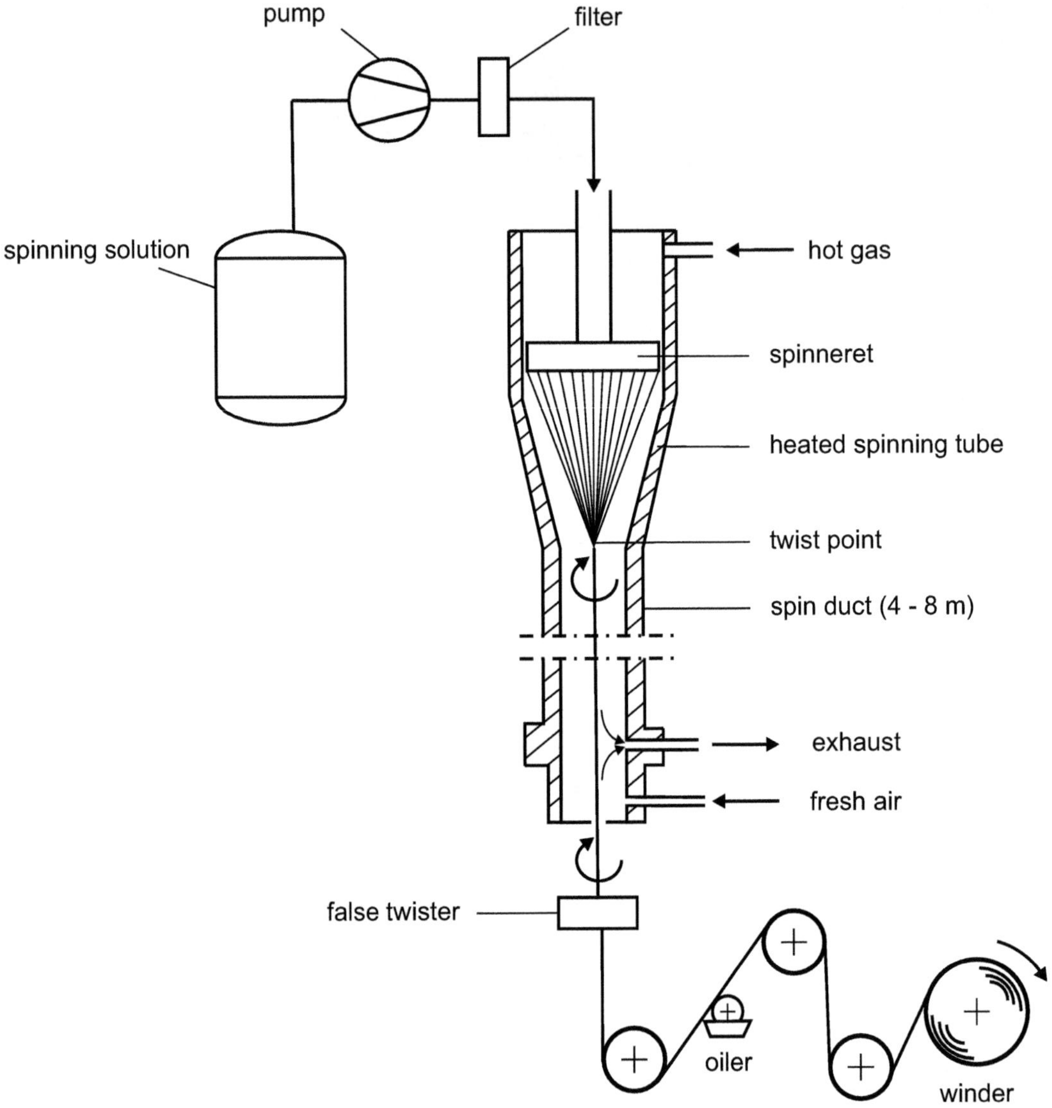

Fig. 36.9 Dry spinning process for elastane fibers

only very low values, which are still below cotton. The modulus of elasticity is correspondingly low (Table 36.1).

The typical force-elongation behavior of elastane fibers compared to other materials is shown in Fig. 36.12. The course of the curve lies between that of a rubber thread and a textured yarn.

The glass temperature is extremely low due to the soft segments, while the melting temperature is high due to the good cohesion and low mobility of the hard segments. The moisture absorption is low, the shrinkage relatively high (Table 36.2).

When elastane fibers are heated under tension, hard segments can slip away from each other and reform in a more sterically favorable shape while reducing internal tension. The higher the temperature and tension, the stronger the thermal rearrangement process. Thermofusibility is of great importance for the finishing and shaping of textiles. Only at temperatures above 170 °C does a noticeable thermal degradation occur.

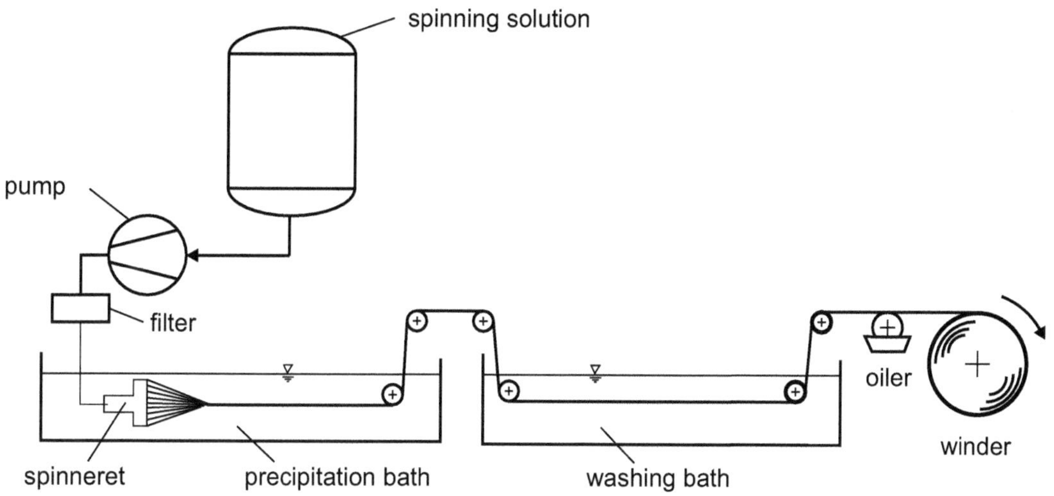

Fig. 36.10 Wet spinning process for elastane fibers

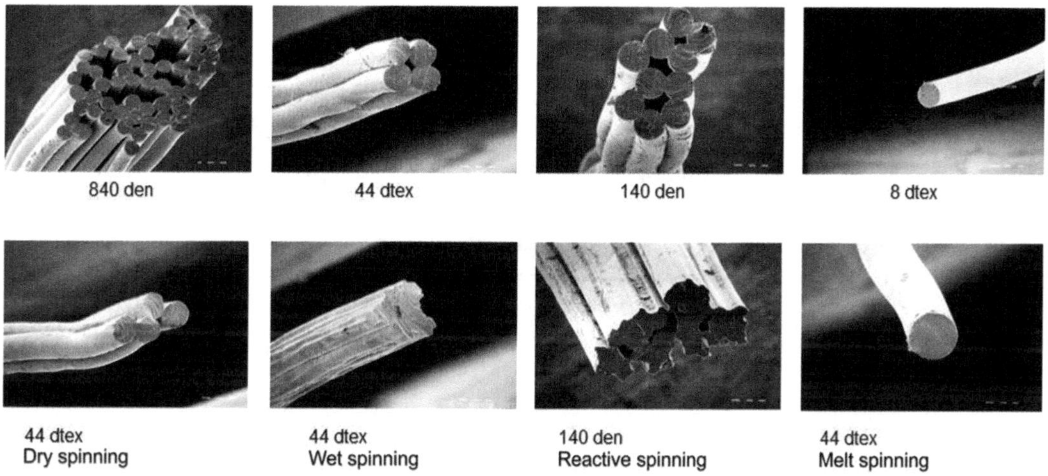

Fig. 36.11 Cross-sectional images of typical elastane fibers

Elastane fibers burn with a luminous flame and develop a pungent odour of isocyanate. A hard, dark residue is formed. Elastane fibers are soluble in highly polar solvents, e.g. dimethylformamide (DMF) and dimethylacetamide (DMAC). Elastanes with polyether soft segments show less degradation by hydrolysis, while elastanes with polyester soft segments are less sensitive to oxidation. Elastane is resistant to some extent to acids, alkalis, oxidizing and reducing agents as well as oxygen and ozone. Exposure to highly concentrated acids and alkalis leads to a loss of elastic properties, especially when heat is applied at the same time. Elastane is insensitive to washing, dry cleaning and chlorinated water. Its resistance to ageing and abrasion is greater than that of rubber threads. However, it is more sensitive to prolonged exposure to light than other polymers.

Elastane fibers can be dyed with disperse, acid and metal complex dyes. If elastane in blends with other fibers would color unevenly, the luster can be increased by reducing the TiO_2 content

Table 36.1 Typical mechanical properties of elastane fibers

Density (g/cm^3)		1.1–1.3
Maximum elongation:	in normal climate [%]	400–800
	"wet" [rel. humidity]	100
Maximum strength:	in normal climate	
	rel. to fineness [cN/tex]	5–12
	rel. to cross-section [MPa]	60–150
	"wet" [rel. humidity]	75–100
Elastic modulus:	rel. to mass [cN/tex]	0.5–1
	rel. to cross-section [MPa]	6–12
Torsion modulus	[cN/tex]	0.4

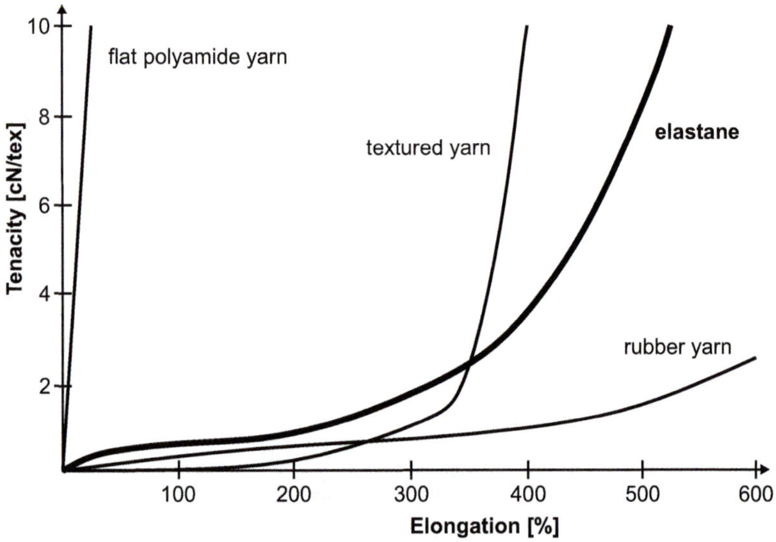

Fig. 36.12 Force-elongation curve of elastane compared to other fibers

whereas the undyed elastane fibers are less noticeable semi-dull or dull.

36.5 Products and Applications

Elastane fibers are used in textiles that are to have a high elastic stretch and at the same time a high restoring force (stretch). Typical applications are underwear, swimwear and sportswear, tights, hygiene products and diapers (cuffs). Elastane fibers are used in varying proportions in blends with other fibers. Elastane fibers are also standard in denim fabrics today (Table 36.3).

The elastane fibers are either added during textile production (knitting, weaving) or combination yarns are produced beforehand (e.g. ring and OE rotor yarns with elastane cores or wrap yarns) (Figs. 36.13 and 36.14).

36.6 Market

Table 36.4 gives an overview of important elastane fiber manufacturers.

Global production in 2020 is around 1 mt, with Hyosung alone as the largest manufacturer with its Creora™ fiber family having a market share of almost 40%. In total, more than 80% of elastane fibers are produced by Chinese companies, with production sites located around the world (e.g. Brazil, India, Turkey) (Fig. 36.15).

Table 36.2 Typical thermal properties of elastane fibers

Glass temperature [°C]	Polyether EL	–60
	Polyester EL	−40 to −20
Softening temperature [°C]		170–230
Melt temperature [°C]		170–290
Heat conductivity [W/mK]		0.15
Moisture absorption [%]	at 21 °C / 65% r. h.	0.5–1.5
	at 24 °C / 95% r. h.	0.5–1.5
Water retention [%]		7–11
Shrinkage [%]	in water 95 °C	3–15

Table 36.3 Typical proportions of elastane in textiles

	Share of elastane [%]
Woven fabrics	2–8
Underwear	2–5
Bath and sports goods	12–20
Fine stockings	2–12
Lingerie	10–45
Medical stockings	35–50

Fig. 36.13 Knitwear with elastane (left: unstretched; right: stretched)

Fig. 36.14 Typical products with elastane content for leisure and work (right)

Table 36.4 Selected manufacturers of elastane fibers

Producer	Trade name	Website	Country
Asahi Kasei Europa GmbH	Dorlastan™	https://asahi-kasei.eu/de/start/	Germany/Japan
Ashai Kasei, Japan	Roica™	http://www.asahi-kasei.co.jp	Japan
Huafon Group	Qianxi™	http://www.spandex.com.cn	China
Hyosung TNC Inc.	Creora™	http://www.creora.com/en/index.do	South Korea
LDZ New Aoshen Spandex Co., Ltd.	Aoshen™	http://de.ldzspandex.com	China
Nisshinbo Ind. Ltd.	Mobilon™	https://www.nisshinbo-textile.co.jp	Japan
The Lycra Company	Lycra™	https://www.lycra.com	USA
The Taekwang & Daehan Company	Acepora™	http://www.acepora-fiber.com	South Korea
Toyobo Co., Ltd.	Espa™	https://www.toyobo-global.com	Japan
Yantai Tayho Advanced Materials Co., Ltd.	Newstar™	http://en.tayho.com.cn	China

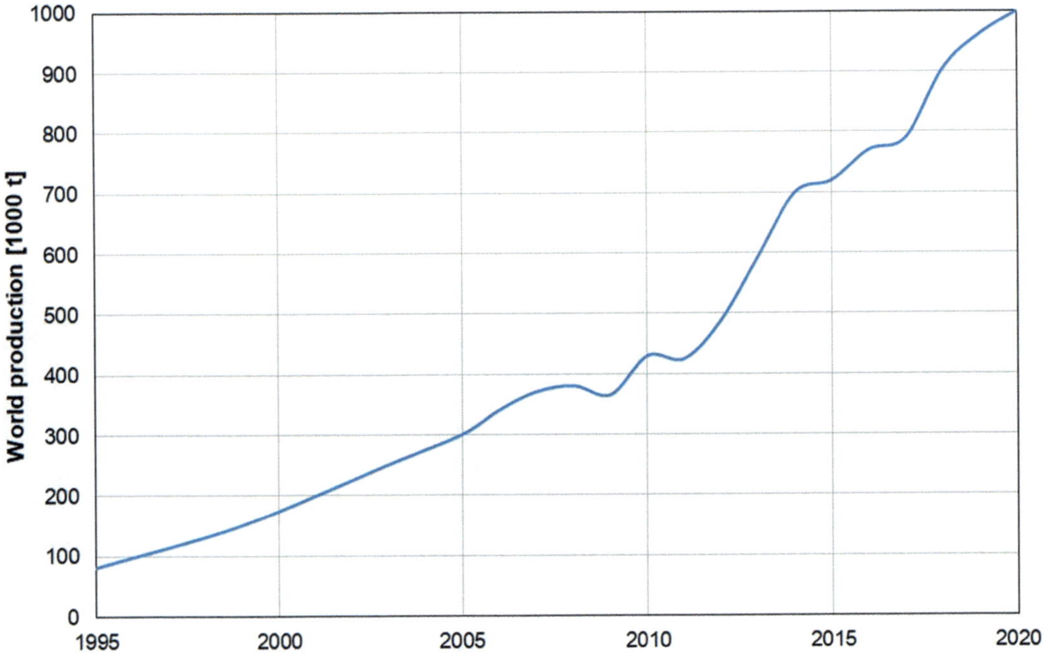

Fig. 36.15 Development of world production (data: The Fiber Year, 2021)

The development of market shares by region shows Fig. 36.16. Since about 2000, most elastane fibers have been produced in China; before that, Europe, Japan, Korea and the USA were leading producers.

The largest exporters are China (70,000 t), Korea (25,000 t) and Taiwan (13,000 t). The price for elastane fibers is approx. 4–8 €/kg for 40 dtex yarns.

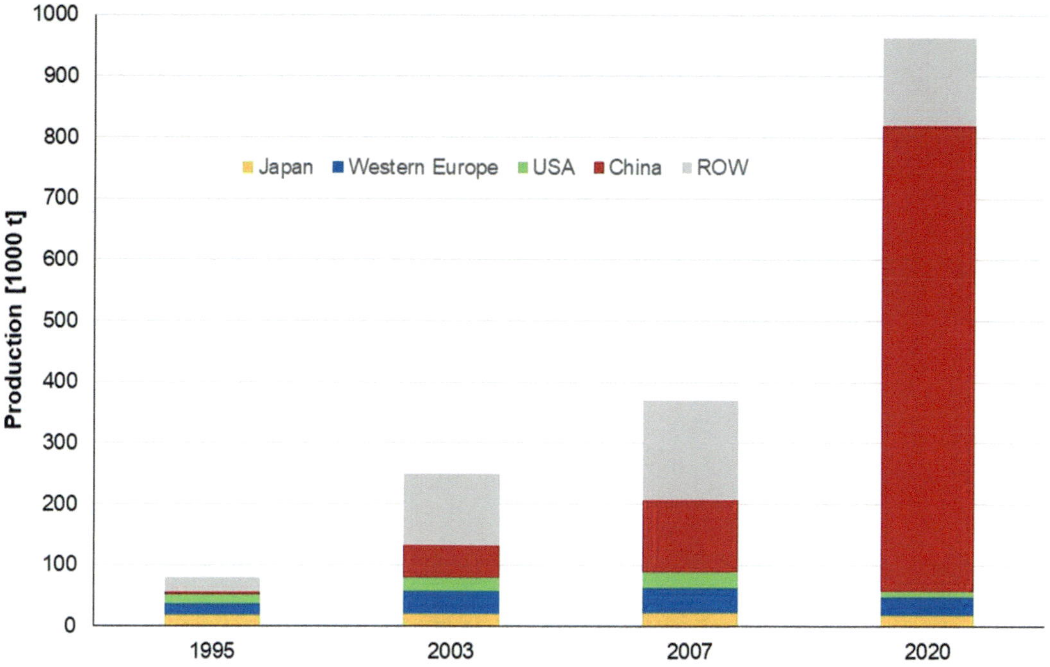

Fig. 36.16 Development of elastane fiber production by region (data: The Fiber Year, 2021)

References

Bayer, O. (1963). Das Diisocyanat-Polydadditionsverfaren. Historische Entwicklung und chemische Grundlagen. In: Vieweg, R. (Hrsg.), *Kunststoff-Handbuch*, Bd. 7: Polyurethane. Hanser, München.

Charch, W. H., & Shivers, J. C. (1959). Structure-properly relationships in synthetic fibers, Part II: Elastomeric condensation copolymers. *Textile Research Journal, 29*, 536–540.

Fabricius, M., Gries, T., & Wulfhorst, B. (1995). Fiber table acc. to P.-A. Koch: Elastan fibers. *Melliand Textilberichte, 76*, 980–990.

Hicks, E. M. (1963). Lycra Spandex fiber, I. Structure and properties. *American Dyestuff Report, 52*(1), 33–35.

Oertel, H. (1965). Elastomere Fäden auf Polyurethanasis, ihr Aufbau, ihre Eigenschaften, ihre Verwendung. *Melliand Textilberichte, 46*, 51–59.

Rinke, H. (1968). Polyurethane als Grundlage für elastomere Fäden. *Chimia, 22*, 164–175.

The Fiber Year. (2021). *The Fiber Year GmbH, Roggwil.* https://www.thefiberyear.com

High-performance Fibers 37

37.1 Definition

Compared to standard polymers, high-performance polymers are characterized by an extraordinary property profile even under extreme external conditions. For example, they have a high thermal and/or chemical resistance as well as good mechanical properties. Some of the polymers in this group can be spun thermoplastically or in solution using standard processes, while special processes have been developed for a few, e.g. PTFE. Due to these special properties and the associated complex manufacturing processes at partly very high temperatures, the fiber prices are 25–100 €/kg and more.

Most polymers of this group have phenyl rings (Fig. 37.1), which makes them rigid on the one hand and extremely chemically stable on the other, because the corresponding chemical bonds can hardly be influenced by external energy input, which explains the good thermal properties.

Aramids also belong to this type of fibers, but they are dealt with in a separate chapter because of their great market importance.

37.2 Polytetrafluoroethylene Fibers (PTFE)

PTFE was first synthesized in 1938 by the American chemist Roy Plunkett. Colette Grégoire came up with the idea of coating pans with PTFE and was granted a patent for it together with Georgette Wamant. In 1954, DuPont, USA, launched the first filament yarns and staple fibers made of PTFE (Fig. 37.2).

PTFE fibers have an extremely low coefficient of friction (0.01–0.04), so that static and sliding friction are approximately equal and the transition from rest to movement is jerk-free.

37.2.1 Production Method

PTFE fibers can be produced in various ways. Mostly the matrix wet spinning process is used. For this, a solution of 60% PTFE and 8% viscose is suspended together with alkyl or acrylic polyglycol ether, filtered, spun out at 20 °C through the spinneret into a spinning bath. This consists of 10% sulfuric acid, 16% sodium sulfate, 10% zinc sulfate and water. After coagulation, the filaments are washed and dried. This is followed by drawing on hot godets (360–390 °C) by a factor of 7 (Fig. 37.3). In the process, the cellulose burns and the PTFE is sintered to form a multifilament yarn. The resulting yarns are yellow-brown in color, which can be bleached by treatment in boiling sulfuric acid with the addition of nitric acid (Fourné, 1995).

Fibers produced in this way have a strength of 10–20 cN/tex, an elongation at break of approx. 18%, shrink relatively strongly at high

D. Veit, *Fibers*, https://doi.org/10.1007/978-3-031-15309-9_37

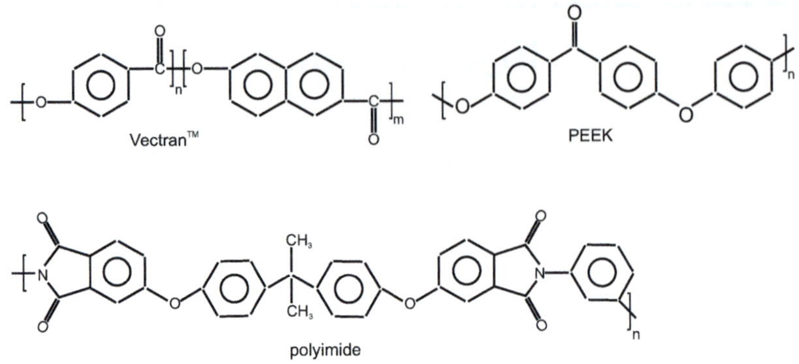

Fig. 37.1 Structural formulas of selected high-performance fibers

temperatures (10–20%) and have a continuous service temperature of 250 °C.

Alternatively, PTFE fibers can also be paste-extruded. For this purpose, PTFE is combined with a lubricant (15–25% by mass). Through the combination of the process steps extrusion and rapid hot stretching, the fiber expands and cracks and a fine network of fibrils and pores is formed. The pore size is 300–800 μm depending on the stretching temperature (250–300 °C). The proportion of pores can be up to 96%. The strength of these so-called ePTFE fibers is significantly higher than that of standard PTFE fibers and ranges from 10–60 cN/tex, the elongation at break is 3–5%, as is the shrinkage. The density of ePTFE is about 0.2–2.2 g/cm^3, depending on the pore size, and the thermal conductivity is 0.1–0.3 W/mK. The continuous service temperature of 260 °C is somewhat higher than for PTFE. The LOI is 95, which is one of the highest values ever for polymers. Both PTFE staple fibers and PTFE filaments are available. Gore-Tex™ membranes are also produced using this process. Water vapor can escape through the pores from the inside, but they are so small that no water in liquid form can penetrate from the outside. A detailed description of the manufacture and properties of ePTFE fibers is given in (Ebnesajjad, 2017).

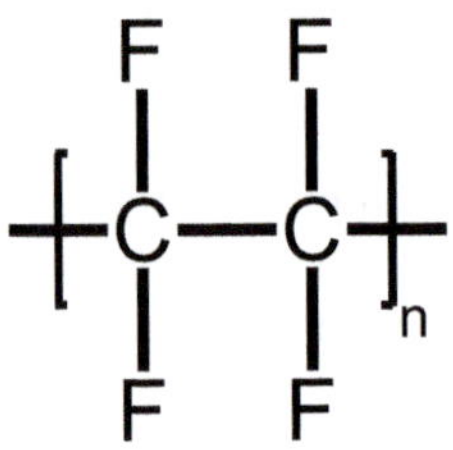

Fig. 37.2 Repeat unit of PTFE

The Fig. 37.4 shows a PTFE ribbon on the left and an ePTFE monofilament on the right.

Due to the inherent stability of PTFE, its resistance to chemicals, heat and UV, as well as its good abrasion resistance and low coefficients of friction, both matrix-spun and paste-extruded PTFE fibers are used for many different products. These include filters, plain bearings, seals, electrostatic shields, dental floss and implants.

37.3 Polyimide Fibers (PI)

The basic structure of polyimides shows the Fig. 37.5. In addition to this basic form, there are also polyamide imides (PAI) and polyetherimides (PEI), which also contain amide and ether groups, respectively.

The first polyimides were developed by DuPont as early as 1953, the first polyamide imides by Amoco in 1964, and the polyimides by General Electric in the early 1980s.

Polyimides are usually formed in a two-stage process from the monomers by polyaddition (thermosets) or polycondensation (thermoplastics). The fibers are usually produced by wet or dry spinning with common solvents; melt spinning is also possible.

Common to all polyimides are

- Chemical resistance to dilute acids and alkalis, organic solvents, fuels, greases and oils,

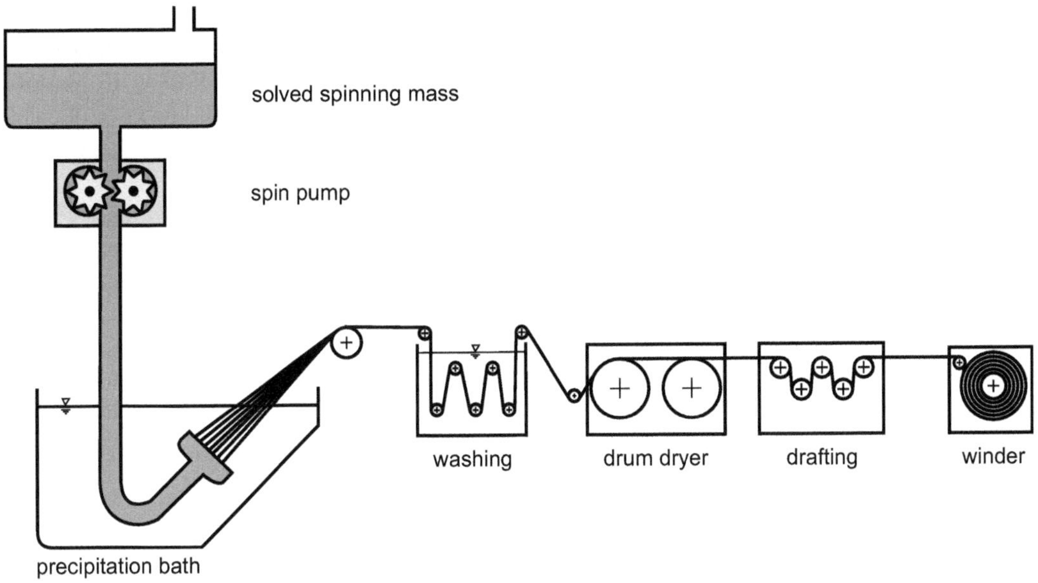

Fig. 37.3 Design of a matrix wet spinning unit for PTFE fibers

- low resistance to concentrated acids and alkalis, amines, hot water and water vapour as well as oxidising agents,
- high radiation resistance even to high-energy radiation,
- medium moisture absorption (1–4%),
- low flammability, self-extinguishing and low smoke emission, LOI value 38,
- moderate weather resistance (deterioration of mechanical values) of PI,
- very good weather resistance, excellent resistance to UV radiation of PAI and PEI.

The Fig. 37.6 shows as a typical example the structure of the polyimide P84™, which is a copolymer of 2 components and is manufactured by Evonik Industries AG. Typical fiber titers are 0.8–6 dtex for staple fibers (fiber length 38–76 mm) and 1.67–3.3 dtex for filaments,

The structure of a typical polyetherimide is shown in Fig. 37.7 by the example of Ultem™. These fibers have a much lower melting point than polyimides at 215–225 °C. Strength (19–27 cN/tex) and Young's modulus are slightly lower than PI, and elongation at break is slightly higher (60–80%). The LOI is 44–45.

The structure of PEI is irregular and largely prevents crystal formation. Therefore, PEI fibers are predominantly amorphous. The weight of the repeat unit of PEI, 591 g/mol, is the highest of all polymers used for fiber production. The mechanical properties change little within a temperature

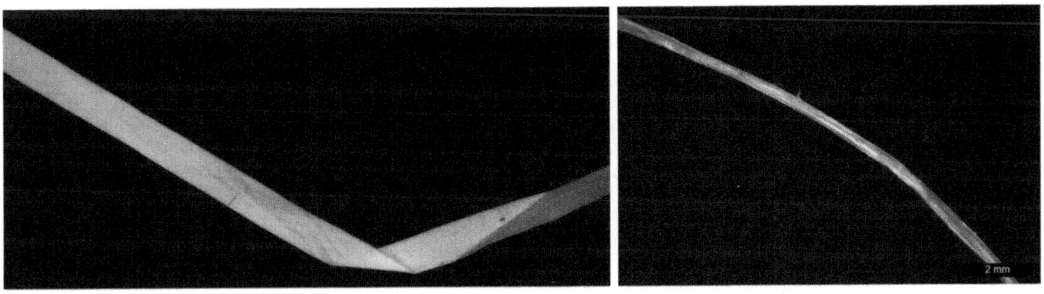

Fig. 37.4 PTFE and ePTFE fibers

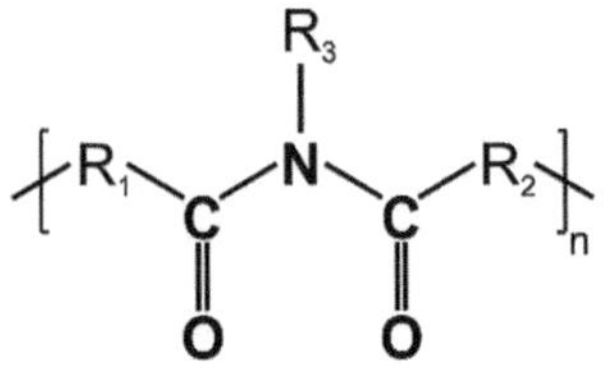

Fig. 37.5 Structure of the repeat unit of polyimides (imide group bold)

range from −240 °C to 260 °C. Polyimides can also withstand temperatures up to 1000 °C for a few seconds. Typical strength values are 25–40 cN/tex, elongation at break is 30–38% for staple fibers and approx. 20% for filaments, modulus of elasticity is 350–500 cN/tex, hot water shrinkage is 1–3%. The glass transition temperature is 315 °C, the density is 1.41 g/cm^3.

Typical filaments made from PI have a fineness of 2–11 dtex, while 0.6–8 dtex are common for staple fibers. Often PI fibers have a multilobal cross-section, which makes them particularly suitable for filter applications. PI fibers are also frequently combined with other materials, e.g. PTFE and glass fibers. PI fibers have good knot and loop strength, which simplifies textile processing (Fig. 37.8).

PEI fibers are produced, for example, by Kuraray, Japan, and marketed under the Kurakiss™ brand. Trofil PEI™ of Monofil-Technik GmbH is also a well-known trade mark. PAI fibers are offered under the brand name Kermel™. They are explained in the chapter "Aramid fibers" because they are counted among the meta-aramids.

Polyimide and related fibers are used to produce technical textiles that are exposed to particularly high temperatures. These include pilot clothing, fireproof underwear, filters for cleaning hot gases, e.g. in power stations and dust extraction plants.

37.4 Polyetherketone Fibers (PEK, PEEK)

This class of polymers is composed of phenyl rings alternately linked by a ketone bridge and an ether bridge. This results in a very rigid structure, similar to PET and the aramids, leading to high glass transition and melting temperatures. The most commonly used PEK for fiber production is polyetheretherketone (PEEK) (Fig. 37.9).

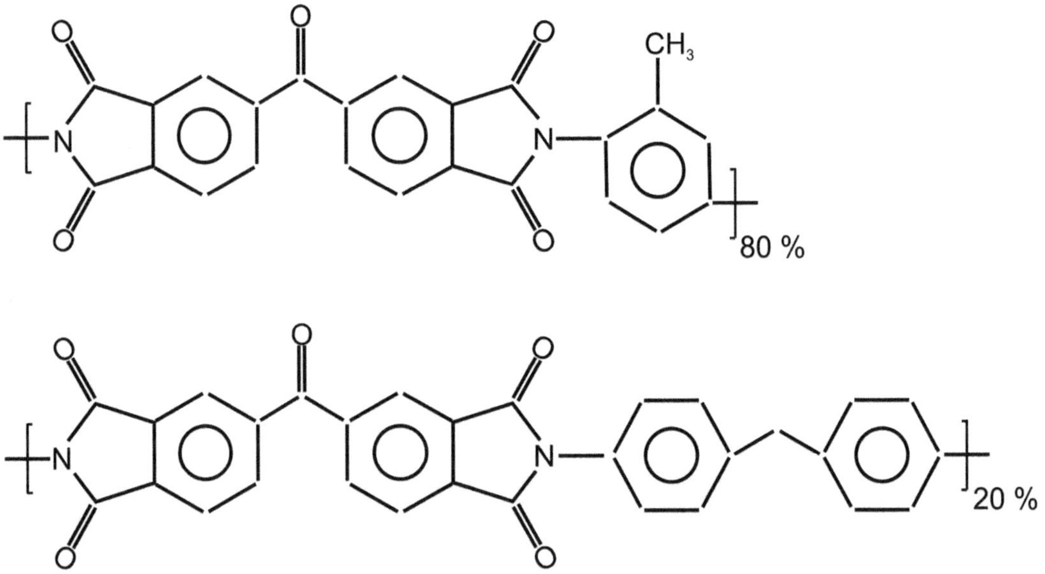

Fig. 37.6 Structure of the repeat units of polyimide (P84™)

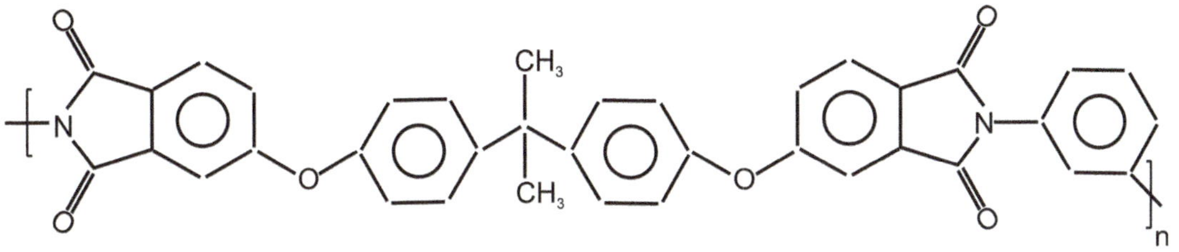

Fig. 37.7 Structure of the repeat unit of PEI (Ultem™)

37.4.1 History

In 1962, DuPont, USA, developed the first PEK polymers, followed shortly afterwards by ICI, UK, under the brand name Victrex™, which still exists today. In the meantime, several manufacturers exist, mainly in the USA and Great Britain.

37.4.2 Structure

PEEK are semi-crystalline (30–50%) thermoplastics. The ratio of ether to ketone bridges essentially determines the physical properties. More ether groups increase the flexibility of the macromolecules and thus facilitate spinning out via better flow behavior. A higher proportion of ketone groups increases the temperature resistance. Ratios of 2:1 to 2:3 are common.

37.4.3 Production

PEEK fibers are melt spun. The principle corresponds to the common melt spinning processes, only the spinning components must be able to withstand significantly higher temperatures compared to conventional polymers. At 380 °C the viscosity is 300–500 Pa·s, at 400 °C still 350 Pa·s. If the process comes to a standstill, the polymer decomposes rapidly (Fourné, 1995).

After spinning out, the filaments are cooled in a water bath; air cooling can also be installed upstream. The water bath is located approx. 0.5–0.8 m below the spinneret and has a water temperature of approx. 60 °C. The amorphous structure remains in the melt due to the rapid cooling. Due to the rapid cooling, the amorphous structure from the melt is largely retained. Subsequently, the filaments are hot drawn in several

Fig. 37.8 Cross-sections of P84 fibers and creels (Evonik Industries AG)

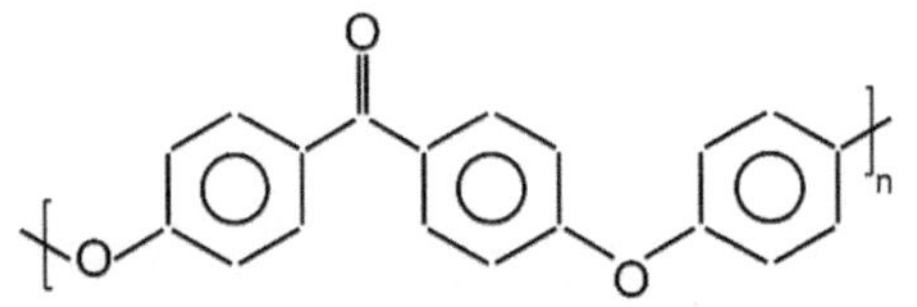

Fig. 37.9 Structural formula of the repeat unit of PEEK

stages up to approx. 4 times their original length at approx. 200 °C. The inner structure of the filaments is then treated by heat treatment. The inner structure is fixed by a heat treatment just below the melting temperature at approx. 280 °C. Finally, the filaments are relaxed. Finally, the filaments are relaxed (5%) and wound up (Fig. 37.10).

Multifilament yarns and monofilaments can be produced and, by means of a downstream cutting process, also staple fibers and short fibers.

37.4.4 Features

Depending on the composition, the glass transition temperature is 139–153 °C, the melting temperature 334–355 °C. PEEK fibers retain their mechanical properties up to approx. 250 °C, and at 300 °C, they still retain 80% of their strength. The heat capacity is 1.34 kJ/kgK and the thermal conductivity is 0.25–0.29 W/mK and the LOI is 38.

PEEK fibers are biocompatible, chemically resistant to almost all chemicals except oxidizing agents and concentrated sulfuric acid, in which they are soluble. They are also resistant to high energy gamma and X-rays. Because they absorb very little water (< 0.1%), they are resistant to hydrolysis up to 280 °C and hot water shrinkage is only 0.4–0.5%. They are flame retardant, self-extinguishing and burn with very low smoke emission.

The density of PEEK is 1.27–1.30 g/cm^3. The strength is 20–100 cN/tex, the ultimate tensile elongation is 15–60% and the Young's modulus is 400–1200 cN/tex (Ullmann's, 2008). Impact strength and flexural fatigue strength are high, PEEK fibers have good sliding properties and wear little.

Typical titres are 75dex/f15 and 1230dtex/f72 for multifilament yarns, 63 dtex - 5 ktex (70–2000 μm) for monofilaments, 3–13 dtex for staple fibers (40–80 mm) and 2–4 dtex for short fibers (3–12 mm).

37.4.5 Applications and Manufacturers

PEEK fibers can be processed into textiles using the usual methods. Typical applications include conveyor belts, filters, abrasion-resistant and high-temperature-resistant cable sheathing, strings for musical instruments and tennis rackets, implants as metal substitutes (e.g. artificial hip joints), sewing threads for shoes and clothing, and the reinforcement of fiber composite parts for high-temperature applications.

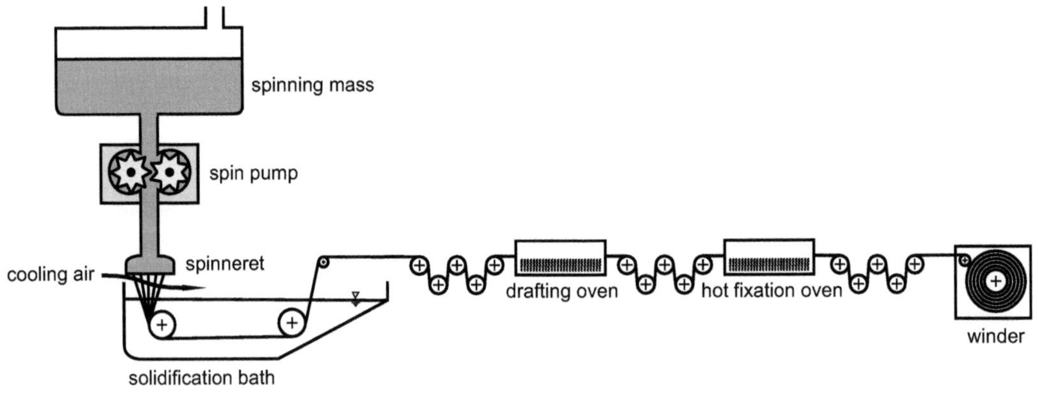

Fig. 37.10 Design of a spinning unit for PEEK fibers

PEEK is beige by nature due to the many phenyl rings. PEEK fibers are often spun-dyed or dyed with disperse dyes.

Until a few years ago, there were numerous manufacturers of PEEK fibers, including BASF, Amoco and Monsanto. Common trade names were Arotone™, Doctalex™, Kadel™, Mindel™, Satolite™, Staber™ and Ultrapek™. Probably the most important manufacturer of PEEK fibers is Zyex, Ltd., Stonehouse, UK, a subsidiary of Victrex plc. In addition, Solvay AS produces PEEK fibers in significant quantities and Evonik supplies PEEK granules. In Germany, Trofil PEEK™ is produced by Monofil-Technik GmbH. There are no other major producers.

37.5 Polysulfone Fibers (PSU, PES)

Polysulfones are amorphous or semi-crystalline, thermoplastic and still stable at high temperatures. The typical structure of polysulfones with the characteristic diphenylsulfone group is shown in Fig. 37.11. This group is very rigid and gives the polymer its high heat resistance, a glass transition temperature of 190–230 °C and a melting point of about 520 °C.

However, the rigid chain makes it difficult to make fibers from long macromolecules because the polymer decomposes before it melts. Therefore, fibers are either made by wet spinning, using DMAC as a solvent, or ether groups are incorporated into the chain (polyether sulfones, PES, Fig. 37.12), which provide better mobility of the macromolecules, allowing melt spinning.

Alternatively, the chain length can be reduced during polymer formation by the use of chain terminators, which reduces the viscosity to such an extent that melt spinning becomes possible.

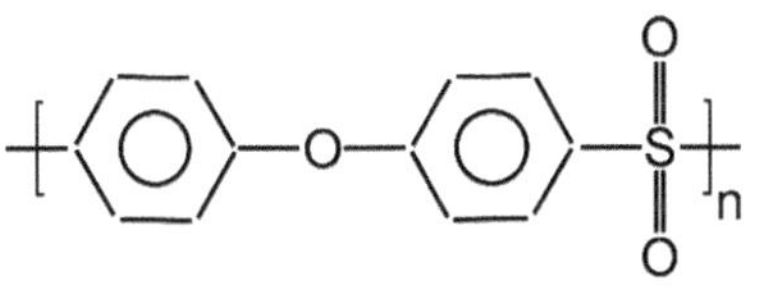

Fig. 37.11 Structure of the repeat unit of polysulfone

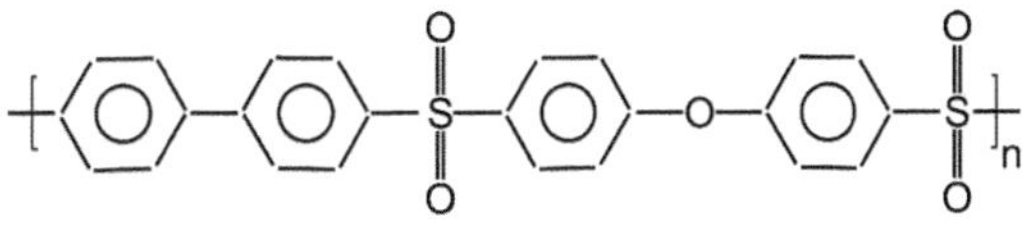

Fig. 37.12 Structure of the repeat unit of polyethersulfone (PES)

However, this reduces the temperature resistance of the polymer.

Polysulfones were first produced by Union Carbide, USA, in 1965. Other early producers included 3 M and ICI, which also produced polyether sulfones.

The most important properties of polysulfone fibers are summarized in Table 37.1.

Compared to the polyimides, which are outstanding in this respect, the polysulfones can only be used in a significantly smaller temperature range. However, the mechanical properties remain largely unchanged between −100 °C and 200 °C (PESU) or 150 °C (PSU). The dimensional stability under the influence of heat is also very high. The inherent color of PSU is pale yellow. PSU fibers are significantly cheaper than other, high-temperature-resistant fibers, e.g. PEEK.

Hollow fiber membranes are a typical product. Porosities of up to 80% are achieved with pore sizes of up to 0.02 μm. Since PSU does not hydrolyze and withstands high temperatures, it is generally suitable for medical products that must be sterilized in autoclaves. PSU and PESU fibers are also processed in blends with other fibers because they not only provide good flame retardancy, but also have little or no degradation of mechanical properties. Often, for high-temperature applications, PSU resins are reinforced with other fibers, e.g. glass.

37.6 Polyphenylene Sulfide Fibers (PPS)

PPS fibers are melt-spinnable and were synthesized as early as 1888 by C. Friedel and J.M. Crafts. Large-scale production did not begin until 1967. Figure 37.13 shows the structure of PPS.

Table 37.1 Typical properties of polysulfone fibers acc. to manufacturer's specifications

Chemical resistance	
Good	Alcohol, fuel, detergents, acids, bases, fats, oils
Bad	Hot water (permanent exposure)
	Aromatic and halogen hydrocarbons
	Ester, ketones
	Concentrated sulphiric and saltpeter acid
Radiation resistance	
Good	Energetic beta-, gamma- and x-ray-radiation
Bad	UV-radiation
Hydrolysis resistance	
– Medium water take-up and high resistance (PSU)	
– High water take-up (PESU)	
Flammability	
– Self-extinguishing	
– Low smoke development	
Other properties	
– Non-continuous weather resistance	
– Good electric insulation properties	
– Very good sterilizable	

PPS is used in linear form is highly crystalline, but there are also cross-linked forms used for resins.

PPS is usually mixed with fillers (up to 70% content) or reinforced as a resin with other fibers, e.g. glass.

Chemical resistance at 150–190 °C, radiation and hydrolysis resistance as well as flammability and electrical properties are similar to those of PSU fibers. PPS fibers are also slightly yellowish. In mechanical properties, PPS fibers are similar to PET, but both the glass transition temperature and the melting temperature are higher (Table 37.2). PPS fibers can be used permanently up to 240 °C and for short periods up to 300 °C.

PPS has a high hardness and stiffness, which makes textile processing more difficult.

PPS fibers are spun as multifilament yarns and staple fibers and are also produced as nonwovens. Typical applications are high temperature filters. The most important manufacturers of PPS fibers are Celanese, Evonik, Toray, Huvis and EMS-Griltech. In Germany, Trofil PPS™ is produced by Monofil-Technik GmbH.

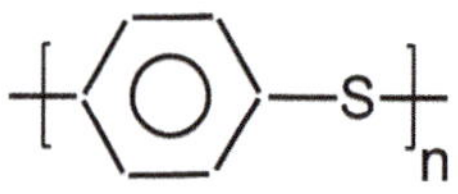

Fig. 37.13 Structure of the repeat unit of polyphenylene sulfide (PPS)

37.7 Melamine Resin Fibers (MF)

Melamine was first synthesized by Justus von Liebig in 1834, and industrial production began in 1930. Melamine fibers (MF) are copolymers and consist of at least 50% crosslinked melamine groups (Fig. 37.14, right). These are grafted onto long alkyl chains, because melamine in its pure form can hardly be spun. The network is formed in a polycondensation reaction of melamine and melamine derivatives with formaldehyde. The melamine groups are then linked together by various ether bridges (Maity & Singha, 2012).

Melamine fibers are produced by the dry spinning process. Basofil™, developed by BASF SA in 1989 and whose production facility was sold in 2002, is the only melamine-based fiber currently marketed. The most important fiber properties are summarized in Table 37.3.

The fiber diameter of melanin fibers is uneven (8–20 μm), therefore no filaments can be produced, only staple fibers. These are combined

Table 37.2 Properties of PPS fibers acc. to manufacturer's specifications compared to PET fibers

	PPS	PET
Typical fiber fineness [dtex]	2–8	0.5–44
Tenacity [cN/tex]	27–47	40–85
Breaking elongation [%]	25–50	8–50
E-modulus [cN/tex]	270–370	
Moisture absorption [%]	0.2–0.6	0.2–0.5
Hot water shrinkage [%]	3.5–5.0	0.5–8
Glass temperature [°C]	90	70
Melting point [°C]	285	255–260
Self-ignition temperature [°C]	590	
LOI	34–35	22
Density [g/cm^3]	1.34–1.37	1.38

with cotton or synthetic man-made fibers, e.g. polyester, viscose, modacrylic and aramid, and processed into fire-resistant textiles using conventional methods (Fig. 37.15).

Basofil fibers do not melt, but begin to decompose above a temperature of approx. 370 °C. Only above approx. 200 °C does the strength gradually decrease (Fig. 37.16).

When the melamine fibers decompose, nitrogen is released, which actively inhibits fire spread. The decomposition of Basofil™ fibers does not lead to toxic degradation products. At 0.028 W/mK, the thermal conductivity of Basofil™ fibers is even lower than that of m-aramid. Therefore, textiles with Basofil™ content are also used as heat shields, e.g. for fire protection suits. These often consist of an aluminum layer (reflection of heat radiation), a dense fabric made of yarns in a core-sheath structure (core: glass filament; sheath: Basofil™ or aramid) and, on the inside, a dense nonwoven fabric made of Basofil™.

37.8 Novoloid Fibers

These fibers are melt spun from phenolic resins (Novolak, Fig. 37.17) and produced by cross-linking with aldehydes. Novoloid fibers must contain at least 85% novolak.

The typical network structure is depicted in Fig. 37.18.

Novoloid fibers are resistant to most acids, dilute alkalis and organic solvents and are only attacked by nitric and sulfuric acid. They are infusible and can be used up to 150 °C in air

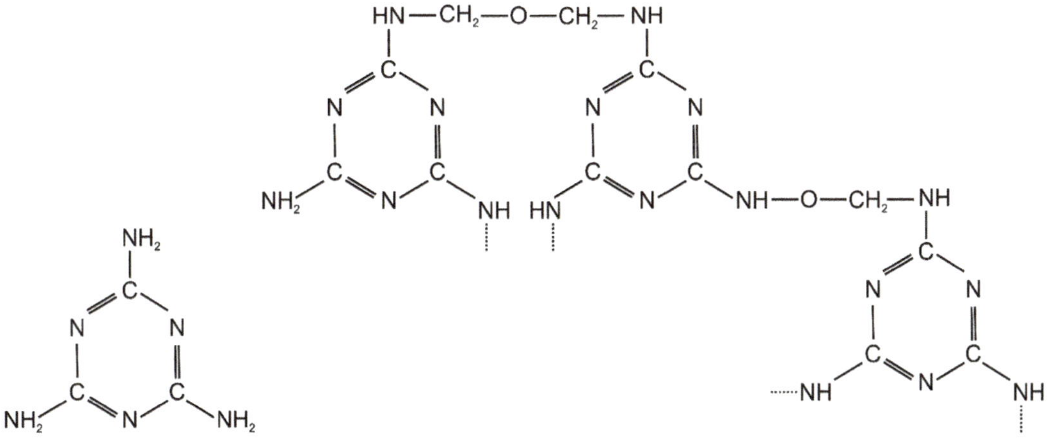

Fig. 37.14 Structure of melamine (left) and melamine network (right)

Table 37.3 Properties of Basofil™ fibers (acc. to manufacturer)

	Basofil™
Typical fiber fineness [dtex]	1.5–3
Tenacity [cN/tex]	20–25
Breaking elongation [%]	15–25
E-modulus [cN/tex]	50
Moisture absorption [%]	5
Hot water shrinkage [%]	< 1
Longterm temperature resistance up to [°C]	200
Disintegration temperature [°C]	400
Heat conductivity [W/mK]	0.028
LOI	30–32
Density [g/cm^3]	14

Fig. 37.15 Basofil™ fibers in cross section

and up to 250 °C in an oxygen-free atmosphere without significant damage. They can withstand temperatures up to 1000 °C for short periods. When decomposing, the outer fiber layer carbonises without losing its inner structure. Hardly any fumes are produced.

The typical properties are summarized in Table 37.4 summarized.

Novoloid fibers have a good moisture absorption capacity and a soft handle combined with acceptable mechanical properties. Therefore, they are well suited as a blend component especially for fire resistant clothing and other high temperature applications.

Novoloid fibers were developed in the late 1960s by Gunei Chemical Industry Co. This company is the world's only manufacturer of this type of fiber and markets it under the brand name Kynol™.

37.9 Polyazole Fibers (PBO)

Polyazole fibers made from poly(p-phenylene-2,6-benzobisoxazole) (PBO) have been commercially produced since 1998 by Toyobo, Japan, the developer and to date the only manufacturer, and marketed under the brand name Zylon™. Due to their high content of aromatic groups, they are relatively rigid (Fig. 37.19). They are

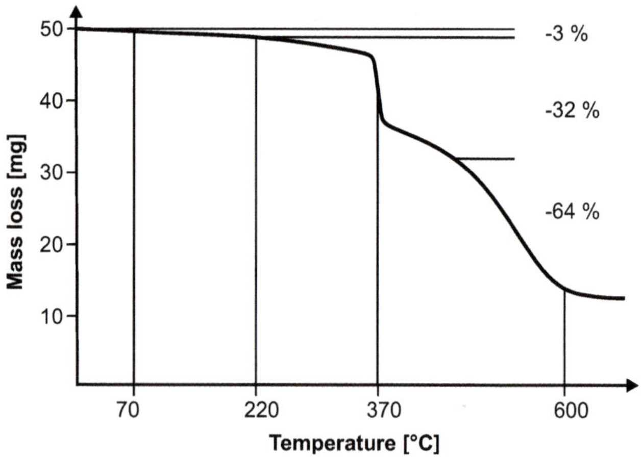

Fig. 37.16 Loss of mass of basofil fibers due to decomposition (acc. to manufacturer's data)

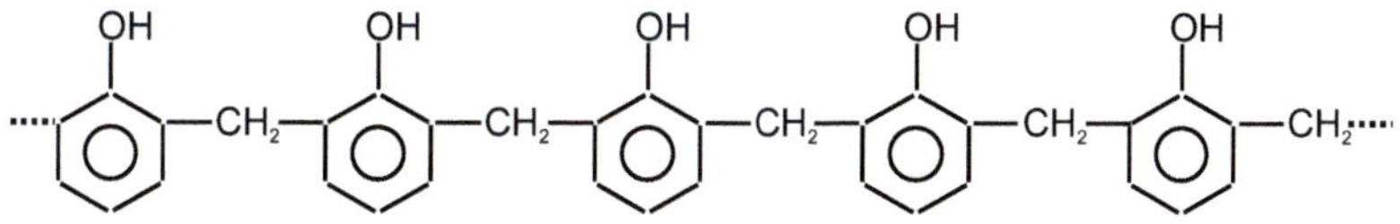

Fig. 37.17 Basic structure of novolak phenolic resin

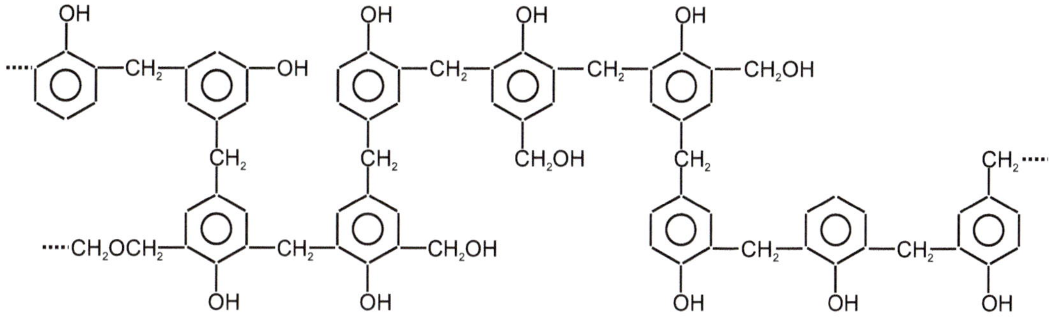

Fig. 37.18 Network structure of Kynol

characterized by extremely good temperature resistance and very good mechanical properties.

PBO fibers are not soluble in organic solvents without modification and are therefore dissolved in strong acids and usually spun out in liquid crystalline form in the air gap spinning process. Polyphosphoric acid or methanesulfonic acid are frequently used for this purpose. Usual spinning speeds are about 100 m/min. PBO fibers are largely isotropic and the macromolecules have a degree of orientation of more than 95%. The fibers have a core-sheath structure. In the core, the macromolecules form microfibrils (Ø 10–50 nm) with a length of about 10 nm, which are separated from each other by microvoids. There are no microvoids in the sheath (Kitagawa, 2001), Fig. 37.20.

PBO fibers still exhibit a very high tensile strength and an extremely high modulus of elasticity even at temperatures above 550 °C. The decomposition of PBO fibers only starts at 650 ° C. Decomposition does not begin until 650 °C. A distinction is made between standard and high modulus types (Table 37.5). Due to their high stiffness, they are difficult to process in textiles. Their LOI value of 68 is also exceptionally high and makes them practically inflammable.

Zylon™ fibers have a golden-yellow color, shrink only slightly even when exposed to heat, are moderately resistant to acids and alkalis, and are very abrasion-resistant. However, despite their low moisture absorption, they are sensitive to water and lose their good strength on prolonged exposure to UV radiation.

Table 37.4 Typical properties of novoloid fibers acc. to manufacturer‘s specifications

	Kynol™
Typical fiber fineness [dtex]	2–11
Tenacity [cN/tex]	12–18
Breaking elongation [%]	10–60
E-modulus [cN/tex]	260–350
Loop strength [cN/tex]	19–27
Knot strength [cN/tex]	10–13
Moisture absorption [%]	6
Hot water shrinkage [%]	0
Longterm temperature stability up to [°C]	150–250
Heat conductivity [W/mK]	0.0002
LOI	30–34
Density [g/cm³]	1.27

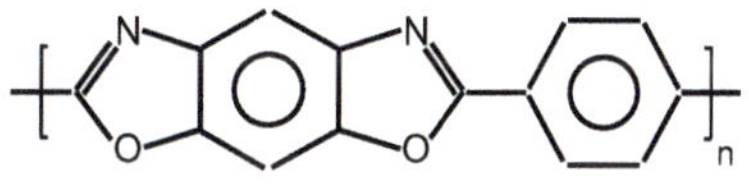

Fig. 37.19 Structural formula of the repeat unit of Zylon™

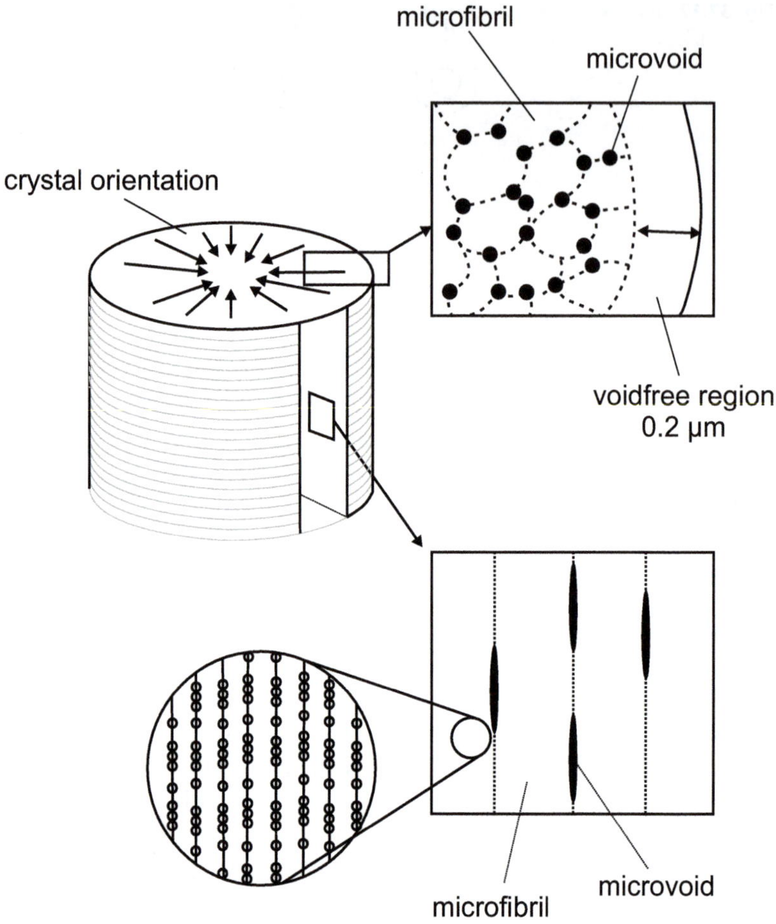

Fig. 37.20 Internal structure of PBO fibers acc. to (Kitagawa et al., 1998)

Typical applications are conveyor belts and roller coverings. Braids made of Zylon™ fibers are used for so-called super balloons, which rise to the upper atmosphere and develop extremely high internal pressures. PBO fibers react sensitively to compressive loads, which is why they are primarily used to absorb tensile forces in the axial direction.

Table 37.5 Typical properties of Zylon™ fibers acc. to manufacturer's specifications

	Zylon™ Standard	Zylon™ Hochmodul
Typical yarn fineness [den]	250, 500, 1000, 1500, 3000	250, 500, 1000, 1500, 3000
Tenacity [cN/tex]	370–530	370–530
Breaking elongation [%]	3.5	2.5
E-modulus [cN/tex]	11,500	17,200
Moisture absporption [%]	2	0.6
Hot water shrinkage [%]	< 1	< 1
Longterm temperature resistance bis [°C]	550	550
Melting temperature [°C]	650	650
LOI	68	68
Density [g/cm^3]	1.54	1.58

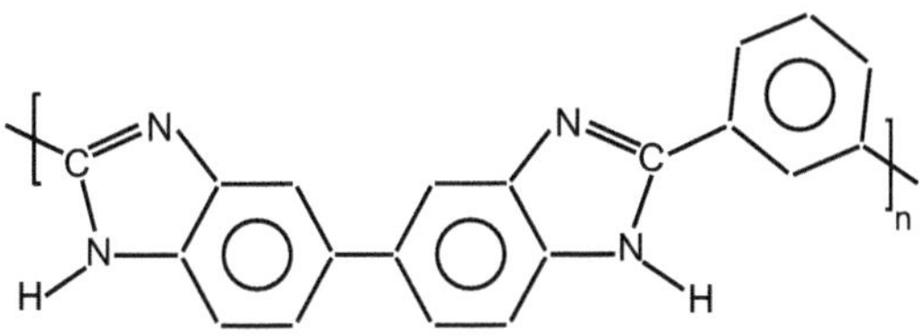

Fig. 37.21 Structural formula of the repeat unit of PBI

37.10 Polybenzimidazole Fibers (PBI)

PBI fibers are extremely fire resistant even at high temperatures. They were developed in the 1950s by C.S. Marvel and H. Vogel and have been used since 1967, after the fire in the Apollo 1 capsule, as a fiber material for astronaut and pilot suits. Since 1983, they have been produced on a larger scale by Celanese Corp, USA.

The Fig. 37.21 shows the structure of PBI, which is very similar to that of PBO.

To increase the fire resistance of the fibers, PBI fibers are usually treated with sulfurous acid (Fig. 37.22). This reduces shrinkage to less than 10% when exposed to heat. The effect can be further increased if phosphoric acid is used instead of sulfurous acid.

PBI is formed by a polycondensation reaction of diphenyl isopthalate and tetraaminobiphenyl in a two-step process. First, the monomers react with each other at approx. 300 °C, splitting off water and phenol and forming long chains. This forms a foam, which is cooled and pulverized. This powder is then heated to about 400 °C for 2–3 hours, further increasing the chain length and forming a fine-grained golden-brown powder. The resulting PBI is dissolved in dimethylacetamide (DMAC) at a polymer concentration of 25%. By adding lithium chloride, the shelf life of the solution can be increased to several months.

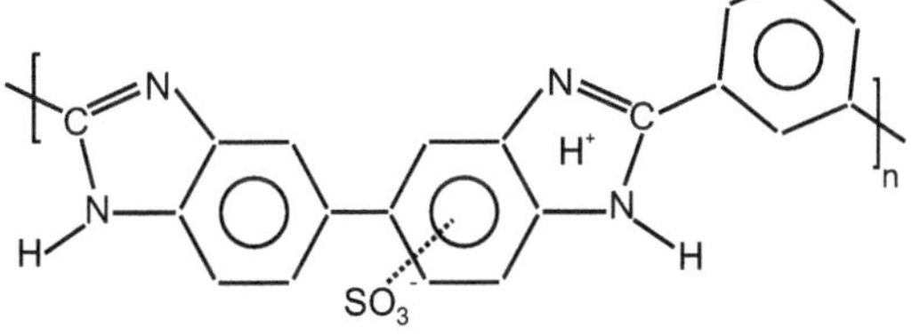

Fig. 37.22 PBI after treatment with sulfurous acid

The filtered solution is then dry spun out into a hot atmosphere, largely evaporating the DMAC. The spun fibers are combined into tows, washed and stretched. Sulfonation is then carried out in aqueous medium. The fibers are optionally crimped and processed into staple fibers or short fibers (Hearle, 2001) (Fig. 37.23).

The glass transition temperature of PBI fibers is 425 °C, above 550 °C they start to decompose. They are therefore temperature resistant at over 300 °C in continuous use and at over 1000 °C for a few seconds. PBI fibers do not burn in air, but carbonize and extinguish automatically after removal of the flame. The LOI value is 38–43. PBI is resistant to most inorganic acids and alkalis as well as many organic chemicals. The Table 37.6 summarizes typical properties of PBI fibers. They can be processed very well into textiles and are used both in their pure form and in blends with other refractory fibers (e.g. aramide). Their color is golden-brown.

Typical applications of PBI fibers are outer fabrics for fire protection suits, fireproof hoods and boots as well as gloves and insulation material (nonwovens).

The sole producer of PBI fibers is PBI Performance Products, Inc.

37.11 Polyoxadiazole Fibers (POD)

Polyoxadiazole fibers (POD) are particularly resistant to heat and dilute inorganic acids and alkalis. The Fig. 37.24 shows the repeat unit of poly-p-phenylene-1,3,4-oxadiazole. It is the only form of POD that is commercially available.

POD is produced by polycondensation of terephthalic acid with hydrazine sulfate. The fibers are produced by solution spinning and have no melting point.

Acc. to the manufacturer, typical filament yarn counts [dtex] are 284f200 and 1000f1000. Staple fibers have single fiber titres of 1.7, 2.2 and 3.3 dtex. Depending on the yarn type, the strength is 27–35 cN/tex (breaking twist: 3%) or 30–57 cN/tex (breaking elongation: 8–21%). The density is 1.44 g/cm^3 and the hot water or hot air shrinkage is 0.2%. Moisture absorption is

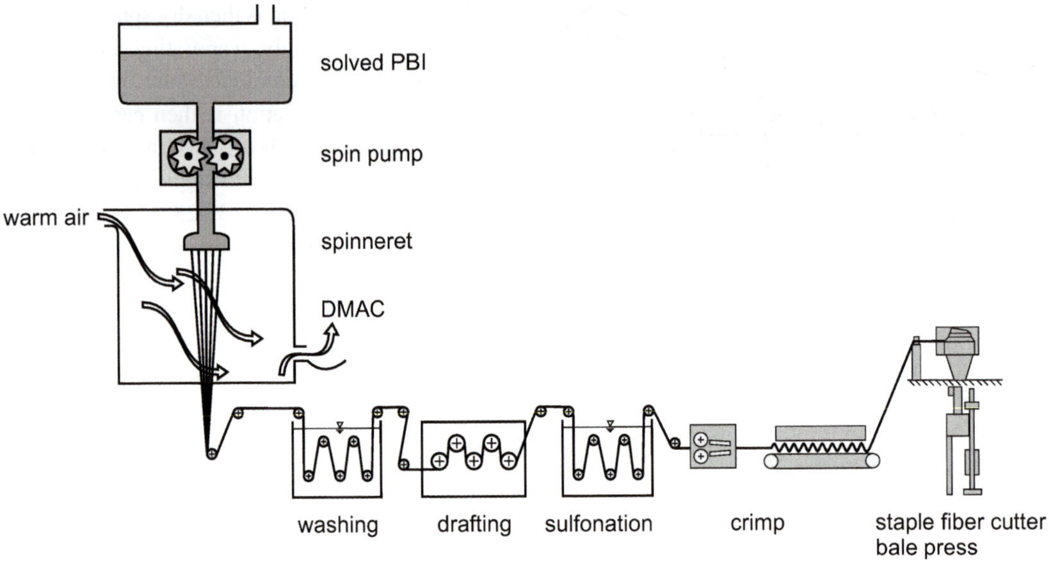

Fig. 37.23 Process for the production of PBI fibers

10–12% and the long-term service temperature is 250 °C, decomposition starts at 460–480 °C and the LOI is 30. These properties are comparable to those of m-aramid. Above 550 °C in an inert atmosphere, the fiber carbonizes, increasing density, tensile strength, and Young's modulus (Kriskovets et al., 2018). The fibers are usually golden-orange in color, but can also be spun-dyed black, grey and olive (Fig. 37.25).

Typical applications include face fabrics and sewing silk (Getadur™) for fire protective clothing, safety gloves, reinforcements for rubber and plastic parts, and filters for high-temperature applications.

The only producer is Khimvolokno Svetlogorsk in Belarus, which markets these fibers under the brand name Arselon™. There are staple fibers and filament yarns, which are processed into woven and non-woven fabrics as well as needle felts.

Table 37.6 Typical properties of PBI fibers acc. to (Hearle, 2001)

	PBI
Typical fiber fineness [dtex]	1.7
Tenacity [cN/tex]	24
Breaking elongation [%]	28
E-modulus [cN/tex]	270
Moisture absorption [%]	15
Hot water shrinkage [%]	< 1
Glass temperature [°C]	425
Disintegration temperature [°C]	450
Heat conductivity [W/mK]	0.038
LOI	41
Density [g/cm^3]	1.43

37.12 Liquid Crystal Polymers (LCP)

In liquid crystal polymers (LCP), crystalline structures are already present in the liquid phase (solution, melt), but without preferred orientation. Aramids and also UHMW polyethylene belong to this group of materials. Both are only solution-spinnable, while the Vectran™ discussed here is obtained from the melt.

Vectran™ was developed by Celanese Corp. in 1990 and is now produced and marketed by Kuraray Co, Ltd. It is spun out as a copolymer.

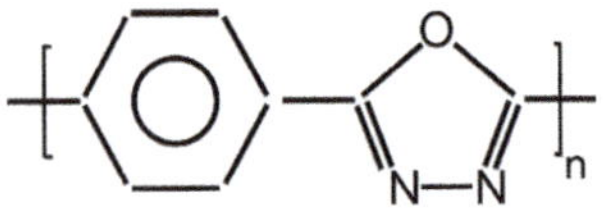

Fig. 37.24 Structural formula of the repeat unit of POD

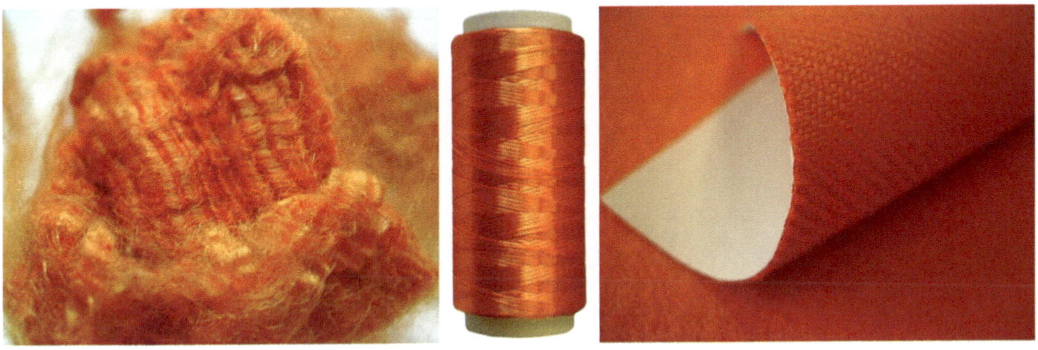

Fig. 37.25 Staple fibers, filament yarns and fabrics from POD (Khimvolokno Svetlogorsk)

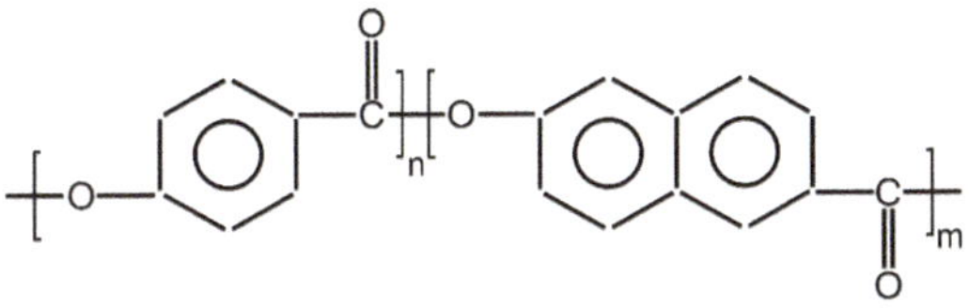

Fig. 37.26 Structural formula of the repeat unit of Vectran™

Because of its phenyl rings, it has a high stiffness (Fig. 37.26).

There are 3 different Vectran™ grades with normal and high strength as well as with high modulus of elasticity, whose typical properties are compiled in Table 37.7.

At 145–150 °C, Vectran™ fibers still retain 50% of their strength. At low mechanical loads, they even retain 90% of their strength up to 300 ° C. Unlike most other fibers, the strength of these fibers increases at very low temperatures. At around −70 °C, the strength increases by around 17%.

Vectran™ fibers have very high abrasion resistance, high fatigue bending strength and high impact strength compared to aramid. Their UV resistance is moderate, but they are resistant to microwave and X-ray radiation. They have very good resistance to highly concentrated acids, bases (conc. < 30%) and organic solvents.

Vectran™ fibers are used for ropes, sails and to reinforce bicycle tires. Because of their excellent properties at low temperatures, they were used for the airbags of the NASA Mars Pathfinder.

Table 37.7 Typical properties of Vectran™ fibers acc. to manufacturer's specifications

	Vectran™ NT	Vectran™ HT	Vectran™ UM
Typical yarn fineness [den]	200, 400, 750, 1000, 1420, 1500, 2250		
Tenacity [cN/tex]	78	220	210
Breaking elongation [%]	2.6	3.8	2.8
E-modulus [cN/tex]	3700	5350	7350
Moisture absorption [%]	< 0.1		
Hot water shrinkage [%]	< 0.2		
Longterm temperature resistance up to [°C]	145	145	150
Melting temperature [°C]	350		
Heat capacity [kJ/kg K]		1100–1420	
Heat conductivity [W/mK]		1.5–2.0	
LOI	28	28	30
Density [g/cm^3]	1.4		

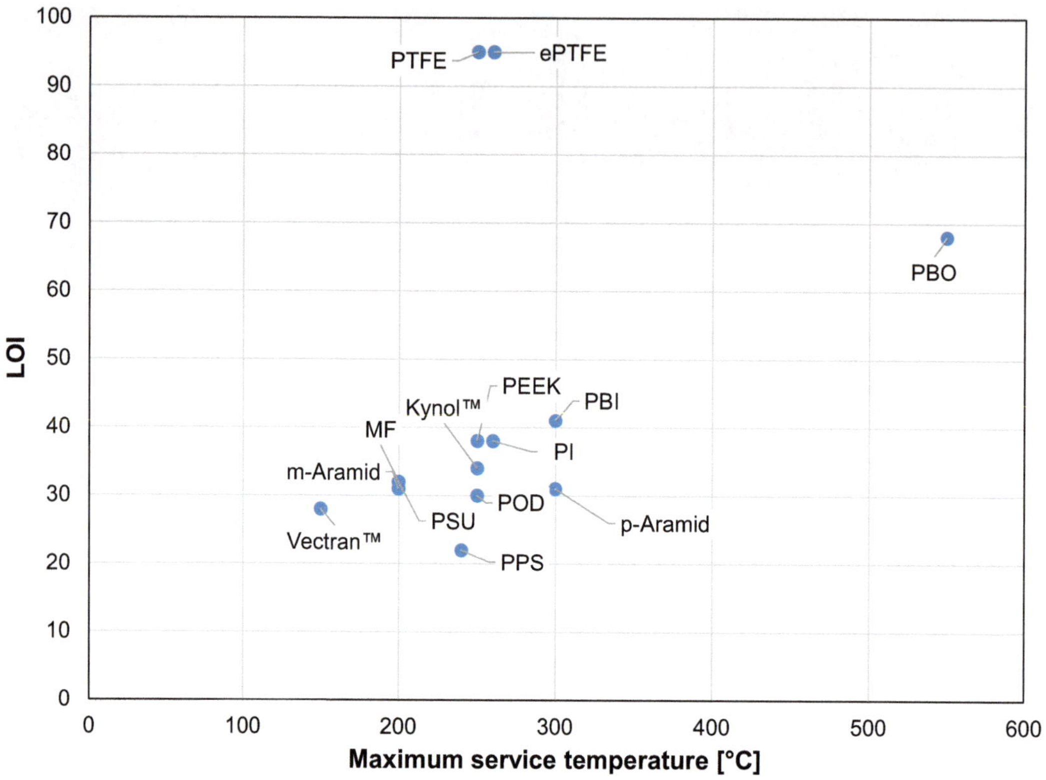

Fig. 37.27 LOI and service temperature for selected fibers

37.13 Properties

Figure 37.27 shows the most important thermal properties of the fibers described. The LOI values are mostly 28–38, the service temperature 200–300 °C. Only PTFE with its very high LOI value and Zylon™ with the highest service temperature of all polymers in this group achieve significantly different values.

37.14 Manufacturers

Most manufacturers of high-performance fibers have their headquarters in Japan or the USA. There are also some manufacturers in Europe. Some of the companies mentioned only produce the polymer, which they sell to other companies that spin it to fibers (Table 37.8).

Table 37.8 Selected manufacturers of high-performance fibers

Producer	Trade mark	Fiber type	Website	Country
3 M	Dyneon™	PTFE	http://www.3m.com	USA
BASF SA	Ultrason S™	PSU	http://www.basf.com	German
	Ultrason E™	PSU		
Bieglo GmbH	Dexnyl™	PI	http://www.bieglo.com	German
Celanese Corp.	Fortron™	PPS	http://www.celanese.com	USA
DuPont de Nemours, Inc.	Teflon™	PTFE	http://www.dupont.com	USA
EMS GRILTECH	Nexylene™	PPS	http://www.emsgriltech.com	Switzerland
Engineered Fibers Technology, LLC	WF Basofil™	MF	http://eftfibers.com	USA
Evonik Resource Efficiency GmbH	VESTAKEEP™	PEEK	https://industrial.vestakeep.com	Austria
	P84™	PI	http://www.p84.com	
Fluoron Chemicals, Ltd.	Fluoron™	PTFE	http://www.fluoronchem.com	China
W. L. Gore & Associates, Inc.	Gore-Tex™	PTFE ePTFE	http://www.gore.com	USA
Gunei Ei Chemical Industry	Kynol™	Novoloid	https://www.gunei-chemical.co.jp	Japan
Huvis Corp.	ZETAONE™	PPS	http://www.huvis.com	South Korea
Kermel SAS	Kermel™	PAI	http://www.kermel.com	France
Kuraray Co., Ltd.	Kurakiss™	PEI	http://www.kuraray.co.jp	Japan
	Vectran™	LPC		
LATI Industria Termoplastici S.p. A	Lasulf™	PSU	http://www.lati.com	Italy
	Laptex A™	PESU		
PBI Performance Products, Inc.		PBI	http://www.pbiproducts.com	USA
Solvay SA	KetaSpire™	PEEK	https://www.solvay.com	Belgium
	Veradel™	PESU		
Svetlogorsk Khimvolokno	Arselon™	POD	http://www.sohim.by	Belarus
Swicofil AG		PI	http://www.swicofil.com	Switzerland
Toray Industries, Inc.	Teflon™	PTFE	http://www.toray.com	Japan
	Torcon™	PPS		
	Toyoflon™	PTFE		
Toyobo Co., Ltd.	Procon™	PPS	http://www.toyobo-global.com	Japan
	Zylon™	PBO		
Victrex plc.	ZYEX™	PEEK	http://www.victrex.com	Great Britain

References

Ebnesajjad, S. (2017). *Expanded PTFE applications handbook*. Elsevier.

Fourné, F. (1995). *Synthetic fibers*. Hanser.

Hearle, J. W. S. (2001). *High-performance fibers*. Woodhead Publishing.

Kitagawa, T. (2001). Polymer fiber processing: High-performance fibers. In *Encyclopedia of materials: Science and technology*. Elsevier, Amsterdam.

Kitagawa, T., Ishitobi, M., & Yabuki, K. (1998). Morphological study on poly-p-phenylenebenzobisoxazole (PBO) fiber. *Journal of Polymer Science, Part B: Polymer Physics, 36*(1), 39–48.

Kriskovets, M. V., Lysenko, V. A., Sazanov, Y. N., Gubanova, G. N., Kulikova, E. M., Ugolkov, V. L., & Lavrent'ev, V. K. (2018). Thermal transformations of polyoxadiazoles. *Russian Journal of Applied Chemistry, 91*(1), 23–30.

Maity, S., & Singha, K. (2012). Melamine fiber – Synthesis, features and applications. *Chemical Fibers International, 62*(4), 183–186.

Ullmann's. (2008). Ullmann's Fibers – Volume 1. Wiley-VCH Verlag, Weinheim.

Victrex. (2021). https://www.victrex.com/~/media/literature/en/vic018-005-zyex-brochure-240518-singles.pdf.

38 Carbon Fibers

Carbon fibers consist of at least 92% carbon atoms, the rest are nitrogen and hydrogen atoms. Their abbreviation is CF. In their internal structure they resemble graphite (Fig. 38.1).

38.1 Development and History

Thomas Edison used carbon filaments for electric light bulbs as early as 1879. The starting material was cellulose fibers, e.g. from bamboo and cotton, which he pyrolysed under controlled conditions. The resulting fibers were resistant to high temperatures, but did not yet have the strength of today's carbon fibers; instead, they were brittle. In the late 1950s, fine carbon staple fibers were first used to reinforce plastics for aircraft because of their significantly higher modulus of elasticity compared to glass fibers. However, the production of such "whiskers" was very costly (Bacon, 1960). The processes for producing carbon filaments were first published in 1959 (Anonymous, 1959; Mock, 1959). Very soon it was recognized that polyacrylonitrile (PAN) was more suitable as a starting material (precursor) than cellulose, which had been most commonly used until then (Shindo, 1961; Standage & Prescott, 1966; Watt et al., 1966). Today, 90% of all carbon fibers are produced from PAN, Morgan (2005).

38.2 Raw Materials

Besides polyacrylonitrile (PAN), the starting materials for carbon fibers are mainly pitch and cellulose (Fig. 38.2).

Pitch is a mixture of a variety of different polyaromatics. It is also known as asphalt or bitumen. Its molecular weight is 180–600 g/mol (Ozel & Bartle, 2002). Advantageous for the production of carbon fibers is the lower amount of nitrogen, hydrogen and other atoms that have to be removed during carbon fiber production compared to PAN and cellulose (Fig. 38.3).

Depending on the starting material, carbon fiber yields of 25% for viscose, 40–50% for PAN and 75–85% for pitch can be achieved. Elemental carbon, such as occurs naturally as graphite or diamond, cannot be used directly due to its unmeltability and insolubility. The prerequisite for a suitable precursor polymer fiber is therefore unmeltability during thermolysis, so that a C-atom arrangement with a high preferred direction parallel to the fiber axis can be formed.

38.3 Fiber and Yarn Production

The objective in the production of carbon fibers is to create graphitic layers and align them in the fiber direction. Starting from PAN or pitch, the

D. Veit, *Fibers*, https://doi.org/10.1007/978-3-031-15309-9_38

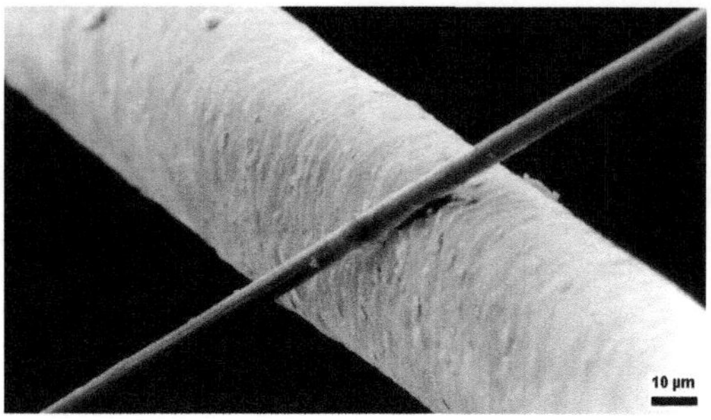

Fig. 38.1 Comparison of a carbon fiber with a human hair (WIKI, 2005)

process stages are similar for both processes. Spinning is followed by drawing, stabilization (oxidation) and carbonization, and for high modulus fibers also graphitization. During stabilization, the precursor is thermally stabilized so that it can withstand the high process temperatures without uncontrolled decomposition. During carbonization, graphite layers are formed, which are oriented in the fiber axial direction during graphitization, which significantly increases the Young's modulus by reducing elongation. This process step is associated with high energy input and is therefore only carried out if it is absolutely necessary for the mechanical properties, i.e. for high modulus fibers (Fig. 38.4).

In the PAN route, the linear polymer is cyclized in an oxidizing atmosphere and rendered unmeltable by cross-linking at a temperature of approx. 200 °C. Cyclization takes place with the elimination of decomposition products, such as hydrocyanic acid and ammonia. It therefore leads to a weight reduction of approx. 50%. If, on the other hand, mesophase pitch is used as precursor material, the aromatic structures only have to be aligned parallel to the fiber axis during spinning. During subsequent carbonization, the weight loss is therefore only approx. 10% (Spengler, 1988).

Depending on their mechanical properties, a distinction is made between HT (high strength), IM (medium modulus), HM (high modulus) and UHM (ultra high modulus) carbon fibers (Fig. 38.5).

38.3.1 Base PAN

PAN dominates in carbon fiber production because it is available in large quantities and at low cost and the quality of the carbon fibers can be adjusted depending on the precursor. Here, the requirements for a precursor fiber are different from those for PAN fibers for use in textiles (Müller, 1973). The morphology of the PAN fibers is preserved in the carbon fiber, only the diameter is reduced as a result of cross-sectional shrinkage. Also, any primary defects in the PAN fibers are transferred to the later carbon fibers and can then lead to fiber breakage in them. Therefore, even the production of PAN fibers is of great importance for the quality of the carbon fibers produced from them. PAN fibers can be produced by the dry or wet spinning process, which leads to specific cross-sectional shapes in each case (Fig. 38.6).

The manufacturing process of carbon fibers takes place in three stages, optionally followed by a surface treatment.

1. Stabilization or oxidation:

The white PAN fiber (precursor fiber) is subjected to temperatures of 200–300 °C under tension. This makes the fiber black, incombustible and non-melting, but only moderately tensile (20 cN/tex). The density increases to approx. 1.36 g/cm^3

Fig. 38.2 Structural formula of PAN (left) and cellulose (right)

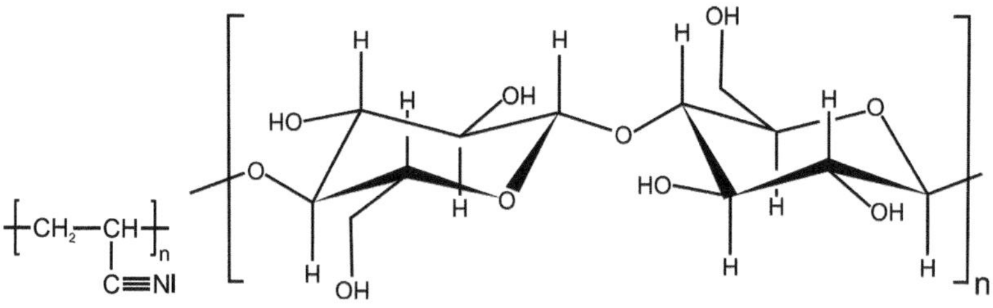

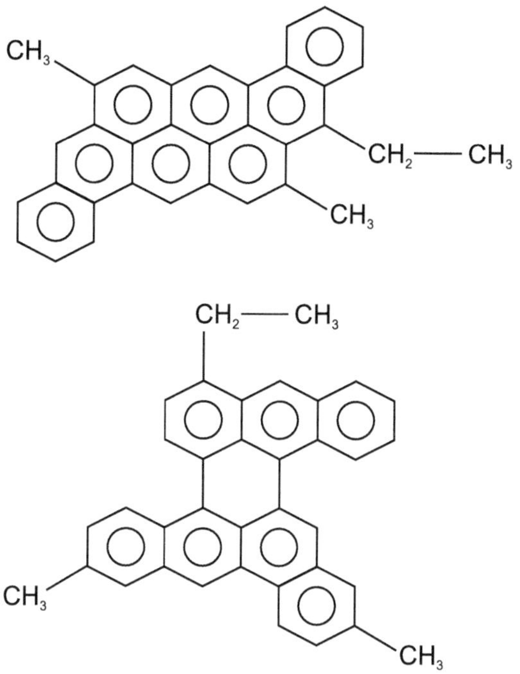

Fig. 38.3 Typical aromatic hydrocarbons in pitch acc. to Aldosari et al. (2020)

and longitudinal shrinkage occurs from approx. 240 °C when no tension is applied. The reaction is strongly exothermic and the carbon content is subsequently 60–70%, the nitrogen content is approx. 20% and the oxygen content approx. 10%.

Two processes take place simultaneously: One is the cyclization of the nitrile groups and the other is the dehydrogenation of the C-C chains by atmospheric oxygen. The superposition of both reactions leads to heteroaromatics with pridine structure. Oxygen is still present, in the form of hydroxyl and carboxyl groups and in acrylone structures. It also has a cross-linking effect, forming ether bridges between the chains (Fitzer et al., 1986), Fig. 38.7.

Figure 38.8 shows the principle of a corresponding oxidation plant. A lot of time is needed to remove the accompanying substances, therefore the residence times are very high (1.5 h) and the process speeds of the individual processing steps are correspondingly very low.

The stabilization reaction with oxygen leads directly to a color change of the carbon fibers from white to yellow to brown to black (Fig. 38.9).

Oxidized PAN fibers (filaments and staple fibers) with a carbon content of approx. 60% are also processed directly into high-temperature resistant products. They have LOI values of

Fig. 38.4 Production of carbon fibers from PAN and from pitch

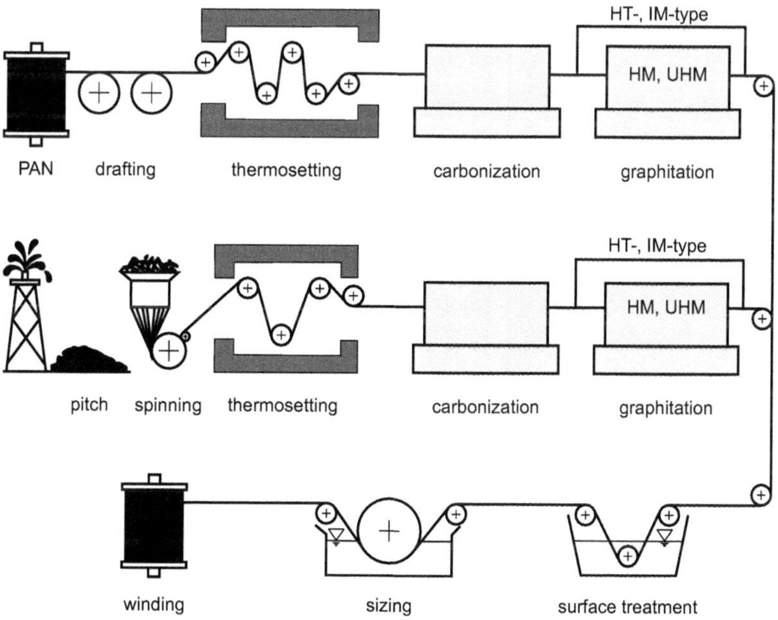

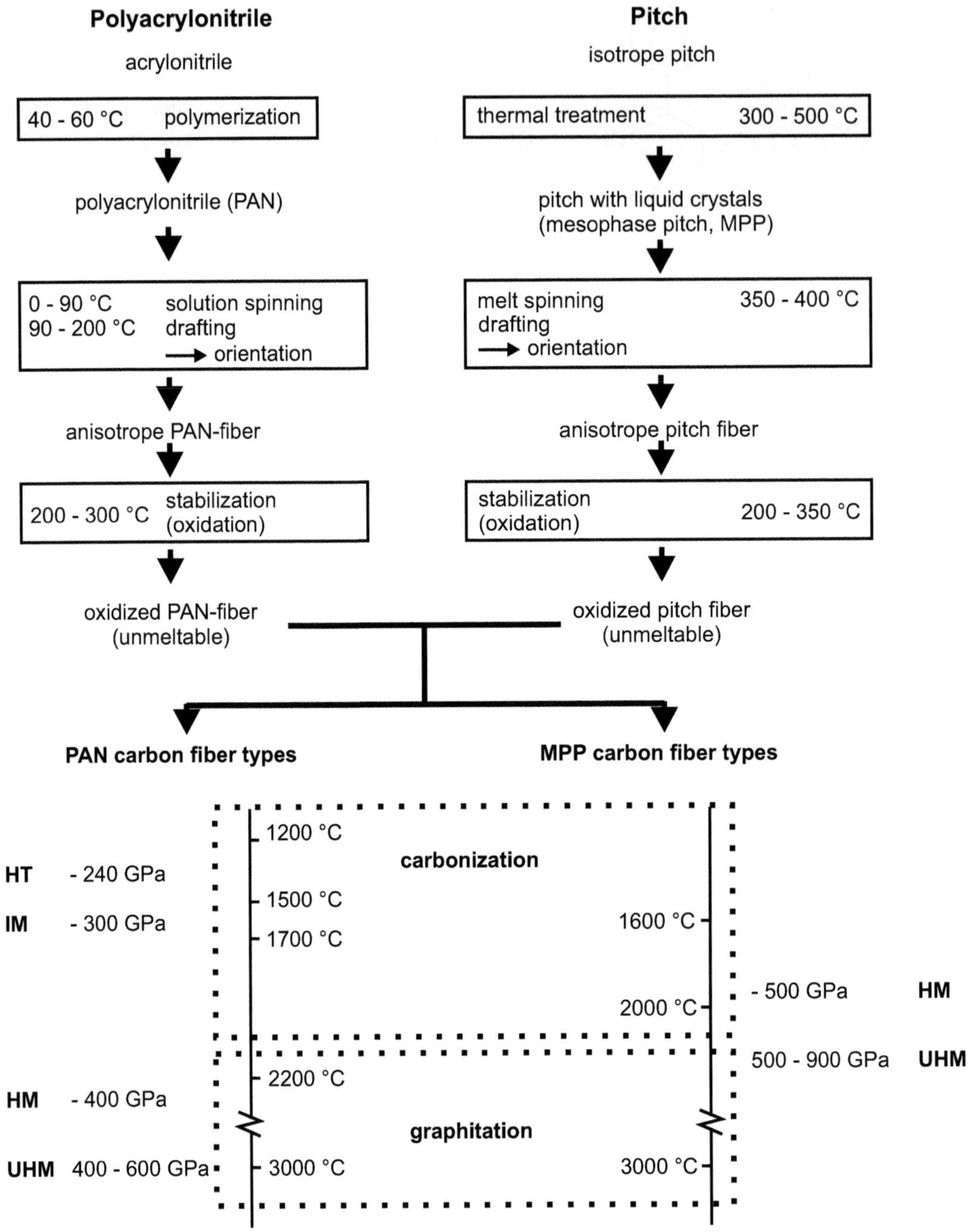

Fig. 38.5 Process routes for the production of carbon fibers acc. to Fitzer et al. (1986)

50–60 and are thus largely fire-resistant. Their strength corresponds to that of cotton with approx. 20 cN/tex, their elongation is 17–28%. A typical application is brake linings for aircraft and racing cars, which have to withstand temperatures of up to 1100 °C. PAN-Ox fibers are also used for heat-protective clothing and in noise protection at high external temperatures.

Fig. 38.6 Spinneret for the production of PAN fibers as carbon precursor (Sossna GmbH, n.d.)

2. Carbonization:

Further thermal treatment in an inert atmosphere (nitrogen) increases the carbon content up to 92%. The exact composition depends on the pretreatment and the maximum carbonization temperature reached or the reaction time. Furthermore, the density increases due to the splitting off of non-carbon atoms (N, O) and the cross-sectional area decreases by approx. 40%. Due to the treatment in a nitrogen stream at up to 1700 ° C, the thermal degradation is then practically completed. The mass loss of approx. 50% results from the splitting off of HCN (hydrocyanic acid), NH_3 (ammonia), H_2O and CO_2. The splitting off of the non-carbon atoms results in planar molecular complexes and the Young's modulus increases sharply. The strength reaches values up to 285 cN/tex (Fig. 38.10).

3. Graphitization:

If the thermal treatment is continued, graphitization occurs at 2400–3000 °C. This increases the degree of orientation of the graphite

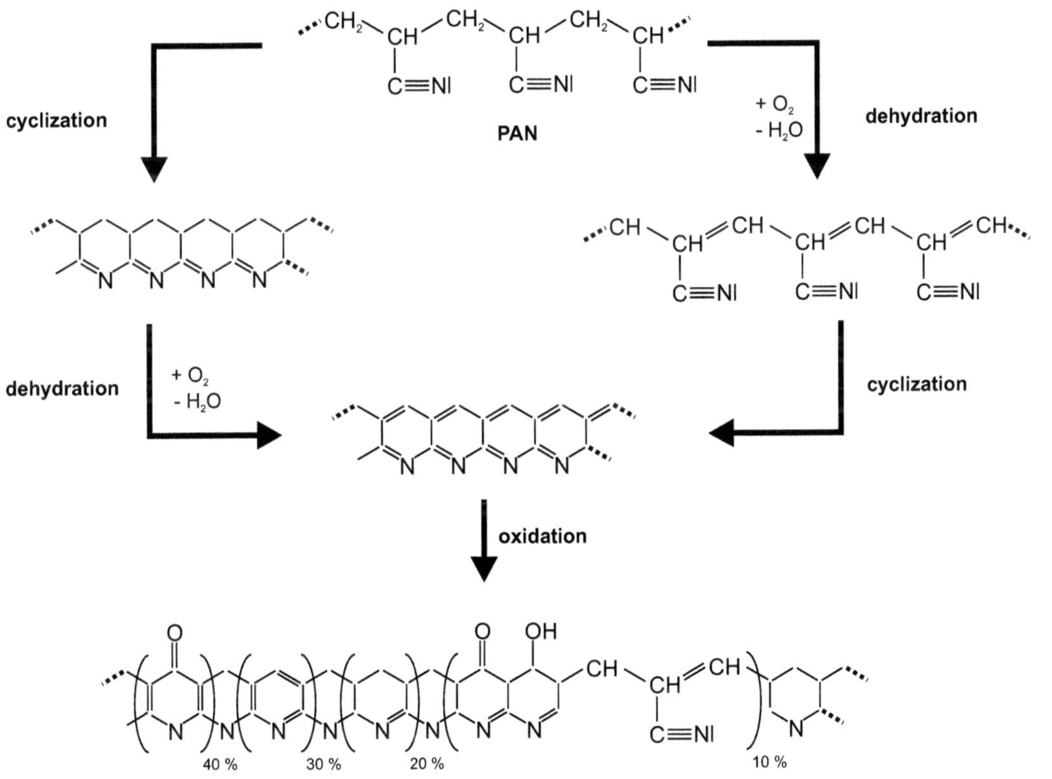

Fig. 38.7 Stabilization reactions acc. to Fitzer et al. (1986)

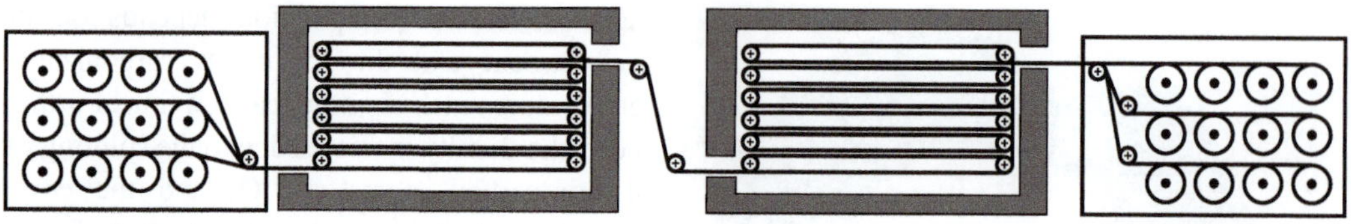

Fig. 38.8 Design of a plant for the oxidation of PAN

layers. If the graphite layers are displaced or twisted against each other, they are referred to as turbostratic layers. The temperature resistance is further increased. This process stage is only necessary for high-modulus fibers and requires a lot of energy due to the high temperatures, despite a dwell time in the thermal furnace of a few seconds or minutes.

The mechanical properties are determined by the final annealing temperature. The steep increase in Young's modulus up to 1000 °C flattens out, but continues up to the highest annealing temperature. This is due to the fact that the graphite layers continue to grow, structural defects are eliminated and recrystallization-like rearrangement processes occur. The tensile strength passes through a maximum during this process (Fitzer et al., 1986), but decreases slightly compared to non-graphitized fibers. The mechanical properties can be improved if the carbon fibers are stretched in the plastic range of carbon above 2400 °C. Young's modulus of up to 500 GPa are possible. The density after graphitization is in a range of 1.75–1.91 g/cm^3 (Fig. 38.11).

4. Surface treatment:

The good mechanical properties of carbon fibers can only be used in composites if there is sufficient fiber-matrix bonding. Therefore, it must be ensured that bonding possibilities for a matrix (e.g. an epoxy resin) are created on the fiber surface. This is done by generating surface oxides that can later react with polymer matrices.

The amount of absorbed oxygen directly influences the density of the stabilized fiber. An oxygen content of 10–14% by weight results in fibers with a density of approx. 1.4 g/cm^3, which have particularly good mechanical properties.

Fig. 38.9 Color change of the carbon fibers depending on the stabilization stage (SGLTechnologies)

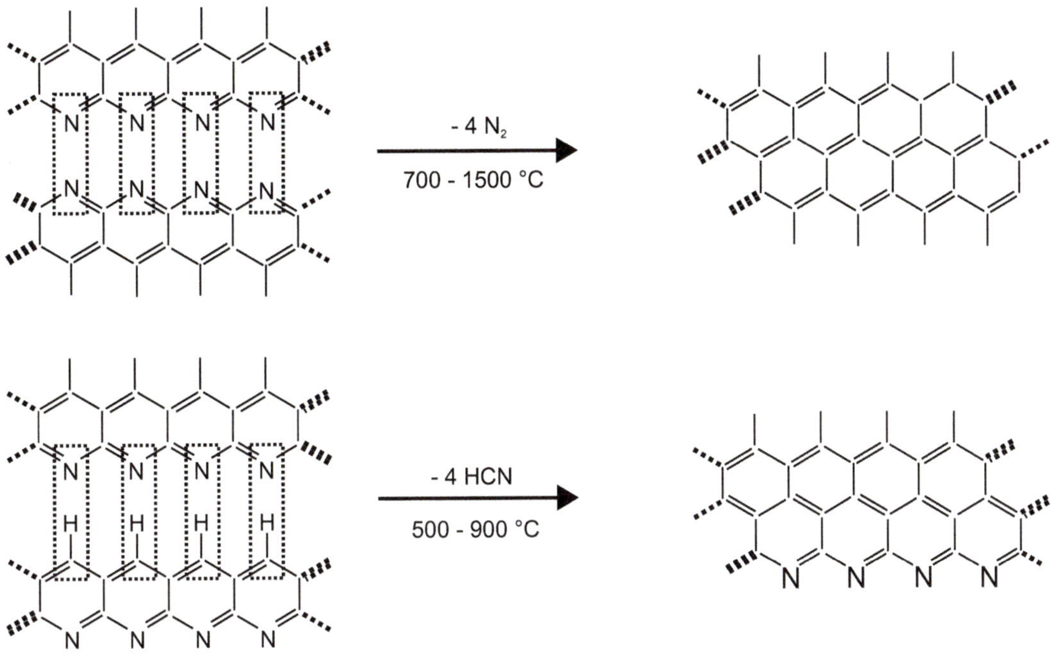

Fig. 38.10 Carbonization processes

38.3.2 Base Pitch

Brooks and Taylor (1965) discovered that pitch forms a liquid crystal-like mesophase from which grapitic carbon can be produced. Carbon fibers have thus been produced from mesophase pitch since the 1970s. Kureha Corp. in Japan was the first to develop an industrial process. In the mesophase state, polyaromatic groups exist in a partially ordered state between solid and liquid, similar to gelatin. In pitch, the mesophase is formed at a temperature of 350–500 °C. Via alkyl bridges, the polyaromatic structures assemble into larger units, forming lamellae and finally nematic liquid crystals. These are spun out to the precursor with the melt spinning process (Fig. 38.12). The spinneret is heated to approx. 350 °C, the filament formation takes place in a nitrogen atmosphere and typical take-off speeds are around 500 m/min.

Fig. 38.11 Output of graphitization with heavy tow (SGL Technologies, n.d.)

Within the spinneret, shear forces cause a pre-orientation of the polyaromatics and after the spinneret a jet expansion ("spinning bulb") is formed, as with other melt-spun polymers.

After stabilization (oxidation) and carbonization of the pitch precursor fibers (Fig. 38.5), the internal structures of the carbon fibers is shown in Fig. 38.13. A very short graphitization process step is also possible (process time <1 s).

Due to the often radial structure of these carbon fibers, micro-cracks occur, which cause these

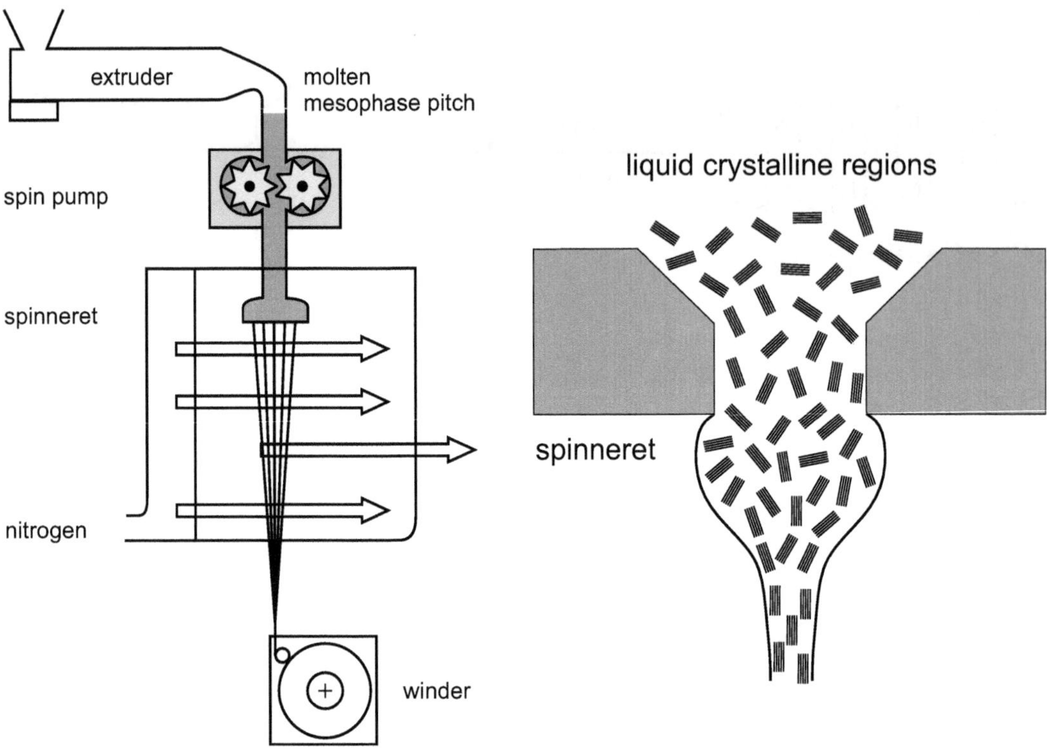

Fig. 38.12 Melt spinning process for pitch precursor fibers and pre-orientation of the nematic areas in the spinneret

fibers to be very brittle and break easily when they are loaded transversely to the direction of the fibers.

In Fig. 38.14, the influence of the temperature of the heat treatment on the orientation of the graphite layers of pitch-based fibers is shown. With increasing temperature, first small (1), then larger (2) groups of graphite pieces form. Finally, these arrange themselves into, at first disordered (3), later planar planes (4) and carbon fibers are formed.

Carbon fibers made from mesophase pitch can be produced inexpensively, for example, fiber mats, as shown schematically in Fig. 38.15.

This process is only of minor market importance at present (approx. 10%), because the mechanical properties are not as good as those of fibers made from PAN. However, capacities have increased significantly in recent years to currently approx. 12,000 t/year, partly because pitch-based precursor fibers are approx. 40% cheaper than those made from PAN.

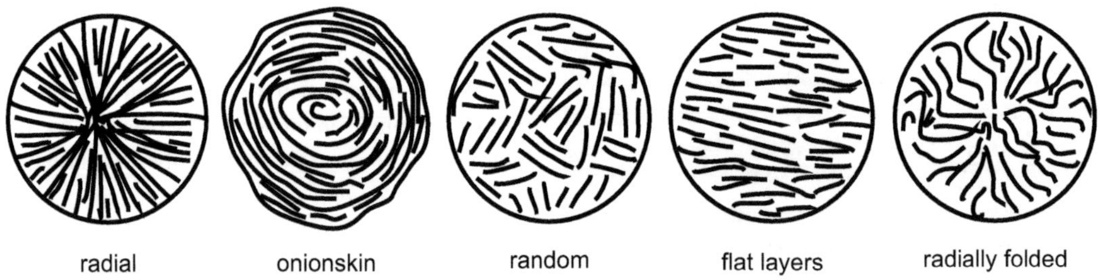

Fig. 38.13 Internal structure of carbon fibers made from mesophase pitch after Aldosari et al. (2020)

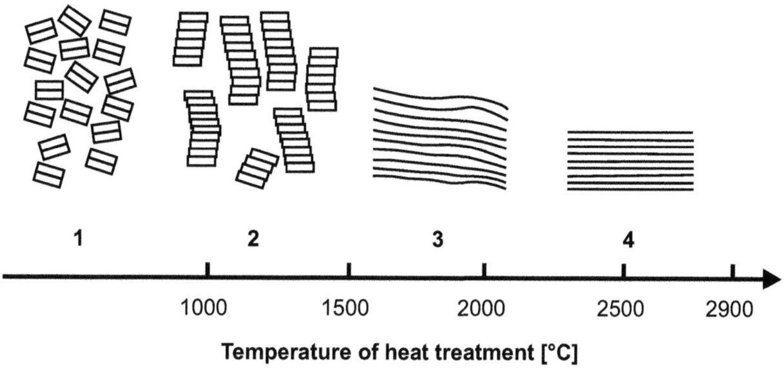

Fig. 38.14 Structural change during heat treatment

38.3.3 Other Precursors

In addition to PAN and mesophase pitch, other polymers can also be used as starting materials for carbon fiber production. All of them have not yet progressed beyond the experimental stage, but some have great potential.

38.3.3.1 Polyolefins

As early as 1978, Horikiri et al. (1978) obtained a patent for a process to produce carbon fibers from polyethylene (PE). In addition to considerable cost advantages in the production of the precursor compared to PAN, the main advantage of PE and PP precursors is that no nitrogen has to be removed. This results in theoretical yields of 85% carbon content relative to the starting polymer. Basically, four different types of PE can be used: LDPE (low density PE), LLDPE (linear low density PE), HDPE (high density PE) and UHMW-PE (ultra high molecular weight PE).

LLDPE can be produced by melt spinning at temperatures of 170 °C and with very low winding speeds of less than 20 m/min. For HDPE, spinning temperatures are 200 °C and winding speeds are up to 3000 m/min (De Palmenaer et al., 2015), which allows for an economical spinning process. Polypropylene is also suitable

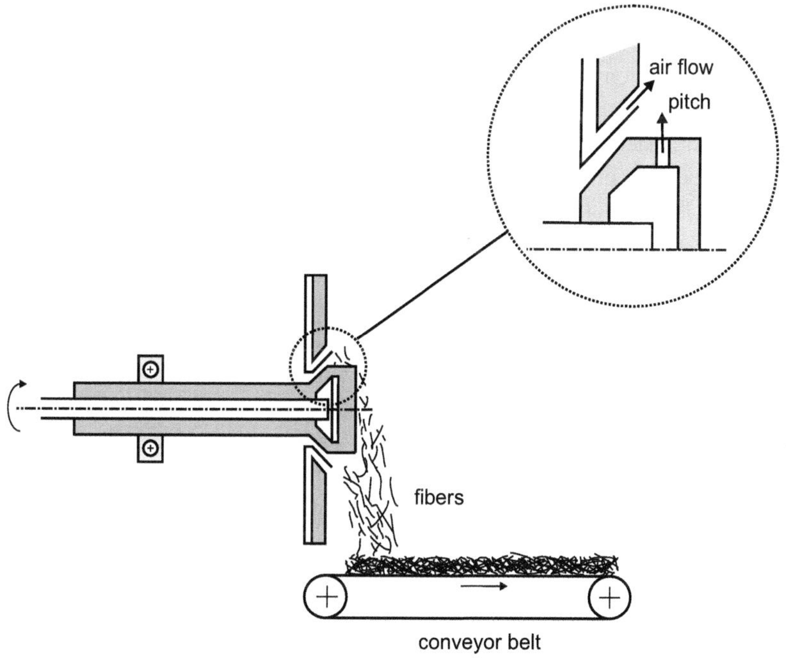

Fig. 38.15 Production of carbon fiber mats

Table 38.1 Properties of carbon fibers with polyolefin precursors acc. to de Palmenaer et al. (2015)

	LLDPE	HDPE	UHWMPE	PP
Tenacity [MPa]	<2160	<2670	<2100	<200
E-modulus [GPa]	<148	<139	<210	<26
Elongation [%]	0.7–3		<1	
Diameter [μm]	9–17	8		
Density [g/cm^3]	2–2.1	0.97	0.97	0.91
Carbon yield [%]	72–75	70–76	75–80	62–88

in principle as a precursor, but the carbon fibers produced from it have significantly poorer mechanical properties compared with those made from PE (Table 38.1).

Precursor fibers made of PE (with the exception of UHWMPE) and PP are thermoplastic and therefore cannot be carbonized and graphitized at high temperatures. Therefore, they are first thermally stabilized by a treatment with concentrated sulfuric acid, the so-called sulfonation. This is done in one or two baths and leads to sulfur bridges between the polymer chains, making the fibers unmeltable. At the same time, their color changes from white to brown to black during the process. This process takes place at temperatures of 130–200 °C and lasts up to 2 h. Sulfonation is an expensive process step if the process takes a long time; if the process takes a short time, the polymer may degrade. These are the main disadvantages of this precursor material so far.

Carbonization of PE precursor fibers occurs at 900–1300 °C, shrinking the fibers by 30–50% (de Palmenaer et al., 2015). Graphitization increases the Young's modulus considerably up to 200 GPa and takes place at temperatures up to 2400 °C (Barton et al., 2017).

38.3.3.2 Cellulose

As early as 1880, Th. Edison received a patent for an incandescent filament made of carbonized cotton (Edison, 1880). In the 1960s, processes were developed at Union Carbide to produce carbon fiber fabrics from rayon. Until the 1970s, attempts were made to produce cellulose-based carbon fibers. However, the yields were too low and the manufacturing costs too high, so that this development direction was largely discontinued. Only in recent years has more work been done on this topic again, because PAN is only available in limited quantities and, against the background of an increasing striving for sustainability, petroleum-based polymers are to be replaced by renewable raw materials. Natural cellulose is too impure and the molecular chains have insufficient orientation, so that it cannot be used as a starting material for carbon fibers. In contrast, cellulose man-made fibers, such as viscose and lyocell, do not have these disadvantages. Instead, their internal structure can be specifically adjusted via the process parameters, which in turn directly determines the properties of the carbon fibers produced from them.

The theoretical carbon fiber yield when using a cellulosic precursor is max. 40%. This results from the consideration that water must be split off from each glucose unit if only carbon is to remain. This dehydration is associated with a mass loss of 60%. If one assumes that there are also other degradation reactions in which CO and CO_2 as well as alcohols, ketones, etc. are formed, the yield is reduced to 10–30% (Donnet, 1984).

For the production of carbon fibers, the cellulose is first stabilized or oxidized. The aim of pyrolysis in combination with stabilization of the precursor is to keep the additional losses due to undesired degradation reactions as low as possible. For this purpose, low heating rates of up to 1/100 K/h are used, for example. However, this is extremely uneconomical and only useful on an experimental scale. Alternatively, the cellulose can also be pre-oxidized. The oxidative decomposition of the cellulose chains is faster than pyrolysis at low temperatures and the yield is therefore higher. The use of flame retardants can also improve pyrolysis and thus lead to a higher yield. Combinations of these processes are also

Table 38.2 Classification of carbon fibers acc. to strength and modulus of elasticity

	E-modulus [GPa]	Tenacity [MPa]
Normal modulus—high tenacity (HT)	200–250	3000–5000
Intermediate modulus (IM)	251–350	4000–5000
High modulus (HM)	351–450	2000–4500
Ultra high modulus (UHM)	451–950	3800–4100

often used (Frank et al., 2015). After pyrolysis, the residue has a carbon mass fraction of 60–70%.

During carbonization, almost all non-carbon atoms are removed in an inert atmosphere at a temperature of 900–1500 °C, thus increasing the carbon content to up to 95%. Polycyclic rings, aromatic structures and graphite-like layers are formed. At the same time, the fiber shrinks, increasing its density and strength.

Graphitization is carried out at a temperature of 1500–3000 °C and results in a carbon content of more than 99%. If the growth of the graphite layers takes place under tensile stress, this is referred to as hot stretching.

The strength of cellulose-based carbon fibers is 500–1500 MPa, elongation is 1–3% and Young's modulus is 40–100 GPa (Frank et al., 2015). Thus, they are not competitive with PAN-based carbon fibers (Table 38.2).

For the production of activated carbon fibers, e.g. as filter material, precursor fibers on a cellulosic basis, e.g. viscose or bamboo, are often carbonized as a finished textile. On the one hand, this is done physically by carbonization at 600–900 °C in an inert atmosphere and subsequent activation by CO_2 and steam at 600–1200 °C. On the other hand, the precursor fibers can be carbonized in a process called "carbonization". On the other hand, the precursor fibers can also be impregnated with the oxidizing substance, e.g. $ZnCl_2$, K_2CO_2, $AlCl_3$, Na_2HPO_4 or H_3PO_4, and then heated in argon or nitrogen (Hina et al., 2018). This method can also be used to generate products for medical textiles and for fiber composites for high temperature applications.

38.3.3.3 Lignin

Lignins are aromatic macromolecules and the second most common plant-based biomaterial after cellulose. Around 20–30% of the dry mass of wood consists of lignin. Therefore, lignin is available everywhere in huge quantities and comparatively cheaply.

Depending on the origin (hardwood, softwood, annual plants), there are different types of lignin. However, because many different lignins are also produced during lignin extraction, e.g. using the sulfate process (85% market share), and these have very different property profiles, industrial use is difficult. However, the carbon content of lignin is over 60% and, in the absence of nitrogen, no toxic by-products (e.g. HCN) can be formed during carbonization. Therefore, attempts have been made for some years to produce lignin-based carbon fibers.

Lignin fibers can be produced by wet, dry and melt spinning. In the case of melt-spun fibers, however, care must be taken to ensure that the lignin fibers do not melt during carbonization. This is achieved by controlled crosslinking of the lignins at temperatures up to 300 °C. The lignin molecules are relatively large and therefore much less mobile than, for example, PAN macromolecules. Therefore, the graphite layers formed during carbonization are also less mobile and thus they are less oriented in the axial direction in the finished carbon fiber. As a result, the strength of lignin-based carbon fibers is significantly lower than that of, for example, PAN-based ones.

An overview of the manufacturing process and possible applications is given by Chatterjee et al. (2015).

38.3.3.4 Other Materials

In principle, all materials are suitable as precursors of carbon fibers that have aromatic components and a high carbon content with as few other atomic species as possible. Therefore,

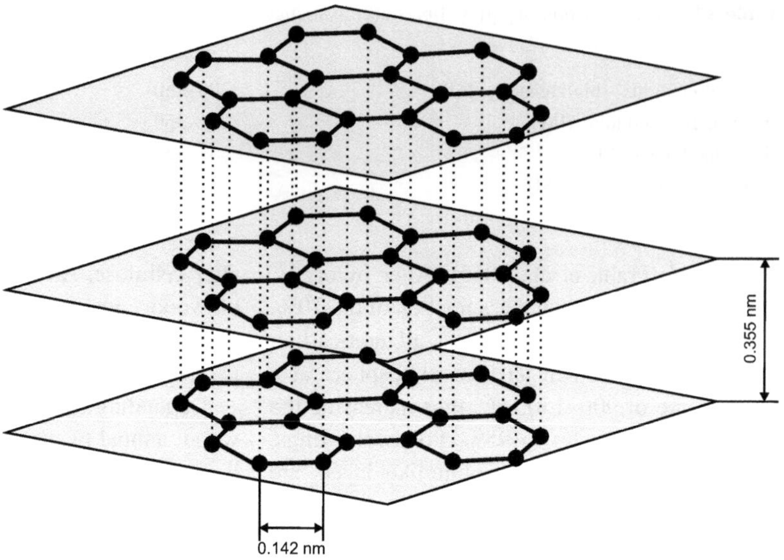

Fig. 38.16 Structure of hexagonal graphite

aromatic polyesters and aromatic polyamides are also suitable in principle. However, it has not yet been possible to develop an economical production process on this basis.

38.4 Inner Structure

Depending on the type, commercially available carbon fibers consist of approx. 95% carbon with a residual content of nitrogen and hydrogen. The structure of carbon fibers is similar to that of hexagonal graphite. Within the layers, the carbon atoms are linked by strong covalent bonds (Fig. 38.16). Perpendicular to this, the bonding forces are much lower. Therefore, carbon fibers can absorb large forces in the axial direction of the fiber, but break brittle when subjected to transverse loads.

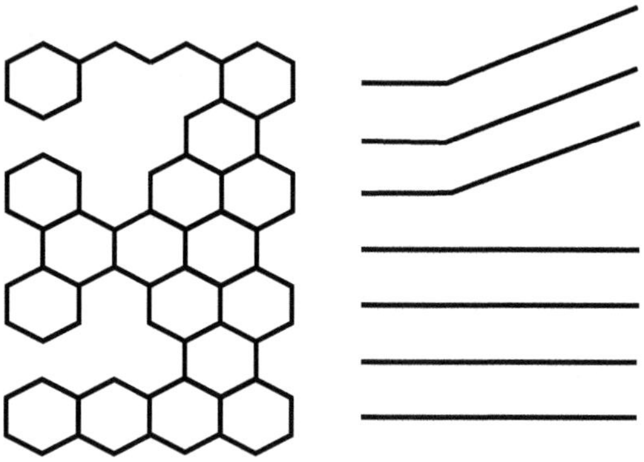

Fig. 38.17 Graphite structure defects in carbon fibers acc. to Fitzer et al. (1986)

Carbon fibers have a turbostratic structure, which is formed by twisting the parallel layers around the crystalline C-axis. This results in the formation of an enlarged layer plane spacing in which lattice defects such as imperfections or stacking faults occur (Fig. 38.17).

The alignment of the graphitic layers is of crucial importance for the magnitude of the Young's modulus. High tensile strengths and Young's moduli can only be achieved with sufficient alignment of the layers in the axial fiber direction. In addition to the anisotropic C-fibers for reinforcement purposes, there are also isotropic fibers which can be produced more economically (Fig. 38.15) and are used, for example, as fillers or sealing materials.

The structure of the graphite layers in the fiber cross-section is also of great importance for the mechanical properties. Different areas and phenomena are distinguished, which the Fig. 38.18 schematically shown: A = surface region, B = core region, C = hairpin defect, D = disclination (defect in rotational symmetry), Fig. 38.18.

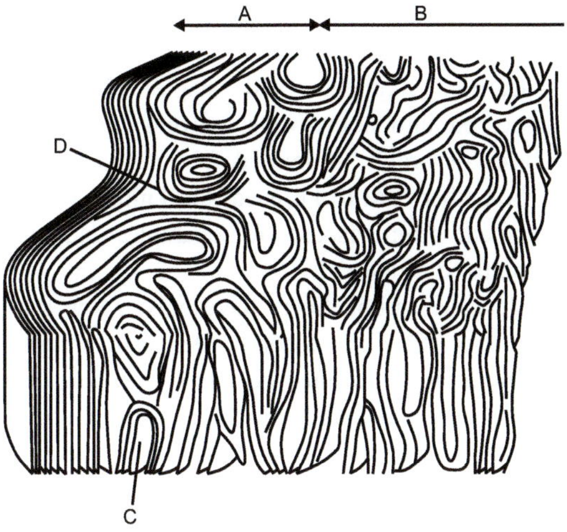

Fig. 38.18 Structure of HM carbon fibers made of PAN in fiber cross-section acc. to Bennett and Johnson (1978)

38.5 Features

Carbon fibers are always first spun as filaments. They are then either combined into rovings (small tow, heavy tow) or cut or milled into staple fibers. Non-carbonized PAN fibers, so-called PAN-Ox fibers, are also processed directly into textile products (Sect. 38.3.1). DIN 65184 (1993) regulates the designation of carbon fibers depending on fineness, strength, elongation, etc. The following table shows the designation of carbon fibers.

Carbon fibers are mainly used as composite materials in combination with matrix materials. Therefore, the embedded textile or composite should always be considered and not only the carbon fiber.

Depending on the modulus of elasticity or tensile strength, a distinction is made between four different fiber types (Table 38.2).

The high modulus of elasticity in combination with the low density leads to very good damping properties, which are greater with pitch-based carbon fibers than with PAN-based ones.

The special properties of carbon fibers include

- high strength, low elongation and high modulus of elasticity (Table 38.3),
- high brittleness,
- low density (1.76–1.93 g/cm^3 (HM)),
- low creep tendency,
- chemical resistance to acids and alkalis,
- good thermal conductivity (thermal conductivity 5 W/mK for normal type to 115 W/mK for HM type),
- good electrical conductivity ($0.78–1.5 \times 10^{-5}$),
- negative thermal expansion (-0.1 to -1.3×10^{-6}/K),
- no moisture absorption.

Carbon fibers made of PAN have a round cross-section. Typical filament diameters are 5–10 µm. The fracture surface is brittle (Fig. 38.19).

Typical mechanical properties of PAN-based carbon rovings summarizes Table 38.3.

The differences in the strengths and Young's modulus values of PAN and pitch-based carbon fiber filaments shown in the Fig. 38.20. Pitch-based carbon fibers have a higher Young's modulus, but their strength is significantly lower than that of PAN-based carbon fibers.

In Table 38.4, typical mechanical properties of carbon staple fibers are listed. A distinction is made between standard types with low strength and low modulus of elasticity and high-strength fibers. Common fiber lengths are between 3 and 50 mm.

Table 38.3 Typical mechanical properties of PAN-based carbon rovings acc. to manufacturer's data

Type	Tenacity [MPa]	E-modulus [GPa]	Elongation [%]	Density [g/cm^3]
HT	3530–5100	200–250	1.5–2.1	1.76–1.82
IM	4400–7000	251–350	1.3–2.2	1.78–1.81
HM	4400	351–450	1.1–1.2	1.82
UHM	3800–4100	451–600	0.7–0.9	1.88–1.93

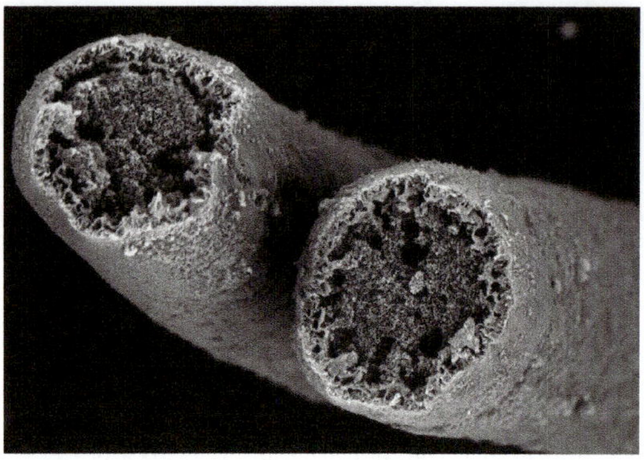

Fig. 38.19 Carbon fiber filaments with fracture

38.6 Further Processing

Carbon fibers are available as filaments (roving), as staple fibers (3–50 mm staple length) and as milled fibers (50–150 μm length) for reinforcing and increasing the conductivity of polymer materials (Fig. 38.21).

Carbon filaments and staple fibers are processed into nonwovens, wovens, non-crimp fabrics or braids. If knitted fabrics made from carbon fibers are to be used, the knitted fabrics are made from viscose, for example, and then carbonized as a textile.

Because carbon fibers have only a low kink resistance, they must not be deflected sharply. Furthermore, care must be taken to ensure that carbon fibers do not get into the machine controls, as they could otherwise cause short circuits. Therefore, machines on which carbon fibers are processed must be encapsulated and care must be taken to ensure that carbon fibers from rooms with such machines do not get into other rooms (e.g. through floor mats).

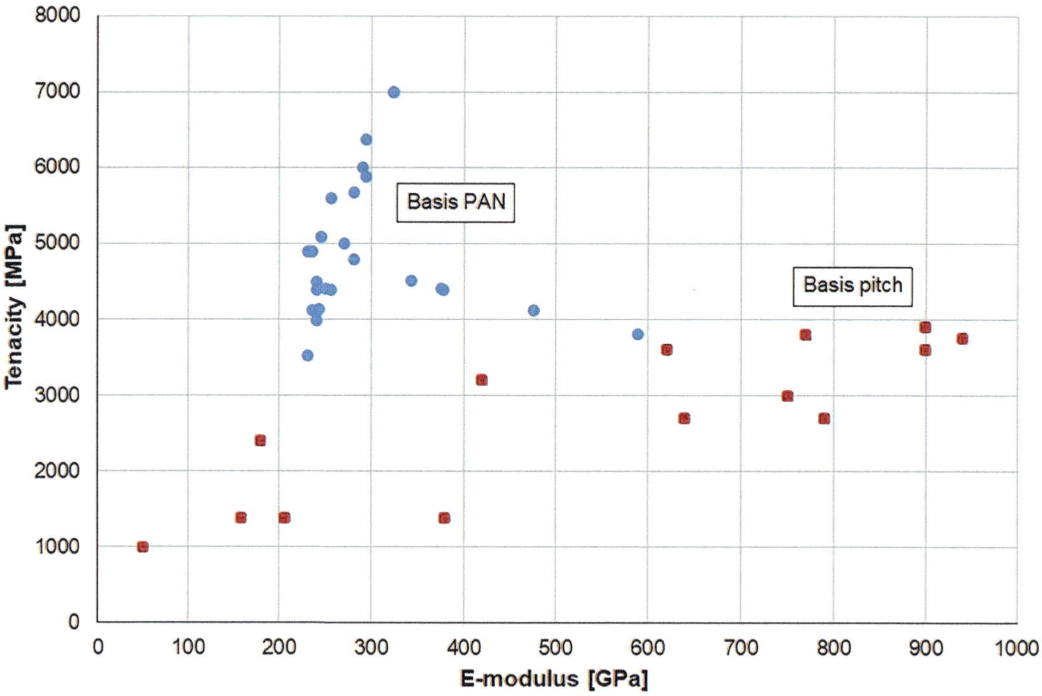

Fig. 38.20 Strength and modulus of elasticity of PAN and pitch-based carbon fibers acc. to manufacturer's data

Table 38.4 Typical mechanical properties of PAN-based carbon staple fibers acc. to manufacturer's data

Type	Tenacity [MPa]	E-modulus [GPa]	Bending strength [MPa]	Bending modulusl [GPa]
Normal type	90–290	14–35	150–500	10–32
High strength type	3300–4400	220–265		

38.7 Areas of Application

Due to their special properties and their high price compared to other fiber materials, carbon fibers are used exclusively for high-performance applications.

The most important areas are aerospace (tail units, wing parts, fuselage), the automotive industry (racing cars and luxury cars) and, increasingly, wind turbines (large wingspan with minimum weight and maximum strength) and sports equipment because of the low specific weight combined with high strength. Concrete is also reinforced with carbon textiles. A combination of carbon fibers with negative coefficients of thermal expansion (CTE) with common resin systems and positive CTE can be used to produce composites that remain dimensionally stable even with large temperature gradients. This is particularly advantageous for applications in space (Fig. 38.22).

It is estimated that 75% of carbon fibers are used for industrial applications, 15% in aerospace and 10% for consumer goods. It is assumed that about 80,000 t/year of carbon fibers will be needed for the production of wind turbines in 2030 and about 50,000 t for the automotive industry. This is an amount higher than the current world production of 120,000 t (2019). However,

Fig. 38.21 Carbon rovings on bobbins (SGL Technologies, n.d.)

Fig. 38.22 Typical applications of carbon in automobiles (left: BMW i3; right: wheel suspension in Formula 1 racing car: Julian Eichhoff)

the capacity of PAN precursor fibers for carbon fibers is about 170,000 t (The Fiber Year, 2021), so the production could be slightly increased.

The price for carbon fibers is 15–50 €/kg depending on the quality.

38.8 Market and Manufacturers

The most important producing regions are the USA and Japan as well as China. The most important markets (2020) for carbon fiber products based on PAN precursor are Europe (40% market share), North America and Asia (25% each), and Central and South America with 5%. For pitch-based fibers, Asia dominates (40%), ahead of North America (35%) and Europe (25%). Sales of pitch-based carbon fibers account for 5–10% of the total market (Statista, 2021), Fig. 38.23.

In 1992, world production was 6500 t, in 1997 it was already 13,000 t and since 2010 production has been growing by 10–15% annually (Fig. 38.24). Further growth in this order of magnitude is also expected in the next 10 years. By 2025, total sales of carbon fibers are expected to reach € 6–7bn.

Around 70% of the carbon fibers produced are so-called small tow fibers (3–24 k), around 30% heavy tow fibers (<24 k), where k stands for the number of filaments in [1000]. Small tow fibers are up to four times more expensive than heavy tow fibers.

Precise information on the production volumes of individual carbon fiber manufacturers is difficult to obtain. There is only the so-called “nameplate capacity”, i.e. the maximum possible production volume, which is approx. 150,000 t in 2020. However, only about 70% of this capacity is currently being utilized. Figure 38.25 shows the estimated production volumes of the most important manufacturers.

Since 2014, Zoltec, one of the largest carbon fiber producers until then, has been part of the Toray Group. Toho Tenax, also one of the major producers, became part of Teijin Ltd. in 2018. The data in Figs. 38.24 and 38.25 do not match exactly. This is due to the process-related inaccuracy of the manufacturers’ data.

Table 38.5 is a list of important manufacturers of carbon fibers. Unless otherwise stated, the trade names refer to PAN-based fibers.

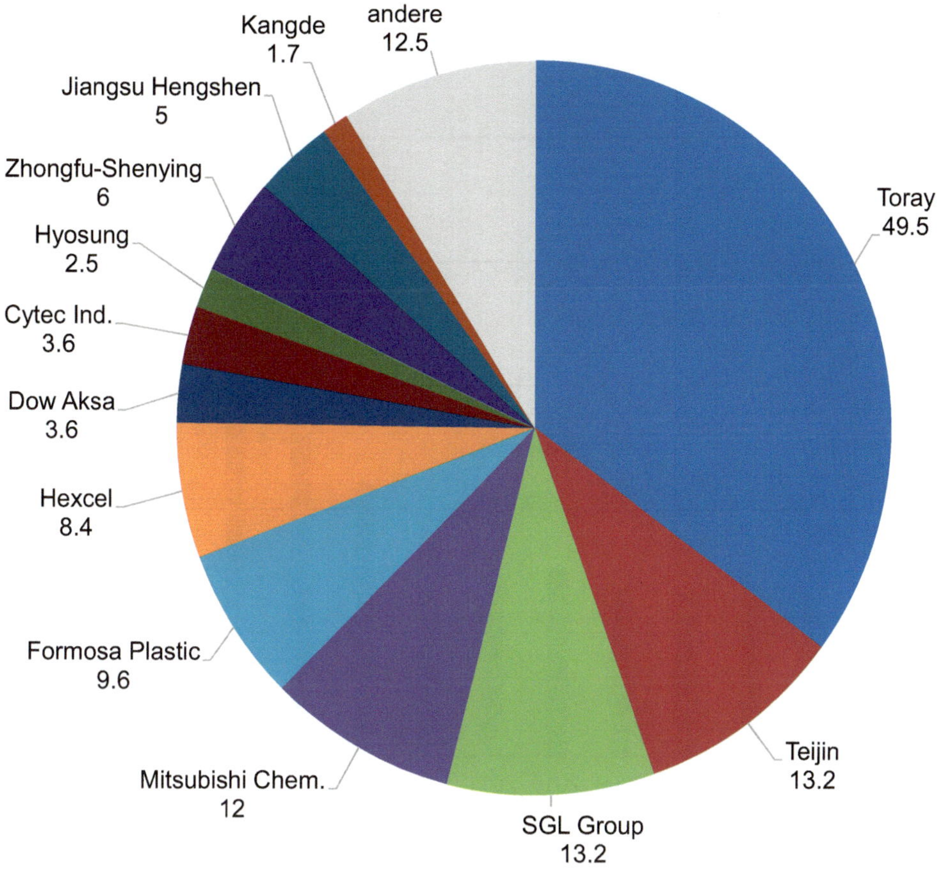

Fig. 38.23 Important regions for the production of carbon fibers (data: statista, 2021)

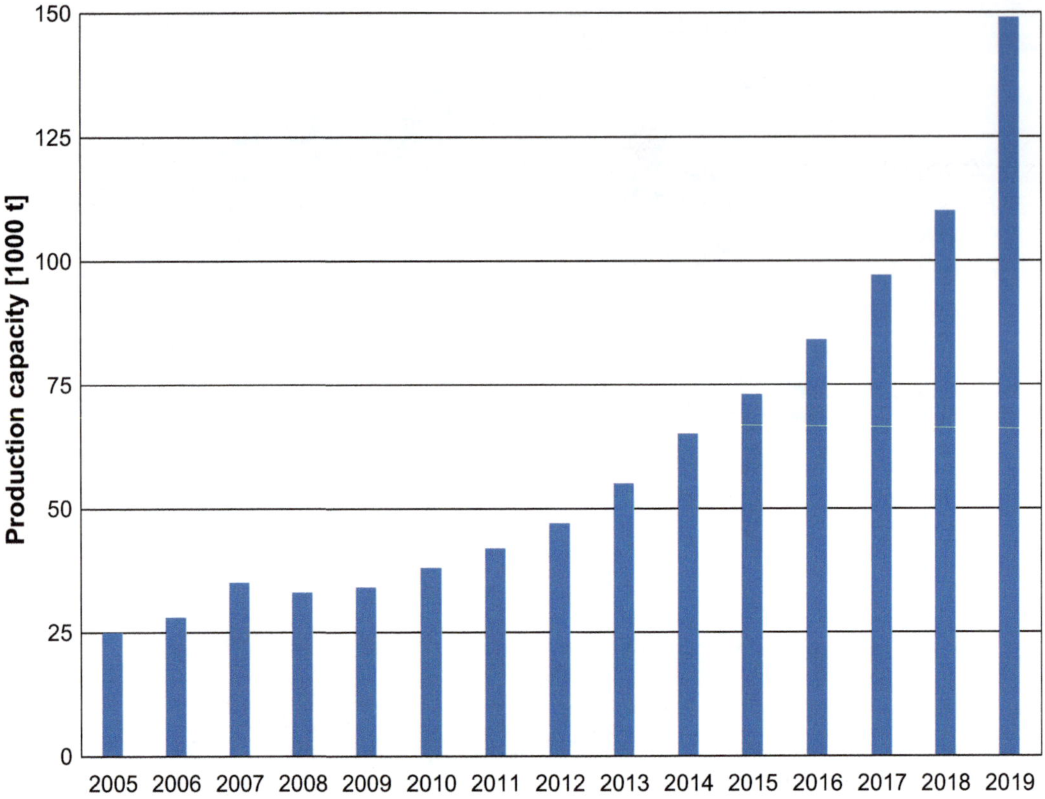

Fig. 38.24 Development of carbon fiber production capacity (data: The Fiber Year, 2021; statista, 2021)

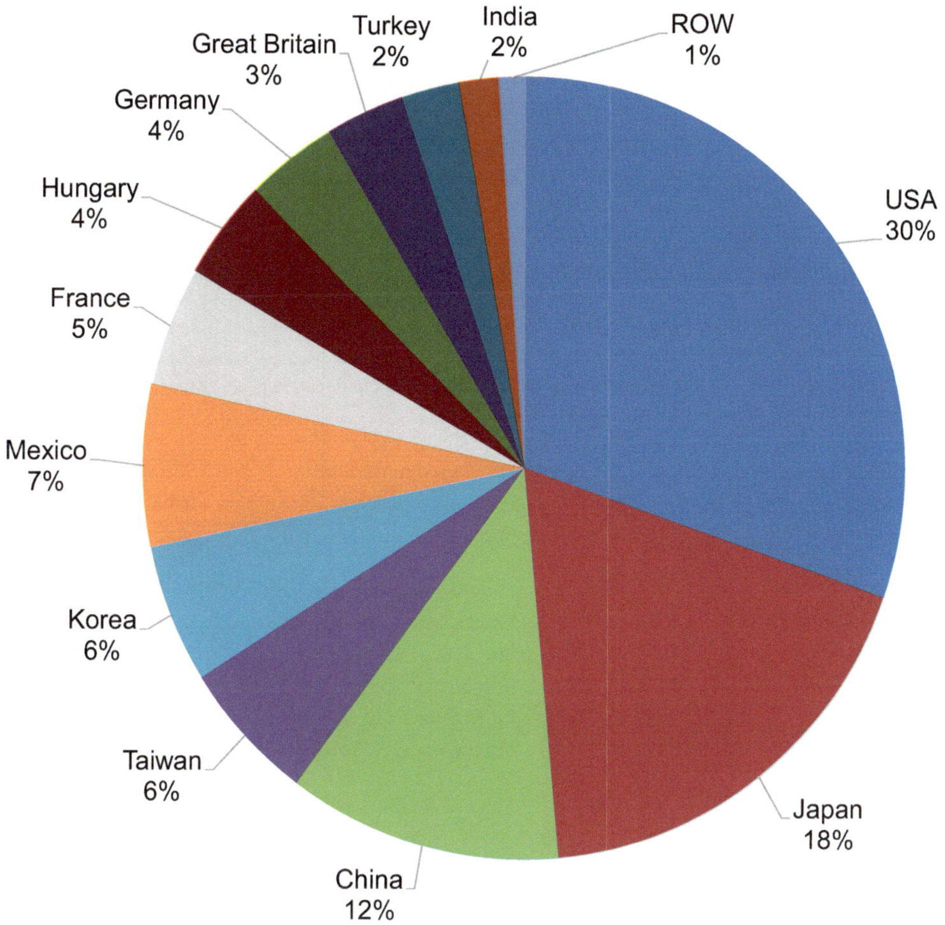

Fig. 38.25 Production capacity of carbon fibers of the most important manufacturers [1000 t] (data: statista, 2021)

Table 38.5 Selected manufacturers of carbon fibers

Producer	Trade mark	Website	Country
Alabuga-Fibre LCC		https://en.umatex.com/production/fiber/	Russia
DowAksa	AKSACA™	http://www.dowaksa.com	Turkey
Formosa Materials	Tairyfil™	http://www.formosam.com	South Korea
Hexcel Corp.	HexTow™	http://www.hexcel.com	USA, China, France
Holding Company Composite		http://www.hccomposite.com/en/	Russia
Hyosung Advanced Materials	TANSOME™	http://www.hyosungadvancedmaterials.com	South Korea
Kureha Group	KRECA™ (Pech)	https://www.kureha.co.jp	Japan
Mitsubishi Chemical	Pyrofil™ DIALEAD™ (Pech)	https://www.m-chemical.co.jp	Japan
Montefibre Carbon		http://www.montefibre.es/	Spain
Reliance Industries		http://www.ril.com	India
SGL Carbon	Sigrafil™	http://www.sglcarbon.com	Germany, China, India, Japan, USA
Sinopec Corp.		http://www.sinopecgroup.com	China
Solvay Cytec Carbon Fibers	Thornel™	http://www.solvay.com	USA, Germany
Taekwang Industrial Co., Ltd.		http://www.taekwang.co.kr	South Korea
Teijin Carbon	TENAX-X™	https://www.teijincarbon.com/de/	Japan, Germany, USA
Toray Group	TORAYCA™	http://www.toray.eu	Japan
Zhongfu Shenying Carbon Fiber Co., Ltd.		http://www.zfsycf.com.cn	China
Zoltek	ZOLTEK PX™	http://www.zoltek.com	USA, Hungary, Mexico

References

Aldosari, S. M., Khan, M., & Rahatekar, S. (2020). Manufacturing carbon fibres from pitch andpolyethylene blend precursors: A review. *Journal of Materials Research and Technology, 9*(4), 7786–7806. Elsevier.

Anonym. (1959). *Metal progress* (Vol. 75, p. 115). Metals Park.

Bacon, R. (1960). Growth, structure, and properties of graphite whiskers. *Journal of Applied Physics, 31*, 283ff. AIP Publishing.

Barton, B. E., Behr, M. J., Patton, J. T., Hukkanen, E. J., Landes, B. G., Wang, W., Horstman, N., Rix, J. E., Keane, D., Weigand, S., Spalding, M., & Derstine, C. (2017). High-modulus low-cost carbon fibers from polyethylene enabled by boron catalyzed graphitization. *Small, 13*. Wiley.

Bennett, S. C., & Johnson, D. J. (1978). Structure-property relationships in carbon fibres. *Journal of Physics D: Applied Physics, 20*, 286–292. AIP Publishing.

Brooks, J. D., & Taylor, G. H. (1965). The formation of praphitizing carbons from the liquid phase. *Carbon, 3*(2), 185–193. Elsevier.

Chatterjee, S., Saito, T., & Bhattacharya, P. (2015). Lignin-derived carbon fibers and their composites. In *Lignin in polymer composites* (1st ed.). Elsevier.

De Palmenaer, A., Wortberg, G., Drissen, F., Seide, G., & Gries, T. (2015). Production of polyethylen based carbon fibres. *Chemical Engineering Transactions, 43*, 1699–1704. AIDIC.

DIN 65184. (1993). *Kohlenstoffasern*. Beuth-Verlag.

Donnet, J. B. (1984). *Carbon fibers*. Marcel Dekker.

Edison, T. A. (1880). *Electric lamp*. US patent 223,898.

Fitzer, E., Heine, M., & Jacobsen, G. (1986). Kohlenstoffasern. In H. Heissler (Ed.), *Verstärkte Kunststoffe in der Luft- und Raumfahrttechnik*. Verlag W. Kohlhammer.

Frank, E., Steudle, L. M., Ingildeev, D., Spörl, J. M., & Buchmeiser, M. R. (2015). *Carbonfasern: Präkursor-Systeme, Verarbeitung, Struktur und Eigenschaften*. Wiley.

Hina, K., Zou, H., Qian, W., Zuo, D., & Yi, C. (2018). Preparation and performance comparison of cellulose-based activated carbon fibres. *Cellulose, 25*, 607–617. Springer.

Horikiri, S., Amagasaki, J., & Minobe, M. (1978). *Process for production of carbon fiber*. US patent 4,070,446.

Mock, J. (1959). *Materials in design engineering* (Vol. 49, p. 149). Elsevier.
Morgan, P. E. (2005). *Carbon fibers and their composites*. Taylor & Francis.
Müller, D. J. (1973). *Zur Herstellung von Kohlenstoffasern aus Polyacrylnitril über die katalytische Cyclisierung des Polymeren*. Dissertation an der TU Karlsruhe, Karlsruhe.
Ozel, M. Z., & Bartle, K. D. (2002). Production of mesophase pitch from coal tar and petroleum pitches using supercritical fluid extraction. *Turkish Journal of Chemistry, 26*, 417–424. Tübitak.
SGL Technologies GmbH. (n.d.). Meitingen. https://www.sglcarbon.com
Shindo, A. (1961). Studies on graphite fiber. *Journal of the Ceramics Association, 69*, 195.
Sossna GmbH. (n.d.). Dorsten. https://sossna.de/
Spengler, H. (1988). *Herstellung von Kohlenstoffasern aus Steinkohlenteerpech*. Sympos. Materialforschung d. BMFT, Hamm.
Standage, A. E., & Prescott, R. (1966). High elastic modulus carbon fibre. *Nature, 211*, 169.
Statista. (2021). http://www.statista.de
The Fiber Year. (2021). *The Fiber Year GmbH*. Roggwil. https://www.thefiberyear.com
Watt, W., Phillips, L. N., & Johnson, W. (1966). Carbon fibres with average tensile strengths. *Thermal Engineering, 27*.
WIKI. (2005). https://commons.wikimedia.org/wiki/File:Cfaser_haarrp.jpg

39 Carbon Nanotubes and Graphene

39.1 Definition and Types

Carbon Nanotubes

Carbon nanotubes (CNTs) are microscopically small tubular structures made of carbon atoms. Their walls consist only of carbon, whereby the carbon atoms adopt a honeycomb-like structure with hexagons and three bonding partners each (specified by the sp^2 hybridization). They therefore resemble graphite in structure and can thus be considered as a different geometrical form of this material (Fig. 39.1). On the other hand, their mechanical properties are very different from graphite, so they are often considered as an independent modification of carbon.

The diameter of the tubes is usually 1–50 nm; theoretically, minimum tube diameters of 0.4 nm are possible. The length of CNTs is usually a few millimeters. A distinction is made between single-wall and multi-wall tubes, between open and closed tubes, and between empty and filled tubes.

Graphene

Graphene also consists only of carbon atoms, which form honeycomb-like structures of hexagons. In contrast to carbon nanotubes, the layers of graphene are planar and ideally consist of only one layer of graphite. Their production and properties are explained in Sect. 39.12.

39.2 History

Carbon nanotubes are not an invention of modern times. Kokarneswaran et al. (2020) discovered that pottery with coatings containing carbon nanotubes was made in India as early as the sixth century BC. Carbon nanotubes have also been discovered in Damascus steel and in ancient Chinese ink. They are formed naturally during incomplete combustion of wood. The first to publish on carbon nanotubes were the Russian scientists (Radushkevich & Lukyanovich, 1952). However, their discovery remained unknown outside Russia because most scientists do not understand Russian. Oberlin et al. (1976) fabricated carbon nanotubes as single- and multi-wall structures by deposition from the gas phase. To honor M. Endo's contributions to this development, the corresponding process by which most multi-wall CNTs are produced today is called the Endo process. Further publications by other researchers followed.

In 1991, Iijima (1991) published his seminal work on the fabrication of carbon nanotubes, which led to many researchers around the world becoming involved in the creation of carbon nanotubes. Iijima and others developed, among other things, the arc process, which is still used today. Since then, CNTs have been researched and new applications developed in a large number of research laboratories worldwide (cf. Sect. 39.9).

D. Veit, *Fibers*, https://doi.org/10.1007/978-3-031-15309-9_39

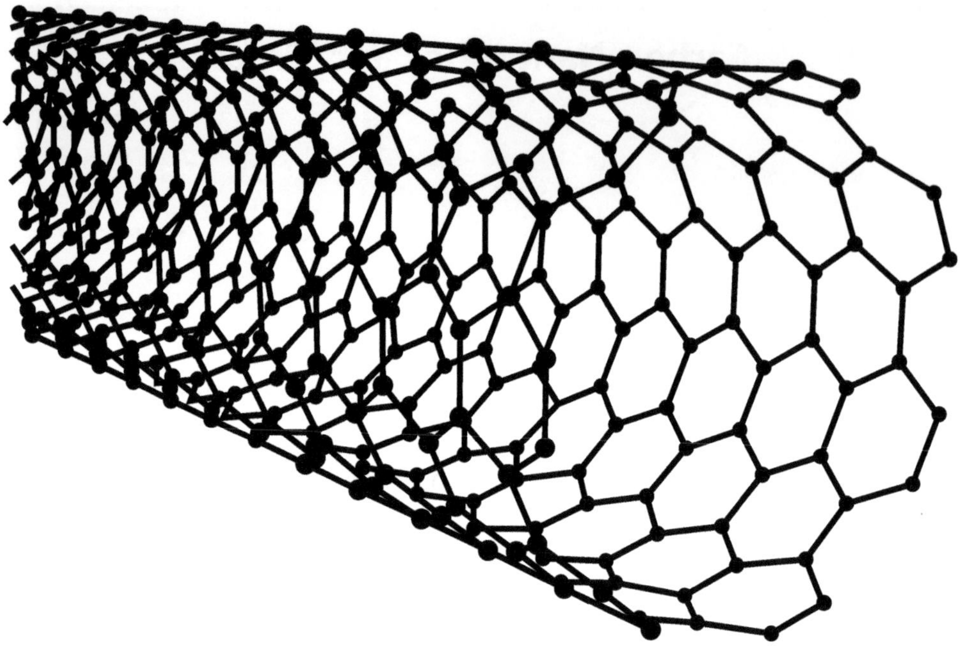

Fig. 39.1 Structure of a single wall carbon nanotube

39.3 Structure

Depending on how the honeycomb network of graphite is rolled into a tube, different structures are formed. A distinction is made between helical (helically wound) and non-mirror symmetric (chiral) structures. The designation of the respective type is done with the help of a pair of numbers (n, m), which indicates the exact winding direction and the size of the nanotube (Fig. 39.2). In general, three types are distinguished:

- Zigzag nanotubes: This non-helical structure is formed when the graphene wall is rolled up in such a way that the ends of an open tube show a zigzag edge as a perfect termination. Thus, the unwinding occurs along the unit vector (1,0) of the lattice.

Fig. 39.2 Geometry and classification of nanotubes

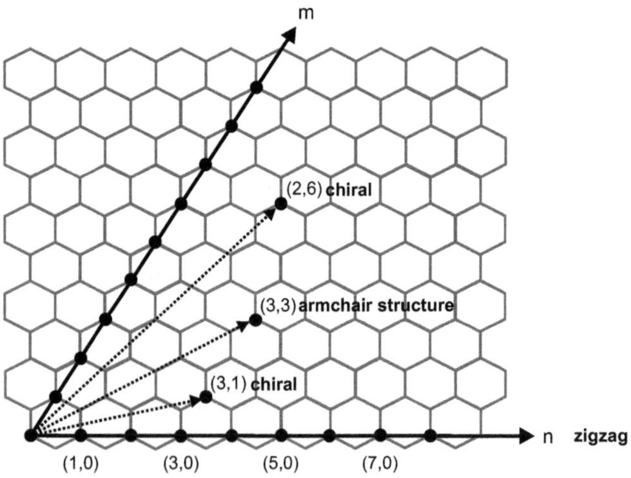

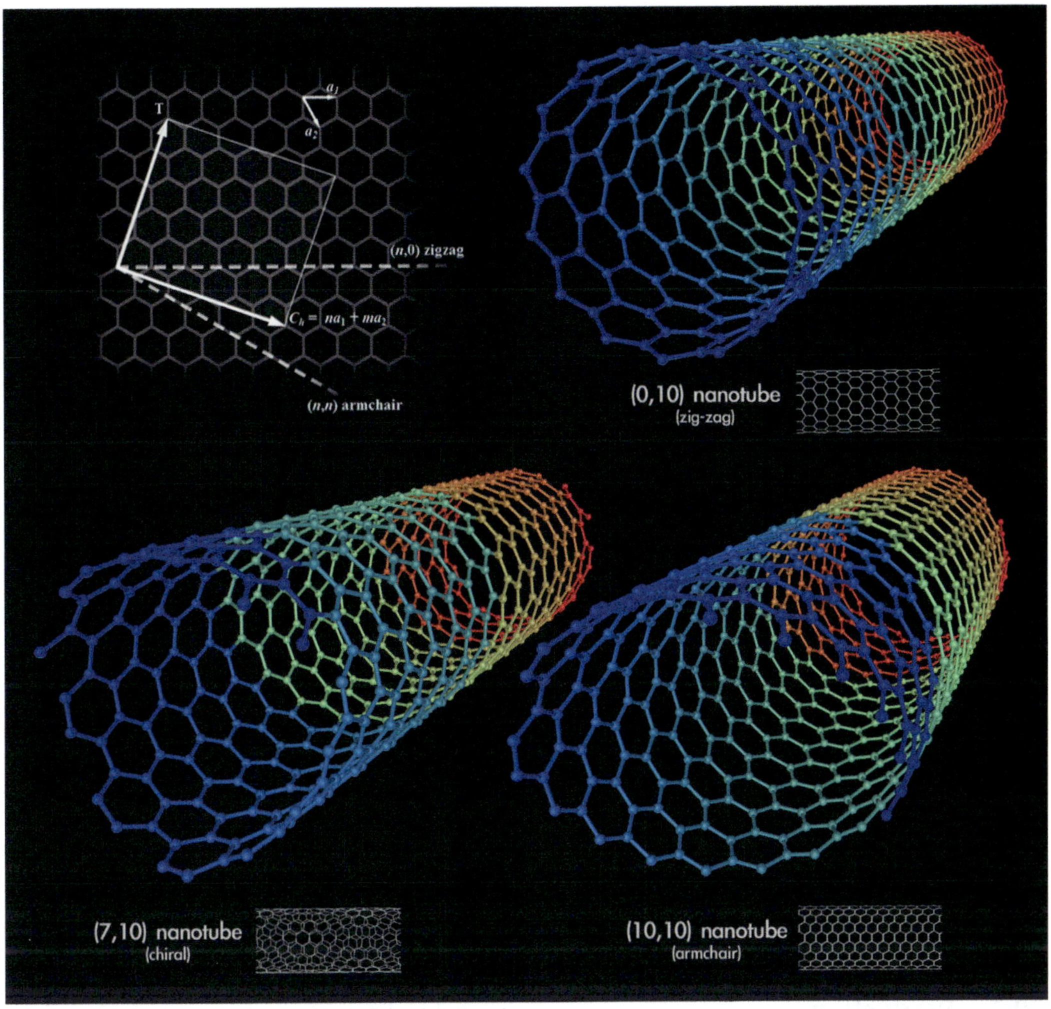

Fig. 39.3 Shapes of carbon nanotubes (Stroeck, 2006)

- Armchair nanotubes: In this helical type, the graph layer is rotated by 30° before rolling up compared to the zigzag tubes, so that an edge is formed from the sides of the last six-ring row as a perfect termination. Thus n = m is valid.
- Chiral nanotubes: If the angle through which the graphene layer is rotated before being rolled up is between 0° and 30°, chiral helical nanotubes are formed. They are characterized by the fact that a line running along the unit vector (1,0) spirals around the nanotube. These nanotubes therefore occur in two enantiomeric forms, e.g. (3,1) and (1,3) or (2,6) in Fig. 39.2.

Figure 39.3 shows the most important types and their naming in the (n,m)-coordinate system.

The diameter of carbon nanotubes can be approximately calculated from

$$d[nm] \approx 0{,}0783 \cdot \sqrt{(n+m)^2 - n \cdot m}$$

In particular, for small values of n and m, this equation is inaccurate because of distortion effects due to strong internal stresses within the carbon nanotube.

The thinnest conceivable nanotube would have the configuration (2,2) and a diameter of about 0.27 nm. The thinnest nanotubes actually produced to date had the configuration (5,1) or

Table 39.1 Diameter of carbon nanotubes in [pm] as a function of (n, m)

n\m	2	3	4	5	6	7	8	9	10
2	271	341	414	489	565	641	718	795	872
3	341	407	476	548	621	696	771	847	923
4	414	476	542	612	683	755	829	903	978
5	489	548	612	678	747	817	889	962	1036
6	565	621	683	747	814	882	953	1024	1096
7	641	696	755	817	882	949	1018	1088	1159
8	718	771	829	889	953	1018	1085	1153	1223
9	795	847	903	962	1024	1088	1153	1221	1289
10	872	923	978	1036	1096	1159	1223	1289	1356

(4,2) and thus a diameter of around 0.4 nm (Torres-Dias et al., 2017), Table 39.1.

In addition to single-wall carbon nanotubes (SWNTs), there are also multi-wall carbon nanotubes (MWNTs) consisting of 2–50 layers. There are two models for their description: Either they are considered as single-wall CNTs pushed into each other (Fig. 39.4, top) or as a rolled-up graphite layer, similar to a parchment roll (Fig. 39.4, bottom).

With more than 50 layers, the distinction from "normal" carbon fibers becomes difficult. The minimum diameter of MWNTs is approx. 1 nm. The distance between the individual layers is 0.34 nm. This corresponds to the distance between the layers in turbostratically arranged graphite.

39.4 Production

There are very different processes for the production of carbon nanotubes. The most important ones are briefly described below. Many of them can also be used for the generation of fullerenes.

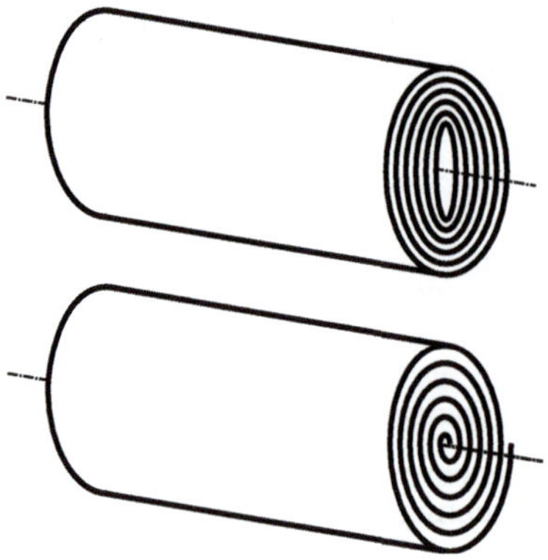

Fig. 39.4 Structure of multi-walled carbon nanotubes

39.4.1 Arc Process

In this process, carbon is vaporized in the form of graphite rods and deposited on cool parts of the apparatus. The graphite electrodes do not touch each other during the conversion and thus an electric arc is formed. Production takes place in a steel container filled with an inert gas (helium, argon) at approx. 700 mbar. The current is 50–100 A, the DC voltage is approx. 20 V. The arc that jumps between the two graphite electrodes when the voltage is applied causes the anode material to be transported into the gas phase, so that it can be deposited in the form of nanotubes. Some of it is deposited as a deposit on the cathode, and some of it condenses on the walls of the vacuum chamber. Because the graphite anode is consumed during the conversion, it must be constantly replenished (approx. 1 mm/min). Due to the large amount of heat generated during the process, the cathode must be cooled (Fig. 39.5).

The Fig. 39.6 shows the formation of the nanotubes in detail. Individual, already formed nanotubes protrude from the columns (zone A) and emit electrons into the plasma in the applied electric field. These electrons ionize carbon clusters or atoms, which then generate a carbon ion flux above zone A. At the same time, the helium is sucked in. At the same time, the helium is sucked in and pushed to the side, so that a

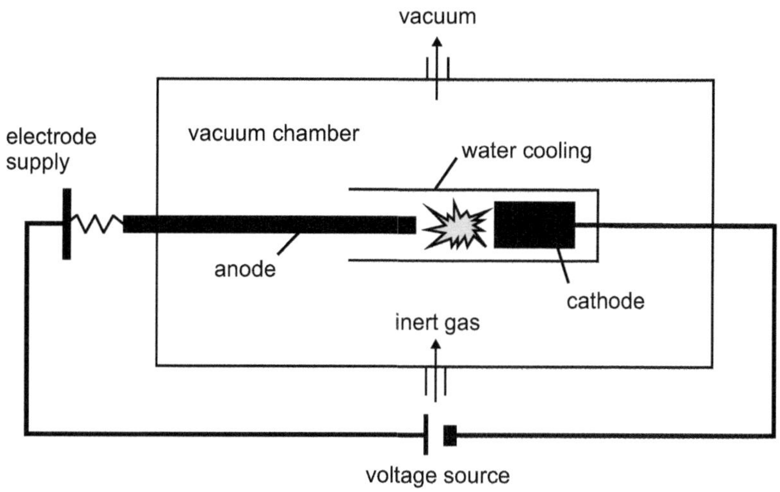

Fig. 39.5 Arc process

circulation is created and new carbon atoms and clusters are constantly supplied. These attach themselves to the already existing nanotube pieces and thus extend them. These nanotubes are mainly located in zone B. If the cathode is not cooled sufficiently, nanotubes already formed evaporate again due to the high heat. The graphite particles evaporating from the anode are only a few nanometers in size and are positively charged. As they pass through the plasma of helium, they are broken down into small C clusters (mostly C_3) and singular carbon atoms. This is only successful for sufficiently small crystals, otherwise the time of plasma passage is insufficient and they deposit as graphitic particles on the cathode. Atomization is therefore one of the limiting parameters of CNT production with this process.

The length of the CNTs produced in this way is about 1 μm, the inner diameter is 1–3 nm and both SWNTs and MWNTs can be produced.

39.4.2 Laser Ablation Process

In this process, a focused laser beam is used to provide the energy needed locally for atomization.

The Fig. 39.7 shows the principle of this process. A quartz tube is placed in a furnace heated to

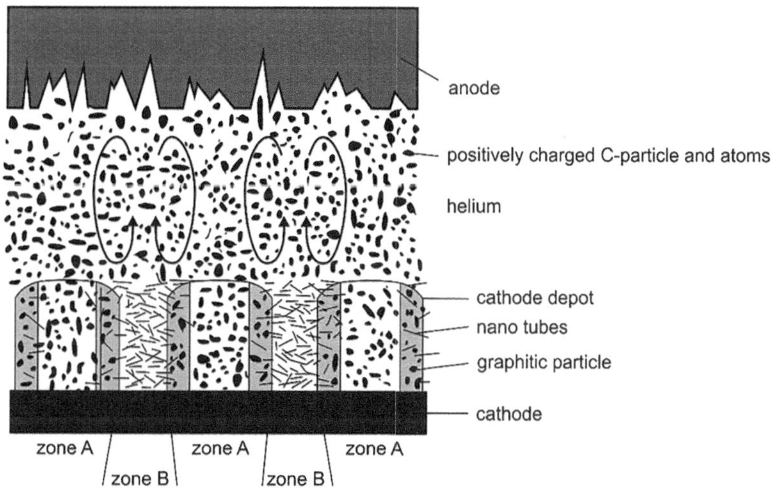

Fig. 39.6 Sequence of nanotube production using the electric arc process

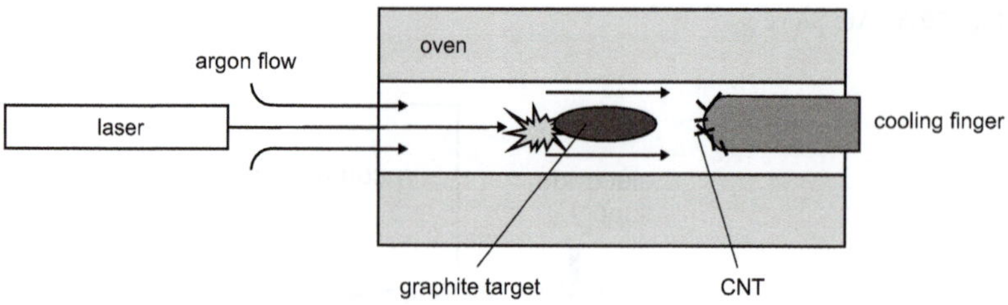

Fig. 39.7 Schematic of laser ablation process

about 1200 °C, at one end of which is a strongly cooled element (often made of copper) for separating the carbon nanotubes. The laser beam vaporizes the graphite, which is passed through the inert carrier gas (usually argon) to the cooling finger where the nanotubes are formed.

Both, SWNTs and MWNTs, can be produced with this process. The former are preferentially produced when a catalyst is added. The length of the individual tubes can reach several hundred nanometers, their diameter is usually 1.2–1.7 nm. A special feature of this process is that whole bundles of CNTs are often formed, the diameter of which is then in the range of 5–20 nm and the length of which is several hundred micrometers. One can therefore almost speak of CNT fibers.

One disadvantage of this process is that it must be operated discontinuously because the graphite electrode consumed during the process cannot be replenished. The production is around 5 g/h SWNTs.

39.4.3 Chemical Vapour Deposition (CVD) Process

This process, which has been known since the 1990s, can be used to produce both carbon fibers and CNTs.

The apparatus consists of a furnace and an evacuable, temperature-stable container containing a carrier material with catalyst particles distributed on it. The gas mixture is passed over the catalyst. After completion of the reaction, the substrate with the CNTs formed thereon is removed from the reaction chamber and the nanotubes are isolated. Typical process temperatures are between 550 and 770 °C (Fig. 39.8).

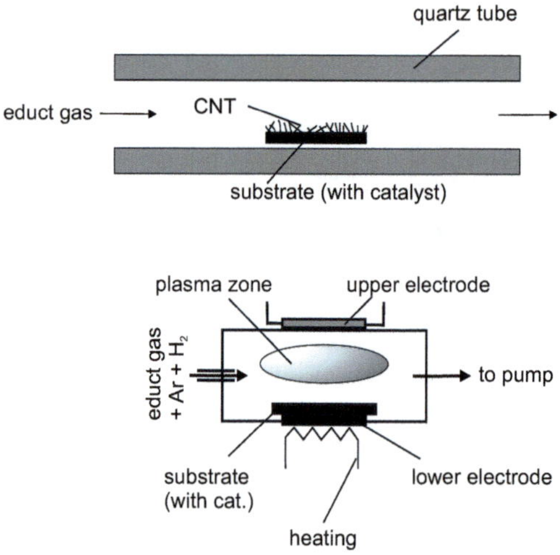

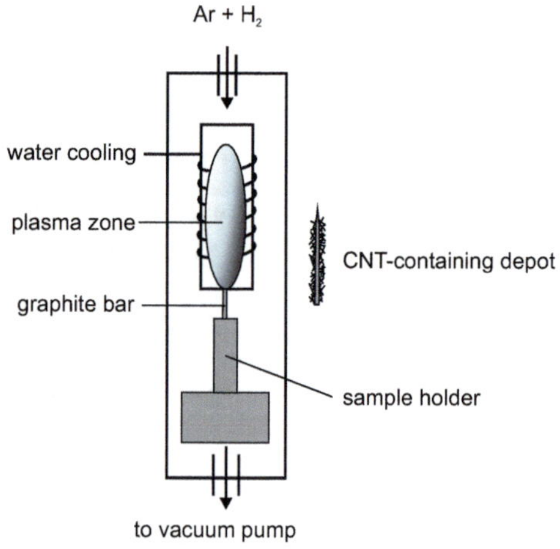

Fig. 39.8 Different possibilities of chemical vapor deposition (top left: classical system; bottom left and right: plasma-assisted systems)

The carbon source is usually hydrocarbons, e.g. methane and ethylene, with admixtures of hydrogen and inert gas. The high process temperature limits the choice because the hydrocarbons must not decompose during the process. The catalyst (e.g. based on iron, cobalt, nickel) is finely distributed on the carrier material (e.g. aluminum or silicon oxide) and must adhere very well to it in order to ensure constant product quality. A large surface area of the catalyst favors the growth process of the nanotubes.

During the process, the catalyst can dissolve carbon to a certain extent (analogous to steel production). Once the saturation state is reached, the actual growth of the nanotubes on the catalyst surface begins. The catalyst also has the task of catalyzing the thermal decomposition of the hydrocarbon present, so that carbon atoms and small clusters are formed in the gas phase, which serve as building blocks for the CNTs.

Particularly advantageous in this process is the ability to control the diameter of the SWNTs by the size of the catalyst particles. For example, iron particles with sizes of 3, 9 and 13 nm result in SWNTs with diameters of 3, 7 and 12 nm. Therefore, the uniform size of the catalyst particles is critical for high quality CNTs. A disadvantage of the process is that it cannot be operated continuously because the substrates have to be removed from the system when the nanotubes have reached the desired length.

A method that is significantly improved in terms of yield involves decomposing ethylene in the presence of a catalyst at a precisely controlled water content of the reactant stream. The oxidizing effect of the water then removes amorphous carbon from the product mixture. This improves the purity of the product and, in particular, also significantly extends the life of the catalyst particles because they are no longer covered by a layer of amorphous carbon. This results in dense, vertically parallel nanotubes that can reach a height of several millimeters after only 10 min. Their diameters are 1–3 nm (Hata et al., 2004).

39.5 Properties

SWNTs have a density of 1.3–1.4 g/cm^3, while for MWNTs it is 1.8 g/cm^3 (Kim et al., 2009) because the voids are filled accordingly.

Mechanical Properties

Carbon nanotubes have a tensile strength of up to 63 GPa in a multi-walled design (Yu et al., 2000). Under optimal conditions, even strengths up to 100 GPa and Young's modulus values up to 1000 GPa are possible (Peng et al., 2008). Thus, 80% of the theoretically possible strength for defect-free tubular structures is achieved. This corresponds to approx. 5500 cN/tex, i.e. 15–25 times the strength of high-strength carbon fibers or 250 times that of high-strength steel. In practical use, however, only strengths in the range of a few GPa are measured, because the cohesion between the tube layers is poor and structural defects can lead to fracture. Thus, the strength is in the range of common carbon fibers. The elongation of CNTs is up to 18%.

The strength in the radial direction is considerably lower, some authors even describe CNTs as "soft". When subjected to pressure in the axial direction, CNTs react by buckling (Jensen et al., 2007).

The Young's modulus ranges between 400 and 1200 GPa (steel: 210 GPa), extreme values for SWNT are up to 4000 GPa (Salvetat et al., 1999). The values for MWNT are generally somewhat higher than for SWNT because, in addition to the strength of the strongest nanotube in the structure, the van der Waals forces between the individual tubes also contribute to the strength.

The mechanical properties are usually measured by clamping a CNT between two force microscope tips and then stretching it. Exact values are very difficult to determine. This is why there are sometimes large deviations in the manufacturers'specifications.

Electrical Properties

Of particular interest to the electronics industry are the current-carrying capacity and thermal

Table 39.2 Electrical properties of carbon nanotubes as a function of (n, m)

n\m	2	3	4	5	6	7	8	9	10
2	■			■			■		
3		■			■			■	
4			■			■			■
5	■			■			■		
6		■			■			■	
7			■			■			■
8	■			■			■		
9					■			■	
10			■			■			■

■ metallic properties

conductivity: the former is estimated to be 1000 times that of copper wires, while the latter, at 6000 W/(m-K) at room temperature, is almost twice as high as that of diamond at 3320 W/(m-K), the best naturally occurring thermal conductor.

Depending on the structure, the electrical conductivity within the tube is metallic or semiconducting (cf. Sect. 39.2).

The index pair (n, m) can be used to determine whether the tube is more like metals in its electrical properties or a semiconductor. If

$$(\mathrm{n}=\mathrm{m}) \text{ or } \frac{n-m}{3} \text{ is an integer,}$$

the carbon nanotube has metallic properties, otherwise it is a semiconductor. Hence, all zigzag nanotubes of appropriate length, all armchair and some chiral nanotubes have metallic properties (Table 39.2).

If individual carbon atoms are replaced by boron or nitrogen, semiconductors of the p- or n-type are obtained. Transistors and simple circuits have also been produced with semiconducting carbon nanotubes.

Thermal Properties

Carbon nanotubes have an extremely good thermal conductivity in the axial direction (up to 3500 W/mK) (Pop et al., 2006), which corresponds to about ten times the value of copper. In the radial direction, on the other hand, the corresponding value is very low at about 1.5 W/mK. Carbon tubes have also been produced which are superconducting at low temperatures (Ferrier et al., 2009).

39.6 Functionalization

Carbon nanotubes are often functionalized to give them specific properties. Chemical reactions can occur at three sites: At the tube ends (particularly reactive), at the outer or inner wall, and at defect sites. A carbon nanotube is essentially equivalent to a graphite layer wound around the z-axis. Therefore, its chemical behavior should resemble that of graphite. However, this is only partially the case. Unlike planar graphite, the cylindrical shape of the tube also produces a curvature in the orientation of the—orbitals. As a result, a mixture of—and—orbitals occurs, so that the—orbitals protrude radially from the tube surface. This effect depends on the diameter of the tubes and decreases with increasing diameter. In addition to the possibility of functionalizing CNTs via covalent bonds, they can also be "wrapped" or have their properties modified by the addition and insertion of atoms and molecules. The Fig. 39.9 gives an overview of the principal possibilities.

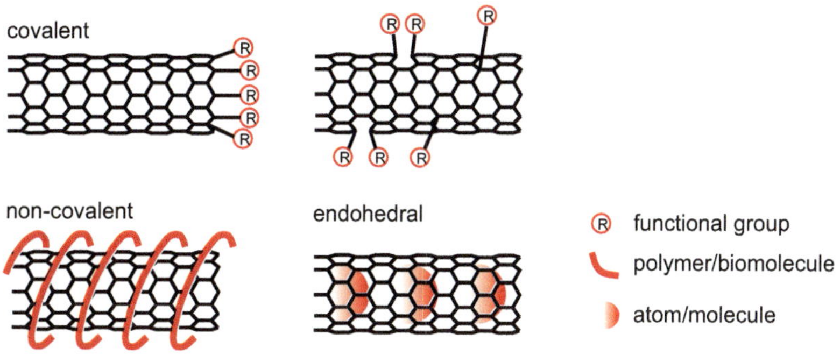

Fig. 39.9 Possibilities for the functionalization of nanotubes

39.7 Production of Yarns from Carbon Nanotubes

Yarns from CNTs are currently mostly produced using the wet spinning process. For this purpose, the nanotubes are dispersed in sodium lauryl sulfate (sodium dodecyl sulfate) and spun out in a liquid crystalline state into a spinning bath of 5% polyvinyl alcohol solution (Vigolo et al., 2000), Fig. 39.10.

The Fig. 39.11 shows the basic principle. After leaving the spinneret and entering the precipitation bath, the carbon nanotubes align themselves in the direction of the fibers due to shear forces. This process was first used to produce CNT tapes and later filament yarns (Mercader et al., 2011). When sodium is replaced by lithium, the strength of the spun-out yarns increases significantly (Dalton et al., 2002). Fibers containing 60% CNTs and 50 μm in diameter possessed a strength of 1.8 GPa. However, this falls far short of the strength values of pure carbon fibers. However, a very large elongation was observed (>100%), resulting in a high energy absorption rate. Subsequent hot stretching via godets led to a high degree of crystallinity of the PVA and thus to a high orientation of the CNTs in the axial direction of the fibers.

A fundamental disadvantage of this spinning method is the high proportion of polymer in the yarn. This significantly reduces the electrical conductivity of the CNTs (Dalton et al., 2002). If the polymer is removed, e.g. by thermal treatment, the electrical conductivity is improved, but the mechanical properties are significantly deteriorated.

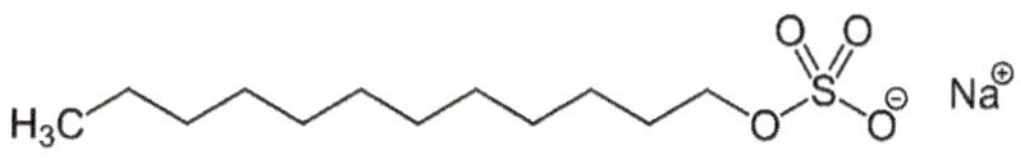

Fig. 39.10 Sodium lauryl sulfate

Néri et al. (2006) developed a spinning process that does not require a polymer-based precipitation bath. The CNTs are first dispersed and then added to the polymer spinning solution. In this way, CNT fibers of good strength can be produced, but the distribution of the CNTs in the spinning solution must be very homogeneous.

Organic solvents were also used, but the maximum concentration of CNTs is very low, so that an industrial process is prohibited from the outset. CNTs are also soluble in strong acids, and the precipitation bath can then consist of water. However, the corresponding processes have so far all only been realized on a laboratory scale.

To avoid the problem of solvents, yarns have also been spun from 100% CNTs by pulling the CNTs from a forest of nanotubes, twisting them and winding them up. The resulting structure is ring yarn-like, typical yarn diameters are 1–60 μm, and the twist height is about 20,000 1/m. The strength of such yarns is 400–750 MPa, the elongation is 3–9% and the Young's modulus reaches values up to 28 GPa (Zhang et al., 2004).

39.8 Production of Composite Materials with Nanotubes

Carbon nanotubes are often used as reinforcing material for composite plastics due to their

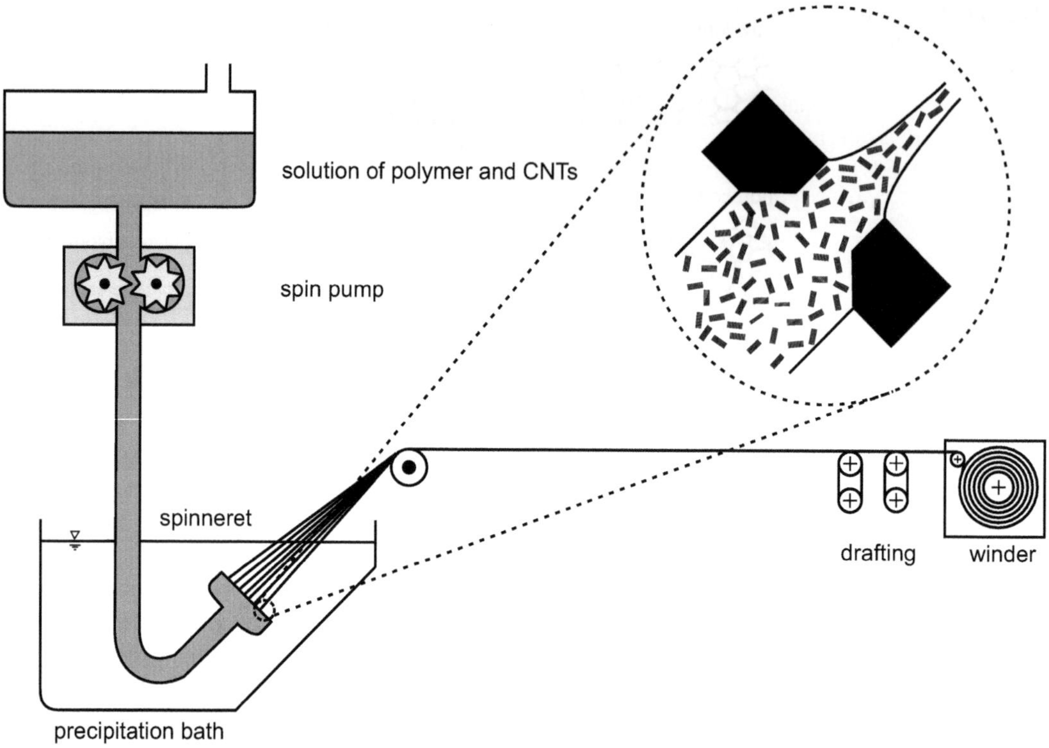

Fig. 39.11 Principle of the CNT wet spinning process

outstanding mechanical properties. However, this has had only very limited success to date. This is due, among other things, to the fact that the carbon tubes easily cluster together and are thus not evenly distributed in the plastic matrix. In addition, the CNTs interfere with the formation of larger crystalline domains. Therefore, the addition of CNTs can even significantly worsen the mechanical properties. A comprehensive overview of numerous material combinations with CNT fillers is given by Hiremath and Bhat (2017).

Attempts have also been made to produce electrically conductive plastics by adding carbon nanotubes. However, this is only successful if the CNTs have direct contact with each other and this means that a considerable amount of CNTs must be added, which has a direct negative effect on the mechanical properties. Therefore, this approach is rather theoretical and has not been successfully implemented in practice so far.

Composites with nanotubes can be produced acc. to different processes. These differ in the time of composite formation and the type of attachment to the nanotubes.

During the Polymerization Process

In this process, the nanotubes react with the active end of a nascent polymer chain. However, the loss of entropy reduces the mobility of the macromolecules and thus also the chance that a carbon nanotube can attach to a polymer chain end. The achievable conversions are therefore low. This process is only rarely used today.

Polymerization Starting from Carbon Nanotubes

Chemical functionalization first binds initiator molecules to the nanotube surface. The chemical reaction then starts from these. With this process, several polymer chains can be attached to a single nanotube. However, the resulting chain length cannot be controlled.

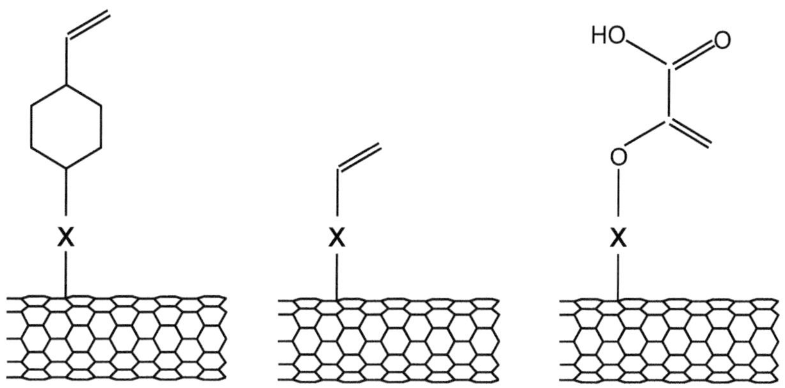

Fig. 39.12 Copolymerization with nanotubes

Alternatively, the nanotubes can be equipped with functional groups that can simultaneously serve as the monomer unit of a polymer (Fig. 39.12). In this way, copolymers can be prepared in which the stoichiometry can be relatively precisely adjusted. However, the handling of nanotubes doped in this way is difficult because the functional groups react easily with each other and are thus no longer available for polymerization.

Direct Chemical Linkage

If polymer chains carry reactive groups (e.g. amino or hydroxyl groups) at their ends, functionalized nanotubes can be covalently bonded to them. This also results in composite materials, but in which the polymer strands are linked to the nanotubes at only a few positions. Reactive groups in the side chains create more docking sites and thus binding can be achieved along the entire length of the nanotube. This was tested, for example, in the production of PA 6 and PA 66 by adding carboxylated nanotubes to the monomer in low concentrations (Fig. 39.12, right). The CNTs were uniformly incorporated into the fiber and resulted in an increase in strength and Young's modulus (Meng et al., 2019). This method has only been successful for polyamide fibers when extensive crystalline regions are not to be formed, which would be hindered in the presence of CNTs. The strength values obtained are therefore far below the strengths of high-strength polyamide.

Non-covalent Connection

Nanotubes can also be integrated into polymers via non-covalent bonding. However, the corresponding adhesive force is usually very low.

Epoxy Resins

>The embedding of nanotubes in epoxy resins has already been carried out successfully for several years. In this process, the CNTs are dispersed in a metastable suspension and then the epoxy resin is cured. This embeds them uniformly and largely singly in the polymer. If the curing reagents have terminal amino groups, then appropriately functionalized nanotubes can also form covalent bonds with the resin directly during curing (Fig. 39.13).

Other Possibilities

In addition to the processes already described, carbon nanotubes can also be incorporated into polymethacrylates, polyanilines, polystyrenes and polyphenylenevinylidene and other polymers. In principle, they could also be added to polyolefins, but the uniform and stable incorporation of the nanotubes is then difficult due to the lack of functional groups in PE and PP.

39.9 Areas of Application

Nanotubes are already used in the production of antistatic films to make them conductive. Antistatic plastic hoses can also be equipped with them to

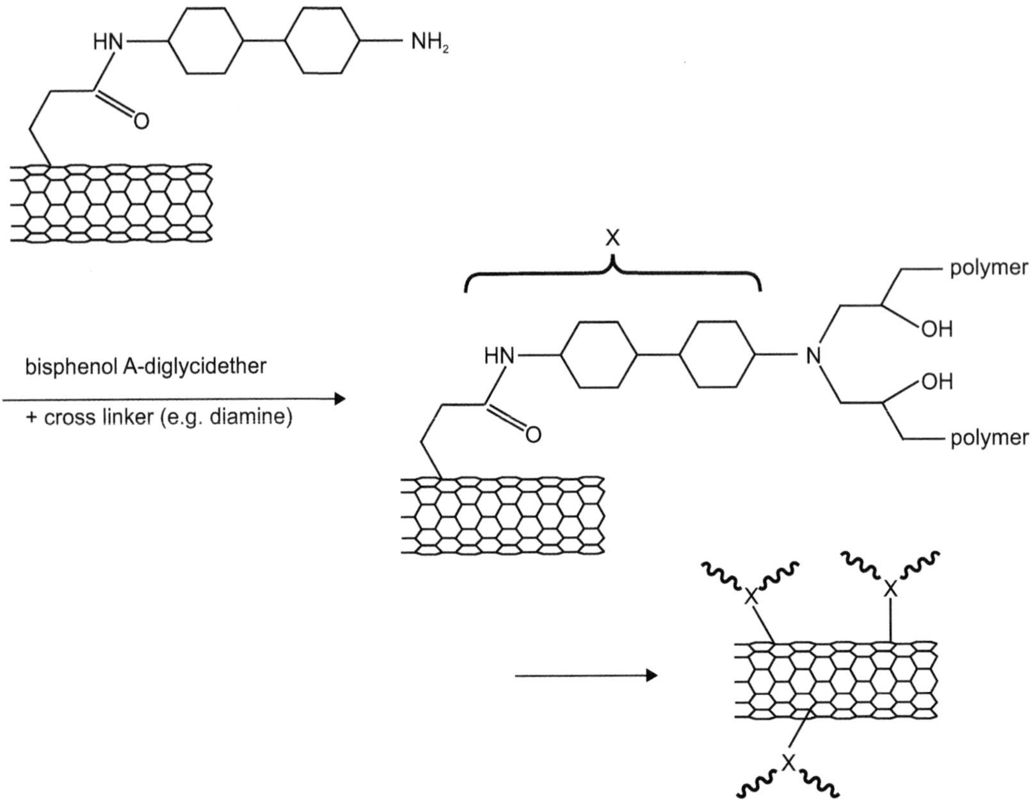

Fig. 39.13 Preparation of CNT-reinforced epoxy composites

prevent dust explosions during the transport of fine particles. In the production of fiber composites, nanotubes are grown on the fibers to improve their bonding to the surrounding resin, the matrix (see above). This results in significant improvements in mechanical properties, but so far only on a laboratory scale. Whole bundles of nanotubes have also already been processed on a laboratory scale into threads or mats to be used as materials. Another field of application is space travel: space elevators are to leave the earth on ropes made of nanotubes. However, this consideration is still very speculative and a realization is not in sight.

Another application is the use of nanotubes for the tip of force microscopes, allowing much finer resolutions (Fig. 39.14).

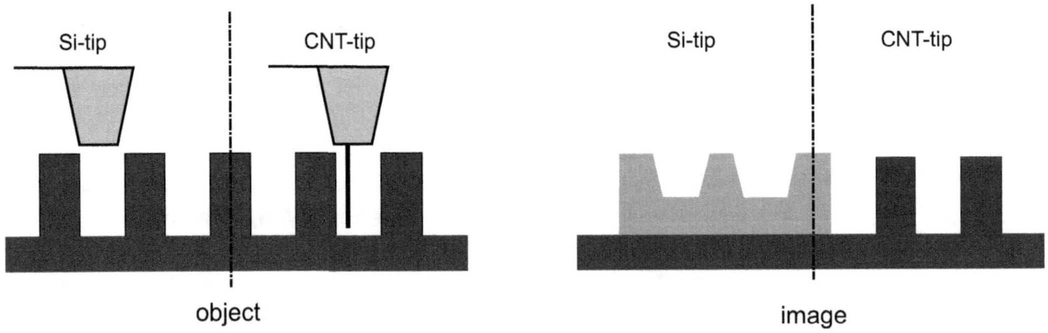

Fig. 39.14 Application of nanotubes as tip of force microscopes

Another field that will gain considerable importance in the future is the use of nanotubes as physical sensors. For example, nanotubes react to a mechanical stress by changing the signal positions in the Raman spectrum, proportional to the applied pressure. This could be used, for example, to monitor stress states in fiber composite materials. Nanotubes can also be used as actuators. They already react to weak electrical signals with a movement, so that one could create controlled-moving objects on the molecular level. Furthermore, it is conceivable that by injecting charge into a bundle of nanotubes in a suitable environment, one could bend them in a targeted manner. In this way, one could, for example, mimic the function of a muscle.

Nanotubes can also be used as chemical sensors. Due to their large surface area, a change in the charge carrier concentration is generated under the influence of electron-donating or—accepting substances. This can be detected by changing the resonance frequency, the conductivity or the electrical resistance and thus the amount of a certain substance can be measured.

The storage of hydrogen for car engines is a problem that has not yet been solved satisfactorily. Conventional carbon structures are too open-pored, so that the hydrogen is not bound. Nanotubes, on the other hand, can store up to 7% of their own weight in hydrogen, both on the outside and inside. This corresponds approximately to the required minimum storage capacity. Desorption only sets in at temperatures above 400 °C (outside) or 600 °C.

The English company Surrey NanoSystems has succeeded in producing an extremely dark material using directed nanotubes. Vantablack absorbs up to 99.965% of incident radiation and has 1 billion nanotubes per square centimetre (diameter 20 nm, length 14–50 μm). Applications for this material could include solar panels and the shielding of radio waves in a very wide frequency range.

Other potential applications include transistors, data storage, displays and electronic switches. However, most of the potential applications mentioned here exist only in theory or on a laboratory scale to date.

39.10 Health Effects

The health effects of human exposure to carbon nanotubes have not yet been conclusively clarified. On the one hand, it is to be expected that due to the asbestos-like dimensions of nanotubes there is a similar hazard. On the other hand, the "incubation period" is several years or even decades, so that no definitive statements can yet be made here. CNTs were toxic in animal experiments under certain conditions. However, this was mostly attributed to contamination of the nanotubes by atoms of the catalysts used in their manufacture (e.g. cobalt, nickel, molybdenum).

In the USA, the National Institute for Occupational Safety and Health (NIOSH, 2013) published a comprehensive study describing the health hazards. Acc. to this, it is recommended that exposure to CNTs should not exceed 1 μg/m^3 over a working lifetime of 45 years (8 h daily).

Because of the unresolved health risks associated with the handling and use of CNTs, the same safety precautions should generally be taken as when handling asbestos (respiratory protection, etc.).

39.11 Market and Manufacturers

Industrial production of carbon nanotubes began in 2004 and by 2014, 5000 t of CNTs had been produced worldwide. A total production of 20,000 t is expected for 2022. Around 60–70% of all carbon nanotubes are added to polymers, and around 10% are used for electrical applications (Statista, 2021). While SWNT production has been increasing recently, some large plants for the production of MWNTs have been shut down again in recent years. This is partly because the demand for carbon nanotubes is not as great as predicted and partly because the commissioning of several large plants led to a drop in prices, making production no longer economically viable for many manufacturers. Nevertheless, 80% of all CNTs continue to be produced as multiwall carbon tubes.

The market for CNTs is currently worth around €5bn and CNT sales are expected to reach around €15bn by 2026 (Statista, 2021). The largest markets are Asia (50%) and Europe and North America (25% each). Acc. to forecasts, this is not expected to change in the next few years (Statista, 2021).

However, both the production quantities and the sales figures are only estimates. Acc. to these official figures, 1 g of carbon nanotubes costs 1 €. However, the average price is more like 25 € per gram. So either the production volume is overstated by a factor of 25 or the sales volume is understated by the same factor.

The purity of industrially produced carbon nanotubes is usually 90–99%, whereby the purity has a great influence on the price. Typical prices per gram for SWNTs are € 1–100, for MWNTs € 25–500. Functionalized SWNTs cost up to 1000 €/g, functionalized MWNTs are priced at up to 300 €/g.

In the Table 39.3, typical characteristic values for selected CNT manufacturers are compiled. Most manufacturers with market relevance have their headquarters in the USA or India.

39.12 Graphene

The term "graphene" was introduced in 1962 by Boehm et al. (1962). Graphene was first created by Novoselov (2004), Konstantin Novoselov and Andre Geim received the Nobel Prize in Physics in 2010 for this and for deciphering the properties of graphene.

There are various methods for the production of graphene. On the one hand, it can be extracted from a graphite block with the help of adhesive strips, but the yield is very low. In another process, graphene layers are created from a mixture of water, dishwashing detergent and graphite. Much higher yields and larger graphene pieces can be obtained by reducing graphene oxide. Graphene can also be produced from polycyclic aromatics in a multi-step process. Heating of graphite oxide to temperatures up to 1000 °C is also used. Growth on metallic substrates has also been successfully tested (WIKI, 2021a). Nevertheless, all these processes produce only small amounts of graphene in each case, which leads to a very high price.

The measured strength of graphene is 125 GPa, the Young's modulus is 1020 GPa (Lee et al., 2008). Thus, it has the highest strength of all known materials. Its density is 2.26 g/cm^3. Thus, one square kilometer of graphene, consisting of one layer of graphite, has a mass of about 750 g. Graphene is electrically conductive, has a high thermal conductivity and is optically almost transparent to almost all wavelengths from UV to IT (97.7% light penetration). It can withstand temperatures of up to 400 ° C, and even up to 1000 °C in the absence of air.

Spinning out graphene in pure form has not yet been successful. However, graphene-added fibers have been melt- and solution-spun. For example, PA 6 and PVA fibers with 1 and 10% graphene, respectively, have been produced (Weise et al., 2017). As expected, the Young's modulus of such fibers is increased because their crystal structure is changed, making it more difficult for the matrix macromolecules to slip. An increase in strength is also observed. Although the electrical conductivity was several orders of magnitude higher than for unadditivated polymers, it was also several orders of magnitude lower than for conductive polymers or metal fibers. In addition, there was undesirable agglomeration of graphene in the polymer. The haul-off speed of up to 1400 m/min would be sufficient for economic production if all other process engineering difficulties are solved.

Gao et al. (2020) added 0.075% graphene to a PAN cursor and were thus able to produce carbon fibers with a strength of 1916 MPa and an elastic modulus of 233 GPa. In absolute values, this corresponds to those of low-quality carbon fibers, but the authors write that they were able to more than double the strength and Young's modulus compared to self-generated carbon fibers without graphene. It is unclear whether this doubling also occurs in high-strength carbon fibers.

The idea of a space elevator was first formulated by Artsutanov (1960). Acc. to this, a cable would be stretched from the earth to a geostationary satellite and thus transport payloads

Table 39.3 Selected manufacturers of carbon nanotubes acc. to Nanowerk (2021)

Producer	Type	Properties	Website	Country
Ad Nano Technologies	SWNT	Ø 0.8 nm L 400–2000 nm	www.ad-nanotech.com	India
	MWNT	Ø 5–15 nm L 10,000 nm		
Beyond Pacific Holdings, LLC	MWNT	Ø 8–13 nmL 50,000–150,000 nm	www.nanophite.com	USA
Cheap Tubes Inc.	MWNT	Ø 10–20 nm L 50,000–150,000 nm	www.cheaptubes.com	USA
He Ji Company Ltd.	SWNT	Ø 8–50	www.nanotubeseu.com	China
	MWNT	Ø 8–50		
M K Impex Corp.	SWNT	Ø 1–2 nm L 5000–30,000 nm	www.mknano.com	Canada
	MWNT	Ø 8–15 nm L 10,000–30,000 nm		
Nanografi	MWNT	Ø 40–80 nm L 10,000–30,000 nm	www.nanografi.com	USA
NanoIntegris Technologies, Inc.	SWNT	Ø 0.8–1.2 nm L 100–1000 nm	www.nanointegris.com	USA
	MWNT	Ø 8 nm L 10,000–30,000 nm		
Nano Research Elements	SWNT	Ø 1.1 nm L 100–300 nm	www.nanorh.com	India
	MWNT	Ø 8–15 nm L 30,000–50,000 nm		
NanoLab, Inc.	SWNT	Ø 1–1.5 nm L 1000–5000 nm	www.nano-lab.com	USA
	MWNT	Ø 10–30 nm L 5000–20,000 nm		
Nanoshel UK Limited	SWNT	Ø 1–2 nm L 3000–8000 nm	www.nanoshel.com	Great Britain
	MWNT	Ø 10–20 nm L 3000–8000 nm		
NoPo Nanotechnologies India Prv. Ltd.	SWNT	Ø 0.8 nm L 1000 nm	www.nopo.in	India
Plasmachem GmbH	SWNT	Ø 1.5–2.1 nm L 5000 nm	www.plasmachem.com	Germany
	MWNT	Ø 5–20 nm L 1000–10,000 nm		
Reinste Nano Ventures Pvt. Ltd.	SWNT	Ø 1.4 nm L 10,000 nm	www.reinste.com	India
	MWNT	Ø 5–20 nm L 1000–10,000 nm		
SabiNano	MWNT	Ø 20–30 nm L 200,000 nm	www.sabinano.co.za	South Africa
SES Research Inc.	SWNT	Ø 2 nm L 5000–30,000 nm	www.sesres.com	USA
Sisco Research Laboratories Pvt. Ltd.	SWNT	Ø 1–2 nm L 200–30,000 nm	www.slrchem.com	India
	MWNT	Ø 8–50 nm L 10,000–30,000 nm		
XinNano Materials Inc.	MWNT	Ø 3–10 nm L 10,000 nm	www.xinnanomaterials.com	Taiwan

Table 39.4 Selected manufacturers of graphene (Weise et al., 2016)

Producer	Website
ACS Materials LLC	http://www.acsmaterial.com
Applied Graphene Materials PLC	http://www.appliedgraphenematerials.com
Directa plus S.L.R.	http://www.directa-plus.com
Graphene Square Inc.	http://www.graphenesq.com
Graphenea S.A.	http://www.graphenea.com
Grapheneall Co., Ltd.	http://www.grapheneall.com
Haydale, Ltd.	http://www.haydale.com
NanoGrafen, Ltd. Sti	http://www.nanografen.com.tr
TheGrapheneBox	http://www.thegraphenebox.com

into space. A corresponding fiber material would have to have an extremely high strength and a low dead weight. At the same time, it would have to be able to withstand the enormous temperature differences in space. Of all the fiber materials known to date, only graphene meets these requirements, at least in theory. Acc. to calculations, a graphene rope with a length of 35,786 km (geostationary altitude of the satellite) would only be subjected to 87% of its tensile strength by its own weight. Accordingly, it would be possible to transport loads along the rope. However, no process is yet known with which such a rope could be produced (WIKI, 2021b).

Other possible applications of graphene, in particular for the reinforcement of composite materials, have so far been of a purely academic nature due to the high production costs of graphene. Prices of several hundred €/g are still far too high for a broad textile application.

A good overview of the patents on graphene as well as the most important manufacturers can be found in Weise et al. (2016). Table 39.4 shows a list of selected graphene manufacturers.

References

Artsutanov, Y. (1960). *To the cosmos by electric train.* Pravda.

Boehm, H. P., Clauss, A., Fischer, G. O., & Hofmann, U. (1962). Das Adsorptionsverhalten sehr dünner Kohlenstoffolien. *Zeitschrift für Anorganische und Allgemeine Chemie, 316*(3–4), 119–127. Wiley.

Dalton, A. B., Collins, S., Muñoz, E., RAzal, J. M., Ebron, V. H., Ferraris, J. P., Coleman, J. N., Kim, B. G., & Baughman, R. H. (2002). Super-tough carbon-nanotube fibres. *Nature, 886*, 706. Springer.

Ferrier, N., Kasumov, A., Deblock, R., Guéron, S., & Bouchiat, H. (2009). Superconducting properties of carbon nanotubes. *Comptes Rendus Physique, 10*, 252–267. Elsevier.

Gao, Z., Zhu, J., Rajabpour, S., Joshi, K., Kowalik, M., Croom, B., Schwab, Y., Zhang, L., Bumgardner, C., Brown, K. R., Burden, D., Klett, J. W., van Duin, A. C. T., Zhigilei, L. V., & Li, X. (2020). Graphene reinforced carbon fibers. *Science Advances, 6*(17) AAAS.

Hata, K., Futaba, D. N., Mizuno, K., Namai, T., Yumura, M., & Iijima, S. (2004). Water-assisted highly efficient synthesis of impurity-free single-walled carbon nanotubes. *Science, 306*(5700), 1362–1364. AAAS.

Hiremath, N., & Bhat, G. (2017). High-performance carbon nanofibers and nanotubes. In *Structure and properties of high performance fibers* (pp. 79–109). Elsevier.

Iijima, S. (1991). Helical microtubules of graphitic carbon. *Nature, 354*, 56–58.

Jensen, K., Mickelson, W., Kis, A., & Zettl, A. (2007). Buckling and kinking force measurements on individual multiwalled carbon nanotubes. *Physics Review B, 76*(195436) American Physical Society.

Kim, S. H., Mulholland, G. W., & Zachariah, M. R. (2009). Density measurement of size selected multiwalled carbon nanotubes by mobility-mass characterization. *Carbon, 47*(5), 1297–1302. Elsevier.

Kokarneswaran, M., Selvaraj, P., Ashokan, T., et al. (2020). Discovery of carbon nanotubes in sixth century BC potteries from Keeladi, India. *Science Reporter, 10*(1), 19786. https://doi.org/10.1038/s41598-020-76720-z

Lee, C., Wei, X., Kysar, J. W., & Hone, J. (2008). Measurement of the elastic properties and intrinsic strength of monolayer graphene. *Science, 321*, 385–388. Springer.

Meng, Q. J., Wang, Z. M., Zhang, X. X., Wang, X. C., & Bai, S. H. (2019). Fabrication and properties of polyamide-6,6-functionalized carboxylic multi-walled carbon nanotube composite fibers. *High Performance Polymers, 22*(7) Sage.

Mercader, C., Denis-Lutard, V., Jestin, S., Maugey, M., Derre, A., Zakri, C., & Poulin, P. (2011). Scalable process for the spinning of PVA-carbon nanotube composite fibers. *Journal of Applied Polymer Science, 125*, 191–196. Wiley.

Nanowerk. (2021). https://www.nanowerk.com/carbon_nanotube_manufacturers_and_suppliers.php

Néri, W., Maugey, M., Miaudet, P., Derré, A., Zakri, C., & Poulin, P. (2006). Surfactant-free spinning of composite carbon nanotube fibers. *Macromolecular Rapid Communications*. Wiley.

NIOSH. (2013). *Occupational exposure to carbon nanotubes and nanofibers*. https://www.cdc.gov/niosh/docs/2013-145/

Novoselov, K. S., Geim, A. K., Morozov, S. V., Jiang, D., Zhang, Y., Dubonos, S. V., Grigorieva, I. V., & Firsov, A. A. (2004). Electric field effect in atomically thin carbon films. *Science, 306*(5696), 666–669. AAAS.

Oberlin, A., Endo, M., & Koyama, T. (1976). Filamentous growth of carbon through benzene decomposition. *Journal of Crystal Growth, 32*(3), 335–349. Elsevier.

Peng, B., Locascio, M., Zapol, P., Li, S., Mielke, S., Schatz, G. C., & Espinosa, H. D. (2008). Measurements of near-ultimate strength for multiwalled carbon nanotubes and irradiation-induced crosslinking improvements. *Nature Nanotechnology, 3*(10), 626–631. Springer.

Pop, E., Mann, D., Wang, Q., Goodson, K., & Dai, H. (2006). Thermal conductance of an individual single-wall carbon nanotube above room temperature. *Nano Letters, 6*(1), 96–100. ACS Publications.

Radushkevich, L. V., & Lukyanovich, V. M. (1952). O Strukture Ugleroda, Obrazujuŝegosja Pri Termičeskom Razloženii Okisi Ugleroda Na Železnom. *Journal of Physical Chemistry, 26*, 88–95. ACS Publications.

Salvetat, J.-P., Bonard, J.-M., Thomson, N. H., Kulik, A. J., Forro, L., Benoit, W., & Zuppiroli, L. (1999). Mechanical properties of carbon nanotubes. *Applied Physics A, 69*, 255–260. Springer.

Statista. (2021). http://www.statista.de

Stroeck, M. (2006). https://commons.wikimedia.org/wiki/File:Types_of_Carbon_Nanotubes.png

Torres-Dias, A. C., Cerqueira, T. F. T., Cui, W., Marques, M. A. L., Bottic, S., Machona, D., Hartmann, M. A., Sun, Y., Dunstan, D. J., & San-Miguel, A. (2017). From mesoscale to nanoscale mechanics in single-wall carbon nanotubes. *Carbon, 123*, 145–150. Elsevier.

Vigolo, B., Pénicaud, A., Coulon, C., Sauder, C., Pailler, R., Jounet, C., Bernier, P., & Poulin, P. (2000). Macroscopif fibers and ribbbons of oriented carbon nanotubes. *Science, 290*, 1331–1334. Science Publishing Group.

Weise, B., Völkel, L., Köppe, G., Schriever, S., Mroszczok, J., Köhler, J., Scheffler, P., & Wegener, M. (2016). Melt- and wet-spinning of graphene-polymer nano-composite fibres for multifunctional textile applications. *Materials Today: Proceedings, 4*, 135–145. Elsevier.

Weise, B., Schneider, J., Wloka, P., & Gries, T. (2017). *Graphenbasierte Faserwerkstoffe*. Faserstofftabelle nach P.-A. Koch des Instituts für Textiltechnik der RWTH Aachen, Shaker, Aachen.

WIKI. (2021a). https://de.wikipedia.org/wiki/Graphen

WIKI. (2021b). https://de.wikipedia.org/wiki/Weltraumlift

Yu, M., Lourie, O., Dyer, M. J., Moloni, K., Kelly, T. F., & Ruoff, R. S. (2000). Strength and breaking mechanism of multiwalled carbon nanotubes under tensile load. *Science, 287*(5453), 637–640. Science Publishing.

Zhang, M., Atkinson, K. R., & Baughman, R. H. (2004). Multifunctional carbon nanotube yarns by downsizing an ancient technology. *Science, 306*, 1358–1362. Science Publishing.

Optical Fibers

40

Optical fibers are fibers that conduct light waves and are primarily used for data transmission. In addition to the long-established glass fibers, polymer optical fibers are also used, especially for short distances (Fig. 40.1).

40.1 History

The idea of transmitting signals over long distances using optical methods was already successfully implemented in antiquity, as shown, for example, in depictions of light signal transmissions using torches from watchtower to watchtower on Trajan's Column (second century). In 1791, Claude Chappe (1767–1805) invented the Télégraphe Chappe, which was able to transmit optical signals at up to 135 km/min. by means of levers, both during the day and at night. In 1880, Alexander Graham Bell developed what was known as the photophone. It converted audio signals into light waves, which were converted back into audio signals after being transmitted through optical lenses (Bell, 1880). However, this was not commercially successful. Starting in 1970, Corning, Inc. produced the first fiber optic cables, which quickly became popular. In 1987, Heraeus GmbH developed high-purity glass, which significantly reduced the content of metallic impurities and thus transmission losses.

40.2 Principle

Optical fibers have a core/sheath structure. The difference in the refractive indicesΔ_{KM} between core and sheath is very small, at approx. 0.003. Depending on the geometry and material, vibrations can only propagate in certain modes. A distinction is made between single-mode fibers (SMF) and multi-mode fibers (MMF). In the case of SMF, only the LP_{01} fundamental mode (LP: linearly polarized) can propagate at certain wavelengths, whereas MMF have a large number of modes.

In multimode fibers, a distinction is made between step index and gradient index fibers. In step index fibers, the refractive index changes in steps from the core to the sheath, while in graded index fibers the change in refractive index is continuous. In multimode fibers, the optical transmission can be described by the total reflection of the wave at the interface between the core and sheath (Fig. 40.2).

Depending on the refractive indices of the core (n_1) and the sheath (n_2), the angle Θ_M for total reflection of the light wave is given by

$$\theta_M = arcsin\left(\frac{n_2}{n_1}\right)$$

From this, the so-called maximum acceptance angle Θ_{Max} between light wave and optical axis of the fiber can be calculated from

D. Veit, *Fibers*, https://doi.org/10.1007/978-3-031-15309-9_40

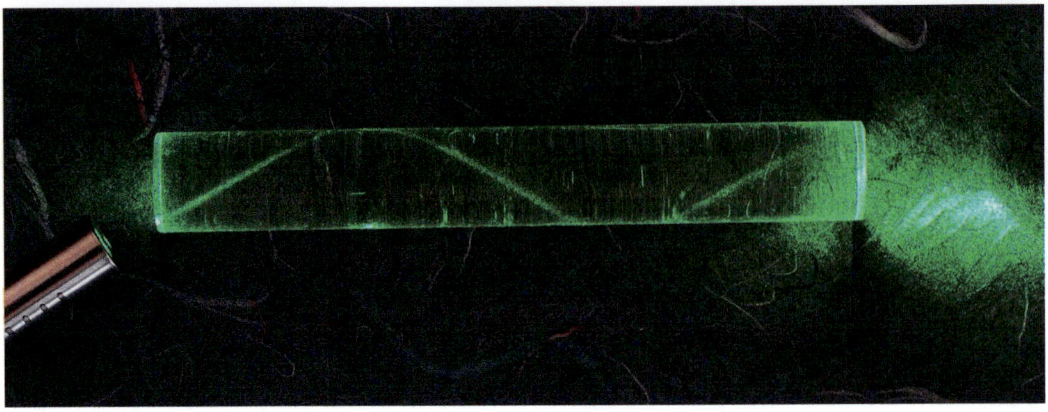

Fig. 40.1 Light conduction through an optical fiber (Timwether, 2005)

$$\theta_{max} = arcsin\left(\frac{1}{n_0}\cdot\sqrt{n_1^2 - n_2^2}\right),$$

where n_0 is the refractive index of the surrounding medium. In air, $n_0 = 1$. Then the so-called numerical aperture NA of an optical fiber can be calculated from:

$$NA = sin\,\theta_{max} = \sqrt{n_1^2 - n_2^2}$$

For single mode fibers, NA is about 0.1 and for multimode fibers it is about 0.2–0.3, depending on the material combination of core and sheath. Figure 40.3 shows the propagation of a light wave in a multimode step index fiber: Incident rays outside the so-called acceptance cone cannot penetrate the fiber and acc.ly cannot be propagated. Typical cones have an angular width of 100–120°.

In multimode index fibers, optical effects can cause wavelength-dependent propagation time differences, which result in a more or less diffuse output signal. Therefore, gradient index fibers have been developed in which these propagation time differences are much less pronounced.

The interrelationships are considerably more complex in the case of monomode fibers and are therefore not dealt with here.

Depending on the material of the core and sheath, losses occur in the light conduction, so that the maximum length over which signals can be transmitted is limited.

Strong bending of an optical fiber can cause part of the light wave to leave the fiber via the sheath, which in turn reduces the maximum possible transmission length. The rule is that a small bending radius leads to greater losses than a large bending radius (Bunge & Beckers, 2016).

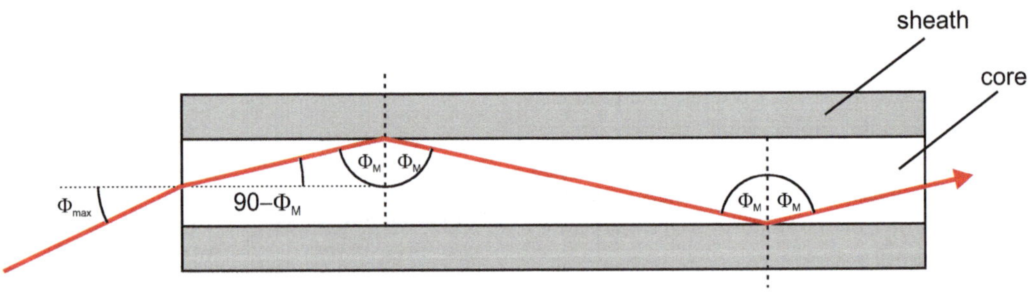

Fig. 40.2 Model of light conduction in an optical fiber

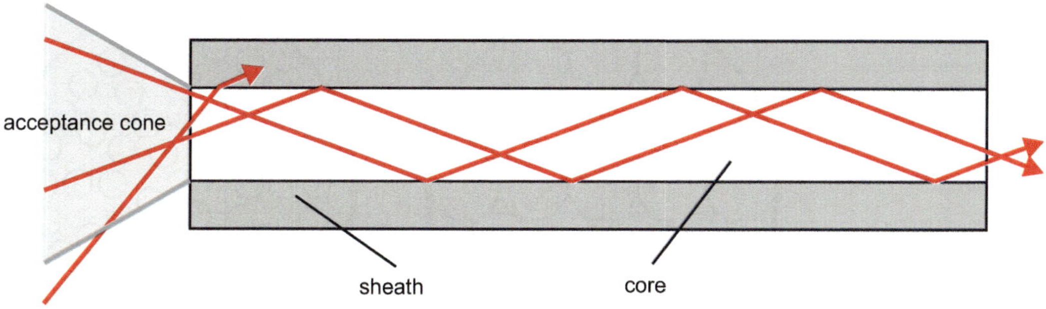

Fig. 40.3 Acceptance cone of a light conducting fiber and light waveguide in a fiber

40.3 Manufacture of Optical Fibers

Optical glass fibers are produced in a two-step process. First, the preform is produced, e.g. by vapour deposition onto a cylindrical rod of Al_2O_3 or graphite. In this way, up to several hundred layers of high-purity glass are applied. At the same time, the subsequent refractive index is adjusted. After removal of the core, an approximately 1 m long hollow rod of glass with a diameter of 10–50 mm is obtained (Mitschke, 2005). In the second process stage, the rod is heated to approx. 2000 °C and then drawn out into a fiber with a draw factor of approx. 40,000, producing hollow glass fibers with a diameter of 5–500 μm or 0.65–6500 dtex. After reaching the final fineness, the fiber is fed through an extruder and coated with plastic (Fig. 40.4). The plastic layer is cured with the aid of UV radiation. The fiber take-up speed is up to 2000 m/min.

The different processes are explained in Nakahara et al. (2008).

40.4 Types of Polymer Optical Fibers

A detailed description of the different POF types, their production and the resulting properties can be found in Bunge and Beckers (2016). In the following, only selected fiber types and the basic principles of the essential production processes are explained.

40.4.1 Multi-core Fibers

When light conducting fibers are bent strongly, large losses can occur. The effect is particularly large with thick fibers. One possible solution is to use many fine fibers arranged in a bundle of up to several hundred filaments in a dense packing. On

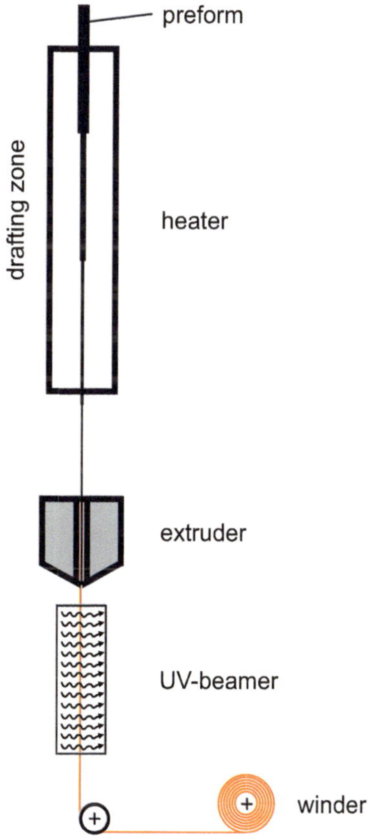

Fig. 40.4 Fabrication of optical glass fibers

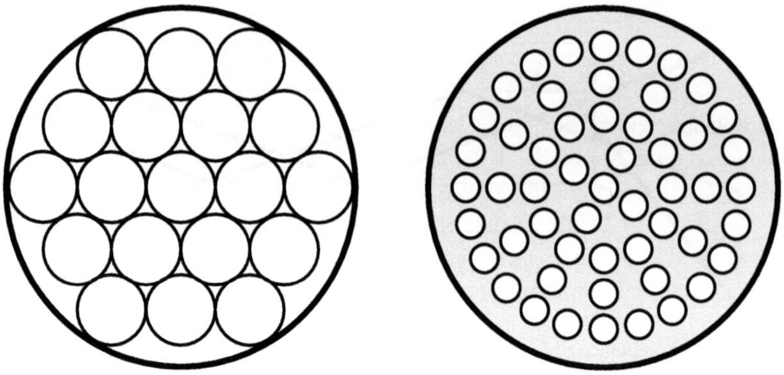

Fig. 40.5 Multi-core fibers (left) and microstructured fibers (right)

the one hand, this reduces the losses and, on the other hand, the light signal is distributed to all filaments by internal scattering between the fibers, which randomly change their position within the conductor, and is thus homogenized. These fibers are also used as a replacement for step index fibers (Bunge & Beckers, 2016).

40.4.2 Monomode Fibers

This type of fiber can transmit large amounts of light, especially coherent light (only one wavelength) over longer distances. It is therefore particularly suitable for sensor applications. However, the fiber diameter and numerical aperture must be very small. This type of POF was first generated by Koike (1992), but is still difficult to fabricate today. Typical core diameters are 5–20 μm, the sheath diameter is 100–800 μm.

40.4.3 Microstructured Fibers

Very fine channels (Ø approx. 1 μm) run longitudinally along the entire length of the mPOF fiber, which are generated from PMMO, for example. These capillaries can also be used to identify chemicals. They have a large surface:volume ratio and therefore surface effects in particular can be exploited. A major advantage of these fibers is their comparative ease of fabrication from a monopolymer, e.g., by melt spinning or injection molding (Fig. 40.5). Also, the cross-sections of the capillaries can be varied, which is not possible, for example, with glass mPOF because of the resulting internal stress (van Eijkelenborg et al., 2001).

40.4.4 Lateral Light Emitting Fibers

Normally, light guides should transport light with as little loss as possible. For applications in lighting technology, however, it may be desired that a light guide emits light uniformly along its length. However, this is not technically easy to achieve because the light emission intensity is proportional to the light energy at the point of emission. However, this decreases along the fiber, so that the light effect becomes weaker and weaker. This can be partially compensated, for example, by sending light in two directions through the light guide (Fig. 40.6).

Technically, lateral light emission can be realized e.g. by partial damage of the sheath, scoring in the sheath, microstructured surfaces in the fiber core, and physical or chemical roughening of the sheath (Bunge & Beckers, 2016).

40.5 Production of Polymer Optical Fibers

In principle, POF can be produced in a discontinuous process or in a continuous process.

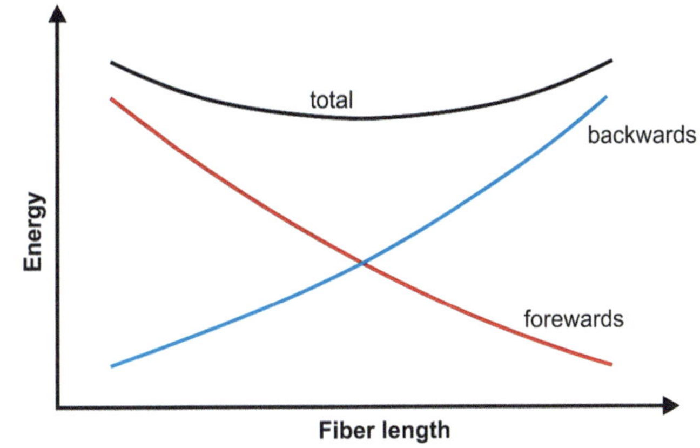

Fig. 40.6 Compensation of the unequal light emission by light beams directed against each other in an optical waveguide acc. to Bunge and Beckers (2016)

40.5.1 Discontinuous Processes

In this manufacturing principle, a preform with specific optical properties (refractive index, etc.) is first created and then stretched to the desired diameter of the light guide. In the process, all geometric relationships are essentially retained and thus also the optical properties. The fibers have a finite length.

Step Index Fiber

Core and sheath consist of different polymers. Either the sheath is filled with the core polymer or the core polymer is coated with the sheath before the in-situ polymerization process begins (wet process). In the dry process, the core and sheath are produced separately and then combined. In both processes, it is essential to avoid contamination of the core-shell surface.

Gradient Index Fiber

Here the core consists of at least two monomers with different molecular weights. A larger molar mass leads to a higher refractive index. In interfacial gel polymerization, the monomers are filled into the rotating shell of PMMA. This leads to a rotationally symmetrical distribution of the monomers in the core. The PMMA sheath is heated and a gel layer is formed on its inner surface. The monomers diffuse into this layer and are incorporated. The diffusion rate of the low molecular weight monomer is higher and it is preferentially incorporated. This leads to a concentration gradient corresponding to the refractive index profile. Finally, both monomers are polymerized out. This process takes about 24 h.

An alternative process is vapor deposition. In this process, a polymer and an additive, both of which have different refractive indices, are combined. Both materials are vaporized and deposited in a rotating cylinder on its surface. The resulting layer structure, which directly determines the gradient of the refractive index, can be very precisely adjusted via the concentration ratios of the components.

Technically simpler is the centrifuge process. Monomers of different densities are filled into a rotating cylinder and pressed outwards by centrifugal force. This creates a concentration profile, which directly reflects the refractive index. This process takes approx. 36 h.

Other manufacturing options include the diffusion process and photochemical polymerization (Bunge & Beckers, 2016).

The fibers are then heated above their glass temperature and stretched in a controlled manner at 10–30 m/min to the desired fineness.

40.5.2 Continuous Processes

In bicomponent melt spinning, core and sheath polymer are melted in two extruders and combined in the spinneret. The POF filaments are then cooled with air (step index profile) or in a water

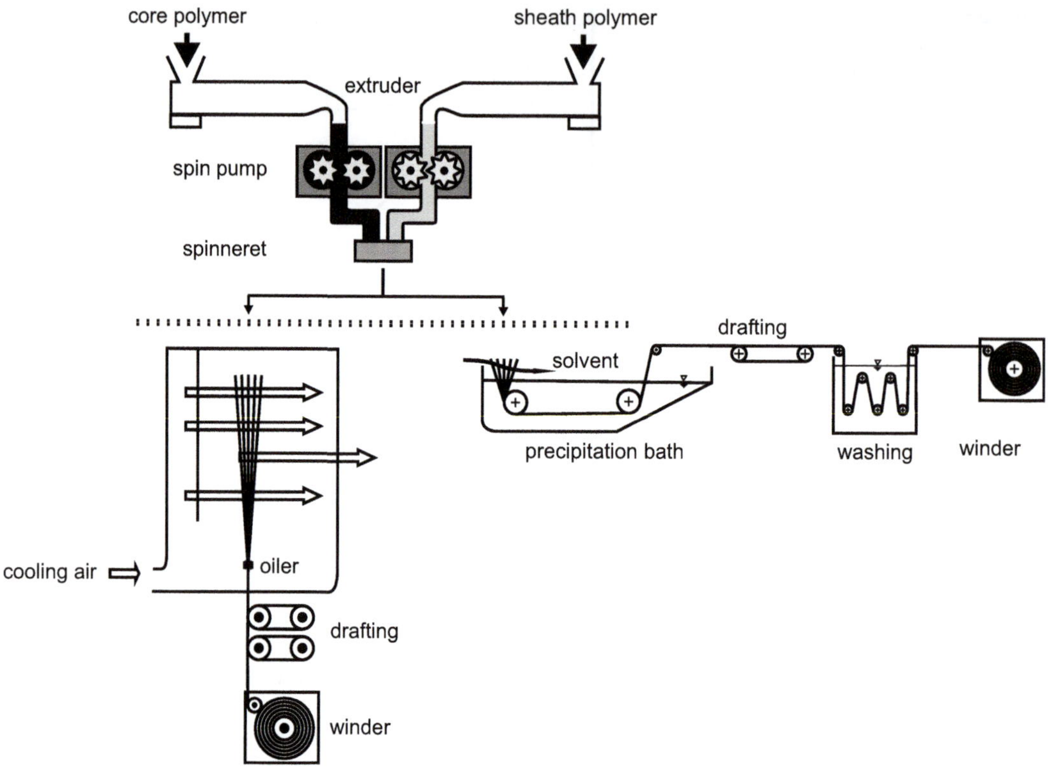

Fig. 40.7 Production of POF in the melt spinning process with air and water cooling

bath (gradient index profile), drawn and wound at a speed of approx. 10–70 m/min (Fig. 40.7).

The advantage of this process is its high productivity; however, POF filaments produced in this way exhibit increased attenuation because the purity of the polymers is not as high as in a manufacturing process in which the core and sheath are polymerized during fiber generation (Marcou, 1997).

In photochemical polymerization, the two materials are embedded in a photoactive resin and combined in core/shell configuration in a mixing chamber. After leaving the spinneret, they are irradiated with UV light, which is absorbed by the photoactive resins, thereby starting the polymerization. This process requires less heat than melt spinning and therefore results in less thermal damage to the polymers.

40.6 Properties

Glass fibers consist mainly of silicate and a polymer sheath, e.g. polyamide, polyimide, polyacrylonitrile. Polymer optical fibers are mostly made of PMMA as core material, but there are also products made of polystyrene and polycarbonate.

The Table 40.1 gives an overview of the typically used materials for core and sheath as well as—for glass fibers—the outer protective sheath

Table 40.1 Typical material combinations, characteristics and ranges of optical fibers

	Outer layer	Core	Sheath	Transmission loss	Range
Glass fiber		SiO_2	SiO_2	0.5 dB/km (@1310 nm)	>10 km
	PCF	SiO_2	Silicone	10 dB/km (@850 nm)	Up to 500 m
	HPCF	SiO_2	Fluor polymer	5 dB/km (@850 nm)	Up to 2 km
Polymeroptical fiber		PMMA	Fluor polymer	150 db/km (@650 nm)	Up to 100 m

Table 40.2 Typical properties of glass- and polymer-based optical fibers acc. to WIKI (2021)

	Glass fibers	POF
Typical filament diameter [µm]		
– Core – Sheath – Outer protective layer	50, 105, 200, 400, 600, 980 125, 125, 230, 440, 660, 1000 245, 245, 500, 480, 760, –, –	240–1960 250–2000
Bandwidth-length-product	500–3500 MHz × km @ 850 nm	10 MHz × 100 m @ 650 nm
Attenuation	0.2–2.3 dB/km @ 850 nm	160 dB/km @ 650 nm
Numerical aperture	0.15–0.22	0.3–0.6
Link length	Up to 10 km	70–100 m
Service temperature [°C]	Up to 500 °C	−40 to 85 °C

and important characteristics. Polymer optical fibers are only suitable for data transmission over short distances of up to 100 m due to the high attenuation losses compared to glass fibers. In laboratory tests, distances of 150 m have also been achieved.

The Table 40.2 shows a comparison of important properties of optical fibers.

Figure 40.8 shows typical diameter ratios for common optical fibers. The data are taken from Table 40.2.

A detailed overview of the optical properties of currently commercially available products is given in Eberlein (2013).

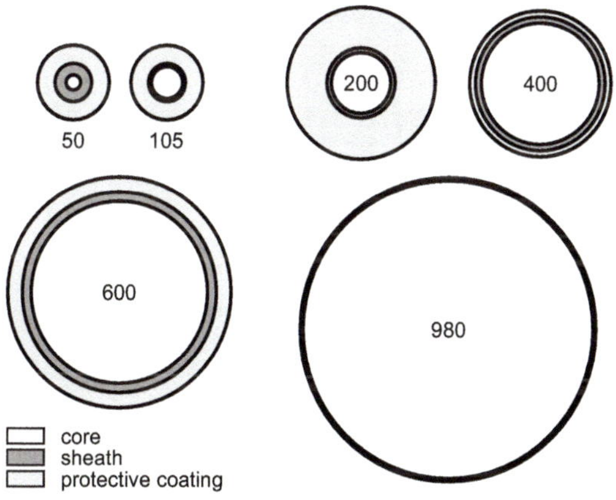

Fig. 40.8 Typical structure of optical glasses as a function of the core diameter [µm] (data from WIKI, 2021)

40.7 Products and Applications

Optical fibers are used for a variety of products. The most important application is data transmission, both over short distances and in the form of so-called deep-sea cables. Since 1988, the first deep-sea cable made of optical fibers has connected France with the USA. By 2020, there are already more than 400 deep-sea cables worldwide (Fig. 40.9).

Light-guiding fibers are also used in the construction industry to achieve extraordinary lighting effects. Here, glass fibers are integrated into concrete components and light effects are created with the aid of LED or other light sources. A detailed account of this is given by Roye et al. (2009). Figure 40.10 shows components made of so-called translucent/illuminated concrete, which are used indoors and outdoors.

Optical fibers are also used in sensor technology and optical metrology. For example, the composition of gases, liquids and solids can be determined by analyzing the optical properties. In energy supply, they are used for passive smart grids and in automotive technology for indirect lighting (cf. Sect. 40.4.4).

Optical fibers are still hardly used in clothing. This should change in the near future, because they are suitable for both, fashionable lighting effects and for data transmission.

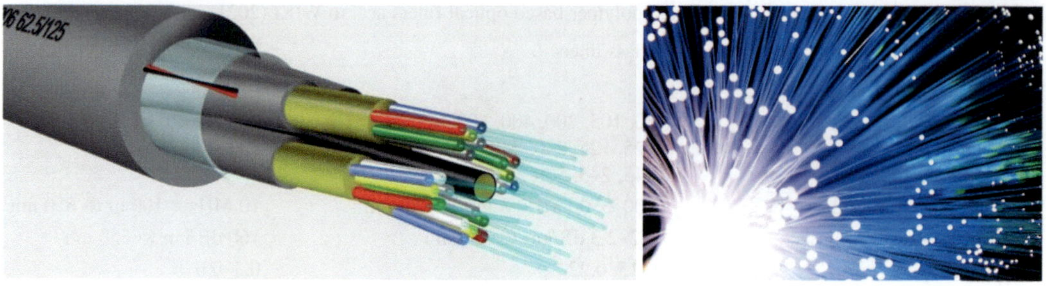

Fig. 40.9 Structure of an optical fiber cable for data transmission (Srleffler, 2009) and optical fibers as a decorative item

Fig. 40.10 Translucent/illuminated concrete (Lucem GmbH, n.d.)

40.8 Market

Table 40.3 is a list of selected manufacturers of optical fibers made of glass and polymers. Most producers come from the USA and Japan, followed by Germany and Great Britain. Chinese manufacturers do not yet play a major role on the world market.

Table 40.3 Selected manufacturers of optical fibers

Producer	Product	Website	Country
AFL	Glass fiber	http://www.aflglobal.com	USA
Asahi Kasai Corp., Ltd.	SingleCore™ (POF) MultiCore™ (POF)	https://www.asahi-kasei.co.jp	Japan
Bones Electronics Co., Ltd	POF	http://www.bones-electronics.com	China
Comcore Co., Ltd.	Glass fiber	http://www.comcore.com	USA
Corning Inc.	ClearCurve™	http://www.corning.com	USA
Eluxi Ltd.	ZLUV™ (glass fiber)	http://www.eluxi.co.uk	Great Britain
FiberLogix Intl. Ltd.	Glass fiber	http://www.fiberlogix.com	Great Britain
Hitachi Cable America Inc.	POF	https://www.hca.hitachi-cable.com	USA
Incom, Inc.	Glass fiber	http://www.incomusa.com	USA
Laser Components GmbH		http://www.lasercomponents.com	Germany
LEONI Fiber Optics GmbH	FiberConnect™ (POF) OptiGrade™ (glass fiber)	http://www.leoni.com	Germany
Mitsubishi Chemical	Eska™ (POF)	https://www.m-chemical.co.jp	Japan
NUFERN (Coherent Inc.)	Glass fiber	http://www.nufern.com	USA
Polymicro Technologies	POF	http://www.molex.com	USA
Toray Industries, Inc.	Toray Raytele™ (POF)	http://www.toray.com	Japan
Vacom GmbH		http://www.vacom.de	Germany

References

Bell, A. G. (1880). On the production and reproduction of sound by light. *American Journal of Science*, (118), 305–324.
Bunge, C.-A., & Beckers, M. (2016). *Polymer optical fibres: Fibre types, materials, fabrication, characterisation and applications*. Woodhead Publishing, Elsevier.
Eberlein, D. (2013). Fasertypen, Parameter, Anwendungen – Überblick über den aktuellen Stand und Trends bei LWL. *Praxis Optische Netze, 4*(13) http://net-im-web.de/freedocs/1304_s18_eberlein_lwl.pdf
Koike, Y. (1992). High bandwidth and low-loss polymer optical fiber. *Wire Industry, 69*, 699.
Lucem GmbH. (n.d.). Aachen. https://lucem.com
Marcou, J. (1997). *Plastic optical fibers – Practical applications*. Wiley.
Mitschke, F. (2005). *Glasfasern - Physik und Technologie*. Spektrum Akademischer Verlag.
Nakahara, T., Hoshikawa, M., Tanaka, S., & Yokota, H. (2008). Fiber optics. In *Ullmann's fibers* (Vol. 2). Wiley.
Roye, A., Barlé, M., & Gries, T. (2009). *Faser- und Textilbasierte Lichtleitung in Betonbauteilen – Lichtleitender Beton*. Shaker.
Srleffler. (2009). *CC BY-SA 3.0*. https://commons.wikimedia.org/wiki/File:Optical_breakout_cable.jpg
Timwether. (2005). *CC BY-SA 3.0*. https://commons.wikimedia.org/wiki/File:Laser_in_fibre.jpg
van Eijkelenborg, M., Large, M. C. J., Argyros, A., Zagari, J., Manos, S., Issa, N. A., Bassett, I., Fleming, S., McPhedran, R. C., de Sterke, C. M., & Nicorovici, N. A. P. (2001). Microstructured polymer optical fibre. *Optics Express, 9*(7), 319–327. The Optical Society.
WIKI. (2021). https://de.wikipedia.org/wiki/Lichtwellenleiter#Vergleich_zwischen_Monomode-_und_Multimodefasern

Part V

Cellulosic Chemical Fibers: Production and Properties

41 Cellulosic Man-Made Fibers

41.1 Basics

In the nineteenth century, cellulosic man-made fibers were the first synthetically produced fibers to be used for textile purposes. In this process, the cellulose was first made soluble by a chemical process, spun out in solution and then recovered ("regenerated") by a chemical reaction in the spinning bath. Following this process, viscose and cupro are produced and modal fibers are produced by modifying the viscose process. Using an alternative process, cellulose can be dissolved directly without the addition of chemicals, resulting in lyocell fibers. These fibers differ from viscose externally primarily in their round cross-sectional shape, and they also fibrillate more. Viscose, modal and cupro are referred to as regenerated fibers because their chemical structure is identical to that of the raw material cellulose (Fig. 41.1).

In contrast, acetate fibers, which are also based on cellulose, are called cellulose-derivative fibers because their chemical structure is permanently altered during the dissolving process and therefore differs from cellulose. This group includes acetate, triacetate and 2½-acetate fibers. Their morphological structure is very similar to that of viscose and cupro.

The abbreviations for the most important cellulosic man-made fibers are:

- Viscose: CV
- Modal: CMD
- Cupro: CUP
- Lyocell: CLY
- Acetate: CA
- Triacetate: CTA

41.1.1 History

Mulberry silk is generally regarded as a model for the manufacture of artificial threads, especially filament yarns. Thus, Hooke (1665) describes a process to "artificially produce a glue-like mass, similar to the substance from which the silk worm spins its balls of thread". R.A.F. de Réaumur also formulated such an idea in 1734. After attempts to produce glass and metal threads were successful, in 1838 Anselm Payen succeeded in isolating cellulose from wood with the help of nitric acid and caustic soda. He gave it its present name ("les cellules"). Payen also discovered that cotton consists of cellulose and broke it down into glucose with the help of sulfuric acid.

However, the path to the production of cellulose-based artificial fibers was only made possible by the discovery of how to dissolve cellulose. In 1845, for example, C.F. Schönbein invented a process for producing collodion (cellulose nitrate), the starting material for nitrate silk, from nitrated cotton (gun cotton) by dissolving it in alcohol ether.

G.P. Audemars used collodion to make fibers in 1855, and J.W. Swan made incandescent lamp filaments from nitrocellulose dissolved in acetic

D. Veit, *Fibers*, https://doi.org/10.1007/978-3-031-15309-9_41

Fig. 41.1 Viscose, modal and lyocell (Lenzing AG n.d.)

acid in 1883. He called the filaments "artificial silk," and his wife made them into home textiles. In 1857, M. E. Schweizer succeeded in dissolving cotton or cellulose with copper oxide ammonia (Schweizer's reagent). M.P. Schützenberger produced a cellulose acetate for the first time in 1865. In 1892, C.F. Cross, E.J. Bevan, and C. Beadle developed viscose, a spinning solution of cellulose sodium xanthogenate. In 1894 they founded the Viscose Syndicate Ltd.

For the industrial production of artificial fibers, however, it was crucial that cellulose could be isolated from lignified fiber cells. Only then was one independent of the comparatively expensive cotton as a starting substance. B.C. Tilgman succeeded in doing this on a technical scale in 1866 by pulping with sulfurous acid. Since 1875, pulp has been produced industrially using A. Mitscherlich's process.

Count Hilaire de Chardonnet (Fig. 41.2) received a patent for the production of artificial silk from nitrocellulose in 1885. Fabrics made from it and also his artificial silk spinning machine were presented at the World's Fair in Paris in 1889. In 1891, he began producing nitrate silk in his first factory in Besançon. By drawing out the still plastic thread, F. Lehner succeeded in refining the fibers and increasing their strength. In the first year, 50 kg were produced daily; by 1907, daily production had risen to 2000 kg. The selling price for 1 kg was 25–40 francs, the cost price was about 20 francs. Due to further increasing costs, the factory was therefore soon unprofitable and was closed again (Becker, 1912).

Fig. 41.2 Joseph Wilson Swan (Wiki1) and Count Hilaire de Chardonnet (Wiki2)

Other factories were set up in Europe, but most were not successful and from about 1905 it became apparent that the nitrate process could not compete with the copper and viscose processes, either economically or in terms of quality.

In 1891, M. Fremery and J. Urban founded the Rheinische Glühlampenfabrik in Aachen, and in 1892 they began manufacturing cellulose filaments for carbon filament incandescent lamps in Oberbruch. These were made from a solution of cellulose in copper oxide ammonia. The process was protected by a patent from (Pauly, 1897). The company became Vereinigte Glanzstoff-Fabriken AG in 1899, with its headquarters in Aachen, which was moved to Elberfeld, now Wuppertal, in 1901. In 1903 a factory was built in Oberbruch, district Heinsberg (Fig. 41.3).

This was followed by the founding of Glanzstoff in St. Pölten (1904) and in Liverpool (1908). These and other companies later converted their production to the viscose process (Glanzstoff 1916). In contrast to the nitrocellulose production companies, the Vereinigte Glanzstoff-fabriken soon made good profits as a result. In

Fig. 41.3 Glanzstoff factory in Oberbruch, Germany (Wiki, 1903)

1908, for example, the net profit was over two million marks, with a share capital of 2.5 million marks (Becker, 1912). J.P. Bemberg AG in Barmen, founded in 1900, on the other hand, continued to produce copper artificial silk for decades. In 1901, the copper process was significantly improved by the development of drawframe spinning with the Thiele spinning funnel.

The viscose process was also modified several times. As early as 1898, C.H. Stearn was granted a patent for a spinning process in which viscose coagulates in solutions of ammonium salts, especially ammonium chloride (Stearn, 1898). In 1900, viscose artificial silk (yarns and fabrics) is shown for the first time at the World‘s Fair in Paris.

C. F. Topham came up with further improvements for the viscose process, including a multi-hole nozzle made of platinum and, in 1901, a regulating device for evenly feeding the spinning mass to the nozzle, the forerunner of the spinning pump (Topham, 1900). The patent registered in 1905 by P. Koppe—a spinning bath (Müllerbad I), consisting of sulfuric acid and sodium sulfate, which goes back to M. Müller—made it possible to produce viscose artificial silk economically. It was improved in 1912 by Müllerbad II. From 1929/30 onwards, the production of semi-matt and low-matt and from 1934 onwards, spun-dyed products began. For technical applications, e.g. fabric inserts for car tires, a special, high-strength artificial cord (tire cord) was developed in 1937.

The term artificial silk was followed by the term rayon in the USA in 1924 (rayonne in France) and "rayon" in Germany in 1948. Initially, they were used as generic terms for all three types of artificial silk (viscose, copper and acetate rayon), later only for filament yarns produced by the viscose process.

E. Bronnert (Fig. 41.4) tested a small experimental plant for the production of staple fibers in 1905 (Bronnert, 1899). In 1916, a first staple fiber spinning machine went into operation in Oberbruch, and in 1917 Vereinigte Glanzstoff-Fabriken AG built a plant with a capacity of 10 t/day in Sydowsaue near Stettin. In Premnitz, Köln-Rottweil AG started production of the viscose staple fiber Vistra® in 1920. The brand name goes back to "Vis" (viscose) and "tra" (trame: designation for particularly lustrous silk yarns). In 1935, the name rayon was introduced for all cellulose-based chemical staple fibers (in reference to "cotton"), after many consumers were unfamiliar with the term "artificial silk" for cellulose-based fibers. In 1937, wool-like special grades were developed for cloth and dress fabrics, and in 1938, high-wet-strength grades for cotton spinning. This was followed by spun-dyed

Fig. 41.4 M. Fremery, J. Urban and E. Bronnert (from left) (Wiki3)

cellular wools, special types for upholstery fabrics and carpets, as well as coarse, highly crimped sheath/core fibers and sheathless round fibers. This meant that man-made fibers were available for almost all types of textiles for the first time.

Süvern (1907) explains all the essential early patents on cellulosic man-made fibers. A comprehensive account of the development of viscose production up to the 1940s and 1960s is given by Bodenbender (1943) and Götze (1967).

41.1.2 Raw Materials, Pulp Production

Cellulose occurs in all plants, but not in pure form. It is accompanied by other substances which interfere with further processing and must therefore be removed (Fig. 41.5).

Wood and, to a lesser extent, cotton linters and other plants are of major industrial importance for the production of cellulose, which is also called pulp. In wood, the soft lignin is reinforced by firm and stiff cellulose fibers, while the soft, non-crystalline hemicellulose (also composed of saccharide groups) acts as a "bonding agent" between the lignin and the cellulose. The cellulose in plants has a fibrous shape and a length of 1–5 mm.

The process for obtaining pulp from wood is shown in Figs. 41.6, 41.7 and 41.8. The tree trunks are first sawn into smaller pieces, barked and then wood chips (2–4 cm) are produced. These can be stored or used directly for pulp production. For this purpose, they are first boiled in lye at 150 °C and 8 bar for approx. 8 h, for example, after which the pulp is extracted and separated from the water. This is followed by a bleaching and dewatering process. After this, the pulp can either be dried to produce pulp sheets or, after dewatering, the pulp is ground and processed into flakes.

Around 30% of the world's land area consists of forest, so the raw material base for man-made cellulosic fibers is almost inexhaustible. Worldwide, around 33mt of cellulose are produced from wood and 12mt from other raw materials each year, primarily for the production of paper and cardboard (FAO, 2021). This is only a vanishingly small fraction of the new cellulose formed each year, the quantity of which is estimated at 1.5 trillion tons (Klemm et al., 2005).

The quality of the pulp is of great importance for the production and properties of man-made

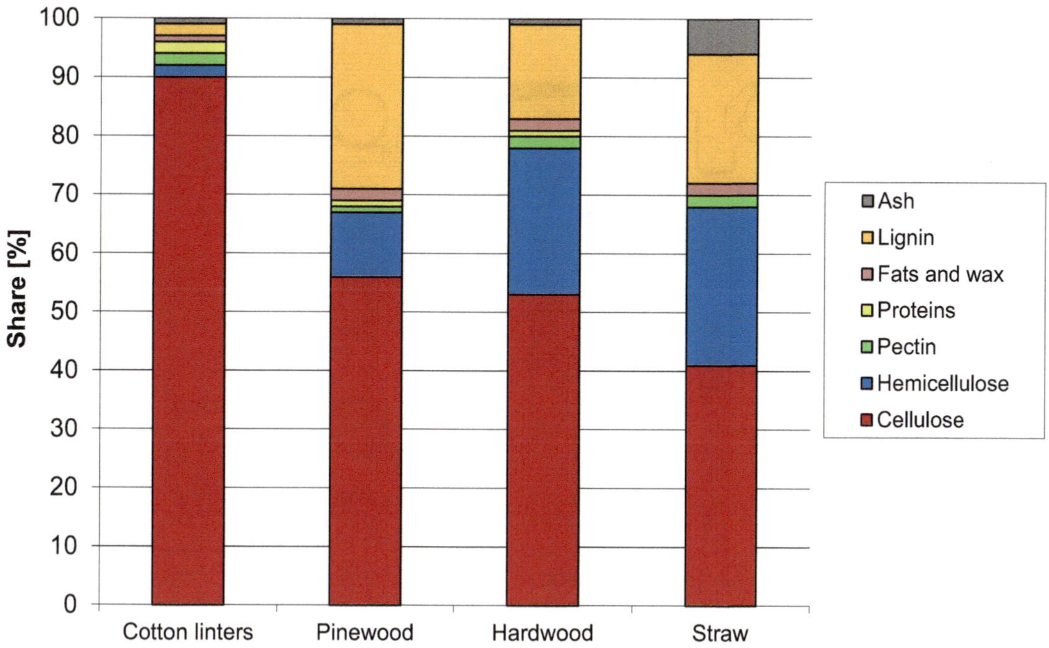

Fig. 41.5 Composition of some plants acc. to (Rogowin, 1982)

cellulosic fibers. Pulp for the production of cellulose fibers is still refined. In this process, accompanying substances and low-molecular hemicellulose are removed and the proportion of pure cellulose (α-cellulose) is increased. Depending on the number of cleaning stages and the conditions, pulps with a purity content of up to approx. 99% are produced.

After bleaching or finishing, the pulp is dewatered and usually dried. Moist pulps are more reactive, but their use requires short distances between the pulp mill and the spinning mill. The pulp is usually supplied as sheets, frequently also as flakes and rarely in roll form.

The pulp quality and the possible applications in man-made fiber production are determined by its chemical composition (α-cellulose and hemicellulose content as well as accompanying substances), its physicochemical properties (degree of polymerization, absorbency, swellability) and its technological properties (degree of soiling, moisture, degree of bleaching, sheet density). Because wood treatment processes pollute air and water, more environmentally friendly processes have been sought for many years. An important result of these efforts is the development of the Lyocell process, which is discussed in detail in Sect. 41.3.

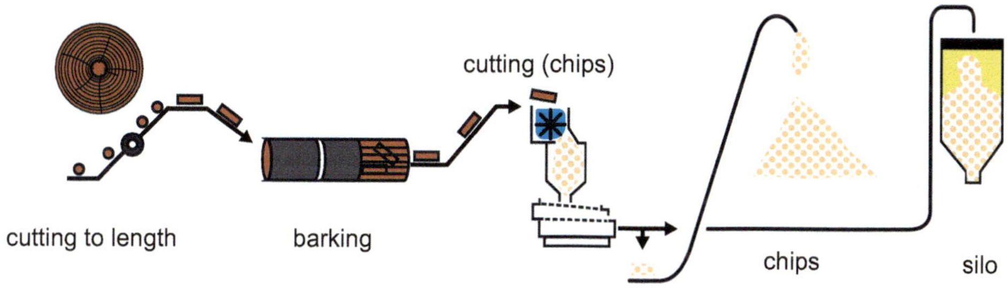

Fig. 41.6 Principle of the production of chips from wood acc. to (Lenzing AG)

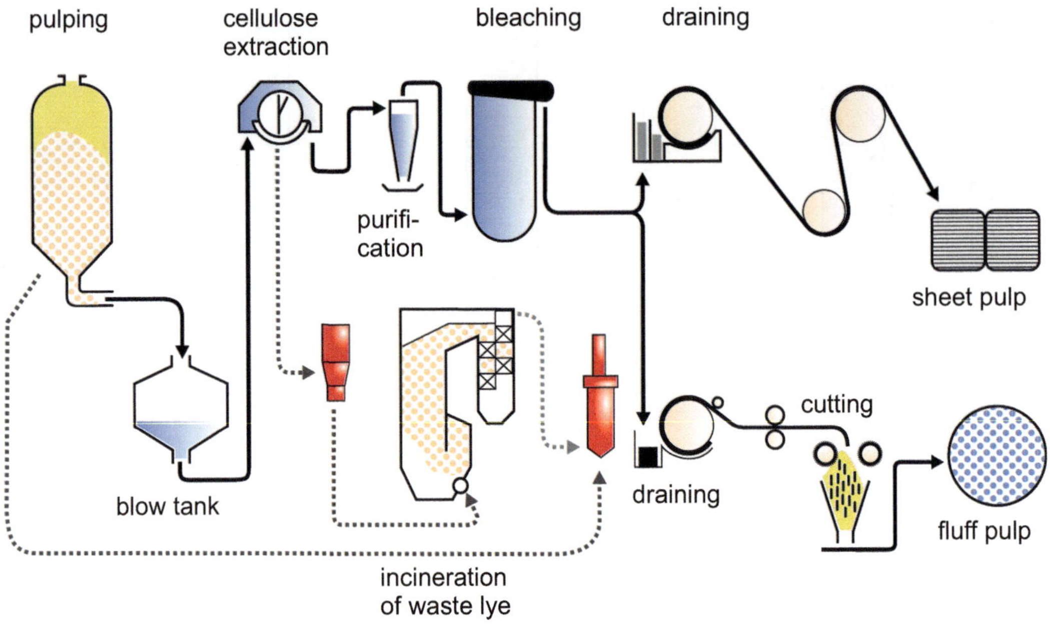

Fig. 41.7 Principle of pulp production from wood chips acc. to (Lenzing AG)

Fig. 41.8 Extraction of the pulp (Franz Neumayr, Lenzing AG)

So far, straw, grass, bamboo, sugar cane, and reeds have not been of great market importance for cellulose production. However, efforts are being made to use these for cellulose production.

41.2 Regenerated Fibers

The cellulosic fibers in the pulp are infusible and too short for direct spinning into yarns. They are therefore dissolved and regenerated in filament form. If their chemical structure is not permanently altered in the process, they are referred to as regenerated fibers or regenerated cellulose. Viscose, Modal and Cupro, among others, are produced using this process.

41.2.1 Viscose Process

Of all the processes for the production of regenerated cellulose fibers, the viscose process is the most important. It is versatile and can be used for the production of textile and technical filament yarns as well as staple fibers.

The fiber is produced in three stages:

- Preparation of the spinning solution from viscose,
- spinning out the fibers,
- after-treatment of the fibers.

The chemical process of derivatizing cellulose to make it soluble can be described in its simplest form by:

The prerequisite for a relatively uniform derivatization of cellulose is the accessibility of the hydroxyl groups. This is achieved by swelling or defibrillation or dissolution of the solid starting material. The more the cellulose is loosened, the easier the access for the solvent and the more uniform and complete the derivatization (Fig. 41.9).

The diagram in Fig. 41.10 shows the viscose spinning process. First, alkali cellulose is formed by treating the pulp with caustic soda ($NaOH_{aq}$). It is a swollen, strongly alkaline, crumbly mass. After dewatering, defibration and pre-maturation, it reacts with carbon disulfide to form sodium cellulose xanthate, which is soluble in further caustic soda. This unstable solution is called viscose. From it, cellulose is regenerated in an aqueous solution of acid and salts. Xanthogenation is followed by the process steps of filtering and ageing as well as deaeration. During deaeration, gases are removed, because otherwise fiber breakage would occur if gas bubbles contained in the polymer burst when the spinning mass leaves the

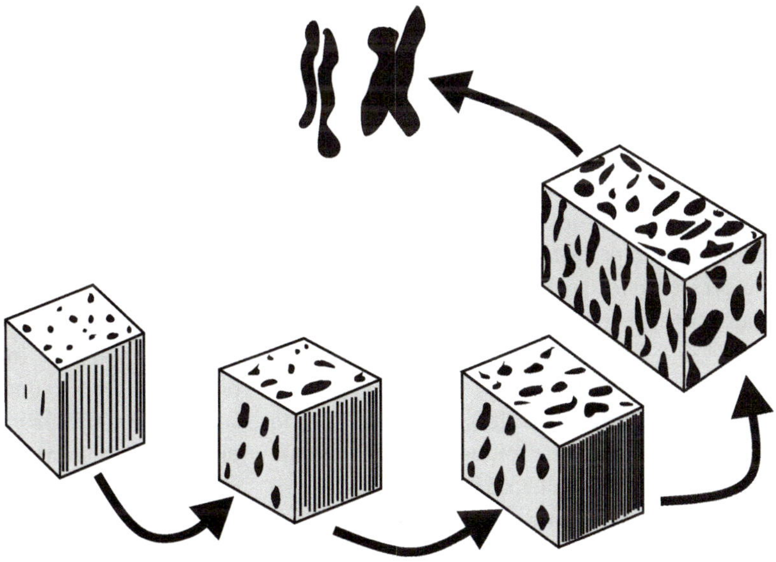

Fig. 41.9 Laminar defibrillation during cellulose swelling

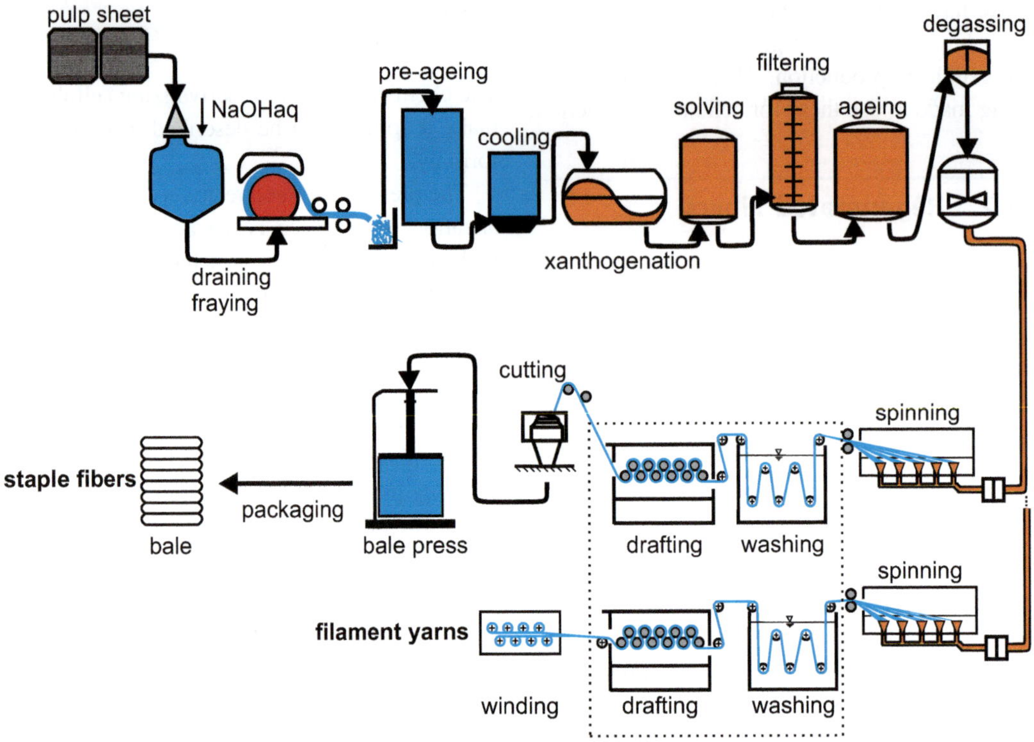

Fig. 41.10 Viscose spinning process acc. to (Lenzing AG)

spinneret. In the so-called spinning bath, the spinning solution is neutralized by the acid and the sodium cellulose xanthate is decomposed. The cellulose coagulates as a filament, which is preformed through fine openings of the spinneret. During the regeneration of the cellulose, sodium sulfate and carbon disulfide are formed, which is released in gaseous form. This is partially condensed and recovered. The non-condensed part is washed out. This also removes the unpleasant smelling components. Spinning is followed by drawing, after-treatment and drying, and a cutting unit to produce staple fibers. Staple fibers are pressed into large bales with a press, while filament yarns are wound directly onto bobbins.

41.2.1.1 Alkalization

To alkalize the cellulose, it is treated with 18 to 22% sodium hydroxide solution at a temperature of 17 to 45 °C. The cellulose is then treated with a water-based alkaline solution. Depolymerization catalysts (e.g. manganese salts) are also added. A content of approx. 30% alkali cellulose is adjusted by pressing. Its composition and properties influence the subsequent process stages (grinding, pre-ageing, i.e. reduction of the degree of polymerization of the cellulose, sulfidation and filtration). Therefore, it is necessary to accurately adjust the temperature and composition of the alkali solution (sodium hydroxide, hemicellulose, sodium carbonate, alkaline earth and heavy metal ions).

If the alkali treatment of the pulp is carried out in sheet form, the alkali cellulose is then defibered. This produces white, voluminous crumbs which are pre-ripened at a temperature of approx. 40 °C for a certain time. During pre-ripening, the crumbs must not dry out on the surface. Pre-ripening of alkali cellulose takes place either discontinuously in closed containers or continuously in rotating drums and lasts for up to 24 h.

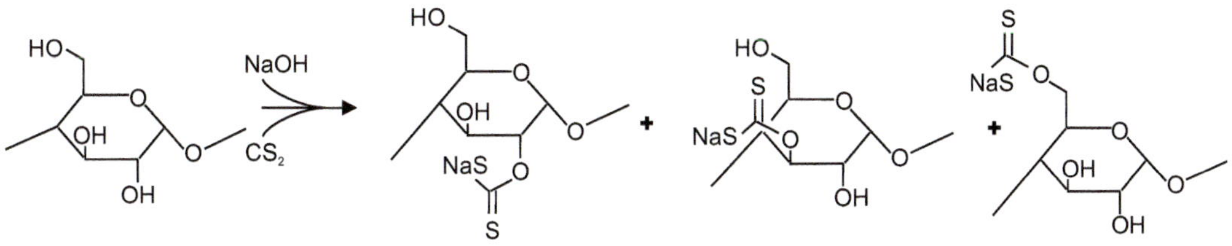

Fig. 41.11 Principle of xanthogenation

41.2.1.2 Xanthogenation

Sulfidation or xanthogenation is carried out by treating the pre-matured alkali cellulose with carbon disulfide (CS_2). The reaction of the alkali cellulose with the carbon disulfide starts very rapidly, initially at the surface of the crumb and slowly penetrates into the forming lumps.

The xanthate group (NaS_2) initially attaches to C atom 2 (kinetically favored). After some time, the xanthate group migrates to the thermodynamically more stable form at C atom 6, resulting in a mixture of different derivatives (Fig. 41.11).

In production, γ-values (number of xanthate groups per 100 glucose units) of 50 to 60 are set. High-strength filament yarns require higher γ-values.

The rate of sulfidation depends on the temperature, composition and properties of the alkali cellulose, and also on the amount of carbon disulfide and the pressure of the carbon disulfide vapors formed during the process. The sulfidation of alkali cellulose is accompanied by many side reactions. They are responsible for the characteristic orange color of xanthate. At high temperatures, xanthogenation proceeds more rapidly, but the side reactions also increase. Therefore, the temperature is usually 25–37 °C over a period of 30–90 min (Fig. 41.12). Depending on the type of fiber and the composition of the alkali cellulose, 220 to 400 kg of carbon disulfide are required to produce 1000 kg of fibers (see also Table 41.1).

The resulting orange-colored cellulose xanthate is dissolved with dilute sodium hydroxide solution at approx. 10 °C to form the viscose (carbon disulfide-air mixtures are explosive, so special safety regulations apply). The solution is assisted by stirring and triturating the lumps

Fig. 41.12 Heated vessel for xanthogenation (Ing. A. Maurer SA, n.d.)

Table 41.1 Consumption of raw and auxiliary materials for the production of 1 t viscose fibers acc. to (Rogowin, 1982)

Raw materials and additives	Textile filament yarns	Staple fibers
Pulp (90% α-cellulose) [kg]	1011–1026	990–1200
Sodium hydroxide [kg]	817–830	850–880
Carbon disulfide [kg]	305–396	171–176
Sulfuric acid		
(100%) [kg][b]	1207–1344	1100–1130
Zinc sulfate		
(heptahydrate) [kg][b]	80–94	145–146
Steam [MJ]		
for the process	167	75
for heating and quenching	226	88
Electric energy [kWh]		
for the process	5700	1200
for heating and quenching	2500	220
Refrigeration [W/h]	140,000	93,000
Water (total consumption of		
fresh water [m^3]	1000	800

[a]Maximum recovery incl. Emission control with activated carbon
[b]Complete spinning bath recovery

present. The viscose contains 7 to 10% cellulose and 5 to 8% sodium hydroxide. Its viscosity varies from 3.5 to over 10 Pas. It depends on the concentration of xanthate, the degree of polymerization of cellulose, the degree of solution and the temperature. To facilitate their spinnability, chemical auxiliaries such as fatty amines or polyglycol mixtures are added in very small quantities.

During post-ripening, the viscose is filtered several times and dissolved and dispersed gases are removed by deaeration. In the course of post-ripening, the sulfidation state of the cellulose becomes uniform and part of the cellulose xanthate already decomposes. The desired maturity (state of the viscose for the regeneration of the cellulose in the regeneration or spinning bath) is set by temperature and dwell time. The state of maturity influences the fiber properties: with increasing maturity, the tendency to coagulation increases and the strength of the fiber decreases. Highly matured viscose is used for the production of textile filament yarns and less matured viscose for technical filament yarns.

41.2.1.3 Spinning of Viscose into Fibers

The spinning of the viscose into fibers in the spinning bath, i.e. the regeneration of the cellulose in fiber form from the viscose, and the further treatment of the fibers take place in several stages (Fig. 41.13). These process steps partly run simultaneously and essentially determine the structure and properties of the fibers.

To spin the fibers, the matured viscose is pressed through the openings (holes) of the spinneret using spinning pumps. The spinnerets are usually made of precious metal alloys; tantalum or glass are also used. Their quality is of decisive importance for the yarn quality. The shape of the capillary openings and their machining play an important role. In the direction of flow, the openings are conical for pre-orientation of the macromolecules and then become cylindrical. At the exit, the capillaries have a diameter between 50 and 250 μm. In the production of filament yarns, the number of openings corresponds to the desired number of filaments in the yarn (e.g. 18 to 1000); in the production of staple fibers, it can reach 90,000. Figure 41.14 shows typical geometries of such spinnerets (Fig. 41.15).

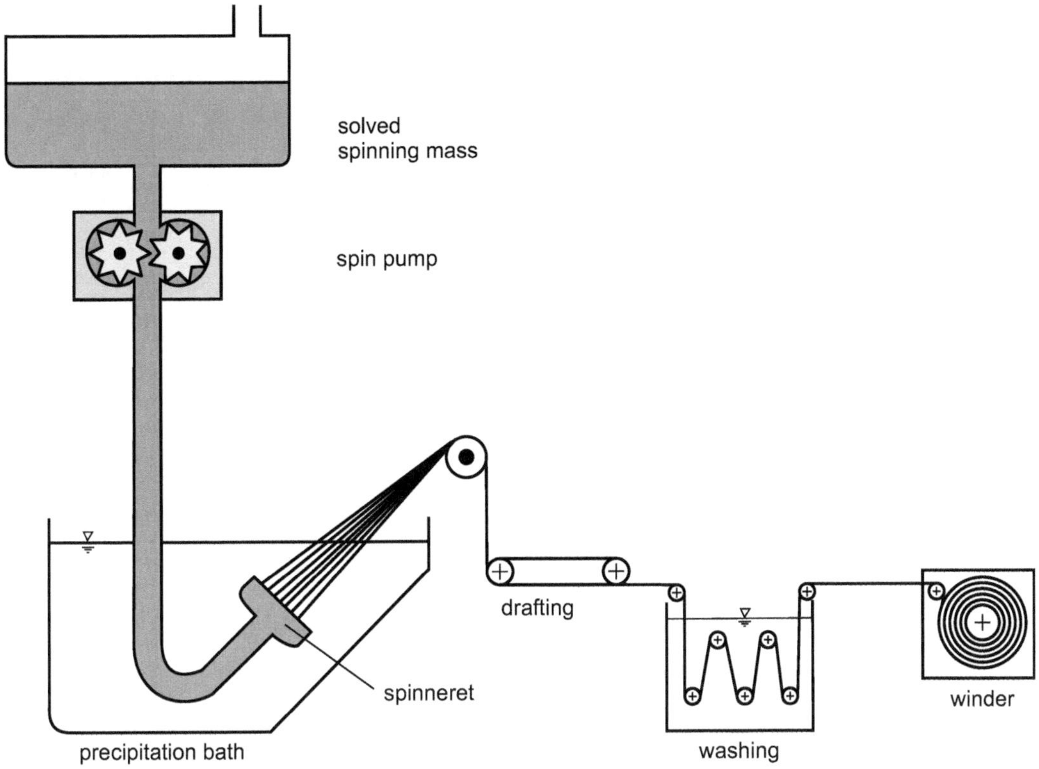

Fig. 41.13 Principle of the wet spinning process

Various chemical reactions take place during filament formation in the spinning bath. The most important is the regeneration of cellulose with sulfuric acid by "splitting" the sodium cellulose xanthate into cellulose, carbon disulfide and sodium sulfate. In addition, secondary reactions take place, resulting, for example, in the unpleasant smelling H_2S.

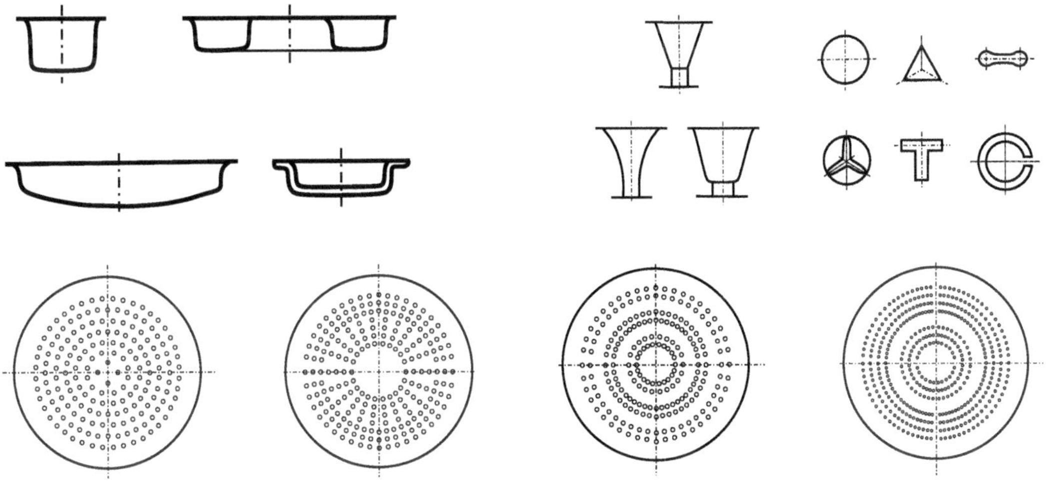

Fig. 41.14 Typical spinneret geometries for filament yarns acc. to (Fourné, 1995)

Fig. 41.15 Drawing off the filaments from the spinneret

$$\text{Cell–O–CSSNa} + H_2SO_4 \rightarrow \text{Cellulose} + CS_2 + Na_2SO_4$$

$$Na_2CS_3 + H_2SO_4 \rightarrow Na_2SO_4 + H_2S + CS_2$$

Furthermore, carbon dioxide is formed by the reaction of alkali lye with carbon disulfide and collodial sulfur. The sodium sulfate present in the spinning bath promotes coagulation of the viscose, and the acid neutralizes the viscose more quickly. The Zn^{2+} ions present in the precipitation bath form a bridge between two cellulose molecules with the sulfur atoms of the xanthate groups, which hinders the regeneration and thus slows it down. However, as a result, the internal structure of cellulose becomes more uniform and the macromolecules become better oriented in the fiber direction. Because the Zn^{2+} ions, due to their size, only penetrate into the outer area of the resulting fibers, a core-sheath structure is created.

The Fig. 41.16 shows the process of coagulation of cellulose xanthate in the precipitation bath. Initially, a thin skin of cellulose (cell) is formed when the cellulosexanthate (cell-X) reacts with the aqueous sulfuric acid $(H_2\ SO_4)_{aq}$. Gradually, the sulfuric acid penetrates deeper inside, reacts with the cellulose xanthate, the cellulose coagulates, and CS_2 diffuses outward into the precipitation bath. As a result, small voids form and the fiber gradually collapses. It cannot shrink uniformly because the originally formed skin of cellulose is solid and does not shrink. This results in a lapped cross-section that resembles a cauliflower.

Figure 41.17 shows light microscopic images of viscose fibers. The irregular cross-section and the grooves are clearly visible in the longitudinal view.

The spinning or precipitation bath composition is adapted to the desired fiber properties (e.g. fineness, strength, crimp), as are the yarn path in the spinning bath, the spinning bath temperature and the spinning speed. Therefore, the spinning conditions for textile and technical filament yarns as well as for staple fibers are different. Generally, the spinning bath contains 7–12% sulfuric acid, 12–24% sodium sulfate and 0.5–4% zinc sulfate. This reduces the swelling of the filaments in the spinning bath. In addition, auxiliary substances can be added to the spinning bath,

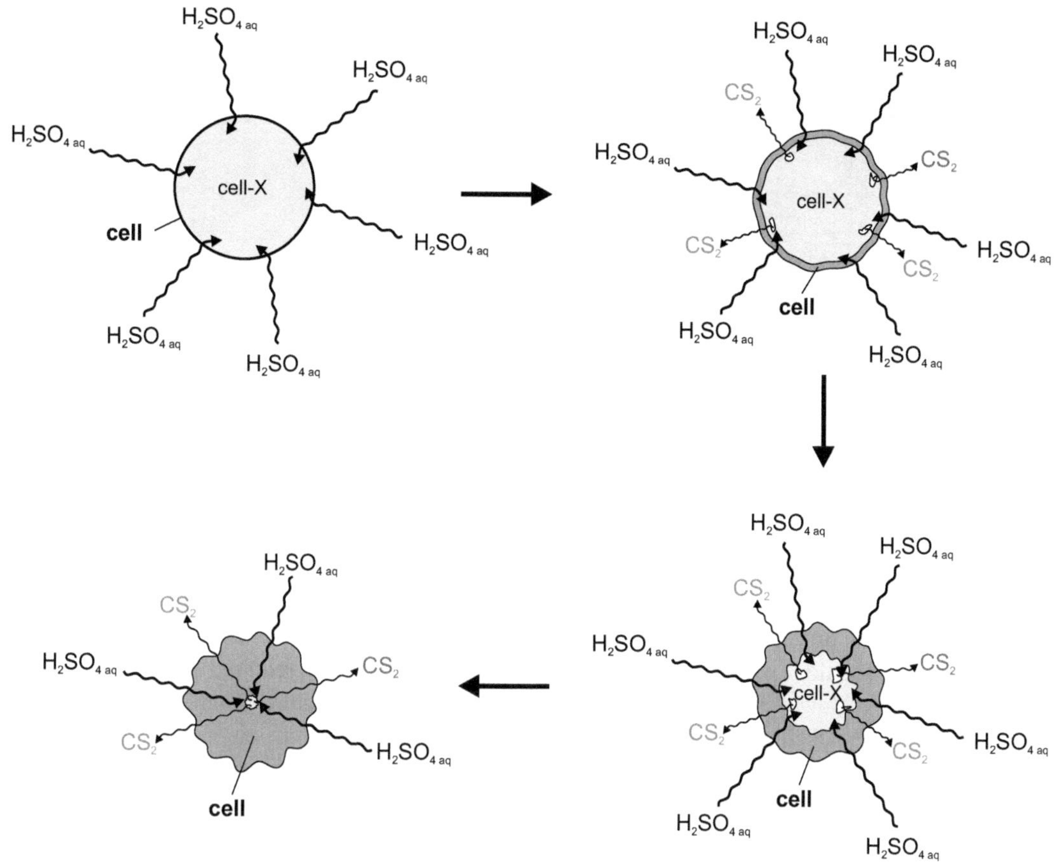

Fig. 41.16 Coagulation of the cellulose xanthate in the precipitation bath

Fig. 41.17 Cross-sectional and longitudinal view of viscose fibers (Lenzing AG)

Fig. 41.18 Spinning machine (left: staple fibers; right: filament yarns, Ing. A. Maurer SA)

which increase the drawability of the freshly spun filaments and the service life of the nozzles.

The wet spinning process is generally also well suited for the addition of substances to the spinning solution. However, the additives must be resistant to the alkaline and acidic process conditions.

Most viscose fibers have a core-sheath structure. The sheath differs from the core by:

- smaller and more crystallites,
- higher strength and elongation,
- greater fatigue resistance.

The core/sheath ratio is about 1:2 for classic viscose fibers and about 1:5 for high tenacity fibers. It depends on the spinning conditions. By adding larger quantities of zinc sulfate to the spinning bath, adjusting the viscose and spinning bath composition, short ageing times and using auxiliaries (amines), fibers with a full sheath structure can also be spun (Fig. 41.18).

41.2.1.4 Degree of Polymerization and Chemical Consumption

While the average degree of polymerization (number of basic building blocks per macromolecule) is still 800–1000 at the beginning, it decreases in the course of viscose production to 300–430 for normal fibers and to 400–500 for high-strength fibers (e.g. tire cord) and modal. For fibers with a high wet modulus, values up to 700 are reached (Fig. 41.19).

The consumption of chemicals and the amount of water and energy required to produce viscose fibers is considerable. Therefore, efforts are always made to use as many auxiliary materials as possible in cycles. The pollution of the waste water by sulfur compounds is particularly problematic.

The fiber properties can be varied within wide limits by changing the process conditions. These include:

- the composition of the viscose and the spinning bath,
- chemical additives,
- modification of the spinning technology (spinning speed, degree of drafting and geometry, number and type of after-treatment baths, etc.).

41.2.1.5 Ancillary Facilities

The excess caustic soda used in alkalization (recycled caustic soda) and the caustic soda pressed out of the alkali cellulose (pressed caustic soda) are largely freed from the hemicellulose by dialysis and used again for alkalization or in the after-treatment for washing the freshly spun fibers.

The used spinning bath is filtered, vacuum aerated, evaporated to remove H_2 S and the

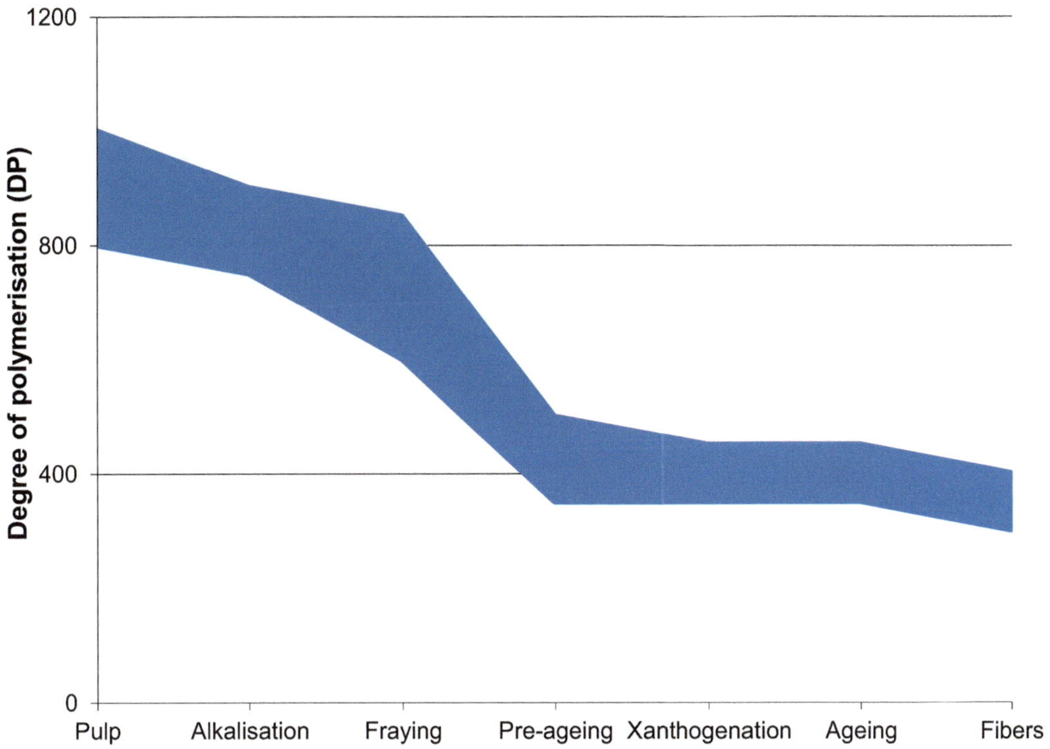

Fig. 41.19 Change in the degree of polymerization during viscose production

remaining solution is reused. The sodium sulfate filtered out in the process is centrifuged and the water content is usually removed by calcination. Around 70% of the carbon disulfide recovered during fiber formation and after-treatment is collected, distilled and also reused.

41.2.1.6 Tape Formation and Drawing

The crystalline content (28–50%), their orientation, the lateral order and the size of the crystallites are influenced by the stretching of the filaments. Textile filament yarns are drawn in an advanced state of regeneration and technical filament yarns in a less advanced state of regeneration (Fig. 41.20).

For sliver formation, the filaments of several spinnerets are combined. This happens immediately after leaving the spinning bath (Fig. 41.21).

The drawing of the filament fiber sliver takes place in different stages. A small amount of drawing already takes place immediately after the spinneret, while the actual higher drawing takes place after leaving the spinning bath between godets. The total draw is approx. 20% for standard fibers and up to 150% for high-strength fibers.

41.2.1.7 After-Treatment of the Filaments

The post-treatment of the spun filaments can be carried out in bobbin form, on the filament yarn or in cable or flake form. The freshly spun filament yarns or staple fibers are treated with the following objectives:

- Removal of impurities,
- equalization and increase of whiteness,
- adjustment of adhesion and slip properties for further processing.

For this purpose, the after-treatment comprises the following process steps:

- Washing with carbon disulfide recovery,
- desulfurization,
- bleaching,

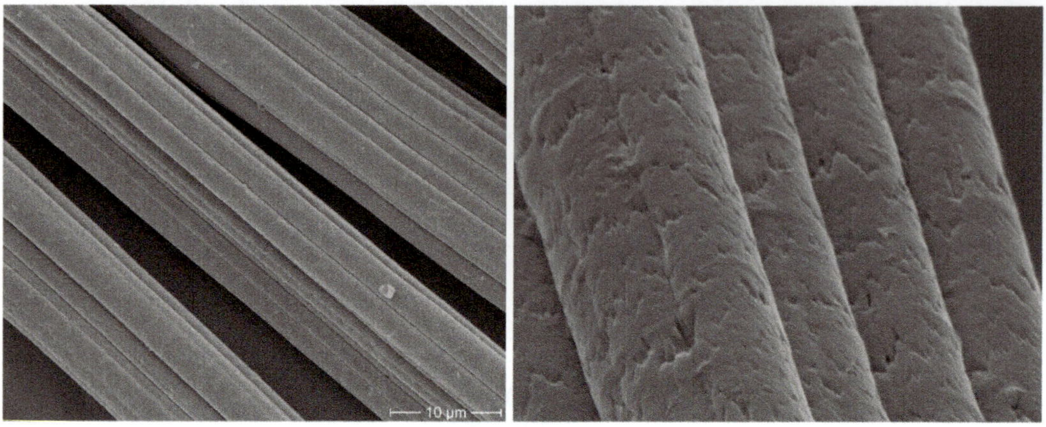

Fig. 41.20 Viscose filaments before (left) and after drawing (right) (Lenzing AG)

Fig. 41.21 Tape formation by combining several fiber strands and drawing over calendar rolls (left: Franz Neumayr, Lenzing AG; right: Ing. A. Maurer SA)

- finish application,
- drying.

The washing of the filament yarns in spool form is carried out under vacuum (suction washing) or under pressure (pressure washing). Solutions of caustic soda with sodium sulfite or sodium sulfide are used for desulfurization. The sulfur content of the filament yarns on leaving the spinning bath is approx. 1%, after washing approx. 0.25%. These values are sufficient for technical filament yarns, while for textile filament yarns and staple fibers the sulfur content is reduced to 0.05%.

Bleaching equalizes and increases the whiteness of textile fibers. As a result, the desired color shades can be better adjusted in subsequent dyeing processes. Sodium hypochlorite, chlorine dioxide or hydrogen or sodium peroxide are used for bleaching. Technical filament yarns are not bleached.

Depending on the type of further processing, lubricants and adhesives that can be easily washed out are mainly used for lubrication in the appropriate concentration. Finish is mainly used to adjust the friction values against fibers, metals and plastics.

The moisture content of the filament yarns after post-treatment is 250–270%, after centrifuging 200–210%. They are dried at temperatures of 70 to 80 °C in the first drying zone and 45–50 °C in the last zone. To ensure uniform shrinkage behavior, drying in the

Fig. 41.22 After-treatment (Ing. A. Maurer SA)

discontinuous process is very slow, usually over many hours. On continuous machines, all post-treatment steps are usually carried out within 3 to 5 min. Due to the coarser yarns, technical filament yarns are dried at higher temperatures (85–90 °C).

After leaving the spinning machine, the filament cable still contains 45–60% of the carbon disulfide used. To remove it, the cable is treated in a closed hot water bath. The released carbon disulfide vapor is then condensed and returned for reuse. The cable can also be simultaneously stretched in this passage. To produce staple fibers, the drawn cables are cut to the desired fiber length and the resulting staple sections are then further processed in post-treatment lines. The staple sections resulting from the cutting are opened, desulfurized, bleached and finished is applied by the baths. In accordance with the high production rate of such lines, the fibers are washed with water between each of the above passages. In the after-treatment the desired crimp can be adjusted (Fig. 41.22).

Finally, the fibers are dried, filament yarns are wound onto tubes, sorted and packed, and staple fibers are pressed into bales (Fig. 41.23).

41.2.1.8 Viscose Fiber Types

The viscose process is characterized by the fact that it is possible to vary the properties of the spun fibers within wide limits by adjusting the chemical or physical manufacturing conditions.

41.2.1.8.1 High Wet Strength Viscose Staple Fibers

Such fibers are produced for conveyor belts and other technical articles under similar conditions as

Fig. 41.23 Baler (left, Lenzing AG), filament winder (right, Ing. A. Maurer SA)

technical filament yarns. Compared to normal viscose fibers, the viscose has a higher concentration of caustic soda, other modifiers (polyethylene glycol, alkylamines, phenol) and is less matured. The addition of the modifiers slows down the regeneration of the cellulose, thus increasing the orientation of the macromolecules under tensile stress and thus significantly increasing their maximum tensile strength and wet modulus compared to viscose. The spinning bath also contains considerably more zinc sulfate (around 40 g/l). The spinning speed is lower and the drafting considerably higher. Typical fineness values are 1.0–3.1 dtex and the fiber cross-section is round to kidney-shaped. The dye affinity is often higher than that of normal viscose. These fibers are also called HWM fibers (High Wet Modulus).

41.2.1.8.2 Modal

Modal fibers are a subgroup of HWM fibers and are produced using a modified viscose process. The cellulose xanthate solution is spun out in liquid crystalline form into a spinning bath that is only slightly acidic and does not contain any Zn^{2+} ions. During the short residence time of the resulting filaments in the precipitation bath, they are simultaneously stretched, forming a fibrillar structure. At the same time, additional bonds are formed between the crystalline domains, which are resistant in water, resulting in a high wet modulus. However, they dissolve quickly in alkaline environments (Fig. 41.24).

For the purpose of differentiation from viscose, the following definition shall apply: the maximum tensile force F_H in the conditioned state and the force $F_{5,N}$ required to cause an elongation of 5% in the wet state are:

$$F_H[cN] \geq 1{,}3 \cdot \sqrt{Tt} + 2 \cdot Tt \, und \, F_{5,N} \geq 0{,}5 \cdot \sqrt{Tt}$$

Tt is the average fineness in dtex.

Modal fibers have a very soft handle compared to other cellulosic man-made fibers.

41.2.1.8.3 Polynosic Fibers

To produce this type of fiber, the ageing time is considerably reduced, resulting in a higher degree of polymerization of up to 800. The sulfuric acid concentration of the spinning bath is relatively low, so that coagulation takes place very slowly. Because the spinning bath temperature as well as the spinning speed are also lower than for normal fibers, drawing can exceed 150% and take place in several stages. This increases orientation and crystallinity. The latter reaches values similar to those of natural cellulose. Polynosic fibers, which consist of a pure sheath structure, can be mercerized, for example, in a blend with cotton. This gives them a round cross-section and a different sheen. The typical fineness ranges from 1.0–3.6 dtex. Their elongation is approx. 10% and they are brittle. Therefore, their textile processing is difficult. Moreover, because the manufacturing conditions of polynosic fibers require adapted plant components and the spinning speed is lower than for standard viscose,

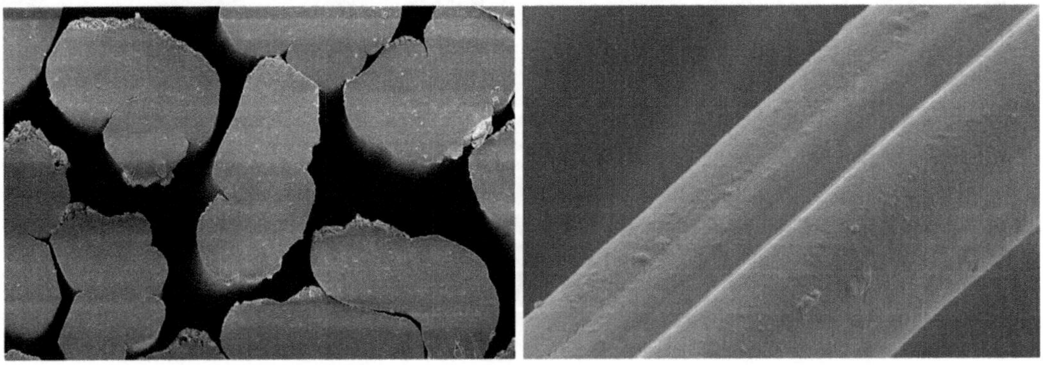

Fig. 41.24 Modal fibers (Lenzing AG)

Fig. 41.25 Longitudinal and cross-sectional view of polynosic fibers

these fibers have not been an economic success (Fig. 41.25).

41.2.1.8.4 Technical Filament Yarns

These include, for example, cord yarns spun from high quality viscose (high-cellulose content). Long-fiber cellulose with a high α-cellulose content (≥ 96%) and a high degree of polymerization is used for their production. By increasing the Zn^{2+} content to 4%, the regeneration of the cellulose is slowed down and thus made more uniform. Supercord yarns consist of filaments with a pure sheath structure. The fibers are then fully drawn in a water bath at 90 °C. The viscose fibers swell during this process. During this process, the viscose fibers swell and the high temperature makes the macromolecules mobile, which facilitates drawing. The fineness of technical filament yarns is between 1000 and 2500 dtex, the filament fineness between 1.2 and 2.0 dtex. The spinnerets have 800 to 1500 holes with a diameter of 50 to 60 μm.

Figure 41.26 shows the reinforcement of car tires with several layers of tire cord fabric made of high-strength viscose at the top and the corresponding open-pore fabric at the bottom. . . .

Figure 41.27 shows typical force-elongation curves of selected viscose types.

41.2.1.8.5 Highly Crimped Viscose Staple Fibers

They are processed alone or in blends with wool, polyester or polyacrylonitrile into clothing and home textiles and are based on a more mature viscose. Due to the higher drawing in the second bath compared to normal viscose fibers, the initially formed sheath breaks up and a new, irregular sheath is formed. This is why the fibers crimp. The number of crimping arcs can be varied within limits. A 3 dtex spun fiber has 100 to 125 crimps/10 cm, which are resistant in the dry state. However, the crimp resistance of cellulosic man-made fibers is lower than that of synthetic fibers. This is particularly true in the wet state. Typical linear densities of such fibers are 2.2–3 dtex.

41.2.1.8.6 Hollow Fibers

There are various technologies for the production of hollow fibers. By adding blowing agents to the viscose and selecting suitable process conditions, hollow fibers with different cross-sectional shapes can be spun out. The outer diameter of a hollow fiber is much larger for the same fineness compared to normal viscose fibers and modal fibers. The wall thickness can be adjusted, in the case of a 1.7 dtex fiber, for example, it is 2–5 μm. Typical applications are oxygenators for blood purification (Fig. 41.28).

41.2.1.8.7 Flame Retardant Fibers

By incorporating organic substances containing phosphorus or sulfur already in the spinning solution, flame-retardant fibers can be produced. Alternatively, $Si(OH)_4$ can also be incorporated into the fibers.

Fig. 41.26 Use of high-strength viscose filament yarns as tire cord fabric

Fig. 41.27 Force-elongation curves of important viscose types

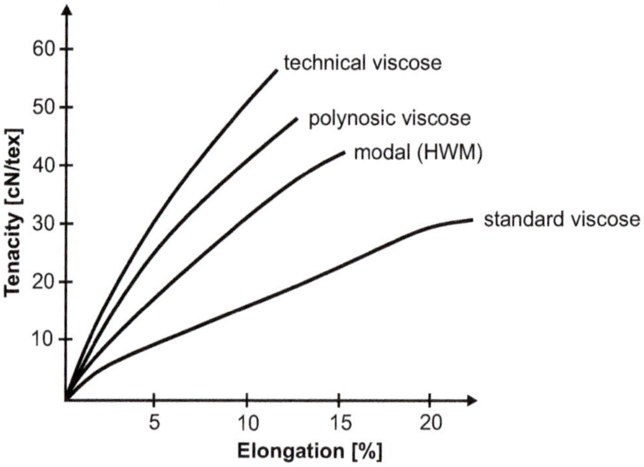

Fig. 41.28 Hollow fibers of viscose and oxygenator

41.2.1.8.8 Water Absorbent Fibers

Cellulosic man-made fibers always absorb approx. 10% water due to the numerous OH groups and their internal structure of large and small pores within the fiber. This amount can be increased by changing the fiber cross-section (e.g. trilobal), increasing the crimp and by incorporating hydrophilic substances into the fiber already during the spinning process (e.g. carboxymethyl cellulose, guaran, alginate). This increases the water retention capacity up to approx. 150%. The use of hollow fibers increases the value up to 400%.

Many water-absorbent textiles made of viscose are disposable products. It is therefore particularly important that the textile structures dissolve well in water after use, for example, when they are thrown down the toilet.

In Fig. 41.29, selected fiber types are compiled to show the diversity of viscose fibers.

41.2.1.9 Chemical Structure, Fine Structure

The molecular formula of cellulose is $(C_6H_{10}O_5)_n$, where n is the number of building blocks (repeat unit) in the chain and is also called the average degree of polymerization (DP). The basic building block of cellulose is glucose or dextrose (Fig. 41.30). Taking into account the atomic weights, cellulose therefore consists of 44.4 wt.% carbon (C), 6.2 wt.% hydrogen (H) and 49.4 wt.% oxygen (O).

Every second glucose building block is rotated by 180° (syndiotactic arrangement), resulting in the structure shown in the Fig. 41.31 structure shown in the figure. The repeat unit is therefore the cellobiose, but the DP usually corresponds to the number of glucose units.

The molar weight of the basic glucose unit is $m_o = 162.14$ g/mol and the molar weight of the OH group or the H atom on the right and left of the square bracket in Fig. 41.31 is 18. This gives the molar weight M_p of the polymer:

$$M_P = m_0 \cdot DP + 18 \approx 162 \cdot DP$$

With Table 41.2, the molar weight of the corresponding cellulose macromolecules can be calculated from the DP value.

Figure 41.32 shows the unit cell of natural cellulose with the chain molecules running from bottom to top.

The treatment of the cellulose with the solvent (usually sodium hydroxide solution) slightly changes the lattice structure, because during alkalization one sodium ion per glucose building

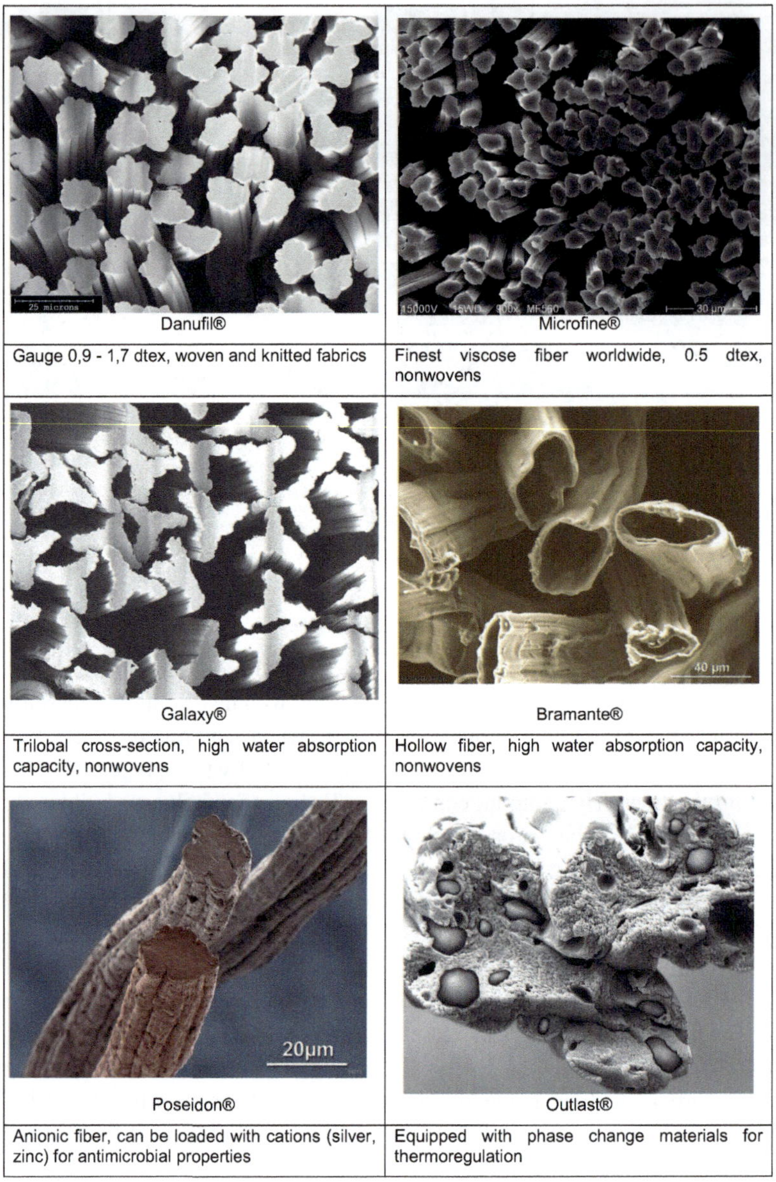

Fig. 41.29 Selected viscose fiber types (Kelheim Fibers GmbH, 2021)

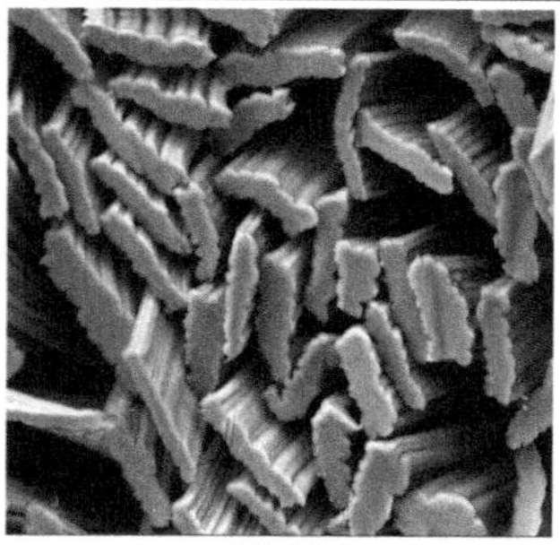	
Viloft®	Leonardo ®
Particularly high absorbency, short pile for wipes	Particularly flat fiber (width : thickness = 40 : 1) for papers

Fig. 41.29 (continued)

block enters the lattice of the cellulose and expands it. After removal of the sodium ions in the precipitation bath, the building blocks of the cell approach each other again, but without assuming their original position.

The regenerated cellulose fibers and cotton are 50% or more amorphous. In contrast, other cellulose fibers such as ramie, hemp and flax have considerably higher degrees of crystallinity.

41.2.1.10 Properties

The fineness and length of viscose staple fibers can be adjusted to the desired further processing conditions (less than 1 to more than 100 dtex, 30 to 200 mm). The density is 1.50–1.52 g/cm^3 (Table 41.3).

The mechanical properties are also variable within wide limits (Fig. 41.33).

The Table 41.4 contains information on the properties of technical viscose filament yarns.

The chemical and thermal behavior of viscose and the other regenerated cellulose fibers is similar to that of cotton.

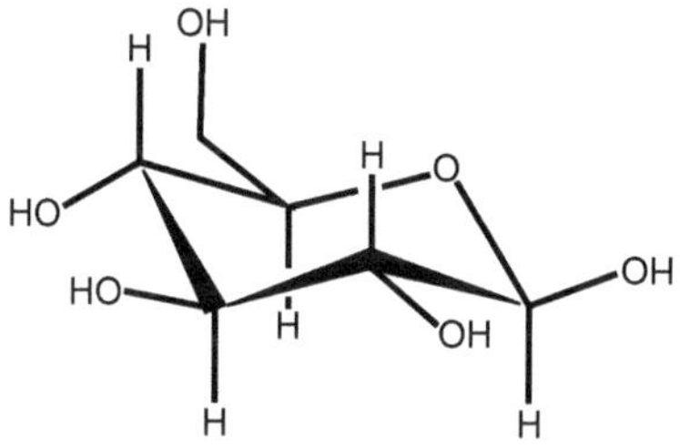

Fig. 41.30 Structural formula of a glucose molecule

41.2.1.11 Processing and Finishing

Viscose staple fibers and viscose filament yarns can be processed into yarns, twisted yarns and textile structures using all standard textile processing methods. Unlike most other man-made fibers, viscose fibers are particularly well suited for air-jet spinning.

Fibers as well as yarns and textile fabrics made of and containing cellulose regenerated fibers are dyed and finished in a similar way to textiles made of natural cellulose fibers (cotton, flax, ramie, etc.). The process differences result from the lower wet strength of cellulose regenerated fibers as well as their sensitivity to alkalis and their higher swelling values. Washing and bleaching processes are much faster than with natural cellulose fibers because cleaning steps are not necessary. Viscose fibers can be dyed with all dyes commonly used for natural cellulose fibers. In order to increase the dimensional stability of clothing and home textiles made from regenerated cellulose fibers, they are often highly refined. This reduces their tendency to swell and makes the textiles more dimensionally stable and changes their feel. Blends with cellulosic natural fibers or synthetic fibers are common to combine the properties of the materials.

41.2.1.12 Carbamate Process

In this process, the pulp is treated with a solution of ammonia and urea $(NH_2)_2$-O and heated. This causes the ammonia to escape and cellulose carbamate to be formed at approx. 140 °C. It is

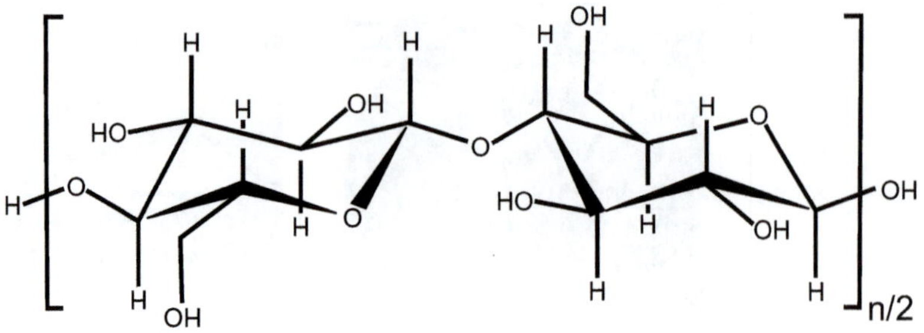

Fig. 41.31 Structural formula of cellulose (n = DP)

Table 41.2 Average degree of polymerization of various materials acc. to (Treiber, 1957)

	DP
Raw cotton	≥ 7000
Cotton linters	6500
Cotton, cleaned	1500–3000
Flax	8000
Ramie	6500
α-cellulose, generated from wood fibers	800–1100
Spruce	3300
Beech	3050
Aspen, fir	2500
Bacterial cellulose	2700

soluble in sodium hydroxide solution and can be regenerated to form cellulosic fibers with the elimination of ammonia (Hill & Jacobson, 1938). If carbamate production is successful, fiber production is not problematic. Difficulties arise in the utilization of the by-product ammonia and in the temperature control during fiber production. The corresponding reaction equations are shown in Fig. 41.34.

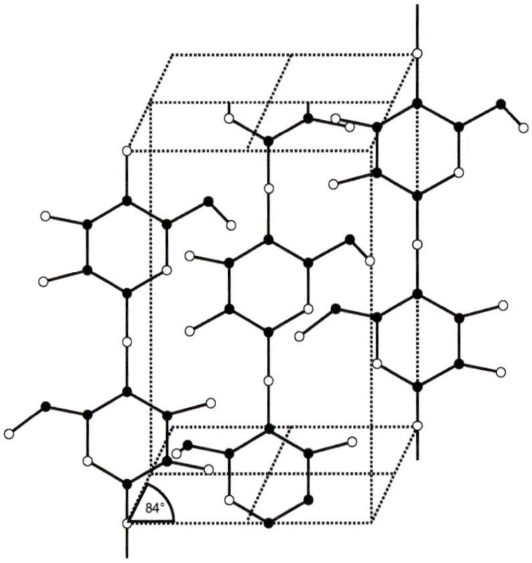

Fig. 41.32 Unit cell of cellulose

The tenacity of the fibers is 12–22 cN/tex, the elongation 16–22%, the loop strength 3.8–5.5 cN/tex (Urbanowski, 1996). The dye uptake corresponds to that of viscose. CarbaCell® was registered as a brand name by Lurgi Zimmer AG, but production on a larger scale was not started.

Fink et al. (2006) produced fibers in a combination of carbamate and NMMO solutions using the funnel spinning process. In this way, they achieved fiber strengths of up to 65 cN/tex and Young's modulus values of up to 3300 cN/tex.

41.2.2 Modified Viscose Fibers

With the aid of pigments added to the viscose, spun-dyed fibers can be produced which are particularly colorfast. An addition of approx. 40% barium sulfate (based on the fiber weight) results in X-ray contrast fibers. Flame retardant fibers are

Table 41.3 Properties of important viscose staple fibers acc. to W. Albrecht and company data

	Normal	Modal fibers	
	Viscose fibers	Polynosic	HWM
Tenacity [cN/tex]	23–47	40–65	40–55
Wet strength ratio [%]	50–80	70–80	
Maximum elongation [%]	5–40	8–12	14–18
Wet maximum elongation ratio [%]	115–140	120–150	
E-modulus, ε = 100% [cN/tex]	200–300	650–750	
Wet E-modulus, ε = 100% [cN/tex]	50–100	140–180	100–200
Moisture absorption in normal climate [%]	13	11–13	11–13
Water retention [%]	90–120	55–70	65–90
Hot water shrinkage [%]	6–15%	5–7	
Mercerizing	No	Very good	Medium
Electrostatic charge	Low		
LOI	26		

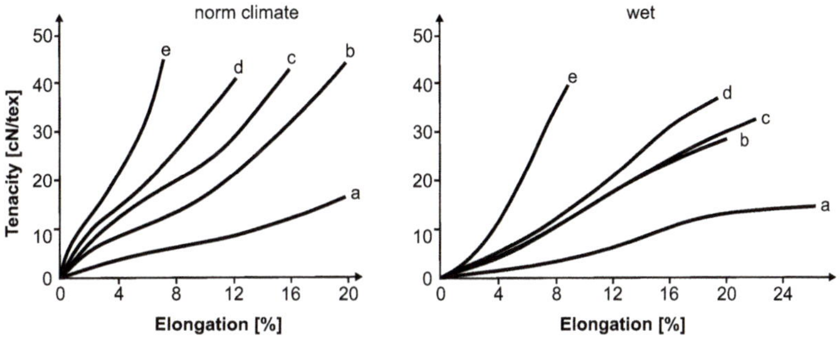

Fig. 41.33 Fineness-related tensile force-elongation curves for various viscose fibers (left: a viscose, textile; b—HWM fiber, type I, c—viscose, technical, d—HWM fiber, type II, e—polynosic fiber)

Table 41.4 Properties of technical viscose yarns

Tenacity, dry [cN/tex]	40–75
Maximum elongation [%]	7–15
Tenacity loss [%], wet	25–30
Tenacity loss [%], after twisting	15–20
Thickness of sheath layer in % of filament cross-sectional area	90–100
Density [g/cm^3]	1.505

produced by spinning in flame retardant compounds, usually containing halogen, phosphorus or nitrogen compounds. Basic additives can modify the dye affinity of regenerated cellulose fibers so that they can be dyed with acid dyes (such as wool). Special color effects can be achieved with blends of such and normal regenerated cellulose fibers. Various manufacturers also offer fibers that are supposed to have antimicrobial properties due to additives that are added during spinning (Table 41.5).

41.2.3 Cupro Process (Cuoxam Process)

Cupro fibers are fibers made of cellulose produced by the copper oxide ammonia process. They are not chemically altered during the process, in contrast to the viscose process. Fibers made from “copper silk” have a soft, supple handle, a silky drape, a good sheen and are easy to wash. Typical applications are outerwear, linings, accessories and underwear.

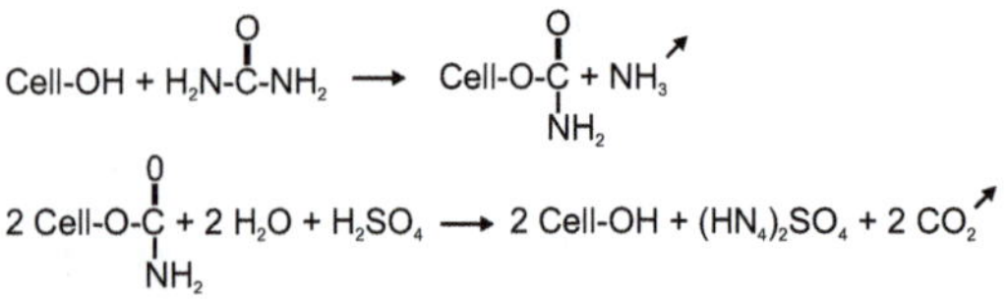

Fig. 41.34 Reactions in the carbamate process

Compared to the viscose process, the cupro process offers several advantages:

- the amount of harmful accompanying substances is lower,
- no harmful gases are formed and
- the waste water is less polluted.

The previously high copper consumption has now been reduced by the recovery process and the waste water contamination by copper ions has been brought to such low levels that there is no longer a problem for bacteriological waste water treatment plants.

Bleached cotton linters are used as the starting material. These are usually dissolved in Cuoxam at 20 °C. $Cu(NH_3)_4^{2+}$ ions form in the solution, making the cellulose soluble.

Depending on the spinning conditions, the spinning solution then contains 4–11% cellulose, 4–6% Cu and 6 to 10% NH_3. Excess NH_3 is removed by deaeration. The viscosity is approx. 200 Pas at 20 °C. Approx. 400 g copper and 650–800 g ammonia are required per kg cellulose. The solubility of the cellulose and the process time can be influenced by the amount of ammonia used. To reduce the degradation of the cellulose, reducing agents such as sodium sulfite and glucose are added. Typical degrees of polymerization are 400–500 in the fiber.

The filaments are spun using the funnel spinning method (Fig. 41.35). In this process, the solution is forced through the holes of the spinneret and coagulation begins immediately below the spinneret when the complex of cellulose and $[Cu(NH_3)_4](OH)_2$ (amine copper (II) hydroxide hydrate) is precipitated on contact with water. The resulting fine filaments are entrained by the softened water (temperature: 35 to 45 °C) flowing from top to bottom. Due to its conical shape, the precipitation bath flows through the funnel at a constantly increasing speed. The greater the difference between the funnel diameter at the top (section A–B) and at the bottom (section C–D), the greater the flow velocity of the water, whereby the resulting filaments are stretched by 10,000–15,000%. The spinneret openings and thus the throughput are correspondingly large. If necessary, the filaments emerging from the funnel are fed to a second bath where the cellulose is regenerated with dilute sulfuric acid (1.5 to 2% at 20 to 24 °C).

The vertically arranged spinning funnels are made of glass or transparent plastic. The spinning speed is approx. 150 m/min.

The after-treatment of the cupro fibers includes:

- the washing out of the acid and salts,
- the removal of the copper and
- the recovery as well as the finish application and drying.

Table 41.5 Substances that change the properties of the fibers

Substances	Effects
Delustering agent	Dye-absorption, shine, luster
Light protection agent	Fade resistance
Bariumsulfate	X-ray contrast effect
Dyestuffs	Color, fastness
Brightener	Degree of whiteness, fluorescence
Flame protection agent	Fire protection
Modifier	Physical and chemical properties
Second component	Bicomponent influence (crimp, physical and chemical properties)
Special dyestuff coupling agent	Color, optical fabric effects

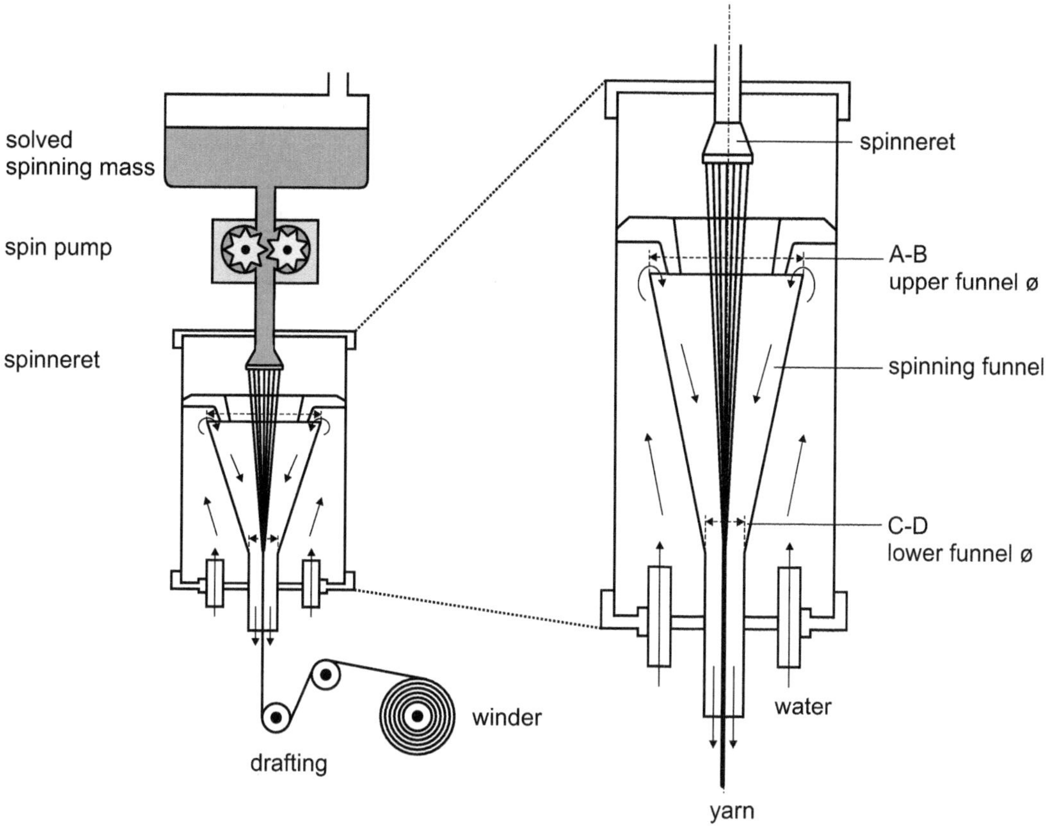

Fig. 41.35 Funnel spinning process for cupro acc. to (Rogowin, 1982)

Wide spinnerets with up to 90,000 holes are used to produce nonwovens from filaments. The filaments formed are deposited on a partially permeable conveyor belt, which traverses, thus achieving an isotropic arrangement of the filaments. The spinning speed is then 35 m/min.

For the production of staple fibers, a 1 to 2% weaker ammonia concentration is used in order to avoid the filament sticking which is a problem here. The spinning funnels are much larger, the spinning nozzles (diameter 80 to 100 mm) have 2000 to 3000 openings (diameter 0.5 to 1 mm). The filaments from 50 to 100 spinning funnels are combined to form a sliver. Post-treatment is carried out in the sliver or in flake form. When the tapes are alternately stretched and relaxed, the sheath breaks open and crimped filaments are produced.

Unlike other fibers made from regenerated cellulose, cupro fibers have a round cross-section because coagulation is very rapid and the fiber therefore shrinks uniformly during coagulation (Fig. 41.36).

The recovery rate of copper is more than 99% when ion exchangers are used. Ammonia is recovered at a rate of 40 to 50% by distillation. Special properties of cupro fibers are their high individual fiber fineness, silk-like appearance and silky feel. In this respect they surpass viscose fibers. The fiber structure makes them particularly suitable for blood dialysis. However, other fibers are now also used for this application.

Typical strength values are 15–20 cN/tex (dry) and 9–12 cN/tex (wet), elongation is 10–20% dry and 16–35% wet. The water absorption capacity is high at 100–120%, the wet modulus at 5% elongation is very low at 1.5–2 cN/tex.

The production of cupro fibers using the copper oxide ammonia process (Cuoxam process) declined sharply in the 1960s and is no longer

Fig. 41.36 Longitudinal and cross-sectional view of cupro fibers

carried out in Germany today. The main reasons are the outdated production facilities, the environmental pollution, especially by copper ions, and the penetration of synthetic man-made fibers into the typical cupro fiber applications. One of the largest manufacturers of cupro fibers, Bemberg AG, converted its production to membranes. Here, the Cuoxam solution is used to produce flat membranes and tubular hollow fibers for artificial kidneys. Cupro fibers for textile use are now only produced in Japan (Asahi Kasei) under the brand name Bemberg™. Nonwovens made from cupro fibers are marketed under the name Benliese™.

41.3 Lyocell

Lyocell fibers are those cellulosic man-made fibers obtained by regenerating cellulose in fiber form from a solution of cellulose in an organic solvent ("Lyo" derived from Greek: lyein = to dissolve, "Cell" from cellulose). The abbreviation is CLY.

Since the start of large-scale cellulose fiber production, attempts have been made to replace the viscose process with other processes. The idea of dissolving cellulose directly and without chemical modification was first patented in 1939 by Ch. Gränacher and R. Sallmann; they used amine oxide. A process was developed by Eastman Kodak in 1969 to produce amine oxide from N-methyl morphine. The process to produce fibers in this way was patented by AKZO in 1979, which led to the development of the lyocell process by Lenzing AG in 1985. Courtaulds started up the first pilot plant in England in 1986, and Lenzing AG in 1992, producing 12,000 t at Heiligenkreuz (Austria) by 1997. Since 1993 Courtaulds also produced lyocell fibers in the USA under the brand name Tencel™. In 2000, the Tencel business unit was acquired by Lenzing AG.

41.3.1 Principle

The main difference between the viscose and lyocell fiber production processes is that the viscose process involves a chemical reaction, whereas the lyocell fiber production process uses only physical processes (Fig. 41.37).

41.3.2 Solvent

For the production of cellulosic fibers based on a purely physical reaction, various solvents were investigated for their suitability. These include N-methylmorpholine oxide (NMMO), dimethylacetami/lithium chloride and diammonium thiocyanate. Of these, only NMMO was able to establish itself on the market. Corresponding products are marketed under the brand names Lyocell™, Tencel™ SeaCell™ and Oricell™, for example.

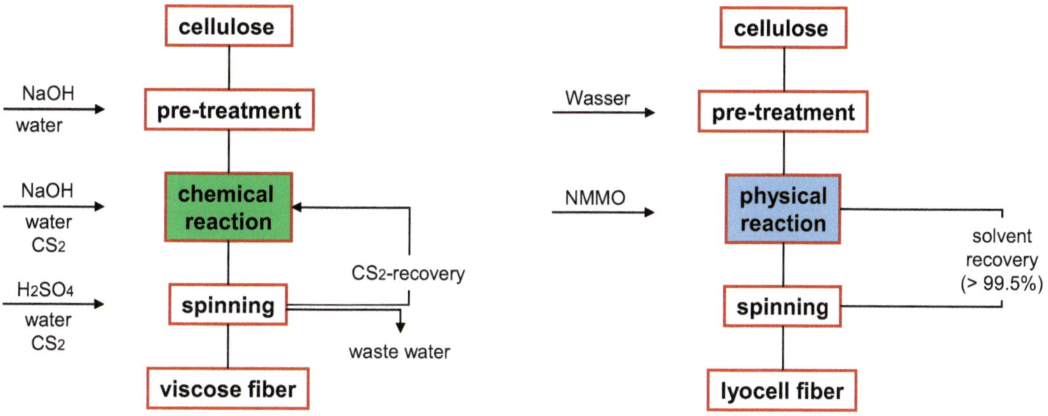

Fig. 41.37 Comparison of viscose and lyocell processes

41.3.3 NMMO Process

The NMMO process is based on the ability of amine oxide to solve cellulose under certain conditions. From this solution, the cellulose can be regenerated. N-methyl-morpholine-N-oxide (NMMO) has proven to be particularly suitable (Fig. 41.38).

The main problems in introducing the new process were initially the high investment costs and the difficulty in recovering the expensive solvent as completely as possible. Now that this has been achieved, the process is being carried out successfully and is being used in more and more plants for the large-scale production of lyocell fibers. In the NMMO process, three process stages are distinguished: The production of the spinning solution, the spinning out of the cellulose fibers and the recovery of the solvent NMMO (Fig. 41.39).

The narrow process window in which cellulose is soluble in NMMO is particularly difficult to control. Initially, many attempts to implement the lyocell process on an industrial scale failed because of this.

Fig. 41.38 Structure of NMMO

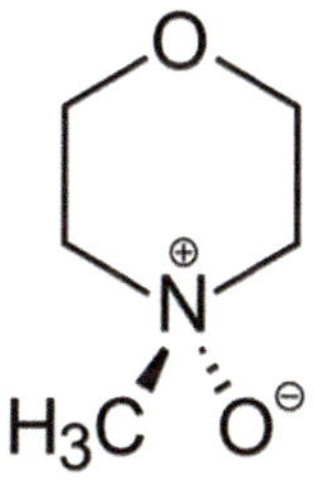

41.3.4 Preparation of a Homogeneous Solution of Pulp, NMMO and Water

The prepared pulp is introduced into an NMMO-water mixture (Fig. 41.40). Initially, the mixture consists of 20% cellulose and 80% water. By adding NMMO and simultaneously extracting water, water is removed from the suspension along the "extraction" line. When the grey marked state area is reached, there is then a complete solution of NMMO, pulp and water, which is stabilized by suitable chemicals. Such a solution state then consists of, for example, 8% water, 75% NMMO and 17% cellulose. By removing the solvent NMMO, the cellulose solidifies and the lyocell fibers are formed (Fig. 41.40).

41.3.5 Fiber Production

The solution is spun out through suitable nozzles at a temperature just above 100 °C using the air gap spinning process (Fig. 41.41). The cellulose is regenerated in the precipitation bath as in the wet spinning process. The fibers are drawn off, washed and wound up. Spinning speeds of approx. 90 m/min are achieved for staple fibers

Fig. 41.39 Principle of the NMMO process

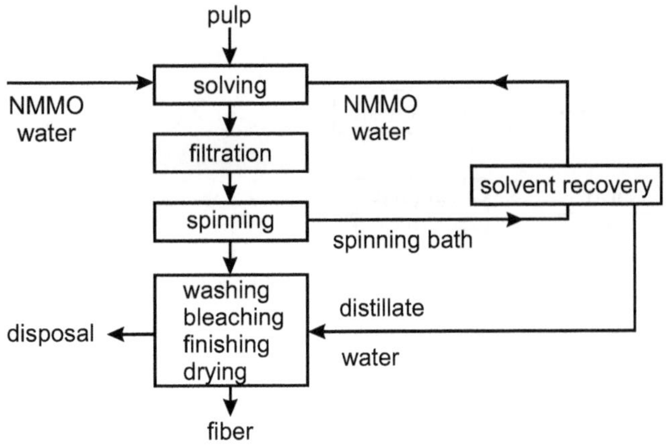

and up to 500 m/min for filament yarns (Michels & Kosan, n.d.).

41.3.6 Possibilities for Modifying the Procedure

Because the NMMO process starts from a cellulose solution, the possibilities for its modification are less than in processes where the cellulose is regenerated from derivatives. Nevertheless, the process allows the production of different types of fibers, including those in which the thickness of the fiber sheath is varied and substances are spun into the cellulose. The best known modifications are the SeaCell® and Tencel® processes.

Fig. 41.40 Ternary mixing ratios for the NMMO process

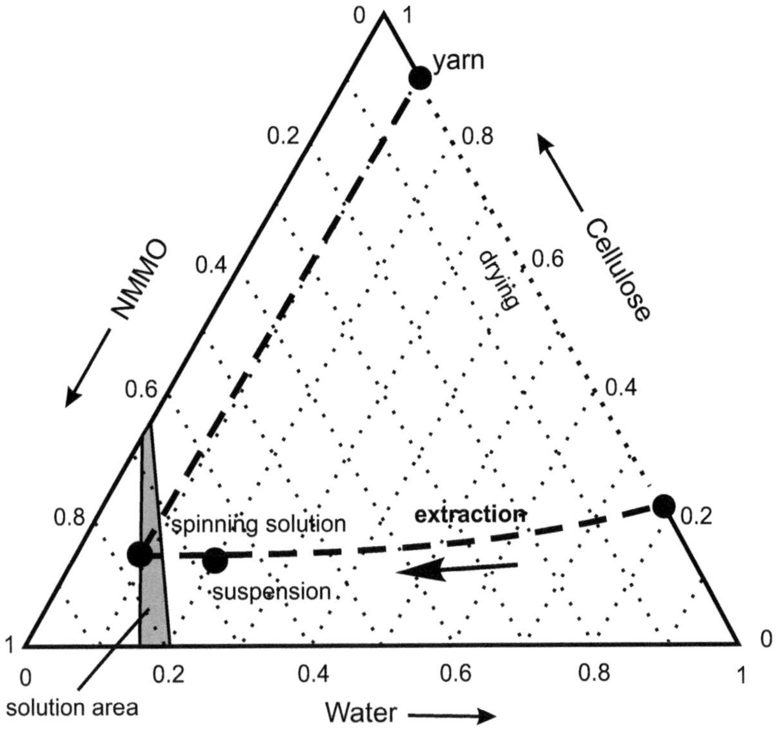

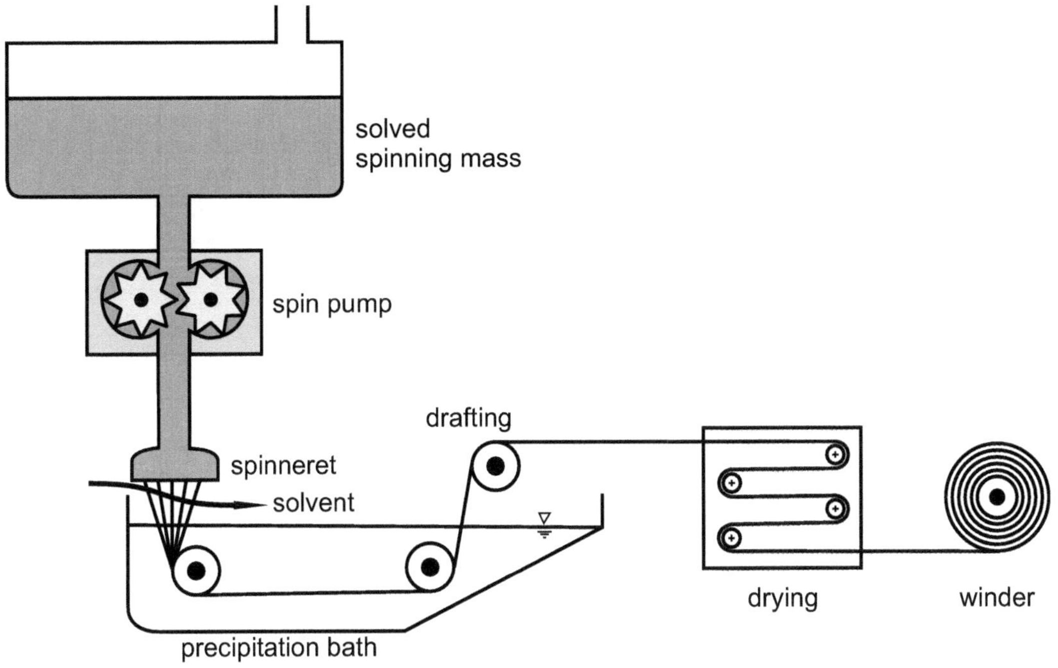

Fig. 41.41 Air-gap spinning process

41.3.7 SeaCell®

This type of cellulose fiber is obtained from brown, red, green and blue algae using the NMMA process (Fig. 41.42). Because marine algae can absorb minerals, this improves the absorption of metal ions by the fiber. This in turn leads to an antibacterial effect, as zinc oxides and silver ions, for example, are incorporated into the fiber. At the same time, the formation of perspiration odor is reduced and thus the wearing comfort is increased.

The metal ions are strongly bound by the phenols contained in the algae of the cellulose and are therefore hardly washed out (long-term effect). The metal ions are not only located on the fiber surface, but also penetrate into the fiber interior and are thus distributed over the entire fiber cross-section. The Fig. 41.43 diagram shows the process schematically.

SeaCell® fibers consist of 84% cellulose, 4% algae substance and 12% water (Fig. 41.44).

During the production of the smartcel™ fiber of Smartfiber AG, up to 16% zinc oxide is incorporated into the fiber. This leads to an antibacterial and odour-inhibiting effect and protects against UV radiation. Typical properties of SeaCell™ and smartcel™ fibers are summarized in Table 41.6.

41.3.8 Structure and Fine Structure of Lyocell Fibers

The special properties of lyocell fibers—high strength and comparatively low elongation—led to investigations of their structure early on. It was found that the fibers are characterized by a high crystallinity (17%) and orientation (26%) of the molecules. The ratio of crystalline to amorphous areas is about 9:1, while the values for viscose fibers are about 6:1. Even in the amorphous areas there is still some kind of orientation. This special property compared to viscose is reflected in the high wet strength of the fibers and ensures the dimensional stability of sheet structures produced from them.

On the other hand, these values are also the cause of the pronounced fibrillation of the fibers, which occurs due to spontaneous coagulation in

Fig. 41.42 Brown algae and harvester (Smartfiber AG n.d.)

the spinning bath (no sufficient interfibrillar interaction). Although fibrillation behavior can be used in textile design to set attractive effects, it is troublesome for most applications. Therefore, attempts are made to suppress fibrillation, e.g. by adding cross-linking substances to the spinning solution or to the bleaching bath, by influencing the spinning conditions in the air gap (concentration of water vapor, temperature, air flow) and by adding additives to the spinning solution (e.g. ε-caprolactam).

41.3.9 Properties

As Table 41.7 shows, the values of lyocell fibers in various parameters deviate significantly from other cellulosic man-made fibers:

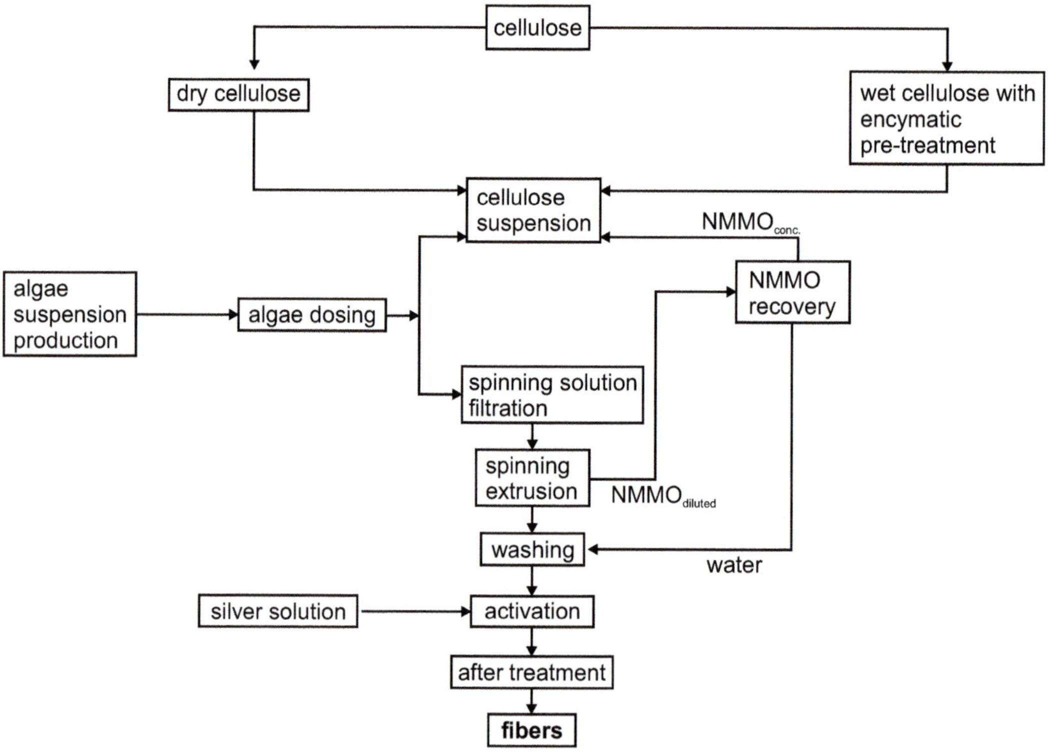

Fig. 41.43 Principle of the SeaCell® process

Fig. 41.44 Dried algae (left) and combed sliver of Seacell® (right) (Smartfiber AG)

The higher strength of lyocell fibers compared to viscose fibers offers advantages in processing into yarns and fabrics. It also allows lighter constructions and is useful in the manufacture of yarns from blends with natural and synthetic fibers (Fig. 41.45).

Here, too, attention must be paid to the fibrillation of the fibers, which occurs under stress along the fiber axis in the wet state. It is used to create textile surfaces that cannot be achieved with conventional fibers. Such surfaces are also known as "peach skin". This effect can be enhanced or reduced, for example, by enzyme treatment of the fabrics or selection of very specific dyestuffs. Such measures can also influence the air permeability of textile fabrics made of lyocell.

41.3.10 Processing

Lyocell fibers can be processed as filament yarns and staple fibers by all known methods into textile fabrics. Lyocell staple fibers can be spun out into fine yarns by the cotton spinning process, both ring and OE rotor spinning. Blended yarns with other fibers are also possible. The properties of lyocell fibers with regard to substance utilization and dimensional stability are particularly advantageous. Lyocell fibers are also used for technical textiles. The strength, the modulus of elasticity and the surface of the fibers play an important role here. The high substance utilization of textiles made of lyocell fibers is also noticeable in nonwovens. In special cases, it can be further enhanced by hydroentanglement at high pressure.

41.3.11 Dyeing, Finishing

Lyocell fibers behave similarly to the other cellulosic fibers when dyed. With certain dyes, it is found that the color yield is somewhat higher compared with viscose, but lower compared with mercerized cotton and modal. The dyeing

Table 41.6 Typical mechanical properties of SeaCell™ and smartcel™ fibers acc. to manufacturer's data

		SeaCell®LT	smartcel®
Cellulose content	[%]	> 85%	> 66
Fineness	[dtex]	1.7–6.7	2.5–6.7
Breaking force, cond.	[cN/tex]	22–29	23–25
Breaking force, wet	[$\%_{trocken}$]	77–86	74–80
Breaking elongation, cond.	[%]	9–11	11
Breaking elongation, wet	[$\%_{trocken}$]	145–155	124–127
Wet E-modulus (BISFA)	[cN/tex]	5.5–7.6	4.8–6.4
Color		Light brown	White, deep dull

Table 41.7 Typical characteristic values of cellulose fibers acc. to manufacturer's data

		Lyocell	Normal viscose	High tenacity viscose	Polynosic	HWM	Cupro
Breaking force	cN/tex	40–50	20–27	40–75	40–65	40–55	15–20
Breaking force, wet	[%$_{dry}$]	65–85	50–80	70–90	70–80	70–80	60
Maximum elongation, cond.	%	10–18	5–40	7–15	8–12	14–18	10–20
Maximum elongation	[%$_{dry}$]	110–130	115–140	150–200	120–150	120–150	16–35
Initial E-modulus, dry [5%]	cN/tex	400	200–300	650–750			200–300
Initial E-modulus, wet [5%]	cN/tex	11–13	40–50		200–350	180–250	30–50
E-modulus, wet	cN/tex	200–350	50		230	120	
Loop breaking force	cN/tex	18–20	20–60		15–20	12–16	
Moisture absorption	[%]		12–14				11–12
Water retention	[%]	65–70	90–120	65–75	55–70	75–90	100–200
Shrinkage	[%]	6–9	6–15		5–7	5–7	0.5–10
Spec. heat capacity	[kJ/kg K]		1.3–1.5				1.4
Thermal conductivity	[W/m K]		0.3–0.6				
Density	[g/cm^3]	1.5	1.52–1.54	1.51			1.52
LOI	[%]	19–20	19–20				
Cellulose	DP	550–600	250–350	400–500	450–500	400–450	300–400

processes also affect the fibrillation effect. Therefore, it is important to adjust the agitation and water flow in the dyeing apparatus accordingly.

41.3.12 Trade Names, Production, Areas of Application

Originally, Tencel™ was the brand name of the American-British company Courtaulds, whose "Tencel" business was taken over by Lenzing AG in 2000. The name Tencel™ is derived from "tenacity" and "cellulose". The process is essentially the same as the classic lyocell process. In addition to Tencel™ fibers (staple, filament), nonwovens are also produced by Lenzing AG and marketed under the name Veocel™ (Fig. 41.46).

By far the largest manufacturer is Lenzing AG, Austria. smartfiber AG, Rudolstadt, with its Seacell™ product, is a newcomer in recent years, but uses a production site of Lenzing. Other suppliers have emerged in Asia, especially

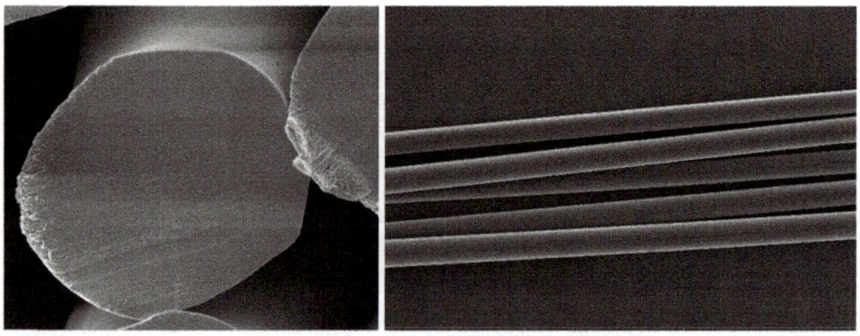

Fig. 41.45 Cross-sectional and longitudinal view of lyocell fibers (Lenzing AG)

Fig. 41.46 Trademarks of Tencel™ and Veocel™ (Lenzing AG) and Seacell™ and smartcel™ (Smartfiber AG)

in China, e.g. Hi-Tech Fiber Group Corporation with its Oricell™ fiber.

Lyocell fibers have been produced in the USA since 1993 under the brand name Tencel™. The quantity produced is about 50,000 t/year. Lenzing produces about 200,000 t in Austria, about 45,000 t in England and a plant under construction in Thailand is expected to produce about 100,000 t of lyocell fibers from 2021. Annual global production is around 420,000 t (2020) and is expected to grow to about 750,000 t by 2030 (Fig. 41.47).

Lyocell is marketed by all manufacturers as a particularly environmentally friendly fiber. For example, the yield per hectare is around 4x as high as cotton and the amount of water required is less than 1000 l/kg. Furthermore, less fertilizer and pesticides have to be used for the cultivation of wood than for the cultivation of cotton. In addition to these ecological advantages, the price for 1 kg of lyocell is also very low at around €1.5. It is therefore to be expected that the market significance of lyocell fibers will continue to increase.

41.3.13 Areas of Application

Due to their good moisture transport, lyocell fibers are mainly used for clothing, bed linen, mattress tops and corresponding fillings as well as hygiene products. In the field of technical textiles, nonwovens, filters and special papers are possible because of the fibrillation tendency. There are also high-strength grades (Fig. 41.48).

41.3.14 Environmental Aspects

The spinning baths are filtered, cleaned, concentrated and then returned to the process. The recovery rate of the NMMO used is over 99.5%. This is imperative because NMMO is already toxic to fish in concentrations of just a few 1000 ppm.

The NMMO process thus enables the production of cellulose man-made fibers without the use of substances that pollute the environment. This makes a circular economy "from cradle to cradle" possible, which is already largely implemented by Lenzing AG, for example (Fig. 41.49).

41.3.15 Comparison of Lyocell and Viscose Processes

In Table 41.8, the viscose and lyocell processes are compared. The Lycocell process is considerably faster, both in terms of solution production and the actual fiber spinning. However, the

Fig. 41.47 Staple fibers and filament yarns made from Lyocell (Lenzing AG)

Fig. 41.48 Typical applications of lyocell fibers (clothing, automotive interiors, diapers, fire protection clothing) (Lenzing AG)

smaller quantities of waste materials and the low energy consumption of the lyocell process are offset by significantly higher investment costs.

41.4 Other Direct Solution Methods

41.4.1 Biocelsol Process

Another process for direct dissolution of cellulose in dilute sodium hydroxide solution was developed by Ciechańska et al. (2007). In this process, cellulose is enzymatically treated in aqueous solution, which completely changes the internal structure of cellulose macromolecules. The average degree of polymerization is decreased and so is the crystallinity by reducing the number of OH-bonds between the chains. At the same time, the internal surface area of cellulose is increased, which increases its solubility in $NaOH_{aq}$. After completion of this pretreatment, excess water is squeezed off, the enzymes are deactivated and the cellulose slurry is filtered and wet spun out in a mixture of caustic soda with small amounts of zinc oxide. The precipitation bath consists of an aqueous solution of sulfuric acid and sodium sulfate.

In pilot scale, strengths of 17–19 cN/tex and elongation of 12–13% were obtained (Ciechańska et al., 2005). Multifilament yarns and hydroentangled nonwovens can thus be produced. An industrial-scale implementation is still pending.

41.4.2 Other

Other direct solvents are mixtures of N, N-dimethylacetamide with 7.8% lithium chloride in aqueous solution (Bianchi et al., 1989) and phosphoric acid in aqueous solution (Northolt et al., 2001). In the latter process, laboratory-scale strengths of up to 113 cN/tex and a Young's modulus of 2860 cN/tex were achieved, which

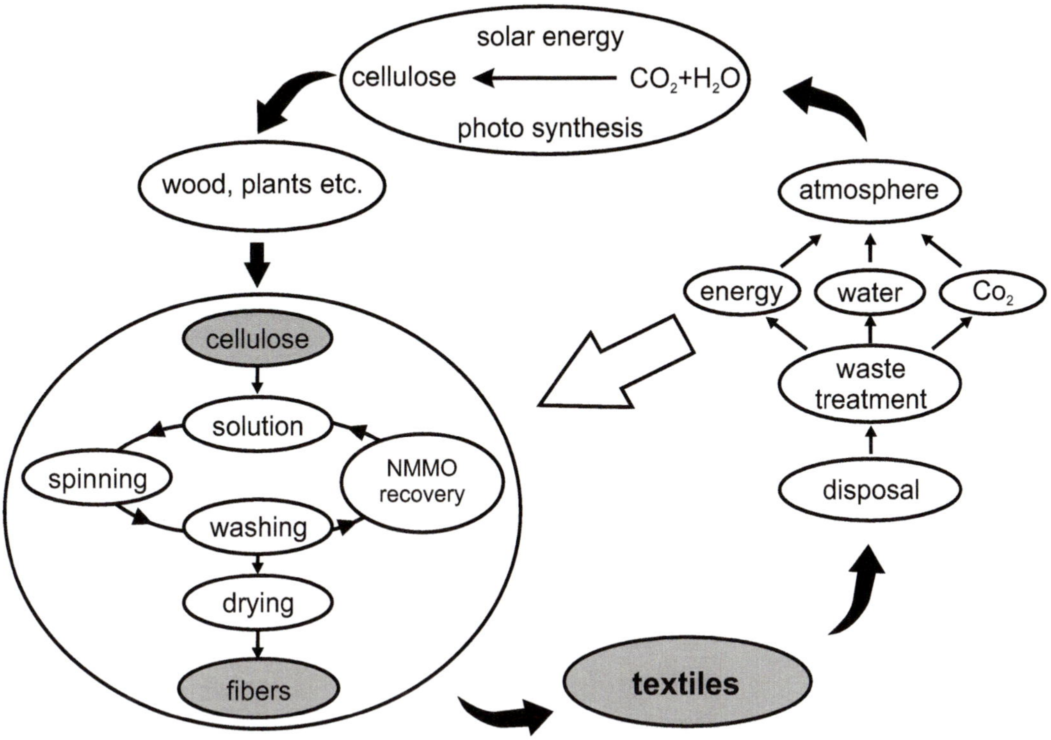

Fig. 41.49 Formation, use, utilization of cellulose acc. to (Lenzing AG)

are extremely high values. The elongation was only 4.6%, which makes textile processing difficult. The length of the crystallites was determined to be up to 17.8 nm, which is twice as long as for viscose.

Laus et al. (2005) gives an overview of the use of so-called ionic liquids for the production of cellulosic fibers. These are liquids which consist almost exclusively of ions and are in part already liquid at room temperature. Values for elongation of 5–17% and for tensile strength of 10–43 cN/tex were achieved with a single filament fineness of approx. 1 dtex.

Further spinning tests with ionic liquids and the funnel spinning process were carried out by Kosan et al. (2008). They achieved fiber strengths of 43–53 cN/tex at elongations of 13–16% and Young's modulus values of 700–900 cN/tex with cellulose concentrations of 13–20%.

41.5 Production

In 2020, approx. 6mt of viscose-based man-made cellulosic fibers were produced worldwide. While the share of cellulosic man-made fibers in world

Table 41.8 Comparison of important characteristics of viscose and lyocell process acc. to (Laszkiewicz et al., 2006)

	Viscose process	Lyocell process
Cellulose concentration in solution [%]	7.5–9.5	10–25
Duration of spinning dope production [h]	35–48	2–4
Spinning speed [m/min]	12–90	60–300
Waste	gases, salts	none
Energy consumption [%]	100	35–40
Investment [%]	30–35	100

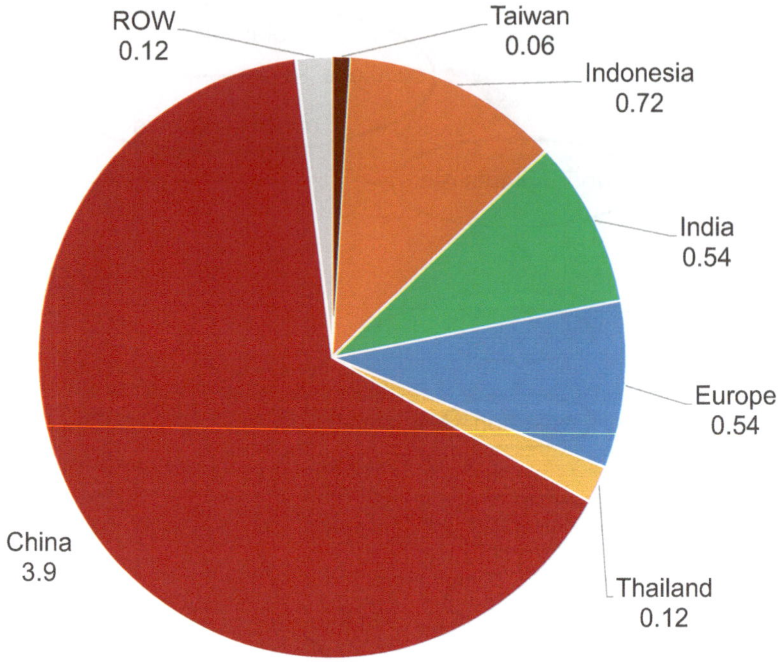

Fig. 41.50 Major producers of viscose staple fibers (The Fiber Year, 2021)

fiber production was still over 95% in 1950 (the rest was accounted for by polyamide), the share has since fallen to around 5% despite a doubling of the production volume. In the case of technical filament yarns, regenerated cellulose fibers are increasingly being displaced by polyester and polyamide; in the clothing sector, demand is very much dependent on fashion. The share of staple fibers in all viscose fibers is currently approx. 94%, the share of filament yarns correspondingly 6% with a downward trend.

There are 120 plants worldwide using the viscose process, 27 using the acetate process, only a few using the lyocell process and one using the cupro process (Fig. 41.50).

The development of viscose fiber production shows Fig. 41.51. While staple fiber production is growing by 2–15% per year, filament yarn production is stagnating or declining. This is mainly due to the fact that staple fibers are used for fashionable clothing, while filament yarns compete with the much cheaper polyester yarns.

With a total production capacity of around 8.5mt, average capacity utilization is therefore around 82%. There is considerable overcapacity in China in particular.

41.5.1 Manufacturers

Table 41.9 shows selected producers of viscose, lyocell and cupro fibers. The market for viscose production is dominated by China (60% market share). However, there are also important producers in Germany and Austria, especially for special fibers.

41.5.2 Areas of Application

Regenerated cellulose fibers are used in pure form and in blends for the production of linings, outerwear and underwear, decorative fabrics, nonwovens, floor coverings, sewing and embroidery yarns, hygiene textiles, car seat covers and interiors, sealing materials, base fabrics and nonwovens for rubberizing and coating, tire cord and interlinings (Fig. 41.52).

Viscose is also used for velvet of film cartridges and case liners as well as jewellery boxes, ballet shoe satin, suitcase linings, cleanroom clothing, wound dressings and much more.

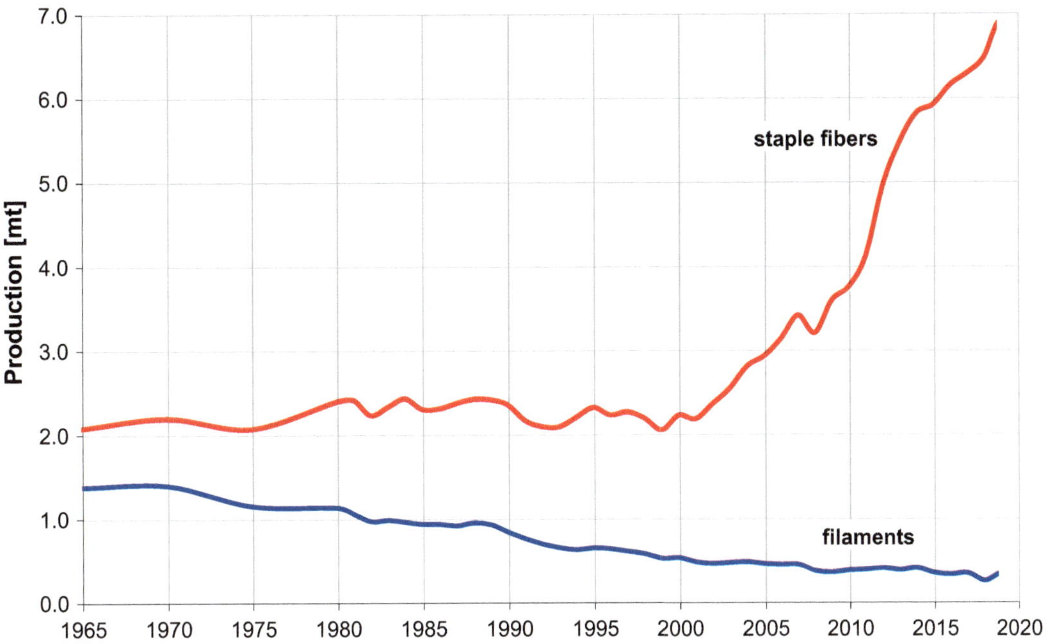

Fig. 41.51 Production of viscose and lyocell fibers (filaments and staple fibers, data: The Fiber Year, 2021)

41.5.3 Cost Trends

The cost of producing regenerated cellulose fibers varies widely between companies and producing countries. Both the availability of cheap raw material and the operating costs (energy, water) play a major role. The world market price for 1 t of pulp in 2020 is around €800. By comparison, 4 t of crude oil are needed to produce 1 t of polyester, which currently costs around €1000. Thus, the raw material basis for viscose fibers is cheaper than for polyester, but due to the higher process costs, the prices for viscose staple fibers are about € 1.20/kg, which is about 25% higher than for polyester staple fibers. For this reason, cellulosic man-made fibers are often marketed as particularly sustainable and environmentally friendly, although this is questionable, at least for viscose fibers.

The higher personnel costs due to the wet-spinning process compared with other man-made fibers can also be significant. Due to the high wastewater load in processes that do not recycle the auxiliary materials, the environmental regulations in Western European countries are also very high.

Nevertheless, viscose is still produced in Germany and 150,000 t of cellulosic man-made fibers were produced in 2019. That is 30% of the total production of man-made fibers. The most important producer is Enka GmbH & Co KG, which is based in Wuppertal and has a production site in Obernburg. Other producers in Germany are Cordenka GmbH & Co KG, Glanzstoff Industries, Kelheim Fibers and Lenzing AG in Austria (Lyocell).

Because the raw material cellulose is available in almost unlimited quantities and relatively inexpensively, and because viscose yarns have advantages over yarns made of polyester and polyamide, particularly with regard to gloss and uniformity of dye absorption as well as color brilliance, cellulosic regenerated fibers have so far held their own against the cheaper competition from polyester and polyamide due to the more economical manufacturing process.

Against the background of demands for greater sustainability and a circular economy

Table 41.9 Selected producers of viscose, modal, cupro and lyocell fibers

Producer	Trade mark	Fiber type	Website	Country
Asahi Kasai	Bemberg™	Cupro	https://www.asahi-kasei.co.jp/fibers/en/bemberg/	Japan
Asia Pacific Rayon Limited		Viscose	https://www.aprayon.com/en/	Indonesia
Birla Cellulose		Viscose	http://www.birlacellulose.com	India
Chengdu Grace Fiber Co.,Ltd		Viscose	https://cdhm.en.ecplaza.net/	China
Cordenka GmbH & Co. KG	Cordenka™		http://www.cordenka.com	Germany
Enka International GmbH & Co. KG	Enka™		http://www.enka.de	Germany
Glanzstoff Industries GmbH	Viscord™	Viscose	http://www.glanzstoff.com	Germany
Hi-Tech Fiber Group Corp.	Oricell™	Lyocell	http://www.htfiber.com	China
Jiangsu Aoyang Technology Co., Ltd.		Viscose	http://www.aoyang.com	China
Kelheim Fibres GmbH	Danufil™	Viscose	http://www.kelheim-fibres.com	Germany
	Galaxy™			
	Viloft™			
	Viseta™			
	Outlast™			
Lenzing AG	Tencel™ Lyocell Tencel™ Modal Veocel™	Lyocell Modal Lyocell	https://www.lenzing.com	Austria
Sateri	EcoCosy™	Viscose	http://www.sateri.com	China
		Lyocell		
Shandong Silver Hawk Chemical Fibre Co., Ltd.		Viscose	http://www.silverhawkfiber.com	China
Smartfiber AG	SeaCell™ LT SeaCell™ MT smartcel™	Lyocell Modal Lyocell	http://www.smartfiber.de	Germany
Tangshan Sanyou GroupXingda Chemical Fibre Co.Ltd		Viscose	http://www.ts-sanyou.com.cn	China
Xinjiang Zhongtai Chemical Co., Ltd.		Viscose	http://www.zhongtaichem.cn	China

that does not require petroleum, cellulosic man-made fibers are an interesting alternative to synthetic man-made fibers.

41.6 Derivative Fibers

If the structure of the cellulose is altered by derivatization and it is not possible to reverse this, the fibers are referred to as derivative fibers. Acetate fibers and nitrocellulose belong to this group.

41.6.1 Acetate Fibers

Acc. to DIN 60001, acetate fibers consist of cellulose acetate with less than 92% but at least 74% by mass of acetylated hydroxyl groups. Triacetate fibers consist of cellulose acetate with more than 92% mass fractions of acetylated hydroxyl groups.

By saponification, the so-called 2½-acetate (alternating 2 or 3 acetate groups per glucose unit) and diacetate (2 acetate groups per glucose

Fig. 41.52 Typical applications of cellulose fibers (Lenzing AG)

unit) with slightly different properties can be produced from triacetate. All types of fibers are produced mainly by the dry-spinning process.

Widely used solvents for acetate fibers are acetone and methylene chloride/alcohol mixtures.

41.6.1.1 History

The first person to produce acetylated cellulose was Paul Schützenberger in 1865. At the beginning of the twentieth century, acetylated cellulose was made soluble in common solvents by partial hydrolysis. It was then used on a large scale as a varnish. At the end of the First World War, there was overcapacity for acetyl cellulose in England. They formed the basis for the industrial production of corresponding fibers (Fig. 41.53).

As early as 1914, there was a small production of filament yarns from completely acetylated cellulose in the USA, but it was discontinued soon after. Chloroform was used as the solvent. In the 1950s, the production of acetate fibers was started again, because a usable solvent was available in the form of methylene chloride.

However, with the advent of synthetic fibers, which can be produced more cheaply and have similar properties, the short-term boom quickly came to an end. Today, acetate fibers are mainly used in the form of fiber cables for cigarette filters.

41.6.1.2 Chemical Basics

Acetate fibers represent a transition between regenerated fibers and synthetic fibers, both chemically and in terms of their properties. In them, the cellulose is permanently altered, because it is also present in the final product in an acetylated state.

If all three hydroxyl groups of each glucose residue of cellulose are acetylated, it is called cellulose triacetate. It is only soluble in chlorinated hydrocarbons. If on average only 2.4–2.6 hydroxyl groups are esterified per

Fig. 41.53 Acetate production (ca. 1950) and spinneret

glucose unit, the acetate dissolves even in simple polar solvents such as acetone. This polymer, known as 2½-acetate, is the raw material for most acetate fibers. In the production of 2½--acetate, the cellulose is first completely acetylated and then partially hydrolyzed.

41.6.1.3 Manufacture of Raw Materials

The starting products for the acetate fibers are cotton linters and pulps containing 99% cellulose. This is necessary because the further process does not include a stage in which hemicelluloses can be removed, which would affect the fiber properties. If necessary, the comminuted pulp is first boiled ("digested") under pressure in caustic soda solution. This corresponds to pre-maturation in the viscose process and serves to make the cellulose more accessible for subsequent acetylation by swelling.

The cellulose pre-treated in this way is placed in a well-cooled stirred vessel containing the acetylation mixture of glacial acetic acid, acetic anhydride and sulfuric acid (as catalyst). Initially, sulfuric acid esters are formed, which gradually change into acetester groups by transesterification (Fig. 41.54).

Fig. 41.54 Formation of triacetate from cellulose by addition of acetic acid

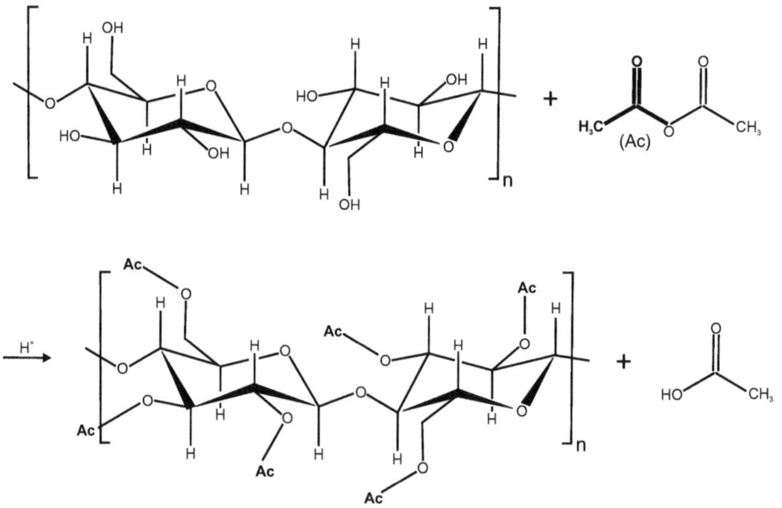

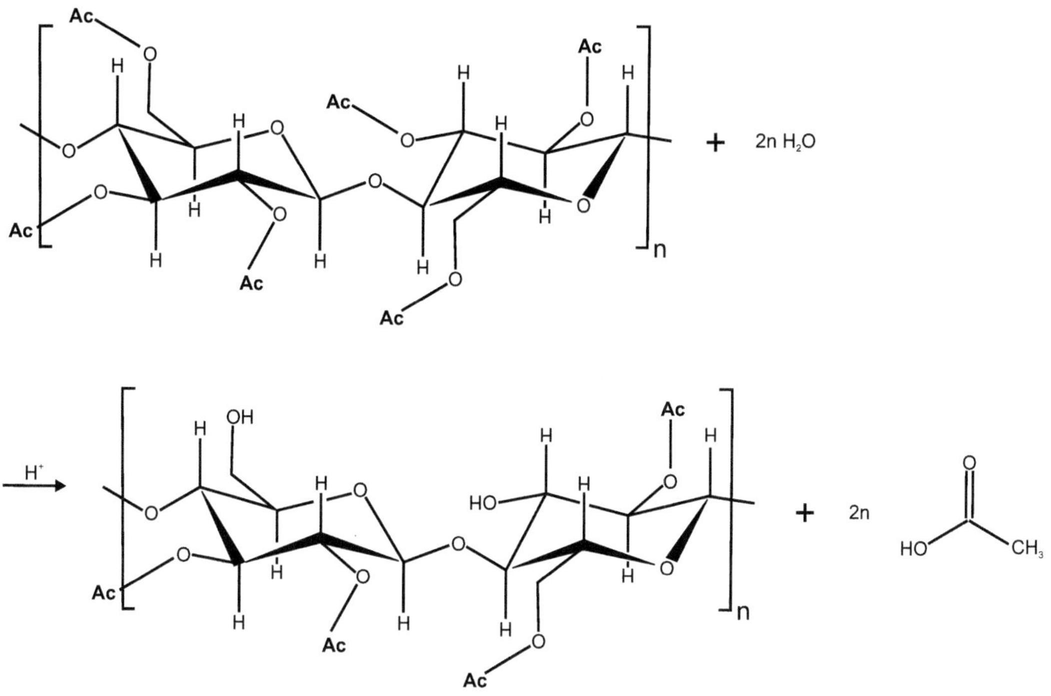

Fig. 41.55 Formation of diacetate from triacetate

The highly exothermic reaction under the action of catalysts produces large quantities of acetic acid. Methylene chloride can be added for cooling, the heat of evaporation of which cools the mass uniformly. When all the pulp is dissolved after approx. 8 h, the rest of the acetic anhydride is destroyed with water.

If the triester is to be converted to 2½-acetate, a precisely calculated amount of water is added, which reduces the acetyl group content by hydrolysis to an average of 2.5 per glucose unit. This process takes up to 20 h at a temperature of 60–80 °C. In the next process step, the reaction product is intensively mixed with diluted acetic acid in a very rapidly stirred vessel. The cellulose acetate precipitates as flakes. These are washed and dried after a squeezing process (Fig. 41.55).

During pretreatment and acetylation of the cellulose and during hydrolysis, its degree of polymerization (DP) is considerably reduced, in the spinning polymers it is only 200–350. An increase of the DP to values of 400–450 leads to a higher strength as well as better abrasion and fatigue bending strength. However, this is associated with an increase in viscosity, which makes spinning more difficult. Possible solutions are a reduction of the polymer concentration in the solvent from 25% to 20% or an increase of pressure and temperature during spinning out.

41.6.1.4 Fiber Production

Filament yarns and staple fibers made of cellulose acetate are generally produced by the dry spinning process (Fig. 41.56). 2½-acetate is dissolved in acetone, to which about 5% water is added to reduce viscosity. Triacetate is dissolved in a mixture of about 90% methylene chloride and 10% methanol. The concentration of acetyl cellulose in the spinning solution is 20–30%, depending on the degree of polymerization. If required, TiO_2 for matting, colorants or carbon black are added to the spinning solution. For spinning, the spinning solution must be filtered very carefully. Before and in the spinning head, it is heated to just below the boiling point of the solvent. Spinning speeds in modern plants reach up to 1000 m/min. Typical filament linear densities are 1.7–8 dtex.

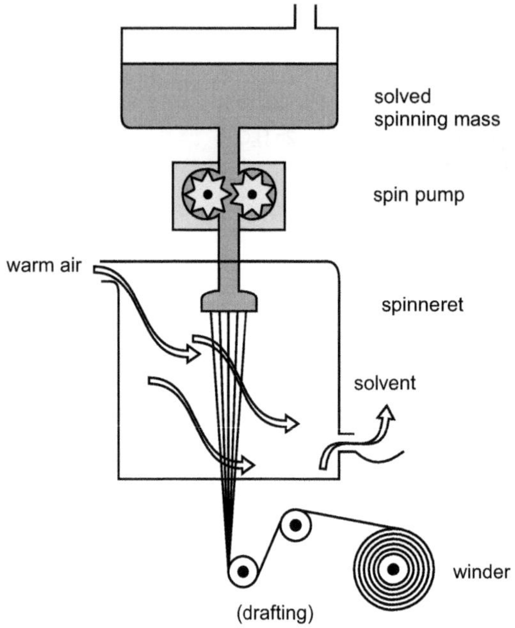

Fig. 41.56 Dry spinning process

Filament yarns are lubricated and wound up below the spinning chutes. In most cases, no further drawing is carried out, as the spinning draft already provides sufficient orientation. If the yarns are made plastically deformable by a temperature treatment, further drafting is possible (Fig. 41.56).

Spun fibers or cables made of triacetate can also be produced directly from the acetylation product using the wet spinning process. The precipitation bath then usually contains diluted acetic acid (Fig. 41.57).

41.6.1.5 Properties and Areas of Application

Table 41.10 contains typical properties of commercially available acetate and triacetate fibers. Compared with other cellulosic fibers, their strength is very low both dry and wet, while elongation is high. The water retention capacity and moisture absorption are low.

The low abrasion resistance and fatigue bending strength of acetate fibers are problematic for use as a clothing material. Their electrostatic charge is also very high due to the polar acetate groups, which makes further processing difficult.

Acetate fibers are sensitive to alkalis, but triacetate much less so. Hot strong alkalis dissolve both. Dilute acids have little effect, strong ones decompose the fibers. Disperse dyes are used to dye acetate products, although dyeing aids may have to be added.

The macromolecules in acetate fibers are not very oriented and also not very crystalline. In the case of triacetate, the degree of order of the crystalline regions can be considerably increased by heat treatment at 230 °C under tension. This is associated with better heat resistance and greater color fastness, provided that dyeing has been carried out prior to heat treatment. Above all, this means that triacetate sheet structures can be fixed and are therefore easy to maintain (Fig. 41.58).

In the clothing sector, they are mainly used for ladies' outerwear, linings and in narrow weaving. In home textiles, they are used for curtain fabrics

Fig. 41.57 Longitudinal and cross-sectional view of 2–1/2 acetate fibers

Table 41.10 Typical properties of acetate and triacetate fibers acc. to manufacturer's specifications

		Acetate	Triacetate
Tenacity	[cN/tex]	10–15	10–15
Wet strength ratio	[%]	50–80	50–80
Maximum elongation	[%]	20–45	20–45
Wet maximum elongation ratio	[%]	120–150	120–150
E-modulus, $\varepsilon = 100\%$	[cN/tex]	200–350	200–350
Wet E-modulus, $\varepsilon = 100\%$	[cN/tex]		
Moisture absorption in normal climate	[%]	6–7	2–5
Water retention	[%]	20–28	10–18
Hot water shrinkage	[%]		5–20
LOI		18–19	18–19
Spec. heat capacity	[kJ/kg K]	1.3–1.5	1.5
Heat conductivity	[W/m K]	0.3	0.3
Glass temperature	[°C]		170–180
Density	[g/cm^3]	1.29–1.33	1.29–1.33

and bedspreads. Where ease of care is important, triacetate is preferred. Cellulose acetate almost completely dominates the cigarette filter market because of its good adsorption capacity and low swelling.

41.6.1.6 Manufacturers and Market

The annual production of acetate fibers is 900,000 t. The main producer since 2009 has been China, after the USA and the EU had previously been the leaders. Production in Mexico was completely discontinued in 2018. Production in Germany, on the other hand, is to be further expanded for the European market by 2025 (Fig. 41.59).

Selected manufacturers and their brand names are compiled in Table 41.11.

41.6.2 Nitrocellulose

Cellulose is soluble in a mixture of concentrated nitric and sulfuric acid. The resulting nitrocellulose is suitable as an explosive and is also known as gun cotton (Fig. 41.60). It can be spun into films or fibers by the dry-spinning process or by introduction into an aqueous precipitation bath. Count Hilaire de Chardonnet first produced a synthetic fiber in this way in 1884. Chardonnet silk, which was initially very popular, lost its appeal, however, after a number of involuntary self-immolations of ballerinas wrapped in Chardonnet robes.

Jules Verne used nitrocellulose as fuel for the moon rocket in his novel "De la Terre à la Lune"

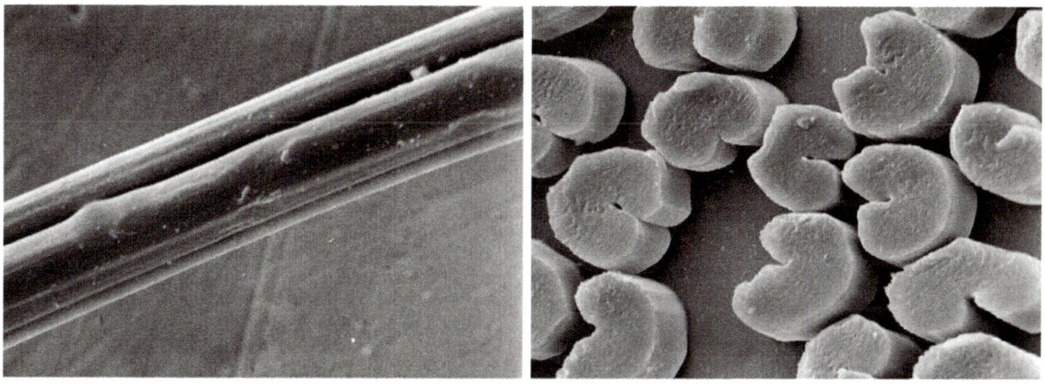

Fig. 41.58 Triacetate fibers in longitudinal and transversal view

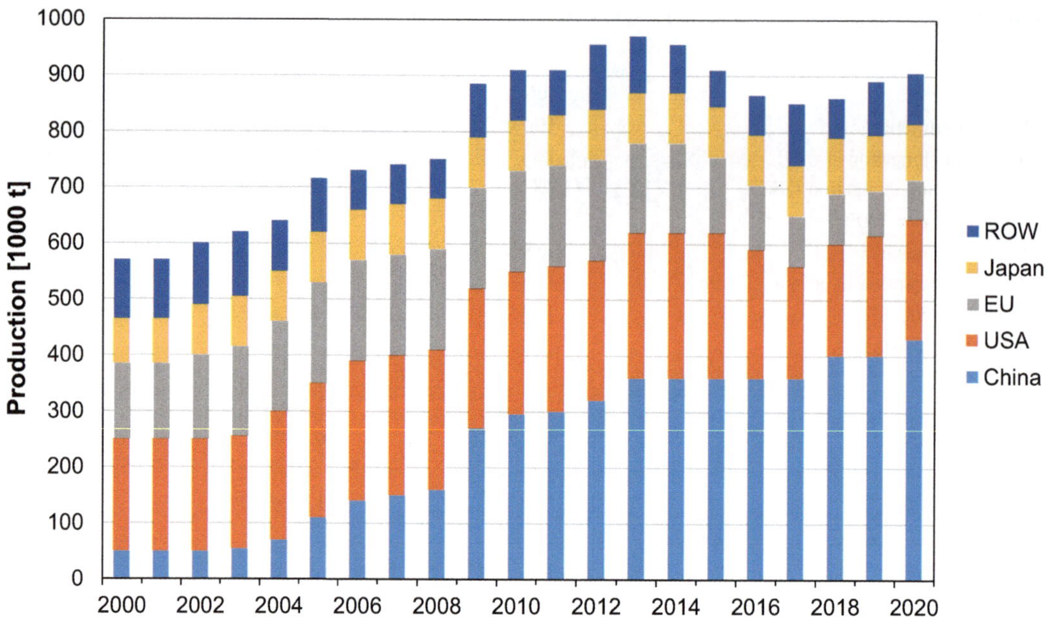

Fig. 41.59 Development of acetate fiber production (data: The Fiber Year, 2021)

Table 41.11 Selected manufacturers of acetate fibers

Producer	Trade mark	Fiber type	Website	Country
Celanese	BlueRidge™	Acetat	http://www.celanese.com	USA
Cerdia International GmbH	Cerdia™	Acetat	http://www.cerdia.com	Switzerland
Daicel Corp.	CAFBLO™	Acetat	http://www.daicel.com	Japan
Eastman Chemical Company	Estron™	Acetat	http://www.eastman.com	USA
Mitsubishi Chemical	Lynda™ Soalon™	DiacetatTriacetat	https://www.m-chemical.co.jp/	Japan

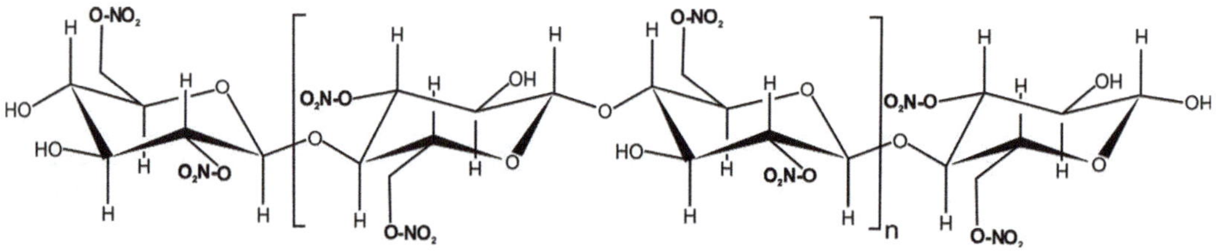

Fig. 41.60 Structure of nitrocellulose

(Fig. 41.61). However, because of the large volume required for this, only one outward flight was possible.

Another product made from nitrocellulose is cellulose dinitrate, which was marketed under the name celluloid. It was developed by Hyat as early as 1869 and used for the manufacture of a wide variety of objects because of its good formability and dyeability. Its use as a photographic base material is noteworthy. However, it was highly flammable, which led to numerous cinema fires. Today it is still used to a small extent for spectacle frames and artificial horn objects.

Fig. 41.61 Moon rocket with nitrocellulose as propellant (Montaut, 1865)

41.7 Nanocellulose

Wood and other plant structures can be regarded as biological fiber composites. They consist of almost equal parts of a largely amorphous matrix, which is mainly composed of hemicellulose and lignin, and cellulose, half of which is in crystalline or amorphous form. The cellulosic microfibrils consist of several cellulose chains and have a diameter of a few nanometers. Therefore, they are also called CNF (cellulose nano fibrils). Because of the many hydroxyl groups, cellulose fibrils tend to agglomerate, which can have a negative effect on the mechanical properties (Dufresne, 2013). As the size of the microfibrils decreases, this effect increases due to the relatively larger surface area.

Cellulose microfibrils can be obtained, among others, by a purely mechanical or by a chemical treatment, electrospinning of a cellulosic solution is also possible. As a result of mechanical shearing processes through a treatment of wood under high pressure (30–150 MPa), the so-called microfibrillated cellulose (MFC) is formed. In order to reduce the amount of energy required for extraction, which ranges from 12–70 kWh/kg (Blanco et al., 2018), depending on the process, the raw material is often pre-treated, e.g. by shredding (Herrick et al., 1983), acid hydrolysis, enzymatic treatment, etc. Microfibrillated cellulose is usually present in an aqueous suspension in a gel-like form with a fiber content of about 2%. Higher fiber contents lead to an increased viscosity, which makes it much more difficult to transport the suspension by pumps. Usually, the MFC have a length of approx. 1 μm and a diameter of 3–100 nm (Dufresne, 2013; Klemm, 2011).

Alternatively, acid hydrolysis with hydrochloric or sulfuric acid dissolves the amorphous regions of cellulosic fibers and exposes and separates the microfibrils in the crystalline regions. The cellulosic nanocrystals (CNC) thus produced are then present in an aqueous solution. Their length is usually 100–250 nm with diameters of 5–70 nm. Nanocellulose can also be produced from low molecular weight sugars using bacteria, e.g. cluconacetobacter. They are referred to as BNC (bacterial nanocellulose) (Blanco et al., 2018), Fig. 41.62.

The mechanical properties of cellulosic microfibrils are exceptional. For example, the modulus of elasticity is approx. 130–250 GPa with a strength of approx. 7500 MPa, which corresponds to approx. 500 cN/tex at a density

Fig. 41.62 Nanocellulose (Wiki, 2021)

of approx.1.5 g/cm^3. If fibers are produced from CNF by gel spinning, typical values for strength are 600–1400 MPa (40–90 cN/tex) and for Young's modulus 60–80 GPa. The maximum fiber length was 840 nm (Mittal et al., 2018). They could be used as reinforcing material of composites (Zimmermann et al., 2010).

Due to the high number of hydroxyl groups in combination with a relatively large surface area of the microfibrils, they can be easily chemically modified. In experimental scale, nanocellulose fibers have been mainly used as reinforcement of other materials, such as collagen, which has good biocompatibility but rather poor mechanical properties (Rodrigues et al., 2017). Reinforcement of composites with CNF is explained in detail in Omran et al. (2021) and in Newman and Staiger (2008). Films for packaging (Duchemin et al., 2009) and medical implants have also been reinforced with CNF. The preparation of aerogels for biomedical applications using CNF are described in Khalil et al. (2020). Applications for thermal insulation products investigated in Gupta et al. (2018). A comprehensive review of the current state of research is given in Kumar et al. (2021).

The Fig. 41.63 shows the number of publications in the Web of Science on the topic of nanocellulose. Since 2006, when there were 2 publications on the topic, the number has increased exponentially to almost 1400 in 2020, showing the extremely high interest in this "new" material.

The price per kg of CNF is about 7–12 € (Chauve and Bras, 2014), which makes the material still too expensive for many applications. World production is estimated at around 500 t (Isogai, 2020).

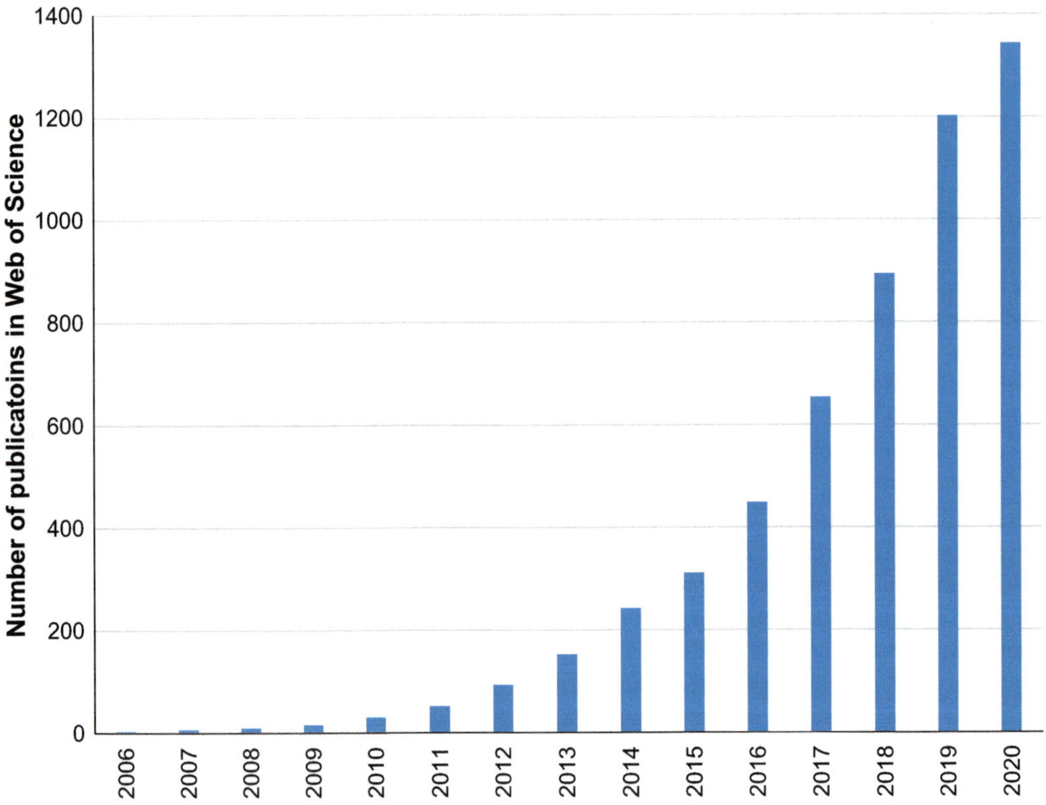

Fig. 41.63 Number of publications on nanocellulose in the Web of Science

References

Becker, F. (1912). *Die Kunstseide*. Reprint by Salzwasser-Verlag GmbH, Paderborn.

Bianchi, E., Ciferri, A., Conio, G., & Tealdi, A. (1989). Fiber formation from liquid-crystalline precursors. II. Cellulose in N.N-dimethylacetamide-lithium chloride. *Journal of Polymer Science B. Polymer Physics, 27*, 1477–1484.

Blanco, A., Concepcion Monte, M., Campano, C., Balea, A., Merayo, N., & Negro, C. (2018). Nanocellulose for industrial use: Cellulose Nanofibers (CNF), Cellulose Nanocrystals (CND), and Bacterial Cellulose (BC). In *Handbook of nanomaterials for industrial applications* (pp. 74–126). Elsevier, Amsterdam.

Bodenbender, H. G. (1943). *Zellwolle – Kunstspinnfasern*. Chemisch-technischer Verlag Dr. Gustav Bodenbender, Berlin.

Bronnert, E. (1899). *Verfahren zur Herstellung von hochprozentigen Lösungen von Zellulose in konzentrierter Chlorzinklösung*. Deutsches Reichspatent 118836, Berlin.

Chauve, G., & Bras, J. (2014). Industrial point of view of nanocellulose materials and their possible applications. In *Handbook of green materials* (pp. 233–252). World Scientific, Singapur.

Ciechańska, D., Wawro, D., Stęplewski, W., Kaszimierczak, J., & Struszczyk, H. (2005). Formation of fibres from bio-modified cellulose pulp. *Fibres and Textiles in Eastern Europe, 13*(6), 19.

Ciechańska, D., Wawro, D., Stęplewski, W., Wesolowska, E., Vehviläinen, M., Nousiainen, P., Kamppuri, T., Hroch, Z., Sandak, J., Janicki, J., Włochowicz, A., Rom, M., & Kovalainen, A. (2007). Ecological method of manufacture of the zellulose fibres for advanced technical products. In *Edana Conference*, Nonwovens Research Academy, Leeds.

Cordenka. http://www.cordenka.com

Duchemin, B. J. C., Newman, R. H., & Staiger, M. P. (2009). Structure–property relationship of all-cellulose composites. *Composites Science and Technology, 69*, 1225–1230.

Dufresne, A. (2013). Nanozellulose: A new ageless bionanomaterial. *Materials Today, 16*, 220–227.

FAO. (2021). http://fao.org

Fink, H.-P., Ebeling, H., & Rihm, R. (2006). Fibre formation from liquid crystalline solutions of cellulose carbamate in N-methylmorpholine-N-oxide. *Proceedings of the 7th International Symposium* "Alternative Zellulose – Manufacturing, Forming, Properties", Rudolstadt.

Götze, K. (1967). *Chemiefasern nach dem Viskoseverfahren*. Springer.

Gupta, P., Singht, B., Agrawal, A. K., & Maji, P. K. (2018). Low density and high strength nanofibrillated cellulose aerogel for thermal insulation application. *Materials & Design, 158*, 224–236.

Herrick, F. W., Casebier, R. L., Hamilton, J. K., & Sandberg, K. R. (1983). Microfibrillated cellulose: Morphology and accessibility. *Journal of Applied Polymer Science, 37*, 797–813.

Hill, J. W., & Jacobson, R. A. (1938). US Patent 2,134,825.

Hooke, R. (1665). *Micrographia*. Royal Society.

Ing. A. Maurer SA, Ittigen, Schweiz. https://maurer-sa.ch

Isogai, A. l. (2020). Cellulose nanofibers: Recent progress and future prospects. *Journal of Fiber Science and Technology, 76*(10), 310–326.

Kelheim Fibres. (2021). http://www.kelheim-fibres.com

Khalil, A., Adnan, A. S., Yahya, E. B., Olaiya, N. G., Safrida, S., Hossain S., Balakrishnan, V., Gopakumar, D. A., Abdullar, C. K., Oyekanmi, A. A., & Pasquini, D. (2020). A review on plant cellulose nanofibre-based aerogels for biomedical applications. *Polymers, 12*, 1759–1784.

Klemm, D., Heublein, B., Fink, H.-P., & Bohn, A. (2005). Cellulose: Fascinating biopolymer and sustainable raw material. *Angewandte Chemie Int. Ed, 44*, 3358–3393.

Klemm, D., Kramer, F., Moritz, S., Lindström, T., Ankerforts, M., Gray, D., & Dorris, A. (2011). Nanocelluloses: A new family of nature-based materials. *Angewandte Chemie Int. Ed, 50*, 5438–5466.

Kosan, B., Michels, C., & Meister, F. (2008). Dissolution and forming of cellulose with ionic liquids. *Zellulose, 15*, 59–66.

Kumar, R., Rai, B., Gahlyan, S., & Kumar, G. (2021). A comprehensive review on production, surface modification and characterization of nanocellulose derived from biomass and its commercial applications. *eXPRESS Polymer Letters, 15*(2), 104–120.

Laszkiewicz, B., Czarnecki, P., Kulpinski, P., Niekraszewicz, B., & Rubacha, M. (2006). Three generations of the Lyocell type fibres. In: *Proceedings of the 7th international symposium alternative cellulose – Manufacturing, forming, properties*, Rudolstadt.

Laus, G., Bentivoglio, G., Schottenberger, H., Kahlenberg, V., Kopacka, H., Röder, T., & Sixta, H. (2005). Ionic liquids: Current developments, potential and drawbacks for industrial applications. *Lenzinger Berichte, 84*, 71–85.

Lenzing AG, Lenzing, Österreich. https://www.lenzing.com

Michels, C., & Kosan, B. (n.d.). *Das Lyocellverfahren – Gegenwärtige Leistungsgrenzen aus stofflicher und technologischer Sicht*, TITK, Greiz. https://www.yumpu.com/de/document/view/13650926/das-lyocell-verfahren-gegenwartige-leistungsgrenzen-lenzing

Mittal, N., Ansari, F., Gowda, K., Bourzet, V. C., Chen, P., Larsson, P. T., Roth, S. V., Lundell, F., Wagberg, L., Kotov, N. A., & Söderberg, L. D. (2018). Multiscale control of nanocellulose assembly: Transferring remarkable nanoscale fibril mechanics to macroscale fibers. *ACS Nano, 12*, 6378–6388.

Montaut, H. (1865). https://commons.wikimedia.org/wiki/File:%27From_the_Earth_to_the_Moon%27_by_Henri_de_Montaut_38.jpg

Newman, N. H., & Staiger, M. P. (2008). Cellulose nanocomposites. In *Properties and performance of natural-fibre composites*. Woodhead Publishing, Cambridge.

Northolt, M. G., Boerstoel, H., Maatman, H., Huisman, R., Veurink, J., & Elzerman, H. (2001). The structure and properties of cellulose fibres spun from an anisotropic phosphoric acid solution. *Polymer, 42*(19), 8249–8266.

Omran, A. A. B., Mohmmed, A. B. A., Sapuan, S., Ilyas, R. A., Asyraf, M. R. M., Koloor, S. S. R., & Petru, M. (2021). Micro- and Nanocellulose in polymer composite materials: A review. *Polymers, 13*, 231–265.

Pauly, H. (1897). *Verfahren zur Herstellung künstlicher Seide aus in Kupferoxydammoniak gelöster Zellulose*. Deutsches Reichspatent 98642, Reichspatentamt, Berlin.

Rodrigues, A. P. H., de Souza, S. D., Gil, C. S. B., Pereira, F. V., de Oliveira, L. C. A., & Patricio, P. S. D. (2017). Biobased nanocomposites based on collagen, cellulose nanocrystals, and plasticizers. *Journal of Applied Polymer Science, 134*(25), 44954.

Rogowin, Z. (1982). *Chemiefasern*. Thieme, Stuttgart.

Smartfiber AG, Rudolstadt. http://www.smartfiber.de

Stearn, C. H. (1898). *Verfahren zur Herstellung von Fäden, Bogen, Filmen u. dgl. aus Viskose*. Deutsches Reichspatent 108511, Reichspatentamt, Berlin.

Süvern, K. (1907). *Die künstliche Seide. Ihre Herstellung, Eigenschaften und Verwendung unter besonderer Berücksichtigung der Patent –Literatur*. Springer.

The Fiber Year. (2021). *The Fiber Year GmbH, Roggwil.* https://www.thefiberyear.com

Topham, C. J. (1900). *Vorrichtung zur Herstellung künstlicher Fäden aus Zelluloselösung oder dgl. und zur Anordnung der gesponnenen Fäden in Strähnform.* Deutsches Reichspatent 125947, Reichspatentamt, Berlin.

Treiber, E. (1957). *Die Chemie der Pflanzenzellwand.* Springer.

Urbanowski, A. (1996). Zellulose carbamate: A new raw material for the manufacture of zellulosic fibres. *Chemical Fibers International, 46*(4).

Wiki. (1903). https://de.wikipedia.org/wiki/Datei:Oberbruch_glanzstoff_1903.jpg

Wiki. (2021). https://commons.wikimedia.org/wiki/File:Nanozellulose.JPG

Wiki1. https://commons.wikimedia.org/wiki/File:Joseph_Wilson_Swan_(HeidICON_33578).jpg

Wiki2. https://commons.wikimedia.org/wiki/File:Hilaire_de_Chardonnet.jpg

Wiki3. https://de.wikipedia.org/wiki/Datei:Fremery,_Urban,_Bronnert.jpg

Zimmermann, T., Tingaut, P., Eyholzer, C., & Richter, K. (2010). Applications of nanofibrillated cellulose in polymer composites. *Proceedings of the Int. Conference on Nanotechnology for the Forest Products Industry*, Espoo, Finnland.

Part VI

Biopolymers

42 Biopolymers

42.1 Basics

Very different materials are referred to as biopolymers. On the one hand, those that are produced from renewable raw materials, and on the other hand, those that are biodegradable. Thus, all natural fibers except asbestos belong to the biopolymers, they even fulfil both requirements. However, there are also other fibrous materials in the group of biopolymers, e.g. special polyesters that are not produced from petroleum but from sugar cane and potatoes, for example, and yet are not biodegradable. A good overview is provided by Blackburn (2006).

Until the 1930s, all man-made fibers were biopolymers because they were produced from cellulose- or protein-based raw materials. Therefore, society's desire for increased use of biopolymers is nothing "new", but rather a throwback to the recent past. The Fig. 42.1 shows the usual classification of the different types of biopolymers in comparison to standard polymers.

Biopolymers from renewable raw materials are usually produced from proteins, polysaccharides (cellulose, starch) or naturally occurring polyesters.

42.2 Protein Fibers

Protein fibers (azlon fibers) are produced from naturally occurring proteins. These include fibers from casein and milk, soy and corn. They are all abbreviated internationally as PROT. A good overview is given by Zhang et al. (2020).

From 1930–1949 in particular, intensive research was carried out on regenerated protein fibers, As Fig. 42.2 for the number of patent applications shows. Initially, the focus was on the development of substitutes for silk, then for wool. From about 1960 onwards, due to the emergence of synthetic man-made fibers, especially polyamide, and from 1970 onwards also polyester, interest very quickly waned and hardly any patents were applied for.

42.2.1 Casein and Milk Fibers

These fibers are precipitated with the aid of an acid from skimmed milk (low fat content) or the rennet ferment and dissolved in caustic soda. The fibers are produced by wet spinning (Fig. 42.3). As early as 1904, F.E. Todtenhaupt was granted a patent for such a process (Todtenhaupt, 1904). About 4 kg of fibers can be produced from 100 kg of skimmed milk (Whittier & Gould, 1940). Dry spinning was also tried, but was not successful.

In 1937, Snia Viscose began production of Lanital™ in Italy (Anon, 1937), and in the United States, production of Aralac™ began in the 1940s (Fig. 42.4). The chemical composition is naturally similar to that of wool, the main constituents being glumatic acid (22%) and proline and leucine with 11% each. The main difference is the

D. Veit, *Fibers*, https://doi.org/10.1007/978-3-031-15309-9_42

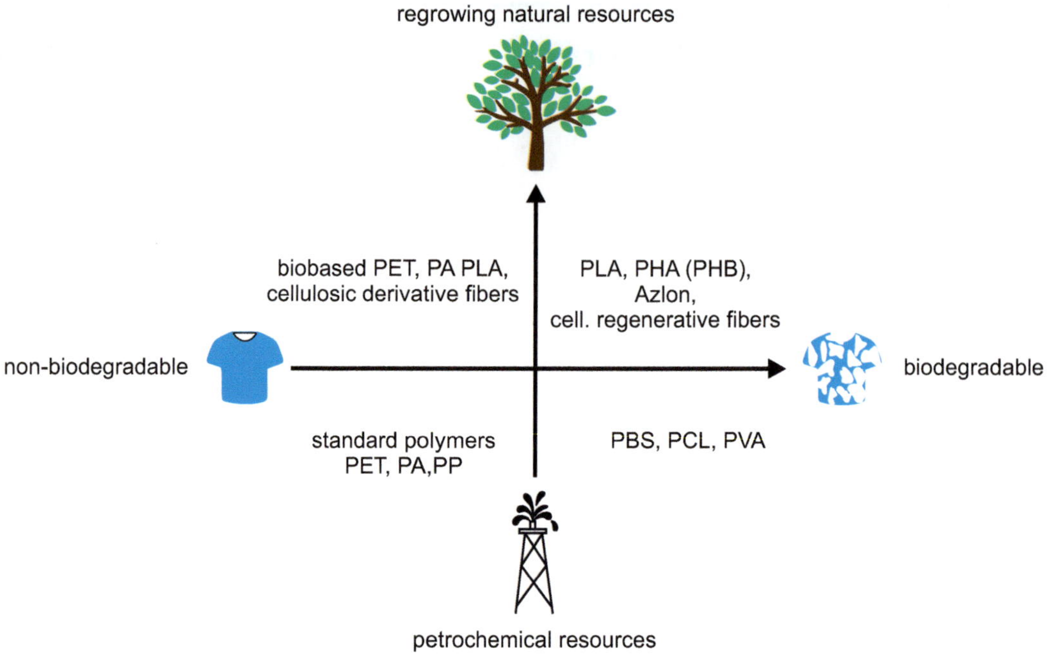

Fig. 42.1 Classification of biopolymers

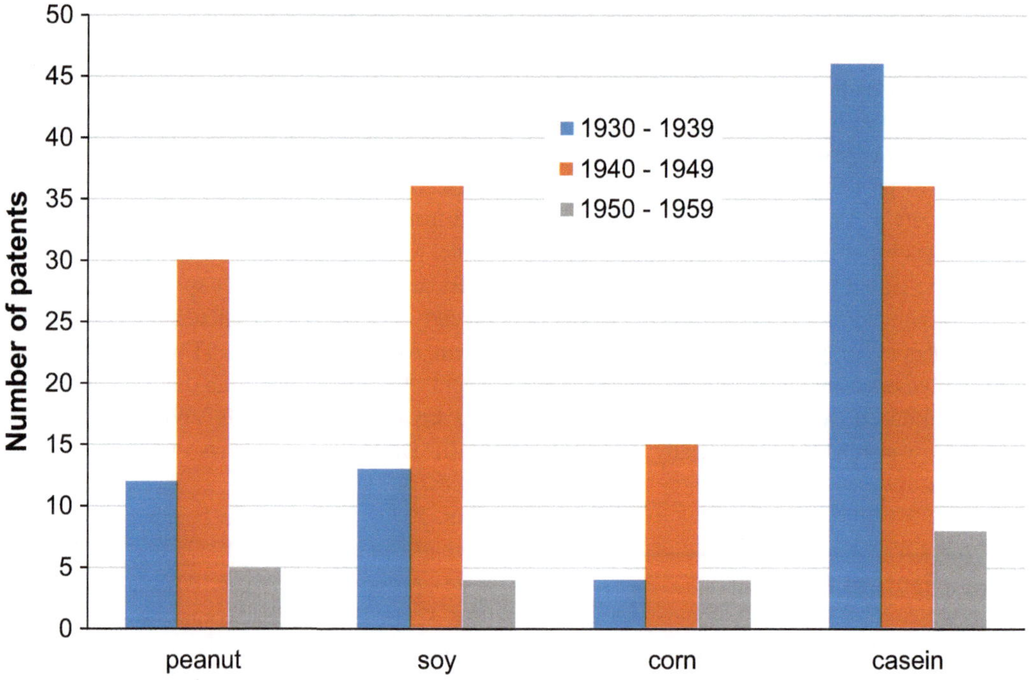

Fig. 42.2 Development of patent applications for protein fibers acc. to (Brooks, 2017)

Fig. 42.3 Take-up of casein fibers from the spinning bath

lower sulfur content compared to wool (0.7% versus 3.6%). Therefore, fewer cystine bridges are formed, which negatively affects the chemical resistance.

With 25–35 cN/tex the strength of wool and silk is achieved. The wet strength is 50–90%, the elongation at break is 25–60% when dry, about twice as high when wet, and the modulus of elasticity is 60–80 cN/tex, which is roughly equivalent to silk. The moisture absorption of 5–8% is somewhat lower than that of wool and silk. The density is 1.22–1.31 g/cm^3. Casein fibers can be washed at 40–60 °C. They have good acid resistance. They have good acid resistance, but dissolve in alkalis, which makes it difficult to use common washing processes. Textiles made from casein fibers have a pleasant feel and a high tendency to crease.

In Australia, casein fibers were blended with wool as a substitute for rabbit hair after the rabbit population was decimated from the 1940s with the help of the myxomatosis virus. This simultaneously reduced production costs (Penfold, 1956). In the USA, Aralac™ was mainly used to make felt hats, and later also clothing textiles (Anon, 1941), Fig. 42.4.

In the 1950s the production of Lanital™ was discontinued, but corresponding fibers were produced and marketed by various companies, e.g. under the brand names Merinova™, Fibrolane™, Casolana™, Caslen™, Cargan™, Lactofil™, Thiolan™ and Silkool™.

Casein fibers are said to have antimicrobial properties, therefore they could be used in the form of nonwoven fabric for hygiene articles. Casein fibers, like all protein fibers, are biocompatible and are therefore suitable in principle for medical textiles (Yang & Reddy, 2012). However, due to their chemical structure, they are attacked by mold and moth larvae (Dirks, 2000).

Assuming an average milk price of 30 ct/kg and that approx. 25 kg of milk are required to produce 1 kg of casein fibers, this results in a raw material price of 7.50 €/kg. If the production costs are added to this, it becomes clear that casein fibers are as expensive as the finest wool.

In China, Zhengjia Milkfiber Sci & Tech Co., Ltd., produces casein-acrylic copolymer fibers with a fineness of 1.5–2.5 dtex and markets them as staple fibers (ring spinning) and filaments (up to 6 dtex) under the name Milkfiber™ (Milkfashion, 2021). They consist of 25–30% casein and 70–75% acrylic. Toyobo (Japan)

Fig. 42.4 Advertising for Aralac™ shirts in the 1950s

produced a similar fiber and the name Chinon™ in the 1970s. QNature UG is currently the only company in Europe producing pure casein fibers on a small scale and markets them under the name QMILK™.

42.2.2 Soy Fibers

Soybeans consist of approx. 36% protein, which is made up of 15–18 different amino acids. Glutamic acid is the most common, accounting for 20% (Brooks, 2005). However, the proteins are spherical and therefore need to be converted into an elongated structure after extraction by adding excipients and enzymes. Then they are spun out wet, stabilized by acetylation, dried, wound up or cut into staple fibers (1–2 dtex). Cross-linking, e.g. with glutaraldehyde, after spinning improves the mechanical properties.

The first patents for this type of process were granted in the 1940s (Kajita & Inoue, 1940, 1946; Huppert, 1945; Boyer, 1940). The Ford Motor Company manufactured seat covers and seat fillings from it from 1939 onwards (Prolon™) with single fiber counts of 1.5–5 dtex, but they were not a commercial success. The same applies to the Silkool™ fiber of the Japanese company Showa Sangyo Kaisha. After the end of World War II, no more soy fibers were produced on a large scale for several decades.

Huang et al. (1995) developed a dry spinning process combining soy protein with maize protein. However, the strength values were far below 10 cN/tex and therefore the fibers were not textile processable. Li (2007) combined soybean protein with PVA and produced fibers with acceptable mechanical properties. Shanghai Winshow Soybean Fiber Industry Co., Ltd., produces about 1500 t of fibers per year under the brand name Winshow™, and Southwest Trading Company produces under the brand name SoySilk™.

The density of soy fiber is 1.29–1.31 g/cm^3, the strength is about 40 cN/tex dry, 25–30 cN/tex wet, elongation at break is about 18–21%, Young's modulus is 500–970 cN/tex, moisture absorption is about 8%. The hot water or hot air shrinkage is 2.2% (Rijavec & Zupin, 2011). They have good acid resistance, moderate alkali resistance and good UV resistance. The stainability depends on the dye, but is generally moderate to good. Above a temperature of 120 °C, the fiber becomes yellow and sticky (Swicofil, 2021). Soy fibers are said to have antibacterial properties and therefore prevent perspiration odor. So far, they have been used in very small quantities and mostly in blends with wool, silk, cotton and viscose (Fig. 42.5).

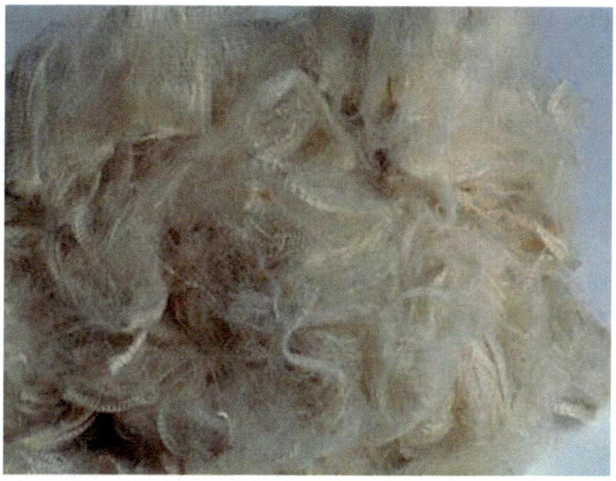

Fig. 42.5 Crimped soy fibers

With a soy price of approx. 30 ct/kg, this results in a raw material price of approx. 83 ct/kg, which is economically competitive.

42.2.3 Corn Fiber

Corn consists of approx. 3% protein. After precipitation of the maize protein, it is solved in caustic soda (protein content 13–16%) and extracted. In the process, the cross-linked macromolecules are denatured so that they can be better oriented later. The fibers, also known as zein fibers, are then spun out in a wet spinning process at a viscosity of 100–200 Pas and coagulate in a precipitation bath (diluted sulfuric acid, acetic acid and zinc sulfate) with the addition of formaldehyde. After a possible hot drawing at 200–300 °C and a washing and drying process, they are wound up. This process was first developed in the USA in 1939 and from 1948–1957 such fibers were marketed under the names Vicara™ (Virginian-Carolina Chemical) and Zycon™ respectively. In Italy and Belgium, zein fibers were sold as Lanital™, in England as Fibrolane™ and in the former GDR as Marinova®.

The fiber strength is approx. 10–12 cN/tex, the elongation at break is 30–45% dry and 25–35% wet, the density is 1.25 g/cm^3 and the moisture absorption reaches values of up to 10%. By subsequent acetylation the strength can be increased by approx. 30%. Zein fibers are

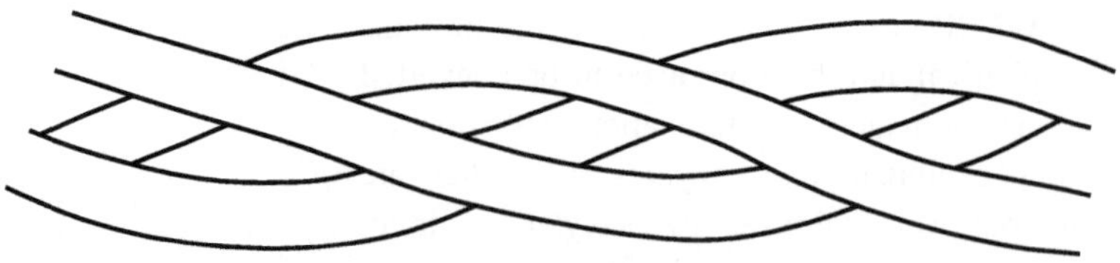

Fig. 42.6 Structure of the collagen triple helix

resistant to both acid and alkali and therefore rot poorly. However, they can be easily dyed with a wide variety of dyestuffs and they have good UV resistance and are stable up to 140 °C. They are usually processed in blends with cotton or viscose fibers. Products in combination with wool and polyamide fibers are also known, e.g. for knitwear (underwear, stockings).

42.2.4 Peanut Fiber

Peanuts consist of about a quarter protein. Therefore, in 1937 ICI in Ardeer (Scotland) developed a synthetic protein fiber from it, which was marketed under the name Ardil™. In parallel, Sarelon™ (Merrifield & Pomes, 1946) was developed in the USA and Fibrolane™ in the UK. Peanut protein was solved in urea or dilute sodium hydroxide solution and spun out wet into a precipitation bath consisting of weakly conc. Sulfuric acid (5%) and sodium sulfate (20%). Several washing and stretching processes follow. Typical strengths of the fibers thus produced were 6–9 cN/tex (wet strength: 25–30%), elongation was 40–60% (dry) and 80% (wet), moisture absorption was good at 12–15% and density was 1.3 g/cm^3. Swelling, at about 50%, was equivalent to that of wool and silk. Because of its low strength, the fiber was processed into textiles in blends with wool, cotton and viscose. Due to the shortage of wool and the great demand for uniform fabrics during World War II, production was quickly expanded. However, production was only economical as long as the price of wool was high and the supply of cheap peanuts was plentiful. When the wool price fell and there were supply shortages of peanuts, production was finally stopped in 1957.

42.2.5 Collagen Fibers

Collagens are long-chain, so-called structural proteins and serve as strength carriers in the connective tissue of most animals. In the human body, they make up 30% of all proteins. Collagens consist of 3 intertwined polypeptide chains and form a triple helix (Fig. 42.6). The chains are essentially composed of 70–300 tripeptides Gly-X-Y. Glycine is the smallest of all amino acids and therefore fits very well as a link in the triple helix, which has very tight turns. At position X there is usually proline, which has a rigid ring structure, forming tight turns in the helix. At position Y there is often 4-hydroxyproline, which forms hydrogen bonds to neighboring polypeptide chains and thus stabilizes the overall structure.

As early as 1895, A. Millar received a patent for the production of the so-called Vandura artificial silk™ from gelatine (Millar, 1895). However, the production was not successful because the fiber completely lost its mechanical strength when it got wet, which made washing very difficult.

However, it has now been possible to produce synthetic collagen fibers on an experimental scale using the wet-spinning process (Siriwardane et al., 2014). Tonndorf et al. (2020) produced wet-spun yarns from 6 filaments with a single fiber fineness of 5 tex and achieved strengths of 12 cN/tex and a Young's modulus of 230 cN/tex. Their wet strength is still very low. The density is 1.32 g/cm^3. Due to their biocompatibility, they could be used for medical purposes, e.g. for the regeneration of nerve cells.

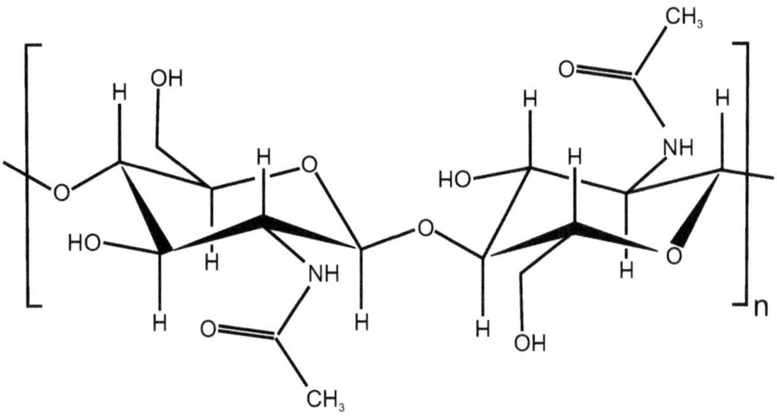

Fig. 42.7 Structural formula of the repeat unit of chitin

42.2.6 Chitin Fibers

Chitin (from Gr.: chitón, engl.: shell, carapace) is the most widely distributed polysaccharide after cellulose. The only difference to cellulose is the acetamide group at C-atom 2 (Fig. 42.7). Chitin serves as a reinforcing material and is found both in fungi (cell wall) and in arthropods, e.g. crustaceans and shrimps.

Attempts to use chitin for fiber production were made as early as the 1920s and 1930s, but without much success. More recent efforts in the 1970s and 1980s were also unsuccessful (Agboh & Qin, 1997).

In early processes, chitin was xanthogenated analogously to viscose and spun out wet in corresponding precipitation baths consisting of sulfuric acid and zinc chloride. Strengths of 6–10 cN/tex were thus obtained. Chitin xanthate can be spun out with viscose xanthate in any mixing ratio as a relatively homogeneous bicomponent fiber, e.g. for medical applications (Noguchi et al., 1978).

A dry spinning process was patented by Capozza (1975). In this process, chitin is solved in hexafluoroisopropyl alcohol with a concentration of 3% polymer. The spinning solution is heated to 55 °C and spun out into a nitrogen atmosphere.

More recent approaches dissolved chitin in dimethylacetamide and LiCl and wet-stretched the solution 5–9% polymer content into a precipitation bath of sulfuric acid and other components. Typical filament fineness was 20–80 μm, strength reached 7–22 cN/tex, Young's modulus 100–550 cN/tex and elongation was 7–75% (Agboh, 1986). The wide variation in these characteristic values shows that the process still has considerable potential for improvement.

The wet strength of chiton fibers decreases rapidly with increasing humidity, dropping to less than 10% of the dry value at 100% humidity. Chitin fibers are biocompatible, non-toxic and are said to support the wound healing process. Therefore, medical sutures and wound dressings have been developed from them.

42.2.7 Chitosan Fibers

Chitosan, also known as poly-D-glucosamine, is a polyaminosaccharide and can be produced from chitin by deacetylase using hot concentrated sodium hydroxide solution at 100–150 °C in the absence of air or enzymatically. The chain length can be adjusted by the reaction conditions and foreign atoms can be introduced into the structure. This leads to a very broad property profile of chitosan. Its structure is very similar to that of cellulose, the only difference being the NH_2 group at C atom 2 (Fig. 42.8).

Fibers made of chitosan are produced for larger quantities by the wet-spinning process at take-off speeds of 20–25 m/min and a chitosan concentration of 6–9%, for smaller quantities,

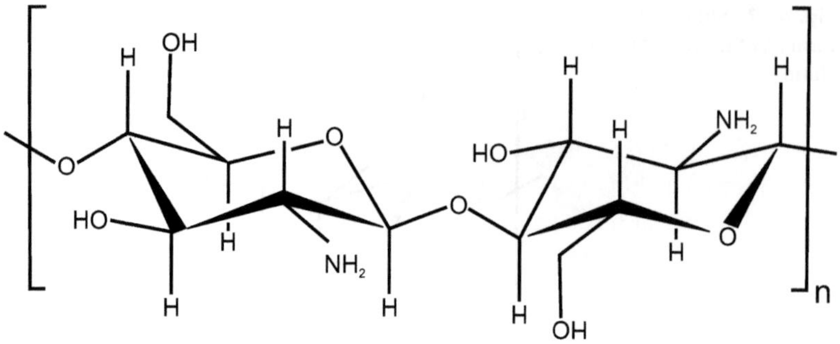

Fig. 42.8 Structural formula of the repeat unit of chitosan

especially for medical textiles, by the electrospinning process. Gel spinning is also possible. Dry spinning is not possible because chitosan can only be dissolved in organic solvents with a high boiling point. The removal of unwanted dyes is achieved by adding $KMnO_4$ or NaOCl (sodium hypochlorite) during the washing process.

Typical strengths are 6–25 cN/tex dry and 5–15 cN/tex wet, elongation reaches 6–20%, Young's modulus is about 820 cN/tex, density is 1.39 g/cm^3 (Agboh & Qin, 1997). Long chain chitosan has low acid resistance and good alkali resistance. As the degree of polymerization decreases, the solubility in alkalis increases. At the same time, the strength and Young's modulus decrease and elongation increases (Maevskaia et al., 2018). Chitosan fibers have a degree of crystallinity of 35–50%.

Due to its free amino groups, it can be positively charged in an acidic environment and thus has an antibacterial and antiviral effect. In addition, chitosan can bind fat very well and could be used for technical filter applications, e.g. in wastewater treatment. Research is also being carried out into wound dressings made from chitosan. There is currently no large-scale production of pure chitosan fibers. However, Chipro GmbH in Bremen offers consulting services for products made of chitosan.

Omikenshi Co, Ltd. (Japan) marketed a fiber blend of viscose and chitosan under the name Crabyon™ (crab: crab shell; yon: rayon) for patients with skin diseases. However, production was discontinued in 2020.

Chitosan is particularly interesting due to its almost inexhaustible raw material base. It is estimated that several billion tons of chitin are produced every year, of which only around 150,000 tons are currently used, mainly from crab and shrimp shells (Swicofil, 2020). Nevertheless, the price of 1 kg of chitin is still far too high for economic use, at around €50–250. Fungi would possibly be a cheaper raw material base, but this is still largely unexplored.

42.2.8 Spider Silk

Spider silk is a natural protein fiber and attempts have been made for some years to produce it synthetically. Information on this can be found in the chapter on silk (cf. Chap. 11).

42.2.9 Fish Protein

The so-called Wikilana fibers were first produced from the dried muscle protein of sea fish by O. Mecheels. However, they were too expensive and the raw material base too small. A combination of 80% cellulose and 20% wikilana resulted in fibers that were resistant to crimping and—compared to other protein-based man-made fibers—had low wear. The tensile strength was 14–17 cN/tex, which was at the lower limit of textile processability. They can be dyed well with acid dyes. However, this "animalized fish cell wool"remained in the experimental stage.

Fig. 42.9 Brown algae (Smartfiber GmbH) and alginate fiber in cross section

42.2.10 Feathers

Feathers have also been investigated for their suitability as a source of keratin for fibers. In the 1940s, tests were carried out at the Western Regional Research Laboratory of the US Department of Agriculture (Traill, 1951). Spinning trials were successful on a small scale, but the unpleasant odor of the fibers prevented a resounding market success (Woodbury, 1945). The keratin obtained from feathers is finer than wool keratin, so it could conceivably be used as a filter material. Water absorption could also be correspondingly high.

42.3 Polysaccharide Fibers

This chapter deals with special polymers on a cellulosic basis. The spinning processes essentially correspond to those of cellulosic regenerated fibers, the composition of the spinning solution and the precipitation baths are material-specific.

42.3.1 Alginate

Alginate for fiber production (Fig. 42.9) can be obtained from the alginic acid of marine algae, especially brown algae. It was discovered by E.C.C. Stanford in 1881 and named "algic cellulose" by him because of its similarity to cellulose. From alginic acid (Fig. 42.10), which builds up the cell walls, is first dissolved in dilute sodium hydroxide solution for 1–2 h at 50–90 °C to obtain sodium alginate and then the fiber is produced with a polymer content of 4–10% in the wet spinning process. Stabilization takes place in metal salt baths, e.g. with common salt (NaCl) or calcium chloride ($CaCl_2$). By adding NaClO (sodium hypochlorite), the fiber can already be bleached during the initial spinning. This spinning process was first used in the 1940s.

Fig. 42.10 Structural formula of alginic acid (left: guluronic acid, right: mannuronic acid)

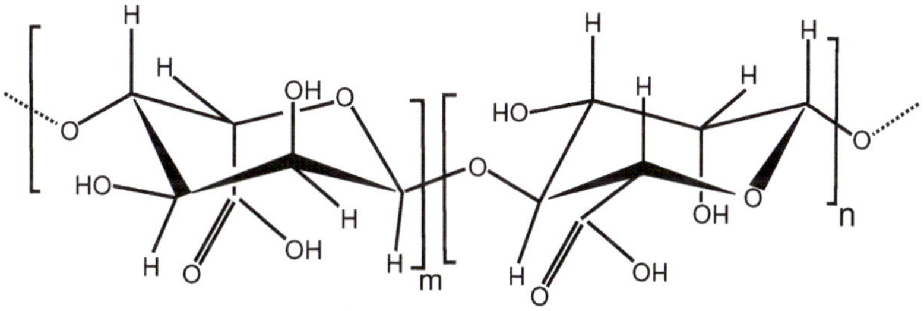

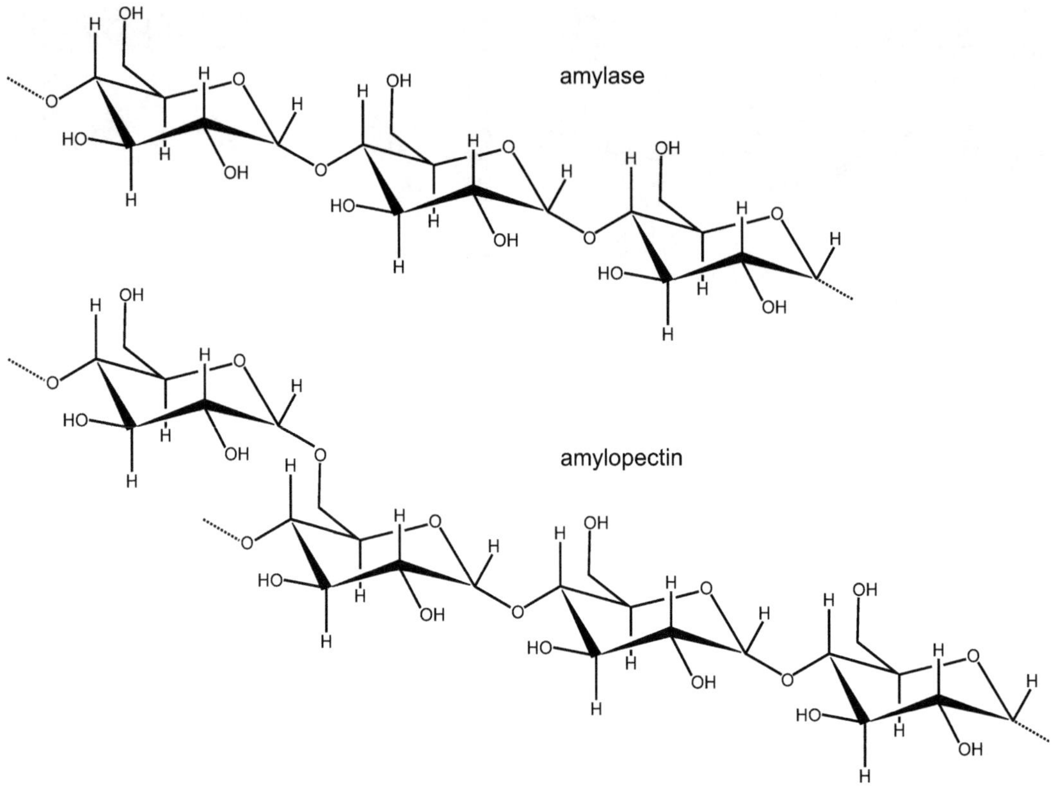

Fig. 42.11 Structural formula of amylose (top) and amylopectin (bottom)

Because they are spun wet, their cross-sectional shape is similar to that of viscose fibers.

Alginic acid is a copolymer of α-L-guluronic acid (G) and β-D-mannuronic acid (M), which form linear chains in 1,4-conformation via a glycosidic linkage. This may also result in homopolymeric regions in which the two components are present as a block. In particular, longer segments of guluronic acid form a zigzag structure. Depending on the part of the alga from which the alginic acid is obtained, different proportions of guluronic and mannuronic acid result. Ratios of 1:2 and 2:3 are common. Extracts from the leaves contain more M-blocks in which calcium is less strongly bound and can therefore be replaced by sodium, e.g. from wound secretions. Therefore, its absorption capacity is more pronounced. If the alginic acid is produced from the algae roots, G-blocks predominate in which the calcium atoms are more strongly bound and the absorption capacity is correspondingly lower. The calcium atoms can be replaced by silver ions in aqueous solution which significantly increases the antimicrobial effect.

Alginate fibers (abbreviation: ALG) have a density of 1.78 g/cm^3, a strength of 11–18 cN/tex with an elongation at break of approx. 5%. Their wet strength is 20–30% of the dry value and the modulus of elasticity is around 700 cN/tex. The moisture absorption is 20–35%. They are soluble in soap solutions and alkaline media. Therefore they can be used e.g. as self-dissolving medical sewing threads or as auxiliary threads for special effects in embroidery.

42.3.2 Starch Fibers

Starch consists of long chain molecules built up from α-D-glucose. A distinction is made between long-chain, unbranched amylose (20–30%) and branched amylopectin with a proportion of 70–80% (Fig. 42.11).

Amylose forms a helical structure, while amylopectin is present as a branched network in the form of granules. Both structures are thus not suitable for fiber formation, which requires long-chain, unbranched molecules. Therefore, attempts were made to modify starch physically or chemically and to melt it. However, this only resulted in sticky filaments that could not be processed into textiles. If PLA is grafted with maleic anhydride to obtain functional groups that improve the interfacial adhesion between PLA and the starch molecules, and then melt-spun in a mixture with up to 70% starch content with the addition of glycerol as a plasticizer, filaments are produced at winding speeds of up to 500 m/min that can be processed into textiles. However, their strength is very low at approx. 3.5 cN/tex, the maximum tensile elongation is 6% and the hot air shrinkage is 35% (Manvi et al., 2017).

It is estimated that around 50mt starch are produced annually, 80% from maize (Globe, 2020) and 1 kg costs 0.3–1.1 €/kg, depending on quality. This would make it an economically interesting alternative raw material base for fibers compared to other biopolymers in terms of the quantity available and the raw material price.

42.3.3 Bamboo

Bamboo plants belong to the family of sweet grasses (Gramineae). It is estimated that there are over 1000 different species. Bamboo is evergreen and has slender, very long culms that can grow up to 40 m long and 80 cm thick. Bamboo grows very fast, up to 1 m within one day. Thus, bamboo plants produce much more wood in the same time than, for example, deciduous trees (Fig. 42.12).

Like many grass species, bamboo has fibers to support the stem. They reach a length of up to 10 mm, are hollow and have a very low lignin content. They are not spinnable, but are used to produce cellulose for regenerated fibers, which are made by the viscose spinning process. Such articles are sold as "made of bamboo", which is misleading.

42.4 Other Biopolymers

In addition to protein- and polysaccharide-based biopolymers, there are others, some of which have recently gained in importance and are summarized in this chapter.

42.4.1 Rubber Threads

Rubber threads can be produced from natural rubber via various conversion processes (mastification, vulcanization). The first usable elastic rubber threads were produced in 1850 by the Mac Intosh company in Great Britain. From 1930 onwards, round, square and grooved rubber threads were produced by spinning out the

Fig. 42.12 Bamboo plant (Heyde, 2007) and bamboo stem fibers

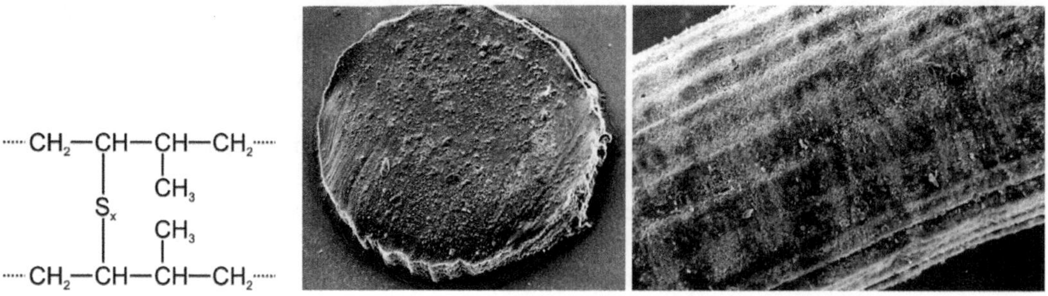

Fig. 42.13 Structural formula of rubber (detail), transverse and longitudinal view of a rubber fiber

masticated rubber mass through spinnerets. After removing the solvent and drying the threads, they were vulcanized. The typical density of such rubber threads is 0.93–1.0 g/cm^3, tensile strength is 2–4 cN/tex, and elongation is 700–900%. These threads have now been almost completely replaced by polyurethane fibers (Fig. 42.13).

42.4.2 Polyvinyl Alcohol Fibers (PVA)

Polyvinyl alcohol fibers (PVA or PVAL) are one of the few thermoplastic polymers that are water soluble. They are often combined with biopolymers and are therefore discussed in this chapter. Their generic name is vinal fibers, but it is rarely used. They were first synthesized from polyvinyl esters by W. Hermann in 1924. Nowadays, they are formed by hydrolysis

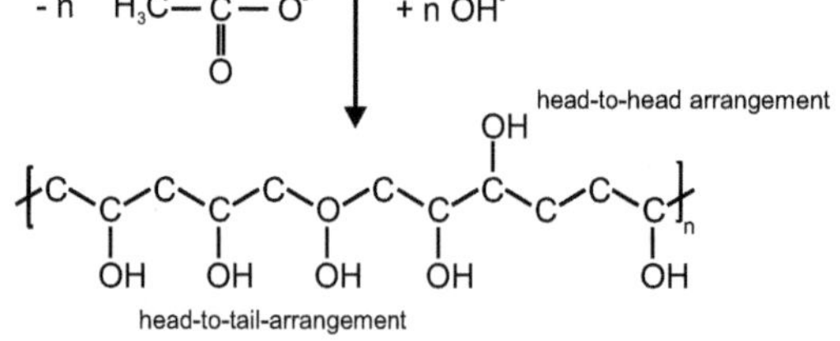

Fig. 42.14 PVA synthesis by saponification of polyvinyl acetate

(saponification) of polyvinyl acetate (Fig. 42.14). A direct synthesis from vinyl alcohol monomers is not possible because of their tautomerism. The DP is 500–2500.

In the PVA chain, 98–99% of the monomers are in head-to-tail arrangement and only 1–2% in head-to-head arrangement. By changing these proportions during hydrolysis, the water solubility of PVA can be altered.

The fibers are spun in a wet-spinning process. The spinneret is made of gold or a gold-platinum alloy and has 2000–60,000 capillary holes with a diameter of approx. 0.1 mm. The take-off speed from the spinneret is approx. 5 m/min. The precipitation bath contains 15–16% sodium sulfate (Na_2SO_4), the bath temperature is 40–50 °C. Often 2 or 3 coagulation baths are used in succession. The PVA filament is then hot-stretched 4–15 times and thermally post-treated at 220–230 °C for 40–180 s. The heat is usually applied via heated coagulation baths. The heat is normally transferred via heated godets or hot air. During this process, the filaments crystallize and are then wound up at a speed of approx. 50 m/min (Fournè, 1995). PVA fibers can be made insoluble in water by cross-linking the hydoxyl groups. In rare cases, PVA fibers are also dry-spun.

The density of PVA fibers is 1.26–1.29 g/cm^3. The typical fiber fineness is 1.5–10 dtex, the strength 20–65 cN/tex (normal type) or 25–80 cN/tex (water insoluble type). The wet strength is 65–85%, the maximum tensile elongation is 10–30% dry and 12–42% wet, the modulus of elasticity is 300–600 cN/tex. The glass transition temperature is in the range of 75–130 °C, the melting temperature is 240–260 °C (water

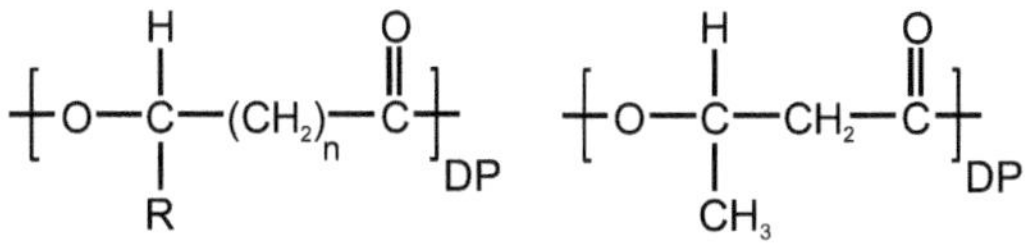

Fig. 42.15 PHA basic structure (left) and PHB structural formula (right)

insoluble grade). For the water insoluble grade, the hot water shrinkage is 2–3%. The LOI is 20.

PVA fibers (normal type) are highly soluble in water, but insoluble in almost all anhydrous organic solvents.

The main textile applications are plastic and cement reinforcement, binder fibers and ropes. The price for 1 kg of PVA fibers is 1.5–3 €, depending on the quality.

In North Korea, PVA fibers are produced under the name Vinalon™. They are also called "Juche fiber" in reference to the Juche ideology. Kuraray, Japan, markets PVA fibers under the brand name Kuralon™.

42.4.3 Polyhydroxyalkanoate Fibers (PHA, PHB)

Polyhydroxyalkanoates (PHA) are naturally occurring, so-called biopolyesters. They are formed by more than 40 different strains of bacteria (Chodák & Blackburn, 2009), are insoluble in water and biodegradable. Figure 42.15 shows the basic structure. Depending on the residue R and number of methylene groups in the repeat unit, different properties result. With $n = 1$ and $R = CH_3$, polyhydroxybutyrate (PHB) is obtained, which is also known as polyhydroxybutyric acid and is used for fiber production. PHB was first described by Lemoigne (1925).

In principle, fibers can be produced from PHB by melt spinning. However, the low thermal stability of PHB means that the resulting fibers degrade quickly. PHB fibers also age quickly and thus become brittle. Crystallization of PHB is very slow, resulting in the formation of large crystallites, which also makes the material very brittle. The addition of nucleating agents, such as boron nitride (Liu et al., 2002), can reduce the crystal size. This reduces the brittleness and increases the drawability. Hot drawing immediately below the spinneret, before large crystals form, can produce fibers with a strength of about 10 cN/tex, Young's modulus of 290 cN/tex and elongation of 95% (Gordeyev & Nekrasov, 1999). Schmack et al. (2000) developed a spinning process in which the spun filaments with a draw of up to 6.9 were wound over two heated godet duos at up to 3500 m/min. This resulted in a strength of 27 cN/tex, an elongation of 37% and a Young's modulus of 630 cN/tex. Their density is 1.2–1.25 g/cm^3. Such fibers can be processed into textiles, but the glass transition temperature of PHB is 4 °C, so that they are already plastically deformable at room temperature.

At present, PHB fibers for textile applications are not produced on a large scale. This is due, on the one hand, to the difficulty of controlling the process and, on the other hand, to the low storability of PHB fibers. Therefore, PHB is only interesting for a few products, e.g. for medical applications. Due to the small quantities, the raw material price of approx. 2–8 €/kg is therefore significantly higher than for standard polyester (PET).

42.4.4 Polyglycolic Acid Fibers (PGA)

Polyglycolic acid is the simplest aliphatic polyester. It is completely degraded hydrolytically in the human body within 2–3 months without harmful degradation products. PGA is produced from glycolide (Fig. 42.16, left), which polymerizes by ring opening.

Its density is 1.53 g/cm^3, its modulus of elasticity approx. 450 cN/tex, its melting temperature is 225–230 °C. Due to its low glass transition

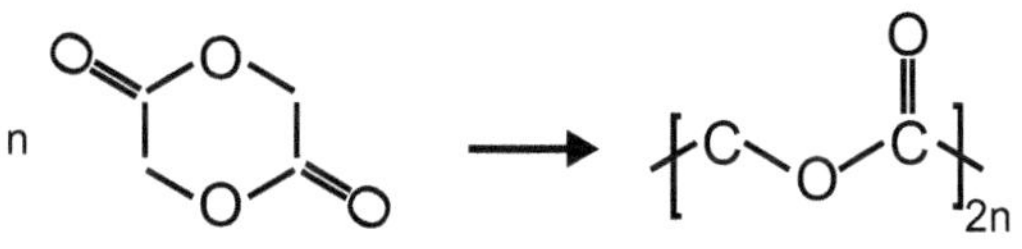

Fig. 42.16 Structural formula of glycolide (left) and repeat unit of PGA (right)

temperature of 35–40 °C, it cannot be used for most textile products. As a copolymer with PLA or PCL, it is used for medical applications, e.g. as absorbable suture material. Dexon™ was developed back in the 1960s and is still sold today.

42.4.5 Polycaprolactone Fibers (PCL)

The polyester polycaprolactone is formed by ring-opening polymerization from ε-caprolactone at 140–150 °C (Fig. 42.17). It is biodegradable anaerobically, but more slowly than PHA. The melting point of PCL is about 60 °C, the glass transition temperature is −72 °C, its density is 1.14 g/cm^3.

Charuchinda et al. (2003) developed a piston melt spinning process in which the polymer is spun out at 85–90 °C and then immediately cooled in a water bath at 5–10 °C. Although the polymer consists of about 50% crystalline regions, it can still be drafted up to 25 times. This indicates that no orientation of the macromolecules takes place during spinning. The drawing increases the strength up to 25 cN/tex, and the elongation reaches 450% at a filament fineness of 5 dtex. The take-up speed is extremely low at 2 m/min. Solution and electrospinning is also possible with PCL.

PCL fibers are characterized by their easy formability and they can be mixed or combined well with other polymers, e.g. with starch and lignin. Due to its low glass transition temperature, it is already soft at room temperature, which makes it unsuitable for textile purposes, but can be advantageous when PCL fibers are used in medical textiles. For such products, it can also be electrospun, resulting in a fiber fineness of 300–900 nm (Zeng et al., 2003). Ethicon, for example, markets sutures made from PCL under the trade name Monocryl™, and Indian manufacturers include Suruglyde™. A comprehensive overview of the manufacture and possible applications of PCL fibers is given in (Azimi et al., 2014).

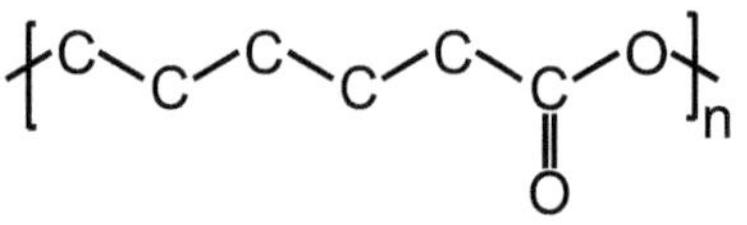

Fig. 42.17 Structural formula of polycaprolactone

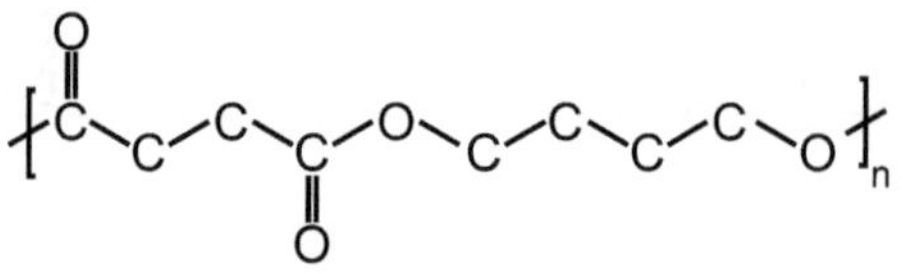

Fig. 42.18 Structural formula of the repeat unit of polybutylene succinate

42.4.6 Other Polymers

There are many other polymers that have been investigated to determine whether they are suitable for textile production. A typical representative of these is PBS (polybutylene succinate), Fig. 42.18.

It has just acceptable mechanical properties in fiber form, e.g. a strength of 6–15 cN/tex with an elongation at break of 60%. The density is about 1.25 g/cm^3 (IFG, 2021). PBS fibers are often advertised as suitable for "high temperature applications". However, their melting temperature is 110–120 °C and their glass transition temperature is −32 °C, hence they are soft and sticky at room temperature. PBS is therefore unsuitable for textile applications, although the material can be used as a binder fiber (Fig. 42.19).

42.5 Interim Summary

For many polymers from which fibers can in principle be produced, the mechanical properties are just about in the same processable range as PBS, but the glass temperatures are frequently much too low and do not reach the minimum required 75–100 °C, which are absolutely essential for heat-setting, washability and general usability in a room climate (Table 42.1).

New biopolymers can only be successful for textile production, if the following conditions are met:

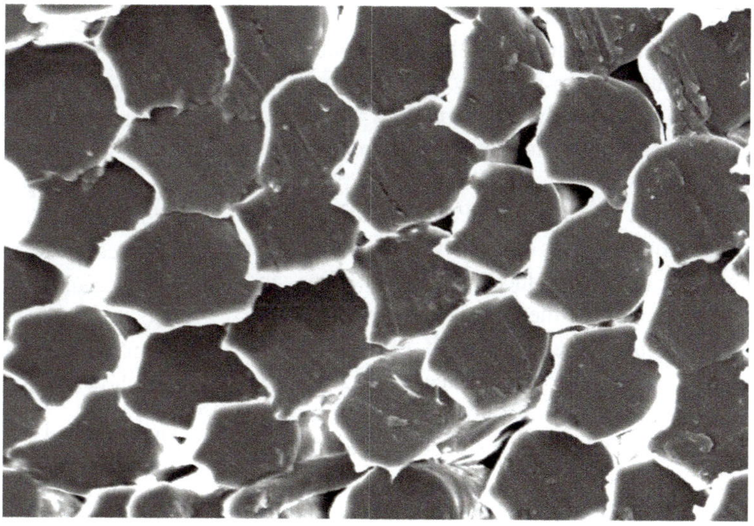

Fig. 42.19 PBS fibers in cross-section (Ø approx. 30 μm)

- raw material base is available in large quantities and cheaply,
- fiber strength > 20 cN/tex,
- elongation at break >8%,
- E-modulus >250 cN/tex,
- glass transition temperature > 75 °C.

The biopolymer should ideally be melt-spinnable with standard equipment and take-up speeds above 3000 m/min. Solution spinning processes are usually too expensive for mass applications because of the comparatively low spinning speed and the low throughput due to the use of solvents.

42.6 BioPET and BioPA Fibers

If synthetic fibers are actually obtained in whole or in part from renewable raw materials, they are also often counted among the biopolymers. Their properties are identical to those of the "original polymers". Typical examples are BioPET and BioPA6, where ethylene glycol, terephthalic acid and ε-caprolactam, respectively, are produced from sugar cane or starch by various fermentation processes. In (IfBB, 2020), the corresponding processes are described in detail,

Table 42.1 Typical properties of fibers from biopolymers

	Fineness [dtex]	Tenacity [cN/tex]	Elongation [%]	Density [g/cm^3]
Casein	1.5–6	25–35	25–60	1.22–1.31
Soy	1–2	25–40	18–21	1.29–1.31
Corn		10–12	30–45	1.25
Peanut		6–9	40–60	1.3
Collagen	5	12	5	1.32
Chitin	5–80	7–22	7–75	1.3
Chitosan		6–25	6–20	1.39
Fish protein		14–17		
Alginate		11–18	5	1.78
Starch		3.5	6	1.5
Rubber		2.4	700–900	0.93–1.0
Polyvinyl alcohol	1.5–10	20–65	10–30	1.26–1.29
Polyhydroxyalcanoate (PHA, PHB)		10–27	37	1.2–1.25
Polyglycolic acid (PGA)				1.53
Polycaprolactone (PCL)	5	25	450	1.14

also for other biopolymers. In particular, the land and water requirements for the production of the basic materials are discussed. For example, the production of 1 t of BioPET requires a minimum of 1 ha of arable land and 10,000 m^3 of water. Accordingly, 0.3 ha of land and 2500 m^3 of water are required to produce BioPA6. In particular, the water requirement is extremely high compared to synthetically produced polyesters and polyamides. But the land requirement is also large: if the entire quantity of PET fibers produced today were to be produced from renewable raw materials, this would require an area of around 400,000 km^2, more than the area of Germany. In addition, polymer production would then be in competition with food production. For these reasons, such biopolymers are "sustainable", but their production is not necessarily sensible.

An exception is PA11, which has been produced from castor oil since 1955 and marketed under the brand name Rilsan™, e.g. for brushes.

BioPET and BioPA are not chemically different from petroleum-based PET and PA, but they can be distinguished using the ^{14}C dating method. This is based on the fact that all living organisms take up carbon in two isotopes, ^{12}C and ^{14}C. After death, the slightly radioactive ^{14}C decays with a half-life of 5730 years, while ^{12}C remains. Petroleum-based PET and PA fibers no longer contain ^{14}C, while PET and PA fibers produced from renewable raw materials still contain ^{14}C.

42.7 Degradability

A material is biodegradable if it can be converted into metabolic end products by biological activity in nature or under artificially created conditions, e.g. in composting plants. Substances are considered to be compostable if they are biodegraded during a composting process. In Table 42.2, the most important terms are defined acc. to DIN.

Acc. to (DIN SPEC 1165), the degradation behavior of polymers is mainly influenced by factors shown in Fig. 42.20:

- The chemical structure of the polymer,
- the manufacture and the nature (thickness, shape, porosity) of the product,
- the product properties and the addition of additives (dyes, plasticizers, etc.),
- the medium in which the plastic is located and the environmental conditions (including UV radiation, temperature, time, humidity and presence/absence of oxygen, number, activity and type of microorganisms).

Biopolymers are usually degraded in two stages (Fig. 42.21). First, abiotic (chemical) degradation occurs. Peroxidation (in the case of saturated hydrocarbon polymers) or hydrolysis (in the case of hydrolytically degradable polymers) splits the macromolecules into smaller molecules. Subsequently, as a result of microbial activity, these fragments are further decomposed by biochemical (cellular) degradation. Natural biopolymers, such as polysaccharides or proteins

Table 42.2 Definition of selected terms on degradability (Becker, 2019)

Term	Explanation	DIN
Biodegradation	Degradation caused by biological activity, especially enzymatic processes, which leads to a significant change in the chemical structure of the material.	DIN EN ISO 16929
Disintegration	Physical breakdown of a plastic material into very small fragments.	DIN EN 14995
Compostability	Property of a material to be biodegradable in a composting process.	DIN EN ISO 16929
Compost	Organic soil improver resulting from the biodegradation of a mixture consisting mainly of various plant residues, occasionally mixed with other organic matter, and having a limited mineral content.	DIN EN ISO 16929
Complete aerobic biodegradation	Decomposition of an organic compound by microorganisms in the presence of oxygen into carbon dioxide, water and mineral salts of all other elements present (mineralisation) with the formation of new biomass.	DIN EN ISO 17556

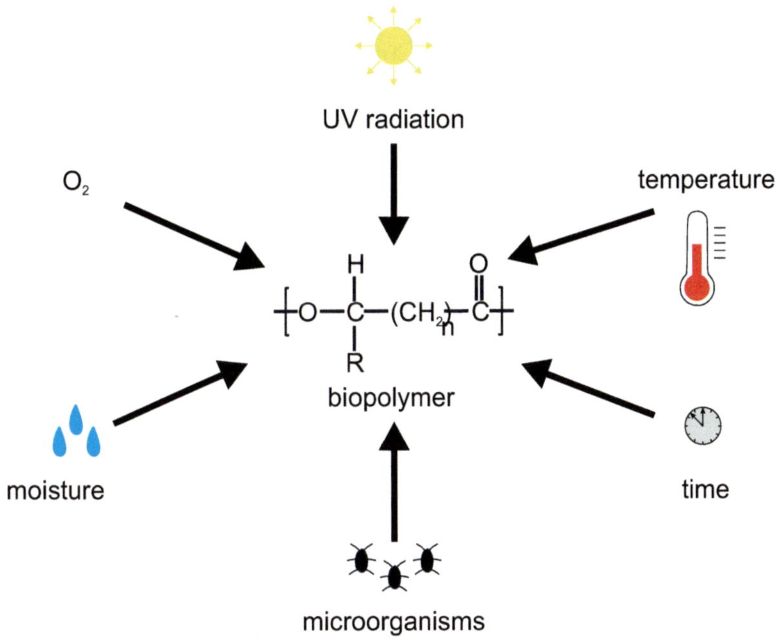

Fig. 42.20 Factors influencing the biodegradability of polymers acc. to (DIN SPEC 1165 , n.d.)

synthesized in living organisms, can also be degraded in a single step by microbial activity.

In principle, easy and fast degradability of a polymer is not necessarily desirable. Especially for clothing, the owner expects a long life and no rotting of his textiles in the wardrobe. For this reason, "biodegradable"usually means that the fibers can only be degraded in a controlled manner, i.e. under high temperature and usually by adding auxiliary substances over a longer period of time. In the interest of resource conservation, it could even be advantageous if the

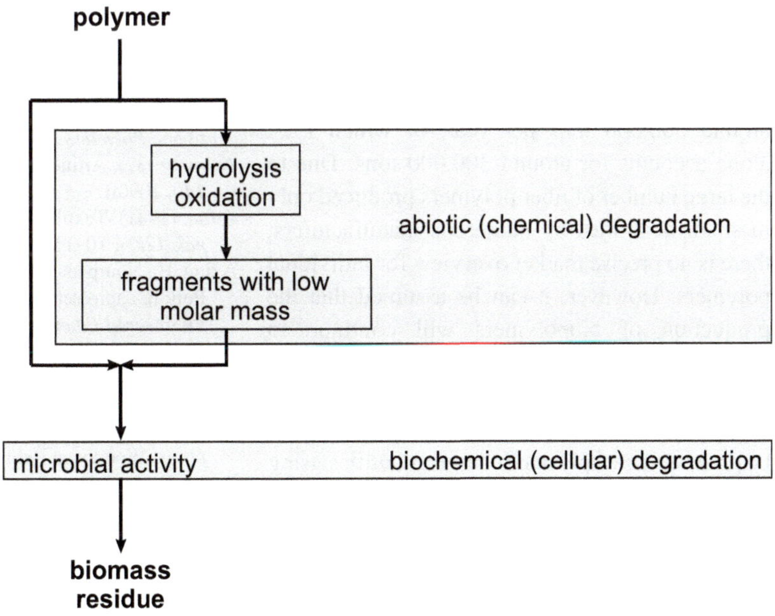

Fig. 42.21 Process of degradation of polymers acc. to (DIN SPEC 1165)

Table 42.3 Selected manufacturers and brand names of biomaterials

Producer	Trade name	Material	Website	Country
Briture Co., Ltd.		PVA	https://briture.en.made-in-china.com/	China
Changzhou Tian Yi Eng. Fiber Co., Ltd.		PVA	https://tianyifiber.en.made-in-china.com	China
ChiPro GmbH		Chitosan	http://chipro.de	Germany
Ethicon Inc.	Monocryl™	PCL	http://www.ethicon.com	USA
Kuraray Europa GmbH	Kuralon™	PVA	http://www.kuraray.eu	Japan Germany
Nantong Wanhe Chemical Co., Ltd.		PVA	https://63ed33d443bd3dd7.en.made-in-china.com	China
Nitivy Co., Ltd.	Solvron™	PVA	https://www.nitivy.co.jp	Japan
Omikenshi Co., Ltd.	Crabyon™	Chitosan	https://omikenshi.co.jp	Japan
Q-Nature UG	QMILK™	Kasein	https://www.qmilkfiber.eu	Germany
Shanghai Winshow Soybean Fibre Industry Co., Ltd.	Winshow™	Sojafaser	https://swsficl.en.china.cn	China
Shenzhen Xiangu High-Tech. Co., Ltd.		PVA	https://xian-gu.en.made-in-china.com/	China
Southwest Trading Company	SoySilk™	Sojafaser	http://www.swtcyarn.com	China
Zhengjia Milkfiber Sci & Tech Co., Ltd.	Milkfiber™	Kasein	http://www.milkfashion.com	China

biodegradability is not given at all and a textile can thus be used for a longer period of time or also be recycled in a high-quality material way.

42.8 Market and Manufacturers

Acc. to a study published by (Nova Institute, 2021), about 13% of all biopolymers produced are used for fiber production. This equates to around 500,000 tons per year, of which PLA alone accounts for around 300,000 tons. Due to the large number of fiber polymers produced only in small quantities by numerous manufacturers, there is no precise market overview for individual polymers. However, it can be assumed that the production of biopolymers will continue to increase, with BioPET and BioPA in particular, in addition to PLA, having the greatest chances of being successful in the market. These fibers can be processed cheaply and in large quantities using established processes and have excellent textile properties. The production volume of lyocell fibers will also continue to increase for the same reasons. All other biopolymers will continue to be of only marginal importance because they are either too expensive to produce or have poor textile properties (cf. Sect. 42.4.5).

Selected manufacturers of biomaterials are listed in Table 42.3.

References

Agboh, O. C. (1986). *The production of fibres from chitin*. PhD Thesis, University of Leeds.

Agboh, O. C., & Qin, Y. (1997). Chitin and chitosan fibers. *Polymers Advanced Technologies, 8*, 355–365.

Anon. (1937). Artificial wool production in Italy. *Nature, 140*, 1090ff.

Anon. (1941). Textiles from Milk. *Textile Research Journal, 12*(2), 10–13.

Azimi, B., Nourpanah, P., Rabiee, M., & Arbab, S. (2014). Poly (ε-caprolactone) Fiber: An overview. *Journal of Engineered Fibers and Fabrics, 9*(3), 74–90.

Becker, A. (2019). *Biopolymere in der Textilindustrie*. Institut für Textiltechnik der RWTH Aachen University.

Blackburn, R. S. (2006). *Biodegradable and sustainable fibres*. Woodhead Publishing.

Boyer, R. A. (1940). Soybean protein Fibers. Experimental production. *Industrial and Engineering Chemistry, 32*, 1549–1551.

Brooks, M. M. (2005). Soya bean protein fibres – Past, presentand future. In *Biodegradable and sustainable*

fibres. Woodhead Publishing Series in Textiles No. 47, Cambridge.
Brooks, M. M. (2017). Astonish the world with… Your new fiber mixture: Producing, promoting and forgetting man-made protein fibers. In *The age of plastic: Ingenuity and responsibility, proceedings of the 2012 MCI symposium*, Smithsonian Museum, Washington.
Capozza, R. C. (1975). *Spinning and shaping poly-(N-acetyl-D-glucosamine)*. US Patent 3,988,411.
Charuchinda, A., Molloy, R., Siripitayananon, J., Molloy, N., & Sriyai, M. (2003). Factors influencing the small-scale melt spinning of poly(ε-caprolactone) monofilament fibres. *Polymer International, 52*, 1175–1181.
Chipro GmbH, Bremen. http://www.chipro.de
Chodák, K., & Blackburn, R. S. (2009). Sustainable synthetic fibres: The case of poly(hydroxyalkanoates) (PHA) and other fibres. In: *Sustainable Textiles – Life cycle and environmental impact*. Woodhead Publishing, Cambridge.
DIN EN 14995, DIN EN 14995, Kunststoffe – Bewertung der Kompostierbarkeit – Prüfschema und Spezifikationen, Beuth, Berlin (n.d.).
DIN EN ISO 16929, DIN EN ISO 16929 (Entwurf), Kunststoffe – Bestimmung des Zersetzungsgrades von Kunststoffmaterialien unter festgelegten Bedingungen der Kompostierung mittels einer Technikums-Maßstab-Prüfung (ISO/DIS 16929: 2018), Beuth, Berlin (n.d.).
DIN EN ISO 17556, DIN EN ISO 17556, Kunststoffe – Bestimmung der vollständigen aeroben biologischen Abbau-barkeit von Kunststoffmaterialien im Boden durch Messung des Sauerstoffbedarfs in einem Respirometer oder der Menge des entstandenen Kohlendioxids (ISO 17556:2012), Beuth, Berlin (n.d.).
DIN SPEC 1165, DIN SPEC 1165, DIN-Fachbericht CEN/TR 15822. Kunststoffe – Bioabbaubare Kunststoffe in oder auf Böden – Verwertung, Entsorgung und verwandte Umweltthemen. Beuth, Berlin (n.d.).
Dirks, K. (2000). Aralac: The cow, the milkmaid and the chemist. *Ars Textrina, 33*, 75–85.
Fournè, F. (1995). *Synthetische Fasern*. Hanser, München.
Globe. (2020). https://www.globenewswire.com/news-release/2020/04/02/2010833/0/en/Global-Starch-Industry.html
Gordeyev, S. A., & Nekrasov, Y. U. P. (1999). Processing and mechanical properties of oriented poly(-β-hydroxybutyrate) fibers. *Journal of Material Science Letters, 18*, 1691–1692.
Heyde, M. (2007). https://commons.wikimedia.org/wiki/File:Bamboo_Richelieu.jpg, CC BY-SA 3.0.
Huang, H. C., Hammond, E. G., Reitmeier, C. A., & Myers, D. J. (1995). Properties of fibers produced from soy protein isolate by extrusion and wet-spinning. *Journal of the American Oil Chemists' Society, 72*(12), 1453–1460.
Huppert, O. (1945). *Process of manufacture of synthetic wool from soya bean protein*. US Patent 2,377,885.
IFG. (2021). https://www.ifbb-hannover.de/files/IfBB/downloads/faltblaetter_broschueren/f+s/Biopolymers-Facts-Statistics-2020.pdf
Kajita, T., & Inoue, R. (1940). *Process for manufacturing artificial fiber from protein contained in soya bean*. US Patent 2,191,194.
Kajita, T., & Inoue, R. (1946). *Process for manufacturing artificial fiber from protein contained in soya bean*. US Patent 2,394,309.
Lemoigne, M. (1925). Etudes sur l'autolyse microbienne — Acidification par formation d'acide β oxybutyrique. *Annales de l'Institut Pasteur, 39*, 144–173.
Li, G. (2007). *Phytoprotein synthetic fibre and method of manufacture therof*. US Patent 7,271,217.
Liu, W. J., Yang, H. L., Wang, Z., Dong, L. S., & Liu, J. J. (2002). Effect of nucleating agents on the crystallization of poly(3-hydroxybutyrate-co-3-hydroxyvalerate). *Journal of Applied Polymer Science, 86*, 2145–2152.
Maevskaia, E. N., Dresvyanina, E. N., Yudenko, A. N., & Yudin, V. E. (2018). The mechanical properties of chitosan fibers obtained in different spinning conditions by coagulation method. *Technical Physics, 63*(9), 1323–1327.
Manvi, P., Balakrishnan, N., Pratiwi, J., Gries, T., & Seide, G. (2017). Starch: A low cost textile raw material. *Future Textiles, 2*, H. 4, 36–38.
Merrifield, A. L., & Pomes, A. F. (1946). Fiber from peanut protein: I. The production and properties of Sarelon. *Textile Research Journal, 16*(8), 369–377.
Milkfashion. (2021). http://www.milkfashion.com
Millar, A. (1895). *Herstellung von für Textilzwecke geeigneten und in Wasser unlöslichen Fäden und Gespinsten aus Gelatine*. Deutsches Reichspatent 88225, Reichspatentamt, Berlin.
Noguchi, J., Tokura, S., & Nishi, N. (1978). Studies on Chitin I. In R. A. A. Muzzarelli & E. R. Pariser (Eds.), *Proceedings of the 1st international conference on chitin and chitosan*. MIT.
Nova Institut. (2021). *Nova-Institut für politische und ökologische Innovation GmbH*, Hürth. http://www.bio-based.eu
Penfold, A. R. (1956). *Plastics and synthetic fibres*. A.H. Pettifer, Government Printer, Sydney.
Qnature UG, Kümmersbruck. http://www.qmilkfiber.eu
Rijavec, T., & Zupin, Z. (2011). *Soybean protein fibres*. University of Ljubljana.
Schmack, G., Jehnichen, D., Vogel, R., & Tändler, B. (2000). Biodegradable fibers of poly (3-hydroxybutyrate) produced by high-speed melt spinning and spin drawing. *Journal of Polymer Science, B Polymer Physics, 38*, 2841–2850.
Shanghai Winshow Soybean Fibre Industry Co., Ltd. (2021). https://swsficl.en.china.cn/
Siriwardane, M. L., de Rosa, K., Collins, G., & Pfister, B. J. (2014). Controlled formation of cross-linked collagen fibers for neural tissue engineering applications. *Biofabrication, 6*, 015012.
Smartfiber AG, Rudolstadt. http://www.smartfiber.de

Swicofil. (2020). http://old.swicofil.com/products/055chitosan.html

Swicofil. (2021). https://www.swicofil.com/commerce/products/soybean/152/introduction

Todtenhaupt, F. (1904). *Verfahren zur Herstellung künstlicher Fäden für Haare und Gewebe*. Deutsches Reichspatent 170051, Reichspatentamt, Berlin.

Tonndorf, R., Aibubu, D., & Cherif, C. (2020). Collage multifilament spinning. *Materials Science & Engineering C, 106*, 110105.

Traill, D. (1951). Some trials by ingenious inquisitive persons: Regenerated protein fibres. *Journal of the Society of Dyers and Colourists, 67*, 257–270.

Whittier, E. O., & Gould, S. P. (1940). Making casein fiber. *Industrial & Engineering Chemistry, 32*(7), 906–907.

Woodbury, C. (1945). Dresses from chicken feathers. *American Magazine, 10*, 32–33, 94–95.

Yang, Y., & Reddy, N. (2012). Properties and potential medical applications of regenerated casein fibers crosslinked with citric acid. *International Journal of Biological Macromolecules*, (51, ½), 37–44.

Zeng, J., Chen, X., Xu, X., Liang, Q., Bian, X., Yang, L., & Jing, X. (2003). Ultrafine fibers electrospun from biodegradable polymers. *Journal of Applied Polymer Science, 89*(4), 1085–1092.

Zhang, J., Liu, Y., Sun, J., Gu, R., Ma, C., & Liu, K. (2020). Biological fibers based on naturally sourced proteins: Mechanical investigation and applications. *Materials Today Advances, 8*, 100095.

Part VII

Inorganic Fibers

Glass Fibers

43

43.1 Development and History

Already more than 3500 years ago, glass threads were used in Egypt to decorate vessels. The glass was pulled out of the liquid melt with the help of tongs, thus creating a thread (Weiss, 1966). The idea of spinning glass fibers had been around since the eighteenth century, and in the mid-nineteenth century, it was put into practice and glass fibers were produced industrially on a larger scale for the first time. From the middle of the twentieth century onwards, glass fiber reinforced composites were produced and in the meantime glass fibers have become an important material for a multitude of technical products. A detailed account of the historical development can be found in the Chap. 2 of this book.

Glass fibers are available as staple fibers, as filaments or rovings and also in the form of so-called glass wool (Fig. 43.1).

43.2 Raw Materials

The basic materials for the production of glass fibers are finely ground quartz sand, limestone, colemantite and kaolin.

Quartz powder is the name for finely ground, mostly iron-free and high-purity quartz sand. Limestones are natural minerals from the group of carbonate rocks that consist predominantly of calcium carbonate ($CaCO_3$). Colemantite with the chemical formula $Ca[B_3O_4(OH)_3]$-H_2O is a borate mineral. Kaolin comprises the group of clay rocks with a varying proportion of elutriable components (grain diameter < 2 μm). It contains the natural silicate kaolinite and has the chemical formula $Al_2[OH)_4/SiO_5]$ or Al_2O_3-$2SiO_2$-$2H_2O$. Kaolinite consists of 40% Al_2O_3, 46% SiO_2, and 14% H_2O. Its internal structure is built up of two layers, a tetrahedral and an octahedral layer (Fig. 43.2). The silicon ions are arranged in a tetrahedral layer, with the oxygen atoms located at the four corners (Fig. 43.3). The aluminum ions are located in an octahedral layer, the building blocks of which are formed by oxygen and hydroxy ions (Falbe & Regitz, 1990).

Depending on the type of glass, precisely dosed quantities of certain oxides can be added to the textile glasses. The mixture created in this way forms the starting material of the subsequent melting process and is fed to the melting units as a so-called "batch". The technological properties of the textile glass strongly depend on the composition of the melt. Therefore, it is imperative to carry out the dosing of the batch components very precisely, to avoid impurities and to adjust the particle size distribution acc. to the desired glass properties (Table 43.1).

43.3 Chemical Composition and Structure

Glass differs significantly from the crystalline, i.e. thermodynamically stable, state of most

D. Veit, *Fibers*, https://doi.org/10.1007/978-3-031-15309-9_43

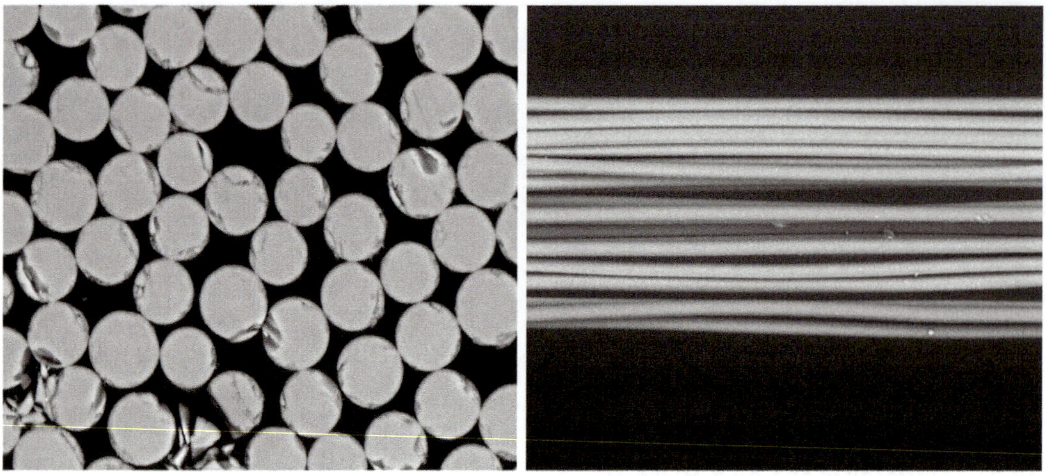

Fig. 43.1 Cross-sectional and longitudinal view of glass filaments

inorganic non-metallic solids in its ground state, which is referred to as vitreous or glassy.

The glassy state is generally defined as the frozen state of a supercooled and thus solidified liquid. It is the result of the suppression of crystallization of a melt. In the Fig. 43.4 it can be seen that the crystallizing melt changes to the crystalline state at melting temperature (T_s) with a sudden change in properties. However, if crystallization is prevented, the melt initially supercools briefly and is then frozen below the so-called transformation or glass formation temperature (T_g). This can be achieved by an extremely short cooling cycle or a quenching process, as is common, for example, in the production of textile glass (Fig. 43.4).

For the structural composition of the glasses, this means that an ordering of their building blocks into a regular crystal lattice cannot take place due to the sudden increase in viscosity. A continuous symmetry or periodicity of the structural composition is therefore not present. The result of the missing crystal order is isotropic, i.e. direction-independent mechanical properties. The frozen state of disorder corresponding to a higher temperature is thermodynamically metastable. It depends essentially on the cooling rate, but also on the temperature range from which cooling has occurred. The properties of the glasses are therefore influenced not only by their chemical composition, but also significantly by the type of thermal pretreatment.

Fig. 43.2 Structure of kaolinite

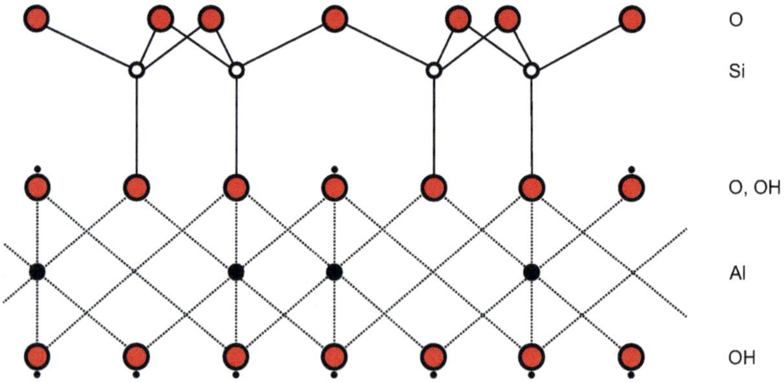

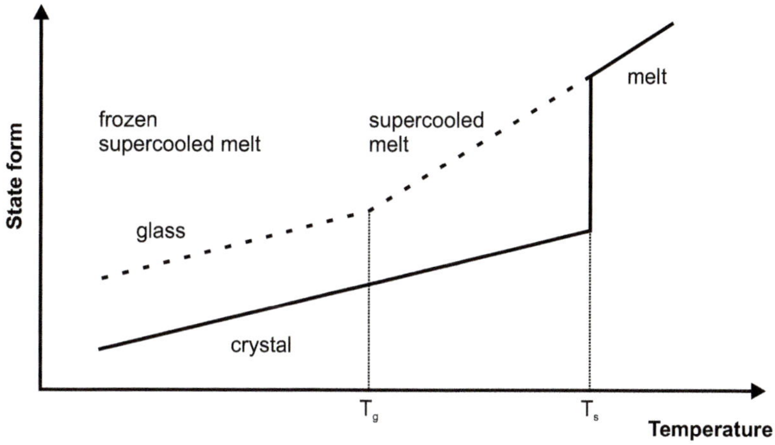

Fig. 43.3 Kinetics of glass formation acc. to Petzold and Hinz (1979)

As an example of the other glass systems, we will now consider the structure of silicate glasses. They consist of two components: a network of SiO_2 and silicates or borates, which fill this framework. Here, the SiO_2 network forms the elastic component, which has no plastic deformation and no elastic aftereffect. In contrast, the silicates and borates constitute the tough component of the glass. It is responsible for the elastic aftereffect and viscous flow observed with increasing temperature. The basic building element of the network is the SiO_4-tetrahedron.

Because the silicon atom wants to complete its electron shell, four oxygen atoms are arranged around each Si atom in a tetrahedron configuration. In networks, several SiO_4-tetrahedra share one oxygen atom ("bridging oxygen"). The formation of these Si-O-Si bridges leads to chain or network formation and thus to high molecular weight silicates (Petzold & Hinz, 1979). Boron oxide B_2O_3 is also a network former which improves the acid and alkali resistance of the glass.

In the field of fiber production, multicomponent glasses predominate. In them, the quartz glass network is present in a somewhat modified form. The cations present, which are incorporated into the gaps of the network, cannot themselves form a regular, spatial lattice. They must therefore modify the existing SiO_4 network to the extent that they are incorporated into the glass ("network converter"). They do this by splitting the Si-O-Si

Table 43.1 Chemical composition of typical source minerals for glass production acc. to Pico (2016)

Oxide	SiO_2	Al_2O_3	B_2O_3	MgO	CaO	Fluor
	sand	china earth	colemantite	dolomite	limestone	
SiO_2	98	44	4–5	< 1	44	< 1
Al_2O_3	0.85	37	0.5	< 0,6		< 1
CaO		0,6	27–28	31	55.4	70.2
Na_2O+K_2O	0.5	2	0.2–0.3	< 0.6		
Fe_2O_3	0.1	< 1	< 0,1	< 0.4	< 0.05	< 0.25
H_2O	0.1	< 1			< 0.4	< 0.4
B_2O_3			40–42			
MgO			2–3	20		
P_2O_5					< 0.1	
MnO_2					< 0.1	
F_2						47.3

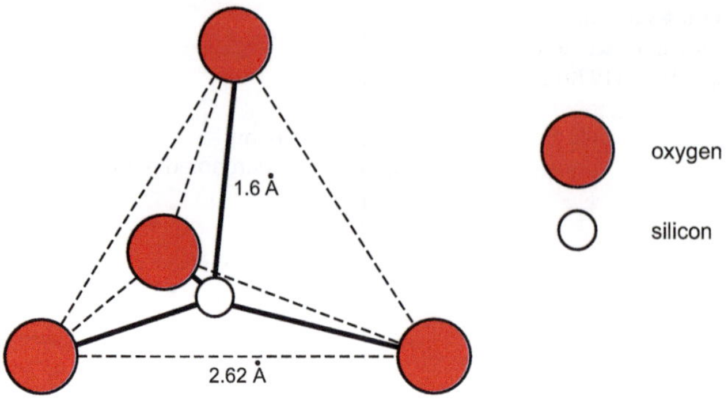

Fig. 43.4 SiO_4 -tetrahedron

bonds with the oxygen they bring along, thus creating separation spots (Fig. 43.5).

The smaller the radius and the larger the charge of the cations, the more they have a splitting effect on the network. In the order K, Na, Li, Ba, Sr, Ca, Mg, Ti, the number of separation sites increases. At the same time, this reduces the viscosity, i.e. more liquid glasses are obtained in the melting range. The addition of CaO increases tensile and flexural strength, MgO improves glass hardness and reduces thermal expansion, ZnO improves acid resistance, Na_2O reduces tensile and flexural strength, K_2O deteriorates chemical resistance and tensile strength, while Li_2O increases acid and alkali resistance.

The network transformers (e.g. K_2O, CaO, BaO, Na_2O, PbO) are thus not glass-forming oxides, but oxides that influence the properties of the glass. Other, so-called intermediate oxides (e.g. MgO, BeO, Al_2O_3 and TiO_2), can assume the functions of both network formers and network transducers.

43.4 Glass Production

In the first process step, the basic materials are melted in furnaces. This is usually done in a unit melter or, for small batches, in a vertical super melter. The essential process steps are described in (Wulfhorst et al., 1993).

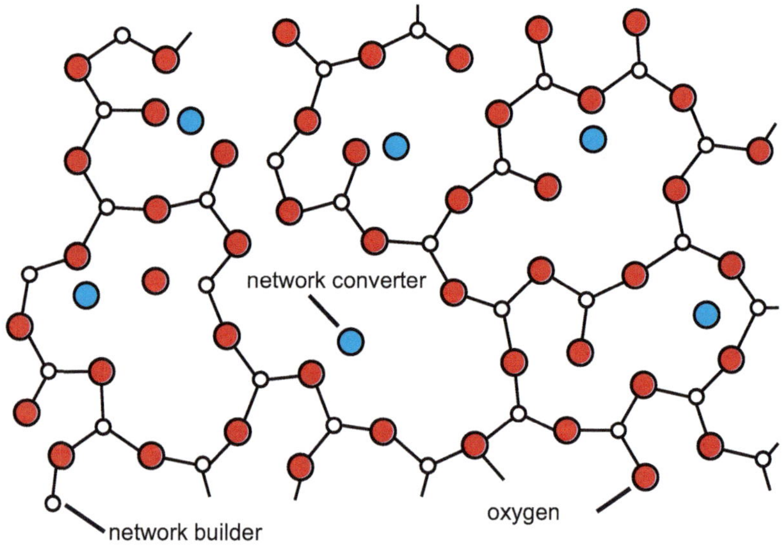

Fig. 43.5 Structural model of glass fibers

43.4.1 Unit Melter

The unit melter usually has a length-to-width ratio of 3:1. It has three different zones through which the glass flows horizontally: the melting zone, the conditioning zone and the working zone. In the melting zone, the batch transported from a silo via screw conveyors into the brick melting chamber is heated and melted. This is done with the aid of 2–20 gas or oil burners arranged on the long side of the furnace. The combustion air required for this is provided by a fan and is usually preheated to approx. 760 °C in a recuperator. In order to counteract the different flows occurring in the melt and to achieve better mixing, one or more rows of air nozzles can be used in the bottom of the melting area. In addition to homogenizing the melt, these nozzles also prevent insufficiently melted particles from flowing out. The melting zone is followed by the conditioning zone. The molten glass is fed to the spinneret via a narrowing channel that retains heavy components. The temperature of the melt in the melting chamber is approx. 1400 °C and decreases to approx. 1260 °C towards the spinneret.

Figure 43.6 schematically shows the design of such a continuously operating plant for the production of glass filaments.

Typical melt throughputs of such plants for fiber production are 20–100 t/day, but there are also small plants with a production of 200 kg/day. Approx. 250 m^3 of air and 120 m^3 of fuel gas are required per ton of glass. The energy input is estimated to be about 340 kWh/t (Worrell et al., 2008). Emissions are mainly NOx in the order of 1 g/kg of glass. Mainly E-glass is produced this way (Fig. 43.7).

43.4.2 Electric Ovens

A frequently used electric furnace is the so-called “Vertical Super Melter” (VSM). This is a shaft furnace with a fully electrically heated melting tank. The round or 6- or 12-cornered melting tank has an average diameter of approx. 5 m and a height of 2–3 m. The furnace is equipped with an electric heating system. Typical for this type of furnace is the long passage with the adjoining riser. Heating takes place via electrodes arranged in three different levels (Fig. 43.8).

The throughput of a VSM is 3–200 t/day, depending on the design. This technology is used, for example, to produce C-glass.

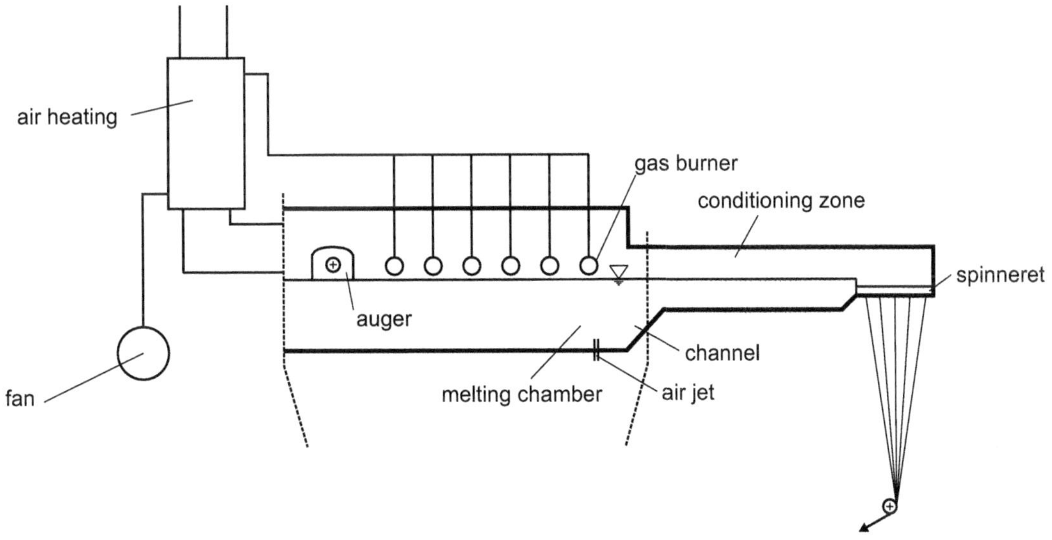

Fig. 43.6 Unit melter

Fig. 43.7 Melting furnace (P-D Glasseiden GmbH Oschatz n.d.)

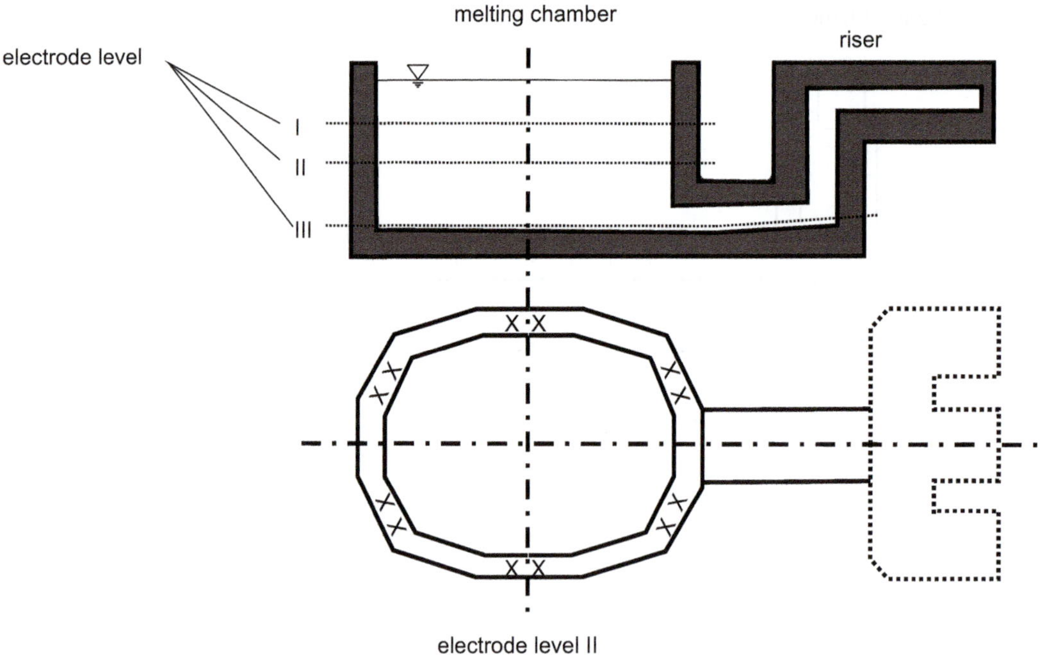

Fig. 43.8 Electric furnace for glass production

43.4.3 Intermediates

As an alternative to the direct processes, in which the melt is conveyed directly to the spinnerets, production can also take place via an intermediate product, e.g. spherules or pellets. Machines with a daily output of up to 4.5 t are used to produce the spherules. The molten glass flows at a temperature of approx. 1150 °C from the furnaces through special channels into a glass cutting device. Spherules or pellets are produced from the molten glass, which are melted and spun out again in a later step.

The production of intermediates as a preliminary stage of glass production is mainly used today for staple fiber production. The direct process dominates in filament production.

43.5 Fiber and Yarn Production

The two main spinning principles used to produce glass fibers and glass yarns are the jet drawing process for filaments and the drum drawing process for staple fibers.

43.5.1 Nozzle Drawing Method for Filaments

Of all the current manufacturing processes for glass filaments, the nozzle drawing process is the most important. About 90% of all glass fibers are produced with this process, for E-glass fibers even 97%. It was first used in the USA in 1930 and in Germany in 1939.

The mixture of basic materials is melted at approx. 1400 °C acc. to one of the processes described above and kept at temperatures between 1250 °C and 1350 °C. The mixture of basic materials is melted at approx. 1400 °C and kept at temperatures between 1250 °C and 1350 °C acc. to one of the processes described above. It enters the drawing nozzles made of a platinum/rhodium alloy, which have a temperature of approx. 1200 °C, via channels. Here it emerges from the bottom of the nozzle by gravity. The viscosity of the molten glass during filament formation is usually 30–100 Pas. Spinnerets (bushings) made of platinum have a high resistance to oxidation and corrosion by the molten glass, rhodium improves the thermal stability of the platinum. Usually a spinneret consists of 10–20% rhodium and 80–90% platinum. The number of holes in the nozzle varies between 800 and 8000 holes in the direct process, capillary hole diameters are usually between 1 and 2.5 mm. The production capacity is approx. 300–1800 kg/day (Fig. 43.9).

Fig. 43.9 Nozzle (spinneret) with glass filaments (P-D Glasseiden GmbH Oschatz, n.d.)

After exiting the spinneret, the glass filaments are continuously drawn off mechanically at high speed (1200–4800 m/min). Cooling fins are located directly below the nozzle to accelerate the solidification of the glass filaments. After cooling at approx. 200 K/cm, the glass filaments are moistened with water to facilitate the subsequent application of sizing (reduction of filament damage). Afterwards, they are combined into a roving with a collecting roller and wound onto a spool (Fig. 43.10). Excessive humidity during glass fiber cooling leads to the incorporation of OH groups into the glass

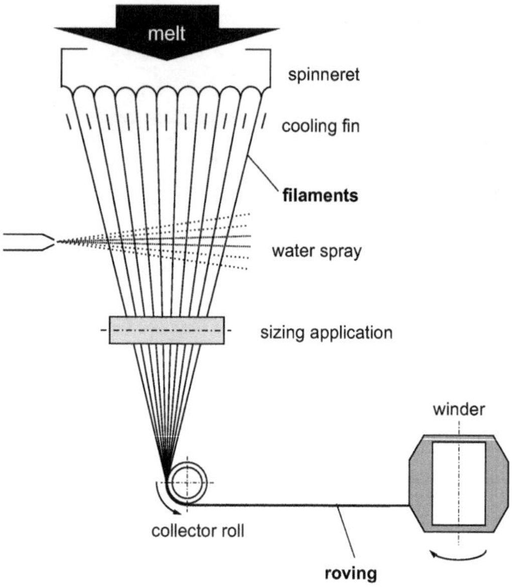

Fig. 43.10 Nozzle drawing process

structure, which reduces fiber strength. Typical single filament fineness is in the range of 0.5–15 dtex.

Figure 43.11 shows a typical take-up unit and the package bearing. The rovings are pulled out of the bobbins from the inside to the outside for further processing.

43.5.2 Drum Drawing Process for Staple Fibers

Approximately 95% of all glass staple fibers are produced today using the drum drawing process

Fig. 43.11 Winder and bobbin bearing (P-D Glasseiden GmbH Oschatz, n.d.)

due to its great economic efficiency. The glass staple fiber rovings are produced in a two-stage process from C-glass (alkali lime glass) or from a special glass adapted to certain applications. In the first stage, the batch is melted in electric tank furnaces and processed into balls or pellets. The actual staple fiber production then takes place in the second process stage (Fig. 43.12).

The pellets are melted and the melt is then pressed through the spinnerets. At temperatures between 1000 and 1200 °C, the melt emerges from the holes in the nozzle. Depending on the product and the desired fineness, approx. 250–1000 holes are arranged in a bar (often made of platinum). In a self-spinning process, the glass drops emerging from the holes are drawn out into filaments by a rotating drum.

The filament fineness is 1.5–3 dtex depending on the take-off speed (up to 3500 m/min). The filaments are mechanically lifted off the drum (e.g. with a scraper or knife edge) and collected in a spinning funnel. Due to the process, fibers with an undefined staple length (approx. 0.02–1 m) are produced. After interlacing in the spinning funnel, the fibers are combined into a sliver in an OE process and drawn off. This glass staple fiber roving is then wound onto cross-wound bobbins at approx. 800 m/min. In this form it forms the starting product for subsequent textile processing methods or can be used directly as weft yarn in the weaving mill. Its fineness is usually 125–2000 tex.

As an alternative to this process, glass rovings can also be cut to a uniform length. Rotating knives are used for this purpose.

43.5.3 Glass Wool

Glass wool consists of amorphous fibers produced from silicate melts using mechanical or pneumatic processes. Such products are also called rock wool and mineral wool.

While glass filament yarns have been produced for centuries, it was not until 1840 that a process for the production of glass wool was developed. This was intended on the one hand to put glass slag to further use and at the same

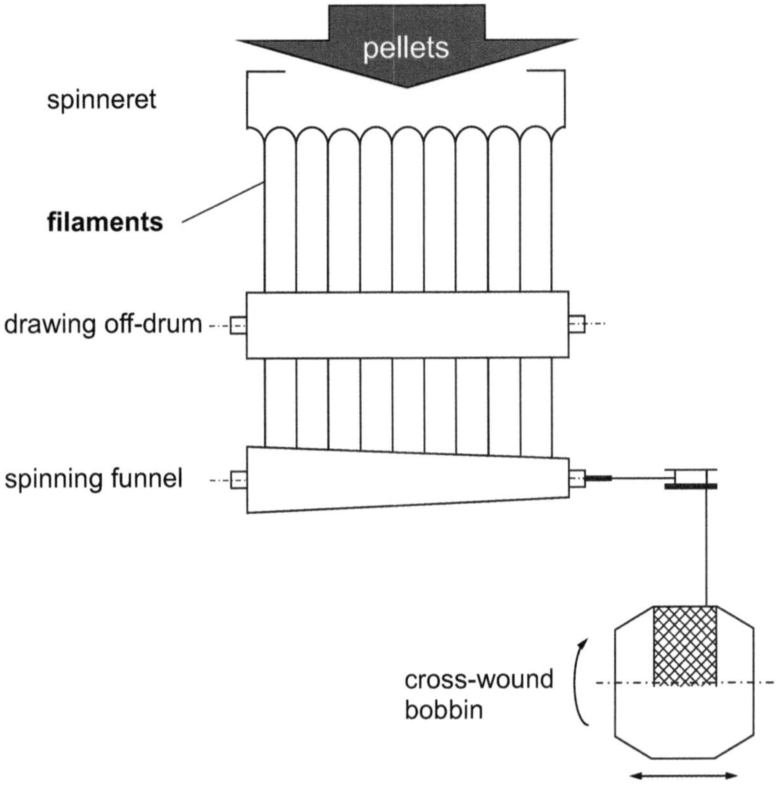

Fig. 43.12 Drum drawing process

time to replace asbestos. It was not until the 1930s that such processes were commercially successful. G. Slayter at the Owens-Illinois Glass Corp. was one of the pioneers of this development. However, this was not initially used to replace asbestos, but such glass wool mats were used for insulation. With the development of appropriate binders in the 1950s, large-scale production of glass wool for insulating buildings and machinery began.

Depending on the process, the fibers of the glass wool have a minimum diameter of approx. 0.18–0.30 μm. This results in specific surface areas of approx. 0.1–0.2 m^2/g. However, the insulating effect of glass wool depends less on the fiber properties than on the overall glass wool/binder system and the installation conditions. The maximum temperature at which glass wool can still be used as an insulation material is between 550 and 750 °C, depending on the material and binder.

Glass wool is inert to organic solvents, but can be attacked by water or aqueous solutions. This can loosen the bond between fibers and binder in particular. Mineral wool containing lime is sensitive to acids.

At present, the manufacturing processes shown schematically in the Fig. 43.13 schematically shown manufacturing processes are used:

Owens Process

In this process, the melt first emerges from the platinum crucible in several strands through nozzle holes. These are then entrained by a downward gas flow and drawn into fibers. The average fiber diameter is 3–6 μm, the fiber length is 3–10 cm. The glass wool produced in this way may still contain glass crumbs that have not yet dissolved into fibers.

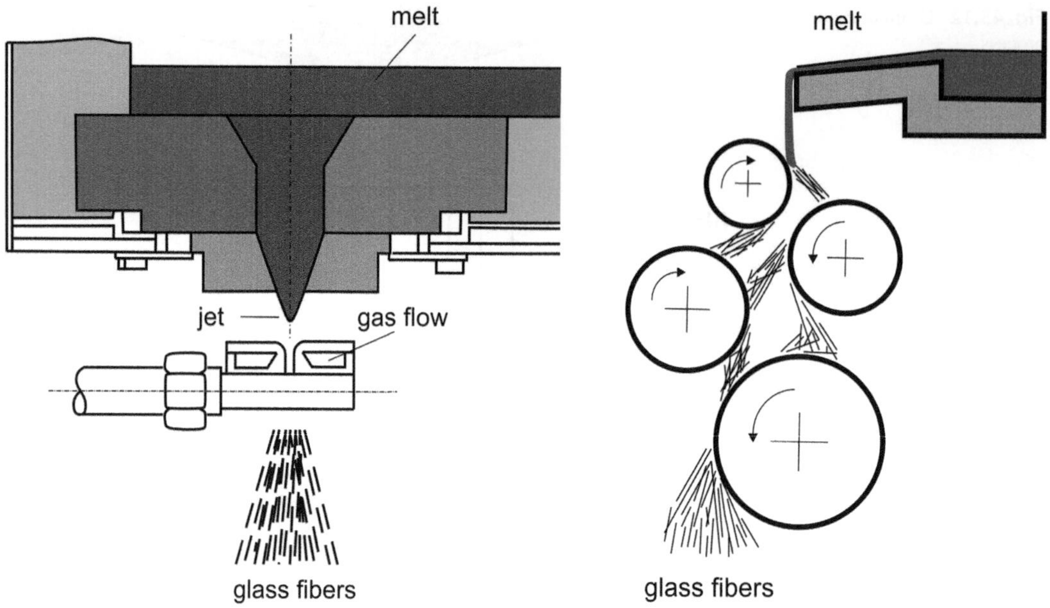

Fig. 43.13 Owens process (left, acc. to Giegerich & Trier, 1964) and Rock Wool process (right, acc. to Skolnikov et al., 1969)

Rock Wool Process
The melt flow is directed onto rotating rollers. The centrifugal forces acting on the melt produce fibers that become finer and finer from roller to roller. Typical fiber lengths are in the range of 3–10 cm.

Rotary Process
This process is a two-stage process. First, the melt is formed into a thin, still liquid strand via a nozzle and passed through the hollow shaft of a rotating centrifuge. From there, the still-liquid glass is evenly distributed on the inside of a rotating container. It then emerges through holes to the outside in fiber form (centrifugal force) and is drawn off downwards by a hot gas stream. The fiber diameter is 2–6 μm. Only special glass types with good flow properties are suitable for this process.

Centrifuge Blowing Process
The melt stream is fed via an outflow channel to a rapidly rotating rotor. From there, the still liquid mass emerges via holes and is stretched into fibers by the centrifugal forces. To produce glass wool blocks, the glass fibers are sprayed with a binder mist and consolidated in a furnace. During this process, the drying of the binder gives glass wool its yellow color (Fig. 43.14).

43.6 Fiber Optic Types

Depending on the area of application, there is a wide range of glasses of different compositions with correspondingly different properties (Table 43.2).

A-Glass
A-glass belongs to the group of alkali lime glasses. Characteristic is a high content of silicon oxide and smaller contents of aluminum trioxide, magnesium oxide, calcium oxide and alkali oxides. The proportion of alkali oxides is less than 1%. In addition, it contains no or only very little boron trioxide.

C-Glass
C-glass is also an alkali lime glass, but has a higher boron trioxide content. Its characteristic feature is its good chemical resistance.

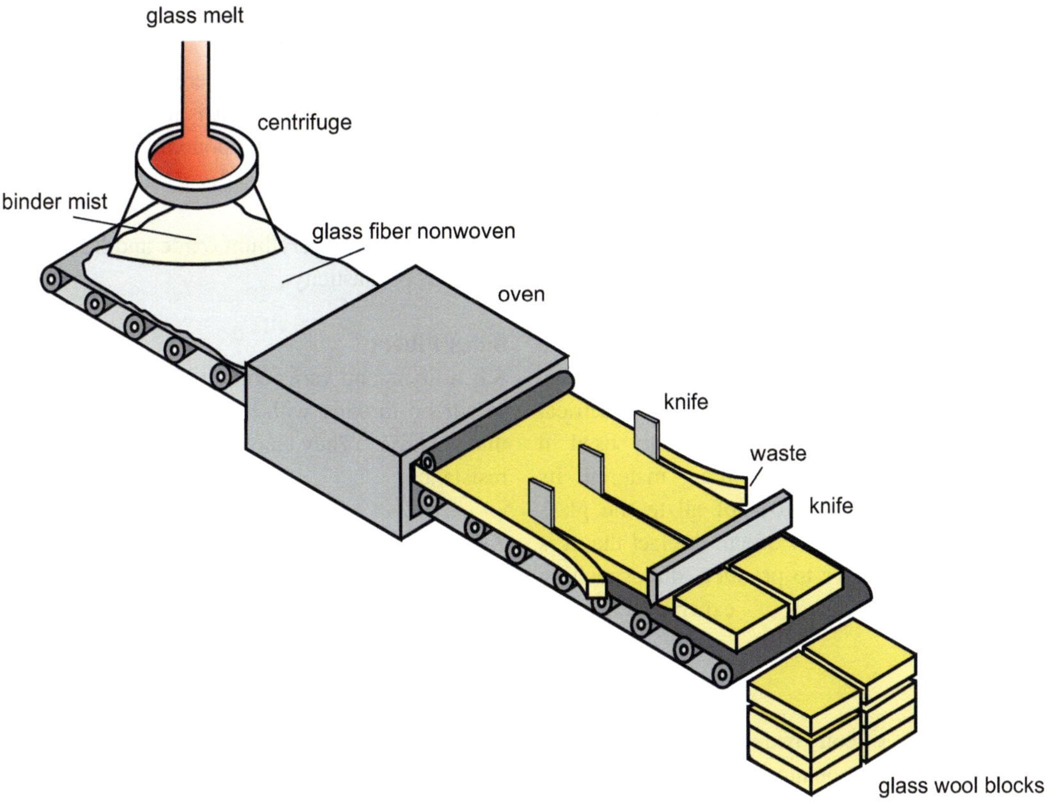

Fig. 43.14 Production scheme of glass wool insulation mats acc. to Eurima (2020)

Table 43.2 Composition of various types of glass acc. to Mohr and Rowe (1978), von Falkai (1981), Loewenstein (1983), Ehrenstein (1981), Liebscher (1983)

Components	Standard types						Alkali resistant		
[Gew.-%]	A-glass	C-glass	D-glass	E-glass	R-glass	S-glass	AR-glass	Z-glass	Silica
SiO_2	70–72	60–66	72–75	50–56	60	62–65	60.9	71	99.8
Al_2O_3	≤ 2,5	≤ 6	≤ 0.5	12–16	24–25	20–26		1	
CaO	5–10	≤ 14		16–25	6–9		4.8		
MgO	0.9–4	≤ 3	0.2	≤ 6	6–9	10–15	0.1		
B_2O_3	≤ 0.5	2–7	22–23	6–13		≤ 1.2			
F				≤ 0.7					
Na_2O	12–15	8–15	1.3	0.3–2.0	0.4	≤ 1,1	14.3	11	
ZrO_2							10.2	16	
K_2O	≤ 1.5	1–8	≤ 1.5	0.2–0.5	0.1		2.7		
Fe_2O_3	≤ 0.5	≤ 0.3	0.2	0.3	0.3			≤ 2.7	
TiO_2					0.2		6.5		
LiO_2								1	
SO_3	0.7						0.2		

D-Glass
D glass has a high proportion of silicon dioxide and approx. 8–13% B_2O_3. It has a high chemical resistance and a good thermal shock resistance and good dielectric properties. It is therefore particularly suitable for applications in insulation technology and for reinforcing computer boards.

E-Glass
E-glasses essentially contain silicon oxide and boron trioxide as well as aluminum trioxide. The mass fraction of alkali oxides is usually less than 1%. It is particularly suitable for electrical applications. Furthermore, it is mainly used in fiber composites as a reinforcing material. Its market share is about 90% of all textile glass filaments. This is mainly due to the fact that it is easy and inexpensive to produce (direct melting process) and decades of experience in its processing.

ECR Glass
In addition to silicon oxide and calcium oxide, this type of glass contains a high proportion of usually more than 17% aluminum trioxide. The content of alkali oxides is usually less than 1%. It has a high chemical resistance, especially to acids. ECR glass is used to reinforce plastics.

R-Glass
R-glass has, in addition to the usual essential components for aluminosilicate glasses, aluminum trioxide and silicon dioxide, additives of calcium and magnesium oxide. It has high strength and high fatigue, temperature and moisture resistance.

S-Glass
This glass belongs to the group of aluminosilicate glasses, but it does not contain calcium oxide. The mass fraction of magnesium oxide is approx. 10%. Due to its high strength, it is mainly used as a strength member, especially in military applications.

AR Glass
This type of glass contains a high addition of zirconium oxide and is therefore particularly alkali-resistant. Therefore, it is used for the reinforcement of concrete components.

M-Glass
M-glass contains beryllium oxide and has a high modulus of elasticity.

Silica Fibers
Silica fibers and yarns consist almost exclusively of silicon dioxide with a mass fraction of more than 99.8%. They have a high temperature resistance.

Z-Glass
Z-glass is comparable to AR-glass, but has a higher proportion of silicon dioxide and zirconium oxide. Therefore, its resistance in cement is higher than that of AR glass.

Figure 43.15 shows the composition of typical glass fibers in comparison to basalt and ceramic fibers. While the SiO_2 contents differ only slightly, glass and basalt fibers, in contrast to ceramic fibers, have significantly higher contents of various other oxides. Accordingly, the properties are more variable.

43.7 Properties

In general, glass fibers are characterized by high tensile strength, low elongation, high weight-related modulus of elasticity, low density, high fineness (5–24 μm) and relatively low price. The properties depend strongly on the glass type.

Acc. to Winkelmann and Schott (1894), the Young's modulus can be calculated from the mass fractions p_i of the oxides that make up the glass. The values E_i and F_i are compiled in Table 43.3.

Thus the modulus of elasticity [GPa] of a glass can be calculated acc. to

Fig. 43.15 Composition of glass, basalt and ceramic fibers

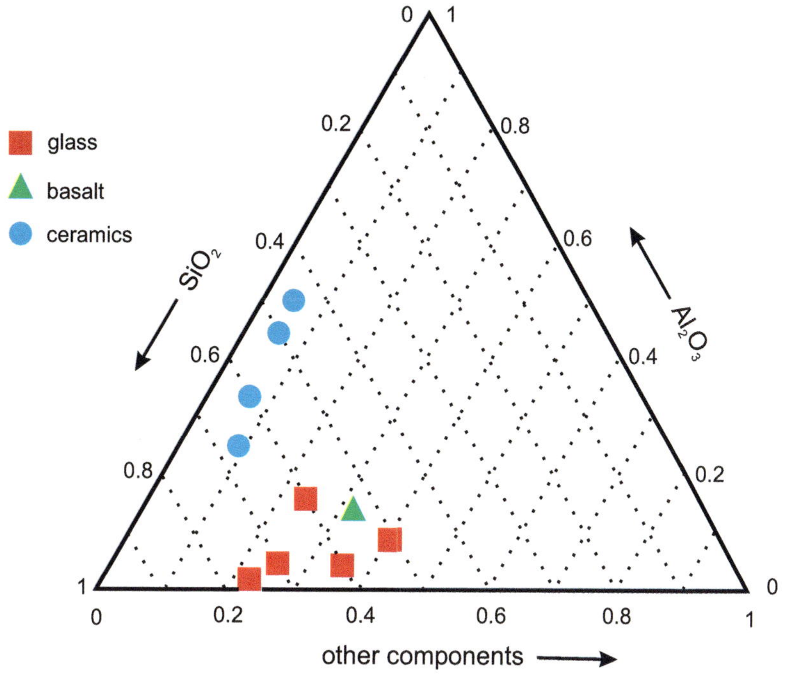

$$E = \sum E_i \cdot p_i$$

The values calculated in this way agree well with the values given in the Table 43.4 given values.

This approach can also be used to estimate the tensile strength of glass fibers in [N/mm^2 or MPa] from the composition acc. to

$$F = \frac{1}{\kappa} \sum F_i \cdot p_i$$

The correction factor for glass fibers should be set at 3.3 for A, C, D and E glass. This is in very good agreement with the values given in Table 43.4.

Glass fibers have lower strengths than carbon and aramid fibers, but have greater elongation (Fig. 43.16).

In principle, it is also true for glass fibers that the theoretically possible strength is far from being achieved in practice (Jones & Huff, 2009). However, the strength values of glass filaments come closer to the theoretically possible values than those of solid glasses. This is mainly due to the very high cooling rate during filament generation, which is about 200 K/cm or 1000 K/ms. As a result, fewer microscopic cracks form on the glass surface ("Griffith's pockets") than in solid glass and the strength is correspondingly higher.

The strength of glass fibers depends on the fiber fineness. With increasing diameter, the probability of a structural defect increases and thus the strength decreases (Table 43.5).

Under the influence of alkalis, oxygen bridges are split in the glass network and the network is gradually dissolved. Alkali resistance is improved by the incorporation of Na_2O and ZrO_2. The effect is based on the formation of a poorly soluble, dense layer of zirconium hydroxide and hydrated ZrO_2 on the glass surface. This barrier

Table 43.3 Factors E_i and F_i for the calculation of Young's modulus and strength of glass fibers acc. to Winkelmann and Schott (1894)

	SiO_2	Al_2O_3	B_2O_3	CaO	MgO	BaO	Na_2O	K_2O	Fe_2O_3
E_i	0.7	1.5	0.6	0.7	0.4	0.7	1.0	0.7	0.2
F_i	90	50	65	200	10	50	20	10	

Table 43.4 Typical mechanical properties of glass filament fibers acc. to Loewenstein (1983), Wallenberger et al. (2001), Xanthos (2010)

	A-glass	C-glass	D-glass	E-glass	R-glass	AR-glass
Density [g/cm^3]	2.46	2.46	2.14	2.6	2.53	2.68
Tenacity [MPa]:						
Filament alone	2,450	2,750	2,500	3,400	4,400	3,000
Filament in composite			1,650	2,400	3,600	1,800
Tenacity [cN/tex]:						
Filament alone			117	131	174	112
Filament in composite			77	92	142	67
Breaking elongation [%]			4.5	4.5	5.2	4.3
E-modulus [GPa]	72	69	55	73	86	73
Moisture absorption [%]			< 0.1	< 0.1	< 0.1	0.1
Softening temp. [°C]			775	846	985	773
Fiber building temp. [°C]	1,280			1,180		

impedes the diffusion processes that lead to the dissolution of the glass. Such AR glasses (alkali-resistant) contain approx. 17% ZrO_2 and 15% Na_2O (Teschner, 2013).

43.8 Further Processing

The most important processes of further processing are shown in Fig. 43.17. Glass filament yarns are processed into fabrics and non-crimp fabrics, glass staple fibers into nonwovens and mats.

Glass fibers can generally be further processed on conventional textile machines, depending on the form in which they are supplied and their properties, which are tailored to the intended application. However, the specific material properties of glass fibers must be taken into account. For example, glass fibers (e.g. in the weaving process) must be cut with special shears which are coated, have diamond discs or are made of a material combination (e.g. steel/plastic). As a result, they are low-closing and produce a clean cut.

To improve processability, the fibers are usually provided with a sizing. This reduces fiber friction and improves the cohesion of the yarn or roving. Of particular importance in the processing of glass fibers is their very low knot strength or sensitivity to kinking. This must be taken into account in the design of the

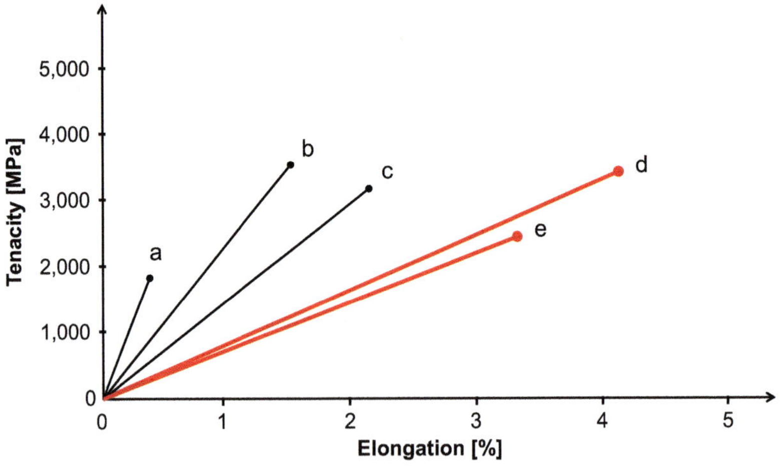

Fig. 43.16 Typical stress-strain diagram for (a,b—carbon fibers; c—aramid; d—R-glass fibers; e—E-glass fibers)

Table 43.5 Dependency of glass fiber tenacity on diameter acc. to Anderegg (1939)

Diameter [μm]	Tenacity [GPa]
2.5	6.0
5.1	2.8
8.6	1.7
10.2	1.3
12.7	1.0
15.2	0.9
19.0	0.7

corresponding processes (e.g. by special yarn guides, few deflection points, etc.). The Si oxide can be "pre-damaged" by a caustic treatment, so that it has a higher temperature resistance.

Often glass fiber rovings are ring twisted (Fig. 43.18) to structure their surface, which increases the fiber-matrix adhesion in fiber composites. Typical spindle speeds are 1500–9000 1/min, common twist heights are 10–250/m (Fig. 43.18).

Glass fibers are usually named with a Latin letter which is a measure of their fineness (Table 43.6). Yarns are characterized acc. to a special system of letters and numbers. The first letter designates the type of glass (A, C, E etc.), the second letter the make-up (S—staple fiber, C—filament, T—air textured), the third letter the fiber fineness in accordance with Table 43.6. The first number indicates the yarn count in [100 yards/American pound] and the fraction indicates the structure of the yarn. The numerator indicates the number of rovings twisted together and the denominator indicates how many of those twists were combined. For example, the specification ECG 75 2/4 means that it is E-glass yarn made of filaments (C) with a single fiber fineness of 9–10.2 μm. The fineness of a roving is 7500 yards/454 g = 60 tex, two rovings are twisted together and the finished yarn consists of four such twists which have been twisted together.

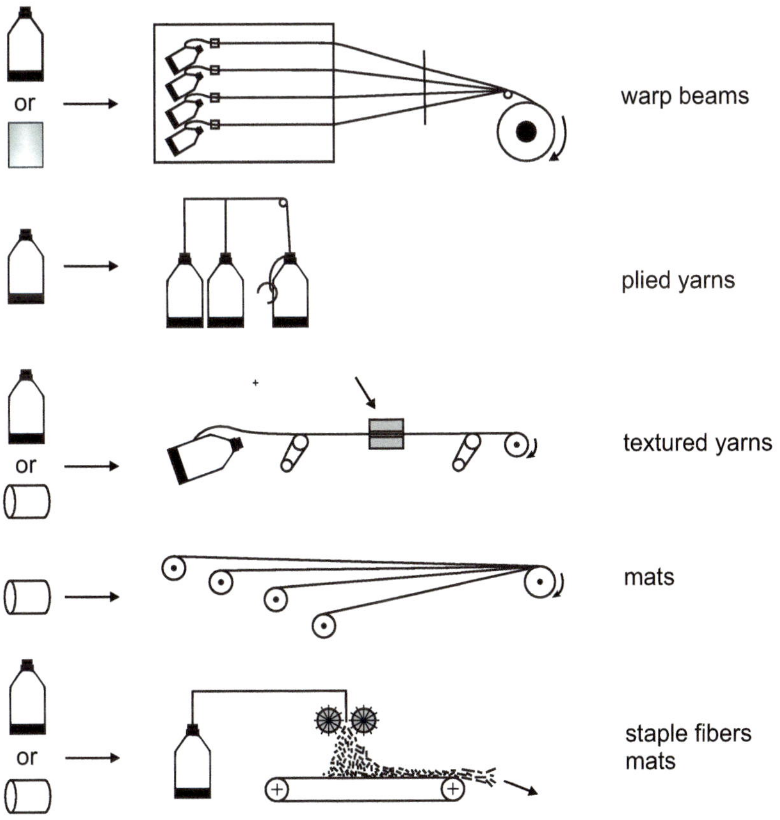

Fig. 43.17 Processes of further processing of glass rovings and fibers

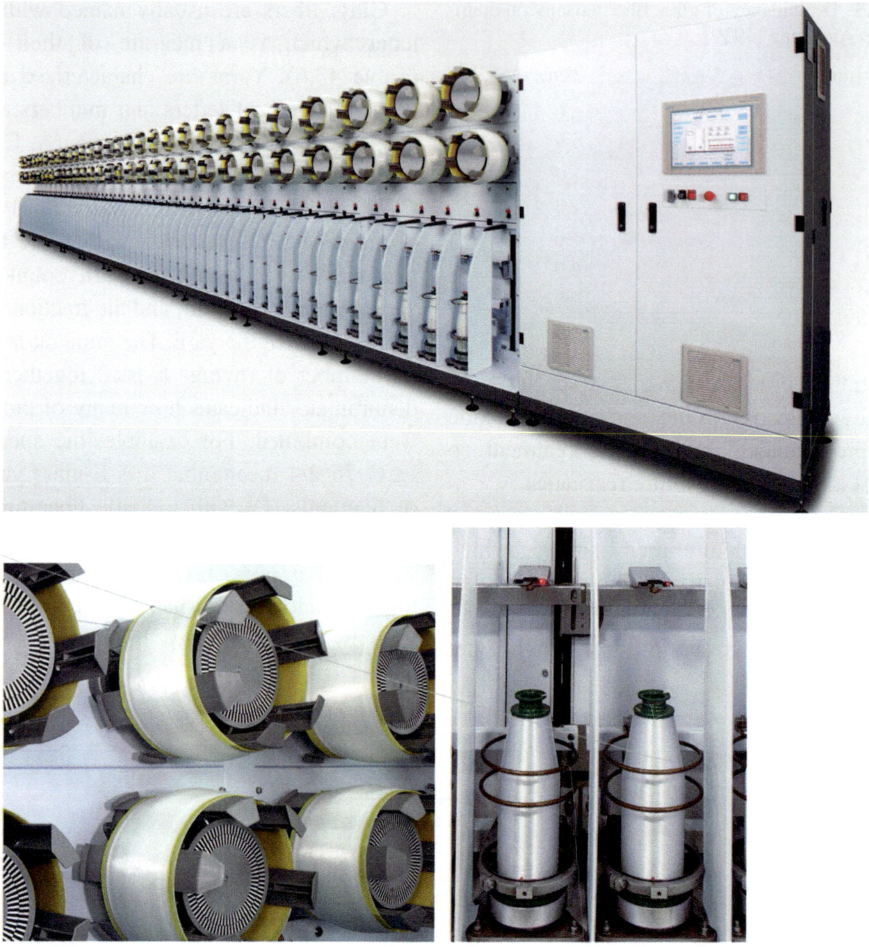

Fig. 43.18 Ring twisting machine for glass roving (top), creel (left) and ring twisting positions with balloon constriction rings (right) (Saurer Volkmann, Krefeld, n.d.)

The twisting direction of the yarn is normally opposite to that of the individual components.

43.9 Products

Glass filament yarns are mostly used to reinforce fiber composites and for products that must withstand high temperatures. Typical examples are (Fig. 43.19)

- sailboat and aircraft components,
- wind power rotor blades,
- filter (hot gases),
- electrical insulation (cable),
- thermal insulation (gaskets, fire blankets),
- fire protection (wallpapers, decorative fabrics),
- reinforcement of printed circuit boards (computers),
- reinforcement of concrete (facade elements).

Glass staple fibers are used as weft and warp material for fabrics and also in braids, e.g. for filters and gaskets. Mats made of glass staple fibers are used for thermal and acoustic insulation. In Germany, for example, a total of 770,000 t of insulation materials were produced in 2019, 80% of which were made from mineral or rock wool, and 20% from glass fibers (Statista, 2021).

Table 43.6 Naming of glass fibers acc. to their fineness (Wallenberger et al., 2001)

Abbreviation	Diameter [μm]
AA	0.8–1.2
A	1.2–2.5
B	2.5–3.8
C	3.8–5.0
D	5.0–6.4
E	6.4–7.6
F	7.6–9.0
G	9.0–10.2
H	10.2–11.4
J	11.4–12.7
Abbreviation	Diameter [μm]
K	12.7–14.0
L	14.0–15.2
M	15.2–16.5
N	16.5–17.8
P	17.8–19.0
Q	19.0–20.3
R	20.3–21.6
S	21.6–22.9
T	22.9–24.1
U	24.1–25.4

Binder-bonded nonwovens are used as dimensionally stable backing material for PVC floor coverings, sealing films, the reinforcement of bitumen as well as insulation boards and battery separators and for plaster reinforcement.

43.10 Market and Manufacturers

The capacity of all fiberglass manufacturers worldwide is currently estimated at 5.8 mt (Statista, 2021), while demand is around 5.3 mt. The annual growth is about 5–8% worldwide. Statistics usually distinguish between glass fibers (insulating materials, filters), glass yarns (<300 tex) and glass rovings (>300 tex). Around 80% of glass fibers are produced as rovings.

Table 43.7 shows selected manufacturers of glass fibers as well as their brand names and main products.

Fig. 43.19 Typical products made of glass fibers (Roving (Noise, 2005); computer board reinforced with glass fiber fabric (Smial, 2011); rotor blade of a wind turbine (Z22 (2013))

Table 43.7 Selected manufacturers of glass fibers

Producer	Trade name	Rovings	Staple fibers	Short fibers	Website	Country
3B-the fibreglass company	HiPer-tex®	x		x	https://www.3b-fibreglass.com	Belgium
China Jushi Co., Ltd.		x		x	https://www.jushi.com	China
Chongqing Polycomp Int. Corp.		x		x	https://en.cpicfiber.com	China
Johns Manville	StarRov®	x			https://www.jm.com/	USA
	ThermoFlow®			x		
	StarStran®			x		
Nippon Electric Glass	Lamion™	x			https://www.neg.co.jp	Japan
	ChopVantage™			x		
Owens Corning Fiberglass	THERMAFIBER®		x		https://www.owenscorning.com	USA
P-D Glasseiden GmbH, Oschatz	PD PowerStrand®	x		x	https://www.oschatz-glas.com	Germany
Rockwool International A.S.				x	https://www.rockwool.com	Denmark
Saint-Gobain Vetrotex	Vetrotex Zero Twist®	x	x	x	https://www.vetrotextextiles.com/	France
Taishan Fiberglass Inc.		x			http://www.ctgf.com	China

References

Anderegg, F.O. (1939). Strength of glass fibers. In *Industrial engineering chemistry* (Vol. 31, pp. 290–298). ACS, .

Ehrenstein, G. W. (1981). *Glasfaserverstärkte Kunststoffe*. Expert Verlag.

Eurima. (2020). https://www.eurima.org/about-mineral-wool/production-process.html

Falbe, J., & Regitz, M. (1990). *RÖMPP–Chemie Lexikon, Bd. 2, 3 u. 5*. Georg Thieme Verlag.

von Falkai, B. (1981). *Synthesefasern–Grundlagen, Technologie*. Verlag Chemie.

Giegerich, W., & Trier, W. (1964). *Glasmaschinen*. Springer.

Jones, F. R., Huff, N. T. (2009). The structure and properties of glass fibres. In *Handbook of textile fibre structure*. Woodhead Publishing.

Liebscher, V. (1983). *Faserstoffe–Begriffe, Eigenschaften, Bezeichnungen, Technische Textilien 26/1* (pp. 1–44). Deutscher Fachverlag.

Loewenstein, K. L. (1983). The manufacturing technology of continuous glass fibres, In *Glass science and technology* (Vol. 6). Elsevier.

Mohr, J. G., & Rowe, W. P. (1978). *Fiber Glass*. Van Nostrand Reinhold.

Noise, D. (2005). *CC BY-SA 3.0*. https://commons.wikimedia.org/wiki/File:Glasfaser_Roving.jpg

P-D Glasseiden GmbH Oschatz. (n.d.). https://www.oschatz-glas.com/de/index.php

Petzold, A., & Hinz, W. (1979). *Silikatchemie–Einführung in die Grundlagen*. Ferdinand Enke Verlag.

Pico D. (2016). Glass fibers. In *Ullmann's polymers and plastics: Products and processes* (1st ed). Wiley VCH.

Saurer Volkmann, Krefeld. (n.d.). http://www.saurer.com

Skolnikov, J. A., Polik, B. M., Kocarov, E. P., & Nigin, E. R. (1969). *Steklannoe stapel'noe volokno*. Isd.

Smial. (2011). *CC BY-SA 3.0*. https://upload.wikimedia.org/wikipedia/commons/8/82/Supermicro_dual_opteron_server_board_cpu_socket_IMGP7338_wp.jpg

Statista. (2021). http://www.statista.de

Teschner, R. (2013). *Glasfasern*. Springer Nature.

Wallenberger, F.T., Watson, J.C., Li, H. (2001), Glass Fibers In: *ASM handbook*, 21, ASM International, Materials Park.

Weiss, G. (1966). *Ullstein Gläserbuch*. Ullstein.

Winkelmann, A., & Schott, O. (1894). Über thermische Widerstandskoeffizienten verschiedener Gläser in ihrer Abhängigkeit von der chemischen Zusammensetzung. *Annalen der Physik, 51*.

Worrell, E., Galitsky, C., Masanet, E., & Graus, W. (2008). *Energy efficiency improvement and cost saving opportunities for the glass industry*. Berkley Lab.

Wulfhorst, B., Kaldenhoff, R., & Hörsting, K.-H. (1993). Glass fibers, Fiber table acc. To P.-a. Koch. In *Technische Textilien* (Vol. 36, pp. 68–85). Deutscher Fachverlag, .

Xanthos, M. (2010). *Functional fillers for plastics*. Wiley-VCH.

Z22. (2013). *CC BY-SA 3.0*. https://commons.wikimedia.org/wiki/File:Fiberglass-reinforced_epoxy_blades_of_Siemens_SWT-2.3-101_wind_turbines.jpg?uselang=de

44 Basalt Fibers

44.1 Basalt Occurrence

When the Earth's mantle melts, thin, SiO_2-poor magma is formed, which solidifies into basalt lava when it emerges on the Earth's surface. Basalt is grey to black in color (Fig. 44.1).

Acc. to estimates, about 13% of the earth's crust consists of basalt, so the deposits are almost inexhaustible. Even the moon consists partly of basalt.

44.2 Chemical Composition and Structure

The chemical composition of basalt fibers is very similar to that of glass fibers. The main difference lies in the absence of boron oxide. Depending on the provenance of the basalt, the content of the respective metal oxides also varies. Table 44.1 summarizes typical values.

The internal structure of basalt fibers is essentially the same as that of glass fibers. Basalt fibers are also present as a solidified melt and do not contain any crystalline areas.

44.3 Fiber and Yarn Production

Yarn production is similar to that of glass fibers. However, the melting temperature of basalt, at approx. 1450 °C, is somewhat higher than that of glass. The equipment and corresponding components are the same as for glass fiber production (Fig. 44.2).

As with glass production, the main production processes for rock wool are centrifugal blowing and centrifugal rolling (rock wool process). Basalt is also used to produce rovings, which are processed in a similar way to glass rovings.

44.4 Fiber Types

Depending on the composition of the basalt used, different properties and fiber types result. In contrast to glass fibers, these cannot be specifically adjusted. On the one hand, this eliminates the need to mix the necessary components as with glass, but the mechanical properties in particular vary depending on the provenance of the basalt (Fig. 44.3).

44.5 Properties

The basalt fiber has a greenish-brown color. The physical properties and thus the areas of application are similar to those of glass fiber.

Typical fiber diameters are 8–20 μm, usual lengths of the staple fibers are 0.1–100 mm. In the form of insulating wool, they are thus similar in thickness to glass fibers and very brittle. They are heat resistant up to 800 °C, depending on the rock composition. Their thermal conductivity coefficient is around 1.7 W/m·K, their specific

D. Veit, *Fibers*, https://doi.org/10.1007/978-3-031-15309-9_44

Fig. 44.1 Basalt columns (Radu, 2007) and basalt stones (DBF Deutsche Basalt Faser GmbH, n.d.)

heat capacity is 0.86 kJ/kg·K and their glass transition temperature is around 1000 °C. Continuous basalt fibers are currently used mainly in fiber composite applications (e.g., automotive). Concrete reinforcement applications are also conceivable because the alkali resistance of basalt fibers is quite good. In general, basalt fibers can in principle be used almost anywhere where glass fibers are used today.

Superfine basalt fibers are now available with fiber diameters of 2–3 μm. Their density is around 2–2.8 g/cm^3, their strength reaches 70–92 cN/tex and their modulus of elasticity is 80–110 GPa. Their moisture absorption is around 0.1–2%.

Especially for applications under high temperature or high humidity, basalt fibers may even be superior to glass fibers, as Table 44.2 for typical basalt and glass fibers shows.

Table 44.1 Composition of basalt fibers (weight percent)

	E-glass	Basalt
SiO_2	52–56	52–58
Al_2O_3	12–16	17–18
CaO	16–25	5–8
MgO	0–5	1–4
B_2O_3	5–10	-
Na_2O	0.8	2.5–6.4
K_2O	0.2–0.8	0.8–4.5
Fe_2O_3	< 0.3	4–10

44.6 Further Processing

Basalt fibers are processed on the same machinery and by the same processes as glass fibers, mainly into woven, non-crimp fabrics and nonwovens (Fig. 44.4).

44.7 Products

Basalt fibers are used as reinforcing fibers in fiber-plastic composites or as heat protection material.

The main problem standing in the way of wider use is the very different quality (composition) of the basalt from which the fibers are obtained. Thus, it is almost impossible to produce a consistent quality of fibers, because even basalts of the same deposit can have different compositions. On the other hand, their colour is similar to that of carbon and their price is much lower, so they still have a market.

44.8 Market and Manufacturers

Basalt fibers have been produced since the 1950s, mainly in Russia and Georgia. In the meantime, there are also producers in China, the USA and Germany. The price is around 3–5 €/kg for

Fig. 44.2 Spinneret, take-up and bobbin (DBF Deutsche Basalt Faser GmbH, n.d.)

Fig. 44.3 Basalt fibers in longitudinal view

Table 44.2 Comparison of the properties of typical basalt and glass fibers

	Basalt	Glass
Density [g/cm^3]	2.8	2.54
Water absorption [%]	0.5	1.0
E-modulus [MPa]	100,000	70,000
Breaking force [MPa]	1,850–2,150	2,450–3,400
Breaking force [%] at rel. humidity of 100 %		
starting point	100	100
after 1 month	92	74
after 2 months	91	72
Breaking force [%]		
at 20 °C	100	100
at 200 °C	95	92
at 400 °C	82	52

Fig. 44.4 Typical products made from basalt fibers (DBF Deutsche Basalt Faser GmbH, n.d.)

Table 44.3 Selected manufacturers of basalt fibers

Producer	Trade name	Website	Country
BASALTEX NV		http://www.basaltex.com	Belgium
Basalt fibertec		https://www.basalt-fibertec.ch	Switzerland
Basalt Fiber Tech		http://www.basaltft.com/	Australia
DBF Deutsche Basalt Faser GmbH		https://www.deutsche-basalt-faser.de	Germany
Hengdian Group		http://www.hg-gbfbasalt.com	China
Kamenny Vek	Basfiber®	https://basfiber.com	Russia
Mafic		https://www.mafic.com/	USA
Sudaglass Fiber Technology, Inc.		http://www.sudaglass.com	USA
Technobasalt-Invest LCC	Technobasalt®	https://technobasalt.com	Ukraine

rovings. The world production quantity is estimated at 20,000 t/year (Table 44.3).

References

DBF Deutsche Basalt Faser GmbH. (n.d.). *Sangerhausen*. https://www.deutsche-basalt-faser.de/

Radu, C. (2007). https://commons.wikimedia.org/wiki/File:DetunataGoala.jpg

Ceramic Fibers

45

Ceramic fibers consist of metal oxides, carbides or nitrides or corresponding mixtures. Accordingly, ceramic fibers are often divided into oxidic and non-oxidic fibers (Fig. 45.1). Ceramic fibers consist of similar basic materials as glass fibers. However, glass fibers are produced by melt spinning and ceramic fibers are produced by pyrolysis. Their main distinguishing feature from all other fibers is an extremely high melting point of usually over 2000 °C.

45.1 Development and History

Research on and development of ceramic fibers began in the 1960s. From the early 1970s, alumina fibers were first produced for insulation material that could withstand high temperatures. Since the 1990s, many new ceramic fibers have been developed and produced. This was mainly due to the increased use of technical textiles, especially in high temperature applications (filters, insulation, etc.).

45.2 Raw Materials and Structure

The starting materials for ceramic fibers are usually silicon oxide, aluminum oxide and possibly various additives (e.g. boron oxide, zirconium oxide). However, there are also special fibers made of other metal oxides. Ceramic fibers are usually highly crystalline or polycrystalline, but there are also amorphous types.

A major challenge in the production of ceramic fibers is their very low elongation. It is a major problem in textile processing and easily leads to fiber breakage. To circumvent this, either very fine fibers are produced or a mixture of several oxides (e.g. SiO_2 and Al_2O_3) is used, which increases the elongation.

Due to their oxidic structure, ceramic fibers cannot oxidize further when heated, which explains their high temperature resistance.

45.2.1 Oxide Ceramics

Silicon Oxide

Silica fibers consist of silicon and oxygen. One silicon atom is connected with fours oxygen atoms and forms a tetrahedron (Fig. 45.2, left). Silicon atoms share the oxygen atoms (Fig. 45.2, right), resulting in the empirical formula SiO_2.

SiO_2 occurs in nature in amorphous and crystalline form and is therefore available in large quantities and relatively cheaply. It is soluble in water, with solubility increasing with increasing temperature and pH. Apart from hydrofluoric acid (HF), SiO_2 has a very high acid resistance. Its density is 2.2–2.7 g/cm^3, depending on the configuration, and its melting point is approx. 1700 °C.

D. Veit, *Fibers*, https://doi.org/10.1007/978-3-031-15309-9_45

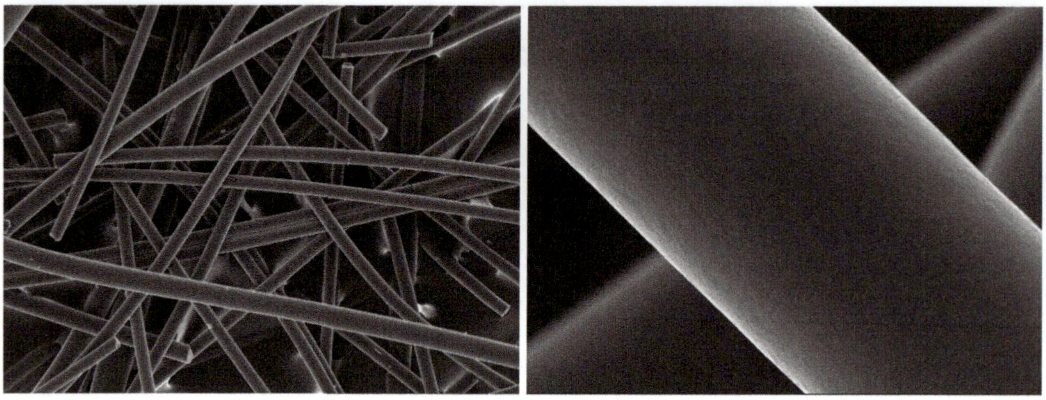

Fig. 45.1 PCW ceramic fibers (Mitsubishi Chemical Corporation, n.d.)

Alumina

Aluminum oxide has the empirical formula Al_2O_3 and is also known as corundum. It is usually extracted from bauxite and is therefore available in large quantities at low cost (approx. 80 ct/kg). The crystal structure is mostly cubic or rhombohedral. The density is 3.7–4.0 g/cm^3, depending on purity, and the melting point is about 2050 °C. Therefore, these fibers cannot be melt spun, but are usually produced by the sol-gel process. For special products, the thermal conversion of organic textiles impregnated with aluminum oxide salt by pyrolysis is also used. The gemstone sapphire consists of Al_2O_3, therefore products made of alumina ceramics are often advertised with the addition "sapphire" for advertising reasons.

Cubic γ-Al_2O_3 is not soluble in water, but is soluble in acids and bases. At temperatures above 800 °C the crystal structure changes to the rhombohedral-form and is then well resistant to acids and bases.

Silica-Alumina (Mullite)

In order to combine the property profiles of silicon and aluminum oxide, mixtures of both

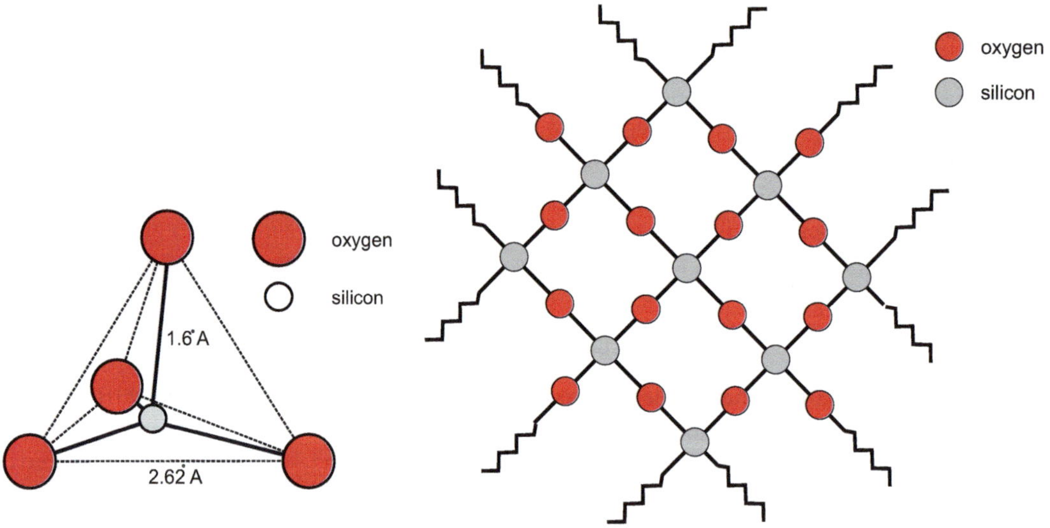

Fig. 45.2 Structure of silicon dioxide (tetrahedron) and internal structure of SiO_2

Fig. 45.3 Insulating fleece made of Al_2O_3 ceramic fibers

materials are often used. Often the proportions are 1:2, in extreme cases only 2–4% SiO_2 are added to the Al_2O_3. Up to an Al_2O_3 content of 45–60%, such fibers can be produced by melt spinning. With higher contents of alumina, the viscosity becomes too high and the sol-gel process is used. With higher contents of Al_2O_3, both, the thermal resistance and the modulus of elasticity increase.

Staple fiber mats for insulation can be directly fabricated from aluminosilicates (Chatterjee et al., 2002), Fig. 45.3.

Zirconia

ZrO_2 occurs naturally (baddeleyite), but is usually obtained from $ZrSiO_4$ by washing, purification and calcination processes for cost reasons. It exists in monoclinic, tetragonal, and cubic forms, with the cubic form having the highest melting point at 2680 °C. It is highly resistant to acids and alkalis and has a density of about 6 g/cm^3. Its price is about 1.50 €/kg. The fibers of ZrO_2 are mostly produced in the sol-gel process.

Magnesium Oxide

Magnesium oxide with the molecular formula MgO occurs naturally as the mineral periclase and crystallizes in cubic form. For cost reasons, it is usually obtained by burning magnesium in air. Its density is 3.6 g/cm^3, its melting point is 2800 °C, which is very high. The price of magnesium is about 2.50 €/kg. The fibers are produced in a sol-gel process.

Fig. 45.4 Structural formula of yttrium oxide

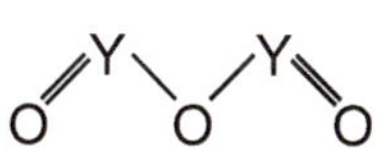

Yttria

Yttrium oxide has the molecular formula Y_2O_3 and crystallizes monoclinic or tetragonal (Fig. 45.4). Its density is very high at 5 g/cm^3 and its melting point is 2410 °C. The corresponding fibers are stable up to 1400 °C. The world market price is about 3 €/kg. Yttrium oxide fibers are spun out of YCl_3 from a hot solution under high pressure and then heat treated.

Titanium Dioxide

Titanium dioxide, TiO_2, occurs naturally in tetragonal and orthorombic form. It is mainly used as a white pigment. Its density is 3.9–4.2 g/cm^3, depending on the crystal structure, and its melting point is 1855 °C. The world market price is approx. 3 €/kg. TiO_2 fibers are produced in the sol-gel process.

TiO_2 is particularly interesting for textile applications because of its property as a photocatalyst. In the presence of TiO_2, a so-called photooxidation takes place under the influence of UV radiation and water. In this process atmospheric oxygen is activated, which can be used, for example, to reduce the exhaust gas load in a filter. If TiO_2 is applied to a textile, this reaction may also destroy the textile. Therefore, it makes sense to produce the textile directly from titanium dioxide fibers (Fig. 45.5).

Fig. 45.5 Textile building shell with integrated TiO_2 (Jan Serode)

45.2.2 Non-oxide Ceramics

Silicon Carbide
SiC crystallizes rhombohedral, has a density of 3.2 g/cm^3 and decomposes above 2300 °C. It does not occur in nature and must therefore be produced artificially. One way to do this is to react SiO_2 with carbon at approx. 1600–2000 °C acc. to.

$$SiO_2 + 3\,C \rightarrow SiC + 2\,CO.$$

However, industrially produced silicon carbide does not consist of equal parts of carbon and silicon. Typical C/Si weight ratios acc. to a compilation by Mahltig and Kyosev (2018) are (57/32), (62/37) and (69/31). Correspondingly, atomic number ratios C/S of 1.31 and 1.39 and 1.05. In addition, there may be impurities from oxygen as well as SiO_2 and graphite. Typical crystal sizes are 0.1–0.5 μm.

SiC fibers are prepared either by vapor deposition or by spinning out a precursor polymer, which is then heat treated. Typically, the precursor polymers contain alkyl and aromatic segments in which Si atoms are incorporated. Some examples are shown in Fig. 45.6.

Boron Nitride
Boron nitride (BN) usually has a cubic (c-BN) or a hexagonal (h-BN) crystal structure. Its density is 2.25 (cubic) to 3.45 (hexagonal) g/cm^3, its melting point is about 2970 °C. c-BN has a thermal conductivity of about 400 W/mK, which is approximately that of pure copper. Nanotubes (BNNTs) can be made from h-BN (Golberg et al., 2010), with diameters of a few nanometers and lengths of a few micrometers. BNNTs act as wide bandgap semiconductors. They also possess inherent magnetic properties that can be induced by external effects. Bansal et al. (2005) report that the addition of as little as 4 wt% BNNT to a glass melt nearly doubles the strength of the resulting composite. However, BN nanotubes are very brittle, so attempts have been made to produce them in fiber form, with similar properties but less brittle.

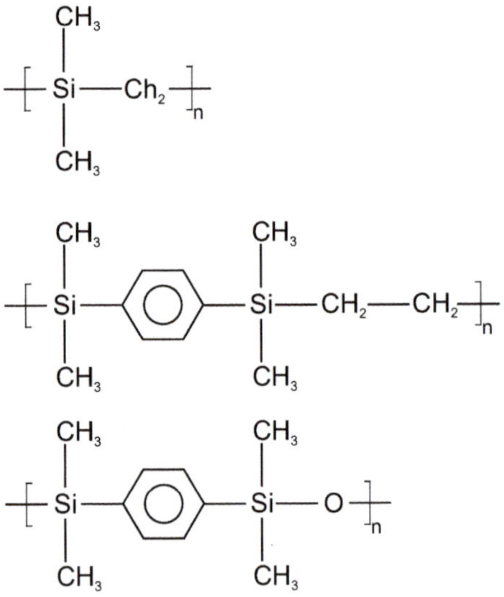

Fig. 45.6 Examples of precursor materials for SiC fiber production

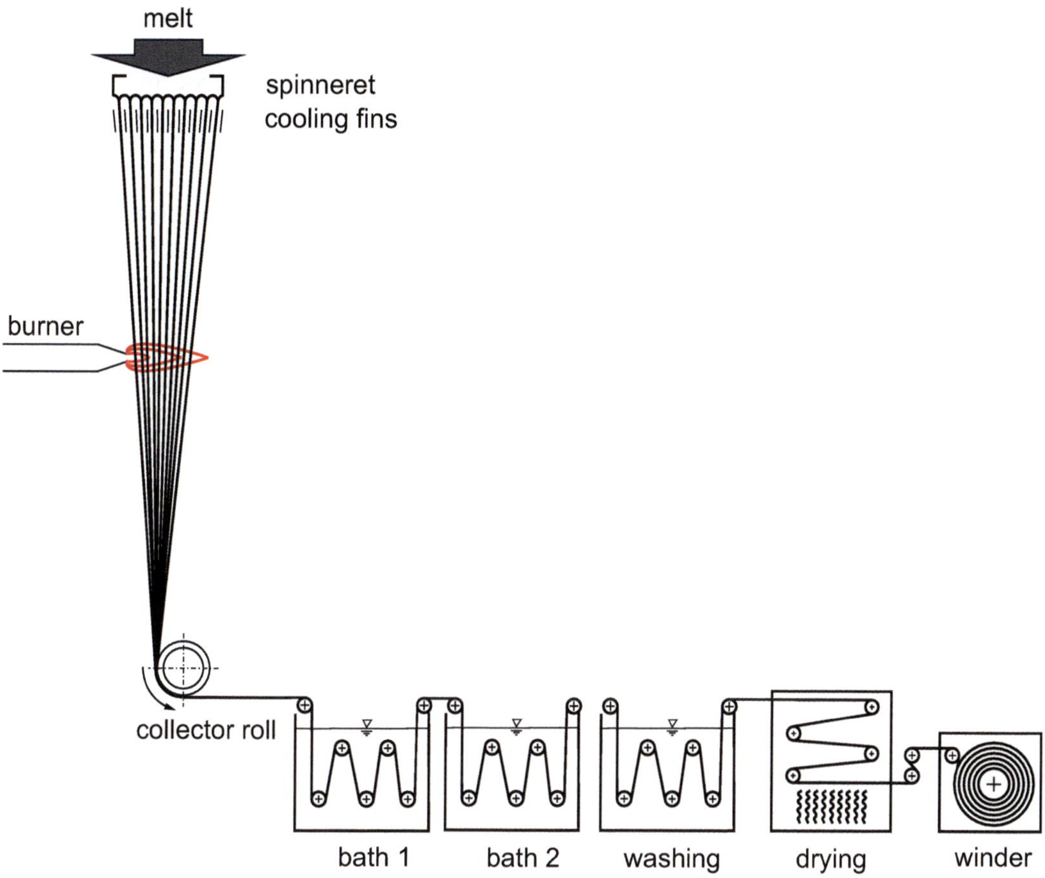

Fig. 45.7 Melt spinning process for filaments

45.3 Fiber and Yarn Production

Silicon and alumina-based ceramic fibers can usually be produced by the melt-spinning process, analogous to glass spinning (Cooke, 1991). Other ceramics are often produced by the sol-gel process and in special cases other processes are also used.

45.3.1 Melt Spinning of Staple Fibers

This process is used to produce ceramic staple fibers. It is essentially the same as the process used to produce glass staple fibers. The melt is spun out through a spinneret and the droplets formed are then stretched into short fibers in an air stream. Alternatively, rotating discs are used, which draw out the melt into short and longer fibers by centrifugal force. Aluminum silicate fibers, for example, are produced in this way. Typical fiber diameters are 1–10 μm, the fiber length several centimeters (Cooke, 1991).

45.3.2 Melt Spinning of Filaments

To produce filaments from SiO_2, a mixture of 75% SiO_2 and 25% Na_2O is melted and spun out in a typical glass melting furnace at approx. 1100 °C. The resulting filaments are then passed by a gas flame and simultaneously stretched. The resulting filaments are then passed over a gas flame and simultaneously stretched. The sodium oxide is then removed in two acid baths. The resulting filaments are washed and dried at 315 ° C. Fiber diameters of 1.5 μm are achievable and the SiO_2 content reaches 99% (Fig. 45.7).

45.3.3 Chemical Vapor Deposition

In this process (chemical vapor deposition–CVD) for the production of ceramic fibers, a chemical reaction takes place on the precursor fiber made of carbon or metal with boron, silicon carbide, etc. The precursor fibers are usually first surface-treated in the first furnace and then coated with a material from the gas phase in the second furnace at temperatures of approx. 1200 °C by means of a chemical reaction. Boron carbide, boron nitride and silicon carbide fibers with a corresponding core can be produced in this way. The finished fibers have diameters of 100–200 μm and are therefore difficult to process in textiles due to their high bending stiffness.

A typical example of this process is the deposition of SiC from CH_3 $SiCl_3$ on a tungsten wire (Ø 20 mm) or a carbon filament at 1000–1300 °C with the release of hydrochloric acid (Hollemann & Wiberg, 1995).

45.3.4 Thermal Conversion by Pyrolysis of Precursor-Impregnated Fibers

In this process, organic polymers are first spun out and processed into textiles. These are then impregnated with a metal salt solution (precursor) and dried. They are then pyrolyzed at 500–800 °C in the presence of air. This removes the organic material and oxidizes the metal salts, forming ceramic fibers. Table 45.1 gives an overview of typical material combinations that are processed in this way.

45.3.5 Sol-Gel Process

This process is a special case of the thermal conversion of polymer-based precursor fibers described above. Usually, low-molecular metal alcoholates are brought into solution (hence "sol") and then combined in a polymer formation reaction to form long chains (aggregation). With increasing chain length and decreasing solvent concentration, networks of macromolecules are formed, the so-called gels. The viscosity increases in the process. Fibers can then be produced from this solution, e.g. using the dry spinning process. After spinning, the solvent is removed by drying and organic residues by pyrolysis, and the resulting fibers are sintered under pressure and temperature. As a result, they become denser and the fiber diameter decreases (Fig. 45.8).

Due to the evaporation of the solvent and the shrinkage of the fiber during sintering, it is difficult to produce fibers with a round cross-section using this process. At sufficiently high temperatures during the sintering process, crystallization occurs, which can improve the mechanical properties (Fig. 45.9).

By adding suitable chemicals to the spinning solution, the range of properties can be varied within wide limits. This process is mainly used to produce ceramic fibers based on silicate and aluminum. Usual spinning speeds are 100–250 m/min.

A typical example of this process is the production of aluminum silicate fibers (Fig. 45.10). For this purpose, aluminum nitrate is first converted into aluminum hydroxide in aqueous solution. Tetraethoxysilane (TEOS) as precursor is acid hydrolyzed in ethanol with the addition of

Table 45.1 Textiles from metal oxides as well as precursors and process temperatures acc. to Mahltig and Kyosev (2018)

Metal oxide	Precursor	Temperature of of pyrolisis [°C]
Al_2O_3	$AlCl_3$	800
ZrO_2	$ZrOCl_2$	500
$ZrO_2/Y2O_3$	$ZrOCl_2/YCl_3$	650
Ta_2O_5	Tantaloxalate	600
Nb_2O_5	Nioboxalate	500
TiO_2	$TiCl_3$	600

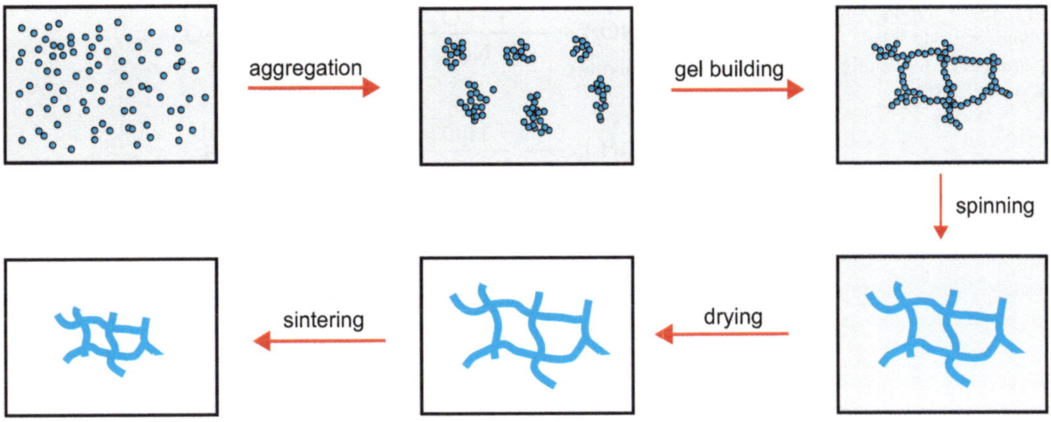

Fig. 45.8 Process steps in the sol-gel spinning process

water in the presence of aluminum nitrate. This forms an aluminum silicate sol. It is then mixed with, for example, polyvinyl butyral (PVB) and the polycondensation of the polymer forms the spinning solution. After spinning out, the fibers are sintered and the polymer is completely removed.

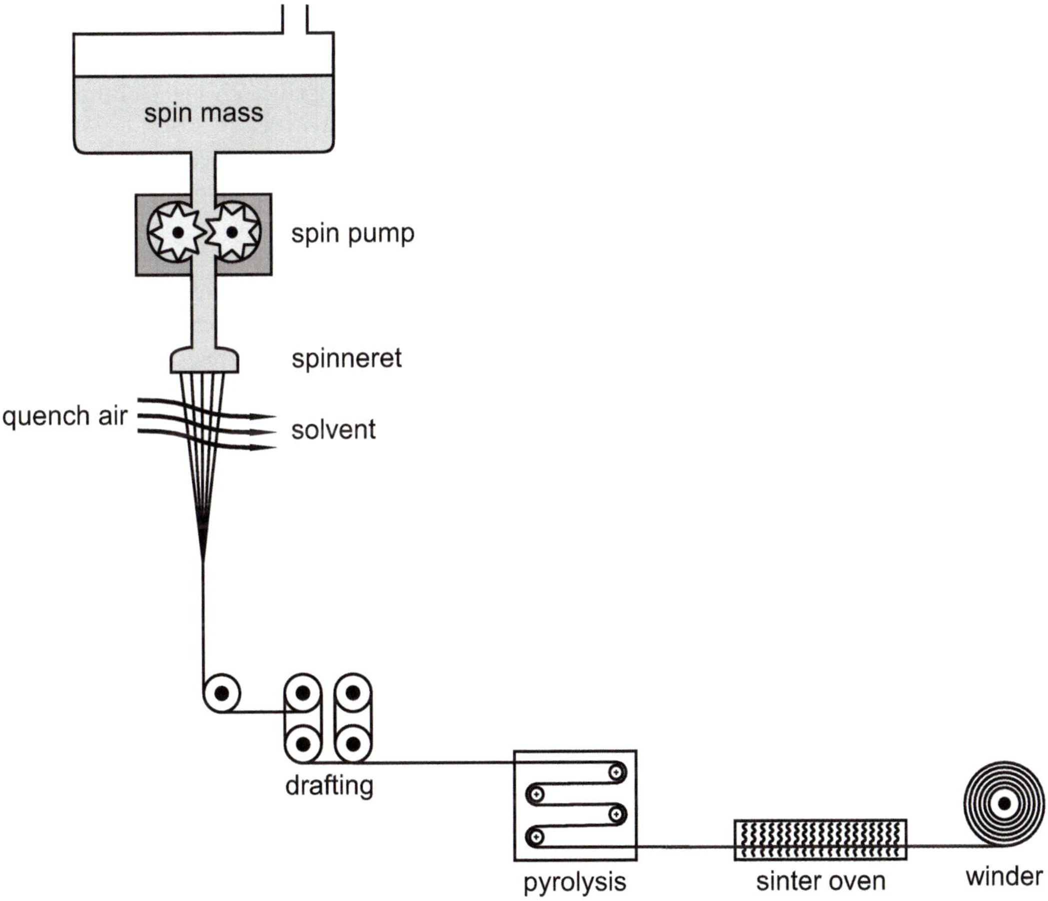

Fig. 45.9 Principle of the sol-gel process incl. heat treatment

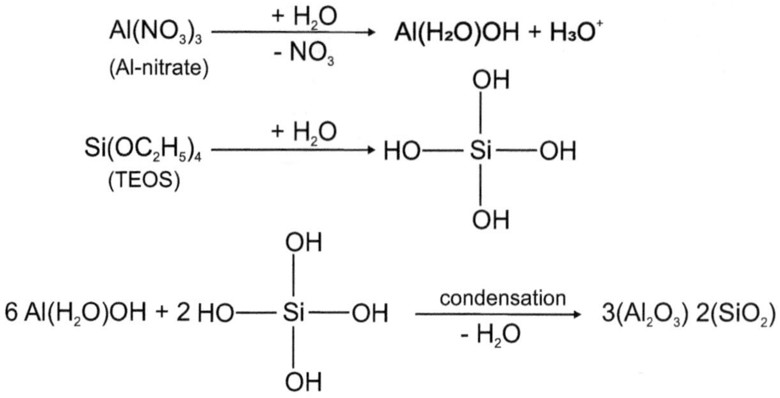

Fig. 45.10 Production of aluminosilicate solar particles acc. to Mahltig and Kyosev (2018)

45.3.6 Other Processes

In addition to the above, there are also special processes for producing ceramic fibers. These include spinning from a dispersion of crystalline ceramic particles using melt or dry spinning processes. The deposition of ceramic particles onto a heated core (e.g. a wire or carbon filament) is used to produce, for example, boron fibers. Ceramic fibers can also be produced by direct crystal growth. This makes them extremely pure and anisotropic due to the completely crystalline orientation of the atoms. However, such polycrystalline fibers can also exhibit isotropic properties. A typical application is optical fibers.

45.4 Properties

The most extraordinary property of ceramic fibers is their extremely high temperature stability. The mechanical properties only change at temperatures above 1200–1400 °C, in an inert atmosphere even often above 1600 °C. Ceramic fibers also have good tensile strength (1000–6000 MPa) and a high modulus of elasticity of mostly 200–400 GPa. They are therefore particularly suitable as reinforcement material for metal and other ceramic matrices. Their elongation at break is very low at 1%, furthermore they are very brittle and therefore difficult to process. The usual diameter of ceramic fibers is between 10 and 20 μm, with special types 1–2 μm.

Ceramic fibers are also very resistant to most acids, especially after heat treatment. However, phosphoric acid still attacks silicate and aluminum silicate ceramics. Resistance to alkalis is usually poor.

In Fig. 45.11, the mechanical properties of selected ceramic fibers are shown. Ceramic fibers made of Al_2O_3/SiO_2 have average strengths of up to approx. 2000 MPa, while the values for fibers made of SiC are significantly higher. Boron fibers have very high strengths.

The density of ceramic fibers is 2.5–6.1 g/cm^3, depending on the composition (Table 45.2).

Ceramic fibers have a good electrical insulating effect and are non-combustible.

Table 45.3 contains the melting temperature and the maximum service temperature in oxidizing and non-oxidizing atmospheres for selected ceramic fibers.

The heat capacity of SiC fibers is 0.7–1.2 kJ/kgK (25 °C) and 1.15–1.17 kJ/kgK (500 °C), respectively, for Al_2O_3 the values are 0.8–1.2 kJ/kgK. The thermal conductivity of SiC fibers is 8–20 W/mK.

45.5 Further Processing

Ceramic staple fibers are often processed into nonwovens, rarely into yarns. Filaments can be woven and braided. However, care must be taken not to use bending radii that are too tight, otherwise there is a risk that the ceramic fibers will break due to their brittleness.

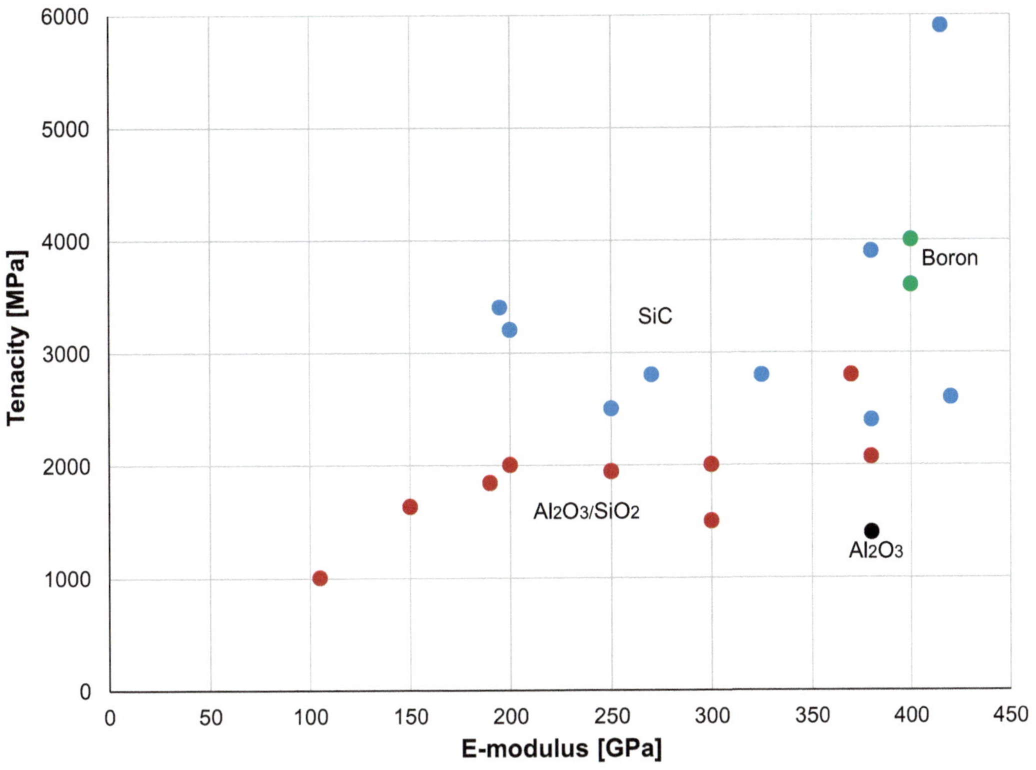

Fig. 45.11 Typical values for strength and modulus of elasticity of ceramic fibers

45.6 Products

Ceramic fibers are produced in very different forms. Very short fibers (a few μm to a few mm; whisker) are used to fill and seal expansion joints or in the form of yarns (e.g. with a steel core). Longer fibers are processed into nonwovens and filament yarns into woven or braided fabrics.

Products made of ceramic fibers are mainly used where very high temperatures prevail permanently and other fiber materials fail. Typical examples are high-temperature dust filters, insulating materials at temperatures above 1000 °C and textiles for fire protection. They are also used in space travel, e.g. in components designed to protect against micro-meteorite impacts.

Table 45.2 Density of selected ceramic fibers

	[g/cm^3]
SiC	2.48–3.1
Al_2O_3	3.4–4.1
Al_2O_3 SiO_2	2.7–3.5
ZrO_2	5.68–6.1
B	2.5–2.6

Ceramic fibers that are used for thermal insulation are also referred to as high-temperature wool. A distinction is made between three groups: The Aluminum Silicate Wools (ASW), which consist of a mixture of Al_2O_3 and SiO_2 and are considered carcinogenic. The alkaline earth silicate wools (AES), which consist of a mixture of CaO, MgO, SiO_2 and ZrO_2 and the polycrystalline wools (PCW), which consist of at least 70% Al_2O_3 . They are not considered carcinogenic.

Due to their complicated manufacturing process, ceramic fibers are usually considerably more expensive than other materials and usually cost 100–800 €/kg, depending on the material and purity. Very special ceramic fibers cost 4000 €/kg and more.

Table 45.3 Characteristic temperatures of inorganic fibers acc. to Cooke (1991)

Maximum use temperature [°C]			
Composition	Melting temp. [°C]	Oxidizing atmosphere	Non-oxidizing atmosphere
B	1,260	560	1,200
SiO_2	1,660	1,060	1,060
Al_2O_3-SiO_2-B_2O_3	1,740	1,427	1,427
Al_2O_3-SiO_2	1,760	1,300	1,300
Al_2O_3-SiO_2-Cr_2O_3	1,760	1,427	1,427
Si_3N_4	1,900	1,300	1,800
Al_2O_3	2,040	1,540	1,600
ZrO_2	2,650	1,650	1,650
SiC	2,690	1,800	1,800
BN	2,980	700	1,650
C	3,650	400	2,500

45.7 Market and Manufacturers

Ceramic fibers are currently produced mainly in Japan and the USA. Table 45.4 lists important manufacturers of ceramic fibers and some trade names.

DuPont (Fiber FP™) and ICI (Saffil™), pioneers in ceramic fiber production, have since sold their ceramic fiber businesses or no longer exist.

Table 45.4 Selected manufacturers of ceramic fibers and trade names

Producer	Trade name	Ceramic type	Website	Country
3M Co.	Nextel 312™	Al_2O_3 SiO_2	https://www.3mdeutschland.de	USA/ Germany
	Nextel 440™	Al_2O_3 SiO_2		
	Nextel 610™	Al_2O_3		
	Nextel 720™	Al_2O_3 SiO_2		
American Elements		Al_2O_3	https://www.americanelements.com	USA
COI Ceramics, Inc.	Sylramic™	SiC	http://www.coiceramics.com	USA
Mitsubishi Chemical Corp.	MAFTEC™	Al_2O_3	https://www.m-chemical.co.jp	Japan
NGS Advanced Fibers Co., Ltd.	Nicalon™ CG	SiC	http://www.ngs-advanced-fibers.com	Japan
Nippon Carbon Co., Ltd.	Nicalon™ SiC	SiC	https://www.carbon.co.jp	Japan
	Nicalon™ Type S	SiC		
	Nicalon™ Hi SiC Low O	SiC		
Specialty Materials Inc.	SCS-6™	SiC	http://www.specmaterials.com	USA
	SCS-Ultra™	SiC		
	4-mil Boron™	B		
	5,6-mil Boron™	B		
UBE Industries Ltd.	Tyranno Fiber™ ZMI	SiC	https://www.ube.com	Japan
	Tyranno Fiber™ SA	SiC		
ZIRCAR Ceramics, Inc.	ALBF™	Al_2O_3	https://www.zircarceramics.com	USA

References

Bansal, N. P., Hurst, J. B., & Choi, S. R. (2005). Boron nitride nanotubes-reinforced glass composites. *Journal of the American Ceramic Society, 89*(1), 388–390.

Chatterjee, M., Naskar, M. K., Chakrabarty, P. K., & Ganguli, D. (2002). Mullite fiber mats by a sol-gel spinning technique. *Journal of Sol-Gel Science and Technology, 25*, 169–174.

Cooke, T. F. (1991). Inorganic fibers–a literature review. *Journal of the American ceramic society, 74*(12), 2959–2978.

Golberg, D., Bando, Y., Huang, Y., Terao, T., Mitome, M., Tang, C., & Zhi, C. (2010). Boron nitride nanotubes and nanosheets. *ACS Nano, 4*, 2979–2993.

Hollemann, A. F., & Wiberg, W. (1995). *Lehrbuch der anorganischen Chemie*. Walter de Gruyter.

Mahltig, B., & Kyosev, Y. (2018). *Inorganic and composite fibers*. Woodhead Publishing.

Mitsubishi Chemical Corporation. (n.d.). https://www.m-chemical.co.jp

Part VIII

Metal Fibers

Metal Fibers

46

46.1 Basics

Metal fibers have been manufactured for more than 5000 years. However, there are no standards which clearly characterize the term metal fiber or deal with the production and use of metal fiber. Furthermore, there is usually little distinction between metal fibers, fine metal wire and metal wire. Whether it is a wire or a fiber depends on the diameter and also on the manufacturing process. Acc. to the German standard DIN 60001 E sheet 2, semi-finished products of unlimited length with mostly round cross-section with diameters below 100 µm are called filament. If this diameter is exceeded, the semi-finished product is called monofilament or wire.

Metal staple fibers (Fig. 46.1) can be produced from such filaments or monofilaments by breaking, among other things. Acc. to diameter, a distinction is often made between fine wire (30 µm to 1.4 mm), ultra-fine wire (10–30 µm) and metal fibers (2–40 µm).

The abbreviation for metal fibers is MTF acc. to DIN 60001 Part 4.

A detailed description of all important aspects of metal fiber production and properties is given in (Mac et al., 2004). This chapter is largely based on this publication.

46.2 Development and History

The history of metal fibers goes back to about 3000 BC. At that time, for example, thin metal wires were used to decorate textiles. In Egypt and the Levant, threads of flax and shell silk were decorated with flattened gold strips or gilded intestines. Wire threads were forged by hammering or cut from thin metal sheets. In the fifth century BC, the drawing iron was invented. This is used to draw a forged, coarse wire through a tapering opening in the drawing tool–the hol. As the wire is drawn, it becomes longer and thinner, and by repeating this process, the desired diameter can be achieved. However, by just drawing without any other aids, the wire is not stretched evenly and therefore breaks at the weakest point. Therefore, aids are used. In Greece, drawn gold wires are used for weaving and embroidery from the fifth century BC. The Romans use textiles made of silver as wound dressings because of its aseptic effect. In the Middle Ages, metal threads from Egypt and China, known as "Cyprian gold", were used in ceremonial robes. They consist of gilded intestinal skins wound around a silk or linen thread, or of gilded paper strips (Wagner, 1981; Haudek & Viti, 1980). From the twelfth century onwards, gold and silver webs are produced in Europe (mainly in France, Spain and Italy). A drawn gold wire is rolled flat and

D. Veit, *Fibers*, https://doi.org/10.1007/978-3-031-15309-9_46

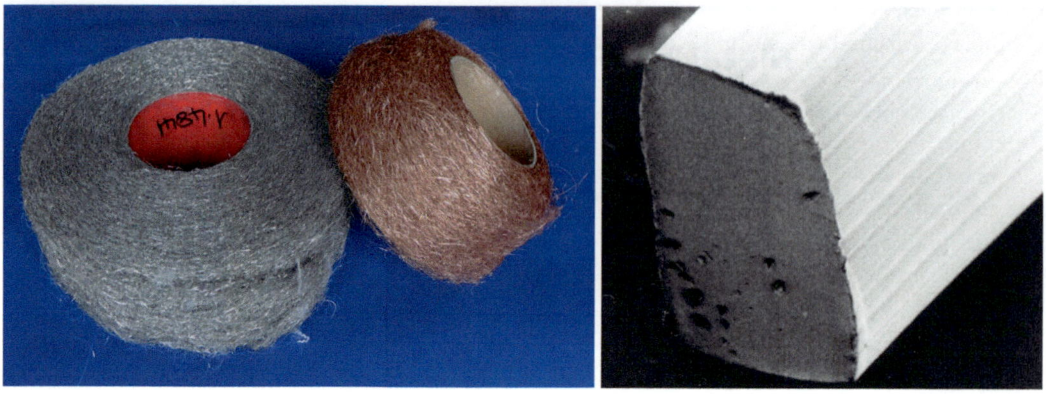

Fig. 46.1 Metal (steel, copper) staple fibers and cross-section of a metal fiber

wound around a silk thread. For price reasons, gold-plated silver or silver-plated or gold-plated copper is often used instead of pure gold. In Spain and France, "faux" metallic threads are also made, consisting of a cotton core thread wrapped in gold or silver. They are also called "leonic threads" after one of the places of production, Lyon. At the same time, lyre drawing benches were developed for the production of fine wire, and from the fourteenth century, shock drawing benches and later water-powered drawing benches were used for coarser wires. In the fifteenth century, pure gold and silver threads were woven into heavy brocade fabrics and posa men ten. In 1569, the first German production facility for the manufacture of "leonic threads" was established in Nuremberg by Anthoni Fournier. In the following years, numerous metalworking workshops were established and in the nineteenth century the first wire rod mills were opened in Germany. Wire drawing develops from a craft to an industry. At the end of the nineteenth century, the drawing iron was replaced by diamond drawing dies. They are still used today for drawing wires. From 1960 onwards, steel becomes established as a fiber material for tyre reinforcement. A pioneer of this development is the tyre manufacturer Michelin (Bruyne & Van Houtte, 1973). In 1962, the shape memory effect of nickel-titanium alloys is discovered (Yahia, 2000), which opens up completely new applications for metal fibers.

Today, metal fibers are used as reinforcing material, especially in tires and other rubber products, as well as for filters. A growing market is applications in architecture, especially for facades.

46.3 Metal Extraction

46.3.1 Occurrence and Extraction of Metals

Metals rarely occur in pure form, but mostly in chemically bound form, e.g. as oxide, sulfide or carbonate ores and as metal salts. Depending on the composition of the ores and their chemical properties, there are different processes for extracting the pure metals from these ores.

Reduction with Carbon

Oxygen can be removed from metal oxides by heating with carbon. The elementary metal and carbon dioxide are formed. Simplified, the following general reaction equation of the redox reaction applies, which proceeds via several partial reactions:

$$2\,\mathrm{MeO} + \mathrm{C} \rightarrow 2\,\mathrm{Me} + \mathrm{CO_2}\ (\mathit{MeO}\ \text{metal oxide},\ \mathit{Me}\ \text{metal}).$$

Strong Heating

Oxidic ores are reduced by strong heating. In this process, oxygen is removed with the aid of reducing agents and the metal is thus obtained. Examples of naturally occurring oxidic ores are Fe_2O_3, $MgCO_3$, $CaCO_3$, $SrCO_3$ and $ZnCO_3$.

Roasting
Sulphide ores are first heated strongly with the addition of oxygen, which is known as "roasting". The sulphidic ore is thereby converted into oxidic ore and can then be further processed as such. Examples of naturally occurring sulphide ores are $CuFeS_2$, ZnS, HgS, MoS_2 and PbS.

Electrolysis
The ores of some metals, if they are not present as oxides, are strongly heated and oxidized and then converted into their salts with acids. The pure metals are then obtained from these salt solutions by electrolysis. A typical example of a naturally occurring metal salt is $CuSO_4$. Metals, such as copper, which occur elementally with other metals, can also be obtained directly by electrolysis (e.g. copper refining).

Calcination
Carbonate ores are calcined without additives. In this process, the ore is heated to such a high temperature that it converts to its oxidic form. The oxide is then reduced again. Examples of naturally occurring carbonate ores are $MgCO_3$, $CaCO_3$, $SrCO_3$ and $ZnCO_3$.

46.3.2 Thermal Annealing Process

During the forming of metals, its microstructure is changed by phase transformations. This makes the material brittle. For this reason, it must subsequently undergo an annealing process. Annealing is the treatment of a workpiece at a certain temperature with a certain holding time and subsequent cooling to achieve certain material properties (Berger et al., 1997). In recrystallization annealing, a new formation of the microstructure is achieved with mechanical properties such as those that existed before deformation. This process is mainly used for cold drawn wire. Other annealing processes used in metal processing are normal, coarse grain, diffusion, soft and stress relief annealing.

46.3.3 Material Numbers According to the German Standard DIN 17007

The systematic designation of iron and steel in accordance with DIN 17006 provides brief information on the characteristic features of a material. This standard specifies the bond, the abbreviated name, the manufacturing process and the type of casting for the respective steel and iron grades.

As a supplement to the short names acc. to DIN 17006, DIN 17007 defines a numbering system for the identification of materials of all kinds. In addition to DIN 17006 for iron and steel, material numbers are also defined for the non-ferrous metals acc. to DIN acc. to main groups.

46.4 Metal Fiber Types

Metal fibers can be made from pure metals, alloys or semi-metals. In addition, plastic fibers or, for cost reasons and to change the properties, metal fibers are also metallized.

46.4.1 Pure Metals

Because of the better properties during further processing, soft and ductile metals, e.g. gold, silver, copper and aluminum, are often used. Depending on the application, nickel, zinc, tin, titanium and iron are also processed, as well as–rarely–molybdenum and platinum (Haudek & Viti, 1980; Heyder & Frommeyer, 1992).

46.4.2 Alloys

Metal fibers can be made from many different alloys, which are created by fusing two or more metals together. Alloys usually have different properties than the pure metal. For example, they often have a different color, a different luster, a different hardness and castability. The melting point of alloys is usually lower than the average melting temperature of the starting materials, while the hardness is often higher. The elementary states of alloys can be taken from state diagrams as a function of the percentages. The microstructure of the alloys is also important.

Important alloys are steel in any form, copper, aluminum and iron alloys, brass, bronze and combinations of nickel with titanium or titanium with chromium.

46.4.3 Semi-metals

Semi-metals are elements and stand between the metals and the non-metals in the periodic table. Only boron and silicon in oxidized form are important for fiber production.

46.4.4 Metallized Fibers

Fibers can be metallized in order to specifically adjust their properties. The basic materials include, for example, polyester, polyamide, glass, aramid and carbon as well as wires, e.g. made of tungsten. Nickel, copper, silver, aluminum, magnesium, zinc, iron, chromium and gold are used for coating. The processes used depend on the material to be coated. The coating can be applied in the form of gas or steam or by electrolytic or chemical deposition (e.g. electroplating). Another possibility is the coating of textile fabrics, e.g. nonwovens and wovens (Wagner, 1981; Marchini & Massa, 1998).

46.5 Production

Metal fibers and wires are produced by mechanical and thermal processes.

46.5.1 Mechanical Manufacturing Process

The dominant processes here are those in which initially thick wires are drawn into ever finer fibers (Fig. 46.2).

Conventional Process: Wire Drawing

Wire drawing is one of the oldest processes for the production of metal fibers. In this process, the wire passes through several drawing stages in which it becomes thinner and longer. The drawing stages are called coarse, medium, fine and scratch drawing. The tool with which the wire is drawn is called a drawing die. It usually consists

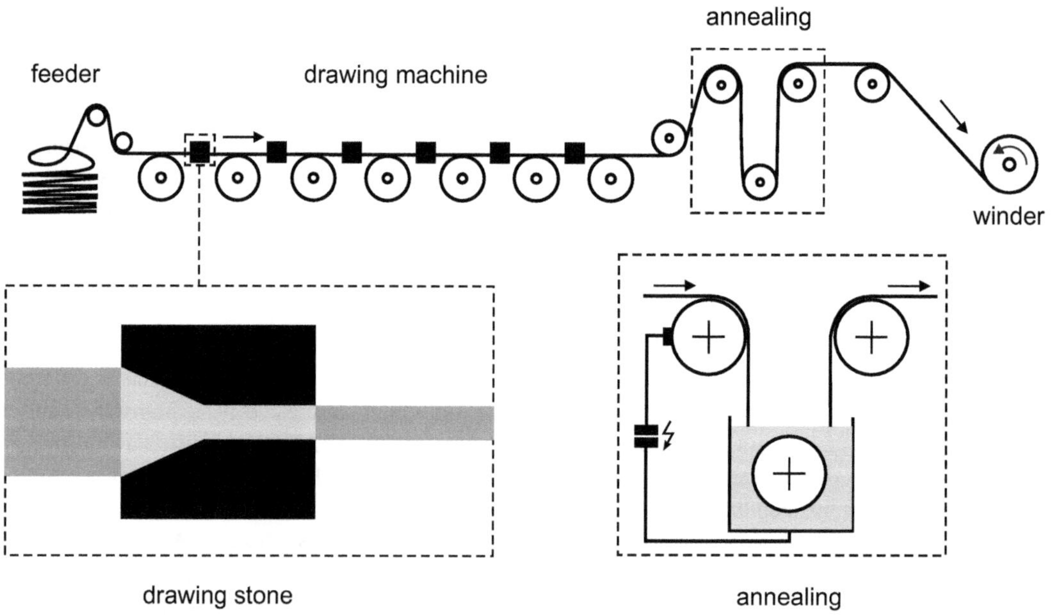

Fig. 46.2 Wire drawing process (single fibers)

of a steel frame with a ceramic, carbide or diamond core and is available in a wide variety of profiles. The drawing die is therefore more stable and does not need to be reworked. High dimensional accuracy and a very good surface quality can be achieved with drawing dies.

The initial average of the wires varies depending on the material. For example, iron begins from 5 mm in diameter, and copper–from 8 mm. Iron wire rod passes through the following stages: first the coarse draw from 5 to 3.8 mm, then the medium draw to 2.9 mm, the fine draw to 2.4 mm and finally the scratch draw to 2 mm. After drawing, the wire is annealed at temperatures between 600 and 900 °C and then quenched. The finished fine metal wire is wound up with up to 600 m/min. Typical fiber fineness values are between 50 µm and several millimeters (Fig. 46.3).

A distinction is made between drawing wires for technical products (wire cloth and wire mesh, steel cord as an insert for car tires, etc.) and leonic threads for textile purposes.

A special process in which a copper rod is exposed to zinc vapors, the precipitate of which combines with the copper surface to form gold-like brass, was developed in France in the sixteenth century. The wire drawn from this process was called or faux (false gold), later called cement in Germany. Because this process was developed in Lyon, the goods made from it, such as platelets, webs, tinsel, braids or braids, were called leonic.

Today, leonic threads include flat wires (Lahn) and screw wires (Bouillon). Metallic webs are created by spinning leonic wires (round or flat) or cut ribbons of cellometal, metal transparencies, lurex and the like around a textile support, core or soul thread (made of silk, viscose, cotton, linen or wool).

Wire Pull in Bundle

Conventionally drawn wires with small diameters are combined into bundles of 1000–2000 strands and sheathed with a thin metal tube. The material of the tube must be chemically more unstable than the wires. The sheathing enables further processing of the complete bundle by drawing.

Fig. 46.3 Drawn copper wire

In several working steps, the initial cross-section of the wire bundle decreases from 150 mm^2 to 0.2 mm in diameter. Then the material of the metal tube has to be removed, e.g. by an acid bath. In the end, individual, endless and parallel fibers with diameters between 4 and 25 µm are obtained. These can either be spun as multifilaments or broken into staple fibers of 50–150 mm length (Fig. 46.4).

The production of stainless steel fibers by conventional drawing of wires and subsequent drawing by the bundle drawing process is time consuming. The fibers produced by the bundle drawing process can be blended with other fibers during the spinning process and the resulting yarn can be knitted or woven. This process is used to produce titanium fibers as well as steel fibers. Typical fiber diameters are 1.5–100 µm, for special shapes 200 nm are also possible.

Metal Cutting Manufacturing Process

Metal fibers are produced endlessly by cutting from a rolled-up metal foil. In this process, several fibers are obtained simultaneously per cutting

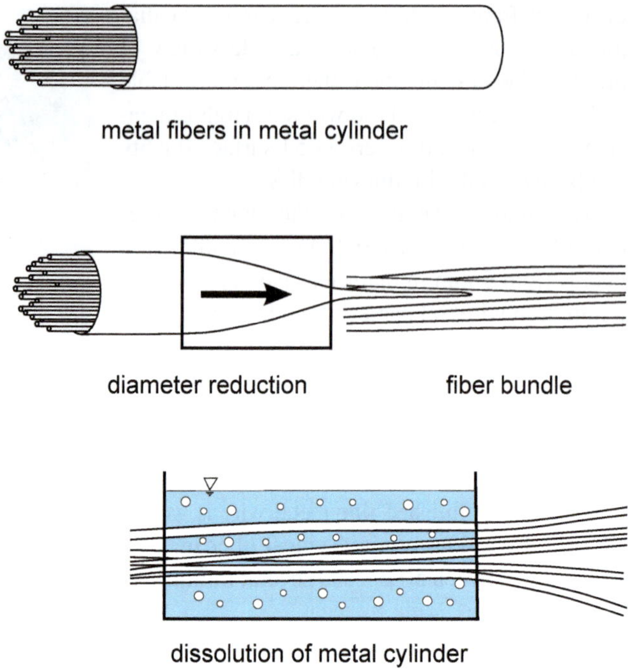

Fig. 46.4 Wire drawing method in the bundle acc. to Bekaert (2003)

tool and removal point. The metal fibers are then bundled into fiber strands and wound onto a roll. Typical fiber diameters are 10–100 μm (Fig. 46.5).

46.5.2 Thermal Manufacturing Processes

The thermal processes are all based on producing the fibers by direct extrusion from the molten phase. The rapidly quenched metals are characterized by unusual properties and combinations of properties. Thus, high hardness and tensile strength can be achieved in amorphous structures, and the yield strengths and flexural fatigue strengths are also very high.

The challenge compared to melt spinning glass or polymers is that the melt viscosity of metals is about 100 times smaller and only a few Pas. Therefore, melt fracture occurs very easily, forming drops instead of fibers.

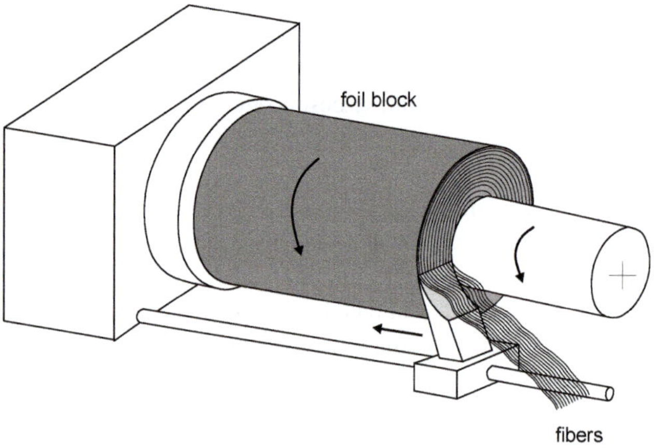

Fig. 46.5 Cut fibers removal process acc. to Bekaert (2020)

The most commonly used processes for metal fiber production are Taylor process, melt spinning into a rotating liquid, melt drag process, Kavesh downpipe method, melt spinning in a gaseous environment, conventional casting, upward strand casting, melt spinning onto a metal wheel and splat quenching. Some of them are explained below.

Taylor Process

The starting material is a metal rod in a glass tube. The material to be processed and the glass must have approximately the same melting point in order to be drawn simultaneously. By heating, the wire melts and the glass tube softens. Now the soft glass tube is drawn out into a very thin hollow glass thread and with it the metallic material. The glass sheath is removed from the filaments by squeezing by means of rollers, ultrasound or by chemical detachment. Very fine fibers (<50 μm) can be produced in this way. However, production is limited to certain temperature intervals, which are determined by the viscosity range. Also, only materials that do not bond with glass can be formed. Another disadvantage of this process is the low production speed due to the small cross-sections (Fig. 46.6).

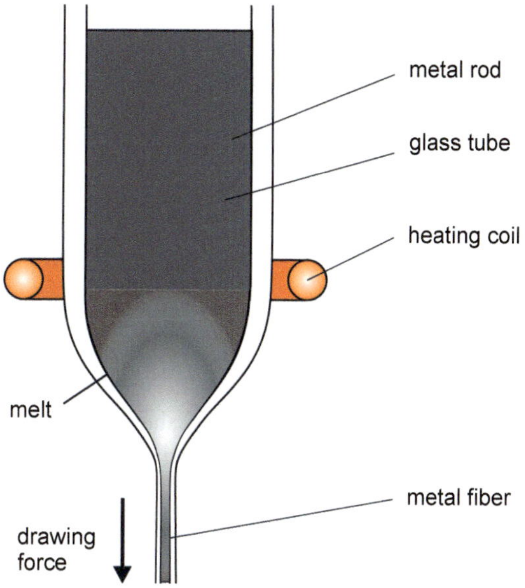

Fig. 46.6 Taylor method acc. to Butler et al. (1972)

Melt Spinning into a Rotating Liquid

This process enables the continuous production of metallic fibers and wires directly from the melt. Pure metals or metallic alloys are melted and superheated in a high-temperature and pressure-resistant crucible under an inert gas atmosphere.

An inert gas overpressure is applied to the melt surface. The resulting overpressure forces the melt through a round nozzle. It leaves the nozzle as a free jet and after a few millimeters dips at a defined angle into a laminar flowing coolant. The solidifying fiber is then accelerated by the rotation (centrifugal forces) and by the difference in density between the metal and the coolant on the inside of the drum, where it solidifies due to the coolant. Removal takes place either continuously or discontinuously, depending on the batch size. The fibers reach diameters of 50–500 μm. Melt spinning into a rotating liquid has several advantages over conventional wire rod drawing. The most important of these are lower equipment costs, the ability to produce smaller batches, improved material properties, less scrap and higher productivity (Fig. 46.7).

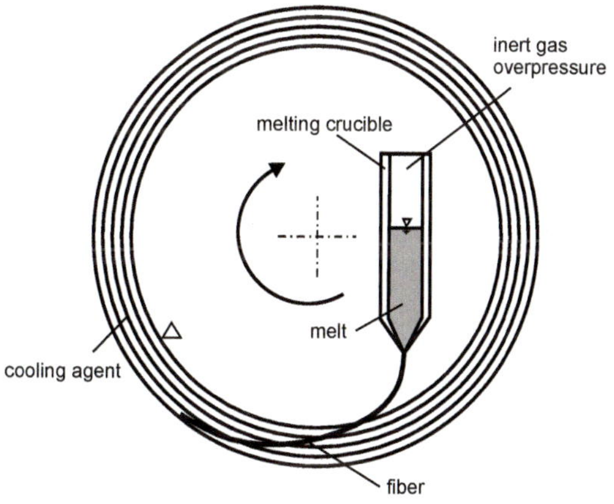

Fig. 46.7 Melt spinning into a rotating liquid acc. to Heyder (1992)

Melt Extraction

In melt extraction, a water-cooled extraction roller rotates in the melt bath. This roller is usually multi-bladed and has V-shaped edges on which the melt filaments are extracted from the melt. The melt solidifies at the edges and detaches

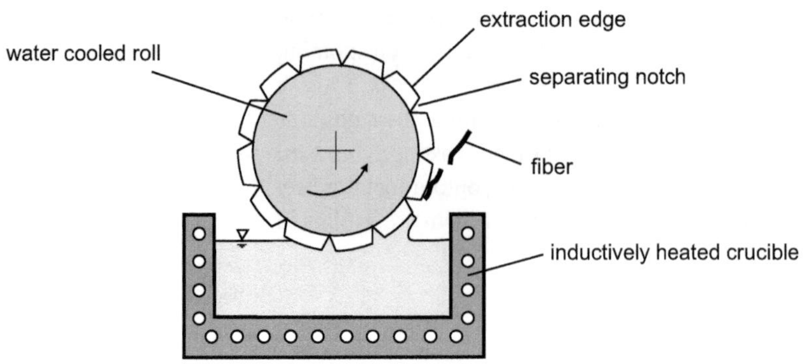

Fig. 46.8 Melt extraction process acc. to Lehnert et al. (1991)

from the roller in the form of short fibers as a result of the heat contraction and the centrifugal forces acting (Fig. 46.8).

Cooling rates of 10^3 to over 10^4 K/s are achieved during melt extraction. To dampen the wave motion at the melt surface, a magnetic field is generated above the melt bath. This method does not allow a continuous process with predefined fiber qualities. Typical fiber fineness values are 40–250 μm.

Aluminum, copper and tin fibers as well as fibers made of aluminum and copper alloys and materials that cannot be processed with conventional methods due to their brittleness (e.g. NiSnB) can be produced in this way. This process is used on a large scale for the production of CrNi steel fibers as a reinforcing component in refractory concretes.

Modified Forms of Melt Extraction

Other ways of obtaining fibers by modified melt extraction have also been investigated. One method is to melt the material by an electron beam or by an acetylene flame and to draw the droplets into fiber form by means of a roller with a rough surface. However, the smallest diameter that can be achieved in this way is only 20 μm.

In another method, a 6 mm thick metal rod is first heated by an induction coil and then warped. This method can process almost all metals and produce fibers of a few micrometers in diameter. However, the production speed is very low.

Melt-Drag Process

In the melt-drag process, a bead stream of liquid metal is held at the end of a nozzle at a constant level in the melting vessel. A water-cooled cylinder rotates immediately adjacent to the bead and drags it across the surface. This process produces large fiber and wire diameters, ribbons and sheets (Fig. 46.9).

46.5.3 Special Forms

Production of Shiny Metal Ribbons: Lurex

Metallic yarns for decorative purposes (lurex) are made from metal foils which are glued between two plastic films and then cut into yarn size. Standard lurex, for example for example, is made from aluminum foil with acetate foils glued on both sides. This is done to protect metal threads from oxidation. Instead of acetate film, more heat-resistant films made of PP, PET, PA or triacetate are also often used (Haudek & Viti, 1980).

Production of Metal Coated Fibers

Electrically conductive fibrous materials can be produced by applying a thin metal layer. The metal layer is applied by electrochemical precipitation, vacuum metallization or spraying

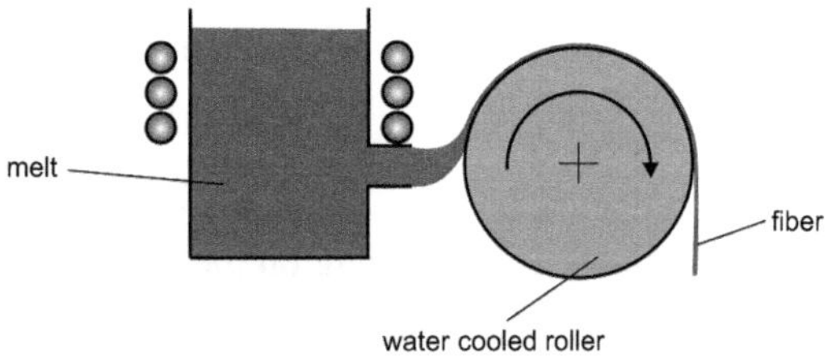

Fig. 46.9 Melt-drag method acc. to Butler et al. (1972)

(Perpepelkin et al., 1974). In electrochemical precipitation, the metallized filaments are created by electroplating gold, silver, or copper, among other methods. In this process, for example, a cotton yarn is coated as a core with a cellulose acetate sheath containing fine metal powder (Wagner, 1981). This process results in a very good adhesion of the metal layer to the fiber, which is much higher than, for example, in processes under vacuum. Moreover, the original textile properties, such as flexibility and handle, are hardly changed (Marchini & Massa, 1998). In vacuum metallization, a thin metal layer is applied to the textile material in a high vacuum by vapor deposition (e.g. physical vapor deposition, PVD) or spraying, e.g. with aluminum (Wagner, 1981).

Production of Semimetal Fibers

In principle, semimetal fibers can be produced using the same processes as metal fibers. However, due to the lack of ductility of some semimetals, they must be obtained by other methods. Boron fibers, for example, cannot be produced by the wire drawing method. And the Taylor method is also ruled out due to reactions of boron with oxygen, aluminum and silicon at higher temperatures. The most suitable process is the gas separation process. Here, for example, a 12 μm thick continuous tungsten wire is drawn through a reaction chamber and brought to annealing by resistance heating. Due to the gas mixture of boron trichloride and hydrogen at temperatures below 1200 °C, pure boron is deposited on the tungsten filament and the filament diameter is thus expanded to 100 μm. The boron diffuses into the tungsten core faster than the tungsten can diffuse outwards, whereby tungsten then forms compounds with the boron. Instead of the tungsten core, a cheaper carbon fiber can also be used, whereby tensile strengths, E-modulus and density are not significantly changed.

46.5.4 Metal Alloys with Shape Memory Effect

In 1932, the Swedish physicist A. Oelander discovered a previously unknown phenomenon while working with an alloy of gold (Au) and cadmium (Cd). The Au-Cd composite could be plastically deformed when cold and returned to its original state when heated. This phenomenon became known as the shape memory effect (SME). Metals that exhibit this behavior are called shape memory alloys. In 1962, while searching for new materials for the U.-S. submarine fleet that would be antimagnetic, hard, and particularly corrosion resistant, the U.-S. Naval Ordnance Laboratory discovered the Shape Memory Effect in an equiatomic alloy of nickel and titanium. The metal was named Nitinol (Nickel-Titanium Naval Ordnance Laboratory) and was a breakthrough in the field of shape memory materials. Other important memory alloys are Cu-Zn-X (X: Si, Sn, Al) and Cu-Al-Ni.

In the case of memory effects, a fundamental distinction is made between the thermal and the mechanical memory effect. Both effects can be adjusted in one and the same alloy by a specific material treatment during production.

Thermal Shape Memory Effect

The thermal shape memory effect is achieved by rapid cooling of the material. In this process, the microstructure of the austenite present at the beginning is converted into martensite. Any subsequent deformation can be reversed by heating. The original austenitic structure is restored and an appropriately treated component resumes its previous shape (Fig. 46.10). This is the one-way memory effect.

A further development of the thermal one-way memory effect is the two-way memory effect. One speaks of a two-way effect when the alloy can "remember" two forms, one form in the low temperature phase and another form in the high temperature phase. The alloy oscillates back and forth between two states due to the change in temperature without any external force. Japanese scientists even discovered a three-way memory effect, also called the all-round memory effect, with the help of which strips of a nitinol alloy can "remember" three different shapes.

Mechanical Shape Memory Effect

The mechanical shape memory effect (superelasticity) is set by deformation of austenite

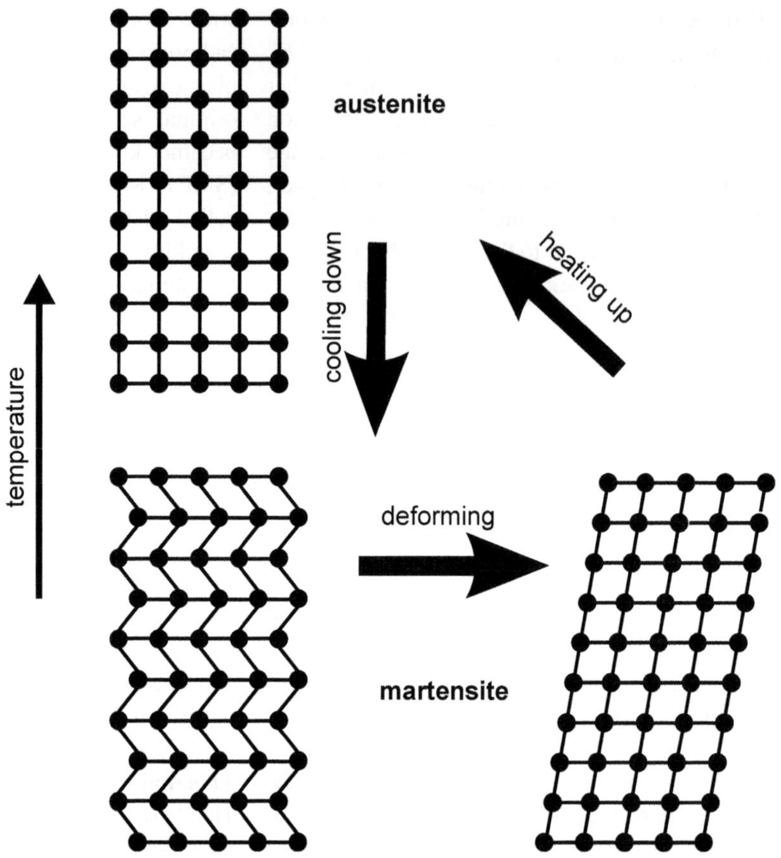

Fig. 46.10 Thermal shape memory effect

in a certain temperature range. As shown in Fig. 46.11, below the temperature A_f, the material is in the austenitic phase. Now, if an external stress σ_{Ms} is applied above a temperature A_s, the transformation of austenite to martensite begins. As a result, some of the external stress is removed and the total free energy of the system is thus less than if the material existed as austenite. Initially a plateau phase is undergone and when the stress σ_{Mf} is reached, the transformation to martensite is complete. If the external stress is reduced, the martensite gradually transforms back to austenite until the initial state is reached. The strain occurring during this reversible, elastic material deformation is up to 10%, which is a very high value for metals and is higher by a factor of 10 than for conventional metallic spring materials. Superelastic components open up new possibilities lich because they remain extremely stable in terms of buckling and shape with the smallest cross-sections.

The economic use of both effects is currently particularly in the fields of medical technology, microsystems technology, measurement and control technology, electrical engineering and automotive engineering. The application of

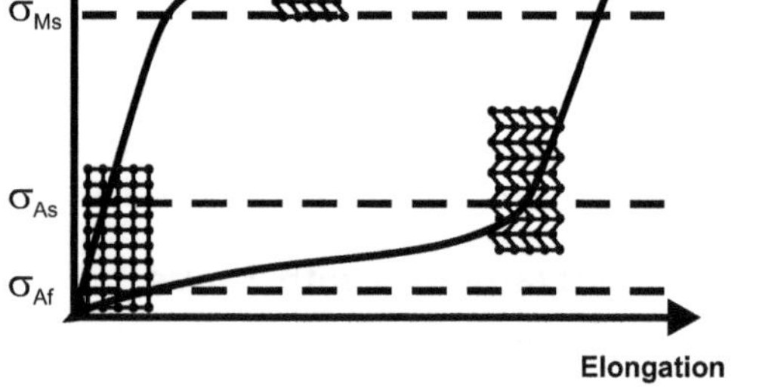

Fig. 46.11 Mechanical shape memory effect and superelasticity

instruments and components made of NiTi in medical technology is additionally promoted by their good body compatibility (biocompatibility).

The properties of NiTi alloys depend on the following parameters:

- Degree of deformation during cold forming
- Temperature and duration of heat treatment (−200 to +100 °C)
- Initial condition of the alloy before the last deformation step
- Sequence of the last heat treatment and forming stage

To achieve a mold imprint, annealing at 400 °C for at least 1–2 min is required. However, temperatures are usually around 500 °C for 5–10 min.

46.6 Processing of Metal Fibers

46.6.1 Yarn Production

Metal staple fibers are basically processed with the same processes as other staple fibers. In spinning preparation, care must be taken to ensure correct mixing when adding metallic fibers so that no accumulations of metallic fibers occur.

Metal filaments are first cut to staple length, formed into slivers of parallel fibers and then spun. Instead of cards, mostly cards are used. Both short and long staple fibers are added as sliver on the draw frame. In order to be able to draw the metal fiber sliver more regularly and to provide better protection during the further processing steps, it should be brought into a sandwich form (Fig. 46.12).

For yarn manufacturing, the ring spinning process is predominant. The friction spinning process is suitable for producing a fiber blend with a core of metal filaments or metal fibers. In this process, the metal core (diameter 8 μm to 50 mm) is given a sheath of usually man-made fibers. Hybrid yarns produced in this way are used, e.g., for cut-resistant textiles.

Today, shielding yarns are mostly manufactured on the basis of continuous copper filaments (diameter 0.02–0.04 mm). These are coated either with a fine polyurethane lacquer layer (surface insulation) or with a special silver alloy (surface conductivity). The filament is then spun with cotton, wool or polyester fibers. In the process, the metal filament is often wrapped in a spiral and thus comes to the surface at specific intervals.

Filaments made of metal fibers can also be twisted. The resulting multifilaments have a very uniform resistance and are therefore well suited for heat transfer, e.g. for seat heaters in cars or in clothing.

46.6.2 Fabric

Weaving wires have a diameter of 20 μm to 6 mm. They are made of various metals, alloyed and unalloyed, annealed or unannealed, hard or

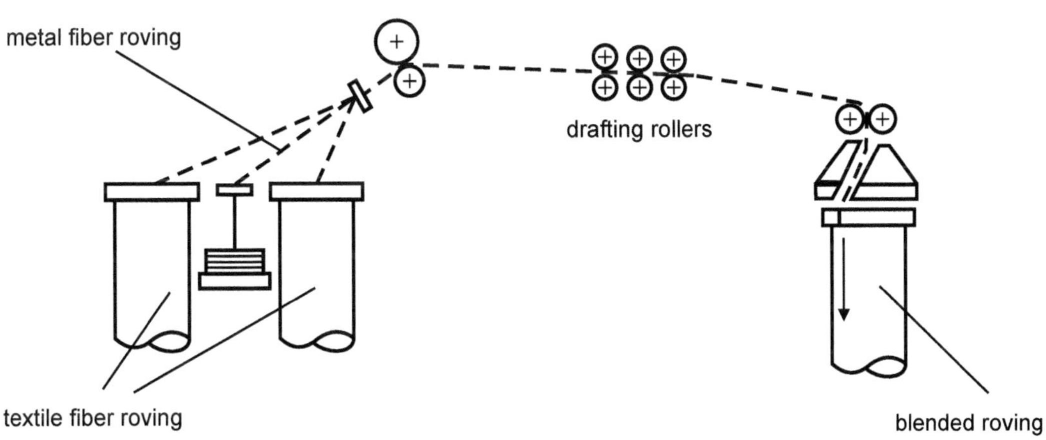

Fig. 46.12 Blending of metal fiber tapes in the short-staple process acc. to N.N. (1974)

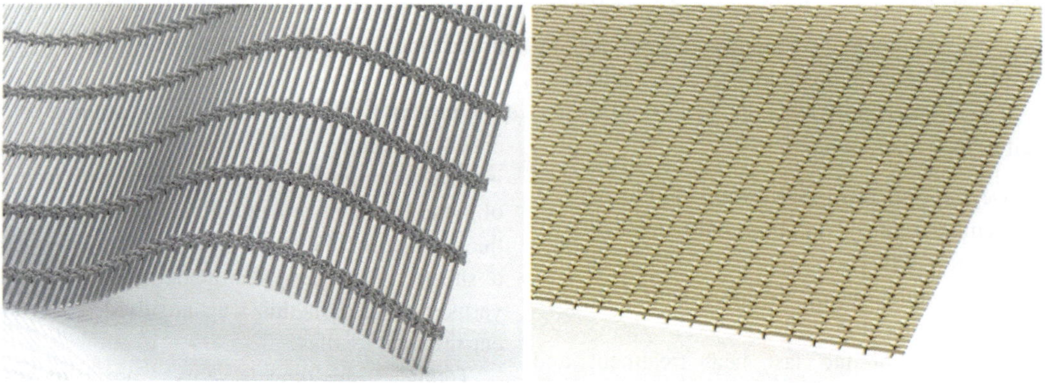

Fig. 46.13 Decorative fabric for facades (GKD Gebr. Kufferath AG, n.d.)

soft, raw or bright polished, with or without corrosion protection by tinning or galvanizing, and have a round, semicircular, elliptical, triangular, square, rectangular or trapezoidal cross-section depending on the intended use.

The main materials for woven wire cloth are steel wires of various strengths, stainless steel wires made of brass wire, tombac wire (brass wire with higher copper content), bronze wire, wires made of nickel alloys as well as aluminum and aluminum alloys, non-ferrous metals, titanium, tungsten, molybdenum and platinum. Weaving strong wires requires good pliability and a minimum elongation of 15–20% (Fig. 46.13).

On heavy modified wire looms, e.g. single-sided rapier or projectile looms, wire cloth can be woven in plain weave or twill weave and produced endless by sewing, soldering or welding. They can also be woven already in tubular form. In the case of woven wire cloth, the mesh size (0.03–50 mm) and the shape of the meshes (square or rectangular) vary acc. to the grain size and shape of the screenings. Wires that are difficult to bend are cranked (bent) before weaving.

46.6.3 Other Sheet Materials of Metal Fibers

Fibers and wires with diameters between 0.05 and 0.6 mm are used for knitted fabrics. There are different versions, as round or flat wire, single or multi-stranded.

Common metals are galvanized steel, aluminum, pure copper, copper alloys, brass, stainless and acid-resistant stainless steels, high-temperature and heat-resistant stainless steels, nickel, nickel alloys and titanium. The characteristic parameters of knitted fabrics are the weight per unit area, the stitch length (1.5–15 mm), the weave (plain knit, bias or arrow), the stitch pitch (2–30 mm), the material and the diameter of the fibers and wires. On flat knitting machines, metal and precious metal wires are made into electric blankets, fencing vests (Fig. 46.14) or catalytic filters for flue gas desulfurization. These are manufactured completely in one piece. Knitted wire meshes can also be produced as hoses using the circular knitting process. The hose is usually further processed as a flat-laid fabric.

Mechanical web bonding (needling) is best suited for processing metal fibers. In the netting of mechanical web formation, the metal fibers are stacked in several layers and then needled. However, the tensile strength of such nonwovens and they may curl or tear due to chafing. To make nonwovens tear resistant, a latex layer can be applied. Another option is to embed the nonwoven in a foam layer, which is then flame cured. Hydrodynamic nonwoven bonding is only possible with very fine fibers that have low bending stiffness. The metal nonwovens produced in this way are used, for example, for shielding electromagnetic waves and radar beams, especially in

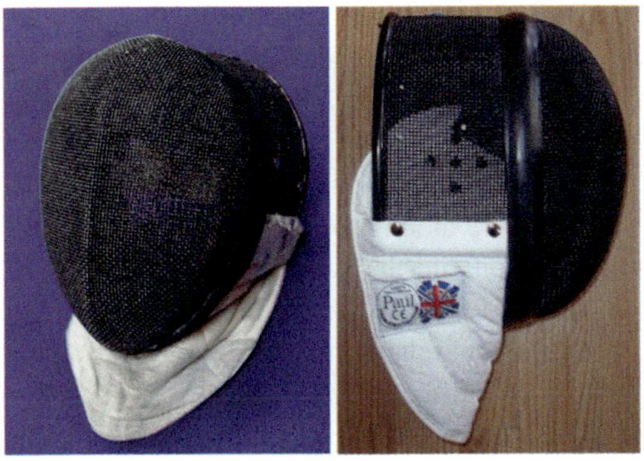

Fig. 46.14 Fencing mask (woven fabric) and neck guard (knitted fabric) (Nguyen, 2013) and details (Schemp, 2007)

aircraft construction and aerospace. Chemical bonding is also conceivable, but is hardly ever used.

Metal fiber nonwovens can be processed into porous mats in the sintering process. The sintered metal fiber nonwovens are used, for example, on a large scale for applications in hydraulics, aeronautics and also in the plastics and synthetic fiber industry, e.g. for filtration.

The production of knitted fabrics from metal fibers and threads (wires) is still difficult due to the extreme mechanical stresses during processing. Open warp-knitted fabrics, voluminous R/R warp-knitted fabrics and spacer fabrics have fewer processing problems due to their textile structure and thus have the greatest potential to become established on the market first, e.g. for antennas. Stitch-bonded knitted fabrics can be produced at a much higher surface formation rate than, for example, woven fabrics. With the modification of the processing technology, the use of steel fibers, steel filament yarns and wires for the production on stitch-bonding machines is conceivable. Warp knitted fabrics based on metallized polymers are a standard product. They are also increasingly being used in medical technology for minimally invasive surgery. The main demands on the product are biocompatibility and low weight per unit area.

46.6.4 Composite Material

A promising development is metal fiber reinforced metals (Metal Matrix Composites, MMC). Combinations of metal with ceramics and carbon, both as matrix and as reinforcing fiber, are also being produced. Often, short fibers are used in metal matrix composites because suitable, high-strength continuous fibers are still lacking. Reasons that speak for the use of MMC are

- high weight-related strength and stiffness at high temperatures,
- high impact strength and ductility at low temperatures,
- good impact behavior,
- higher fatigue strength and better dimensional stability etc.

The following three processes are most commonly used to reinforce metals

- Coating of the metal fibers with matrix material and subsequent compaction,
- matrix infiltration,
- in-situ production of directionally solidified eutectics.

46.7 Properties

Metal fibers are available as wires, filaments and staple fibers both in pure form and in blends, e.g. with cotton (antistatic clothing) and wool (antistatic carpets). Diameters and staple lengths vary within large ranges.

Metal fibers almost all have the following general properties

- high density,
- high tensile strength,
- heat conductive and good electrical conductivity,
- magnetizable and antistatic,
- corrosion resistant, chemically resistant,
- resistant to high temperatures and non-flammable,
- washable and water repellent,

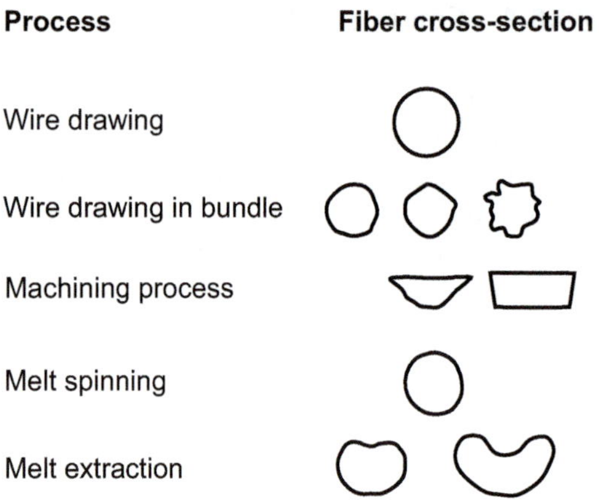

Fig. 46.15 Typical fiber cross sections depending on the manufacturing process

- microwave reflection and radiation shielding,
- metallic luster.

The density of metal fibers is usually 7.9–8.9 g/cm^3 and the elongation at break is usually 1–10%. The tensile strength is typically 200–900 MPa, which is a high value in absolute terms, but only corresponds to approx. 2.5–11 cN/tex due to their high density.

Depending on the manufacturing process, the fibers differ, among other things, in the fiber cross-section and the surface finish. Metal fibers with a rough surface (machining process) have a lower fineness-related maximum tensile strength due to notch effects. For all processes, however, the tensile strength increases with decreasing wire diameter (fewer microstructural defects) and the flexibility of the wires improves (Fig. 46.15).

46.8 Applications

Metal fibers are used in a wide range of product groups. These include

- Automotive: tire cord, filters, heat protection and reinforcement of hoses and belts.

Fig. 46.16 Selected products made of metal fibers and yarns (top: Façade of the FIFA headquarters in Geneva, candle filters; GKD Gebr. Kufferath AG; bottom: steel cables of a suspension bridge)

Table 46.1 Selected manufacturers of metal fibers

Producer	Products	Website	Country
Arcelor Mittal	Metal fibers	https://corporate.arcelormittal.com	LUX
Bekaert	Steel wire, metal fibers and rovings Yarns, intermediate products	https://www.bekaert.com	BEL
Berkenhoff GmbH	Soldering, welding and EDM wire	http://www.bedra.com	D
Biologa Danell GmbH	Metal fabrics	http://www.swiss-shield.de	D
Créafibres S.A.S.	Steel fibers	http://www.creafibres.fr	F
Deutsches Metallfaserwerk Dr. Schwabbauer GmbH & Co. KG	Metal fibers	http://www.stax.de	D
Elektrisola Dr. Gerd Schildbach GmbH & Co. KG	Enameled copper wire	http://www.elektrisola.com	D
Fiber Technology Ltd	Steel fibers	http://www.fibretech.com	UK
SC Fibro Meals SRL	Metal fibers	http://www.fibrometals.ro/	ROU
Frenzelit GmbH	Metal fabrics	http://www.frenzelit.com	D
GKD – Gebr. Kufferath AG	Metal fabrics	http://www.gkd.de	D
Green Steel Group	Metal fibers, metal wools	http://www.greensteelgroup.com	ITA
Haver & Boecker OHG	Metal fabrics	http://www.haverboecker.com	D
Hebei Swiit Metallic Fiber Co.,Ltd	Metal fibers	http://www.swiit-fiber.com	China
Hunan Huitong Advanced Materials Co., Ltd.	Metal fibers	http://en.ht-metalfiber.com/	China
Hunan Sunshine Steel Fiber Co.,Ltd.	Metal fibers	https://sunshinefiber.en.china.cn/	China
IntraMicron, Inc	Metal fibers, nonwovens	http://www.intramicron.com	USA
King's Metal Fiber Technologies CO., LTD	Metal fibers and yarns	http://www.kingsmetalfiber.com	UK
KOOLON ADV. MAT. TECHN. CO., LTD.	Metal fibers	http://www.koolon.com	Taiwan
Leoni Ag	Metal fibers and yarns	http://www.leoni.com	D
Metlon Corp.	Metal fibers and yarns	http://www.metlon.com	USA
Nikko Techno, Ltd.	Metal fibers, nonwovens	http://www.nikkotechno.jp	JPN
Nippon Seisen Co., Ltd.	Metal fibers and yarns	https://www.n-seisen.co.jp	JPN
Nobel Biomaterials, Inc.	Metal fibers for medical products	http://www.noblebiomaterials.com	USA
R.Stat SAS	Metal fibers	http://www.r-stat.com	F
Rhodius GmbH	Metal textiles	http://www.rhodius.com	D
SPRINT METAL GmbH	Metal wires	https://sprintmetal.swisssteel-group.com	CH
Statex GmbH	Metal yarns	http://www.statex.de	D
Stewols India (P), Ltd.	Metal fibers, steel wool	http://www.stewols.com	India
Teijin Ltd.	Metallized fibers and yarns	http://www.teijin.com	JPN
Ugitech SA	Metal fibers	http://www.ugitech.com	F

- Aerospace: noise insulation and fiber composites, antistatic components.
- Textiles: Antistatic carpets and clothing, conductive yarns for smart textiles, heatable textiles, cut protection, decorations of luxurious garments (e.g. brocade), polymer filters.
- Power generation: Filter.

- Building and living: Reinforcement of concrete components and decorative facades, upholstery fabrics.

Of steel fibers, the largest group of metal fibers and wires, about 60% are currently used in construction, about 20% for fiber composites (e.g. tyres), 10% in fire protection and another 10% for all other applications. Figure 46.16 shows the wide range of applications for textiles made of metal fibers.

46.9 Market and Manufacturers

About 60% of all steel fibers are produced by wire drawing, about 25% by foil slitting, 10% by melt spinning and 5% by all other processes. Exact market figures are not recorded. Acc. to estimates, the production of metal fibers per year is approx. 1 m tons (2020) with a growth rate of approx. 4%.

An overview of important manufacturers and processors of metal fibers and their main products is provided in Table 46.1.

References

Bekaert, S. A. (2003). Manufacturing of metal fibres. *Zwevegem*. https://www.bekaert.com

Bekaert, S. A. (2020). Metal fiber technology. *Zwevegem*. https://www.bekaert.com

Berger, C., Burr, A., Habig, K.H., Harsch, G., Kloos, K. H., & Speckhardt, H. (1997). Glühbehandlung. In *DUBBEL, Taschenbuch für den Maschinenbau*. Springer.

Bruyne, D. E. A., & Van Houtte, J. S. (1973). Current inorganic fibers for tires: Steel. *Journal of Applied Polymer Science, 17*, 2033ff.

Butler, I. G., Kurz, W., Gillot, J., & Lux, B. (1972). The production of metal fibres and wires directly from the melt. *Fibre Science and Technology, 5*(4), 243–262.

GKD Gebr. Kufferath AG. (n.d.). *Düren*. https://www.gkd-group.com

Haudek, H. W., & Viti, E. (1980). *Textilfasern*. Bondi Verlag.

Heyder, B., Frommeyer, G. (1992). Herstellung metallischer Verstärkungsfasern für Metallmatrix-Verbundwerkstoffe durch Schmelzspinnen in eine rotierende Flüssigkeit. In *Metall-Wirtschaft, Wissenschaft, Technik* 46, pp. 134–138, GDMB Verlag.

Heyder, B. (1992). *Kontinuierliches Gießen metallischer Fasern und Drähte durch Schmelzspinnen in eine rotierende Flüssigkeit*. Dissertation, RWTH Aachen University.

Lehnert, F., Lotze, G., & Stephani, G. (1991). Herstellung, Eigenschaften und Anwendungen metallischer Kurzfasern. *Materialwissenschaften und Werkstofftechnik, 22*(9), 355–358.

Mac, T., Houis, S., & Gries, T. (2004). "Metallfasern", Faserstofftabelle nach P-A. Koch. In *Technische Textilien* (Vol. 47, pp. 17–32). Deutscher Fachverlag.

Marchini, F., & Massa, V. (1998). Metallbeschichtete Fasern: Eigenschaften und Einsatzgebiete. *Chemiefasern in der Textilindustrie, 38*, T5–T7.

Nguyen, M. L. (2013). *CC BY 2.5*. https://commons.wikimedia.org/wiki/File:Mask_French_Fencing_Championship_2013_n02.jpg

N.N. (1974). Metallfasern für Industrietextilien. *Text, 92*(10), 33ff.

Perpepelkin, K. E., Muchin, B. A., & Smirnov, V. S. (1974). Chemiefaserstoffe für spezielle Zwecke, ihre Eigenschaften und Anwendungen. *Faserforschung und Textiltechnik, 25*(2), 72ff.

Schemp, P. (2007). *CC BY 2.5*. https://commons.wikimedia.org/wiki/File:Fencingmask.jpg

Wagner, E. (1981). *Die textilen Rohstoffe*. Deutscher Fachverlag.

Yahia, L. (2000). *Shape memory implants*. Springer.

Part IX

Test Methods

47 Test Methods

There are numerous test methods for determining the properties of fibers, most of which are standardized. Typical methods are explained below. For further details, please refer to the relevant standards.

47.1 Determination of the Fiber Content

There are a variety of methods for determining the material of a pulp. The most important ones are briefly presented below.

47.1.1 Microscope

Fibers differ in their microscopic transverse and longitudinal views. Cotton is often kidney-shaped in transverse view, while bast fibers often have an angular cross-section and possess striations in longitudinal view. Wool, on the other hand, is usually round and, like almost all animal fibers, has scales. The number of scales along the circumference and the scale height can be used to distinguish different animal fibers. Man-made fibers are usually round in cross-section, although trilobal shapes, snowflakes, etc., do exist. In longitudinal view, melt-spun fibers are usually smooth, while solution-spun ones often have longitudinal grooves (Fig. 47.1).

47.1.2 Density Method

The density of polymer fibers varies from 0.9 g/cm^3 (PP) to 2.3 (PTFE) and can therefore be used for material determination. For this purpose, the fibers are dispersed in a liquid of known density. If the fiber floats on the water surface, its density is lower and thus the density can be determined in several steps (Table 47.1).

47.1.3 Chemical Determination

Another method for determining the fiber material, which has been used for a long time, is the solubility method. In this method, commercially available solvents are used and the fiber materials are chemically determined in the exclusion method (soluble, non-soluble) (Fig. 47.2).

47.1.4 Dyeing Method

Natural and man-made fibers react differently to dyes. Therefore, dyeing tests can also be carried out with corresponding dyes, analogous to the solubility method (staining, not staining). For example, cellulosic fibers can be distinguished from polyester.

Indicator solutions are also used because fiber materials sometimes show characteristic color

D. Veit, *Fibers*, https://doi.org/10.1007/978-3-031-15309-9_47

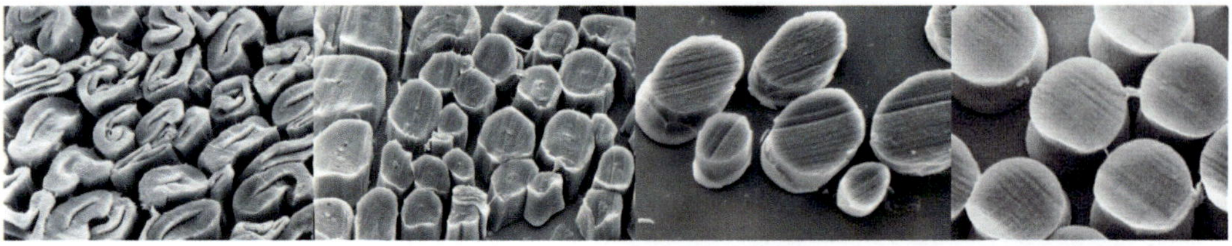

Fig. 47.1 Selected natural fiber cross sections

Table 47.1 Density of fibrous materials and typical reference liquids acc. to Falkai (1981)

	Density [g/cm^3]
PP	0.90–0.92
PE	0.95–0.96
H_2O	*1.0*
PUE	1.0–1.32
PA 11	1.05
Chlorobenzene (C_6H_5Cl)	*1.11*
PA 6, PA 6.6	1.14–1.20
PAN	1.17–1.19
Saturated, aqueous $MgCl_2$-solution	*1.34*
PET, PVC, aramid	1.36–1.45
Carbon tetrachloride (CCl_4)	1.59
PTFE	2.1–2.3

reactions by which they can be identified. In these processes, it must be taken into account that accompanying fiber materials, e.g. lubricants and additives, can also show color reactions.

47.1.5 Thermal Methods

Differential Scanning Calorimetry (DSC)
In this method, the sample is placed in a sealed crucible and this, as well as a second, empty crucible (reference), are heated in a controlled manner in a heat bath. The lid can be porous to avoid an undesired pressure build-up. The temperature of the sample and reference crucibles are kept constant during the measurement. When the glass transition temperature and the melting temperature of the sample are reached, the heat capacity of the sample changes in a material-specific manner due to exo- or endothermic processes. Thus, the heat flow required to maintain the desired temperature differs and from this the glass and melting temperature of the fiber material can be calculated.

In Fig. 47.3, a typical curve for a semi-crystalline polymer is shown. The glass transition region is followed by exothermic recrystallization, then endothermic melting and finally exothermic decomposition.

ISO 11357 describes the method and the evaluation.

Thermogravimetric Analysis (TGA)
This method is mainly used to determine the decomposition temperature and the filler or additive content of a polymer. The sample with a mass of 10–100 mg is heated in a controlled manner, whereby the rate of temperature increase is usually between 0.5 and 20 K/min. A very accurate balance is used to determine the loss of mass as a function of temperature. When the polymer has completely decomposed, the weight remaining in the test crucible corresponds to the proportion of fillers or additives in the polymer. ASTM E1131 and ISO 11358 and regulate the exact test procedure and evaluation (Fig. 47.4).

Melting Point Method
In the melting point method, which is much simpler in terms of equipment, the fiber samples are heated uniformly at approx. 3 K/min under a microscope with polarized light. When the melting point is reached, the birefringence disappears.

Burn Sample
In the so-called burning test, the fibers are burned in a controlled manner with an open flame and

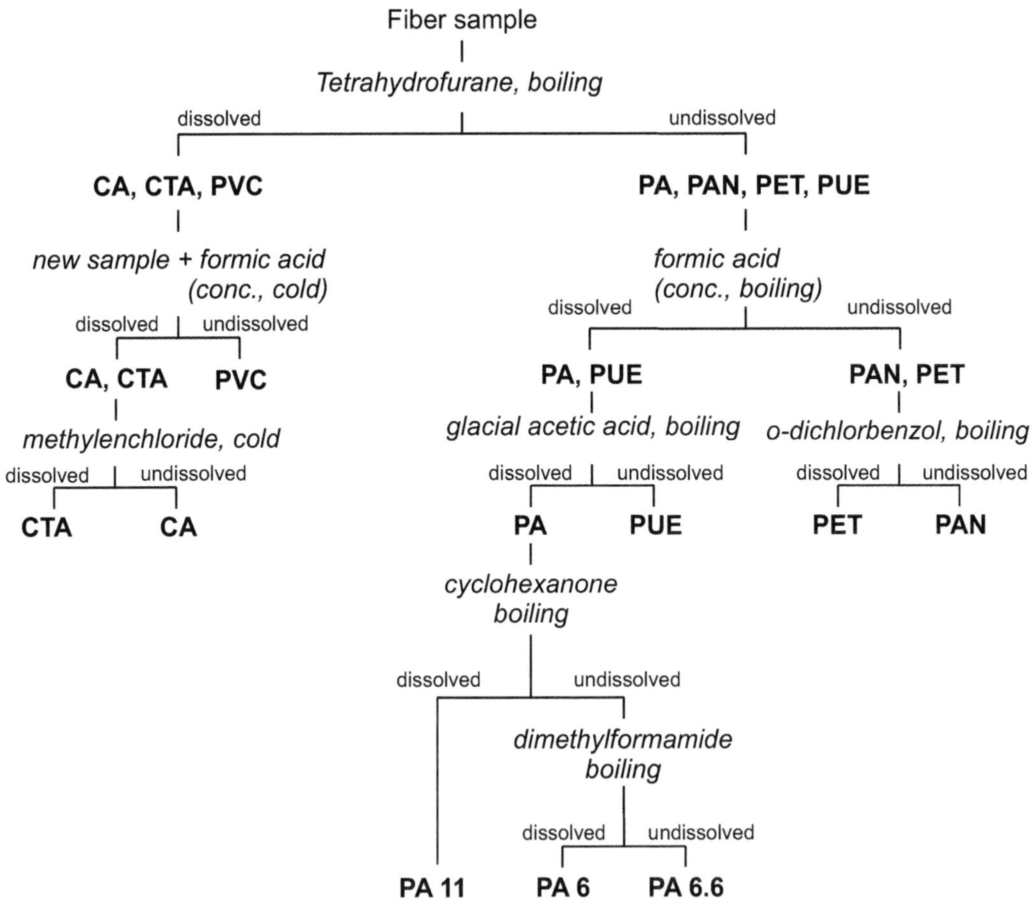

Fig. 47.2 Determination of the fiber material with the solubility method acc. to Agster (1967)

both the development of odor and the deformation of the sample as well as the resulting residue are observed (Table 47.2).

47.1.6 Chromatographic Methods

Polymers can be broken down into their components by acids or alkalis. These are then determined with the aid of gas or thin-layer chromatography, thus allowing conclusions to be drawn about the chemical structure of the polymers.

47.2 Mechanical Properties

Mechanical properties include tensile strength, elongation, Young's modulus, and loop and knot strength.

47.2.1 Basics

Figure 47.5 shows the basic structure of a cellulosic fiber with the orientation of the macrofibrils, which are composed of cellulose chains. The primary wall provides protection against external influences, the secondary walls absorb external forces and the lumen ensures that the fiber is supplied with water and nutrients. The secondary wall S_2 is usually the thickest of the secondary walls.

The orientation of the macrofibrils relative to the fiber axis (Macrofibril Angle, MFA) is decisive for the mechanical properties of a fiber, as the example in Fig. 47.6 shows for Young's modulus. There is no such relationship for the tensile strength.

Various models have been developed to describe the relationships between MFA and

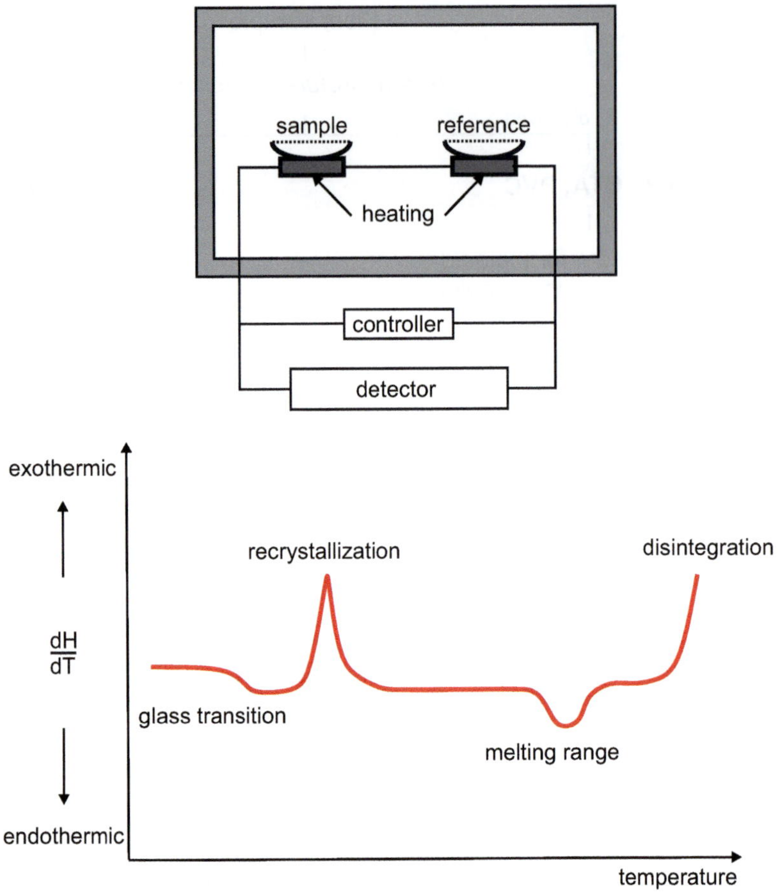

Fig. 47.3 Structure of the DSC apparatus (top) and typical DSC curve (bottom)

mechanical properties, but none of them are universally applicable. Mwaikambo (2006) gives a brief overview.

47.2.2 Strength, Elongation, Modulus of Elasticity

The fiber strength is one of the most important parameters for textile processing. It is a measure of how much stress a fiber can be subjected to in the process stages of spinning preparation and spinning without breaking. Particularly during cleaning, carding and on draw frames and spinning machines, large forces occur which can lead to fiber breakage. The fiber strength is usually determined by single or repeated tensile tests.

The most important terms used to describe fiber strength acc. to EN ISO 5079:2020 are shown in the diagram in Fig. 47.7. The maximum tensile force F_H is the highest value of the tensile force in the tensile test. The breaking force F_B is the force at which the fiber breaks. Acc.ly, the maximum tensile strength and the elongation at break also result.

The fiber strength is usually given in terms of fineness in [cN/tex] by dividing the absolute value of the force by the fiber fineness (present at the beginning). It can be easily and with good accuracy converted into [N/mm^2] or MPa with:

$$\left[\frac{N}{mm^2}\right] = 10 \cdot \rho \left[\frac{g}{cm^3}\right] \cdot \left[\frac{cN}{tex}\right],$$

Where ρ is the density of the material. Aramid, for example, has a strength of 190 cN/tex and a density ρ of 1.44 g/cm^3, giving a strength value of 2736 N/mm^2.

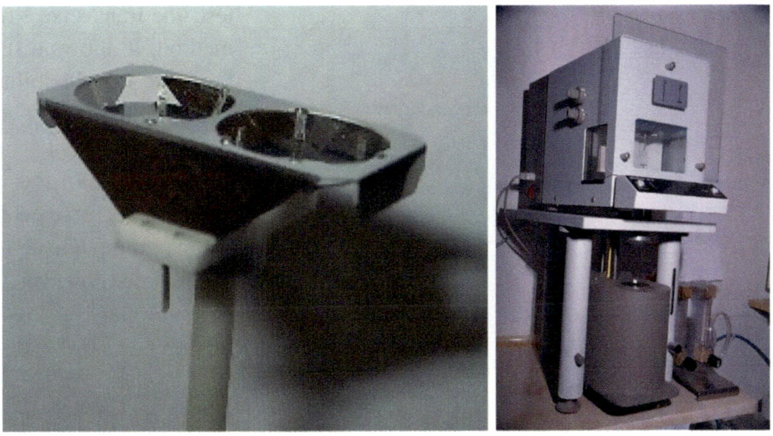

Fig. 47.4 DSC crucible (left, Wiki, 2005) and TGA device with furnace extended downwards (left, Düren, 2005)

Table 47.2 Typical behavior of fibers in the burning test acc. to Agster (1967), Koch and Wagner (1953)

	Melting range [°C]	Odor	Residue
PP	160–175	After burning paraffins	Light brown, round pearls
PA 6	215–220	Slightly after burning hair	Hard, tough, grey-brown round pearls
PA 6.6	255–260	Non-stinging	
PET	250–260	Sweet, aromatic, slightly stinging	Hard, tough, brown-black, round pearls
PAN	Charred	Sweet, aromatic	Hard, brittle, black, uneven formed pearls
PVC	160–200	Stinging, cough irritant	Hard, black, uneven formed pearls
PVA	Charred	After incinerated sugar	Brown, brittle
CO	–	After burning paper	White-grey to yellowish (small amount)
FL	–		
HA	–		
WO	–	After burning horn	Blistered, coaly, molten spherules
CV	–	After burning paper	White-grey (small amount)
CUP	–		
CA	–	Stinging sour	Blistered, coaly, molten spherules

Sometimes the unit [g/den] or [g/dtex] is also used. In this case, the following applies as a good approximation:

$$\left[\frac{cN}{tex}\right] = 10 \cdot \left[\frac{g}{dtex}\right] = 11 \cdot \left[\frac{g}{den}\right]$$

A value of 15 cN/tex therefore corresponds to 1.5 cN/dtex and 1.65 g/den.

The conversion for the Young's modulus, which is often given either in [cN/tex] or in [MPa] = [N/mm^2], is analogous. Sometimes the "initial E-modulus" is given, it then corresponds to the slope of the force-strain curve at the beginning of the loading (Fig. 47.7).

If the tensile test is carried out on both, a dry and a wet sample, the relative wet strength can be determined as a percentage in relation to the dry strength. Depending on the fiber material, it is above (e.g. flax) or below (e.g. viscose) 100%. Thus, if the dry strength is 40 cN/tex and the wet strength is 110%, this corresponds to 44 cN/tex.

In general, a low pretensioning force F_v is applied to the fiber before the test begins in order to create defined test conditions. This must not yet lead to substance elongation.

47.2.3 Loop Strength

The loop strength is tested acc. to ASTM D3217/D3217M-20. While the fiber is only stressed in the longitudinal direction in the tensile strength

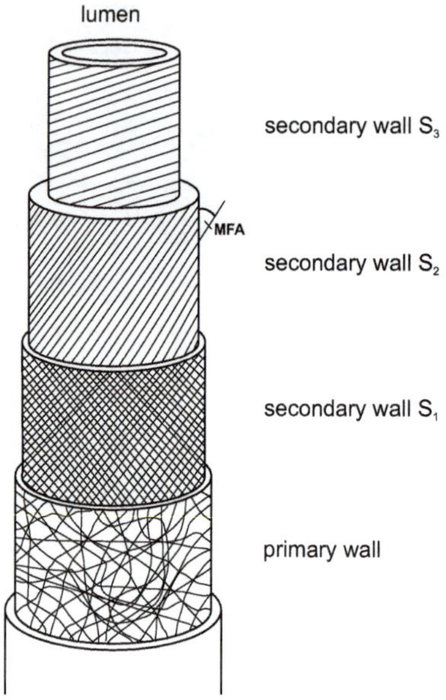

Fig. 47.5 Model of the structure of a cellulosic fiber

test, the transverse strength is included with this method. It allows a statement to be made on the brittleness or flexibility of the fiber material. As shown in Fig. 47.8 two fibers are intertwined and subjected to tensile stress.

In the case of ideally supple fibers, the maximum tensile force determined in this way is twice the maximum tensile force of the individual fiber. In practice, however, the loop maximum tensile force is always lower. Its value is given in [cN/tex].

47.2.4 Mechanical Utility Value

Acc. to Böhringer (1957), strength increases and elongation decreases with increasing stretching, but not linearly. Correspondingly, the total working capacity initially increases until it reaches an optimum and then falls again (Fig. 47.9).

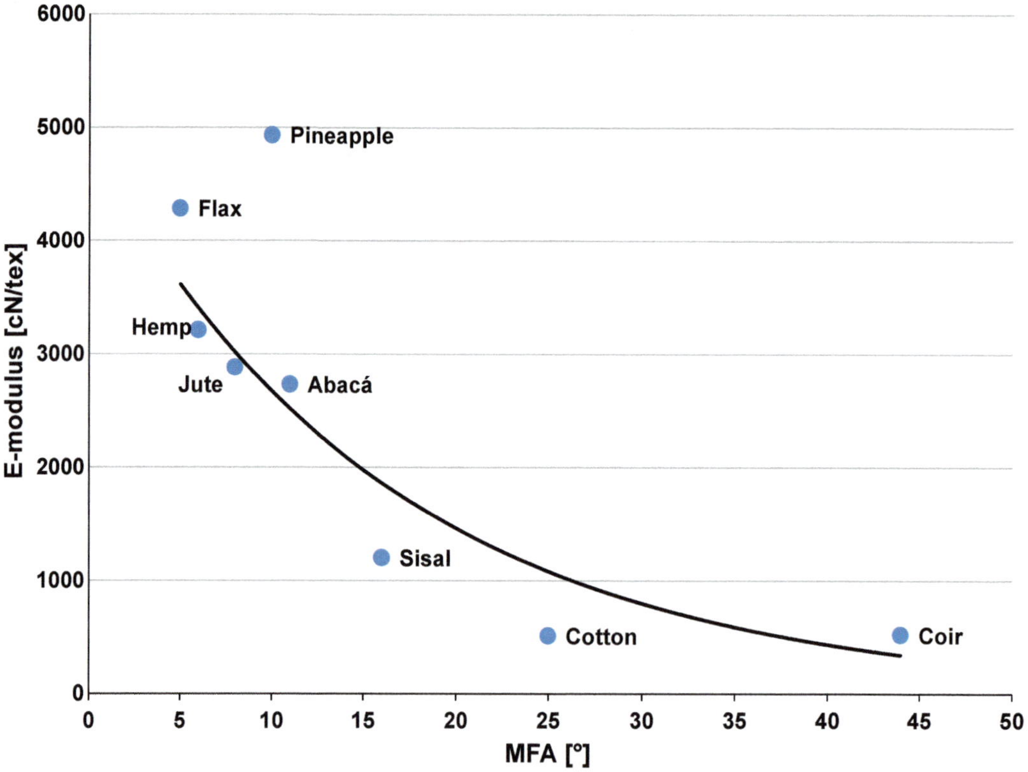

Fig. 47.6 Dependence of the Young's modulus on the macrofibril angle MFA acc. to Mwaikambo (2006)

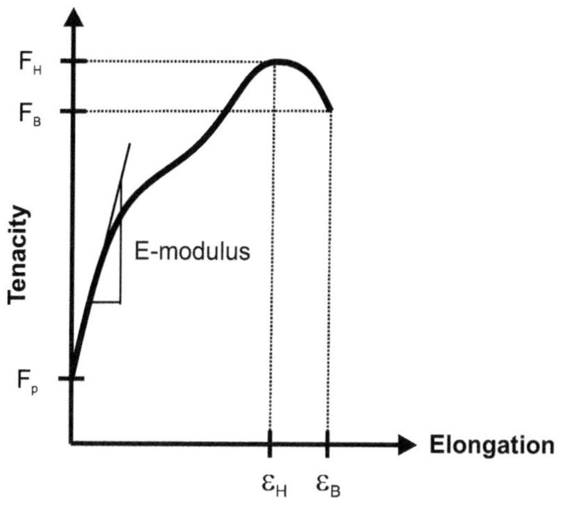

Fig. 47.7 Typical force-elongation diagram of a polymeric fiber

47.2.5 Optical Methods for Determining the Internal Structure

47.2.5.1 Birefringence

Birefringence is a measure of the orientation of macromolecules. The process is based on the fact that anisotropic materials divide an incident light beam into two mutually perpendicular beams. Isotropic materials also exhibit this behavior when an external voltage is applied. This effect results from different refractive indices $n_{\|}$ and $n_{\perp}$ of the polymer depending on the propagation direction and polarization of the light beam. The refractive index is the ratio of the wavelength of light in a vacuum to the wavelength in the material.

Birefringence is defined as $D = n_{\perp} - n_{\|}$. Depending on the sign, one speaks of optically positive and optically negative materials.

47.2.5.2 Wide Angle X-Ray Diffraction (WAXD)

Wide Angle X-Ray Diffraction (WAXD) is a technique for determining the crystal structure of polymers in the nanometer range and below. Diffraction of an X-ray wave occurs when the spacing of the grating lines of the diffraction grating is of the order of the wavelength of the incident waves. The wavelength of X-rays is 1 pm to 10 nm, which is also approximately the spacing of atoms in crystal structures of polymers. Therefore, polymer crystals act like a three-dimensional diffraction grating on X-rays.

Fig. 47.8 Experimental setup of a loop strength test

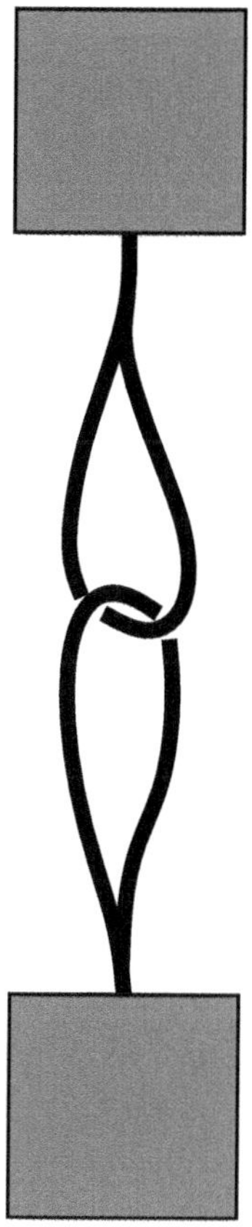

The X-rays are diffracted at the electron shell of the irradiated atoms and the diffracted waves emanating from the individual atoms generate interference. Depending on the distance between the atoms, different path differences are obtained for the diffracted waves. Whether constructive or destructive interference occurs at a fixed angle depends on the distance between the atoms.

Because crystals consist of three-dimensional and mostly periodically repeating units, constructive interference occurs only for very specific angles. From these angles, the distance of the planes of the crystal can be calculated with the help of the Bragg equation:

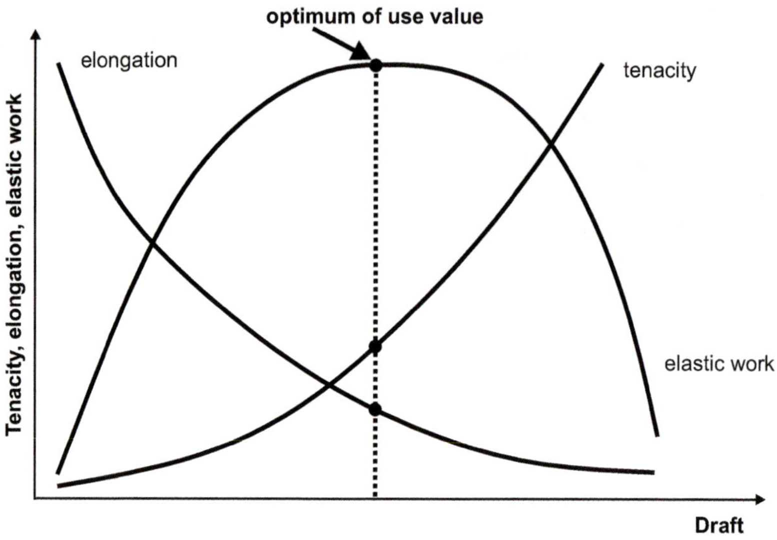

Fig. 47.9 Optimization of the utility value of fibers acc. to Böhringer (1957)

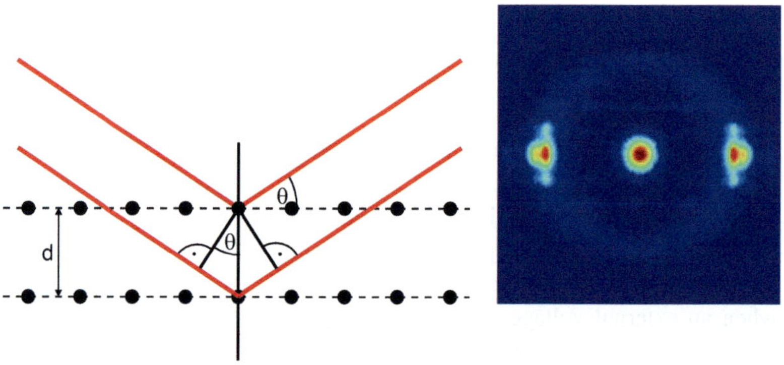

Fig. 47.10 Principle of the Bragg grating (left) and typical WAXD image of a polymer

$$n \cdot \lambda = 2d \cdot \sin(\Theta)$$

Where λ stands for the wavelength of the monochromatic X-ray beam used to irradiate the sample, d is the spacing of the grid planes, the angle between the angle of incidence of the X-ray beam and the grid plane, and n is the degree of the examined maximum from the center, counted in the form of a natural number. Figure 47.10 shows the principle of the Bragg grating (left) and a typical WAXD image with the interference maxima (right).

Mo and Zhang (1996) describe in detail the procedure and also others for determining the crystal structure.

47.2.5.3 SAXS

Small angle scattering (SAS) can be used to investigate the morphology of fibers in the micro- and nanometer range. In this method, the size of the structures to be analyzed is significantly larger than the wavelength of the radiation used. If X-rays are used, this is referred to as the SAXS method. The basic principle corresponds to the WAXD method.

47.2.5.4 FTIR

An FTIR spectrometer (Fourier transform infrared spectrometer) is a variant of a spectrometer. With the help of a fixed and a movable mirror, light waves are separated from each other,

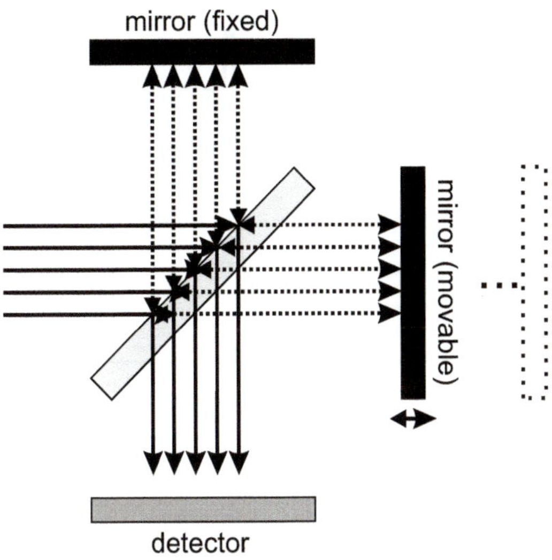

Fig. 47.11 Setup of an FTIR spectrometer

spatially shifted and then combined again. The basic principle is depicted in Fig. 47.11. Light is refracted at a lens which is arranged in the middle. A part of the light waves (ideally half) is deflected to the fixed mirror, the other part reaches the movable mirror. From the mirrors, the light waves are reflected back to the lens, combined there and then enter the detector.

Depending on the distance of the moving mirror to the measuring field, an interference pattern is created from the reflected light beams. A stationary detector picks up the intensity signal. If the optical path length difference is equal to zero, the waves constructively overlap and the maximum intensity is produced. If the optical path length difference corresponds to exactly half a wavelength of light (mirror shift of a quarter wavelength), the superposition is destructive and the waves cancel each other out.

As the mirror shift increases, the phase differences of the individual wavelengths differ to varying degrees; they no longer add up to the maximum value. If all the wavelengths of the light involved in the measurement are superimposed, the intensity detected by the detector results from the integral over the bandwidth of the array. In the case of large mirror shifts, the phases of the individual wave trains are almost uniformly distributed and add up to an average value. This mean value does not contain any spectroscopically relevant information. It is therefore subtracted from the intensity signal and only the deviation of the detected signal from the mean value is analyzed. Thus, the incident intensity is obtained at the detector as a function of the mirror shift as an interferogram (Fig. 47.12).

With the help of a Fourier transformation, the IR spectrum (Fig. 47.13). From this, information about the internal orientation of polymers can be derived.

47.2.5.5 Raman Spectroscopy

Raman spectroscopy is based on a similar principle to FTIR spectroscopy. Here, the sample to be examined is irradiated with monochromatic light, usually with the aid of a laser. The incident light is scattered by the sample, resulting in frequency differences between the incident and emitted light

Fig. 47.12 Interference intensity as a function of the displacement of the moving mirror

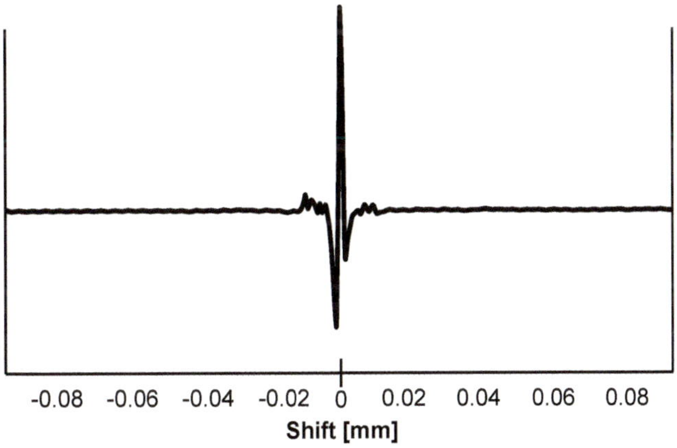

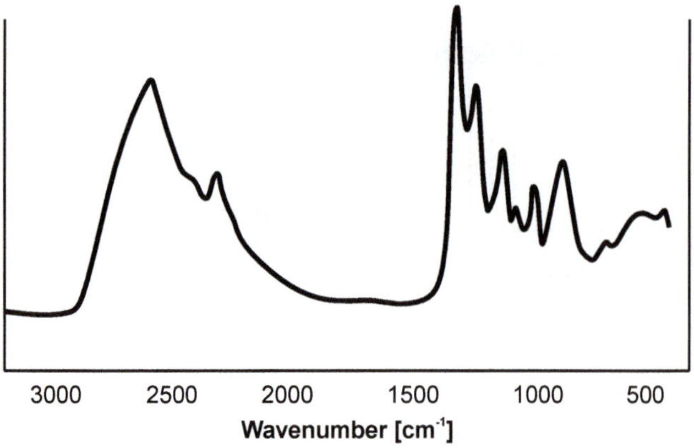

Fig. 47.13 IR spectrum of an interferogram after Fourier transform

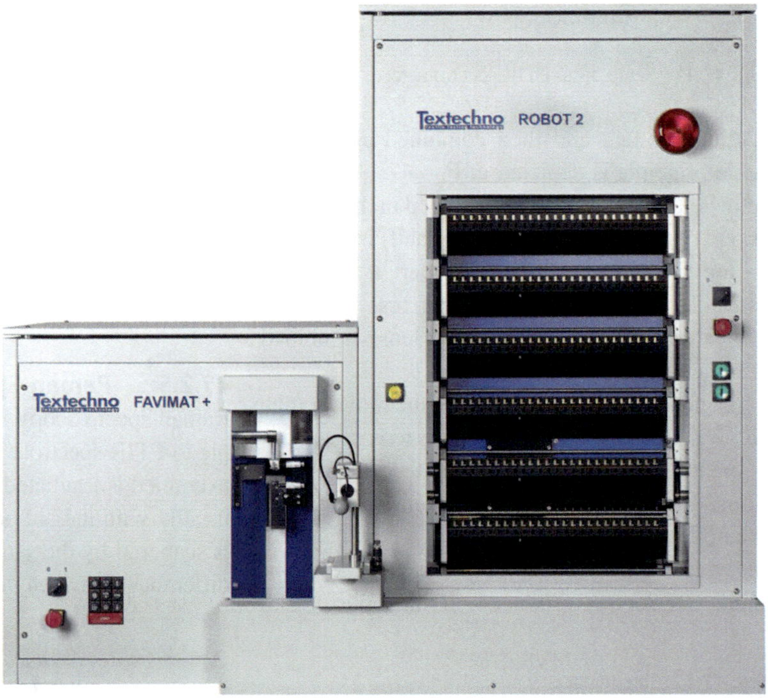

Fig. 47.14 Single fiber fineness tester (Textechno, n.d.)

waves. This allows conclusions to be drawn about the internal structure of a fiber, e.g. the crystallinity and crystal orientation.

47.3 Geometric Properties

The geometric properties of fibers include fiber fineness or its diameter, fiber length and crimp.

47.3.1 Fiber Fineness

The fiber fineness is often also referred to as the titer. It is one of the most important quantities for characterizing fibers. Depending on the type of fiber, it is given in different units. These, in turn, can differ from country to country.

The single fiber fineness is often determined by means of an acoustic resonance method, a corresponding test device shows Fig. 47.14.

The fiber fineness has a decisive influence on the stiffness, drape and handle of textile fabrics. Furthermore, their appearance and dyeing behavior depend on the fineness. The number of twists required in the yarn to achieve a certain strength also depends on the fiber fineness and is lower for finer fibers (i.e. more fibers in the yarn cross-section for the same yarn fineness). The yarn uniformity is also directly determined by the number of fibers in the yarn cross-section, i.e. the fiber fineness.

For wool and other fibers which have an almost round cross-section, the diameter is given in μm, e.g. for superfine wool: 16 μm.

In the case of man-made fibers, despite their often round cross-section, the fineness is given in [dtex] or [den]:

$$1 dtex = \frac{1\,g}{10{,}000\,m}$$

$$\text{respectively } 1\ den = \frac{1\,g}{9{,}000\,m}$$

The fineness is then abbreviated as Tt ("titer in tex") or Td ("titer in denier"). Although the unit [den] was replaced by [dtex] in 1970, it is still used because this usually results in "even" numerical values. A typical fineness of man-made fiber yarns, for example, is 150 den, which corresponds to 167 dtex.

47.3.2 Fiber Length

The fiber length is only determined for natural fibers (exception: silk) as well as cut man-made fibers. For–in principle–endless filament yarns, this test makes no sense.

Definitions and Terms

Normally applies

Fiber length: Length of the uncrimped, undrawn fiber.

Staple length: Often used but not standardized term for fiber length.

Mean fiber length: Arithmetic mean of the lengths of all fibers in a sample.

Maximum fiber length: Largest occurring fiber length.

For *cotton* are also common

Classifying staple: (fiber length that 16% of all fibers reach at least).

Short fiber content: z. e.g. proportion of fibers with a length < 6 mm

For *man-made staple fibers* is common

Nominal length: The theoretical length to which man-made fibers can be cut. It can deviate from the actual length and is usually used in the shipping documents.

For simplification, the fiber length is often divided into classes. The individual class covers a certain length range. The width of the class range is called the class width, and the arithmetic mean of the two class boundaries is called the class center.

Frequency Chart

The frequency chart is obtained by plotting the class frequencies on the ordinate and the class centers on the abscissa. This makes it immediately clear which class occurs particularly frequently, i.e. contains the most fibers. In this way, e.g. cutting errors in the production of synthetic staple fibers can be easily identified (Fig. 47.15).

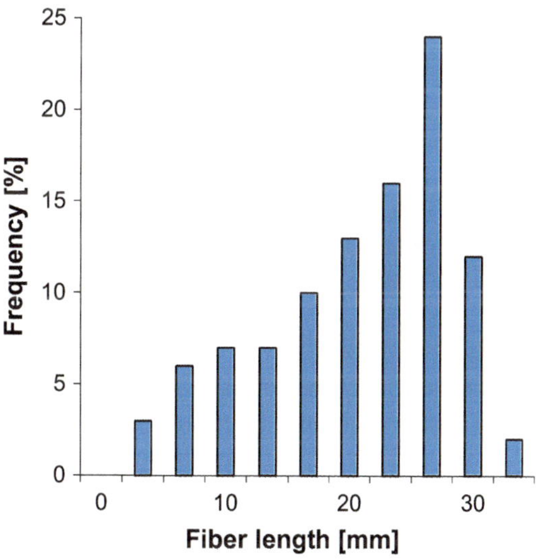

Fig. 47.15 Frequency chart

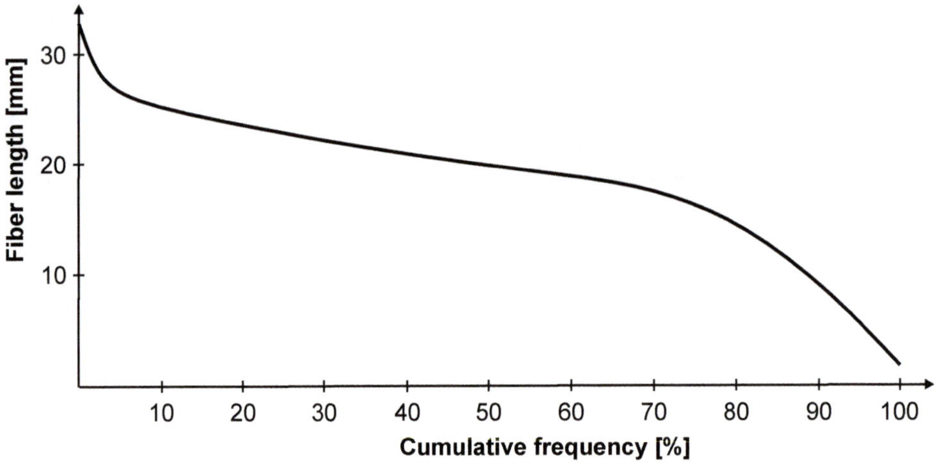

Fig. 47.16 Staple length diagram using cotton as an example

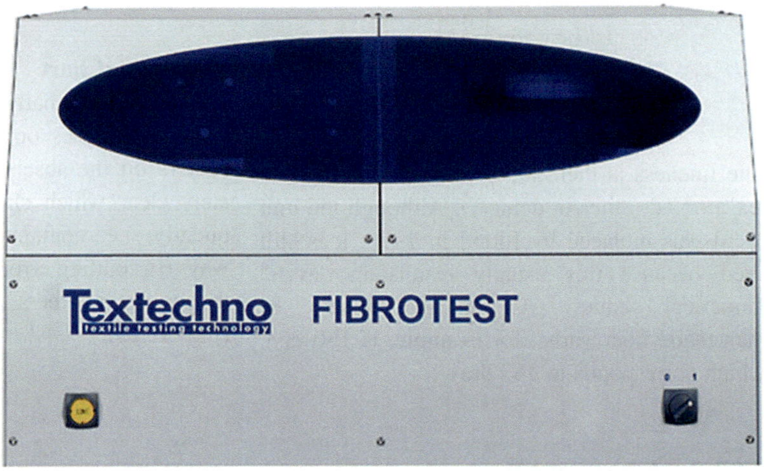

Fig. 47.17 Testing device for determining fiber fineness and bundle strength (Textechno, n.d.)

Depending on the test method, either the unstressed, the length-stressed, the weight-stressed or the cross-section-stressed fiber length is determined. Details are regulated by ASTM D5103–07(2018).

Staple Length Diagram

The staple length diagram is the cumulative line of the continuously added frequencies. It serves to illustrate the fiber length distribution. It largely corresponds to an experimentally produced fiber sample ordered by fiber length. The lower class limit is plotted on the ordinate and the corresponding cumulative frequency on the abscissa. The points obtained are connected by a smooth line (Fig. 47.16).

A testing device for determining fiber length and fiber bundle strength shows Fig. 47.17. It can be used for the analysis of cotton and man-made fibers.

47.3.3 Fiber Crimp

The term "crimp"refers to an undulating course of the fiber along its length. Due to the fineness-dependent area moment of inertia, the crimp is more pronounced in fine fibers (e.g. wool) than in coarse fibers. The crimp improves the cohesion of the fibers in the sliver ("sliver adhesion") and also influences the usage properties of the textiles produced from them. A strong fiber crimp gives the textile greater volume, better temperature insulation, better moisture transport and also the elongation is increased.

Natural fibers are usually present in a largely stretched form. Although twisting of the fiber about the longitudinal axis may also occur in cotton, for example, no true crimped bends are formed. Some fibers, such as wool, are crimped (due to their ortho- and paracortex structure).

Man-made fibers are also often crimped after the spinning process. This makes them easier to blend with other crimped fibers. In addition, they become more voluminous.

The crimp can be described by different characteristic values. The three most important are:

- The number of crimping arcs per unit length.
- The ratio between crimped and stretched length (crimp contraction).
- The crimp that returns after a load (crimp stability).

47.4 Moisture and Water Absorption

If fibers, yarns or textiles are brought into direct contact with steam or water (e.g. during finishing), they usually absorb water accordingly. Their mechanical properties change accordingly, and this may also play a major role in dyeing.

47.4.1 Moisture Absorption

The absolute moisture content of a fiber sample is determined from the difference in weight of the sample in its initial state and in its dried state. The moisture content is related to the dry weight and given as a percentage. ASTM D2654–22 describes standard test methods for moisture in textiles and ASTM D1909–13(2020) lists commercial moisture regain values and commercial allowances for textile fibers.

47.4.2 Water Absorption and Water Retention Capacity

Water absorption is understood to be the behavior of fibers under the influence of liquid water. In addition to the actual water absorption capacity, the wettability of the fibers also plays a major role, e.g. in dyeing processes.

47.4.3 Wettability

Good wettability of fibers can be both a desirable and an undesirable property. In textile finishing processes, good wettability is usually beneficial to allow water to penetrate the fiber, yarn or textile quickly and evenly. Good wettability is also important for incontinence products, e.g. diapers. In other cases, however, e.g. for rainwear, a poor wettability is desired.

One of the most important factors influencing wettability is the surface structure of the fiber. In addition to the chemical structure, the macroscopic structure (e.g. "lotus effect") also plays a major role. Furthermore, the capillary effect between the fibers can also be important.

There are numerous test methods for assessing wettability, most of which are carried out on the finished textiles.

47.4.4 Water Retention Capacity

The water retention capacity acc. to ASTM D2402–07(2018) indicates as a percentage value how much water a fiber can absorb including any surface water adhering to it. Before measurement, excess surface water is removed by defined centrifuging.

The characteristic value itself can be related to the dry weight of the sample or to the weight of the fiber sample adjusted in the standard climate.

47.5 Standards

There are a large number of standards for determining the properties of fibers. Table 47.3 contains selected examples. Depending on the country, there are different standards, e.g. in the USA the standards of ASTM (American Society for Testing and Materials), in Great Britain the standards of BSI (British Standards Institution), in France the standards of AFNOR (Association française de normalization) etc. apply.

The ISO standards, which are often based on DIN standards, are valid worldwide. Accordingly, they are then abbreviated as DIN ISO. When applying the standards, it should be noted in particular that lengths and masses may have to be specified in units other than the SI units. For example, especially in Anglo-Saxon countries, measurements are still often made in feet, stones, bushel and other non-SI units.

References

Agster, A. (1967). *Färberei- und textilchemische Untersuchungen*. Springer.
Böhringer, H. (1957). *Gebrauchswertoptimale Erspinnung von Chemiefasern*. Akademie-Verlag.
Düren, R. (2005). *CC BY-SA 3.0*. https://commons.wikimedia.org/wiki/File:TGA-Messung.jpg
Falkai, B. v. (1981). *Synthesefasern*. Verlag Chemie.
Koch, P.-A., & Wagner, E. (1953). *Textile Prüfungen*. Dr. Spohr-Verlag.
Mwaikambo, L. Y. (2006). Review of the history, properties and application of plant fibres. *African Journal of Science and Technology, 7*(2), 120–133.

Table 47.3 Selected textile standards (DIN and ASTM)

	German standard	International standard
Textiles, basic terms	DIN 60 000	ASTM D123-19
Fibers, basic terms, fineness	DIN 60 910	ASTM D276-12
Fibers–technological testing processes	DIN 53801-3	
Fiber length of spun fibers	DIN 53 805DIN 53 806DIN 53 808-1	ISO 6989ASTM D5103
Maturity grade of cotton	DIN 53 943	ASTM D1442-06
Length and evenness of cotton	DIN 53 944	ASTM D1447-07
Length of wool		ISO 920ASTM D519
Fiber fineness	DIN EN ISO 1973	ASTM D 1577
Fineness of monofilaments	DIN EN 13 392	
Fineness of spun fibers–terms, processes	DIN 53 810	
Fiber diameter in microprojection	DIN 53 811	ASTM D2130-13
Fiber fineness, gravimetric processes	DIN 53 812	
Glass fiber diameter		ISO 1888
Maximum tenacity and elongation	DIN EN ISO 5079	ASTM D3822
Cotton, bundle strength	DIN ISO 3060	ASTM D1455
Monofilaments, strength	DIN EN 13 895	
Fibers (monofilaments)–shrinkage	DIN EN 13 844	ASTM D2102-02
Fibers–elasticity	DIN EN 15 930	ASTM D2591-07
Loop strength	DIN 53 843-2	ASTM D3217
Standard terminology for cotton fibers		ASTM D7139–20
Standard terminology relating to wool		ASTM D4845-10
Cashmere, fine animal hair–quantitative analysis	DIN EN ISO 17751-1	ASTM D2252-18
Light transmitting fibers	DIN 58 143-1DIN 53 143-2DIN 53 143-3DIN 53 143-4	ASTM F3092-14
Textured yarns–crimp	DIN 53 840-1DIN 53 840-2	ASTM D6774-12

Mo, Z., & Zhang, H. (1996). The degree of crystallinity in polymers by WideAngle X-ray diffraction (Waxd). *Journal of Macromolecular Science, Part C: Polymer Reviews*, 555–580.

Textechno. (n.d.). *Herbert Stein GmbH & Co. KG Textile Mess- und Prüftechnik*. Mönchengladbach. https://www.textechno.com

Wiki. (2005). *CC BY-SA 3.0*. https://commons.wikimedia.org/wiki/File:DSC-Probentr%C3%A4ger.jpg

Part X

Properties of Natural and Chemical Fibers

48 Properties

The values of typical properties of natural fibers are taken from various works. Common values are given, in` some cases with wide ranges, reflecting the sometimes unclear data situation.

The values for man-made fibers compiled in this chapter come for the most part from the fiber producers themselves. The corresponding links to the websites of the fiber producers are compiled in the next chapter. The values for "fiber length"refer to staple fibers, for filaments the length is technically endless.

The conversion of important units is carried out in accordance with:

$$Fiber\ diameter\ [\mu m] = 11.3 \cdot \sqrt{\frac{Titer[dtex]}{density[g/cm^3]}}$$

For a fiber with a fiber fineness of 3 dtex and a density of 1.5 g/cm^3, the fiber diameter is therefore $d = 16$ µm. The value 10 is often used instead of the factor 11.3, which strictly speaking is only correct for square fibers. However, fibers are rarely ideally round, so the value 11.3 is also only an approximation.

Strength and Young's modulus are expressed in the SI system in [cN/tex], [MPa] (= [N/mm^2]) or in the imperial system in [lbf/in^2] (pounds-force per square inch), abbreviated psi, with 1000 psi = 1 ksi and 1000,000 psi = 1 Msi. A unit still occasionally used is [grams per denier], abbreviated [gpd] or [g/den]. In very old works, strength is given in breaking length in kilometers [Rkm]. This value corresponds exactly to the value in [cN/tex].

It applies

$$Tenacity[MPa] = 10 \cdot \rho[g/cm^3] \cdot Tenacity[cN/tex]$$
$$= 1.45 \cdot \rho[g/cm^3] \cdot Tenacity[ksi]$$

Example
2400 MPa at a density of the fiber of 1.5 g/cm^3 corresponds to 160 cN/tex or 16 cN/dtex and 18 gpd and 1100 ksi and 1.1 Msi.

48.1 Fineness

Depending on the material, different systems are used to specify the fiber fineness which is also called linear density and titer. In addition to specifications in µm and dtex (g/10,000 m), values in the (g/9000 m) and other units are also common. In Fig. 48.1, the fineness of selected fiber materials is compared.

Figure 48.2 shows the comparison of the cross-sectional size of different fibers. For human hair, 80 µm were taken as a basis, whereby blond and children's hair can be finer (up to 40 µm). Wool was assumed to be 19 µm, cotton 15 µm, flax (hemp, jute) 22 µm, the PET microfiber 0.2 dtex (4 µm) and glass or carbon 10 µm.

D. Veit, *Fibers*, https://doi.org/10.1007/978-3-031-15309-9_48

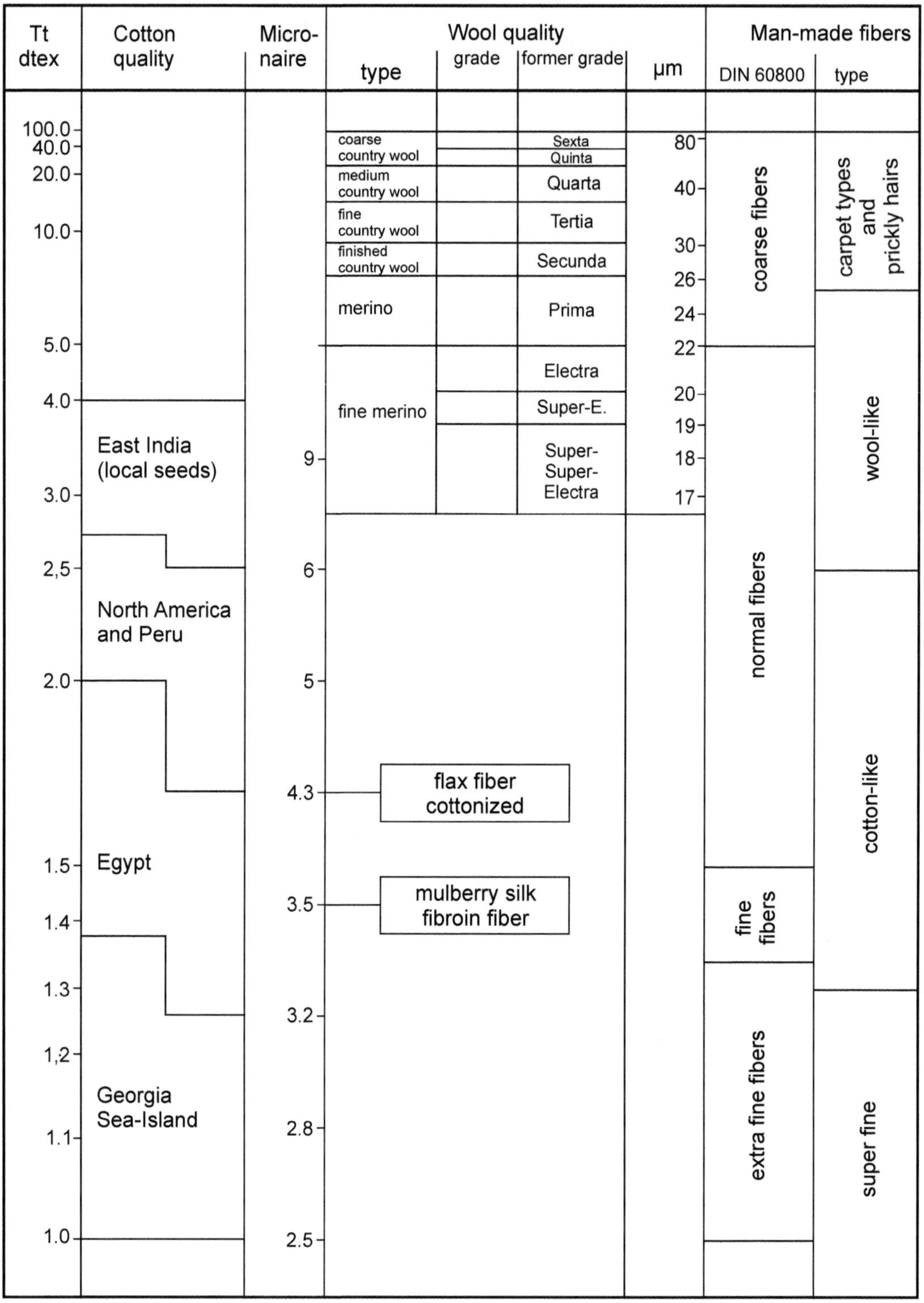

Fig. 48.1 Comparison of fineness data in different fineness systems

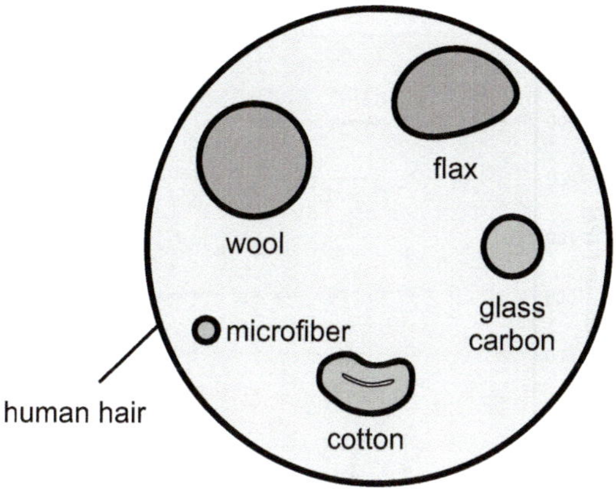

Fig. 48.2 Comparison of the fineness of important fibers

48.2 Strength and Modulus of Elasticity

Figure 48.3 shows the values of strength and Young's modulus for fibers used mainly for clothing. It is obvious that the bast fibers have a much larger Young's modulus than all other fibers, which partly explains their tendency to crease.

Strength and Young's modulus for selected fibers for technical textiles are summarized in Fig. 48.4. The density of the fiber material is included in the diagram on the left, while the absolute strengths are shown on the right, which

Fig. 48.3 Strength and modulus of elasticity of selected fibers for clothing

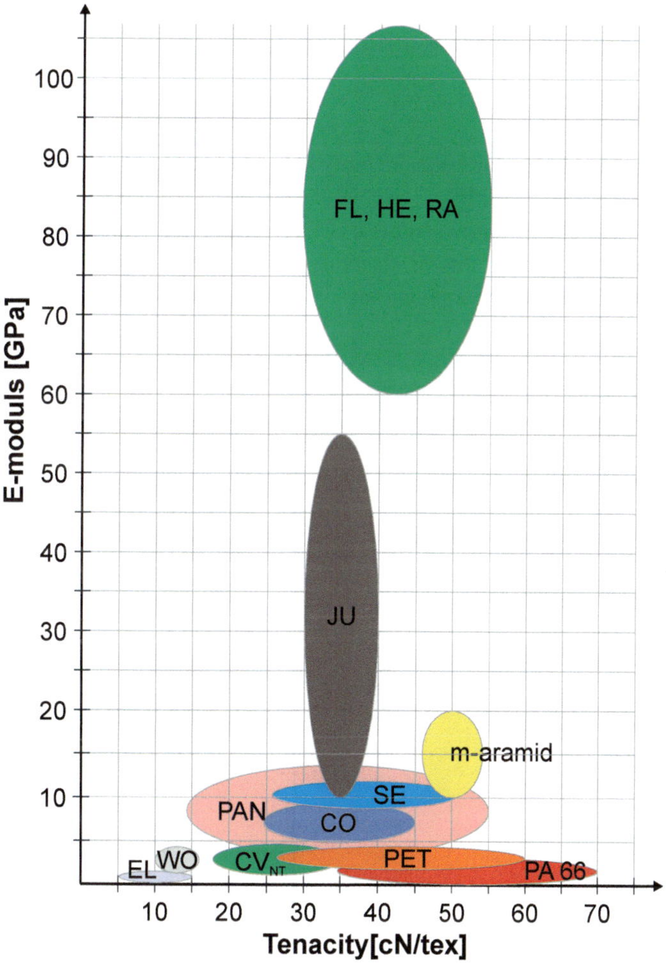

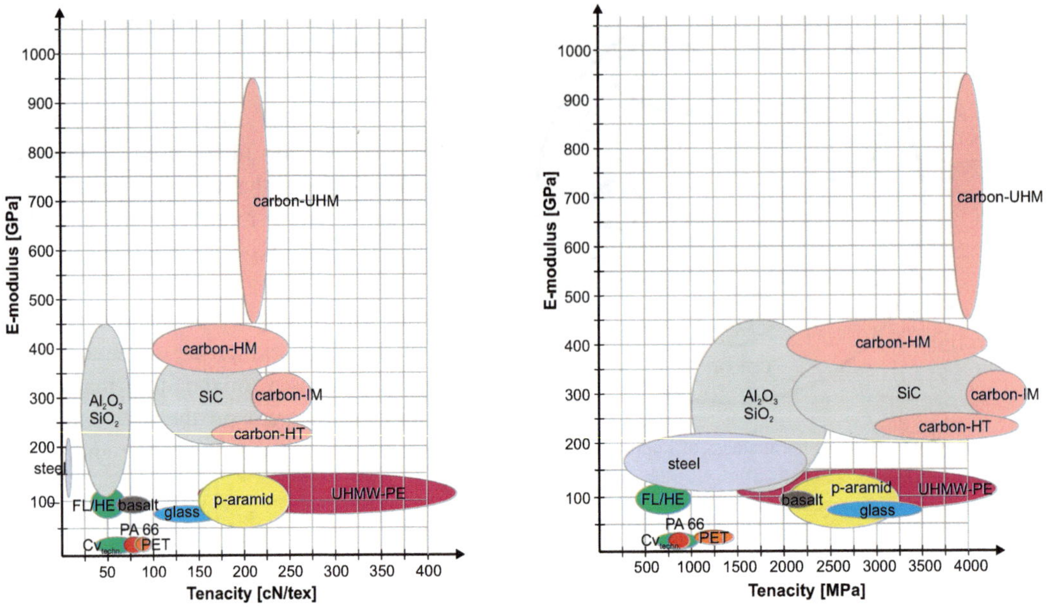

Fig. 48.4 Strength and modulus of elasticity of selected technical fibers in [cN/tex] and [MPa] respectively

changes the picture considerably in some cases (steel, glass, basalt).

Figure 48.5 shows typical value ranges for strength and modulus of elasticity of man-made fibers for different applications.

Typical force-elongation curves of natural fibers shown in Fig. 48.6. The very high Young's modulus and great strength of all bast fibers compared with all other natural fibers is apparent. Wool has by far the largest elongation, cotton

Fig. 48.5 Typical strength and Young's modulus values for man-made fibers acc. to Hearle (2008)

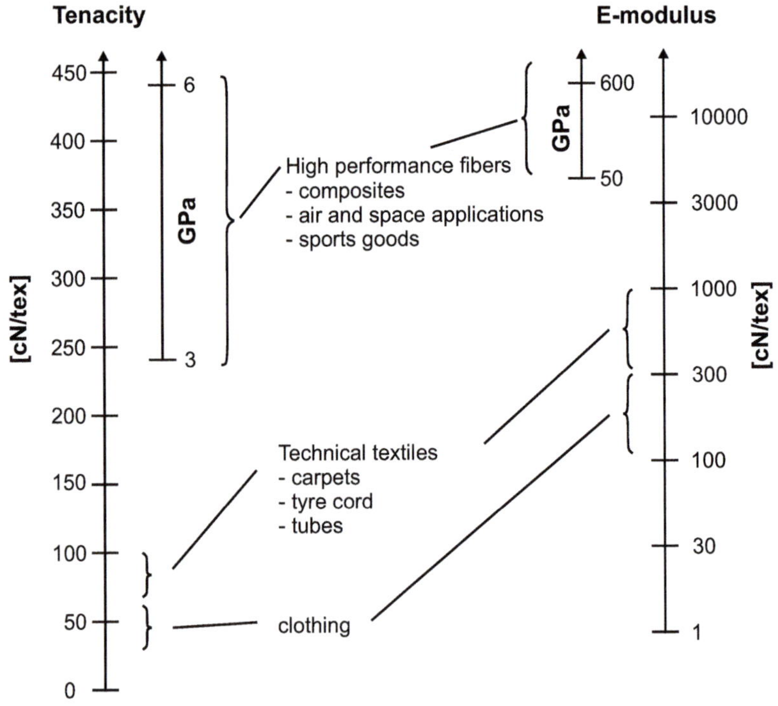

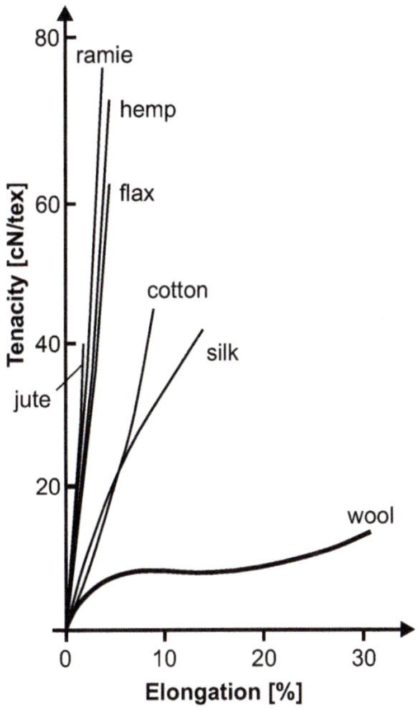

Fig. 48.6 Typical force-elongation curves of natural fibers

and silk have a similar initial modulus and strength.

The mechanical properties of fibers are usually determined in the fiber axis direction. Even with a small deviation between fiber axis direction and external load F, the strength decreases rapidly, as Fig. 48.7 shows.

In Table 48.1, typical degrees of elasticity of selected fibers are listed. The degree of elasticity corresponds to the reversible portion at an elongation of the fiber of 5%. In the case of PU (elastane), the degree of elasticity is almost 100% even at an elongation of around 800%. Due to their usually significantly lower crystalline content, man-made fibers are normally more elastic than natural fibers, which form much more slowly and are therefore usually more highly crystalline.

Table 48.2 shows the torsional brittleness of various fiber materials. It is a measure of how elastically a fiber reacts to twisting, e.g. during yarn formation. The angle of breakage torsion is determined by twisting the fibers until they break. An imaginary line on the fiber sheath in the direction of the fiber axis takes on a helical shape when twisted. The pitch angle of the helix when the fiber breaks relative to the fiber axis is then the break twist angle. The larger the angle, the more the fiber could be twisted without breaking.

With d = fiber diameter, l = fiber length, and T = number of turns on length l, we have:

Fig. 48.7 Dependence of fiber strength on the angle between fiber axis and force

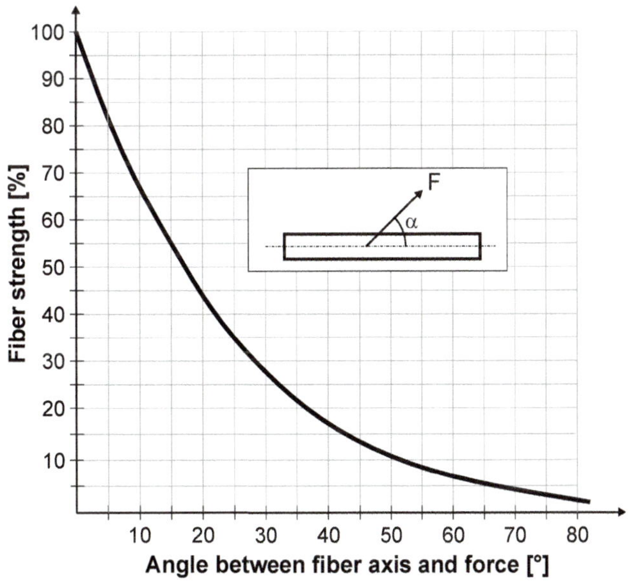

Table 48.1 Elasticity grades of selected fibrous materials

Fiber	Degree of elasticity at total elongation E = 5 %
PA	98
PU	98
Modacrylic	92
PP	88
PET	80
PVC	77
PAN	75
Silk	70
Wool	65
CT	63
CV	50
CA	50
PVA	50
Cotton	45
Fl	40

$$\tan(\alpha) = \frac{l}{\pi \cdot d \cdot T}$$

48.3 Degree of Polymerization and Length of the Polymers

Figure 48.8 and Table 48.3 show the molecular weights of the repeat unit and the chain length of the macromolecules of selected fibrous materials. Most polymers have a length of 200–300 nm. Cotton has the largest chain length among cellulosic fibers, but is surpassed by wool. UHMW-PE molecules are even up to 61,000 nm long. The shortest polymer of technical importance is PLA with around 120 nm.

48.4 Composition

Table 48.4 gives an overview of the typical chemical composition of cellulosic fibers. While seed, bast and hard fibers have a high cellulose content, almost half of coir fibers consist of lignin, which can hardly be removed. Therefore, only coarse yarns can be produced from coir fibers. In principle, a large number of different values for the chemical composition of plant fibers can be found in the relevant literature. Therefore, such data are always only approximate values and these are highly dependent on the region of cultivation.

48.5 Density

The density of selected fibrous materials summarized in Table 48.5. Cellulose-based fibers reach values around 1.5 g/cm^3, fibers based on protein fibers are around 1.3–1.4 g/cm^3, as are many synthetic fibers. Carbon, glass and basalt reach values around 1.7–2.8 g/cm^3, ceramics and steel fibers even significantly higher. Further values can be found in the chapters "High-performance fibers" and "Biopolymers", among others.

Table 48.2 Torsional brittleness of selected natural and man-made fibers acc. to Koch and Wagner (1953)

	Fractional twist angle [°]	Range [°]	Labelling
Casein	28–31.5	<40	Very smooth
Polyamide	42–42.5	40–50	Smooth
Acetate	44–49.5		
Cupro	48–50.5		Normal
Wool	48.5–51.5		
Silk	51	50–60	
Angora	51.5		
Viscose	50.5–54.5		Supple
Cotton	53–56		
Flax	60	60–70	Brittle
Viscose, high strength	67		
Glass	85–87.5	> 70	Very brittle

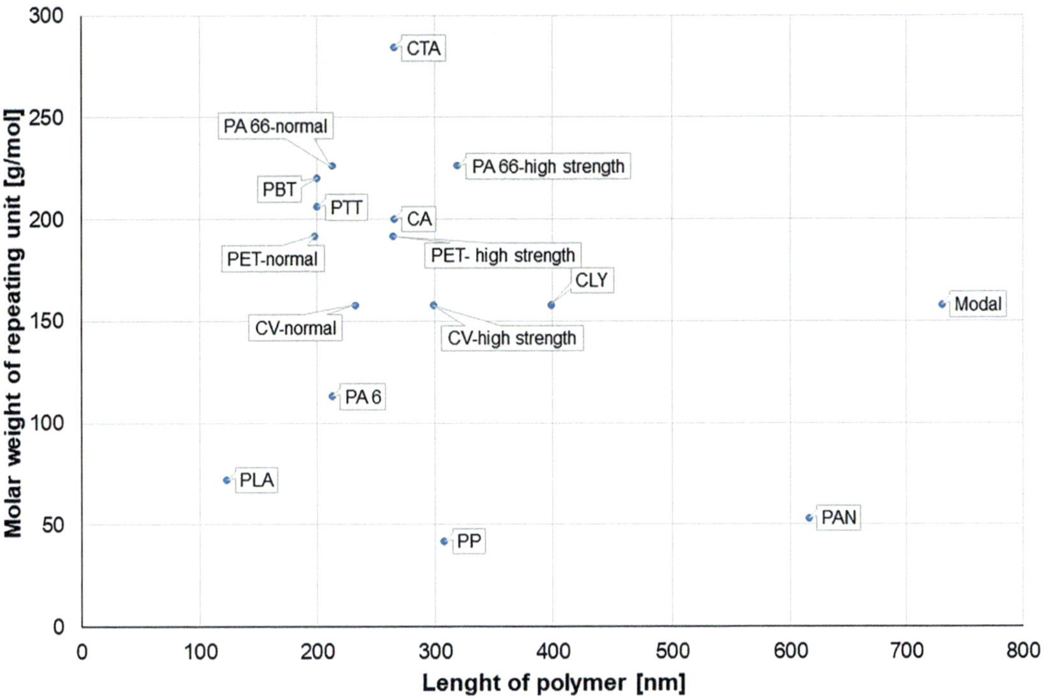

Fig. 48.8 Molar weight of the repeating unit and chain length of the polymers of selected fibrous materials

Table 48.3 Degrees of polymerization and length of macromolecules for selected fibrous materials

Polymer	Length of repeat unit [pm]	DP		Lenght of macromolecules [nm]	
		Min	Max	Min	Max
PE	308	100,000	200,000	30,800	61,600
Wool	448	60,000	100,000	26,880	44,800
Cotton	664	3000	10,000	1992	6640
Flax	664	5000	8000	3320	5312
Ramie	664	600	3000	398	1992
Silk	2688	500	700	1344	1882
Modal	664	700	1100	465	730
PAN	308	1000	2000	308	616
CLY	664	550	600	365	398
PA 66-high strength	2128	80	150	170	319
PP	308	800	1000	246	308
CV-high strength	664	400	450	266	299
CA	664	300	400	199	266
CTA	664	300	400	199	266
PET-high strength	1322	150	200	198	264
CV-normal	664	300	350	199	232
PA 6	1064	150	200	160	213
PA 66-normal	2128	50	100	106	213
PET-normal	1322	100	150	132	198
PLA	440	140	280	62	123

Table 48.4 Composition of selected cellulosic fibers acc. to Gurunathan et al. (2015)

	Cellulose	Hemicellulose	Lignin	Pectin	Water	Wax
Seed fibers						
Cotton	85–90	6	1–2	0–1	1	0.6
Bast fibers						
Nettle	86	4	5	1	2	3
Flax	71–78	19–21	2.2	2.3	4–11	1.7
Hemp	70–74	18–22	4–6	1	2	0.8
Jute	61–72	14–20	12–13	0,2	1	0.5
Kenaf	45 -57	22	8–13	3–5		
Ramie	69–76	13–17	0.6	6–13	6	0.3
Hard fibers						
Abacá	56–63	22	12–13	1	1.6	0.2
Pineapple	80–83	15–20	8–12	2–4	1–3	4–7
Curaua	74	10	8			
Henequen	78	4–8	13			
Sisal	67–78	10–14	8–11	10	1,3	2
Fruit fibers						
Coir	36–43	0.2	41–45	3–4	5–16	

48.6 Flammability and Use Temperature

Figure 48.9 shows the use temperature and the flammability of selected fiber materials. The LOI value (Limiting Oxygen Index) is determined, for example, acc. to DIN EN ISO 4589 or ASTM D 2863–77. It indicates the oxygen content in the air required for combustion. However, the test method is not practical, because the fire test is carried out with a flame acting from above, whereas normally in the case of fire the flame comes from below. Therefore, fibers can burn in an atmosphere with an oxygen content below the official LOI value. All fibers below an LOI value of 21 are combustible in air. However, when assessing the fire behavior, it must be taken into account whether the fiber melts when burning (skin burns!) or decomposes directly. The service temperature indicates the ambient temperature up to which the fiber can be used, at least for a short time, without self-igniting or melting. As can be seen, a textile made of wool is very suitable for barbecues, partly because sparks extinguish on contact with wool, whereas in cotton textiles holes are created by burning of the cellulosic fibers. Safety-oriented barbecue masters prefer to wear aprons made of carbon or PTFE.

In Table 48.6, typical temperature ranges for selected fibers are summarized. The values depend to a large extent on the grade and the manufacturing process and therefore vary considerably in some cases.

Table 48.5 Density of selected fibrous materials

Fiber	Density [g/cm³]
Seed fibers	
Cotton	1.50–1.54
Bast fibers	
Flax	1.43–1.52
Hemp	1.48–1.5
Jute	1.44–1.49
Ramie	1.51–1.55
Nettle	1.50
Kenaf	1.50
Hard fibers	
Sisal	1.16–1.45
Abacá	1.35–1.50
Pineapple	1.44–1.54
Fruit fibers	
Coir	1.40
Kapok	0.2–0.3
Animal fibers	
Wool	1.32
Fine animal hair	1.23–1.35
Silk	1.25–1.37
Cellulosic man-made fibers	
Acetate	1.29–1.33
Lyocell	1.50
Viscose	1.52
Synthetic man-made fibers	
Carbon	1.7–1.9
ePTFE	0.2–2.2
m-aramid	1.37–1.38
Modacryl	1.30–1.42
PA 6, PA 66	1.14
PAN	1.14–1.18
p-aramid	1.44–1.45
PE (UHMW)	0.95–0.98
PET	1.36–1.41
PP	0.99
PU	1.14–1.30
PVA	1.26–1.31
Other fibers	
Basalt	2.6–2.8
Glass	2.14–2.68
Ceramics	2.8–4.1
Steel	7.9–8.9
Kapok	0.2–0.3
ePTFE	0.2–2.2
PE (UHMW)	0.95–0.98
PP	0.99
PA 6, PA 66	1.14
PAN	1.14–1.18

(continued)

Table 48.5 (continued)

Fiber	Density [g/cm³]
PU	1.14–1.30
Sisal	1.16–1.45
Feine Tierhaare	1.23–1.35
Seide	1.25–1.37
PVA	1.26–1.31
Acetat	1.29–1.33
Modacryl	1.30–1.42
Wolle	1.32
Abacá	1.35–1.50
PET	1.36–1.41
m-Aramid	1.37–1.38
Kokos	1.40
Flachs	1.43–1.52
p-Aramid	1.44–1.45
Jute	1.44–1.49
Ananas	1.44–1.54
Hanf	1.48–1.5
Brennnessel	1.50
Kenaf	1.50
Lyocell	1.50
Baumwolle	1.50–1.54
Ramie	1.51–1.55
Viskose	1.52
Carbon	1.7–1.9
Glas	2.14–2.68
Basalt	2.6–2.8
Keramik	2.8–4.1
Stahl	7.9–8.9

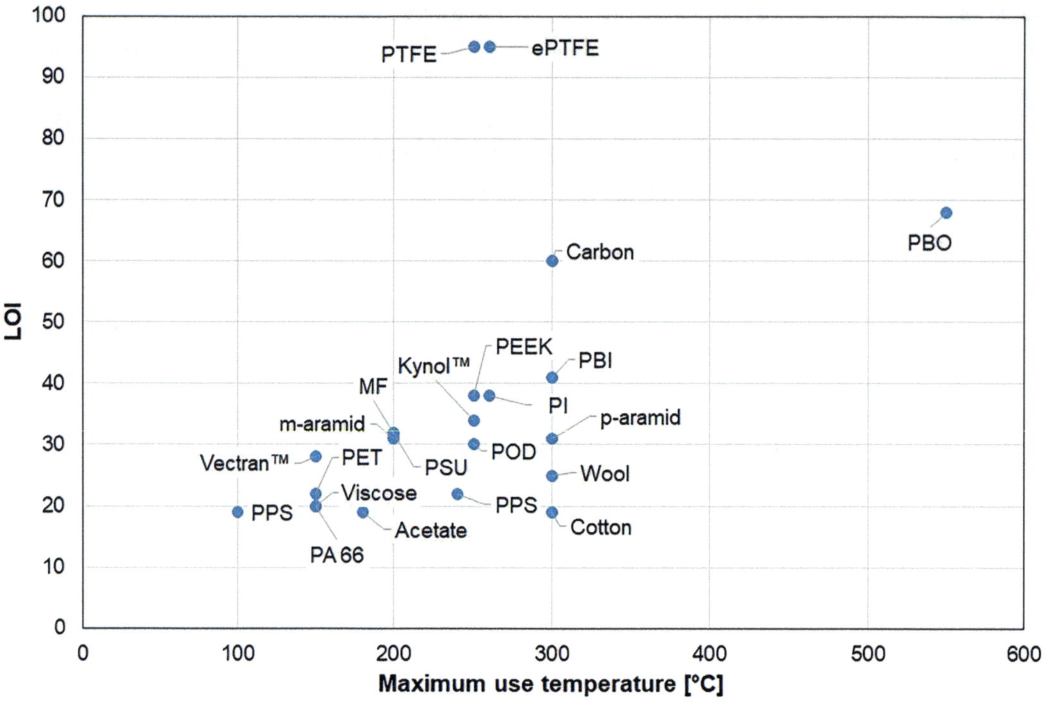

Fig. 48.9 Maximum use temperature and flammability of selected fibers

Table 48.6 Characteristic temperatures of selected fibrous materials acc. to Haudek and Viti (1980)

	Temperature range [°C]						Max. temp. in [°C] for	
Fiber	Softening	Melting	Disintegration	Charring	Flashpoint	Self-ignition	Washing	Ironing
Cotton	–	–	190–210	300–400	288	400	95–100	180–220
Flax	–	–	260–320	320	–	–	95–100	215–240
Wool	–	–	130–240	210–400	224	590	30–40	160–170
Silk	–	–	170–180	300–400	–	–	30–40	140–160
CV	–	–	175–205	250–400	–	–	60	150–180
CA	180–190	255–260	–	–	–	500	30–50	170–180
CTA	190–250	290–300	–	–	–	500	30	180–190
PA 6	170–200	215–220	–	–	354	510	30–60	110–120
PA 66	220–235	255–260	–	–	–	–	30–60	120–150
PET	230–240	250–260	–	–	372	508	30–60	140–150
PAN	190–250	–	250–350	8	245	515	30–40	120–130
PVC	125–130		–	–	182	–	30	70–100
PP	150–160		–	–	242	440	30	110–130

48.7 Resistance to Light and Weather

Especially for outdoor applications, the resistance to weather and light is crucial (Table 48.7).

Table 48.7 Resistance to light and weather acc. to Haudek and Viti (1980)

Fiber	Light fastness	Weather fastness
PAN	Very high	Very high
PVC		
PET		High
CTA	High	Medium
CA		
Cotton	Medium	
PA 6, PA 66	Low(without UV protection)	
Viscose	Low	Low
Flax		
Wool		
Silk		
PP	Very low(without UV protection)	Very low(without UV protection)

48.8 Resistance to Acids and Alkalis

Table 48.8 Resistance to acids and alkalis acc. to Haudek and Viti (1980)

Fiber	Acid resistance		Alkali resistance	
PVC	Very high	Only attacked by hot, conc. saltpeter acid	Very high	Not attacked by lyes
PP		Only attacked by hot, conc. saltpeter acid		Attacked by hot, conc. lyes
PAN	High	Soluble only in conc. acid (except hydrochloric acid)	High–medium	Attacked by conc. Lyes(especially caustic soda)
PET		Only attacked by conc. acids, soluble in conc. Sulphuric acid	High–medium	Destroyed by conc. Lyes, very high resistance to weak lyes
Wool		Attacked by hot, conc. acids, in weak acids tendency to felt	Very low	Destroyed by lyesin weak lyes tendency to felt
Silk		Resistance slightly lower than wool	Low	Resistance better than wool
PA 6, PA 66	Medium	Attacked by cold, conc. acids	High	Attacked by 10% caustig soda
Triacetate	Low	Destroyed by strong acids	Sufficient	Slightly more resistant than acetate
Acetate		Destroyed by strong acids		Strong lyes lead to saponification
Flax		Resistance higher than cotton		Lower resistance than cotton
Viscose	Very low	Similar to cotton	High	Dissolved in 10% caustic soda
Cotton		Destroyed in hot, diluted or cold, conc. acids		10%-caustic soda leads to swelling (no destruction), mercerisation

48.9 Resistance to Microorganisms

Table 48.9 Resistance to microorganisms and moths acc. to Haudek and Viti (1980)

Fiber	Resistance to		
	Microorganisms	Moth bite	Moth damage
PVC	Very high	Medium	Very high
PP			
PAN		High	
PET		Very high	
PA 6, PA 66	Very high–high	High	
CTA	High	Low	
CA	Medium		
Silk			
Wool		Very low	Very low
Flax	Low (unbleached)	Very high	Very high
Cotton			
CV	Very low	Low	

48.10 General Overview

Table 48.10 Selected properties of important fibrous materials (data from Fourne, 1995, among others)

		Cotton (CO)	Flax (FL)	Hemp (HE)	Jute (JU)	Kenaf (KF)	Ramie (RA)	Nettle	Wool (WO)	Silk (SE)	Viscose (CV)		
Density	[g/cm^3]	1.50–1.54	1.43–1.52	1.48–1.5	1.44–1.49	1.5	1.51–1.55	1.5	1.32	r: 1.37. e: 1.25	1.52		
Fineness	[µm] [dtex]	10–12 1.6–2	E: 8–22. T: 26–53 E: 1–7. T: 10–40	12–20 2–6	12–20 2–6	10–40 2–25	18–30 5–13	15–25 3–9	16–40 3–20	10–18 1–4	10–40 1.3–22		
Length	[mm]	25–32 ELS: 34–40	E: 10–50 T: 100–1300	E: 15–25 T: 1000–2250	E: 1–6 T: 1500–2500	E: 1–12 T: 900–1800	E: 60–250	E: 5–50 T: 19–80	25–400		38–200		
Mechanical properties											normal type	HWM type	techn. type
Tenacity in normal climate	[cN/tex] [MPa]	25–45 375–675	30–55 430–825	35–70 525–1050	30–40 450–600	40–50 600–750	40–75 600–1100	25–35 375–525	10–16 130–210	25–50 300–600	18–35 270–550	35–45 550–685	40–75 600–1100
Tenacity (wet)	[% des Trockenwerts]	100–110	105–120 (techn. f.)	102–106	85–110		116–125	100–104	70–90	75–95	40–70	70–80	70–80
Maximum elongation	[%]	6–10	1.5–4.0	1–6	1–2	3–4	2–3	3–4	25–50	10–30	15–30	8–18	7–15
Maximum elongation (wet)	[% des Trockenwerts]	100–110	110–125	100–110	100–110	100–110	100–110	110–125	110–140	120–200	100–130	120–150	150–200
Breaking elongation	[%]	6–10	1.5–4.0	1–6	1–2	3–4	2–3	3–4	25–50	10–30	15–30		
Loop strength	[% of tenacity]	65–70	20–40						75–85	60–80	20–60	15–20	30–70
Knot strength	[% of tenacityt]	60–100							80–85	80–85	30–70	25–65	25–65
E-modulus	[cN/tex] [GPa]	300–600 4.5–9	4000–6700 60–100	4700–8000 70–120	670–3700 10–55		2900–3900 44–59		150–300 2–4	700–1000 9–12	200–300 3–4.5	650–750 10–12	240 3.6
Torsion modulus	[cN/tex] [MPa]	160 2400	80–100 1100–1500						80–100 1000–1300	150–230 1900–2900	50–100		
Transverse brittleness	[°]	53–56	48–52						48–52	51	50–60		
Degree of elasticity	[% of total elongation]												
– at ε = 2% – at ε = 5% – at ε = 10%		75 45 –							95–99 60–70 40–50	11–14 20–22 30	70–95 40–60		
Moisture absorption													
at 20/65 at 24/95	[%] [%]	7–11 14–18	8–10 bis 20	8 24	10–15 34		12–17 31		15–17 25–30	9–11 e: 20–40	11–14 26–28		
Water retention	[%]	45–50	50–55						40–45	e: 40–45	85–120	65–75	

(continued)

Table 48.10 (continued)

		Cotton (CO)	Flax (FL)	Hemp (HE)	Jute (JU)	Kenaf (KF)	Ramie (RA)	Nettle	Wool (WO)	Silk (SE)	Viscose (CV)
Spec. electrical resistance	Ω cm	10^6–10^8	10^8–10^{11}						10^8–10^{11}	10^8–10^{10}	10^6–10^9
Electrostatic charge											
at 20/65 at 24/95		low low	low low	low low	low low	low low	low low		low medium	e: high e: high	low low
Thermal properties											
Glass temperature	[°C]	–									
Fixation temperature	[°C]	–									
Ironing temperature	[°C]	180 -320	215–240						160–170	140–165	150–180
Softening temperature	[°C]	–							140		
Melting temperature	[°C]	–									
Disintegration temperature	[°C]	400							130–240	170–180	175–205
Self-ignition temperature	[°C]	400							590–600		420
Shrinkage											
– in water (95 °C) – in hot air (150 °C) – in hot air (190 °C)	[%]									1	0.5–10
Specific heat	[kJ/kg K]	1.3	1.4						1.3–1.6	1.4–1.5	1.3–1.5
Heat of fusion	[kJ/kg K]	–									
Thermal conductivity	[W/m K]	0.3–0.5	0.3						0.2	0.2–0.4	0.3–0.6
Burning behavior		flammable burnes rapidly	flammable burnes rapidly	flammable burnes rapidly	flammable burnes rapidly	flammable burnes rapidly	flammable burnes rapidly		flammable burns slowly	flammable burns slowly	flammable burnes rapidly
Limiting Oxygen Index (LOI)		19							25	24	20
Usage properties											
Wrinkling											
– dry – hot wash		high high	high high						low	e: high r: medium	high high

		Cotton (CO)	Flax (FL)	Hemp (HE)	Jute (JU)	Kenaf (KF)	Ramie (RA)	Nettle	Wool (WO)	Silk (SE)	Viscose (CV)		
Fibrillation		low	high						low				
Pilling		low	low						low				
Abrasion resistance		very high	high–medium	high	medium	high	high		low–medium	very high			
Resistance to													
acids (20 °C–1000 h/100 °C–10 h)	[%]												
– phosphoric acid (10 %) – saltpeter acid (1 %) – hydrochloric (1 %) – sulphuric acid (1 %)		60–80/0–20 60–80/0–20 60–80/0–20 60–80/0–20							90–100/20–60 90–100/0–20 90–100/90–100 90–100/20–60				
Alkali (20 °C–1000 h/100 °C–10 h)	[%]												
– ammonia (1 %) – caustic soda (1 %) – soda solution (1 %)		90–100/ 90–100 80–90/ 80–90 90–100/ 90–100							90–100/90–100 0–20/0–20 80–90/0–20				
UV resistance													
Thermal treatment	[°C/time in h/ breaking strength in %]												
– hot air		120/100/ 40–60							120/100/60	120/100/ 60–80	120/ 100/ 20–40		165/8/ 70–85
– steam		120/1/ 80–100							120/1/50–80	120/1/70–90	120/1/ 80–95		
Biological resistance													
– bacteria and fungi		unbleached: low	unbleached: low bleached: high						not moth proof rot resistance good	unbleached: low	low		

(continued)

		Lyocell (CLY)	2-1/2 Acetate (CA)	Triacetate (CTA)	PA 6. PA 66			PA 11	PET			m-aramid (AR)	p-aramid (AR)
Density	[g/cm^3]	1.5	1.29–1.33		1.14			1.04	1.36–1.41			1.37–1.38	1.44–1.45
Fineness	[μm] [dtex]	8–16 0.9–3.3	12–28 2–10		11–45 1.4–22			17–26 3–7	5–60 0.5–45			10–30 1–14	8–12 1–2
Length	[mm]	25–51 (114)	40–120				19–150				38–200	38–102	38–50
Mechanical properties		standard type			normal type	techn. type	staple fibers		normal type	techn. type	staple fibers		
Tenacity in normal climate	[cN/tex] [MPa]	38–50	10–15 130–200		36–68 410–775	68–85 775–970	32–63 365–720	45–68	25–60 340–820	70–95 960–1300	25–55 340–770	43–53 600–730	145–250 2100–3600
Tenacity (wet)	[% des Trockenwerts]	65–85	50–80		80–90			96–98	95–100			75–100	100
Maximum elongation	[%]	10–18	20–40		25–46	14–22	40–70	15–40	20–30	8–20	25–55	15–45	2.0–4.5
Maximum elongation (wet)	[% des Trockenwerts]	110–130	120–150		104–122	107–150	103–125		100–105			60–80	
Breaking elongation	[%]		31–42	30–40	55–63							15–45	2.0–4.5
Loop strength	[% of tenacity]	15–20	70–90	70–95	65–85	70–80			30–95	70–95		65–95	45–80
Knot strength	[% of tenacityt]		80–90	80–90	80–90	60–70		75	70–80	40–70		80–85	30–50
E-modulus	[cN/tex] [GPa]		200–350 2.6–4.6		50–300 0.6–3.5	600–900 7–10	100 1.1	45–50 0.47–0.53	250–400 3.4–5.5	700–1200 9.5–16	250–400 3.4–5.5	140–280 10–20 (filament)	3400–9900 50–140 (filament)
Torsion modulus	[cN/tex] [MPa]		60–80 800–1000		340–520 3900–5900	800 9100			65 880	120 1600			
Transverse brittleness	[°]		44–50		35–43		27–41	36–44	41–48	47–48	30–49	50	
Degree of elasticity	[% of total elongation]												
– at ε = 2% – at ε = 5% – at ε = 10%			90–95 40–60	90–95 40–60	95–100 95–100 90–95	90–95 83–90	95–100 95–100 90–95	100 100	90–98 70–90 50–80				
Moisture absorption													
at 20/65 at 24/95	[%] [%]	12–13	CA: 6–7 CA: 13–15	CTA: 2–5 CTA: 8–10	3.5–4.5 6–9			0.9–1.3	0.2–0.5 0.8–1.0			4.5–5.5 6.5–7	2–4 4–7
Water retention	[%]	65–70	CA: 20–80	CTA: 10–17	10–15			3–7	3–5			12–17	4–8

		Lyocell (CLY)	2-1/2 Acetate (CA)	Triacetate (CTA)	PA 6. PA 66			PA 11	PET			m-aramid (AR)	p-aramid (AR)
Spec. electrical resistance	Ω cm		10^9–10^{12}	10^{14}	10^9–10^{12}				10^{11}–10^{14}				10^{15}
Electrostatic charge													
at 20/65 at 24/95			high	high									
Thermal properties													
Glass temperature	[°C]				40–42			40–50	72–75			270–290	300
Fixation temperature	[°C]		180–220	180–220	185–190	220–225			180–230				
Ironing temperature	[°C]		180	220–250	150	180–200			150–200				
Softening temperature	[°C]		190	190–250	185–190	200–225		120	230–240				
Melting temperature	[°C]		250–255	300	215–220	255–260		190	254–260				
Disintegration temperature	[°C]		280	300								370–430	500
Self-ignition temperature	[°C]		475	480	510–530				510			675	
Shrinkage													
– in water (95 °C) – in hot air (150 °C) – in hot air (190 °C)	[%]	6–9	4–5	5–20 0.5–2	1–15 1–15 2–18		5–9		5–10	7–8	0.5–1	0.5–4 0.1 0.5	0.1 0.1 0.1
Specific heat	[kJ/kg K]		1.3–1.5	1.5	1.5–1.9				1.1–1.4			1.2	1.4
Heat of fusion	[kJ/kg K]												
Thermal conductivity	[W/m K]		0.3	0.3	0.2–0.3			0.29	0.2–0.3			0.13–0.25	0.04

(continued)

Table 48.10 (continued)

		Lyocell (CLY)	2-1/2 Acetate (CA)	Triacetate (CTA)	PA 6. PA 66	PA 11	PET	m-aramid (AR)	p-aramid (AR)
Burning behavior			melts. flammable. burns		melts. flammable. goes out		melts. catches fire and burns sooty. goes out	hardly flammable. goes out	hardly flammable. goes out
Limiting Oxygen Index (LOI)		19–20	18–19		20	18–19	21–22	28	29–40
Usage properties									
Wrinkling									
– dry – hot wash			high medium	medium medium	low high	low high	low high		
Fibrillation			low	low	low		low		
Pilling			low	low	low		high		
Abrasion resistance									
Resistance to									
acids (20 °C–1000 h/ 100 °C–10 h)	[%]								
– phosphoric acid (10 %) – saltpeter acid (1 %) – hydrochloric (1 %) – sulphuric acid (1 %)		medium	90–100/90–100 80–90/0–20 60–80/90–100 80–90/80–90		90–100/90–100 90–100/90–100 90–100/90–100 90–100/90–100	similar to PA 6 high	90–100/90–100 90–100/90–100 90–100/90–100 90–100/90–100	90–100/ 90–100 90–100/80–90 80–90/90–100 90–100/80–90	low
Alkali (20 °C–1000 h/ 100 °C–10 h)	[%]								
– ammonia (1 %) – caustic soda (1 %)		medium	60–80/70–90 0–20/40–60 60–80/60–80		90–100/90–100 90–100/90–100 90–100/90–100	similar to PA 6 high	90–100/90–100 90–100/90–100 90–100/90–100	90–100/ 90–100 - 90–100 90–100/ 90–100	low–medium

		Lyocell (CLY)	2-1/2 Acetate (CA)	Triacetate (CTA)	PA 6. PA 66	PA 11	PET	m-aramid (AR)	p-aramid (AR)
– soda solution (1 %)									
UV resistance								low	
Thermal treatment	[°C/time in h/ breaking strength in %]								
– hot air			120/100/ 55–70	130/250/ 70–80	120/100/60–80		120/1000/90–100	180/1000/ 90–100	high
– steam			120/1/40–60		120/100/70–90		120/100/70–90	150/1000/ 20–40	strength loss at 200–300 ° C
Biological resistance									
– bacteria and fungi			high		high–very high	high	very high	very high	very high

(continued)

		PBT	PAN	Modacrylic (MOD)	PP	PE (UHMW)	PVC	PVA	PTFE	Polyurethane (PU)	Carbon	Glass	Basalt	Ceramic	Steel
Density	[g/cm^3]	1.35	1.14–1.18	1.30–1.42	0.90	0.95–0.98	1.35–1.42	1.26–1.31	2.1	1.14–1.30	1.7–1.9	2.14–2.68	2.6–2.8	Al_2O_3: 2.8–4.1	7.9–8.9
Fineness	[µm] [dtex]	15–38 3–20	8–45 0.6–25	11–70 1.5–60	13–70 1.5–40	14–20 2–4	12–70 1.5–60	12–30 1.5–10	15–35 5–25	30–500 50–210	5–12 0.5–2.5	1–25 0.02–15	8–13 1–4	1–20 0.04–16	4–12 1–12
Length	[mm]	38–200	38–200	38–200	38–200		38–200				3–80 mm		0.1–100		20–100
Mechanical properties															
Tenacity in normal climate	[cN/tex] [MPa]	25–35 335–470	14–55 160–625	10–25 130–325	15–60 135–540	140–380 (430) 1445–3725 (4200)	20–35 300–525	20–65 340–820	8–14 170–300	4–15 45–170	110–290 2000–5000	100–140 2450–3400	75–85 1850–2150	25–150 1000–6000	2.5–30 200–2400
Tenacity (wet)	[% des Trockenwerts]	95–100	75–100	80–100	100	100	100	65–85	100	75–100	100	100	100	100	100
Maximum elongation	[%]	24–40	15–70	25–50	15–200	3–10	10–35	10–30	18–75	400–800	0.7–2.2	4.3–5.8		1–3	1–2 (10)
Maximum elongation (wet)	[% des Trockenwerts]		22–110	80–100	100	100		120–140	100	100	100	100		100	100
Breaking elongation	[%]		20–70	25–50	23	3–10	45			400–800	0.7–222	4.3–5.8			1–2 (10)
Loop strength	[% of tenacity]		30–70	50–70	85–95		35–70	35–80	60–90			15–30			65–75
Knot strength	[% of tenacityt]		75–80	80	70–90		65	45–75	75–90			10–25			60–90
E-modulus	[cN/tex] [GPa]		250–1150 2.8–13	100–450 1.3–5.9	300–500 2.7–4.4	7750–16.000 73–152	200–400 2.7–5.4	300–600 3.8–7.6	35–460 0.7–10	0.3–1.0 0.003–0.01	12.000–55.000 200–950	2200–3440 55–86	3500–4400 91–110	2500–10500 100–420	1500–2600 120–210
Torsion modulus	[cN/tex] [MPa]		100 1140				60–70 810–950	90–150 1150–1920	20–120 420–2520	0.4 5		1600			750–950
Transverse brittleness	[°]		45–60				27–55					85–88		87–89	
Degree of elasticity	[% of total elongation]														
– at ε = 2% – at ε = 5% – at ε = 10%			90–95 50–90 55–80	95–99 85–98 55–95	90–95 85–90 80–85		70–90 60–65	60–80 40–60 30–50		100 100 100		100			100

		PBT	PAN	Modacrylic (MOD)	PP	PE (UHMW)	PVC	PVA	PTFE	Polyurethane (PU)	Carbon	Glass	Basalt	Ceramic	Steel
Moisture absorption															
at 20/65 at 24/95	[%] [%]	0.2–0.5	1–2.5 2–5	0.4–4 1–6	0 0	0 0	0–0.2 0–2	3.5–5		0.5–1.5 0.5–1.5	1 1	0 0.3	0.1–0.5	0	0 0
Water retention	[%]		5–30	10–20			25–35	4–6		7–11					
Spec. electrical resistance	Ω cm		10^8–10^{14}	10^{12}–10^{13}			10^{12}–10^{14}		10^{18}		10^{-3}	10^{10}–10^{15}			10^{-6}–10^{-4}
Electrostatic charge															
at 20/65 at 24/95		medium			medium medium		medium high	low medium				medium high			
Thermal properties															
Glass temperature	[°C]	20–40	30–104	85–95	-5–10		70–100	75–130	30	-60–20					
Fixation temperature	[°C]			100–110				90–98	180	180–200					
Ironing temperature	[°C]		120–180		130		65–70	210		150–180					
Softening temperature	[°C]		200–320	130–200	150–160		90–150		327	170–230		773–960	800		
Melting temperature	[°C]	225	320	130–170	160–175	144–152				170–290			1450	1260–3650	1400–1450
Disintegration temperature	[°C]		250–320	170		149–152	160–200	180–260		230	400				
Self-ignition temperature	[°C]		510–560	650–700	430–450										
Shrinkage															
– in water (95 °C) – in hot air (150 °C) – in hot air (190 °C)	[%]	18	1–40	0.2–16 5–30	0.5 30–50			2–3	2	3–15 5–10				0–1 0–1 0–1	
Specific heat	[kJ/kg K]		1.2–1.5		1.6		0.8–1.3		1.0–1.1		0.7	0.7–0.8		0.7–1.2	0.46
Heat of fusion	[kJ/kg K]														

(continued)

Table 48.10 (continued)

		PBT	PAN	Modacrylic (MOD)	PP	PE (UHMW)	PVC	PVA	PTFE	Polyurethane (PU)	Carbon	Glass	Basalt	Ceramic	Steel
Thermal conductivity	[W/m K]		0.2		0.1–0.3	0.5	0.2		0.23	0.15	5–115	0.8–1.0	1.7–2.0	8–20 (SiC)	15
Burning behavior			strong shrinkage, flammable, burns sooty		shrinks, melts, flammable, burns slowly		shrinks quickly, does not melt, inflammable			burns, melts, glowing flame, stingy odour		inflammable	inflammable	inflammable	
Limiting Oxygen Index (LOI)			17–18	26–31	19	17–18	37–45	20	95		40–60				
Usage properties															
Wrinkling															
– dry – hot wash			high	medium–high	high low		low	high very high							
Fibrillation					high										
Pilling			medium–high	medium–high			medium	high							
Abrasion resistance						very high	low								
Resistance to															
acids (20 °C–1000 h/ 100 °C–10 h)	[%]														
– phosphoric acid (10 %) – saltpeter acid (1 %) – hydrochloric (1 %) – sulphuric acid (1 %)		similar to PET	90–100/80–90 90–100/80–90 90–100/90–100 90–100/80–90	gut	very high	very high	high–very high	high	very high	90–100/- 90–100/0–20 90–100/90–100 90–100/90–100	very high	only attacked by concentrated anorganic acids (e.g. hydrofluoric acid–HF)	high	very high	high
Alkali (20 °C–1000 h/ 100 °C–10 h)	[%]														
– ammonia (1 %) – caustic soda (1 %)		similar to PET	90–100/90–100 90–100/60–80	sehr gut	very high	very high	high	high	very high		very high	very high	high	very high	high

		PBT	PAN	Modacrylic (MOD)	PP	PE (UHMW)	PVC	PVA	PTFE	Polyurethane (PU)	Carbon	Glass	Basalt	Ceramic	Steel
– soda solution (1 %)			90–100/90–100												
UV resistance						very high									
Thermal treatment	[°C/time in h/ breaking strength in %]														
– hot air		similar to PET	120/1000/ 70–100	wie PAN					177/1000/ 90–100	high resistance					
– steam			120/1000/ 70–90	wie PAN					288/1000/ 65–75	high resistance					
Biological resistance															
– bacteria and fungi		similar to PET	very high	high–very high	very high	high	very high	very high		high	very high			very high	

E: elementary fiber
T: technical fiber
r: raw
e: degummed

References[1]

Afshari, M. (2016). *Electrospun nanofibers*. Woodhead Publishing.

Autorenkollektiv. (1963). *Textile Faserstoffe*. VEB Fachbuchverlag.

Babu, K. M. (2013). *Silk: Processing, properties and applications*. Woodhead Publishing.

Bhat, G. (2017). *Structure and properties of high-performance fibers*. Elsevier.

Blackburn, R. S. (2005). *Bodegradable and sustainable fibres*. Woodhead Publishing.

Bobeth, W. (Ed.). (1993). *Textile Faserstoffe*. Springer.

Bunsell, A. R. (2009). *Handbook of tensile properties of textile and technical fibres*. Woodhead Publishing.

Cook, J. G. (1984a). Handbook of textile fibres. In *Volume 1: Natural fibres*. Woodhead Publishing.

Cook, J. G. (1984b). Handbook of textile fibres. In *Volume 2: Man-made fibres*. Woodhead Publishing.

Crawshaw, G., & Simpson, W. S. (2002). *Wool: Science and technology*. Woodhead Publishing.

Deopura, B. L., Alagirusamy, R., Joshi, M., & Gupta, B. S. (2008). *Polyesters and polyamides*. Woodhead Publishing.

Eichhorn, S. J., Hearle, J. W. S., Jaffe, M., & Kikutani, T. (2008). *Handbook of textile fibre structure* (Vol. 1, 2). Woodhead Publishing.

von Falkai, B. (1981). *Synthesefasern*. Verlag Chemie.

Fourne, F. (1995). *Synthetic Fibers*. Hanser Verlag.

Franck, R. R. (2001). *Silk, Mohair*. Woodhead Publishing.

Franck, R. R. (2005). *Bast and other plant fibres*. Woodhead Publishing.

Gordon, S., & Hsieh, Y.-L. (2005). *Cotton: Science and technology*. Woodhead Publishing.

Gurunathan, T., Mohanty, S., & Nayak, S. K. (2015). A review of the recent developments in biocomposites based on natural fibres and their application perspectives. *Composites: Part A, 77*, 1–25.

Haudek, W., & Viti, E. (1980). *Textilfasern*. Verlag Johann L. Bondi & Sohn.

Hearle, J. W. S. (Ed.). (2008). *High-performance fibres*. Woodhead Publishing.

Koch, P.-A., & Wagner, E. (1953). *Textile Prüfungen*. Spohr-Verlag.

Koslowski, H.-J. (2008). *Chemiefaser-Lexikon*. Deutscher Fachverlag.

Kozlowski, R. (2012a). *Handbook of natural fibres. Vol. 1: Types, properties and factors affecting breeding and cultivatoin*. Woodhead Publishing.

Kozlowski, R. (2012b). *Handbook of natural fibres. Vol. 2: Processing and applications*. Woodhead Publishing.

Massy, C. (2007). *The Australian merino*. Random House.

Mather, R. R., & Wardman, R. H. (2015). *The chemistry of textile fibres*. The Royal Society of Chemistry.

McIntyre, J. E. (2004). *Synthetic fibres: Nylon, polyester, acrylic, polyolefin*. Woodhead Publishing.

Morton, W. E., & Hearle, J. W. S. (2008). *Physical properties of textile fibres*. Woodhead Publishing.

Rogowin, Z. (1982). *Chemiefasern*. Thieme.

Wagner, E. (1981). *Die textilen Rohstoffe*. Deutscher Fachverlag.

Woodings, C. (2001). *Regenerated cellulose fibres*. Woodhead Publishing.

[1] The following books contain numerous characteristic data on natural and man-made fibers. Even though some of them are older works, the characteristic values, especially for natural fibers, are still valid to a large extent.

Part XI

Producers and Trade Names

49 Producers and Trade Names

49.1 Producers

In the following table, selected producers of chemical fibers on cellulosic, synthetic and inorganic basis are listed. It is recommend to visit the respective websites to obtain further and often detailed information (Tables 49.1 and 49.2).

The following maps show the locations of selected manufacturers for important fiber groups (Figs. 49.1, 49.2, 49.3, 49.4, 49.5 and 49.6).

49.2 Trade names

Many fiber manufacturers have registered trade names of which a selection is listed in the following table (Table 49.2).

D. Veit, *Fibers*, https://doi.org/10.1007/978-3-031-15309-9_49

Table 49.1 Selected producers of chemical fibers

Company	Fiber type	Trade name	Website
DuPont de Nemours, Inc.	Aramid (meta)	Nomex™	https://www.dupont.com/brands/nomex.html
Teijin Limited	Aramid (meta)	Teijinconex™	https://www.teijin.com
Threads India, Ltd.	Aramid (meta)	M-Tech™M-Bonded™	http://www.threadsindia.com
DuPont de Nemours, Inc.	Aramid (para)	Kevlar™	https://www.dupont.com/brands/kevlar.html
PROTEX new advanced textiles GMBH	Aramid (para)		http://www.protexonline.com
Teijin Limited	Aramid (para)	Twaron™ Technora™	https://www.teijin.com
Threads India, Ltd.	Aramid (para)	P-Advance™P-Steel™ P-Tech™P-Bonded™	http://www.threadsindia.com
Basalt Fiber Tech	Basalt		http://www.basaltft.com/
BASALTEX NV	Basalt		http://www.basaltex.com
DBF Deutsche Basalt Faser GmbH	Basalt		https://www.deutsche-basalt-faser.de
Hengdian Group	Basalt		http://www.hg-gbfbasalt.com
Kamenny Vek	Basalt	Basfiber™	https://basfiber.com
Mafic	Basalt		https://www.mafic.com/
Sudaglass Fiber Technology, Inc.	Basalt		http://www.sudaglass.com
Technobasalt-Invest LCC	Basalt	Technobasalt™	https://technobasalt.com
Celanese	CA	BlueRidge™	http://www.celanese.com
Cerdia International GmbH	CA	Cerdia™	http://www.cerdia.com
Daicel Corp.	CA	CAFBLO™	http://www.daicel.com
Eastman Chemical Company	CA	Estron™	http://www.eastman.com
Mitsubishi Chemical	CA	Lynda™	https://www.m-chemical.co.jp/
Alabuga-Fibre LCC	Carbon		https://en.umatex.com/production/fiber/
DowAksa	Carbon	AKSACA™	http://www.dowaksa.com
Formosa Materials	Carbon	Tairyfil™	http://www.formosam.com
Hexcel Corp.	Carbon	HexTow™	http://www.hexcel.com
Holding Company Composite	Carbon		http://www.hccomposite.com/en/
Hyosung Advanced Materials	Carbon	TANSOME™	http://www.hyosungadvancedmaterials.com
Kureha Group	Carbon	KRECA™ (Pech)	https://www.kureha.co.jp
Mitsubishi Chemical	Carbon	Pyrofil™DIALEAD™ (Pech)	https://www.m-chemical.co.jp
Montefibre Carbon	Carbon		http://www.montefibre.es/
Reliance Industries	Carbon		http://www.ril.com
SGL Carbon	Carbon	Sigrafil™	http://www.sglcarbon.com
Sinopec Corp.	Carbon		http://www.sinopecgroup.com
Solvay Cytec Carbon Fibers	Carbon	Thornel™	http://www.solvay.com
Taekwang Industrial Co., Ltd.	Carbon		http://www.taekwang.co.kr

(continued)

Table 49.1 (continued)

Company	Fiber type	Trade name	Website
Teijin Carbon	Carbon	TENAX-X™	https://www.teijincarbon.com/de/
Teijin Limited	Carbon	Tenax™	https://www.teijin.com
Toray Group	Carbon	TORAYCA™	http://www.toray.eu
Zhongfu Shenying Carbon Fiber Co., Ltd.	Carbon		http://www.zfsycf.com.cn
Zoltek	Carbon	ZOLTEK PX™	http://www.zoltek.com
Omikenshi Co., Ltd.	Chitosan	Crabyon™	https://omikenshi.co.jp
Lenzing Aktiengesellschaft	CLY	TENCEL™	https://www.lenzing.com/de
Lenzing Aktiengesellschaft	CLY	VEOCEL™ Lyocell	https://www.lenzing.com
Mitsubishi Chemical	CTA	Soalon™	https://www.m-chemical.co.jp/
Asahi Kasei Corporation	Cupro	Bemberg™	https://www.asahi-kasei.co.jp/fibers/en/bemberg/
Asia Pacific Rayon Limited	CV		https://www.aprayon.com/en/
Birla Cellulose	CV		http://www.birlacellulose.com
Chengdu Grace Fiber Co.,Ltd	CV		https://cdhm.en.ecplaza.net/
Cordenka GmbH & Co. KG	CV	Cordenka™	http://www.cordenka.com
Enka International GmbH & Co. KG	CV	Enka™	http://www.enka.de
Glanzstoff Industries GmbH	CV	Viscord™	http://www.glanzstoff.com
Jiangsu Aoyang Technology Co., Ltd.	CV		http://www.aoyang.com
Kelheim Fibres GmbH	CV	Danufil™ Galaxy™ Viloft™ Viseta™ Outlast™	http://www.kelheim-fibres.com
Lenzing Aktiengesellschaft	CV	EcoVero™	https://www.ecovero.com/
Lenzing Aktiengesellschaft	CV	VEOCEL™	https://www.veocel.com/en-us/about-us
Sateri	CV	EcoCosy™	http://www.sateri.com
Shandong Silver Hawk Chemical Fibre Co., Ltd.	CV		http://www.silverhawkfiber.com
Tangshan Sanyou Group Xingda Chemical Fibre Co. Ltd	CV		http://www.ts-sanyou.com.cn
Xinjiang Zhongtai Chemical Co., Ltd.	CV		http://www.zhongtaichem.cn
W. L. Gore & Associates, Inc.	ePTFE	Gore-Tex™	http://www.gore.com
3B-the fibreglass company	Glass	HiPer-tex™	https://www.3b-fibreglass.com
China Jushi Co., Ltd.	Glass		https://www.jushi.com
Chongqing Polycomp Int. Corp.	Glass		https://en.cpicfiber.com
Johns Manville	Glass	StarRov™ Thermoflow™ StarStran™	https://www.jm.com/
Owens Corning Fiberglass	Glass	THERMAFIBER™	https://www.owenscorning.com
P-D Glasseiden GmbH, Oschatz	Glass	PD PowerStrand™	https://www.oschatz-glas.com
Rockwool International A.S.	Glass		https://www.rockwool.com

(continued)

Table 49.1 (continued)

Company	Fiber type	Trade name	Website
Saint-Gobain Vetrotex	Glass	Vetrotex Zero Twist™	https://www.vetrotextextiles.com/
Taishan Fiberglass Inc.	Glass		http://www.ctgf.com
DuPont de Nemours, Inc.	HDPE	Tyvek™	https://www.dupont.com/brands/tyvek.html
Q-Nature UG	Casein	QMILK™	https://www.qmilkfiber.eu
Zhengjia Milkfiber Sci & Tech Co., Ltd.	Casein	Milkfiber™	http://www.milkfashion.com
3 M Co.	Ceramics	Nextel™	https://www.3mdeutschland.de
American Elements	Ceramics		https://www.americanelements.com
COI Ceramics, Inc.	Ceramics	Sylramic™	http://www.coiceramics.com
Mitsubishi Chemical Corp.	Ceramics	MAFTEC™	https://www.m-chemical.co.jp
NGS Advanced Fibers Co., Ltd.	Ceramics	Nicalon™ CG	http://www.ngs-advanced-fibers.com
Nippon Carbon Co., Ltd.	Ceramics	Nicalon™	https://www.carbon.co.jp
Specialty Materials Inc.	Ceramics	Boron™	http://www.specmaterials.com
Specialty Materials Inc.	Ceramics	SCS-6™	http://www.specmaterials.com
Specialty Materials Inc.	Ceramics	SCS-Ultra™	http://www.specmaterials.com
UBE Industries Ltd.	Ceramics	Tyranno Fiber™	https://www.ube.com
ZIRCAR Ceramics, Inc.	Ceramics	ALBF™	https://www.zircarceramics.com
Kuraray Co., Ltd.	LCP	Vectran™	http://www.kuraray.co.jp
Toray Industries, Inc.	LCP	SIVERAS™	https://www.toray.com
Hi-Tech Fiber Group Corp.	Lyocell	Oricell™	http://www.htfiber.com
Lenzing AG	Lyocell	Tencel®	http://www.lenzing.com
Sateri	Lyocell	EcoCosy™	http://www.sateri.com
Smartfiber AG	Lyocell	Seacell®	http://www.smartfiber.de
Arcelor Mittal	Metal		https://corporate.arcelormittal.com
Bekaert	Metal	Dramix™Cercas Pro™	https://www.bekaert.com
Berkenhoff GmbH	Metal	gapstar™	http://www.bedra.com
Biologa Danell GmbH	Metal	Swiss Shield™	http://www.swiss-shield.de
Créafibres S.A.S.	Metal		http://www.creafibres.fr
Deutsches MetallfaserwerkDr. Schwabbauer GmbH & Co. KG	Metal		http://www.stax.de
Elektrisola Dr. Gerd Schildbach GmbH & Co. KG	Metal		http://www.elektrisola.com
Fiber Technology Ltd	Metal		http://www.fibretech.com
Frenzelit GmbH	Metal	hicoTEC™ hicoFORM™	http://www.frenzelit.com
GKD–Gebr. Kufferath AG	Metal		http://www.gkd.de
Green Steel Group	Metal		http://www.greensteelgroup.com
Haver & Boecker OHG	Metal		http://www.haverboecker.com

(continued)

Table 49.1 (continued)

Company	Fiber type	Trade name	Website
Hebei Swiit Metallic Fiber Co., Ltd	Metal		http://www.swiit-fiber.com
Hunan Huitong Advanced Materials Co., Ltd.	Metal		http://en.ht-metalfiber.com/
Hunan Sunshine Steel Fiber Co., Ltd.	Metal		https://sunshinefiber.en.china.cn/
IntraMicron, Inc	Metal	MicroLobal™	http://www.intramicron.com
King's Metal Fiber Technologies Co., Ltd	Metal		http://www.kingsmetalfiber.com
Koolon Adv. Mat. Techn. Co., Ltd.	Metal		http://www.koolon.com
Leoni Ag	Metal		http://www.leoni.com
Metlon Corp.	Metal		http://www.metlon.com
Nikko Techno, Ltd.	Metal		http://www.nikkotechno.jp
Nippon Seisen Co., Ltd.	Metal	ELEMETAL™ HYBREM™ TouPH Stainless™	https://www.n-seisen.co.jp
Nobel Biomaterials, Inc.	Metal	circuitex™IONIC+™NOBLE DFENCE™X-STATIC™	http://www.noblebiomaterials.com
R.Stat SAS	Metal	silveR.STAT™ R.STAT/S™ R.STAT/N™	http://www.r-stat.com
Rhodius GmbH	Metal		http://www.rhodius.com
SC Fibro Meals SRL	Metal		http://www.fibrometals.ro
Sprint METAL GmbH	Metal		https://sprintmetal.swisssteel-group.com
Statex GmbH	Metal	Shieldex™	http://www.statex.de
Teijin Ltd.	Metal		http://www.teijin.com
Ugitech SA	Metal		http://www.ugitech.com
Engineered Fibers Technology, LLC	MF	WF Basofil™	http://eftfibers.com
Gunei Ei Chemical Industry	Novoloid	Kynol™	https://www.gunei-chemical.co.jp
beti d.d.	PA 6		https://www.beti.si
EMS-Griltech AG	PA 6	Grilon™	http://www.emsgriltech.com
Franke Chemiefasern GmbH & Co. KG	PA 6		https://www.franke-fibers.de
Hangzhou Dikai Industrial Fabrics Co.,Ltd	PA 6		http://www.dikai-fabrics.com
Jiangyin Huacai Textile Co., Ltd.	PA 6		https://www.huacaifiber.com
Ladylon Kft.	PA 6		http://www.ladylon.hu
Neshin Spinning	PA 6		http://www.neshin.com.tw
Threads India, Ltd.	PA 6	Nybond	http://www.threadsindia.com
Asahi Kasei Corporation	PA 66	Leona™	https://www.asahi-kasei.co.jp
beti d.d.	PA 66		https://www.beti.si
EMS-Griltech AG	PA 66	Nexylon™	http://www.emsgriltech.com
Franke Chemiefasern GmbH & Co. KG	PA 66		https://www.franke-fibers.de
Hangzhou Dikai Industrial Fabrics Co.,Ltd	PA 66		http://www.dikai-fabrics.com
Ladylon Kft.	PA 66		http://www.ladylon.hu

(continued)

Table 49.1 (continued)

Company	Fiber type	Trade name	Website
Minifibers, Inc.	PA 66		https://www.minifibers.com
Neshin Spinning	PA 66		http://www.neshin.com.tw
PHP Fibers GmbH (Indorama Ventures)	PA 66		https://www.indoramaventures.com
PROTEX new advanced textiles GMBH	PA 66		http://www.protexonline.com
Threads India, Ltd.	PA 66	Opel™Bondex™Bond-S™	http://www.threadsindia.com
Toray Industries, Inc.	PA 66	NYLON™	https://www.toray.com
TWD Fibres GmbH	PA 66	TIMBRELLE®	http://www.twd-fibres.de
Kermel SAS	PAI	Kermel™	http://www.kermel.com
Aksa AS	PAN	Armora™	http://www.aksa.com
Asahi Kasei Corporation	PAN	Lastan™	https://www.asahi-kasei.co.jp
Companhia Sudamericana de Fibras S.A.	PAN	Drytex™Drymix™ pimacryl™	http://sdef.com/
Dewan Textile Mills Limited	PAN		https://yousufdewan.com
Dolan GmbH (Lenzing AG)	PAN	Dolan™Dolanit™	http://www.dolan-gmbh.de
Dow Aksa AS	PAN	Acryluna™Acrysole™ Acrylusion™Acryterna™	http://www.aksa.com
Dralon GmbH	PAN	Dralon™	https://www.dralon.com/
Drytex technical yarns	PAN	Drytex™	http://drytex.es
Formosa Plastics Corp.	PAN	Tairylan™	http://www.fpc.com.tw
Indian Acrylics, Ltd.	PAN		https://www.indianacrylics.com
Jiangsu Zhongxin Resources Group Co., Ltd.	PAN		http://www.china-acrylicyarn.com
Jilin Qifeng Chemical Fiber	PAN		http://www.jlhxjt.com
Kaltex	PAN	Kaltex™	http://kaltex.com
Kaneka Corp.	PAN	Kanecaron™Protex™	https://www.kanecaron.comhttps://www.modacrylic.com
Pasupati Acrylon	PAN		https://pasupatiacrylon.com
Polyacryl Iran Corp.	PAN		http://www.polyacryl.ir
Polymir	PAN	Nitron D™	https://polymir.by
Sterling Fibers, Inc.	PAN	Conductrol™	http://www.sterlingfibers.com
Thai Acrylic Fiber Co., Ltd.	PAN	Amicor™Birlacril™	http://www.amicorpure.co.uk
The Taekwang & Daehan Company	PAN	Acepora™	http://www.acepora-fiber.com
Tong Hwa Synthetic Fiber Co.	PAN		www.tonghwa.com.tw
Toray Industries, Inc.	PAN	TORAYLON™	https://www.toray.com
Toyobo Co., Ltd.	PAN	Exlan™	https://www.toyobo-global.com
Vardhman Acrylics, Ltd.	PAN		https://www.vardhmanacrylics.com
PBI Performance Products, Inc.	PBI		http://www.pbiproducts.com
Toyobo Co., Ltd.	PBO	Zylon™	http://www.toyobo-global.com
Jiangsu Yinyu Chemical Fiber Co., Ltd.	PBT		http://www.pbtyarn.cn

(continued)

Table 49.1 (continued)

Company	Fiber type	Trade name	Website
Minifibers, Inc.	PBT		https://www.minifibers.com
Taşdelen Group	PBT	POLY-tera™	http://www.tasdelengroup.com
Yinyuda	PBT		http://www.yinyuda.com
Ethicon Inc.	PCL	Monocryl™	http://www.ethicon.com
PROTEX new advanced textiles GMBH	PE (UHMW)		http://www.protexonline.com
Teijin Limited	PE (UHMW)	Endumax™	https://www.teijin.com
Evonik Resource Efficiency GmbH	PEEK	VESTAKEEP™	https://industrial.vestakeep.com
Solvay SA	PEEK	KetaSpire™	https://www.solvay.com
Victrex plc.	PEEK	ZYEX™	http://www.victrex.com
Kuraray Co., Ltd.	PEI	Kurakiss™	http://www.kuraray.co.jp
LATI Industria Termoplastici S.p.A	PESU	Laptex A™	http://www.lati.com
Solvay SA	PESU	Veradel™	https://www.solvay.com
DT Group	PET		https://dtcompany.com
Fujian JinLun Fiber Shareholding Co, Ltd	PET		http://www.fjsjlgx.com
HEDA Chemical Fiber	PET		https://www.cnfibers.com
Huvis Corp.	PET		http://www.huvis.com
JB Ecotex LLB	PET		https://www.jbecotex.com
Märkische Faser GmbH	PET	GRISUTEN™	https://maerkische-faser.com
Nan Ya Plastics Corporation	PET		https://www.npc.com.tw
PPH Legs Sp. z o.o.	PET		https://legs.com.pl
PROTEX new advanced textiles GMBH	PET	Gualon™	http://www.protexonline.com
Recron (Reliance Industries, Ltd.)	PET	Recron™	http://www.recron.com
Swicofil AG	PET		https://www.swicofil.com
Taşdelen Group	PET	POLY-set™ POLY-eco™POLY-FR™ POLY-UVRESIST™POLY-melt™ POLY-thermo™POLY- glow™ POLY-tristar™POLY-micro™ POLY-mats™OCEANLINE™	http://www.tasdelengroup.com
Teijin Limited	PET	EcoPET™	https://www.teijin.com
Teijin Limited	PET	Nanofront™	https://www.teijin.com
Thai Polyester	PET		http://www.thaipolyester.com
Threads India, Ltd.	PET	Ultimo	http://www.threadsindia.com
Trevira GmbH (Indorama Ventures)	PET	Bioactive	https://www.trevira.de/
Trevira GmbH (Indorama Ventures)	PET	Classixx™CS™Micro™Perfomr™ Sinfineco™Xpand™Bioactive™	https://www.trevira.de/
Trevira GmbH (Indorama Ventures)	PET	Xpand	https://www.trevira.de/
TWD Fibres GmbH	PET	DIOLEN®	http://www.twd-fibres.de
Wuxi Longshine Textile Co., Ltd	PET		http://www.wxlongshine.com

(continued)

Table 49.1 (continued)

Company	Fiber type	Trade name	Website
Bieglo GmbH	PI	Dexnyl™	http://www.bieglo.com
Evonik Resource Efficiency GmbH	PI	P84™	http://www.p84.com
Swicofil AG	PI		http://www.swicofil.com
Mitsui Chemicals Group	PLA	Lacea™	https://www.mitsuichemicals.com/
NatureWorks LLC	PLA	Ingeo™	https://www.natureworksllc.com
Shimadzu Corp.	PLA	Lacty™	https://www.shimadzu.com
Toray Industries, Inc.	PLA	ecodear™	https://www.toray.com
Svetlogorsk Khimvolokno	POD	Arselon™	http://www.sohim.by
Leoni Ag	POF		http://www.leoni.com
PROTEX new advanced textiles GMBH	PP		http://www.protexonline.com
Celanese Corp.	PPS	Fortron™	http://www.celanese.com
Ems Griltech AG	PPS	Nexylene™	http://www.emsgriltech.com
Huvis Corp.	PPS	ZETAONE™	http://www.huvis.com
Toray Industries, Inc.	PPS	Torcon™	https://www.toray.com
Toyobo Co., Ltd.	PPS	Procon™	http://www.toyobo-global.com
Teijin Limited	PPT	Solotex™	https://www.teijin.com
BASF SA	PSU	Ultrason E™	http://www.basf.com
BASF SA	PSU	Ultrason S™	http://www.basf.com
LATI Industria Termoplastici S.p.A	PSU	Lasulf™	http://www.lati.com
3 M	PTFE	Dyneon™	http://www.3m.com
DuPont de Nemours, Inc.	PTFE	Teflon™	http://www.dupont.com
Fluoron Chemicals, Ltd.	PTFE	Fluoron™	http://www.fluoronchem.com
Toray Industries, Inc.	PTFE	Teflon™Toyoflon™	https://www.toray.com
W. L. Gore & Associates, Inc.	PTFE	Gore-Tex™	http://www.gore.com
DuPont de Nemours, Inc.	PTT	Sorona™	http://www.sorona.com
HaiXin Kaisheng Technology Co.,Ltd.	PTT	Sustans™	http://www.haixingtec.com
Asahi Kasei Corporation	PU	ROICA™	https://www.asahi-kasei.co.jp
Asahi Kasei Europa GmbH	PU	Dorlastan™	https://asahi-kasei.eu/de/start/
Huafon Group	PU	Qianxi™	http://www.spandex.com.cn
Hyosung TNC Inc.	PU	Creora™	http://www.creora.com/en/index.do
LDZ New Aoshen Spandex Co., Ltd.	PU	Aoshen™	http://de.ldzspandex.com
Nisshinbo Ind. Ltd.	PU	Mobilon™	https://www.nisshinbo-textile.co.jp
The Lycra Company	PU	Lycra™	https://www.lycra.com
The Taekwang & Daehan Company	PU	Acepora™	http://www.acepora-fiber.com
Toray Industries, Inc.	PU	LYCRA™	https://www.toray.com
Toyobo Co., Ltd.	PU	Espa™	https://www.toyobo-global.com
Yantai Tayho Advanced Materials Co., Ltd.	PU	Newstar™	http://en.tayho.com.cn

(continued)

Table 49.1 (continued)

Company	Fiber type	Trade name	Website
Bricture Co., Ltd.	PVA		https://briture.en.made-in-china.com/
Changzhou Tian Yi Engineering Fiber Co., Ltd.	PVA		https://tianyifiber.en.made-in-china.com
Kuraray Europa GmbH	PVA	Kuralon™	http://www.kuraray.eu
Nantong Wanhe Chemical Co., Ltd.	PVA		https://63ed33d443bd3dd7.en.made-in-china.com
Nitivy Co., Ltd.	PVA	Solvron	https://www.nitivy.co.jp
Shenzhen Xiangu High-Tech. Co., Ltd.	PVA		https://xian-gu.en.made-in-china.com
Asahi Kasei Corporation	PVDC	Saran™	http://www.asahi-kasei.co.jp
Minifibers, Inc.	PVDF	Kynar™	https://www.minifibers.com
Shanghai Winshow Soybean Fibre Ind. Co., Ltd.	Soy fiber	Winshow™	https://swsficl.en.china.cn
Southwest Trading Company	Soy fiber	SoySilk™	http://www.swtcyarn.com
Minifibers, Inc.	TPC-ET	Hytrel™	https://www.minifibers.com

Table 49.2 Trade names of selected chemical fibers

Company	Fiber type	Trade name	Website
The Taekwang & Daehan Company	PAN	Acepora™	http://www.acepora-fiber.com
The Taekwang & Daehan Company	PU	Acepora™	http://www.acepora-fiber.com
Dow Aksa AS	PAN	Acryluna™Acrysole™ Acrylusion™Acryterna™	http://www.aksa.com
DowAksa	Carbon	AKSACA™	http://www.dowaksa.com
ZIRCAR Ceramics, Inc.	Ceramics	ALBF™	https://www.zircarceramics.com
Thai Acrylic Fiber Co., Ltd.	PAN	Amicor™	http://www.amicorpure.co.uk
LDZ New Aoshen Spandex Co., Ltd.	PU	Aoshen™	http://de.ldzspandex.com
Aksa AS	PAN	Armora™	http://www.aksa.com
Svetlogorsk Khimvolokno	POD	Arselon™	http://www.sohim.by
Kamenny Vek	Basalt	Basfiber™	https://basfiber.com
Asahi Kasei Corporation	CUP	Bemberg™	https://www.asahi-kasei.co.jp/fibers/en/bemberg/
Trevira GmbH (Indorama Ventures)	PET	Bioactive	https://www.trevira.de/
Trevira GmbH (Indorama Ventures)	PET	Bioactive™	https://www.trevira.de/
Thai Acrylic Fiber Co., Ltd.	PAN	Birlacril™	http://www.amicorpure.co.uk
Celanese	CA	BlueRidge™	http://www.celanese.com
Threads India, Ltd.	PA 66	Bondex™	http://www.threadsindia.com
Threads India, Ltd.	PA 66	Bond-S™	http://www.threadsindia.com
Specialty Materials Inc.	Ceramics	Boron™	http://www.specmaterials.com
Daicel Corp.	CA	CAFBLO™	http://www.daicel.com
Bekaert	Metal	Cercas Pro™	https://www.bekaert.com

(continued)

Table 49.2 (continued)

Company	Fiber type	Trade name	Website
Cerdia International GmbH	CA	Cerdia™	http://www.cerdia.com
Nobel Biomaterials, Inc.	Metal	circuitex™	http://www.noblebiomaterials.com
Trevira GmbH (Indorama Ventures)	PET	Classixx™	https://www.trevira.de/
Sterling Fibers, Inc.	PAN	Conductrol™	http://www.sterlingfibers.com
Cordenka GmbH & Co. KG	CV	Cordenka™	http://www.cordenka.com
Omikenshi Co., Ltd.	Chitosan	Crabyon™	https://omikenshi.co.jp
Hyosung TNC Inc.	PU	Creora™	http://www.creora.com/en/index.do
Trevira GmbH (Indorama Ventures)	PET	CS™	https://www.trevira.de/
Kelheim Fibres GmbH	CV	Danufil™	http://www.kelheim-fibres.com
Bieglo GmbH	PI	Dexnyl™	http://www.bieglo.com
Mitsubishi Chemical	Carbon	DIALEAD™ (Pech)	https://www.m-chemical.co.jp
TWD Fibres GmbH	PET	DIOLEN®	http://www.twd-fibres.de
Dolan GmbH (Lenzing AG)	PAN	Dolanit™	http://www.dolan-gmbh.de
Dolan GmbH (Lenzing AG)	PAN	Dolan™	http://www.dolan-gmbh.de
Asahi Kasei Europa GmbH	PU	Dorlastan™	https://asahi-kasei.eu/de/start/
Dralon GmbH	PAN	Dralon™	https://www.dralon.com/
Bekaert	Metal	Dramix™	https://www.bekaert.com
Companhia Sudamericana de Fibras S.A.	PAN	Drymix™	http://sdef.com/
Companhia Sudamericana de Fibras S.A.	PAN	Drytex™	http://sdef.com/
Drytex technical yarns	PAN	Drytex™	http://drytex.es
3 M	PTFE	Dyneon™	http://www.3m.com
Sateri	Lyocell	EcoCosy™	http://www.sateri.com
Sateri	CV	EcoCosy™	http://www.sateri.com
Toray Industries, Inc.	PLA	ecodear™	https://www.toray.com
Teijin Limited	PET	EcoPET™	https://www.teijin.com
Lenzing Aktiengesellschaft	CV	EcoVero™	https://www.ecovero.com/
Nippon Seisen Co., Ltd.	Metal	ELEMETAL™	https://www.n-seisen.co.jp
Teijin Limited	PE (UHMW)	Endumax™	https://www.teijin.com
Enka International GmbH & Co. KG	CV	Enka™	http://www.enka.de
Toyobo Co., Ltd.	PU	Espa™	https://www.toyobo-global.com
Eastman Chemical Company	CA	Estron™	http://www.eastman.com
Toyobo Co., Ltd.	PAN	Exlan™	https://www.toyobo-global.com
Fluoron Chemicals, Ltd.	PTFE	Fluoron™	http://www.fluoronchem.com
Celanese Corp.	PPS	Fortron™	http://www.celanese.com
Kelheim Fibres GmbH	CV	Galaxy™	http://www.kelheim-fibres.com
Berkenhoff GmbH	Metal	gapstar™	http://www.bedra.com
W. L. Gore & Associates, Inc.	ePTFE	Gore-Tex™	http://www.gore.com
W. L. Gore & Associates, Inc.	PTFE	Gore-Tex™	http://www.gore.com
Ems-Griltech AG	PA 6	Grilon™	http://www.emsgriltech.com
Märkische Faser GmbH	PET	GRISUTEN™	https://maerkische-faser.com
PROTEX new advanced textiles GMBH	PET	Gualon™	http://www.protexonline.com

(continued)

Table 49.2 (continued)

Company	Fiber type	Trade name	Website
Hexcel Corp.	Carbon	HexTow™	http://www.hexcel.com
Frenzelit GmbH	Metal	hicoFORM™ hicoTEC™	http://www.frenzelit.com
3B-the fibreglass company	Glass	HiPer-tex™	https://www.3b-fibreglass.com
Nippon Seisen Co., Ltd.	Metal	HYBREM™	https://www.n-seisen.co.jp
Minifibers, Inc.	TPC-ET	Hytrel™ (DuPont)	https://www.minifibers.com
NatureWorks LLC	PLA	Ingeo™	https://www.natureworksllc.com
Nobel Biomaterials, Inc.	Metal	IONIC+™	http://www.noblebiomaterials.com
Kaltex	PAN	Kaltex™	http://kaltex.com
Kaneka Corp.	PAN	Kanecaron™	https://www.kanecaron.com
Kermel SAS	PAI	Kermel™	http://www.kermel.com
Solvay SA	PEEK	KetaSpire™	https://www.solvay.com
DuPont de Nemours, Inc.	Aramid (para)	Kevlar™	https://www.dupont.com/brands/kevlar.html
Kureha Group	Carbon	KRECA™ (Pech)	https://www.kureha.co.jp
Kuraray Co., Ltd.	PEI	Kurakiss™	http://www.kuraray.co.jp
Kuraray Europa GmbH	PVA	Kuralon™	http://www.kuraray.eu
Minifibers, Inc.	PVDF	Kynar™	https://www.minifibers.com
Gunei Ei Chemical Industry	Novoloid	Kynol™	https://www.gunei-chemical.co.jp
Mitsui Chemicals Group	PLA	Lacea™	https://www.mitsuichemicals.com/
Shimadzu Corp.	PLA	Lacty™	https://www.shimadzu.com
LATI Industria Termoplastici S.p.A	PESU	Laptex A™	http://www.lati.com
Asahi Kasei Corporation	PAN	Lastan™	https://www.asahi-kasei.co.jp
LATI Industria Termoplastici S.p.A	PSU	Lasulf™	http://www.lati.com
Asahi Kasei Corporation	PA 66	Leona™	https://www.asahi-kasei.co.jp
The Lycra Company	PU	Lycra™	https://www.lycra.com
Toray Industries, Inc.	PU	LYCRA™	https://www.toray.com
Mitsubishi Chemical	CA	Lynda™	https://www.m-chemical.co.jp/
Hi-Tech Fiber Group Corp.	Oricell™	Lyocell	http://www.htfiber.com
Mitsubishi Chemical Corp.	Ceramics	MAFTEC™	https://www.m-chemical.co.jp
Threads India, Ltd.	Aramid (meta)	M-Bonded™M-Tech™	http://www.threadsindia.com
IntraMicron, Inc	Metal	MicroLobal™	http://www.intramicron.com
Trevira GmbH (Indorama Ventures)	PET	Micro™	https://www.trevira.de/
Zhengjia Milkfiber Sci & Tech Co., Ltd.	Casein	Milkfiber™	http://www.milkfashion.com
Nisshinbo Ind. Ltd.	PU	Mobilon™	https://www.nisshinbo-textile.co.jp
Ethicon Inc.	PCL	Monocryl™	http://www.ethicon.com
Teijin Limited	PET	Nanofront™	https://www.teijin.com
Yantai Tayho Advanced Materials Co., Ltd.	PU	Newstar™	http://en.tayho.com.cn
3 M Co.	Ceramics	Nextel™	https://www.3mdeutschland.de
Ems Griltech AG	PPS	Nexylene™	http://www.emsgriltech.com

(continued)

Table 49.2 (continued)

Company	Fiber type	Trade name	Website
Ems Griltech AG	PA 66	Nexylon™	http://www.emsgriltech.com
Nippon Carbon Co., Ltd.	Ceramics	Nicalon™	https://www.carbon.co.jp
NGS Advanced Fibers Co., Ltd.	Ceramics	Nicalon™ CG	http://www.ngs-advanced-fibers.com
Polymir	PAN	Nitron D™	https://polymir.by
Nobel Biomaterials, Inc.	Metal	NOBLE DFENCE™	http://www.noblebiomaterials.com
DuPont de Nemours, Inc.	Aramid (meta)	Nomex™	https://www.dupont.com/brands/nomex.html
Threads India, Ltd.	PA 6	Nybond	http://www.threadsindia.com
Toray Industries, Inc.	PA 66	NYLON™	https://www.toray.com
Taşdelen Group	PET	OCEANLINE™	http://www.tasdelengroup.com
Threads India, Ltd.	PA 66	Opel™	http://www.threadsindia.com
Kelheim Fibres GmbH	CV	Outlast™	http://www.kelheim-fibres.com
Evonik Resource Efficiency GmbH	PI	P84™	http://www.p84.com
Threads India, Ltd.	Aramid (para)	P-Advance™P-Bonded™ P-Steel™P-Tech™	http://www.threadsindia.com
P-D Glasseiden GmbH, Oschatz	Glass	PD PowerStrand™	https://www.oschatz-glas.com
Trevira GmbH (Indorama Ventures)	PET	Perfomr™	https://www.trevira.de/
Companhia Sudamericana de Fibras S.A.	PAN	pimacryl™	http://sdef.com/
Taşdelen Group	PET	POLY-eco™POLY-FR™POLY- glow™POLY-mats™POLY-melt™POLY-micro™POLY-set™ POLY-thermo™POLY-tristar™POLY-UVRESIST™	http://www.tasdelengroup.com
Taşdelen Group	PBT	POLY-tera™	http://www.tasdelengroup.com
Toyobo Co., Ltd.	PPS	Procon™	http://www.toyobo-global.com
Kaneka Corp.	PAN	Protex™	https://www.modacrylic.com
Mitsubishi Chemical	Carbon	Pyrofil™	https://www.m-chemical.co.jp
Huafon Group	PU	Qianxi™	http://www.spandex.com.cn
Q-Nature UG	Casein	QMILK™	https://www.qmilkfiber.eu
R.Stat SAS	Metal	R.STAT/S™ R.STAT/N™	http://www.r-stat.com
Recron (Reliance Industries, Ltd.)	PET	Recron™	http://www.recron.com
Asahi Kasei Corporation	PU	ROICA™	https://www.asahi-kasei.co.jp
Asahi Kasei Corporation	PVDC	Saran™	http://www.asahi-kasei.co.jp
Specialty Materials Inc.	Ceramics	SCS-6™	http://www.specmaterials.com
Specialty Materials Inc.	Ceramics	SCS-Ultra™	http://www.specmaterials.com
Smartfiber AG	CLY	Seacell®	http://www.smartfiber.de
Statex GmbH	Metal	Shieldex™	http://www.statex.de
SGL Carbon	Carbon	Sigrafil™	http://www.sglcarbon.com
R.Stat SAS	Metal	silveR.STAT™	http://www.r-stat.com
Trevira GmbH (Indorama Ventures)	PET	Sinfineco™	https://www.trevira.de/
Toray Industries, Inc.	LCP	SIVERAS™	https://www.toray.com
Mitsubishi Chemical	CTA	Soalon™	https://www.m-chemical.co.jp/
Teijin Limited	PPT	Solotex™	https://www.teijin.com

(continued)

Table 49.2 (continued)

Company	Fiber type	Trade name	Website
Nitivy Co., Ltd.	PVA	Solvron	https://www.nitivy.co.jp
DuPont de Nemours, Inc.	PTT	Sorona™	http://www.sorona.com
Southwest Trading Company	Soy fiber	SoySilk™	http://www.swtcyarn.com
Johns Manville	Glass	StarRov™	https://www.jm.com/
Johns Manville	Glass	StarStran™	https://www.jm.com/
HaiXin Kaisheng Technology Co., Ltd.	PTT	Sustans™	http://www.haixingtec.com
Biologa Danell GmbH	Metal	Swiss Shield™	http://www.swiss-shield.de
COI Ceramics, Inc.	Ceramics	Sylramic™	http://www.coiceramics.com
Formosa Materials	Carbon	Tairyfil™	http://www.formosam.com
Formosa Plastics Corp.	PAN	Tairylan™	http://www.fpc.com.tw
Hyosung Advanced Materials	Carbon	TANSOME™	http://www.hyosungadvancedmaterials.com
Technobasalt-Invest LCC	Basalt	Technobasalt™	https://technobasalt.com
Teijin Limited	Aramid (para)	Technora™	https://www.teijin.com
DuPont de Nemours, Inc.	PTFE	Teflon™	http://www.dupont.com
Toray Industries, Inc.	PTFE	Teflon™	https://www.toray.com
Teijin Limited	Aramid (meta)	Teijinconex™	https://www.teijin.com
Teijin Limited	Carbon	Tenax™	https://www.teijin.com
Teijin Carbon	Carbon	TENAX-X™	https://www.teijincarbon.com/de/
Lenzing AG	CLY	Tencel®	http://www.lenzing.com
Lenzing Aktiengesellschaft	CLY	TENCEL™	https://www.lenzing.com/de/produkte/tenceltm
Owens Corning Fiberglass	Glass	THERMAFIBER™	https://www.owenscorning.com
Johns Manville	Glass	Thermoflow™	https://www.jm.com/
Solvay Cytec Carbon Fibers	Carbon	Thornel™	http://www.solvay.com
TWD Fibres GmbH	PA 66	TIMBRELLE®	http://www.twd-fibres.de
Toray Group	Carbon	TORAYCA™	http://www.toray.eu
Toray Industries, Inc.	PAN	TORAYLON™	https://www.toray.com
Toray Industries, Inc.	PPS	Torcon™	https://www.toray.com
Nippon Seisen Co., Ltd.	Metal	TouPH Stainless™	https://www.n-seisen.co.jp
Toray Industries, Inc.	PTFE	Toyoflon™	https://www.toray.com
Teijin Limited	Aramid (para)	Twaron™	https://www.teijin.com
UBE Industries Ltd.	Ceramics	Tyranno Fiber™	https://www.ube.com
DuPont de Nemours, Inc.	HDPE	Tyvek™	https://www.dupont.com/brands/tyvek.html
Threads India, Ltd.	PET	Ultimo	http://www.threadsindia.com
BASF SA	PSU	Ultrason E™	http://www.basf.com
BASF SA	PSU	Ultrason S™	http://www.basf.com
Kuraray Co., Ltd.	LCP	Vectran™	http://www.kuraray.co.jp
Lenzing Aktiengesellschaft	CV	VEOCEL™	https://www.veocel.com/en-us/about-us
Lenzing Aktiengesellschaft	CLY	VEOCEL™ Lyocell	https://www.lenzing.com/de/produkte/veoceltm
Solvay SA	PESU	Veradel™	https://www.solvay.com

(continued)

Table 49.2 (continued)

Company	Fiber type	Trade name	Website
Evonik Resource Efficiency GmbH	PEEK	VESTAKEEP™	https://industrial.vestakeep.com
Saint-Gobain Vetrotex	Glas	Vetrotex Zero Twist™	https://www.vetrotextextiles.com/
Kelheim Fibres GmbH	CV	Viloft™	http://www.kelheim-fibres.com
Glanzstoff Industries GmbH	CV	Viscord™	http://www.glanzstoff.com
Kelheim Fibres GmbH	CV	Viseta™	http://www.kelheim-fibres.com
Engineered Fibers Technology, LLC	MF	WF Basofil™	http://eftfibers.com
Shanghai Winshow Soybean Fibre Ind. Co., Ltd.	Soy fiber	Winshow™	https://swsficl.en.china.cn
Trevira GmbH (Indorama Ventures)	PET	Xpand	https://www.trevira.de/
Trevira GmbH (Indorama Ventures)	PET	Xpand™	https://www.trevira.de/
Nobel Biomaterials, Inc.	Metal	X-STATIC™	http://www.noblebiomaterials.com
Huvis Corp.	PPS	ZETAONE™	http://www.huvis.com
Zoltek	Carbon	ZOLTEK PX™	http://www.zoltek.com
Victrex plc.	PEEK	ZYEX™	http://www.victrex.com
Toyobo Co., Ltd.	PBO	Zylon™	http://www.toyobo-global.com

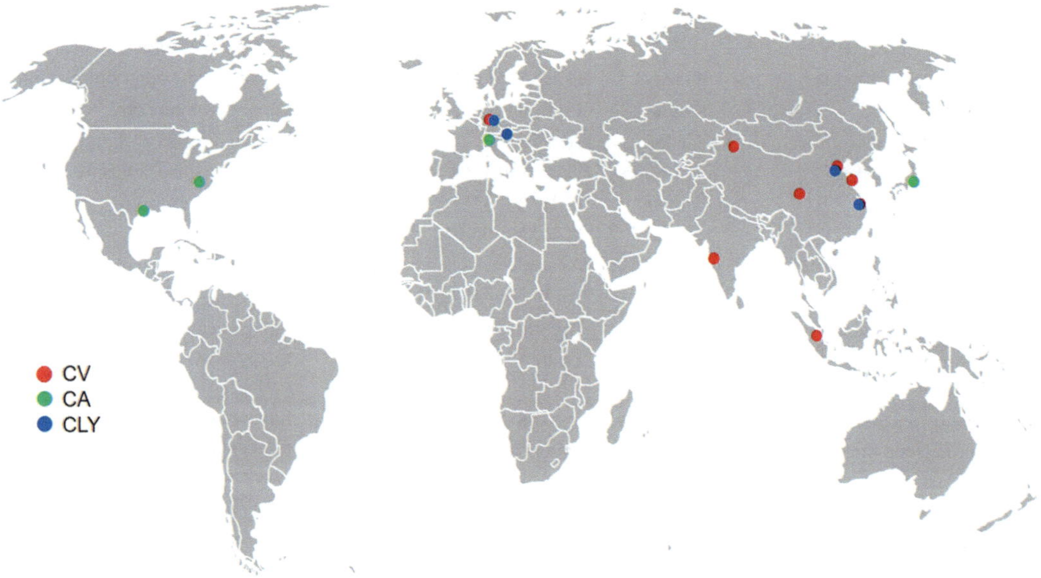

Fig. 49.1 Locations of selected manufacturers of cellulosic man-made fibers

Fig. 49.2 Locations of PET and PBT fiber manufacturers

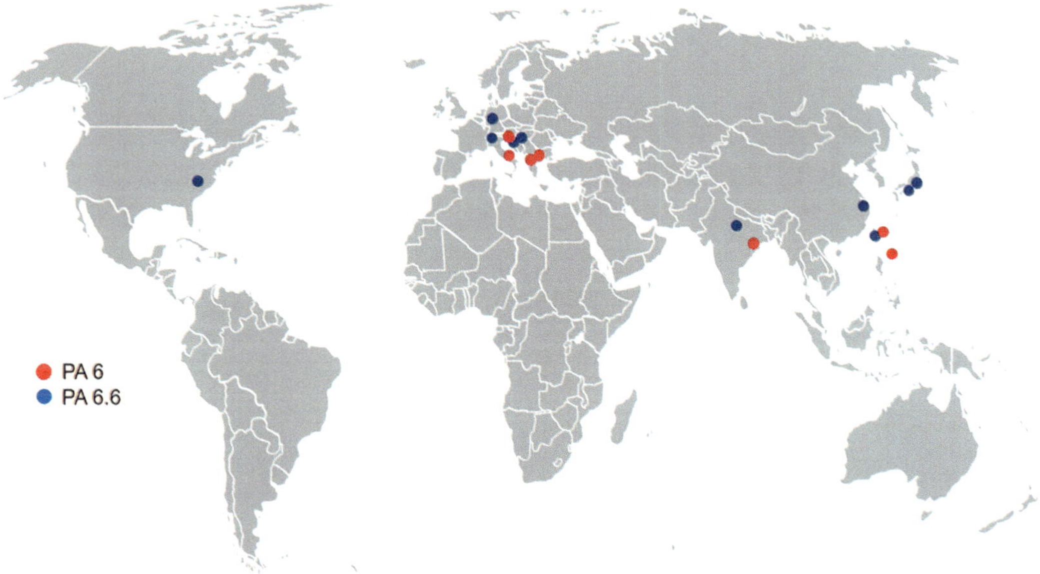

Fig. 49.3 Locations of PA fiber manufacturers

Fig. 49.4 Locations of PAN fiber manufacturers

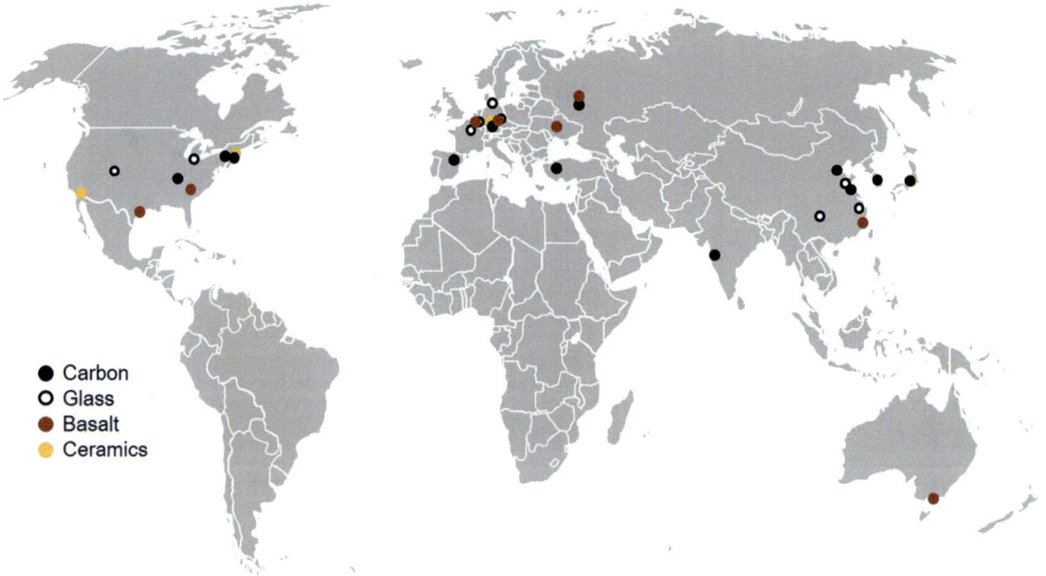

Fig. 49.5 Locations of carbon, glass, basalt and ceramic fiber manufacturers

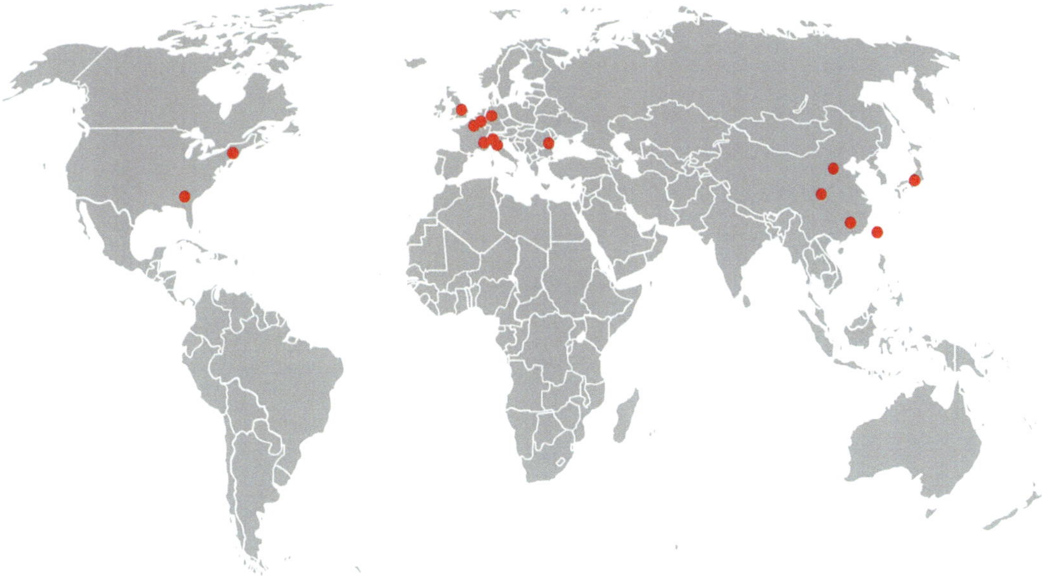

Fig. 49.6 Locations of metal fiber manufacturers

Part XII

Outlook and Trends

Outlook and Trends

50

A good overview of the current market developments for natural fibers, especially against the background of calls for greater sustainability and improved environmental protection, is provided in a study by Textile Exchange (2021).

50.1 Natural Fibers

50.1.1 Cotton

The cultivation of cotton requires large amounts of water and the use of fertilizers, herbicides, pesticides and fungicides. The use of more efficient irrigation methods, e.g. in Australia, and the development of genetically modified cotton (Bt cotton) have solved some of the resulting environmental problems. However, new problems arose, particularly with the cultivation of Bt cotton, which have not yet been solved. These include the development of resistance in pests due to large monocultures and also the necessity for farmers to buy new seeds every year because Bt cotton cannot be self-propagated. Also, cotton cultivation is poorly paid due to low market prices, so many farmers in developing countries continue to live in poverty.

One possible way out is the sustainable cultivation of cotton. Several labels have been developed for this purpose, with which different stages of cotton cultivation are certified acc. to certain criteria. However, compliance with the standards is not always strictly monitored and therefore leaves room for abuse. It is also disputed whether these standards are at all suitable for reducing the environmental impact of cotton cultivation and improving social conditions in the cultivation and processing of cotton. Nevertheless, the share of cotton produced in this way has risen from 5% (2012) to around 25% today (2022), and a further increase is to be expected. Around 90% of this is currently certified by the Better Cotton Initiative (BCI) (Better Cotton Initiative, n.d.).

So-called organic cotton, e.g. in combination with the Cotton made in Africa (CmiA) initiative, appears to be a sensible way of both protecting the environment as far as possible and at the same time improving the working conditions of local cotton farmers. However, the market share of cotton grown in this way is still very low at 0.02% and is also subject to high price pressure from consumers.

It is up to the consumers in the industrialized countries, how cotton production will develop in the future. If they are prepared to pay higher prices and prefer to buy cotton textiles that are certified, then the conditions for cotton farmers and in textile production will also improve. The trend towards recycled clothing, which is now also offered by large retail chains, could lead to fewer new textiles being produced, but a higher price being paid for them. Here, too, there is still a lot of "room for improvement", as only around 1% of textiles worldwide are reused as clothing. A detailed account of this can be found in Thomas (2020).

D. Veit, *Fibers*, https://doi.org/10.1007/978-3-031-15309-9_50

50.1.2 Bast and Hard Fibers

For many years, attempts have been made to re-establish flax and hemp fibers, which were widely used until the twentieth century, in the market for clothing textiles. Subsidies for cultivation led to short-lived renaissances of these fiber materials, but they were unable to establish themselves on the market. This was due on the one hand to their still higher price compared to cotton and on the other hand to their processing and care properties. The production of yarns from bast fibers requires different machinery than for processing cotton, and the corresponding investments are therefore expensive. Also, the further processing, in particular the finishing, differs significantly from that of cotton, so that massive investments have to be made also in later stages of the textile production. Therefore, these two and also other bast fibers (e.g. nettle fibers) have only been able to conquer niche markets in the clothing sector. However, these are stable and it is expected that their market shares will continue to increase moderately because bast fibers can be produced locally in Europe, which is advantageous for sustainable production and also saves transport costs if they are processed locally. Buyers of textiles made from bast fibers are usually prepared to pay a higher price for them because, compared to cotton and man-made fibers, they are considered to be "natural" and, in some cases, "noble". The latter is at least questionable in view of their tendency to crease and the coarse fibers which can irritate the skin on direct contact.

Hard fibers are increasingly being used in automotive construction, e.g. coir fibers in seats and abaca fibers in structural components. This does not make automobiles significantly more "sustainable", but it is a selling point in the industrialized countries of Europe, especially in the luxury car segment. This trend will continue, even if the quantities processed in this way will remain small compared to other fibers.

50.1.3 Wool

Textiles made of wool are largely fashion-independent and have had a stable, absolute market share of approx. 1 m tons for decades. For a long time, they were regarded as a model natural product, because their production does not harm the environment. Recently, however, reports of the cruel practice of mulesing have tarnished this image, resulting in increasing pressure on wool farmers to avoid it. Nevertheless, mulesing is widespread and some farmers are still unwilling to bow to pressure from consumers and large retail chains to use different methods or even breed different breeds of sheep. Since there are hardly any corresponding labels as in the case of cotton, consumers do not recognize how the wool of the products they buy was produced. It is therefore to be expected that nothing will change in the medium term unless a large-scale campaign, similar to that against whaling by Greenpeace, is launched to force all wool farmers around the world to rethink.

50.1.4 Other Natural Fibers

Other natural fibers, such as cashmere, camel hair and yak will gain in importance if they are produced sustainably and under good social conditions. Buyers of such textiles are willing to pay a higher price for them, which can make processing the fibers in the country of extraction attractive. This in turn is a purchasing argument for consumers with ethical standards.

Silk is not currently the focus of animal rights activists, but could become more difficult to market if potential buyers become aware of how many animals have to be killed for 1 kg of silk (approx. 10,000). Silks from other animals and also spider silk will continue to have no market significance.

Numerous, mainly locally endemic fibrous materials have already been investigated for fiber composites, in particular bast and hard fibers. However, their properties are too inconsistent, the quantities that can be produced too low and their price too high for them to ever achieve market significance.

50.2 Man-Made Fibers

50.2.1 Synthetic Fibers

The production of synthetic man-made fibers has increased continuously in recent decades and an end to this growth is not expected in the next 15–20 years. Their properties can be precisely adjusted within wide limits and their production and the subsequent care of the textiles have little impact on the environment compared to cotton. However, the production of synthetic man-made fibers has recently been called into question, on the one hand because of their extraction from petroleum and on the other because of the problem of microplastics polluting the environment.

Man-made fibers are indispensable because of their properties and it is not possible to switch the entire fiber production to natural fibers owing to the enormous amount of land required. For this reason, increased efforts have recently been made to change the raw material base away from petroleum to renewable raw materials. For example, it is technically possible to produce polyester from cellulose and solutions are also being developed for polyamides. The extent to which this makes sense from an environmental and energy point of view remains a question that has not yet been conclusively answered.

According to a study by Fraunhofer UMSICHT (2021), the share of fibers in the problem of microplastics is 2–4% and is therefore only of comparatively minor importance (abrasion of tyres: 30%). Nevertheless, this issue is playing an increasingly important role among consumers, so an answer must be found. This could be filters in washing machines or alternative yarn constructions. The addition of additives which dissolve the fibers under defined conditions without leaving residues is also being discussed.

In the medium term, the volume of synthetic man-made fibers produced annually will continue to increase due to global population growth, increasing prosperity and the associated greater demand for textiles per capita, and because of their comparatively low price. However, more sustainably produced fibers based on polyester and polyamide will gain market share in industrialized countries, especially in Europe, and this trend will intensify in the coming decades.

50.2.2 Cellulosic Fibers

The relative growth of cellulosic fibers has been stronger than that of synthetic fibers in recent years. This trend will continue because cellulosic man-made fibers are seen as an environmentally friendly and sustainable alternative to synthetic man-made fibers and are therefore attractive to many consumers. However, scaling up production involves large investments and will therefore take a long time. Nevertheless, growth will continue to intensify and it is expected that by mid-century approximately 30% of all man-made fibers will be produced on a cellulosic basis. If the oil price rises again to the record levels of 2008–2015, cellulosic man-made fibers will also be price competitive, which has not been the case so far in the mass market. Then the development could accelerate considerably.

50.3 Inorganic and Carbon Fibers

The production of glass fibers is currently growing by 5–8% per year, mainly for the production of fiber composites in lightweight construction. If this growth were to continue, glass fiber production in 2050 would be over 20 m tons, which is unlikely. A more realistic assumption is 10 m tons, which is still double the current level. The production of basalt fibers will also continue to increase, competing with both, glass fibers and carbon fibers, the latter being similar mainly because of their color rather than their mechanical properties. Ceramic matrix composites (CMC) will also gain importance, especially for high temperature applications in power plants and engines, both for filters and structural components. The current growth of carbon fibers is 10–15% and is expected to continue at this rate until 2030. Main applications for carbon fibers will be wind power rotor blades, followed by lightweight products, possibly in automotive. In order to achieve the climate targets, it will be essential to further expand the use of wind power, not only building new turbines but also renewing existing ones. For this reason, it is expected that the production of carbon fibers will increase to 500,000–800,000 tons per year by 2050. It is difficult to make a forecast for metal fibers because production volumes are already not accurately recorded today. However, it can be assumed that their quantity will continue to increase in the course of general economic development. The production of asbestos will have ceased completely by 2050, because the countries in which asbestos is still produced today will also switch to alternative fiber materials due to pressure from their populations.

50.4 The World in 2050

In 2050, a large proportion of our clothing will still be made of cotton, although man-made fibers will gain considerably in importance. The proportion of recycled fibers will increase significantly, provided that it is possible to separate them by type. Appropriate processes have already been developed so that, for example, cotton can be separated from polyester and new fibers can be produced from the individual components (Thomas, 2020). However, such processes are not yet economical under current market conditions. A better approach is a different design of textiles than is common today (difficult or inseparable fiber blends). The question of collecting and sorting used textiles also needs to be addressed for the large quantities that will then be produced.

Figure 50.1 shows a forecast of the development for the world fiber production until 2050. According to estimates by the UN and other organizations, the world population is expected to increase to approximately 10 bn by 2050 (Statista, 2021). This increase will mainly take place in Asia and Africa, where the standard of living will also continue to rise. As a result, the demand for fibers will increase more strongly there than in the industrialized countries, where the population is even expected to decline. World fiber demand will thus be around 200 m tons, if we assume that fiber demand per capita is around 20 kg/year. Growth in cotton production will be moderate because the available arable land competes with food production and an increase in production can only be achieved through higher-yielding varieties. Wool production will remain constant because new breeding of sheep breeds with higher wool yields is not expected and there is no increased demand for wool clothing in either Africa or Asia. Bast and hard fiber production will increase slightly because it is unlikely to compete with food production. The production of synthetic fibers will continue to increase, but growth will slow down in favor of man-made fibers based on cellulose. For these, annual growth rates of 6–8% are expected from 2030 onwards, accompanied by a reduction in oil production to meet climate targets. Towards the middle of the century, the production of synthetic fibers will gradually decline and by 2070 it will be surpassed by the volume of cellulosic man-made fibers. This does not mean that viscose and lyocell fibers will then dominate, but rather that even the polyester and polyamide fibers that predominate today will gradually be produced from renewable raw materials, i.e. presumably from cellulose. Nevertheless, the share of viscose and especially lyocell fibers will continue to increase strongly (Fig. 50.1).

The political situation will change in the coming decades due to China's growing economic power. If there are trade wars with punitive tariffs between China, the USA and the EU, this will also affect fiber production in the medium term. It is to be expected that the production of man-made fibers will continue to take place to a large extent in Asia, especially in China. However, Southeast Asian countries will gain in importance, e.g.

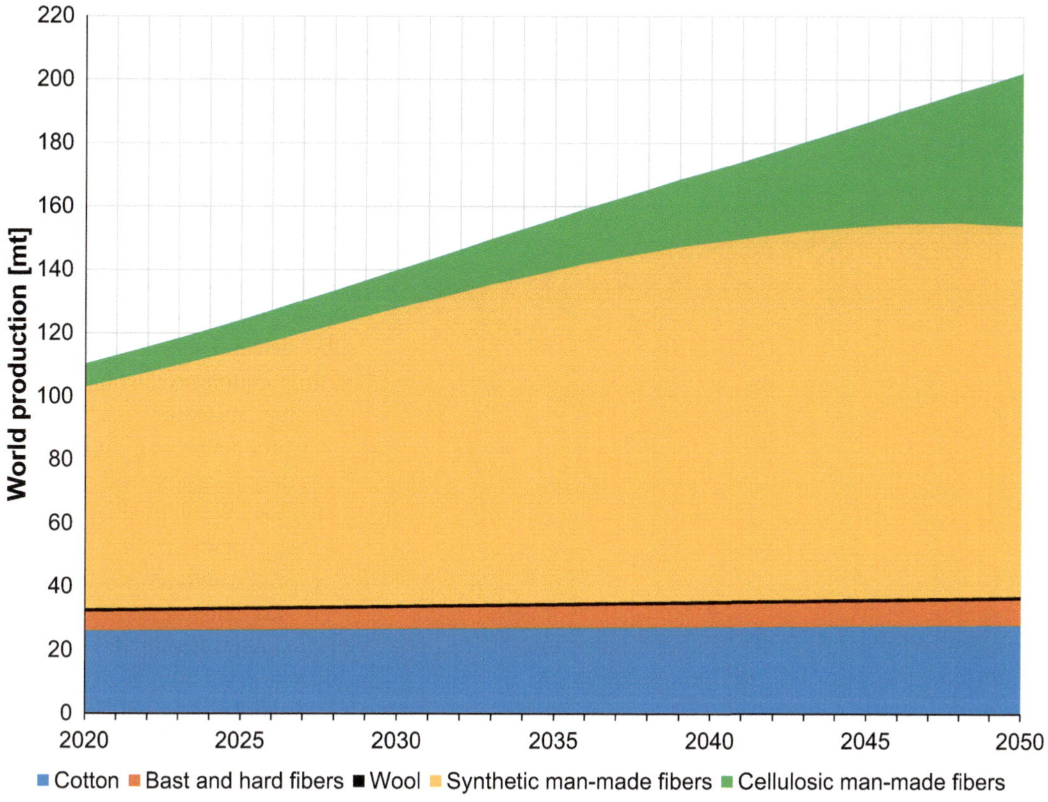

Fig. 50.1 Forecast of the development of world fiber production

Thailand and Indonesia, where large new plants are already being built, and India will also play a major role.

Due to the Covid 19 pandemic 2020–2022, the weaknesses of increasing globalization have been generally recognized and there will therefore be a partial renaissance of regional production of fibers and textiles. This will not only affect breathing masks and air filters made of polypropylene, but also other textiles that need to be available at short notice, e.g. for medical devices. The increasing demand of many consumers in the industrialized countries for sustainable production will also strengthen this trend for other textiles. However, it is not to be expected that textile production will return to the industrialized countries to a large extent, because labor costs there are too high for economic production, especially in ready-made clothing.

This forecast is based on various assumptions and model calculations that are plausible from today's perspective. However, the Danish saying that "forecasts are difficult, especially when they concern the future" also applies here.

References

Better Cotton Initiative. (n.d.). https://bettercotton.org/

Fraunhofer UMSICHT. (2021). https://www.umsicht.fraunhofer.de/content/dam/umsicht/de/dokumente/publikationen/2018/kunststoffe-id-umwelt-konsortialstudie-mikroplastik.pdf

Statista. (2021). https://de.statista.com/statistik/daten/studie/1717/umfrage/prognose-zur-entwicklung-der-weltbevoelkerung/

Textile Exchange. (2021). *Preferred fiber & materials*. Market Report 2020. http://textileexchange.org/wp-content/uploads/2020/06/Textile-Exchange_Preferred-Fiber-Material-Market-Report_2020.pdf

Thomas, D. (2020). *Unfair fashion*. Riva Verlag.

Index

A
Aachen, 16
Abacá, 255–257
Abbreviations, 7
Abitz, W., 69
Abott, W.F., 33
Acetate, 831
Acetate fibers, 80, 870–875
α-crystals, 75, 664
Additivation, 428
Adipic acid, 23
Aerogel fibers, 605–607
Aerogels, 605–608
Africa, 161–163, 238
Agave cantala, 250
Agave fibers, 249–255
Agave fourcroydes, 250
Agave sisalana, 249
A-glass, 914
α-helix, 295
AH salt, 654, 661
Air-gap spinning process, 686, 861
Air gap wet spinning process, 684
Air-jet texturing, 617–621
Air-jet texturing machine, 620
Air texturing jet, 619
α-keratin, 83, 84
Akon, 403
AKZO, 626
Alexander the Great, 351
Algae, 862
Alginate, 891
Alginate fibers (ALG), 456, 891, 892
Aliphatic chains, 64
Alkalization, 838–839
Alloys, 943–944
Alpaca, 331–334
Alpaca suri, 331
American Society for Testing and Materials (ASTM), 972
Amino acids, 81, 291
Amorphous, 64, 465, 471
Amorphous areas, 77
Amorphous regions, 86, 88
Amphibole, 386, 388, 389
Amylopectin, 892
Amylose, 892
Angora goat, 339
Angora rabbit, 344
Annealing, 65
Antistatic, 669
Antistatic behavior, 709
Antron, 669
Apollo 1, 773
Apron, 612
Aralac, 883, 885
Aral Sea, 99
Aramid, 27, 76–78, 681–691
Aramina, 404
Arc process, 805
Ardil, 888
Argentina, 313–314
AR glass, 916
Aristotle, 351
Artificial silk, 19, 832, 833
Artificial turf, 541
Arvin, J.A., 625
Asahi Kasai, 17
Asbestos, 12, 381–401
Asclepias fiber, 405
Atactic PP, 699
Audemars, G.P., 13, 831
Austenite, 950
Australia, 151–157, 277, 304–307
Austria, 22
Autoclave, 433–434, 662
Avrami equation, 65
Azlon fibers, 883

B
Babylonia, 11
Bacillus thuringiensis, 174
Back-scoured wool, 273
Bactrian camel, 327–330
Bale press, 117–118
Baler, 554
Bale size, 118
Bamboo, 243–246, 893
Bangladesh, 229, 231

D. Veit, *Fibers*, https://doi.org/10.1007/978-3-031-15309-9

Barus effect, 487
Basalt fibers, 37, 923–926
Basofil, 768, 770
Batch stirred tank, 432
Bayer AG, 722, 749
Bayer, O., 29, 721, 749
BCF, 502, 531, 534
β-crystals, 664
Beadle, C., 14, 832
Belgium, 192
Bell, A.G., 819
Bemberg AG, 17, 833
Bending instabilities, 589
Benin, 163
Betel nut fibers, 269–270
Bevan, E.J., 14, 832
Bicomponent fibers, 569–582
Bicomponent spinning processes, 574
Bingham, 438
Biocelsol process, 866
Biodegradability, 899
BioPA, 897–898
BioPET, 897–898
Biopolymers, 883–900
Birefringence, 86, 965
Bi-rotor, 502
β-keratin, 84
Blades, H., 681
Boehmeria nivea, 232
Boehmeria utilis, 232
Boehm, H.P., 814
Bolivia, 329
Boll weevil, 169
Bollworm, 170
Bombax ceiba, 267
Border Leicester, 282
Boron nitride, 930–931
Bourette spinning, 363–364
Brazil, 157–158, 253
Break conversion, 551
Breaking, 203
Breaking converter, 550–551
Bronner, E., 20
Bronnert, E., 18, 833
Brooks, J.D., 785
Broom fiber, 405
Bt resistance, 176
Bt toxin, 174, 175
Bulk continuous filament (BCF), 706
Burkina Faso, 163
Bushings, 911
Butterfly silk, 374–375
BXN cotton, 174
Byssus, 375–376
Byzantium, 353

C
CA, 831
Cabling, 523–525
Cactus fibers, 270–271
Camel, 327–338
Capillary holes, 488
Carbamate process, 853
Carbon fibers, 33–36, 779–798
Carbonization, 783, 785
Carbonized wool, 273
Carbon nanotubes (CNT), 801–816
Cargan, 885
Carothers, 625
Carothers equation, 422, 423
Carothers, W.H., 22, 23, 27, 31, 649, 651
Carpasian flax, 382
Carpet yarns, 531–535
Carrier-free dyeing, 637
Cartwright, E., 96
Casein and milk fibers, 883–886
Cashmere, 338–339
Caslen, 885
Casolana, 885
Catalyst, 629, 701
Cationic dyes, 637–638
Cauliflower effect, 468
Cayven, G., 37, 38
C-curve, 466
Ceiba pentandra, 267
Celanese Corp., 774
Cellulose, 413
Cellulose I, 78, 82
Cellulose II, 79
Cellulosic fibers, 78–81
Cellulosic man-made fibers, 831–878
Central Asia, 151
Centrifugal spinning, 601–602
Centrifuge blowing, 914
Ceramic fibers, 37–38, 927–936
C-glass, 914
Chain growth, 701
Chain growth reaction, 424
Chain length distribution, 427
Chain lengths, 69, 84, 85, 423, 431
Chain reaction, 418
Chappe, C., 819
Charlemagne, 382
Chemical constitution, 652–660
Chemical structure, 632, 851–853
Chemical vapour deposition (CVD) process, 806–807, 932
Cheviot, 279, 282
China, 9, 11, 12, 31, 32, 38, 146–147, 180, 184, 189, 213, 353, 359
Chips, 445
Chiru, 407
Chitin fibers, 889
Chitosan fibers, 889–890
Chrysotile, 381, 386–388
Churka, 110
Classification, 132–133, 209, 296–301, 359–360, 385–386, 586

Classification of biopolymers, 884
Classification of man-made fibers, 5
Classification of nanotubes, 802
Classification of natural fibers, 4
CLY, 831
CMD, 831
Coarse wool, 274
Coat of arms, 192, 215
Cocoon, 355
Co-current flow, 442
Coextrusion, 428
Coir fibers, 263–267
Collagen fibers, 888
Color code diagram, 131
Colored cotton, 180
Combing machines, 276
Comeback, 274, 280
Comonomers, 659
Conex, 27
Congo Jute, 404
Consumption, 43–45
Conversion, 547–553
Cooley, J.F., 585
Cooling drum, 532
Cooling rates, 465
Cooling zone, 613–614
Copolyamide, 658–660
Copolymerization, 725
Copolymerization with nanotubes, 811
Copolymers, 427–428
Copper process, 13–14
Core/sheath (C/S) fiber, 573–577, 580
Core/sheath (C/S) type, 569
Corn fiber, 887–888
Corriedale, 282
Cortex, 288, 289
Côte d'Ivoire, 163
Cotlook, 182–183
Cotton, 11, 93–187
Cotton belt, 97
Cotton bug, 170
Cotton louse, 170
Cotton Made in Africa, 161
Cotton seed, 121
Cottonseed oil, 120
Cotton trade, 181
Cotton varieties, 106
Coulomb forces, 589
Counterflow, 442
Courtoulds, 14, 20
Covid-19, 42
Crafts, J.M., 767
Crepe, 362
Crimp, 615
Crimped staple fibers, 549
Crookes, W., 13
Crossbreds, 274, 280
Cross, C.F., 832
Cross, F.C., 14
Cross-section, 633
Crude asbestos, 385
Crystal length, 73
Crystalline, 358, 465, 468, 471
Crystalline regions, 71, 85
Crystallinity, 63–69, 85, 465, 467
Crystallites, 63
Crystallization, 67–68, 450–451, 465–467
Crystallization rate, 64, 423
Crystals, 66, 87, 88, 460
Crystal structures, 70–83, 122, 683, 685, 704
CTA, 831
Cuoxam, 13
Cuoxam process, 855–858
Cuprama, 14
Cupro (CUP), 831
Cupro process, 855–858
Cutting conversion, 548–550
CV, 831
Cyprian gold, 38, 941

D

Dacron, 28, 626
Date palm fibers, 261
de Chardonnet, H., 15, 832, 875
Dederon, 25, 650
Deep dyeing, 637
Defibrillation, 837
Definitions, 3–7
Degradability, 898–900
Degree of crystallinity, 68, 466
Degree of order, 71
Degree of polymerization, 419, 422
Degumming, 236, 368–369
Density, 959
De Réaumur, F., 13
Derivative fibers, 870–877
Despaissis, H., 14
D-glass, 916
Dhé, P., 37
Diamines, 661
Dicarboxylic acids, 661
Dickson, J.T., 27, 625
Differential dyeing, 669
Differential scanning calorimetry (DSC), 960, 962
Dilactide, 743
Dimensionally stable (DST), 521
Dimethylformamide (DMF), 721
Dimethyl terephthalate (DMT), 630
Diolen, 28, 626
Diphyl, 443
Direct cabling, 524
Disk ring reactors (DRR), 434–435
Dispersion spinning, 460
Distillation tower, 414
Doffing, 504
Dog hair, 347
Doppi-trame, 362
Dorlastan, 749

Dosing additives, 632
Drafting, 498–500, 548
Drafting point, 499
Dralon, 722
Drawing, 68, 468–470
Drawing process, 86–88
Draw ratio, 76, 474
Draw test, 471
Draw twisting, 521–523
Draw winding process, 667
Dromedary, 327, 330
Drum drawing, 912
Dryer, 446–452
Drying process, 451
Drying speed, 451
Dry spinning, 455–457, 684, 715, 725–726, 752, 754
DSM, 30
Dunova, 722
DuPont, 22, 24, 27, 28, 31, 650, 681, 722, 739, 749, 765
DuPont de Nemours, 626
Duraron, 696
Dwarf palm fibers, 261
Dye affinity, 640–641
Dyeing defects, 614
Dyneema, 30, 696
Dynel, 723

E
Eastman Kodak, 626
ε-Caprolactam, 422, 657
ECR glass, 916
Edison, T.A., 14, 33
Edison, Th., 779, 788
E-glass, 916
Egypt, 11, 38, 158–161, 178, 190, 230
Eichengrün, A., 14
Eilfeld, F., 17
Elastane (EL), 749–759
Elastomers, 751–752
Electoral, 281
Electric field, 592–593
Electric heating tapes, 442–443
Electric ovens, 909–910
Electrospinning, 585–599, 606–607
Electrowriting, 601
Eli Whitney, 96
Emballage, 118–120
Emulsion polymerization, 420, 723
Ending Stocks, 138
Engelshaar, 36
Engineered cotton, 173–178
England, 36
English eicester, 282
English Leicester, 279
Enkalure, 669
Entangling, 500
Epoxy composites, 812
Eriodendron anfractuosum, 267
Escurial, 281
Esparto fibers, 260
Esterification, 427
Eternit, 382
Ethicon Inc., 739
Ethylene glycol, 629
Europe, 31, 32, 165–167
Evonik Industries AG, 763
Extra long staple cotton (ELS), 139–143
Extruder, 477–478

F
Fabric filter, 480
False twist, 611
False twister, 613
False twist texturing, 609–617
False twist texturing machine, 610, 616
FDY, 502, 514–516, 521
Feathers, 891
Ferretti, A., 22
Fertile Crescent, 274
Fiber, 3
Fiber A, 722
Fiber classifying system, 133
Fiber fineness, 280
Fiber optic types, 914–916
Fiber orientation, 594–599
Fiber V, 626
Fibroin, 355, 358–359
Fibrolane, 885, 887, 888
Fikentscher, H., 29, 721
Filament fineness, 698
Filament numbers, 616
Filament yarn types, 636, 674
Film yarns, 537–541
Filter, 478–482, 507–508
Fine animal hair, 327–349
Fine fibers, 576, 585–602
Fineness, 299, 555, 573, 616, 636, 674, 968
Fish protein, 890
Fixing, 615
Flame retardant (FR), 669, 709
Flame retardant (FR) fibers, 849–851
Flammability, 639–640
Flash dryer, 447–448
Flash spinning, 600–601
Flax, 9, 189–213
Fleece, 284
Flock fibers, 551–553
Flocking, 553
Floss silk tree, 403
Flow rate, 441
Flow-through vessel, 432
Flow tube, 433
Fluidized bed dryer, 449–450
Follicle, 286–288
Force-elongation curves, 87, 470–472, 756
Force-elongation diagrams, 472
Force microscopes, 812
Forces, 464

Ford Motor Company, 887
Formhals, A., 585
Fourier transform infrared spectrometer (FTIR), 966–967
Fournier, A., 942
FoxFiber, 180
FOY, 473, 497, 516–519
France, 192
Freidlina, R.CH., 26
Fremery, M., 14, 16, 832
Frequency chart, 969–970
Friction discs, 612
Friedel, C., 767
Functionalization, 808–809
Funnel spinning, 456–458, 857
Fur hair, 347–349
Futures, 182

G

Gandhi, 149
Gay-Lussac, J.L., 31
Gazza, 105
γ-crystals, 664
Geim, A., 814
Gel spinning, 458, 694–695, 726–727
Genas, M., 25
Genetically modified cotton, 164
Genetic engineering, 173–178
Germany, 14, 19, 24, 28, 37, 50–61, 167–169, 193, 242, 316–317, 354, 650, 722
Gerngroß, O., 69
Gilbert, W., 585
Gin, 114–115
Ginning, 110–121
Glanzstoff, 832
Glanzstoff-Fabriken, V., 17
Glass fibers, 36–37, 905–922
Glass transition temperature, 450
Glass wool, 912–915
Glühlampenfabrik, R., 832
Goat, 338–344
Godets, 68, 498, 499, 533
Gordon, D., 36
Gossypium, 93
Gossypium arboreum, 93, 101, 102, 104
Gossypium barbadense, 93, 103, 104
Gossypium brasiliense, 103
Gossypium herbaceum, 93, 101, 102, 105
Gossypium hirsutum, 93, 104
Gossypium nangking, 101
Gossypium peruvianum, 103
Gossypium purpurascens, 103
Gradient index fiber, 823
Grafting, 427–428
Granulate, 445
Granulate drying, 663–664
Granulator, 445–446
Graphene, 801–816
Graphitization, 783–785
Grass fibers, 260–261
Greece, 165, 167
Green Flax, 201
Grenadine, 362
Grilon, 26, 669
Guanaco, 334–335

H

Hackling, 204–205
Hagen-Poiseuille, 439, 443
Hagiwaba, K., 585
Hall, C.C., 37
Hand picking, 106–107
Harappa, 147
Hargreaves, J., 95
Harvest times, 112
Hearle, J.W.S., 70, 71
Heater, 613–614
Heat exchange, 440–442
Heat exchanger, 439–443
Heating media, 444
Heat transfer media (HTM), 491
Heat transfer oil, 443
Heavy tow, 785
Hemp, 9–11, 213–223
Henequen, 249–255
Hermann, W., 894
Herodotus, 11, 93, 147
Herrmann, K., 69
Heuck, C., 721
Heuvel, H.M., 75
Hexamethylene diamine, 23
High-modulus low-shrinkage (HMLS), 521
High-performance fibers, 761–777
High-shrink fibers, 637
High tenacity filament yarns, 637
High Volume Instruments, 133
Hill, J.W., 23, 649
HM 50, 682
Hoechst, 626
Hole, 485
Hole patterns, 509
Hollow fibers, 489, 490, 849
Holmes, H.O., 111
Honeywell, 30
Hooke, R., 12, 831
Hop fiber, 405
Horsehair, 345–347
Hottenroth, V., 18
Houtz, R.C., 29, 721
Huacaya alpaca, 333
Hubert, E., 22, 715
Huisman, R., 75

I

ICI, 28, 626, 693, 765, 888
IG Farben AG, 24, 25, 650, 715, 721
Iijima, S., 801
India, 9, 11, 32, 102, 147–150, 178–179, 189, 213, 229, 232, 238

Indonesia, 32, 263
Industrial yarns, 521–525
Inner structure, 753
Inorganic man-made fFibers, 36–38
Insertion polymerization, 701
Interlacing, 501
Internal structure, 694
Internal structure of carbon fibers, 786
Intrinsic viscosity (IV), 424
Ionic polymerization, 420
Irrigated cotton, 99
Irrigation, 101
Isis, 190
Islands-in-the-sea, 576
ISO, 972
Isotactic polypropylene, 699
IZANAS, 696

J
Japan, 12, 17, 20, 29, 31, 32, 35, 722
Java Jute, 404
Jumel, 103
Junk, 191
Jute, 223–230

K
Kanatnik, 405
Kanebo Gohsen Ltd., 739
Kaolinite, 906
Kapok, 267–268
Kashgora, 341–342
Kazakhstan, 151
Kenaf, 238–241
Kendyr fiber, 405
Kermel, 688
Kevlar, 27, 681
Kinetics of crystal formation, 63–66
Kistler, S.S., 605
Klatte, F., 22, 29, 715, 721
Knit fixation process, 621–622
Kobayashi, K., 87, 88
Kodel, 28, 626
Koppe, P., 18
Korea, 33
Kreca, 35
Kuraray Co, Ltd., 774
Kureha, 34
Kureha Corp., 785
Kurtz, P., 29, 721
Küttner, H., 16
Kwolek, S.L., 27, 681
Kynol, 771
Kyrgyzstan, 151, 180

L
Lactam, 657–658
Lactic acid, 742
Lactofil, 885
Lamellae, 66
Lanital, 22, 883, 885, 887
Laser ablation process, 805–806
Leaflet structure, 684
Lehner, F., 16, 832
Length of the macromolecules, 83–85
Lenzing AG, 22
Leonic threads, 38, 942
Libbey, D., 37
Liebig, J. von, 768
Light conduction, 820
Light emitting fibers, 822
Lignin, 789
Lilienfeld, L., 15
Lily fibers, 260
Lime bast fiber, 405
Lincoln, 279, 282
Linseed, 189
Lint, 93
Linters, 93, 120
Linum angustifolium, 190
Linum usitatissimum Linné, 189
Liquid crystal polymers (LCP), 774–776
Liquid crystals, 76, 785
Liquid crystal spinning, 458–460
Livius, T., 191
Llama, 329–331
Loading, 369–370
LOI, 776
Lotus silk, 405
Low pilling types, 732–733
Low-pill staple fibers for use in apparel textiles, 637
LOY, 473, 496, 513
Ludewig, H., 27, 625
Lukyanovich, V.M., 801
Lupine fiber, 405
Lurex, 948
Lycra, 749
Lyocell, 20–22, 41, 831, 832, 858–866
Lyotropic, 461

M
MAC, 721
Macrofibril angle, 964
Macromolecules, 63, 465–468
Macromonomers, 427–428
Magnesium oxide, 929
Magnetic spindle, 612
Mako cotton, 158
Malaysia, 263
Mali, 163
Mandeville, 95
Mangalica, 407–408
Manila Hemp, 255–257
Manufacture, 731
Manufacturers, 691, 737, 766–767, 813–814, 875
Manufacturers and brand names of biomaterials, 900
Manufacturers of acetate fibers, 876
Manufacturers of basalt fibers, 926
Manufacturers of bicomponent fibers, 582

Manufacturers of carbon fibers, 798
Manufacturers of carbon nanotubes, 815
Manufacturers of ceramic fibers and trade names, 936
Manufacturers of elastane fibers, 758
Manufacturers of glass fibers, 922
Manufacturers of graphene, 816
Manufacturers of high-performance fibers, 777
Manufacturers of metal fibers, 955
Manufacturers of optical fibers, 826
Manufacturers of PA 6 and PA 66, 674
Manufacturers of PET and PBT fibers, 647
Manufacturers of PVA fibers, 719
Marco Polo, 382
Marinova, 887
Market, 41–61
Mark, H., 681
Martensite, 950
Marvel, C.S., 773
Mass and gas phase polymerization, 703–704
Masterbatch, 475
Matrix/fibril (M/F) fibers, 576, 578, 580–582
Matrix/fibril (M/F) type, 569–570
Matrix wet spinning, 763
Maturity, 130–131
Maximum acceptance angle, 819
McArthur, John, 306
Mechanical harvesting, 107–108
Mechanical solubilizing, 220–221
Melamine, 769
Melamine resin (MF) fibers, 768–769
Meltblown, 557
Meltblown process, 560–561, 563
Melt electrospinning, 599–600
Melt extraction, 947–948
Melting temperature during, 428
Melting temperatures of aliphatic polyamides, 656–657
Melting temperatures of polyamides from lactams, 658
Melt spinneret, 484–489
Melt spinning, 455, 461, 462
Melt spinning into a rotating liquid, 947
Merino, 274, 275, 277, 279–282
Merinova, 885
Metal coated fibers, 948–949
Metal cutting manufacturing process, 945–946
Metal fibers, 38, 941–956
Metal filters, 480
Metallized fibers, 944
Metallocene, 701
Metallocene catalysts, 699–703
Mexico, 11, 103, 232, 249, 261
Meyer, K.H., 69, 78
M-glass, 916
Michelin, 942
Microfibers, 668
Micronaire, 105, 125
Microscope, 959
Microstructured fibers, 822
Milkfiber, 885
Millar. A., 22
Misch, L., 78
Mitscherlich, A., 15, 832
Mmulberry moth, 356
Mobilon, 749
Modacrylic fibers, 721, 730
Modal, 15, 832
Model, 831, 848
Modification, 517–518, 631–632, 708–710
Modifier, 519
Mohair, 339–341
Mohamed Ali, K., 93
Mohamed Ali, P., 158
Mohenjo Daro, 147
Moisture absorption, 992
Molar mass distribution, 426
Molar ratio, 423, 629
Molecular weight, 591
Molecular weight distribution, 424–427
Monofilaments, 527–529
Monomode fibers, 822
Monsanto, 27, 681
Montecatini, 30
Morphology, 704
Morton, W.E., 70, 71
Morton, W.J., 585
Mouflon, 274
Moureau, M.C., 721
Mulesing, 284–285
Müllerbad I, 18, 833
Müllerbad II, 18, 833
Mullite, 928–929
Multi-core fibers, 821–822
Multi-mode fibers (MMF), 819
Multi-wall carbon nanotubes (MWNTs), 804
Musa textilis, 255–257
Musk oxen, 342
Mycelia, 73

N

Nanocellulose, 877–878
Nanofiber nonwoven, 588
Napoleon, 36, 96, 281
Natta, G., 30, 697
Natural draw ratio, 87
NatureWorks, 31, 740
Needleless spinning, 590–591
Negretti, 281
Neptune grass, 261
Nettle fiber, 241–243
Network transformers, 908
New Zealand, 307–310
New Zealand flax, 260
Nitinol, 38
Nitrocellulose, 13, 875–877
N-methylmorpholine-N-oxide (NMMO), 20
NMMO process, 859, 860
Nomex, 27, 681
Nonwoven bonding, 559–560
Nonwovens, 557–566

Nonwovens filter, 480–481
Novolak, 769
Novoloid, 771
Novoloid fibers, 769–770
Novoselov, K.S., 814
Nozzle drawing, 911–912
Number of filaments, 516, 636, 674
Nylon, 24, 25, 422, 650
Nyltest, 25, 651, 652

O
Oberlin, A., 801
Oiler, 497
Oil flax, 197
Oil palm fibers, 261
Okra fiber, 405
Oligomers, 422
Optical fibers, 819–826
Options, 182
Organic cotton, 178–181
Organzine, 362
Orientation, 85–86, 462, 465–468, 694
Otani, S., 34
Otto, 161
Owens, 37
Owens process, 913

P
P84, 764
PA 4.6, 655
PA 6, 23–25, 65, 72, 74, 88, 424, 650
PA 6.6, 650, 653–655
PA 6.10, 655
PA 7, 26, 651, 658
PA 11, 651, 657, 898
PA 12, 26, 658
PA 66, 23–25, 65, 72, 74, 88, 433
Packaging, 504
PACM-12, 659–660
Paczinsky. W.v., 37
Paddle dryer, 449–450
Pakistan, 11, 103, 150–151, 231
Pako vicuña, 337–338
Palm fibers, 261–262
Papin, D., 433
Parry, E., 37
Partially crystalline, 64, 77
Pastoral zones, 307
Pauly, H., 14, 832
Payen, A., 13, 831
Peanut fiber, 888
PeCe, 22, 715
Pechmann, 693
Pelletizer, 445, 447, 448
Pellets, 447, 477
Pelouze, T.-J., 739
Pennings, A., 693
Perendale, 279, 282
Perlagon, 27
Perlon, 25, 422, 650, 669
Perluran, 24
Peru, 11, 329, 331
PE-Shish-Kebab structure, 697
Pest control, 171–173
Pests, 169
PESU, 767
Peterlin, 88
Petryanov-Sokolov, I.V., 585
Philippines, 255, 257
Phormium, 260
Pile dwellings, 9
Pima, 103
Pima cotton, 105
Pineapple, 257–259
Pipe flows, 440
Pitch, 779, 785–787
Plant, 504
Plastic deformation, 88
Player, J., 37
Pleated filter, 483
Pliny, 351, 381
Plutarch, 381
POF, 824
Pohl, H.A., 749
Polar groups, 64
Polwarth, 279, 282
Polyacrylonitrile (PAN), 29, 50, 58, 77–78, 81, 721–737, 779, 814
Polyaddition, 424, 749
Polyamide (PA), 47, 55, 58, 72–76, 86, 422, 448, 551, 649–680
Polyamide imides (PAI), 762
Polyazole (PBO), 770–773
Polyazole fibers, 770–773
Polybenzimidazole (PBI), 774
Polybenzimidazole (PBI) fibers, 773
Polybutylene succinate (PBS), 896, 897
Polybutylene terephthalate (PBT), 28, 626, 628, 633
Polycaprolactone fibers (PCL), 896
Polycondensation, 420–424, 629, 652, 660, 661
Polyester, 27–29, 76, 477, 550, 551, 625–648
Polyesterification, 629
Polyester terephthalate (PET), 27–28, 41, 47, 55, 58, 65, 76, 86, 88, 448, 450, 472, 490, 518, 522, 633
Polyetherimides (PEI), 762
Polyetherketone (PEK, PEEK) fibers, 764–767
Polyethylene (PE), 30, 693–696
Polyethylene terephthalate (PET), 625–627
Polyglycolic acid fibers (PGA), 895–896
Polyhydroxyalkanoate fibers, 895
Polyhydroxyalkanoates (PHA), 895
Polyhydroxybutyrate (PHB), 895
Polyimide, 764
Polyimide fibers (PI), 762–764
Polylactic acid (PLA), 739–747
Polylactide (PLA), 31
Polymer formation, 413–429, 431–452
Polymerization, 418–420

Polymer optical fibers, 821–822
Polymethyl methycrylate (PMMA), 518, 824
Polynosic fibers, 15, 848–849
Polyolefins, 77, 693–719
Polyoxadiazole (POD) fibers, 773–774
Polyphenylene sulfide (PPS) fibers, 767–768
Polypropylene (PP), 29–30, 80, 538, 696–714
Polysaccharide fibers, 891–893
Polysulfone (PSU, PES) fibers, 767
Polytetrafluoroethylene fibers (PTFE), 761–762
Polytrimethylene terephthalate (PTT), 28, 626–627, 633
Polyurethane, 749
Polyvinyl alcohol (PVA), 718–719
Polyvinyl alcohol (PVA) fibers, 894–895
Polyvinyl chloride (PVC), 22, 714–716
Polyvinyl dichloride (PVDC), 716–717
Polyvinyl difluoride (PVDF), 717–718
Poplar fibers, 268–269
POY, 472, 474, 497, 503, 513, 515, 518
PP, 30, 65, 77, 86
Precipitation bath, 843
Precipitation polymerization, 420
Premnitz, 20
Preston, J., 681
Prices, 61
Process lines, 545, 727
Process routes, 700, 724, 782
Prolon, 887
Protein, 417–418
Protein fibers, 81–83, 883–891
Pulp, 415, 417, 836
Pumps, 436–437

Q
Qiana, 659–660
Qiviut, 342–344
Q-Milk, 22, 886
Quench, 496
Quench duct, 492–496

R
Radical polymerization, 418–419
Radushkevich, L.V., 801
Rain-grown cotton, 99
Raman spectroscopy, 967–968
Ramie, 230–238
Raw silk, 359–360
Raw wool, 273
Rayon, 19, 833
Reaction rate, 421–423
Reaction spinning, 460–461
Reactive spinning process, 752–753
Reactor, 431–433, 507, 631
Réaumur, R.-A.F. de, 36, 831
Rectangular spinneret, 545, 559
Rectangular spinning package, 487
Recycling, 428–429, 712–714
Redon, 29, 722
Reeling, 360–362
Refractive indices, 819
Regenerated, 831
Regenerated fibers, 837–858
Rein, H., 29, 721
Relaxation time, 462
Renewable raw materials, 745
Repeat units of m-and p-aramide, 683
Resistance, 174, 176
Resistances, 175–176
Retting, 202, 219–220
Revolver filter, 481–482
R-glass, 916
Rilsan, 26, 651, 657, 898
Ring-opening polymerization, 652, 741
Ring spinnerets, 546
Ring spinning package, 487
Ring twisting, 920
Rinke, H., 749
Rippling, 202, 203
Rock wool process, 914, 923
Roller gin, 114–115
Romney, 279, 282
Rosella, 404
Rotary process, 914
Round spinning package, 487
Rozenblum, N.D., 585
Rubber threads, 893–894
Russia, 192, 384

S
Sand filter, 479
Sarelon, 888
Saw gin, 115
Scale structure, 297
Schappe spinning, 363
Scheele, C.W., 739
Schlack, P., 24, 27, 625, 650, 651, 749
Schönbein, C.F., 13, 831
Schuller, W., 37
Schützenberger, M.P., 832
Schützenberger, P., 14, 871
Schwabe, L., 36
Schweizer, M.E., 13, 14
Scoured wool, 273
Screw, 477–479
Scutching, 202–204
SeaCell, 861, 864
Seagrass, 260–261
Sea island cotton, 103
Seed oil, 189
Segmented pie, 576
Seifriz, W., 69
Sericin, 358, 359
Service temperature, 776
Shaft dryer, 447–448
Shape memory effect, 942, 949
Shatush, 407
Shearing, 283–285
Shearing robots, 287

Shear rate, 438
Shear stress, 438
Shimadzu, 31
Shivers, J.C., 749
Siberian ibex, 406
Side-by-side (S/S) fiber, 573, 577, 579–580
Side-by-side (S/S) type, 569
Side stream extruder, 428
Silica-alumina, 928–929
Silica fibers, 916
Silicon carbide, 930
Silicon oxide, 927
Silk, 12, 81, 351–379
Silkool, 885, 887
Silk Road, 353
Single-mode fibers (SMF), 819
Single-wall carbon nanotubes (SWNTs), 802, 804
Sinsen, 29, 722
SiO_4, 907
Sisal, 249–255
Small angle scattering (SAS), 966
Snowflake, 487
Snyder, M.D., 749
SOHIO process, 723
Sol-gel process, 606, 932–934
Sol-gel spinning, 933
Solidification, 466
Solid state extrusion process, 695
Solid state polycondensation (SSP), 423–424
Soltes, W.T., 33
Solution polymerization, 420, 724
Solution spinning, 453–461
Solution spun man-made fibers, 507–511
Solution-spun yarns, 87
South Africa, 277, 310–311, 382
Soy fibers, 886–887
Spain, 165, 167, 260, 316
Spandex, 749
Spanish broom, 405
Spantel, 749
Spectra, 30, 696
Spherolites, 66–67
Spider silk, 376–379, 890
Spindle picker, 108
Spin-draw processes, 667
Spin-draw texturing, 547
Spin dyeing, 474–475
Spinnability, 461–464
Spinneret, 455, 461–463, 486, 490, 508–509, 531, 574, 783
Spinneret cross-sections, 509
Spinnerets, 488
Spinning Beam, 491–492
Spinning Jenny, 95
Spinning orientation, 472–474
Spinning packages, 489–491
Spinning pump, 484
Spinning solutions, 508, 725
Spin preparation, 496–498
Spin pumps, 482–484
Split Film, 707
Spunbond, 557–560
Spunbond line, 558, 564
Spun dyed fibers, 668
Sri Lanka, 263
Standards, 972
Staple diagrams, 552
Staple fibers, 543–556
Staple fiber types, 636
Staple length diagram, 970
Starch, 897
Starch fibers, 892–893
Static mixers, 443–445
Staudinger, H., 22
Stearn, C.H., 18, 833
Step growth reaction, 421, 424
Step index fiber, 823
Steppe wild sheep, 274
Stick machine, 113–114
Stink bug, 170
Stoichiometric ratio, 423
Strabon, 191
Straw and peat fiber, 405
Strepicheev, A.A., 26
Stress-strain curves, 638, 688
Stretching point, 472
Stripper, 108–109
Structural formula, 626, 633, 655
Structural models, 63
Structure, 63–89, 199, 218–219, 226–227, 233, 286–293, 357–359, 386, 418, 427, 461
Subsidies, 184
Substance or mass polymerization, 420
Sudan, 163
Suffolk, 279
Sugar cane, 897
Sulfite process, 415
Sun, 404–405
Supercritical drying, 606
Superpolyesters, 649
Surging, 614
Suri, 333
Suspension or dispersion polymerization, 420, 723–724
Suspension polymerization, 703
Swan, J.W., 13, 831
Syndiotactic PP, 699

T
Tacticity, 699, 700
Taiwan, 32
Tajikistan, 151
Take-up speed, 89
Tanzania, 163
Target, 598–599
Taslan, 617–621
Taxonomy, 4–5
Taylor cone, 585
Taylor, G.H., 785

Taylor, G.I., 585
Taylor process, 947
Technical filament yarns, 523, 666–668, 849
Technora, 688
Teijin, 27
Temperature, 64, 65, 87
Tenacity-elongation curves, 554
Tencel, 22, 864
Terephthalic acid (TPA), 630, 631
Test methods, 959–972
Tetrahydrocannabinol (THC), 217
Textile filament yarns, 665–666
Texturing, 609–622
Thermal manufacturing processes, 946–948
Thermogravimetric analysis (TGA), 960
The total cultivated, 167
Thiele spinning funnel, 459
Thiolan, 885
Thornel, 34
Tiberius, 351
Tibetan antelope, 407
Tie molecules, 73
Tilghman, B.C., 15
Tire cord, 25
Titanium dioxide, 929
Todtenhaupt, F., 22
Todtenhaupt, F.E., 883
Topham box, 19
Topham, C.F., 14, 17, 833
Toray, 31
Tower dryer, 111–113
Trade names, 641–643, 716, 719, 737, 747, 864–865
Trajan's Column, 819
Trame, 362
Translucent/illuminated concrete, 826
Trash, 127–129
Trevira, 28, 626
Triacetate, 831, 872
Triangular trade, 94
Tricacetate fibers, 80
Triclinic unit cell, 635
Tri-component fibers, 578–579
Trofil PPS, 768
Tsunooga, 696
Tube bundle heat exchanger, 442
Tube flows, 437–439
Tumble dryer, 448–449
Turka, 405
Turkey, 163–165, 180, 184
Turkmenistan, 151
Turnover, 421–423
Tussah silk, 373–374
Twaron, 27, 686
Twist insertion, 609–612
Typha, 404

U
UHMWPE, 30, 693, 695
Ultem, 765
Uniformity index, 71
Union Carbide, 33, 788
United Kingdom, 314–316
Unit melter, 909
Upland cotton, 105, 141
Urban, J., 14, 16, 832
Urena-Ur, 404
Urtica dioica, 241
Uruguay, 311–313
USA, 19, 20, 31, 32, 37, 143–146, 180, 184, 651, 721
US Universal Cotton Standards, 133, 134
Uzbekistan, 99, 151

V
van der Waals forces, 63
Vandura, 22
Vectran, 774, 775
Veocel, 864
Vereinigte Glanzstoff, 25
Vereinigte Glanzstoff-Fabriken AG, 832
Verne, J., 875
Vestan, 28
Vicara, 887
Victre, 765
Vicuñas, 335–337
Viscose, 31, 42, 453, 456, 468, 779, 831, 832
Viscose fiber types, 847–852
Viscose spinning process, 838
Viscosities, 508
Viscosity, 461
Vistra, 20, 833
VK tube, 435–437, 661–662
Vogel, H., 773
von Pechmann, H., 30
Vulkollan, 749
Vyrene, 749

W
Water retention, 971
Watt, W., 34
Weight fraction, 426
Weighting, 369–370
Wet spinning, 456, 458, 726, 752, 755
Whale whiskers, 408
Whinfield, J.R., 27, 625
White fly, 170–171
White ramie, 234
Whitney, 111
Wide-angle X-ray, 86
Wide Angle X-Ray Diffraction (WAXD), 965–966
Wikilana, 890
Wilhelm Pieck, 25
Willow, 403–404
Windemuth, E., 749
Winder, 500–504, 509–511, 533
Wipolan, 22
Wire drawing, 944–945
Wire pull in bundle, 945
Wolcrylon, 722

Wolfen, 715
Wolpryla, 722
Wool, 11, 82, 83, 273–325
Woolgrass, 404
Woolly pig, 407
Wool scouring, 277
Wool seals, 325
Wool top, 273
Wool trade, 318
Wool tree, 267–268
Wormell, R., 22

X
X-101, 681
X-400, 681
X-500, 681
Xanthate, 838
Xanthogenation, 839–840
Xinjiang, 146

Y
Yak, 344–345
Yangir, 406
Yarn counts, 516
Yields, 120–121, 138
Young's modulus, 964
Yttria, 929
Yucca fibers, 260
Yucca filamentosa, 260

Z
Zeltner, W., 25
Z-glass, 916
Ziegler, K., 30, 697
Ziegler-Natta catalyst systems, 702
Zirconia, 929
Zycon, 887
Zylon, 771

GPSR Compliance

The European Union's (EU) General Product Safety Regulation (GPSR) is a set of rules that requires consumer products to be safe and our obligations to ensure this.

If you have any concerns about our products, you can contact us on ProductSafety@springernature.com

In case Publisher is established outside the EU, the EU authorized representative is:

Springer Nature Customer Service Center GmbH
Europaplatz 3
69115 Heidelberg, Germany

Batch number: 10740024

Printed by Printforce, the Netherlands